German Dictionary of Environmental Technology
Wörterbuch Umwelttechnologie
Deutsch–Englisch/Englisch–Deutsch

Routledge German Dictionary of Environmental Technology
Wörterbuch Umwelttechnologie
Deutsch–Englisch/Englisch–Deutsch

ROUTLEDGE
London and New York

First published 1997
by Routledge
11 New Fetter Lane, London EC4P 4EE

Simultaneously published in the USA and Canada
by Routledge
29 West 35th Street, New York, NY 10001

Typeset in Monotype Times, Helvetica Neue and Bauer Bodoni
by Routledge

Printed in England by TJ Press (Padstow) Ltd., Cornwall

Printed on acid-free paper

British Library Cataloguing-in-Publication Data
A catalogue record for this book is available from the British Library

Library of Congress Cataloging-in-Publication Data
Applied for

ISBN: 0–415–11243–5

German Dictionary of Environmental Technology
Wörterbuch Umwelttechnologie Deutsch–Englisch/Englisch–Deutsch

Project Managers/Projektmanagement
Sabine Bohnacker-Bruce Susanne Jordans

Managing Editor/Redaktionsleitung
Sinda López

Editorial/Lektorat
Justine Bird Lisa Carden Janice McNeillie Robert Timms

Programme Manager/Programmleitung
Elizabeth White

Marketing
Vanessa Markey Rachel Miller

Systems/Datenbanksystem
Omar Raman Simon Thompson

Administration/Verwaltung
Amanda Brindley

Production/Herstellung
Michelle Draycott Maureen James Nigel Marsh Joanne Tinson

Compilers/Verfasser
Dr Jan C. Bongaerts Amy R. Newland

Terminologists/Terminologie
Michael Hippler Jörn Sickerling

Lexicographers & Proofreaders/Lexikographie & Korrekturlesen
Stephen Curtis Erica Fawcus-Gibbs Anke Kornmüller Lisa Williams Jill Williams Helen Wyler

Keyboarders/Datenerfassung
Emmanuelle Bels Sara Fenby Antonio Fernández Entrena Rosa Gálvez López
Christiane Grosskopf Michael Jopling Ute Krebs Ilona Lehmann Geir Moulson
Fabienne Rangeard Beate Schmitt Debbie Thomas Robert Timms

Acknowledgements

We are particularly grateful to Dr Jan C. Bongaerts and Amy R. Newland for providing us with the German subject area names.

Danksagung

Wir bedanken uns besonders bei Dr Jan C. Bongaerts und Amy R. Newland, die die deutschen Fachgebietsnamen auswählten.

Contents/Inhalt

Foreword	ix–x	Vorwort	ix–x
Introduction	xi–xii	Einleitung	xi–xii
Features of the dictionary	xiii–xv	Aufbau und Anordnung der Einträge	xiii–xv
Using the dictionary	xvii–xviii	Hinweise für die Benutzung des Wörterbuches	xix–xxi
Abbreviations used in this dictionary	xxiii–xxiv	Im Wörterbuch verwendete Abkürzungen	xxiii–xxiv
GERMAN–ENGLISH DICTIONARY	1–140	WÖRTERBUCH DEUTSCH–ENGLISCH	1–140
ENGLISH–GERMAN DICTIONARY	141–272	WÖRTERBUCH ENGLISCH–DEUTSCH	141–272

Foreword/Vorwort

Environmental technology is as much a part of everyday economic life as process engineering, logistics or sales. All branches of the economy are affected by environmental regulations which apply not only to production processes, but also, and to an increasing extent, to products themselves.

There now exists a body of environmental legislation that lays down where and how environmental technology should be applied in the form of post-production techniques – otherwise known as end-of-pipe technology – for the collection and conversion of harmful substances within, as far as possible, closed systems.

Experience has taught us that post-production techniques merely transferred environmental problems elsewhere (from the air, via the water to the soil, say) and were also expensive in terms of energy and raw materials. As a result, they were gradually supplemented, and we moved towards the age of the reuse or recycling of raw materials and the recovery of energy.

An integrated environmental technology followed, whereby it was believed that a more appropriate choice of materials, together with the optimization of energy use, could avoid pollution and provide cheaper solutions to environmental problems. The emphasis was now on developing products that would have as little impact as possible on the environment throughout the whole of their lifetime or life cycle.

The need to work out the process for developing such a product has brought together product developers, process and systems engineers, logisticians and waste-management and recycling experts and is generally referred to as life-cycle analysis, the overall view or the holistic approach. In the long term, such methods are intended to uncouple economic growth from increased pollution. The model for long-term, sustainable future development established by the United Nations Conference on the Environment and Development held in Rio de Janeiro in 1992, is that economic activity will not

Die Umwelttechnologie gehört zum Wirtschaftsalltag wie die Verfahrenstechnik, die Logistik und der Vertrieb. Alle Wirtschaftszweige sind von Umweltvorschriften betroffen, die nicht nur für die Produktionsverfahren, sondern auch - und in zunehmendem Umfang - für die Erzeugnisse gelten.

Eine Umweltgesetzgebung ist entstanden, die den Einsatz von Umwelttechnologie vorschreibt, anfänglich in Form von nachgeschalteten Techniken – auch als 'End-of-pipe-Technik' bezeichnet – zur Einsammlung und Umwandlung von Schadstoffen in möglichst geschlossenen Verfahren. Die Erfahrung lehrte jedoch, daß nachgeschaltete Techniken Umweltprobleme lediglich verlagern (etwa von der Luft über Gewässer zum Boden hin) und wegen des Rohstoff- und Energiebedarfs teuer sind. Sie wurden deshalb nach und nach ergänzt und das Zeitalter der Wiederverwendung bzw. -verwertung von Rohstoffen und der Rückgewinnung von Energie nahm seinen Anfang.

Eine Weiterentwicklung dieses umwelttechnischen Gedankenguts führte zur integrierten Umwelttechnologie. Aus dem Grundsatz, daß die Vermeidung von Umweltbelastungen durch geeignete Werkstoffauswahl und durch Optimierung des Energieverbrauchs alternative, kostengünstigere Lösungsansätze für Umweltprobleme liefern kann, entstand die Vorgabe, daß ein Erzeugnis während seiner gesamten Lebensdauer (Lebenszyklus) möglichst geringe Umweltauswirkungen verursachen sollte. Das dazu passende Verfahren, das Produktentwickler, Verfahrenstechniker, Logistiker, Abfallbehandlungs- und Abfallverwertungsexperten und Systemtechniker zusammenführt, wird mit Begriffen wie Lebenszyklusanalyse bzw. Gesamtbetrachtung und holistische Vorgehensweise bezeichnet.

Langfristig wird eine sogenannte Entkopplung des Wirtschaftswachstums und der Umweltbelastungen beabsichtigt: Wirtschaftswachstum und Umweltbeeinträchtigungen verlaufen nicht parallel, sondern gehen getrennte Wege. Im Endeffekt ver-

cause environmental problems but will, instead contribute to their solution.

The compilation of a dictionary in a subject area which is as continuously and rapidly developing as environmental technology presents an interesting challenge. As a starting point, the scientific – that is the physical, biological, chemical and ecosystemic – foundations for the development of environmental technology need to be covered. Then, the broad range of technical terms associated with end-of-pipe technologies and techniques for recycling and reutilization needs to be included. Another important aspect to be taken into account is the measurement and evaluation of the quality of the environment, whether in the air, water or soil. These have led to the development of extremely sophisticated forms of environmental analysis and, in turn, have given rise to concepts associated with examining the effectiveness of depollution measures.

With the integration of environmental protection into the overall process of product and production planning and manufacturing, it is very difficult to draw a line between environmental technology and technology in general. The environmental component can no longer be isolated from the whole and, consequently, circumscribed within a subject-specific terminology. It has required a very conscious effort to keep coverage targeted to the specific field of environmental technology whilst also including technical phrases to illustrate these terms in the wider context in which they operate.

A dictionary is not simply the work of its compilers, it is a collaborative venture and many people contribute to its success. Among these must be mentioned the editors at Routledge, who oversaw its progress, and the consultants who checked each entry critically in the light of their technical knowledge. This is the place to thank them for their efforts and their commitment and to hope that users can now look forward to finding their work just that little bit easier.

Translation by Stephen Curtis

ursacht das Wirtschaften keine Umweltprobleme, sondern trägt zu deren Lösung bei. Diese Vorstellung entspricht dem Leitbild der nachhaltigen Entwicklung, das die Konferenz der Vereinten Nationen über Umwelt und Entwicklung 1992 in Rio de Janeiro geprägt hat.

Die Zusammenstellung eines Wörterbuches in einem Gebiet, das sich so kontinuierlich und rapide weiterentwickelt wie die Umwelttechnologie, stellt eine interessante Herausforderung dar. Es kommt zunächst darauf an, die physikalischen, biologischen, chemischen und ökosystemischen wissenschaftlichen Grundlagen, welche die Entwicklung der Umwelttechnologie kennzeichnen, abzudecken. Sodann soll ein möglichst vollständiger Überblick über die Fachausdrücke nachgeschalteter Umwelt, Verwertungs- und Recyclingtechniken enthalten sein. Ein weiterer wichtiger Aspekt, der in Betracht gezogen werden muß, ist die Messung und Bewertung der Umweltqualität, ob Luft, Gewässer oder Boden. Diese haben zur Entwicklung einer äußerst anspruchsvollen Umweltanalytik geführt, welche wiederum Begriffsbildungen im Zusammenhang mit der Überprüfung der erreichten Umweltentlastungen mit sich bringt.

Bei der Betrachtung der Integration der Umweltvorsorge in die gesamte Produkt- und Herstellungsplanung und -durchführung wird die Abgrenzung von der Technik äußerst schwierig. Die umwelttechnische Komponente kann nicht mehr vom Herstellungsverfahren abgetrennt und folgerichtig nicht mehr durch fachspezifische Begriffsbildungen umschrieben werden. Wir haben uns bewußt darum bemüht, den Wortschatz auf das Gebiet der Umwelttechnologie einzugrenzen, jedoch gleichzeitig auch technische Redewendungen und Ausdrucksweisen aufzunehmen, die verdeutlichen, wie diese Fachbegriffe in der Praxis Anwendung finden.

Ein Wörterbuch entsteht kaum aus der alleinigen Arbeit der Autoren, sondern aus dem Zusammenwirken vieler, die sich am Gelingen beteiligen. Dazu gehören die Lektoren von Routledge, die den Fortgang begleiteten und die Gutachter, die jede Eintragung kritisch und mit Sachverstand überprüften. Ihnen sei an dieser Stelle für ihre Einsatzbereitschaft gedankt, und den Benutzern sei eine Erleichterung ihrer Arbeit in Aussicht gestellt.

Dr Jan C. Bongaerts & Amy R. Newland, Bonn und Leiden

Introduction/Einleitung

Following on from the programme of general technical and business dictionaries launched in October 1994, the *German Dictionary of Environmental Technology* is the latest in the series of German specialist dictionaries to be published by Routledge. Given the growing importance of environmental technology in everyday life and the fact that English and German as a language combination has become increasingly important over recent years, this dictionary will be a valuable resource for specialist translators, terminologists and experts working in the field. Approximately 25,000 entries in each language cover terminology used in areas such as Air Pollution & Air Quality Control, Soil Contamination & Remediation and Renewable Energy Sources.

It would not have been possible to compile this dictionary within a realistic timescale, and to the standard achieved, without the use of a custom-designed database. The database used is a relational one: term records for each language are held in separate files, with further files consisting only of link records. Links between terms in different language files represent translations which enable us to handle various types of one-to-many and many-to-one translation equivalences. Links between terms within a single-language file represent cross-references, themselves of a wide variety of types: spelling variants, geographical variants and abbreviations.

The content of the database for this dictionary was compiled and vetted by English- and German-speaking terminologists with current practical experience of environmental technology, in order to ensure its currency, the accuracy of explanations, and the adequacy of coverage. Finally, each language file was reviewed by editors to ensure coverage of North American terms and spelling variants; these are clearly labelled and distinguished.

The compilation and editing of the database of terms was, however, only the first stage in the

Das vorliegende *Wörterbuch Umwelttechnologie Deutsch–Englisch/Englisch–Deutsch* ist das neueste Werk in einer zweisprachigen Fachwörterbuchreihe von Routledge, die sich den seit Oktober 1994 erscheinenden Wörterbuchreihen Technik sowie Wirtschaft, Handel und Finanzen anschließt. Umwelttechnologie spielt eine wachsende Rolle im Alltagsleben, und gleichzeitig hat die Sprachkombination Englisch–Deutsch in den letzten Jahren an Bedeutung gewonnen. Auf diesem Hintergrund ist das vorliegende Wörterbuch ein wertvolles Mittel für alle Fachübersetzer und Fachübersetzerinnen, Terminologen und Terminologinnen sowie Fachleute in diesem Bereich. Etwa 25 000 Einträge in jeder Sprache decken Gebiete wie zum Beispiel Luftverschmutzung & Luftreinhaltung, Bodenkontaminierung & Bodensanierung und erneubare Energiequellen.

Es wäre ohne den Einsatz einer für unsere Bedürfnisse speziell zugeschnittenen Datenbank unmöglich gewesen, dieses Wörterbuch innerhalb einer realistischen Zeitspanne und in der vorliegenden Qualität zusammenzustellen. Die Datenbank ist relational: Die Datensätze zu den einzelnen Stichwörtern sind für jede Sprache in einer separaten Datei untergebracht, während gesonderte Dateien nur Datensätze enthalten, die die Verbindungen zwischen Ausgangs- und Zielsprache herstellen. Diese Verbindungen zwischen den Sprachen stellen die Übersetzungen dar, sie schaffen die Möglichkeit, unterschiedliche Verknüpfungen herzustellen: von einem Ausgangswort zu mehreren Übersetzungen oder auch von mehreren Ausgangswörtern zu einer gemeinsamen Übersetzung. Innerhalb der einsprachigen Dateien lassen sich ebenfalls Verknüpfungen schaffen, nämlich Querverweise verschiedener Art wie auf abweichende Schreibweisen, geographische Varianten und Abkürzungen.

Der Inhalt der Datenbank für dieses Wörterbuch wurde von einer Gruppe von deutsch- und englischsprachigen Terminologen mit praktischer Erfahrung auf dem Gebiet Umwelttechnologie

making of the dictionary. Within the database, the distinction between source and target languages is not meaningful, but for this printed dictionary it was necessary to format the data to produce separate German–English and English–German sections. The data was processed by a further software module to produce two alphabetic sequences, of German headwords with English translations and vice versa, each displaying the nesting of compounds, ordering of translations, style for cross-references of different types, and other features according to the algorithm.

At this stage the formatted text was edited by a team of experienced English and German lexicographers whose task it was to eliminate duplication or inconsistency; edit the contextual information; and to remove terms that were too general for inclusion in a specialist dictionary.

The editorial team

zusammengestellt und überprüft. So wurde sichergestellt, daß die Wortliste höchste Aktualität hat, daß die Worterklärungen präzise sind und daß wirklich das in Fachkreisen geläufigste Vokabular behandelt wird. Schließlich wurde noch jede Sprachdatei von Redakteuren überprüft, um eine internationale Terminologieabdeckung zu gewährleisten. Varianten in britischem und nordamerikanischem Englisch werden unterschieden und sind entsprechend gekennzeichnet.

Die Erstellung und Überprüfung der Datenbankgrundlage war jedoch nur der erste Schritt bei der Herstellung dieses Wörterbuches. Innerhalb einer Datenbank erübrigt sich im Grunde die Unterscheidung zwischen Ausgangs- und Zielsprache, aber für das vorliegende Wörterbuch mußten natürlich die Daten so bearbeitet werden, daß separate Abschnitte für Deutsch–Englisch und Englisch–Deutsch entstehen konnten. Mit Hilfe eines weiteren Softwaremoduls wurden zwei alphabetische Listen erstellt, eine deutsche Stichwortliste mit englischen Übersetzungen und umgekehrt. Dabei wurden mit Hilfe eines Algorithmus zusammengesetzte Begriffe in Blöcken aufgeführt, Übersetzungen in sinnvoller Reihenfolge angeordnet und verschiedene Arten von Querverweisen hergestellt.

Dann wurde der formatierte Text von einem Team erfahrener deutscher und englischer Lexikographen redigiert; ihre Aufgabe war es, Doppelnennungen und Unstimmigkeiten auszumerzen, die zum Wortzusammenhang gegebenen Informationen zu bearbeiten und all die Stichwörter zu streichen, die zu allgemeinsprachlich für ein Fachwörterbuch sind.

Das Redaktionsteam

Features of the dictionary/ Aufbau und Anordnung der Einträge

The following text extracts illustrate the principal features of the dictionary. For a more detailed explanation of each of these features and a full guide to using the dictionary, see pages xvii–xviii.

Die folgenden Textbeispiele illustrieren Aufbau und Anordnung der Einträge. Weitere Erläuterungen und Hinweise zur Benutzung des Wörterbuches befinden sich auf den Seiten xix–xxi.

Brennkammer *f* ABFALL, ABWASSER, BODEN, LUFT combustion chamber; **Brennkammeraustritt** *m* LUFT combustion chamber outlet; **Brennkammerendtemperatur** *f* LUFT combustion chamber final temperature

Brenn: **Brennlinie** *f* ERNEU ENER focal line

Subject area labels given in alphabetical order show appropriate translation — **Brennstoff** *m* ABFALL, ERNEU ENER fuel, SICHERHEIT combustible material; **Brennstoff aus Biomasse** ERNEU ENER fuel from biomass; **Brennstoffe** *m pl* ABFALL, ABWASSER, LUFT, SICHERHEIT combustibles; **~ aus Müll** (*BRAM*) WASSER refuse-derived fuel (*RDF*); — *Sachgebietskürzel in alphabetischer Reihenfolge helfen beim Finden der korrekten Übersetzung*

Genders are indicated at German nouns — **Brennstoffmangel** *m* LUFT lack of fuel; **Brennstoffschicht** *f* ABFALL fuel layer; **Brennstoffüberschuß** *m* LUFT excess fuel; **Brennstoffzelle** *f* ERNEU ENER, LUFT fuel cell; **Brennstoffzusammensetzung** *f* ABFALL fuel composition — *Bei deutschen Substantiven wird das grammatikalische Geschlecht angegeben*

Brenn: **Brenntrommel** *f* BODEN firing drum

Brigg: **~scher Logarithmus** *m* LUFT Brigg's logarithm

Brikett *nt* ABFALL *vorherrschende BRAM-Form*, ERNEU ENER *Bioenergienausgangsstoff* briquette

Brikettierpresse *f* ABFALL briquetting press

German dummy entries in bold italics indicate the continuation of an earlier nest of compounds — ***Brikett***: **Brikettpreßstempel** *m* ABFALL briquette press piston — *Deutscher Blindeintrag in Fettkursivschrift zeigt die Fortsetzung eines vorhergehenden Nests mit zusammengesetzten Formen an*

Bringbehältersystem *nt* ABFALL, WASSER stationary container system (*AmE*) (*SCS*)

Bringsystem *nt* ABFALL, WASSER collection point system

Britisch: **~es Institut** *nt* **für Normung** (*BSI*) ANALYTIK British Standards Institute (*BSI*)

BROM *abbr* (*BRAM aus organischer Masse*) ABFALL RDF from organic mass

Base form of adjective is replaced by a swung dash in adjective–noun combinations — **bromidhaltig** *adj* ABFALL, WASSER bromide-containing; **~es Wasser** *nt* WASSER bromide-containing water — *Die Grundform des Adjektivs wird in flektierten Formen durch die Tilde ersetzt*

Brown: **~sche Molekularbewegung** *f* ABWASSER, LUFT *stochastische Bewegung von Partikeln* Brownian movement

Bruch: **Bruchbelastung** *f* ABFALL, WASSER breaking load;

Contexts give supplementary information to help locate the right — **Bruchlast** *f* ABFALL *Chlor*, WASSER *Stickstoffoxid* breaking load; **Bruchzone** *f* **der Kontinentalplatten** ERNEU ENER *Stelle des geothermischen Energiepotentials* fracture zone of continental plates — *Die Angabe von Zusammenhängen ergänzt die gegebenen Informationen und unterstützt die Suche nach dem passenden Übersetzungsäquivalent*

Brückenmolekül *nt* ABWASSER *Flockulation* bridging molecule

Brüden *m pl* ABWASSER exhaust vapors (*AmE*), exhaust vapours (*BrE*); **Brüdenbehandlung** *f* ABFALL, WASSER

British English and American English spelling variants are given in full and labelled accordingly — vapor treatment (*AmE*), vapour treatment (*BrE*); **Brüdenkompressor** *m* LUFT vapor compressor (*AmE*), vapour compressor (*BrE*); **Brüdenkondensator** *m* ABWASSER vent condenser — *Britische und amerikanische Schreibvarianten werden voll ausgeschrieben und entsprechend gekennzeichnet*

Brunnen *m* BODEN, WASSER well; **Brunnenabsenkung** *f* WASSER well sinking; **Brunnenausbau** *m* WASSER well construction; **Brunnenbetrieb** *m* WASSER well operation; **Brunnenentwicklung** *f* WASSER well development; **Brunnenergiebigkeit** *f* WASSER well yield;

German terms are ordered in strict letter-by-letter order — **Brunnenfassung** *f* WASSER well catchment; **Brunnenfilter** *nt* WASSER well filter; **Brunnenregeneration** *f* WASSER well regeneration; **Brunnenrohr** *nt* WASSER well pipe; **Brunnenschacht** *m* WASSER well shaft; **Brunnenschaft** *m* WASSER well shaft; **Brunnenwasser** *nt* WASSER well water — *Deutsche Einträge sind streng alphabetisch geordnet*

Cross-references from abbreviations to their full forms are shown for both the German and the English translation — **BSB** *abbr* ABFALL, ABWASSER, ANALYTIK, WASSER (*biochemischer Sauerstoffbedarf*) BOD (*biochemical oxygen demand*), (*biologischer Sauerstoffbedarf*) BOD (*biological oxygen demand*) — *Sowohl für den deutschen Eintrag wie für die englische Übersetzung werden Querverweise von Abkürzungen auf ihre Vollformen gegeben*

Sachgebietskürzel in alphabetischer Reihenfolge helfen beim Finden der korrekten Übersetzung

Subject area labels given in alphabetical order show appropriate translation

atomize *vt* AIR zerstäuben, SOIL versprühen, WASTE, WATER SUPP verdüsen

atomizer *n* AIR Einspritzdüse *f*, Zerstäuber *m*, SOIL Versprüher *m*, WATER SUPP Verdüser *m*; **~ nozzle** *n* AIR Zerstäuberluftdüse *f*

atomspectrometric *adj* ANALYSIS atomspektrometrisch

Sowohl für den deutschen wie den englischen Eintrag werden Querverweise auf Abkürzungen gegeben

Cross-references to abbreviations are shown for both the English and the German terms

ATP *abbr* (*adenosine triphosphate*) WATER POLL ATP (*Adenosintriphosphat*)

attached: **~ shell** *n* NOISE *surface lining* Vorsatzschale *f*

attemperation *n* AIR Temperaturregelung *f*

attenuate *vt* AIR, ANALYSIS schwächen, NOISE dämpfen, schwächen

attenuation *n* AIR, ANALYSIS Schwächung *f*, NOISE Dämpfung *f*, Schwächung *f*

Die Angabe von Zusammenhängen ergänzt die gegebenen Informationen und unterstützt die Suche nach dem passenden Übersetzungsäquivalent

Contexts give supplementary information to help locate the right translation

attenuator *n* NOISE *silencer* Dämpfungsregler *m*, Lautstärkeregler *m*; **~ pad** *n* NOISE Dämpfungsglied *nt*

attraction *n* AIR, WATER POLL Anziehung *f*

attrition *n* AIR Abrieb *m*; **~ resistance** *n* AIR Abriebfestigkeit *f*

auctioning *n* POLICY *of tradeable emissions rights* Versteigerung *f*

Zusammengesetzte Begriffe werden alphabetisch hinter dem ersten Element angeordnet

Compound forms are nested alphabetically at the first element

audibility *n* NOISE Schalleistung *f*; **~ level** *n* ANALYSIS Schallleistungspegel *m*; **~ reference value** *n* ANALYSIS Bezugsschalleistung *f*

audible *adj* NOISE hörbar; **~ sound** *n* NOISE hörbarer Schall *m*, Hörschall *m*

audiometer *n* ANALYSIS Audiometer *nt*

auger *n* ANALYSIS, SOIL *bioremediation* Bodenbohrer *m*

Auger: **~ electron spectroscopy** *n* (*AES*) ANALYSIS Auger-Elektronspektroskopie *f* (*AES*)

Bei deutschen Substantiven wird das grammatikalische Geschlecht angegeben

Genders are indicated at German noun translations

authority *n* POLICY Behörde *f*; **~ decision** *n* POLICY Behördenentscheidung *f*

authorization *n* POLICY Erlaubnis *f*, Genehmigung *f*, Zulassung *f*; **~ application** *n* POLICY Genehmigungsantrag *m*; **~ impact** *n* POLICY Genehmigungswirkung *f*; **~ process** *n* POLICY Genehmigungsverfahren *nt*

autochthonous: **~ matter** *n* SOIL eigenständige Substanz *f*

automate *vt* GEN automatisieren

Englische Einträge sind streng alphabetisch geordnet

English terms are ordered in strict letter-by-letter order

automated *adj* GEN automatisiert

automatic *adj* GEN automatisch

automatically: **~ injected** *adj* ANALYSIS automatisch eingespritzt

Englischer Blindeintrag in Fettkursivschrift zeigt die Fortsetzung eines vorhergehenden Nests mit zusammengesetzten Formen

English dummy entries in bold italics indicate the continuation of an earlier nest of compounds

automatic: **~ detector tube** *n* ANALYSIS automatisches Prüf-röhrchen *nt*; **~ injection blow molder** *AmE*, **~ injection blow moulder** *BrE* *n* WASTE Spritzblasautomat *m*; **~ recognition unit** *n* WASTE automatische Erkennungseinrichtung *f*; **~ sampling** *n* ANALYSIS automatische Probenahme *f*; **~ sorting** *n* WASTE automatische Sortierung *f*, automatische Trennung *f*, mechanische Sortierung *f*, mechanische Trennung *f*

Britische und amerikanische Schreibvarianten werden voll ausgeschrieben und entsprechend gekennzeichnet

British English and American English spelling variants are given in full and are labelled accordingly

automobile: **~ battery** *n* *AmE* (*cf car battery BrE*) WASTE Autobatterie *f*; **~ tax** *n* *AmE* (*cf road tax BrE*) POLICY Kfz-Steuer *f*

autosampler *n* ANALYSIS Autosampler *m*, automatischer Probengeber *m*

autosampling *n* ANALYSIS Autosampling *nt*

autothermic *adj* AIR autotherm

autotrophic *adj* WATER POLL *bacteria* autotroph; **~ bacteria** *n pl* WATER POLL autotrophe Bakterien *f pl*; **~ denitrification** *n* WATER POLL autotrophe Denitrifikation *f*

Using the dictionary

Range of coverage

This is a specialist German–English/English–German dictionary that covers the whole range of environmental technology and the scientific knowledge that underlies it. It contains a broad base of terminology drawn from fundamental areas of technology such as Waste Treatment & Management, Water Supply & Drinking Water and Air Pollution & Air Quality Control.

Selection of terms

We have aimed to include the essential vocabulary of each specialist area, and the material has been checked by leading subject experts to ensure that both the English and the German terms are accurate and current, that the translations are valid equivalents, and that there are no gaps in coverage.

Placement of terms

All terms are ordered alphabetically. This is also the policy for hyphenated compounds.

Stoplists

All terms are entered at their first element except where that element is an article, preposition, conjunction, pronoun or other delexicalized word. In such cases, the term is entered at its next valid element.

Compound terms are listed at their first element. When this first element is itself a headword with a technical sense of its own, compound forms follow the simple form. In the case of noun nests, open forms are listed first with the headword replaced by a swung dash.

If the first element is not itself translated, a colon precedes the compounds. For example:

Emscher: ~ **tank** *n* WATER POLL Emscherbrunnen *m*

German compounds are spelled out in full and directly follow the nest of open forms. For example:

Becken *nt* ABFALL, ABWASSER, ANALYTIK, WASSER basin; ~ **für Rückstände** ABWASSER tailings pond; **Beckenauskleidung** *f* ABFALL, ABWASSER, WASSER reservoir lining

Compounds are entered in alphabetical sequence. When a nest is interrupted by other entries, the run of compounds is picked up again later in the correct alphabetical sequence. A dummy headword in bold italics indicates the continuation of a run of compounds. For example:

Partikel *f* ALLG particle; **Partikelabscheidung** *f* LUFT particle separation; **Partikelaufladung** *f* LUFT particle charging

partikelbeladen *adj* LUFT particulate-loaded; **~es Abgas** *nt* LUFT particulate-loaded waste gas; **~er Gasstrom** *m* LUFT particulate-loaded gas stream

Partikel: **Partikelbenetzung** *f* LUFT particle wetting; **Partikeldichte** *f* LUFT particle thickness

Adjective–noun combinations are entered at the base form of the adjective. In the nest, the base form is replaced by a swung dash and followed by the relevant inflection. For example:

belüftet *adj* ABFALL, ABWASSER, WASSER aerated; **~er Abwasserteich** *m* ABWASSER aerated oxidation pond; **~es Becken** *nt* ABWASSER aerated lagoon; **~e Kompostmiete** *f* ABWASSER aerated compost pile

Within nests, articles, pronouns, prepositions, conjunctions, and in German some adjectives and adverbs (e.g. *ober, etwa, nur*), are ignored in determining the sequence of nested open forms. For example:

Sammlung *f* ALLG collection; ~ **durch beauftragte Dritte** ABFALL collection by contracted third party; ~ **an zentralen Orten** ABFALL collection at central sites; **Sammlungszeitraum** *m* ANALYTIK collection period

enthalpy *n* AIR, ANALYSIS Enthalpie *f*; ~ **of fusion** *n* AIR Schmelzenthalpie *f*; ~ **of vaporization** *n* AIR Verdampfungsenthalpie *f*

Homographs

In general, terms are accompanied by a label indicating their part of speech. (For a complete list of these labels and their expansions, please see pages xxvii–xxviii.) When terms beginning with the same element fall into two or more part-of-speech categories, the different categories are distinguished by numbers. The sequence is adjective, adverb, noun and verb, followed by less frequent parts of speech. For example:

capillary 1. *adj* ANALYSIS kapillar; **2.** *n* ANALYSIS Kapillare *f*

Ordering of translations

Every term is accompanied by one or more labels indicating the technological area in which it is used. (For a complete list of these labels and their expansions, please see pages xxvi–xxvii.)

Where the same term is used in more than one technological area, multiple labels are given as appropriate. These labels appear in alphabetical order. Where a term has the same translation in more than one technological area, this translation is given after the sequence of labels. For example:

emitted *adj* AIR, NOISE, WATER SUPP emittiert

When a term has different translations according to the technological area in which it is used, the appropriate translation is given after each label or set of labels. For example:

scoop *n* ANALYSIS Schöpfgerat *nt*, MARINE Behälter *m*, Container *m*, Frachtbehälter *m*, WASTE Schrappergefäß *nt*

Supplementary information

In many cases additional data is given about a term in order to show how it is used. Such contextual information can be:

(a) the typical subject or object of a verb, for example:

scavenge *vt* AIR *gases* ausspülen

(b) typical nouns used with an adjective, for example:

drahtbeheizt *adj* ANALYTIK *Röhrenofen* wire-heated

(c) words indicating the reference of a noun, for example:

Dacron *nt* LUFT *Filtermaterial* dacron

(d) information which supplements the subject area label, for example:

saisonal: **~er Wärmespeicher** *m* ERNEU ENER *Solarenergie* seasonal heat storage

(e) a paraphrase or broad equivalent, for example:

sweeper *n* WASTE *machine* Kehrmaschine *f*

When various different translations apply in the same subject area, contextual information is also used to show which translation is appropriate in different circumstances. For example:

Niederschlag *m* BODEN, LUFT precipitate, *Regen, Schnee* precipitation, *radioaktiv* fallout

Cross-references

Geographical variants, both spelling and lexical, are given in full in the German–English section when they are translations:

Muldenkippwagen *m* MEER skip lorry (*BrE*), skip truck (*AmE*)

In the English–German section both British English and North American English terms are covered, and these are differentiated by regional labels. In the case of lexical variants, full information – including translations and cross-references to the other form – is given at each entry. For example:

muffler *n AmE* (*cf silencer BrE*) NOISE Dämpfer *m*, Schalldämpfer *m*

silencer *n BrE* (*cf muffler AmE*) NOISE Dämpfer *m*, Schalldämpfer *m*

Both abbreviations and their full forms are entered in the main body of the dictionary. Full information – including translations and cross-references to the full form or abbreviation – is given.

Where an abbreviation in the source language translates to an abbreviation in the target language, the full form in each case is given after the abbreviation in brackets and italics. For example:

MAC *abbr* (*maximum allowable concentration*) AIR, ANALYSIS, SAFETY, WATER SUPP HZK (*höchstzulässige Konzentration*)

If a source language abbreviation does not translate to a target language abbreviation, the translation of the abbreviation is in roman. For example:

SOFC *abbr* (*solid oxide fuel cell*) AIR Oxidkeramik-Brennstoffzelle *f*

Hinweise für die Benutzung des Wörterbuches

Umfang des Wörterbuches

Sie haben ein Fachwörterbuch Deutsch–Englisch/Englisch–Deutsch vor sich, das alle Bereiche Umwelttechnologie und des zugrundeliegenden technisch-naturwissenschaftlichen Wissens abdeckt. Dieses Wörterbuch enthält einer breiten Grundlage technischer Ausdrücke aus grundsätzlichen Bereichen wie etwa Abfallbehandlung, Wasserwirtschaft & Trinkwasserbereitsellung und Luftverschmutzung & Luftreinhaltung.

Auswahl der Stichwörter

Wir haben uns bemüht, aus jedem Gebiet das wesentliche Fachvokabular aufzunehmen; hierzu wurde die Stichwortliste sowohl im Englischen wie im Deutschen von führenden Experten in den einzelnen Fachgebieten auf Genauigkeit, Richtigkeit und Vollständigkeit überprüft.

Reihenfolge der Anordnung

Alle Einträge sind alphabetisch angeordnet. Zusammengesetzte Einträge erscheinen unter ihrem Basiswort. Dies gilt auch für Komposita mit Bindestrich.

Stopplisten

Alle Stichwörter sind nach ihrem ersten Element, dem Basiswort, angeordnet, es sei denn, dieses Element ist ein Artikel, eine Präposition, eine Konjunktion, ein Pronomen oder ein anderes ausgegliedertes Wort. In solchen Fällen erscheint das Stichwort unter seinem ersten gültigen Element.

Zusammengesetzte Einträge sind in Nestform unter ihrem Basiswort aufgeführt. Wenn das Basiswort ein eigenes Stichwort mit einer technischen Bedeutung darstellt, folgen alle weiteren Zusammensetzungen in alphabetischer Reihenfolge. Hier ersetzt die Tilde das Stichwort. Zusammengesetzten Einträgen mit nicht übersetztem Basiswort geht ein Doppelpunkt voraus. Zum Beispiel:

Emscher: **~ tank** *n* WATER POLL Emscherbrunnen *m*

Deutsche Komposita werden voll ausgeschrieben und folgen auf Einträge mit mehreren Wörtern. Zum Beispiel:

Becken *nt* ABFALL, ABWASSER, ANALYTIK, WASSER basin; **~ für Rückstände** ABWASSER tailings pond; **Beckenauskleidung** *f* ABFALL, ABWASSER, WASSER reservoir lining

Deutsche Komposita sowie englische zusammengesetzte Formen sind in alphabetischer Reihenfolge angeordnet. Nester, die aufgrund anderer Stichwörter unterbrochen werden müssen, werden entsprechend alphabetisch wieder aufgenommen. Das Basiswort als Blindeintrag in Fettkursivdruck weist auf die Fortführung des Artikels hin. Zum Beispiel:

Partikel *f* ALLG particle; **Partikelabscheidung** *f* LUFT particle separation; **Partikelaufladung** *f* LUFT particle charging
partikelbeladen *adj* LUFT particulate-loaded; **~es Abgas** *nt* LUFT particulate-loaded waste gas; **~er Gasstrom** *m* LUFT particulate-loaded gas stream
Partikel: **Partikelbenetzung** *f* LUFT particle wetting; **Partikeldichte** *f* LUFT particle thickness

Zusammensetzungen aus Adjektiv und Substantiv sind unter der Grundform des Adjektivs aufgeführt. Im sich anschließenden Artikel wird die Grundform des Adjektives durch die Tilde ersetzt. Die entsprechende Adjektivendung schließt sich der Tilde an. Zum Beispiel:

belüftet *adj* ABFALL, ABWASSER, WASSER aerated; **~er Abwasserteich** *m* ABWASSER aerated oxidation pond; **~es Becken** *nt* ABWASSER aerated lagoon; **~e Kompostmiete** *f* ABWASSER aerated compost pile

Innerhalb eines Nests werden Artikel, Pronomen, Präpositionen, Konjunktionen und im Deutschen auch manche Adjektive und Adverbien (z.B. *ober, etwa, nur*) bei der Reihenfolge der eingenesteten

Stichwörter und Phrasen ignoriert. Zum Beispiel:

Sammlung *f* ALLG collection; ~ **durch beauftragte Dritte** ABFALL collection by contracted third party; ~ **an zentralen Orten** ABFALL collection at central sites; **Sammlung-szeitraum** *m* ANALYTIK collection period

enthalpy *n* AIR, ANALYSIS Enthalpie *f*; ~ **of fusion** *n* AIR Schmelzenthalpie *f*; ~ **of vaporization** *n* AIR Verdampfungsenthalpie *f*

Homographen

Im allgemeinen sind Einträge mit einem Label versehen, das die Wortklassenzugehörigkeit angibt. (Eine vollständige Liste dieser Labels befindet sich auf Seiten xxvi–xxvii.) Einträge mit gemeinsamem Basiswort jedoch unterschiedlicher Wortklassenzugehörigkeit erhalten für jede Wortklasse eine eigene Numerierung. Die Einträge sind in der Reihenfolge Adjektiv, Adverb, Substantiv, Verb aufgeführt. Dem schließen sich weniger häufig vorkommende grammatikalische Einheiten an. Zum Beispiel:

capillary 1. *adj* ANALYSIS kapillar; **2.** *n* ANALYSIS Kapillare *f*

Reihenfolge der Übersetzungen

Auf jedes Stichwort der Ausgangssprache folgen ein oder mehrere Labels, die das Fachgebiet angeben, in dem das Wort benutzt wird. (Eine vollständige Liste dieser Labels mit Erklärungen befindet sich auf den Seiten xxvii–xxviii.)

Wenn dasselbe Stichwort in mehr als einem Fachgebiet benutzt wird, so werden entsprechend mehrere Labels angegeben. Sie werden stets in alphabetischer Reihenfolge aufgeführt. Falls die angegebene Übersetzung in mehr als einem Fachgebiet benutzt wird, so folgt die Übersetzung jeweils im Anschluß an die entsprechenden Labels. Zum Beispiel:

emitted *adj* AIR, NOISE, WATER SUPP emittiert

Hat ein Stichwort verschiedene, von seinen jeweiligen Fachgebieten abhängige Übersetzungen, so ist die zutreffende Übersetzung jeweils nach dem entsprechenden Kürzel zu finden. Zum Beispiel:

scoop *n* ANALYSIS Schöpfgerat *nt*, MARINE Behälter m, Container *m*, Frachtbehälter *m*, WASTE Schrappergefäß *nt*

Zusatzinformationen

Deutsche Substantive sind mit Geschlechtsangabe versehen. In vielen Fällen wird über das Stichwort noch zusätzliche Information gegeben, die über die Benutzung des Wortes Aufschluß gibt. Solche Informationen zum Benutzungszusammenhang können verschiedene Formen annehmen:

(a) bei einem Verb ein typisches Subjekt oder Objekt:

scavenge *vt* AIR *gases* ausspülen

(b) typische Substantive, die mit einem bestimmten Adjektiv verwendet werden:

drahtbeheizt *adj* ANALYTIK *Röhrenofen* wire-heated

(c) Wörter, die für Substantive einen typischen Bezug erläutern:

Dacron *nt* LUFT *Filtermaterial* dacron

(d) Informationen, die das Fachgebiet noch weiter eingrenzen:

saisonal: **~er Wärmespeicher** *m* ERNEU ENER *Solarenergie* seasonal heat storage

(e) Umschreibungen oder ungefähre Äquivalente:

sweeper *n* WASTE *machine* Kehrmaschine *f*

Wenn in einem Fachgebiet verschiedene Übersetzungen für ein Stichwort möglich sind, so soll hier die zusätzliche Information anzeigen, welches im jeweiligen Zusammenhang die korrekte Übersetzung ist. Zum Beispiel:

Niederschlag *m* BODEN, LUFT precipitate, *Regen, Schnee* precipitation, *radioaktiv* fallout

Querverweise

Geographische Varianten der Übersetzungen, sowohl orthographischer als auch lexikalischer Art, werden im deutsch–englischen Teil des Wörterbuches immer angegeben und voll ausgeschrieben. Zum Beispiel:

Muldenkippwagen *m* MEER skip lorry (*BrE*), skip truck (*AmE*)

Im englisch–deutschen Teil des Wörterbuches werden Einträge sowohl in britischem als auch in amerikanischem Englisch aufgeführt und sind entsprechend mit geographischen Kürzeln gekennzeichnet. Bei allen lexikalischen Unterscheidungen werden Übersetzungen und Querverweise für beide Englischvarianten gegeben. Zum Beispiel:

muffler *n AmE* (*cf silencer BrE*) NOISE Dämpfer *m*, Schalldämpfer *m*
silencer *n BrE* (*cf muffler AmE*) NOISE Dämpfer *m*, Schalldämpfer *m*

Im Hauptteil des Wörterbuches sind auch Abkürzungen und ihre ausgeschriebenen Formen angeordnet. Bei jedem Eintrag wird die vollständige Information – inklusive Übersetzungen und Querverweisen zur Vollform respektive Abkürzung in Kursivschrift – aufgeführt. Zum Beispiel:

MAC *abbr* (*maximum allowable concentration*) AIR, ANALYSIS, SAFETY, WATER SUPP HZK (*höchstzulässige Konzentration*)

Handelt es sich um Abkürzungen, die in der jeweiligen Zielsprache kein direktes Äquivalent besitzen, bleibt der Übersetzungsvorschlag in Magerschrift. Zum Beispiel:

SOFC *abbr* (*solid oxide fuel cell*) AIR Oxidkeramik-Brennstoffzelle *f*

Abbreviations used in this dictionary/ Im Wörterbuch verwendete Abkürzungen

Parts of speech/Wortarten

abbr	abbreviation	Abkürzung
adj	adjective	Adjektiv
adv	adverb	Adverb
f	feminine	Femininum
f pl	feminine plural	Femininum Plural
in cpds	base form of compounds	Grundwort bei Komposita
m	masculine	Maskulinum
m pl	masculine plural	Maskulinum Plural
n	noun	Substantiv
n pl	noun plural	Substantiv Plural
nt	neuter	Neutrum
nt pl	neuter plural	Neutrum Plural
phr	phrase	fachsprachliche Redewendung
pl	plural	Plural
pref	prefix	Präfix
prep	preposition	Präposition
vi	intransitive verb	intransitives Verb
v refl	reflexive verb	reflexives Verb
vt	transitive verb	transitives Verb

Geographic codes/Geographische Kürzel

AmE	American English	amerikanisches Englisch
BrE	British English	britisches Englisch

Subject area labels/Fachgebietskürzel

AIR	Air Pollution & Air Quality Control	Luftverschmutzung & Luftreinhaltung
ANALYSIS	Sampling, Measurement & Analysis	Probenahme, Messung & Bewertung
GEN	General	Allgemein
MANAGE	Management	Management
MARINE	Marine Pollution	Meeresverschmutzung
NOISE	Noise Pollution	Lärmbelästigung
POLICY	Policy & Legal Instruments	Politik & Rechtsinstrumente
RENEW ENERG	Renewable Energy Sources	Erneuerbare Energiequellen
SAFETY	Safety Engineering	Sicherheitstechnik
SOIL	Soil Contamination & Remediation	Bodenkontaminierung & Bodensanierung
WASTE	Waste Treatment & Management	Abfallbehandlung
WATER POLL	Water Pollution & Wastewater Treatment	Gewässerverschmutzung & Abwasserbehandlung

WATER SUPP	Water Supply & Drinking Water	Wasserwirtschaft & Trinkwasserbereitstellung
ALLG	Allgemein	General
ABFALL	Abfallbehandlung	Waste Treatment & Management
ABWASSER	Gewässerverschmutzung & Abwasserbehandlung	Water Pollution & Wastewater Treatment
ANALYTIK	Probenahme, Messung & Bewertung	Sampling, Measurement & Analysis
BODEN	Bodenkontaminierung & Bodensanierung	Soil Contamination & Remediation
ERNEU ENER	Erneuerbare Energiequellen	Renewable Energy Sources
LÄRM	Lärmbelästigung	Noise Pollution
LUFT	Luftverschmutzung & Luftreinhaltung	Air Pollution & Air Quality Control
MANAG	Management	Management
MEER	Meeresverschmutzung	Marine Pollution
POLITIK	Politik & Rechtsinstrumente	Policy & Legal Instruments
SICHERHEIT	Sicherheitstechnik	Safety Engineering
WASSER	Wasserwirtschaft & Trinkwasserbereitstellung	Water Supply & Drinking Water

Trademarks/Warenzeichen

Note that neither the presence nor the absence of the symbol ® should be regarded as affecting the legal status of any trademark.

Bitte beachten Sie, daß weder die Anwesenheit noch die Abwesenheit des Symbols ® den gesetzlichen Status irgendeines Warenzeichens beeinflußt.

German–English
Deutsch–Englisch

A

aaRdT *abbr* (*allgemein anerkannte Regeln der Technik*) POLITIK, WASSER generally accepted technical procedures
AAS *abbr* ANALYTIK (*Atomabsorptionsspektrometer*) AAS (*atomic absorption spectrometer*), (*Atomabsorptionsspektrometrie*) AAS (*atomic absorption spectrometry*)
Abbau *m* ALLG decomposition, degradation, depletion; **~ der Ozonschicht in der Stratosphäre** LUFT depletion of stratospheric ozone layer; **~ von Schadstoffen** ABFALL, ABWASSER, WASSER decomposition of harmful substances; **~ in der unteren Atmosphäre** LUFT degradation in lower atmosphere
abbaubar *adj* ABFALL, ABWASSER, WASSER degradable; **~e organische Substanz** *f* (*AOS*) ABFALL, ABWASSER, WASSER biodegradable matter, degradable organic matter
Abbaubarkeit *f* ABFALL, ABWASSER, WASSER degradability; **Abbaubarkeitstest** *m* ANALYTIK degradability test
abbauen *vt* ALLG decompose, degrade, deplate
Abbau: **Abbaugeschwindigkeit** *f* ABFALL, ABWASSER, WASSER *Kompostierung* rate of decomposition, speed of degradation; **Abbaugrad** *m* ABFALL, ABWASSER, WASSER degree of degradability, degree of degradation; **Abbauleistung** *f* ABWASSER decomposition capacity, degradation performance; **Abbaupotential** *nt* BODEN decomposition potential; **Abbauprozeß** *m* ABFALL, WASSER destruction process; **Abbaustufe** *f* ABFALL stage of decompostion
Abbindezeit *f* ABFALL setting time
abbrechen *vt* ABFALL demolish
Abbrennrate *f* LUFT burn-off rate
Abbruch *m* ABFALL demolition; **Abbruchabfall** *m* ABFALL demolition waste
Abdampf *m* ABFALL waste steam; **Abdampfrückstand** *m* ABFALL, ANALYTIK, ERNEU ENER evaporation residue
Abdeck- *in cpds* ALLG covering; **Abdeckboden** *m* ABFALL, BODEN covering soil
abdecken *vt* ALLG cover, seal
Abdecken *nt* ALLG covering, sealing
Abdeck-: **Abdeckfolie** *f* LÄRM covering sheet; **Abdeckmaterial** *nt* ABFALL, WASSER covering material; **Abdeckrohr** *nt* BODEN *Erdfilter* cover tube; **Abdeckschicht** *f* ABFALL, WASSER covering layer, sealing layer
Abdeckung *f* ALLG covering, sealing; **Abdeckungsschicht** *f* ABFALL, BODEN, LÄRM, WASSER sealing layer, covering layer
abdichten *vt* ABFALL *Deponie*, ABWASSER, BODEN, LÄRM, WASSER seal, line
Abdichtung *f* ABFALL *einer Deponie*, BODEN, LUFT, WASSER sealing, lining; **Abdichtungsmaßnahmen** *f pl* ABFALL, WASSER sealing measures; **Abdichtungswand** *f* BODEN retention wall
Abdrift *f* MEER drift
A-bewertet: **~er Schallpegel** *m* LÄRM A-weighted sound level
abfackeln *vt* ABFALL, ABWASSER burn off, LUFT, WASSER flare
Abfackeln *nt* ABFALL, ABWASSER burning off LUFT, WASSER flaring
Abfackelung *f* ABFALL, ABWASSER, LUFT flaring; **Abfackelungsanlage** *f* ABFALL flaring installation
Abfall *m* ABFALL garbage (*AmE*), junk, refuse, rubbish (*BrE*), waste; **~ aus der Landwirtschaft** ABFALL agricultural waste; **~ der Pharmaindustrie** ABFALL pharmaceutical industry waste; **Abfallabfuhr** *f* ABFALL waste collection; **Abfallabgabe** *f* POLITIK waste charge; **Abfallagerung** *f* ABFALL waste storage; **Abfallanfall** *m* ABFALL waste production
abfallarm: **~es Produktionsverfahren** *nt* ABFALL, LUFT low-waste production method; **~e Technologie** *f* ABFALL, LUFT, WASSER low-waste technology
Abfall: **Abfallaufkommen** *nt* ABFALL, WASSER appearance of waste; **Abfallausgrabung** *f* ABFALL disinterment of waste; **Abfallbehälter** *m* ABFALL refuse can, waste container; **Abfallbehandlung** *f* ABFALL waste management, waste processing, waste treatment, MANAG, POLITIK waste management; **Abfallbehandlungsanlage** *f* ABFALL waste treatment plant; **Abfallbeschickung** *f* ABFALL waste injection
Abfallbeseitigung *f* ABFALL garbage disposal (*AmE*), refuse disposal, waste disposal, waste removal, waste treatment; **Abfallbeseitigungsanlage** *f* ABFALL waste treatment plant; **Abfallbeseitigungssteuer** *f* POLITIK waste disposal tax; **Abfallbeseitigungsunternehmen** *nt* ABFALL, MANAG waste disposal company, waste disposal firm, waste management company, waste management firm
Abfall: **Abfallbewirtschaftung** *f* ABFALL, MANAG, POLITIK waste management; **Abfallbörse** *f* ABFALL Waste Exchange; **Abfalldeponie** *f* ABFALL waste dump; **Abfalldeponierung** *f* ABFALL, WASSER waste landfilling; **Abfalldesinfektion** *f* ABFALL waste disinfection
Abfallentsorgung *f* ABFALL, MANAG, POLITIK refuse disposal, waste disposal; **Abfallentsorgungsanlage** *f* ABFALL, MANAG waste management plant; **Abfallentsorgungsplan** *m* ABFALL, MANAG, POLITIK waste disposal plan; **Abfallentsorgungsverordnung** *f* POLITIK waste disposal regulation
abfallerzeugend *adj* ABFALL, ABWASSER, LUFT waste-generating; **~er Betrieb** *m* ABFALL, ABWASSER, LUFT waste-generating factory
Abfall: **Abfallerzeuger** *m* ABFALL, ABWASSER, LUFT waste generator, waste producer; **Abfallerzeugung** *f* ABFALL, ABWASSER, LUFT waste generation, waste production; **Abfallextraktionssystem** *nt* ABFALL, ABWASSER, LUFT waste extraction system
abfallfrei *adj* ABFALL, WASSER waste-free
Abfall: **Abfallgärung** *f* ABFALL fermentation of refuse
abfallgefeuert *adj* ABFALL waste-fired; **~e Anlage** *f* ABFALL waste-fired plant
Abfall: **Abfallgemisch** *nt* ABFALL waste mixture; **Abfallgesetz** *nt* ABFALL, ABWASSER, LUFT, POLITIK, WASSER Waste Avoidance and Management Act, Waste Management Act; **Abfallkataster** *m* ABFALL, POLITIK waste inventory; **Abfall-Klärschlamm-Gemisch** *m* ABFALL, WASSER waste/sludge mixture; **Abfall-Kohle-Gemisch** *m* ABFALL *Brennstoff aus Müll* waste/coal mixture; **Abfallmenge** *f* ABFALL garbage volume (*AmE*), refuse volume, volume of waste
abfallos: **~e Technologie** *f* ALLG non-waste-technology (*NWT*)
Abfall: **Abfallqualität** *f* ABFALL waste quality; **Abfallrecht** *nt* ABFALL, POLITIK law on waste
abfallrechtlich *adj* ABFALL, POLITIK according to the law on waste; **~e Anlagengenehmigung** *f* ABFALL, POLITIK plant permit according to the law on waste
Abfall: **Abfallrohstoff** *m* ERNEU ENER *Bioenergienausgangsstoff* waste raw material; **Abfallrückholung** *f* ABFALL *Atommüll* , WASSER disinterment of waste; **Abfallsäure** *f* ABFALL waste acid; **Abfallschicht** *f* ABFALL waste layer; **Abfallsortieranlage** *f* ABFALL garbage separation plant

(*AmE*), refuse separation plant, waste sorting plant; **Abfallsortierung** *f* **am Anfallsort** ABFALL source separation; **Abfallsteuer** *f* POLITIK waste tax; **Abfallstoff** *m* ABFALL waste product, waste substance; **Abfallstrom** *m* ABFALL waste stream; **Abfalltourismus** *m* POLITIK waste tourism; **Abfallverbrennungsanlage** *f* ABFALL, LUFT garbage incinerator (*AmE*), refuse incinerator; **Abfallverbrennungsofen** *m* ABFALL, ERNEU ENER, LUFT incinerator; **Abfallvermeidung** *f* ABFALL waste avoidance; **Abfallverursacher** *m* ABFALL waste producer; **Abfallverwertung** *f* ABFALL waste processing, waste treatment, waste utilization; **Abfallverwertungsanlage** *f* ABFALL waste processing plant, waste treatment plant, waste utilization plant; **Abfallvorbehandlung** *f* ABFALL waste pretreatment

Abfallwirtschaft *f* ABFALL, MANAG waste management

abfallwirtschaftlich *adj* ABFALL, MANAG waste-economical; **~es Konzept** *nt* ABFALL, MANAG waste-management concept; **~e Planung** *f* ABFALL, MANAG waste-management planning, waste-economical planning

Abfallwirtschaft: **Abfallwirtschaftskonzept** *nt* ABFALL, MANAG waste-management concept; **Abfallwirtschaftsmarkt** *m* ABFALL, MANAG waste-management market; **Abfallwirtschaftsplanung** *f* ABFALL, MANAG waste-management planning

Abfall: **Abfallzellstoff** *m* ABFALL waste pulp; **Abfallzerkleinerer** *m* ABFALL waste disintegrator; **Abfallzusammensetzung** *f* ABFALL waste composition

abfangen *vt* ABWASSER collect, intercept

Abfangen *nt* ABWASSER collection, interception

Abfangleitung *f* ABWASSER intercepting sewer, collection sewer

abfiltern *vt* ABFALL, ANALYTIK, LUFT, WASSER filter out

Abfließen *nt* ABWASSER, WASSER drainage, outflow, runoff

Abfluß *m* ABWASSER *Ablauf*, MEER, discharge, drainage, effluent, runoff, POLITIK effluent, WASSER drainage, effluent, runoff; **~ im Oberflächenbereich** ABFALL, ABWASSER, WASSER surface runoff; **Abflußkanal** *m* ABWASSER effluent channel, effluent sewer; **Abflußkoeffizient** *m* ABWASSER, WASSER runoff coefficient; **Abflußkreislauf** *m* WASSER runoff cycle; **Abflußleistung** *f* ABWASSER, WASSER runoff capacity; **Abflußleitung** *f* ABFALL, ABWASSER, WASSER discharge pipe, effluent sewer

abflußlos: **~es Einzugsgebiet** *nt* WASSER blind drainage area

Abfluß: **Abflußmenge** *f* ABWASSER, ANALYTIK, WASSER rate of flow; **Abflußmengendauerlinie** *f* ABWASSER, ANALYTIK, WASSER flow duration curve; **Abflußmessung** *f* ANALYTIK, WASSER flow measurement; **Abflußregelung** *f* WASSER flow regulation; **Abflußrinne** *f* ABFALL, ABWASSER, WASSER effluent trough; **Abflußverhältnis** *nt* ABWASSER, WASSER runoff coefficient; **Abflußvermögen** *nt* ABFALL, WASSER runoff capacity

Abfuhr *f* ABFALL collection; **Abfuhrwagen** *m* ABFALL tipper truck, skip

Abfüll: **Abfüllbetrieb** *m* ABFALL, WASSER filling plant; **Abfüllmaschine** *f* ABFALL, WASSER dosing machine

Abfüllung *f* ANALYTIK *von Probenahmeflaschen* filling

Abgabe *f* MANAG, POLITIK *Qualitätssicherung* charge, levy; **Abgabenbefreiung** *f* MANAG, POLITIK charge exemption; **Abgabenkürzung** *f* MANAG, POLITIK charge reduction; **Abgabenquote** *f* MANAG, POLITIK charges quota; **Abgabensatz** *m* MANAG, POLITIK rate of charge; **Abgabentarif** *m* MANAG, POLITIK charges tariff

Abgas *nt* ABFALL tail gas, exhaust gas, waste gas, BODEN off-gas, LUFT *von Verkehr* exhaust emissions, exhaust gas, *einer Raffinerie* off-gas, tail gas, *schwebstaubhaltig* fumes, *einer Turbine* combustion gas, waste gas, WASSER exhaust gas, waste gas; **~ als Brennstoff** LUFT fuel gas; **Abgasabsaugung** *f* LUFT waste gas extraction; **Abgasanalyse** *f* ANALYTIK analysis of exhaust gas; **Abgasdruck** *m* LUFT *Schornstein* waste gas pressure; **Abgasdurchsatz** *m* LUFT waste gas throughput, waste gas thruput (*AmE*); **Abgasemissionen** *f pl* LUFT waste gas emissions; **Abgasentschwefelung** *f* ABFALL waste gas desulfurization (*AmE*), waste gas desulphurization (*BrE*); **Abgasfahne** *f* LUFT waste gas plume; **Abgasgebläse** *nt* LUFT waste gas blower; **Abgasgegendruck** *m* LUFT waste gas back pressure; **Abgasgehäuse** *nt* LUFT exhaust housing, exhaust casing; **Abgasgeschwindigkeit** *f* LUFT *Schornstein* waste gas velocity; **Abgasgrenzwert** *m* LUFT, POLITIK, WASSER exhaust gas limit value; **Abgasinhaltsstoff** *m* LUFT waste gas component; **Abgaskomponente** *f* LUFT waste gas component; **Abgasrauch** *m* LUFT exhaust smoke; **Abgasreiniger** *m* LUFT waste gas purifier

Abgasreinigung *f* ABFALL, LUFT, WASSER waste gas cleaning, waste gas cleansing; **~ nach Müllverbrennung** ABFALL, LUFT exhaust gas cleaning following waste incineration; **Abgasreinigungsanlage** *f* LUFT exhaust emissions cleaning system

Abgas: **Abgasrohr** *nt* LUFT tail pipe; **Abgasrohremissionen** *f pl* LUFT *Kohlenmonoxid, Stickstoffoxid* tail pipe emissions; **Abgasrückführung** *f* (*AGR*) LUFT exhaust gas recirculation (*EGR*); **Abgassonde** *f* LUFT waste gas probe; **Abgasstrom** *m* ANALYTIK, LUFT exhaust gas flow, waste gas flow; **Abgastemperatur** *f* LUFT waste gas temperature; **Abgastrübung** *f* ANALYTIK *Staub, Rauchdichte* waste gas turbidity; **Abgasüberwachung** *f* LUFT exhaust emissions control; **Abgasüberwachungsanlage** *f* LUFT exhaust emissions cleaning system; **Abgasverbrennung** *f* LUFT waste gas incineration; **Abgasvolumen** *nt* LUFT waste gas volume; **Abgasvorwärmung** *f* LUFT waste gas preheating; **Abgaswärme** *f* LUFT exhaust heat; **Abgaswolke** *f* LUFT waste gas plume; **Abgaszusammensetzung** *f* LUFT waste gas composition

abgebaut *adj* ALLG degraded

abgedeckt *adj* ABFALL *Deponie* covered, BODEN sealed

abgedichtet *adj* ABFALL, BODEN sealed

abgehängt: **~e Decke** *f* LÄRM suspended ceiling

abgehend: **~e Anlage** *f* ABFALL scrapped plant

abgelagert *adj* ABFALL, ABWASSER, WASSER deposited

abgeschätzt *adj* ALLG assessed

abgeschieden *adj* LUFT collected; **~er Staub** *m* LUFT collected dust; **~e Staubpartikel** *f* LUFT collected dust particle

abgeschlossen *adj* ABFALL sealed; **~e Deponie** *f* ABFALL complete fill

abgesetzt *adj* ABFALL, ABWASSER, WASSER settled; **~es Abwasser** *nt* ABWASSER settled wastewater; **~er Niederschlag** *m* ABFALL, WASSER direct precipitation; **~er Schlamm** *m* ABWASSER settled sludge

abgestuft: **~e Lüftung** *f* ABWASSER tapered aeration

abgießen *vt* ABFALL, WASSER decant

abhängbar: **~es Schallschluckelement** *nt* LÄRM overhead sound absorber

abhängig: **~es Ökosystem** *nt* ABWASSER, BODEN, LUFT, WASSER dependent ecosystem

Abhitze *f* ABFALL, ABWASSER, LUFT waste heat; **Abhitzedampfkessel** *m* ABFALL, ABWASSER, LUFT waste heat boiler; **Abhitzekessel** *m* ABFALL, ABWASSER, LUFT waste heat boiler; **Abhitzekesselsystem** *nt* ABFALL, ABWASSER, LUFT waste heat boiler system; **Abhitzesystem** *nt* ABFALL, LUFT waste heat system

Abholcontainersystem *nt* ABFALL, WASSER hauled container system (*AmE*) (*HCS*)

abholen *vt* ABFALL collect, pick up, ABWASSER, ANALYTIK, LUFT, MEER, WASSER collect

Abholen *nt* ABFALL collection

Abholzung *f* LUFT deforestation

abiotisch *adj* ABFALL, ABWASSER, WASSER abiotic

abkippen *vt* ABFALL, ABWASSER, WASSER dump

Abkippstelle *f* ABFALL, ABWASSER, WASSER dumping location

abkochen *vt* WASSER boil off

Abkommen 1. *nt* POLITIK convention; **2. ein ~ entwerfen** *vi* POLITIK draft a convention
abkühlen 1. *vt* ALLG cool, BODEN quench; **2.** *vi* ANALYTIK, LUFT, WASSER cool down
Abkühlung *f* ALLG cooling, cooling down
Abkühlzone *f* LUFT cooling zone
ablagern *vt* ALLG *Müll* deposit, dump, tip (*BrE*)
Ablagerung *f* ALLG accretion, build-up, deposit, deposition, sediment; **Ablagerungsbereich** *m* ABFALL dumping area
ablagerungsfähig *adj* ABFALL dumpable
***Ablagerung*: Ablagerungsgeschwindigkeit** *f* ABFALL, LUFT, WASSER deposition velocity
ablagerungsorientiert *adj* ABFALL disposal-oriented; **~e Abfallbehandlung** *f* ABFALL disposal-oriented treatment of waste
***Ablagerung*: Ablagerungsplan** *m* ABFALL dumping plan; **Ablagerungsplatz** *m* ABFALL storage area; **Ablagerungsverfahren** *nt* ABFALL dumping method; **Ablagerungswert** *m* ABFALL, BODEN deposition value
Ablaß *m* ABWASSER outlet; **Ablaßgraben** *m* ABWASSER outlet channel; **Ablaßrohr** *nt* ABWASSER outlet pipe
Ablauf *m* ABFALL, ABWASSER, MEER, WASSER *Senkloch* sink, *Abfluß* effluent, outflow, *Vorgang* runoff; **Ablaufdauer** *f* ABWASSER runoff duration
ablaufen *vi* ABFALL run off, ABWASSER drain, WASSER run off
***Ablauf*: Ablaufkanal** *m* ABFALL, ABWASSER, WASSER effluent channel, effluent sewer; **Ablaufkonzentration** *f* ABFALL, ABWASSER, WASSER effluent concentration; **Ablaufleitung** *f* ABFALL, ABWASSER, WASSER discharge pipe, effluent sewer; **Ablaufrinne** *f* ABFALL, ABWASSER, WASSER effluent side wall; **Ablaufrohr** *nt* ABFALL, ABWASSER, WASSER delivery pipe; **Ablauftrichter** *m* ABFALL, ABWASSER, WASSER discharge hopper; **Ablaufwehr** *nt* ABFALL, ABWASSER, LUFT, WASSER overflow weir
Ablauge *f* ABFALL, ABWASSER, WASSER spent lye, waste lye
Ablegetisch *m* ABFALL tipping table
ableichtern *vt* MEER lighten
ableiten *vt* ABWASSER, LUFT discharge, WASSER divert
Ableitung *f* ABFALL offtake, ABWASSER discharge, LUFT *des Abgases über Schornstein* discharge, WASSER offtake; **Ableitungskanal** *m* ABWASSER discharge canal
Ablenker *m* LUFT deflector
ablesbar *adj* ANALYTIK readable
Ablese- *in cpds* ANALYTIK reading; **Ablesefehler** *m* ANALYTIK reading error; **Ablesegenauigkeit** *f* ANALYTIK reading accuracy, reading precision
ablesen *vt* ANALYTIK read
Ablesung *f* ANALYTIK reading
ablösen *vt* ABFALL, BODEN, separate, loosen, detach, WASSER separate, detach
Abluft *f* LUFT waste gas; **Abluftdurchsatz** *m* LUFT waste air throughput, waste air thruput (*AmE*); **Abluftkomponente** *f* LUFT waste air constituent; **Abluftkonzentration** *f* LUFT waste air concentration; **Abluft-reiniger** *m* LUFT exhaust cleaning installation; **Abluftreinigung** *f* LUFT waste gas cleaning, waste gas cleansing; **Abluftstrom** *m* LUFT waste air flow; **Abluftsystem** *nt* LUFT, SICHERHEIT exhaust vent installation; **Ablufttemperatur** *f* LUFT waste air temperature; **Abluftventilator** *m* LUFT waste air fan
abmessen *vt* ALLG gage (*AmE*), gauge (*BrE*), measure
Abmessung *f* ANALYTIK measurement
Abnutzung *f* MEER wear
Abperleffekt *m* LUFT repellency effect
abpipettieren *vt* ANALYTIK withdraw by pipette
abprallen *vt* LUFT *Teilchen auf Faser* rebound
abpumpen *vt* ABFALL, ABWASSER, BODEN, WASSER pump out
Abpumpen *nt* WASSER *Brunnen* pumping-out
Abrasionsschutz *m* ABFALL abrasion protection
abrasiv *adj* ABFALL abrasive
Abraum *m* ABFALL excavated material
abreinigen *vt* ALLG clean
Abreinigung *f*: **~ mit Druckluft** LUFT cleaning with compressed air; **Abreinigungsdauer** *m* LUFT duration of cleaning; **Abreinigungsdruck** *m* LUFT cleaning pressure; **Abreinigungsimpuls** *m* LUFT *einer Schlauchfilter* cleaning pulse; **Abreinigungsstellung** *f* LUFT *einer Schlauchfilter* cleaning position; **Abreinigungsvorgang** *m* LUFT cleaning operation
Abrieb *m* ABFALL abrasion, LUFT attrition, abrasion; **Abriebfestigkeit** *f* LUFT abrasion resistance, attrition resistance
Abriegelung *f* ABFALL, BODEN, WASSER blocking
Abrollcontainer *m* ABFALL, WASSER transport container
ABS *abbr* ABFALL (*Acrylnitril-Butadien-Styrol*) ABS (*acrylonitrile butadiene styrene*), (*Alkylbenzolsulfonat*) ABS (*alkylbenzenesulfonate AmE, alkylbenzenesulphonate BrE*)
absatzfähig *adj* MANAG marketable
Absäuerung *f* LUFT acid formation
Absaug- *in cpds* ABFALL, BODEN, LUFT, SICHERHEIT suction
absaugen *vt* ALLG suck, remove by suction
Absaugen *nt* ALLG suction; **~ der Bodenluft** BODEN soil vapor extraction (*AmE*), soil vapour extraction (*BrE*)
***Absaug-*: Absaugfilter** *nt* LUFT exhauster filter; **Absaug- und Filtervorrichtung** *f* **für Staub und Späne** ABFALL, BODEN, LUFT, SICHERHEIT suction and filter installation for dust and chippings; **Absaugleitung** *f* ABFALL suction extraction line; **Absaugpegel** *m* BODEN suction level; **Absaugpumpe** *f* ANALYTIK *für Schnellbestimmung* suction pump
Absaugung *f* ABWASSER removal by suction, suction, BODEN, LUFT, MEER, SICHERHEIT, WASSER suction
***Absaug-*: Absaugversuch** *m* ANALYTIK extraction test
Abschälen *nt* ABWASSER *Langsamfilter* flaking
Abschätzung *f* ANALYTIK estimation, MANAG, POLITIK, WASSER assessment, estimation
abscheidbar *adj* ABFALL, ABWASSER, LUFT separable
Abscheidbarkeitstest *m* ABWASSER, ANALYTIK separability test
Abscheide- *in cpds* LUFT separation ; **Abscheideeigenschaften** *f pl* LUFT separation properties; **Abscheideeinheit** *f* LUFT precipitation unit; **Abscheidegrad** *m* ABFALL, ABWASSER, LUFT degree of separation; **Abscheidegüte** *f* LUFT removal quality; **Abscheideleistung** *f* LUFT separation capacity
abscheiden *vt* ABFALL, ABWASSER, LUFT, WASSER separate, *Staubpartikel* remove, precipitate
Abscheider *m* ABFALL, ABWASSER, MEER, WASSER separator
***Abscheide-*: Abscheideverhalten** *nt* LUFT removal behavior (*AmE*), removal behaviour (*BrE*), collection behavior (*AmE*), collection behaviour (*BrE*); **Abscheidevorgang** *m* LUFT collection operation, removal operation; **Abscheidewirkungsgrad** *m* LUFT collection efficiency, removal efficiency, separating efficiency
Abscheidung *f* ABFALL, ABWASSER, LUFT, WASSER segregation, separation
abscheren *vt* ABFALL shear off
Abschirmung *f* LÄRM screen; **~ im Freien** LÄRM outdoor shielding; **~ in Räumen** LÄRM indoor shielding; **~ von Schall** LÄRM acoustic shielding
Abschirmwand *f* LÄRM shielding wall, screening wall
abschlämmen *vt* ABFALL, ABWASSER blow down, clarify
abschleißend *adj* LUFT abrasive; **~er Staub** *m* LUFT abrasive dust
abschließen *vt* MEER seal
abschnittweise *adj* BODEN batch-bulk; **~s Einpressen** *nt* BODEN batch-bulk injection
Abschöpf- *in cpds* MEER skimming; **Abschöpfbarke** *f* MEER skimming barge; **Abschöpfeinrichtung** *f* MEER skimmer
abschöpfen *vt* ABFALL, ABWASSER, WASSER siphon off, skim off
Abschöpfen *nt* ABWASSER, MEER skimming off, siphoning off

Abschöpfer *m* ABWASSER skimmer

Abschöpf-: **Abschöpfkopf** *m* MEER *eines Abschöpfgerätes* skimming head; **Abschöpfölsperre** *f* MEER skimming barrier

Abschottung *f* BODEN *durch Dichtwände* cutting off

abschwemmen *vt* ABFALL, LUFT, WASSER scour, wash away

Abschwemmung *f* ABFALL, LUFT, WASSER scouring, washing

absenken *vi* WASSER lower

Absetz- *in cpds* ABFALL, LUFT settling

absetzbar *adj* ABFALL, WASSER settleable; **~e Feststoffe** *m pl* ABWASSER settleable solids; **~er Schlamm** *m* ABWASSER settleable sludge; **~e Schwebstoffe** *m pl* ABWASSER settleable suspended solids

Absetzbarkeit *f* ABWASSER settleability

Absetzbecken *nt* ABFALL, ABWASSER, WASSER sedimentation tank, settling basin, settling tank; **~ mit peripherem Einlauf** ABWASSER settling tank with peripheral inlet; **~ mit zentralem Einlauf** ABWASSER settling tank with central inlet

Absetz-: **Absetzbedingungen** *f pl* ABWASSER settling conditions; **Absetzbehälter** *m* ABWASSER decay tank, LUFT *biologische Gaswäsche* settling tank; **Absetzeigenschaften** *f pl* ABFALL, ABWASSER, WASSER settling characteristics, settling properties

absetzen *vt* ABFALL, ABWASSER settle, WASSER settle out, decant

Absetzen *nt* ABFALL, ABWASSER, LUFT, WASSER decantation, settling, settling out

absetzfähig *adj* ABFALL, ABWASSER settleable; **~e Makroflocken** *f pl* ABWASSER settleable macroflocs

Absetz-: **Absetzgefäß** *nt* ABWASSER settling vessel; **Absetzgeschwindigkeit** *f* ABFALL, ABWASSER settling velocity; **Absetzglas** *nt* ANALYTIK sedimentation glass; **Absetzgrube** *f* ABWASSER cesspit, cesspool; **Absetzkammer** *f* ABFALL, WASSER settling chamber; **Absetzklärung** *f* ABFALL, WASSER decantation; **Absetzkurve** *f* ABWASSER sedimentation curve; **Absetzleistung** *f* BODEN deposition rate; **Absetzprobe** *f* ANALYTIK settling test; **Absetzteich** *m* ABFALL sedimentation pond, ABWASSER settling lagoon, WASSER sedimentation pond; **Absetzzeit** *f* ABFALL, ABWASSER, WASSER settling time; **Absetzzone** *f* ABWASSER settling zone

absieben *vt* ABFALL, BODEN, POLITIK, WASSER classify, screen, sift

absinken 1. *vt* LUFT *Temperatur* lower; **2.** *vi* WASSER drop

Absinken *nt* ABFALL, WASSER drawdown; **~ des pH-Wertes** *nt* ABWASSER pH drop

Absinkstoff *m* ABWASSER sinking matter

Absolutdruck *m* ABWASSER, WASSER *Pumpe* absolute pressure

Absorbat *nt* ABWASSER, ANALYTIK, LUFT absorbate, solute

Absorbens *nt* LUFT absorbent; **Absorbensreinigung** *f* LUFT absorbent cleaning

Absorber *m* ERNEU ENER absorber plate, LUFT absorber; **Absorberbehälter** *m* LUFT absorber tank; **Absorberkopf** *m* LUFT absorber head; **Absorbermotor** *m* LUFT absorber motor; **Absorberplatte** *f* LÄRM sound-absorbing panel, acoustical tile, acoustical board, acoustic tile; **Absorbersumpf** *m* LUFT absorber sump; **Absorbersuspension** *f* LUFT absorber suspension; **Absorberwand** *f* LUFT absorber wall

absorbierbar *adj* ALLG absorbable

Absorbierbarkeit *f* ALLG absorbency

absorbieren *vt* ALLG absorb

absorbierend *adj* ALLG absorbing; **~ ausgekleidet** *adj* LÄRM absorbent-lined

absorbiert *adj* ALLG absorbed; **~er Zustand** *m* LUFT absorbed state

Absorption *f* ALLG absorption; **Absorption/Filtration/Adsorption-Verfahren** *nt* (*AFA-Verfahren*) LUFT absorption/filtration/adsorption process (*AFA process*); **Absorptionsapparat** *m* **ohne Einbauten** LUFT absorption apparatus without internal fittings; **Absorptionsband** *nt* ANALYTIK absorption band; **Absorptionseinrichtung** *f* **zur Trockenabtrennung giftiger Gase** SICHERHEIT absorption plant for the dry separation of noxious gases; **Absorptionsfähigkeit** *f* ABFALL, BODEN, LUFT, WASSER absorbency, absorption capacity, absorptivity; **Absorptionsfaktor** *m* ABFALL, LUFT, WASSER absorption coefficient; **Absorptionsflüssigkeit** *f* LUFT absorption liquid; **Absorptionsgefäß** *nt* ANALYTIK absorption vessel; **Absorptionsgrad** *m* LÄRM absorption coefficient, LUFT degree of absorption; **Absorptionsisotherme** *f* LUFT absorption isotherm; **Absorptionskoeffizient** *m* LUFT absorption coefficient; **Absorptionskolonne** *f* LUFT absorption column; **Absorptionslösung** *f* ANALYTIK absorption solution; **Absorptionsmaterial** *nt* LÄRM absorbent; **Absorptionsmittel** *nt* LUFT absorbent; **Absorptionspipette** *f* ANALYTIK absorption pipette; **Absorptionsrate** *f* ABFALL, WASSER rate of absorption; **Absorptionsschalldämpfer** *m* LÄRM, SICHERHEIT absorption silencer, absorptive attenuator, absorptive silencer, sound-absorbent damper; **Absorptionsspektrum** *nt* ANALYTIK *eines IR-Spektralphotometers* absorption spectrum; **Absorptionsstufe** *f* LUFT absorption stage; **Absorptionstemperatur** *f* LUFT absorption temperature; **Absorptionsturm** *m* LUFT absorption tower; **Absorptionsvermögen** *nt* ABFALL, ABWASSER, BODEN, WASSER absorption capacity

Absorptiv *nt* ABWASSER, ANALYTIK solute, LUFT solute, absorbent

Abspaltung *f* ABFALL splitting off

Absperr- *in cpds* WASSER closing, shutoff; **Absperrklappe** *f* LUFT butterfly valve; **Absperrorgan** *nt* ABFALL, WASSER shutoff device; **Absperrventil** *nt* ABFALL, WASSER check valve

abspülen *vt* ABWASSER wash away

Abstimmung *f* POLITIK negotiation

Abstoßung *f* ABFALL, WASSER *Physik* repulsion; **Abstoßungskraft** *f* ABWASSER, WASSER *Kolloidteilchen* electrostatic repulsive force, repulsive power; **Abstoßungsvermögen** *nt* ABWASSER, WASSER electrostatic repulsive force, repulsive power

abstrahlen *vt* ABFALL, LÄRM, LUFT radiate

Abstrahlungsverluste *m pl* ABFALL heat radiation losses

Abteilung *f* POLITIK *Verwaltungseinheit* department, division

abteufen *vt* ABFALL sink

Abtragung *f* MEER wear

abtransportieren *vt* BODEN *Boden bei Off-site Verfahren* transport

abtrennbar *adj* ABFALL, ABWASSER, BODEN, WASSER separable

abtrennen *vt* ABFALL *schädliche oder gefährliche Substanzen in Abfällen*, ABWASSER, BODEN, WASSER separate

Abtrennung *f* ABFALL separation; **~ des Extraktionsmittels** ABFALL, WASSER desolventizing

Abwärme *f* ABFALL, ABWASSER, LUFT, WASSER thermal discharge, waste heat; **Abwärmenutzung** *f* ABFALL, LUFT, WASSER waste heat use; **Abwärmeverwertung** *f* ABFALL, LUFT cogeneration, waste heat recovery, waste heat utilization

Abwärtsflächeneinbau *m* ABFALL, WASSER downward surface installation

Abwasser *nt* ABFALL *von Deponie* effluent, wastewater, ABWASSER, WASSER wastewater, sewage, effluent; **~ in der Entaschung** ABFALL, WASSER wastewater in ash removal; **~ aus Haushaltungen** ABWASSER domestic wastewater, household wastewater; **~ aus der Müllverbrennungsanlage** ABFALL, WASSER wastewater from waste incineration plant; **~ aus der Rauchgaswäsche** ABFALL, WASSER wastewater from flue-gas scrubbing; **~ aus Sanitäranlagen** ABWASSER sanitary wastewater; **~ aus der Trinkwasseraufbereitung** ABFALL, WASSER wastewater resulting from drinking water treatment; **Abwasser-**

abgabe *f* ABWASSER wastewater load, POLITIK effluent charge; **Abwasseranalysegerät** *nt* ANALYTIK wastewater analyser (*BrE*), wastewater analyzer (*AmE*); **Abwasseranfall** *m* ABWASSER wastewater flow, wastewater volume, wastewater yield; **Abwasseraufbereitung** *f* ABWASSER wastewater treatment; **~ der ersten Stufe** ABWASSER primary wastewater treatment; **~ der zweiten Stufe** ABWASSER secondary wastewater treatment; **Abwasseraufbereitungsanlage** *f* ABWASSER wastewater treatment plant; **Abwasserausfluß** *m* ABWASSER wastewater discharge, wastewater outfall, wastewater outflow; **Abwasserbecken** *nt* ABWASSER stabilization pond; **Abwasserbehandlung** *f* ABWASSER wastewater clarification, wastewater treatment; **Abwasserbehandlungsanlage** *f* ABWASSER clarification plant, wastewater treatment plant; **Abwasserbeseitigung** *f* ABWASSER effluent disposal, wastewater disposal; **Abwasserdekontaminierung** *f* ABWASSER wastewater decontamination; **Abwassereinleitung** *f* **in den Untergrund** ABWASSER underground wastewater disposal; **Abwasserentgiftung** *f* ABWASSER wastewater decontamination; **Abwasserentsorgung** *f* ABWASSER wastewater disposal; **Abwasserentstehung** *f* ABWASSER wastewater production; **Abwassererfassung** *f* ABWASSER wastewater collection; **Abwasserfahne** *f* ABWASSER wastewater trail; **Abwasserfärbung** *f* ANALYTIK wastewater coloration (*AmE*), wastewater colouration (*BrE*); **Abwasserfischteich** *m* ABWASSER wastewater fish pond

abwasserfrei *adj* LUFT *Sprühabsorptionsverfahren* wastewater-free; **~e Rauchgasreinigung** *f* LUFT wastewater-free flue-gas cleaning

***Abwasser*: Abwassergebühr** *f* ABWASSER wastewater treatment fee; **Abwasserkanal** *m* ABWASSER sewer; **Abwasserkanalreinigung** *f* ABWASSER sewer cleaning; **Abwasserkläranlage** *f* ABWASSER wastewater clarification plant, wastewater purification plant; **Abwasserklärung** *f* ABWASSER wastewater purification, wastewater treatment; **Abwasserkontrolle** *f* ABWASSER wastewater monitoring, wastewater control; **Abwasserkontrollstation** *f* ABWASSER *für automatische Untersuchung der Abwasserbeschaffenheit* wastewater monitoring station; **Abwasserlagerung** *f* ABWASSER wastewater storage; **Abwasserlast** *f* ABWASSER wastewater load; **Abwasserleitung** *f* WASSER canalization; **Abwassermenge** *f* ABWASSER wastewater flow, wastewater volume; **Abwasserorganismen** *m pl* ABWASSER wastewater organisms; **Abwasserpilz** *m* ABWASSER wastewater fungus; **Abwasserprobe** *f* ANALYTIK wastewater sample; **Abwasserqualität** *f* ABWASSER wastewater quality

Abwasserreinigung *f* ABWASSER wastewater purification, wastewater treatment, *mechanisches Abwasseraufbereitungsverfahren* wastewater clarification; **Abwasserreinigungsverfahren** *nt* ABWASSER wastewater purification process; **Abwasserreinigungsanlage** *f* **in Containern** ABWASSER container wastewater treatment plant; **Abwasserreinigungsstufe** *f* ABWASSER wastewater treatment stage; **Abwasserreinigungsverfahren** *nt* ABFALL, ABWASSER, WASSER wastewater purification process

***Abwasser*: Abwassersammelsystem** *nt* ABWASSER wastewater collection system; **Abwassersammeltank** *m* ABFALL, WASSER wastewater collection tank; **Abwassersammler** *m* ABWASSER wastewater collector; **Abwassersanierung** *f* ABWASSER wastewater renovation; **Abwasserschlamm** *m* ABWASSER effluent sludge, municipal waste stream (*MWS*); **Abwasserspeicherung** *f* ABWASSER wastewater storage; **Abwasserstrom** *m* ABWASSER wastewater stream; **Abwassertechnik** *f* ABWASSER wastewater technology

abwassertechnisch *adj* ABWASSER disposal-related, effluent-related

***Abwasser*: Abwasserteich** *m* ABWASSER sewage lagoon, oxidation pond, stabilization pond; **Abwassertemperatur** *f* ABWASSER wastewater temperature; **Abwassertoxizität** *f* ABWASSER wastewater toxicity; **Abwasserüberwachung** *f* ABWASSER wastewater control; **Abwasserverregnung** *f* ABWASSER wastewater sprinkling; **Abwasserverrieselung** *f* ABWASSER broad irrigation with wastewater; **Abwasserversenkung** *f* ABWASSER underground wastewater disposal; **Abwasserverwertung** *f* ABWASSER wastewater utilization; **Abwasserwiederverwertung** *f* ABWASSER wastewater reuse; **Abwasserzusammensetzung** *f* ABWASSER wastewater composition

abwechseln: sich ~ *v refl* LUFT alternate

abwerfen *vt* ABFALL discard, throw away

abziehen *vt* BODEN, LUFT draw off

Abzug *m* ANALYTIK fumehood, BODEN *der Bodenluft* venting, LUFT conduit, flue, hood; **Abzugsgraben** *m* ABFALL, WASSER drainage ditch; **Abzugskanal** *m* ABFALL, WASSER delivery channel

Abzweig *m* ABFALL, WASSER branch pipe; **Abzweigrohr** *nt* ABFALL, WASSER branch pipe

Abzweigung *f* WASSER *Kanal* branching

***Abzweig*: Abzweigwiderstand** *m* WASSER branch pipe resistance

ACCR-Verfahren *nt* LUFT activated carbon catalytic reduction (*ACCR*)

Acetat *nt* ABFALL acetate; **Acetatbildung** *f* ABFALL, WASSER acetate build-up, acetate formation

acetogen *adj* ABWASSER acetogenic; **~e Bakterien** *f pl* ABWASSER acetogenic bacteria

Acetogenese *f* ABWASSER acetogenesis

Aceton *nt* ABFALL, LUFT acetone

Achse *f* ANALYTIK, ERNEU ENER *Biokonversion* axis

Acidogenese *f* ABWASSER acidogenesis

Acrylglas *nt* ABFALL acrylic plastic

Acrylnitril-Butadien-Styrol *nt* (*ABS*) ABFALL acrylonitrile butadiene styrene (*ABS*)

acyclisch *adj* LUFT acyclic

adaptiv *adj* ABWASSER adaptive; **~es Enzym** *nt* ABWASSER adaptive enzyme

Additiv *nt* ALLG *gebräuchliche Sorbents* additive

additiv: ~es System *nt* ABFALL *Hausmüllsystem*, WASSER additive system

***Additiv*: Additivsilo** *m* LUFT additives silo; **Additivsuspension** *f* LUFT additive suspension

Adenosintriphosphat *nt* (*ATP*) ABWASSER adenosine triphosphate (*ATP*)

Adhäsion *f* ABFALL, WASSER adherence

Adhäsionskraft *f* ABFALL, WASSER adhesive force

adhäsiv: ~ wirkendes Ölaufnahmeband *nt* MEER *endlos* oil mop

adiabatisch *adj* LUFT adiabatic

ADI-Wert *m* (*tolerierbare tägliche Aufnahme*) LÄRM, SICHERHEIT *Schallemissionsanalyse* ADI (*acceptable daily intake*)

Adsorbat *nt* ABWASSER, LUFT adsorbate

Adsorbens *nt* ALLG adsorbent, adsorbing agent; **Adsorbensaustritt** *m* LUFT adsorbent outlet; **Adsorbensbett** *nt* LUFT adsorbent bed; **Adsorbensdichte** *f* LUFT adsorbent density; **Adsorbensdosierung** *f* LUFT adsorbent dosage; **Adsorbenseintritt** *m* LUFT adsorbent inlet; **Adsorbensoberfläche** *f* LUFT adsorbent surface; **Adsorbenspartikel** *f* LUFT adsorbent particle

Adsorber *m* LUFT adsorber; **Adsorberfilter** *nt* LUFT adsorber filter; **Adsorberharz** *nt* ABWASSER adsorber resin

adsorbierbar *adj* ALLG adsorbable; **~e organische Halogenverbindungen** *f pl* (*AOX*) ABFALL, ABWASSER, ANALYTIK, WASSER adsorbable organic halogen compounds (*AOX*)

Adsorbierbarkeit *f* ALLG adsorbability

adsorbieren *vt* ALLG adsorb

adsorbierend *adj* ALLG adsorptive

adsorbiert *adj* ALLG adsorbed; **~es Gas** *nt* LUFT adsorbed gas;

~e Komponente *f* LUFT absorbed compound, adsorbed component; **~er Stoff** *m* LUFT adsorbed substance

Adsorption *f* ALLG *Sorption von Gasen in Flüssigkeiten* adsorption; **Adsorptionschromatographie** *f* ANALYTIK adsorption chromatography; **Adsorptionsdauer** *m* LUFT adsorption duration; **Adsorptionseffizienz** *f* ABFALL, ABWASSER, BODEN, LUFT adsorbability, WASSER adsorption efficiency; **Adsorptionsfähigkeit** *f* LUFT adsorption power; **Adsorptionsfilter** *nt* ABFALL, WASSER adsorption filter; **Adsorptionsgeschwindigkeit** *f* LUFT adsorption velocity; **Adsorptionsisotherme** *f* ABFALL, ABWASSER, LUFT, WASSER adsorption isotherm; **Adsorptionskapazität** *f* ABFALL, WASSER adsorption capacity; **Adsorptionsleistung** *f* LUFT adsorption capacity; **Adsorptionsmittel** *nt* ABFALL, ABWASSER, LUFT, WASSER adsorbent; **Adsorptionspotential** *nt* LUFT adsorption potential; **Adsorptionsrohr** *nt* ANALYTIK adsorption tube; **Adsorptionssäule** *f* ANALYTIK adsorption column; **Adsorptionsstufe** *f* LUFT adsorption stage; **Adsorptionstemperatur** *f* LUFT adsorption temperature; **Adsorptionstest** *m* ANALYTIK adsorption test; **Adsorptionsvermögen** *nt* ABFALL adsorption capacity, LUFT adsorption power, WASSER adsorption capacity; **Adsorptionswirkungsgrad** *m* ABFALL, BODEN, LUFT, WASSER adsorption efficiency; **Adsorptionszone** *f* LUFT adsorption zone

adsorptiv *adj* ABWASSER, BODEN adsorbing, adsorptive

Adsorptiv *nt* ALLG adsorptive agent

ADSOX-Verfahren *nt* LUFT *Adsorptionsverfahren* ADSOX process

AED *abbr* (*Atomemissionsdetektion*) ANALYTIK *gute Laborpraxis* AED (*atomic emission detection*)

aerob *adj* ABFALL, ABWASSER, ERNEU ENER, WASSER *auf Sauerstoff angewiesen* aerobic; **~er Abbau** *m* ABFALL aerobic degradation, ABWASSER aerobic decomposition, WASSER aerobic degradation; **~er Abbauprozeß** *m* ABFALL, WASSER aerobic decomposition process; **~e Ausfaulung** *f* ABWASSER *biologische Schlammstabilisierung* aerobic digestion, aerobic fermentation; **~e Bakterien** *f pl* ABFALL, ABWASSER, WASSER aerobic bacteria; **~es Becken** *nt* ABWASSER aerobic pond, aerobic lagoon; **~e Bedingungen** *f pl* ABWASSER aerobic conditions; **~e Faulung** *f* ABWASSER, WASSER *biologische Schlammstabilisierung* aerobic digestion, aerobic fermentation; **~e Gärung** *f* ABFALL, ABWASSER, WASSER aerobic fermentation; **~e Mikrobe** *f* ABWASSER aerobic microbe, aerobic microorganism; **~er Mikroorganismus** *m* ABWASSER aerobic microorganism; **~e Oxidation** *f* ABFALL, WASSER aerobic oxidation; **~ stabilisiert** *adj* ABWASSER aerobically stabilized, aerobically digested; **~ stabilisierter Schlamm** *m* ABWASSER aerobically-stabilized sludge, aerobically-digested sludge; **~e Stabilisierung** *f* ABWASSER aerobic stabilization; **~er Stoffwechsel** *m* ABWASSER aerobic metabolism; **~er Teich** *m* ABWASSER aerobic lagoon; **~e Vergärung** *f* ERNEU ENER *Biokonversion* aerobic fermentation; **~e Zersetzung** *f* ABFALL, WASSER aerobic decomposition; **~e Zone** *f* ABWASSER *des Biofilms* aerobic zone

aerobisch: **~er Bioreaktor** *m* ABWASSER aerobic bioreactor; **~es Heterotroph** *nt* ABWASSER aerobic heterotroph

aerob-thermophil *adj* ABWASSER aerobic-thermophilic; **~e Stabilisierung** *f* (*ATS*) ABWASSER aerobic-thermophilic stabilization; **~e Stabilisierung** *f* **mit anschließender Faulung** ABWASSER aerobic-thermophilic stabilization with digestion

aerodynamisch *adj* ERNEU ENER aerodynamic; **~er Auftrieb** *m* ERNEU ENER *Windturbine* aerodynamic lift; **~er Durchmesser** *m* ANALYTIK, LUFT aerodynamic diameter; **~e Strömungsabriß** *f* ERNEU ENER *Windturbine* stall

Aerometer *nt* ANALYTIK *zur Bestimmung der Luftdichte oder zum Wägen von Luft* aerometer

Aerosol *nt* ABFALL, LUFT *schwebstoffhaltige Luft* aerosol; **Aerosolabscheidung** *f* LUFT aerosol precipitation; **Aerosolpackung** *f* ABFALL, LUFT aerosol dispenser

AES *abbr* (*Atomemissionsspektrometrie*) ANALYTIK AES (*atomic emission spectrometry*)

AFA-Verfahren *nt* (*Absorption/Filtration/Adsorption-Verfahren*) LUFT AFA process (*absorption/filtration/adsorption process*)

Affinität *f* WASSER *Ionenaustauscher* affinity

AFS *abbr* (*Atomfluoreszenzspektrometrie*) ANALYTIK AFS (*atomic fluorescence spectrometry*)

Agglomerat *nt* LUFT agglomerate

Agglomeration *f* ABFALL, ABWASSER, LUFT, WASSER agglomeration; **Agglomerationsfähigkeit** *f* LUFT agglomeration capability

agglomerieren *vt* ALLG agglomerate, nodulize

Agglomerieren *nt* LUFT agglomeration, nodulizing

agglomerierend: **~er Staub** *m* LUFT *Schlauchfilter* agglomerating dust

Aggradation *f* ABFALL, WASSER aggradation

Aggregation *f* ABWASSER aggregation

Aggregatzustand *m* LUFT condition of matter, physical condition, state of aggregation

aggressiv *adj* ABFALL, LUFT, WASSER aggressive; **~er Gasbestandteil** *m* LUFT aggressive gas component; **~e Gaskomponente** *f* LUFT aggressive gas component; **~es Grubenwasser** *nt* WASSER aggressive pit water; **~es Grundwasser** *nt* WASSER aggressive groundwater; **~es Wasser** *nt* WASSER aggressive water

AGR *abbr* (*Abgasrückführung*) LUFT EGR (*exhaust gas recirculation*)

A-Horizont *m* BODEN A-horizon

Airmould-Verfahren *nt* ABFALL, WASSER air mold process (*AmE*), air mould process (*BrE*)

AK *abbr* ABFALL, ABWASSER, BODEN, LUFT (*Aktivkohle*) activated carbon, (*Aktivkoks*) activated coke; **AK-Filter** *nt* ABFALL, ABWASSER, WASSER *Volumenfilter* activated carbon filter

Akku *abbr* (*Akkumulator*) ABFALL storage battery

Akkumulation *f* ABFALL, ANALYTIK, MEER, WASSER accumulation

Akkumulator *m* (*Akku*) ABFALL storage battery

Akkumulierbarkeit *f* ABFALL, LUFT, WASSER accumulability

akkumulieren **1.** *vt* ABFALL, ANALYTIK, MEER, WASSER accumulate; **2. biologisch ~** *vi* ABWASSER, BODEN, LUFT bioaccumulate

Aktinometer *nt* ANALYTIK actinometer

Aktiv- *in cpds* ANALYTIK, BODEN, WASSER activated

aktiv *adj* ALLG active; **~es Lösungsmittel** *nt* ABFALL active solvent; **~er Pellistor** *m* ANALYTIK active pellistor; **~e thermische Sonnenenergienutzung** *f* ERNEU ENER active thermal solar energy use; **~es Wasser** *nt* WASSER active water; **~e Zone** *f* LUFT charging zone

Aktiv-: **Aktiventgasung** *f* ABFALL *Basis-Oberflächen Abdichtung* active degassing

aktivieren *vt* ABFALL, ABWASSER, BODEN, LUFT activate

aktiviert *adj* ABFALL, ABWASSER, BODEN, LUFT activated; **~es Aluminiumoxid** *nt* ABFALL, LUFT *Adsorptionsmittel* activated aluminium oxide (*BrE*), activated aluminum oxide (*AmE*)

Aktivierung *f* ABWASSER, ANALYTIK activation; **Aktivierungsanalyse** *f* ANALYTIK *Schwermetallbestimmung* activation analysis; **Aktivierungsenergie** *f* LUFT activation energy

Aktivkohle *f* (*AK*) ABFALL, ABWASSER, BODEN, LUFT activated carbon; **Aktivkohleadsorber** *m* ABWASSER, LUFT activated carbon adsorber; **Aktivkohleadsorption** *f* ABFALL, ABWASSER, LUFT, WASSER activated carbon adsorption; **Aktivkohlebehandlung** *f* LUFT activated carbon treatment; **Aktivkohlefestbettreaktor** *m* LUFT activated carbon fixed-bed reactor; **Aktivkohlefilter** *nt* ABFALL *Volumenfilter*, ABWASSER activated carbon filter, LUFT activated carbon canister, WASSER *Volumenfilter* activated carbon filter; **Aktivkohlefiltration** *f* ABFALL, ABWASSER, LUFT, WASSER activated carbon filtration; **Aktivkohlesäule** *f* ANALYTIK activated carbon column; **Aktivkohlesuspension** *f* ANALYTIK activated carbon suspension; **Aktivkohleverfahren** *nt* LUFT activated carbon process; **Aktivkohlewanderbett** *nt* LUFT activated carbon moving bed; **Aktivkohlewirbelbett** *nt* LUFT activated carbon fluidized bed

Aktivkoks *m* (*AK*) ABFALL, ABWASSER, BODEN, LUFT activated coke; **Aktivkoksadsorber** *m* ABFALL, LUFT, WASSER activated coke adsorber; **Aktivkoksverfahren** *nt* LUFT activated coke process

Aktiv-: **Aktivtonerde** *f* ABWASSER activated alumina

Akustik *f* LÄRM, MEER, SICHERHEIT acoustics; **Akustikdecke** *f* LÄRM acoustical ceiling; **Akustikkoppler** *m* ANALYTIK acoustic coupler; **Akustikplatte** *f* LÄRM acoustic panel, *Material für die Luftschalldämpfung* acoustic plate; **Akustiktest** *m* LÄRM, MEER, SICHERHEIT acoustical test

akustisch *adj* ANALYTIK, LÄRM, SICHERHEIT acoustic, acoustical; **~e Abschirmung** *f* LÄRM acoustic shielding; **~e Emissionen** *f pl* ANALYTIK, LÄRM acoustical emissions; **~er Koppler** *m* ANALYTIK, LÄRM acoustic coupler; **~es Leck** *nt* LÄRM acoustical leak; **~e Schwachstelle** *f* LÄRM acoustical leak

akut: **~e Vergiftung** *f* SICHERHEIT acute poisoning; **~e Wirkung** *f* ABFALL, LUFT acute effect

Al *abbr* (*Aluminium*) ABFALL, ABWASSER, BODEN Al (*aluminium BrE, aluminum AmE*)

Alarmsignal *nt* ANALYTIK emergency signal

Alaun *m* ABFALL, WASSER alum

Albedo *f* LUFT albedo; **Albedo-Veränderung** *f* ERNEU ENER *Auswirkung der erneuerbaren Energietechnik* albedo change

Algen *f pl* ABFALL, ABWASSER algae, MEER seaweed, WASSER algae; **Algenbekämpfung** *f* ABFALL, ABWASSER, WASSER algae abatement; **Algenbiomasse** *f* ABWASSER algal biomass; **Algenphotosynthese** *f* ABWASSER algal photosynthesis; **Algenteich** *m* ABFALL, WASSER algae pond; **Algentest** *m* ANALYTIK algae test; **Algenwachstum** *nt* ABWASSER algae growth; **Algenwachstumspotential** *nt* ABWASSER algae growth potential (*AGP*)

alicyclisch *adj* LUFT alicyclic; **~e Verbindungen** *f pl* LUFT alicyclic compounds

aliphatisch *adj* ABFALL, ABWASSER, LUFT aliphatic; **~e Kohlenwasserstoffe** *m pl* ABFALL, ABWASSER, LUFT *Alkane, Alkene* aliphatic hydrocarbons; **~es Polyamid** *nt* LUFT *Filter-Fasermaterial* aliphatic polyamid; **~e Verbindungen** *f pl* LUFT aliphatic compounds

Alkali *nt* ALLG alkali

Alkalinität *f* ALLG alkalinity; **Alkalinitätsmessung** *f* ANALYTIK alkalinity measurement

alkalisch *adj* ALLG alkaline; **~es Abwasser** *nt* ABWASSER alkaline wastewater; **~e Gärung** *f* ABWASSER alkaline fermentation; **~-oxidativ** *adj* LUFT alkaline-oxidative

Alkalität *f* ALLG *Laugengrad* alkalinity; **~ vor der Ansäuerung** LUFT preacidification alkalinity

Alkane *nt* LUFT *gesättigte Kohlenwasserstoffe* alkane

Alkene *nt* LUFT *Olefine, ungesättigte Kohlenwasserstoffe* alkene

Alkohol *m* ERNEU ENER, LUFT alcohol

Alkoholyse *f* ABFALL, WASSER alcoholysis

Alkylbenzolsulfonat *nt* (*ABS*) ABFALL alkylbenzenesulfonate (*AmE*) (*ABS*), alkylbenzenesulphonate (*BrE*) (*ABS*)

Alleinkompostierung *f*: **~ entwässerter Klärschlämme** ABWASSER separate composting of dewatered sludges

Alluvialschicht *f* ABFALL, WASSER alluvial bed

Alluvium *nt* ABFALL, WASSER alluvial deposit

alpha-mesosaprob *adj* ABWASSER alpha-mesosaprobic

Alt- *in cpds* ABFALL, LUFT, WASSER used; **Altakkumulator** *m* ABFALL used storage battery; **Altanlage** *f* ABFALL existing plant, LUFT antiquated plant, POLITIK existing plant; **Altanlagenmodernisierung** *f* ABWASSER modernization of existing plants; **Altbatterie** *f* ABFALL used battery; **Altblei** *nt* ABFALL scrap lead; **Altdeponie** *f* ABFALL existing landfill site

Alterung *f* ABFALL, LUFT, WASSER ageing

alterungsbeständig *adj* LUFT, WASSER age-resistant

Alterung: **Alterungsbeständigkeit** *f* ABFALL, LUFT, WASSER ageing resistance, ageing stability; **Alterungseigenschaften** *f pl* ABFALL, WASSER ageing characteristics; **Alterungstest** *m* ABFALL, WASSER ageing test

Altfahrzeug *nt* ABFALL, WASSER end-of-life vehicle; **Altfahrzeugdeponie** *f* ABFALL used car dump; **Altfahrzeugschrott** *m* ABFALL car scrap; **Altfahrzeugverschrottungsanlage** *f* ABFALL car fragmentation plant

Altglas *nt* ABFALL waste glass, recycled glass, used glass; **Altglasbehälter** *m* ABFALL bottle bank, waste glass container; **Altglaseinsatz** *m* ABFALL recycled glass application; **Altglasscherben** *f pl* ABFALL recycled glass cullet; **Altglasverwertung** *f* ABFALL, WASSER glass recycling

Alt-: **Altkleider** *nt pl* ABFALL recycled clothing; **Altkunststoff** *m* ABFALL, WASSER recycled plastic

Altlast *f* ABFALL, BODEN contaminated site; **Altlastaufarbeitung** *f* BODEN clean-up of contaminated site; **Altlastenentstehung** *f* BODEN contaminated site generation; **Altlastenkataster** *m* BODEN contaminated site register; **Altlastensanierung** *f* BODEN clean-up of contaminated site; **Altlastpyrolyse** *f* BODEN contaminated site pyrolysis

Altmaterial: **aus ~ gewinnen** *vt* ABFALL recover

Altmetall *nt* ABFALL scrap metal; **Altmetallhändler** *m* ABFALL scrap metal dealer, scrap metal merchant; **Altmetallverwertungsanlage** *f* ABFALL scrap yard

Altöl *nt* ABFALL residual oil, spent oil, used oil, waste oil, ABWASSER spent oil; **Altölaufbereitungsbetrieb** *m* ABFALL oil regeneration plant; **Altöleinsammlung** *f* ABFALL waste oil collection; **Altölgesetz** *nt* POLITIK, WASSER Waste Oil Act; **Altölprodukt** *nt* BODEN old oil product; **Altölrückgewinnung** *f* ABFALL waste oil recovery; **Altölsammlung** *f* ABFALL waste oil collection; **Altöltank** *m* ABFALL slop tank; **Altölwiederverwertung** *f* ABFALL waste oil recovery

Altpapier *nt* ABFALL recycled paper, waste paper; **Altpapieraufbereitung** *f* ABFALL, WASSER waste paper preparation; **Altpapiereinsatz** *m* ABFALL recycled paper application; **Altpapiereinsatzquote** *f* ABFALL recycled paper application rate; **Altpapierrecycling** *nt* ABFALL, WASSER waste paper recycling; **Altpapiersammlung** *f* ABFALL paper collection, waste paper collection; **Altpapierverwertung** *f* ABFALL recycled paper use

Altreifen *m* ABFALL used tire (*AmE*), used tyre (*BrE*); **Altreifenbeseitigung** *f* ABFALL used tire disposal (*AmE*), used tyre disposal (*BrE*); **Altreifenkarkasse** *f* ABFALL used tire carcass (*AmE*), used tyre carcass (*BrE*); **Altreifenpyrolyse** *f* ABFALL used tire pyrolysis (*AmE*), used tyre pyrolysis (*BrE*)

Altstoff *m* ABFALL used material, waste material; **Altstoffeinsatzquote** *f* ABFALL waste material use rate; **Altstoffrücklaufquote** *f* ABFALL waste material return rate

Alufolie *f* ABFALL aluminium foil (*BrE*), aluminum foil (*AmE*)

Aluminat *nt* ABFALL aluminate

Aluminium *nt* (*Al*) ABFALL, ABWASSER, BODEN aluminium (*BrE*) (*Al*), aluminum (*AmE*) (*Al*); **Aluminiumchlorid** *nt* ABFALL, WASSER aluminium chloride (*BrE*), aluminum chloride (*AmE*); **Aluminiumfolie** *f* ABFALL aluminium foil (*BrE*), aluminum foil (*AmE*)

aluminiumhaltig *adj* ABFALL aluminium-containing (*BrE*), aluminum-containing (*AmE*); **~es Material** *nt* ABFALL aluminium-containing material (*BrE*), aluminum-containing material (*AmE*)

Aluminium: **Aluminiumoxid** *nt* ABFALL, ABWASSER aluminium oxide (*BrE*), aluminum oxide (*AmE*); **Aluminiumrecycling**

nt ABFALL, ABWASSER aluminium recycling (*BrE*), aluminum recycling (*AmE*); **Aluminiumsalze** *nt pl* ABWASSER *Fällmittel, Flockungsmittel* aluminium salts (*BrE*), aluminum salts (*AmE*); **Aluminiumschrott** *m* ABFALL, WASSER aluminium scrap (*BrE*), aluminum scrap (*AmE*); **Aluminiumsulfat** *nt* ABFALL, ABWASSER, WASSER *Hauptkomponent reiner Luft* aluminium sulphate (*BrE*), aluminum sulfate (*AmE*); **Aluminiumverpackung** *f* ABFALL aluminium packaging (*BrE*), aluminum packaging (*AmE*)

Aluverpackung *f* ABFALL aluminium packaging (*BrE*), aluminum packaging (*AmE*)

Ammoniak *nt* ALLG ammonia; **Ammoniak-Strippen** *nt* ABWASSER ammonia stripping; **Ammoniak-Verfahren** *nt* LUFT ammonia process

Ammonium *nt* ABFALL, ABWASSER, WASSER ammonium; **Ammoniumfracht** *f* ABWASSER ammonium load

ammoniumhaltig *adj* ABWASSER ammonium-containing; **~es Abwasser** *nt* ABWASSER ammonium-containing wastewater

amorph *adj* ABWASSER, ERNEU ENER amorphous, LUFT amorphous, noncrystalline; **~es Adsorbens** *nt* ABWASSER *wie Aktivkohle, Aktivkoks, Kieselgel,* LUFT amorphous adsorbent; **~es Adsorptionsmittel** *nt* ABWASSER *wie Aktivkohle, Aktivkoks, Kieselgel,* LUFT amorphous adsorbent; **~e Siliziumzelle** *f* ERNEU ENER amorphous silicon cell

Amortisationszeit *f* ERNEU ENER pay-back time

Amperemeter *nt* ANALYTIK ammeter, amperometer

Amperometrie *f* ANALYTIK *Verfahren der Voltammetrie* amperometry

amperometrisch *adj* ANALYTIK amperometric; **~e Bestimmungsmethoden** *f pl* ANALYTIK amperometric determination methods; **~e Meßsonde** *f* ANALYTIK amperometric measuring head; **~e Titration** *f* ANALYTIK amperometric titration

Amts- *in cpds* POLITIK official; **Amtsblatt** *nt* POLITIK *der EU* official journal; **Amtshilfe** *f* POLITIK cooperation among administrations; **Amtssprache** *f* POLITIK official language

Anabolismus *m* ABWASSER anabolism

anaerob *adj* ABFALL, ABWASSER, ERNEU ENER, WASSER *nicht auf Sauerstoff angewiesen* anaerobic; **~er Abbau** *m* ABFALL, ABWASSER, WASSER anaerobic degradation; **~er Atmungsprozeß** *m* ABFALL, WASSER anaerobic respiration process; **~e Ausfaulung** *f* ABFALL, ABWASSER, WASSER anaerobic digestion; **~e Bakterien** *f pl* ABWASSER anaerobic bacteria; **~es Becken** *nt* ABWASSER anaerobic lagoon, anaerobic pond; **~e Bedingungen** *f pl* ABWASSER anaerobic conditions; **~e Faulung** *f* ABFALL, ABWASSER, WASSER anaerobic digestion, anaerobic fermentation; **~e Gärung** *f* ABFALL, ABWASSER, WASSER anaerobic digestion, anaerobic fermentation; **~e Mikrobe** *f* ABWASSER anaerobic microbe, anaerobic microorganism; **~er Mikroorganismus** *m* ABWASSER anaerobic microorganism; **~ stabilisiert** *adj* ABWASSER anaerobically stabilized; **~ stabilisierter Schlamm** *m* ABWASSER anaerobically-stabilized sludge; **~er Teich** *m* ABWASSER anaerobic lagoon, anaerobic pond; **~e Vergärung** *f* ERNEU ENER *Biokonversion* anaerobic fermentation; **~e Zone** *f* ABWASSER *des Biofilms* anaerobic zone

anaerobisch: **~er Bioreaktor** *m* ABWASSER anaerobic bioreactor; **~es Heterotroph** *nt* ABWASSER anaerobic heterotroph

analoggesteuert *adj* ANALYTIK *Programmierung* analog-controlled

Analogkanal *m* ANALYTIK analog channel

Analysator *f* ANALYTIK analyser (*BrE*), analyzer (*AmE*)

Analyse *f* ANALYTIK analysis; **Analysedaten** *pl* ANALYTIK analysis data; **Analysenergebnis** *nt* ANALYTIK analytical finding, analytical result, analysis result; **Analysenfehler** *m* ANALYTIK analytical error; **Analysengerät** *nt* ANALYTIK analyser (*BrE*), analyzer (*AmE*); **Analysenkalibrierfunktion** *f* ANALYTIK analysis calibration function; **Analysenmeßtechniken** *f pl* ANALYTIK analysis measuring techniques; **Analysenort** *m* ANALYTIK analysis location; **Analysenparameter** *f* ANALYTIK analytical parameter

analysenrein *adj* ANALYTIK analytically pure

Analyse: **Analysenumfang** *m* ANALYTIK analysis extent; **Analysenvorschrift** *f* ANALYTIK analytical procedure; **Analysenwaage** *f* ANALYTIK analytical scale; **Analyseverfahren** *nt* ANALYTIK analytical procedure, analytical process; **Analysezeit** *f* ANALYTIK analysis time

analysierbar *adj* ANALYTIK analysable (*BrE*), analyzable (*AmE*)

analysieren *vt* ANALYTIK analyse (*BrE*), analyze (*AmE*)

Analyt *m* ANALYTIK analyte; **Analyt-Abscheidung** *f* ANALYTIK analyte separation

analytisch *adj* ANALYTIK analytical; **~e Qualitätssicherung** *f* (*AQS*) ANALYTIK analytical quality control

anbacken *vt* ABFALL, ABWASSER, LUFT, WASSER cake

anbackend: **~er Staub** *m* LUFT caked dust

Anbackung *f* ABFALL, ABWASSER, LUFT, WASSER caking

Andreasen-Pipette *f* ANALYTIK *Messung der Sedimentation von Schwebstoffen in Flüssigkeiten* Andreasen pipette

Anemometer *nt* ANALYTIK anemometer; **Anemometerhöhe** *f* LUFT *Schornstein* anemometer height

anerkannt: **allgemein ~e Regeln der Technik** *f pl* (*aaRdT*) POLITIK, WASSER generally accepted technical procedures; **~e Technik** *f* ABWASSER, ERNEU ENER, LUFT, WASSER established technologies

Anfahrbetrieb *m* LUFT starting operation

Anfahrzeit *f* ABWASSER initiation time

Anfallspitze *f* WASSER peak water flow

Anfallstelle *f* ABFALL source of waste generation

Anfangs- *in cpds* ALLG initial, preliminary; **Anfangsadsorption** *f* **bei optimalem Flockulantenzusatz** ABWASSER initial adsorption at optimum polymer dosage; **Anfangssetzung** *f* ABFALL, WASSER preliminary settlement

Anforderungen *f pl* MANAG demands

Angebotspotential *nt* ERNEU ENER *der erneuerbaren Energiequellen* supply potential

angesäuert: **~e Bodenfläche** *f* BODEN acidified area

angewandt: **~e Forschung** *f* ANALYTIK applied research; **~e Umweltforschung** *f* ANALYTIK applied environmental research

angezapft: **~er Fluß** *m* WASSER beheaded river

anhaften 1. *vt* ABWASSER *Partikeln* adhere; **2.** *vi* LUFT adhere

anhaftend *adj* ALLG adherent

Anhaftung *f* ABWASSER, LUFT *der Partikeln* adherence

Anhäufung *f* LUFT agglomeration

Anhydrit *m* LUFT *Gipsart* anhydrite

An-IA *abbr* (*Anionenaustauscher*) ABFALL, ABWASSER, WASSER anion exchanger

Anion *nt* ABFALL, ABWASSER, WASSER anion

anionenaktiv *adj* ABWASSER anionic-active

Anion: **Anionenaustausch** *m* ABFALL, BODEN, WASSER anion exchange; **Anionenaustausch-Chromatographie** *f* ANALYTIK anion-exchange chromatography; **Anionenaustauscher** *m* (*An-IA*) ABFALL, ABWASSER, WASSER anion exchanger

anionendurchlässig: **~e Membran** *f* WASSER anion-permeable membrane

anionenselektiv *adj* ABWASSER anion-selective; **~e Membran** *f* ABWASSER anion-selective membrane

anisokinetisch *adj* ANALYTIK anisokinetic; **~e Probenahme** *f* ANALYTIK anisokinetic sampling

Ankerboje *f* MEER mooring buoy

Ankergruppe *f* ABWASSER *Ionenaustauscher* anchor group, fixed group

ankuppeln *vt* ABFALL, WASSER connect

Anlage *f* ALLG facility, installation, plant

anlagebezogen *adj* ALLG installation-related, plant-related

Anlage: **Anlagengenehmigung** *f* POLITIK plant permit, plant

authorization, plant licence (*BrE*), plant license (*AmE*); **Anlagensicherheit** *f* SICHERHEIT plant safety
anlagenspezifisch *adj* ALLG installation-specific, plant-specific
anlagentechnisch *adj* ALLG plant-related, plant-specific, installation-related
anlagern *vt* LUFT *Staubteilchen an die Flüssigkeittröpfchen* add on
Anlandung *f* ABFALL, WASSER alluvial deposit
Anlieferung *f* ABFALL, ABWASSER, WASSER delivery; **Anlieferungsfahrzeug** *nt* ABFALL delivery vehicle; **Anlieferungszeit** *f* ABFALL delivery time
Annäherung *f* ANALYTIK approximation; **Annäherungssensor** *m* ANALYTIK proximity pick-up, proximity sensor; **Annäherungstest** *m* ANALYTIK approximation test
Annahmeerklärung *f* POLITIK acceptance certificate
Annahmestelle *f* ABFALL collection station
Anode *f* BODEN, LUFT anode, *Elektrofilter* positive electrode
anorganisch *adj* ABFALL, ABWASSER, LUFT, WASSER inorganic; **~er Ballaststoff** *m* ABFALL, WASSER inorganic roughage; **~e Chemikalie** *f* ABWASSER, WASSER inorganic chemical; **~e Feststoffe** *m pl* ABFALL, WASSER inorganic solids; **~es Flockungsmittel** *nt* ABFALL, ABWASSER, ANALYTIK, LUFT, WASSER coagulant; **~es Gas** *nt* LUFT inorganic gas; **~e Kontaminanten** *nt pl* ABWASSER, LUFT inorganic contaminants; **~e Salze** *nt pl* ABFALL, WASSER inorganic salts; **~es Säuregas** *nt* LUFT inorganic acid gas; **~e Schadstoffe** *m pl* ABWASSER, LUFT inorganic pollutants; **~er Stickstoffdünger** *m* BODEN *Bodenverbesserungsmittel* anorganic nitrogen fertilizer; **~es Trägermaterial** *nt* LUFT *Biofilter* inorganic base material
anoxisch *adj* ABWASSER anoxic; **~e Bedingungen** *f pl* ABWASSER anoxic conditions; **~e Denitrifikationszone** *f* ABWASSER anoxic denitrification zone
anreichern *vt* ALLG accumulate, recharge, replenish, enrich
Anreicherung *f* ABFALL accumulation, ANALYTIK accumulation, enrichment, LUFT *Staubablagerung*, MEER, WASSER accumulation; **~ von Schadstoffen** ABFALL, ABWASSER, BODEN, LUFT, WASSER accumulation of contaminants, accumulation of harmful substances, accumulation of pollutants; **Anreicherungsbecken** *nt* ABFALL, WASSER recharge basin
Anreiz *m* MANAG, POLITIK incentive
***Anreiz*: ~ setzen** *vi* MANAG *Recycling* offer incentives
Ansammlung *f* BODEN accumulation, agglomeration
ansäuern *vt* ABFALL, ABWASSER, BODEN, LUFT acidify
ansäuernd *adj* ABFALL, ABWASSER, BODEN, LUFT acidifying
Ansäuerung *f* ABFALL, ABWASSER, BODEN, LUFT acidification, acid formation
Ansaug- *in cpds* ABWASSER, LUFT intake; **Ansaugdruck** *m* ABWASSER intake pressure
ansaugen: sich ~ *v refl* LUFT adhere by suction
Ansaugen: zum ~ bringen *vt* ABFALL, ABWASSER, WASSER *Pumpe* prime
Ansaugenlassen *nt* ABWASSER, WASSER *einer Pumpe* priming
***Ansaug-*: Ansaugluft** *f* LUFT intake air; **Ansaugöffnung** *f* ABFALL, ABWASSER, LUFT, WASSER pump inlet; **Ansaugrohr** *nt* ABFALL priming pipe, suction pipe, ABWASSER priming pipe, WASSER suction pipe, *einer Pumpe* priming pipe; **Ansaugvolumen** *nt* ABWASSER, LUFT intake volume
anschließen *vt* WASSER *Gebäude an die Trinkwasserversorgung* connect
Anschluß *m* ABFALL, ABWASSER, WASSER supply, connection; **Anschlußgebühr** *f* POLITIK sewerage connection duty; **Anschlußleitung** *f* ABFALL, WASSER connecting pipe, service mains; **Anschlußrohr** *nt* ABFALL, WASSER connecting pipe
Ansetzbehälter *m* ABWASSER batching tank
ansprengen *vt* MEER wring
Anspringtemperatur *f* LUFT *Katalysator* light-off temperature
Anströmgeschwindigkeit *f* ABWASSER face velocity, LUFT face velocity, *Katalysator* blower stream velocity
Anströmungsgeschwindigkeit *f* LUFT laminar flow velocity
anthropogen *adj* ALLG anthropogenic; **~ bedingter Treibhauseffekt** *m* LUFT anthropogenic greenhouse effect; **~ bedingte Übersäuerung** *f* BODEN anthropogenic acidification; **~ bedingte Wasserverschmutzung** *f* ABWASSER, WASSER anthropogenic water pollution; **~ bedingte Wasserverunreinigung** *f* ABWASSER, WASSER anthropogenic water pollution; **~e Belastung** *f* ABWASSER, BODEN, LUFT anthropogenic pollution; **~e Klimaveränderungen** *f pl* LUFT anthropogenic climate changes; **~es Ökosystem** *nt* ABWASSER, BODEN, LUFT anthropogenic ecosystem; **~e Quellen** *f pl* LUFT, WASSER anthropogenic sources; **~er Schadstoff** *m* ABWASSER, BODEN, LUFT anthropogenic pollutant; **~ verursacht** *adj* ABWASSER, BODEN, LUFT anthropogenically caused, manmade, WASSER manmade; **~ verursachte Bodenkontamination** *f* BODEN anthropogenically-caused soil contamination; **~ verursachtes Erdbeben** *nt* BODEN manmade earthquake; **~e Wasserinhaltsstoffe** *m pl* WASSER anthropogenic water components
Anthroposphäre *f* BODEN anthroposphere
Antikatalysator *m* LUFT inhibitor
Antikatalyse *f* LUFT negative catalysis
Antiklopfbrennstoff *m* ABFALL, LUFT antiknock fuel
Antiklopfmittel *nt* ABFALL, LUFT antiknock additive, antiknock agent
Antimon *nt* (*Sb*) WASSER antimony (*Sb*)
antimonhaltig *adj* ABFALL, WASSER antimonial
antistatisch: ~e Schutzkleidung *f* gegen aggressive Chemikalien SICHERHEIT antistatic protective clothing against chemical aggression
Antriebselemente *nt pl* ABFALL, WASSER drive units
Anwendbarkeit *f* ALLG *von Verfahren* applicability
Anwendung *f* ABWASSER *von Verfahren*, ANALYTIK, ERNEU ENER, LUFT application, POLITIK instrumentation; **~ der Technik** ANALYTIK, MANAG, SICHERHEIT technology application; **Anwendungsmöglichkeiten** *f pl* ABWASSER, ANALYTIK, ERNEU ENER, LÄRM, LUFT, MEER, SICHERHEIT, WASSER *von Verfahren* application possibilities; **Anwendungssicherheit** *f* SICHERHEIT use safety; **Anwendungstechnik** *f* ABWASSER, ANALYTIK, BODEN, LUFT, SICHERHEIT, WASSER application technology
anwendungstechnisch *adj* ALLG application-technological
***Anwendung*: Anwendungsverfahren** *nt* ABWASSER, ANALYTIK, BODEN, LUFT, SICHERHEIT, WASSER method of application
anzapfen *vt* ABFALL, WASSER draw water from
Anzeigeinstrument *nt* ANALYTIK indicating instrument
Anzeigevorrichtung *f* ANALYTIK indicating instrument
Anziehung *f* ABWASSER, ERNEU ENER attraction; **Anziehungskraft** *f* ERNEU ENER force of attraction; **Anziehungsvermögen** *nt* ABWASSER, ERNEU ENER force of attraction
AOS *abbr* (*abbaubare organische Substanz*) ABFALL, ABWASSER, WASSER degradable organic matter
AOX *abbr* (*adsorbierbare organische Halogenverbindungen*) ABFALL, ABWASSER, ANALYTIK, WASSER AOX (*adsorbable organic halogen compounds*); **AOX-Analyse** *f* ANALYTIK AOX analysis
AQS *abbr* (*analytische Qualitätssicherung*) ANALYTIK analytical quality control
Aquädukt *m* WASSER aqueduct
Aquakultur *f* ABFALL, WASSER aquaculture
aquatisch *adj* MEER, WASSER aquatic; **~e Lebensform** *f* MEER, WASSER aquatic life; **~es Ökosystem** *nt* MEER, WASSER aquatic ecosystem

Aquifer *m* ABFALL, ABWASSER, WASSER aquifer; **Aquiferverunreinigung** *f* ABFALL, ABWASSER, WASSER *durch Abwasser* aquifer contamination
Äquipotentialfläche *f* ABFALL, WASSER equipotential surface
Äquivalent: **~ je Million** *phr* (*EPM*) LUFT equivalent per million (*EPM*); **Äquivalentdurchmesser** *m* LUFT equivalent diameter
Äquivalenznormalruß *m* LUFT equivalent standard smoke
Ar *abbr* (*Argon*) LUFT Ar (*argon*)
aramidfaserverstärkt: **~es Duromer** *nt* ABFALL, WASSER aramide fiber reinforced thermoset (*AmE*), aramide fibre reinforced thermoset (*BrE*)
Aräometer *nt* ANALYTIK *Dichtebestimmung von Flüssigkeiten* densimeter
Arbeit *f*: **~ mit Gefahrstoffen** SICHERHEIT working with dangerous materials, working with hazardous substances
arbeitsintensiv *adj* ALLG labor-intensive (*AmE*), labour-intensive (*BrE*); **~e Methode** *f* ALLG labor-intensive method (*AmE*), labour-intensive method (*BrE*)
Arbeit: **Arbeitsmedium** *nt* ERNEU ENER *Wärmepumpe* working medium; **Arbeitsmittel** *nt* ERNEU ENER *Wärmepumpe* working medium; **Arbeitsplatz** *m* SICHERHEIT workplace; **am Arbeitsplatz** *phr* SICHERHEIT at the workplace
arbeitsplatzbezogen *adj* SICHERHEIT workplace-related; **~er Wasserbedarf** *m* WASSER workplace-related water demand
Arbeit: **Arbeitsplatzgrenzschwellenwert** *m* SICHERHEIT TLV at the workplace; **Arbeitssicherheit** *f* SICHERHEIT work safety; **Arbeitsweise** *f* ABWASSER, ANALYTIK, LUFT, WASSER procedure
archimedisch: **~es Prinzip** *nt* ABFALL, ABWASSER, WASSER Archimedes' principle; **~e Wasserschnecke** *f* ABWASSER, WASSER Archimedean screw
Argon *nt* (*Ar*) ABFALL, LUFT argon (*Ar*)
Armatur *f* ABFALL, ABWASSER, ANALYTIK, WASSER armature, fitting
armieren *vt* ABFALL, WASSER reinforce
aromatisch *adj* ABFALL, ABWASSER, LUFT, WASSER aromatic; **~e Kohlenwasserstoffe** *m pl* ABFALL, ABWASSER, ANALYTIK, LUFT, WASSER aromatic hydrocarbons; **~es Polyamid** *nt* LUFT *Filter-Fasermaterial* aromatic polyamid; **~e Verbindungen** *f pl* ABFALL, ABWASSER, LUFT, WASSER aromatic compounds
Arsen *nt* (*As*) ABFALL, ABWASSER, BODEN arsenic (*As*)
artesisch *adj* WASSER artesian; **~er Brunnen** *m* WASSER artesian well; **~es Grundwasser** *nt* ABFALL, WASSER artesian groundwater; **~e Quelle** *f* WASSER artesian spring
Artikel *m* POLITIK *im Grundsatz* article
As *abbr* (*Arsen*) ABFALL, ABWASSER, BODEN *ICP-OES* As (*arsenic*)
Asbest *nt* ABWASSER, LUFT, SICHERHEIT asbestos; **Asbestbelastung** *f* ABWASSER, SICHERHEIT asbestos exposure; **Asbestbelastungsgrenzwert** *m* ABWASSER, SICHERHEIT asbestos-exposure limit
asbestfrei *adj* SICHERHEIT asbestos-free; **~e Isolierplatte** *f* SICHERHEIT asbestos-free insulating plate
Asbest: **Asbestzementrohr** *nt* ABFALL, ABWASSER, WASSER asbestos cement pipe
Asche *f* ABFALL, ANALYTIK, LUFT ash
aschefrei *adj* ABFALL, LUFT ash-free
Asche: **Aschegehalt** *m* ABFALL, LUFT ash content; **Aschenabzug** *m* ABFALL clinker pit; **Aschenaustrag** *m* ABFALL ash discharge; **Aschenschlamm** *m* ABWASSER sludge containing ash; **Aschensinterung** *f* ABFALL ash vitrification; **Asche- und Verbrennungsrückstand** *m* ABFALL, ANALYTIK, LUFT ash and combustion residue
Asphalt *m* ABFALL, BODEN asphalt; **Asphaltbetondichtung** *f* ABFALL, BODEN asphalt sealing; **Asphaltrecycling** *nt* ABFALL, BODEN asphalt recycling
Aspirationspsychrometer *nt* ANALYTIK aspiration psychrometer
Assimilation *f* ABWASSER, ANALYTIK *biologisches Denitrifikationsverfahren* assimilation
assimilieren *vt* ABWASSER, ANALYTIK assimilate
Assoziationskolloid *nt* ABFALL, ABWASSER, WASSER association colloid
Asymptote *f* LUFT *Schornstein* asymptote
Asynchrongenerator *m* ERNEU ENER *Windturbine* induction generator
Atem *m* SICHERHEIT breath; **Atemgerät** *nt* SICHERHEIT breathing apparatus; **Atemgerät** *nt* **mit externer Luftversorgung** SICHERHEIT supplied-air breathing apparatus; **Atemgrenzwert** *m* LUFT breathing capacity; **Atemluft** *f* SICHERHEIT breathable air
Atemschutz *m* SICHERHEIT breathing protection; **Atemschutzfilter** *nt* SICHERHEIT respiratory filter; **Atemschutzgerät** *nt* SICHERHEIT respiratory protection apparatus; **Atemschutzmaske** *f* SICHERHEIT respiratory mask; **Atemschutzsystem** *nt* SICHERHEIT breathing protection system, respiratory protection system
Atemzwecken: **zu ~ ungeeignete Luft** *f* SICHERHEIT air unfit for respiration
Atmosphäre *f* LUFT atmosphere; **Atmosphärendruck** *m* LUFT atmospheric pressure; **Atmosphärengewicht** *nt* LUFT atmospheric mass; **Atmosphärenmasse** *f* LUFT atmospheric mass; **Atmosphärentemperatur** *f* LUFT atmospheric temperature; **Atmosphärentrübung** *f* LUFT air turbidity
atmosphärisch *adj* LUFT atmospheric; **~e Aufwärmung** *f* LUFT atmospheric heat trapping; **~er Auswaschvorgang** *m* LUFT atmospheric scrubbing; **~e Erscheinung** *f* LUFT atmospheric phenomenon; **~e Gase** *nt pl* LUFT atmospheric gases; **~e Luftbelastung** *f* LUFT atmospheric loading; **~er Niederschlag** *m* LUFT atmospheric fallout, atmospheric precipitation, precipitation, WASSER atmospheric precipitation; **~es Ozon** *nt* LUFT atmospheric ozone; **~e Ozonkonzentration** *f* LUFT atmospheric ozone concentration; **~e Rückstrahlung** *f* LUFT atmospheric reflection; **~er Sauerstoff** *m* LUFT atmospheric oxygen; **~e Säurekapazität** *f* LUFT atmospheric acidity; **~er Schadstoff** *m* LUFT atmospheric pollutant; **~er Schwefel** *m* LUFT atmospheric sulfur (*AmE*), atmospheric sulphur (*BrE*); **~e Sperrschicht** *f* LUFT atmospheric barrier layer; **~e Spurengase** *nt pl* LUFT *Wasserdampf, Kohlendioxid, Methan, Distickstoffoxid, Ozon* atmospheric trace gases; **~e Verdunklung** *f* LUFT atmospheric obscurity
Atmung: **~ beeinflussende Umgebungsluft** *f* LUFT, SICHERHEIT atmosphere which affects breathing
Atom *nt* ANALYTIK atom; **Atomabsorptionsspektrometer** *nt* (*AAS*) ANALYTIK atomic absorption spectrometer (*AAS*); **Atomabsorptionsspektrometrie** *f* (*AAS*) ANALYTIK *flammenphotometrische Detektion* atomic absorption spectrometry (*AAS*); **Atomemissionsdetektion** *f* (*AED*) ANALYTIK atomic emission detection (*AED*); **Atomemissionsdetektor** *m* ANALYTIK atomic emission detector; **Atomemissionsspektrometrie** *f* (*AES*) ANALYTIK atomic emission spectrometry (*AES*); **Atomfluoreszenz** *f* ANALYTIK atomic fluorescence; **Atomfluoreszenzspektrometrie** *f* (*AFS*) ANALYTIK atomic fluorescence spectrometry (*AFS*); **Atommüll** *m* ABFALL, SICHERHEIT radioactive waste; **Atommüll** *m* **mit langer Halbwertzeit** ABFALL long-life radioactive waste
atomrechtlich *adj* POLITIK according to the law on nuclear power; **~e Anlagengenehmigung** *f* POLITIK plant permit according to the law on nuclear power
atomspektrometrisch *adj* ANALYTIK atomspectrometric
ATP *abbr* (*Adenosintriphosphat*) ABWASSER ATP (*adenosine triphosphate*)
ATS *abbr* (*aerob-thermophile Stabilisierung*) ABWASSER aerobic-thermophilic stabilization
ätzend *adj* SICHERHEIT *als Gefahrsymbol* irritant
Ätzkalk *m* BODEN, MEER quicklime

Audiometer *nt* ANALYTIK audiometer
Auenböden *m pl* BODEN fluvial alluvium
Aufarbeitung *f* ABFALL, LUFT recovery; **Aufarbeitungsvorrichtung** *f* ABFALL, LUFT, WASSER recovery device; **Aufarbeitungszentrum** *nt* **für wiederverwertbare feste Altstoffe** ABFALL processing center for recyclable solid waste materials (*AmE*), processing centre for recyclable solid waste materials (*BrE*)
Aufbau *m* ABFALL structure
aufbauen *vt* ABFALL *Deponiekörper* construct
aufbereiten *vt* ALLG treat, process
aufbereitet: **~e Abfälle** *m pl* ABFALL processed wastes; **~e Schlacke** *f* ABFALL processed slag
Aufbereitung *f* ALLG treatment, processing; **Aufbereitungsaggregat** *nt* ABFALL processing unit; **Aufbereitungsanlage** *f* ABFALL preparation plant, processing plant; **Aufbereitungseinheit** *f* ABFALL treatment unit, processing unit, WASSER treatment unit; **Aufbereitungskosten** *pl* ABFALL, WASSER reprocessing costs; **Aufbereitungsverfahren** *nt* ABFALL treatment process, conditioning process, preparation process, WASSER treatment process, conditioning process
aufbewahren *vt* ANALYTIK *Probe* store
aufblähen *vt* LUFT, MEER inflate
Aufblasanschluß *m* MEER *einer Ölsperre* inflation cuff
aufblasbar *adj* ALLG inflatable
aufblasen *vt* LUFT, MEER inflate
Aufblasen *nt* LUFT, MEER inflation
aufbrechen *vt* ABFALL break open, ABWASSER rupture, SICHERHEIT, WASSER break open
Aufbrechen *nt*: **~ der Flocke** ABWASSER rupture of floc
Aufbringfähigkeit *f* ABWASSER possibility of wastewater deposition
Aufbringung *f* ABWASSER deposit; **~ von Klärschlamm auf Grünland** ABWASSER deposition of wastewater sludge on grassland; **Aufbringungsverbot** *nt* **für Klärschlamm** ABWASSER wastewater sludge deposition ban
Aufenthaltszeit *f* ABFALL, ABWASSER, LUFT, WASSER detention time, retention time
Auffangeinrichtung *f* ABFALL, MEER, WASSER *für Öl* collection device
auffangen *vt* ABFALL, ABWASSER, ANALYTIK collect, ALLG *Schadstoffe* capture, LUFT intercept, capture, collect, MEER *Schadstoffe* capture, *Öl* collect, WASSER collect
Auffangen *nt* ABFALL collection, ABWASSER, BODEN, LÄRM capture, LUFT capture, *von Partikeln* interception, MANAG capture, MEER *von Öl* collection, *von Schadstoffen* capture
Auffanggefäß *nt* ANALYTIK receptacle
Auffangtrichter *m* ABFALL, MEER, WASSER collection hopper
aufflammen *vi* ABFALL, ABWASSER, LUFT, WASSER flare
auffüllen *vt* ABFALL, BODEN fill up
Auffüllung *f* ABFALL, BODEN *angefahrener oder aufgeschütteter Boden* backfill
Aufgabe *f* ABFALL, ANALYTIK, BODEN task; **Aufgabegut** *nt* LUFT charging material; **Aufgabenstellung** *f* ANALYTIK problem definition; **Aufgabeschacht** *m* ABFALL, WASSER feed shaft; **Aufgabetrommel** *f* ABFALL, WASSER feed drum; **Aufgabevorrichtung** *f* ABFALL, WASSER feeder
aufgeladen: **~e Partikel** *f* LUFT charged particle
aufgelöst *adj* ABWASSER, BODEN solubilized; **~e organische Verbindungen** *f pl* ABWASSER solubilized organics
aufgenommen *adj* ABFALL, WASSER ingested, absorbed
aufgeschäumt: **~es Thermoplast** *nt* ABFALL expandable thermoplastics
Aufgrundlaufen *nt* MEER *eines Schiffes* grounding
aufhaspeln *vt* ABFALL spin off
aufheizen *vt* ANALYTIK, LUFT heat up
Aufheizung *f* ANALYTIK, LUFT heating up
Aufkocher *m* LUFT boiling vessel
Aufkommen *nt* POLITIK revenue, yield
aufkommensneutral *adj* POLITIK revenue-neutral
Aufkommen: **Aufkommensneutralität** *f* POLITIK revenue neutrality
Auflagerost *m* LUFT supporting grate
auflastbedingt: **~e Beschädigung** *f* ABFALL load-induced damage
Auflast *f*: **~ der Deponie** ABFALL load of landfill
auflockern *vt* ABFALL, BODEN loosen, loosen up
Auflockerung *f* ABFALL, ANALYTIK breaking-up, loosening up, WASSER breaking-up
auflösen *vt* ABFALL, ABWASSER, ANALYTIK, LUFT *Gas in Flüssigkeit* dissolve, solubilize
Auflösung *f* ABFALL, ABWASSER dissolution, ANALYTIK, BODEN decomposition, WASSER dissolution; **Auflösungsanalyse** *f* ANALYTIK electrochemical stripping analysis (*ESA*), ESA (*electrochemical stripping analysis*)
Aufnahme *f* ABFALL absorption, ingestion, ABWASSER, ANALYTIK, LUFT absorption, WASSER absorption, ingestion; **Aufnahmebunker** *m* ABFALL receiving bunker, refuse bunker, waste bunker; **Aufnahmefähigkeit** *f* ABFALL absorption capacity, holding capacity, LUFT absorptivity, absorption capacity, SICHERHEIT absorbency, WASSER absorption capacity, holding capacity; **Aufnahmekapazität** *f* ABFALL, WASSER absorption capacity
aufnehmen *vt* ABFALL ingest, SICHERHEIT absorb, WASSER ingest
Aufprall *m* ABFALL collision, LUFT impact, impaction, *Elektrofilter* impingement
aufprallen: **1. ~ auf** *vt* LUFT collide with, impact; **2.** *vi* LUFT impinge
Aufprall: **Aufprallmethode** *f* ANALYTIK *Messung der Luftqualität* impinge method
aufpumpen *vt* LUFT, MEER inflate
Aufpumpen *nt* LUFT, MEER inflation
Aufsalzung *f* ABFALL, WASSER salination
Aufschlag-Abscheider *m* LUFT impingement separator
aufschlagen: **1. ~ auf** *vt* LUFT collide with, impact; **2.** *vi* LUFT impinge
aufschließen *vi* BODEN, WASSER *Schadstoffe* digest
aufschmelzen 1. *vt* ABFALL, ABWASSER, BODEN melt; **2.** *vi* ABFALL, ABWASSER, BODEN melt
aufschütten *vt* ABFALL, BODEN, WASSER fill up
Aufschüttung *f* ABFALL, BODEN, WASSER *Haldendeponie* earth bank
Aufschwimmen *nt* ALLG floating
aufschwimmend: **~e Stoffe** *m pl* ABWASSER floating solids
Aufsicht *f* ALLG supervision; **Aufsichtsamt** *nt* ABFALL, MANAG, WASSER river authority; **Aufsichtsorgan** *nt* MEER supervisor
Aufstiegsgeschwindigkeit *f* LUFT *von Gasen* rate of ascent
aufströmen *vt* ABWASSER, LUFT entrain
Aufstromfilter *nt* WASSER upflow filter
Aufstromklassierer *m* WASSER up-current classifier
Auftanken *nt* MEER refueling (*AmE*), refuelling (*BrE*)
auftragen *vt* ANALYTIK *Chromatogramm* plot
Auftragung *f*: **~ von Klärschlamm auf Flächen** ABWASSER *in der Landwirtschaft* surface application
auftreffen: **1. ~ auf** *vt* LUFT collide with, impact; **2.** *vi* LUFT impinge
Auftrieb *m* ERNEU ENER *Windturbine* lift
Aufwand-Nutzen-Relation *f* MANAG cost-benefit ratio
Aufwärmofen *m* ABFALL, WASSER heating furnace
Aufwärts: **Aufwärtsfiltration** *f* ABFALL, WASSER reverse filtration; **Aufwärtsflächeneinbau** *m* ABFALL, WASSER upward surface installation
aufweichen *vt* ABFALL, WASSER macerate
Aufweichen *nt* ABFALL, WASSER *von Abfällen* maceration
Aufwindkraftwerk *nt* ERNEU ENER upwind power plant
Augenscheinanalyse *f* ANALYTIK visual analysis

Augerelektronspektroskopie *f* ANALYTIK Auger electron spectroscopy
Ausbeutung *f*: **~ der geothermalen Energie** LUFT geothermal energy exploitation
ausblasbar: **~e organische Halogenverbindungen** *f pl* (*POX*) BODEN, WASSER purgeable organic halogen compounds (*POX*)
ausblasen *vt* ABFALL, WASSER blow out
Ausbrand *m* LUFT burnout, burning-out; **Ausbrandrost** *m* ABFALL burnout grate
ausbrechen *vi* SICHERHEIT break out
ausbreiten: **sich ~** *v refl* ABFALL, LUFT, MEER diffuse, disperse, propagate, spread
Ausbreitung *f* ABFALL dispersion, diffusion, LUFT propagation, diffusion, MEER spreading, WASSER dispersion, diffusion; **~ von Schadstoffen** ABFALL, ABWASSER, BODEN, LUFT, WASSER propagation of pollutants; **Ausbreitungsdämpfung** *f* LÄRM propagation deadening; **Ausbreitungsklasse** *f* LUFT *Abgasfahne von Schornstein* propagation class; **Ausbreitungsverhalten** *nt* ABFALL, WASSER dispersion behavior (*AmE*), dispersion behaviour (*BrE*); **Ausbreitungsweg** *m* BODEN *der Schadstoffe im Boden* propagation path
Ausbrennaggregat *nt* ABFALL, WASSER burn-out aggregate
ausbrennen *vt* ABFALL, BODEN, LUFT *Rauchgase* burn out
Ausdämpfen *nt* LUFT steam out
ausdehnen: **mechanisch ~** *vt* BODEN *Injektion* mechanically expand
Ausdehnungsthermometer *nt* ANALYTIK expansion thermometer
Ausdünnung *f* LUFT *der Ozonschicht* thinning
Ausdünstungen *f pl* SICHERHEIT *von Farben, Lacken, Klebstoffen* fumes
ausfällen *vt* ALLG precipitate, coagulate
Ausfällung *f* ABWASSER, ANALYTIK, LUFT precipitation
ausfaulen: **~ lassen** *vt* ABFALL putrefy
Ausfaulgrad *m* ABWASSER degree of stabilization
ausfiltern *vt* ABFALL, ANALYTIK, LUFT, WASSER filter out
ausfließen *vi* ABFALL, WASSER discharge
Ausfließen *nt* ABFALL, WASSER discharge
ausflockbar *adj* ABWASSER, ANALYTIK, LUFT, WASSER coagulable
ausflocken *vi* ALLG coagulate, flocculate
Ausflockung *f* ALLG coagulation, flocculation
Ausfluß *m* ABFALL discharge, ABWASSER discharge, outflow, MEER *Vorgang* runoff, WASSER discharge; **Ausflußrohr** *nt* ABFALL, WASSER delivery pipe; **Ausflußviskosimeter** *nt* ANALYTIK flowing-out viscosimeter
Ausfüllungsgrad *m* ERNEU ENER *Windturbine* solidity
Ausgang *m* ANALYTIK output; **Ausgangsgrößen** *f pl* MANAG output; **Ausgangsmaterial** *nt* ABFALL, ABWASSER base material, compost raw material, WASSER base material; **Ausgangsstoff** *m* ABFALL, ABWASSER, ERNEU ENER, WASSER base material
Ausgasen *nt* ABFALL, WASSER gas evolution
Ausgasung *f* ABFALL gas evolution, ANALYTIK *von Proben* gas release, WASSER *der Deponie* gas evolution
ausgeflossen: **~es Öl** *nt* MEER oil spill, spill
ausgegeben: **~e Meßwerte** *m pl* ANALYTIK read-out of measured values
ausgeglichen: **~e CO_2-Bilanz** *f* LUFT zero CO_2 emissions
ausgehoben *adj* ABFALL, BODEN excavated
ausgekoffert *adj* BODEN excavated; **~er Boden** *m* BODEN excavated soil; **~es Material** *nt* BODEN excavated material
ausgelöst *adj* ALLG dissolved
Ausgleich *m* ABFALL, BODEN, WASSER adjustment, compensation; **Ausgleichbecken** *nt* ABFALL, WASSER equalizing tank
ausgleichen *vt* POLITIK, SICHERHEIT compensate
Ausgleich: **Ausgleichsbehälter** *m* ABFALL, WASSER distribution reservoir; **Ausgleichsbunker** *m* BODEN surge hopper; **Ausgleichsschicht** *f* ABFALL equalizing layer; **Ausgleichszahlung** *f* POLITIK compensation
Ausgleichung *f* ABWASSER equalization
ausgraben *vt* BODEN *kontaminiertes Material* dig out, dig up
Ausguß *m* ABFALL, WASSER *einer Pumpe* spout, sink; **Ausgußrohr** *nt* ABFALL, WASSER delivery pipe
ausheben *vt* ABFALL, BODEN excavate
Aushub *m* ABFALL, BODEN excavation; **Aushubmasse** *f* BODEN excavated mass; **Aushubstation** *f* ABFALL excavation system
auskleiden *vt* ALLG line
Auskleidung *f* ABFALL, ABWASSER, LÄRM, WASSER *eines Beckens, Behälters* lining
auskoffern *vt* BODEN excavate
Auskolkung *f* ABFALL underwashing
auskondensieren *vt* LUFT condense
auskristallisieren *vt* LUFT crystallize
Auslage *f*: **~ von Unterlagen** POLITIK presentation of documents for public inspection
Auslaß *m* ABWASSER outlet; **Auslaßdurchflußregelung** *f* ABWASSER outlet flow control; **Auslaßventil** *nt* LUFT exhaust valve
Auslastungsgrad *m* ABFALL, ABWASSER, WASSER rate of utilization
auslaufen *vi* ABFALL drain off, run out, MEER leak
Auslaufen *nt* ALLG leakage
Auslaug- *in cpds* ABWASSER leaching
Auslaugbarkeit *f* BODEN leachability
auslaugen *vt* ABFALL leach, ABWASSER elutriate, BODEN, MEER, WASSER leach
Auslaugen *nt* ABFALL leaching, ABWASSER elutriation, leaching, lixiviation, BODEN, MEER, WASSER leaching
auslaugsicher *adj* ABFALL, BODEN, WASSER safe against leaching
Auslaug-: **Auslaugtest** *m* BODEN leaching test
Auslaugung *f* ABFALL leaching, ABWASSER elutriation, leaching, lixiviation, WASSER leaching; **Auslaugungsgraben** *m* ABFALL, WASSER leaching trench
Auslaug-: **Auslaugverfahren** *nt* ABFALL, WASSER leaching process; **Auslaugverhalten** *nt* BODEN leaching behavior (*AmE*), leaching behaviour (*BrE*), leaching property
Auslegen *nt*: **~ von Unterlagen** POLITIK presentation of documents for public inspection
auslesbar *adj* ABFALL separable; **~er Ballaststoff** *m* ABFALL separable roughage
auslöschen *vt* BODEN quench
Ausmaß *nt* BODEN *einer Kontamination* extent; **~ des Ozonabbaus** LUFT degree of ozone depletion
Ausmauerung *f* LUFT *des Katalysators* lining
Ausmündungsstück *nt* ABFALL, WASSER scouring outlet with grate
Auspuff *m* LÄRM, LUFT exhaust; **Auspuffgas** *nt* LUFT exhaust emissions; **Auspuffgeräusch** *nt* LÄRM exhaust noise
Auspuffkrümmer *m* LUFT exhaust manifold; **Auspuffkrümmerkatalysator** *m* LUFT exhaust manifold catalyst; **Auspuffkrümmerreaktor** *m* LUFT exhaust manifold reactor
Auspuff: **Auspuffrohr** *nt* LUFT tail pipe; **Auspuffschalldämpfer** *m* LÄRM exhaust silencer, exhaust muffler; **Auspuffschalldämpfung** *f* LÄRM exhaust silencing; **Auspuffventil** *nt* LUFT exhaust valve
auspumpen *vt* ABFALL, ABWASSER, BODEN, WASSER pump out
Auspumpen *nt* ABFALL, ABWASSER, BODEN, WASSER pumping, pumping-out
ausrüsten *vt* ANALYTIK *Meßgeräte*, SICHERHEIT equip
ausschlämmen *vt* ABWASSER *Fluß* cleanse
Ausschluß *m* POLITIK *der Öffentlichkeit* exclusion
Ausschmelzen *nt* ABWASSER melting
Ausschuß *m* ABFALL industrial scrap, MANAG, POLITIK committee
Ausschwemmung *nt* ABWASSER sloughing; **~ von Biomassen** ABWASSER sloughing of biomass, sloughed biomass

Außen- *in cpds* ANALYTIK, LÄRM, LUFT outer; **Außenelektrode** *f* ANALYTIK outer electrode; **Außenfilter** *nt* ANALYTIK outer filter; **Außenhaut** *m* LÄRM *in der Kapselung* outer shell; **Außenlärm** *m* LÄRM outdoor noise; **Außenluft** *f* LUFT outer air; **Außentrombe** *f* LUFT outer vortex; **Außenwirbel** *m* ABFALL, WASSER *Zentrifugalfeld* outer eddy
außerbetrieblich: **~e Verwertung** *f* ABFALL third-party recycling
äußerlich *adj* ABFALL, ABWASSER, LUFT, WASSER external
aussickern *vt* ABFALL, MEER, WASSER seep
Aussickern *nt* MEER seepage
Aussickerung *f* ABFALL, WASSER exit percolation
aussieben *vt* MEER screen
aussondern *vt* ABFALL screen
Aussonderungsmagnet *m* ABFALL, WASSER screening magnet
Aussortierung *f*: **~ von Hand** ABFALL manual sorting
ausspülen *vt* ABFALL scour, clean out, flush, ABWASSER clean out, LUFT *Gase* scavenge, MANAG clean out, WASSER flush, clean out, scour
Ausspülen *nt* ABFALL, ABWASSER scouring, flushing, cleaning out, LUFT scavenging, WASSER cleaning out, flushing, scouring
Ausspülung *f* ABFALL, ABWASSER cleaning out, scouring, LUFT scavenging, WASSER cleaning out, scouring
ausstoßen *vt* LUFT *Emissionen* discharge, emit
Ausströmraum *m* ABFALL, WASSER volute chamber
Austausch *m* ABFALL, ABWASSER, LUFT, WASSER exchange
austauschbar *adj* ABWASSER exchangeable, ANALYTIK interchangeable; **~er Detektor** *m* ANALYTIK interchangeable detector; **~e Lampe** *f* ANALYTIK interchangeable lamp
austauschen *vt* ABFALL, ABWASSER, LUFT, WASSER exchange
Austauscher: **Austauscherharz** *nt* ABWASSER exchange resin; **Austauscherkapazität** *f* ABWASSER, WASSER exchanger capacity; **Austauschermaterial** *nt* ABFALL, ABWASSER, WASSER exchanger material; **Austauschersäule** *f* ABWASSER exchanger column
Austausch: **Austauschfähigkeit** *f* ABFALL, WASSER exchange capacity; **Austauschgrad** *m* ABFALL, WASSER degree of exchange; **Austauschvorgang** *m* ABFALL, WASSER exchange operation
Austrag: **Austragregler** *m* ABFALL refuse extractor; **Austragschnecke** *f* ABFALL refuse worm, LUFT discharge worm; **Austragseinheit** *f* ABFALL, WASSER discharger
Austritt *m* LUFT outlet; **Austrittsbedingungen** *f pl* LUFT *Schornstein* discharge conditions
Austrocknung *f* ABFALL, WASSER *der Deponiekörper* drying up
austrommeln *vt* MEER *Sperre* reel out
auswaschen *vt* ABFALL, ANALYTIK, BODEN, WASSER elutriate, wash out
Auswaschen *nt* ABFALL, ANALYTIK *Wasserbestimmung in Gasen* washing-out, BODEN washing-out, elutriation, WASSER *Filterfasern* washing-out
Auswaschung *f* ABFALL, LUFT, WASSER washout
auswechselbar: **~es Kation** *nt* LUFT exchangeable cation
auswerten *vt* ANALYTIK evaluate
Auswertesoftware *f* ANALYTIK evaluation software
Auswertung *f* ANALYTIK, SICHERHEIT evaluation; **~ der Effekte auf Menschen** SICHERHEIT evaluation of effects on human beings
auswiegen *vt* ANALYTIK weigh out; **analytisch ~** *vt* ANALYTIK weigh out analytically
Auswirkung *f* MANAG impact; **Auswirkungsanalyse** *f* MANAG impact analysis
Auto- *in cpds* ANALYTIK auto; **Autobatterie** *f* ABFALL automobile battery (*AmE*), car battery (*BrE*)
automatisch 1. *adj* ANALYTIK automatic; **2. ~ eine Probe nehmen** *vt* ANALYTIK take a sample automatically
automatisch: **~ eingespritzt** *adj* ANALYTIK automatically injected; **~e Erkennungseinrichtung** *f* ABFALL automatic recognition unit; **~ ladendes Müllfahrzeug** *nt* ABFALL self-loading refuse vehicle; **~ ladendes Müllfahrzeug** *nt* **mit internem Kompaktor** ABFALL self-loading vehicle with internal compactor; **~e Müllsortierung** *f* ABFALL mechanical waste separation; **~e Probenahme** *f* ANALYTIK automatic sampling; **~er Probengeber** *m* ANALYTIK autosampler; **~es Prüfröhrchen** *nt* ANALYTIK automatic detector tube; **~e Sortierung** *f* ABFALL automatic sorting; **~e Trennung** *f* ABFALL automatic sorting
automatisieren *vt* ALLG automate
automatisiert *adj* ANALYTIK automated
Auto-: **Autosampler** *m* ANALYTIK autosampler; **Autosampling** *nt* ANALYTIK autosampling; **Autoschrott** *m* ABFALL car scrap
autotherm *adj* LUFT autothermic
autotroph *adj* ABWASSER *Bakterien* autotrophic; **~e Bakterien** *f pl* ABWASSER autotrophic bacteria; **~e Denitrifikation** *f* ABWASSER autotrophic denitrification
Auto-: **Autowrack** *nt* ABFALL scrap motorcar
Avogadro-Konstante *f* LUFT Avogadro's constant
axial *adj* LUFT axial; **~ angeströmter Zyklon** *m* LUFT axial-entry cyclone; **~e Verformung** *f* ABFALL *Deponie* axial deformation
Axial: **Axialeinlauf-Zyklon** *m* LUFT axial-entry cyclone; **Axialkreiselpumpe** *f* ABWASSER axial-flow rotary pump; **Axialüberdruckventilator** *m* LUFT reaction-type axial fan; **Axialventilator** *m* LUFT axial fan
Azotometer *nt* ANALYTIK nitrometer

B

Babcock-Verfahren *nt* ABWASSER *enzymatische Stabilisierung* Babcock method
Bacharach-Skala *f* ANALYTIK *Staubmessung* Bacharach scale
Bachbildung *f* LUFT channeling (*AmE*), channelling (*BrE*)
Backenbrecher *m* ABFALL jaw crusher
Bagger *m* BODEN excavator
bakteriell *adj* ABFALL, ABWASSER, WASSER bacterial
Bakterien *f pl* ALLG bacteria; **Bakterieneliminierung** *f* ABFALL, WASSER bacteria elimination; **Bakterienfilter** *nt* LUFT bacteria filter; **Bakteriengehalt** *m* ABFALL, WASSER bacteria content; **Bakteriengift** *nt* ABWASSER bacterial poison
bakterienhaltig *adj* ABWASSER bacteria-containing; **~er Schlamm** *m* ABWASSER bacteria-containing sludge
Bakterien: **Bakterienmasse** *f* ABWASSER bacteria mass; **Bakterienpopulation** *f* ABFALL, ABWASSER, WASSER bacteria population; **Bakterientätigkeit** *f* ABFALL, LUFT *Biofilter*, WASSER *Kompostierung* bacterial activity; **Bakterientest** *m* ANALYTIK bacteria test; **Bakterienverseuchung** *f* ABFALL, WASSER *Kompostierung* bacterial contamination; **Bakterienwachstum** *nt* ABFALL, ABWASSER, WASSER bacterial growth; **Bakterienwachstumsrate** *f* ABFALL, ABWASSER, WASSER bacterial growth rate; **Bakterienzahl** *f* ABFALL, ABWASSER, WASSER bacteria count, bacterial count
bakteriologisch: **~e Reinigung** *f* ABFALL, WASSER bacteriological purification, bacteriological treatment
Ballast *f* ABFALL, WASSER ballast; **Ballastmaterial** *nt* ABFALL, WASSER ballast material, inert material; **Ballaststoff** *m* ABFALL, WASSER roughage
ballaststofffrei *adj* ABFALL, WASSER roughage-free
Ballen *m* ABFALL bale; **Ballenauflöser** *m* ABFALL, WASSER bale disintegrator; **Ballendeponie** *f* ABFALL bale landfill; **Ballenpresse** *f* ABFALL baling press; **Ballenspeicherband** *nt* ABFALL bale storage belt; **Ballenverschnürung** *f* ABFALL bale cording
ballistisch *adj* ABFALL ballistic; **~e Auslesevorrichtung** *f* ABFALL ballistic sorter; **~er Sichter** *m* ABFALL ballistic sorter; **~e Sichtung** *f* ABFALL ballistic separation, ballistic sorting; **~e Sortierung** *f* ABFALL ballistic separation, ballistic sorting
Ballonsonde *f* ANALYTIK balloon sonde
Ballungsgebiet *nt* ABFALL, MANAG, POLITIK, WASSER *Wasserbedarf* densely populated area, overcrowded area
Band *nt* ABWASSER band, belt; **Bandabschöpfgerät** *nt* MEER rope skimmer
Banderoliermaschine *f* ABFALL strapping machine
Band: **Bandfilter** *nt* ABWASSER band filter, belt filter, ANALYTIK bandpass filter; **Bandfilterpresse** *f* ABWASSER *Entwässerung* band filter press; **Bandpresse** *f* ABWASSER *bei Entwässerung* belt press; **Bandräumer** *m* ABWASSER flight scraper; **Bandtrockner** *m* ABWASSER belt dryer
Bar *nt* ANALYTIK bar
Barometer *nt* ANALYTIK barometer
Barriere *f* ABFALL, BODEN, WASSER barrier; **Barrierensystem** *nt* ABFALL, WASSER barrier system
Basenkation *nt* LUFT base cation
Basisabdichtung *f* ABFALL, WASSER bottom sealing, *einer Deponie* base sealing
basisabgedichtet: **~e Deponie** *f* ABFALL base-sealed landfill
basisch *adj* ALLG basic; **~es Absorbens** *nt* LUFT basic absorbent; **~er Anionenaustauscher** *m* ABWASSER basic anionic exchanger; **~er Kationenaustauscher** *m* ABWASSER basic cationic exchanger
Basisdichtung *f* BODEN base sealing
Basis-Schutz-Niveau *nt* POLITIK basic protection level
Basizität *f* ABFALL, ABWASSER, WASSER basic capacity, basicity
BAT *abbr* (*beste verfügbare Technik*) ALLG *integrierte Verminderung der Umweltverschmutzung* BAT (*best available technology*)
Batch *m* ABWASSER, ANALYTIK, BODEN, batch; **Batchdestillation** *f* BODEN batch distillation; **Batchdestillationstank** *m* BODEN batch distillation tank; **Batchverfahren** *nt* ABWASSER batch process; **Batchversuch** *m* ANALYTIK batch test
BATNEEC *abbr* (*wirtschaftlich vertretbare beste verfügbare Technik*) ALLG BATNEEC (*best available technology not entailing excessive costs*)
Batterie *f* ABFALL battery
batteriebetrieben *adj* ANALYTIK battery-operated, battery-powered; **~es Fahrzeug** *nt* LUFT battery-operated vehicle; **~er Gaschromatograph** *m* ANALYTIK battery-powered gas chromatograph
Batterie: **Batterierecycling** *nt* ABFALL battery recycling; **Batteriereststoff** *m* ABFALL battery residue
Bau- *in cpds* POLITIK building; **Baudichte** *f* POLITIK building density; **Baugenehmigung** *f* MANAG, POLITIK construction licence (*BrE*), construction license (*AmE*), building permit, construction permission; **Baulärm** *m* LÄRM noise associated with construction; **Baumaschinen** *f pl* BODEN earth-moving machinery
Bau-: **Bauschutt** *m* ABFALL construction waste, demolition waste, *Abfallart* building rubble; **Bauschuttrecycling** *nt* ABFALL building rubble recycling
Baustelle *f* ABFALL, BODEN, LÄRM building site, construction site; **Baustellenabfall** *m* ABFALL construction waste, building site waste
Bau-: **Bautechnik** *f* ABWASSER, ERNEU ENER, WASSER structural engineering
bautechnisch *adj* ABWASSER, ERNEU ENER, WASSER structurally engineered; **~e Barriere** *f* ABFALL, WASSER structurally-engineered barrier; **~e Verfahrenstechnik** *f* ABWASSER, ERNEU ENER, WASSER structural engineering technology
Bau-: **Bauwerkschwingung** *f* LÄRM building vibration
Bauxit *m* LUFT *Adsorbens* bauxite
BB *abbr* (*Belebtbecken*) ABWASSER activated sludge tank
Be *abbr* (*Beryllium*) LUFT *Elementaranalyse* Be (*beryllium*)
Beanspruchung *f* ABFALL, MEER, WASSER stress
Bebauung *f* MANAG building density; **Bebauungsdichte** *f* POLITIK, WASSER building density; **Bebauungsplan** *m* POLITIK, WASSER building development plan
Becher *m* ABFALL, WASSER cup; **Becherglas** *nt* ANALYTIK beaker glass; **Bechersieb** *nt* ABFALL, WASSER *bei Sortieranlage* screen for cups; **Becherwerk** *nt* LUFT *Schüttschichtfilter* chain conveyor
Becken *nt* ABFALL, ABWASSER, ANALYTIK, WASSER basin; **~ für Rückstände** ABWASSER tailings pond; **Beckenauskleidung** *f* ABFALL, ABWASSER, WASSER reservoir lining; **Beckeninfiltration** *f* ABFALL, WASSER tank infiltration; **Beckensohle** *f* ABFALL basin floor, ABWASSER tank bottom, basin bottom, WASSER basin floor
Bedeckungsgrad *m* ANALYTIK coverage rate, LUFT *Schornstein* cloud amount
Bedingungen *f pl* ABWASSER, ANALYTIK, BODEN, LUFT conditions
Bedüsung *f* ALLG spraying

beeinträchtigen *vt* ALLG impair
Beeinträchtigung *f* ALLG impairment
befestigen *vt* ANALYTIK fix, MEER moor, WASSER clip
Befestigungsgeschirr *nt* MEER mooring bracket
Beförderung *f* ABFALL, WASSER transport, transportation; **~ über die Schiene** ABFALL rail transport, railroad transportation (*AmE*), railway transportation (*BrE*); **~ über die Straße** ABFALL road transportation; **~ zu Wasser** ABFALL, WASSER *von Abfällen* water transport
Begleitschein *m* POLITIK accompanying document
begrenzen *vt* ABFALL *Schadstoffausbreitung*, BODEN, LUFT *Schadstoffausbreitung* limit
Begrenzungsfläche *f* LÄRM boundary, periphery
Behälter *m* ALLG container, ABWASSER cistern, ANALYTIK vessel, LUFT vessel, MEER scoop, WASSER tank; **Behälteradsorber** *m* LUFT tank adsorber; **Behälterglas** *nt* ABFALL container glass; **Behälterreinigung** *f* ABFALL, WASSER reservoir cleaning; **Behältersystem** *nt* ABWASSER container system; **Behälterwand** *f* LUFT *des Zyklons* vessel wall
behandeln *vt* BODEN *katalytisch* catalyse (*BrE*), catalyze (*AmE*), ALLG treat
behandelt *adj* ALLG treated; **~es Abwasser** *nt* ABWASSER treated effluent; **~er Schlamm** *m* ABWASSER treated sludge
Behandlung *f* ALLG treatment; **Behandlungsanlage** *f* ABFALL treatment plant, BODEN treatment cell; **Behandlungsdauer** *f* BODEN treatment duration, treatment time; **Behandlungsintensität** *f* BODEN treatment intensity
beheizt *adj* ABWASSER, ANALYTIK heated; **~er Faulraum** *m* ABWASSER heated digestion chamber, heated digestor; **~er Injektor** *m* ANALYTIK *Flüssigkeits- oder Gasprobenahme* heated injector; **~er Ofen** *m* ANALYTIK *Gaschromatograph* heated kiln
Behinderung *f* SICHERHEIT hindrance
Behörde *f* POLITIK authority; **Behördenentscheidung** *f* POLITIK authority decision; **Behördenpflicht** *f* POLITIK duty of authorities
behördlich: **~e Überwachung** *f* POLITIK *von Anlagen* official monitoring, official supervision
beimischen *vt* ABFALL *Kunststoffe* blend into
beißend: **~er Geruch** *m* ABFALL, ABWASSER, LUFT, WASSER acrid odor (*AmE*), acrid odour (*BrE*)
Bekämpfung *f* LUFT abatement; **Bekämpfungsmaßnahmen** *f pl* LUFT abatement measures
Beladefähigkeit *f* LUFT loading capacity
Beladeluft *f* LUFT air charge; **Beladelufttemperatur** *f* LUFT air charge temperature
beladen 1. *adj* ALLG laden, loaded, saturated; **2.** *vt* ALLG load, saturate
beladen: **~es Adsorbens** *nt* LUFT loaded adsorbent, saturated adsorbent; **~e Aktivkohle** *f* ABWASSER saturated activated carbon; **~er Aktivkoks** *m* LUFT loaded activated coke; **~es Filter** *nt* ABFALL, LUFT, WASSER loaded filter; **~es Harz** *nt* ABWASSER saturated resin; **~es Lösungsmittel** *nt* LUFT loaded solvent; **~e Suspension** *f* LUFT loaded suspension; **~es Waschmittel** *nt* LUFT loaded washing agent
Beladung *f* ABFALL, BODEN, LUFT, WASSER loading
Belastbarkeit *f* BODEN stress capacity, LUFT burden, loadability, loading capacity
belasten *vt* ALLG *mit Schadstoffen* contaminate, pollute; **etwas chronisch ~** *vt* SICHERHEIT pose a chronic health risk
belastend: **~er Stoff** *m* ALLG polluting agent
belastet *adj* ALLG contaminated, polluted; **~e Gewässer** *nt pl* ABWASSER polluted water body; **~es Regenwasser** *nt* ABWASSER polluted rainwater; **~er Staub** *m* LUFT polluted dust, contaminated dust
Belastung *f* ABFALL, ABWASSER impact, pollution, contamination, ANALYTIK load, loading, BODEN, LUFT, MANAG, MEER, POLITIK pollution, SICHERHEIT, WASSER impact, pollution; **~ durch Dämpfe** SICHERHEIT exposure to fumes; **Belastungspfad** *m* SICHERHEIT exposure path; **Belastungsprobe** *f* ANALYTIK load capacity test; **Belastungsversuch** *m* ANALYTIK load capacity test
beleben *vt* ABWASSER *Schlamm* activate
belebt *adj* ABWASSER *Schlamm* activated
Belebtbecken *nt* (*BB*) ABWASSER activated sludge tank
Belebtschlamm *m* ABWASSER activated sludge; **Belebtschlammanlage** *f* ABWASSER activated sludge plant; **Belebtschlammbecken** *nt* ABWASSER activated sludge tank; **Belebtschlammflocke** *f* ABWASSER activated sludge floc; **Belebtschlammgehalt** *m* ABWASSER activated sludge content; **Belebtschlammrücklauf** *m* ABWASSER activated sludge process, activated sludge return flow; **Belebtschlammverfahren** *nt* LUFT *Biowäscher*, WASSER activated sludge process
Belebung *f* ABWASSER activation; **Belebungsanlage** *f* ABWASSER activated sludge plant; **Belebungsbecken** *nt* ABWASSER, LUFT *Biowäscher* activated sludge tank; **Belebungsverfahren** *nt* ABWASSER activated sludge process
Beleuchtungsanlage *f* ERNEU ENER *mit Sonnenenergie* lighting installation
belüften *vt* ABFALL, ABWASSER, WASSER aerate
Belüfter *m* ABFALL, ABWASSER, WASSER aerator
belüftet *adj* ABFALL, ABWASSER, WASSER aerated; **~er Abwasserteich** *m* ABWASSER aerated oxidation pond; **~es Becken** *nt* ABWASSER aerated lagoon; **~e Kompostmiete** *f* ABWASSER aerated compost pile; **~e Mietenkompostierung** *f* ABFALL, ABWASSER, BODEN aerated bed composting, aerated pile composting (*APC*); **~er Sandfang** *m* ABWASSER aerated grit chamber; **~er vertunnelter Komposthaufen** *m* ABFALL aerated tunneled compost heap (*AmE*), aerated tunnelled compost heap (*BrE*); **~e Wasserlamelle** *f* WASSER ventilated nappe
Belüftung *f* ABFALL, ABWASSER, WASSER *Abwasserkanal* ventilation, *Grundwasseraufbereitung* aeration; **Belüftungsaggregate** *nt pl* ABWASSER aerator; **Belüftungsbecken** *nt* ABWASSER aeration basin, aeration tank; **Belüftungsbehälter** *m* ABWASSER aeration basin, aeration tank; **Belüftungsdauer** *f* ABWASSER aeration time; **Belüftungseinrichtungen** *f pl* ABWASSER aerator; **Belüftungslochdämpfer** *m* LÄRM vent silencer; **Belüftungspaddel** *nt* ABWASSER aeration paddle; **Belüftungsrohr** *nt* ABWASSER aerator pipe; **Belüftungssystem** *nt* ABWASSER aeration system; **Belüftungsversuch** *m* ABWASSER aeration test; **Belüftungszeit** *f* ABWASSER aeration time
Bemessungswassermenge *f* ABFALL, WASSER design flow
Benachrichtigung *f* MEER notification
Benennung *f* ANALYTIK designation
benetzbar *adj* ABWASSER, BODEN, LUFT wettable; **~er Staub** *m* LUFT wettable dust
Benetzbarkeit *f* ABWASSER, BODEN, LUFT wettability
benetzt *adj* ABWASSER, BODEN, LUFT wetted; **~es Filter** *nt* LUFT *Art des Naßwäschers* wetted-filter-type scrubber
Benetzungsmittel *nt* LUFT, MEER wetting agent
Benzindampfrückgewinnungsanlage *f* ABFALL, LUFT gasoline vapor recovery plant (*AmE*), petrol vapour recovery plant (*BrE*)
Benzofuran *nt* BODEN benzofurane
Benzol *nt* ABFALL *Deponiegaskomponente*, LUFT benzene; **Benzol, Toluol, Ethylbenzol, Xylol** *nt* (*BTEX*) ABWASSER benzene, toluol, ethyl benzene, xylene (*BTEX*); **Benzol, Toluol, Xylol** *nt* (*BTX*) ABWASSER benzene, toluol, xylene (*BTX*)
beobachtbar *adj* ANALYTIK observable
beobachten *vt* ABFALL, ANALYTIK, WASSER monitor
Beobachtung *f* ABFALL, ANALYTIK, WASSER monitoring, observation; **Beobachtungsbrunnen** *m* ABFALL, ANALYTIK, WASSER monitoring well, observation well; **Beobachtungszeit** *f* ANALYTIK monitoring time, observation time

Beplankung *f* LÄRM planking
Beprobung *f* ANALYTIK sampling
beprobungsfrei *adv* ANALYTIK without taking samples
Beprobungskapazität *f* ANALYTIK sampling capacity
berechenbar *adj* ANALYTIK computable, calculable
berechnen *vt* ANALYTIK compute, calculate
Berechnung *f* ANALYTIK calculation, computation
Beregnung *f* ABWASSER sprinkling, BODEN overhead irrigation; **Beregnungsanlage** *f* ABWASSER sprinkler
Bereitschaft- *in cpds* MEER stand-by; **Bereitschaftsdienst** *m* MEER stand-by duty
bergmännisch: **~ hergestellter Hohlraum** *m* ABFALL mined space
Berichterstattung *f* MANAG reporting
berieseln *vt* LUFT scrub
Berieselung *f* ABFALL irrigation, ABWASSER wastewater irrigation, sprinkling, irrigation, BODEN surface irrigation, LUFT *des Filters* watering, WASSER irrigation; **~ von Flächen mit Abwasser** ABWASSER land application; **Berieselungsbrunnen** *m* ABWASSER, WASSER irrigation well; **Berieselungsgraben** *m* ABFALL, WASSER irrigation ditch, catch feeder; **Berieselungskanal** *m* ABFALL, WASSER irrigation ditch, catch feeder
Berl-Füllkörper *m pl* LUFT *Füllkörper des Füllkörperkolonnenabscheiders* berl saddles
Berme *f* ABFALL, WASSER segregation berm
Bernoulli: **~-Gleichung** *f* ANALYTIK Bernoulli's equation; **Bernoullische Gleichung** *f* ANALYTIK Bernoulli's equation
Berst- *in cpds* LUFT bursting; **Berstdruck** *m* ABFALL, WASSER bursting pressure; **Berstfestigkeit** *f* ABFALL, LUFT, WASSER bursting strength; **Berstfestigkeitsprüfgerät** *nt* ABFALL, LUFT, WASSER bursting strength tester
beruflich *adj* SICHERHEIT occupational
berufsbedingt: **~e Belastungsgrenzen** *f pl* (*OEL*) SICHERHEIT occupational exposure limits (*OEL*)
Berufskrankheit *f* SICHERHEIT occupational disease
Beryllium *nt* (*Be*) LUFT beryllium (*Be*)
beschichten *vt* LUFT coat
beschichtet *adj* ABFALL, LUFT coated; **~e Fläche** *f* ABFALL, LUFT coated surface; **~er Keramikkörper** *m* LUFT *Katalysator* coated ceramic body
Beschichtung *f* ABFALL enameling (*AmE*), enamelling (*BrE*), ALLG coating
beschicken *vt* ALLG charge, dose, feed
Beschicken *nt* ALLG charging, dosing, feeding
Beschickung *f* ALLG charging, dosage, dosing, feeding; **Beschickungseinrichtung** *f* ABFALL, WASSER charging facility, loading mechanism; **Beschickungshöhe** *f* ABFALL, WASSER application rate; **Beschickungskammer** *f* ABFALL, WASSER *Atommüllagerung* charging chamber; **Beschickungspumpe** *f* ABWASSER metering pump, dosing pump; **Beschickungstrichter** *m* BODEN feeding hopper; **Beschickungsvorrichtung** *f* ABFALL, WASSER feeder
beschleunigen *vt* ABFALL, ANALYTIK, LUFT accelerate
beschleunigt *adj* ABFALL, ANALYTIK, LUFT accelerated; **~es Elektron** *nt* LUFT accelerated electron; **~e Kompostierung** *f* ABFALL mechanical composting, rapid fermentation
Beschleunigung *f* ABFALL, ANALYTIK, LUFT, WASSER acceleration; **Beschleunigungsaufnehmer** *m* ANALYTIK accelerometer; **Beschleunigungsmesser** *m* ANALYTIK accelerometer
Beschränkung *f*: **~ der Giftigkeit am Arbeitsplatz** SICHERHEIT restriction of toxicity at the workplace
beseitigen *vt* ALLG *Recycling, Schadstoffe* remove, dispose of
Beseitigen *nt* ABFALL, BODEN, MEER, WASSER removal, disposal
beseitigt *adj* ALLG disposed of, removed
Beseitigung *f* ABWASSER, BODEN disposal, LUFT *abgeschiedener Staub* removal, MEER disposal; **~ der Altlasten** BODEN removal of contaminated sites; **~ von kontaminierten Böden** BODEN removal of contaminated soils; **~ von Lösemittelresten** BODEN residual solvent removal; **~ von organischen Bestandteilen** ABFALL, LUFT, WASSER removal of organic matter; **~ von Schwebstoffen durch Sedimentablagerung** LUFT removal of suspended solids by sedimentation; **~ von Schwebstoffen durch Sedimentation** ABFALL, WASSER removal of suspended solids by sedimentation
besprengen *vt* ABFALL, WASSER sprinkle
besprühen *vt* MEER spray
beständig *adj* ABFALL, ANALYTIK, WASSER resistant
Bestandteil *m* ANALYTIK, BODEN, LUFT component, constituent
Bestaubung *f* LUFT dusting
beste: **~ verfügbare Technik** *f* (*BAT*) ALLG *objektive Bewertung aller Umweltauswirkungen eines Produktes* best available technology (*BAT*)
besteuern *vt* POLITIK tax
bestimmbar *adj* ANALYTIK estimable
bestimmen *vt* ANALYTIK determine; **quantitativ ~** *vt* ANALYTIK assess quantitatively, quantify
Bestimmung *f* ANALYTIK, SICHERHEIT analysis, determination; **~ der absetzbaren und ungelösten Stoffe von Abwasser** ABWASSER, ANALYTIK, WASSER determination of settleable and undissolved matter in wastewater; **~ der Abwasserdichte** ABWASSER, ANALYTIK, WASSER determination of wastewater density; **~ der Abwassertemperatur** ABWASSER, ANALYTIK, WASSER determination of wastewater temperature; **~ des Ammoniumgehaltes in Abwasser** ABWASSER, ANALYTIK, WASSER determination of ammonium content in wastewater; **~ des BSB von Abwasser** ABWASSER, ANALYTIK, WASSER BOD determination in wastewater; **~ des Chloridgehaltes in Abwasser** ABWASSER, ANALYTIK, WASSER determination of chloride content in wastewater; **~ des Chromitgehaltes in Abwasser** ABWASSER, ANALYTIK, WASSER determination of chromite content in wastewater; **~ der elektrischen Leitfähigkeit von Abwasser** ABWASSER, ANALYTIK, WASSER determination of conductivity of wastewater; **~ des Gehaltes an Blei und Bleiverbindungen** ABWASSER, SICHERHEIT, WASSER determination of particulate lead and lead compounds; **~ des Gesamt-, Abdampf- und Glührückstandes von Abwasser** ABWASSER, ANALYTIK determination of total, evaporation and ignition residue in wastewater; **~ des Kaliumgehaltes in Abwasser** ABWASSER, ANALYTIK, WASSER determination of potassium content in wastewater; **~ der Luftverschmutzung durch saure Gase** SICHERHEIT determination of gaseous acid air pollution; **~ des Magnesiumgehaltes in Abwasser** ABWASSER, ANALYTIK, WASSER determination of magnesium content in wastewater; **~ des Natriumgehaltes in Abwasser** ABWASSER, ANALYTIK, WASSER determination of sodium content in wastewater; **~ des Nitratgehaltes in Abwasser** ABWASSER, ANALYTIK, WASSER determination of nitrate content in wastewater; **~ des Nitritgehaltes in Abwasser** ABWASSER, ANALYTIK, WASSER determination of nitrite content in wastewater; **~ des pH-Wertes von Abwasser** ABWASSER, ANALYTIK, WASSER determination of wastewater pH-value; **~ nach Probendestillation** ANALYTIK determination following sample distillation; **~ der Säurekapazität von Abwasser** ANALYTIK determination of acid capacity in wastewater; **~ der Stickstoffdioxidkonzentration** SICHERHEIT determination of concentration of nitrogen dioxide; **~ des Sulfatgehaltes in Abwasser** ABWASSER, ANALYTIK, WASSER determination of sulfate content in wastewater (*AmE*), determination of sulphate content in wastewater (*BrE*); **~ der Trübung von Abwasser** ABWASSER, ANALYTIK, WASSER determination of wastewater turbidity; **Bestimmungsergebnis** *nt* ANALYTIK analytical finding, analytical result; **Bestimmungsgrenze** *f* ANALYTIK determination limit; **Bestimmungskomponente** *f* ANALYTIK

determination component; **Bestimmungsmethode** *f* ANALYTIK method of determination

bestmöglich: **~e Option** *f* **für die Umwelt** (*BPEO*) POLITIK best practical environmental option (*BPEO*)

Bestrahlung *f* ABFALL, ANALYTIK, ERNEU ENER, LUFT, SICHERHEIT radiation, SICHERHEIT exposure; **Bestrahlungsdauer** *f* ABFALL, WASSER irradiation time; **Bestrahlungsdosis** *f* ABFALL, WASSER irradiation dosage; **Bestrahlungsstärke** *f* ABFALL, WASSER *der Mikroorganismen* irradiation intensity

Beta-Absorption-Probenahmemeßgerät *nt* ANALYTIK betaray absorption sampler

beta-mesosaprob *adj* ABWASSER beta-mesosaprobic

Beton: **Betonrohr** *nt* ABFALL, WASSER concrete pipe; **Betonschachtring** *m* ABFALL, WASSER concrete ring of tubbings

betreiben *vt* ERNEU ENER, MANAG, POLITIK, WASSER operate

Betreiber *m* MANAG, POLITIK, WASSER operator; **Betreiberpflichten** *f pl* MANAG, POLITIK, WASSER duties of operator

Betrieb *m* ALLG operation

betrieben: **~er Deponieabschnitt** *m* ABFALL operational landfill section

Betrieb: **Betriebsablauf** *m* ABFALL, WASSER industrial process

Betriebsbeauftragter *m* MANAG, POLITIK, WASSER plant representative; **~ für Abfall** ABFALL, MANAG, POLITIK, WASSER waste management official representative; **~ für Gewässerschutz** ABFALL, MANAG, POLITIK, WASSER water pollution control official representative; **~ für Immissionsschutz** LUFT, MANAG, POLITIK air quality protection official representative

betriebsbedingt: **~er Fehler** *m* ABFALL, MANAG, WASSER operational error

Betrieb: **Betriebsdauer** *f* ERNEU ENER operation time; **Betriebsdrehzahl** *f* ERNEU ENER *Windturbine* operating speed; **Betriebsfehler** *m* ABFALL, MANAG, WASSER operational error; **Betriebsgenehmigung** *f* POLITIK *Erlaubnis* operating permission, operating authorization, *Zulassung* operating license (*AmE*), operating licence (*BrE*)

betriebsintern: **~es Recycling** *nt* ABFALL in-house recycling

Betrieb: **Betriebsmittel** *nt* LUFT operating material, resource; **Betriebsnormen** *f pl* POLITIK operating standards; **Betriebssicherheit** *f* SICHERHEIT operations safety; **Betriebsspannung** *f* LUFT operating voltage

betriebsspezifisch *adj* POLITIK company-specific, firm-specific; **~e Informationen** *f pl* POLITIK company-specific information, firm-specific information; **~e Unterlagen** *f pl* POLITIK company-specific documents, firm-specific documents

Betrieb: **Betriebstemperatur** *f* ANALYTIK *von Detektoren* working temperature; **Betriebswasser** *nt* ABFALL process water, production water, ABWASSER production water, industrial water, process water, BODEN, LUFT process water, WASSER process water, production water; **Betriebszuverlässigkeit** *f* MEER reliability

Beugung *f* ANALYTIK diffraction

beurteilt *adj* ALLG assessed; **~e Probe** *f* ANALYTIK assessed sample

Beurteilungspegel *m* ANALYTIK assessment level, LÄRM assessment level, noise rating level, MEER assessment level

Beutel *m* ABFALL bag, dustbin liner (*BrE*), garbage can bag (*AmE*); **Beutelpapier** *nt* ABFALL bag paper; **Beutelverpackung** *f* ABFALL bag packaging

Bevölkerung *f* ANALYTIK, LUFT, SICHERHEIT, WASSER population; **Bevölkerungsgesamtdosis** *f* LUFT, SICHERHEIT population dose; **Bevölkerungsteildosis** *f* LUFT, SICHERHEIT subpopulation collective dose; **Bevölkerungswachstum** *nt* BODEN *massiver Druck auf die Bodenfunktion* population growth

Bewässerung *f* ABFALL, ABWASSER irrigation, LUFT *des Filters* watering, WASSER irrigation; **Bewässerungsbrunnen** *m* ABFALL, WASSER irrigation well; **Bewässerungsgraben** *m* ABFALL, WASSER irrigation ditch, catch feeder

beweglich *adj* ABFALL mobile, ANALYTIK *Gasanalysengerät* portable, LUFT mobile; **~e Einbauten** *m pl* LUFT movable fittings; **~e Emissionsquellen** *f pl* LUFT mobile emission sources; **~e Komponente** *f* ABFALL, LUFT mobile component; **~e Pumpanlage** *f* ABWASSER mobile pumping unit; **~e Trennwand** *f* ABFALL movable bulkhead

Bewegung *f*: **~ des Grundwassers** WASSER movement of groundwater; **Bewegungsenergie** *f* ERNEU ENER kinetic energy

bewerten *vt* ALLG assess, evaluate

bewertet: **~er Schallpegel** *m* LÄRM weighted sound level

Bewertung *f* ALLG assessment, evaluation; **~ von Auswirkungen** MANAG impact analysis; **Bewertungskriterien** *nt pl* ABFALL, BODEN, LÄRM, LUFT, MANAG, MEER assessment criteria, POLITIK evaluation criteria, WASSER assessment criteria; **Bewertungsmaßstäbe** *m pl* POLITIK evaluation criteria

Bewuchs *m* ABFALL, MEER, WASSER fouling

bewuchshindernd *adj* ABFALL, WASSER antifouling

Bezug: **Bezugselektrode** *f* ANALYTIK *pH-Messung* reference electrode; **Bezugsgröße** *f* ANALYTIK reference value; **Bezugsschalldruck** *m* ANALYTIK sound pressure reference value, LÄRM sound pressure reference value, reference sound pressure; **Bezugsschalleistung** *f* ANALYTIK audibility reference value

BHKW *abbr* (*Blockheizkraftwerk*) ERNEU ENER, LUFT CHP station (*combined heat and power station*); **BHKW mit Brennstoffzellen** *nt* LUFT CHP station with fuel cell plant

B-Horizont *m* BODEN B-horizon

biegesteif: **~er Plattenabsorber** *m* LÄRM rigid plate absorber

biegeweich *adj* LÄRM flexible; **~e Platte** *f* LÄRM flexible plate; **~er Plattenabsorber** *m* LÄRM flexible plate absorber; **~e Schale** *f* LÄRM flexible shell

Biegewellen *f pl* LÄRM bending waves

bilateral *adj* POLITIK bilateral; **~e Umweltvereinbarung** *f* POLITIK bilateral environmental agreement; **~e Verhandlung** *f* POLITIK bilateral negotiation

Bildanalyse *f* ANALYTIK image analysis

Bilge *f* MEER bilge; **Bilgenpumpe** *f* MEER bilge pump, gully sucker; **Bilgenwasser** *nt* MEER *aus Schiffen* bilge water

binär *adj* ANALYTIK binary

Binde: **Bindefähigkeit** *f* ABWASSER *Ionen* binding strength; **Bindemittel** *nt* ABFALL, BODEN, MEER, WASSER binder, binding agent

binden *vt* LUFT *Partikeln mit einer Waschflüssigkeit* bind

Binder *m* BODEN binding agent, *Verfestigung* binder, MEER binding agent; **Binderschicht** *f* ABFALL *von Recyclingmaterial* binding layer

Bindung *f* LUFT *Partikeln an der Oberfläche* bonding

Binnen: **Binnenemissionen** *f pl* LUFT domestic emissions, **Binnengewässer** *nt pl* WASSER inland waters; **Binnenquelle** *f* WASSER internal source

Bio- *in cpds* ALLG bio-; **Bioabfall** *m* ABFALL biological waste, putrescible waste, organic waste, biowaste, WASSER biowaste, biological waste, organic waste; **Bioabfallfraktion** *f* ABFALL, WASSER organic waste fraction

Bioakkumulation *f* ABWASSER, BODEN, LUFT bioaccumulation; **Bioakkumulationsfaktor** *m* ABWASSER, BODEN, LUFT bioaccumulation factor; **Bioakkumulationspotential** *nt* ABWASSER, BODEN, LUFT bioaccumulation potential; **Bioakkumulationsvermögen** *nt* ABWASSER, BODEN, LUFT bioaccumulation capacity

bioakkumulativ *adj* ABWASSER, BODEN, LUFT bioaccumulative

Bio: **Biobett** *nt* BODEN *Bett ohne künstliche Düngung und chemische Schädlingsbekämpfung* biological bed, soil zone

biochemisch *adj* ABFALL, ABWASSER, ANALYTIK, WASSER biochemical; **~er Indikator** *m* ABFALL, ANALYTIK, WASSER biochemical tracer; **~er Sauerstoffbedarf** *m* (*BSB*) ABFALL,

ABWASSER, ANALYTIK, WASSER biochemical oxygen demand (*BOD*); **~er Sauerstoffbedarf** *m* **in fünf Tagen** (*BSB_5*) ABFALL, ABWASSER, ANALYTIK, LUFT, WASSER biochemical oxygen demand in five days (*BOD_5*); **~e Selbstreinigung** *f* ABFALL, ABWASSER, WASSER biochemical self-purification; **~e Umwandlung** *f* ERNEU ENER *Biomasse* biochemical conversion

Biodenipho-Verfahren *nt* ABWASSER *Phosphorelimination* Biodenipho method

***Bio*: Biodiesel** *m* ERNEU ENER, LUFT biodiesel; **Bioenergie** *f* ERNEU ENER *Biokonversion* bioenergy; **Bioenergieanlage** *f* ERNEU ENER bioenergy plant

Biofilm *m* ABWASSER biofilm; **Biofilmoberfläche** *f* ABWASSER biofilm surface; **Biofilmwachstum** *nt* ABWASSER biofilm growth

***Bio*: Biofilter** *nt* ABFALL, ABWASSER biofilter, LUFT biological filter, *biologisches Sorptionsverfahren* biofilter, WASSER biofilter; **Biofilterung** *f* ABFALL, ABWASSER, LUFT biological filtration; **Biofiltration** *f* LUFT *gasförmiges Oxidationsmittel* biofiltration; **Bioflockung** *f* ABWASSER bioflocculation

Biogas *nt* ABFALL biogas, digester gas, fermentation gas, ABWASSER, WASSER digester gas, fermentation gas; **Biogasgewinnung** *f* ABFALL biogas recovery, ERNEU ENER biogas generation; **Biogasreaktor** *m* ABFALL biogas reactor; **Biogaszusammensetzung** *f* ABFALL biogas composition

biogen: ~es Sediment *nt* ABFALL biogenic sediment

***Bio*: Bioindikator** *m* ABFALL, ANALYTIK, WASSER bioindicator, biological indicator; **Biokatalysator** *m* LUFT biological catalyst

biokatalytisch *adj* LUFT biologically catalytic; **~e Abluftreinigung** *f* LUFT biologically catalytic waste air cleaning; **~e Regeneration** *f* LUFT biologically catalytic regeneration

***Bio*: Biokonversion** *f* ERNEU ENER biomass conversion; **Biokonversionsanlage** *f* ERNEU ENER bioenergy plant; **Biokraftstoff** *m* ERNEU ENER biofuel

biologisch *adj* ALLG bio-, biological; **~er Abbau** *m* ABFALL, ABWASSER, BODEN, MEER, WASSER biodegradation, biological degradation; **~ abbaubar** *adj* ABFALL, ABWASSER, BODEN, MEER, WASSER biodegradable; **~ nicht abbaubar** *adj* ABFALL, ABWASSER, BODEN, MEER, WASSER nonbiodegradable; **~ abbaubarer Abfall** *m* ABFALL, WASSER biodegradable waste; **~ nicht abbaubarer Abfall** *m* ABFALL, WASSER nonbiodegradable waste; **~ abbaubare Kunststoffe** *m pl* ABFALL biodegradable plastics; **~ abbaubare organische Verbindungen** *f pl* ABWASSER biodegradable organics; **~ nicht abbaubare organische Verbindungen** *f pl* ABWASSER, BODEN, LUFT nonbiodegradable organics; **~ abbaubarer Packstoff** *m* ABFALL, WASSER biodegradable packaging material; **~ nicht abbaubarer Packstoff** *m* ABFALL nonbiodegradable packaging material; **~ abbaubarer Stoff** *m* ABFALL, ABWASSER, WASSER biodegradable matter; **~ nicht abbaubarer Stoff** *m* ABFALL, ABWASSER, WASSER nonbiodegradable matter; **~ abbaubare Substanz** *f* ABFALL, WASSER biodegradable substance; **~ nicht abbaubare Substanz** *f* ABFALL, WASSER nonbiodegradable substance; **~e Abbaubarkeit** *f* ABFALL, ABWASSER, BODEN biodegradability, LUFT *biologische Gaswäsche* biodegradation, MEER, WASSER biodegradability; **~e Abluftreinigung** *f* LUFT biological waste air treatment; **~e Abwasseraufbereitung** *f* ABWASSER biological wastewater treatment; **~es Agens** *nt* ABFALL, WASSER biological agent; **~e Altlastensanierung** *f* BODEN bioremediation; **~e Behandlung** *f* ABFALL, WASSER biological treatment; **~e Belastung** *f* ABFALL *Deponie* biological strain; **~e Belebungsanlage** *f* ABWASSER biological activated sludge plant; **~e Denitrifikation** *f* ABFALL, WASSER biological denitrification; **~e Eliminierung** *f* BODEN biological elimination; **~es Erdfilter** *nt* LUFT *Biofilter* biological ground filter; **~es Filter** *nt* ABFALL, WASSER biological filter; **~e Gaswäsche** *f* LUFT *Sorptionsverfahren* biological gas scrubbing; **~e Gefahr** *f* SICHERHEIT biological hazard; **~e Gewässergüte** *f* WASSER biological water quality; **~ katalytisch** *adj* LUFT biologically catalytic; **~ katalytische Abluftreinigung** *f* LUFT biologically catalytic waste air cleaning; **~ katalytische Regeneration** *f* LUFT biologically catalytic regeneration; **~e Kläranlage** *f* ABWASSER biological clarification plant; **~er Körper** *m* ABWASSER bacteria bed; **~e Nachreinigung** *f* ABWASSER secondary wastewater treatment; **~e Nährstoffentfernung** *f* ABWASSER biological nutrient removal; **~er Nährstoffentzug** *m* ABWASSER biological nutrient removal; **~e Phosphoreliminierung** *f* ABWASSER biological phosphorus removal; **~er Rasen** *m* ABWASSER filter film, biofilm, fixed film, LUFT *biowasher* filter film; **~e Reinigung** *f* ABWASSER biological purification; **~er Sauerstoffbedarf** *m* (*BSB*) ABFALL, ABWASSER, ANALYTIK, WASSER *bei Meßgeräten* biological oxygen demand (*BOD*); **~er Sauerstoffbedarf** *m* **in fünf Tagen** (*BSB_5*) ABFALL, ABWASSER, ANALYTIK, LUFT, WASSER biological oxygen demand in five days (*BOD_5*); **~ schwer abbaubar** *adj* ABWASSER biologically difficult to decompose; **~es Sorptionsverfahren** *nt* LUFT biological sorption process; **~e Stabilisierung** *f* BODEN biological stabilization; **~ tot** *adj* BODEN biologically dead; **~e Umwandlung** *f* ABFALL, WASSER biological energy conversion; **~es Waschpulver** *nt* ABWASSER biological washing powder; **~e Wassergütebestimmung** *f* ANALYTIK biological water quality determination; **~e Zersetzung** *f* ABFALL, BODEN, MEER biodegradation, biological degradation

Biomasse *f* ALLG biomass; **Biomassekonzentration** *f* ABWASSER biomass concentration; **Biomassekurve** *f* ABWASSER biomass curve; **Biomassepotential** *nt* ERNEU ENER biomass potential; **Biomasseproduktion** *f* ERNEU ENER biomass production

Biomüll *m* ABFALL, WASSER biowaste, organic waste, putrescible waste; **Biomüllerfassung** *f* ABFALL organic waste recovery; **Biomüllfraktion** *f* ABFALL, WASSER organic waste fraction; **Biomüllverwertung** *f* ABFALL organic waste use

***Bio*: Bio-Öl** *nt* ERNEU ENER bio-oil; **Bioreaktor** *m* ABWASSER bioreactor; **Biorestauration** *f* BODEN bioaugmentation; **Biosensor** *m* ANALYTIK biosensor; **Biostabilisator** *m* ABFALL, LUFT, WASSER biostabilizer

biotechnologisch *adj* BODEN, LUFT biotechnological; **~e Behandlung** *f* LUFT biotechnological treatment

***Bio*: Biotest** *m* ANALYTIK biological test

biotisch *adj* ABFALL, ABWASSER, WASSER biotic; **~e Akkumulation** *f* ABFALL, WASSER biotic accumulation

***Bio*: Biotonne** *f* ABFALL collection container for biological waste, Green Container; **Biowäsche** *f* LUFT biological gas scrubbing; **Biowäscher** *m* LUFT *biologische Sorptionsverfahren* biowasher

Biozid *nt* ABWASSER, BODEN biocide

Biozönose *f* ABWASSER, LUFT biocenosis (*AmE*), biocoenosis (*BrE*)

biozönotisch *adj* ABWASSER, LUFT biocenotic (*AmE*), biocoenotic (*BrE*)

Blähschlamm *m* ABWASSER bulking sludge; **Blähschlammbekämpfung** *f* ABWASSER bulking sludge control; **Blähschlammbildung** *f* ABWASSER bulking sludge formation

Blase *f* LUFT bubble

blasen *vt* LUFT blow

***Blase*: Blasensäule** *f* LUFT bubble column; **Blasenverpackung** *f* ABFALL bubble pack

Blatt *nt* ERNEU ENER *Windturbine* blade

Blättchen *nt* ABWASSER lamella

***Blatt*: Blatteinstellungswinkel** *m* ERNEU ENER *Windturbine* blade angle; **Blatteinstellungswinkelregelung** *f* ERNEU ENER *Windturbine* blade angle control; **Blattindex** *m* ABFALL, LUFT leaf-area index; **Blattoberfläche** *f* LUFT foliar

surface; **Blattumdrehung** *f* ERNEU ENER *Windturbine* blade rotation
blau: **~er Engel** *m* POLITIK *Umweltzeichen* blue angel; **~e Tonne** *f* ABFALL collection container for paper and cardboard
Blaualgen *f pl* ABFALL, WASSER blue algae
Blech *nt* LÄRM *wird in Schichten oder Schalen zur Kapselung benutzt* sheet metal; **Blechtrommel** *f* ABFALL metal drum
Blei *nt* (*Pb*) ALLG lead (*Pb*); **Bleiakkumulator** *m* ABFALL *Batterie* lead accumulator
bleibend: **~e Härte** *f* ABFALL, ABWASSER, WASSER noncarbonate hardness (*NCH*), permanent hardness
Bleichereiabwasser *nt* ABWASSER bleaching plant wastewater
bleifrei: **~es Benzin** *nt* ABFALL, LUFT lead-free gasoline (*AmE*), lead-free petrol (*BrE*)
bleihaltig *adj* ABFALL, LUFT lead-containing, plumbiferous; **~er Staub** *m* ABFALL, LUFT lead-containing dust
Blei: **Bleivergiftung** *f* SICHERHEIT *Berufskrankheit* lead poisoning; **Bleizusatzstoffe** *m pl* ABFALL, LUFT lead additives
Blendengehäuse *nt* LUFT baffle chamber
Blind- *in cpds* ANALYTIK blank; **Blindprobe** *f* ANALYTIK blind test, blank test; **Blindwert** *m* ANALYTIK blank value
Blisterpackung *f* ABFALL blister pack
Blockheizkraftwerk *nt* (*BHKW*) ERNEU ENER, LUFT combined heat and power station (*CHP station*); **~ mit Brennstoffzellen** LUFT combined heat and power station with fuel cell plant
Blockierung *f* LUFT *Katalysator* blockage
Blockpolymer *nt* ABFALL block polymer
Blowby-Gas *nt* LUFT blowby gas
Boden *m* ABFALL, BODEN soil, LUFT tray, WASSER soil; **im ~ eingebundene Schadstoffe** *m pl* BODEN contaminants bound to the soil; **Bodenabsorber** *m* LUFT tray absorber; **Bodenabzugskanal** *m* WASSER bottom culvert
bodenähnlich: **~es Material** *nt* BODEN *von gereinigten Böden* soil-like material
Boden: **Bodenanalyse** *f* ANALYTIK, BODEN soil analysis; **Bodenaufbereitung** *f* BODEN soil processing; **Bodenaushub** *m* ABFALL excavated soil, BODEN excavated soil, extracted soil; **Bodenaustausch** *m* BODEN soil exchange; **Bodenbedeckung** *f* BODEN soil covering; **Bodenbereich** *m* BODEN soil area; **Bodenbestandteil** *m* BODEN soil component, soil constituent; **Bodenblech** *nt* LUFT *Glockenboden* tray, plate; **Bodenbohrer** *m* ANALYTIK *Probenehmen* , BODEN auger
Bodendegradation *f*: **~ durch Bergbau und Prospektierung** BODEN soil degradation through mining and prospecting; **~ durch industrielle Landwirtschaft** BODEN soil degradation through mechanized farming; **~ infolge militärischer Einwirkungen** BODEN soil degradation following military action; **~ durch Tourismus** BODEN soil degradation through tourism; **~ durch ungeregelte Urbanisierung** BODEN soil degradation through uncontrolled urbanization
Boden: **Bodendurchlässigkeit** *f* ABFALL, BODEN, WASSER soil permeability; **Bodendurchlüftung** *f* BODEN soil aeration; **Bodeneigenschaften** *f pl* ABFALL, BODEN, WASSER soil characteristics, soil properties; **Bodenentseuchung** *f* BODEN *radioaktiv* soil decontamination; **Bodenfeuchtemessung** *f* ANALYTIK soil humidity measuring; **Bodenfilter** *nt* ABFALL, WASSER soil filter; **Bodenfilterschicht** *f* ABFALL, WASSER soil filter layer
bodenfremd: **~e Substanz** *f* BODEN allochthonous matter
Boden: **Bodenfunktion** *f* BODEN soil function; **Bodengas** *nt* BODEN soil gas; **Bodengehalt** *m* BODEN soil content; **Bodenhorizont** *m* BODEN soil horizon; **Bodenkolonne** *f* LUFT plate column; **Bodenkontamination** *f* BODEN soil contamination
Bodenluft *f* BODEN soil vapor (*AmE*), soil vapour (*BrE*); **Bodenluftanalyse** *f* ANALYTIK soil vapor analysis (*AmE*), soil vapour analysis (*BrE*); **Bodenluftkonzentration** *f* BODEN soil vapor concentration (*AmE*), soil vapour concentration (*BrE*); **Bodenluftmessung** *f* ANALYTIK soil vapor measurement (*AmE*), soil vapour measurement (*BrE*); **Bodenluftprobenehmen** *nt* ANALYTIK soil vapor sampling (*AmE*), soil vapour sampling (*BrE*); **Bodenluftsensor** *m* ANALYTIK soil vapor probe (*AmE*), soil vapour probe (*BrE*)
Boden: **Bodenmatrix** *f* ABFALL, BODEN, WASSER soil matrix; **Bodenmatrixprobe** *f* ANALYTIK soil matrix sample; **Bodenmechanik** *f* BODEN soil mechanics
bodenmechanisch *adj* ABFALL soil-mechanical
Boden: **Bodenmilieu** *nt* BODEN soil environment
bodennah: **~e Schadstoffe** *m pl* LUFT ground-level contaminants
Boden: **Bodennutzung** *f* BODEN soil use, land use; **Bodenpassage** *f* ABFALL, WASSER soil passage
bodenphysikalisch *adj* BODEN soil-physical; **~er Parameter** *m* BODEN soil-physical parameter
Boden: **Bodenprobe** *f* ANALYTIK, BODEN soil sample; **Bodenprobenahme** *nt* ANALYTIK, BODEN soil sampling; **Bodenräumerschild** *nt* ABWASSER bottom scraper blade; **Bodensanierung** *f* BODEN soil clean-up; **Bodensättigung** *f* BODEN soil saturation; **Bodensatz** *m* BODEN deposit; **Bodensäule** *f* BODEN soil column; **Bodensäulenabscheider** *m* LUFT *Naßwäscher* plate-column scrubber, tray-tower, plate tower; **Bodensäure** *f* BODEN soil acidity; **Bodenschicht** *f* ABFALL bottom layer, BODEN soil layer, WASSER bottom layer; **Bodenschlamm** *m* ABWASSER bottom deposit, BODEN soil slurry; **Bodenschlammgreifer** *m* ANALYTIK soil deposit sampler
bodenschützend *adj* BODEN soil-protecting
bodenschutzrechtlich *adj* BODEN, POLITIK according to the law on soil protection
Boden: **Boden- und Sedimentationswäsche** *f* BODEN soil and sediment washing; **Bodensonde** *f* ANALYTIK soil probe
bodenspezifisch *adj* BODEN soil-specific
Boden: **Bodenspülung** *f* BODEN soil flushing; **Bodenstaub** *m* BODEN soil dust; **Bodenströmung** *f* ABFALL, WASSER bottom current; **Bodenteilchen** *nt* BODEN soil particle; **Bodentextur** *f* BODEN soil texture; **Bodenumbruch** *m* BODEN soil break-up; **Bodenuntersuchung** *f* ANALYTIK, BODEN soil analysis; **Bodenverbesserer** *m* BODEN soil ameliorant; **Bodenverbesserung** *f* BODEN soil conditioning; **Bodenverbesserungsmittel** *nt* ABWASSER *Klärschlamm* soil conditioner, BODEN soil ameliorant; **Bodenvereisung** *f* BODEN soil freezing
bodenverschmutzend: **~e Stoffe** *m pl* ABFALL land pollutants, BODEN land pollutants, soil pollutants
bodenverunreinigend: **~e Stoffe** *m pl* ABFALL land pollutants, BODEN land pollutants, soil pollutants
Boden: **Bodenverunreinigung** *f* BODEN soil pollution; **Bodenwäscher** *m* BODEN soil washer
Bodenwasser *nt* ABFALL soil water; **Bodenwasserbilanz** *f* WASSER soil water balance; **Bodenwasserprobenahme** *f* ANALYTIK soil water sampling
Boden: **Bodenzone** *f* BODEN soil zone; **Bodenzustand** *m* BODEN soil condition
Bogen *m* WASSER *im Kanal* bend; **Bogenrechen** *m* ABWASSER curved screen; **Bogensieb** *nt* ABWASSER curved sieve
Bohr-: **Bohrbrunnen** *m* ABFALL, WASSER bored well
bohren *vt* ANALYTIK, BODEN, ERNEU ENER drill
Bohren *nt* ANALYTIK *Grundwasserprobenahme*, BODEN, ERNEU ENER well drilling
Bohrer *m* BODEN drill
Bohr-: **Bohrgerät** *nt* ANALYTIK, WASSER bore machine; **Bohrgestänge** *nt* BODEN drilling rod assembly; **Bohrkernuntersuchung** *f* ANALYTIK drilling core analysis
Bohrloch *nt* ALLG borehole; **Bohrlochgreifer** *m* BODEN drilling grab; **Bohrlochpumpe** *f* ABFALL, WASSER borehole pump; **Bohrlochwand** *f* BODEN borehole wall

Bohr-: **Bohrrohr** *nt* WASSER boring pipe
Bohrung *f* ANALYTIK, BODEN drilling, WASSER boring
Bohr-: **Bohrwagen** *m* ANALYTIK wagon drill; **Bohrwelle** *f* BODEN drilling shaft; **Bohrwerkzeug** *nt* WASSER boring tool
Boosterpumpe *f* ABWASSER, WASSER booster pump
Böschung *f* ABFALL, WASSER slope, *von Deponien* ramp landfill, slope landfill; **Böschungsdichtung** *f* BODEN slope sealing
Boxenkompostierung *f* ABFALL rotting chamber composting process
Boyle-Mariotte: **~sches Gesetz** *nt* ANALYTIK Boyle's law
BPEO *abbr* (*bestmögliche Option für die Umwelt*) POLITIK BPEO (*best practical environmental option*)
brackig *adj* WASSER brackish
Brackwasser *nt* WASSER brackish water
Bragg: **~sches Spektrometer** *nt* ANALYTIK ionization spectrometer; **Bragg-Spektrometer** *nt* ANALYTIK ionization spectrometer
BRAM *abbr* (*Brennstoff aus Müll*) ABFALL RDF (*refuse-derived fuel*); **~ aus organischer Masse** *m* (*BROM*) ABFALL RDF from organic mass; **~ aus Papier** *m* (*BRAP*) ABFALL RDF from paper; **~-Verbrennung** *f* ABFALL RDF incineration
Brand *m* ALLG fire; **Brandgefahr** *f* SICHERHEIT fire risk, fire hazard; **Brandrodungslandwirtschaft** *f* LUFT *unmittelbare Auswirkungen auf den Kohlendioxidhaushalt der Welt* slash-and-burn agriculture; **Brandschutz** *m* SICHERHEIT fireproofing, protection against fire; **Brandsicherheit** *f* SICHERHEIT fire safety
Branntkalk *m* ABWASSER, MEER quicklime
BRAP *abbr* (*BRAM aus Papier*) ABFALL RDF from paper
brauchbar *adj* ANALYTIK usable
Brauchbarkeit *f* ANALYTIK usability
Brauchwasser *nt* ABFALL, ABWASSER industrial water, process water, production water, BODEN process water, ERNEU ENER *Wärmepumpe* service water, LUFT process water, WASSER process water, industrial water, production water
Brauereiabwasser *nt* ABWASSER brewery wastewater
Braunglas *nt* ABFALL brown glass
Braunkohle *f* ABFALL, BODEN, LUFT brown coal, lignite; **Braunkohlenflugasche** *f* ABFALL *industrielles Nebenprodukt* lignite ash; **Braunkohlenkraftwerk** *nt* LUFT brown coal power station, lignite power station
Brecher *m* BODEN breaker
Brechungsindex *m* ANALYTIK *für Detektor* refractive index
Brei *m* ABFALL pulp
brennbar *adj* ABFALL, ABWASSER, LUFT, SICHERHEIT combustible; **~er Abfall** *m* ABFALL combustible waste; **~e gasförmige Schadstoffe** *m pl* LUFT combustible gaseous pollutants; **~es Lösemittel** *nt* LUFT combustible solvent; **~es Material** *nt* ABFALL, ABWASSER, LUFT, SICHERHEIT combustible material; **~er Staub** *m* LUFT combustible dust
Brennbarkeit *f* ABFALL, ERNEU ENERG, LUFT, SICHERHEIT combustibility
brennen *vt* ABFALL, LUFT burn
Brenner *m* ABFALL, LUFT burner
Brennereiabwasser *nt* ABWASSER distillery wastewater
Brenner: **Brennersystem** *nt* LUFT combustion system
Brenn: **Brenngas** *nt* LUFT burnable gas; **Brenngut** *nt* ABFALL incinerator charge; **Brennholz** *nt* ERNEU ENER *Bioenergienausgangsstoff* firewood
Brennkammer *f* ABFALL, ABWASSER, BODEN, LUFT combustion chamber; **Brennkammeraustritt** *m* LUFT combustion chamber outlet; **Brennkammerendtemperatur** *f* LUFT combustion chamber final temperature
Brenn: **Brennlinie** *f* ERNEU ENER focal line
Brennstoff *m* ABFALL, ERNEU ENER fuel, SICHERHEIT combustible material; **Brennstoff aus Biomasse** ERNEU ENER fuel from biomass; **Brennstoffe** *m pl* ABFALL, ABWASSER, LUFT, SICHERHEIT combustibles; **~ aus Müll** (*BRAM*) WASSER refuse-derived fuel (*RDF*); **Brennstoffmangel** *m* LUFT lack of fuel; **Brennstoffschicht** *f* ABFALL fuel layer; **Brennstoffüberschuß** *m* LUFT excess fuel; **Brennstoffzelle** *f* ERNEU ENER, LUFT fuel cell; **Brennstoffzusammensetzung** *f* ABFALL fuel composition
Brenn: **Brenntrommel** *f* BODEN firing drum
Brigg: **~scher Logarithmus** *m* LUFT Brigg's logarithm
Brikett *nt* ABFALL *vorherrschende BRAM-Form*, ERNEU ENER *Bioenergienausgangsstoff* briquette
Brikettierpresse *f* ABFALL briquetting press
Brikett: **Brikettpreßstempel** *m* ABFALL briquette press piston
Bringbehältersystem *nt* ABFALL, WASSER stationary container system (*AmE*) (*SCS*)
Bringsystem *nt* ABFALL, WASSER collection point system
Britisch: **~es Institut** *nt* **für Normung** (*BSI*) ANALYTIK British Standards Institute (*BSI*)
BROM *abbr* (*BRAM aus organischer Masse*) ABFALL RDF from organic mass
bromidhaltig *adj* ABFALL, WASSER bromide-containing; **~es Wasser** *nt* WASSER bromide-containing water
Brown: **~sche Molekularbewegung** *f* ABWASSER, LUFT *stochastische Bewegung von Partikeln* Brownian movement
Bruch: **Bruchbelastung** *f* ABFALL, WASSER breaking load; **Bruchlast** *f* ABFALL *Chlor*, WASSER *Stickstoffoxid* breaking load; **Bruchzone** *f* **der Kontinentalplatten** ERNEU ENER *Stelle des geothermischen Energiepotentials* fracture zone of continental plates
Brückenmolekül *nt* ABWASSER *Flockulation* bridging molecule
Brüden *m pl* ABWASSER exhaust vapors (*AmE*), exhaust vapours (*BrE*); **Brüdenbehandlung** *f* ABFALL, WASSER vapor treatment (*AmE*), vapour treatment (*BrE*); **Brüdenkompressor** *m* LUFT vapor compressor (*AmE*), vapour compressor (*BrE*); **Brüdenkondensator** *m* ABWASSER vent condenser
Brunnen *m* BODEN, WASSER well; **Brunnenabsenkung** *f* WASSER well sinking; **Brunnenausbau** *m* WASSER well construction; **Brunnenbetrieb** *m* WASSER well operation; **Brunnenentwicklung** *f* WASSER well development; **Brunnenergiebigkeit** *f* WASSER well yield; **Brunnenfassung** *f* WASSER well catchment; **Brunnenfilter** *nt* WASSER well filter; **Brunnenregeneration** *f* WASSER well regeneration; **Brunnenrohr** *nt* WASSER well pipe; **Brunnenschacht** *m* WASSER well shaft; **Brunnenschaft** *m* WASSER well shaft; **Brunnenwasser** *nt* WASSER well water
BSB *abbr* ABFALL, ABWASSER, ANALYTIK, WASSER (*biochemischer Sauerstoffbedarf*) BOD (*biochemical oxygen demand*), (*biologischer Sauerstoffbedarf*) BOD (*biological oxygen demand*)
BSB$_5$ *abbr* (*biochemischer Sauerstoffbedarf in fünf Tagen, biologischer Sauerstoffbedarf in fünf Tagen*) ABFALL, ABWASSER, ANALYTIK, WASSER BOD$_5$ (*biochemical oxygen demand in five days, biological oxygen demand in five days*); **BSB$_5$-Abbau** *m* ABFALL, ABWASSER, WASSER BOD$_5$ decomposition, BOD$_5$ degradation; **BSB$_5$-Bestimmung** *f* ANALYTIK BOD$_5$ determination
BSB: **BSB-Bestimmung** *f* ANALYTIK BOD determination; **BSB-Fracht** *f* ABWASSER BOD load
BSI *abbr* (*Britisches Institut für Normung*) ANALYTIK BSI (*British Standards Institute*)
BTEX *abbr* (*Benzol, Toluol, Ethylbenzol, Xylol*) ABWASSER *EA* BTEX (*benzene, toluol, ethyl benzene, xylene*)
BTX *abbr* (*Benzol, Toluol, Xylol*) ABWASSER *PB GC-MS* BTX (*benzene, toluol, xylene*)
Büchse *f* ABFALL can
Bucht *f* ERNEU ENER *Gezeitenkraftwerk* bay
Bügel *m* MEER shackle
Bündel *nt* ABFALL bundle; **Bündelmaschine** *f* ABFALL bundle-tying machine
bündeln *vt* ERNEU ENER concentrate
Bunkertank *m* ABFALL, LUFT, WASSER bunker tank

Bunsen: **~scher Verteilungskoeffizient** *m* ABFALL, WASSER Bunsen distribution coefficient
Bunsenbrenner *m* ANALYTIK Bunsen burner
Bürette *f* ANALYTIK burette
Büroabfall *m* ABFALL office waste
Bürstenbelüftung *f* ABWASSER brush aeration
Bußgeld *nt* POLITIK fine, punitive fee, compensation fee
Bußgeld: **~ verhängen gegen** *vt* POLITIK fine
Butadien *nt* LUFT butadiene
Butan *nt* LUFT butane
Bypass *m* LUFT *vor Überhitzung zu schützen* bypass; **Bypass-Absperrklappe** *f* LUFT bypass stop valve; **Bypass-Drucksensor** *m* LUFT bypass pressure sensor; **Bypass-Filter** *nt* LUFT bypass filter; **Bypass-Kamin** *m* LUFT bypass chimney, bypass stack; **Bypass-Leitung** *f* LUFT bypass duct; **Bypass-Pumpe** *f* LUFT bypass pump; **Bypass-Strom** *m* LUFT bypass flow; **Bypass-System** *nt* ANALYTIK *direkter Massestrom* , LUFT bypass system

C

CA *abbr* (*Zelluloseacetat*) ABFALL, ABWASSER CA (*cellulose acetate*)
Cadmium *nt* (*Cd*) ABFALL, ABWASSER, BODEN, WASSER cadmium (*Cd*)
Caisson *m* ERNEU ENER caisson
Calcium *nt* ABFALL, ABWASSER, WASSER calcium; **Calciumcarbonat** *nt* ABFALL, WASSER calcium carbonate; **Calciumhydroxid** *nt* ABFALL, ABWASSER, WASSER calcium hydroxide; **Calciumion** *nt* ABFALL, WASSER calcium ion; **Calciumoxid** *nt* (*CaO*) LUFT calcium oxide (*CaO*); **Calciumphosphat** *nt* ABWASSER calcium phosphate, phosphate of calcium; **Calcium-Sulfat-Dihydrat** *nt* LUFT *Gipsart* calcium sulfate hemihydrated (*AmE*), calcium sulphate hemihydrated (*BrE*)
CA: **CA-Membran** *f* WASSER CA membrane
CaO *abbr* (*Calciumoxid*) LUFT CaO (*calcium oxide*)
Carbonathärte *f* (*CH*) ABWASSER, WASSER *Dosierung von Flüssigkeiten* carbonate hardness (*CH*); **Carbonathärtebestimmung** *f* ANALYTIK carbonate hardness determination in wastewater
Carbonylsulfid *nt* ABFALL, LUFT, WASSER carbonyl sulfide (*AmE*), carbonyl sulphide (*BrE*)
CaZ *abbr* (*Cetanzahl*) LUFT cetane number
CCS-System *nt* LUFT controlled combustion system (*CCS*)
Cd *abbr* (*Cadmium*) ABFALL, ABWASSER, BODEN, WASSER Cd (*cadmium*)
CEN *abbr* (*Europäisches Normungsinstitut*) ANALYTIK CEN (*European Standardization Committee*)
Cetan *m* LUFT cetane; **Cetanwert** *m* LUFT cetane number; **Cetanzahl** *f* (*CaZ*) LUFT cetane number
CH *abbr* (*Carbonathärte*) ABWASSER *gesamter anorganisch gebundener Kohlenstoff*, WASSER CH (*carbonate hardness*)
Charge *f* ABWASSER, ANALYTIK, BODEN batch; **Chargenbetrieb** *m* ABWASSER, LUFT batch operation
chargenmäßig: **~ betrieben** *adj* ABWASSER, LUFT batch-processed
Charge: **Chargenneutralisation** *f* ABWASSER batch neutralization; **Chargenprozeß** *m* LUFT batch process; **Chargenverfahren** *nt* LUFT batch process
chargenweise *adv* ABFALL, ABWASSER, BODEN, LUFT in batch quantities
Chelat *nt* ABWASSER, BODEN *Zusatzstoff* chelate
chelatbildend *adj* ABWASSER chelating; **~er Ionenaustauscher** *m* ABWASSER chelating ion exchanger
Chelat: **Chelatbildner** *m* ABWASSER, BODEN chelate former
CHEMFIX-Verfahren® *nt* ABFALL, BODEN, WASSER *Verfestigungsverfahren für Sonderabfälle* CHEMFIX process®
Chemieabwasser *nt* ABWASSER chemical wastewater
chemiebeständig *adj* ABFALL, ABWASSER, LUFT, SICHERHEIT chemically resistant
Chemikalie *f* ABWASSER, LUFT, WASSER chemical reagent, chemical
chemikalienbeständig *adj* ABFALL, ABWASSER, LUFT, SICHERHEIT chemically resistant; **~er Schutzanzug** *m* SICHERHEIT chemically resistant protective suit
Chemikalie: **Chemikalienreste** *m pl* ABFALL chemical residue
Chemilumineszenz *f* ANALYTIK chemiluminescence; **Chemilumineszenzdetektion** *f* (*CLD*) ANALYTIK chemiluminescence detection (*CLD*); **Chemilumineszenzmeßgerät** *nt* ANALYTIK *zum Nachweis von Ozon* chemiluminescence measuring instrument
chemisch *adj* ALLG chemical; **~er Abfall** *m* ABFALL chemical waste; **~e Abwasseraufbereitung** *f* ABWASSER chemical wastewater treatment; **~e Adsorption** *f* LUFT chemisorption, chemical adsorption; **~ aggressiv** *adj* LUFT chemically aggressive; **~ aggressiver Staub** *m* LUFT chemically aggressive dust; **~e Aggressivität** *f* ABFALL, WASSER chemical aggressivity; **~e Atmosphärendruck-Ionisation** *f* (*APCI*) ANALYTIK atmospheric pressure chemical ionization; **~e Belastung** *f* ABFALL chemical strain; **~e Beständigkeit** *f* LUFT *Filtermaterialen* chemical resistance; **~e Desinfektion** *f* ABFALL, WASSER chemical disinfection; **~es Dosimeter** *nt* ANALYTIK chemical dosimeter; **~e Fällung** *f* ABFALL *Wasserenthärtung*, ABWASSER, LUFT chemical precipitation; **~e Gefahr** *f* SICHERHEIT chemical hazard; **~e Gewässergüte** *f* WASSER chemical water quality; **~e Ionisation-Massenspektrometrie** *f* (*CI-MS*) ANALYTIK chemical ionization mass spectrometry (*CI-MS*); **~er Kampfstoff** *m* ABFALL, LUFT *Militär* toxic agent; **~e Konditionierung** *f* ABFALL, WASSER chemical conditioning; **~e Oxidation** *f* ABFALL, ABWASSER, WASSER chemical oxidation; **~ oxidative Gaswäsche** *f* LUFT chemical-oxidative gas scrubbing; **~er Parameter** *m* ABWASSER, ANALYTIK, BODEN chemical parameter; **~er Prozeß** *m* **in der Atmosphäre** LUFT atmospheric chemical process; **~ qualitativ** *adj* ANALYTIK chemically qualitative; **~ quantitativ** *adj* ANALYTIK chemically quantitative; **~es Recycling** *nt* ABFALL tertiary plastic recycling, feedstock recycling; **~e Regeneration** *f* ABWASSER chemical regeneration; **~ resistent** *adj* ABFALL, LUFT, SICHERHEIT chemically resistant; **~er Sauerstoffbedarf** *m* (*CSB*) ABFALL, ABWASSER, LUFT, WASSER chemical oxygen demand (*COD*); **~es Sediment** *nt* ABFALL, ABWASSER, WASSER chemical sediment; **~e Selbstreinigung** *f* ABFALL, ABWASSER, WASSER chemical self-purification; **~ stabilisiert** *adj* ABWASSER chemically stabilized; **~e Stabilisierung** *f* ABWASSER chemical stabilization; **~e Verwertung** *f* ABFALL feedstock recycling
chemisch-physikalisch *adj* ALLG chemical-physical; **~e Vorbehandlung** *f* ABFALL, WASSER chemical-physical pretreatment; **~e Phosphoreliminierung** *f* ABWASSER physiochemical phosphorus removal
chemisorbieren *vt* LUFT chemisorb
Chemisorptionsfilter *nt* LUFT chemisorption filter
Chemolyse *f* ABWASSER chemolysis
Chemosorption *f* LUFT chemisorption, chemical adsorption
Chevron-Baffle *nt* LUFT chevron baffle
Chlor *nt* (*Cl*) ABWASSER, BODEN, LUFT, WASSER chlorine (*Cl*); **Chlor-Alkali-Elektrolyse** *f* ABFALL, ABWASSER, WASSER chlor-alkali electrolysis
chlorarm: **~ gebleichtes Papier** *nt* ABFALL low-chlorine bleached paper
Chlorat *nt* ABFALL, BODEN, LUFT, WASSER chlorate
Chlor: **Chlorbedarf** *m* ABFALL, WASSER chlorine demand (*CD*); **Chlorbenzol** *nt* LUFT chlorobenzene; **Chlorbleichlauge** *f* ABFALL, WASSER chlorine bleaching; **Chlordosierung** *f* WASSER chlorine feeding; **Chlorethen** *nt* ABFALL *Vinylchlorid* chloroethylene
chlorfrei *adj* ABFALL chlorine-free; **~es Papier** *nt* ABFALL chlorine-free paper
Chlor: **Chlorgas** *nt* ABFALL, LUFT chlorine gas; **Chlorgasüberschuß** *m* ABFALL, LUFT excess chlorine gas
Chlorid *nt* ABFALL, BODEN, LUFT, WASSER chloride
chlorieren *vt* ABWASSER, BODEN, LUFT, WASSER chlorinate

chloriert *adj* ABWASSER, BODEN, LUFT, WASSER chlorinated; **~er Kohlenwasserstoff** *m* LUFT chlorinated hydrocarbon
***Chlor*: Chlorkohlenwasserstoff** *m* (*CKW*) LUFT chlorinated hydrocarbon; **Chlorphenol** *nt* LUFT chlorophenol; **Chlorüberschuß** *m* WASSER excess chlorine
Chlorung *f* ABWASSER, BODEN, LUFT, WASSER chlorination
***Chlor*: Chlorzusatzgerät** *nt* ABWASSER, BODEN, LUFT, WASSER chlorinator
C-Horizont *m* BODEN C-horizon
Chrom *nt* (*Cr*) ABWASSER, BODEN, LUFT chromium (*Cr*)
chromathaltig *adj* ABWASSER chromate-containing; **~er Abfall** *m* ABFALL chromate-containing waste; **~es Abwasser** *nt* ABWASSER chromate-containing wastewater
Chromatogramm *nt* ANALYTIK chromatogram
Chromatographiedatensystem *nt* ANALYTIK chromatographic data system
chromatographierbar *adj* ANALYTIK able to be chromatographed
chromatographieren *vt* ANALYTIK chromatograph
chromatographisch *adj* ANALYTIK chromatographic; **~e Säule** *f* ANALYTIK chromatographic column
Chromatreduktion *f* ABFALL chromate reduction
chromhaltig *adj* ABWASSER chromium-containing; **~er Schlamm** *m* ABWASSER chromium-containing sludge
Chromogen *nt* ANALYTIK chromogen
chronisch *adj* SICHERHEIT chronic
***chronisch*: ~e Vergiftung** *f* SICHERHEIT chronic poisoning; **~e Wirkung** *f* ABFALL, SICHERHEIT chronic effect
Chronopotentiometrie *f* ANALYTIK chronopotentiometry
CIEF *abbr* (*kapillarisoelektrische Fokussierung*) ANALYTIK CIEF (*capillary isoelectric focusing*)
CI-MS *abbr* (*chemische Ionisation-Massenspektrometrie*) ANALYTIK CI-MS (*chemical ionization mass spectrometry*)
CKW *abbr* (*Chlorkohlenwasserstoff*) LUFT chlorinated hydrocarbon
Cl *abbr* (*Chlor*) ABFALL, LUFT, WASSER *Gaschromatographie* Cl (*chlorine*)
Claus-Anlage *f* LUFT, WASSER *Schwefelrückgewinnungsanlage* Claus plant
Claus-Verfahren *nt* LUFT, WASSER *Schwefelrückgewinnung* Claus process
CLD *abbr* (*Chemilumineszenzdetektion*) ANALYTIK CLD (*chemiluminescence detection*)
C/N/P-Verhältnis *nt* (*Kohlenstoff-Stickstoff-Phosphor-Verhältnis*) ANALYTIK C/N/P ratio (*carbon-nitrogen-phosphorus ratio*)
C/N-Verhältnis *nt* (*Kohlenstoff-Stickstoff-Verhältnis*) ANALYTIK C/N ratio (*carbon-nitrogen ratio*)
Co 1. *abbr* (*Kobalt*) BODEN Co (*cobalt*); **2.** *in cpds* co-
CO *abbr* (*Kohlenmonoxid*) LUFT CO (*carbon monoxide*)
CO_2 *abbr* (*Kohlendioxid*) ABFALL, LUFT, WASSER CO_2 (*carbon dioxide*)
***CO*: CO-Emissionen** *f pl* LUFT CO emissions
***CO_2*: CO_2-Emissionsverminderung** *f* LUFT CO_2 emissions reduction
***Co-*: Co-Extrudierung** *f* ABFALL co-extrusion; **Co-Extrusion** *f* ABFALL co-extrusion; **Co-Injektion** *f* ABFALL co-injection
***CO_2*: CO_2-Konzentration** *f* LUFT CO_2 concentration
coliform: **~e Bakterien** *f pl* ANALYTIK *Indikator* coliform bacteria; **~e Keime** *m pl* ABFALL, WASSER coliforms
Colititer *m* ANALYTIK coli test
Combicontainer *m* ABFALL multiple material container
computergesteuert *adj* ANALYTIK computerized; **~er Analysator** *m* ANALYTIK computerized analyser (*BrE*), computerized analyzer (*AmE*)
Container *m* ALLG container, MEER scoop
containerisierbar *adj* ABFALL, BODEN containerizable
containerisieren *vt* ABFALL, BODEN containerize
containerisiert *adj* ABFALL, BODEN containerized
Containerisierung *f* ABFALL, BODEN containerization
***Container*: Containerkläranlage** *f* ABWASSER container wastewater treatment plant; **Containerpresse** *f* ABFALL container press; **Container-Sanierungsanlage** *f* BODEN container remediation installation; **Containerschlitten** *m* ABFALL container slide; **Containersystem** *nt* ABFALL container system
Controlling-Kreislauf *m* ABFALL controlling loop
Copolymer *nt* ABFALL copolymer
Copolymerisation *f* ABFALL copolymerization
Cordierit *nt* LUFT cordierite
Coriolis-Kraft *f* LUFT Coriolis force
Cottrell-Möller-Verfahren *nt* LUFT *zur elektrostatischen Gasreinigung* Cottrel process
Coulometer *nt* ANALYTIK coulometer
Coulometrie *f* ANALYTIK coulometry
coulometrisch *adj* ANALYTIK coulometric; **~er Analysator** *m* ANALYTIK coulometric analyser (*BrE*), coulometric analyzer (*AmE*); **~e Titration** *f* ANALYTIK coulometric titration
***CO*: CO-Wert** *m* LUFT CO-value
Cr *abbr* (*Chrom*) ABWASSER *GPT*, BODEN *Schwermetalle*, LUFT Cr (*chromium*)
cracken *vt* ABFALL *Pyrolyse* crack
Cradle-to-Grave-Analyse *f* MANAG, POLITIK, WASSER cradle-to-grave analysis
Crossflow-Filtration *f* ABWASSER cross-flow filtration; **~ mit Membranen** ABWASSER cross-flow filtration with membranes
Crossflow-Mikrofiltration *f*: **~ mit überlagertem elektrischem Feld** ABWASSER cross-flow microfiltration with underlying electrical field
CSB *abbr* (*chemischer Sauerstoffbedarf*) ABFALL, ABWASSER, LUFT, WASSER COD (*chemical oxygen demand*); **CSB-Bestimmung** *f* ANALYTIK COD determination; **CSB-Reduktion** *f* ABWASSER COD reduction
CSD: **CSD-Kommission** *f* POLITIK Commission on Sustainable Development (*CSD*)
Cu *abbr* (*Kupfer*) ALLG Cu (*copper*)
CVS-Test *m* LUFT *amerikanisches Prüfverfahren mit Teilstromentnahme* constant volume sampling (*CVS*)
Cyanid *nt* ABWASSER, BODEN cyanide; **Cyanideliminierung** *f* ABWASSER cyanide removal
cyanidhaltig *adj* ABWASSER cyanide-containing; **~er Abfall** *m* ABFALL, ABWASSER cyanide-containing waste; **~er Schlamm** *m* ABFALL, ABWASSER, WASSER cyanide-containing sludge

D

dachintegriert: **~er Flachkollektor** *m* ERNEU ENER built-in roof surface collector
Dacron *nt* LUFT *Filtermaterial* dacron
Dalton: **~sches Gesetz** *nt* LUFT Dalton's law of partial pressure
Damm *m* ERNEU ENER *Gezeitenkraftwerk* dike
Dämm: **Dämmplatte** *f* LÄRM insulating board, acoustical slab, acoustic tile; **Dämmschicht** *f* LÄRM insulating layer; **Dämmstein** *m* LÄRM insulating brick; **Dämmzahl** *f* LÄRM sound absorption coefficient
Dampf *m* ALLG steam, vapor (*AmE*), vapour (*BrE*)
Dampfdruck *m* ABFALL, ABWASSER *Pumpe* ANALYTIK, BODEN *Bodenluftabsaugung*, LUFT, WASSER *Pumpe* steam pressure, vapor pressure (*AmE*), vapour pressure (*BrE*); **Dampfdruckbestimmung** *f* ANALYTIK vapor pressure analysis (*AmE*), vapour pressure analysis (*BrE*), vapor pressure determination (*AmE*), vapour pressure determination (*BrE*); **Dampfdruckfederthermometer** *nt* ANALYTIK steam pressure thermometer; **Dampfdruckmessung** *f* ANALYTIK vapour pressure measurement (*BrE*), vapor pressure measurement (*AmE*); **Dampfdruckwaage** *f* ANALYTIK vapor pressure scale (*AmE*), vapour pressure scale (*BrE*)
Dampf: **Dampfemissionen** *f pl* LUFT vapor emissions (*AmE*), vapour emissions (*BrE*), vapor-laden emissions (*AmE*), vapour-laden emissions (*BrE*)
dämpfen *vt* LÄRM *Elektrotechnik* attenuate, *Geräusch* deaden, muffle, *mechanisch* cushion, damp
Dämpfer *m* LÄRM muffler (*AmE*), silencer (*BrE*)
Dampf: **Dampferzeuger** *m* ABWASSER steam generator; **Dampfextraktion** *f* BODEN vapor extraction (*AmE*), vapour extraction (*BrE*); **Dampffahne** *f* LUFT, WASSER vapor plume (*AmE*), vapour plume (*BrE*)
dampfförmig *adj* LUFT vaporous; **~er Brennstoff** *m* LUFT vaporous fuel; **~es Lösungsmittel** *nt* LUFT vaporous solvent; **~e Schadstoffe** *m pl* LUFT vaporous pollutants
Dampf: **Dampfkessel** *m* ABFALL boiler; **Dampfluftvorwärmer** *m* LUFT steam air preheater; **Dampfluvo** *m* LUFT steam air preheater; **Dampfprozeß** *m* ERNEU ENER steam process; **Dampfraumanalyse** *f* ANALYTIK headspace analysis; **Dampfstrahl** *m* WASSER steam jet; **Dampfstrahlen** *nt* ABFALL steam cleaning; **Dampfstrahlpumpe** *f* ABFALL steam jet pump; **Dampfstrippen** *nt* ABFALL, BODEN, WASSER steam stripping; **Dampfstripping** *f* ABWASSER steam stripping; **Dampfturbine** *f* ERNEU ENER steam turbine
Dämpfung *f* LÄRM sound deadening, attenuation; **Dämpfungsbelag** *m* LÄRM damped layer; **Dämpfungsglied** *nt* LÄRM isolation pad, attenuator pad; **Dämpfungsregler** *m* LÄRM attenuator
Dampf: **Dampfvakuum Strippen** *nt* BODEN steam-vacuum stripping
Daphnientest *m* ANALYTIK daphnia test
Darrieus-Rotor *m* ERNEU ENER *Biokonversion* Darrieus rotor
Daten *pl* ANALYTIK data; **Datenanalyse** *f* ANALYTIK data analysis; **Datenauswertung** *f* ANALYTIK data interpretation; **Datenbank** *f* ANALYTIK data base; **Datenerfassung** *f* ANALYTIK data logging, data acquisition; **Datenlogger** *m* ANALYTIK data logger; **Datennetz** *nt* ANALYTIK data network; **Datensicherheit** *f* ANALYTIK data security; **Datenspeicherung** *f* ANALYTIK data storage; **Datensystem** *nt* ANALYTIK data system; **Datenübertragung** *f* ANALYTIK data transfer; **Datenverarbeitung** *f* ANALYTIK data processing
Dauerbetrieb *m* MANAG, POLITIK *einer Anlage*, WASSER commercial operation
dauerhaft: **~e Hörschwellenverschiebung** *f* LÄRM permanent audibility threshold shift (*PTS*)
dB *abbr* (*Dezibel*) ANALYTIK dB (*decibel*)
DC *abbr* (*Dünnschichtchromatographie*) ANALYTIK TLC (*thin-layer chromatography*)
DCR *abbr* (*Dispersion durch chemische Reaktion*) ABFALL, WASSER DCR (*dispersion by chemical reaction*); **DCR-Verfahren** *nt* ABFALL *Sonderabfallbehandlung* DCR process
DDT *abbr* (*Dichlordiphenyltrichlorethan*) BODEN DDT (*dichlorodiphenyltrichloroethane*)
Deck- *in cpds* ABWASSER, BODEN, LÄRM, LUFT covering; **Deckband** *nt* BODEN shroud
Deckel *f* LUFT top
Deckenplatte *f* LÄRM ceiling panel
Deck-: **Deckplatte** *f* LUFT *Schlauchfilter* cover plate; **Deckschicht** *f* ABFALL *einer Deponie* covering layer
defekt: **~e Kanalsysteme** *nt pl* BODEN *Quelle von Bodenverunreinigung* defective canal systems
definierbar *adj* ANALYTIK definable
degradieren *vt* ALLG degrade
degradiert *adj* ALLG degraded
Degussa-Plinke-Verfahren *nt* LUFT Degussa-Plinke process
dehalogenieren *vt* BODEN dehalogenate
Dehydrogenasetest *m* ANALYTIK dehydrogenase test
De-Inking *nt* ABFALL, ABWASSER de-inking; **~ durch Flotation** ABWASSER de-inking by flotation; **De-Inking-Anlage** *f* ABFALL, ABWASSER de-inking plant, de-inking unit; **De-Inking-Schlamm** *m* ABWASSER de-inking sludge
Deionat *nt* ABWASSER, WASSER deionized water, demineralized water
Deionisation *f* ABFALL, ABWASSER, WASSER deionization, demineralization
deionisieren *vt* ABFALL, ABWASSER, WASSER deionize, demineralize
deionisiert *adj* ABFALL, ABWASSER, WASSER deionized, demineralized
Deionisierung *f* ABFALL, ABWASSER, WASSER deionization, demineralization
Dekanter *m* ABWASSER decanter
dekantieren *vt* ALLG decant, elutriate
Dekantieren *nt* ALLG decantation, elutriation
Dekantierzentrifuge *f* ABFALL, WASSER decanting centrifuge
dekompaktieren *vt* ABFALL, WASSER decompact
Dekontamination *f* ALLG decontamination; **Dekontaminationsreaktor** *m* BODEN decontamination reactor
Dekontaminierbarkeit *f* ALLG decontaminability
dekontaminieren *vt* ABFALL, ABWASSER decontaminate, BODEN remediate, LUFT, SICHERHEIT, WASSER decontaminate
dekontaminiert *adj* BODEN *Boden* decontaminated, remediated; **~er Boden** *m* BODEN decontaminated soil
demineralisieren *vt* ABFALL, ABWASSER, WASSER deionize, demineralize
demineralisiert *adj* ABFALL, ABWASSER, WASSER deionized, demineralized
Demineralisierung *f* ABFALL, ABWASSER, WASSER deionization, demineralization
Demister *m* LUFT demister
Demonstrationskraftwerk *nt* ERNEU ENER demonstration power plant
Demulgator *m* ABFALL demulsifier, emulsion breaker, ABWAS-

SER demulsifier, demulsifying agent, MEER emulsion breaker, WASSER demulsifier, demulsifying agent, emulsion breaker

Demulgierprodukt *nt* ABWASSER, WASSER demulsifying product

Denitrierung *f*: ~ **des Abfalls** ABFALL waste denitrification

Denitrifikant *m* ABFALL, ABWASSER, LUFT denitrifying bacteria

Denitrifikation *f* ALLG *Umwandlung des Nitratstickstoffs in gasförmigen Stickstoff* denitrification; **Denitrifikationszone** *f* ABFALL, ABWASSER, LUFT, WASSER denitrification zone

denitrifizieren *vt* ALLG denitrify

Denitrifizierung *f* ALLG denitrification

Denox *nt* (*NO_x-Entstickung*) BODEN, LUFT denox (*NO_x removal*); **Denoxanlage** *f* LUFT *selektiv-katalytische Reduktion* denoxer; **Denox-Katalysator** *m* LUFT denox-catalyst

DEP *abbr* (*Direktentschwefelungsprozeß*) LUFT direct desulfurization process (*AmE*), direct desulphurization process (*BrE*)

Depolarisation *f* ANALYTIK depolarization

Depolarisator *m* ANALYTIK depolarizer

Depolymerisation *f* ABFALL depolymerization

Deponie *f* ABFALL, SICHERHEIT disposal site, garbage disposal site (*AmE*), refuse disposal site, repository, storage site, tip (*BrE*), waste dump, waste tip (*BrE*), rubbish tip, ABFALL, ANALYTIK, BODEN, LUFT, POLITIK landfill; **Deponieabdichtung** *f* ABFALL landfill sealing; **Deponieabgabe** *f* POLITIK landfill charge; **Deponieabschnitt** *m* ABFALL landfill section; **Deponiealter** *nt* ABFALL landfill age; **Deponieart** *f* ABFALL landfill design, landfill type; **Deponieaufbau** *m* ABFALL landfill structure; **Deponiebasis** *f* ABFALL landfill foundation; **Deponiebasisabdichtung** *f* ABFALL landfill base sealing; **Deponiebauwerk** *nt* ABFALL landfill building; **Deponiebetreiber** *m* ABFALL landfill operator; **Deponiebetrieb** *m* ABFALL site operation, landfill operation; **Deponiebinder** *m* ABFALL landfill binder; **Deponiedichte** *f* ABFALL landfill thickness; **Deponieeinbau** *m* ABFALL landfill installation; **Deponieentsorgung** *f* ABFALL landfill disposal; **Deponieerfordernis** *nt* ABFALL landfill requirement

deponiefähig *adj* ABFALL landfill-suitable; **~es Fahrzeug** *nt* ABFALL landfill-suitable vehicle

***Deponie*: Deponiefläche** *f* ABFALL landfill area

Deponiegas *nt* ABFALL, LUFT landfill gas; **Deponiegasableitung** *f* ABFALL, LUFT landfill gas drainage; **Deponiegasanfall** *m* ABFALL, LUFT landfill gas yield; **Deponiegasbehandlung** *f* ABFALL landfill gas treatment; **Deponiegasentwicklung** *f* ABFALL landfill gas development; **Deponiegaserfassung** *f* ABFALL, LUFT landfill gas collection; **Deponiegasfassung** *f* ABFALL, LUFT landfill gas collector; **Deponiegasgrenzwert** *m* ABFALL landfill gas limit; **Deponiegasmenge** *f* ABFALL landfill gas quantity; **Deponiegasmessung** *f* ANALYTIK landfill gas measurement; **Deponiegasmigration** *f* ABFALL, LUFT landfill gas migration; **Deponiegasnutzung** *f* ABFALL landfill gas use; **Deponiegasproblem** *nt* ABFALL landfill gas problem; **Deponiegasproduktion** *f* ABFALL landfill gas production; **Deponiegasuntersuchung** *f* ANALYTIK landfill gas analysis; **Deponiegaszusammensetzung** *f* ABFALL landfill gas composition

***Deponie*: Deponiegebühr** *f* ABFALL landfill fee

deponiegeeignet *adj* ABFALL landfill-suitable

***Deponie*: Deponiegelände** *nt* ABFALL landfill site; **Deponiegrenze** *f* ABFALL landfill perimeter; **Deponiegut** *nt* ABFALL waste mass, fill mass; **Deponiehang** *m* ABFALL landfill slope; **Deponiekapazität** *f* ABFALL landfill capacity; **Deponiekataster** *m* POLITIK landfill register; **Deponiekörper** *m* BODEN landfill body; **Deponiematerial** *nt* ABFALL landfill material; **Deponieoberfläche** *f* ABFALL working face, operating face; **Deponiephase** *f* ABFALL landfill phase; **Deponieplanum** *nt* ABFALL landfill support; **Deponieplanung** *f* ABFALL waste landfill planning; **Deponieraum** *m* ABFALL landfill area; **Deponieraumschonung** *f* ABFALL landfill area conservation

deponierbar *adj* ABFALL landfill-suitable

Deponierbarkeit *f* ABFALL, ABWASSER *Klärschlammasche* landfilling suitability

Deponierfähigkeit *f* ABFALL landfilling suitability

deponiert *adj* ABFALL landfilled

***Deponie*: Deponierückbau** *m* ABFALL recovery of landfill space

Deponierung *f* ABFALL landfilling; **Deponierungspraxis** *nt* ABFALL landfilling practice; **Deponierungsverfahren** *nt* ABFALL landfilling method, tip filling method (*BrE*), waste disposal method

***Deponie*: Deponiesicherheit** *f* ABFALL, SICHERHEIT landfill location security

Deponiesickerwasser *nt* ABFALL, ABWASSER landfill leachate; **Deponiesickerwasserbecken** *nt* ABFALL, ABWASSER landfill leachate tank; **Deponiesickerwasserbehandlung** *f* ABFALL, ABWASSER landfill leachate treatment; **Deponiesickerwasserreinigung** *f* ABFALL, ABWASSER, WASSER landfill leachate purification

***Deponie*: Deponiesohle** *f* ABFALL landfill bottom; **Deponiestandort** *m* ABFALL landfill location, MEER landfill site; **Deponiestandortstechnik** *f* ABFALL landfill site technique; **Deponiestraßen** *f pl* ABFALL landfill roads; **Deponietechnik** *f* ABFALL landfill technique; **Deponieuntergrund** *m* ABFALL landfill underground; **Deponieverhalten** *nt* ABFALL landfill behavior (*AmE*), landfill behaviour (*BrE*); **Deponieverschluß** *m* ABFALL site closure; **Deponievolumen** *nt* ABFALL landfill volume

Depotcontainer *m* ABFALL depot container

Derrick *m* ERNEU ENER *Gezeitenkraftwerk* derrick

desaktivieren *vt* ABFALL, LUFT deactivate

desaktiviert *adj* ABFALL, LUFT deactivated

Desaktivierung *f* ABFALL, LUFT deactivation

Desinfektion *f* ABFALL ABWASSER, SICHERHEIT, WASSER disinfection; **~ von Trinkwasser** WASSER drinking water disinfection; **Desinfektionsmittel** *nt* ABWASSER, SICHERHEIT, WASSER disinfectant; **Desinfektionsschutz** *m* WASSER disinfection protection; **Desinfektionswirkung** *f* ABWASSER, WASSER disinfection effect, disinfection efficiency

desinfizieren *vt* ABFALL, ABWASSER, SICHERHEIT, WASSER disinfect

desinfiziert *adj* ABFALL, ABWASSER, SICHERHEIT, WASSER disinfected

Desintegrator *m* LUFT disintegrator

desodorieren *vt* ABFALL, ABWASSER, LUFT, WASSER deodorize

desodoriert *adj* ABFALL, ABWASSER, WASSER deodorized

Desonox *nt* (*SO_2/NO_x-Entstickung*) BODEN, LUFT desonox (*SO_2/NO_x removal*); **Desonox-Verfahren** *nt* LUFT desonox process; **Desonox-Verfahren** *nt* **in der zirkulierenden Wirbelschicht** LUFT desonox in circulating fluidized bed

Desorberzulauf *m* LUFT desorber feed

desorbieren *vt* ANALYTIK *thermisch*, BODEN, LUFT desorb

desorbiert *adj* ANALYTIK, BODEN, LUFT desorbed; **~es Absorbens** *nt* LUFT desorbed absorbent; **~er Aktivkoks** *m* LUFT desorbed activated coke; **~es Reichgas** *nt* LUFT desorbed rich gas

Desorption *f* ABWASSER, ANALYTIK, LUFT *Rückgewinnung der Gaskomponente aus der Wäschlösung* desorption; **Desorptionskolonne** *f* LUFT desorption column; **Desorptionsluft** *f* LUFT desorption air; **Desorptionsluftwärmetauscher** *m* LUFT desorption air heat exchanger; **Desorptionsmedium** *nt* LUFT desorbent; **Desorptionsmittel** *nt* LUFT desorbent; **Desorptionszeit** *f* LUFT desorption time; **Desorptionszone** *f* LUFT desorption zone

destabilisieren *vt* ABFALL, ABWASSER, WASSER destabilize

destabilisiert *adj* ABFALL, ABWASSER, WASSER destabilized; **~es Teilchen** *nt* ABWASSER *Flockulation* destabilized particle

Destabilisierung *f* ABFALL, ABWASSER, WASSER destabilization
Destillat *nt* ANALYTIK distillate
Destillation *f* ABFALL, BODEN, WASSER distillation; **Destillationskolonne** *f* LUFT distilling column, WASSER *Pyrolyse* distillation column; **Destillationsrückstand** *m* ABFALL distillation residue; **Destillationswasser** *nt* ABFALL distillation water
Destillatkolben *m* ANALYTIK distillate flask
destillieren *vt* ABFALL, ANALYTIK, LUFT distil (*BrE*), distill (*AmE*)
destilliert: **~es Wasser** *nt* ANALYTIK distilled water
Detektion *f* ANALYTIK detection; **Detektionsprinzip** *nt* ANALYTIK detection principle
Detektor *m* ANALYTIK detector; **~ für Flüssigkeitschromatographie** ANALYTIK *Chlor* liquid chromatography detector; **Detektorlampe** *f* ANALYTIK detector lamp
Detergens *nt* ABWASSER MEER, WASSER detergent, surface-active agent
Detritus *m* ABWASSER detritus, suspended and settable solids
Deutsch: **~e Einheitsverfahren** *nt pl* **zur Wasser-, Abwasser- und Schlammanalyse** (*DEV*) ANALYTIK German unit processes for water, wastewater and sludge analysis; **~es Institut** *nt* **für Normung** (*DIN*) ANALYTIK German Institute for Standards (*DIN*)
DEV *abbr* (*Deutsche Einheitsverfahren zur Wasser-, Abwasser- und Schlammanalyse*) ANALYTIK German unit processes for water, wastewater and sludge analysis
Devastierung *f* ABFALL land degradation, land disturbance
Dewar-Gefäß *nt* ANALYTIK Dewar vessel
dezentral: **~e photoelektrische Nutzung** *f* ERNEU ENER decentral photoelectric use; **~e thermische Nutzung** *f* ERNEU ENER isolated thermal use
Dezibel *nt* (*dB*) ANALYTIK decibel (*dB*)
DHS *abbr* (*dynamische Headspace-Technik*) ANALYTIK DHS (*dynamic headspace technique*)
Diagnose *f* ANALYTIK diagnosis
Dialoganzeige *f* ANALYTIK *bei Meßgeräten* dialog display (*AmE*), dialogue display (*BrE*)
dialoggesteuert *adj* ANALYTIK *Programming* dialog-controlled (*AmE*), dialogue-controlled (*BrE*)
Dialyse *f* ABWASSER, ANALYTIK, WASSER dialysis
diamagnetisch *adj* ANALYTIK diamagnetic
Diatomeenerde *f* ABFALL, WASSER diatomaceous earth; **Diatomeenerdefiltration** *f* ABWASSER diatomaceous earth filtration
Dichlordifluormethan *nt* (*Freon 12*) LUFT dichlorodifluoromethane (*Freon 12*)
Dichlordiphenyltrichlorethan *nt* (*DDT*) BODEN dichlorodiphenyltrichloroethane (*DDT*)
Dichlormethan *nt* LUFT *halogenierter Kohlenwasserstoff* dichloromethane
dicht: **~e Atmosphärenschicht** *f* LUFT *Meteorologie* friction layer
Dichte *f* ABFALL *Deponieschichten* thickness, ANALYTIK, WASSER *des Wassers* density; **Dichtebestimmung** *f* ANALYTIK density determination
dichten *vt* MEER seal
Dichte: **Dichtesortierung** *f* ABFALL sorting based on density; **Dichteströmung** *f* WASSER density current
dichthalten *vt* ANALYTIK *Probe* keep sealed; **dichtmachen** *vt* BODEN densify
Dichtung *f* ABFALL, BODEN seal; **Dichtungsauflager** *nt* ABFALL seal-bearing surface; **Dichtungsbahn** *f* ABFALL liner sheet; **Dichtungsfolie** *f* BODEN sealing sheeting; **Dichtungssystem** *nt* ABFALL barrier system; **Dichtungswand** *f* BODEN retention wall
Dichtwand *f* BODEN retention wall
Dichtwirkung *f* ABFALL sealing properties
Dickschlamm *m* ABWASSER thickened sludge
Dickstoffstrom *m* ABWASSER thick matter flow
Diesel *m* ERNEU ENER, LUFT *Brennstoff* diesel fuel, diesel, *Motor* diesel engine; **Dieselabgas** *nt* LUFT diesel exhaust gas; **Dieselgenerator** *m* ERNEU ENER diesel generator; **Dieselmotor** *m* LUFT diesel engine
Differential *nt* ANALYTIK differential; **Differential-Gaschromatogramm** *nt* ANALYTIK differential gas chromatogram; **Differentialmanometer** *nt* LUFT differential manometer
Differenzkalorimetrie *f* ANALYTIK differential calorimetry
Differenzthermoanalyse *f* (*DTA*) ANALYTIK differential thermal analysis (*DTA*)
Diffraktion *f* ANALYTIK diffraction
diffundieren *vt* LUFT diffuse
diffus: **~e Einstrahlung** *f* ERNEU ENER diffuse radiation; **~es Schallfeld** *nt* LÄRM *wenn zu viele Schallreflektoren im Raum sind* diffuse field
Diffusion *f* LUFT, WASSER diffusion; **Diffusionsaufladung** *f* LUFT diffusion charging; **Diffusionsbeiwert** *m* WASSER coefficient of diffusion; **Diffusionskoeffizient** *m* WASSER coefficient of diffusion; **Diffusionsuntersuchung** *f* ANALYTIK diffusion analysis; **Diffusionsversuch** *m* ANALYTIK diffusion test
Diffusor *m* ABFALL diffuser, *einer Kreiselpumpe, Zentrifugalpumpe* volute chamber, ANALYTIK, LÄRM diffuser, LUFT diffuser, *Venturidüse* expander, WASSER diffuser
digital: **~ anzeigen** *vt* ANALYTIK read out digitally
Digitalanzeige *f* ANALYTIK digital readout
Digitalkanal *m* ANALYTIK digital channel
Dilution *f* ABWASSER dilution; **Dilutionskapazität** *f* ABWASSER *der Gewässer* dilution capacity; **Dilutionsvermögen** *nt* ABWASSER *der Gewässer* dilution capacity
DIN *abbr* (*Deutsches Institut für Normung*) ANALYTIK DIN (*German Institute for Standards*)
Diode *f* ANALYTIK diode; **Dioden-Array-Detektor** *m* ANALYTIK diode array detector; **Diodenlaser** *m* ANALYTIK *Messung von gasförmigen Schadstoffen* diode laser
Dioxine *nt pl* BODEN dioxins
direkt *adj* GESCHÄFT direct; **mit ~er Ablesung** *phr* ANALYTIK *Meßmethode* direct-reading; **~es Auffangen** *nt* LUFT direct interception; **~ beheizter Drehrohrofen** *m* BODEN direct-heated rotary kiln; **~e Beheizung** *f* BODEN direct heating; **~e Einstrahlung** *f* ERNEU ENER direct radiation; **~er Flüssigkeitseinlaß** *m* ANALYTIK direct liquid introduction; **~e Grundwasseranreicherung** *f* ABFALL, WASSER direct groundwater recharge; **~e Industrieabwassereinleitung** *f* ABWASSER direct industrial wastewater discharge; **~er Luftprobenehmer** *m* ANALYTIK direct air sampler; **~e Meßmethode** *f* ANALYTIK direct measurement method; **~er Schallpegel** *m* LÄRM direct sound level; **~e Trocknung** *f* ABWASSER *Schwebe-, Trommel- oder Fließbetttrockner* direct drying; **~e Verseuchung** *f* LUFT direct poisoning
Direkt: **Direktablesung** *f* ANALYTIK direct reading; **Direkteinleiter** *m* ABWASSER direct discharger; **Direkteinspritzung** *f* BODEN direct injection; **Direktentschwefelungsprozeß** *m* (*DEP*) LUFT direct desulfurization process (*AmE*), direct desulphurization process (*BrE*); **Direktextrusion** *f* ABFALL direct extrusion
direktgekoppelt *adj* BODEN close-coupled
Direkt: **Direktimpulsfilter** *nt* LUFT direct pulse filter; **Direktinjektion** *f* ANALYTIK direct injection
Direktorat *nt* POLITIK *Verwaltungseinheit* directorate
Direkt: **Direktschall** *m* LÄRM direct sound, primary sound
diskontinuierlich *adj* ALLG discontinuous, intermittent; **~er Festbettadsorber** *m* LUFT discontinuous fixed-bed adsorber; **~e Hydrolyse** *f* BODEN discontinuous hydrolysis; **~e Messung** *f* ANALYTIK discontinuous measuring; **~e Reinigung** *f* LUFT *Filterschlauch* intermittent cleaning; **~e UF** *f* ABWASSER discontinuous ultrafiltration; **~e**

Ultrafiltration *f* ABWASSER discontinuous ultrafiltration; **~es Verfahren** *nt* ABWASSER batch process
diskret: **~e Messung** *f* ANALYTIK discontinuous measuring
Dispergator *m* BODEN, MEER dispersant, dispersing agent
Dispergens *nt* ANALYTIK *grenzflächenaktiver chemischer Stoff*, LUFT, MEER dispersant
dispergieren *vt* ABWASSER, BODEN, LUFT, MEER disperse
Dispergiermittel *nt* ABWASSER, BODEN, LUFT, MEER dispersing agent
dispergiert *adj* ABWASSER, BODEN dispersed, LUFT dispersed, suspended; **~es Feststoffteilchen** *nt* LUFT suspended solid particle; **~e Flüssigphase** *f* LUFT suspended liquid phase
Dispergierung *f* ABWASSER, BODEN, LUFT, MEER dispersion
dispers: **~e Phase** *f* ANALYTIK, LUFT disperse phase
Dispersant *nt* ABWASSER, BODEN, LUFT, MEER dispersant; **Dispersantbehälter** *m* MEER *unter einem Hubschrauber* bucket
Dispersion *f* ALLG dispersion; **~ durch chemische Reaktion** (*DCR*) ABFALL dispersion by chemical reaction (*DCR*); **Dispersionsaerosol** *nt* LUFT dispersion aerosol; **Dispersionseinrichtung** *f* MEER *zur Bekämpfung von Ölverschmutzungen* dispersing equipment; **Dispersionskolloid** *nt* ABFALL, ABWASSER, WASSER dispersion colloid; **Dispersionskraft** *f* LUFT *van der Waalssche Kräfte* dispersion force; **Dispersionsmittel** *nt* MEER dispersant
Dissimilation *f* ABWASSER *biologisches Denitrifikationsverfahren* dissimilation
Dissoziation *f* ABWASSER, ANALYTIK dissociation; **Dissoziationsgrad** *m* ANALYTIK degree of dissociation
dissoziierbar *adj* ABWASSER, ANALYTIK dissociable; **~er Stoff** *m* ABWASSER, ANALYTIK dissociable substance
dissoziieren *vt* ABWASSER, ANALYTIK dissociate
Distickstoffoxid *nt* (N_2O) ABFALL, LUFT, WASSER nitrous oxide (N_2O)
DOC *abbr* (*gelöster organischer Kohlenstoff*) ANALYTIK, WASSER DOC (*dissolved organic carbon*)
Dokumentation *f* ABFALL, WASSER documentation
Dolomitmehl *nt* LUFT *Additiv* dolomite meal;
Doppel- *in cpds* ANALYTIK double, dual; **Doppelalkaliwäsche** *f* LUFT *nasse Rauchgasentschwefelung* dual alkaline scrubbing; **Doppelbandpresse** *f* ABFALL double-belt press
Doppelbett: **Doppelbettkatalysator** *m* LUFT dual-bed catalyst; **Doppelbettmonolithkatalysator** *m* LUFT dual-bed monolithic catalyst; **Doppelbettschüttgutkatalysator** *m* LUFT dual-bed bulk material catalyst
Doppel-: **Doppeldetektor-GC** *m* ANALYTIK double detector GC
Doppelrohr *nt* LUFT double pipe; **Doppelrohrkondensator** *m* LUFT double-pipe condenser; **Doppelrohrwärmetauscher** *m* LUFT double-pipe heat exchanger
Doppel-: **Doppelschaufelabscheider** *m* ABFALL, WASSER double-bucket collector; **Doppelschneckenextruder** *m* ABFALL twin-screw extruder; **Doppelstrahlphotometer** *nt* ANALYTIK double-beam photometer; **Doppelwand** *f* LÄRM *zur Pegelminderung* double wall
Dorr-Sandfang *m* ABWASSER Dorr grit chamber
Dortmundbecken *nt* ABWASSER Dortmund tank
Dortmundbrunnen *m* ABWASSER Dortmund tank
Dose *f* ABFALL can; **Dosenschrott** *m* ABFALL can scrap
Dosier: **Dosierbehälter** *m* ABWASSER feed tank; **Dosiereinheit** *f* ABFALL, ANALYTIK, WASSER metering unit
dosieren *vt* ALLG dose, feed
Dosierer *m* ABFALL, WASSER dosing feeder
Dosier: **Dosiergerät** *nt* ANALYTIK *Auftragen von fließfähigen Medien* dosage device; **Dosiermaschine** *f* ABFALL, WASSER metering machine, dosing machine; **Dosierpumpe** *f* ABFALL dosing pump, ABWASSER metering pump, dosing pump, WASSER dosing pump; **Dosierschnecke** *f* ABWASSER metering worm
Dosierung *f* ALLG dosage, dosing, feeding
Dosier: **Dosiervorrichtung** *f* ABFALL, ANALYTIK, WASSER dosing apparatus, metering equipment, metering unit; **Dosierwaage** *f* LUFT dosing scales
Dosis *f* ABFALL, ABWASSER, ANALYTIK, SICHERHEIT dose; **Dosisgrenzwert** *m* **für berufliche Bestrahlung** SICHERHEIT occupational dose limit; **Dosisleistungsmesser** *m* ANALYTIK dosimeter; **Dosismeßgerät** *nt* ANALYTIK dosimeter; **Dosiswirkung** *f* ABFALL, ANALYTIK, WASSER dose response; **Dosis-Wirkungs-Beziehung** *f* ABFALL, ANALYTIK, WASSER dose-response relationship
dotieren *vt* ERNEU ENER, LUFT dope
DPP *abbr* (*orientiertes Polypropylen*) ABFALL DPP (*directed polypropylene*); **DPP-Film** *m* ABFALL oriented polypropylene film
drahtbeheizt *adj* ANALYTIK *Röhrenofen* wire-heated
Draht: **Drahtgewebe** *nt* ABWASSER woven-wire cloth; **Drahtkäfig** *m* ABFALL, WASSER wire cage; **Drahtstützkorb** *m* LUFT *Schlauchfilter* support wire cage
Drallregler *m* LUFT torque adjuster
Dralltrockner *m* ABWASSER spin dryer
Drän *m* ABFALL, BODEN, WASSER drain
Dränage *f* ABFALL, BODEN, WASSER drainage; **Dränagesystem** *nt* ABFALL, WASSER drainage system; **Dränagewand** *f* ABFALL drainage wall
Drän: **Dränleitung** *f* ABFALL drainage pipe, drainage line; **Dränmaterial** *nt* ABFALL drainage material; **Dränrohr** *nt* ABFALL drainage pipe; **Dränschicht** *f* ABFALL, WASSER drainage layer
Drechsel-Gaswaschflasche *f* ANALYTIK Drechsel gas-washing bottle
Dreh- *in cpds* WASSER rotating; **Dreharm** *m* MEER rotatable arm
drehbar: **~er Flügel** *m* LUFT movable blade
Dreh-: **Drehbewegung** *f* LUFT *Zyklon* rotation movement; **Drehfilter** *nt* ABFALL, WASSER rotary filter; **Drehgasvorwärmer** *m* LUFT rotating gas preheater; **Dreh-Gavo** *m* LUFT rotating gas preheater; **Drehkolbengebläse** *nt* LUFT rotary piston compressor; **Drehkolbenpumpe** *f* ABFALL, ABWASSER rotary pump; **Drehkolbenverdichter** *m* LUFT rotary piston compressor; **Drehluftvorwärmer** *m* LUFT rotary air heater; **Drehluvo** *m* LUFT rotary air heater; **Drehmoment** *nt* ERNEU ENER torque; **Drehofen** *m* ABFALL, BODEN, LUFT rotary furnace, rotary kiln; **Drehofenverbrenner** *m* LUFT rotary kiln incinerator; **Drehrohrofen** *m* ABFALL, ABWASSER rotary furnace, rotary kiln, BODEN *Verbrennung* rotary kiln, rotary furnace, LUFT rotary furnace, rotary kiln; **Drehsieb** *nt* ABWASSER drum screen, rotary screen; **Drehsprenger** *m* ABWASSER rotary distributor; **Drehstromfrequenz** *f* ERNEU ENER *Windturbine* rotary current frequency; **Drehtrommel** *f* ABFALL, BODEN *direkt oder indirekt beheizt für thermische Bodenbehandlung* revolving drum; **Drehtrommelofen** *m* ABWASSER rotary drum furnace; **Drehvakuumfilter** *nt* ABWASSER rotary vacuum filter; **Drehzahl** *f* ERNEU ENER *Windturbine* number of revolutions
drehzahlfest: **~e Rotoren** *m pl* ERNEU ENER *Windturbine* rotors with fixed number of revolutions
drehzahlvariabel: **drehzahlvariable Rotoren** *m pl* ERNEU ENER *Windturbine* rotors with variable number of revolutions
Dreiecksmiete *f* ABFALL, WASSER triangular compost heap
Drei: **Drei-Liter-Auto** *nt* LUFT three-liter automobile (*AmE*), three-litre car (*BrE*); **Drei-Wege-Katalysator** *m* LUFT three-way catalyst (*TWC*)
Driftgeschwindigkeit *f* LUFT drift velocity
dritte: **~ Reinigungsstufe** *f* ABWASSER advanced wastewater treatment, tertiary wastewater treatment
Drossel *f* WASSER throttle; **Drosselklappe** *f* LUFT, WASSER butterfly valve

Druck *m* ABFALL compression, pressure, ABWASSER pressure, ANALYTIK compression, pressure, BODEN pressure, compression, ERNEU ENER pressure, LUFT compression, pressure, MEER thrust, WASSER compression; **Druckabfall** *m* LUFT pressure drop; **Druckausgleich** *m* ABFALL, WASSER pressure compensation; **Druckausgleichskammer-Schalldämpfer** *m* LÄRM expansion chamber silencer; **Druckbeanspruchung** *f* ABFALL, WASSER compression load; **Druckdifferenz** *f* ABFALL, ANALYTIK pressure difference, LUFT *in der Filterschicht* differential pressure, WASSER pressure difference; **Druckdifferenzanzeige** *f* LUFT differential pressure gage (*AmE*), differential pressure gauge (*BrE*); **Druckenergie** *f* LUFT pressure energy; **Druckentlastung** *f* LUFT *des Filterapparats* pressure release; **Druckentwässerung** *f* ABWASSER pressure drainage

Druckeranschluß *m* ANALYTIK *an Analysatorgeräte* printer connection

Druck: **Druck-Erdgas** *nt* LUFT compressed natural gas (*CNG*)

Druckerhöhung *f* LUFT pressure increase; **Druckerhöhungspumpe** *f* ABWASSER booster pump; **Druckerhöhungspumpwerk** *nt* WASSER booster pump works; **Druckerhöhungsventilator** *m* LUFT booster fan

Drückerkopf *m* ABFALL, WASSER siphon head

Druckfarbe: **Druckfarben von den Fasern lösen** *phr* ABFALL, WASSER *De-Inking-Prozeß* dissolve printed color from fiber (*AmE*), dissolve printed colour from fibre (*BrE*)

druckfest *adj* LUFT compression-proof

Druck: **Druckfestigkeit** *f* ABFALL, WASSER compression strength; **Druckfilter** *nt* ABFALL, ABWASSER, LUFT, WASSER pressure filter; **Druckfiltration** *f* ANALYTIK compression filtration; **Druckflotation** *f* ABWASSER dissolved air flotation; **Druckflüssigkeit** *f* LUFT pressurized liquid; **Druckflüssigkeitsstrom** *m* LUFT pressurized liquid stream; **Druckgebläse** *nt* LUFT forced draft fan (*AmE*), forced draught fan (*BrE*); **Druckgradient** *m* BODEN pressure gradient; **Druckhöhe** *f* ABWASSER discharge head, WASSER pressure head; **Druckkammer** *f* LUFT *Erdfilter* pressure chamber; **Druckleitung** *f* ABFALL delivery pipe, penstock, delivery main, discharge piping, WASSER discharge piping, penstock, delivery main, *Abzugskanal* delivery pipe

drucklos *adj* ABFALL, WASSER unpressurized; **~e Versorgung** *f* WASSER unpressurized supply

Druckluft *f* ABWASSER, BODEN, LUFT, SICHERHEIT compressed air, diffused air

druckluftabgereinigt *adj* LUFT compressed-air-cleaned

Druckluft: **Druckluftatemgerät** *nt* **für industrielle Einsätze** SICHERHEIT compressed-air industrial breathing apparatus; **Druckluftbecken** *nt* ABWASSER diffused air tank; **Druckluftbehälter** *m* LUFT compressed-air receiver; **Druckluftbelüftung** *f* ABWASSER diffused-air aeration

druckluftgereinigt: **~es Jet-Filter** *nt* LUFT compressed-air-cleaned jet filter

Druckluft: **Druckluftimpuls** *m* LUFT compressed-air impulse; **Druckluftinjektor** *m* ABWASSER pneumatic injector; **Druckluftlanze** *f* BODEN *Bodenluftabsaugung* compressed-air lance; **Druckluftreinigung** *f* LUFT cleaning with compressed air; **Druckluftspeicher** *m* LUFT *Puls-Jet Filter* compressed-air storage tank; **Druckluftstoß** *m* LUFT pulse stroke, *Schlauchfilter* compressed-air blast; **Druckluftverbrauch** *m* LUFT compressed-air consumption

Druck: **Druckmesser** *m* LUFT pressure gage (*AmE*), pressure gauge (*BrE*); **Druckprüfmaschine** *f* ANALYTIK, WASSER compression test machine; **Druckpumpe** *f* SAMPLING pressure pump; **Druckregler** *m* ANALYTIK, LUFT pressure governor (*BrE*), pressure regulator (*AmE*); **Druckrohr** *nt* ABWASSER, WASSER pressure pipe; **Druckrohrleitung** *f* ABFALL, WASSER penstock, delivery main; **Druckschacht** *m* WASSER pressure well; **Druckschalter** *m* LUFT manometric switch

druckschwach *adj* ALLG low-pressure; **~e Versorgung** *f* WASSER low-pressure supply

Druck: **Druckschwingung** *f* WASSER pressure oscillation; **Drucksensor** *m* LUFT pressure sensor; **Drucksondierung** *f* ANALYTIK pressure sounding; **Druckstoßfilter** *nt* LUFT jet filter, reverse-jet filter

druckstoßgereinigt: **~es Filter** *nt* LUFT pulse-jet-cleaned filter

Druck: **Druckstoßreinigung** *f* LUFT pulse-jet cleaning; **Druckverformung** *f* ABFALL, WASSER compression damage

Druckverlust *m* ABWASSER, LUFT *Filterschlauch*, WASSER head loss, loss of head, pressure loss; **~ von Rohrleitungen** WASSER pressure loss of pipeline; **~ in Verteilerrohrnetzen** WASSER pressure loss in distribution network

Druck: **Druckversuch** *m* ABFALL, WASSER compression test; **Druckzone** *f* WASSER pressure zone

DSD *abbr* (*Duales System Deutschland GmbH*) ABFALL, MANAG *abfallwirtschaftliches Konzept* Dual System Germany Ltd

DSR *abbr* (*Dünnsäurerückgewinnung*) ABFALL spent acid recovery

DTA *abbr* (*Differenzthermoanalyse*) ANALYTIK DTA (*differential thermal analysis*)

Dual: **~es System** *nt* **Deutschland GmbH** (*DSD*) ABFALL, MANAG *abfallwirtschaftliches Konzept*

Dualsystem *nt* ABFALL, MANAG *abfallwirtschaftliches Konzept* dual system

duldbar *adj* BODEN *Schadstoffkonzentration* tolerable

Duldbarkeit *f* BODEN *Schadstoffkonzentration* tolerance

Dünger *m* ABWASSER, BODEN fertilizer

Düngewert *m* ABWASSER, BODEN fertilizing value

Düngung *f* BODEN fertilization

dunkel: **~ halten** *vt* ANALYTIK *Probe* keep in dark

dunkel: **dunkle Absorberfolie** *f* ERNEU ENER *Solarteich* dark heat-absorbing material

Dünnsäurerückgewinnung *f* (*DSR*) ABFALL spent acid recovery

Dünnschicht- *in cpds* ABWASSER thin-film, thin-layer; **Dünnschichtchromatographie** *f* (*DC*) ANALYTIK *Chrom* thin-layer chromatography (*TLC*); **Dünnschichteinbau** *m* ABFALL, WASSER *Flächeneinbau* thin-layer installation; **Dünnschichtelektrophorese** *f* ANALYTIK thin-layer electrophoresis; **Dünnschichtmodul** *nt* ERNEU ENER *Solarzelle* thin-film module; **Dünnschichtphotovoltaikzelle** *f* ERNEU ENER thin-film photovoltaic cell; **Dünnschichttrocknung** *f* ABWASSER thin-layer drying; **Dünnschichtverdampfer** *m* ABWASSER, WASSER thin-film evaporator

dünnwandig *adj* ANALYTIK *Probenahme* thin-walled

Dunst *m* LUFT haze; **Dunstkoeffizient** *m* LUFT coefficient of haze

Durchblasen *nt* LUFT blowing-through

durchbohren *vt* ABFALL, WASSER bore through

Durchbruch *m* ABFALL, WASSER *Filterschlauch* rupture

durchdringen *vt* ALLG permeate

Durchdrückverpackung *f* ABFALL blister pack

Durchfall *m* BODEN *direkt in den Boden eingehender Niederschlag* throughput, thruput (*AmE*), LUFT throughfall

durchfiltern *vt* ABWASSER percolate

Durchfluß: **Durchflußmenge** *f* ABWASSER, ANALYTIK, WASSER flow rate, rate of flow; **Durchflußmengenmeßgerät** *nt* ANALYTIK flow rate measuring device; WASSER instant flowmeter, flowmeter; **Durchflußmesser** *m* ANALYTIK flowmeter; **Durchflußquerschnitt** *m* ABWASSER, WASSER cross-sectional area of flow; **Durchflußrate** *f* ABWASSER, ANALYTIK, WASSER flow rate, rate of flow; **Durchflußschwimmermesser** *m* ANALYTIK rotameter; **Durchflußzähler** *m* ANALYTIK flowmeter

Durchführbarkeit *f* ALLG feasibility

durchführen *vt* ALLG *Versuche* conduct, carry out

Durchgang *m* LUFT heat transition

durchlässig *adj* ABFALL porous, ABWASSER permeable, porous,

ANALYTIK permeable, BODEN permeable, pervious, porous, LUFT permeable, WASSER permeable, porous; **~e Fläche** *f* ABWASSER pervious area
Durchlässigkeit *f* ABFALL porosity, ABWASSER, BODEN permeability, WASSER porosity; **Durchlässigkeitskoeffizient** *m* ABFALL, BODEN, LUFT, WASSER coefficient of permeability; **Durchlässigkeitsmesser** *m* ANALYTIK transmittance meter; **Durchlässigkeitsmeter** *nt* LUFT transmittance meter; **Durchlässigkeitswert** *m* ABFALL, BODEN, LUFT, WASSER coefficient of permeability
Durchlaufbecken *nt* WASSER rainwater tank
durchlaufen *vt* LUFT pass through
Durchlaufneutralisation *f* ABWASSER continuous neutralization
Durchmesser *m* ABWASSER diameter
durchmischen *vt* ANALYTIK mix thoroughly
Durchmischung *f* ABFALL, ABWASSER, LUFT, WASSER complete mixing, thorough mixture
Durchsatz *m* ALLG throughput, thruput (*AmE*); **Durchsatzschwankung** *f* LUFT throughput fluctuation, thruput fluctuation (*AmE*)
durchschlagen *vt* LUFT fuse
Durchschlagen *nt* LUFT electrical breakdown, *Sicherung* blow-out
Durchschnitt *m* ANALYTIK average
durchschnittlich *adj* ANALYTIK average
***Durchschnitt*: Durchschnittsprobe** *f* ANALYTIK average sample
Durchsichtigkeit *f* ABFALL, WASSER transparency; **Durchsichtigkeitsmessung** *f* ABFALL, WASSER transparency measurement
durchsickern *vi* ABWASSER, MEER percolate, seep
Durchsickern *nt* ABWASSER, MEER percolation, seepage
durchsieben *vt* ABWASSER, MEER screen, sift
Durchsieben *nt* ABWASSER, MEER screening, sifting
durchspülen *vt* ABFALL, WASSER flush
Durchstoßfestigkeit *f* ABFALL, LUFT, SICHERHEIT puncturability, puncture resistance
durchstrahlen *vt* ANALYTIK penetrate by radiation, radiograph
Durchstrahlungselektronenmikroskopie *f* ANALYTIK transmission electron microscopy (*TEM*)
durchströmen *vt* ABFALL, ABWASSER flow through, ANALYTIK run through, LUFT, WASSER flow through
Duromer *nt* ABFALL, WASSER thermoset
Duroplast *m* ABFALL, WASSER thermoset
duroplastisch *adj* ABFALL thermosetting
Düse *f* ABWASSER, LUFT, WASSER jet, nozzle; **Düsenflügel** *m* LUFT *Patronenfilter* nozzle blade; **Düsenrohr** *nt* LUFT nozzle tube; **Düsenstock** *m* LUFT blast connection; **Düsenstrahl** *m* ABWASSER, LUFT, WASSER jet; **Düsenstrahlinjektion** *f* BODEN jet injection; **Düsenzerstäuber** *m* LUFT *Sprühabsorber* nozzle atomizer; **Düsenzerstäubung** *f* LUFT *Sprühabsorption* nozzle atomization
dynamisch *adj* ANALYTIK, MEER, WASSER dynamic; **~es Abschöpfgerät** *nt* MEER dynamic skimmer; **~er Auslaugtest** *m* ANALYTIK dynamic leachability test; **~er Druck** *m* ANALYTIK dynamic pressure; **~e Druckhöhe** *f* ABFALL, ABWASSER, WASSER velocity head; **~er Eluationstest** *m* ANALYTIK dynamic leachability test; **~es Headspace** *nt* ANALYTIK dynamic headspace; **~e Headspace-Technik** *f* (*DHS*) ANALYTIK dynamic headspace technique (*DHS*); **~e Viskosität** *f* ABFALL, WASSER dynamic viscosity; **~es Vorrotteverfahren** *nt* BODEN windrow method; **~e Zähigkeit** *f* ABFALL, WASSER dynamic viscosity
Dynamometer *nt* ANALYTIK dynamometer

E

EA *abbr* (*Elementaranalyse*) ANALYTIK EA (*elemental analysis, elementary analysis*)
Ebbe *f* ERNEU ENER *Gezeitenkraftwerk* ebb-tide; **~ und Flut** *f* ERNEU ENER *Gezeitenenergie* ebb and flow
ECD *abbr* (*Elektroneneinfangdetektor*) ANALYTIK ECD (*electron capture detector*)
Echtzeit- *in cpds* ANALYTIK real-time; **Echtzeit-Analysator** *m* ANALYTIK real-time analyser (*BrE*), real-time analyzer (*AmE*); **Echtzeitbetrieb** *m* ANALYTIK real-time operation; **Echtzeit-Chromatogramm** *nt* ANALYTIK real-time chromatogram; **Echtzeit-Signalverarbeitung** *f* ANALYTIK real-time signal processing; **Echtzeit-Stichprobenahme** *f* ANALYTIK real-time sampling
ED *abbr* (*Elektrodialyse*) ABWASSER *von kulturfähigem unbelastetem Boden*, ANALYTIK, WASSER ED (*electrodialysis*)
Edelmetall *nt* ABFALL, LUFT, WASSER noble metal; **Edelmetallkatalysator** *m* LUFT noble metal catalyst
Edelstahlgehäuse *nt* LUFT special steel housing
E-Einfang *abbr* ANALYTIK electron capture
effektiv: **~e Energienutzung** *f* POLITIK effective energy use; **~e Evaporation** *f* ABFALL, WASSER effective evaporation; **~e Kaminhöhe** *f* LUFT effective stack height; **~er Schalldruck** *m* LÄRM effective sound pressure; **~e Schornsteinhöhe** *f* LUFT effective stack height
Effektivität *f*: **~ der Verfahren** ALLG process effectiveness
E-Filter *nt* (*Elektrofilter*) ABFALL, LUFT, WASSER electrostatic precipitator (*ESP*)
EGW *abbr* (*Einwohnergleichwert*) ABFALL, ABWASSER inhabitant equivalent
E-Horizont *m* BODEN eluvial horizon
Eich- *in cpds* ANALYTIK calibration
eichbedürftig *adj* ANALYTIK subject to calibration
eichen *vt* ANALYTIK calibrate, gage (*AmE*), gauge (*BrE*)
Eich-: **Eichgas** *nt* ANALYTIK calibration gas; **Eichimpuls** *m* ANALYTIK calibration pulse; **Eichkurve** *f* ANALYTIK calibration curve; **Eichlösung** *f* ANALYTIK calibration solution; **Eichmarke** *f* ANALYTIK calibration mark; **Eichsubstanz** *f* ANALYTIK calibrated substance
Eichung *f* ANALYTIK calibration
Eich-: **Eichwert** *m* ANALYTIK calibrated value
eigenständig: **~e Substanz** *f* BODEN autochthonous matter
einaxial: **~e Druckfestigkeit** *f* ABFALL *Deponie* uniaxial compressive strength
Einbau *m* ALLG installation; **~ der Abfälle** ABFALL *in einer Deponie* placement of wastes
einbauen *vt* ABWASSER install, ALLG build in, BODEN, ERNEU ENER, LÄRM, LUFT, MEER, SICHERHEIT, WASSER install
Einbau: **Einbaufeuchte** *f* ABFALL moisture of installation; **Einbaufläche** *f* ABFALL *einer Deponie* constructed area; **Einbauschichten** *f pl* ABFALL *einer Deponie* constructed layers
Einbauten: **ohne ~** *phr* LUFT *Waschraum* without built-in structure
einbetten *vt* ABFALL embed
Einbindung *f* ABFALL binding, *von Sondermüll* grain encapsulation, WASSER binding
einblasen *vt* ABWASSER blow into, LUFT inject, blow into
Einblasen *nt* LUFT injection
einbringen *vt* ABFALL, LUFT, WASSER feed, load
EINCS *abbr* (*Europäisches Neustoffverzeichnis*) POLITIK, WASSER EINCS (*European inventory of new chemical substances*)
eindampfen *vt* BODEN vaporize
Eindampfen *nt*: **~ einer Säure** LUFT acid concentration
eindeutig: **~e Gesundheitsgefahr** *f* SICHERHEIT clear health risk
eindicken *vt* ABFALL, ABWASSER thicken
Eindicker *m* ABFALL, ABWASSER, WASSER thickener; **Eindicker-Überlaufbehälter** *m* ABFALL, ABWASSER thickener overflow tank; **Eindicker-Vorlagebehälter** *m* ABFALL, ABWASSER thickener prestorage tank
Eindickgrad *m* ABWASSER *Entwässerung* degree of thickening
Eindickung *f* ABFALL, ABWASSER thickening
eindringen *vi* ABFALL, LUFT, WASSER infiltrate, penetrate; **~ in** *vi* ABFALL percolate through, LUFT penetrate, SICHERHEIT intrude, WASSER percolate through
Eindringen *nt* ABFALL infiltrate, LUFT *der Staubpartikel* penetration, SICHERHEIT *von giftiger Luft* intrusion, WASSER infiltrate
eindrücken *vt* ABFALL press in
eindüsen *vt* ABFALL, BODEN, LUFT, WASSER inject, spray
Eindüsung *f* ANALYTIK, BODEN, LUFT injection
EINECS *abbr* (*Europäisches Altstoffverzeichnis*) POLITIK, WASSER EINECS (*European inventory of existing chemical substances*)
einfach: **~es Verfahren** *nt* POLITIK simplified procedure
Einfassung *f* MEER skirt
Einflügler *m* ERNEU ENER *Windturbine* single-blade machine
Einflußgebiet *nt* WASSER area of influence
Einfügungsdämpfung *f* LÄRM insertion loss
einführen *vt* POLITIK *beste verfügbare Technik* introduce
Einführung *f* POLITIK introduction
Eingang *m* ANALYTIK input; **Eingangsbereich** *m* ABFALL entrance area; **Eingangsgrößen** *f pl* MANAG inputs; **Eingangskontrolle** *f* ABFALL *einer Deponie* weigh office
eingebaut *adj* ABFALL, ANALYTIK, BODEN, WASSER built-in; **~e Alterung** *f* ABFALL, WASSER built-in obsolescence; **~e Meßeinrichtung** *f* ANALYTIK, BODEN built-in measuring equipment; **~e Regeleinrichtung** *f* ANALYTIK, BODEN built-in regulating equipment; **~e Überwachungseinrichtung** *f* ANALYTIK, BODEN built-in monitoring equipment
eingedampft: **~e Probenmenge** *f* ANALYTIK concentrated amount of sample
eingekapselt *adj* BODEN, LÄRM encapsulated
eingelagert: **~er Abfall** *m* ABFALL emplaced waste
eingestrahlt: **~e Energie** *f* ERNEU ENER irradiated energy; **~e Energiemenge** *f* ERNEU ENER amount of irradiated energy
eingezwängt: **~er Dämpfungsbelag** *m* LÄRM *Schwingungsdämpfung* constrained layer
eingrenzen *vt* ABWASSER, MEER *Ölteppich* corral
Einheit *f* ANALYTIK unity; **Einheitsoperation** *f* ABWASSER *mechanische Abtrennung von Schadstoffen* unit operation; **Einheitsverfahren** *nt pl* ABFALL, ABWASSER, WASSER *biologische bzw. chemische Abtrennung von Schadstoffen* unit processes
Einimpfung *f* ANALYTIK, BODEN injection
einkapseln *vt* ABFALL, BODEN, LÄRM encapsulate, cap
Einkapselung *f* ABFALL, *gefährlicher Abfallstoffe*, BODEN, LÄRM encapsulation, capping; **~ durch Inertmaterial** ABFALL *Sonderabfallbehandlung* encapsulation using inert materials; **Einkapselungsmaßnahmen** *f pl* BODEN encapsulation measures
Einkaufsrichtlinie *f* MANAG ecological buying guideline
Einkaufsvorgabe *f* MANAG ecological buying guideline
einkommensneutral *adj* POLITIK income-neutral; **~e ökolo-**

gische Steuerreform *f* POLITIK income-neutral eco-tax reform
Einkreislaufwäscher *m* LUFT single-cycle scrubber
einlagern *vt* ABFALL, WASSER store; **wieder ~** *vt* BODEN *nach Behandlung* redeposit
Einlagerung *f* ALLG storage; **Einlagerungskammer** *f* ABFALL *für radioaktive Abfälle* storage chamber; **Einlagerungsverhalten** *nt* ABFALL storage behavior (*AmE*), storage behaviour (*BrE*)
Einlaß *m* ABWASSER inlet, LUFT inlet, intake, WASSER inlet; **Einlaßemissionen** *f pl* LUFT inlet emissions; **Einlaßkrümmer** *m* LUFT intake manifold
Einlauf *m* ABWASSER, LUFT, WASSER inlet, intake, influent; **Einlaufbauwerk** *nt* ABFALL, ABWASSER, WASSER inlet structure, intake structure; **Einlaufhebewerk** *nt* ABWASSER influent pump works; **Einlaufschacht** *m* ABFALL sink, LUFT entry inlet, WASSER sink; **Einlaufschwelle** *f* WASSER intake sill
einleiten *vt* ABWASSER, MEER discharge
Einleiten *nt* ABWASSER, MEER discharge
Einleiter *m* ABWASSER, MEER discharger
Einleitung *f* ABWASSER, MEER *von Abwässern* discharge; **Einleitungsanlage** *f* ABWASSER *für Abwässer*, MEER discharge system
Einmündung *f* WASSER *eines Kanals* junction
Einnahme *f* POLITIK revenue; **~ aus Ökosteuern** POLITIK revenue from eco-taxes; **Einnahmequelle** *f* POLITIK source of revenue
Einpreß- *in cpds* BODEN injection, grouting; **Einpreßabschnitt** *m* BODEN grouting section; **Einpreßdruck** *m* BODEN grouting pressure, injection pressure
einpressen *vt* ABFALL press in
Einpreß-: **Einpreßschnitt** *m* BODEN injection section; **Einpreßverfahren** *nt* BODEN grouting
einrammen *vt* ABFALL pile drive
einrichten *vt* ANALYTIK *Meßstelle* install
Einrichtung *f* ALLG device, equipment
einrütteln *vt* BODEN shake in
Einsammellogistik *f* ABFALL, ANALYTIK collection logistics
einsammeln *vt* ALLG collect
Einsammeln *nt* ABFALL, ANALYTIK *von Daten*, LUFT, WASSER collection; **~ von Müllsäcken** ABFALL garbage bag collection (*AmE*), refuse sack collection
Einsammelquote *f* ABFALL, WASSER collection rate
Einsammlungsrunde *f* ABFALL collection route
Einsammlungsverfahren: **von ~** *phr* ABFALL door to door
Einsatz *m* ABFALL insert, ABWASSER, ANALYTIK, ERNEU ENER, LUFT application, WASSER insert; **Einsatzbereich** *m* BODEN operative range; **Einsatzgebiet** *nt* ERNEU ENER *der Solarfarmen* installation area; **Einsatzmöglichkeiten** *f pl* ABWASSER, ANALYTIK, ERNEU ENER, LÄRM, LUFT, MEER, SICHERHEIT, WASSER application possibilities
einschalig *adj* LÄRM single-leaf, *Kapselung* single-sheet
einschließen *vt* MEER *Ölteppich* entrap
Einschluß *m* ABFALL *von Sondermüll* grain encapsulation; **Einschlußflockung** *f* ABWASSER sweep coagulation
einschmelzen *vt* ABFALL, ANALYTIK melt down
Einschneckenextruder *m* ABFALL single-screw extruder
einsetzbar *adj* LUFT adjustable
Einsetzbarkeit *f* ABWASSER, ANALYTIK *von Verfahren*, ERNEU ENER *von Verfahren*, LUFT, WASSER *von Verfahren* applicability
einsetzen *vt* ANALYTIK adjust
einsickern 1. *vt* ABFALL infiltrate; **~ in** WASSER percolate through; **2.** *vi* WASSER infiltrate; **~ in** ABFALL percolate through
einsinken *vi* LUFT subside
Einsinken *nt* LUFT subsidence
Einspritzdüse *f* LUFT atomizer, injection nozzle
einspritzen *vt* BODEN inject
Einspritzen *nt* ANALYTIK, BODEN injection
Einstabmeßkette *f* ANALYTIK *pH-Meter* combination pH-electrode
einstampfbar: **~er Klebstoff** *m* ABFALL repulpable adhesive
Einstampfgerät *nt* ABFALL repulping equipment
Einstaufelder *nt pl* ABWASSER spreading grounds
Einstauflächen *f pl* ABWASSER spreading grounds
einstellbar *adj* ALLG adjustable
einstellen *vt* ALLG set, adjust
Einstellung *f* ALLG setting
Einstiegschacht *m* ABFALL, WASSER entrance shaft
Einstoffdüse *f* LUFT *Strahlwäscher* single-phase jet
Einstoffsystem *nt* ABFALL single-material system
einströmen *vi* LUFT stream in
einstufig *adj* ABFALL, ABWASSER, LUFT, WASSER single-stage; **~es Belebungsverfahren** *nt* ABWASSER single-stage activated sludge process; **~er Faulraum** *m* ABWASSER single-stage digestor; **~e Peripheralpumpe** *f* ABFALL, ABWASSER, WASSER single-stage peripheral pump; **~er Wirbelschichtadsorber** *m* LUFT single-stage fluidized-bed adsorber
Eintauchrührer *m* ANALYTIK *pH-Meter* immersion stirrer
eintragen *vt* ABFALL, BODEN, WASSER introduce
Eintragschnecke *f* BODEN feeder worm
Eintritt *m* LUFT inlet; **~ des Wärmeübertragermediums** ERNEU ENER *Solarkollektor* heat transfer medium in; **Eintrittsgeschwindigkeit** *f* LUFT inlet velocity; **Eintrittstaubbelastung** *f* LUFT inlet dust burden
eintrommeln *vt* MEER *eine Sperre* reel in
Einweg- *in cpds* ABFALL disposable, nonreturnable, one-way; **Einwegbehälter** *m* ABFALL one-way container; **Einwegflasche** *f* ABFALL nonreturnable bottle, one-way bottle; **Einwegkunststofftüte** *f* ABFALL disposable plastic bag; **Einwegpackung** *f* ABFALL one-way pack; **Einwegpalette** *f* ABFALL expendable pallet, nonreturnable pallet, one-way pallet; **Einwegpapiertüte** *f* ABFALL disposable paper bag; **Einwegprodukt** *nt* ABFALL disposable product, throw-away product; **Einweg-PT-Verpackung** *f* ABFALL expanded polythene packaging; **Einwegverpackung** *f* ABFALL nonreturnable container, nonreturnable packaging, one-way packaging, throw-away pack
einwohnerbezogen *adj* WASSER population-related, population-specific; **~er Wasserverbrauch** *m* WASSER population-related water consumption, population-specific water demand
Einwohnergleichwert *m* (*EGW*) ABFALL, ABWASSER inhabitant equivalent
einwohnerspezifisch *adj* ALLG population-specific
Einzel- *in cpds* ABFALL, LUFT individual, single; **Einzelkläranlage** *f* ABWASSER separate sewerage system
einzellig: **~er Zyklon** *m* LUFT single-unit cyclone
Einzel-: **Einzelmessung** *f* ANALYTIK spot measurement, single measurement; **Einzelprobe** *f* ANALYTIK individual sample; **Einzelschichtwellfaserplatte** *f* ABFALL single-wall corrugated fibreboard (*BrE*), single-wall corrugated fiberboard (*AmE*); **Einzelstaubabscheider** *m* SICHERHEIT individual dust removal apparatus; **Einzelstoffanalyse** *f* ANALYTIK individual substance analysis; **Einzeltröpfchen** *nt* LUFT individual droplet; **Einzeluntersuchung** *f* ANALYTIK spot analysis; **Einzelzyklon** *m* LUFT single-unit cyclone
Einzugsgebiet *nt* WASSER catchment area, drainage area, drainage basin; **~ mit Binnenentwässerung** WASSER blind drainage area
Eisen *nt* ABFALL, WASSER iron; **Eisenbestimmung** *f* **von Abwasser** ANALYTIK analysis of iron in wastewater; **Eisenchlorid** *nt* ABFALL, WASSER ferric chloride
eisenhaltig *adj* ABFALL, WASSER iron-containing; **~es Material** *nt* ABFALL iron-containing material
Eisen: **Eisenmetalle** *nt pl* ABFALL ferrous metals; **Eisen(III)-Salz** *nt* ABWASSER *Flockungsmittel* ferric salt; **Eisenschrott**

m ABFALL ferrous scrap, scrap iron, junk iron, iron scrap; **Eisenschrottpresse** *f* ABFALL scrap-baling press, junk press; **Eisensulfat** *nt* ABFALL, ABWASSER, WASSER iron sulfate (*AmE*), iron sulphate (*BrE*); **Eisen(III)-Sulfat** *nt* ABFALL, WASSER ferric sulfate (*AmE*), ferric sulphate (*BrE*)

Eiweiß *nt* ALLG *Enzyme mit höchster Wirkungsspezifität* albumen

Ejektorbelüfter *m* ABWASSER sludge siphon, jet aerator

Ejektorbelüftung *f* ABWASSER jet aeration, sludge syphoning

EKC *abbr* (*elektrokinetische Kapillarchromatographie*) ANALYTIK EKC (*electrokinetic capillary chromatography*)

elastisch: **~e Zwischenlage** *f* LUFT *Katalysator* elastic temporary layer

Elastomer *nt* ABFALL, LÄRM *federndes Material* elastomer

Elektretmikrofon *nt* ANALYTIK *Lärmmessung* electret microphone

elektrisch *adj* ALLG electric, electrical; **~ betrieben** *adj* ERNEU ENER, LUFT, SICHERHEIT electrically driven; **~ betriebene Wärmepumpe** *f* ERNEU ENER electrically driven heat pump; **~e Energie** *f* ERNEU ENER electrical energy, electric energy; **~e Entstaubungsanlage** *f* LUFT, SICHERHEIT electrical dust removal installation; **~es Feld** *nt* ERNEU ENER electrical field, LUFT electric field, *Elektrofilter* electrical field; **~e Feuchtemessung** *f* ANALYTIK electrical moisture measurement; **~e Gefahr** *f* SICHERHEIT electrical hazard; **~e Kraft** *f* LUFT electrical force; **~es Meßgerät** *nt* SICHERHEIT electric measuring apparatus; **~er Strom** *m* ERNEU ENER electrical power; **~er Widerstand** *m* ANALYTIK electric resistance

Elektro- *in cpds* ABFALL, LUFT electric

elektroakustisch *adj* BODEN electroacoustic

elektroanalytisch *adj* ANALYTIK electroanalytic

Elektro-: **Elektroantrieb** *m* LUFT electric drive; **Elektroauto** *nt* LUFT electric car

elektrochemisch *adj* ANALYTIK electrochemical; **~e Meßzellen** *f pl* ANALYTIK *als Gasmeßgerät* electrochemical detector; **~e Methoden** *f pl* ANALYTIK electrochemical methods; **~e Ozonmessung** *f* ANALYTIK electrochemical ozone measurement; **~e Ozonsonde** *f* ANALYTIK electrochemical ozone sonde

Elektrode *f* ABFALL *Batterie, Elektrolyse*, ANALYTIK electrode; **Elektrodensensor** *m* ANALYTIK electrode sensor

Elektro-: **Elektrodialyse** *f* (*ED*) ABFALL, ABWASSER, ANALYTIK electrodialysis (*ED*); **Elektrofilter** *nt* (*E-Filter*) ABFALL, LUFT, WASSER electrostatic precipitator (*ESP*); **Elektroflotation** *f* ABWASSER electroflotation

elektrokinetisch *adj* BODEN electrokinetic; **~e Kapillarchromatrographie** *f* (*EKC*) ANALYTIK electrokinetic capillary chromatography (*EKC*); **~e Sanierung** *f* BODEN electrokinetic remediation

Elektrolyse *f* ABFALL, ABWASSER, BODEN, WASSER electrolysis; **Elektrolyseanlage** *f* ERNEU ENER electrolysis installation; **Elektrolyserückstand** *m* ABFALL electrolysis residue; **Elektrolysezelle** *f* ANALYTIK electrolysis cell

Elektrolyt *m* ABFALL, ANALYTIK, WASSER electrolyte

elektrolytisch: **~ abscheiden** *vt* ABFALL *gelöste Metalle in der Batterie* separate electrolytically

elektrolytisch: **~e Zelle** *f* ABFALL electrolytic cell

Elektrolyt: **Elektrolytkonzentration** *f* ABWASSER electrolytic concentration; **Elektrolytmatrix** *f* LUFT electrolyte matrix

Elektro-: **Elektromagnet** *m* ABFALL, ABWASSER, LUFT electromagnet

elektromagnetisch *adj* ABFALL, ABWASSER, LUFT, WASSER electromagnetic; **~e Abstrahlung** *f* LUFT electromagnetic radiation; **~e Pumpe** *f* ABFALL, ABWASSER, WASSER electromagnetic pump

elektromechanisch: **~e Trennung** *f* ABFALL electromechanical separation

Elektro-: **Elektromelioration** *f* BODEN electron amelioration

elektrometrisch *adj* ANALYTIK potentiometric, electrometric; **~e pH-Wert-Bestimmung** *f* ANALYTIK electrometric pH-value determination; **~e Titration** *f* ANALYTIK potentiometric titration

Elektro-: **Elektromotor** *m* ABWASSER electrical motor

Elektron *nt* ANALYTIK, LUFT, WASSER electron; **Elektronbeugung** *f* ANALYTIK electron diffraction; **Elektronbeugungsbild** *nt* ANALYTIK electron diffraction pattern; **Elektronenakzeptor** *m* WASSER electron acceptor; **Elektronenbeschleuniger** *m* LUFT electron accelerator; **Elektronenbestrahlung** *f* WASSER electron irradiation; **Elektronendonator** *m* WASSER electron donator; **Elektroneneinfang** *m* ANALYTIK electron capture; **Elektroneneinfangdetektor** *m* (*ECD*) ANALYTIK *speziell* electron capture detector (*ECD*); **Elektronenlinse** *f* ANALYTIK electron lens; **Elektronenmikroskop** *nt* ANALYTIK *Staubanalyse* electron microscope; **Elektronenmikroskopie** *f* ANALYTIK electron microscopy

elektronenoptisch *adj* ABFALL optoelectric; **~es Verfahren** *nt* ABFALL *Art der Einbindung* optoelectric process

Elektron: **Elektronenspinresonanz** *f* (*ESR*) ANALYTIK electron spin resonance (*ESR*)

Elektronenstoß *m* ANALYTIK electron impact; **Elektronenstoß-Ionisation** *f* ANALYTIK electron impact ionization; **Elektronenstoß-Massenspektrometer** *nt* ANALYTIK electron impact mass spectrometer (*EI-MS*); **Elektronenstoß-Massenspektrometrie** *f* ANALYTIK electron impact mass spectrometry (*EI-MS*); **Elektronenstoßmethode** *f* ANALYTIK electron impact method

Elektronenstrahl *m* LUFT *ESV* electron beam; **Elektronenstrahlmikrosonde** *f* ANALYTIK electron microprobe (*EM*); **Elektronenstrahlverfahren** *nt* (*ESV*) LUFT electron beam process

Elektron: **Elektronenvolt** *nt* (*eV*) ANALYTIK electronvolt (*eV*)

elektronisch *adj* ANALYTIK *Meßgeräte* electronic; **~e Sortierung** *f* ABFALL electronic sorting

Elektro-: **Elektroosmose** *f* BODEN electroosmosis

elektroosmotisch *adj* BODEN electroosmotic

Elektro-: **Elektrophorese** *f* ANALYTIK, BODEN electrophoresis

elektrophoretisch *adj* ANALYTIK, BODEN electrophoretic

Elektro-: **Elektrostatik** *f* LUFT electrostatics

elektrostatisch *adj* ABFALL, ABWASSER, LUFT, WASSER electrostatic; **~er Abscheider** *m* ABFALL, LUFT, WASSER electrostatic precipitator (*ESP*); **~e Abscheidung** *f* ABFALL, WASSER electrostatic separation; **~e Abstoßung** *f* ABWASSER electrostatic repulsion; **~e Anziehung** *f* LUFT electrostatic attraction; **~er Naßabscheider** *m* LUFT wet electrostatic precipitator (*WEP*); **~er Staubabscheider** *m* ABFALL, LUFT, WASSER electrostatic precipitator (*ESP*)

Element *nt* ANALYTIK element

elementar: **~es Platin** *nt* LUFT platinum metal

Elementaranalyse *f* (*EA*) ANALYTIK elemental analysis (*EA*), elementary analysis (*EA*)

Elementarzusammensetzung *f* ANALYTIK elemental composition

elementbezogen *adj* ANALYTIK element-specific; **~e Analytik** *f* ANALYTIK element-specific analysis

elementspezifisch *adj* ANALYTIK element-specific; **~e Analytik** *f* ANALYTIK element-specific analysis

Eluat *nt* ABWASSER, ANALYTIK eluate; **Eluatherstellung** *f* ANALYTIK eluate preparation

Eluation *f* BODEN leaching; **Eluationsversuch** *m* BODEN leaching test

Eluat: **Eluatkontrolle** *f* ANALYTIK eluate control

eluierbar *adj* ABFALL, ABWASSER, ANALYTIK, WASSER suitable for elution

eluieren *vt* ABFALL, ABWASSER, ANALYTIK elute

Eluierung *f* ABFALL, ABWASSER, ANALYTIK, WASSER elution

Elutionschromatographie *f* ANALYTIK elution chromatography

Elutionstest *m* ANALYTIK elution test

Elutriation *f* ANALYTIK elutriation

Emballagenpresse *f* ABFALL packing press

Emission *f* ABFALL, ABWASSER, BODEN, LUFT *Luftverunreinigung* emission; **Emission-Banking** *nt* POLITIK emissions banking; **Emissionenpfad** *m* ABWASSER, LUFT emissions pathway; **Emissionen** *f pl* **während der Wartung** LUFT emissions during maintenance; **Emissionsabgabe** *f* POLITIK emissions charge; **Emissionsanalyse** *f* ANALYTIK emissions analysis; **Emissionsauflagen** *f pl* POLITIK emissions requirements; **Emissionsbegrenzung** *f* LUFT emissions limitation, emissions limit, POLITIK emissions limit, emissions limitation; **Emissionsbegrenzung** *f* **für Kraftfahrzeuge** LUFT, POLITIK emissions limit for cars

emissionsbezogen *adj* LUFT emissions-related; **~e Wartung** *f* LUFT emissions-related maintenance

***Emission*: Emissionsdaten** *pl* ANALYTIK, LUFT emissions data; **Emissionsdetektion mit Plasma** *f* ANALYTIK plasma emission detection; **Emissionserklärung** *f* POLITIK, WASSER emissions statement; **Emissionsfähigkeit** *f* LUFT emissivity; **Emissionsgrenzwert** *m* ANALYTIK emissions standard, LUFT emissions standard, emissions limit value, POLITIK emissions standard; **Emissionsguthaben** *nt* POLITIK emissions credit; **Emissionskataster** *m* ANALYTIK emissions inventory, POLITIK, WASSER emissions register; **Emissionskontrolle** *f* BODEN, LUFT emissions control; **Emissionskontrollsystem** *nt* LUFT emissions control system; **Emissionsmessung** *f* ANALYTIK emissions measurement; **Emissionsminimierung** *f* ABWASSER, LUFT, POLITIK, WASSER emissions minimalization

emissionsorientiert *adj* POLITIK emissions-targeted; **~e Kfz-Steuer** *f* POLITIK emissions-targeted automobile tax (*AmE*), emissions-targeted road tax (*BrE*)

***Emission*: Emissionsort** *m* LUFT emissions point; **Emissionspegel** *m* LÄRM *von Schall* emissions level; **Emissionspotential** *nt* ABFALL emissions potential; **Emissionsquelle** *f* ABFALL, BODEN, LUFT, POLITIK, WASSER emission source, pollution emitter; **Emissionsquelle** *f* **in einem Gebiet** LUFT area emission source; **Emissionsrate** *f* LUFT emissions rate; **Emissionsraum** *m* BODEN *Untergrund, Grundwasser* environmental medium compartment; **Emissionsspektrometer** *nt* ANALYTIK emissions spectrometer; **Emissionsspektrometrie** *f* **mit induktiv gekoppeltem Plasma** (*ICP-OES*) ANALYTIK *QS* inductively coupled plasma with optical emission spectrometry (*ICP-OES*); **Emissionsspektrum** *nt* ANALYTIK, LUFT emissions spectrum; **Emissionssteuer** *f* POLITIK emissions tax; **Emissionsüberwachung** *f* ANALYTIK emissions monitoring; **Emissionsvermeidung** *f* BODEN, LUFT emissions avoidance; **Emissionsverminderung** *f* ABWASSER, BODEN, LUFT emissions reduction; **Emissionsvermögen** *nt* LUFT emissivity; **Emissionsverzeichnis** *nt* ANALYTIK *Messung der Luftqualität* emissions inventory; **Emissionswert** *m* LUFT emissions value

Emittent *m* LUFT, POLITIK, WASSER emission source, emitter

emittieren *vt* LÄRM, LUFT, WASSER emit

emittiert *adj* LÄRM, LUFT, WASSER emitted; **~e Partikel** *f* LUFT emitted particle; **~er Stoff** *m* LUFT emitted substance

Empfänger *m* LUFT receptor; **Empfängerbereich** *m* LUFT receptor region; **Empfängerröhre** *f* ERNEU ENER receiver; **Empfängerturm** *m* ERNEU ENER receiving tower

Empfangsteleskop *nt* ANALYTIK *Streulichtfernmessung* receiving telescope

Empfehlung *f* POLITIK recommendation

empfindlich *adj* ANALYTIK *Methode*, LUFT sensitive; **~er Nachweis** *m* ANALYTIK sensitive detection

Empfindlichkeit *f* ANALYTIK *von Detektoren*, LUFT sensitivity

Emscherbecken *nt* ABWASSER Emscher-type settling tank

Emscherbrunnen *m* ABWASSER Emscher tank

Emulgator *m* ABWASSER, BODEN, SICHERHEIT, WASSER emulsifying agent; **Emulgatorflüssigkeit** *f* ABFALL, ABWASSER, BODEN, SICHERHEIT, WASSER emulsifying liquid

emulgieren *vt* ABFALL, ABWASSER, WASSER emulsify

emulgiert *adj* ABFALL, ABWASSER, WASSER emulsified; **~es Öl** *nt* ABWASSER emulsified oil

Emulgierungsmittel *nt* ABFALL, ABWASSER, BODEN, WASSER emulsifying agent

Emulsion *f* ABFALL, ABWASSER, MEER, WASSER emulsion

emulsionbeschichtet *adj* ABFALL, WASSER emulsion-coated

***Emulsion*: Emulsionsspalter** *m* ABFALL emulsion breaker, MEER demulsifier, emulsion breaker, WASSER emulsion breaker; **Emulsionsspaltung** *f* ABFALL, ABWASSER, WASSER emulsion separation, demulsification; **Emulsionstrennung** *f* ABFALL, WASSER emulsion separation; **Emulsionszusammensetzung** *f* ABWASSER emulsion composition

End- *in cpds* BODEN final; **Endabdeckung** *f* ABFALL final covering, *einer Deponie* final cover

endabgedichtet: **~e Deponie** *f* ABFALL finally-sealed landfill; **~er Deponieabschnitt** *m* ABFALL finally-sealed landfill section

Endenergie *f* ERNEU ENER final energy; **Endenergiesubstitution** *f* ERNEU ENER final energy substitution; **Endenergieverbrauch** *m* ERNEU ENER final energy consumption

***End-*: Endgas** *nt* ABFALL, LUFT tail gas; **Endgasreinheit** *f* LUFT end gas purity

endgültig: **~e Sanierung** *f* BODEN final remediation

***End-*: Endhaltung** *f* ABFALL final reach; **Endlager** *nt* ABFALL disposal zone; **Endlagerstätte** *f* **für radioaktiven Abfall** ABFALL repository for radioactive waste; **Endlagerung** *f* ABFALL *von Müll* final storage, permanent storage; **Endlagerungsstätte** *nt* ABFALL disposal zone; **End-of-pipe-Technologie** *f* ABFALL, POLITIK end-of-pipe technology

endogen: **~er Katabolismus** *m* ABWASSER endogenous catabolism

endothermisch *adj* ABFALL, LUFT, WASSER endothermic; **~e Reaktion** *f* ABFALL, LUFT, WASSER endothermic reaction

***End-*: Endprodukt** *nt* ABFALL *Wiederaufbereitung* finished product, LUFT *Gips aus Rauchgaswäsche* end product; **Endprodukterzeugung** *f* ABFALL end product generation; **Endserienverpackung** *f* ABFALL end-of-line packaging; **Endverbraucher** *m* ABFALL end user

energetisch *adj* ALLG energy-related; **~er Stoffstrom** *m* ERNEU ENER energy material stream; **~er Wirkungsgrad** *m* BODEN energy efficiency

Energie *f* ALLG energy; **~ aus Biomasse** ERNEU ENER energy from biomass; **~ der Meeresströmung** ERNEU ENER ocean current power; **Energieabsorption** *f* ERNEU ENER energy absorption; **Energieamortisationszeit** *f* ERNEU ENER energy pay-back time; **Energieanwendungsmöglichkeiten** *f pl* ERNEU ENER *von Verfahren* energy application possibilities

energiearm *adj* LUFT low-energy

***Energie*: Energieaufwand** *m* ABWASSER, ERNEU ENER, LUFT, WASSER *von Verfahren* energy expenditure; **Energieausbeute** *nt* ERNEU ENER energy yield; **Energiebarriere** *f* ABWASSER energy barrier; **Energiebedarf** *m* ERNEU ENER energy demand

***Energie*: Energiebedarf decken** *vi* ERNEU ENER cover energy demand

***Energie*: Energiebilanz** *f* ABFALL, ABWASSER, ERNEU ENER, WASSER energy balance; **Energiedichte** *f* ABFALL, ERNEU ENER energy density; **Energieeffizienz** *f* ERNEU ENER, LUFT energy efficiency; **Energieeinsatzmöglichkeiten** *f pl* ERNEU ENER energy application possibilities; **Energieeinsparung** *f* ABFALL, LUFT energy saving; **Energieempfindlichkeit** *f* ANALYTIK energy sensitivity; **Energieerzeugung** *f* ERNEU ENER power generation; **Energie-Farming** *nt* ERNEU ENER *Gewinnung nachwachsender Energieträger aus Biomasse* energy farming; **Energiegehalt** *m* ERNEU ENER energy content; **Energiegewinn** *m* ERNEU ENER energy gain; **Energiegewinnung** *f* ABWASSER *Schlammverbrennung* energy recovery, ERNEU ENER energy extraction, energy generation; **Energiehaushalt** *m* LUFT energy budget; **Energieimport** *m* ERNEU ENER energy import; **Energieinhalt** *m* ERNEU ENER energy content; **Energieintensität** *f* ERNEU ENER energy intensity

energieintensiv *adj* ABWASSER, ERNEU ENER, LUFT, WASSER energy-intensive; **~es Verfahren** *nt* ABWASSER, BODEN, ERNEU ENER, LUFT, WASSER energy-intensive method

Energie: **Energiekosten** *pl* ERNEU ENER energy costs; **Energieleistung** *f* ERNEU ENER energy output; **Energiemenge** *f* ERNEU ENER amount of energy; **Energienutzung** *f* ABFALL, ERNEU ENER, LUFT, WASSER energy use; **Energiepotential** *nt* ERNEU ENER energy potential; **Energieproduktivität** *f* ERNEU ENER energy productivity; **Energie-Pro-Kopf-Verbrauch** *m* ERNEU ENER per capita energy consumption; **Energiequelle** *f* ERNEU ENER energy source; **Energiereserven** *f pl* ERNEU ENER energy reserves; **Energieressourcen** *f pl* ABFALL, LUFT, WASSER energy resources; **Energierohstoffe** *m pl* ERNEU ENER energy raw materials; **Energierückgewinnung** *f* ABFALL, ERNEU ENER, LUFT energy recovery; **Energieschwelle** *f* ABFALL energy barrier

energiesparend *adj* LUFT energy-saving; **~e Technologie** *f* ABFALL, LUFT, WASSER energy-saving technology; **~es Verfahren** *nt* LUFT energy-saving method

Energie: **Energiespeicherung** *f* ERNEU ENER energy storage; **Energiespender** *m* ERNEU ENER energy source; **Energiesteuer** *f* POLITIK energy tax; **Energiesteueraufkommen** *nt* POLITIK energy tax revenue; **Energietechnologie** *f* ERNEU ENER energy technology; **Energieträger** *m* ABFALL energy carrier, ERNEU ENER energy source, WASSER energy carrier; **Energietransport** *m* ERNEU ENER energy transport; **Energieumwandlung** *f* ERNEU ENER energy conversion; **Energieverbrauch** *m* ABWASSER, ERNEU ENER, LUFT, WASSER *von Verfahren* energy consumption; **unter Energieverbrauch** *phr* ABWASSER *aerobe Faulung* energy-consumptive; **Energieverlust** *m* ERNEU ENER energy loss; **Energieverschwendung** *f* ABFALL, LUFT waste of energy

Energieversorgung *f* ERNEU ENER energy supply; **Energieversorgungsnetz** *nt* ERNEU ENER energy supply grid

Energie: **Energiewirtschaft** *f* ERNEU ENER energy industry, *Bewirtschaftung* energy management; **Energiezufuhr** *m* LUFT energy supply; **Energiezweck** *m* ERNEU ENER energy purpose

engmaschig *adj* ABFALL, ABWASSER, LUFT, WASSER *Rost* close-meshed; **~er Siebboden** *m* LUFT close-meshed sieve plate

engporig *adj* LUFT porous; **~es Absorbens** *nt* LUFT porous adsorbent; **~e Aktivkohle** *f* LUFT porous activated carbon

Entascher *m* ABFALL, ABWASSER ash remover

Entaschung *f* ABFALL, ABWASSER ash removal

entcarbonisieren *vt* ABFALL, WASSER decarbonize

Entcarbonisieren *nt* ABFALL, WASSER decarbonization

entcarbonisiert *adj* ABFALL, WASSER decarbonized; **~es Grundwasser** *nt* ABFALL, WASSER decarbonized groundwater

Entcarbonisierung *f* ABFALL, WASSER decarbonization; **~ mit Kalkhydrat** ABFALL, WASSER decarbonization with slaked lime; **~ mit Kalkwasser** ABFALL, WASSER decarbonization with lime water; **Entcarbonisierungsanlage** *f* ABFALL, WASSER, ERNEU ENER decarbonization plant

entchloren *vt* ABFALL, BODEN, WASSER dechlorinate

Entchlorung *f* ABFALL, BODEN, WASSER dechlorination

Entdröhnung *f* LÄRM sound deadening; **Entdröhnungsmaterial** *nt* LÄRM sound deadener material; **Entdröhnungsmittel** *nt* LÄRM sound deadener material

Enteisenung *f* WASSER *Grundwasseraufbereitung* deferrization

entfärben *vt* ABFALL, ABWASSER, ANALYTIK, WASSER remove color (*AmE*), remove colour (*BrE*)

Entfärbung *f* ABFALL, ABWASSER, ANALYTIK, WASSER decoloration (*AmE*), decolouration (*BrE*)

entfernen *vt* ALLG remove

Entfernen *nt* ALLG removal

entfernt: **~e Quelle** *f* ABFALL, LUFT, WASSER distant source

Entfernung *f* ALLG removal; **~ gelöster organischer Stoffe** WASSER removal of dissolved organic substances

entfeuchten *vt* ANALYTIK dehumidify

Entfeuchtung *f* ANALYTIK dehumidification; **Entfeuchtungswirkung** *f* ANALYTIK dehumidification effect

entflammbar *adj* ABFALL, LUFT, SICHERHEIT flammable, inflammable; **nicht ~** *adj* SICHERHEIT flameproof, noninflammable, nonflammable; **~er Abfall** *m* ABFALL *Sondermüll* flammable waste; **~er Dampf** *m* SICHERHEIT flammable vapor (*AmE*), flammable vapour (*BrE*); **~e Flüssigkeit** *f* SICHERHEIT flammable liquid; **~er Werkstoff** *m* SICHERHEIT flammable material

Entflammbarkeit *f* ABFALL, LUFT, SICHERHEIT flammability

entgasen *vt* ALLG degas, degasify

Entgaser *m* ALLG degasser

Entgasung *f* ABFALL *einer Deponie*, ABWASSER, BODEN, LUFT degasification, degassing, gas drainage; **~ über Entgasungskammer** ABFALL degassing with degassing chamber; **~ über Gabionen** ABFALL degassing with gabion; **~ über Gasbrunnen** ABFALL degassing with gas wells; **~ über Schottersäule** ABFALL degassing with ballasted column; **~ über Sonden** ABFALL degassing with sonde; **Entgasungseinrichtung** *f* ABFALL degassing facility; **Entgasungsfilterrohr** *nt* BODEN degasification filter tube; **Entgasungskammer** *f* ABFALL degassing chamber; **Entgasungsrohr** *nt* ABFALL degassing pipe; **Entgasungsschacht** *m* ABFALL degassing shaft; **Entgasungsschicht** *f* ABFALL degassing layer

entgiften *vt* ABFALL, BODEN, LUFT, WASSER detoxify

Entgiftung *f* ABFALL, BODEN, LUFT, WASSER detoxification

Enthalpie *f* ANALYTIK, LUFT enthalpy

enthalten: **~e Energie** *f* ERNEU ENER contained energy

enthärten *vt* ABFALL, WASSER soften

Enthärter *m* ABFALL, WASSER softener

enthärtet *adj* ABFALL, WASSER softened; **~es Wasser** *nt* WASSER soft water, softened water

Enthärtung *f* ABFALL, WASSER softening; **~ des Trinkwassers** WASSER drinking water softening; **Enthärtungsanlage** *f* WASSER softening plant; **Enthärtungsmittel** *nt* WASSER softener

entkeimen *vt* ABFALL, ABWASSER, WASSER disinfect, sterilize

entkeimt *adj* ABFALL, ABWASSER, WASSER disinfected, sterilized

Entkeimung *f* ABWASSER, WASSER sterilization, disinfection

entkoppelt: **~er Biowäscher** *m* ABWASSER, LUFT decoupling biological scrubber

entkrusten *vt* ABFALL, WASSER wash out

entladbar: **~e Batterie** *f* ABFALL dischargeable battery

Entlade- *in cpds* ABFALL unloading; **Entladebunker** *m* ABFALL unloading hopper; **Entladehalle** *f* ABFALL unloading hall

entladen *vt* ABFALL, WASSER unload

Entladung *f* ABFALL, WASSER unloading; **Entladungslampe** *f* ABFALL gas discharge lamp

Entlastung: **Entlastungsbrunnen** *m* ABFALL, WASSER relief well; **Entlastungsgerinne** *nt* ABWASSER discharge flume; **Entlastungsschleuse** *f* ABWASSER discharge sluice

entleeren *vt* ABFALL, ABWASSER, WASSER blow off, empty, drain, drain off

Entleeren *nt* ABFALL, ABWASSER, WASSER *von Abwasserkanälen* blowing-off, draining, emptying

Entleerung *f* ABFALL, ABWASSER blowing-off, drainage, emptying

entlüften *vt* ABFALL, WASSER ventilate, ABWASSER deaerate, ventilate

Entlüftung *f* ABFALL, LUFT ventilation; **Entlüftungsrohr** *nt*

ABFALL, LUFT vent; **Entlüftungsventil** *nt* ABFALL, LUFT air relief valve
Entmanganung *f* WASSER *Grundwasseraufbereitung* demanganization; **Entmanganungsfilter** *nt* WASSER demanganization filter
Entmischung *f* ABWASSER demixing
Entnahme *f* ABWASSER, LUFT, WASSER intake, WASSER withdrawal, *von Wasser* tapping; **Entnahmebauwerk** *nt* ABFALL, ABWASSER, WASSER inlet structure, intake structure, ABWASSER inlet structure; **Entnahmebrunnen** *m* ABFALL intake well, BODEN excavation well, WASSER intake well; **Entnahmegefäß** *nt* ANALYTIK sampling vessel; **Entnahmegerät** *nt* ABFALL extractor, ANALYTIK sampling instrument, WASSER extractor; **Entnahmehilfsmittel** *nt* ANALYTIK sampling aid; **Entnahmeleitung** *f* ABFALL, WASSER intake main; **Entnahmemethode** *f* ANALYTIK sampling method; **Entnahmestelle** *f* **von Proben** ABWASSER, ANALYTIK, BODEN sampling point; **Entnahmestollen** *m* ABFALL, WASSER intake tunnel; **Entnahmetrichter** *m* ABFALL, WASSER cone of depression; **Entnahmeversuch** *m* ABFALL, ABWASSER, WASSER pumping test; **Entnahmevorrichtung** *f* ANALYTIK sampling instrument
Entölen *nt* ABWASSER oil removal, oil separation
Entpalettisiermaschine *f* ABFALL depalletizer
Entphenolung *f* ABWASSER dephenolizing, dephenolating
entsalzen *vt* ABFALL, WASSER desalinate
Entsalzung *f* ABFALL, WASSER desalination; **Entsalzungsanlage** *f* ABFALL, WASSER desalination plant, salt water plant; **Entsalzungsverfahren** *nt* ABFALL, WASSER desalination technique
entsanden *vt* ABWASSER degrit, WASSER *Brunnen* develop
Entsandung *f* WASSER *eines Brunnens* development
entsäuern *vt* ABFALL, WASSER neutralize
entsäuert *adj* ALLG neutralized
Entsäuerung *f* ABFALL, WASSER neutralization
entschädigen *vt* POLITIK, SICHERHEIT compensate
Entschädigung *f* POLITIK, SICHERHEIT compensation
Entschlackung *f* ABFALL, LUFT slag removal; **Entschlackungseinrichtung** *f* ABFALL slag removal equipment, ABWASSER slag-removing plant
entschlammen *vt* ABFALL, ABWASSER, WASSER desludge
Entschlämmer *m* ABFALL desludger, scourer, ABWASSER, WASSER desludger
Entschlammung *f* ABFALL, ABWASSER, WASSER desludging
Entschrottung *f* ABFALL scrap metal separation
entschwefeln *vt* ABFALL, LUFT, WASSER desulfurize (*AmE*), desulphurize (*BrE*)
Entschwefelung *f* ABFALL *Regen, Niederschlag*, LUFT *höchstzulässige Konzentration*, WASSER desulfurization (*AmE*), desulphurization (*BrE*); **~ mit Magnesiumoxid** LUFT desulfurization with magnesium oxide (*AmE*), desulphurization with magnesium oxide (*BrE*); **~ von Prozeßgasen** LUFT desulfurization of process gases (*AmE*), desulphurization of process gases (*BrE*); **Entschwefelungsgrad** *m* LUFT degree of desulfurization (*AmE*), degree of desulphurization (*BrE*); **Entschwefelungssorbens** *nt* LUFT desulfurization sorbent (*AmE*), desulphurization sorbent (*BrE*)
entseuchen *vt* ALLG decontaminate, disinfect
entseucht *adj* ALLG decontaminated, disinfected
Entseuchung *f* ALLG decontamination, disinfection
entsorgen *vt* ABFALL, ABWASSER, BODEN, MEER dispose of
entsorgend: **zu ~e Rückstände** *m pl* ABFALL residues to be disposed of
entsorgt *adj* ABFALL, ABWASSER, BODEN, MEER disposed of; **~e Abfälle** *m pl* ABFALL disposed-of wastes; **~e Rückstände** *m pl* ABFALL, ABWASSER disposed-of residues
Entsorgung *f* ABWASSER, BODEN disposal, MANAG waste disposal, MEER disposal; **Entsorgungsanlage** *f* **an Land** MEER shore reception facility; **Entsorgungsgebühr** *f* ABFALL disposal fee; **Entsorgungskosten** *pl* MANAG disposal costs; **Entsorgungslogistik** *f* ABFALL logistics of disposal; **Entsorgungsnachweis** *m* POLITIK disposal proof
entsorgungspflichtig *adj* ABFALL subject to obligatory disposal
Entsorgung: **Entsorgungssicherheit** *f* ABFALL guaranteed disposal; **Entsorgungsweg** *m* ABFALL disposal channel, disposal route
entspannen *vt* ABWASSER, WASSER *Druck* reduce, relax
entspannt *adj* ABFALL, ABWASSER, WASSER reduced, relaxed
Entspannung *f* ABFALL, ABWASSER *Druck* reduction, relaxation; **Entspannungsflotation** *f* ABWASSER dissolved air flotation, pressure flotation; **Entspannungsverdampfer** *m* ABWASSER flash evaporator; **Entspannungsverdampfung** *f* ABFALL flash distillation
entstabilisieren *vt* ABFALL, ABWASSER *Kolloide* destabilize
entstabilisiert *adj* ABFALL, ABWASSER *Kolloide* destabilized
Entstabilisierung *f* ABFALL, ABWASSER *der Kolloide* destabilization
entstauben *vt* LUFT dust, free from dust, *Staub* precipitate
Entstauber *m* ABFALL dust separator, LUFT deduster, *Zyklon, Filter, E-Filter* dust remover, SICHERHEIT *Luftreinigung* dust separator
entstaubt *adj* LUFT dust-removed; **~es Rohgas** *nt* LUFT dust-removed dirty gas
Entstaubung *f* LUFT dedusting, dust removal, dust extraction; **Entstaubungsanlage** *f* LUFT deduster, dust removal plant, dust extraction plant, SICHERHEIT dust removal installation; **Entstaubungsgrad** *m* LUFT degree of dust removal; **Entstaubungstechnik** *f* LUFT dust removal technology; **Entstaubungswirkungsgrad** *m* LUFT dust removal efficiency
entstehen *vi* ABFALL, ERNEU ENER, WASSER generate
entsticken *vt* LUFT remove nitrogen from
Entstickungsgrad *m* LUFT degree of nitrogen oxides removal
Entwaldung *f* BODEN, LUFT deforestation
entwässerbar *adj* ABWASSER, LUFT drainable; **~er Kristall** *m* LUFT drainable crystal
Entwässerbarkeit *f* ABWASSER, LUFT drainability
entwässern *vt* ALLG *Teich* concentrate
entwässert *adj* ALLG concentrated, thickened; **~er Schlamm** *m* ABWASSER drained sludge, concentrated sludge
Entwässerung *f* ABFALL drainage, ABWASSER drainage, thickening, BODEN, LUFT drainage; **~ durch Gefrieren** ABWASSER drainage by freezing, thickening by freezing; **Entwässerungsgraben** *m* ABFALL, WASSER drainage canal, drainage ditch; **Entwässerungsgraben** *m* **mit Böschung** ABFALL, WASSER berm ditch; **Entwässerungsgrad** *m* ABWASSER degree of drainage; **Entwässerungskanal** *m* ABFALL drainage canal, drainage channel, ABWASSER sewer, WASSER drainage canal, drainage channel; **Entwässerungsmaschine** *f* ABFALL draining engine; **Entwässerungsmaterial** *nt* ABFALL drainage material; **Entwässerungsmöglichkeit** *f* ABFALL drainage possibility; **Entwässerungspumpe** *f* ABFALL, WASSER drainage pump, draining pump; **Entwässerungsschacht** *m* ABFALL *einer Deponie* drainage shaft; **Entwässerungsschicht** *f* ABFALL, WASSER *einer Deponie* drainage layer
entwickeln *vt* ALLG develop
Entwicklung *f* ALLG development
entzinken *vt* ABFALL deplate
entzinnen *vt* ABFALL detin
Entzinnung *f* ABFALL detinning
entzünden *vt* ABFALL, ABWASSER, ANALYTIK, LUFT ignite
entzündlich: **~er Staub** *m* LUFT flammable dust
Entzündungsgefahr *f* SICHERHEIT fire hazard, fire risk
Enzym *nt* ABWASSER, LUFT *Biowäscher* enzyme
enzymatisch *adj* ABWASSER enzymatic; **~er Immunoassaytest** *m* ANALYTIK enzyme-mediated immunoassay test (*EMIT*); **~e Stabilisierung** *f* ABWASSER enzymatic stabilization

Enzym: **Enzymtätigkeit** *f* ABWASSER enzyme activity

EOCL *abbr* (*extrahierbare organische Chloride*) ABWASSER, ANALYTIK, BODEN, LUFT EOCL (*extractable organic chlorides*)

EOX *abbr* (*extrahierbare organische Halogenverbindungen*) ABFALL, ANALYTIK, WASSER EOX (*extractable organic halogen compounds*)

Epilimnion *nt* ABWASSER epilimnion

EPM *abbr* (*Äquivalent je Million*) LUFT EPM (*equivalent per million*)

Epoxidharz *nt* ABFALL epoxy resin

Erd- *in cpds* LUFT earth; **Erdalkalien** *f pl* ABFALL, WASSER alkaline earths; **Erdatmosphäre** *f* LUFT earth's atmosphere; **Erdaushub** *m* ABFALL excavated soil; **Erdbehälter** *m* ABFALL earth reservoir; **Erdbeschleunigung** *f* (*g*) LUFT acceleration due to gravity (*g*); **Erdboden** *m* ABFALL soil, ground, BODEN soil; **Erdbohrer** *m* ANALYTIK *Bodenprobenehmen*, BODEN *biologische Altlastensanierung* soil sample tube; **Erdgasaustauschgas** *nt* (*SNG*) ERNEU ENER synthetic natural gas (*SNG*); **Erdinnere** *nt* ERNEU ENER *Geothermie* earth's interior; **Erdkruste** *f* ERNEU ENER *Geothermie* earth's crust; **Erdoberfläche** *f* ABFALL, ERNEU ENER, LUFT earth's surface; **Erdöl** *nt* ABFALL petroleum; **Erdölerzeugnis** *nt* ABFALL petroleum product; **Erdreich** *nt* ABFALL soil, ground, BODEN soil; **Erdstrahlung** *f* LUFT earth radiation; **Erdumdrehung** *f* ERNEU ENER *Ursprung der Gezeitenenergie* earth's rotation; **Erdwärme** *f* ERNEU ENER geothermal energy

ereignisabhängig *adj* ANALYTIK incident-dependent

Ereignispegel *m* LÄRM single-event noise level

erfaßbar *adj* ABFALL recoverable, ANALYTIK collectable, WASSER recoverable

erfassen *vt* ABFALL recover, ABWASSER recover, *Schadstoffe* collect, capture, WASSER recover

Erfassung *f* ABFALL recovery, ABWASSER, BODEN, LUFT, MANAG, MEER collection, capture; **Erfassungsgrad** *m* ABFALL degree of collection; **Erfassungsquote** *f* ABFALL, ABWASSER capture rate, recovery rate

Ergiebigkeit *f* WASSER yield

erhalten *vt* ERNEU ENER, MANAG, POLITIK conserve

Erhaltung *f* ERNEU ENER, MANAG, POLITIK conservation

erhärten *vt* BODEN harden, set

Erhärtung *f* BODEN *Dichtwände* hardening, setting

erhitzen *vt* ALLG heat

Erhitzen *nt* ALLG heating

Erhitzer *m* LUFT, WASSER fired heater

Erhitzung *f* LUFT, WASSER heating

Erhöhung: ~ **des pH-Wertes** *f* ABWASSER rise of pH-value

Eriochromschwarz *nt* ANALYTIK *Indikator* eriochrome black (*EBT*)

erkunden *vt* ANALYTIK, SICHERHEIT investigate

erkündet *adj* ANALYTIK, SICHERHEIT investigated

Erkundung *f* ANALYTIK investigation

Erlaubnis *f* ABFALL, LUFT permission, POLITIK permit, licence (*BrE*), license (*AmE*), *Wasser* licensing, permission, authorization; **Erlaubnisvorbehalt** *m* POLITIK reservation of permission

erleiden *vt* SICHERHEIT sustain

Erlenmeyerkolben *m* ANALYTIK Erlenmeyer flask

Ermächtigung *f* POLITIK *Kraft eines Gesetzes* statutory powers; **Ermächtigungsgrundlage** *f* POLITIK, WASSER statutory power

Ermessen *nt* POLITIK administrative discretion; **Ermessensentscheidung** *f* POLITIK discretionary decision; **Ermessensfehler** *m* POLITIK discretionary error

ermessensfehlerfrei *adj* ANALYTIK free of discretionary error

ermitteln *vt* ANALYTIK detect, determine

Ermittlung *f* ANALYTIK detection, determination; ~ **des Wasserbedarfs** WASSER determination of water demand

ermüden *vi* ALLG fatigue

Ermüdung *f* ALLG fatigue

erneuerbar *adj* ERNEU ENER renewable; **~e Energie** *f* ERNEU ENER renewable energy; **~e Energieform** *f* ERNEU ENER renewable energy form; **~e Energiequelle** *f* ERNEU ENER renewable energy source; **~e Energietechnik** *f* ERNEU ENER renewable energy technology; **~e Energietechnologie** *f* ERNEU ENER renewable energy technology; **~er Energieträger** *m* ERNEU ENER renewable energy source; **~e Energieversorgung** *f* ERNEU ENER renewable energy supply; **~es Energieversorgungssystem** *nt* ERNEU ENER renewable energy supply system; **~e Ressourcen** *f pl* ERNEU ENER renewable resources; **~e Technologie** *f* ERNEU ENER renewable technology

erneuern *vt* ERNEU ENER renew

Ernte *f* ERNEU ENER harvest; **Ernteabfall** *m* ABFALL crop waste

ernten *vt* ERNEU ENER *Energie* harvest

Erörterung *f* POLITIK public hearing; **Erörterungstermin** *m* POLITIK public hearing date

Erosion *f* BODEN, WASSER *Kunststoffe* erosion; **Erosionsschutz** *m* BODEN erosion protection

erosiv *adj* BODEN, LUFT, WASSER *Staubeigenschaften* erosive

erschöpfen *vt* ERNEU ENER, WASSER *Filteraufnahmekapazität* exhaust

erschöpflich *adj* ERNEU ENER, WASSER exhaustible

Erschöpflichkeit *f* ERNEU ENER, WASSER *der Energiequellen* exhaustibility

erschütterungsfrei: **~e Aufstellung** *f* LÄRM antivibration mounting

erstarren *vi* ABFALL, WASSER set

Erstarren *nt* ABFALL setting, solidification

Erstbewertung *f* ANALYTIK *für Bodensanierung* preliminary evaluation, BODEN *vorläufige Gefährdungsabschätzung* first assessment

Ertüchtigung *f* LUFT upgrading

erwärmen *vt* ALLG heat

Erwärmung *f* ALLG *von Schlammen* heating; ~ **der Erde** LUFT global warming

Erweiterung *f*: ~ **einer Steuerbemessungsgrundlage** POLITIK extension of tax base

Erz *nt* ABFALL ore

erzeugen *vt* ABFALL, ERNEU ENER *Heißdampf*, LÄRM, WASSER generate

erzeugt *adj* ABFALL, ERNEU ENER, LÄRM, WASSER generated; **~er Strom** *m* ERNEU ENER generated electricity, generated power

ESA *abbr* ABFALL, LUFT, WASSER ESP (*electrostatic precipitator*)

Escherichia: ~ **coli** *f* WASSER *Indikatorkeim* coli bacillus

ESR *abbr* (*Elektronenspinresonanz*) ANALYTIK ESR (*electron spin resonance*)

Essengas *nt* LUFT stack gas

Essigsäure *f* ABWASSER acetic acid; **Essigsäurebildung** *f* ABWASSER acetic acid formation

ESV *abbr* (*Elektronenstrahlverfahren*) LUFT electron beam process

Etagen- *in cpds* ABWASSER, LUFT tiered; **Etagenbiofilter** *nt* LUFT *biologisches Sorptionsverfahren* tiered biofilter; **Etagenofen** *m* ABWASSER tiered furnace; **Etagentrockner** *m* ABWASSER tiered dryer

Ethanolerzeugung *f* ERNEU ENER *Biokonversion* ethanol generation

Etherverdunstung *f* ANALYTIK *Taupunkthygrometer* ether evaporation

Ethylen *nt* LUFT ethylene; **Ethylendiamintetraessigsäure** *f* ANALYTIK ethylene diamine-tetracetic acid (*EDTA*)

Ethylmercaptan *nt* ABFALL, WASSER ethylmercaptane

Europäisch: **~es Altstoffverzeichnis** *nt* (*EINECS*) POLITIK European inventory of existing chemical substances (*EINECS*); **~es Neustoffverzeichnis** *nt* (*EINCS*) POLITIK European inventory of new chemical substances (*EINCS*);

~es Normungsinstitut *nt* (*CEN*) ANALYTIK European Standardization Committee (*CEN*)
eutroph *adj* ABFALL, ABWASSER, WASSER eutrophic, nutrient-rich
Eutrophierung *f* ABFALL, ABWASSER, WASSER eutrophication
eV *abbr* (*Elektronenvolt*) ANALYTIK eV (*electronvolt*)
evakuieren *vt* ABFALL, WASSER evacuate
Evapotranspiration *f* WASSER evapotranspiration
Exoelektronendosimeter *nt* ANALYTIK exoelectron dosimeter
Exo-Enzym *nt* LUFT exoenzyme
exotherm 1. *adj* ABFALL, ABWASSER, LUFT, WASSER exothermic; **2. ~ reagieren** *vi* ABFALL react exothermally
exotherm: **~er Prozeß** *m* ABWASSER *Abwasserverbrennung* exothermic process
exothermisch *adj* ABFALL, ABWASSER, LUFT, WASSER exothermic; **~e Reaktion** *f* ABFALL, LUFT, WASSER exothermic reaction
explosibel: **explosible Stäube** *m pl* LUFT explosive particles
Explosibilität *f* LUFT explosibility
Explosion *f* ANALYTIK, BODEN, LUFT, SICHERHEIT explosion
explosionsfähig *adj* LUFT, WASSER explosive; **~es Gas** *nt* LUFT explosive gas; **~es Gemisch** *nt* LUFT *Gase* explosive mixture
Explosion: **Explosionsgefahr** *f* ABFALL *durch Inhaltsstoffe im Deponiegas* , ANALYTIK, SICHERHEIT explosion risk
explosionsgeschützt *adj* ABFALL, ANALYTIK, LUFT, SICHERHEIT explosion-proof
Explosion: **Explosionsgrenze** *f* ANALYTIK, LUFT explosibility limit; **Explosionsschutz** *m* ANALYTIK, LUFT explosion protection
explosionssicher *adj* ABFALL, ANALYTIK, LUFT, SICHERHEIT explosion-proof
Explosion: **Explosionssicherheit** *f* ANALYTIK, SICHERHEIT explosion safety; **Explosionsunterdrückung** *f* LUFT *des Filterapparats* explosion suppression; **Explosionswarngerät** *nt* ANALYTIK, SICHERHEIT explosion warning device
explosiv *adj* ABFALL, LUFT, SICHERHEIT explosive; **~er Abfall** *m* ABFALL, SICHERHEIT explosive waste; **~e Luft** *f* SICHERHEIT explosive atmosphere
Exposition *f* SICHERHEIT exposure
Exsikkator *m* ABFALL, ANALYTIK, BODEN, LUFT desiccator
extern: **~e Begutachtung** *f* MANAG external validation; **~er Effekt** *m* POLITIK external effect, externality; **~e Kosten** *pl* POLITIK external costs; **~e Validierung** *f* MANAG external validation
Extinktion *f* ANALYTIK extinction; **Extinktionskoeffizient** *m* ANALYTIK linear absorption coefficient, *Atomspektrometrie* extinction coefficient, LUFT extinction coefficient
extrahierbar: **~e organische Chloride** *nt pl* (*EOCL*) ABWASSER, ANALYTIK, BODEN, LUFT extractable organic chlorides (*EOCL*); **~e organische Halogenverbindungen** *f pl* (*EOX*) ABFALL, ANALYTIK, WASSER extractable organic halogen compounds (*EOX*)
extrahieren *vt* ALLG extract
extrahierend *adj* ALLG extracted; **~e Komponente** *f* ABWASSER extracted component
extrahiert *adj* BODEN extracted; **~er Dampf** *m* BODEN extracted vapor (*AmE*), extracted vapour (*BrE*)
Extrakt *m* ABFALL, ABWASSER, ANALYTIK, WASSER extract
Extraktion *f* ALLG extraction; **Extraktionsgas** *nt* ANALYTIK extraction gas; **Extraktionslösungsmittel** *nt* ANALYTIK, BODEN extraction solvent
Extraktionsmittel *nt* ABWASSER, ANALYTIK, BODEN eluant, extractant, extracting agent, extraction agent, extractive agent; **Extraktionsmittelaufbereitung** *f* BODEN extractant processing, extracting agent processing, extraction agent processing; **Extraktionsmittelrest** *m* ABFALL, WASSER extract residue
Extraktion: **Extraktionstank** *m* BODEN extraction tank; **Extraktionsturm** *m* ABWASSER extraction vessel; **Extraktionsvorgang** *m* ABWASSER extraction operation; **Extraktionszentrifuge** *f* ABWASSER extraction centrifuge
Extraktor *m* ABWASSER, BODEN extractor
extraterrestrisch: **~ operierender Satellit** *m* ANALYTIK extra-terrestrially-operating satellite
Extrudieren *nt* ABFALL extruding
Extrusion *f* ABFALL extrusion; **~ mit Regranulierung** ABFALL extrusion with regranulation; **Extrusionsanlage** *f* ABFALL extrusion line, *Verarbeitung von gebrauchten Kunststoffen* extrusion plant; **Extrusionsblasen** *nt* ABFALL extrusion blow molding (*AmE*), extrusion blow moulding (*BrE*); **Extrusionsblasverfahren** *nt* ABFALL extrusion blow molding (*AmE*), extrusion blow moulding (*BrE*); **Extrusionsspritzkopf** *m* ABFALL extrusion head; **Extrusionsstreckblasformmaschine** *f* ABFALL extrusion stretch-flow molding machine (*AmE*), extrusion stretch-flow moulding machine (*BrE*)

F

F *abbr* (*Fluor*) BODEN F (*fluorine*)
Fach *nt* POLITIK sector; **Fachabteilung** *f* POLITIK sectoral division; **Fachbehörde** *f* POLITIK sectoral authority, sectoral unit
fächerförmig: **~e Rauchfahne** *f* LUFT fanning smoke plume
Fächermethode *f* ABFALL, WASSER cell method
Fähigkeit *f* ALLG capacity
Fahr: **Fahrsprenger** *m* WASSER traveling distributor (*AmE*), travelling distributor (*BrE*)
Fahrzeug: **~ mit Vergasermotor** ABFALL gasoline engine vehicle (*AmE*), petrol engine vehicle (*BrE*); **Fahrzeugkapazität** *f* ABFALL *Abfalltransport* vehicle capacity
Fäkal- *in cpds* ABWASSER faecal (*BrE*), fecal (*AmE*); **Fäkalabwasser** *nt* ABWASSER faecal wastewater (*BrE*), fecal wastewater (*AmE*)
Fäkalienpumpe *f* ABWASSER wastewater pump
Fäkalschlamm *m* ABWASSER faecal sludge (*BrE*), fecal sludge (*AmE*), ABWASSER faecal matter (*BrE*), fecal matter (*AmE*); **~ aus Hauskläranlagen** ABWASSER faecal solids sludge from domestic clarification plants (*BrE*), fecal solids sludge from domestic clarification plants (*AmE*); **Fäkalschlammbehandlung** *f* ABWASSER faecal solids sludge treatment (*BrE*), fecal solids sludge treatment (*AmE*); **Fäkalschlammzusammensetzung** *f* ABWASSER faecal solids sludge composition (*BrE*), fecal solids sludge composition (*AmE*)
Faktor *m* ANALYTIK volumetric factor
fakultativ *adj* ABWASSER facultative; **~ aerob** *adj* ABWASSER facultative-aerobic; **~ anaerobe Bakterien** *f pl* ABWASSER facultative anaerobic bacteria; **~es Heterotroph** *nt* ABWASSER facultative heterotroph; **~e Mikroben** *f pl* ABWASSER facultative microorganisms; **~er Teich** *m* ABWASSER facultative lagoon; **~e Zone** *f* ABWASSER facultative zone
fällbar: **~e organische Verbindungen** *f pl* ABWASSER precipitable organics
fällen *vt* ABFALL, ABWASSER, BODEN, WASSER precipitate
Fall: **Fallfilmabsorber** *m* LUFT falling film absorber; **Fallkörperviskosimeter** *nt* ANALYTIK falling sphere viscosimeter
Fällmittel *nt* ALLG coagulant, precipitant, precipitate; **Fällmittelaufwand** *m* ABWASSER quantity of precipitate; **Fällmitteldosierung** *f* ABWASSER precipitate feeding
Fallout *m* BODEN, LUFT fallout
Fall: **Fallrohr** *nt* ABWASSER downpipe, soil pipe, LUFT standpipe, WASSER soil pipe; **Fallschacht** *m* LUFT *Glockenboden* downspout, WASSER pressure well
Fällung *f* ABWASSER, WASSER precipitation; **~ mit Aluminiumsalzen** ABWASSER precipitation with aluminium salts (*BrE*), precipitation with aluminum salts (*AmE*); **~ mit Eisen(III)-Salzen** ABWASSER precipitation with ferric salt; **Fällungsmittel** *nt* ABWASSER precipitant, precipitate; **Fällungsschlamm** *m* ABWASSER precipitation sludge; **Fällungsvorgang** *m* ABWASSER precipitation operation
Fall: **Fallwindkraftwerk** *nt* ERNEU ENER downwind power plant
Faltenfilter *nt* ABWASSER folded filter
Fangsieb *nt* ABWASSER dry screen
Faraday-Gesetz *nt*: **~ der elektromagnetischen Induktion** ANALYTIK Faraday's law of electromagnetic induction
Farb- *in cpds* WASSER color (AmE), colour (BrE); **Farbabfälle** *m pl* ABFALL *Sonderabfall* waste paint
Farbe *f* ABFALL color (*AmE*), colour (*BrE*), paint, ANALYTIK color (*AmE*), colour (*BrE*)
farbecht *adj* ABFALL colorfast (*AmE*), colourfast (*BrE*)
Farb-: **Farbechtheit** *f* ABFALL colorfastness (*AmE*), colourfastness (*BrE*)
farbgebend *adj* ABWASSER chromophoric; **~e Wasserinhaltsstoffe** *m pl* ABWASSER chromophoric water components
Farb-: **Farbindikator** *m* ANALYTIK color indicator (*AmE*), colour indicator (*BrE*); **Farbkennzeichnung** *f* ANALYTIK color designation (*AmE*), colour designation (*BrE*)
farblich: **~ gemischt** *adj* ABFALL mixed-color (*AmE*), mixed-colour (*BrE*); **~ gemischtes Bruchglas** *nt* ABFALL mixed-color glass cullet (*AmE*), mixed-colour glass cullet (*BrE*); **~e Sortierung** *f* ABFALL *Altglas* separation by color (*AmE*), separation by colour (*BrE*); **~e Trennung** *f* ABFALL *Altglas* separation by color (*AmE*), separation by colour (*BrE*)
farblos *adj* LUFT colorless (*AmE*), colourless (*BrE*); **~es Gas** *nt* LUFT colorless gas (*AmE*), colourless gas (*BrE*); **~es Produkt** *nt* MEER white product
Farb-: **Farbreaktion** *f* ANALYTIK *Absorptionslösung* color reaction (*AmE*), colour reaction (*BrE*); **Farbschlamm** *m* ABFALL waste paint sludge; **Farbtiefe** *f* ANALYTIK color depth (*AmE*), colour depth (*BrE*); **Farbton** *m* ANALYTIK shade of color (*AmE*), shade of colour (*BrE*); **Farbumschlag** *m* ANALYTIK color change (*AmE*), colour change (*BrE*)
Färbung *f* ABFALL, ANALYTIK coloring (*AmE*), colouring (*BrE*); **Färbungsbestimmung** *f* ANALYTIK color analysis (*AmE*), colour analysis (*BrE*)
Faser *f* ABFALL, LUFT fiber (*AmE*), fibre (*BrE*)
faserartig *adj* LUFT *Teilchenbeschreibung* fibrous
Faser: **Faserdurchmesser** *m* LUFT fibre diameter (*BrE*), fiber diameter (*AmE*); **Faserfilter** *nt* LUFT fabric filter
faserförmig *adj* LUFT *Teilchenbeschreibung* fibrous
Faser: **Faserholz** *nt* ABFALL pulp wood
faserig *adj* ABFALL fibrous; **~e Inhaltsstoffe** *m pl* ABFALL fibrous constituents
Faser: **Fasermaterial** *nt* LUFT *für Filter* fiber material (*AmE*), fibre material (*BrE*); **Fasermatte** *f* LUFT fibrous mat; **Fasermattenfilter** *nt* LUFT fibrous mat filter; **Fasermischung** *f* LUFT fibre mixture (*BrE*), fiber mixture (*AmE*); **Faseroberfläche** *f* LUFT *filternde Abscheider* fibre surface (*BrE*), fiber surface (*AmE*); **Faserschicht** *f* LUFT *Filtermedium* fiber layer (*AmE*), fibre layer (*BrE*); **Faserschichtfilter** *nt* LUFT fibre layer filter (*BrE*), fiber layer filter (*AmE*); **Faserstoff** *m* ABFALL, LUFT fiber (*AmE*), fibre (*BrE*); **Faserstoffzusammensetzung** *f* ABFALL, LUFT fiber composition (*AmE*), fibre composition (*BrE*); **Fasersuspension** *f* WASSER *De-Inking-Prozeß* fibre suspension (*BrE*), fiber suspension (*AmE*); **Faservlies** *nt* LUFT *Filterfaser* nonwoven fabric; **Faserzusammensetzung** *f* ABFALL, LUFT fibre composition (*BrE*), fiber composition (*AmE*)
Faßpumpe *f* WASSER drain pump
Fassung *f* WASSER water catchment; **Fassungselement** *nt* ABFALL *Kollektor* collection element; **Fassungsrohr** *nt* WASSER tapping pipe
Faul: **Faulbecken** *nt* ABWASSER digestion tank, sludge digester; **Faulbehälter** *m* ABWASSER digestion tank, sludge digester, digester
Faulen *nt* ABFALL, ABWASSER putrefaction, rot

Faul: **Faulfähigkeit** *f* ABFALL putrescibility, rot

Faulgas *nt* ABFALL, ABWASSER, WASSERdigester gas, fermentation gas, sludge gas; **Faulgasausbeute** *f* ABWASSER digester gas quantity; **Faulgasertrag** *m* ABWASSER digester gas yield; **Faulgasmenge** *f* ABWASSER digester gas quantity; **Faulgasverwertung** *f* ABWASSER digester gas utilization

Faul: **Faulgrube** *f* ABWASSER cesspit, cesspool

Fäulnis *f* ABFALL, ABWASSER putrefaction

fäulniserregend *adj* ABFALL, ABWASSER putrefactive, putrescible

fäulnisfähig *adj* ABFALL, ABWASSER putrefactive, putrescible; **~er Schlamm** *m* ABWASSER putrescible sludge; **~er Stoff** *m* ABFALL putrescible matter

fäulnisunfähig *adj* ABFALL, ABWASSER nonputrefactive, nonputrescible

Faulraum *m* ABWASSER digestion tank, sludge digester, digester; **Faulraumbeheizung** *f* ABWASSER digestion tank heating; **Faulraumheizung** *f* ABWASSER digestion chamber heating; **Faulraumstörung** *f* ABWASSER digester disruption; **Faulraumwasser** *nt* ABWASSER supernatant liquor

Faul: **Faulschlamm** *m* ABWASSER digested sludge; **Faulschlammwasser** *nt* ABWASSER sludge liquor; **Faulturm** *m* ABWASSER digester

Faulung *f* ABFALL, ABWASSER, WASSER digestion; **~ mit anschließender Kompostierung** ABWASSER digestion with added composting

Faul: **Faulzeit** *f* ABWASSER digestion time, digestion period

FBK *abbr* (*Festbettkaskadenreaktor*) ABWASSER fixed-bed cascade reactor

FCKW *abbr* (*Fluorchlorkohlenwasserstoff*) LUFT CFC (*chlorofluorocarbon*)

FD *abbr* (*Felddesorption*) ANALYTIK FD (*field desorption*)

federführend: **~e Behörde** *f* POLITIK responsible authority

federnd *adj* LÄRM resilient; **~e Aufstellung** *f* LÄRM *von Maschinen für die Begrenzung von Schallausbreitung* resilient mounting; **~es Lager** *nt* LÄRM resilient bearing; **~e Schicht** *f* LÄRM resilient layer

Feedstock-Recycling *nt* ABFALL tertiary plastic recycling, feedstock recycling

Fehlanzeige *f* ANALYTIK indication error

Fehlbedienung *f* ANALYTIK operating error

fehlend: **~e Angaben** *f pl* MANAG *bei Ökobilanz* missing data

Fehlermöglichkeit *f* ANALYTIK possibility of error

Fehlmessung *f* ANALYTIK faulty measurement

fein *adj* LUFT fine; **~e Partikeln** *f pl* LUFT fine particles

feinblasig: **~e Belüftung** *f* ABWASSER fine-bubble aeration

Fein: **Feinfaser-Nadelfilz** *m* LUFT fine-fibered needled felt (*AmE*), fine-fibred needled felt (*BrE*); **Feinfilter** *nt* ABWASSER microfilter; **Feinflocken** *f pl* ABWASSER fine flocs

feinkörnig *adj* ABFALL fine-grained

Fein: **Feinmüll** *m* ABFALL *Korngrößen 0 bis 8 mm* fine refuse

feinporig *adj* ABFALL, ABWASSER, LUFT, WASSER microporous

Fein: **Feinrechen** *m* ABFALL, ABWASSER, LUFT fine screen

Feinst- *in cpds* LUFT finest

Fein: **Feinstaub** *m* LUFT fine dust; **Feinstaubabscheidung** *f* LUFT fine dust precipitation

Feinst-: **Feinststaub** *m* LUFT finest dust; **Feinsttropfenabscheider** *m* LUFT finest drop precipitator

feinstverteilt *adj* ABWASSER dispersed

Feinst-: **Feinstzerteilung** *f* ABFALL, WASSER *der Tropfen* dispersion

feinverteilt *adj* ABFALL, ABWASSER, WASSER suspended

Fein: **Feinzuschlagstoff** *m* WASSER fine-grained additive

FEL *abbr* (*Frank effect level*) SICHERHEIT FEL (*Frank effect level*)

Feld- *in cpds* ANALYTIK, LUFT, MEER, WASSER field; **Feldanalyse** *f* ANALYTIK field analysis; **Feldaufladung** *f* LUFT field charging; **Felddaten** *pl* ANALYTIK field data; **Felddesorption** *f* (*FD*) ANALYTIK field desorption (*FD*); **Feldionisation** *f* (*FI*) ANALYTIK *FEL* field ionization (*FI*); **Feldkapazität** *f* **der Deponie** ABFALL field capacity of landfill; **Feldstärke** *f* ANALYTIK, LUFT *Elektrofilter* field strength; **Feldversuch** *m* ANALYTIK field test, field trial; **Feldwirksamkeit** *f* ANALYTIK field performance

Fender *m* MEER fender

fermentieren *vt* ABFALL, ABWASSER, ERNEU ENER, WASSER ferment

Fern- *in cpds* ANALYTIK remote; **Fernerkundung** *f* ANALYTIK, MEER remote sensing; **Fernmessung** *f* MEER remote sensing; **Fernsteueranlage** *f* ABWASSER, WASSER remote control system; **Fernsteuereinheit** *f* ABFALL, ABWASSER remote control unit; **Fernsteuersystem** *nt* ABWASSER, WASSER telecontrol system; **Fernwasserleitung** *f* WASSER long-distance water main; **Fernwasserversorgung** *f* WASSER distant water supply, long-distance water supply

Ferrisalz *nt* WASSER *Koagulans* ferric salt

ferromagnetisch *adj* ANALYTIK ferromagnetic

Fertigkompost *m* ABFALL readied compost

Fertigprodukt *nt* ABFALL *Wiederaufbereitung* finished product

Fertigungsstraße *f* ABFALL production line

Fest- *in cpds* LUFT solid

fest: **~er Abfall** *m* ABFALL solid waste; **~er Aggregatzustand** *m* LUFT solid state of aggregation; **~er Brennstoff** *m* LUFT solid fuel; **~er Kohlenstoff** *m* ERNEU ENER *Biomasse* fixed carbon; **~e Partikel** *f* ABWASSER solid particle; **~er Reaktionspartner** *m* ANALYTIK solid-reacting agent; **~es Rotorblatt** *nt* ERNEU ENER *Windturbine* fixed rotor blade; **~er Rückstand** *m* ABFALL, WASSER solid residue

Festbett *nt* ABWASSER, LUFT fixed bed; **Festbettadsorber** *m* ABWASSER, LUFT fixed-bed adsorber, *Adsorptionsverfahren* stationary-bed adsorber; **Festbettelektrolyse** *f* ABWASSER fixed-bed electrolysis; **Festbettionenaustauscher** *m* ABWASSER fixed-bed ion exchanger; **Festbettkaskadenreaktor** *m* (*FBK*) ABWASSER fixed-bed cascade reactor; **Festbettkatalysator** *m* LUFT fixed-bed catalyst; **Festbettofen** *m* ABWASSER *Klärschlammverbrennung* fixed-bed furnace; **Festbettreaktor** *m* ABFALL, BODEN, WASSER fixed-bed reactor

Fest-: **Festbrennstoff** *m* LUFT solid fuel; **Fest-Flüssig-Extraktion** *f* BODEN solid-liquid extraction

Festigkeit *f* ABFALL *einer Deponie* strength

Fest-: **Festphasenextraktion** *f* (*SPE*) ANALYTIK solid-phase extraction (*SPE*)

Feststoff *m* ABFALL, ABWASSER, LUFT, WASSER solid material, solid matter; **Feststoffe und Gase voneinander trennen** *phr* LUFT separate solid matter from gas; **Feststoffabfall** *m* ABFALL solid waste; **Feststoffabscheider** *m* ABFALL, ABWASSER solids separator; **Feststoffabtrennung** *f* LUFT solid-matter separation; **Feststoffausstoß** *m* LUFT particulate emissions; **Feststoffbeladung** *f* LUFT solid-matter loading; **Feststoffgehalt** *m* ABFALL solids contents, ABWASSER dry mass, solids contents

feststoffhaltig: **~e Abluft** *f* LUFT solid-containing waste air

Feststoff: **Feststoffkatalysator** *m* LUFT fixed catalyst; **Feststoffteilchen** *nt* ABFALL, WASSER solid particles; **Feststoffverweildauer** *m* WASSER residence time of solids; **Feststoffverweilzeit** *f* WASSER residence time of solids

Fett *nt* ABFALL, ABWASSER grease, LUFT fat, WASSER grease; **Fettabscheider** *m* ABWASSER grease trap, oil trap, MEER skimming tank, WASSER grease trap, grease separator; **Fettalkohol** *m* ABWASSER fatty alcohol; **Fettfang** *m* ABWASSER grease trap, oil trap; **Fettfänger** *m* MEER skimming tank

fettfrei *adj* ABFALL grease-free, LUFT fat-free, WASSER grease-free; **~e Abluft** *f* LUFT fat-free waste air

fetthaltig: **~es Abwasser** *nt* ABWASSER oily wastewater

Fett: **Fettlöslichkeit** *f* ABWASSER liposolubility; **Fettsäure** *f* ABWASSER fatty acid

Feuchte *f* ANALYTIK moisture, humidity; **Feuchtemessung** *f* ANALYTIK humidity measurement; **Feuchtesensor** *m* ANALYTIK moisture sensor

Feuchtgehalt *m* ABFALL, ABWASSER, ANALYTIK *von Gasen*, ERNEU ENER, LUFT moisture content; **Feuchtgehaltsbestimmung** *f* ANALYTIK humidity content determination, moisture content determination

Feuchtigkeit *f* ABFALL, LUFT humidity, moisture

feuchtigkeitsbeständig *adj* ABFALL, LUFT, WASSER moisture-proof, humidity-proof

Feuchtigkeit: **Feuchtigkeitsbestimmung** *f* ABFALL moisture determination

feuchtigkeitsfest *adj* ABFALL, WASSER moisture-resistant

Feuchtigkeit: **Feuchtigkeitsgehalt** *m* ABFALL, ABWASSER, ERNEU ENER, LUFT moisture content; **Feuchtigkeitsprobe** *nt* ABFALL moisture test

Feuer *nt* ALLG fire

feuerbeständig *adj* SICHERHEIT fire-resistant

feuerfest 1. *adj* SICHERHEIT fireproof; **2. ~ machen** *vt* SICHERHEIT fireproof

feuerhemmend *adj* SICHERHEIT fire-retardant

Feuerlöschmonitor *m* MEER fire monitor

Feuerraum *m* ABFALL, ABWASSER, BODEN, LUFT combustion chamber; **Feuerraumanbackung** *f* ABFALL combustion chamber slagging; **Feuerraumaschesilo** *m* LUFT combustion chamber ash silo; **Feuerraumtemperatur** *f* ABFALL combustion chamber temperature

Feuerung *f* ABFALL, ERNEU ENER, LUFT, SICHERHEIT combustion; **Feuerungsabgas** *nt* LUFT combustion waste gas; **Feuerungsanlage** *f* ABFALL, ERNEU ENER, LUFT boiler plant

feuerungstechnisch *adj* ABFALL, LUFT fuel engineering-related

feuerverzinkt *adj* ABFALL hot-dipped, hot-galvanized; **~er Stahl** *m* ABFALL hot-galvanized steel; **~es Stahlrohr** *nt* ABFALL hot-galvanized steel pipe

FI *abbr* (*Feldionisation*) ANALYTIK *LOAEL* FI (*field ionization*)

FIA *abbr* (*Fließinjektionsanalyse*) ANALYTIK FIA (*flow injection analysis*)

FIAS *abbr* (*Fließinjektionsanalysensystem*) ANALYTIK FIAS (*flow injection analysis system*)

Fibel *f* ANALYTIK handbook

Fick: **~sches Gesetz** *nt* LUFT *der Diffusion* Fick's law

FID *abbr* (*Flammenionisationsdetektor*) ANALYTIK FID (*flame ionization detector*)

Film *m* WASSER film; **Film-Coating** *nt* WASSER film coating; **Filmplakette** *f* ANALYTIK *Strahlungsdosimeter* film badge

Filter *nt* ALLG filter; **Filteranlage** *f* LUFT, SICHERHEIT, WASSER filtration plant, filtering plant, filtering unit; **Filteranlage** *f* **gegen Staub und Fasern** SICHERHEIT *Textilindustrie* filter plant for dust and fibers (*AmE*), filter plant for dust and fibres (*BrE*); **Filterart** *f* ABWASSER, LUFT filter type; **Filterasche** *f* ABFALL filter ash; **Filterauslegung** *f* LUFT filter layout; **Filteraustausch** *m* ABWASSER, LUFT filter replacement; **Filterband** *nt* ANALYTIK filter belt; **Filterbecken** *nt* WASSER filtering tank; **Filterbetrieb** *m* ABFALL, WASSER filter operation

Filterbett *nt* ABFALL, ABWASSER, LUFT, WASSER filter bed; **Filterbetthöhe** *f* LUFT, WASSER filter bed height; **Filterbettporosität** *f* LUFT filter bed porosity

Filter: **Filterbrunnen** *m* ABWASSER, WASSER filtering well; **Filterdeckel** *m* LUFT cover plate; **Filterdimensionierung** *f* LUFT filter dimensioning; **Filterdüse** *f* ABWASSER, WASSER filter nozzle; **Filtereinheit** *f* ABFALL, ABWASSER, LUFT, WASSER filter unit; **Filterelement** *nt* ABFALL, ABWASSER, LUFT, SICHERHEIT, WASSER filter element

filterfähig *adj* ALLG filterable

Filter: **Filterfilz** *m* LUFT filter felt; **Filterfläche** *f* ABWASSER filter area; **Filterflächenbelegung** *f* LUFT *Gewebefiltration* air-to-cloth ratio; **Filtergehäuse** *nt* LUFT, WASSER filter housing; **Filtergeschwindigkeit** *f* WASSER filter velocity; **Filtergewebe** *nt* LUFT filter fabric; **Filterhilfsmittel** *nt* ABFALL, ABWASSER, LUFT, WASSER filter aid; **Filterhilfsstoff** *m* ABFALL, ABWASSER, LUFT, WASSER filter aid; **Filterhilfsstoffwirkungsgrad** *m* ABFALL, ABWASSER, LUFT, WASSER filter aid efficiency; **Filterkammer** *f* ABWASSER, WASSER filter gallery; **Filterkies** *m* ABWASSER filter gravel; **Filterkuchen** *m* ABWASSER, LUFT filter cake; **Filterkühler** *m* LUFT filter cooler; **Filtermaterial** *nt* ABFALL, ABWASSER, LUFT, WASSER filter material, filter medium; **Filtermaterialkreislauf** *m* LUFT filter material circulation; **Filtermedium** *nt* ABFALL, ABWASSER, LUFT, WASSER filter material, filter medium; **Filtermediumoberfläche** *f* LUFT filter medium surface; **Filtermittel** *nt* LUFT filtering agent

filtern *vt* ALLG filter, filtrate

Filtern *nt* ALLG filtering, filtration

filternd: **~er Abscheider** *m* LUFT filtering precipitator

Filter: **Filteroberfläche** *f* ABWASSER, LUFT, WASSER filter surface; **Filterpapier** *nt* ANALYTIK filter paper; **Filterpore** *f* ABWASSER filter pore; **Filterpresse** *f* ABWASSER, WASSER *Oberflächenfilter* filter press; **Filterraum** *m* ABWASSER, LUFT filter compartment; **Filterregenierung** *f* ABWASSER, LUFT filter regeneration; **Filterrückstände** *m pl* ABWASSER, LUFT filter residues; **Filtersäule** *f* ABFALL, WASSER filter column; **Filterschicht** *f* ABFALL, ABWASSER filter bed, LUFT filter bed, filter layer, WASSER filter bed; **Filterschichthöhe** *f* ABWASSER, LUFT filter bed depth; **Filterschlauch** *m* LUFT filter bag, tubular filter bag; **Filterschnecke** *f* LUFT filter worm; **Filterspülung** *f* ABWASSER, LUFT, WASSER filter washing; **Filterstaub** *m* ABFALL, LUFT filter dust; **Filterstoff** *m* ABFALL, ABWASSER, LUFT, WASSER filter cloth; **Filterstufe** *f* LUFT filter stage; **Filtertasche** *f* LUFT filter bag; **Filtertest** *m* ANALYTIK filter test; **Filtertuch** *nt* ABFALL, ABWASSER, LUFT, WASSER filter cloth; **Filtertyp** *m* ABWASSER, LUFT filter type

Filterung *f* ALLG filtration

Filter: **Filterwandung** *f* LUFT filter wall; **Filterwiderstand** *m* ABWASSER, LUFT, WASSER filter resistance; **Filterwirksamkeit** *f* LUFT filter efficacy; **Filterwirkungsgrad** *m* ABWASSER, LUFT filter efficiency; **Filterzelle** *f* LUFT feed tank

Filtrat *nt* ALLG filtrate; **Filtratbehälter** *m* ABFALL, WASSER filtrate tank; **Filtratgut** *nt* ABFALL, ABWASSER, WASSER filtered matter

Filtration *f* ALLG filtering, filtration; **Filtrationsanlage** *f* LUFT, WASSER filtration plant; **Filtrationsdruck** *m* LUFT filtration pressure; **Filtrationseffekt** *m* ABWASSER, LUFT, WASSER filtration effect; **Filtrationsgeschwindigkeit** *f* ABWASSER, ANALYTIK, LUFT, WASSER filtering rate, filtration rate, filtration velocity; **Filtrationskurve** *f* ABWASSER, LUFT, WASSER filtration curve; **Filtrationsleistung** *f* ABWASSER, LUFT, WASSER filtration capacity; **Filtrationsrichtung** *f* LUFT filtration direction; **Filtrationsrückstand** *m* ANALYTIK filtration residue; **Filtrationsvorgang** *m* ABWASSER, LUFT filtering operation; **Filtrationszyklus** *m* LUFT filtration cycle

Filtrat: **Filtratkontamination** *f* ABWASSER, WASSER filtrate contamination; **Filtratnachchlorung** *f* WASSER filtrate post-chlorination; **Filtratqualität** *f* ABWASSER, WASSER filtrate quality; **Filtrat- und Restwasserrücklauf** *m* ABWASSER filtration and residue water return; **Filtratstrom** *m* ABWASSER, WASSER filtrate flow; **Filtratverschmutzung** *f* ABWASSER, WASSER filtrate contamination

Filtrier- *in cpds* ABWASSER, ANALYTIK, LUFT filtering

filtrierbar *adj* ALLG filterable; **~e Feststoffe** *m pl* ABFALL, ABWASSER, WASSER filterable solids; **~e Flocken** *f pl* ABWASSER filterable flocs

Filtrier-: **Filtrierdauer** *f* ABFALL, LUFT, WASSER filtering time; **Filtriereffekt** *m* ABWASSER, LUFT, WASSER filtration effect

filtrieren *vt* ALLG filter

Filtrier-: **Filtrierstellung** *f* LUFT *Schlauchfilter* filtering position

filtriert *adj* ALLG filtered; **~e Probe** *f* ANALYTIK filtered sample

Filtrierung *f* ALLG filtering, filtration

Filtrier-: **Filtriervorgang** *m* ABWASSER, LUFT filtering operation; **Filtrierzeit** *f* ABFALL, LUFT, WASSER filtering time

Filz *m* LÄRM *Dämmung* felt
finanzieren *vt* MANAG, POLITIK finance
Fisch: **Fischgiftigkeit** *f* ABWASSER fish toxicity; **Fischtest** *m* ANALYTIK fish test; **Fischtoxizität** *f* ABWASSER fish toxicity
Fitting *nt* ABFALL, ABWASSER, ANALYTIK, WASSER armature, fitting
Fixierung *f* ABFALL, BODEN fixation, fixing
flach *adj* ABFALL, BODEN, WASSER flat; **~e Fraktion** *f* ABFALL flat fraction; **~er Untergrund** *m* BODEN flat subsurface
Flach: **Flachbettbiofilter** *nt* LUFT *biologisches Sorptionsverfahren* flat-bed biofilter; **Flachbrunnen** *m* WASSER shallow well; **Flachbunker** *m* ABFALL flat bunker
Fläche *f* ALLG surface; **Flächenbedarf** *m* ERNEU ENER surface requirement; **Flächenbelastung** *f* ABFALL surface loading, LUFT surface load; **Flächenbelüftung** *f* ABWASSER area aeration
flächenbezogen *adj* ALLG area-related; **~er Lärm** *m* LÄRM, MEER area-related noise; **~er Wasserbedarf** *m* WASSER area-related water demand
Fläche: **Flächeneinbau** *m* ABFALL, WASSER surface installation; **Flächenfilter** *nt* WASSER surface filter; **Flächenfiltration** *f* WASSER surface filtration; **Flächennutzungsplan** *m* ERNEU ENER, MANAG, POLITIK, WASSER zoning plan; **Flächenquelle** *f* ABWASSER, BODEN, LUFT area source; **Flächenverkleidung** *f* LÄRM *schallabsorbierend* surface lining
Flach: **Flachglas** *nt* ABFALL sheet glass; **Flachkollektor** *m* ERNEU ENER flat-plate collector, surface collector; **Flachsieb** *nt* ABFALL, WASSER flat sieve
Flamm- *in cpds* ANALYTIK, LUFT flame
flammbeständig *adj* SICHERHEIT flame-resistant
Flamme *f* ANALYTIK, LUFT flame; **Flammen-AAS** *f* ANALYTIK flame-AAS; **Flammenabkühlung** *f* LUFT flame cooling; **Flammenfärbung** *f* ANALYTIK flame coloration (*AmE*), flame colouration (*BrE*)
flammenfest *adj* SICHERHEIT flameproof
Flamme: **Flammenionisation** *f* ANALYTIK flame ionization; **Flammenionisationsdetektor** *m* (*FID*) ANALYTIK flame ionization detector (*FID*); **Flammenphotometer** *nt* ANALYTIK flame photometer; **Flammenphotometrie** *f* ANALYTIK *Schwermetallbestimmung* flame photometry
flammenphotometrisch *adj* ANALYTIK flame-photometric; **~e Bestimmung** *f* ANALYTIK flame-photometric determination; **~er Detektor** *m* (*FPD*) ANALYTIK flame-photometric detector (*FPD*)
flammensicher *adj* SICHERHEIT flameproof
Flamme: **Flammenspektrometer** *nt* ANALYTIK flame spectrometry; **Flammenspektrometrie** *f* ANALYTIK flame spectrometry
flammenspektrometrisch *adj* ANALYTIK flame-spectrometric; **~e Bestimmung** *f* ANALYTIK flame-spectrometric determination
Flamm-: **Flammgas** *nt* LUFT flame gas
Flammpunkt *m* LUFT flame point; **Flammpunktbestimmung** *f* ANALYTIK flash point determination; **Flammpunktprüfer** *m* ANALYTIK flash point tester
Flanken- *in cpds* BODEN, LÄRM flanking; **Flankenabdichtung** *f* BODEN flanking sealing
flankengedichtet: **~e Deponie** *f* ABFALL edge-sealed landfill
Flanken-: **Flankenübertragung** *f* LÄRM flanking transmission; **Flankenweg** *m* LÄRM *um eine Schallwand herum* flanking path; **Flankenwegübertragung** *f* LÄRM flanking transmission
Flansch *m* LUFT, MEER flange
Flasche *f* ANALYTIK bottle; **~ ohne Pfand** ABFALL nonreturnable bottle; **Flaschenhülle** *f* ABFALL bottle jacket; **Flaschenhülse** *f* ABFALL bottle sleeve; **Flaschenkasten** *m* ABFALL bottle crate; **Flaschenlecksensor** *m* ABFALL bottle leak detector; **Flaschenmethode** *f* **nach Wagner** ANALYTIK Wagner's bottle method; **Flaschenpfand** *nt* ABFALL bottle deposit, deposit; **Flaschenverschlußkappe** *f* ABFALL bottle cap; **Flaschenversuch** *m* ANALYTIK bottle test; **Flaschenwaschmaschine** *f* ABFALL bottle-rinsing machine; **Flaschenzug** *m* SICHERHEIT hoist
flaumig *adj* LUFT fluffy
flexibel: **flexibler Schlauch** *m* MEER flexible hose; **flexibler Schwimmtank** *m* MEER *zur Aufnahme von Öl* floating flexible tank
Fliehkraft *f* ERNEU ENER *Gezeitenenergie*, LUFT centrifugal force; **Fliehkraftabscheider** *m* LUFT cyclone; **Fliehkraftabscheidung** *f* LUFT centrifugal precipitation; **Fliehkraftsedimentation** *f* ANALYTIK *Meßverfahren für Staube* centrifugal force sedimentation; **Fliehkraftwirkung** *f* LUFT centrifugal force activity
Fließ- *in cpds* ABWASSER, ANALYTIK, WASSER flow, flowing
Fließbett *nt* ABWASSER moving text, LUFT, WASSER moving bed; **Fließbettadsorber** *m* LUFT moving-bed adsorber; **Fließbettkatalysator** *m* LUFT fluid catalyst; **Fließbettreaktor** *m* ABFALL, BODEN moving-bed reactor; **Fließbettsandfilter** *nt* ABWASSER, WASSER moving-bed sand filter; **Fließbetttrockner** *m* ABWASSER moving-bed dryer
fließen *vi* ABWASSER, ERNEU ENER, WASSER flow
fließend: **~es Gewässer** *nt* ABFALL running water; **~es Wasser** *nt* WASSER flowing water
Fließ-: **Fließgeschwindigkeit** *f* ABWASSER, WASSER flow velocity
Fließinjektionsanalyse *f* (*FIA*) ANALYTIK flow injection analysis (*FIA*); **~ für Atomspektrometrie** ANALYTIK flow injection analysis spectrometry; **Fließinjektionsanalysensystem** *nt* (*FIAS*) ANALYTIK flow injection analysis system (*FIAS*)
Fließ-: **Fließsand** *m* WASSER fluidized sand; **Fließwasser** *nt* WASSER flowing water; **Fließwiderstand** *m* WASSER flow resistance
flockbar *adj* ABWASSER, ANALYTIK, LUFT, WASSER coagulable
Flockbarkeit *f* ABWASSER, ANALYTIK, LUFT, WASSER coagulability
Flocke *f* ALLG floc
flocken *vi* ALLG coagulate
Flocken *f pl* WASSER clusters; **Flockenbildung** *f* ABWASSER floc formation; **Flockenfragmente** *nt pl* ABWASSER floc fragments
flockig *adj* ABFALL, ABWASSER, ANALYTIK, LUFT, WASSER flocculent, LUFT *Staubeigenschaften* fluffy
Flockung *f* ALLG coagulation, flocculation; **Flockungsbecken** *nt* ABFALL, ABWASSER, ANALYTIK, LUFT, WASSER flocculation tank, flocculator basin, flocculator; **Flockungsfiltration** *f* ABWASSER contact filtration, WASSER floc filtration; **Flockungsgeschwindigkeit** *f* ABWASSER flocculation rate; **Flockungshilfsmittel** *nt* ABWASSER, WASSER coagulation aid; **Flockungsmittel** *nt* ABFALL, ABWASSER, ANALYTIK, LUFT, WASSER coagulant, flocculant, flocculator; **Flockungsreaktor** *m* ABFALL, ABWASSER, ANALYTIK, LUFT, WASSER flocculator; **Flockungsreaktor** *m* **mit Zylinderrührer** ABWASSER flocculator with cylindrical mixer; **Flockungstest** *m* ANALYTIK flocculation test; **Flockungszone** *f* ABWASSER flocculation zone
Flokkulant *m* ALLG flocculant
Flotation *f* ABWASSER *mechanisches Abwasseraufbereitungsverfahren*, ANALYTIK, MEER, WASSER *Naßtrennungsverfahren für Kunststoffe* flotation; **~ mit dispergiertem Gas** ABWASSER flotation with dispersed gas; **~ mit elektrolytisch erzeugtem Gas** ABWASSER flotation with electrolytically generated gas; **~ mit integrierter Filtration** ABWASSER flotation with integrated filtration; **~ durch Lufteinblasen** WASSER pneumatic flotation; **~ mit Tauchbelüfter** ABWASSER flotation with submerged aerator; **Flotationsanlage** *f* ABWASSER flotation plant; **Flotationsbecken** *nt* ABWASSER flotation tank; **Flotationseindickung** *f* ABWASSER flotation thickening;

Flotationskammer *f* MEER *einer Ölsperre* flotation chamber; **Flotationsmittel** *nt* ABWASSER, BODEN flotation agent; **Flotationszelle** *f* BODEN flotation cell
Flowmeter *nt* ANALYTIK flowmeter
flüchtig *adj* ABWASSER, BODEN, LUFT volatile; **~e anorganische Verbindungen** *f pl* BODEN, LUFT volatile inorganic compounds; **~e Emissionen** *f pl* LUFT fugitive emissions, volatile emissions; **~e organische Verbindungen** *f pl* (*VOC*) BODEN, LUFT volatile organic compounds (*VOC*); **~e Schwermetalle** *nt pl* LUFT volatile heavy metals
Flugasche *f* LUFT fly ash; **Flugascheanfall** *m* WASSER fly ash yield; **Flugaschesilo** *m* LUFT fly ash silo; **Flugaschezement** *m* WASSER cement with fly ash
Flügel *m* ABFALL, ABWASSER vane; **Flügelpumpe** *f* WASSER semirotary pump; **Flügelradanemometer** *nt* ANALYTIK impeller anemometer; **Flügelscherfestigkeit** *f* ABFALL, BODEN *Deponie* vane shear strength; **Flügelzellenpumpe** *f* ABWASSER vane pump
Fluglärmsteuer *f* LÄRM, POLITIK aircraft noise tax
Flugstaub *m* LUFT fly ash; **Flugstaubkonditionierung** *f* LUFT fly ash conditioning
Flugstromverfahren *nt* BODEN fly stream method
Flugzeugabgas *nt* LUFT aircraft waste gas
Fluor *nt* (*F*) BODEN fluorine (*F*); **Fluorchlorkohlenwasserstoff** *m* (*FCKW*) LUFT chlorofluorocarbon (*CFC*)
Fluoreszenz *f* ANALYTIK fluorescence; **Fluoreszenzanalyse** *f* ANALYTIK fluorescence analysis; **Fluoreszenzdetektion** *f* ANALYTIK fluorescence detection; **Fluoreszenzindikatoradsorption** *f* ANALYTIK fluorescent indicator adsorption; **Fluoreszenzpolarisation** *f* ANALYTIK fluorescence polarization; **Fluoreszenzspektrometrie** *f* ANALYTIK *Röntgen-Fluoreszenz-Analysen* fluorescence spectrometry; **Fluoreszenzspektroskopie** *f* ANALYTIK *Röntgen-Fluoreszenz-Analysen* fluorescent spectroscopy; **Fluoreszenzspektrum** *nt* ANALYTIK fluorescence spectrum
fluoreszierend: **~er Stoff** *m* LUFT fluorescent compound
Fluorid *nt* WASSER fluoride; **Fluoridbestimmung** *f* **in Abwässern** ANALYTIK fluoride analysis in wastewater
Fluorimetrie *f* ANALYTIK *Schwermetallbestimmung* fluorimetry
fluorimetrisch *adj* ANALYTIK fluorimetric
Fluor: **Fluorwasserstoff** *m* LUFT fluorohydrogen, hydrogen fluoride
Fluß *m* ABWASSER, POLITIK, WASSER river; **Flußablagerung** *f* WASSER aggradational deposit; **Flußaufsichtsbehörde** *f* ABFALL, MANAG, WASSER river authority; **Flußbau** *m* WASSER river training, river works; **Flußbecken** *f* ABFALL, WASSER river basin; **Flußbelüftung** *f* ABWASSER stream aeration; **Flußbett** *nt* WASSER river bed; **Flußeinzugsgebiet** *nt* WASSER river basin; **Flußeinzugsgebietsbehörde** *f* POLITIK river basin authority; **Flußfassung** *f* WASSER river tapping; **Flußgebiet** *nt* WASSER river basin; **Flußgröße** *f* ANALYTIK flow parameter
flüssig *adj* ALLG liquid; **~er Abfall** *m* ABFALL liquid waste; **~er Abfallstoff** *m* ABWASSER, LUFT liquid waste substance; **~er Aggregatzustand** *m* LUFT liquid state of aggregation; **~es Arbeitsmittel** *nt* ERNEU ENER *Wärmepumpe* liquid working medium; **~er Reaktionspartner** *m* ANALYTIK liquid-reacting agent; **~er Rückstand** *m* LUFT liquid residue; **~er Sonderabfall** *m* ABFALL hazardous liquid waste
Flüssig: **Flüssigerdgas** *nt* LUFT liquefied natural gas (*LNG*); **Flüssig-Fest-Extraktion** *f* ANALYTIK liquid-solid extraction; **Flüssig-Flüssig-Extraktion** *f* ABWASSER, ANALYTIK liquid-liquid extraction; **Flüssiggas** *nt* (*LPG*) ERNEU ENER, LUFT liquid petroleum gas (*LPG*)
Flüssigkeit *f* ALLG liquid; **Flüssigkeit-Flüssigkeitschromatographie** *f* ANALYTIK *Glasrecycling* liquid chromatography-liquid chromatography; **Flüssigkeitsabzug** *m* LUFT *Absorber* liquid discharge; **Flüssigkeitsaufgabe** *f* LUFT *Absorber* feeding-in of liquid
Flüssigkeitschromatographie *f* (*LC*) ANALYTIK liquid chromatography; **~ mit elektrochemischer Detektion** ANALYTIK *Ultraviolett* liquid chromatography-electrochemical detection; **~ mit Fluoreszenzdetektion** ANALYTIK liquid chromatography-fluorescence detection; **~ mit Ultraviolettdetektion** ANALYTIK liquid chromatography-ultraviolet detection
Flüssigkeit: **Flüssigkeitsfederthermometer** *nt* ANALYTIK liquid spring thermometer; **Flüssigkeitsfilm** *m* LUFT liquid film; **Flüssigkeitsglasthermometer** *nt* ANALYTIK liquid glass thermometer; **Flüssigkeitsoberfläche** *f* LUFT liquid surface; **Flüssigkeitsprobenahme** *f* ANALYTIK liquid sampling; **Flüssigkeitssäule** *f* ANALYTIK liquid column; **Flüssigkeitsschicht** *f* LUFT liquid layer; **Flüssigkeitströpfchen** *nt* LUFT droplet, liquid droplet; **Flüssigkeitsverteiler** *m* LUFT liquid distributor
Flüssig: **Flüssigkompostierung** *f* ABWASSER liquid composting; **Flüssigkristallsichtanzeige** *f* (*LCD*) ANALYTIK liquid crystal display (*LCD*); **Flüssigphase** *f* LUFT liquid phase
Flüssigschlamm *m* ABWASSER liquid sludge, slurry; **Flüssigschlammkonditionierung** *f* ABWASSER liquid sludge conditioning; **Flüssigschlammverwertung** *f* ABWASSER wet sludge utilization
Fluß: **Flußkanalisierung** *f* WASSER river training; **Flußmittel** *nt* WASSER fluxing agent; **Flußregulierung** *f* WASSER river training; **Flußsohle** *f* WASSER river floor; **Flußufermauer** *f* WASSER river wall; **Flußverwaltungsplan** *m* POLITIK river basin management plan
Flußwasser *nt* WASSER river water; **Flußwasserentnahme** *f* WASSER river water tapping; **Flußwasserfassung** *f* WASSER river water tapping; **Flußwasserqualität** *f* WASSER river water quality; **Flußwasserreinigung** *f* WASSER river water purification
Flut *f* ERNEU ENER flood tide
Flutung *f* ABWASSER, WASSER flooding
Fokussierung *f* ERNEU ENER focusing
Folgenutzung *f* ABFALL, WASSER after-use
Folie *f* BODEN sheeting; **Folienabsorber** *m* LÄRM sheet absorber; **Folienextrudieranlage** *f* ABFALL film extrusion equipment; **Folienpartikel** *f* ABFALL, WASSER foil particle
Förder- *in cpds* ABWASSER carrying, conveying, delivery; **Förderanlage** *f* ABFALL, WASSER conveying equipment
Förderband *nt* ABFALL belt conveyor, conveying belt; **~ zur Ölaufnahme** MEER *Bekämpfung der Ölverschmutzung* belt skimmer; **Förderbandskimmer** *m* ABFALL, WASSER conveyor-belt skimmer
Förder-: **Förderbrunnen** *m* BODEN conveying well; **Förderdruck** *m* WASSER delivery pressure
Förderer *m* ABFALL, WASSER conveyor
Förder-: **Förderhöhe** *f* ABWASSER, WASSER delivery head, *Pumpen* discharge head; **Förderleistung** *f* ABFALL delivery, ABWASSER hauling capacity, transport capacity, delivery, WASSER *durch Pumpe* delivery, *einer Pumpe* hauling capacity; **Förderleitung** *f* WASSER penstock, delivery main, discharge piping; **Fördermenge** *f* ABFALL *einer Pumpe* pump capacity, ABWASSER, WASSER *einer Pumpe* hauling capacity, pump capacity; **Fördermöglichkeit** *f* ABWASSER transport capacity
fördern *vt* ABFALL discharge, ABWASSER *Fäkalienpumpe*, WASSER *durch Pumpe* deliver, *Pumpe* discharge
Förder-: **Förderpumpe** *f* BODEN conveyor pump; **Förderschnecke** *f* BODEN, LUFT screw conveyor
Förderung *f* ABFALL, ABWASSER offtake, MANAG output, POLITIK support, WASSER offtake
Form *f* ABFALL form, shape, *Gußform* mold (*AmE*), mould (*BrE*), WASSER form, shape; **Formaktivkohle** *f* LUFT chunk-form activated carbon
Formaldehyd *m* LUFT formaldehyde

formen *vt* ABFALL mold (*AmE*), mould (*BrE*)
Form: **Formerkennungsverfahren** *nt* ABFALL *Sortieren von Kunststoffen* shape recognition methods; **Formfülltechnik** *f* ABFALL mold-filling technique (*AmE*), mould-filling technique (*BrE*); **Formpressen** *nt* ABFALL compression molding (*AmE*), compression moulding (*BrE*); **Formstabilität** *f* LUFT *von Filtern* form stability; **Formstück** *nt* ABFALL, WASSER pipe fitting; **Formteile** *nt pl* **unter Verwendung von Recyclingrohstoffen** ABFALL molded parts using recycled plastics for core material (*AmE*), moulded parts using recycled plastics for core material (*BrE*)
Formungstemperatur *f* ABFALL forming temperature
fortgeschritten: **~e Abwasserbehandlung** *f* ABWASSER advanced wastewater treatment, tertiary wastewater treatment
fortleiten *vt* LÄRM *Körperschall* conduct
Fortpflanzung *f*: **~ von Schall** LÄRM noise propagation
fossil: **~ befeuert** *adj* ERNEU ENER fossil-fuel-fired; **~ befeuerter Dampferzeuger** *m* ERNEU ENER fossil-fuel-fired steam generator; **~ betrieben** *adj* ERNEU ENER fossil-fuel-driven; **~ betriebene Wärmepumpe** *f* ERNEU ENER fossil-fuel-driven heat pump; **~er Brennstoff** *m* ERNEU ENER fossil fuel, LUFT combustible fossil fuel; **~er Energieträger** *m* ERNEU ENER fossil energy source; **~e Verbrennung** *f* LUFT fossil fuel combustion
Fourier-Transformation *f* ANALYTIK Fourier-transform
FPD *abbr* (*flammenphotometrischer Detektor*) ANALYTIK FPD (*flame-photometric detector*)
Fracht *f* ABWASSER load; **Frachtbehälter** *m* MEER scoop; **Frachtkosten** *pl* ABFALL freight rate; **Frachttarif** *m* ABFALL freight rate
Fraktion *f* ANALYTIK, BODEN, LUFT fraction
fraktionieren *vt* ANALYTIK, BODEN, LUFT fractionate
fraktionierend: **~e Destillation** *f* ANALYTIK fractional distillation
Fraktionierung *f* BODEN, LUFT fractionation
Fraktion: **Fraktionsabscheidegrad** *m* LUFT degree of fraction collection, degree of fraction removal; **Fraktionsabscheider** *m* ABFALL fraction separator; **Fraktionsabscheidung** *f* ABFALL fraction separation; **Fraktionstrennung** *f* ABFALL fraction separation
Frank: **~ effect level** *nt* (*FEL*) SICHERHEIT *NOAEL* Frank effect level (*FEL*)
Fraunhofer: **~sche Linien** *f pl* ANALYTIK Fraunhofer lines
frei: **~es Cyanid** *nt* ABWASSER free cyanide; **~es Elektron** *nt* LUFT free electron; **~es Grundwasser** *nt* WASSER unconfined groundwater, unconfined water; **~es Radikal** *nt* LUFT free radical; **~er Zufluß** *m* WASSER free inflow; **~er Zufluß** *m* **über einen Hochbehälter** WASSER free inflow via an elevated reservoir; **~er Zufluß** *m* **über einen Wasserturm** WASSER free inflow via a water tower
Frei: **Freifeld** *nt* LÄRM free field; **Freilandmessung** *f* ANALYTIK open air measurement; **Freiräumer** *m* ABFALL clearance slide; **Freiraumschieber** *m* ABFALL, WASSER clearance slide
freisetzen *vt* ALLG release
Freisetzung *f* ALLG *von Schadstoffen* release
Frei: **Freispiegelleitung** *f* ABWASSER gravity sewer
freistehend: **~er Turm** *m* ERNEU ENER *Windturbine* free-standing tower
Frei: **Freiwasser** *nt* WASSER surplus water
freiwillig: **~e Selbstverpflichtung** *f* POLITIK voluntary commitment; **~e Vereinbarung** *f* POLITIK voluntary agreement
Fremd- *in cpds* ABFALL, ABWASSER, LUFT, WASSER external, foreign; **Fremdemissionen** *f pl* LUFT foreign emissions; **Fremdfarbanteil** *m* ABFALL *Altglaseinsammlung* share of different colored glass (*AmE*), share of different coloured glass (*BrE*); **Fremdgeräusch** *nt* LÄRM extraneous noise; **Fremdquelle** *f* LUFT foreign source; **Fremdstoff** *m* ABFALL *Recycling* foreign substance; **Fremdstoffniederschlag** *m* LUFT contamination fallout; **Fremdteil** *m* ABFALL *Recycling* foreign material; **Fremdwasser** *nt* ABWASSER external water; **Fremdzufuhr** *f* ABFALL, LUFT, WASSER external input
Freon: **~ 12** *nt* (*Dichlordifluormethan*) LUFT Freon 12 (*dichlorodifluoromethane*)
Frequenz *f* ANALYTIK, LÄRM frequency; **Frequenzanalysator** *m* ANALYTIK frequency analyser (*BrE*), frequency analyzer (*AmE*); **Frequenzbereich** *m* LÄRM, MEER frequency range; **Frequenzspektrum** *nt* LÄRM frequency spectrum
Fresnel: **Fresnellinse** *f* ERNEU ENER Fresnel lens; **~sche Linse** *f* ERNEU ENER Fresnel lens
Freunde: **~ der Erde** *m pl* POLITIK Friends Of The Earth
Friedrichs-Gaswaschflasche *f* ANALYTIK Friedrichs's gas-washing bottle
Frisch- *in cpds* LUFT fresh; **Frischkoks** *m* LUFT fresh coke; **Frischkompost** *m* ABFALL fresh compost; **Frischschlamm** *m* ABWASSER untreated sludge; **Frischwasser** *nt* WASSER fresh water; **Frischwasserzufuhr** *m* WASSER freshwater supply
Frontschaufellader *m* MEER front-end loader
FTIR *abbr* (*Infrarotspektrometrie mit Fourier-Transformation*) ANALYTIK *Peroxoacetylnitrat* FTIR (*Fourier-transform infrared spectrometry*); **FTIR-Spektrometer** *nt* ANALYTIK FTIR spectrometer; **FTIR-Spektrometrie** *f* ANALYTIK FTIR spectrometry
fühlen *vt* ANALYTIK feel
Füll- *in cpds* ABFALL, LUFT, WASSER filling; **Füllart** *f* **der Deponie** ABFALL tip filling method (*BrE*)
füllen *vt* ABFALL, WASSER fill
Fullererde *f* LUFT *Adsorbens* Fuller's earth
Füllkörper *m* ABFALL, LUFT packing; **Füllkörperhöhe** *f* LUFT packed-bed height; **Füllkörperkolonne** *f* LUFT packed-bed column, WASSER packed column; **Füllkörperkolonnenabscheider** *m* LUFT packed-tower scrubber, *Naßwäscher* packed-column scrubber; **Füllkörpersäulenabscheider** *m* LUFT packed-tower scrubber, *Naßwäscher* packed-column scrubber; **Füllkörperschüttung** *f* LUFT packing; **Füllkörperwäscher** *m* LUFT packed-bed scrubber
Füll-: **Füllstoff** *m* BODEN filler, bulking agent, *von gereinigten Böden* fill material
Füllung *f*: **~ mit schallschluckenden Stoffen** LÄRM sound-absorbent material filler
Füll-: **Füllvorgang** *m* ABFALL filling operation
Fumarole *f* ERNEU ENER *geothermische Energie* fumarole
Funken *m* LUFT spark; **Funken-AAS** *f* ANALYTIK *Schwermetallbestimmung* spark-ablation atomic absorption spectrometry; **Funkenableiter** *m* LUFT spark arrestor; **Funkenfänger** *m* LUFT spark arrestor; **Funkenflug** *m* LUFT flying sparks
Funktion *f* SICHERHEIT function; **Funktionssicherheit** *f* SICHERHEIT function safety; **Funktionstüchtigkeit** *f* ABFALL *der Dichtungsabdichtungssysteme* function reliability; **Funktionsüberwachung** *f* SICHERHEIT function monitoring
Furchentechnik *f* ABWASSER ridge-and-furrow technique
Fusionsprozeß *m* ERNEU ENER *Sonnenenergie* fusion process
Fußröhre *f* WASSER tail pipe

G

g *abbr* (*Erdbeschleunigung*) LUFT g (*acceleration due to gravity*)

Gabel: **Gabelpunkt** *m* WASSER *im Rohr* bifurcation point; **Gabelrohr** *nt* LUFT breeches pipe; **Gabelstapler** *m* ABFALL fork lift truck

GAK *abbr* (*granulierte Aktivkohle*) ABWASSER, LUFT, WASSER GAC (*granulated activated carbon*)

Galvanik- *in cpds* WASSER galvanic; **Galvanikabfälle** *m pl* ABFALL galvanic wastes; **Galvanikschlamm** *m* ABFALL, ABWASSER, WASSER galvanic sludge

Galvanometer *nt* ANALYTIK galvanometer

Gamma- *in cpds* ABWASSER gamma; **Gammabestrahlung** *f* ABWASSER gamma-ray irradiation; **Gammastrahlenbestrahlung** *f* ABFALL gamma-ray irradiation

Gär- *in cpds* ERNEU ENER fermentation

gären 1. *vt* ABFALL, ABWASSER, ERNEU ENER, WASSER ferment; **2.** *vi* ABFALL, ABWASSER, ERNEU ENER, WASSER ferment

gärfähig *adj* ABFALL, ABWASSER, ERNEU ENER fermentable

Gär-: **Gärfaulverfahren** *nt* ABWASSER filtration putrefaction process

Garn *nt* LUFT *Gewebefiltration* yarn

Gär-: **Gärprobe** *f* ANALYTIK fermentation test; **Gärturm** *m* ABWASSER fermentation tower

Gärung *f* ABWASSER digestion

gärungserregend *adj* ABFALL, ABWASSER, ERNEU ENER, WASSER fermentative

Gärung: **Gärungsgas** *nt* ABFALL, ABWASSER, WASSER fermentation gas

gärungshemmend *adj* ABWASSER anti-fermentative

Gas *nt* ALLG gas; **Gase** *nt pl* **und Dämpfe** *m pl* SICHERHEIT gases and fumes; **Gasabkühlung** *f* LUFT gas cooling; **Gasabsaugung** *f* ABFALL gas suction; **Gasabscheidung** *f* LUFT gas scrubbing; **Gasabsorption** *f* LUFT gas absorption; **Gasabsorptionsmittel** *nt* LUFT gas absorbent; **Gasanalysator** *m* ANALYTIK gas analyser (*BrE*), gas analyzer (*AmE*)

Gasanalyse *f* ANALYTIK gas analysis; **Gasanalyseapparat** *m* **nach Orsat** ANALYTIK Orsat apparatus, Orsat gas analyser (*BrE*), Orsat gas analyzer (*AmE*); **Gasanalysengerät** *nt* ANALYTIK gas analyser (*BrE*), gas analyzer (*AmE*)

Gas: **Gasaufkommen** *nt* ABFALL gas formation; **Gasaustausch** *m* ABFALL, ABWASSER, WASSER gas transfer; **Gasaustritt** *m* ABFALL gas discharge, LUFT gas outlet; **Gasbehälter** *m* ABWASSER gas container, ANALYTIK gasometer; **Gasbeschleunigung** *f* LUFT *Venturidüse* gas acceleration

Gasbildung *f* ABFALL *in Deponien* gas build-up; **Gasbildungspotential** *nt* ABFALL gas build-up potential, ANALYTIK *Deponiegas* gas formation potential; **Gasbildungsrate** *f* ABFALL, WASSER gas build-up rate

Gas: **Gasblase** *f* ABWASSER gas bubble; **Gasbrunnen** *m* ABFALL gas well; **Gasbrunnenkette** *f* ABFALL gas well chain; **Gasbürette** *f* ANALYTIK gas burette; **Gaschromatograph** *m* (*GC*) ANALYTIK gas chromatograph (*GC*)

Gaschromatographie *f* (*GC*) ANALYTIK gas chromatography (*GC*); **~ mit Flammenionisationsdetektion** (*GC-FID*) ANALYTIK gas chromatography with flame ionization detection (*GC-FID*); **~ mit flammenphotometrischer Detektion** (*GC-FPD*) ANALYTIK gas chromatography with flame photometric detection (*GC-FPD*); **~ in Kopplung mit Elektroneneinfangsdetektoren** (*GC-ECD*) ANALYTIK gas chromatography coupled with electron capture detection (*GC-ECD*); **~ in Kopplung mit Flammenionisationsdetektoren** ANALYTIK gas chromatography coupled with flame ionization detectors; **~ in Kopplung mit Massenspektrometern** ANALYTIK gas chromatography coupled with mass spectrometers; **~ mit Massenspektrometer** (*GC-MS*) ANALYTIK gas chromatography with mass spectrometer (*GC-MS*); **~ mit Massenspektrometrie** (*GC-MS*) ANALYTIK gas chromatography with mass spectrometry (*GC-MS*); **~ mit Photoionisationsdetektion** (*GC-PID*) ANALYTIK photoionization gas chromatography; **Gaschromatographie-Detektor** *m* (*GCD*) ANALYTIK gas chromatography detector (*GCD*); **Gaschromatographie-Stickstoff/Phosphor-Detektion** *f* (*GC-NPD*) ANALYTIK gas chromatography nitrogen/phosphorus detection (*GC-NPD*)

gaschromatographisch *adj* ANALYTIK gas-chromatographic; **~e Trennung** *f* ANALYTIK gas-chromatographic separation

Gas: **Gas- und Dampfanlage** *f* (*GuD-Anlage*) LUFT combined-cycle gas turbine (*CCGT*); **Gasdispergierung** *f* LUFT gas dispersion; **Gasdränschicht** *f* ABFALL gas drainage layer

Gasdurchfluß *m* LUFT gas flow; **Gasdurchflußgeschwindigkeit** *f* LUFT gas flow velocity; **Gasdurchflußzähler** *m* ANALYTIK gas flow counter

gasdurchlässig *adj* ABFALL, WASSER gas-permeable

Gas: **Gasdurchsatz** *m* LUFT gas throughput, gas thruput (*AmE*); **Gasdurchtritt** *m* ABFALL *bei Deponien* gas permeability; **Gaseinschlüsse** *m pl* LUFT gas entrainment; **Gaseintritt** *m* LUFT gas inlet; **Gasemissionen** *f pl* ABFALL, ERNEU ENER, LUFT gas emissions

gasempfindlich: **~e Elektrode** *f* ANALYTIK gas-sensitive electrode

Gas: **Gasentschwefelung** *f* ABFALL, LUFT gas desulfurization (*AmE*), gas desulphurization (*BrE*); **Gaserfassung** *f* LÄRM, LUFT gas capture; **Gaserfassung** *f* **außerhalb der Deponie** ABFALL gas collection outside of landfill; **Gas-Feststoff-Chromatographie** *f* (*GSC*) ANALYTIK gas-solid chromatography (*GSC*); **Gasflußmesser** *m* ANALYTIK gas flow meter

gasförmig *adj* ABFALL, ERNEU ENER, LUFT, WASSER gaseous; **~er Aggregatzustand** *m* LUFT gaseous state of aggregation; **~es Arbeitsmittel** *nt* ERNEU ENER *Wärmepumpe* gaseous working medium; **~er Brennstoff** *m* LUFT gaseous fuel; **~e Emissionen** *f pl* ABFALL, LUFT gaseous emissions; **~e Luftverunreinigung** *f* LUFT gaseous air pollution; **~es Medium** *nt* ABFALL, LUFT, WASSER gaseous medium; **~e Schadstoffe** *m pl* LUFT gaseous pollutants, gaseous contaminants; **~es Verbrennungsprodukt** *nt* ABFALL, LUFT gaseous combustion product

gasgefüllt *adj* ANALYTIK gas-filled; **~er Detektor** *m* ANALYTIK gas-filled detector

Gas: **Gasgemisch** *nt* ANALYTIK gas mixture; **Gasgeschwindigkeit** *f* LUFT gas velocity; **Gasgewinnung** *f* **aus Mülldeponien** ABFALL landfill gas extraction; **Gasgravimetrie** *f* ANALYTIK gravimetric gas analysis; **Gashaube** *f* ABWASSER gas hood; **Gasinterferometer** *nt* ANALYTIK gas interferometer; **Gaskollektor** *m* ABFALL gas collector; **Gaskomponente** *f* ABFALL *Deponiegas*, LUFT, WASSER gas component; **Gaskonzentration** *f* ABFALL gas concentration; **Gaskühler** *m* LUFT gas cooler; **Gaskühlung** *f* LUFT gas cooling; **Gasleitung** *f* BODEN gas soil pipe; **Gaslöslichkeit** *f* LUFT gas solubility; **Gasmaske** *f* SICHERHEIT gas mask; **Gasmenge** *f* ANALYTIK gas quantity, LUFT amount of gas; **Gasmesser** *m* ANALYTIK gas meter; **Gasmeßgerät** *nt* ANALYTIK gas measuring device, gas measurement device;

Gasmessung *f* ANALYTIK gas measuring, gas measurement; **Gasmigration** *f* ABFALL, BODEN gas migration; **Gasmischer** *m* ANALYTIK gas mixer; **Gasmischkammer** *f* ANALYTIK gas-mixing chamber; **Gasmolekül** *nt* LUFT gas molecule; **Gasnutzungsanlage** *f* ABFALL gas use plant

Gasometer *nt* ABWASSER, ANALYTIK gasometer

***Gas*: Gaspermeation** *f* LUFT gas permeation

Gasphase *f* LUFT gas phase; **Gasphasenoxidation** *f* ABWASSER gas-phase oxidation; **Gasphasentitration** *f* (*GPT*) ANALYTIK gas-phase titration (*GPT*); **Gasphasentitrator** *m* ANALYTIK gas-phase titrator

***Gas*: Gaspipette** *f* ANALYTIK gas sampling pipette

Gasprobe *f* ANALYTIK gas sample, gas sampling; **Gasprobenahmerohr** *nt* ANALYTIK gas sampling tube; **Gasprobenahmeventil** *nt* ANALYTIK gas sampling valve; **Gasprobenehmen** *nt* ANALYTIK gas sampling

***Gas*: Gasprüfer** *m* ANALYTIK, LUFT gas detector; **Gasraum** *m* LUFT gas compartment; **Gasreiniger** *m* LUFT gas purifier; **Gasrückgewinnung** *f* ABFALL, LUFT gas recovery; **Gasrückstand** *m* ABFALL, LUFT residual gas; **Gassammelbeutel** *m* ANALYTIK gas sampling bag; **Gassammelröhre** *f* ANALYTIK gas sampling pipette; **Gassättigung** *f* ANALYTIK gas saturation; **Gassättigungsmethode** *f* ANALYTIK *Dampfdruckbestimmung* gas saturation method; **Gasschacht** *m* ABFALL gas shaft; **Gasschmelze** *f* ABFALL gas melt

gassensitiv *adj* ANALYTIK gas-sensitive; **~e Elektrode** *f* ANALYTIK gas-sensitive electrode; **~e Elektrode *f* mit gaspermeabler Membrane** ANALYTIK gas-sensitive electrode with gas permeable membrane

***Gas*: Gasspeicher** *m* ABWASSER gas storage tank; **Gasspurenanalysator** *m* ANALYTIK gas trace analyser (*BrE*), gas trace analyzer (*AmE*); **Gasspürgerät** *nt* ANALYTIK gas detector; **Gasstrom** *m* ANALYTIK, LUFT gas flow, gas stream; **Gasströmungsgeschwindigkeit** *f* LUFT gas velocity of flow; **Gasströmungsrichtung** *f* LUFT direction of gas stream; **Gastemperatur** *f* LUFT gas temperature; **Gasturbine** *f* ABWASSER gas turbine

gasunspezifisch *adj* ANALYTIK *Meßprinzip* gas non-specific

***Gas*: Gasvermessung** *f* ANALYTIK gasometry; **Gasverteilung** *f* LUFT gas distribution; **Gasverteilungschromatographie** *f* ANALYTIK gas partition chromatography; **Gasverwertungsanlage** *f* ABFALL gas reuse plant; **Gasvolumetrie** *f* ANALYTIK volumetric gas analysis

gasvolumetrisch *adj* ANALYTIK gasometric; **~e Methode** *f* ANALYTIK gasometric method

***Gas*: Gasvorwärmer** *m* (*GAVO*) LUFT gas preheater; **Gaswärmeaustauscher** *m* LUFT gas heat exchanger; **Gaswarngerät** *nt* ANALYTIK gas detector, gas indicator

Gaswaschflasche *f* ANALYTIK gas-wash bottle, gas-washing bottle; **~ nach Drechsel** ANALYTIK Drechsel gas-washing bottle; **~ nach Friedrichs** ANALYTIK Friedrichs's gas-washing bottle; **~ mit Glasfritte** ANALYTIK gas-washing bottle with sintered head

***Gas*: Gaswaschturm** *m* LUFT scrubbing tower, scrubber tower; **Gaswasser** *nt* ABWASSER ammonia water, gas water; **Gaswolke** *f* LUFT gas cloud; **Gaszähler** *m* ANALYTIK gas meter; **Gaszusammensetzung** *f* LUFT gas composition; **Gaszutrittsmöglichkeit** *f* ABFALL, WASSER gas entrance possibility

Gate-Elektrode *f* ANALYTIK gate electrode

GAU *abbr* (*größter anzunehmender Unfall*) ABFALL, SICHERHEIT MCA (*maximum credible accident*)

Gauß- *in cpds* ANALYTIK Gaussian; **Gauß-Filter** *nt* ANALYTIK Gaussian filter; **Gauß-Modell** *nt* ANALYTIK Gaussian model

GAVO *abbr* (*Gasvorwärmer*) LUFT gas preheater

GC *abbr* ANALYTIK (*Gaschromatograph*) GC (*gas chromatograph*), (*Gaschromatographie*) GC (*gas chromatography*)

GCD *abbr* (*Gaschromatographie-Detektor*) ANALYTIK GCD (*gas chromatography detector*)

GC-ECD *abbr* (*Gaschromatographie in Kopplung mit Elektroneneinfangsdetektoren*) ANALYTIK GC-ECD (*gas chromatography coupled with electron capture detection*)

GC-FID *abbr* (*Gaschromatographie mit Flammenionisations-Detektion*) ANALYTIK GC-FID (*gas chromatography with flame ionization detection*)

GC-FPD *abbr* (*Gaschromatographie mit flammenphotometrischer Detektion*) ANALYTIK GC-FPD (*gas chromatography with flame photometric detection*)

***GC*: GC-Injektor** *m* ANALYTIK GC-injector

GC-MS *abbr* ANALYTIK (*Gaschromatographie mit Massenspektrometer*) GC-MS (*gas chromatography with mass spectrometer*), (*Gaschromatographie mit Massenspektrometrie*) GC-MS (*gas chromatography with mass spectrometry*)

GC-NPD *abbr* (*Gaschromatographie-Stickstoff/Phosphor-Detektion*) ANALYTIK GC-NPD (*gas chromatography nitrogen/phosphorus detection*)

GC-PID *abbr* (*Gaschromatographie mit Photoionisationsdetektion*) ANALYTIK photoionization gas chromatography

***GC*: GC-Thermodesorption** *f* ANALYTIK GC-thermodesorption

gealtert: **~er Katalysator** *m* LUFT aged catalyst

gebietsbezogen *adj* ALLG area-related; **~er Lärm** *m* LÄRM, MEER area-related noise

Gebläse *nt* LÄRM, LUFT fan, ventilator

gebogen: **~er Abwasserkanal** *m* ABFALL, ABWASSER, WASSER curved sewer

Gebot *nt* POLITIK obligation

gebrannt: **~er Kalk** *m* ABFALL, WASSER burnt lime

gebraucht *adj* ABFALL, ERNEU ENERWASSER used; **~es Aluminium** *nt* ABFALL used aluminium (*BrE*), used aluminum (*AmE*); **~e Flasche** *f* ABFALL, WASSER used bottle; **~es Produkt** *nt* ABFALL used product; **~e Verkaufsverpackung** *f* ABFALL used commercial packaging

Gebühr *f* MANAG charge

gebündelt *adj* ERNEU ENER concentrated; **~e Sonnenenergie** *f* ERNEU ENER concentrated solar energy

geeicht *adj* ANALYTIK calibrated; **~es Wasserrückhaltebecken** *nt* ANALYTIK, BODEN *Landwirtschaft* calibrated watershed

geerdet: **~e Wandung** *f* LUFT earthed wall (*BrE*), grounded wall (*AmE*)

Gefahr *f* SICHERHEIT danger, hazard; **~ durch Hochfrequenzwellen** SICHERHEIT radiowave hazard; **~ durch Laserstrahlung** SICHERHEIT laser radiation hazard; **~ durch Ultraschall** SICHERHEIT ultrasonic hazard

gefährden *vt* SICHERHEIT endanger

gefährdet *adj* SICHERHEIT endangered

Gefährdung *f* SICHERHEIT danger; **Gefährdungsabschätzung** *f* ABFALL, ABWASSER, ANALYTIK, BODEN, LUFT, MEER, MANAG, POLITIK, SICHERHEIT danger estimate, risk assessment, risk estimate; **Gefährdungshaftung** *f* POLITIK strict liability; **Gefährdungspotential** *nt* MANAG risk potential, SICHERHEIT danger potential

***Gefahr*: Gefahrenkennzeichnung** *f* SICHERHEIT identification of hazards; **Gefahrenverhütung** *f* MANAG, SICHERHEIT *in Gebäuden* hazard prevention; **Gefahrenvorsorge** *f* SICHERHEIT danger prevention; **Gefahrenzone** *f* SICHERHEIT danger zone; **Gefahrgüter** *nt pl* ABFALL dangerous goods; **Gefahrguttransport** *m* ABFALL transport of hazardous substances; **Gefahrgutumschlag** *m* ABFALL transshipment of hazardous goods

gefährlich *adj* ALLG dangerous; **~er Abfall** *m* ABFALL, ABWASSER, WASSER hazardous waste; **~e und anthropogene Störung *f* des Klimasystems** LUFT dangerous anthropogenic interference with the climate system; **~e Güter** *nt pl* ABFALL, WASSER dangerous goods; **~e Ladungen** *f pl* ABFALL, WASSER dangerous loads; **~e Situation** *f* ABFALL, WASSER dangerous situation; **~er Sonderabfall** *m* ABFALL,

WASSER dangerous hazardous waste; **~er Stoff** *m* ABWASSER hazardous substance

Gefährlichkeitsabschätzung *f* SICHERHEIT danger assessment

Gefahrstoff *m* SICHERHEIT dangerous material, dangerous substance, hazardous substance; **Gefahrstoffkataster** *nt* ABFALL, BODEN, LUFT, POLITIK, WASSER inventory of hazardous substances, register of hazardous substances; **Gefahrstofflager** *nt* SICHERHEIT storage of dangerous materials

Gefahr: **Gefahrsymbol** *nt* SICHERHEIT danger symbol

Gefälle *nt* WASSER slope, head, fall of water; **~ des Wasserspiegels** WASSER fall; **Gefälleleitung** *f* ABWASSER gravity sewer

Gefällsverlust *m* WASSER loss of head

Gefäß *nt* ANALYTIK, LUFT container, vessel

gefiltert *adj* ALLG filtered, filtrated; **~er Schlamm** *m* ABWASSER filtration sludge

geflanscht *adj* LUFT, MEER flanged; **~er Wärmeaustauscher** *m* LUFT flanged heat exchanger

Gefriergerät *nt* ABFALL freezer

Gegen- *in cpds* LUFT reverse; **Gegenbehälter** *m* WASSER online storage tank; **Gegendruck** *m* LUFT back pressure; **Gegenion** *nt* ABWASSER, WASSER counter-ion; **Gegenklopfmittel** *nt* LUFT antiknock agent, antiknock additive

Gegenstrom *m* LUFT *Gasströmung* reverse flow; **Gegenstromabscheider** *m* LUFT reverse-flow precipitator; **Gegenstrombelüfter** *m* ABWASSER countercurrent aerator; **Gegenstromextraktion** *f* ABWASSER, BODEN countercurrent extraction; **Gegenstromextraktor** *m* ABWASSER, BODEN countercurrent extractor; **Gegenstromfilter** *nt* ABWASSER reverse-flow filter, LUFT reverse-air-flow filter; **Gegenstromrechen** *m* ABWASSER countercurrent screen; **Gegenstromregeneration** *f* ABWASSER countercurrent regeneration (*CFR*); **Gegenstromröhrenleitung** *f* LUFT *Schlauchfilter* reverse-flow duct; **Gegenstromschüttschichtfilter** *nt* LUFT reverse-flow packed-bed filter; **Gegenstromspülung** *f* ABWASSER *bei Filtern* countercurrent washing, LUFT reverse-flow cleaning; **Gegenstromstripping** *nt* BODEN countercurrent stripping; **Gegenstromwäsche** *f* ABWASSER countercurrent washing; **Gegenstromwäscher** *m* LUFT reverse-flow scrubber; **Gegenstromzentrifuge** *f* ABWASSER countercurrent centrifuge

gegen: **~ den Wind** *adv* ERNEU ENER, LUFT upwind

Gehäuse *nt* ERNEU ENER *Solarkollektor* housing

gehemmt: **~es Absetzen** *nt* ABWASSER hindered settling

Gehör *nt* LÄRM hearing; **Gehörschädlichkeit** *f* LÄRM hearing damage

Gehörschutz *m* LÄRM hearing protection

Gehörschützer *m* LÄRM ear protector

Gehörschutz: **Gehörschutzhelm** *m* LÄRM hearing protection helmet, noise exclusion helmet; **Gehörschutzkappe** *f* LÄRM ear muffs; **Gehörschutzstöpsel** *m pl* LÄRM earplugs, hearing protection plugs; **Gehörschutzwatte** *f* LÄRM hearing protection padding

Geiger: **Geiger-Müllerzähler** *m* ANALYTIK Geiger-Müller counter (*GM counter*), contamination meter, Geiger counter; **Geigerzähler** *m* ANALYTIK Geiger-Müller counter (*GM counter*), contamination meter, Geiger counter

geklärt: **~es Abwasser** *nt* ABWASSER purified wastewater

gekoppelt: **~er Biowäscher** *m* ABWASSER, LUFT coupled biological scrubber

gekürzt: **~er Satz** *m* POLITIK reduced rate of charge

geladen *adj* ABWASSER, LUFT charged; **~er Gasstrom** *m* LUFT charged gas stream; **~es Ion** *nt* ABWASSER charged ion; **~es Polymer** *nt* ABWASSER charged polymer; **~es Teilchen** *nt* LUFT *Elektrofilter* charged particle

gelb: **~er Sack** *m* ABFALL yellow plastic collection bag; **~e Tonne** *f* ABFALL Yellow Container; **~er Wertstoffsack** *m* **für DSD-Abfälle** ABFALL yellow plastic liner bag for DSD wastes

Gelchromatographie *f* ANALYTIK gel chromatography

Gelelektrophorese *f* ANALYTIK gel electrophoresis

gelelektrophoretisch *adj* ANALYTIK gel electrophoretic

Geliermittel *nt* MEER gelling agent

gelocht *adj* ALLG *Rohr* perforated

gelöst *adj* ALLG dissolved; **~er anorganischer Kohlenstoff** *m* ABFALL, ANALYTIK, LUFT, WASSER dissolved inorganic carbon; **~er organischer Kohlenstoff** *m* (*DOC*) ABFALL, ANALYTIK, LUFT, WASSER dissolved organic carbon (*DOC*); **~e organische Substanz** *f* ABFALL, ANALYTIK, LUFT, WASSER dissolved organic matter; **~er Sauerstoff** *m* ABFALL, LUFT, WASSER dissolved oxygen (*DO*); **~e Stoffe** *m pl* ABFALL, ABWASSER, WASSER dissolved solids, dissolved substances

Gelpermeations-Chromatographie *f* (*GPC*) ANALYTIK *LOAEL* gel permeation chromatography (*GPC*)

gemeinschaftlich: **~ stranggepreßte Folie** *f* ABFALL co-extruded film

gemindert *adj* LÄRM reduced

Gemischregler *m* LUFT fuel-air ratio control unit

gemischt *adj* ALLG mixed; **~e Ablagerung** *f* ABFALL co-deposition, co-disposal; **~e Kunststoffe** *m pl* ABFALL mixed plastics; **~er Primärschlamm** *m* ABWASSER mixed primary sludge

Gemischtwasserleitung *f* ABWASSER combined sewer

Genehmigung *f* ABFALL, LUFT, POLITIK, WASSER authorization, licence (*BrE*), license (*AmE*), permission, permit; **Genehmigungsantrag** *m* POLITIK, WASSER authorization application, licensing application, permit application

genehmigungsbedürftig *adj* POLITIK requiring a licence (*BrE*), requiring a license (*AmE*), requiring authorization; **~e Anlage** *f* POLITIK plant requiring permit, plant requiring licence (*BrE*), plant requiring license (*AmE*)

Genehmigung: **Genehmigungsbehörde** *f* POLITIK licensing authority; **Genehmigungserteilung** *f* POLITIK granting of permit; **Genehmigungskonferenz** *f* POLITIK licensing conference; **Genehmigungsrecht** *nt* POLITIK licensing law; **Genehmigungsverfahren** *nt* POLITIK licensing process, authorization process, permit process; **Genehmigungsvoraussetzung** *f* POLITIK conditions for licensing, conditions for authorization; **Genehmigungsvorbehalt** *m* POLITIK reservation of permission; **Genehmigungswiderruf** *m* POLITIK repeal of licence (*BrE*), repeal of license (*AmE*), repeal of authorization, repeal of permit; **Genehmigungswirkung** *f* POLITIK authorization impact, permit impact, licensing impact

Generator *m* ANALYTIK, ERNEU ENER generator; **Generatorgehäuse** *nt* LÄRM generator house

genetisch *adj* ANALYTIK, SICHERHEIT genetic; **~ wichtige Dosis** *f* SICHERHEIT genetically significant dose

geologisch: **~e Barriere** *f* ABFALL geological barrier; **~e Bedingungen** *f pl* ERNEU ENER geological conditions

geophysikalisch: **~e Standortuntersuchung** *f* ANALYTIK geophysical site analysis

geordnet: **~es Ablagern** *nt* ABFALL landfilling; **~e Ablagerung** *f* ABFALL controlled dumping, controlled tipping *BrE*, proper disposal, sound disposal; **~e Beseitigung** *f* ABFALL safe disposal; **~e Deponie** *f* ABFALL sanitary landfill; **~e Deponierung** *f* ABFALL sanitary landfilling; **~e Hochdeponie** *f* ABFALL controlled high-level landfill

geothermal: **~e Energie** *f* ABFALL, ERNEU ENER, LUFT geothermal energy

Geothermie *f* ABFALL, ERNEU ENER, LUFT geothermal energy

Geothermik *f* ERNEU ENER geothermics

geothermisch *adj* ERNEU ENER geothermal, geothermic; **~e Energie** *f* ERNEU ENER geothermal energy; **~es Energiepotential** *nt* ERNEU ENER geothermal energy potential; **~e Energieressourcen** *f pl* ERNEU ENER geothermal energy resources; **~es Heizwerk** *nt* ERNEU ENER geothermal heating plant; **~es Kraftwerk** *nt* ERNEU ENER geothermal power plant; **~e Lagerstätte** *f* ERNEU ENER geothermal deposit; **~es Reservoir** *nt* ERNEU ENER geothermal reservoir; **~e Ressourcen** *f pl* ERNEU ENER geothermal resources; **~e Stromerzeugung** *f* ERNEU ENER geothermal electrical power generation; **~e Tiefenstufe** *f* ERNEU ENER geothermal gradient

gepackt: **~e Säule** *f* ANALYTIK packed column

gepumpt *adj* ABFALL, WASSER pumped; **~er Wasserstrom** *m* ABFALL, WASSER pumped water flow

Gerät *nt* ALLG device, equipment; **Geräteeichung** *f* ANALYTIK apparatus calibration; **Geräteselbsttest** *m* ANALYTIK instrument self test

Geräusch *m* LÄRM, SICHERHEIT noise; **Geräuschanalyse** *f* LÄRM noise analysis; **Geräuschart** *f* LÄRM noise type; **Geräuschbekämpfung** *f* LÄRM noise abatement; **Geräuschemissionen** *f pl* LÄRM noise emissions; **Geräuschemittent** *m* LÄRM noise emitter

geräuschintensiv *adj* LÄRM noise-intensive

geräuschlos *adj* LÄRM noiseless; **~er Motor** *m* LÄRM, SICHERHEIT noiseless motor

Geräusch: **Geräuschminderung** *f* LÄRM noise reduction; **Geräuschpegel** *m* LÄRM noise level; **Geräuschpegelanzeiger** *m* LÄRM noise level indicator

geräuschunterdrückend *adj* LÄRM, MEER, SICHERHEIT antinoise

gereckt *adj* ABFALL oriented; **~e Kunststoffe** *m pl* ABFALL oriented plastics; **~e Polypropylenfolie** *f* ABFALL oriented polypropylene film

geregelt: **~e Ablagerung** *f* ABFALL regulated deposition; **~er Katalysator** *m* LUFT computer-controlled catalytic converter

gereinigt *adj* ALLG decontaminated, cleaned; **~e Abluft** *f* LUFT cleaned waste air; **~er Boden** *m* BODEN decontaminated soil; **~er Faserbrei** *m* ABFALL cleaned pulp fibre (*BrE*), cleaned pulp fiber (*AmE*); **~es Gas** *nt* LUFT clean gas; **~e Scherben** *f pl* ABFALL cleaned glass cullet

gering: **~e Bodendurchlässigkeit** *f* BODEN low-soil permeability

geringpermeabel: **geringpermeabler Grundwasserleiter** *m* ABFALL, WASSER aquiclude, aquitard

Gerinne *nt* WASSER flume

Geruch *m* ALLG odor (*AmE*), odour (*BrE*)

Geruch: **~ beseitigen** *vi* ABFALL, ABWASSER, LUFT, WASSER deodorize

geruchlos *adj* ALLG odorless (*AmE*), odourless (*BrE*); **~es Gas** *nt* LUFT odorless gas (*AmE*), odourless gas (*BrE*)

Geruch: **Geruchsbekämpfung** *f* ABFALL, ABWASSER, LUFT, WASSER odor control (*AmE*), odour control (*BrE*), deodorization

geruchsbeladen *adj* LUFT odor-laden (*AmE*), odour-laden (*BrE*); **~e Abluft** *f* LUFT odor-laden waste air (*AmE*), odour-laden waste air (*BrE*); **~es Rohgas** *nt* LUFT odor-laden dirty gas (*AmE*), odour-laden dirty gas (*BrE*)

geruchsbelästigend: **~es Abwasser** *nt* ABWASSER, LUFT malodorous wastewater; **~er Stoff** *m* LUFT malodorous substance

Geruch: **Geruchsbelästigung** *f* ABFALL, ABWASSER, LUFT, WASSER odor impairment (*AmE*), odour impairment (*BrE*), odor nuisance (*AmE*), odour nuisance (*BrE*), odor pollution (*AmE*), odour pollution (*BrE*); **Geruchsbeseitigung** *f* ABFALL, ABWASSER, WASSER deodorization; **Geruchsbestimmung** *f* ANALYTIK odor analysis (*AmE*), odour analysis (*BrE*), odor determination (*AmE*), odour determination (*BrE*); **Geruchsbestimmung** *f* **bei Abwässern** ANALYTIK odor analysis in wastewater (*AmE*), odour analysis in wastewater (*BrE*)

geruchsbildend *adj* ABFALL, ABWASSER, WASSER odor-forming (*AmE*), odour-forming (*BrE*); **~e Substanz** *f* ABWASSER odor-forming substance (*AmE*), odour-forming substance (*BrE*)

Geruch: **Geruchsbildung** *f* ABFALL, ABWASSER, WASSER odor formation (*AmE*), odour formation (*BrE*); **Geruchsemission** *f* ABFALL, ABWASSER, LUFT, WASSER odor emission (*AmE*), odour emission (*BrE*); **Geruchsentfernung** *f* LUFT odor removal (*AmE*), odour removal (*BrE*); **Geruchsentstehung** *f* ABWASSER, LUFT odor source (*AmE*), odour source (*BrE*)

geruchsfrei *adj* ABFALL, WASSER odor-free (*AmE*), odour-free (*BrE*)

geruchsintensiv *adj* ALLG odor-intensive (*AmE*), odour-intensive (*BrE*); **~es Lösungsmittel** *nt* LUFT odor-intensive solvent (*AmE*), odour-intensive solvent (*BrE*); **~er Stoff** *m* LUFT odor-intensive substance (*AmE*), odour-intensive substance (*BrE*)

Geruch: **Geruchsmessung** *f* ANALYTIK olfactometry, odorimetry; **Geruchsproblem** *nt* ABWASSER odor problem (*AmE*), odour problem (*BrE*); **Geruchsprüfung** *f* ANALYTIK odor test (*AmE*), odour test (*BrE*)

Geruchsschwelle *f* ABFALL, ABWASSER, WASSER threshold odor (*AmE*), threshold odour (*BrE*); **Geruchsschwellenkonzentration** *f* ABFALL, ABWASSER, ANALYTIK, WASSER threshold odor concentration (*AmE*), threshold odour concentration (*BrE*); **Geruchsschwellenwert** *m* (*GSW*) ABFALL, ABWASSER, ANALYTIK threshold odor number (*AmE*) (*TON*), threshold odour number (*BrE*) (*TON*)

Geruch: **Geruchsstoff** *m* ABFALL, LUFT, WASSER odorous substance, odoriferous substance

geruchsstoffbeladen: **~e Luft** *f* LUFT odor-substance-laden air (*AmE*), odour-substance-laden air (*BrE*)

Geruch: **Geruchsstoffemissionen** *f pl* LUFT odoriferous substance emissions; **Geruchsträger** *m* LUFT odoriphore, osmophore; **Geruchszahl** *f* ANALYTIK odor quotient (*AmE*), odour quotient (*BrE*)

gesamt *adj* ALLG total; **~er anorganisch gebundener Kohlenstoff** *m* (*TIC*) ANALYTIK total inorganic carbon (*TIC*); **~er organisch gebundener Kohlenstoff** *m* (*TOC*) ANALYTIK total organic carbon (*TOC*); **~er vorhandener Kohlenstoff** *m* (*TC*) ANALYTIK total carbon (*TC*)

Gesamt: **Gesamtabfallaufkommen** *nt* ABFALL total appearance of waste; **Gesamtablagerung** *f* ABFALL, BODEN, LUFT, WASSER total deposition; **Gesamtabscheidegrad** *m* LUFT degree of total removal; **Gesamtabscheidung** *f* LUFT total collection, total removal; **Gesamtanalyse** *f* ANALYTIK total analysis; **Gesamtbakterienzahl** *f* WASSER total bacterial count; **Gesamtchlor** *nt* ABFALL total chlorine; **Gesamtdurchfluß** *m* ANALYTIK *bei jeder Einzelmessung* total flow; **Gesamtfallhöhe** *f* ABWASSER, WASSER total head; **Gesamtfluor** *nt* ABFALL total fluorine; **Gesamtgasmenge** *f* ABFALL, ANALYTIK total gas quantity; **Gesamtgeräusch** *nt* LÄRM total noise; **Gesamtgewässergüte** *f* WASSER overall water quality; **Gesamthärte** *f* (*GH*) ABWASSER, WASSER total hardness; **Gesamthärtebestimmung** *f* **von Abwasser** ABWASSER, ANALYTIK, WASSER determination of total hardness in wastewater; **Gesamtkeimzahl** *f* WASSER total germ number; **Gesamtkontamination** *f* ABWASSER total contamination; **Gesamtkosten** *pl* MANAG total costs; **Gesamtleistung** *f* ERNEU ENER total capacity; **Gesamtluftqualität** *f* LUFT total air quality; **Gesamtmenge** *f* **ungelöster Schwebstoffe** ABWASSER total undissolved suspended solids (*TSS*); **Gesamtmüll** *m* ABFALL total refuse; **Gesamtpapierverbrauch** *m* ABFALL total paper consumption; **Gesamtrisiko** *nt* SICHERHEIT total risk; **Gesamtrückstand** *m* ANALYTIK total residue; **Gesamtschadstoffmenge** *f* ANALYTIK total pollutant emissions; **Gesamtschlammanfall** *m* ABWASSER total sludge yield

gesamtschuldnerisch: **~e Haftung** *f* ABFALL, POLITIK, WASSER joint and several liability

Gesamt: **Gesamtschwebestoffe** *m pl* WASSER total dissolved solids (*TDS*); **Gesamtschwefel** *m* ABFALL total sulfur (*AmE*), total sulphur (*BrE*); **Gesamtstaub** *m* ANALYTIK, LUFT total dust; **Gesamtstaubgehalt** *m* LUFT total dust content; **Gesamtvolumen** *nt* ANALYTIK total volume; **Gesamtwasserbedarf** *m* WASSER total water requirements; **Gesamtwirkungsgrad** *m* ERNEU ENER total efficiency
gesättigt *adj* ALLG saturated; **~e Bodenzone** *f* BODEN saturated soil zone; **~es Gas** *nt* ANALYTIK saturated gas; **~e Zone** *f* ABFALL, WASSER *Untergrund* saturated zone
geschäumt: **~es Polysterol** *nt* ABFALL expanded polystyrene
geschlossen *adj* MEER sealed; **~e Decke** *f* LÄRM sealed ceiling; **~er Faulbehälter** *m* ABWASSER covered digester; **~e Grubendeponie** *f* ABFALL closed pit landfill; **~er Kanal** *m* ABFALL, ABWASSER, WASSER closed canal; **~e Kompostierung** *f* ABFALL mechanical composting, rapid fermentation; **~er Kreislauf** *m* ABWASSER *Kreislaufführung von Abwasser* closed loop, pipe-to-pipe recycling, ERNEU ENER closed cycle (*CC*); **~er Kreislauf-Naßwäscher** *m* LUFT closed-loop wet scrubbing; **~er Rankine-Kreislauf** *m* ERNEU ENER *Meereswärmeenergie* closed Rankine cycle; **~es System** *nt* ABFALL, WASSER closed system; **~er Tropfkörper** *m* ABWASSER covered trickling filter; **~e Wand** *f* LÄRM sealed wall
geschluckt *adj* ALLG absorbed
geschmacksfrei *adj* LUFT *Gas* tasteless
geschmacksneutral *adj* LUFT *Gas* tasteless
Geschmacksstoff *m* WASSER flavor substance (*AmE*), flavour substance (*BrE*)
geschmolzen: **~es Metall** *nt* SICHERHEIT molten metal
geschreddert *adj* ABFALL shredded; **~e Kunststoffe** *m pl* ABFALL shredded plastics
Geschwindigkeit *f* ALLG velocity; **Geschwindigkeitshöhe** *f* ABFALL, ABWASSER, WASSER velocity head; **Geschwindigkeitsmesser** *m* ANALYTIK speedometer
Gesetz *nt* POLITIK law, statute, act; **~ über Vermeidung und Entsorgung von Abfällen** ABFALL, ABWASSER, POLITIK Waste Avoidance and Management Act; **~ zur Wasserhaushaltung** MEER Clean Water Act (*CWA*); **Gesetzbuch** *nt* POLITIK statute book
gesetzeskonform *adj* POLITIK conforming to the law
Gesetz: **Gesetzgeber** *m* POLITIK legislator
gesetzlich *adj* POLITIK statutory, legal; **~e Auflagen** *f pl* POLITIK legal requirements; **~es Instrument** *nt* POLITIK legal instrument; **~e Vorschrift** *f* POLITIK, SICHERHEIT statutory regulation
gespannt: **~es Grundwasser** *nt* WASSER confined groundwater; **~er Grundwasserleiter** *m* WASSER confined aquifer
gespeichert *adj* ERNEU ENER stored; **~e Erdwärme-Energie** *f* ERNEU ENER stored geothermal energy; **~e Sonnenenergie** *f* ERNEU ENER stored solar energy; **~e Wärme** *f* ERNEU ENER stored heat; **~es Wasser** *nt* ERNEU ENER stored water
gestaltlos *adj* LUFT amorphous, noncrystalline
Gesteinsschicht *f* BODEN rock bed
gestört: **~e Probe** *f* ANALYTIK disturbed sample
gestrichen: **~es Papier** *nt* ABFALL coated paper
gestuft: **~e Brennstoffdosierung** *f* LUFT staged fuel feeding
Gesundheit *f* SICHERHEIT health
gesundheitsbezogen *adj* SICHERHEIT health-related; **~e Probenentnahme** *f* SICHERHEIT health-related sampling
Gesundheit: **Gesundheitsgefahr** *f* SICHERHEIT health risk, health hazard
gesundheitsgefährdend *adj* SICHERHEIT health-threatening
Gesundheit: **Gesundheitsrisiko** *nt* SICHERHEIT health risk, health hazard; **Gesundheitsschaden** *m* SICHERHEIT damage to health
gesundheitsschädlich *adj* SICHERHEIT harmful to health
Gesundheit: **Gesundheits- und Sicherheitsbedingungen** *f pl* SICHERHEIT health and safety requirements
Getränkeflasche *f* ABFALL beverage bottle
Getränkeverpackung *f* ABFALL beverage container, beverage packaging, drinks packaging
getrennt *adj* ABWASSER, ANALYTIK, BODEN, WASSER separate; **~e Entsorgung** *f* ABFALL separate disposal; **~e Klärung** *f* **von Abwässern** ABWASSER separate sewerage system; **~e Lagerung** *f* ABFALL separate storage; **~e Müllabfuhr** *f* ABFALL separate refuse collection, selective refuse collection, separate collection, selective collection; **~e Müllsammlung** *f* ABFALL separate refuse collection, selective refuse collection, separate collection, selective collection
Getrenntsammlung *f* ABFALL separate refuse collection, selective refuse collection
Getriebe *nt* ERNEU ENER drive train
getrocknet *adj* ABFALL, ABWASSER dried; **~er Schlamm** *m* ABWASSER dried sludge
getrübt: **~es Abwasser** *nt* ABWASSER turbid wastewater
Gewässer: **Gewässerbelastung** *f* ABWASSER water pollution; **Gewässerbeschaffenheit** *f* ANALYTIK, WASSER water quality; **Gewässerbiologie** *f* WASSER water biology; **Gewässergüte** *f* WASSER water quality; **Gewässergüteklasse** *f* ANALYTIK, WASSER water quality class; **Gewässerkunde** *f* WASSER hydrology, hydroscience; **Gewässerverdunstung** *f* WASSER evaporation of waters; **Gewässerverschmutzung** *f* ABWASSER water pollution; **Gewässerverunreinigung** *f* ABWASSER water pollution
Gewebe *nt* LUFT filter; **Gewebefilter** *nt* LUFT fabric filter; **Gewebeschlauchfilter** *nt* LUFT fabric filter
Gewerbe: **Gewerbeabfall** *m* ABFALL process waste, industrial waste; **Gewerbemüll** *m* ABFALL industrial waste
gewerblich *adj* ALLG commercial, industrial; **~er Abfall** *m* ABFALL commercial waste, industrial waste, process waste, trade waste; **~es Abwasser** *nt* ABWASSER, WASSER industrial effluent, industrial wastewater; **~er Klärschlamm** *m* ABWASSER industrial wastewater sludge; **~e Quelle** *f* ABFALL, LUFT, WASSER commercial source, industrial source; **~e Sammlung** *f* ABFALL industrial collection; **~er Wasserverbrauch** *m* WASSER industrial water consumption
Gewichtsanalyse *f* ANALYTIK gravimetric analysis
gewichtsanalytisch *adj* ANALYTIK gravimetric
gewinnbar *adj* ABFALL, ABWASSER, WASSER recoverable; **~e Wassermenge** *f* ABFALL, ABWASSER, WASSER recoverable volume of water
gewinnen *vt* ABFALL, ABWASSER, BODEN reclaim, recover, ERNEU ENER generate, WASSER recover, generate
Gewinnung *f* ABFALL, ABWASSER, BODEN *von Land* reclamation, recovery, ERNEU ENER *elektrischer Energie* generation, WASSER recovery, generation; **Gewinnungsgebiet** *nt* WASSER recovery area
Geysir *m* ERNEU ENER *geothermische Energie* geyser
Gezeiten *f pl* ERNEU ENER tide; **Gezeitenenergie** *f* ERNEU ENER tidal power, *erneuerbare Energiequelle* tidal energy; **Gezeitenenergieressourcen** *f pl* ERNEU ENER tidal energy resources; **Gezeitenkraft** *f* ERNEU ENER tidal power; **Gezeitenkraftwerk** *nt* ERNEU ENER tidal power plant, tidal mill; **Gezeitensperrwerk** *nt* ERNEU ENER *Gezeitenkraftwerk* tidal barrage
GF *abbr* (*Graphitrohrofen*) ANALYTIK GF (*graphite furnace*)
GFK *abbr* (*glasfaserverstärkter Kunststoff*) ABFALL, WASSER FRP (*fiberglass-reinforced plastics AmE*, *fibreglass-reinforced plastics BrE*); **GFK-Laminat** *nt* ABFALL, WASSER glass fiber laminate (*AmE*), glass fibre laminate (*BrE*)
GH *abbr* (*Gesamthärte*) WASSER total hardness
Gichtgas *nt* LUFT stack gas; **Gichtgasrückführung** *f* LUFT top gas recycling; **Gichtgasschlamm** *m* ABFALL top gas sludge
Gichtschlamm *m* ABWASSER blast furnace sludge
Gift *nt* ABFALL, LUFT, SICHERHEIT, WASSER poison, toxin
giftig *adj* ABFALL, ABWASSER toxic, ANALYTIK lethal, toxic, BODEN toxic, LUFT, SICHERHEIT *als Gefahrensymbol* poisonous, WASSER toxic; **~er Abfall** *m* ABFALL toxic waste

Giftigkeit *f* ALLG toxicity
Gift: **Giftmüll** *m* ABFALL toxic waste, poisonous waste; **Giftmüllentsorgungsanlage** *f* ABFALL toxic waste disposal plant; **Giftstoff** *m* ABFALL, BODEN, LUFT, WASSER toxic agent, toxicant, toxin
Gips *m* ABFALL, BODEN, LUFT calcium sulfate (*AmE*), calcium sulphate (*BrE*), gypsum, slaked lime; **Gipsabtrennung** *f* LUFT gypsum separation; **Gipsanfall** *m* LUFT gypsum yield; **Gipsaufbereitung** *f* LUFT gypsum processing; **Gipsbehandlung** *f* LUFT gypsum treatment; **Gipsbildung** *f* LUFT gypsum formation; **Gipsentwässerung** *f* LUFT gypsum dewatering
gipserzeugend *adj* LUFT gypsum-generating
Gips: **Gipskartonplatte** *f* LÄRM gypsum plaster board; **Gipskristall** *m* LUFT gypsum crystal; **Gipspartikel** *f* LUFT gypsum particle; **Gipssilo** *m* LUFT gypsum silo; **Gipssuspension** *f* LUFT gypsum suspension; **Gipstrocknung** *f* LUFT gypsum drying; **Gipsverwertung** *f* LUFT gypsum use
Gitter *nt* ABFALL, ANALYTIK, WASSER grid; **Gitterboxpalette** *f* ABFALL box pallet with sidewalls; **Gitterpunkt** *m* ANALYTIK grid point; **Gitterzerkleinerer** *m* ABFALL, WASSER grid crusher
Glas *nt* ABFALL glass container, glass jar; **Glasbehälter** *m* ABFALL glass container; **Glasbruch** *m* ABFALL waste glass cullet, cullet; **Glascontainer** *m* ABFALL bottle bank; **Glasdach** *nt* ERNEU ENER glass roof
Glasfaser *f* ABFALL, LÄRM, LUFT *Filtermedium* fibreglass (*BrE*), fiberglass (*AmE*), *glasfaserverstärkter Kunststoff* glass fibre (*BrE*), glass fiber (*AmE*); **Glasfaserabsorber** *m* LÄRM glass fiber absorber (*AmE*), glass fibre absorber (*BrE*); **Glasfaserfilter** *nt* ANALYTIK glass fiber filter (*AmE*), glass fibre filter (*BrE*); **Glasfaserlaminat** *nt* ABFALL glass fiber laminate (*AmE*), glass fibre laminate (*BrE*)
glasfasermattenverstärkt: **~es Thermoplast** *nt* ABFALL glass fiber-mat-reinforced thermoplast (*AmE*), glass fibre-mat-reinforced thermoplast (*BrE*)
Glasfaser: **Glasfaserschichtkunststoff** *m* ABFALL glass fiber laminate (*AmE*), glass fibre laminate (*BrE*)
glasfaserverstärkt *adj* ABFALL, LÄRM glass-fiber-reinforced (*AmE*), glass-fibre-reinforced (*BrE*), fiberglass-reinforced (*AmE*), fibreglass-reinforced (*BrE*); **~er Kunststoff** *m* (*GFK*) ABFALL fiberglass-reinforced plastic (*AmE*) (*FRP*), fibreglass-reinforced plastic (*BrE*) (*FRP*)
Glasfaser: **Glasfaservlies** *nt* LÄRM *Absorptionsmaterial* fiberglass fleece (*AmE*), fibreglass fleece (*BrE*)
Glas: **Glasflasche** *f* ANALYTIK *Probenehmen* glass bottle; **Glasgefäß** *nt* ABFALL glass jar; **Glasgewebe** *nt* ABFALL glass fabric; **Glaskolben** *m* ANALYTIK flask
glasmattenverstärkt: **~es Thermoplast** *nt* ABFALL glass-mat-reinforced thermoplast
Glas: **Glasrecycling** *nt* ABFALL glass recycling; **Glasrohr** *nt* ANALYTIK glass tube; **Glasscheibe** *f* ERNEU ENER *Chrom* glazing, *Kobalt* glass pane; **Glasscherben** *f pl* ABFALL waste glass cullet; **Glasschliffstopfen** *m* ANALYTIK *Probenflasche* ground glass stopper; **Glasschmelzverfahren** *nt* ABFALL *Behandlung von Müllverbrennungsschlacken* glass melting process; **Glasverbrauch** *m* ABFALL glass consumption; **Glaswolle** *f* ABFALL glass wool; **Glaszerkleinerer** *m* ABFALL glass crusher, glass pulverizer
Gleichgewicht *nt* ABFALL, ERNEU ENER, LUFT, WASSER balance; **Gleichgewichtsbeladung** *f* LUFT *Adsorption* equilibrium load; **Gleichgewichtskonstante** *f* LUFT equilibrium constant; **Gleichgewichtskonzentration** *f* ABFALL equilibrium concentration
gleichlaufend: **~e Zentrifuge** *f* ABWASSER cocurrent centrifuge
Gleichrichter *m* (*GR*) LUFT rectifier
Gleichspannung *f* ANALYTIK direct voltage; **Gleichspannungsfeld** *nt* LUFT DC voltage field; **Gleichspannungsschreiber** *m* ANALYTIK direct voltage recorder
Gleichstrom *m* LUFT *Abgasstrom*, WASSER co-current flow
Gleichstromung *f* LUFT *Abgasstrom*, WASSER co-current flow
Gleichstrom: **Gleichstromwäsche** *f* LUFT co-current washing; **Gleichstromwäscher** *m* LUFT co-current flow scrubber, co-current scrubber
gleichzeitig *adj* ANALYTIK, ERNEU ENER simultaneous, synchronous
global *adj* ERNEU ENER, LUFT global; **~es Klimamodell** *nt* LUFT global climate modeling (*AmE*), global climate modelling (*BrE*); **~es Klimamodellieren** *nt* LUFT global climate modeling (*AmE*), global climate modelling (*BrE*); **~er Umweltfonds** *m* POLITIK Global Environment Facility (*GEF*); **~es Umweltüberwachungssystem** *nt* ANALYTIK, POLITIK Global Environment Monitoring System
Globalstrahlung *f* LUFT global irradiance
Glocke *f* LUFT cap, bubble cap
Glockenboden *m* LUFT bubble-cap tray tower; **Glockenbodenkolonne** *f* LUFT bubble-cap tray column, bubble-cap column; **Glockenbodenwäscher** *m* LUFT bubble washer
Glocke: **Glockenhals** *m* LUFT vapor riser (*AmE*), vapour riser (*BrE*)
GLP *abbr* (*gute Laborpraxis*) ANALYTIK GLP (*good laboratory practice*)
Glüh- *in cpds* ABFALL, ANALYTIK, LUFT, WASSER ignition
glühen *vt* ABFALL, ABWASSER, ANALYTIK LUFT ignite
Glüh-: **Glühfadenlampe** *f* ANALYTIK incandescent lamp; **Glührückstand** *m* ABWASSER, ANALYTIK ignition residue; **Glühverlust** *m* ABFALL, ABWASSER, ANALYTIK, LUFT ignition loss; **Glühzündung** *f* ABFALL, ABWASSER, ANALYTIK, LUFT ignition
GM-Zähler *m* (*Geiger-Müllerzähler*) ANALYTIK GM counter (*Geiger-Müller counter*), contamination meter, Geiger counter
Gondel *f* ERNEU ENER gondola
GPC *abbr* (*Gelpermeations Chromatographie*) ANALYTIK GPC (*gel permeation chromatography*)
GPT *abbr* (*Gasphasentitration*) ANALYTIK GPT (*gas-phase titration*)
GR *abbr* (*Gleichrichter*) LUFT rectifier
graben *vt* ABFALL, WASSER *Sonden* dig
Graben *m* ABFALL, BODEN, MEER, WASSER trench, ditch; **Grabenbewässerung** *f* ABFALL, WASSER ditch irrigation; **Grabenmethode** *f* ABFALL, WASSER trench method
graduieren *vt* ANALYTIK graduate
Grandfathering-Prinzip *nt* POLITIK grandfathering principle
Granulat *nt* LUFT *Schüttschichtfilter* granulate; **Granulatkörper** *m* LUFT *Schüttschichtfilter* granulate body; **Granulatmaterial** *nt* LUFT *Schüttschichtfilter* granulate
Granulator *m* ABFALL granulator, granulating machine
Granulat: **Granulatröhrenfilter** *nt* LUFT granulate tubular filter; **Granulatschicht** *f* LUFT *Schüttschichtfilter* granulate layer
granulieren *vt* ABFALL pelletize, granulate
Granulierkopf *m* ABFALL granulating head
granuliert: **~e Aktivkohle** *f* (*GAK*) ABWASSER, LUFT, WASSER granulated activated carbon (*GAC*)
Granulierung *f* ABFALL pelletizing
Granuliervorrichtung *f* ABFALL pelletizer
Graphit *m* ABFALL *Batterie* graphite; **Graphitrohrofen** *m* (*GF*) ANALYTIK graphite furnace (*GF*); **Graphitrohrofenatomabsorptionspektrometrie** *f* (*GF-AAS*) ANALYTIK graphite furnace atomic absorption spectrometry
grau: **~e Tonne** *f* ABFALL *Einsammlungscontainer für Restmüll* gray container (*AmE*), grey container (*BrE*)
Gravimetrie *f* ANALYTIK gravimetric analysis
gravimetrisch *adj* ANALYTIK gravimetric; **~e Analyse** *f* ANALYTIK gravimetric analysis; **~e Bestimmung** *f* ANALYTIK

gravimetric determination; **~es Verfahren** *nt* ANALYTIK gravimetric method
Gravitation *f* ALLG gravitation, gravity; **Gravitationskräfte** *f pl* ABWASSER, ANALYTIK, ERNEU ENER, LUFT, WASSER *Entwässerung* gravitational force
Gray *nt* (*Gy*) ANALYTIK gray (*Gy*)
Greifarm *m* ABFALL grappler
Greifbagger *m* ABFALL grab
Greifer *m* ABFALL pickup mechanism, grab; **Greiferbohrung** *f* BODEN grab boring; **Greifersystem** *nt* ABFALL grab system
Greifroboter *m* ABFALL *bei der automatischen Sortierung* gripper robot
Grenz- *in cpds* POLITIK, SICHERHEIT limit
grenzaktiv *adj* LUFT surface-active; **~er Stoff** *m* LUFT surface-active agent
Grenze *f* LUFT boundary
Grenz-: **Grenzfläche** *f* ANALYTIK, LUFT, MEER interface
grenzflächenaktiv: **~er Stoff** *m* ABFALL, LUFT, WASSER surfactant
Grenz-: **Grenzflächenerneuerung** *f* LUFT interfacial renewal; **Grenzflächenspannung** *f* ANALYTIK interfacial tension; **Grenzkonzentration** *f* LUFT limiting concentration, SICHERHEIT limit concentration; **Grenzrisiko** *nt* SICHERHEIT risk limit; **Grenzschicht** *f* LUFT boundary layer; **Grenzschichtströmung** *f* LUFT boundary layer flow; **Grenzteilchenbahn** *f* LUFT *Abscheidevorgang* periphery particle orbit
grenzüberschreitend *adj* ABWASSER, LUFT, POLITIK transboundary; **~e Luftverunreinigung** *f* LUFT transboundary air pollution; **~e Wasserverschmutzung** *f* ABWASSER transboundary water pollution; **~e Wasserverunreinigung** *f* ABWASSER transboundary water pollution
Grenzwert *m* WASSER limit value; **~ für Belastung** POLITIK, SICHERHEIT *Gefahrstoffe* exposure limit to hazardous substances; **~ für Gefahrstoffe** POLITIK, SICHERHEIT exposure limit to hazardous substances; **Grenzwertkontrolle** *f* ANALYTIK limit value control; **Grenzwert-Monitoring** *nt* ANALYTIK limit value control
Grieß *m* ABWASSER grit
grob *adj* ABWASSER, BODEN, LUFT coarse; **~e Feststoffe** *m pl* ABWASSER coarse solids; **~e Partikel** *f* LUFT coarse particle
Grob- *in cpds* ABFALL, ABWASSER, LUFT coarse; **Grobascheabscheider** *m* LUFT coarse ash separator
grobblasig: **~e Belüftung** *f* ABWASSER large-bubble aeration
Grob-: **Grobfilter** *nt* LUFT *Speicherfilter* coarse filter; **Grobflocken** *f pl* ABWASSER coarse flocs
grobflockig *adj* ABFALL, WASSER coarse flocculent
Grob-: **Grobklärbecken** *nt* ABWASSER roughing tank
grobkörnig *adj* ABFALL coarse-grained
Grob-: **Grobmüll** *m* ABFALL coarse refuse; **Grobrechen** *m* ABWASSER, WASSER coarse screen; **Grobstaub** *m* LUFT grit; **Grobstaubabscheidung** *f* LUFT grit precipitation; **Grobstoffänger** *m* ABWASSER roughing tank
grobstückig *adj* ABFALL coarse-particled
Groß- *in cpds* ERNEU ENER large; **Großanlage** *f* ERNEU ENER *Wind, Sonne* large-scale installation; **Großbehälter** *m* ABFALL large container
Größe: **nach ~ sortieren** *vt* ABFALL sort according to size
Groß-: **Großfeldberegner** *m* ABWASSER long-throw sprinkler
großflächig *adj* ERNEU ENER large-area; **~e Nutzung** *f* **der Sonnenenergie** ERNEU ENER large-area use of solar energy
Groß-: **Großkläranlage** *f* ABWASSER large-scale clarification plant; **Großsammelstelle** *f* ABFALL large collection station
größt: **~er anzunehmender Unfall** *m* (*GAU*) ABFALL, SICHERHEIT maximum credible accident (*MCA*)
großtechnisch *adj* ALLG commercial-scale, ERNEU ENER utility-scale; **~e Anlage** *f* ALLG commercial-scale plant; **~e Anwendung** *f* ALLG commercial-scale application; **~er Einsatz** *m* ALLG commercial-scale application
großvolumig: **~es Luftprobenahmegerät** *nt* ANALYTIK high-volume sampler
Grube *f* ABFALL pit; **Grubendeponie** *f* ABFALL pit landfill; **Grubengasinterferometer** *nt* ANALYTIK interferential pit gas tester; **Grubenüberlauf** *m* ABWASSER cesspool overflow; **Grubenwand** *f* ABFALL embankment
Grün: **~er Punkt** *m* ABFALL, MANAG environmental labelling mark for packaging
grün: **~e Revolution** *f* LUFT green revolution
Grünabfall *m* ABFALL green waste
Grund- *in cpds* ALLG bottom; **Grundablaß** *m* WASSER bottom outlet; **Grundgeräusch** *nt* LÄRM background noise; **Grundkonzentration** *f* LUFT background concentration, WASSER instantaneous concentration; **Grundlastkraftwerk** *nt* ERNEU ENER base load power station
grundlegend: **~e Verschmutzung** *f* LUFT background pollution
Grund-: **Grundnetz** *nt* MEER trawl net; **Grundschleppnetz** *nt* MEER trawl net; **Grundschwelle** *f* ABFALL, WASSER sill; **Grundstoffen verarbeiten** *f* MANAG process of raw material
Grundwasser *nt* BODEN groundwater, WASSER groundwater, subterranean water, underground water; **Grundwasserabsenkung** *f* WASSER lowering of groundwater; **Grundwasseranreicherung** *f* WASSER groundwater recharge; **Grundwasseraufbereitung** *f* WASSER groundwater treatment; **Grundwasseraufbereitungsanlage** *f* WASSER groundwater treatment plant; **Grundwasserbecken** *nt* WASSER groundwater basin; **Grundwasserbelastung** *f* WASSER impact on groundwater; **Grundwasserbeobachtungsbrunnen** *m* ANALYTIK groundwater monitoring well; **Grundwasserbestand** *m* WASSER groundwater reserve; **Grundwasserbewirtschaftung** *f* WASSER groundwater management; **Grundwasserbildung** *f* WASSER groundwater formation; **Grundwasserbrunnen** *m* WASSER groundwater well; **Grundwasserdargebot** *nt* WASSER groundwater supply; **Grundwassereigenschaften** *f pl* WASSER groundwater properties; **Grundwassereinzugsgebiet** *nt* WASSER groundwater catchment; **Grundwasserentnahmeabgabe** *f* POLITIK groundwater withdrawal charge; **Grundwasserentnahmesteuer** *f* POLITIK groundwater extraction tax, groundwater withdrawal tax; **Grundwasserentnahmesumme** *f* WASSER total amount of groundwater tapping; **Grundwassererfassung** *f* WASSER groundwater capture, groundwater tapping; **Grundwassererschließung** *f* WASSER groundwater capture, groundwater tapping; **Grundwassergefährdung** *f* WASSER groundwater risk; **Grundwasserkontamination** *f* WASSER groundwater contamination; **Grundwasserleiter** *m* ABWASSER, WASSER aquifer; **Grundwassermeßstelle** *f* ANALYTIK measuring point of groundwater, groundwater measuring point; **Grundwassermodell** *nt* ANALYTIK, MANAG groundwater model; **Grundwassernichtleiter** *m* WASSER aquifuge; **Grundwassernutzung** *f* WASSER groundwater use; **Grundwasserprobeentnahme** *f* ANALYTIK groundwater sampling; **Grundwasserproblem** *nt* WASSER groundwater problem; **Grundwasserpumpwerk** *nt* WASSER groundwater pumping station; **Grundwasserquelle** *f* WASSER groundwater source; **Grundwasserschürfung** *f* WASSER groundwater prospecting
Grundwasserschutz *m* WASSER groundwater protection; **Grundwasserschutzgebiet** *nt* WASSER groundwater protection area; **Grundwasserschutzzone** *f* WASSER groundwater protection area
Grundwasser: **Grundwassersohle** *f* WASSER groundwater bottom; **Grundwassersonde** *f* ANALYTIK groundwater probe; **Grundwasserspiegel** *m* WASSER water table, groundwater table; **Grundwasserstand** *m* WASSER groundwater level; **Grundwasserstauer** *m* WASSER aquifuge, barrier; **Grundwasserstockwerk** *nt* WASSER aquifer; **Grundwassertiefe** *f* WASSER groundwater depth; **Grundwasserverfahren**

nt (*GW-Verfahren*) BODEN groundwater method; **Grundwasserverschmutzung** *f* BODEN groundwater contamination; **Grundwasserverunreinigung** *f* BODEN groundwater contamination; **Grundwasservorkommen** *nt* WASSER groundwater resources, presence of groundwater; **Grundwasservorrat** *m* WASSER groundwater reserve

Grund-: **Grundwehr** *nt* WASSER drowned weir

Grünen: **die ~** *pl* POLITIK *politische Partei* the Greens

Grün: **Grünglas** *nt* ABFALL green glass; **Grünsand** *m* ABFALL, WASSER *Zeolith* greensand

Gruppenkläranlage *f* ABWASSER group clarification plant

gruppenkollektiv: **~e Dosis** *f* ANALYTIK, LUFT group collective dose

Gruppenparameter *m* ANALYTIK group parameter

GSC *abbr* (*Gas-Feststoff-Chromatographie*) ANALYTIK GSC (*gas-solid chromatography*)

GSW *abbr* (*Geruchsschwellenwert*) ABFALL, ABWASSER, ANALYTIK TON (*threshold odor number AmE, threshold odour number BrE*)

GuD-Anlage *f* (*Gas- und Dampfanlage*) LUFT *Wirbelschichtfeuerung* CCGT (*combined-cycle gas turbine*)

Gülle *f* ABFALL slurry, ABWASSER liquid manure, WASSER slurry

Gully *m* WASSER sink

Gummiabfall *m* ABFALL waste rubber

Gußeisenrohr *nt* ABWASSER cast-iron pipe

gut: **~ absorbierbar** *adj* LUFT highly absorbable; **~e Laborpraxis** *f* (*GLP*) ANALYTIK good laboratory practice (*GLP*)

Gütenormen *f pl*: **~ für Trinkwasser** WASSER drinking water standards

GW-Verfahren *nt* (*Grundwasserverfahren*) BODEN groundwater method

Gy *abbr* (*Gray*) ANALYTIK Gy (*gray*)

H

Haar: **Haarfeuchtigkeitsmesser** *m* ANALYTIK hair hygrometer; **Haarhygrometer** *nt* ANALYTIK hair hygrometer
häckseln *vt* ABFALL chop, hack up
Häckseln *nt* ABFALL chopping, hacking up
Haft- *in cpds* ANALYTIK, LUFT adhesive; **Haftfähigkeit** *f* ABWASSER adhesiveness; **Haftfestigkeit** *f* ABFALL, ANALYTIK adherence, LUFT adhesive force; **Haftkraft** *f* LUFT *Elektrofilter* adhesive force; **Haftmittel** *nt* LUFT adhesive agent, bonding agent
Haftung *f* ABFALL, ABWASSER, ANALYTIK, LUFT adhesion, POLITIK liability, WASSER adhesion; **~ zur ungeteilten Hand** ABFALL, POLITIK, WASSER joint and several liability; **Haftungsfestigkeit** *f* ABFALL, ABWASSER, ANALYTIK, LUFT, WASSER adhesion
Haft-: **Haftvermögen** *nt* ANALYTIK, LUFT adhesive force
HAK *abbr* (*höchstzulässige Konzentration am Arbeitsplatz*) SICHERHEIT maximum allowable concentration at the workplace
halbautomatisch *adj* ABFALL, WASSER partially automated
halbdurchlässig *adj* ABFALL, ABWASSER, WASSER semipermeable; **~e Membran** *f* ABFALL, ABWASSER, WASSER semipermeable membrane; **~e Schicht** *f* ABFALL, WASSER semi-impermeable layer
halbfest: **~er brennbarer Abfall** *m* ABFALL semisolid combustible waste
Halb: **Halbfeuchttrennung** *f* ABFALL semi-wet sorting
halbkontinuierlich *adj* ABWASSER, ANALYTIK partially continuous
Halbleiter *m* ANALYTIK, ERNEU ENER semiconductor; **Halbleiterbauelement** *nt* ERNEU ENER semiconductor component; **Halbleiterdetektor** *m* ANALYTIK *Messung von Ionisation im Festkörper* semiconductor detector
Halb: **Halbstundenmittelwert** *m* ANALYTIK half-hourly average value, half-hourly mean value; **Halbwertsbreite** *f* ANALYTIK half-width value; **Halbwertzeit** *f* LUFT half-life
Halid *nt* LUFT halide
Hallraum *m* LÄRM reverberation chamber
Haloformverbindungen *f pl* ABFALL, ABWASSER, WASSER haloform compounds
halogenfrei: **~e organische Lösungsmittel** *nt pl* ABFALL, WASSER halogen-free organic solvents
halogenhaltig *adj* ABFALL, WASSER halogenated; **~es Gas** *nt* LUFT halogenated gas; **~e organische Lösungsmittel** *nt pl* ABFALL, WASSER halogenated organic solvents
Halogenid *nt* LUFT halide
halogeniert: **~e Kohlenwasserstoffe** *f pl* (*HKW*) WASSER halogenated hydrocarbons
Halogenkohlenwasserstoff *m* ABFALL, LUFT halogenated hydrocarbon
Halon *nt* LUFT halon
haltbar *adj* ABFALL, ABWASSER nonputrescible, *Probe* durable
halten: **im Gleichgewicht ~** *vt* ALLG balance, maintain
Hammer: **Hammerbrecher** *m* ABFALL hammermill, hammermill crusher; **Hammermühle** *f* ABFALL hammermill, hammermill crusher; **Hammerzerkleinerer** *m* ABFALL hammermill, hammermill crusher
Hand: **von ~** *phr* ABFALL, ANALYTIK, WASSER manual; **Handauslese** *f* ABFALL hand sorting
handbedient *adj* ANALYTIK hand-operated
handbeladen: **~er Kompaktor** *m* ABFALL, WASSER hand-loaded compactor
handbetrieben *adj* ANALYTIK hand-driven; **~es Gasspürgerät** *nt* ANALYTIK hand-driven gas detector
Hand: **Handbohrung** *f* ANALYTIK, BODEN hand augering
handelsüblich *adj* ANALYTIK, ERNEU ENER commercially available; **~e Anlage** *f* ERNEU ENER commercially available installation; **~es Meßgerät** *nt* ANALYTIK commercially available measuring device; **~es pH-Meßgerät** *nt* ANALYTIK commercially available pH-measuring device; **~es Thermometer** *nt* ANALYTIK commercially available thermometer
handgehalten *adj* ANALYTIK hand-held
Händler *m* MANAG retailer
Hand: **Handscanner** *m* ABFALL, WASSER *Mülltrennung* hand scanner; **Handsortierstrecke** *f* ABFALL hand sorting line
handsortiert: **~e Fraktion** *f* ABFALL hand-sorted fraction, manually-sorted fraction
Hand: **Handsortierung** *f* ABFALL hand sorting, manual separation, manual sorting; **Handspektroskop** *nt* ANALYTIK hand-held spectroscope; **Handzentrifuge** *f* ANALYTIK hand-held centrifuge
Hang *m* ABFALL slope; **Hangeinschnitt** *m* ABFALL slope recess
Hängerahmen *m* LUFT *Schlauchfilter* suspended frame
Hang: **Hangquelle** *f* WASSER inclined spring; **Hangsickerwasser** *nt* ABFALL slope leachate; **Hangsickerwasseraustritt** *m* ABFALL, WASSER slope leachate emission; **Hangwasser** *nt* ABFALL, WASSER slope water
harmonisieren *vt* POLITIK harmonize
harmonisiert *adj* POLITIK harmonized; **~er Umweltschutz** *m* POLITIK harmonized environmental protection
Harmonisierung *f* POLITIK harmonization; **~ des Umweltschutzes** POLITIK harmonization of environmental protection
Harnstoff *m* ABWASSER urea
hart: **~es Wasser** *nt* ABFALL, WASSER hard water
Härte *f* WASSER hardness; **Härtebestimmung** *f* ANALYTIK hardness determination; **Härtebildner** *m* WASSER hardness former; **Härtegrad** *m* ABWASSER, ANALYTIK, WASSER degree of hardness; **Härtenachweis** *m* ANALYTIK hardness detection
Hart-PVC *nt* ABFALL rigid PVC
Harz *nt* ABFALL resin
Haube *f* LUFT hood
Häufigkeit *f* ANALYTIK frequency; **~ der Wellen** ERNEU ENER wave frequency; **Häufigkeitskoeffizient** *m* ANALYTIK abundance coefficient; **Häufigkeitsverteilung** *f* ANALYTIK frequency distribution
Haupt: **Hauptabluftgebläse** *nt* LUFT main waste air blower; **Hauptanzeige** *f* ANALYTIK *bei Meßgeräten* main display; **Hauptbrennkammer** *f* ABFALL, LUFT main combustion chamber; **Hauptdichtung** *f* WASSER main sealing layer; **Hauptgasstrom** *m* LUFT main gas stream; **Hauptkatalysator** *m* LUFT main catalyst; **Hauptleitung** *f* WASSER trunk main, *einer Pumpe* head pipe; **Hauptpumpwerk** *nt* WASSER main pumping station; **Hauptquelle** *f* LUFT, WASSER major source; **Hauptrotte** *f* ABFALL principal rotting treatment; **Hauptsammler** *m* ABFALL, WASSER main pipe, main collector; **Hauptstrom** *m* LUFT main stream; **Hauptströmung** *f* LUFT main flow; **Hauptuntersuchung** *f* ANALYTIK main analysis; **Hauptversorgungsgebiet** *nt* WASSER main supply area; **Hauptwaschkreislauf** *m* LUFT *Hauptstromoxidation* main wash circulation, main wash

circulation loop; **Hauptwasserleitung** *f* WASSER water main, mains; **Hauptwasserrohr** *nt* WASSER water main
Hausanschlußleitung *f* WASSER service mains
Haushalts- *in cpds* WASSER domestic; **Haushaltsabfall** *m* ABFALL domestic waste, urban waste, municipal solid waste (*MSW*), municipal waste stream; **Haushaltsabwasser** *nt* ABWASSER domestic wastewater; **Haushaltswasserversorgung** *f* WASSER domestic water supply
häusliches: **~ Abwasser** *nt* ABWASSER domestic wastewater
Hausmüll *m* ABFALL consumer waste, urban waste, domestic waste, municipal solid waste (*MSW*); **Hausmüllabfuhr** *f* ABFALL domestic waste collection, municipal waste collection; **Hausmüllanalyse** *f* ANALYTIK domestic waste analysis; **Hausmülldeponie** *f* ABFALL municipal waste landfill; **Hausmülleinsammlung** *f* ABFALL domestic waste collection, garbage disposal (*AmE*), refuse disposal; **Hausmüllkompostierung** *f* ABFALL municipal waste composting; **Hausmüllpyrolyse** *f* ABFALL domestic waste pyrolysis; **Hausmüllschlacke** *f* ABFALL domestic waste slag; **Hausmüllverbrennung** *f* ABFALL domestic waste incineration; **Hausmüllverbrennungsanlage** *f* ABFALL domestic waste incineration plant; **Hausmüllzusammensetzung** *f* ABFALL domestic waste composition
Haussammlung *f* ABFALL house collection
HCB *abbr* (*Hexachlorbenzol*) BODEN HCB (*hexachlorobenzene*)
HC-Emissionen *f pl* LUFT *unverbrannter Kohlenwasserstoff* HC-emissions
HCH *abbr* (*Hexachlorcyclohexan-Isomer*) BODEN HCH (*hexachlorocyclohexane isomer*)
HC-Wert *m* (*Kohlenwasserstoffwert*) LUFT HC value
HDW-Strahl *m* (*Hochdruckwasserstrahl*) BODEN *Bodenwäsche* high-pressure water jet
He *abbr* (*Helium*) LUFT He (*helium*)
Headspace: **Headspace-Gaschromatographie** *f* (*HSGC*) ANALYTIK headspace gas chromatography (*HSGC*); **Headspace-Sampler** *m* ANALYTIK headspace sampler; **Headspace-Technik** *f* ANALYTIK headspace technique
heben *vt* ABFALL, ABWASSER, WASSER lift
Hebezug *m* SICHERHEIT hoist
Heilbrunnen *m* WASSER mineral spring
Heiß- *in cpds* LUFT, MEER hot
heiß: **~es Abgas** *nt* LUFT hot waste gas; **~es Probegas** *nt* ANALYTIK hot sampling gas
Heiß-: **Heißdampf** *m* ERNEU ENER, LUFT superheated steam; **Heißdampfkessel** *m* ABFALL superheated boiler
Heißgas *nt* BODEN, LUFT hot gas; **Heißgasentstaubung** *f* LUFT *Verbrennungsgas* hot gas dust removal; **Heißgasfilter** *nt* LUFT hot gas filter; **Heißgasinjektion** *f* BODEN hot gas injection; **Heißgasverdichter** *m* LUFT hot gas compressor
Heiß-: **Heißluftheizung** *f* SICHERHEIT hot-air radiation heating system; **Heißtrocknung** *f* ABWASSER *Schlammentsorgung* heat drying; **Heißwassererhitzer** *m* LUFT hot water heater; **Heißwasserquelle** *f* ERNEU ENER hot water source; **Heißwasserwaschen** *nt* MEER hot water washing
Heiz- *in cpds* LUFT heating; **Heizanlage** *f* ERNEU ENER heating installation
Heizer *m* LUFT stoker, *Röhre* heating filament
Heiz-: **Heizleistung** *f* ABFALL, LUFT thermal capacity; **Heizöl** *nt* LUFT *Zusatzbrennstoff* fuel oil, crude oil; **Heizöl** *nt* **S** LUFT heavy fuel; **Heizsystem** *nt* ERNEU ENER heating system
Heizung *f* ERNEU ENER heating
Heiz-: **Heizwert** *m* ABFALL, ABWASSER, ANALYTIK, ERNEU ENER, LUFT calorific value; **Heizzone** *f* LUFT heating zone; **Heizzweck** *m* ERNEU ENER heating purpose; **Heizzylinder** *m* ABWASSER heating cylinder
Heliostat *m* ERNEU ENER heliostat
Helium *nt* (*He*) LUFT helium (*He*); **Helium-Neon-Laser** *m* ANALYTIK helium-neon laser
Henry: **~sches Gesetz** *nt* ABWASSER, BODEN, LUFT Henry's law
Henry-Dalton: **~sches Gesetz** *nt* ABFALL Henry-Dalton law
Heptan *nt* ABFALL heptane
Herabsetzung: **~ des pH-Wertes** *f* ABWASSER, ANALYTIK, LUFT pH depression
herauslösen *vt* ANALYTIK elute, BODEN dissolve away
Herauslösung *f* ANALYTIK elution, BODEN dissolving away
herkömmlich *adj* ALLG conventional; **~e Kohlereinigung** *f* LUFT conventional coal cleaning
herstellen *vt* ABFALL, MANAG produce
Hersteller *m* ABFALL, MANAG producer
Herstellung *f*: **~ des Werkstoffes** MANAG production of raw material
herunterdrücken *vt* ABFALL press down
herunterwalzen *vt* ABFALL break down
Heteroatom *nt* ANALYTIK *wie Stickstoff, Halogen* heteroatom
heterogen: **~e Abfälle** *m pl* ABFALL heterogeneous wastes; **~er Katalysator** *m* LUFT heterogeneous catalyst
heterotroph *adj* ABWASSER *Bakterien* heterotrophic
Heterotroph *nt* ABWASSER heterotroph
heterotroph: **~e Denitrifikation** *f* ABWASSER heterotrophic denitrification
Hexachlor- *in cpds* BODEN hexachloro-; **Hexachlorbenzol** *nt* (*HCB*) BODEN hexachlorobenzene (*HCB*); **Hexachlorcyclohexanisomer** *nt* (*HCH*) BODEN hexachlorocyclohexane isomer (*HCH*)
Hexan *nt* LUFT hexane
HFB *abbr* (*Horizontalfilterbrunnen*) WASSER horizontal filter well, horizontal filtering well
HFCKW *abbr* (*teilhalogenierte Fluorchlorkohlenwasserstoffe*) LUFT HCFC (*hydrochlorofluorocarbon*)
Hg *abbr* (*Quecksilber*) ABFALL, ABWASSER, BODEN, WASSER Hg (*mercury*)
Hg-haltig *adj* ABFALL, ABWASSER mercury-containing; **~e Lampe** *f* ABFALL mercury-containing lamp
High-Speed-Mischer *m* ABFALL, WASSER high-speed mixer
High-Tech-Schrott *m* ABFALL high tech scrap
Hilfs-: **Hilfsbrenner** *m* LUFT auxiliary burner; **Hilfsgas** *nt* ANALYTIK auxiliary gas; **Hilfsstoffe** *m pl* ABFALL, LUFT, WASSER process materials; **Hilfsstoffstrom** *m* LUFT process materials stream
Hintereinanderschaltung *f* LUFT *des Zyklons* tandem mounting
Hintergrund *m* LÄRM, LUFT background; **Hintergrundgeräusch** *nt* LÄRM background noise; **Hintergrundkonzentration** *f* ANALYTIK, BODEN, WASSER background level; **Hintergrundprobe** *f* ANALYTIK background sample
Hitzdrahtanemometer *nt* ANALYTIK hot-wire anemometer
Hitze *f* ALLG heat
hitzeabweisend *adj* SICHERHEIT heat-resistant, heatproof
hitzebeständig *adj* ABFALL, WASSER heat-resistant
HKW *abbr* (*halogenierte Kohlenwasserstoffe*) WASSER halogenated hydrocarbons; **HKW-Eliminierung** *f* WASSER *Trinkwasseraufbereitung* elimination of halogenated hydrocarbons
hobeln *vt* ABFALL plane
Hoch- *in cpds* ABFALL high
hochabsorbierend *adj* LÄRM highly absorbent; **~es Lärmschutzsystem** *nt* LÄRM, MEER highly absorbent noise protection system
hochalkalisch *adj* ABFALL, WASSER highly alkaline
Hoch-: **Hochbehälterabfluß** *m* WASSER water tower runoff
hochbelastet *adj* ABFALL, ABWASSER, WASSER highly polluted, high-rate; **~es Abwasser** *nt* ABWASSER highly polluted wastewater; **~es Belebungsbecken** *nt* ABWASSER high-rate activated sludge tank; **~er Fäkalschlamm** *m* ABWASSER highly polluted faecal solids sludge (*BrE*), highly polluted fecal solids sludge (*AmE*); **~er Faulraum** *m* ABWASSER high-

rate digester; **~e Gewässer** *nt pl* ABWASSER highly polluted waters; **~es kommunales Abwasser** *nt* ABWASSER highly contaminated municipal wastewater; **~er Tropfkörper** *m* ABWASSER high-rate trickling filter; **~es Verfahren** *nt* ABWASSER *Belebungsschlamm* high-rate process

Hochdruck- *in cpds* ABWASSER high-pressure; **Hochdruckfilter** *nt* ABWASSER high-pressure filter; **Hochdruckhydrierung** *f* ABFALL high-pressure hydrogenation; **Hochdruckkamm** *m* ERNEU ENER *Wellenenergie* high-pressure cog; **Hochdruck-Naßoxidation** *f* ABWASSER high-pressure wet oxidation; **Hochdruck-Polyethylen** *nt* ABFALL high-density polyethylene (*HDPE*); **Hochdruckpumpe** *f* ABFALL, ABWASSER, WASSER high-pressure pump; **Hochdruckradialverdichter** *m* BODEN *Bodenluftabsaugung* high-pressure radial flow compressor; **Hochdruckreaktor** *m* ERNEU ENER *Biomasseumwandlung* high-pressure reactor; **Hochdruckreinigung** *f* MEER *mit Wasser* high-pressure water blasting; **Hochdruckseite** *f* LUFT high-pressure side; **Hochdruck-Spritzdüse** *f* LUFT high-pressure spray nozzle; **Hochdruck-Sprühdüse** *f* LUFT high-pressure spray nozzle; **Hochdruckspülen** *nt* MEER high-pressure flushing; **Hochdruckspülung** *f* LUFT high-pressure cleaning; **Hochdruck-Venturiwäscher** *m* LUFT high-pressure Venturi scrubber

Hochdruckwasserstrahl *m* (*HDW-Strahl*) BODEN *Bodenwäsche* high-pressure water jet; **Hochdruckwasserstrahlgerät** *nt* WASSER *Reinigung der Wärmetauscher* high-pressure water jet torch; **Hochdruckwasserstrahlrohr** *nt* BODEN high-pressure water jet pipe; **Hochdruckwasserstrahlverfahren** *nt* BODEN *Bodenwäsche* high-pressure water jet method

hochempfindlich *adj* ANALYTIK highly sensitive; **~er Detektor** *m* ANALYTIK highly sensitive detector; **~es Meßgerät** *nt* ANALYTIK highly sensitive measuring device; **~er Test** *m* ANALYTIK *Wasseranalyse* highly sensitive test; **~e Waage** *f* ANALYTIK *Gasabfüllung* highly sensitive scale

Hoch-: **Hochfrequenzerwärmung** *f* BODEN *thermische Desorption* high-frequency heating, radio frequency heating

Hochgeschwindigkeitsgasstrom *m* LUFT high-velocity gas stream

Hoch-: **Hochglanzanstrich** *m* ABFALL *Recyclinghindernis* high-gloss coating

hochkontaminiert *adj* BODEN highly contaminated

Hochleistungs- *in cpds* ANALYTIK, LUFT, WASSER high efficiency, high performance; **Hochleistungsabscheider** *m* LUFT high-performance precipitator; **Hochleistungsbelüftung** *f* ABFALL, WASSER high-performance aeration; **Hochleistungsbrunnen** *m* WASSER high-performance well; **Hochleistungs-Dünnschichtchromatographie** *f* (*HPDC*) ANALYTIK high-performance thin-layer chromatography; **Hochleistungsentstaubung** *nt* LUFT high-performance dust removal; **Hochleistungsfaulraum** *m* ABWASSER high-rate digester; **Hochleistungsfestbettreaktor** *m* WASSER high-performance fixed-bed reactor; **Hochleistungsfilter** *nt* LUFT high-performance filter; **Hochleistungsfilterung** *f* ABWASSER, LUFT, WASSER high-performance filtration, high-rate filtration; **Hochleistungsfiltration** *f* ABWASSER, LUFT, WASSER high-performance filtration, high-rate filtration; **Hochleistungs-Flüssigkeitschromatographie** *f* (*HPLC*) ANALYTIK *UEG* high-performance liquid chromatography (*HPLC*); **Hochleistungs-Flüssigkeitschromatographie** *f* **in Kopplung mit Massenspektrometrie** (*HPLC-MS*) ANALYTIK high-performance liquid chromatography mass spectrometry (*HPLC-MS*); **Hochleistungskoaleszer** *m* ABWASSER high-rate coalescence oil separator; **Hochleistungspresse** *f* ABFALL, WASSER high-pressure press; **Hochleistungswäscher** *m* LUFT high-performance scrubber; **Hochleistungszyklon** *m* LUFT high-efficiency cyclone, high-performance cyclone

hochmolekular: **~es Polymer** *nt* ABFALL high molecular weight polymer

Hoch-: **Hochofenschlacke** *f* ABFALL *industrielles Nebenprodukt* blast furnace slag

hoch: **~ radioaktiv** *adj* ABFALL high-level radioactive

Hoch-: **Hochseebagger** *m* MEER marine dredge; **Hochseeschiff** *nt* MEER sea-going ship, sea-going vessel; **Hochsicherheit** *f* BODEN *des Standorts*, SICHERHEIT high security; **Hochsicherheitsdeponie** *f* ABFALL high-security landfill

Hochspannungs- *in cpds* WASSER high-voltage; **Hochspannungsaggregat** *nt* LUFT high-voltage aggregate; **Hochspannungs-Ausströmelektrode** *f* LUFT high-voltage discharge electrode; **Hochspannungselektrofilter** *nt* LUFT high-voltage ESP; **Hochspannungs-Entladungselektrode** *f* LUFT high-voltage discharge electrode; **Hochspannungserzeuger** *m* LUFT high-voltage generator; **Hochspannungsleitung** *f* ERNEU ENER high-voltage line

Höchst- *in cpds* SICHERHEIT maximum

Höchstbelastung *f* SICHERHEIT maximum exposure; **Höchstbelastungsgrenze** *f* SICHERHEIT maximum exposure limit; **Höchstbelastungsgrenzen** *f pl* **für Gefahrstoffe** SICHERHEIT maximum exposure limits to hazardous substances

höchste: **~ Dosis** *f*, **die noch nicht zu nachteiligen Wirkungen führt** (*NOAEL*) SICHERHEIT no observed adverse effect level (*NOAEL*); **~ Konzentration** *f* **ohne eine beobachtete nachteilige Wirkung** (*NOEC*) SICHERHEIT no observed effect concentration (*NOEC*)

Höchst-: **Höchstgeschwindigkeit** *f* ABFALL, ANALYTIK, SICHERHEIT, WASSER peak velocity; **Höchstwert** *m* POLITIK, SICHERHEIT limit value

höchstzulässig: **~e Konzentration** *f* (*HZK*) ANALYTIK, LUFT, SICHERHEIT, WASSER maximum allowable concentration (*MAC*); **~e Konzentration** *f* **am Arbeitsplatz** (*HAK*) SICHERHEIT maximum allowable concentration at the workplace; **~er Konzentrationswert** *m* (*HZK*) LÄRM, POLITIK, SICHERHEIT threshold limit value (*TLV*)

hochtemperaturbeständig: **~es Öl** *nt* ERNEU ENER high-temperature-resistant oil

Hochtemperatur: **Hochtemperaturkorrosion** *f* ABFALL high-temperature corrosion; **Hochtemperaturverbrennung** *f* ABFALL, ABWASSER high-temperature incineration; **Hochtemperaturvergasung** *f* (*HTV*) ABFALL, LUFT high-temperature gasification; **Hochtemperaturwärme** *f* ERNEU ENER high-temperature heat

hochtoxisch *adj* ABFALL, BODEN, WASSER highly toxic; **~er Sonderabfall** *m* ABFALL highly toxic hazardous waste

hochturbulent *adj* ABWASSER highly turbulent

hochverschmutzt *adj* ABWASSER highly polluted; **~e Gewässer** *nt pl* ABWASSER highly polluted waters

Hochwasser *nt* MEER, WASSER flood water, high water; **Hochwasserentlastungsanlage** *f* WASSER flood spillway; **Hochwasserschutz** *m* WASSER flood control, flood prevention; **Hochwasserüberlauf** *m* WASSER high-water overflow; **Hochwasserüberschwemmungsgebiet** *nt* BODEN, WASSER flood plain; **Hochwasserüberwachung** *f* WASSER flood control

hochwertig *adj* ABFALL, WASSER high-grade; **~es Recyclingprodukt** *nt* ABFALL high-grade recycling product; **~er Schrott** *m* ABFALL high-grade scrap

hohe: **~ Bodendurchlässigkeit** *f* BODEN high soil permeability; **~ Konzentration** *f* ABFALL, ANALYTIK, LUFT, WASSER high concentration; **~ Rohgasbelastung** *f* ABFALL high dirty gas load

Höhenabsorber *m* LÄRM treble absorber

hohe: **hoher Wasserverbrauch** *m* WASSER high water consumption; **hoher Solarstand** *m* ERNEU ENER *Photovoltaik* high position of the sun; **hoher Sonnenstand** *m* ERNEU ENER *Photovoltaik* high position of the sun

Hohl- *in cpds* ABFALL, LUFT hollow; **Hohlglas** *nt* ABFALL bottle glass

Hohlraum *m* ABFALL, ABWASSER void, *des Tropfkörpers* void space; **Hohlraumabsorber** *m* LÄRM cavity absorber; **Hohlraumvolumen** *nt* LUFT void volume

Hohl-: **Hohlzylinder** *m* LUFT *rotierender Adsorber* hollow cylinder

Holsystem *nt* ABFALL curb-side collection system (*AmE*), kerb-side collection system (*BrE*)

Holz: **Holzabfall** *m* ABFALL wood waste; **Holzfeuerung** *f* ERNEU ENER *Biokonversion* wood firing; **Holzfeuerungsanlage** *f* ERNEU ENER *Biokonversion* wood firing plant; **Holzschnitzel** *m pl* ERNEU ENER *Bioenergienausgangsstoff* wood chips; **Holzspanplatte** *f* ABFALL chipboard

homogenisieren *vt* ABFALL, BODEN, WASSER homogenize

homogen: **~er Abfall** *m* ABFALL homogeneous wastes; **~er Katalysator** *m* LUFT homogeneous catalyst

Homogenisierung *f* ABFALL, BODEN, WASSER homogenization

Homopolymer *nt* ABFALL homopolymer

Homopolymerisat *nt* ABFALL homopolymer

hörbar *adj* LÄRM audible; **~er Schall** *m* LÄRM audible sound

Hörbarkeitsschwelle *f* LÄRM threshold of audibility

Hörbereich *m* LÄRM range of audibility

Horizontal- *in cpds* ALLG horizontal

horizontal *adj* ALLG horizontal; **~er Brunnen** *m* WASSER horizontal well; **~er Dosierstößel** *m* ABFALL, WASSER horizontal feeder ram; **~e Entgasung** *f* ABFALL, WASSER horizontal degassing; **~er Festbettadsorber** *m* LUFT horizontal fixed-bed adsorber; **~er Flächeneinbau** *m* ABFALL, WASSER horizontal surface installation; **~e Gasdränage** *f* ABFALL, BODEN, WASSER horizontal gas drainage; **~e Gaswanderung** *f* ABFALL, BODEN, WASSER horizontal gas migration; **~es linienförmiges Fassungselement** *nt* ABFALL horizontal linear-shaped collection element; **~er Sickerbrunnen** *m* ABFALL, WASSER horizontal percolation well; **~er Wanderbettadsorber** *m* LUFT horizontal moving-bed adsorber

Horizontal-: **Horizontalachsenturbine** *f* ERNEU ENER horizontal axis wind turbine (*HAWT*); **Horizontalfilterbrunnen** *m* (*HFB*) WASSER horizontal filter well, horizontal filtering well

horizontalflächig: **~es Fassungselement** *nt* ABFALL horizontal blanket-type collection element

Hörschall *m* LÄRM audible sound

Hot-dry-rock *nt* ERNEU ENER *geothermische Energieressource* hot dry rock (*HDR*)

Hot-tubing *nt* LUFT hot tubing (*ht*)

HPDC *abbr* (*Hochleistungs-Dünnschichtchromatographie*) ANALYTIK high-performance thin-layer chromatography

HPLC *abbr* (*Hochleistungs-Flüssigkeitschromatographie*) ANALYTIK HPLC (*high-performance liquid chromatography*)

HPLC-MS *abbr* (*Hochleistungs-Flüssigkeitschromatographie in Kopplung mit Massenspektrometrie*) ANALYTIK HPLC-MS (*high-performance liquid chromatography mass spectrometry*)

HSGC *abbr* (*Headspace-Gaschromatographie*) ANALYTIK HSGC (*headspace gas chromatography*)

HTV *abbr* (*Hochtemperaturvergasung*) ABFALL, LUFT high-temperature gasification

Hub *m* ABFALL, ABWASSER, WASSER lifting; **Hubkolbenverdichter** *m* ABFALL, ABWASSER piston compressor; **Hubraum** *m* ABFALL, WASSER *des Pressenkolbens* piston capacity

Humifizierung *f* ABFALL, ABWASSER, BODEN humification; **Humifizierungsgrad** *m* ABFALL, ABWASSER, BODEN degree of humification

Huminsäure *f* ABFALL, ABWASSER, BODEN humic acid

Huminstoff *m* ABFALL, ABWASSER, BODEN humic mass; **Huminstoffbildung** *f* ABFALL, ABWASSER, BODEN humic mass formation

Humus *m* ABFALL, ABWASSER humus; **Humusbehälter** *m* ABWASSER humus tank; **Humusbildung** *f* ABWASSER humus formation

humushaltig: **~er Boden** *m* BODEN humus-containing soils

100-Punkt-Kalibrierung *f* ANALYTIK hundred points calibration

Huygens-Fresnel: **~sches Prinzip** *nt* ANALYTIK Huygen-Fresnel principle

Hydrant *m* WASSER water hydrant

Hydratverfahren *nt* ABWASSER *Entsalzung* hydrate process

Hydraulik- *in cpds* ALLG hydraulic; **Hydraulikflüssigkeit** *f* ERNEU ENER hydraulic fluid; **Hydraulikpumpe** *f* ABFALL, ABWASSER, LUFT, WASSER pump

hydraulisch *adj* ALLG hydraulic; **~er Antrieb** *m* ERNEU ENER hydraulic transmission; **~e Begleitmaßnahmen** *f pl* BODEN accompanying hydraulic measures; **~e Belastung** *f* ABWASSER hydraulic loading; **~ betrieben** *adj* ABFALL, WASSER hydraulically operated; **~ betriebene Kippbühne** *f* ABFALL hydraulically operated tipping ramp; **~ betriebener Dosierstößel** *m* ABFALL, WASSER hydraulically operated feeder ram; **~er Durchmesser** *m* ABFALL, WASSER hydraulic diameter; **~ durchmischter Flockungsreaktor** *m* ABWASSER hydraulically mixed flocculator; **~ gesteuert** *adj* LUFT hydraulically controlled; **~ gesteuerte Klappe** *f* LUFT hydraulically controlled flap; **~er Klassierer** *m* BODEN hydraulic classifier; **~e Presse** *f* ABFALL hydraulic press; **~er Radius** *m* ABFALL, WASSER hydraulic radius; **~er Verlust** *m* ABFALL, ABWASSER, WASSER hydraulic loss; **~er Widder** *m* ABWASSER hydraulic ram

hydrieren *vt* ABFALL, ERNEU ENER, LUFT hydrogenate

Hydrierung *f* ABFALL, ERNEU ENER, LUFT hydrogenation

Hydro- *in cpds* ALLG hydro; **Hydrodynamik** *f* WASSER hydrodynamics

hydrodynamisch *adj* BODEN hydrodynamic; **~e Verhältnisse** *nt pl* **im Untergrund** BODEN hydrodynamic behavior in subsoil (*AmE*), hydrodynamic behaviour in subsoil (*BrE*)

Hydrogencarbonat *nt* WASSER hydrocarbonate

Hydro-: **Hydrogeologie** *f* BODEN, WASSER hydrogeology

hydrogeologisch: **~e Untersuchung** *f* BODEN, WASSER hydrogeological analysis

hydrokinetisch *adj* ABWASSER hydrokinetic

Hydro-: **Hydroliftreaktor** *m* LUFT hydrolift reactor

hydrolisierbar *adj* ABFALL, ABWASSER, ERNEU ENER, WASSER hydrolysable (*BrE*), hydrolyzable (*AmE*)

Hydrologie *f* WASSER hydrology

hydrologisch *adj* ABFALL, WASSER hydrological; **~e Bilanz** *f* ABFALL, WASSER hydrological balance; **~er Zyklus** *m* ABFALL, WASSER hydrological cycle

Hydrolyse *f* ABFALL, ABWASSER, ERNEU ENER, WASSER hydrolysis

hydrolysieren *vt* ABFALL, ABWASSER, ERNEU ENER, WASSER hydrolyse (*BrE*), hydrolyze (*AmE*)

hydrolysiert *adj* ABFALL, ABWASSER, ERNEU ENER, WASSER hydrolysed (*BrE*), hydrolyzed (*AmE*); **~er Rohschlamm** *m* ABWASSER hydrolysed raw sludge (*BrE*), hydrolyzed raw sludge (*AmE*)

hydrolytisch *adj* ABFALL, ABWASSER, ERNEU ENER, WASSER hydrolytic

Hydrometrie *f* WASSER hydrometry

hydrophil *adj* ABWASSER, LUFT hydrophilic

hydrophob *adj* ABWASSER, LUFT hydrophobic

Hydro-: **Hydropumpe** *f* MEER pump; **Hydrosphäre** *f* BODEN, WASSER hydrosphere; **Hydrostatik** *f* WASSER hydrostatics

hydrostatisch *adj* ABWASSER hydrostatic; **~e Barriere** *f* ABWASSER hydrostatic barrier; **~er Druck** *m* ABFALL, ABWASSER, WASSER hydrostatic pressure; **~e Druckdifferenz** *f* ABWASSER hydrostatic pressure difference

hydrothermal *adj* ERNEU ENER *geothermische Energieressource* hydrothermal

Hydroxidschlamm *m* ABWASSER, WASSER hydroxide sludge

Hydro-: **Hydrozyklon** *m* ABWASSER, BODEN, LUFT, WASSER hydrocyclone; **Hydrozyklonbatterie** *f* LUFT hydrocyclone battery
Hygienepapier *nt* ABFALL toilet paper, toilet tissue
Hygienisierung *f* ABFALL sanitation
Hygrometer *nt* ABFALL, ANALYTIK, LUFT, WASSER hygrometer
Hygrometrie *f* ABFALL, ANALYTIK, LUFT, WASSER hygrometry
Hygroskop *nt* ABFALL, ANALYTIK, LUFT, WASSER hygroscope
hygroskopisch *adj* ABFALL, ANALYTIK, LUFT, WASSER hygroscopic; **~es Wasser** *nt* ABFALL, WASSER hygroscopic water
Hygroskopizität *f* ABFALL, ANALYTIK, LUFT, WASSER hygroscopicity
Hyperschall- *in cpds* LÄRM hypersonic; **Hyperschallbereich** *m* LÄRM hypersonic range
Hypochlorung *f* ABWASSER hypochlorination
Hypolimnion *nt* ABWASSER hypolimnion
HZK *abbr* ANALYTIK (*höchstzulässige Konzentration*) MAC (*maximum allowable concentration*), LÄRM (*höchstzulässiger Konzentrationswert*) TLV (*threshold limit value*), LUFT (*höchstzulässige Konzentration*) MAC (*maximum allowable concentration*), POLITIK (*höchstzulässiger Konzentrationswert*) TLV (*threshold limit value*), SICHERHEIT, WASSER (*höchstzulässige Konzentration*) MAC (*maximum allowable concentration*), (*höchstzulässiger Konzentrationswert*) TLV (*threshold limit value*); **~ am Arbeitsplatz** *f* SICHERHEIT MAC at the workplace; **~ in der Umgebung** *m* SICHERHEIT TLV in the free environment

I

IA *abbr* (*Ionenaustausch*) ABWASSER, ANALYTIK, WASSER IE (*ion exchange*)

IC *abbr* (*Ionenchromatographie*) ANALYTIK IC (*ion chromatography*)

ICP-MS *abbr* (*induktiv gekoppeltes Plasma in Kopplung mit Massenspektrometrie*) ANALYTIK ICP-MS (*inductively coupled plasma mass spectrometry*)

ICP-OES *abbr* (*Emissionsspektrometrie mit induktiv gekoppeltem Plasma*) ANALYTIK ICP-OES (*inductively coupled plasma with optical emission spectrometry*)

ideal: **~es Gas** *nt* LUFT ideal gas, perfect gas

identifizieren *vt* ANALYTIK identify

Identifizierung *f* ANALYTIK *einer Probe* identification

illegal *adj* ABFALL, ABWASSER, POLITIK illegal; **~e Dünnsäureverklappung** *f* ABFALL illegal barging of spent acid, MEER illegal dumping of spent acid, WASSER illegal barging of spent acid

Imhoffbrunnen *m* ABWASSER Imhoff tank

Imhofftrichter *m* ABWASSER Imhoff sedimentation cone

Immission *f* ANALYTIK, LUFT ambient air; **Immissionsanalyse** *f* ANALYTIK, LUFT ambient air analysis; **Immissionsgrenzwert** *m* ANALYTIK, LUFT ambient air concentration value; **Immissionsgrenzwert** *m* **der Luft** ANALYTIK, LUFT ambient air quality standard; **Immissionskonzentration** *f* ANALYTIK, LUFT ambient air concentration; **Immissionsmeßgerät** *nt* ANALYTIK, LUFT ambient air quality analyser (*BrE*), ambient air quality analyzer (*AmE*); **Immissionsmessung** *f* ANALYTIK, LUFT ambient air measuring

immissionsneutral *adj* ABFALL, LUFT ambient-air neutral

Immission: **Immissionsrecht** *nt* LUFT, POLITIK law on ambient air

immissionsschutzrechtlich *adj* LUFT, POLITIK according to the law on ambient air protection; **~e Anlagengenehmigung** *f* POLITIK plant permit according to the law on ambient air protection; **~er Widerruf** *m* POLITIK, WASSER repeal according to law on ambient air protection

Immission: **Immissionswert** *m* ANALYTIK, LUFT ambient air quality value

immobilisieren *vt* ABFALL, ABWASSER, ANALYTIK, BODEN immobilize

immobilisiert *adj* ABFALL, ABWASSER, ANALYTIK, BODEN immobilized; **~er Mikroorganismus** *m* ABWASSER immobilized microorganism

Immobilisierung *f* ABFALL, ABWASSER, ANALYTIK, BODEN immobilization

Immunoanalyse *f* ANALYTIK immunoassay test

Immunversuch *m* ANALYTIK immunoassay

Impaktion *f* ANALYTIK *Meßverfahren für Staube* impaction

Impaktor *m* ANALYTIK impactor, impinger, LUFT impactor

impfen *vt* BODEN *Mikroorganismus und Nährstoffe* inject

Impfkompostrückführung *f* ABFALL recycling of inoculated compost

Impfschlamm *m* ABFALL, WASSER seeding sludge

importieren *vt* ERNEU ENER import

imprägnieren *vt* ANALYTIK, LUFT impregnate

imprägniert *adj* ANALYTIK, LUFT impregnated; **~es Filterpapier** *nt* ANALYTIK impregnated filter paper

Imprägnierung *f* ANALYTIK, LUFT impregnation

Impulslärm *m* LÄRM impulse noise

Impulsschallpegelmesser *m* ANALYTIK impulse sound level meter

inaktivieren *vt* ABFALL, WASSER inactivate

Inaktivierung *f* ABFALL, WASSER inactivation

Indikator *m* ABWASSER, ANALYTIK indicator, LUFT tracer, WASSER indicator; **Indikatorelektrode** *f* ANALYTIK indicating electrode; **Indikatorlösung** *f* ANALYTIK indicator solution; **Indikatororganismus** *nt* ABWASSER indicator organism; **Indikatorpapier** *nt* ANALYTIK indicator paper

indirekt *adj* ABWASSER, ANALYTIK, LUFT indirect; **~ beheizter Drehrohrofen** *m* BODEN indirect-heated rotary kiln; **~e Beheizung** *f* BODEN indirect heating; **~e Grundwasseranreicherung** *f* ABFALL, WASSER indirect groundwater recharge; **~e Industrieabwassereinleitung** *f* ABWASSER indirect industrial wastewater discharge; **~e Meßmethode** *f* ANALYTIK indirect measurement method; **~e Trocknung** *f* ABWASSER indirect drying

Indirekteinleiter *m* ABWASSER indirect discharger; **Indirekteinleiterquellen** *f pl* LUFT indirect emission sources

Indirekteinleitung *f* ABWASSER indirect discharge

Induktion *f* ABWASSER, LUFT induction; **Induktionserwärmung** *f* LUFT induction heating; **Induktionsheizung** *f* LUFT induction heating; **Induktionsmotor** *m* **mit gewickeltem Läufer** ABWASSER *Pumpenantrieb* wound-rotor induction motor

induktiv: **~ gekoppeltes Plasma** *nt* **in Kopplung mit Massenspektrometrie** (*ICP-MS*) ANALYTIK inductively coupled plasma mass spectrometry (*ICP-MS*)

Industrie *f* ALLG industry; **Industrieabfall** *m* ABFALL industrial waste; **Industrieabwasser** *nt* ABWASSER, WASSER industrial effluent, industrial wastewater; **Industrieabwasserreinigung** *f* ABWASSER, WASSER industrial wastewater purification

industrieeigen: **~e Kläranlage** *f* ABWASSER company-owned treatment plant

Industrie: **Industriegas** *nt* LUFT manufactured gas

industriell *adj* ALLG industrial; **~e Emissionen** *f pl* LUFT industrial emissions; **~er Klärschlamm** *m* ABWASSER industrial wastewater sludge; **~es Recycling** *nt* ABFALL industrial recycling; **~es Schüttgutcontainersystem** *nt* ABFALL industrial bulk container system

Industrie: **Industriemülldeponie** *f* ABFALL industrial landfill; **Industrienorm** *f* ALLG industrial standard

industriespezifisch *adj* ALLG industry-specific; **~e Reinigung** *f* ALLG industry-specific purification

Industrie: **Industriestandort** *m* BODEN industrial site; **Industrieunfall** *m* SICHERHEIT industrial accident; **Industrieverpackung** *f* ABFALL industrial packing; **Industrieversicherung** *f* POLITIK, SICHERHEIT industrial insurance; **Industriewasserversorgung** *f* WASSER industrial water supply

Inert- *in cpds* ABFALL, LUFT, WASSER inert

inert: **~er Abfall** *m* ABFALL, ABWASSER inert waste; **~er Verschluß** *m* ANALYTIK *Probeflasche* inert closure

Inert-: **Inertdeponie** *f* ABFALL inert landfill; **Inertfraktion** *f* ABFALL inert fraction

Inertgas *nt* LUFT inert gas; **Inertgasbrennkammer** *f* LUFT inert gas combustion chamber; **Inertgasstrom** *m* LUFT inert gas flow

Inertialabscheider *m* LUFT inertial separator

inertisieren *vt* ABWASSER render inert

inertisiert *adj* ABWASSER rendered inert; **~e Rückstände** *m pl* ABFALL *unverbrennbar* incombustible residue

Inertisierung *f* ABFALL *von Schadstoffen* inerting, BODEN *biologische Behandlung* mineralization, LUFT inerting

Inert-: **Inertmaterial** *nt* ABFALL, LUFT inert material; **Inertstoffdeponie** *f* ABFALL material landfill

infektiös: **~er Abfall** *m* ABFALL, ABWASSER, MEER anatomical waste, infectious waste, pathological waste; **~er Krankenhausabfall** *m* ABFALL infectious hospital waste

Infiltration *f* ABWASSER, WASSER infiltration; **Infiltrationsbecken** *nt* ABWASSER infiltration basin, infiltration tank; **Infiltrationsbrunnen** *m* BODEN, WASSER infiltration well; **Infiltrationsfläche** *f* WASSER area of water filtration; **Infiltrationsgebiet** *nt* WASSER area of water filtration; **Infiltrationsgelände** *nt* WASSER *Wasserwerke* infiltration ground; **Infiltrationsgewässer** *nt pl* WASSER infiltration waters; **Infiltrationsrate** *f* WASSER infiltration rate

infiltriert: **~es Rohwasser** *nt* ABFALL, WASSER infiltrated raw water

Informationsinstrumente *nt pl* POLITIK information instruments

Infrarot *nt* (*IR*) ANALYTIK, LUFT infrared (*IR*); **Infrarotbereich** *m* ANALYTIK infrared region; **Infrarot-Durchlaufofen** *m* ABFALL infrared through-type furnace, infrared continuous furnace; **Infrarotofen** *m* ANALYTIK infrared kiln; **Infrarotspektrometrie** *f* ANALYTIK infrared spectrometry; **Infrarotspektrometrie** *f* **mit Fourier-Transformation** (*FTIR*) ANALYTIK Fourier-transform infrared spectrometry (*FTIR*); **Infrarotspektrum** *nt* ANALYTIK infrared spectrum; **Infrarotstrahlung** *f* ABFALL, LUFT infrared radiation

infrarot-thermisch: **~e Zerstörung** *f* BODEN infrared thermal destruction

Inhaltskennzeichnung *f* SICHERHEIT *Industriegasbehälter* identification of contents

Inhaltsstoffe: **~ der Proben** *m pl* ANALYTIK sample components

Inhibitor *m* LUFT, WASSER inhibitor

Injektion *f* ANALYTIK, BODEN injection; **Injektionsanlage** *f* BODEN *Einpressen* injection plant; **Injektionsbohrung** *f* BODEN injection drilling; **Injektionsbrunnen** *m* BODEN injection well; **Injektionsflüssigkeit** *f* BODEN grouting fluid, injection fluid; **Injektionslanze** *f* BODEN injection lance; **Injektionsspritzen** *nt* ANALYTIK injection spraying; **Injektionssystem** *nt* ANALYTIK injection system

Injektor *m* ANALYTIK injector; **Injektorwirkung** *f* LUFT *Schlauchfilter* injector action

injizieren *vt* ANALYTIK, BODEN inject

Inkrustation *f* ABFALL, BODEN, WASSER incrustation

Inkubation *f* ANALYTIK incubation; **Inkubationszeit** *f* ANALYTIK incubation period

inkubieren *vt* ANALYTIK incubate

inkubiert *adj* ANALYTIK incubated

Inlands-: **Inlandsemissionen** *f pl* LUFT domestic emissions; **Inlandsquelle** *f* WASSER internal source

Innen- *in cpds* ABFALL, ANALYTIK, LUFT, WASSER internal

innen: **~ polierter Zyklon** *m* LUFT interior-polished cyclone

Innen-: **Innendruck** *m* LUFT *Schlauchfilter* internal pressure; **Innenelektrode** *f* ANALYTIK interior electrode; **Innenpanzerung** *f* LUFT wear-resistant lining; **Innenraumluftverunreinigung** *f* LUFT indoor air pollution; **Innentrombe** *f* LUFT inner vortex; **Innenwirbel** *m* ABFALL, WASSER *Zentrifugalfeld* internal eddy

innerbetrieblich: **~e Verwertung** *f* ABFALL in-house recycling; **~er Wasserkreislauf** *m* ABWASSER in-house water recirculation system

innovativ: **~e Technik** *f* ABWASSER, ERNEU ENER, LUFT, WASSER innovative technologies

In-situ *in cpds* ALLG in-situ; **~-biologische Altlastensanierung** *f* BODEN *für Kohlenwasserstoffe* in-situ bioremediation; **~-Denitrifikation** *f* WASSER in-situ denitrification; **~-Extraktion** *f* BODEN in-situ extraction; **~ thermische Desorption** *f* BODEN in-situ thermal desorption; **~-Verbrennung** *f* BODEN on-site incineration; **~-Verfahren** *nt* BODEN in-situ method; **~-Verfestigung** *f* BODEN in-situ solidification; **~-Verglasung** *f* BODEN in-situ vitrification (*ISV*)

insolubel *adj* ABFALL, ABWASSER, WASSER insoluble

Inspektion *f* ALLG inspection

Inspektor *m* ALLG supervisor

instabil *adj* ABFALL unstable, LUFT *Abgasfahne-Ausbreitungsklasse* labile, WASSER unstable; **~e Methangärung** *f* ABFALL unstable methane fermentation

Instabilität *f* ABFALL instability

Installation *f* ABFALL erection, installation, ABWASSER, ANALYTIK installation, BODEN, ERNEU ENER, LÄRM, LUFT, MEER, SICHERHEIT, WASSER erection, installation

installieren *vt* ABWASSER install, ALLG erect, BODEN, ERNEU ENER, LÄRM, LUFT, MEER, SICHERHEIT, WASSER install

Instandhaltung *f* ALLG maintenance

Institution *f* POLITIK institution

instrumentell: **~e Analytik** *f* ANALYTIK instrumental analysis

integral *adj* ANALYTIK integral

Integrator *m* ANALYTIK *zur Auswertung* integrator

integriert *adj* LUFT, MANAG, POLITIK integrated; **~e Abfallbewirtschaftung** *f* MANAG integrated waste management (*IWM*); **~er Ansatz** *m* POLITIK integrated approach; **~e Grenzwerte** *m pl* POLITIK integrated limit values; **~es System** *nt* ABFALL *Hausmüllsystem* integrated system; **~er Umweltschutz** *m* MANAG, POLITIK integrated environmental protection; **~e Vermeidung** *f* **der Umweltverschmutzung** (*IVU*) MANAG, POLITIK integrated pollution control (*IPC*); **~e Vorsorge** *f* **der Umweltverschmutzung** MANAG integrated pollution prevention; **~e Vorsorge** *f* **und Vermeidung** *f* **der Umweltverschmutzung** MANAG, POLITIK integrated prevention and pollution control (*IPPC*); **~es Wärmeaustauschersystem** *nt* LUFT integrated heat exchanger system

Intensität *f* ANALYTIK intensity; **~ der UV-Strahlung** LUFT intensity of UV radiation

intensiv: **~er Ackerbau** *m* BODEN intensive farming

Intensivrotte *f* ABFALL intensive rotting, ABWASSER rapid rotting

Interferenz *f* ANALYTIK interference; **Interferenzfilter** *nt* ANALYTIK interference filter; **Interferenzlichtfilter** *nt* ANALYTIK interference filter; **Interferenzmeßverfahren** *nt* ANALYTIK interferometry; **Interferenzrefraktometer** *nt* ANALYTIK interference refractometer

Interferogramm *nt* ANALYTIK interferogram

Interferometer *nt* ANALYTIK interferometer

intermittierend *adj* ANALYTIK, LUFT intermittent; **~e Reinigung** *f* LUFT *Filterschlauch* intermittent cleaning

intern: **~e Rezirkulation** *f* ABWASSER internal recirculation

Internalisierung *f* POLITIK internalization; **~ externer Effekten** POLITIK internalization of external effects, internalization of externality; **~ externer Kosten** POLITIK internalization of external costs

International: **~e Normungsorganisation** *f* (*ISO*) ANALYTIK, POLITIK International Standards Organization (*ISO*); **~es Referenzsystem** *nt* (*IRS*) POLITIK *GAK* International Referral System (*IRS*)

international: **~ verhandelbare Emissionsrechte** *nt pl* POLITIK internationally tradeable emissions rights; **~e Verhandlung** *f* POLITIK international negotiation

International: **~es Verzeichnis** *nt* **für potentiell toxische Chemikalien** (*IRPTC*) POLITIK International Register of Potentially Toxic Chemicals (*IRPTC*)

Intervallwert *m* ANALYTIK threshold value range

Intrusion *f* ABFALL, WASSER intrusion

Inversion: **Inversionskappe** *f* LUFT lid; **Inversionsschicht** *f*

LUFT atmospheric inversion layer, inversion layer; **Inversionswetterlage** *f* LUFT inversion atmospheric conditions

Inversvoltammetrie *f* ANALYTIK inverse voltammetry

involutorisch *adj* LUFT involutional

iodometrisch *adj* ANALYTIK iodometric; **~e Bestimmung** *f* ANALYTIK iodometric determination; **~es Verfahren** *nt* ANALYTIK iodometric method

Ion *nt* ABWASSER, ANALYTIK, WASSER ion; **Ionaustauschermembran** *f* ABWASSER ion exchanger membrane

Ionen- *in cpds* ABWASSER, ANALYTIK, BODEN, WASSER ionic; **Ionenaktivität** *f* ANALYTIK ion activity; **Ionenaustausch** *m* (*IA*) ABWASSER, ANALYTIK, WASSER ion exchange (*IE*); **Ionenaustauschenthärter** *m* ANALYTIK, WASSER ion exchange softener; **Ionenaustauschenthärtung** *f* ABWASSER, WASSER ion exchange softening; **Ionenaustauscher** *m* ABWASSER, WASSER ion exchanger; **Ionenaustauschharz** *nt* BODEN, WASSER ion exchange resin; **Ionenbeweglichkeit** *f* ANALYTIK ionic mobility

Ionenchromatographie *f* (*IC*) ANALYTIK ion chromatography (*IC*)

Ionen-: **Ionenfalle** *f* ANALYTIK ion trap; **Ionenfalle-Detektor** *m* ANALYTIK ion trap detector; **Ionenfluß** *m* LUFT *Elektrofilter* ion current; **Ionenform** *f* ABWASSER, ANALYTIK, WASSER ionic form; **Ionenhaushalt** *m* BODEN, LUFT, WASSER ion budget; **Ionenkonzentration** *f* ABWASSER, WASSER ionic concentration; **Ionenpaar** *nt* ANALYTIK ion pair (*IP*); **Ionenpaarchromatographie** *f* ANALYTIK IP chromatography; **Ionenquelle** *f* ANALYTIK ion source

ionenselektiv *adj* ABFALL, ABWASSER, ANALYTIK, WASSER ion-selective; **~e Elektrode** *f* (*ISE*) ANALYTIK ion-selective electrode (*ISE*); **~e Elektrode** *f* **mit Glasmembran** ANALYTIK ion-selective electrode with glass membrane; **~e Membran** *f* ABFALL, ABWASSER, WASSER ion-selective membrane

ionensensitiv *adj* ANALYTIK ion-sensitive; **~e Elektrode** *f* ANALYTIK ion-sensitive electrode

Ionen-: **Ionenstrom** *m* LUFT *Elektrofilter* ion current; **Ionenwanderung** *f* BODEN ionic migration; **Ionenwertigkeit** *f* ABWASSER ionic valence

Ionisation *f* ANALYTIK, BODEN, LUFT, WASSER ionization; **~ durch Laser** (*LEI*) ANALYTIK laser-enhanced ionization (*LEI*); **Ionisationskammer** *f* ANALYTIK ionization chamber; **Ionisationszone** *f* LUFT ionizing zone

ionisch *adj* ALLG ionic

ionisieren *vt* ALLG ionize

ionisierend *adj* ALLG ionizing; **~er Naßwäscher** *m* LUFT ionizing wet scrubber (*IWS*); **~e Strahlung** *f* ABFALL, BODEN, LUFT, WASSER ionizing radiation

ionisiert *adj* ALLG ionized; **~es Gasmolekül** *nt* LUFT ionized gas molecule

Ionisierung *f* ANALYTIK, BODEN, LUFT, WASSER ionization; **Ionisierungsenergie** *f* ANALYTIK ionization potential

ionogen *adj* ABWASSER, ANALYTIK ionogenic

IR *abbr* (*Infrarot*) ANALYTIK, LUFT, WASSER IR (*infrared*)

Irisieren *nt* MEER iridescence

IR: **IR-Laser-Rückstreuung** *f* ANALYTIK IR laser backscattering; **IR-Lichtabsorption** *f* ANALYTIK IR optical absorption

IRPTC *abbr* (*Internationales Verzeichnis für potentiell toxische Chemikalien*) POLITIK IRPTC (*International Register of Potentially Toxic Chemicals*)

irreversibel *adj* ABWASSER, LUFT irreversible; **irreversibles Verfahren** *nt* ABWASSER, LUFT irreversible process

IRS *abbr* (*Internationales Referenzsystem*) POLITIK IRS (*International Referral System*)

IR: **IR-Spektralphotometer** *nt* ANALYTIK IR spectrophotometer; **IR-Spektrometrie** *f* ANALYTIK IR spectrometry; **IR-Spektrophotometer** *nt* ANALYTIK IR spectrophotometer; **IR-Strahlung** *f* LUFT infrared radiation; **IR-Thermometer** *nt* ANALYTIK IR thermometer; **IR-Trocknung** *f* ANALYTIK IR drying; **IR-VOC Analysator** *m* ANALYTIK *von flüchtigen organischen Verbindungen* IR-VOC analyser (*BrE*)

ISE *abbr* (*ionenselektive Elektrode*) ANALYTIK ISE (*ion-selective electrode*)

ISO *abbr* (*Internationale Normungsorganisation*) ANALYTIK, POLITIK ISO (*International Standards Organization*)

isokinetisch *adj* ANALYTIK isokinetic; **~es Ansaugen** *nt* ANALYTIK isokinetic suction; **~e Probenahme** *f* ANALYTIK isokinetic sampling; **~es Probenehmen** *nt* ANALYTIK isokinetic sampling

Isolation *f* ALLG insulation

Isolator *m* ANALYTIK *Halbleiterdetektor*, LUFT insulator

Isolier- *in cpds* ALLG insulating; **Isoliereigenschaften** *f pl* LÄRM insulating properties

isolieren *vt* ALLG insulate

Isolier-: **Isolierglas** *nt* LÄRM insulating glass; **Isoliermasse** *f* LÄRM insulating compound; **Isolierplatte** *f* LÄRM structural insulation board; **Isolierstoff** *m* LÄRM insulating material

isoliert *adj* ALLG insulated

Isolierung *f* ANALYTIK *einer Probe* isolation, ERNEU ENER, LÄRM, LUFT, MEER, SICHERHEIT insulation

isotherm *adj* ANALYTIK, LUFT isothermal; **~er Gaschromatograph** *m* ANALYTIK isothermal gas chromatograph; **~e Gaschromatographie** *f* ANALYTIK isothermal gas chromatography

Isotherme *f* ANALYTIK, LUFT isotherm

isothermisch *adj* ANALYTIK, LUFT isothermal

Iteration *f* LUFT iteration

IVU *abbr* (*integrierte Vermeidung der Umweltverschmutzung*) MANAG, POLITIK IPC (*integrated pollution control*)

J

Jahres- *in cpds* ABFALL, ANALYTIK, WASSER annual; **Jahresabfluß** *m* WASSER annual runoff; **Jahresdurchfluß** *m* WASSER annual flow; **Jahreshochwasser** *nt* WASSER annual flood; **Jahresmittelwert** *m* ANALYTIK annual mean value; **Jahresmüllmenge** *f* ABFALL annual refuse volume; **Jahresverbrauch** *m* WASSER yearly consumption; **Jahreswasserverbrauch** *m* WASSER annual water consumption; **Jahreswindgeschwindigkeit** *f* ERNEU ENER annual wind velocity

jährlich *adj* ABFALL, ANALYTIK, WASSER annual; **~e Entwässerung** *f* ABFALL, WASSER annual drainage; **~e Last** *f* ABFALL, WASSER annual load; **~er Mittelwasserstand** *m* WASSER annual mean water level; **~e Niederschlagshöhe** *f* WASSER annual precipitation; **~e Verdunstung** *f* ABFALL, WASSER annual evaporation

Jamin-Interferometer *nt* ANALYTIK Jamin interferometer

Jar-Test *m* ANALYTIK, WASSER jar test

Jauche *f* ABFALL, ABWASSER, WASSER slurry, liquid manure

Jigger *m* ABFALL, BODEN, WASSER jigger

Joule-Heizung *f* BODEN *thermische Desorption* joule heating

juristisch *adj* POLITIK juridical

juvenil: **~es Grundwasser** *nt* ABFALL, WASSER juvenile groundwater; **~es Wasser** *nt* ABFALL, WASSER juvenile water

K

Kabeltrommel *f* ABFALL cable drum

Kabine *f* SICHERHEIT *Baumaschinen* cab; **~ mit integrierter Aktivkohlefilteranlage** SICHERHEIT cab with integrated activated carbon filter unit; **Kabinenfilter** *nt* SICHERHEIT cab filter

Käfigläufermotor *m* ERNEU ENER *Windturbine* squirrel-cage induction motor

Kalender *m* LUFT *im Filtermedium* calender

Kalibrationsgerät *nt* ANALYTIK calibration device

Kalibrator *m* ANALYTIK calibrator

Kalibrier- *in cpds* ANALYTIK, calibration; **Kalibrierbereich** *m* ANALYTIK calibration range; **Kalibrierdaten** *pl* ANALYTIK calibration data; **Kalibriereinheit** *f* ANALYTIK calibration unit

kalibrieren *vt* ANALYTIK calibrate, gage (*AmE*), gauge (*BrE*), graduate

Kalibrieren *nt* ANALYTIK calibration

Kalibrier-: **Kalibriergas** *nt* ANALYTIK calibration gas; **Kalibriergasflasche** *f* ANALYTIK calibration gas bottle; **Kalibrierimpuls** *m* ANALYTIK calibration pulse; **Kalibrierkurve** *f* ANALYTIK calibration curve; **Kalibriermodus** *m* ANALYTIK calibration mode; **Kalibrierprotokoll** *nt* ANALYTIK calibration record; **Kalibrierpuffer** *m* ANALYTIK *pH-Meter* calibration buffer; **Kalibrierspeicher** *m* ANALYTIK calibration storer

kalibriert *adj* ANALYTIK calibrated

Kalibrierung *f* ANALYTIK calibration

Kalibrierwert *m* ANALYTIK calibrated value

Kalk *m* ABFALL *Glasrecycling*, BODEN, WASSER lime; **Kalkablagerung** *f* ABFALL furring; **Kalkaggressivität** *f* WASSER aggressivity of lime

kalkangreifend *adj* ABFALL, ABWASSER, WASSER lime-aggressive

kalkarm: **~er Schotter** *m* ABFALL low-lime-content rubble

Kalk: **Kalkfällung** *f* ABWASSER precipitation with lime

kalkhaltig *adj* ABFALL, WASSER calcareous; **~es Wasser** *nt* ABFALL, WASSER hard water

Kalk: **Kalkhärte** *f* WASSER calcium hardness; **Kalkhydrat** *nt* ABWASSER calcium hydroxide, hydrated lime, slaked lime; **Kalkkohlensäuregleichgewicht** *nt* WASSER carbonate balance; **Kalkkonzentration** *f* WASSER lime concentration; **Kalklöschen** *nt* ABFALL lime slaking

kalklöschend *adj* ABFALL lime-slaked

Kalk: **Kalkmilch-Anrührbehälter** *m* LUFT slaked lime mixing tank; **Kalkmilchsuspension** *f* LUFT slaked lime suspension; **Kalkschlamm** *m* ABFALL lime slurry, WASSER lime sludge; **Kalksinter** *m* ABFALL, WASSER calcareous sinter; **Kalksteinmehl** *nt* ABFALL, WASSER *Additiv* limestone meal; **Kalksteinschotter** *m* ABFALL, WASSER limestone ballast; **Kalksteinsuspension** *f* ABFALL, WASSER limestone suspension

Kalkung *f* WASSER liming

Kalk: **Kalkwasser** *nt* WASSER lime water; **Kalkwasserbereitung** *f* WASSER lime water treatment

Kalorimeter *nt* ANALYTIK calorimeter; **Kalorimeterbombe** *f* ANALYTIK bomb calorimeter

Kalorimetrie *f* ANALYTIK calorimetry

kalorimetrisch *adj* ANALYTIK calorimetric; **~e Pararosanilinmethode** *f* ANALYTIK calorimetric pararosaniline method

kalt: **~es Meerwasser** *nt* ERNEU ENER *Meereswärmekraftwerk* cold seawater

Kaltdampf-Atomabsorptionsspektroskopie *f* ANALYTIK cold-vapor AAS (*AmE*), cold-vapour AAS (*BrE*)

Kältemittel *nt* ABFALL, LUFT refrigerant

kalzinieren *vt* ABFALL, LUFT, WASSER calcine

Kalzinierung *f* LUFT calcination

Kamin *m* ABFALL, ERNEU ENER, LUFT chimney, flue, smoke stack, stack; **Kaminaustritt** *m* LUFT plume; **Kaminauswurf** *m* LUFT chimney discharge, chimney emissions; **Kaminbauhöhe** *f* LUFT overall stack height, overall chimney height; **Kamindurchmesser** *m* LUFT stack diameter; **Kaminhöhe** *f* LUFT stack height, chimney height; **Kaminmündung** *f* LUFT stack orifice; **Kaminruß** *m* LUFT soot; **Kaminwand** *f* LUFT stack wall

Kammer *f* ALLG chamber; **Kammerfilterpresse** *f* ABFALL, BODEN, WASSER chamber filter press; **Kammerwasserspiegel** *m* ERNEU ENER *Wellenkraftwerk* chamber water level

Kanal *m* LÄRM duct, LUFT channel, ABFALL, ABWASSER, WASSER canal, channel; **Kanalabzweig** *m* WASSER lateral canal; **Kanalabzweigung** *f* WASSER canal branching; **Kanalauskleidung** *f* LÄRM duct lining; **Kanaleinmündung** *f* WASSER canal junction; **Kanalentwässerung** *f* WASSER canal drainage

Kanalisation *f* ABFALL drainage system, ABWASSER sewerage, sewerage system, sewer system, ABFALL, ABWASSER, WASSER canalization; **Kanalisationsnetz** *nt* WASSER canalization; **Kanalisationssystem** *nt* WASSER canalization

kanalisieren *vt* ABFALL, ABWASSER, WASSER canalize

Kanal: **Kanalnetz** *nt* ABWASSER sewerage system, sewer system, WASSER drainage system; **Kanalradkreiselpumpe** *f* WASSER channel impeller pump; **Kanalradpumpe** *f* ABWASSER nonclogging pump, nonclog pump, WASSER nonclogging pump; **Kanalrekuperator** *m* LUFT channel recuperator; **Kanalverengung** *f* WASSER canal constriction

Kanister *m* ABFALL canister

Kantenpressung *f* ABFALL edge pressure

Kapazität *f* ALLG capacity; **Kapazitätsbelastung** *f* ABFALL *bei Fahrzeug* carrying capacity

Kapillar- *in cpds* ALLG capillary

kapillar *adj* ANALYTIK capillary; **~e Steighöhe** *f* ABFALL, BODEN, WASSER capillary rise

Kapillare *f* ANALYTIK capillary

Kapillar-: **Kapillar-Elektrophorese** *f* ANALYTIK capillary electrophoresis (*CE*); **Kapillar-Elektrophorese** *f* **in Kopplung mit Massenspektrometrie** ANALYTIK capillary electrophoresis mass spectrometry (*CE-MS*); **Kapillar-GC** *f* ANALYTIK capillary GC; **Kapillar-GC-MS** *nt* ANALYTIK capillary GC-MS; **Kapillargel-Elektrophorese** *f* ANALYTIK capillary gel electrophoresis (*CGE*)

kapillarisoelektrisch: **~e Fokussierung** *f* (*CIEF*) ANALYTIK capillary isoelectric focusing (*CIEF*)

Kapillar-: **Kapillarrohr** *nt* ANALYTIK capillary tube; **Kapillarströmungsmesser** *m* ANALYTIK capillary flow meter; **Kapillarviskosimeter** *nt* ANALYTIK capillary viscosimeter; **Kapillarwasser** *nt* ABWASSER, BODEN, WASSER capillary water; **Kapillarzone-Elektrophorese** *f* ANALYTIK capillary zone electrophoresis (*CZE*)

Kaplanturbine *f* ERNEU ENER *Gezeitenkraftwerk* Kaplan turbine

Kapnograph *m* ANALYTIK *Messung der Staubkonzentration* capnograph

Kapseln *nt* LÄRM *um Schallausbreitung zu verhindern* enclosure

Kapselung *f* LÄRM *um Schallausbreitung zu verhindern* enclosure

Karbonatschmelze-Brennstoffzelle *f* LUFT molten carbonate fuel cell (*MCFA*)

karbonisiert: **~er BRAM** *m* ABFALL carbonized RDF

kardiert: **~e Produkte** *nt pl* ABFALL carded packaging

Karst *m* WASSER karst; **Karstgrundwasser** *nt* WASSER karst groundwater; **Karsthydrologie** *f* WASSER karst hydrology; **Karstquelle** *f* WASSER intermittent spring, karstic spring; **Karstwasservorkommen** *nt* WASSER karst water resources

Karton *m* ABFALL cardboard, ALLG *Schachtel* carton

Kartonage *f* ABFALL cardboard packaging

Karton: **Kartonautomat** *m* ABFALL box-making machine; **Kartonröhre** *f* ABFALL cardboard tube; **Kartonverpackung** *f* ABFALL cardboard packaging

Karussellextraktor *m* BODEN carousel extractor

karzinogen *adj* SICHERHEIT carcinogenic

Karzinogen *nt* SICHERHEIT carcinogenic substance

Kaschierung *f*: **~ der Oberflächen** ABFALL lamination of surface

Kaskade *f* ABWASSER, ANALYTIK, BODEN cascade; **Kaskadenbelüftung** *f* ABWASSER cascade aeration; **Kaskadendenitrifikation** *f* ABWASSER cascade denitrification; **Kaskadenimpaktor** *m* ANALYTIK cascade impactor; **Kaskadensäuerung** *f* BODEN cascade acidification

Kassette *f* ABFALL landfill cell, refuse cell, *biochemischer Sauerstoffbedarf* subcell, LÄRM coffer

Kasten *m* ALLG box, carton

Kat *m* LUFT catalytic converter (*BrE*), catalytic muffler (*AmE*)

Katabolismus *m* ABWASSER catabolism

Katalysator *m* BODEN, LUFT, heat exchanger, catalytic converter (*BrE*), catalytic muffler (*AmE*), WASSER catalyst; **~ mit Lambdaregelung** LUFT catalytic converter closed loop; **~ ohne Lambdaregelung** LUFT catalytic converter open loop; **Katalysatorbehälter** *m* LUFT catalyst container; **Katalysatorbett** *nt* LUFT catalyst bed; **Katalysatordesaktivierung** *f* LUFT catalyst deactivation; **Katalysatorebene** *f* LUFT catalyst level; **Katalysatorelement** *nt* LUFT catalyst element; **Katalysatorfüllung** *f* LUFT catalyst charge; **Katalysatorgift** *nt* LUFT catalyst poison; **Katalysatorgranulat** *nt* LUFT granular catalyst; **Katalysator-Krackanlage** *f* LUFT catalytic cracking unit; **Katalysatorlebensdauer** *f* LUFT catalyst lifetime; **Katalysatormaterial** *nt* LUFT catalyst material; **Katalysatormodul** *nt* LUFT catalyst module; **Katalysatoroberfläche** *f* LUFT catalyst surface; **Katalysatorpaket** *nt* LUFT catalyst module; **Katalysatorplatte** *f* LUFT catalyst plate; **Katalysatorregeneration** *f* LUFT catalyst regeneration; **Katalysatorschädigung** *f* LUFT catalyst degradation; **Katalysatorschicht** *f* LUFT catalyst bed; **Katalysatorschüttung** *f* LUFT catalyst charge; **Katalysatorstandzeit** *f* LUFT catalyst lifetime; **Katalysatorstaub** *m* LUFT catalyst dust, catalyst fines; **Katalysatorstufe** *f* LUFT catalyst step; **Katalysatortechnik** *f* LUFT catalyst technology; **Katalysatorträger** *m* LUFT, WASSER catalyst carrier; **Katalysatorturm** *m* LUFT catalytic tower; **Katalysatorvergiftung** *f* LUFT catalyst contamination, catalyst poisoning; **Katalysatorverschlechterung** *f* LUFT catalyst deterioration; **Katalysatorvolumen** *nt* LUFT catalyst volume; **Katalysatorwirksamheit** *f* LUFT catalyst efficiency; **Katalysatorwirkungsgrad** *m* LUFT catalyst efficiency; **Katalysatorwirkzeit** *f* LUFT catalyst lifetime; **Katalysatorzerstörung** *f* LUFT catalyst deterioration

Katalyse *f* BODEN, LUFT, WASSER catalysis

katalysieren *vt* BODEN, LUFT, WASSER catalyse (*BrE*), catalyze (*AmE*)

katalytisch *adj* BODEN, LUFT catalytic; **~er Abgasreiniger** *m* LUFT catalytic waste gas cleaner; **~ aktive Schicht** *f* LUFT *Katalysator* catalytic active layer; **~ aktives Metall** *nt* LUFT catalytic active metal; **~e Aktivität** *f* LUFT catalytic activity; **~e Beschichtung** *f* LUFT *des Filtermaterials* catalytic coating; **~e Beschleunigung** *f* LUFT catalytic acceleration; **~e Imprägnierung** *f* LUFT catalytic impregnation; **~es Kracken** *nt* LUFT catalytic cracking; **~es Krackmittel** *nt* LUFT catalytic cracker; **~e Nachverbrennung** *f* (*KNV*) ABWASSER catalytic afterburning, LUFT catalytic combustion, catalytic incineration; **~e Nachverbrennungsanlage** *f* LUFT catalytic combustion plant, catalytic incineration plant; **~e Oxidation** *f* ABFALL, BODEN, LUFT, WASSER catalytic oxidation; **~e Reaktion** *f* LUFT catalytic reaction; **~er Reaktor** *m* LUFT catalytic reactor; **~es Reduktionsverfahren** *nt* LUFT catalytic reduction process; **~e UV-Oxidation** *f* ABFALL, ABWASSER, WASSER catalytic UV oxidation; **~er Verbrenner** *m* LUFT catalytic incinerator

Kathode *f* LUFT cathode, *Elektrofilter* negative electrode; **Kathodenstrahlröhre** *f* SICHERHEIT cathode ray tube

kathodisch: **~er Schutz** *m* ERNEU ENER *Gezeitenkraftwerk* cathodic protection

Kation *nt* ABWASSER cation, positive ion, WASSER cation, positive cation

kationenaktiv *adj* ABWASSER cationic-active

Kation: **Kationenaustausch** *m* ABWASSER, BODEN, WASSER cation exchange; **Kationenaustauscher** *m* ABWASSER cationic exchanger, BODEN, WASSER cation exchanger; **Kationenaustauschkapazität** *f* BODEN cation exchange capacity (*CEC*)

kationenpermeabel *adj* ABWASSER cation-permeable; **kationenpermeable Membran** *f* ABWASSER cation-permeable membrane

kationenselektiv *adj* ABWASSER cation-selective; **~e Membran** *f* ABWASSER cation-selective membrane

kationisch *adj* ABWASSER, WASSER cationic

Kautschuk *m* ABFALL caoutchouc, unvulcanized rubber

Kavitationsbestimmung *f* ABWASSER, WASSER cavitation determination

Kavitationsempfindlichkeit *f* ABWASSER, WASSER cavitation susceptibility

kegelförmig: **~e Rauchfahne** *f* LUFT coning smoke plume

Kehr: **Kehrfahrzeug** *nt* ABFALL road-sweeping lorry (*BrE*), road-sweeping truck (*AmE*), street cleaner, street-cleaning lorry (*BrE*), street-cleaning truck (*AmE*); **Kehrmaschine** *f* ABFALL sweeper; **Kehrwalze** *f* ABFALL *am Kehrfahrzeug* cylinder broom

Keim *m* ABWASSER, BODEN, WASSER germ

keimfrei *adj* ABFALL, ANALYTIK *Probe* germ-free; **~e Abwasserprobe** *f* ABFALL, ANALYTIK germ-free wastewater sample

Keimverminderung *f* WASSER germ reduction

Kennwert *m* ALLG parameter

kennzeichnen *vt* ABFALL label, SICHERHEIT identify

Kennzeichnung *f* ABFALL labeling (*AmE*), labelling (*BrE*), SICHERHEIT identification; **Kennzeichnungspflicht** *f* ABFALL *der Kunststoffe* labeling obligation (*AmE*), labelling obligation (*BrE*); **Kennzeichnungssystem** *nt* **für wiederverwertbare Plastikcontainer** ABFALL labeling system for recyclable plastic containers (*AmE*), labelling system for recyclable plastic containers (*BrE*)

Keramik *f* LUFT ceramics; **Keramikfilter** *nt* LUFT ceramic filter; **Keramikkatalysator** *m* LUFT ceramic catalyst; **Keramikkörper** *m* LUFT *des Katalysators* ceramic body; **Keramikmonolith** *m* LUFT *als Katalysator* ceramic monolith; **Keramikträger** *m* LUFT *Katalysator* ceramic carrier

keramisch: **~ ausgekleidete Brennkammer** *f* LUFT ceramic-lined combustion chamber; **~es Material** *nt* LUFT *Filtermaterial* ceramic material; **~es Pellet** *nt* BODEN ceramic pellet

Keramisierung *f* BODEN *von Pellets* ceramization

Kern *m* ANALYTIK core; **Kernmaterial** *nt* ANALYTIK core

material; **Kernprobe** *f* ANALYTIK core sample; **Kernrohr** *nt* ANALYTIK *Bodenprobe* core barrel
Kerzenfilter *nt* LUFT, WASSER candle filter
Kessel *m* ABFALL, ABWASSER, ERNEU ENER, LUFT boiler; **Kesselanlage** *f* ABFALL, ERNEU ENER, LUFT boiler plant; **Kesselkorrosion** *f* ABFALL boiler corrosion; **Kesselraumasche** *f* ABFALL boiler room ash; **Kesselspeisewasser** *nt* ABFALL, LUFT boiler feedwater; **Kesselspeisewasserpumpe** *f* LUFT *Vollast und Teillast* boiler feedwater pump; **Kesselstein** *m* ABFALL scale, boiler scale; **Kesselsteinablagerung** *f* ABFALL scale formation; **Kesselsteinbildung** *f* ABFALL scale formation; **Kesselzusatzwasser** *nt* ABFALL, LUFT, WASSER boiler makeup water
Kette *f* ABFALL, ABWASSER, WASSER chain; **Kettenkratzer** *m* ABWASSER chain-and-flight collecting mechanism; **Kettenpumpe** *f* ABWASSER, WASSER paternoster pump; **Kettenräumer** *m* ABWASSER chain-and-flight collecting mechanism; **Kettenstabrechen** *m* ABWASSER chain screen
Kf-Wert *m* ABFALL, BODEN, LUFT, WASSER coefficient of permeability
Kfz *abbr* (*Kraftfahrzeug*) LUFT motor vehicle; **Kfz-Abgas** *nt* LUFT motor vehicle exhaust gas; **Kfz-Steuer** *f* POLITIK automobile tax (*AmE*), road tax (*BrE*)
Kielwasser *nt* MEER wake
Kies *m* ABFALL, ABWASSER, WASSER gravel; **Kiesablagerung** *f* ABFALL, WASSER gravel deposition; **Kiesbettfilter** *nt* LUFT gravel-bed filter
Kiesel- *in cpds* ALLG gravel, silica, pebble, gravel; **Kieselerdestaub** *m* SICHERHEIT silica dust; **Kieselfilterschicht** *f* WASSER *Deponie* gravel filter layer; **Kieselgel** *nt* ABWASSER *Adsorbens*, LUFT silica gel; **Kieselgurfiltration** *f* ABWASSER diatomaceous earth filtration
Kies: **Kiesfilter** *nt* WASSER gravel filter; **Kiessäule** *f* ABFALL gravel column; **Kiesschicht** *f* WASSER *Filterbecken* gravel layer
kiesummantelt: **~e Dränleitung** *f* ABFALL gravel-sheathed drainage line
Kilo: **Kilojoule pro Kilogramm** *phr* (*kJ/kg*) ABFALL kilojoule per kilogram, kilojoule per kilogramme (*kJ/kg*); **Kilonewton pro Quadratmeter** *phr* (kN/m^2) ABFALL kilonewton per square meter (*AmE*), kilonewton per square metre (*BrE*) (kN/m^2)
Kilowatt *nt* (*kW*) ERNEU ENER kilowatt (*kW*); **Kilowattanlage** *f* ERNEU ENER kilowatt plant; **Kilowattstunde** *f* ERNEU ENER *Wärmepumpe* kilowatt hour; **Kilowattstundenwärme** *f* ERNEU ENER *Wärmepumpe* kilowatt-hour heat
kinematisch: **~e Viskosität** *f* ABFALL, WASSER kinematic viscosity; **~e Zähigkeit** *f* ABFALL, WASSER kinematic viscosity
kinetisch: **~e Energie** *f* ERNEU ENER, LUFT kinetic energy; **~ erzeugter Auftrieb** *m* ABFALL, WASSER kinetically induced buoyancy
Kipp- *in cpds* ABFALL, ABWASSER, MEER, WASSER dumping, tipping; **Kippbühne** *f* ABFALL tipping ramp
Kipper *m* ABFALL tipper truck, skip
Kipp-: **Kippkanteneinbau** *m* ABFALL tipped-edge installation; **Kippkübel** *m* MEER skip; **Kipprinne** *f* ABWASSER tipping trough; **Kippstufenrost** *m* ABFALL rocking grate
Kiste *f* ALLG box, carton
kJ/kg *abbr* (*Kilojoule pro Kilogramm*) ABFALL kJ/kg (*kilojoule per kilogram, kilojoule per kilogramme*)
KKA *abbr* (*kommunale Kläranlage*) ABWASSER municipal wastewater plant; **KKA-Zulauf** *m* (*Zulauf von kommunalen Klärschlammanlagen*) ABWASSER discharge of municipal wastewater sludge plant
klar: **~e Abwasserprobe** *f* ANALYTIK clear wastewater sample
Klär: **Kläranlage** *f* ABWASSER purification plant, clarification plant, wastewater treatment plant; **Kläranlagebehälter** *m* ABWASSER filtration vat; **Klärbecken** *nt* ABWASSER clarifier, clarification tank; **Klärbehälter** *m* ABWASSER filtering tank
klären *vt* ABWASSER clarify, purify, settle
Klär: **Klärfilter** *nt* ABWASSER wastewater bed; **Klärgas** *nt* ABWASSER biogas, digester gas; **Klärgrube** *f* ABWASSER cesspit, cesspool
Klarheit *f*: **~ des Wassers** ABWASSER, WASSER water clearness
Klärschlamm *m* (*KS*) ABFALL, ABWASSER, WASSER sewage sludge, wastewater sludge; **Klärschlammabgabe** *f* ABFALL, ABWASSER, WASSER wastewater sludge delivery; **Klärschlammbehandlung** *f* ABFALL, ABWASSER, WASSER wastewater sludge treatment; **Klärschlammeigenschaften** *f pl* ABFALL, ABWASSER, WASSER sludge characteristics; **Klärschlammenge** *f* ABFALL, ABWASSER, WASSER wastewater sludge amount; **Klärschlammentwässerungsanlage** *f* ABFALL, ABWASSER, WASSER wastewater sludge dewatering equipment; **Klärschlammkatalyse** *f* ABFALL, ABWASSER, WASSER wastewater sludge catalysis; **Klärschlammkompostierung** *f* ABFALL, ABWASSER, WASSER wastewater sludge composting; **Klärschlammkonditionierungsanlage** *f* ABFALL, ABWASSER, WASSER wastewater sludge conditioning plant; **Klärschlammverbrennung** *f* ABFALL, ABWASSER, WASSER wastewater sludge incineration; **Klärschlammverbrennung** *f* **in Braunkohlefeuerungen** ABFALL, ABWASSER, WASSER incineration of wastewater sludge in lignite furnaces; **Klärschlammverbrennungsanlage** *f* ABFALL, ABWASSER, WASSER wastewater sludge incineration plant; **Klärschlammverbrennung** *f* **in Steinkohlefeuerungen** ABFALL, ABWASSER, WASSER incineration of wastewater sludge in coal furnaces; **Klärschlammverbrennung** *f* **in Zementöfen** ABFALL, ABWASSER, WASSER incineration of wastewater sludge in cement furnaces; **Klärschlammvortrocknung** *f* ABFALL, ABWASSER, WASSERpredrying of wastewater sludge
Klär: **Klärüberlauf** *m* ABWASSER overflow structure for settled combined water; **Klärwerk** *nt* ABWASSER clarification plant, wastewater treatment plant; **Klärwerkausrüstung** *f* ABWASSER wastewater treatment plant installation; **Klärwerkspumpwerk** *nt* ABWASSER combined aeration; **Klärwirkung** *f* ABWASSER clarification operation
klassieren *vt* ABFALL, BODEN, POLITIK, WASSER classify, screen
Klassierer *m* ABFALL, BODEN, WASSER classifier, screen
klassierfähig *adj* ABFALL, BODEN, POLITIK, WASSER classifiable
klassifizieren *vt* ABFALL, BODEN, POLITIK, WASSER classify, grade
Klassifizierung *f* ABFALL, BODEN, POLITIK, WASSER classification, grading
klassisch: **~e Grundwasseraufbereitung** *f* ABFALL, WASSER conventional groundwater treatment; **~e Recyclingbranchen** *f pl* ABFALL, MANAG *Altmetall-, Schrott- und Altpapier* classical recycling sectors
klebend: **~e Staubbestandteile** *m pl* LUFT tacky particulate components
Kleber *m* ABFALL, LUFT adhesive
klebrig *adj* LUFT glutinous; **~e Ablagerung** *f* LUFT sticky deposit; **~er Staub** *m* LUFT sticky dust
Klebrigkeit *f* ABFALL stickiness, LUFT stickiness, viscidity, WASSER stickiness
Klein- *in cpds* ERNEU ENER small-scale; **Kleinanlage** *f* ERNEU ENER *Wind, Sonne* small-scale installation; **Kleindeponie** *f* ABFALL small landfill; **Kleineinzugsgebiet** *nt* WASSER experimental catchment basin; **Kleinware** *f* **auf Kartonunterlagen** ABFALL carded packaging
Kletterrechen *m* ABWASSER climbing screen
Klima *nt* LUFT climate; **Klimaänderungen** *f pl* LUFT *Erwärmung der Erde* climate changes; **Klimaanlage** *f* LUFT air conditioning installation; **Klimafaktor** *m* LUFT climate factor; **Klimagefahr** *f* LUFT climatic hazard; **Klimakammer** *f* ABFALL, LUFT climatic chamber; **Klimaraum** *m* ABFALL, LUFT climatic chamber

klimarelevant *adj* LUFT climate-relevant; **~e Spurengase** *nt pl* LUFT climate-relevant trace gases
klimaschädlich *adj* LUFT climate-damaging; **~e Gase** *nt pl* LUFT climate-damaging gases
Klima: **Klimaschutz** *m* LUFT, POLITIK climatic protection; **Klimaschutzsteuer** *f* POLITIK climate protection tax; **Klimaschwankungen** *f pl* LUFT climate fluctuations, climatic variations; **Klimasystem** *nt* LUFT climate system
klimatisch *adj* LUFT climatic; **~e Bedingungen** *f pl* ABFALL, LUFT, WASSER climatic conditions; **~er Streß** *m* LUFT climatic stress
Klima: **Klimaverschlechterung** *f* LUFT climate deterioration; **Klimaversuch** *m* ABFALL climatic test; **Klimazone** *f* LUFT climatic zone
Klopf- *in cpds* LUFT knocking, rapping
klopfen *vt* LUFT knock, rap
Klopfen *nt* LUFT rapping, knocking
Klopf-: **Klopffestigkeit** *f* LUFT knock resistance; **Klopffestigkeitswert** *m* LUFT antiknock value; **Klopffilter** *nt* LUFT knock filter; **Klopfgrenze** *f* LUFT knock limit; **Klopfintervall** *nt* LUFT rapping interval
Klopfung *f* LUFT knocking, rapping; **~ des Staubes zur Lösung von der Niederschlagselektrode** LUFT rapping of electrodes
Klopf-: **Klopfvorrichtung** *f* LUFT rapper; **Klopfwert** *m* LUFT octane number, antiknock value
Kn *abbr (Knudsen-Zahl)* LUFT Kn *(Knudsen number)*
knapp *adj* ALLG scarce; **~e Umweltressourcen** *f pl* ABWASSER, BODEN, ERNEU ENER, LUFT, POLITIK scarce environmental resources
Kneter *m* ABFALL kneader
Knickpunktchlorung *f* ABWASSER, WASSER breakpoint chlorination
kN/m² *abbr (Kilonewton pro Quadratmeter)* ABFALL kN/m² *(kilonewton per square meter AmE, kilonewton per square metre BrE)*
Knopfzelle *f* ABFALL mercury battery
Knudsen-Zahl *f (Kn)* LUFT Knudsen number *(Kn)*
KNV *abbr (katalytische Nachverbrennung)* ABWASSER catalytic afterburning, LUFT catalytic combustion, catalytic incineration; **KNV-Anlage** *f* LUFT catalytic combustion plant, catalytic incineration plant
Koagulans *nt* ABWASSER, ANALYTIK, LUFT, WASSER coagulant
Koagulation *f* ALLG coagulation; **Koagulationsabscheidung** *f* ABWASSER coagulation-precipitation; **Koagulationsflockung** *f* ABWASSER coagulation-flocculation; **Koagulationsmittel** *nt* ABWASSER, ANALYTIK, LUFT, WASSER coagulant
koagulieren *vi* ALLG coagulate
Koaleszenz *f* ABFALL, ABWASSER *mechanische Ölabscheidung* coalescence, coalescence method, LUFT, WASSER coalescence; **Koaleszenzabscheider** *m* ABWASSER coalescence oil separator
koaleszieren *vi* ABFALL, ABWASSER, LUFT, WASSER coalesce
Kobalt *nt (Co)* BODEN, LUFT cobalt *(Co)*
Ko-Extrusion *f* ABFALL co-extrusion
Kofferdamm *m* ERNEU ENER *Gezeitenkraftwerk* cofferdam
Kohäsion *f* LUFT cohesion
Kohle *f* LUFT coal; **Kohleentschwefelung** *f* LUFT coal desulfurization *(AmE)*, coal desulphurization *(BrE)*; **Kohlefaserpapier** *nt* LUFT carbon fiber paper *(AmE)*, carbon fibre paper *(BrE)*
Kohlendioxid *nt* (CO_2) ABFALL, LUFT, WASSER carbon dioxide (CO_2); **Kohlendioxidhaushalt** *m* LUFT, POLITIK carbon dioxide budget; **Kohlendioxidtreibhauseffekt** *m* LUFT carbon dioxide greenhouse effect
Kohlenkraftwerk *nt* LUFT coal power station
Kohlenmonoxid *nt (CO)* ABFALL, LUFT carbon monoxide *(CO)*; **Kohlenmonoxidanzeiger** *m* ANALYTIK carbon monoxide detector
Kohlensäure *f* WASSER carbonic acid; **Kohlensäureanhydrid** *nt* WASSER carbonic anhydride
kohlensäurehaltig *adj* ABFALL, WASSER carbonic; **~es Wasser** *nt* ABFALL, WASSER carbonic water
Kohlenstoff *m* ERNEU ENER carbon; **Kohlenstoff-Absorption** *f* BODEN carbon absorption
kohlenstoffhaltig *adj* LUFT carboniferous, carbonic; **~es Adsorbens** *nt* LUFT carbonic adsorbent
Kohlenstoff: **Kohlenstoffkreislauf** *m* ERNEU ENER *Biomass* carbon cycle; **Kohlenstoff-Stickstoff-Phosphor-Verhältnis** *nt (C/N/P-Verhältnis)* ANALYTIK carbon-nitrogen-phosphorus ratio *(C/N/P ratio)*; **Kohlenstoff-Stickstoff-Verhältnis** *nt (C/N-Verhältnis)* ANALYTIK carbon-nitrogen ratio *(C/N ratio)*
Kohlenwasserstoff *m (KW-Stoff)* ABWASSER, LUFT, MEER, WASSER hydrocarbon *(HC)*; **Kohlenwasserstoff-Katalysator** *m* LUFT hydrocarbon catalyst; **Kohlenwasserstoffteppich** *m* WASSER hydrocarbon slick
Kolben *m* ABWASSER, LUFT, WASSER piston; **Kolbenkompressor** *m* ABWASSER, WASSER piston compressor; **Kolbenkraftmaschine** *f* LUFT reciprocating piston engine; **Kolbenpumpe** *f* ABWASSER, WASSER piston pump, reciprocating pump; **Kolbenstange** *f* **der Pumpe** ABWASSER, WASSER pump rod
Kollaps *m* LUFT collapse
Kollektorfeld *nt* ERNEU ENER collector field
Kollektorkopf *m* ABFALL collector head
kollidieren *vi* LUFT *Partikeln* collide
Kollimator *m* ANALYTIK collimator
Kolloid *nt* ABWASSER, BODEN, LUFT, WASSER colloid
kolloidal *adj* ABWASSER, BODEN, LUFT, WASSER colloidal; **~e Lösung** *f* ABWASSER colloidal solution; **~e Stoffe** *m pl* ABWASSER colloidal substance; **~es Teilchen** *nt* ABFALL, ABWASSER, LUFT, WASSER colloidal particle
kolloiddispers *adj* ABWASSER colloid-disperse; **~es System** *nt* LUFT colloid-disperse system
Kolloid: **Kolloidkonzentration** *f* ABWASSER, WASSER colloidal concentration; **Kolloidsuspension** *f* ABWASSER, WASSER colloidal suspension; **Kolloidteilchen** *nt* ABFALL, ABWASSER, LUFT, WASSER colloidal particle
Kolorimeter *nt* ANALYTIK colorimeter
Kolorimetrie *f* ANALYTIK *Beurteilung der Färbung des Abwassers* colorimetry
kolorimetrisch *adj* ANALYTIK colorimetric; **~e Anzeigeröhre** *f* ANALYTIK colorimetric indicator tube; **~e Bestimmung** *f* ANALYTIK colorimetric determination; **~er Nachweis** *m* ANALYTIK colorimetric detection; **~e pH-Wert-Bestimmung** *f* ANALYTIK colorimetric pH-value determination
Kombination *f* ABFALL, LUFT combination; **~ BB-TK** *(Kombination Belebtbecken-Tropfkörper)* ABWASSER combination of trickling filter and activated sludge tank; **~ Belebtbecken-Tropfkörper** *(Kombination BB-TK)* ABWASSER combination of trickling filter and activated sludge tank; **Kombinationsdichtung** *f* ABFALL *aus Kunststoffbahnen und mineralischer Schicht* combination sealing, composite seal; **Kombinationswäscher** *m* LUFT hybrid scrubber
kombiniert *adj* ABWASSER, BODEN combined; **~e Basisdichtung** *f* BODEN combined base sealing; **~es Basisdichtungssystem** *nt* BODEN combined base-sealing system; **~e Belüftung** *f* ABWASSER combined aeration; **~e Verpackung** *f* ABFALL combined packaging
Kombiwäscher *m* LUFT hybrid scrubber
kommerziell *adj* ALLG commercial; **~er Globalenergiebedarf** *m* ERNEU ENER commercial global energy demand; **~ nutzbar** *adj* ERNEU ENER commercially usable
kommunal *adj* ABWASSER, POLITIK, WASSER municipal; **~es Abwasser** *nt* ABWASSER municipal wastewater; **~e Abwassereinleitung** *f* ABWASSER municipal wastewater discharge; **~er Abwasserschlamm** *m* ABWASSER municipal

wastewater sludge (*MWS*); **~e Kläranlage** *f* (*KKA*) ABWASSER municipal wastewater plant; **~er Klärschlamm** *m* ABFALL, ABWASSER municipal wastewater sludge (*MWS*); **~e Müllabfuhr** *f* ABFALL communal waste collection by local authority; **~e Mülleinsammlung** *f* ABFALL communal waste collection by local authority; **~er Umweltschutz** *m* POLITIK municipal environmental protection; **~es Wasser** *nt* WASSER town water

Kompakt- *in cpds* ALLG compact

kompakt *adj* ABWASSER, LUFT, WASSER compact

Kompakt-: **Kompaktabsorber** *m* LÄRM compact absorber; **Kompaktfilter** *nt* LUFT *Containerfilter* compact filter; **Kompaktflockungsanlage** *f* WASSER compact flocculation plant; **Kompaktgaswäscher** *m* LUFT compact gas scrubber

Kompaktierung *f* ABFALL compacting, compaction

Kompakt-: **Kompaktmodul** *nt* ABWASSER *mit Membranen* compact module

Kompaktor *m* ABFALL packer unit, landfill compactor, compactor

Kompakt-: **Kompaktreaktor** *m* ABWASSER, BODEN compact reactor; **Kompaktwärmeaustauscher** *m* LUFT compact heat exchanger

Kompensationsschlauch *m* WASSER compensating tube

kompensieren *vt* ANALYTIK compensate

Komplexbildner *m* ABWASSER, ANALYTIK complexing agent

komplexometrisch *adj* ANALYTIK complexometric; **~e Bestimmung** *f* ANALYTIK complexometric determination; **~e Methode** *f* ANALYTIK complexometric method; **~e Titration** *f* ANALYTIK complexometric titration

komponentenspezifisch *adj* ANALYTIK, BODEN component-specific; **~er Gasanalysator** *m* ANALYTIK component-specific gas analyser (*BrE*), component-specific gas analyzer (*AmE*)

Kompost *m* ABFALL, ABWASSER compost; **Kompostaufbereitung** *f* ABFALL compost reprocessing; **Kompostbehälter** *m* ABFALL compost container; **Kompostbelüftung** *f* ABFALL compost aeration; **Kompostfilter** *nt* LUFT *Biofilter* compost filter; **Kompostgut** *nt* ABFALL composting material; **Komposthalde** *f* ABFALL compost heap

kompostierbar *adj* ABFALL, ABWASSER compostable; **~er Abfall** *nt* ABFALL compostable waste; **~e Fraktion** *f* ABFALL compostable fraction; **~er Schlamm** *m* ABWASSER compostable sludge

kompostieren *vt* ABFALL, ABWASSER compost

kompostierfähig *adj* ABFALL, ABWASSER compostable

Kompostierfähigkeit *f* ABFALL, ABWASSER compostability

Kompostierung *f* ABFALL, ABWASSER composting; **Kompostierungsanlage** *f* ABFALL, BODEN composting plant, composting works; **Kompostierungsrückstand** *m* ABFALL composting residue; **Kompostierungstechnik** *f* ABFALL composting technique; **Kompostierungsvorgang** *m* ABFALL composting operation

Kompost: **Kompostkörper** *m* ABFALL rotting body; **Kompostmaterial** *nt* ABFALL compost material; **Kompostqualität** *f* ABFALL compost quality; **Kompostreifung** *f* ABFALL compost maturing, compost ripening; **Kompostrohstoff** *m* ABFALL compost raw material; **Kompostsortieranlage** *f* ABFALL compost sorting plant; **Komposttonne** *f* ABFALL compost container; **Kompostwerk** *nt* ABFALL, BODEN composting plant, composting works

Kompounder *m* ABFALL compounder

Kompressibilität *f* ABFALL *des Mülls* compressibility

Kompressor *m* ABWASSER, ERNEU ENER, LUFT compressor; **Kompressorenschalldämpfer** *m* SICHERHEIT noise protection for compressors

komprimierbar *adj* ABFALL, ABWASSER, WASSER compressible; **~er Abfall** *m* ABFALL compressible waste

Kondensat *nt* ABWASSER, BODEN condensate, LUFT condensate, precipitate, WASSER condensate; **Kondensatabscheider** *m* ABFALL condensate separator; **Kondensatbildung** *f* LUFT condensate formation; **Kondensatfilm** *m* LUFT condensate film

kondensatführend: **~e Leitung** *f* ABFALL condensate pipe

Kondensation *f* ABWASSER, BODEN, LUFT condensation; **Kondensationsaerosol** *nt* LUFT condensation aerosol; **Kondensations-E-Filter** *nt* LUFT condensation-force ESP; **Kondensationskern** *m* LUFT condensation nucleus; **Kondensationskernzähler** *m* LUFT condensation nucleus counter; **Kondensationswäscher** *m* LUFT condensation scrubber

Kondensator *m* ANALYTIK, BODEN, ERNEU ENER, LUFT condenser

Kondensat: **Kondensatreinigung** *f* ABWASSER condensate polishing; **Kondensatstrom** *m* LUFT condensate flow

kondensieren 1. *vt* LUFT condense; **2.** *vi* LUFT precipitate

Kondenswasser *nt* ABWASSER condensation water; **Kondenswasserabscheider** *m* ABWASSER condensation water separator

konditionieren *vt* ABFALL, ABWASSER, LUFT, WASSER condition

konditioniert *adj* ABFALL, ABWASSER, LUFT, WASSER conditioned; **~er Schlamm** *m* ABWASSER conditioned sludge

Konditionierung *f* ABFALL, ABWASSER, LUFT, WASSER conditioning

Konditionierungsreaktor *m* LUFT conditioning reactor

Konduktometrie *f* ANALYTIK conductimetry

konduktometrisch *adj* ANALYTIK conductometric; **~e Maßanalyse** *f* ANALYTIK conductometric analysis; **~e Methode** *f* ANALYTIK conductometric method; **~es Verfahren** *nt* ANALYTIK conductometric method

Konfektioniereinheit *f* ABFALL assembling unit

konfigurieren *vt* ANALYTIK configure

konglomerieren *vt* ABWASSER conglomerate

Konimeter *nt* ANALYTIK konimeter

konisch *adj* LUFT conical; **~er Schneckenextruder** *m* ABFALL, WASSER conical screw extruder; **~e Staffelbelüftung** *f* ABWASSER tapered aeration; **~es Unterteil** *nt* LUFT *des Zyklons* conical base

konjugiert: **~e Verbindung** *f* ANALYTIK conjugate compound

Konkurrenzfähigkeit *f* ERNEU ENER, MANAG competitiveness

Konservenbüchse *f* ABFALL can, tin

Konsistenz *f* ABFALL consistence, consistency

konstitutiv *adj* ABWASSER constitutive; **~es Enzym** *nt* ABWASSER constitutive enzyme

Konsum *m* ABFALL, MANAG, WASSER consumption

Konsument *m* ABFALL, MANAG, WASSER consumer

Konsum: **Konsumgüter** *nt pl* ABFALL, MANAG consumer goods

konsumieren *vt* ABFALL, MANAG, WASSER consume

Konsum: **Konsumverhalten** *nt* ABFALL, LUFT, MANAG, WASSER consumer behavior (*AmE*), consumer behaviour (*BrE*)

Kontakt *m* ABFALL, ABWASSER, LUFT, WASSER contact; **Kontaktfilterbett** *nt* ABWASSER contact bed; **Kontaktgift** *nt* LUFT catalyst poison; **Kontaktkatalyse** *f* LUFT contact catalysis, surface catalysis

kontaktkatalytisch *adj* ABFALL, WASSER contact-catalytic

Kontakt: **Kontaktkondensator** *m* LUFT contact condenser, direct contact condenser; **Kontaktoberfläche** *f* LUFT contact surface; **Kontaktofen** *m* LUFT contact reactor; **Kontaktstabilisierung** *f* ABWASSER contact stabilization; **Kontakttrockner** *m* ABWASSER contact dryer; **Kontaktturm** *m* LUFT contact tower; **Kontaktverfahren** *nt* WASSER *Biogasgewinnung* contact process; **Kontaktzeit** *f* ABWASSER, WASSER contact time

Kontaminant *nt* ALLG contaminant, pollutant

Kontamination *f* ALLG contamination; **Kontaminationsart** *f* BODEN contamination type; **Kontaminationsgrad** *m* BODEN degree of contamination; **Kontaminationspfad** *m* ABWASSER, WASSER contamination pathway

kontaminieren *vt* ALLG contaminate

kontaminiert *adj* ALLG contaminated; **~es Grundwasser** *nt* ABFALL, ABWASSER, WASSER contaminated groundwater; **~er Kunststoff** *m* ABFALL contaminated plastic; **~er Schlamm** *m* ABWASSER contaminated sludge; **~er Standort** *m* ABFALL contaminated site, BODEN problem site; **~es Wasser** *nt* ABWASSER, WASSER contaminated water

kontinuierlich *adj* ANALYTIK continuous; **~ beschickbar** *adj* ABFALL, WASSER able to feed continuously; **~e Hydrolyse** *f* BODEN continuous hydrolysis; **~e Messung** *f* ANALYTIK continuous measuring; **~e Presse** *f* ABFALL continuous press; **~e Pressenstraßen** *f pl* ABFALL *Abfallmaterialen* continuous press lines; **~e Probenahme** *f* ANALYTIK continuous sampling; **~ regelbar** *adj* ABFALL continuously adjustable; **~e Registrierung** *f* ANALYTIK continuous recording; **~e Reinigung** *f* ABWASSER, LUFT *Filterschlauch* continuous cleaning; **~e Siebbandpresse** *f* ABWASSER continuous screen belt press; **~ strömender Wasserfilm** *m* LUFT continuous flowing film; **~e UF** *f* ABWASSER continuous ultrafiltration; **~e Ultrafiltration** *f* ABWASSER continuous ultrafiltration

Kontrollart *f* ABFALL, ANALYTIK, WASSER indicator species

Kontrollbrunnen *m* ABFALL, WASSER monitoring well, observation well

Kontrolle *f* ABFALL, ANALYTIK check, checking, control; **~ der Bodenauslaugung** BODEN leaching control; **~ der Giftigkeit am Arbeitsplatz** SICHERHEIT control of toxicity at the workplace, monitoring of toxicity at the workplace

Kontrollierbarkeit *f* ALLG controllability

kontrollieren *vt* ABFALL, ABWASSER inspect, ALLG control, ANALYTIK inspect, check, BODEN, ERNEU ENER, LÄRM inspect, LUFT monitor, inspect, MEER monitor, SICHERHEIT, WASSER inspect

kontrolliert: **~e Gärung** *f* ABFALL controlled fermentation; **~e Müllablagerung** *f* ABFALL controlled dumping, controlled tipping *BrE*, sanitary landfill

Kontrollschacht *f* WASSER control shaft

Kontrollvorrichtung *f* ABFALL, WASSER inspection equipment

Konus *m* LUFT *eines Zyklons* cone; **Konusgetriebe** *nt* ABWASSER, WASSER *Pumpe* cone drive

Konvektion *f* ERNEU ENER convection; **Konvektionsteil** *nt* LUFT convection section; **Konvektionstrockner** *m* ABFALL, ABWASSER convection dryer; **Konvektionswärmeaustauscher** *m* LUFT convection heat exchanger

Konvention *f* POLITIK convention

konventionell *adj* ALLG conventional; **~e Abwasserreinigung** *f* ABWASSER conventional wastewater clarification; **~es Dränsystem** *nt* ABFALL, WASSER conventional drainage system; **~e Energietechnik** *f* ERNEU ENER *spezifisch* conventional energy technology; **~e Energietechnologie** *f* ERNEU ENER conventional energy technology; **~e Energieversorgung** *f* ERNEU ENER conventional energy supply; **~es Energieversorgungssystem** *nt* ERNEU ENER conventional energy supply system; **~es Filtersystem** *nt* LUFT conventional filter system; **~e Grundwasseraufbereitung** *f* ABFALL, WASSER conventional groundwater treatment; **~e Heizungstechnik** *f* ERNEU ENER conventional heating technology; **~er Oxidationskatalysator** *m* LUFT conventional oxidation catalyst (*COC*); **~e Technik** *f* ABWASSER, ANALYTIK, BODEN, LÄRM, LUFT conventional technology; **~e Technologie** *f* ABWASSER, ANALYTIK, BODEN, LÄRM, LUFT conventional technology; **~e Transportpalette** *f* ABFALL conventional transportable pallet

Konversion *f* ALLG conversion

Konverter *m* ABFALL, ERNEU ENER converter

Konvertierungsreaktion *f* LUFT *Brennstoffzelle* conversion reaction

Konzentrat *nt* ABWASSER, BODEN, LUFT, WASSER concentrate; **Konzentratgewinnung** *f* ABWASSER concentrate recovery

Konzentration *f* ALLG concentration; **~ in der Atmosphäre** LUFT atmospheric concentration; **~ auf Bodenhöhe** BODEN ground-level concentration (*GLC*); **Konzentrationsschwankung** *f* LUFT *von Schadstoffen* concentration fluctuation

Konzentrat: **Konzentratstrom** *m* ABWASSER, WASSER concentrate flow

konzentrieren *vt* ABFALL, ABWASSER, BODEN, LUFT concentrate

konzentrierend: **~er Kollektor** *m* ERNEU ENER *Parabolkollektoren* concentrating collector

konzentriert *adj* LUFT, WASSER concentrated

Konzentrierung *f*: **~ von Lärm auf einen Bereich** LÄRM, MEER concentration of noise to area

konzipieren *vt* ABFALL, ANALYTIK, WASSER design

Kopf *m* LÄRM, LUFT head; **Kopfhörer** *m* LÄRM headphones; **Kopfplatte** *f* LUFT head plate; **Kopfseite** *f* LUFT *des Reaktors* head

Kopplung *f* ANALYTIK coupling; **in ~ mit** *phr* ANALYTIK *Pyrolyse-Gaschromatographie* coupled with

Korn *nt* ABWASSER, ABFALL particle, ANALYTIK, BODEN grain, particle, LUFT, WASSER particle; **Kornfeinheit** *f* ABFALL fineness of particle size

Korngröße *f* ABFALL particle size, BODEN grain size, LUFT particle size; **Korngrößenanalyse** *f* ANALYTIK grain size analysis; **Korngrößenanalyse** *f* **von Schwebestoffen** ANALYTIK grain size analysis of particulates; **Korngrößenbereich** *m* ABFALL sieve fraction; **Korngrößenklasse** *f* ABFALL particle size class; **Korngrößentrennung** *f* ABFALL particle size separation; **Korngrößenverteilung** *f* ABWASSER particle size distribution, ANALYTIK, BODEN grain size distribution, LUFT particle size distribution; **Korngrößenverteilung** *f* **von Staub** ANALYTIK dust particle size distribution

körnig: **~e Schicht** *f* LUFT *Filtermedium* granulated layer

Korona *f* LUFT corona; **Koronabeständigkeit** *f* LUFT corona resistance; **Koronaeffekt** *m* LUFT corona effect; **Koronaentladung** *f* LUFT corona discharge

Körperschall *m* LÄRM structure-borne noise, solid-borne noise; **Körperschallabsorber** *m* LÄRM structure-borne absorber; **Körperschalldämmung** *f* LÄRM structure-borne-noise proofing

körperschalldämpfend *adj* LÄRM *Schichten* structure-borne-noise-reducing

Körperschall: **Körperschalldämpfung** *f* LÄRM structure-borne-noise reduction; **Körperschallübertragung** *f* LÄRM structure-borne-noise transmission

Korrekturfaktor *m* ANALYTIK volumetric factor

korrespondierend: **~e Probenahme** *f* ANALYTIK corresponding sampling

korrodieren *vt* ABFALL, ABWASSER, LUFT, WASSER corrode

Korrosion *f* ABFALL, ABWASSER, LUFT, WASSER corrosion; **~ an Armaturen** WASSER corrosion in fittings; **~ an Pumpen** WASSER corrosion in pumps; **~ in Pumpwerken** WASSER corrosion in pumping stations; **~ in Wasserinstallation** WASSER corrosion in water installations; **~ von Wasserleitungen** ABWASSER corrosion of water pipelines

korrosionsbeständig *adj* ABFALL, LUFT, WASSER corrosion-resistant, noncorrosive; **~e Sickerwasserpumpe** *f* ABFALL corrosion-resistant leachate pump

Korrosion: **Korrosionsbeständigkeit** *f* LUFT, WASSER *Leitungsnetz* corrosion resistance; **Korrosionsgefahr** *f* LUFT corrosion risk

korrosionshemmend *adj* WASSER corrosion-inhibiting

Korrosion: **Korrosionshemmstoff** *m* WASSER corrosion inhibitor; **Korrosionsinhibitor** *m* WASSER corrosion inhibitor; **Korrosionsschaden** *m* WASSER corrosion damage; **Korrosionsschutz** *m* ABWASSER, WASSER corrosion protection; **Korrosionsschutzauskleidung** *f* WASSER corrosion protection lining; **Korrosionstest** *m* ANALYTIK corrosion test

korrosiv *adj* ABFALL *Deponiegas*, LUFT, SICHERHEIT, WASSER corrosive; **~e Eigenschaften** *f pl* ABFALL, WASSER corrosive properties; **~es Gas** *nt* LUFT corrosive gas; **~er Sondermüll** *m* ABFALL corrosive hazardous waste; **~er Stoff** *m* SICHERHEIT corrosive substance

Korrosivität *f* ABFALL, LUFT, SICHERHEIT, WASSER corrosivity, corrosiveness

kosmisch *adj* LUFT cosmic; **~e Primärstrahlung** *f* LUFT primary cosmic radiation; **~e Sekundärstrahlung** *f* LUFT secondary cosmic radiation; **~e Solarstrahlung** *f* LUFT cosmic solar radiation; **~e Strahlung** *f* LUFT cosmic radiation

Kosten *pl* MANAG cost; **Kostenanalyse** *f* MANAG cost analysis; **Kostenaufwand** *m* MANAG cost; **Kostenbewertung** *f* MANAG cost appraisal

kosteneffektiv *adj* MANAG cost-effective

Kosten: **Kosteneffizienz** *f* MANAG cost-effectiveness; **Kostenfaktor** *m* MANAG cost factor

kostenintensiv *adj* ABWASSER, LUFT, MANAG, WASSER cost-intensive; **~e Behandlungsverfahren** *nt pl* ABWASSER, LUFT, MANAG, WASSER cost-intensive methods of treatment

Kosten: **Kosten-Nutzen-Analyse** *f* MANAG cost-benefit study; **Kosten-Nutzen-Rechnung** *f* MANAG cost-benefit-analysis; **Kosten-Nutzen-Verhältnis** *nt* MANAG cost-benefit ratio; **Kostenplanung** *f* MANAG cost planning; **Kostenrechnung** *f* MANAG cost calculation; **Kosten-Standzeit-Relation** *f* MANAG cost-service life ratio; **Kostenstruktur** *f* MANAG cost structure; **Kostenvorteil** *m* MANAG cost advantage

Kr *abbr* (*Krypton*) LUFT Kr (*krypton*)

Kraft 1. *f* ERNEU ENER power; **2. in ~ treten** *vi* POLITIK *Gesetz, Abkommen* come into force

Kraft: **Kraftfahrzeug** *nt* (*Kfz*) LUFT motor vehicle; **Kraftfahrzeugabgas** *nt* LUFT motor vehicle exhaust gas

Kraftstoff *m* LUFT fuel; **Kraftstoffaufnahme** *f* MEER refueling (*AmE*), refuelling (*BrE*); **Kraftstoffersatz** *m* LUFT fuel replacement; **Kraftstofftank** *m* LUFT fuel tank; **Kraftstofftankemissionen** *f pl* LUFT *Kohlenwasserstoff* fuel tank emissions; **Kraftstoffverbrauch** *m* LUFT fuel consumption; **Kraftstoffverbrennung** *f* ABFALL, LUFT fuel combustion

Kraft: **Kraft-Wärme-Kopplung** *f* (*KWK*) ABFALL, LUFT cogeneration, combined heat and power (*CHP*); **Kraftwerk** *nt* ERNEU ENER power house, power station, LUFT power plant

Krählarm *m* ABFALL rabble arm

Krählwerk *nt* ABWASSER rabble rake

Krankenhausabfall *m* ABFALL health care waste, hospital waste

krankenhausspezifisch: **~er Abfall** *m* ABFALL health care waste, hospital waste

krankheitserregend *adj* ABFALL, ABWASSER, WASSER pathogenic

Krankheitserreger *m* ABWASSER, WASSER pathogen

Kratzer *m* ABWASSER scraper; **Kratzerarm** *m* ABWASSER discharge hopper; **Kratzerkette** *f* ABFALL drag chain; **Kratzerkettenentschlacker** *m* ABFALL drag conveyor ash remover

krebserzeugend *adj* SICHERHEIT cancer-causing, carcinogenic; **~er Stoff** *m* SICHERHEIT carcinogenic substance

Krebsrisiko *nt* SICHERHEIT cancer risk

Kreisbahn *f* LUFT orbit

Kreisbewegung *f* LUFT circular movement

Kreiselpumpe *f* ABWASSER, MEER, WASSER centrifugal pump

Kreislauf *m* ABFALL, ABWASSER, LUFT, WASSER circulation; **Kreislaufflüssigkeit** *f* LUFT circulating liquid; **Kreislaufführung** *f* LUFT closed-loop circulation; **Kreislaufgas** *nt* LUFT recycling gas; **Kreislaufheizung** *f* ABWASSER recirculated heating; **Kreislaufpumpe** *f* LUFT circulation pump; **Kreislaufsystem** *nt* ABFALL closed-loop system, ABWASSER recirculation system, WASSER closed-loop system; **Kreislaufwaschmittel** *nt* LUFT recirculating washing agent; **Kreislaufwasser** *nt* ABWASSER recirculating water, LUFT circulating water; **Kreislaufwasserführung** *f* WASSER recirculating water management; **Kreislaufwirtschaft** *f* POLITIK, WASSER recycling economy; **Kreislaufwirtschaftsgesetz** *nt* POLITIK, WASSER recycling economy law

Kreuzstrom *m* LUFT *Abgasstrom* cross-flow; **Kreuzstromwäscher** *m* LUFT cross-flow scrubber

kristallin *adj* ERNEU ENER, LUFT crystalline; **~e Siliziumzelle** *f* ERNEU ENER *Sonne* crystalline silicon cell

kristallinisch *adj* LUFT crystalline

Kristallisation *f* LUFT crystallization; **Kristallisationskeim** *m* LUFT crystallization germ; **Kristallisationszone** *f* LUFT crystallization zone

Kristallographie *f* ANALYTIK crystallography

Kriterien *nt pl* ANALYTIK criteria; **~ zur Bewertung** POLITIK evaluation criteria

kritisch: **~e Belastung** *f* BODEN critical load; **~e Wasserlinie** *f* ABFALL, WASSER critical water level; **~er Wasserstand** *m* ABFALL, WASSER critical water level

Krume *f* BODEN A-horizon

Krümmer *m* LUFT *Abscheider, Wäscher* elbow

Krypton *nt* (*Kr*) LUFT krypton (*Kr*)

KS *abbr* (*Klärschlamm*) ABFALL sewage sludge, ABWASSER wastewater sludge, WASSER sewage sludge; **KS-Menge** *f* ABWASSER wastewater sludge amount; **KS-Verbrennung** *f* ABWASSER wastewater sludge incineration; **KS-Vortrocknung** *f* ABWASSER predrying of wastewater sludge

Kugelregensystem *nt* LUFT shot cleaning system

Kühl- *in cpds* ALLG cooling

kühl: **~ halten** *vt* ANALYTIK *Probe* keep cool

Kühl-: **Kühlbecken** *nt* ABWASSER cooling pond; **Kühlbehälter** *m* LUFT cooling tank

kühlen *vt* ABFALL, ANALYTIK, ERNEU ENER, WASSER cool

Kühler *m* ABFALL, ANALYTIK, LUFT, WASSER cooler

Kühl-: **Kühlflüssigkeit** *f* LUFT cooling liquid; **Kühlgasgebläse** *nt* LUFT cooling gas fan; **Kühlkreislauf** *m* ERNEU ENER cooling water loop

Kühlluft *f* LUFT cooling air; **Kühlluftaustritt** *m* LUFT cooling air outlet; **Kühllufteintritt** *m* LUFT cooling air inlet

Kühlmittel *nt* ABFALL, LUFT, WASSER coolant, cooling medium; **Kühlmittelaustritt** *m* LUFT coolant out, coolant outlet; **Kühlmitteleintritt** *m* LUFT coolant inlet, coolant in; **Kühlmittelströmungsgeschwindigkeit** *f* LUFT coolant flow rate

Kühl-: **Kühlöl** *nt* ABFALL, WASSER cooling oil; **Kühlsystem** *nt* ABFALL, ERNEU ENER, LUFT cooling system; **Kühlteich** *m* ABWASSER cooling pond; **Kühltrommel** *f* BODEN cooling drum

Kühlturm *m* ERNEU ENER cooling tower; **Kühlturmabluft** *f* LUFT cooling tower exhaust air; **Kühlturmschwaden** *m* LUFT Gaussian plume

Kühlung *f* ALLG cooling

Kühl-: **Kühlwasser** *nt* ABWASSER, LUFT cooling water

Kulturboden *m* BODEN arable land

Kundt: **~sches Rohr** *nt* LÄRM standing wave tube

Kunstfaser *f* ABFALL manmade fiber (*AmE*), manmade fibre (*BrE*)

Kunstharz *nt* ABFALL *jedes Kunststoffmaterial* resin; **Kunstharzbeton** *m* ABFALL *Nachbehandlung von Filterstäuben* polymer-concrete composite material

künstlich *adj* ALLG artificial, manmade, synthetic; **~e Abdichtung** *f* ABFALL, WASSER synthetic lining; **~e Alterung** *f* ABFALL, WASSER artificial ageing; **~ angereichert** *adj* ABFALL, WASSER artificially recharged, artificially replenished; **~ angereichertes Grundwasser** *nt* WASSER artificially recharged groundwater, artificially replenished groundwater; **~e Belüftung** *f* ABWASSER artificial aeration; **~es Grundwasser** *nt* WASSER artificial groundwater; **~e Grundwasseranreicherung** *f* WASSER artificial groundwater recharge; **~er Schleusenkanal** *m* ERNEU ENER *Gezeitenkraftwerk* sluiceway; **~er Staub** *m* LUFT manmade dust; **~e Übersäuerung** *f* BODEN artificial acidification

Kunststoff *m* ABFALL plastic; **Kunststoffabdichtung** *f* ABFALL plastic sealing; **Kunststoffabfall** *m* ABFALL plastics waste; **Kunststoffabfälle** *m pl* ABFALL waste plastics; **Kunststoffabfall** *m* **nach Gebrauch** ABFALL post-consumer plastics waste; **Kunststoffabtrennung** *f* ABFALL plastic separation; **Kunststoffbahn** *f* BODEN plastic liner, plastic sheet; **Kunststoffbehälter** *m* MEER dracone; **Kunststoffdichtungsbahn** *f* BODEN plastic sheet; **Kunststoffeinbauten** *m pl* ABWASSER *Tropfkörper-Füllkörper* plastic module; **Kunststoffeinsatz** *m* ABFALL plastics application; **Kunststoffentsorgung** *f* ABFALL, ABWASSER disposal of plastics; **Kunststoffgebinde** *nt* ABFALL plastic container; **Kunststoffilm** *m* ABFALL plastics sheeting; **Kunststofflasche** *f* ABFALL, ANALYTIK *Probenahme* plastic bottle; **Kunststoffmatrize** *f* ABFALL plastic matrix; **Kunststoffolie** *f* ABFALL plastic sheet, plastics sheeting; **Kunststofformen** *nt* ABFALL plastics molding (*AmE*), plastics moulding (*BrE*); **Kunststoffrecycling** *nt* ABFALL plastics recycling; **Kunststoffrecyclinganlage** *f* ABFALL plastics recycling plant; **Kunststoffrecycling-Wirbelschichtanlage** *f* ABFALL plastics recycling fluidized-bed plant; **Kunststoffrohr** *nt* ERNEU ENER *Wärmepumpe* plastic pipe; **Kunststoffrücknahmeverpflichtung** *f* ABFALL take-back obligation for plastics; **Kunststoffschnitzel** *nt* ABFALL shredded plastics; **Kunststoffsortieranlage** *f* ABFALL plastics sorting plant

kunststoffverarbeitend *adj* ABFALL plastics-processing; **~e Industrie** *f* ABFALL plastics-processing industry

Kunststoff: **Kunststoffverarbeitungsextruder** *m* ABFALL plastics extruder; **Kunststoffverpackung** *f* ABFALL plastic container, plastic packaging, plastic packaging material

Kunststoffverwertung *f* ABFALL plastics recycling; **~ der ersten Stufe** ABFALL primary plastics recycling; **~ der zweiten Stufe** ABFALL secondary plastics recycling; **Kunststoffverwertungsanlage** *f* ABFALL plastics recycling plant

Kupfer *nt* (*Cu*) ALLG copper (*Cu*); **Kupfer-Indium-Diselenid-Zelle** *f* ERNEU ENER *Deponiegaskomponente* copper-indium diselenide cell; **Kupferrohr** *nt* WASSER copper pipe; **Kupfer (II)-Carbonat** *nt* WASSER copper carbonate

Kupolofen-Winderhitzer *m* LUFT cupola-blast heater

Kupplung *f* LUFT coupling

Kurbelgehäuse *nt* LUFT crankcase; **Kurbelgehäuseemissionen** *f pl* LUFT *Abgas* crankcase emissions; **Kurbelgehäuseentlüftung** *f* LUFT crankcase ventilation, positive crankcase ventilation (*PCV*); **Kurbelgehäusegas** *nt* LUFT blowby gas

Kurzanalyse *f* ANALYTIK *zur Bestimmung der wichtigsten Bestandteile* proximate analysis

kurzgebaut: **~er dicker Zyklon** *m* LUFT short and thick cyclone

kurzglasfaserverstärkt: **~es Thermoplast** *nt* ABFALL, WASSER short-glass-fiber reinforced thermoplast (*AmE*), short-glass-fibre reinforced thermoplast (*BrE*)

Kurzschlußläufermotor *m* ABWASSER, WASSER *Pumpenantrieb* squirrel-cage induction motor

Kurzwellenhörer *m* LÄRM short-wave listener (*SWL*)

kurzwellig *adj* LUFT short-wave; **~es Sonnenlicht** *nt* LUFT short-wave sunlight; **~e Sonnenstrahlung** *f* LUFT short-wave incoming radiation

Kurzzeit- *in cpds* ALLG short-term; **Kurzzeit-Exposition** *f* (*STEL*) ANALYTIK short-term exposure limit (*STEL*)

kurzzeitig *adj* WASSER short-term

Kurzzeit-: **Kurzzeitmaßnahmen** *f pl* ALLG short-term measures; **Kurzzeitprobenahme** *f* ANALYTIK short-term sampling; **Kurzzeitspeicher** *m* ERNEU ENER *Energie* short-term storage; **Kurzzeitüberwachung** *f* ANALYTIK short-term monitoring

Küsten- *in cpds* ABFALL, WASSER coastal; **Küstenfluß** *m* WASSER coastal river; **Küstenform** *f* ERNEU ENER coastline form; **Küstenlinie** *f* MEER shoreline; **Küstenschlamm** *m* ABFALL coastal deposit; **Küstenschlick** *m* ABFALL coastal deposit; **Küstenschutz** *m* WASSER shore protection

Kuverwasser *nt* MEER seepage

kW *abbr* (*Kilowatt*) ERNEU ENER kW (*kilowatt*); **kW-Anlage** *f* ERNEU ENER kilowatt plant

KWK *abbr* (*Kraft-Wärme-Kopplung*) ABFALL, LUFT CHP (*combined heat and power*)

KW-Stoff *m* (*Kohlenwasserstoff*) ABWASSER, LUFT, MEER, WASSER HC (*hydrocarbon*)

L

labil *adj* LUFT *Abgasfahne-Ausbreitungsklasse* labile; **~e Temperaturschichtung** *f* LUFT labile temperature layering
Labor *nt* ANALYTIK lab, laboratory
Laboratorium *nt* ANALYTIK lab, laboratory
Labor: **Laborbefunde** *m pl* ANALYTIK laboratory findings, lab results, laboratory results; **Laborexperiment** *nt* ANALYTIK laboratory experiment; **Laborextruder** *m* ANALYTIK laboratory extruder
laborgeprüft *adj* ANALYTIK lab-tested, laboratory-tested
Labor: **Labormaßstab** *m* ANALYTIK laboratory scale; **Labormesokosmen** *nt pl* ANALYTIK laboratory mesocosms; **Labormethode** *f* ANALYTIK laboratory method, lab method; **Laborpraxis** *f* ANALYTIK lab practice, laboratory practice; **Laborprobe** *f* ANALYTIK laboratory sample, lab sample; **Labortechniker** *m* ANALYTIK lab technician, laboratory technician; **Labortest** *m* ANALYTIK lab test, laboratory test; **Laboruntersuchung** *f* ANALYTIK bench-scale study; **Laborversuch** *m* ANALYTIK lab test, laboratory test
Lachgas *nt* LUFT nitrous oxide (N_2O)
Lackmuspapier *nt* ANALYTIK pH-paper strip
Lackschlamm *m* ABFALL varnish waste sludge, ABWASSER paint sludge, WASSER varnish waste sludge
Ladefähigkeit *f* LUFT loading capacity
Ladepumpe *f* MEER cargo pump
Laderaupe *f* BODEN loading tractor
Ladung *f* ABWASSER *Ionenaustauscher*, LUFT loading; **Ladungsträger** *m* LUFT *Elektrofilter* charge carrier; **Ladungsverteilung** *f* LUFT loading distribution
Lage *f* ABFALL, BODEN, LUFT layer
Lager- *in cpds* ALLG storage; **Lagerbeständigkeit** *f* ABFALL storage durability; **Lagerbrücke** *f* LUFT *Schüttschichtfilter* bearing bridge; **Lagerbunker** *m* ABFALL receiving bunker, refuse bunker, waste bunker
lagerfähig *adj* ALLG storable
Lager-: **Lagergut** *nt* ABFALL, BODEN fill mass; **Lagerkammer** *f* ABFALL *für radioaktive Abfälle* storage chamber; **Lagermöglichkeit** *f* ABFALL, BODEN, LUFT, WASSER storage facility; **Lagerraum** *m* ABFALL, BODEN storage area, storage space; **Lagerstätte** *f* ABFALL, BODEN repository, storage facility
Lagerung *f* ALLG storage
Lambda *nt* LUFT lambda; **Lambdafenster** *nt* LUFT lambda window; **Lambdasonde** *f* ANALYTIK lambda probe, LUFT oxygen sensor, *Katalysator* lambda probe
Lambert-Beer: **~sches Gesetz** *nt* ANALYTIK Lambert-Beer law
Lamelle *f* ABWASSER, LUFT, WASSER lamella; **Lamellenabscheider** *m* LUFT lamellar precipitator; **Lamellenbündel** *nt* LUFT lamella bundle
lamellenförmig *adj* ABFALL, LUFT, WASSER lamellar
Lamelle: **Lamellenwärmeaustauscher** *m* LUFT lamellar heat exchanger
laminar *adj* ABFALL, BODEN, LUFT, WASSER laminar; **~es Festbett** *nt* BODEN laminar fixed bed; **~e Grenzschicht** *f* LUFT *eines Zyklons* laminar boundary layer; **~e Grenzschichtströmung** *f* LUFT laminar boundary layer flow
Laminarströmung *f* ABFALL, LUFT, WASSER laminar flow, streamline flow
laminieren *vt* ABFALL laminate
LAMS *abbr* (*Laser-Massenspektrometrie*) ANALYTIK LAMS (*laser mass spectrometry*)
Land: **von ~ aus** *adv* ERNEU ENER land-based
Landfarming-Methode *f* BODEN landfarming method
landwirtschaftlich *adj* ABFALL, ABWASSER agricultural; **~er Abfall** *m* ABFALL agricultural waste, farm waste; **~e Nutzung** *f* ABWASSER land application; **~e Verwertung** *f* ABWASSER *von Abwasser* agricultural reuse
Langlebigkeit *f* BODEN *der Schadstoffe*, LUFT longevity
langsam: **~e Gärung** *f* ABFALL, WASSER slow fermentation; **~e Kompostierung** *f* ABFALL slow composting
Langsam: **Langsamfilter** *nt* WASSER slow-process filter; **Langsamfiltrat** *nt* WASSER slow-process filtrate; **Langsamsandfilter** *nt* ABWASSER slow-sand filter, WASSER slow-process sand filter; **Langsamsandfilterschicht** *f* WASSER slow-process sand filter bed; **Langsamsandfiltration** *f* ABWASSER *biologisches Verfahren* slow-sand filtration
Längsbecken *nt* ABWASSER rectangular basin, rectangular clarifier, rectangular tank
Längsräumer *m* ABWASSER longitudinal scraper
Langweg- *in cpds* ANALYTIK long-path; **Langwegmonitor** *m* ANALYTIK long-path monitor
langwellig *adj* LUFT long-wave; **~e terrestrische Ausstrahlung** *f* LUFT long-wave radiation from the earth; **~e Wärmestrahlung** *f* LUFT long-wave heat radiation
Langzeit- *in cpds* ALLG extended-period, long-running, long-term; **Langzeitabsicherung** *f* ABFALL *einer Deponie* long-term safety; **Langzeitbelebung** *f* ABWASSER extended-period aeration; **Langzeitbelüftung** *f* ABWASSER extended-period aeration; **Langzeitbeständigkeit** *f* ABWASSER, LÄRM, LUFT, MEER, WASSER long-term durability; **Langzeiteffekt** *m* ABWASSER, LÄRM, LUFT, MEER, POLITIK, WASSER long-term effect; **Langzeitgasanalyse** *f* ANALYTIK long-term gas analysis
langzeitig *adj* ALLG long-running, long-term
Langzeit-: **Langzeitkontrolle** *f* ABFALL, ABWASSER, LÄRM, LUFT long-term control; **Langzeitlager** *nt* ABFALL long-term storage; **Langzeitlagerung** *f* ABWASSER extended deposition; **Langzeitmaßnahmen** *f pl* ABFALL, ABWASSER, BODEN, LÄRM, LUFT, MEER, WASSER long-term measures; **Langzeitmeter** *nt* ANALYTIK long-term dosimeter; **Langzeitprobenahme** *nt* ANALYTIK long-term sampling; **Langzeitprüfröhrchen** *nt* ANALYTIK long-term detector tube; **Langzeitsicherheit** *f* ABFALL long-term security; **Langzeitspeicher** *m* ERNEU ENER *Energie* long-term storage; **Langzeitüberwachung** *f* ABFALL, ABWASSER, ANALYTIK, LÄRM, LUFT, MEER, WASSER long-term monitoring; **Langzeitverhalten** *nt* ABFALL *einer Deponie* long-term behavior (*AmE*), long-term behaviour (*BrE*); **Langzeitwirkung** *f* ABWASSER, LÄRM, LUFT, MEER, WASSER long-term effect
Lanze *f* LUFT lance
Lärm *m* LÄRM, SICHERHEIT noise; **~ am Arbeitsplatz** LÄRM noise at the workplace; **~ im Bauwesen** LÄRM construction noise; **durch ~ verursacht** *adj* LÄRM, MEER, SICHERHEIT noise-induced; **durch ~ verursachter Hörschaden** *m* LÄRM, MEER, SICHERHEIT noise-induced hearing impairment; **Lärmabstrahlung** *f* LÄRM noise radiation; **Lärmabwehr** *f* LÄRM noise suppression
lärmarm *adj* LÄRM low-noise; **~e Arbeitsverfahren** *nt pl* LÄRM low-noise working methods; **~e Baumaschinen** *f pl* LÄRM low-noise building machinery; **~e Geräte** *nt pl* LÄRM, MEER low-noise equipment; **~e Konstruktion** *f* LÄRM low-noise construction; **~e Maschine** *f* LÄRM low-noise machine
lärmbedingt *adj* LÄRM noise-induced; **~e dauerhafte**

Hörschwellenverschiebung *f* LÄRM noise-induced permanent threshold shift (*NIPTS*)

Lärm: **Lärmbegrenzung** *f* LÄRM, SICHERHEIT noise control; **Lärmbekämpfung** *f* LÄRM noise abatement; **Lärmbelästigung** *f* LÄRM, SICHERHEIT noise pollution, sound pollution; **Lärmbelastung** *f* LÄRM noise nuisance; **Lärmbeurteilung** *f* LÄRM noise rating, noise weighting

Lärmbewertung *f* ANALYTIK, LÄRM noise rating (*NR*); **Lärmbewertungsindex** *m* ANALYTIK, LÄRM noise rating index (*NR index*); **Lärmbewertungskurve** *f* ANALYTIK, LÄRM noise rating curve (*NR curve*); **Lärmbewertungszahl** *f* ANALYTIK, LÄRM noise rating index (*NR index*)

lärmdämmend *adj* LÄRM noise-excluding

Lärm: **Lärmdosimeter** *nt* ANALYTIK noise dosimeter; **Lärmeinwirkung** *f* LÄRM noise exposure; **Lärmemissionen** *f pl* ERNEU ENER *von Windkraftanlagen*, LÄRM noise emissions; **Lärmemissionskennzeichnung** *f* LÄRM noise labeling (*AmE*), noise labelling (*BrE*); **Lärmempfindlichkeit** *f* LÄRM sensitivity to noise; **Lärmentstehung** *f* LÄRM noise generation, *Quelle* noise source; **Lärmerzeugung** *f* LÄRM noise generation; **Lärmexposimeter** *nt* ANALYTIK noise exposure meter; **Lärmexposition** *f* LÄRM noise exposure

lärmfrei: **~es Feld** *nt* **im Freien** LÄRM outdoor noise-free field

lärmgedämpft *adj* LÄRM noise-reducing

Lärm: **Lärmgrenzwert** *m* ANALYTIK, LÄRM noise limit, noise limit value, noise limiting value; **Lärmgrenzwertkurve** *f* ANALYTIK noise criterion curve (*NC-curve*); **Lärmgrenzwertmelder** *m* ANALYTIK, LÄRM noise limit indicator; **Lärmimmissionen** *f pl* LÄRM noise ambient impairment; **Lärmimmissionen** *m* **am Arbeitsplatz** LÄRM noise at the workplace; **Lärmimmissionsprognosewert** *m* LÄRM noise exposure forecast (*NEF*)

lärmintensiv *adj* LÄRM noise-intensive

Lärm: **Lärmkarte** *f* LÄRM noise map; **Lärmkataster** *m* LÄRM noise register; **Lärmkontrolle** *f* LÄRM, SICHERHEIT noise control; **Lärmkriteriumkurve** *f* ANALYTIK noise criterion curve (*NC-curve*); **Lärmmeßgerät** *nt* ANALYTIK noise meter; **Lärmmessung** *f* ANALYTIK noise measurement

lärmmindernd *adj* LÄRM noise-reducing

Lärm: **Lärmminderung** *f* LÄRM noise reduction; **Lärmpegel** *m* LÄRM noise level; **Lärmquelle** *f* LÄRM noise source; **Lärmrichtwert** *m* LÄRM noise guiding value, noise standard value; **Lärmschaden** *m* LÄRM, SICHERHEIT noise-induced hearing damage; **Lärmschirm** *m* LÄRM noise shield, noise screen

lärmschluckend *adj* LÄRM noise-absorbing; **~es Bauelement** *nt* LÄRM noise-absorbing structural element

Lärmschutz *m* LÄRM, SICHERHEIT noise protection; **Lärmschutzbereich** *m* LÄRM, SICHERHEIT noise abatement zone, noise protection area, noise protection zone; **Lärmschutzgebiet** *nt* LÄRM noise protection area; **Lärmschutzgesetz** *nt* LÄRM, SICHERHEIT noise protection law; **Lärmschutzkabine** *f* LÄRM *deren Innenraum vor außerhalb erzeugtem Schall geschützt wird* acoustic booth; **Lärmschutzmaßnahmen** *f pl* LÄRM noise protection measures; **Lärmschutzmittel** *nt* LÄRM noise guards; **Lärmschutzrecht** *nt* LÄRM, SICHERHEIT noise protection law; **Lärmschutzschirm** *m* LÄRM noise shield; **Lärmschutzsystem** *nt* LÄRM noise protection system; **Lärmschutztunnel** *m* LÄRM noise protection tunnel; **Lärmschutzwall** *m* LÄRM noise insulation wall; **Lärmschutzwand** *f* LÄRM noise protection wall, noise-insulating wall, noise-insulating barrier; **Lärmschutzwand** *f* **aus Abfall** LÄRM noise protection wall made from waste; **Lärmschutzwand** *f* **aus Kunststoffmüll** LÄRM noise protection wall made from plastic waste; **Lärmschutzzone** *f* LÄRM noise protection zone, noise abatement zone

Lärm: **Lärmschwerhörigkeit** *f* LÄRM noise deafness; **Lärmspitze** *f* LÄRM noise peak; **Lärmstörung** *f* LÄRM noise intrusion; **Lärmtaubheit** *f* LÄRM noise-induced deafness; **Lärmteppich** *m* LÄRM noise carpet; **Lärmüberwachung** *f* ANALYTIK noise monitoring, LÄRM noise check, noise monitoring; **Lärmüberwachungsgerät** *nt* ANALYTIK, LÄRM noise monitor; **Lärmverbot** *nt* LÄRM noise ban; **Lärmverhinderung** *f* LÄRM, SICHERHEIT noise prevention

lärmverseucht *adj* LÄRM noise-contaminated

Lärm: **Lärmverseuchung** *f* LÄRM noise contamination; **Lärmvorsorge** *f* LÄRM, SICHERHEIT noise prevention; **Lärmzone** *f* LÄRM influence zone of noise, noise area; **Lärmzonenkarte** *f* LÄRM noise area map

Laser *m* ANALYTIK, SICHERHEIT laser

laserangeregt *adj* ANALYTIK laser-excited; **~e Atomfluoreszenzspektrometrie** *f* (*LEAFS*) ANALYTIK laser-excited atomic fluorescence spectrometry (*LEAFS*)

Laser: **Laserfernmessung** *f* ANALYTIK remote laser measuring; **Laser-Massenspektrometrie** *f* (*LAMS*) ANALYTIK laser mass spectrometry (*LAMS*); **Laser-Mikrospektralanalyse** *f* ANALYTIK laser microprobe mass analysis; **Laser-Radar** *nt* ANALYTIK *Messung von Schadstoff-Konzentration* laser radar; **Laser-Raman-Spektroskopie** *f* ANALYTIK laser Raman spectroscopy; **Laser-Sampler** *m* ANALYTIK *Festprobenanalytik* laser sampler; **Laserspektroskopie** *f* ANALYTIK laser spectroscopy; **Laserstrahl** *m* ABFALL *Trennung nach Farbe*, ANALYTIK *in LIDAR* laser beam; **Laserstrahlung** *f* SICHERHEIT laser radiation

Lästigkeitspegel *m* LÄRM perceived noise level (*PNdB*)

Last: **Lastkraftwagen** *m* (*LKW*) ABFALL heavy goods vehicle (*HGV*), lorry (*BrE*), truck (*AmE*); **Lastwagen** *m* ABFALL truck; **Lastwagenanhänger** *m* ABFALL *Abfalltransport* lorry trailer (*BrE*), truck trailer (*AmE*); **Lastzug** *m* ABFALL *Abfalltransport* lorry with trailer (*BrE*), tractor trailer (*AmE*)

Laufrad *nt* LUFT impeller

Laugen *f pl* BODEN bases

Laugewäsche *f* LUFT caustic scrubbing

Lautheit *f* LÄRM loudness

Lautstärke *f* LÄRM volume; **Lautstärkeempfindung** *f* LÄRM loudness perception; **Lautstärkemesser** *m* ANALYTIK, LÄRM phonometer; **Lautstärkemessung** *f* ANALYTIK, LÄRM loudness measurement; **Lautstärkepegel** *m* LÄRM loudness level; **Lautstärkepegelmesser** *m* ANALYTIK, LÄRM phonometer; **Lautstärkeregler** *m* LÄRM attenuator

LBG *abbr* (*niedrigkaloriges Biogas*) ERNEU ENER LBG (*low thermal value biogas*)

LC_0 *abbr* (*letale Konzentration Null*) SICHERHEIT LC_0 (*lethal concentration zero*)

LC_{50} *abbr* (*letale Konzentration, bei der 50% der Probanden überleben*) SICHERHEIT LC_{50} (*lethal concentration fifty*)

LC_x *abbr* (*letale Konzentration*) SICHERHEIT LC_x (*lethal concentration*)

LCD *abbr* (*Flüssigkristallsichtanzeige*) ANALYTIK LCD (*liquid crystal display*); **LCD-Bildschirm** *m* ANALYTIK LCD screen; **LCD-Monitor** *m* ANALYTIK LCD monitor

LC-UV-Detektion *f* ANALYTIK LC-UV detection

LD *abbr* (*letale Dosis*) ANALYTIK LD (*lethal dose*)

LD_{50} *abbr* (*letale Dosis, bei der 50% der Probanden überleben*) SICHERHEIT LD_{50} (*lethal dose fifty*)

LD_{Lo} *abbr* (*niedrigste letale Dosis*) SICHERHEIT LD_{Lo} (*lethal dose low*)

LD_x *abbr* (*letale Dose*) SICHERHEIT LD_x (*lethal dose*)

LEAFS *abbr* (*laserangeregte Atomfluoreszenzspektrometrie*) ANALYTIK LEAFS (*laser-excited atomic fluorescence spectrometry*)

Leasing *nt* POLITIK leasing; **~ von Zertifikaten** POLITIK leasing of certificates

lebend: **~e Bakterien** *f pl* ABWASSER live bacteria

Lebensdauer *f* ABFALL, BODEN *der Sanierungsmaßnahme* lifespan; **~ der Atmosphäre** LUFT atmospheric lifetime

lebensfähig: **~e Bakterien** *f pl* ABFALL, WASSER viable bacteria

Lebensmittelverpackung *f* ABFALL food packaging
Lebensweg *m*: **~ eines Produktes** MANAG product life cycle; **Lebenswegbilanz** *f* (*LWB*) ABFALL, MANAG life cycle analysis (*LCA*), life cycle assessment (*LCA*)
Lebenszyklusanalyse *f* (*LWB*) MANAG life cycle analysis (*LCA*), ABFALL, MANAG life cycle assessment (*LCA*)
Leck *nt* ALLG leak, leakage
Leckage *f* ALLG leak, leakage
Leck: **Leckalarm** *m* SICHERHEIT leakage warning; **Leckanzeiger** *m* ANALYTIK leak detector; **Leckbehälter** *m* WASSER spillage tank; **Leckbestimmung** *f* ABFALL, WASSER leak detection; **Leckdetektion** *f* ANALYTIK, BODEN leak detection
lecken *vi* SICHERHEIT leak
leckgeworden *adj* BODEN *von Chemikalien in den Boden* leaking
Leck: **Leckstelle** *f* ABFALL leak; **Leckwarnanlage** *f* ABWASSER seepage warning installation; **Leckwarngerät** *nt* ABWASSER seepage warning device; **Leckwarnsystem** *nt* SICHERHEIT leakage indicator system
LED *abbr* (*lichtemittierende Diode*) ANALYTIK LED (*light-emitting diode*); **LED-Anzeige** *f* ANALYTIK LED display
leer *adj* ABFALL, WASSER empty
Leer-: **Leerfahrt** *f* ABFALL empty trip; **Leergut** *nt* ABFALL empties; **Leerlauf** *m* ABFALL, WASSER draining
leerlaufen *vi* ABFALL drain off
leermachen *vt* WASSER drain, drain off
leerpumpen *vt* ABFALL, ABWASSER, BODEN, WASSER pump out
Lehm *m* ABFALL loam; **Lehmwall** *m* ABFALL, WASSER clay barrier
LEI *abbr* (*Ionisation durch Laser*) ANALYTIK LEI (*laser-enhanced ionization*)
leicht: **~er Nebel** *m* LUFT haze, mist
Leichtbauplatte *f* LÄRM lightweight panel
leichtentzündlich *adj* SICHERHEIT *als Gefahrkennzeichnung* highly inflammable
Leichter *m* MEER lighter, lightening vessel
Leichtern *nt* MEER *Ölabgabe von Schiff zu Schiff* lightening
Leichtfaß *nt* ABFALL fiber drum (*AmE*), fibre drum (*BrE*)
leichtflüchtig *adj* ABWASSER, BODEN, LUFT volatile; **~e chlorierte Kohlenwasserstoffe** *m pl* BODEN semivolatile chlorinated hydrocarbons; **~er Chlorkohlenwasserstoff** *m* LUFT volatile chlorinated hydrocarbon; **~er Halogenkohlenwasserstoff** *m* ABFALL, LUFT, WASSER volatile halogenated hydrocarbon
Leichtfraktion *f* ABFALL fine fraction, light-density fraction
Leichtstofffraktion *f* ABFALL fine fraction, light-density fraction
leimig *adj* LUFT *Staubeigenschaften* glutinous
Leistung *f* ALLG capacity; **Leistungsabgabe** *f* ERNEU ENER power output; **Leistungsaufnahme** *f* **des Rotors** ERNEU ENER *Windturbine* input of rotor; **Leistungsbereich** *m* ERNEU ENER range of capacity; **Leistungsfähigkeit** *f* WASSER *Leitungsnetz* efficiency; **Leistungsvermögen** *nt* WASSER *Leitungsnetz* efficiency
Leit- *in cpds* ALLG control, guide; **Leitapparat** *m* ABFALL, WASSER control device
leiten *vt* ANALYTIK, ERNEU ENER *Strom* conduct
Leitfähigkeit *f* ALLG conductibility, conductivity; **Leitfähigkeitsmesser** *m* ANALYTIK conductivity meter; **Leitfähigkeitsmeßgerät** *nt* ANALYTIK conductivity measuring instrument, conductivity meter; **Leitfähigkeitsmessung** *f* ANALYTIK conductimetry
Leit-: **Leitflügel** *m* ERNEU ENER guide vane
leitprobenfrei: **~e Auswertung** *f* ANALYTIK evaluation without standard samples
leitprobengebunden: **~e Auswertung** *f* ANALYTIK evaluation using standard samples
Leit-: **Leitschaufel** *f* LUFT, WASSER *Pumpe* guide vane; **Leitstelle** *f* ABFALL, LUFT, WASSER control room
Leitung *f* LUFT duct; **Leitungsmesser** *m* ANALYTIK conductometer; **Leitungsnetz** *nt* WASSER supply network; **Leitungswasser** *nt* ABWASSER, WASSER tap water
Leit-: **Leitwand** *f* ABWASSER guide wall
Leitwertmesser *m* ANALYTIK conductometer
Lenkungssteuer *f* POLITIK regulatory tax
lesbar *adj* ANALYTIK readable
Leseband *nt* ABFALL sorting belt
letal *adj* SICHERHEIT lethal; **~e Dosis** *f* (LD_x) SICHERHEIT lethal dose (LD_x); **~e Dosis** *f*, **bei der 50% der Probanden überleben** (LD_{50}) SICHERHEIT lethal dose fifty (LD_{50}); **~e Konzentration** *f* (LC_x) SICHERHEIT *PED* lethal concentration (LC_x); **~e Konzentration** *f*, **bei der 50% der Probanden überleben** (LC_{50}) SICHERHEIT lethal concentration fifty (LC_{50}); **~e Konzentration** *f* **Null** (LC_0) SICHERHEIT lethal concentration zero (LC_0)
Letaldosis *f* ANALYTIK, LUFT, WASSER lethal dose (*LD*)
Letalkonzentration *f* ANALYTIK lethal concentration
Leuchtbakterientest *m* ANALYTIK luminescent bacteria test
Leuchtgas *nt* LUFT coal gas
Leuchtstoff *m* ABFALL luminous substance, luminophore; **Leuchtstofflampe** *f* ABFALL fluorescent tube
Licht *nt* ABFALL, ANALYTIK light; **Lichtabsorption** *f* ANALYTIK optical absorption; **Lichtabsorption** *f* **in Gasen** ANALYTIK optical absorption in gases
lichtemittierend *adj* ANALYTIK light-emitting; **~e Diode** *f* (*LED*) ANALYTIK *heavy metal* light-emitting diode (*LED*)
Licht: **Lichtimpuls** *m* ANALYTIK light-induced pulse; **Lichtintensität** *f* ANALYTIK luminous intensity
lichtmaserbetrieben: **~es Radar** *nt* (*LIDAR*) ANALYTIK light detection and ranging (*LIDAR*)
Licht: **Lichtmesser** *m* ANALYTIK photometer; **Lichtmessung** *f* ANALYTIK light measuring; **Lichtmikroskop** *nt* ANALYTIK light-optical microscope; **Lichtstrahl** *m* ABFALL light beam; **Lichtstreuung** *f* ANALYTIK light scattering, diffused light; **Lichtzerlegung** *f* ANALYTIK *Trennung nach Farbe* diffraction of light
LIDAR *abbr* (*lichtmaserbetriebenes Radar*) ANALYTIK LIDAR (*light detection and ranging*)
liefern *vt* ABFALL *Pumpe*, ABWASSER *Fäkalienpumpe*, WASSER *Pumpe* discharge
Lieferprogramm *nt* ANALYTIK delivery program (*AmE*), delivery programme (*BrE*)
Limnologie *f* ANALYTIK limnology
Linderung *f* POLITIK mitigation
Linienquelle *f* LUFT line source
Linse *f* ERNEU ENER *Solarenergieanlage* lens
lipophil *adj* ABWASSER lipophilic
lipophob *adj* ABWASSER lipophobic
lizenziert *adj* ABFALL, POLITIK, WASSER licensed
Lizenzzeichen *nt* MANAG licensing mark
LKW *abbr* (*Lastkraftwagen*) ABFALL HGV (*heavy goods vehicle*), lorry (*BrE*), truck (*AmE*); **LKW-Anhänger** *m* ABFALL *Abfalltransport* lorry trailer (*BrE*), truck trailer (*AmE*)
LOAEL *abbr* (*geringste Dosis, die bereits zu nachteiligen Wirkungen führt*) SICHERHEIT LOAEL (*lowest observed adverse effect level*)
Lobby-Arbeit *f* POLITIK *von Umweltverbänden* lobbying
Loch- *in cpds* ALLG perforated; **Lochband** *nt* MEER sifting belt; **Lochblech** *nt* ABWASSER, LÄRM, LUFT perforated panel, perforated plate; **Lochblechabsorber** *m* LÄRM perforated panel absorber
lochen *vt* ALLG *Rohr* perforate
Loch-: **Lochfraß** *m* WASSER pitting; **Lochplatte** *f* LÄRM, LUFT, WASSER perforated panel, perforated plate; **Lochplattenabsorber** *m* LÄRM perforated panel absorber
LOEC *abbr* (*niedrigste Konzentration eines Schadstoffes bei der Wirkung*) SICHERHEIT LOEC (*lowest observed effect concentration*)
LOEL *abbr* (*geringste Dosis, die bereits zu erkennbaren

Wirkungen führt) SICHERHEIT LOEL *(lowest observed effect level)*

Löffel *m* MEER *beim Löffelbagger* shovel; **Löffelbagger** *m* BODEN *Herstellung von Dichtwanden* ditcher, MEER backhoe; **Löffeltiefbagger** *m* MEER backhoe

logarithmisch *adj* ANALYTIK logarithmic; **~e Eichung** *f* ANALYTIK logarithmic calibration

Lokalisierung *f* BODEN *von Atlasten* localization

Longitudinalwellen *f pl* LÄRM longitudinal waves

los: **~e Ablagerung** *f* ABFALL, WASSER bulk deposition; **~er Transport** *m* ABFALL, WASSER bulk transport

lösbar *adj* ABFALL, ABWASSER, LUFT, WASSER soluble

Löschbehälter *m* LUFT quencher tank

löschen *vt* ABFALL, LUFT quench, *Kalk* slake, MEER *bei Schiffen* unload, WASSER quench

Löscher *m* BODEN quencher

Löschturm *m* LUFT quenching tower

Lösebehälter *m* ABWASSER solution tank, LUFT solvent tank

Lösemittel *nt* ABFALL, LUFT, WASSER solvent; **Lösemittelemissionen** *f pl* LUFT solvent emissions

lösemittelfrei *adj* ABFALL, LUFT, WASSER solvent-free; **~e Abluft** *f* LUFT solvent-free waste air

lösemittelhaltig *adj* LUFT: **~e Abluft** *f* LUFT solvent-containing waste air; **~e Luft** *f* LUFT solvent-laden air

Lösemittel: **Lösemittelrest** *m* ABFALL, BODEN, WASSER solvent residue

lösen *vt* ANALYTIK dissolve

Lösereaktor *m* WASSER solvents reactor

löslich *adj* ALLG soluble; **~es Gas** *nt* LUFT soluble gas

Löslichkeit *f* ALLG *der Schadstoffe* solvability, solubility

Löslichmachung *f* ALLG solubilization

Lösung *f* ABWASSER, ANALYTIK solution; **Lösungsbecken** *nt* ABWASSER solution tank

Lösungsmittel *nt* ALLG solvent; **auf ~ basierte Beschichtung** *f* ABFALL solvent-based coating; **Lösungsmittelanreicherung** *f* LUFT solvent accumulation; **Lösungsmittelbeladung** *f* LUFT solvent loading; **Lösungsmittelemissionen** *f pl* LUFT solvent emissions

lösungsmittelfrei *adj* ABFALL, LUFT, WASSER solvent-free; **~e Abluft** *f* LUFT solvent-free waste air; **~e Lacke** *f* ABFALL solvent-free paint

lösungsmittelhaltig *adj* LUFT solvent-laden; **~e Abluft** *f* LUFT solvent-laden waste air

Lösungsmittel: **Lösungsmittelkonzentration** *f* LUFT solvent concentration; **Lösungsmittelrückgewinnung** *f* ABFALL, LUFT solvent recovery; **Lösungsmittelrückgewinnungsanlage** *f* ABFALL, LUFT solvent recovery plant, solvent recovery unit; **Lösungsmittelzerstäubung** *f* LUFT solvent vaporization

LPG *abbr (Flüssiggas)* ERNEU ENER, LUFT LPG *(liquid petroleum gas)*

Lückengrad *m* LUFT voids fraction

Luft *f* ALLG air; **Luftabschluß** *m* ABFALL exclusion of air; **Luftabsorption** *f* LÄRM *Luftschalldämpfung* air absorption; **Luftabzug** *m* ABFALL, WASSER air excess; **Luftaufklärung** *f* MEER aerial reconnaissance; **Luftaufklärungsflugzeug** *nt* MEER *für Ölverschmutzungen* spotter plane; **Luftausfällung** *f* LUFT air shed; **Luftaustausch** *m* LUFT air exchange; **Luftbefeuchter** *m* SICHERHEIT air humidifier; **Luftbeschaffenheit** *f* LUFT air quality; **Luftbeschaffenheitsnormen** *f pl* LUFT air quality standards; **Luftbildauswertung** *f* ANALYTIK aerial view evaluation; **Luftbildfernerkundung** *f* MEER airborne remote sensing; **Luftblasenflotation** *f* LUFT air flotation; **Luftchemie** *f* LUFT atmospheric chemistry; **Luftdämpfung** *f* ANALYTIK *bei Meßgeräten* air friction damping

luftdicht *adj* ALLG air-tight

Luft: **Luftdiffusor** *m* ABWASSER air diffuser; **Luftdosierung** *f* LUFT air dosage; **Luftdurchlässigkeit** *f* BODEN *bei Bodenluftabsaugung* air permeability; **Luftdüse** *f* LUFT air jet; **Lufteinlaß** *m* LUFT air intake; **Lufteinwirkungszone** *f* BODEN zone of aeration; **Luftemissionen** *f pl* LUFT air emissions; **Luftentfeuchter** *m* SICHERHEIT air dehumidifier

luftentzündlich *adj* LUFT pyrophoric

Lüfter *m* LÄRM, LUFT fan; **Lüfterdämpfung** *f* LÄRM fan silencing; **Lüftergehäuse** *nt* LÄRM fan shroud; **Lüfterlärm** *m* LÄRM fan noise

Luft: **Luftfeder** *f* LÄRM air spring; **Luftfeuchtigkeit** *f* LUFT air humidity; **Luftfilter** *nt* LUFT, SICHERHEIT air filter; **Luftförderer** *m* ABFALL, WASSER aerial conveyor; **Luftführungszylinder** *m* LUFT air conduction cylinder

luftgefährdend *adj* LUFT hazardous to air, harmful to air

luftgekühlt *adj* ABFALL, LUFT air-cooled

Luft: **Luftgeschwindigkeit** *f* LUFT air velocity, air speed; **Luftmangel** *m* ABFALL, LUFT, WASSER deficiency of air; **Luftmasse** *f* ERNEU ENER *Windkraft* air mass; **Luftmengenmesser** *m* LUFT air flow sensor; **Luftpolster** *nt* LÄRM air cushion; **Luftprobennahmegerät** *nt* ANALYTIK air sampler

Luftqualität *f* ANALYTIK, LUFT, SICHERHEIT air quality; **Luftqualitätsdaten** *nt pl* ANALYTIK, LUFT air quality data; **Luftqualitätskriterien** *nt pl* ANALYTIK, LUFT air quality criteria; **Luftqualitätswert** *m* ANALYTIK, LUFT ambient air concentration value

Luft: **Luftreinheit** *f* LUFT air purity; **Luftreinigung** *f* SICHERHEIT *Umweltschutz* clean air device; **Luftreinigungs- und Deodorisierungsgerät** *nt* SICHERHEIT air purification and deodorization equipment; **Luftsauerstoff** *m* ABWASSER atmospheric oxygen; **Luftschadstoff** *m* LUFT air pollutant; **Luftschadstoff-Prognosemodell** *nt* MANAG airborne pollutant observation model

Luftschall *m* LÄRM, SICHERHEIT airborne noise; **Luftschalldämmung** *f* LÄRM airborne-noise proofing; **Luftschalldämpfung** *f* LÄRM *Luftabsorption* airborne-noise reduction; **Luftschallübertragung** *f* LÄRM airborne-noise transmission

Luft: **Luftschicht** *f* LUFT air layer; **Luftschleiertür** *f* SICHERHEIT air curtain installation for open doors; **Luftschleuse** *f* SICHERHEIT airlock; **Luftsperre** *f* MEER air bubble boom; **Luftstrahl** *m* ABWASSER, LUFT *Schlauchfilter* air jet; **Luftstrahlsieb-Verfahren** *nt* ANALYTIK *Staubanalyse* air-jet screening method; **Luftstrippen** *nt* ABFALL, ABWASSER, WASSER air stripping; **Luftstrom** *m* ABFALL, ERNEU ENER, WASSER air flow, air stream; **Luftströmung** *f* ABFALL, LUFT *Massenkraftabscheider*, WASSER air flow, air stream; **Lufttemperatur** *f* ABWASSER air temperature; **Lufttrocknung** *f* ABWASSER air drying; **Lufttrübung** *f* LUFT air turbidity; **Luftüberschuß** *m* ABFALL, LUFT, WASSER excess air; **Luftüberwachung** *f* MEER aerial surveillance; **Luftverdichter** *m* LUFT air compressor; **Luftvergiftung** *f* LUFT, SICHERHEIT air poisoning

Luftverschmutzung *f* LUFT, SICHERHEIT air pollution, atmospheric pollution; **Luftverschmutzungskontrolle** *f* LUFT, SICHERHEIT air pollution control; **Luftverschmutzungsvorhersage** *f* LUFT, SICHERHEIT air pollution forecast

Luft: **Luftverseuchung** *f* LUFT, SICHERHEIT air poisoning; **Luftverteiler** *m* ABWASSER air diffuser; **Luftverteilung** *f* BODEN air sparging, LUFT *Filtern* air distribution

Luftverunreinigung *f* LUFT, SICHERHEIT air pollution; **Luftverunreinigungsemissionen** *f pl* LUFT, SICHERHEIT air pollution emissions

Luft: **Luftvorwärmer** *m (LUVO)* ABFALL, ABWASSER, LUFT, WASSER air preheater; **Luftvorwärmung** *f* ABFALL, ABWASSER, LUFT, WASSER air preheating; **Luftwechsel** *m* SICHERHEIT ventilation; **Luftwiderstand** *m* ERNEU ENER *Windturbine* drag; **Luftzahl** *f* LUFT fuel-air ratio; **Luftzufuhr** *f* ABFALL, LUFT, WASSER air introduction, air supply

lungengängig *adj* SICHERHEIT respirable; **~e Reststoffe** *m pl* SICHERHEIT respirable residues

lungenreizend *adj* SICHERHEIT lung-irritant
LUVO *abbr* (*Luftvorwärmer*) ABFALL, LUFT air preheater
LWB *abbr* ABFALL, MANAG (*Lebenswegbilanz*) LCA (*life cycle assessment*), (*Lebenszyklusanalyse*) LCA (*life cycle analysis*)
Lysimeter *nt* ANALYTIK lysimeter

M

Machbarkeit *f* ALLG feasibility; **Machbarkeitsbewertung** *f* MANAG feasibility assessment; **Machbarkeitsstudie** *f* MANAG feasibility study

Mach-Zehnder-Interferometer *nt* ANALYTIK Mach-Zehnder interferometer

Magermotor *m* LUFT lean-mix engine; **~ mit Oxidationskatalysator** LUFT lean-mix engine with oxidation catalyst

Magma *nt* ERNEU ENER *geothermische Energieressource* magma

Magnesium *nt* ABFALL, WASSER magnesium; **Magnesiumhärte** *f* WASSER magnesium hardness; **Magnesiumhydroxid** *nt* WASSER magnesium hydroxide; **Magnesiumion** *nt* WASSER magnesium ion; **Magnesium-Nichtcarbonathärte** *f* WASSER magnesium noncarbonate hardness; **Magnesiumoxid-Wäsche** *f* LUFT magnesium oxide scrubbing; **Magnesiumsulfat** *nt* ABFALL, WASSER magnesium sulfate (*AmE*), magnesium sulphate (*BrE*)

Magnet *m* ABFALL, ANALYTIK, BODEN magnet

Magnetabscheider *m* ABFALL, BODEN, WASSER magnetic separator; **~ mit hoher Intensität** ABFALL, BODEN high-intensity magnet separator; **~ mit niedriger Intensität** ABFALL, BODEN low-intensity magnet separator

Magnet: **Magnetabscheidung** *f* ABFALL, BODEN magnetic separation; **Magnetfeld** *nt* ANALYTIK magnetic field

magnetisch *adj* ABFALL, ANALYTIK, BODEN magnetic; **~e Abscheidung** *f* ABFALL, BODEN magnetic separation; **~er Sauerstoffanalysator** *m* ANALYTIK magnetic oxygen analyser (*BrE*), magnetic oxygen analyzer (*AmE*); **~e Sortierung** *f* ABFALL, BODEN magnetic separation; **~e Suszeptibilität** *f* ANALYTIK magnetic susceptibility

Magnet: **Magnetkupplung** *f* ABWASSER magnetic coupling; **Magnetrührstab** *m* ABWASSER magnetic stirrer; **Magnetventil** *nt* LUFT solenoid valve

mahlen *vt* ABFALL pulverize

Mahlgut *nt* ABFALL ground stock

Makro- *in cpds* BODEN macro-; **Makroabfall** *m* ABFALL macrowaste; **Makroeinkapselung** *f* BODEN macroencapsulation; **Makroklima** *nt* LUFT macro-climate

makromolekular *adj* ABFALL macromolecular

Makro-: **Makropore** *f* ABWASSER, LUFT macropore

makroskopisch: **~-biologische Wassergütebestimmung** *f* ANALYTIK macroscopic biological water quality determination

MAK-Wert *m* (*maximal zulässige Arbeitsplatzkonzentration*) ABFALL occupational TLV (*occupational threshold limit value*)

Mammutpumpe *f* ABWASSER mammoth pump

Mammutrotor *m* ABWASSER mammoth rotor

Mangan *nt* LUFT manganese; **Manganbakterien** *f pl* WASSER manganese bacteria; **Manganhärte** *f* WASSER manganese hardness; **Manganhydroxid** *nt* ABFALL, WASSER manganese hydroxide

Mannloch *nt* ABWASSER manhole

Manometer *nt* ANALYTIK manometer; **Manometerflüssigkeit** *f* ANALYTIK manometer liquid; **Manometerwert** *m* ANALYTIK manometer value

manometrisch *adj* ABFALL, ABWASSER, ANALYTIK, WASSER manometric; **~e Förderhöhe** *f* ABFALL, ABWASSER, WASSER manometric discharge head; **~e Messung** *f* ANALYTIK manometric measurement; **~e Saughöhe** *f* ABFALL, ABWASSER, WASSER manometric suction head

Mantel *m* LUFT shell

manuell *adj* ABFALL, ANALYTIK manual; **~e Aufbereitung** *f* ABFALL manual processing; **~e Aussortierung** *f* ABFALL manual sorting; **~ beladender Kompaktor** *m* ABFALL manually loaded compactor

manuselektiv: **~er Detektor** *m* (*MSD*) ANALYTIK mass selective detector (*MSD*)

marin *adj* MEER marine; **~es Ökosystem** *nt* MEER marine ecosystem

Markt *m* MANAG market; **Marktabfall** *m* ABFALL market waste

marktabhängig *adj* MANAG market-dependent

marktgängig *adj* MANAG marketable

Markt: **Marktkräfte** *f pl* MANAG, POLITIK market forces; **Marktpreis** *m* MANAG, POLITIK market price

marktwirtschaftlich *adj* MANAG, POLITIK market-economy; **~er Anreiz** *m* MANAG, POLITIK market-economy incentive; **~es Instrument** *nt* MANAG, POLITIK market-economy instrument

Marsch *f* WASSER marsh, marshland

Masche *f* ABWASSER mesh; **Maschengröße** *f* ABWASSER mesh size

maschinell *adj* ALLG mechanical, mechanically assisted; **~e Aufbereitung** *f* ABFALL, WASSER mechanical processing; **~ entwässert** *adj* ABWASSER mechanically dewatered, mechanically thickened; **~ entwässerter Schlamm** *m* ABWASSER mechanically-thickened sludge, mechanically-dewatered sludge; **~ gestützte Handsortierung** *f* ABFALL mechanically-assisted hand separation

Maschinenausfall *m* ABWASSER, BODEN, LÄRM machine failure

Maschinenkapsel *f* LÄRM machine enclosure

Maßanalyse *f* ANALYTIK volumetric analysis

maßanalytisch *adj* ANALYTIK volumetric, titrimetric; **~e Bestimmung** *f* ANALYTIK volumetric determination

Maßeinheit *f* ANALYTIK measuring unit

Massen- *in cpds* ANALYTIK, LUFT, SICHERHEIT mass; **Massenabfall** *m* ABFALL mass waste; **Massenbestimmung** *f* **von Schwefeldioxid in der Umgebungsluft** SICHERHEIT determination of mass concentration of sulfur dioxide in ambient air (*AmE*), determination of mass concentration of sulphur dioxide in ambient air (*BrE*); **Massendurchflußmesser** *m* ANALYTIK mass flow rate meter; **Massenflußregler** *m* ANALYTIK mass flow regulator; **Massenfragmentographie** *f* ANALYTIK mass fragmentography; **Massenkonzentration** *f* LUFT *Gas, Staub* mass concentration

massenselektiv *adj* ANALYTIK mass-selective; **~er Detektor** *m* ANALYTIK mass-selective detector

Massen-: **Massenspektrometer** *nt* ANALYTIK mass spectrometer; **Massenspektrometrie** *f* (*MS*) ANALYTIK mass spectrometry (*MS*); **Massenspektroskop** *nt* ANALYTIK mass spectroscope; **Massenspektroskopie** *f* ANALYTIK *Verfahrensart in der Massenspektrometrie* mass spectroscopy; **Massenspektrum** *nt* ANALYTIK mass spectrum

mäßigen *vt* ANALYTIK temper

mäßig: **~ sauer** *adj* BODEN *Ultraviolett* moderately acidic

Maßstab *m* ANALYTIK measuring rod

Material *nt* ABFALL, POLITIK material; **Materialbilanz** *f* POLITIK materials balance; **Materialeigenschaften** *f pl* ABFALL characteristics of materials; **Materialeinsparung** *f* ABFALL materials saving; **Materialstromverteiler** *m* ABFALL materials stream distributor

mathematisch: **~es Klimamodell** *nt* LUFT mathematical climate model
Matrize *f* ABFALL matrix
Matte *f* LÄRM *Flächenverkleidung* matting
maximal: **~e Emissionskonzentration** *f* (*MEK*) ABFALL, LUFT maximum emissions concentration; **~e Fließgeschwindigkeit** *f* ABFALL, WASSER maximum flow velocity; **~e Immissionskonzentration** *f* ABFALL, LUFT maximum immission concentration; **~er Tagesbedarf** *m* WASSER maximum daily demand; **~ verträgliche Dosis** *f* (*MTD*) SICHERHEIT maximum tolerated dose (*MTD*); **~er Wasserbedarf** *m* WASSER maximum water demand; **~ zulässige Arbeitsplatzkonzentration** *f* (*MAK-Wert*) ABFALL occupational maximum allowable concentration (*occupational MAC*), occupational threshold limit value (*occupational TLV*); **~ zulässige Emissionskonzentration** *f* ANALYTIK maximum allowable emissions concentration; **~e zulässige Immissionskonzentration** *f* (*MIK-Wert*) ANALYTIK, LUFT maximum allowable concentration in the free environment; **~ zulässige Organkonzentration** *f* ANALYTIK maximum allowable organ concentration
MBG *abbr* (*mittelkaloriges Biogas*) ERNEU ENER MBG (*middle thermal value biogas*)
mechanisch *adj* ALLG mechanical; **~er Abscheider** *m* LUFT *wie Zyklonen* mechanical collector; **~e Abwasseraufbereitung** *f* ABWASSER mechanical wastewater treatment; **~e Abwasserreinigung** *f* ABWASSER primary wastewater treatment; **~e Aufbereitung** *f* ABFALL, WASSER mechanical processing; **~er Einschluß** *m* ABFALL *von Schadstoffen* physical stabilization; **~e Energie** *f* ERNEU ENER mechanical energy; **~ entwässert** *adj* ABWASSER mechanically dewatered, mechanically thickened; **~ entwässerter Schlamm** *m* ABWASSER mechanically-thickened sludge, mechanically-dewatered sludge; **~es Entwässerungsverfahren** *nt* ABWASSER *Vakuumfiltration, Zentrifugation* mechanical sludge-thickening method; **~e Filtration** *f* ABFALL, WASSER mechanical filtration; **~e Gefahr** *f* SICHERHEIT mechanical hazard; **~ gereinigt** *adj* ABWASSER mechanically cleaned; **~ gereinigter Rechen** *m* ABWASSER mechanically cleaned bar screen; **~ gerührter Flockungsreaktor** *m* ABWASSER flocculator with mechanically-operated paddles; **~e Müllsortierung** *f* ABFALL mechanical waste separation; **~e Mülltrennung** *f* ABFALL mechanical waste separation; **~er Rechen** *m* ABWASSER manually-operated screen, mechanical raking; **~es Recycling** *nt* ABFALL, ABWASSER mechanical recycling; **~e Reinigung** *f* ABWASSER mechanical purification, *des Rechens* mechanical cleaning; **~e Sortierung** *f* ABFALL automatic sorting; **~e Spannung** *f* WASSER stress; **~e Trennung** *f* ABFALL automatic sorting, mechanical separation; **~er Verlust** *m* ABFALL, ABWASSER, WASSER mechanical loss; **~e Verwertung** *f* ABFALL, ABWASSER mechanical recycling; **~e Wasseraufbereitung** *f* ABFALL, WASSER physical water treatment
medienspezifisch *adj* ANALYTIK medium-specific; **~er Bestandteil** *m* ANALYTIK medium-specific component
Meeres- *in cpds* MEER marine; **Meeresbucht** *f* ERNEU ENER bay, lagoon
meeresbürtig *adj* MEER sea-based; **~e Verschmutzung** *f* MEER sea-based pollution
Meeres-: **Meereseinleitung** *f* ABWASSER *von Abwasser* ocean disposal; **Meeresenergie** *f* ERNEU ENER ocean energy; **Meeresökologie** *f* MEER marine ecology; **Meeresspiegel** *m* MEER sea level
Meeresströmung *f* ERNEU ENER ocean current; **Meeresströmungsenergie** *f* ERNEU ENER ocean current energy; **Meeresströmungskraftwerk** *nt* ERNEU ENER ocean current power plant; **Meeresströmungsressourcen** *f pl* ERNEU ENER ocean current resources
Meeres-: **Meeresverschmutzung** *f* MEER marine pollution
Meereswärme: **Meereswärmeenergie** *f* ERNEU ENER ocean thermal energy; **Meereswärmekraftwerk** *nt* ERNEU ENER ocean thermal energy conversion plant (*OTEC plant*); **Meereswärmekraftwerk** *nt* **am Land** ERNEU ENER land-based OTEC plant; **Meereswärmeumwandlung** *f* ERNEU ENER ocean thermal energy conversion (*OTEC*)
Meeres-: **Meereswassergewinnung** *f* WASSER seawater recovery
Meerwasser *nt* ERNEU ENER, MEER, WASSER ocean water, seawater; **Meerwassereinbruch** *m* MEER seawater intrusion
Megawatt: **Megawatt-Anlage** *f* ERNEU ENER megawatt plant
Mehrbereich- *in cpds* ANALYTIK multirange; **Mehrbereichsinstrument** *nt* ANALYTIK multirange instrument
Mehreinheit-Container *m* ABFALL multiunit container
Mehrfachmeßstellen *f pl* ANALYTIK multiple measuring stations
Mehrkammer- *in cpds* ABFALL, LUFT multichamber, multicompartment; **Mehrkammerabsetzbecken** *nt* ABWASSER mixing worm; **Mehrkammer-Container** *m* ABFALL multicompartment container; **Mehrkammerschlauchfilter** *nt* LUFT multichamber bag filter
Mehrröhrenzyklon *m* LUFT multitube cyclone
mehrschalig *adj* LÄRM *Lärmschutzwände* multisheet; **~e Wand** *f* LÄRM compound wall
Mehrschicht- *in cpds* ABFALL, LÄRM multilayered; **Mehrschichtenkonstruktion** *f* LÄRM sandwich construction; **Mehrschichtfilter** *nt* ABWASSER multilayer filter
mehrschichtig *adj* ABFALL multilayer; **~er Aufbau** *m* ABFALL multilayered construction; **~er Behälter** *m* ABFALL multilayered container; **~e Filtration** *f* ABFALL, ABWASSER, WASSER multilayered filtration; **~e Platte** *f* LÄRM sandwich panel
Mehrschicht-: **Mehrschichtprinzip** *nt* ABFALL multibarrier principle
mehrseitig *adj* POLITIK multilateral; **~e Verhandlung** *f* POLITIK multilateral negotiation
Mehrstoffbrenner *m* ABWASSER compound burner
Mehrstoffsystem *nt* ABFALL multimaterial system
Mehrstufenverdampfer *m* ABWASSER multistage evaporator
mehrstufig *adj* ALLG multistage; **~e Abwasserbehandlungsanlage** *f* ABWASSER comprehensive wastewater treatment plant; **~es Belebungsverfahren** *nt* ABWASSER multistage activated sludge process; **~e Peripheralpumpe** *f* ABFALL, ABWASSER, WASSER multistage peripheral pump; **~e Wirbelschicht** *f* LUFT multistage fluidized bed; **~er Wirbelschichtadsorber** *m* LUFT multistage fluidized-bed adsorber; **~er Zyklon** *m* ANALYTIK multistage cyclone
Mehrweg- *in cpds* ABFALL returnable; **Mehrwegflasche** *f* ABFALL recycled bottle, returnable bottle, deposit bottle; **Mehrweggebinde** *nt* ABFALL returnable container; **Mehrwegkarton** *m* ABFALL reusable box; **Mehrwegquote** *f* ABFALL return rate; **Mehrwegverpackung** *f* ABFALL recyclable packaging, reusable packaging
mehrwertig *adj* ABFALL multivalent; **~es Kation** *nt* ABFALL, WASSER multivalent cation
Meißel *m* BODEN cutter
MEK *abbr* (*maximale Emissionskonzentration*) ABFALL, LUFT maximum emissions concentration; **MEK-Wert** *m* ANALYTIK maximum allowable emissions concentration
Meldung *f* MEER notification
Melioration *f* BODEN soil vapor concentration (*AmE*), soil vapour concentration (*BrE*)
Membran *f* ALLG membrane; **Membranbioreaktor** *m* LUFT membrane biological reactor; **Membrandestillation** *f* ABWASSER membrane distillation; **Membrandurchlässigkeit** *f* ABWASSER, WASSER membrane permeability; **Membranfilter** *nt* ABFALL, ABWASSER, ANALYTIK, WASSER membrane filter; **Membranfilterpresse** *f* ABFALL, ABWASSER, SICHERHEIT, WASSER membrane filter press; **Membranfiltration** *f* ABWAS-

SER, ANALYTIK, WASSER membrane filtration; **Membranoberfläche** *f* ABWASSER, ANALYTIK, WASSER membrane surface; **Membranpolarometer** *nt* ANALYTIK *zur Messung von gelöstem Sauerstoff in Wasser* foil polarometer; **Membranpumpe** *f* ABWASSER, ANALYTIK, MEER, WASSER diaphragm pump; **Membranreaktor** *m* LUFT membrane reactor; **Membranschicht** *f* ABWASSER, WASSER membrane layer; **Membrantrennverfahren** *nt* ABWASSER, WASSER membrane separation process; **Membranventil** *nt* ABWASSER, LUFT, WASSER diaphragm valve; **Membranverfahren** *nt* ABWASSER *beste verfügbare Technik*, SICHERHEIT membrane process; **Membranwiderstand** *m* ABWASSER, WASSER membrane resistance

Menge *f* ALLG amount

mengenabhängig *adj* ANALYTIK *Probenahme* quantity-dependent

Mengenfluß *m* ANALYTIK mass flow; **Mengenflußregelgerät** *nt* ANALYTIK mass flow controller

mengenproportional *adj* ANALYTIK flow-proportional; **~er Probenehmer** *m* ANALYTIK sampler for flow-proportional samples, flow-proportional sampler

Mengenreduktion *f* ABFALL, ANALYTIK volume reduction

menscheninduziert *adj* BODEN human-induced; **~e Bodendegradation** *f* BODEN human-induced soil degradation

menschlich: **~ bedingte Übersäuerung** *f* BODEN anthropogenic acidification

Mensur *f* ANALYTIK measuring jar

Mergel *m* ABFALL marl; **Mergelton** *m* ABFALL marly clay

mesophil *adj* ABWASSER mesophilic; **~e Bakterien** *f pl* ABWASSER mesophilic bacteria; **~e Faulung** *f* ABWASSER mesophilic digestion

Mesopore *f* LUFT *Biowäscher* mesopore

Meso-Scale-Zirkulation *f* LUFT mesoscale circulation

Mesosphäre *f* LUFT mesosphere

Meß- *in cpds* ANALYTIK measuring; **Meßanordnung** *f* ANALYTIK measuring set-up; **Meßapparat** *m* ANALYTIK measuring instrument, measuring device; **Meßaufgabe** *f* ANALYTIK measurement task; **Meßausstattung** *f* ANALYTIK measuring equipment

meßbar *adj* ANALYTIK measurable

Meßbarkeit *f* ANALYTIK measurability

Meß-: **Meßbecher** *m* ANALYTIK measuring cup; **Meßbedingungen** *f pl* ANALYTIK measurement conditions; **Meßbehälter** *m* ANALYTIK measuring container, measuring vessel; **Meßbereich** *m* ANALYTIK measuring range; **Meßbereichsüberschreitung** *f* ANALYTIK measuring range deviation; **Meßblende** *f* ANALYTIK measuring orifice; **Meßbrücke** *f* ANALYTIK measuring bridge

Meßdaten: **Meßdatenabtastung** *f* ANALYTIK measuring data scanning; **Meßdatenerfassung** *f* ANALYTIK data logging; **Meßdatenerfassungsprogramm** *nt* ANALYTIK *Computer* data logging programme; **Meßdatenfilterung** *f* ANALYTIK measuring data filtering

Meß-: **Meßdauer** *f* ANALYTIK measuring time; **Meßdraht** *m* ANALYTIK gage wire (*AmE*), gauge wire (*BrE*); **Meßdüse** *f* ANALYTIK measuring nozzle; **Meßeinrichtung** *f* ANALYTIK measuring equipment; **Meßelektrode** *f* ANALYTIK measuring electrode; **Meßelement** *nt* ANALYTIK measuring element; **Meßempfindlichkeit** *f* ANALYTIK measuring sensitivity; **Meßempfindlichkeitskontrolle** *f* ANALYTIK measurement sensitivity control

messen *vt* ANALYTIK gage (*AmE*), gauge (*BrE*), measure, meter

Messer *m* ANALYTIK meter

Meß-: **Meßergebnis** *nt* ANALYTIK measurement result, measuring result; **Meßfahrzeug** *nt* ANALYTIK measurement vehicle; **Meßfehler** *m* ANALYTIK measurement error, measuring error; **Meßfläche** *f* ANALYTIK measuring surface; **Meßflüssigkeit** *f* ANALYTIK measuring fluid; **Meßfühler** *m* ANALYTIK sensing device, sensor, detector, probe; **Meßgas** *nt* ANALYTIK measuring gas; **Meßgaskühler** *m* ANALYTIK measuring gas cooler; **Meßgaspumpe** *f* ANALYTIK sample pump; **Meßgefäß** *nt* ANALYTIK burette, measuring container, measuring vessel; **Meßgenauigkeit** *f* ANALYTIK measuring precision, measuring accuracy

Meßgerät *nt* ANALYTIK measuring instrument, measuring device, meter, MEER sensing device; **~ zur Analyse von Staub und Aerosolen** SICHERHEIT aerosol and dust measuring and analysis apparatus; **~ auf der Grundlage des Paramagnetismus** ANALYTIK *Messung der Sauerstoffkonzentration* measuring instrument based on paramagnetism; **Meßgerätegehäuse** *nt* ANALYTIK meter housing; **Meßgeräteklemme** *f* ANALYTIK meter terminal

Meß-: **Meßgerinne** *nt* ABWASSER, WASSER measuring flume, flow-measuring flume; **Meßgröße** *f* ANALYTIK measured variable, quantity to be measured, measuring quantity; **Meßinstrument** *nt* ANALYTIK measuring instrument, measuring device; **Meßkammer** *f* ANALYTIK measuring chamber; **Meßkanal** *m* ABWASSER, ANALYTIK, WASSER measuring channel, measuring flume, flow-measuring flume; **Meßkelch** *m* ANALYTIK measuring cup

Meßkette *f* ANALYTIK electrode; **Meßkettenimpedanz** *f* ANALYTIK electrode impedance; **Meßkettenspannung** *f* ANALYTIK electrode potential; **Meßkettenstatistik** *f* ANALYTIK electrode statistics

Meß-: **Meßkolben** *m* ANALYTIK volumetric flask; **Meßkomponente** *f* ANALYTIK measurement component; **Meßkopf** *m* ANALYTIK measuring sonde, measuring head; **Meßküvette** *f* ANALYTIK measuring cuvette; **Meßlehre** *f* ANALYTIK measuring gage (*AmE*), measuring gauge (*BrE*); **Meßlöffel** *m* ANALYTIK measuring scoop; **Meßlösung** *f* ANALYTIK test solution; **Meßmethode** *f* ANALYTIK measuring method; **Meßmittel** *nt* ANALYTIK measuring means; **Meßmotor** *m* ANALYTIK integrating motor; **Meßnetz** *nt* ANALYTIK field of measurement; **Meßort** *m* ANALYTIK measuring point, measuring location; **Meßperiode** *f* ANALYTIK measuring period; **Meßpipette** *f* ANALYTIK measuring pipette; **Meßplan** *m* ANALYTIK measurement plan; **Meßplatte** *f* ANALYTIK measuring panel; **Meßplatz** *m* ANALYTIK measuring position; **Meßprinzip** *nt* ANALYTIK measuring principle; **Meßprogramm** *nt* ANALYTIK measurement program (*AmE*), measurement programme (*BrE*); **Meßprotokoll** *nt* ANALYTIK measuring report; **Meßpunkt** *m* ANALYTIK measuring point; **Meßreihe** *f* ANALYTIK run; **Meßrohr** *nt* ANALYTIK measuring tube; **Meßschacht** *m* ANALYTIK measuring shaft; **Meßschraube** *f* ANALYTIK micrometer; **Meßsignal** *nt* ANALYTIK measuring signal; **Meßsonde** *f* ANALYTIK measuring sonde, measuring probe, measuring head; **Meßsondekabel** *nt* ANALYTIK measuring head cable; **Meßstation** *f* ANALYTIK measuring station, *für Gewässerkontrolle* sampling station

Meßstelle *f* ANALYTIK measuring point, measuring location; **Meßstellenabtaster** *m* ANALYTIK scanner; **Meßstellendichte** *f* ANALYTIK density of measurement points; **Meßstellennetz** *nt* ANALYTIK monitoring network; **Meßstellenumschalter** *m* ANALYTIK scanner

Meß-: **Meßstrecke** *f* ANALYTIK measuring section; **Meßtechnik** *f* ANALYTIK measurement technique

meßtechnisch *adj* ANALYTIK metrology-related

Meß-: **Meßuhr** *f* ANALYTIK dial gage (*AmE*), dial gauge (*BrE*); **Meßumformer** *m* ANALYTIK measuring transducer; **Meßumgebung** *f* ANALYTIK test environment; **Meßumschalter** *m* ANALYTIK measuring switch

Messung *f* ALLG measurement, measuring; **~ von gasförmigen Schadstoffen** ANALYTIK measurement of gaseous pollutants; **~ von Gasmengen** ANALYTIK measurement of gas quantities; **~ der Gasstromgeschwindigkeit** ANALYTIK measurement of gas flow velocity; **~ der Luftqualität** ANALYTIK, LUFT air quality measurement; **~ der**

Luftverschmutzung ANALYTIK, SICHERHEIT air pollution measurement; ~ **der Sauerstoffkonzentration** ANALYTIK measurement of oxygen concentration; ~ **der Streulichtintensität** ANALYTIK measurement of intensity of scattered light; ~ **der Strömungsgeschwindigkeit** ANALYTIK measurement of flow rate; ~ **mit Ultraschall** ANALYTIK measuring with ultrasound; ~ **von Umweltlärm** LÄRM, SICHERHEIT environmental noise measurement

Meß-: **Meßverstärker** *m* ANALYTIK measuring amplifier; **Meßvoraussetzungen** *f pl* ANALYTIK measurement prerequisites; **Meßvorgang** *m* ANALYTIK measuring procedure; **Meßwagen** *m* ANALYTIK measurement vehicle; **Meßwehr** *nt* WASSER measuring weir

Meßwert *m* ANALYTIK reading, measured value; **Meßwertanzeige** *f* ANALYTIK read-out display; **Meßwertausgabe** *f* ANALYTIK read-out display; **Meßwertcodierung** *f* ANALYTIK coding of measured value; **Meßwertdecodierung** *f* ANALYTIK decoding of measured value; **Meßwertfernübertragung** *f* ANALYTIK telemetry; **Meßwertgeber** *m* ANALYTIK sensing device, measuring transducer; **Meßwertkonstanz** *f* ANALYTIK reading constancy; **Meßwertregistriergerät** *nt* ANALYTIK measured value recording device; **Meßwertspeicherung** *f* ANALYTIK data storage; **Meßwertübermittlungsgerät** *nt* ANALYTIK device for measured value transmission; **Meßwertverarbeitung** *f* ANALYTIK measured value processing; **Meßwertverfälschung** *f* ANALYTIK measured value distortion

Meß-: **Meßwiderstand** *m* ANALYTIK measuring resistor, measuring resistance; **Meßzeit** *f* ANALYTIK measuring time; **Meßzelle** *f* ANALYTIK measuring cell; **Meßzweck** *m* ANALYTIK measuring purpose; **Meßzylinder** *m* ANALYTIK cylindrical graduated measuring glass, measuring jar

Metabolitentest *m* ANALYTIK metabolites test

Metalimnion *nt* ABWASSER metalimnion, thermocline

Metall *nt* ABFALL, ABWASSER, BODEN, LUFT metal; **Metallabfall** *m* ABFALL metal waste; **Metallgehäuse** *nt* LUFT metal housing

metallhaltig *adj* ABWASSER metalliferous; **~es Abwasser** *nt* ABWASSER metalliferous wastewater; **~es Pflanzenschutzmittel** *nt* BODEN metal fungicides

metallisch: **~er Staub** *m* LUFT metallic dust

Metallisierung *f* ABFALL metal coating

Metall: **Metallplatte** *f* LUFT metal plate; **Metallrohr** *nt* LUFT metal pipe; **Metallsalze** *nt pl* ABWASSER metallic salts; **Metallschlamm** *m* ABFALL, ABWASSER metallic sludge; **Metallschrott** *m* ABFALL metal scrap, metal waste; **Metallsortieranlage** *f* ABFALL metal separator; **Metallträger** *m* LUFT *Katalysator* metal carrier

metallurgisch: **~e Schlacke** *f* ABFALL metallurgical slag

Metall: **Metallverbindungen** *f pl* BODEN metal complexes

Meter *m* ANALYTIK meter

Methan *nt* ABFALL, ABWASSER, LUFT methane

Methanal *nt* LUFT formaldehyde

Methan: **Methanbildner** *m* ABFALL, ABWASSER methane former; **Methanbildung** *f* ABFALL, ABWASSER methane build-up, methanogenesis; **Methanerzeugung** *f* ABFALL methanogenesis; **Methangärung** *f* ABFALL methane fermentation; **Methangewinnung** *f* ABFALL methane recovery

methanogen *adj* ABWASSER methanogenic; **~e Bakterien** *f pl* ABWASSER methanogenic bacteria

Methanogenese *f* ABWASSER methanogenesis

Methanolyse *f* ABFALL methanolysis

Methan: **Methanreaktor** *m* ABFALL methane reactor

Methode *f* ALLG method

Methylenblauprobe *f* ANALYTIK methylene blue test

Methylenblautest *m* ANALYTIK methylene blue test

Methylenchlorid *nt* LUFT methylene chloride

MGB *abbr* (*Müllgroßbehälter*) ABFALL large refuse container

mg/kg *abbr* (*Milligramm pro Kilogramm*) ANALYTIK mg/kg (*milligram per kilogram, milligramme per kilogramme*)

Michelson-Interferometer *nt* ANALYTIK Michelson interferometer

Mie-Streuung *f* ANALYTIK, LUFT Mie scattering

Miete *f* ABFALL heap, BODEN pit

Migration *f*: ~ **der Schadstoffe** ABFALL, BODEN migration of harmful substances

migrieren *vi* ABFALL, BODEN migrate

migrierend: **~es Gas** *nt* ABFALL, BODEN migrating gas

Mikro- *in cpds* ALLG micro-

Mikrobe *f* ALLG microbe, microorganism; **Mikrobenaktivität** *f* ABWASSER, BODEN microbial activity; **Mikrobenpopulation** *f* ABWASSER, BODEN microbial community, microbial population

mikrobiell *adj* ABWASSER, BODEN microbial; **~er Abbau** *m* BODEN microbial degradation, microbial decomposition; **~e Atmung** *f* ABWASSER microbial respiration; **~e Auslaugung** *f* BODEN microbial leaching

mikrobiologisch *adj* ALLG microbiological; **~er Abbau** *m* ABWASSER microbiological decomposition; **~e Gefahr** *f* SICHERHEIT microbiological hazard; **~er Stoffwechselprozeß** *m* LUFT microbiological metabolic process; **~e Untersuchung** *f* ANALYTIK microbiological testing; **~e Verfahren** *nt pl* BODEN *Bodenbehandlung* microbiological processes

Mikro-: **Mikrocoulometrie** *f* ANALYTIK microcoulometry

mikrocoulometrisch *adj* ANALYTIK microcoulometric; **~e Titration** *f* ANALYTIK microcoulometric titration

Mikro-: **Mikrodipper** *m* ANALYTIK microdipper; **Mikroeinkapselung** *f* ABFALL grain encapsulation, BODEN microencapsulation; **Mikrofiltration** *f* ABWASSER microfiltration; **Mikroflocke** *f* WASSER microfloc; **Mikroflockung** *f* ABWASSER microcoagulation

Mikrofon *nt* ANALYTIK microphone

Mikro-: **Mikroklima** *nt* LUFT microclimate; **Mikroklimaveränderungen** *f pl* ERNEU ENER microclimatic changes; **Mikroluftblasenflotation** *f* ABWASSER dissolved air flotation

μm *abbr* (*Mikrometer*) ANALYTIK μm (*micrometer AmE, micrometre BrE*)

mikromaterialbearbeitet *adj* ANALYTIK micromachined

Mikro-: **Mikrometer** *nt* (*μm*) ANALYTIK *Einheit* micro, micrometer (*AmE, μm*), micrometre (*BrE, μm*), *Instrument* micrometer; **Mikrometerschraube** *f* ANALYTIK micrometer; **Mikronährstoff** *m* ABWASSER micronutrient; **Mikroorganismus** *m* ABWASSER microorganism, microbe, BODEN, LUFT, WASSER microorganism; **Mikropore** *f* ABWASSER, LUFT *Biowäscher* micropore

mikroporös *adj* ABFALL, ABWASSER, LUFT, WASSER microporous; **~e Membran** *f* ABFALL, ABWASSER, WASSER microporous membrane

Mikro-: **Mikroprozessor** *m* ANALYTIK microprocessor

mikroprozessorgesteuert: **~er Analysator** *m* ANALYTIK computerized analyser (*BrE*), computerized analyzer (*AmE*); **~er PID** *m* ANALYTIK microprocessor-controlled PID

Mikro-: **Mikrosiebfilter** *nt* ABWASSER microfilter; **Mikrosiebung** *f* ABWASSER microscreening

Mikroskop *nt* ANALYTIK microscope

mikroskopisch *adj* ANALYTIK microscopic; ~ **qualitativ** *adj* ANALYTIK microscopic qualitative; ~ **qualitative Nachweismethode** *f* ANALYTIK microscopic qualitative detection method; **~er Staub** *m* SICHERHEIT microscopic dust

Mikro-: **Mikrowellenverfahren** *nt* ABFALL microwave process

MIK-Wert *m* (*maximale zulässige Immissionskonzentration*) ANALYTIK, LUFT maximum allowable concentration in the free environment

Milligramm: ~ **pro Kilogramm** *phr* (*mg/kg*) ANALYTIK milligram per kilogram (*AmE*) (*mg/kg*), milligramme per kilogramme (*BrE*) (*mg/kg*)

mindern *vt* LÄRM *Schallpegel* reduce
Minderung *f* POLITIK mitigation
minderwertig: **~es Altpapier** *nt* ABFALL lower-quality recycled paper
Mindestprobemenge *f* ANALYTIK minimum sample quantity
Mineral- *in cpds* ALLG mineral; **Mineralfaser** *f* LÄRM, LUFT mineral fiber (*AmE*), mineral fibre (*BrE*); **Mineralfaserabsorber** *m* LÄRM mineral fiber absorber (*AmE*), mineral fibre absorber (*BrE*)
Mineralisation *f* ABWASSER mineralization
mineralisch: **~e Abdichtung** *f* ABFALL mineral sealing; **~e Dichtung** *f* ABFALL mineral sealing; **~e Dichtungsschicht** *f* ABFALL mineral sealing layer; **~er Staub** *m* LUFT mineral dust
mineralisieren *vt* ALLG mineralize
Mineralisierung *f* ABFALL, ABWASSER, BODEN, LUFT mineralization; **Mineralisierungsgrad** *m* ABFALL mineralization level
mineralogisch: **~ zerstört** *adj* BODEN destroyed in minerals
Mineralöl *nt* ABFALL mineral oil; **Mineralölabfall** *m* ABFALL mineral oil waste
mineralölhaltig: **~es Abwasser** *nt* ABWASSER wastewater containing mineral oil
Mineralöl: **Mineralölprodukt** *nt* ABFALL mineral oil product; **Mineralölsteuer** *f* POLITIK mineral oil tax
Mineral-: **Mineralquelle** *f* WASSER mineral spring; **Mineralstoffdeponie** *f* ABFALL mineral aggregate landfill
mineralstofffrei *adj* ABFALL, WASSER mineral-matter-free
Mineral-: **Mineralwolle** *f* LÄRM mineral fiber (*AmE*), mineral fibre (*BrE*), mineral wool; **Mineralwolleplatte** *f* LÄRM, MEER mineral fiber panel (*AmE*), mineral fibre panel (*BrE*)
minimal: **~e Fließgeschwindigkeit** *f* ABFALL, WASSER minimum flow velocity; **~er Tagesbedarf** *m* WASSER minimal daily demand; **~er Wasserbedarf** *m* WASSER minimal water demand
minimieren *vt* ABFALL, LUFT, MANAG minimize
Minimierung *f* ABFALL, LUFT minimization
Misch- *in cpds* ALLG mixed
mischbar: **~e Substanz** *f* ABFALL, ABWASSER, LUFT, WASSER miscible substance
Mischbarkeit *f* ABWASSER, LUFT miscibility
Misch-: **Mischbecken** *nt* ABWASSER, WASSER mixing tank; **Mischbettionenaustauscher** *m* ABWASSER mixed-bed ion exchanger; **Mischdeponie** *f* ABFALL co-disposal landfill; **Mischeinrichtung** *f* LUFT *KNV* collating device
Mischen *nt* BODEN mixing; **~ mit Zuschlagsstoffen** BODEN mixing with aggregates
Mischer *m* ABWASSER mixing device, blender
Misch-: **Mischgas** *nt* LUFT mixed gas; **Mischglas** *nt* ABFALL mixed glass; **Mischglasscherben** *f pl* ABFALL mixed glass cullet; **Mischkanalisation** *f* ABWASSER mixed canalization; **Mischkultur** *f* ABWASSER mixed culture; **Mischprobe** *f* ANALYTIK composite sample; **Mischsalz** *nt* ABFALL mixed salt; **Mischschlamm** *m* ABWASSER mixed sludge; **Mischschnecke** *f* ABWASSER mixing worm
Mischung *f* ABFALL, ABWASSER, WASSER mixing
Misch-: **Mischverfahren** *nt* ABWASSER *Abwasser* combined treatment; **Mischwäscher** *m* LUFT combination scrubber; **Mischwasser** *nt* WASSER mixed water
Mitbehandlung *f*: **~ von Klärschlamm** ABWASSER treatment of wastewater sludge
mitgerissen *adj* ABWASSER, BODEN, LUFT entrained; **~es Bodenteilchen** *nt* BODEN entrained soil particle; **~es Feststoffteilchen** *nt* LUFT entrained solid particle; **~e Flüssigkeit** *f* LUFT entrained liquid; **~er Flüssigkeitstropfen** *m* LUFT entrained liquid droplet
Mitnahmefaktor *m* LUFT entrainment factor
mitnehmen *vt* LUFT entrain
Mitnehmen *nt* ABWASSER, LUFT entrainment
mitreißen *vt* ABWASSER, BODEN, LUFT entrain
Mitreißen *nt* ABWASSER, BODEN, LUFT entrainment
Mitreißgrenze *f* ABWASSER, BODEN, LUFT entrainment point
Mittel- *in cpds* ERNEU ENER medium-scale; **Mittel** *nt* ABFALL, ABWASSER, MEER medium, modifier; **Mittel** *nt* **zur Modifizierung der Oberflächenspannung** MEER surface tension modifier
mittelbelastet: **~er Tropfkörper** *m* ABWASSER intermediate trickling filter
mittelblasig: **~e Belüftung** *f* ABWASSER medium-size bubble aeration
Mittel-: **Mitteldruckspülung** *f* LUFT medium-pressure cleaning; **Mittelfraktion** *f* ABFALL intermediate fraction
mittelgroß: **~er Müll** *m* ABFALL medium-sized refuse
mittelkalorig: **~es Biogas** *nt* (*MBG*) ERNEU ENER middle thermal value biogas (*MBG*)
mittelkörnig *adj* ABFALL medium-grained
Mitteltemperatur- *in cpds* ERNEU ENER, LUFT mid-temperature; **Mitteltemperaturverbrennung** *f* LUFT mid-temperature incineration; **Mitteltemperaturwärme** *f* ERNEU ENER mid-temperature heat
Mittelungspegel *m* LÄRM average sound level
Mittel-: **Mittelwert** *m* ANALYTIK average value, mean value
mittlere: **mittlere Anlage** *f* ERNEU ENER medium-scale installation; **mittlerer jahreszeitlicher Abfluß** *m* ABFALL, WASSER mean seasonal runoff; **mittlere Letalzeit** *f* ANALYTIK, LUFT median lethal time; **mittlere Monatsdosis** *f* ANALYTIK, SICHERHEIT average monthly dose
Mixer-Settler-Extraktor *m* ABWASSER mixer-settler extractor
mobil *adj* ALLG mobile; **~e Anlage** *f* BODEN *für Bodenbehandlung* mobile unit; **~e Feldmeßtechnik** *f* ANALYTIK mobile field-measuring technique; **~es Labor** *nt* ANALYTIK mobile laboratory, mobile lab; **~e Labormeßtechnik** *f* ANALYTIK mobile laboratory-measuring technique; **~e Meßstation** *f* ANALYTIK mobile measuring unit; **~es Probenahmegerät** *nt* ANALYTIK mobile sampler; **~e Sanierungsanlage** *f* BODEN container remediation installation; **~e Zerkleinerungsanlage** *f* ABFALL mobile crushing unit
mobilisieren *vt* ABFALL, BODEN, WASSER mobilize
mobilisiert *adj* ABFALL, BODEN, WASSER mobilized
Mobilisierung *f* ABFALL, BODEN *Schwermetalle*, WASSER *von Schadstoffen* mobilization
Mobillabor *nt* ANALYTIK mobile laboratory
Modellberechnung *f* LUFT model calculation
Modelleichung *f* ANALYTIK model calibration
Modellierung *f*: **~ der Gewässerqualität** WASSER modeling of water quality (*AmE*), modelling of water quality (*BrE*)
modernisieren *vt* ABFALL, ABWASSER, POLITIK, WASSER modernize
Modernisierung *f* POLITIK modernization
modifiziert: **~e Bodenmaterialien** *nt pl* WASSER *Deponieabdichtung* modified soil materials
modular *adj* ANALYTIK modular
Modulschallpegelmesser *m* ANALYTIK module sound level meter
möglich: **~es Sicherheitsrisiko** *nt* SICHERHEIT potential safety risk
Mohr: **~sche Waage** *f* ANALYTIK Mohr balance
MOK-Wert *m* ANALYTIK maximum allowable organ concentration
Mol *nt* LUFT mole (*gram-molecule*)
Molekül *nt* ALLG molecule
Molekular- *in cpds* ALLG molecular; **Molekulargewicht** *nt* ABFALL, ABWASSER, WASSER molecular weight; **Molekularkraft** *f* LUFT *Adsorption* molecular force; **Molekularsieb** *nt* ABWASSER, ANALYTIK, BODEN, LUFT, WASSER molecular sieve; **Molekularsiebung** *f* ABWASSER, ANALYTIK, BODEN, LUFT, WASSER molecular sieving
Molekül: **Molekülkolloid** *nt* ABWASSER, WASSER molecular colloid; **Molekülspektrometrie** *f* ANALYTIK molecular spectrometry

Molverhältnis *nt* LUFT mol ratio
Molvolumen *nt* LUFT molar volume
Monatsmittelwert *m* ANALYTIK monthly mean value
Monitor *m* ANALYTIK monitor
Monitoring *nt* ANALYTIK, SICHERHEIT monitoring
Monitorröhre *f* ANALYTIK *Massenspektrometrie* monitor tube
Monochlorbenzol *nt* LUFT monochlorobenzene
Monodeponie *f* ABFALL, ABWASSER mono-dump
monokristallin *adj* ERNEU ENER single-crystalline; **~e Siliziumzelle** *f* ERNEU ENER single-crystalline silicon cell
Monolith *m* LUFT monolith
monolithisch *adj* LUFT monolithic
Monomer *nt* ABFALL monomer
Montage *f* ABWASSER, WASSER assembly
montieren *vt* ABFALL, ABWASSER, WASSER assemble
Moor *nt* ABFALL, WASSER moor, moorland, swamp
Mooring *nt* MEER mooring
Moränenfilterschicht *f* ABFALL *Deponie* morainic filter layer
Morast *m* ABFALL, WASSER swamp
Mörtelmischer *m* BODEN pug mill
Motor *m* ABFALL, LUFT, MEER, WASSER motor; **~ mit einstellbarer Drehzahl** ABWASSER variable-speed motor; **~ mit Fremdzündung** LUFT spark ignition engine
motorbetrieben: **~es Abschöpfgerät** *nt* MEER *für Öl* self-propelled skimmer
motorfern: **~er Katalysator** *m* LUFT underfloor catalyst
motornah: **~er Katalysator** *m* LUFT underhood catalyst, exhaust manifold catalyst
Motor: **Motorpumpe** *f* MEER, WASSER motor pump
MS *abbr* (*Massenspektrometrie*) ANALYTIK MS (*mass spectrometry*)
MTD *abbr* (*maximal verträgliche Dosis*) SICHERHEIT MTD (*maximum tolerated dose*)
Muffe *f* ABWASSER *Erweiterung am Ende eines Rohres*, WASSER pipe sleeve, socket
Muldenkippwagen *m* MEER skip lorry (*BrE*), skip truck (*AmE*)
Müll *m* ABFALL, ABWASSER garbage (*AmE*), refuse, rubbish (*BrE*), waste; **Müllabfuhr** *f* ABFALL garbage collection (*AmE*), garbage disposal (*AmE*), refuse collection, refuse collection service, refuse disposal, rubbish collection; **Müllabfuhrlast** *f* ABFALL municipal waste collection load; **Müllabfuhrunternehmen** *nt* ABFALL collector, waste collector; **Müllabladeplatz** *m* ABFALL dump ground; **Müllabwurfschacht** *m* ABFALL garbage chute (*AmE*), garbage slide (*AmE*), refuse chute; **Müllanfall** *m* ABFALL waste production; **Müllaufbereitung** *f* ABFALL conditioning of waste; **Müllaufgabetrichter** *m* ABFALL hopper-furnace feed chute; **Müllaufkommen** *nt* ABFALL appearance of waste; **Müllballen** *m* ABFALL waste bale; **Müllbehälter** *m* ABFALL dustbin (*BrE*), garbage can (*AmE*), refuse can, rubbish bin (*BrE*), trash can (*AmE*), waste container; **Müllbeseitigung** *f* ABFALL waste disposal; **Müllbeutel** *m* ABFALL dustbin liner (*BrE*), garbage bag (*AmE*), garbage can bag (*AmE*), refuse sack, rubbish bag (*BrE*), waste bag; **Müllbunker** *m* ABFALL receiving bunker, waste bunker; **Müllbunkerung** *f* ABFALL waste storage; **Müllcontainer** *m* ABFALL roll-out waste container; **Mülldeponie** *f* ABFALL, ABWASSER, ANALYTIK, BODEN, LUFT, POLITIK, WASSER landfill; **Mülldetektor** *m* ABFALL waste detector; **Mülleimer** *m* ABFALL dustbin (*BrE*), garbage can (*AmE*), rubbish bin (*BrE*), trash can (*AmE*), waste container; **Mülleinsammlung** *f* ABFALL garbage collection (*AmE*), rubbish collection (*BrE*), refuse collection, waste collection; **Müllerfassung** *f* ABFALL garbage recovery (*AmE*), refuse recovery; **Müllerzeugung** *f* ABFALL waste generation; **Müllfahrer** *m* ABFALL waste collector; **Müllfahrzeug** *nt* ABFALL collection body, collection vehicle, dustbin lorry (*BrE*), garbage collection truck (*AmE*), garbage truck (*AmE*), refuse collection vehicle, refuse body; **Müllfeuchte** *f* ABFALL waste dampness; **Müllgroßbehälter** *m* (*MGB*) ABFALL large refuse container; **Müllsammlung** *f* rubbish collection
Mullit *m* LUFT *Keramikträger* mullite
Müll: **Müllkippe** *f* ABFALL garbage disposal site (*AmE*), rubbish tip (*BrE*), refuse disposal site, waste dump, waste tip (*BrE*); **Müll-Klärschlamm-Preßtrocknung** *f* ABWASSER press drying of waste with wastewater sludge; **Müllkompost** *m* ABFALL garbage compost (*AmE*), refuse compost; **Müllkompostierung** *f* ABFALL garbage composting (*AmE*), refuse composting; **Müllkörper** *m* ABFALL waste body; **Müllmann** *m* ABFALL waste collector; **Müllmenge** *f* ABFALL garbage mass (*AmE*), refuse mass; **Müllsack** *m* ABFALL dustbin liner (*BrE*), garbage bag (*AmE*), garbage can bag (*AmE*), refuse sack, rubbish bag (*BrE*), waste bag; **Müllsammelfahrzeug** *nt* ABFALL collection vehicle; **Müllsammlung** *f* ABFALL rubbish collection (*BrE*), garbage collection (*AmE*), refuse collection; **Müll- und Schlammkompostierung** *f* ABWASSER composting of waste and sludge; **Müllschlucker** *m* ABFALL rubbish chute (*BrE*), garbage chute (*AmE*), garbage slide (*AmE*), refuse chute; **Müllsortieranlage** *f* ABFALL garbage separation plant (*AmE*), refuse separation plant, waste sorting plant; **Müllsortierung** *f* ABFALL sorting of refuse, sorting of waste, sorting of garbage (*AmE*), sorting of rubbish (*BrE*); **Mülltonne** *f* ABFALL dustbin (*BrE*), garbage can (*AmE*), refuse can, rubbish bin (*BrE*), trash can (*AmE*), waste container; **Mülltourismus** *m* ABFALL, POLITIK waste tourism
Müllverbrennung *f* ABFALL garbage incineration (*AmE*), refuse incineration, trash burning (*AmE*), waste incineration; **Müllverbrennungsanlage** *f* (*MVA*) ABFALL garbage incineration plant (*AmE*), refuse incineration plant, waste incineration plant; **Müllverbrennungsfilterstaub** *m* ABFALL filter dust from incineration plant, waste incineration ash; **Müllverbrennungsschlacke** *f* ABFALL slag from waste incineration plant, waste incineration slag
Müll: **Müllverdichter** *m* ABFALL landfill compactor; **Müllverweildauer** *m* ABFALL residence time of waste; **Müllverweilzeit** *f* ABFALL residence time of waste; **Müllvolumen** *nt* ABFALL garbage volume (*AmE*), refuse volume; **Müllwagen** *m* ABFALL collection body, collection vehicle, refuse body, dustbin lorry (*BrE*), garbage collection truck (*AmE*), garbage truck (*AmE*), refuse collection vehicle; **Müllzerkleinerer** *m* ABFALL garbage grinder (*AmE*), refuse grinder; **Müllzusammensetzung** *f* ABFALL garbage composition (*AmE*), refuse composition, waste composition
Multizyklon *m* LUFT multiunit cyclone
Mündung *f*: **~ der Ablaufleitung** WASSER drain pipe spout
Mündungshöhe *f* LUFT top height
mutagen *adj* SICHERHEIT mutagenic; **~e Substanz** *f* SICHERHEIT mutagen
Mutterboden *m* ABFALL top soil
MVA *abbr* (*Müllverbrennungsanlage*) ABFALL garbage incineration plant (*AmE*), refuse incineration plant; **MVA-Schlacke** *f* ABFALL slag from waste incineration plant, waste incineration slag
MW-Anlage *f* ERNEU ENER *Windturbine* megawatt plant

N

N *abbr* ABFALL, ABWASSER (*Stickstoff*) N (*nitrogen*), ANALYTIK (*Newton*) N (*newton*), LUFT (*Stickstoff*) N (*nitrogen*)

Na *abbr* (*Natrium*) LUFT, WASSER Na (*sodium*)

Nach- *in cpds* ALLG secondary; **Nachabscheider** *m* LUFT secondary precipitator

Nachbarschaftsschutz *m* POLITIK protection of neighbors (*AmE*), protection of neighbours (*BrE*)

Nachbarschlauch *m* LUFT adjacent bag

nachbarschützend: **~e Rechtsnormen** *f pl* POLITIK legal standards

Nach-: **Nachbearbeitungsvorrichtung** *f* ABFALL, WASSER finishing device

nachbehandeln *vt* ABFALL, LUFT re-treat

Nach-: **Nachbehandlung** *f* ABFALL, BODEN, LUFT, WASSER secondary treatment; **Nachbrenner** *m* LUFT afterburner; **Nachbrennkammer** *f* LUFT afterburner chamber; **Nacheindicker** *m* ABWASSER final thickener; **Nachfällung** *f* ABWASSER post-precipitation; **Nachfaulbecken** *nt* ABWASSER secondary digestion tank; **Nachfaulraum** *m* ABWASSER secondary digestion tank; **Nachfilter** *nt* ABWASSER, LUFT, WASSER final filter; **Nachfiltern** *nt* ABWASSER, LUFT, WASSER final filtering, final filtration; **Nachfilterung** *f* ABWASSER, LUFT, WASSER final filtering, final filtration; **Nachfiltration** *f* ABWASSER post-filtration

Nachführeinrichtung *f* ERNEU ENER tracker, *Solarkollektoren* tracking mechanism

nachgeführt: **~er Spiegel** *m* ERNEU ENER *Solarturm* tracking reflector

nachgeordnet: **~e Behörde** *f* POLITIK lower-level administration

nachgeschaltet *adj* ALLG secondary; **~e Abgasreinigung** *f* ABWASSER, LUFT secondary waste gas treatment, subsequent exhaust gas cleaning; **~es Belebtbecken** *nt* ABWASSER secondary activated sludge tank; **~e Denitrifikation** *f* ABWASSER post-denitrification; **~er Katalysator** *m* LUFT secondary catalyst; **~e Maßnahmen** *f pl* MANAG end-of-pipe measures; **~er Partikelabsorber** *m* LUFT secondary particle separator; **~er Schönungsteich** *m* ABWASSER secondary polishing pond, secondary tertiary pond; **~er Staubabsorber** *m* LUFT secondary particle separator; **~er Tropfkörper** *m* ABWASSER secondary trickling filter

Nach-: **Nachhall** *m* LÄRM reverberation

nachhallend *adj* LÄRM reverberant

Nach-: **Nachhallzeit** *f* LÄRM reverberation time

nachhaltig *adj* BODEN, MANAG, POLITIK sustainable; **~e Bodennutzung** *f* BODEN sustainable soil use; **~e Entwicklung** *f* MANAG, POLITIK sustainable development; **~e Forstwirtschaft** *f* POLITIK sustainable forestry; **~e Kreislaufwirtschaft** *f* POLITIK sustainable recycling economy; **~es Wirtschaften** *nt* MANAG sustainable business, POLITIK sustainable economy

Nachhaltigkeit *f* BODEN, MANAG, POLITIK sustainability

Nach-: **Nachholbedarf** *m* POLITIK back-up requirement; **Nachkalibrierung** *f* ANALYTIK recalibration; **Nachklärbecken** *nt* ABWASSER final settling tank, secondary settling tank; **Nachklärung** *f* ABWASSER final clarification; **Nachkühler** *m* LUFT aftercooler

nachprüfbar *adj* ANALYTIK verifiable

Nach-: **Nachreaktion** *f* LUFT post-reaction; **Nachreaktionszeit** *f* ABWASSER post-reaction time, secondary reaction time; **Nachrotte** *f* ABFALL, ABWASSER post-rotting treatment

nachrüsten *vt* ALLG retrofit

Nach-: **Nachrüstung** *f* ALLG retrofitting

Nachsorge *f* ABFALL, WASSER monitoring, post-closure care; **Nachsorgeaufwand** *f* ABFALL post-closure expenditure; **Nachsorgemaßnahmen** *f pl* ABFALL post-closure measures; **Nachsorgeüberwachung** *f* ABFALL, BODEN post-closure monitoring

nachträglich: **~e Entwässerung** *f* BODEN subsequent dewatering; **~e Flockung** *f* ABFALL post-coagulation

Nach-: **Nachverbrenner** *m* LUFT afterburner; **Nachverbrennergehäuse** *nt* LUFT afterburner housing; **Nachverbrennung** *f* BODEN *bei thermischen Verfahren*, LUFT afterburning; **Nachverbrennungskammer** *f* ABFALL secondary combustion chamber (*SCC*); **Nachverrottung** *f* ABFALL, ABWASSER post-rotting treatment

nachwachsend: **~e Rohstoffe** *m pl* ERNEU ENER renewable raw materials

Nach-: **Nachwäscher** *m* LUFT rescrubber

Nachweis *m* ANALYTIK detection, identification

nachweisbar *adj* ANALYTIK detectable, identifiable

Nachweisbarkeit *f* ANALYTIK detectability, identifiability

Nachweis: **Nachweisempfindlichkeit** *f* ANALYTIK detection sensitivity

nachweisen *vt* ANALYTIK identify, detect

Nachweis: **Nachweisgrenze** *f* ANALYTIK testing limit, detection limit, identification limit; **Nachweishäufigkeit** *f* ANALYTIK detection frequency, identification frequency; **Nachweismethode** *f* ANALYTIK detection method, identification method; **Nachweisreaktion** *f* ANALYTIK detection reaction, identification reaction

Nadelfilz *m* LUFT needled felt

Nah-Infrarot-Spektroskopie *f* ABFALL, WASSER *automatische Kunststoffsortierung* near infrared spectroscopy

Nährlösungsbehälter *m* LUFT *biologische Gaswäsche* nutrient solution tank

Nährstoff *m* ABWASSER, BODEN, MEER, WASSER nutrient

nährstoffarm *adj* ABFALL, BODEN, MEER, WASSER nutrient-poor; **~es Wasser** *nt* WASSER nutrient-poor water

Nährstoff: **Nährstoffentfernung** *f* ABWASSER nutrient removal; **Nährstoffentzug** *m* ABWASSER nutrient removal

nährstoffreich *adj* ALLG nutrient-rich; **~es Wasser** *nt* ABWASSER, WASSER nutrient-rich water

Nährstoff: **Nährstoffversorgung** *f* ABFALL, ABWASSER, BODEN, LUFT, MEER nutrient supply

Nanogramm *nt* (*ng*) ANALYTIK nanogram (*ng*), nanogramme (*ng*)

Nanometer *nt* (*nm*) ANALYTIK nanometer (*AmE*) (*nm*), nanometre (*nm*) (*BrE*)

Naß- *in cpds* ABWASSER, BODEN, LUFT, WASSER wet; **Naßablagerung** *f* ABFALL, LUFT wet deposition; **Naßabscheider** *m* LUFT, WASSER wet scrubber; **Naßabscheidung** *f* LUFT wet precipitation, wet separation; **Naßabsorber** *m* LUFT wet collector; **Naßbagger** *m* MEER dredger

naßbaggern *vt* MEER dredge

naßchemisch: **~e Methoden** *f pl* ANALYTIK wet-chemical methods

nasse: **~ Abgasbehandlung** *f* LUFT wet waste gas treatment; **~ Abreinigung** *f* LUFT wet cleaning process; **~ Entstaubung** *f* LUFT wet dust removal; **~ Korrosion** *f* ABFALL, WASSER wet corrosion

Naß-: **Naßentascher** *m* ABFALL wet ash remover; **Naßentschwefelungsprozeß** *m* ABFALL, LUFT wet desulfurization process (*AmE*), wet desulphurization process (*BrE*)

nasse: **~s Elektrofilter** *nt* LUFT wet electrostatic filter; **~r saurer Niederschlag** *m* LUFT wet acidic fallout

Naß-: **Naßmahlen** *nt* LUFT wet grinding, wet milling; **Naßmischer** *m* ABFALL wet mixer; **Naßoxidation** *f* ABFALL, ABWASSER, BODEN, WASSER wet air oxidation (*WAO*), wet oxidation; **Naßperiode** *f* LUFT wet period; **Naßrauchgasentschwefelungsanlage** *f* (*Naß-REA*) LUFT wet flue-gas desulfurization plant (*AmE*), wet flue-gas desulphurization plant (*BrE*); **Naß-REA** *f* (*Naß-Rauchgasentschwefelungsanlage*) LUFT wet flue-gas desulfurization plant (*AmE*), wet flue-gas desulphurization plant (*BrE*)

naßreinigend: **~es Gichtgas** *nt* LUFT wet-scrubbed waste top gas

Naß-: **Naßreinigung** *f* ABFALL, ABWASSER, LUFT wet cleaning; **Naßscheider** *m* LUFT wet scrubbing device; **Naßschlamm** *m* ABWASSER liquid sludge, slurry, wet sludge; **Naßschlammverwertung** *f* ABWASSER wet sludge utilization; **Naßsiebung** *f* ABWASSER wet screening; **Naßsorption** *f* ABFALL wet sorption; **Naßsortierer** *m* ABWASSER wet classifier; **Naßtrennung** *f* ABFALL wet sorting; **Naßtrennverfahren** *nt* ABFALL wet sorting; **Naßwäscher** *m* ABFALL, LUFT wet scrubber, wet scrubbing device; **Naßzerkleinerung** *f* ABFALL wet grind

national: **~er Umweltschutz** *m* POLITIK national environmental protection

Natrium *nt* (*Na*) LUFT, WASSER sodium (*Na*); **Natriumcarbonatanhydrid** *nt* WASSER soda ash, sodium trioxocarbonate; **Natriumchlorid** *nt* ABWASSER, MEER sodium chloride, sodium trioxocarbonate; **Natriumion** *nt* WASSER sodium ion

Natronlauge *f* WASSER caustic soda solution

Naturgips *m* LUFT natural gypsum

Naturkraft *f* WASSER physical agent

natürlich; **~e Abdichtung** *f* ABFALL *einer Deponie* natural lining; **~e Bodenmaterialien** *nt pl* ABFALL *Deponie-Abdichtung* natural soil materials; **~e Energiebilanz** *f* ERNEU ENER natural energy balance; **~e Energiequelle** *f* ERNEU ENER natural energy source; **~e Entwässerung** *f* ABFALL, WASSER natural drainage; **~es Entwässerungssystem** *nt* ABWASSER natural dewatering system; **~e Grundbelastung** *f* LUFT background level; **~e Grundwasseranreicherung** *f* WASSER natural groundwater recharge; **~e Grundwasserneubildung** *f* WASSER natural groundwater recharge; **~e Infiltration** *f* ABFALL, WASSER natural infiltration; **~e Konvektion** *f* ABWASSER natural convection; **~es Ökosystem** *nt* BODEN, LUFT, WASSER natural ecosystem; **~er Ozongehalt** *m* LUFT natural ozone content; **~e Quellen** *f pl* LUFT natural sources; **~e Säuerung** *f* ABFALL, BODEN, LUFT, WASSER natural acidification; **~er Schadstoff** *m* ABWASSER, BODEN, LUFT natural pollutant; **~e Sedimentation** *f* ABFALL, ABWASSER, WASSER natural sedimentation; **~er Sedimentationsvorgang** *m* ABFALL, ABWASSER, WASSER natural sedimentation process; **~e Selbstreinigung** *f* ABFALL, ABWASSER, WASSER natural self-purification; **~er Treibhauseffekt** *m* LUFT natural greenhouse effect; **~e Umwelt** *f* LUFT, POLITIK, WASSER natural environment; **~e Versickerung** *f* ABFALL, WASSER natural percolation; **~er Wasserkreislauf** *m* ABFALL, WASSER natural hydrological cycle

natursauer: **natursaurer See** *m* WASSER naturally acid lake

NCH *abbr* (*Nichtkarbonathärte*) ABWASSER, WASSER NCH (*noncarbonate hardness*), permanent hardness

NCI *abbr* (*negative chemische Ionisation*) ANALYTIK NCI (*negative chemical ionization*)

NDIR *abbr* (*nichtstreuendes Infrarot*) ANALYTIK NDIR (*nondispersive infrared*)

NDIRA *abbr* (*nichtstreuende Infrarot-Absorption*) ANALYTIK NDIRA (*nondispersive infrared absorption*)

NDIR: **~ Spektrometrie** *f* ANALYTIK NDIR spectrometry

Ne *abbr* (*Neon*) LUFT Ne (*neon*)

Nebel *m* LUFT fog

Nebeltropfen *m* LUFT mist drops

Neben- *in cpds* ALLG additional; **Nebenanzeige** *f* ANALYTIK *bei Meßgeräten* additional display; **Nebengrundwasserspiegel** *m* ABFALL apparent water table; **Nebenprodukt** *nt* ABFALL, LUFT, MANAG by-product; **Nebenproduktverwertung** *f* ABFALL, ERNEU ENER, LUFT, MANAG by-product recovery; **Nebenreaktion** *f* LUFT side reaction; **Nebensammler** *m* ABFALL, ABWASSER branch sewer, lateral sewer, WASSER branch pipe; **Nebensprechen** *nt* LÄRM crosstalk; **Nebenwegübertragung** *f* LÄRM flanking transmission; **Nebenzweigung** *f* WASSER submains

negativ *adj* ABFALL, LUFT negative; **~e chemische Ionisation** *f* (*NCI*) ANALYTIK negative chemical ionization (*NCI*); **~er Inhibitor** *m* LUFT inhibitor; **~e Mantelreibung** *f* ABFALL negative skin friction

Neigung *f* ABFALL, WASSER *Steigung* inclination; **~ zur Schaumbildung** LUFT foaming tendency, frothing tendency

NE-Metalle *nt pl* ABFALL nonferrous metals

Nennlast *f* ABWASSER, ERNEU ENER, LUFT, WASSER rated capacity, WASSER rated load

Nennleistung *f* ABWASSER, ERNEU ENER, LUFT, WASSER power capacity, rated capacity

Neon *nt* (*Ne*) LUFT neon (*Ne*)

Nephelometrie *f* ANALYTIK nephelometry

nephelometrisch *adj* ANALYTIK nephelometric

Neßlers-Reagens *nt* ANALYTIK Nessler's solution

Nettodonator *m* ABFALL, WASSER net donator

Netzanschluß *m* ERNEU ENER grid connection, mains connection

netzbar *adj* LUFT wettable

netzend *adj* ABWASSER, BODEN, LUFT wetting

Netzfrequenz *f* ERNEU ENER power frequency

netzgekoppelt *adj* ERNEU ENER grid-coupled; **~e Photovoltaikmodule** *m pl* ERNEU ENER grid-coupled photovoltaic modules

Netzmittel *nt* MEER wetting agent

neu: **~e Anlage** *f* ABFALL, LUFT, POLITIK, WASSER new plant; **~e Betriebsanlage** *f* ABFALL, LUFT, WASSER new plant; **auf dem ~esten Stand der Technik** *phr* ALLG state-of-the-art; **~ere Technologie** *f* ALLG emerging technology

Neuproduktion *f*: **~ von Papier** ABFALL primary production of paper

neutral *adj* ALLG neutral; **~e Abwasserprobe** *f* ANALYTIK neutral wastewater sample; **~es Gasmolekül** *nt* LUFT neutral gas molecule; **~e Steuer** *f* POLITIK neutral tax; **~e Temperaturschichtung** *f* LUFT neutral temperature layering

Neutralglasfilter *nt* ANALYTIK gray filter (*AmE*), grey filter (*BrE*)

Neutralisation *f* ALLG neutralization; **~ alkalischer Abwässer** ABWASSER neutralization of alkaline wastewater; **~ mit Kalk** ABWASSER neutralization with lime; **~ mit Kohlendioxid** ABWASSER neutralization with carbon dioxide; **~ mit Mineralsäuren** ABWASSER neutralization with mineral acids; **~ saurer Abwässer** ABWASSER neutralization of acid wastewaters; **Neutralisationsmittel** *nt* ABWASSER, LUFT neutralizer, neutralizing agent; **Neutralisationspunkt** *m* ANALYTIK neutralization point

neutralisieren *vt* ALLG neutralize, render harmless

neutralisiert *adj* ALLG neutralized, rendered harmless

Neutralisierung *f* ABWASSER, ANALYTIK, WASSER neutralization, neutralizing; **Neutralisierungsmittel** *nt* SICHERHEIT neutralizing agent

Newton *nt* (*N*) ANALYTIK newton (*N*)

ng *abbr* (*Nanogramm*) ANALYTIK ng (*nanogram, nanogramme*)

Ni *abbr* (*Nickel*) ABWASSER, BODEN, LUFT, WASSER Ni (*nickel*)
nichtaufbereitet *adj* ABFALL, WASSER unprocessed; **~er Abfall** *m* ABFALL unprocessed waste
nichtauslaugbar *adj* BODEN nonleachable
nichtbeladen *adj* MEER unladen
nichtdispersiv *adj* ANALYTIK nondispersive; **~es Absorptionsgerät** *nt* ANALYTIK nondispersive infrared absorber; **~er Infrarot-Analysator** *m* ANALYTIK nondispersive infrared analyser (*BrE*), nondispersive infrared analyzer (*AmE*)
Nichteisenmetalle *nt pl* ABFALL nonferrous metals
Nichteisenschrott *m* ABFALL nonferrous scrap
nichtentflammbar *adj* ABFALL, ANALYTIK, LUFT, SICHERHEIT flameproof, noninflammable
nichtentwässert *adj* ABWASSER unthickened; **~er Schlamm** *m* ABWASSER unthickened sludge
nichtfiltrierbar *adj* ABWASSER nonfilterable; **~e Rückstände** *m pl* ABWASSER nonfilterable residues
nichtflüchtig *adj* LUFT *Absorbens* nonvolatile
nichtgedämpft *adj* LÄRM undamped
nichtgewebt *adj* LUFT nonwoven
nichtionisch *adj* ABWASSER nonionic
nichtionogen *adj* ABWASSER nonionogenic; **~es Polymer** *nt* ABWASSER nonionogenic polymer
Nichtkarbonathärte *f* (*NCH*) ABWASSER, WASSER noncarbonate hardness (*NCH*), permanent hardness
nichtkatalytisch *adj* LUFT noncatalytic
nichtkegelförmig: **~e Rauchfahne** *f* LUFT nonconing smoke plume
nichtkompostierbar *adj* ABFALL noncompostable; **~er Abfall** *m* ABFALL noncompostable waste
nichtkonzentrierend: **~er Kollektor** *m* ERNEU ENER *Flach-, Vakuumkollektoren* nonconcentrating collector
nichtkristallin *adj* LUFT amorphous, noncrystalline
Nichtleiter *m* ANALYTIK insulator
nichtnetzend *adj* LUFT nonwetting
nichtporös *adj* ABFALL, ABWASSER, WASSER nonporous
nichtresonanzaktiv: **~es Ion** *nt* ANALYTIK nonresonant ion
nichtrückgewinnbar *adj* ABFALL nonrecoverable; **~e Abfälle** *m pl* ABFALL nonrecoverable waste
nichtsolar: **~e erneuerbare Energiequellen** *f pl* ERNEU ENER *Gezeiten-, geothermische Energie* non-solar-based renewable energy sources
nichtstreuend *adj* ANALYTIK nondispersive; **~es Infrarot** *nt* (*NDIR*) ANALYTIK nondispersive infrared (*NDIR*); **~e Infrarotabsorption** *f* (*NDIRA*) ANALYTIK nondispersive infrared absorption (*NDIRA*); **~er Infrarotanalysator** *m* ANALYTIK nondispersive infrared analyser (*BrE*), nondispersive infrared analyzer (*AmE*); **~e Infrarotspektrometrie** *f* ANALYTIK nondispersive infrared spectrometry
nichttoxisch *adj* ALLG nontoxic
nichtverdichtet *adj* ABFALL uncompacted; **~e Deponie** *f* ABFALL uncompacted dump, uncompacted landfill, uncompacted tip (*BrE*)
nichtverstellbar: **~es Rotorblatt** *nt* ERNEU ENER *Windturbine* fixed-pitch rotor blade
nichtverwertbar *adj* ABFALL unusable; **~er Reststoff** *m* ABFALL unusable residue; **~er Rückstand** *m* ABFALL waste product
nichtvulkanisiert *adj* ABFALL nonvulcanized
nichtwägbar *adj* ANALYTIK nonweighable; **~es Lysimeter** *nt* ANALYTIK nonweighable lysimeter
nichtwassergefährdend *adj* ABFALL, ABWASSER, WASSER not harmful to water, not hazardous to water
Nickel *nt* (*Ni*) ABWASSER, BODEN, LUFT, WASSER nickel (*Ni*)
Niederdruck- *in cpds* ALLG low-pressure; **Niederdruckdampf** *m* BODEN low-pressure steam; **Niederdruck-Naßoxidation** *f* ABWASSER low-pressure wet oxidation; **Niederdruck-Polyethylen** *nt* ABFALL low-density polyethylene (*LDPE*); **Niederdruckpumpe** *f* ABWASSER, WASSER low-pressure pump; **Niederdruckrückspülung** *f* LUFT low-pressure backwashing; **Niederdruckseite** *f* LUFT low-pressure side; **Niederdruckspülen** *nt* MEER low-pressure flushing; **Niederdruckspülung** *f* LUFT low-pressure cleaning; **Niederdruck-Venturiwäscher** *m* LUFT low-pressure Venturi scrubber
Niederfrequenzabsorber *m* LÄRM low-frequency absorber
niedergeschlagen: **~er Staub** *m* LUFT collected dust
Niederländisch: **~er Leitfaden** *m* BODEN *Grenzwerte für Schadstoffekonzentrationen* Dutch list; **~e Liste** *f* BODEN *Grenzwerte für Schadstoffekonzentrationen* Dutch list
Niederschlag *m* BODEN, LUFT precipitate, *Regen, Schnee* precipitation, *radioaktiv* fallout
niederschlagen *vt* BODEN, LUFT precipitate
Niederschlag: **Niederschlagsabwasser** *nt* WASSER storm wastewater; **Niederschlagselektrode** *f* ANALYTIK, LUFT collecting electrode; **Niederschlagsgebiet** *nt* WASSER catchment area, drainage area, drainage basin, precipitation area, rainfall area; **Niederschlagshäufigkeit** *f* WASSER frequency of rain; **Niederschlagsmenge** *f* WASSER amount of precipitation; **Niederschlagsmesser** *m* WASSER precipitation gage (*AmE*), precipitation gauge (*BrE*), rain gage (*AmE*), rain gauge (*BrE*); **Niederschlagssammler** *m* LUFT precipitation collector; **Niederschlagsvorfall** *m* LUFT precipitation event; **Niederschlagswasser** *nt* ABWASSER, WASSER rainwater, stormwater
Niederspannung *f* ANALYTIK, LUFT low voltage; **Niederspannungselektrofilter** *nt* LUFT low-voltage ESP; **Niederspannungslampe** *f* ANALYTIK low-voltage lamp
Niedertemperatur- *in cpds* ERNEU ENER low-temperature; **Niedertemperaturkollektoranlage** *f* ERNEU ENER low-temperature solar collector plant; **Niedertemperaturkonvertierung** *f* ABWASSER *Klärschlammverbrennung* low-temperature conversion; **Niedertemperaturwärme** *f* ERNEU ENER low-temperature heat; **Niedertemperaturwärmeverteilung** *f* ERNEU ENER lower-temperature heat distribution
niederwertig *adj* ABFALL low-grade; **~es Papier** *nt* ABFALL low-grade paper
Niedrig- *in cpds* ALLG low
niedrig: **~ste erreichbare Emissionsrate** *f* LUFT lowest achievable emission rate; **~e Konzentration** *f* ANALYTIK low concentration; **~ste Konzentration** *f* **eines Schadstoffes bei der Wirkung** (*LOEC*) SICHERHEIT lowest observed effect concentration (*LOEC*); **~ste letale Dosis** *f* (LD_{Lo}) SICHERHEIT lethal dose low (LD_{Lo}); **~ster Niedrigwasserabfluß** *m* WASSER minimum flow; **~ster Niedrigwasserstand** *m* WASSER minimum low water; **~e Rohgasbelastung** *f* ABFALL low dirty gas load; **~e Schüttdichte** *f* ANALYTIK low-settled apparent density; **~er Solarstand** *m* ERNEU ENER *LBG* low position of the sun; **~er Sonnenstand** *m* ERNEU ENER low position of the sun
Niedrigalkalität *f* WASSER low alkalinity
niedrigkalorig: **~es Biogas** *nt* (*LBG*) ERNEU ENER low thermal value biogas (*LBG*)
Niedrig-: **Niedrigtrübung** *f* WASSER low turbidity
Niedrigwasser *nt* MEER, WASSER low water; **Niedrigwasserabfluß** *m* ABFALL low-water discharge; **Niedrigwasserstand** *m* WASSER low-water level; **Niedrigwasserverbrauch** *m* WASSER low water consumption
Nipptiden *f pl* ERNEU ENER *Gezeitenenergie* neap tides
Nitrat *nt* ABWASSER, BODEN, LUFT, WASSER nitrate; **Nitrateliminierung** *f* ABWASSER nitrate removal; **Nitratfracht** *f* ABWASSER nitrate load
nitrathaltig *adj* ABWASSER nitrate-containing; **~es Abwasser** *nt* ABWASSER nitrate-containing effluent; **~er Abwasserstrom** *m* ABWASSER nitrate-containing wastewater stream
Nitrat: **Nitratkonzentration** *f* ABWASSER nitrate concentration

Nitrifikanten *pl* ABWASSER nitric bacteria
Nitrifikation *f* ABWASSER, WASSER nitrification; **Nitrifikation-Denitrifikation** *f* ABWASSER nitrification-denitrification; **Nitrifikationszone** *f* ABWASSER nitrification zone
nitrifizieren *vt* ABFALL, ABWASSER, WASSER nitrify
nitrifizierend *adj* ABFALL, ABWASSER, WASSER nitrifying; **~e Bakterien** *f pl* ABFALL, ABWASSER, WASSER nitrifying bacteria
Nitrit *nt* ABWASSER, BODEN, WASSER nitrite; **Nitritkonzentration** *f* ABWASSER nitrite concentration
nitritreich: **~er Schlamm** *m* ABFALL nitrite rich sludge
Nitrosomas-Bakterien *f pl* ABWASSER nitrosomas bacteria
nm *abbr* (*Nanometer*) ANALYTIK nm (*nanometer AmE*), (*nanometre BrE*)
NO_2 *abbr* (*Stickstoffdioxid*) LUFT NO_2 (*nitrogen peroxide*)
N_2O *abbr* (*Distickstoffoxid*) ABFALL, LUFT, WASSER N_2O (*nitrous oxide*)
NO_x *abbr* (*Stickstoffoxide*) LUFT NO_x (*nitrogen oxides*)
N_2O_5 *abbr* (*Stickstoffpentoxid*) LUFT N_2O_5 (*nitrogen pentoxide*)
NOAEL *abbr* (*höchste Dosis, die noch nicht zu nachteiligen Wirkungen führt*) SICHERHEIT NOAEL (*no observed adverse effect level*)
NO_x: **NO_x-armer Brenner** *m* LUFT low-NO_x burner (*LNB*), low-NO_x combustor
NOEC *abbr* (*höchste Konzentration ohne eine beobachtete nachteilige Wirkung*) SICHERHEIT NOEC (*no observed effect concentration*)
No-Effect: **~-Level** *m* SICHERHEIT no-effect level
NO_x: **NO_x-Emissionen** *f pl* LUFT NO_x-emissions
Nord-Süd-Beziehung *f* POLITIK north-south relationship
Normalbedingungen *f pl* ANALYTIK standard temperature and pressure (*STP*)
Normalbetrieb *m* MANAG, POLITIK normal operation
Normbrandkurve: **~-Zeit-Temperatur** *f* ANALYTIK standard curve-time temperature
Normdruck *m* ANALYTIK normal pressure
Normenkontrolle *f* POLITIK standards implementation monitoring
Normenkontrollverfahren *nt* POLITIK implementation procedure
Normzustand *m* ANALYTIK standard conditions (*SC*)
Not- *in cpds* ALLG emergency; **Notbehandlung** *f* ABWASSER, BODEN, LUFT, SICHERHEIT emergency treatment
Notfall *m* ALLG emergency; **Notfallmaßnahmen** *f pl* ABWASSER, BODEN, LUFT, SICHERHEIT emergency measures
Not-: **Nothochwasserentlastungsanlage** *f* WASSER emergency spillway; **Notüberlauf** *m* ABWASSER emergency overflow; **Notumlauf** *m* ABWASSER emergency by-pass; **Notverschluß** *m* ABWASSER emergency gate
Novalox-Sattelkörper *m* LUFT *Füllkörper des Füllkörpersäulenabscheiders* intalox saddle
Novelle *f* POLITIK amendment
novellieren *vt* POLITIK amend
NO_x: **NO_x-Wert** *m* LUFT NO_x-value
NPD *abbr* (*Stickstoff-Phosphor-Detektor*) ANALYTIK NPD (*nitrogen-phosphorus detector*)
NPSH-Wert *m* ABFALL, ABWASSER NPSH (*net positive suction head*)
nuklear *adj* ABFALL, ERNEU ENER, SICHERHEIT nuclear; **~er Energieträger** *m* ERNEU ENER nuclear energy source; **~ ungefährlicher Niederschlag** *m* LUFT dry deposition
Nuklearsicherheit *f* SICHERHEIT nuclear safety
Nullabgleich *m* ANALYTIK null balance
Nullpunktkorrektur *f* ANALYTIK zero effect compensation
numerisch *adj* ANALYTIK numerical; **~es Gittermodell** *nt* ANALYTIK numerical grid model
Nutz- *in cpds* ALLG usable, useful
nutzbar *adj* ABWASSER, ERNEU ENER, LUFT, WASSER available, usable; **~er Anteil** *m* **der Hydrosphäre** LUFT *für menschliche Nutzung* usable portion of hydrosphere; **~e Energieform** *f* ERNEU ENER usable energy form; **~e Kapazität** *f* ABWASSER available capacity; **~es Potential** *nt* ERNEU ENER usable potential; **~er Speicherraum** *m* ABFALL live storage
Nutzbarkeit *f* ERNEU ENER usability
Nutzbarmachung *f* ERNEU ENER utilization
nutzen *vt* ABFALL, ERNEU ENER use, utilize, harness
Nutz-: **Nutzfläche** *f* ABFALL useful area; **Nutzlast** *f* ABFALL, ABWASSER service load, MEER payload; **Nutzmasse** *f* MEER payload; **Nutzraum** *m* ABWASSER *eines Klärbeckens* working capacity; **Nutzstaub** *m* LUFT reusable dust
Nutzung: **Nutzungsbeschränkung** *f* POLITIK use reduction; **Nutzungsdauer** *f* ABWASSER, LUFT, WASSER *einer Anlage* serviceable life; **Nutzungseinschränkung** *f* POLITIK use reduction; **Nutzungsempfehlung** *f* POLITIK use recommendation; **Nutzungsmöglichkeiten** *f pl* ERNEU ENER use possibilities; **Nutzungsregelung** *f* POLITIK use regulation; **Nutzungsverbot** *nt* POLITIK use prohibition; **Nutzungsvertrag** *m* POLITIK *des Umweltzeichens* user's contract
Nutz-: **Nutzwertanalyse** *f* ANALYTIK efficiency analysis, MANAG use-value analysis
nylonverstärkt *adj* ABFALL, WASSER nylon-reinforced

O

O₂ *abbr* (*Sauerstoff*) ALLG O_2 (*oxygen*)
Oberboden *m* ABFALL top layer, BODEN A-horizon
obere: **oberes Bett** *nt* LUFT *Katalysator* upper bed; **oberer Heizwert** *m* ABFALL gross calorific value
Oberfläche *f* ALLG surface; **~ der Partikeln** LUFT *Naßabscheider* particle surface; **~ der Teilchen** LUFT *Naßabscheider* particle surface; **Oberflächenabdeckung** *f* BODEN surface cover; **Oberflächenabdichtung** *f* ABFALL surface sealing; **Oberflächenabfluß** *m* ABFALL, ABWASSER surface runoff
oberflächenaktiv *adj* BODEN surface-active; **~er Stoff** *m* ABWASSER, MEER surface-active agent; **~e Stoffe** *m pl* ABWASSER surface-active substances; **~e Substanzen** *f pl* BODEN *Zusatzstoffe* surface-active substances
Oberfläche: **Oberflächenausstrahlung** *f* BODEN surface irradiation; **Oberflächenbelüfter** *m* ABFALL surface aerator; **Oberflächenbelüftung** *f* ABFALL, ABWASSER surface aeration; **Oberflächenbereich** *m* BODEN surface area; **Oberflächenbrecher** *m* MEER breaker board; **Oberflächenentlastungsanlage** *f* WASSER flood spillway; **Oberflächenfilm** *m* ABWASSER surface film; **Oberflächenfilter** *nt* LUFT surface filter; **Oberflächenfiltration** *f* ABWASSER, LUFT, WASSER surface filtration; **Oberflächenfiltration** *f* **mit Membranen** ABWASSER surface filtration with membranes; **Oberflächengewässer** *nt* WASSER surface waters; **Oberflächengewässeranteil** *m* WASSER share of surface waters; **Oberflächengewässeraufbereitung** *f* WASSER surface waters treatment; **Ober-flächengewässerbewirtschaftung** *f* WASSER surface waters management; **Oberflächenglättung** *f* BODEN surface smoothing; **Oberflächenkondensator** *m* LUFT surface condenser, indirect condenser; **Oberflächenleitfähigkeit** *f* LUFT *Elektrofilter* surface conductivity; **Oberflächenmeerwasser** *nt* ERNEU ENER surface seawater; **Oberflächenmeßfühler** *m* ANALYTIK surface sensor; **Oberflächenmethode** *f* ABFALL, WASSER surface method; **Oberflächenneubildung** *f* ABFALL, WASSER surface refurbishing, surface renewal; **Oberflächenrauhigkeit** *f* ABFALL surface roughness; **Oberflächensammelbecken** *nt* ABWASSER surface reservoir; **Oberflächenschicht** *f* ABFALL surface layer; **Oberflächenverdunstung** *f* WASSER surface evaporation; **Oberflächenversiegelung** *f* ABFALL final cover; **Oberflächenwasser** *nt* WASSER surface water; **Oberflächenwassertemperatur** *f* ERNEU ENER surface water temperature
oberflächig: **~ abfließen** *vi* ABFALL run off the surface
oberirdisch *adj* ABFALL, WASSER surface, above-ground; **~er Abfluß** *m* ABFALL, ABWASSER, WASSER surface runoff; **~e Deponie** *f* ABFALL surface landfill; **~e Infiltrationsanlage** *f* WASSER above-ground infiltration facility; **~e Sonderabfalldeponie** *f* ABFALL special waste surface landfill, hazardous waste surface landfill
Oberkante *f* ABFALL upper surface
objektbezogen *adj* LÄRM, MEER object-related; **~er Lärm** *m* LÄRM, MEER object-related noise
Octanzahl *f* LUFT octane number
OEL *abbr* (*berufsbedingte Belastungsgrenzen*) SICHERHEIT OEL (*occupational exposure limits*)
offen: **~er Abzugsgraben** *m* ABFALL open drain, ABWASSER open drain, open sewer; **~e Deponie** *f* ABFALL open landfill, open pit landfill; **~er Deponieabschnitt** *m* ABFALL open landfill section; **~er Kreislauf** *m* ERNEU ENER open cycle (*OC*); **~er Rankine-Kreislauf** *m* ERNEU ENER *Meereswärmeenergie* open Rankine cycle; **~es System** *nt* ABFALL open system; **~er Tropfkörper** *m* ABWASSER open trickling filter
Offener-Kreislauf-Meereswärmekraftwerk *nt* ERNEU ENER open-cycle OTEC plant
öffentlich: **~e Gesundheit** *f* POLITIK, WASSER public health; **~e Kanalisation** *f* ABWASSER public sewerage system; **~er Sammelplatz** *m* ABFALL public collection station; **~e Sammelstation** *f* ABFALL public collection station; **~er Trinkwasserverbrauch** *m* WASSER public drinking water consumption; **~e Trinkwasserversorgung** *f* WASSER public drinking water supply; **~e Versorgungsunternehmen** *nt pl* WASSER public utilities; **~e Wasserversorgung** *f* ABWASSER, WASSER public water supply
Öffentlichkeitsbeteiligung *f* POLITIK public participation
off-line *adj* ANALYTIK off-line
Off-line-Betrieb *m* ANALYTIK off-line operation
Off-shore- *in cpds* ERNEU ENER offshore; **Off-shore-Potential** *nt* ERNEU ENER offshore potential; **Off-shore-Windenergiepotential** *nt* ERNEU ENER offshore potential of wind energy; **Off-shore-Windpark** *m* ERNEU ENER offshore wind farm
Off-site- *in cpds* ABFALL, ANALYTIK, BODEN, WASSER off-site: **Off-site-Analyse** *f* ANALYTIK off-site analysis; **Off-site-Untersuchung** *f* ANALYTIK off-site analysis; **Off-site-Verbrennung** *f* BODEN off-site incineration
Ohrstöpsel *m pl* LÄRM earplugs
Öko- *in cpds* LUFT, MANAG, MEER, POLITIK eco-; **Öko-Audit** *nt* MANAG, POLITIK eco-audit; **Öko-Auto** *nt* LUFT eco-car, eco-automobile; **Ökobilanz** *f* MANAG eco-balance, *Kunststoffrecycling* eco-balance; **Ökobilanz** *f* **als Doppelsystem** MANAG double eco-balance system; **Ökobilanzsystematik** *f* MANAG eco-balance design; **Öko-Controlling** *nt* MANAG eco-controlling; **Ökodatenbank** *f* MANAG ecological database; **Ökofilter** *nt* ABFALL ecological filter; **Ökohändler** *m* MANAG eco-retailer; **Öko-Labeling** *nt* POLITIK, WASSER eco-labeling (*AmE*), eco-labelling (*BrE*)
Ökologie *f* ALLG ecology
ökologieorientiert *adj* MANAG ecology-oriented; **~es Management** *nt* MANAG ecology-oriented management; **~es Marketing** *nt* MANAG ecology-oriented marketing
ökologierelevant *adj* POLITIK ecologically relevant
ökologisch *adj* ALLG ecological; **~es Gleichgewicht** *nt* ABFALL, POLITIK, WASSER ecological balance; **~e Pyramide** *f* ABFALL, POLITIK, WASSER ecological pyramid; **~e Randbedingungen** *f pl* POLITIK ecological conditions; **~e Restriktionen** *f pl* POLITIK ecological restrictions; **~er Schaden** *m* ABWASSER, BODEN, LUFT ecological damage; **~es Steuerpaket** *nt* POLITIK eco-tax package; **~e Steuerreform** *f* POLITIK eco-tax reform, green tax reform, ecological tax reform; **~es Steuersystem** *nt* POLITIK eco-tax system, green tax system, ecological tax system; **~er Strukturwandel** *m* POLITIK ecological structural change; **~er Zusammenbruch** *m* BODEN, LUFT environmental collapse, POLITIK ecological collapse
Öko-: **Ökomaske** *f* MANAG eco-window
ökonomisch *adj* MANAG, POLITIK economic; **~e Effizienz** *f* MANAG, POLITIK economic efficiency
Öko-: **Ökoprodukt** *nt* MANAG green product; **Öko-qualitätskriterien** *nt pl* MANAG eco-quality criteria; **Öko-Rating** *nt* MANAG eco-rating; **Ökosphäre** *f* BODEN ecosphere
Ökosteuer *f* MANAG, POLITIK eco-tax; **Ökosteueraufkommen**

nt POLITIK eco-tax revenue; **Ökosteuern** *f pl* MANAG, POLITIK green taxes; **Ökosteuerpaket** *nt* MANAG, POLITIK eco-tax package; **Ökosteuerreform** *f* POLITIK eco-tax reform, green tax reform

Ökosystem *nt* ALLG ecosystem; **~ des Meeres** MEER marine ecosystem; **~ des offenen Ozeans** MEER open-ocean ecosystem; **Ökosystemstreß** *m* ABWASSER, BODEN, LUFT, MEER ecosystem stress

Öko-: **Ökotourismus** *m* MANAG eco-tourism, green tourism; **Ökotoxikologie** *f* ABWASSER, POLITIK, SICHERHEIT, WASSER ecotoxicology; **Ökotoxizität** *f* ABWASSER ecotoxicity

Okular *nt* ANALYTIK ocular

Öl *nt* ABWASSER, ERNEU ENER, MEER oil; **mit ~ gefüllte Empfängerröhre** *f* ERNEU ENER oil-filled receiving tube; **~ aus Schlamm** ABWASSER oil from sludge (*OFS*); **Ölabfall** *m* ABFALL oil waste, residual oil; **Ölabscheider** *m* ABFALL oil separator, ABWASSER degreaser, skimming tank, oil separator, BODEN oil separator; **Ölabscheidung** *f* ABWASSER de-oiling, oil removal, oil separation; **Ölabschöpfgerät** *nt* MEER recovery device; **Ölabschöpfsystem** *nt* MEER recovery system

ölanziehend *adj* ABWASSER, MEER oleophilic

Ölaufbereitungsanlage *f* ABFALL oil regeneration plant

ölaufnehmend *adj* ABWASSER, MEER oleophilic; **~es Band** *nt* MEER oleophilic belt; **~es Förderband-Abschöpfgerät** *nt* MEER oleophilic belt skimmer

Öl: **Ölausbiß** *m* MEER seepage; **Ölaustritt** *m* MEER oil leakage; **Ölbekämpfung** *f* MEER oil-spill response; **Ölbekämpfungsschiff** *nt* MEER oil-pollution combatting ship; **Ölbeseitigung** *f* **durch Trennmittel** ABFALL removal of oil by separators

oleophil *adj* ABWASSER, MEER oleophilic

Olfaktometrie *f* ANALYTIK olfactometry

Öl: **Ölfang** *m* ABFALL oil trap, ABWASSER grease trap, oil trap; **Ölfilm** *m* MEER oil slick, oil film; **Ölfleck** *m* MEER oil patch, oil slick

ölhaltig *adj* ABFALL, ABWASSER, ERNEU ENER oil-containing, oleaginous; **~e Abfälle** *m pl* ABFALL oleaginous wastes; **~es Abwasser** *nt* ABFALL oleaginous wastewater, ABWASSER oily wastewater; **~e Pflanzen** *f pl* ERNEU ENER *Bioenergieausgangsstoff* oil-containing plants

ölig: **~es Wasser** *nt* MEER black water

oligodynamisch *adj* ABFALL oligodynamic

oligosaprob *adj* ABFALL, ABWASSER, WASSER oligosaprobic

oligotroph *adj* ABFALL nutrient-poor, oligotrophic, ABWASSER oligotrophic, WASSER nutrient-poor, oligotrophic

Öl: **Ölkatastrophe** *f* ABFALL, ABWASSER, MEER oil spill disaster; **Ölkonzentration** *f* ABWASSER oil concentration; **Öllache** *f* ABFALL, MEER oil spill; **Ölleck** *nt* ABFALL, MEER oil leakage; **Ölnebelabscheider** *m* ABFALL oil aerosol separator; **Ölpest** *f* ABFALL, ABWASSER oil pollution, MEER black tide; **Ölregenerat** *nt* ABFALL recovered oil; **Ölrückgewinnungsschiff** *nt* ABFALL oil-recovery vessel; **Ölrückgewinnungsschute** *f* ABFALL oil-recovery barge; **Ölrückgewinnungsskimmer** *m* ABFALL oil-recovery skimmer; **Ölrückstände** *m pl* ABFALL, MEER oil waste, residual oil; **Ölschicht** *f* ABFALL, MEER oil slick, oil film; **Ölschlamm** *m* MEER *Öl-Wasser-Emulsion* chocolate mousse; **Ölsperre** *f* MEER oil boom; **Ölsperrengebinde** *nt* MEER boom pack; **Ölspill-Identifikationssystem** *nt* MEER oil spill identification system; **Ölteppich** *m* ABFALL, MEER oil slick, oil film; **Ölteppichbekämpfungsausrüstung** *f* MEER oil spill combatting equipment; **Ölteppichidentifikation** *f* MEER oil spill identification; **Öltröpfchen** *nt* ABFALL, ABWASSER, MEER oil droplet; **Ölverdichtungsmittel** *nt* ABFALL oil-concentrating agent

ölverschmutzt: **~es Abwasser** *nt* ABWASSER oil-polluted wastewater

Öl: **Ölverschmutzung** *f* ABFALL, MEER oil pollution, oil contamination; **Ölverschmutzungsnotfall** *m* ABFALL, MEER oil pollution emergency; **Ölverschüttung** *f* ABFALL, ABWASSER, MEER oil spill, oil spillage; **Öl/Wasser-Abscheider** *m* ABWASSER, BODEN oil/water separator; **Öl/Wasser-Abscheidung** *f* ABWASSER, BODEN oil/water separation; **Öl/Wasser-Berührungsfläche** *f* BODEN, MEER oil/water interface; **Öl/Wasser-Trennung** *f* ABWASSER, BODEN oil/water separation

on-board: **~ Mikroprozessor** *m* ANALYTIK on-board computer

On-line *adj* ANALYTIK on-line; **On-line-Analysengerät** *nt* ANALYTIK on-line analyser (*BrE*), on-line analyzer (*AmE*); **On-line-Betrieb** *m* ANALYTIK on-line operation; **On-line-Messung** *f* ANALYTIK on-line measurement

On-site- *in cpds* ABFALL, ANALYTIK, BODEN, WASSER on-site, on-the-spot; **On-site-Analyse** *f* ANALYTIK on-site analysis, field screening; **On-site-Anwendung** *f* BODEN on-site application; **On-site-Aushub** *m* BODEN on-site excavation, on-site extraction; **On-site-Behandlung** *f* ABFALL on-site treatment; **On-site-Einsatz** *m* BODEN on-site application; **On-site-Lagerung** *f* ABFALL on-site storage; **On-site-Verbrennung** *f* BODEN on-site incineration; **On-site-Vorsortierung** *f* ABFALL on-site presorting

Open-Top-Container *m* ABFALL open-top container

o-Phosphat *nt* ABWASSER orthophosphate

optimal: **~er Flockulantenzusatz** *m* ABWASSER optimum polymer dosage

optisch *adj* ANALYTIK *Analysengeräte* optical; **~er Abscheider** *m* ABFALL optical sorter; **~e Emissionsspektrometrie** *f* ANALYTIK optical emission spectrometry (*OES*); **~e Sortierung** *f* ABFALL optical sorting

Ordnungsrecht *nt* POLITIK administrative law

ordnungswidrig *adj* ABFALL, ABWASSER POLITIK illegal

Ordnungswidrigkeit *f* POLITIK breach of law

organisch *adj* ABFALL, ABWASSER, LUFT, WASSER organic; **~er Abfall** *m* ABFALL putrescible waste, organic waste, biowaste, WASSER organic waste; **~er Ballaststoff** *m* ABFALL organic roughage; **~e Belastung** *f* ABWASSER organic loading; **~e Biomasse** *f* ERNEU ENER *Biokonversion* organic biomass; **~e Chemikalie** *f* ABWASSER, WASSER organic chemical; **~e Chlorverbindungen** *f pl* ABFALL, WASSER organic chlorine compounds; **~e Feststoffe** *m pl* ABFALL, ABWASSER organic solids; **~e Fraktion** *f* ABFALL, ABWASSER organic fraction; **~es Gas** *nt* LUFT organic gas; **~er Kontaminant** *m* ABWASSER, LUFT organic contaminant, organic pollutant; **~es Lösungsmittel** *nt* LUFT organic solvent; **~e Müllfraktion** *f* ABFALL, WASSER organic waste fraction; **~es Sauergas** *nt* LUFT organic acid gas; **~er Schadstoff** *m* ABWASSER, LUFT organic contaminant, organic harmful substance, organic pollutant; **~er Stoff** *m* ABFALL, LUFT, WASSER organic matter; **~es Trägermaterial** *nt* LUFT *Biofilter* organic base material; **~e Trockensubstanz** *f* ABFALL, ABWASSER, WASSER organic dry mass; **~e Verbindungen** *f pl* ABWASSER organic compounds

orientiert *adj* MANAG, POLITIK oriented; **~es Polypropylen** *nt* (*DPP*) ABFALL directed polypropylene (*DPP*)

Originalboden *m* BODEN original soil

Originalprobe *f* ANALYTIK original sample

Orlon *nt* LUFT *Filtermaterial* orlon

Orsatapparat *m* ANALYTIK Orsat apparatus, Orsat gas analyser (*BrE*), Orsat gas analyzer (*AmE*)

Orsatbürette *f* ANALYTIK Orsat burette

orthokinetisch *adj* ABFALL orthokinetic; **~e Flockung** *f* ABFALL, WASSER orthokinetic flocculation

Orthophosphat *nt* ABWASSER orthophosphate

örtlich: **~e Emissionsquelle** *f* ABWASSER, ERNEU ENER, LUFT local emission source; **~ repräsentativ** *adj* ANALYTIK *Probe* locally representative

ortsabhängig *adj* ERNEU ENER site-dependent, stationary

ortsfest: **~e Emissionsquelle** *f* LUFT site-dependent emission source, stationary emission source

Ortskennzeichnung *f* ABFALL, ABWASSER, ERNEU ENER site characterization
Osmose *f* ABFALL, ABWASSER, WASSER osmosis
osmotisch *adj* ABFALL, ABWASSER, WASSER osmotic; **~er Druck** *m* ABFALL, ABWASSER, WASSER osmotic pressure
Oszillator *m* ERNEU ENER, LÄRM oscillator
oszillieren *vt* ERNEU ENER, LÄRM oscillate
oszillierend: **~e Wassersäule** *f* (*OWC*) ERNEU ENER oscillating water column (*OWC*)
Oszilloskop *nt* ANALYTIK oscilloscope
Ottomotor *m* LUFT spark ignition engine
Oversampling *nt* ANALYTIK oversampling
OWC *abbr* (*oszillierende Wassersäule*) ERNEU ENER *Quecksilber* OWC (*oscillating water column*); **OWC-Betrieb** *m* ERNEU ENER *Glasrecycling* OWC operation
Oxidation *f* ALLG oxidation; **Oxidationsbehälter** *m* LUFT oxidation tank; **Oxidationsbeständigkeit** *f* ABFALL, WASSER resistance to oxidation; **Oxidationsgraben** *m* ABFALL, ABWASSER, WASSER oxidation ditch; **Oxidationsluft** *f* LUFT oxidation air; **Oxidationsmittel** *nt* ABWASSER, LUFT, WASSER oxidant, oxidizing agent; **Oxidationsreduktion** *f* ABWASSER oxidation reduction; **Oxidationsteich** *m* ABWASSER aerated lagoon, oxidation pond; **Oxidationszone** *f* LUFT oxidation zone
oxidativ *adj* ABWASSER oxidative; **~er Abbau** *m* ABWASSER oxidative decomposition
Oxidator *m* LUFT oxidator
oxidierbar *adj* ALLG oxidizable; **~er Stoff** *m* LUFT oxidizable substance
Oxidierbarkeit *f* ANALYTIK oxidability
oxidieren 1. *vt* BODEN *von Schadstoffen* oxidize; **2.** *vi* ABFALL, ABWASSER, ANALYTIK, LUFT, WASSER oxidize
oxidierend *adj* ALLG oxidizing; **~er Katalysator** *m* LUFT oxidizing catalyst; **~es Katalysatorbett** *nt* LUFT oxidizing catalyst bed; **~er Stoff** *m* LUFT, SICHERHEIT oxidizing substance
oxidimetrisch *adj* ANALYTIK oxidimetric; **~e Bestimmung** *f* ANALYTIK oxidimetric determination; **~e Titration** *f* ANALYTIK oxidimetric titration
Oxidkeramik-Brennstoffzelle *f* LUFT solid oxide fuel cell (*SOFC*)
oxygenieren *vt* ABWASSER oxygenate
Ozeanbecken *nt* MEER ocean basin
ozeanisch *adj* MEER oceanic
Ozon *nt* ABFALL, ANALYTIK, LUFT, SICHERHEIT ozone; **Ozonabbau** *m* LUFT ozone depletion
ozonabbauend *adj* LUFT ozone-depleting
Ozon: **Ozonabbaupotential** *nt* LUFT ozone depletion potential (*ODP*); **Ozonanalysator** *m* ANALYTIK ozone analyser (*BrE*), ozone analyzer (*AmE*); **Ozonanlage** *f* ABFALL ozone plant; **Ozonbelastung** *f* LUFT ozone impact; **Ozonbestimmung** *f* ANALYTIK ozone detection; **Ozonbildung** *f* LUFT ozone formation; **Ozondefizit** *nt* LUFT ozone deficit
ozonerzeugend *adj* LUFT ozone-generating
ozonfreundlich *adj* LUFT ozone-friendly
Ozon: **Ozongehalt** *m* LUFT ozone content; **Ozongenerator** *m* ANALYTIK, LUFT ozone generator; **Ozonhaushalt** *m* LUFT ozone budget; **Ozonhintergrundkonzentration** *f* LUFT background ozone concentration
Ozonierung *f* ABFALL, ABWASSER, BODEN, WASSER ozonization
Ozon: **Ozonimmissionen** *f pl* LUFT ozone immissions
Ozonisierung *f* ABFALL, ABWASSER, BODEN, WASSER ozonization
Ozon: **Ozonkapillare** *f* LUFT ozone capillary; **Ozonkonzentration** *f* ABFALL, LUFT ozone concentration; **Ozonloch** *nt* LUFT ozone hole; **Ozonmessung** *f* ANALYTIK ozone measuring; **Ozonpegel** *m* LUFT ozone level
ozonresistent *adj* LUFT ozone-resistant
Ozon: **Ozonschaden** *m* LUFT ozone damage
ozonschädlich *adj* LUFT ozone-damaging
Ozon: **Ozonschicht** *f* LUFT ozone layer, ozone shield; **Ozonschichtzerstörung** *f* LUFT ozone layer destruction; **Ozonschutzzerstörungspotential** *nt* LUFT ozone depletion potential (*ODP*); **Ozonschwund** *m* LUFT ozone shrinkage; **Ozonsmog** *m* LUFT ozone smog; **Ozonspitzenbelastung** *f* LUFT peak ozone damage
ozontolerant *adj* LUFT ozone-tolerant
Ozon: **Ozonverlust** *m* LUFT ozone loss; **Ozonwarnung** *f* LUFT, SICHERHEIT ozone warning; **Ozonwert** *m* LUFT ozone value; **Ozonzerstörer** *m* LUFT ozone destroyer; **Ozonzerstörung** *f* LUFT ozone destruction

P

P *abbr* (*Phosphor*) ABWASSER, LUFT P (*phosphorus*)
Pa *abbr* (*Pascal*) ANALYTIK, SICHERHEIT Pa (*pascal*)
PA *abbr* (*Polyamid*) ABFALL, ABWASSER PA (*polyamide*)
Packer *m* ABFALL, MANAG packer
Packstoff *m* ABFALL packaging material; **Packstoffindustrie** *f* MANAG packaging material producing industry
Paddel *nt* ABWASSER paddle; **Paddelbelüfter** *m* ABWASSER aeration paddle; **Paddelmischer** *m* WASSER paddle mixer; **Paddelradbelüftung** *f* ABWASSER paddle-wheel aeration
PAK *abbr* (*polyzyklischer aromatischer Kohlenwasserstoff*) ABFALL, ABWASSER, LUFT, WASSER PAH (*polycyclic aromatic hydrocarbon*)
Paketieranlage *f* ABFALL bundle-tying machine
Paketiermaschine *f* ABFALL bundle-tying machine
Palette *f* ABFALL pallet; **Palettenentnahme** *f* ABFALL depalletization; **Palettenentnehmer** *m* ABFALL depalletizer
Pallring *m* LUFT pall ring
PA-Membran *f* ABWASSER polyamide membrane
PAN *abbr* (*Peroxoacetylnitrat*) LUFT, WASSER PAN (*peroxoacetyl nitrate*)
Papier *nt* ABFALL, ANALYTIK paper; **Papierbeutel** *m* ABFALL paper bag; **Papierchromatographie** *f* (*PC*) ANALYTIK paper chromatography (*pc*); **Papierfaser** *f* ABFALL paper fibre (*BrE*), paper fiber (*AmE*); **Papierfaserstoff** *m* ABFALL paper fiber (*AmE*), paper fibre (*BrE*); **Papierfilter** *nt* ANALYTIK paper filter
papierfremd: **~er Bestandteil** *m* ABFALL nonpaper component
Papier: **Papierholz** *nt* ABFALL pulp wood; **Papierqualität** *f* ABFALL paper quality; **Papierrohstoff** *m* ABFALL paper raw material; **Papiersack** *m* ABFALL paper sack; **Papierschneidemaschine** *f* ABFALL guillotine; **Papiertasche** *f* ABFALL paper bag; **Papiertüte** *f* ABFALL paper bag; **Papierverbrauch** *m* ABFALL paper consumption; **Papierverpackung** *f* ABFALL paper wrapping
Pappe *f* ABFALL cardboard; **Pappeabfall** *m* ABFALL cardboard waste, paper board waste
Pappfaß *nt* ABFALL fibre drum (*BrE*), fiber drum (*AmE*)
Parabol- *in cpds* ERNEU ENER parabolic
parabolisch *adj* ERNEU ENER parabolic; **~ geformte Rinne** *f* ERNEU ENER *Kollektor* parabolically formed gutter
Parabol-: **Parabolschüssel** *f* ERNEU ENER parabolic dish; **Parabolschüsselanlage** *f* ERNEU ENER parabolic dish installation; **Parabolspiegel** *m* ERNEU ENER parabolic reflector
Paragraph *m* POLITIK article
parallel *adj* ANALYTIK *Messungen* parallel
Parallelpumpe *f* ABWASSER, WASSER parallel pump
paramagnetisch *adj* ANALYTIK paramagnetic; **~e Flüssigkeitsanalyse** *f* ANALYTIK paramagnetic fluid analytical method
Paramagnetismus *m* ANALYTIK paramagnetism
Parameter *m* ALLG parameter; **Parameterbestimmung** *f* ANALYTIK parameter determination
Pararosanilin *nt* ANALYTIK pararosaniline; **Pararosanilinmethode** *f* ANALYTIK pararosaniline method
Partialdruck *m* LUFT partial pressure; **Partialdruckdifferenz** *f* ABWASSER partial pressure difference, LUFT partial differential pressure; **Partialdruckgesetz** *nt* **von Dalton** LUFT Dalton's law of partial pressure
Partikel *f* ALLG particle; **Partikelabscheidung** *f* LUFT particle separation; **Partikelaufladung** *f* LUFT particle charging
partikelbeladen *adj* LUFT particulate-loaded; **~es Abgas** *nt* LUFT particulate-loaded waste gas; **~er Gasstrom** *m* LUFT particulate-loaded gas stream
Partikel: **Partikelbenetzung** *f* LUFT particle wetting; **Partikeldichte** *f* LUFT particle thickness; **Partikeldurchmesser** *m* ABWASSER particle diameter; **Partikelemissionen** *f pl* LUFT particle emissions; **Partikelfiltration** *f* ABWASSER particle filtration
partikelförmig *adj* LUFT particulate; **~er Feststoff** *m* LUFT particulate solid; **~e Luftverunreinigung** *f* LUFT particulate air pollution; **~e Rußemissionen** *f pl* LUFT particulate smoke emissions; **~e Schadstoffe** *m pl* LUFT particulate pollutants
Partikel: **Partikelgröße** *f* ABFALL, LUFT particle size; **Partikelgrößenverteilung** *f* LUFT particulate size distribution; **Partikelreibung** *f* LUFT particle friction; **Partikelschadstoffe** *m pl* LUFT particulate pollutants; **Partikelstrahl-Flüssigchromatographie-Massenspektrometrie** *f* ANALYTIK particle beam liquid chromatography-mass spectrometry; **Partikel-Tropfen-Agglomerat** *nt* LUFT particle-drop agglomerate; **Partikelzähler** *m* ANALYTIK particle counter
Partometer *nt* ANALYTIK partometer
PAS *abbr* (*photoakustische Spektroskopie*) ANALYTIK PAS (*photoacoustic spectroscopy*)
Pascal *nt* (*Pa*) ANALYTIK, SICHERHEIT pascal (*Pa*)
passiv *adj* ALLG passive; **~er Pellistor** *m* ANALYTIK passive pellistor; **~er Schallschutz** *m* LÄRM *Kopfhörer, Ohrenstöpsel* passive sound protection; **~e thermische Sonnenenergienutzung** *f* ERNEU ENER passive thermal solar energy use
Passiventgasung *f* ABFALL passive degassing
Pasteurisierung *f* ABWASSER pasteurization
Pastillenpresse *f* ANALYTIK tablet press
pastös *adj* ABFALL pasty; **~er Abfall** *m* ABFALL pasty waste; **~er Sonderabfall** *m* ABFALL pasty hazardous waste
pathogen *adj* ABFALL, ABWASSER, WASSER pathogenic
Pathogen *nt* ABFALL, ABWASSER, WASSER pathogen
pathogen: **~er Abfall** *m* ABFALL, ABWASSER, MEER anatomical waste, pathological waste, infectious waste; **~e Bakterien** *f pl* ABFALL pathogenic bacteria; **~e Gefahr** *f* SICHERHEIT pathogenic hazard; **~e Keime** *m pl* ABFALL, WASSER pathogenic germs; **~er Mikroorganismus** *m* ABFALL, ABWASSER pathogenic microorganism
pathogenfrei *adj* ABFALL pathogen-free
Patronenfilter *nt* LUFT cartridge filter
Patrouillenboot *nt* MEER patrol boat
Pb *abbr* (*Blei*) ALLG Pb (*lead*)
Pb-Akku *m* ABFALL *Batterie* lead accumulator
PC *abbr* (*Papierchromatographie*) ANALYTIK pc (*paper chromatography*)
PC-Anschluß *m* (*Personalcomputer-Anschluß*) ANALYTIK PC connection (*personal computer connection*)
PCB *abbr* (*polychloriertes Biphenyl*) BODEN PCB (*polychlorinated biphenyl*); **PCB-Kondensator** *m* ABFALL, LUFT PCB-capacitor
PCDD *abbr* (*polychloriertes Dibenzo-p-Dioxin*) BODEN PCDD (*polychlorinated dibenzo-p-dioxin*)
PCDF *abbr* (*polychloriertes Dibenzofuran*) BODEN PCDF (*polychlorinated dibenzofurane*)
PCI *abbr* (*positive chemische Ionisation*) ANALYTIK PCI (*positive chemical ionization*)

PCN *abbr* (*polychloriertes Naphthalin*) BODEN, LUFT PCN (*polychlorinated naphthalene*)
PCP *abbr* (*Pentachlorphenol*) BODEN, LUFT PCP (*pentachlorophenol*)
PCT *abbr* BODEN, LUFT (*polychloriertes Terphenyl*) PCT (*polychlorinated terphenyl*)
Pd *abbr* (*Palladium*) LUFT Pd (*palladium*)
Pd II/Cu II-Katalysator *m* LUFT *Nickel* Pd II/Cu II catalyst
PE *abbr* (*Polyethylen*) ABFALL PE (*polyethylene*)
Peak *nt* ABWASSER, ANALYTIK, LUFT peak
***PE*: PE-Container** *m* ABFALL PE container, polyethylene container; **PE-Film** *m* ABFALL PE film; **PE-Flasche** *f* ABFALL PE bottle
Pegel *m* ABWASSER, ANALYTIK, LÄRM level indicator; **Pegelminderung** *f* LÄRM, MEER level reduction; **Pegelschreiber** *m* ABWASSER, ANALYTIK, LÄRM level recorder
Pellet *nt* ABWASSER *Schlammkuchen*, BODEN *Sinterung, Verglasung* pellet; **Pelletbrennen** *nt* BODEN, LUFT pellet incineration
pelletisieren *vt* ABFALL, ABWASSER, BODEN pelletize
Pelletisierung *f* ABFALL, ABWASSER, BODEN pelletization
***Pellet*: Pelletmühle** *f* ABFALL pellet mill; **Pelletpresse** *f* ABFALL pellet mill; **Pelletschicht** *f* BODEN pellet layer
Pellistor *m* ANALYTIK pellistor
Peltier-Element *nt* ANALYTIK *Kühlung* Peltier element
Pendelräumer *m* ABWASSER swinging scraper
Penetrometer *nt* ANALYTIK penetrometer
PentaCDF *abbr* (*Pentachlordibenzofuran*) BODEN pentaCDF (*pentachlorodibenzofurane*)
Pentachlordibenzofuran *nt* (*PentaCDF*) BODEN pentachlorodibenzofurane (*pentaCDF*)
Pentachlorphenol *nt* (*PCP*) BODEN, LUFT pentachlorophenol (*PCP*)
PER *abbr* (*Perchlorethylen*) LUFT PER (*perchloroethylene*)
Perchlorethylen *nt* (*PER*) LUFT perchloroethylene (*PER*)
perfekt: **~es Gas** *nt* LUFT ideal gas, perfect gas
perforiert *adj* ALLG *Rohr* perforated
perikinetisch *adj* ABFALL, WASSER perikinetic; **~e Flockung** *f* ABFALL, WASSER perikinetic flocculation
periodisch *adj* ANALYTIK, LUFT periodic; **~e Reinigung** *f* LUFT periodic cleaning; **~e Spülung** *f* ABFALL, WASSER periodic flushing
Peripheralpumpe *f* ABWASSER, WASSER peripheral pump
peristaltisch: **~e Pumpe** *f* ABFALL, ABWASSER peristaltic pump
Perkolation *f* ABWASSER, ANALYTIK, WASSER percolation; **Perkolationsversuch** *m* ANALYTIK percolation test
perkolativ: **~e Infiltration** *f* ABFALL, WASSER percolative infiltration
perkolieren *vt* ABFALL, ABWASSER, ANALYTIK, WASSER percolate
permanent: **~e Härte** *f* ABWASSER, WASSER noncarbonate hardness (*NCH*), permanent hardness; **~e Infiltration** *f* ABFALL, WASSER permanent infiltration
Permanentmagnet *m* ABFALL, ANALYTIK permanent magnet
permanentmagnetisch *adj* ABFALL permanent magnetic
permeabel *adj* ALLG permeable
Permeabilität *f* ABWASSER, ANALYTIK permeability; **Permeabilitätsmeßgerät** *nt* ANALYTIK permeability measuring instrument, permeameter
Permeameter *nt* ANALYTIK permeability measuring instrument, permeameter
Permeat *nt* ABWASSER, LUFT, WASSER permeate; **Permeatgewinnung** *f* ABWASSER permeate recovery
Permeationsgeschwindigkeit *f* LUFT permeation velocity
Permeationsmembran *f* LUFT *Menbranreaktor* permeation membrane
Peroxoacetylnitrat *nt* (*PAN*) LUFT, WASSER peroxoacetyl nitrate (*PAN*)
persistent: **~es Öl** *nt* ABFALL, BODEN persistent oil
Persistenz *f* BODEN, LUFT persistence; **Persistenzpotential** *nt* BODEN persistence potential
Personalcomputer-Anschluß *m* (*PC-Anschluß*) ANALYTIK personal computer connection (*PC connection*)
Perzentilwert *m* ANALYTIK percentile value
PES *abbr* (*Polyester*) ABFALL PES (*polyester*); **PES-Membran** *f* ABWASSER polyester membrane
Pestizid *nt* ABWASSER pesticide
pestizidhaltig *adj* ABFALL, WASSER: **~es Wasser** *nt* WASSER water containing pesticides
PETRIFIX-Verfahren *nt* ABFALL PETRIFIX process
PF *abbr* (*Phenol-Formaldehyd-Harz*) ABFALL phenol-formaldehyde-resin
pfadspezifisch *adj* BODEN path-specific
Pfand *nt* ABFALL deposit; **Pfandflasche** *f* ABFALL returnable bottle, deposit bottle; **Pfandkasten** *m* ABFALL reusable box; **Pfandregelung** *f* POLITIK deposit fee system
Pflanze *f* ERNEU ENER plant; **Pflanzenklärsystem** *nt* ABFALL, ABWASSER purification system using plants; **Pflanzenöle** *nt pl* ERNEU ENER plant oils; **Pflanzenschutzgesetz** *nt* POLITIK Plant Protection Act
pflanzlich: **~er Abfall** *m* ABFALL vegetable waste; **~er Staub** *m* LUFT plant dust
pH *abbr* (*Wasserstoffexponent*) ALLG pH (*potential of hydrogen*); **pH-Abnahme** *f* ANALYTIK, LUFT pH drop
Phase *f* ANALYTIK, LUFT phase; **Phasenanalyse** *f* ANALYTIK *einer Probe* phase analysis; **Phasengrenzfläche** *f* LUFT interphase
phasenspezifisch *adj* ABFALL phase-specific; **~e sequentielle Extraktion** *f* ABFALL phase-specific sequential extraction
***Phase*: Phasentrenngefäß** *nt* LUFT phase break vessel; **Phasentrennung** *f* ABFALL, ANALYTIK, LUFT phase separation, separation of phases
***pH*: pH-Bereich** *m* ABWASSER pH range; **pH-Elektrode** *f* ANALYTIK pH electrode; **pH-empfindlich** *adj* ABFALL, ABWASSER, WASSER pH-sensitive
Phenol *nt* ABWASSER phenol; **Phenol-Formaldehyd-Harz** *nt* (*PF*) ABFALL phenol-formaldehyde-resin; **Phenolharz** *nt* ABFALL, WASSER phenolic resin
Pheromon *nt* BODEN pheromone
***pH*: pH-Meßgerät** *nt* ANALYTIK pH meter; **pH-Messung** *f* ANALYTIK pH measurement
Phon *nt* ANALYTIK phon; **Phonmeter** *nt* ANALYTIK, LÄRM phonometer
Phoredox-Verfahren *nt* ABWASSER *Phosphorelimination* Phoredox method
Phosphat *nt* ABWASSER phosphate; **Phosphateliminierung** *f* ABWASSER phosphate removal, phosphate elimination; **Phosphatfällung** *f* ABWASSER phosphate precipitation; **Phosphatfällungsmittel** *nt* ABWASSER phosphate precipitate
phosphatfrei *adj* ABWASSER phosphate-free; **~es Waschmittel** *nt* ABWASSER phosphate-free detergent, phosphate-free washing powder
phosphathaltig *adj* ABFALL, ABWASSER, WASSER phosphate-containing
***Phosphat*: Phosphatkonzentration** *f* ABWASSER phosphate concentration; **Phosphatschlamm** *m* ABWASSER phosphate sludge; **Phosphatstrippen** *nt* ABWASSER phosphate stripping
Phosphor *f* (*P*) ABWASSER, LUFT phosphorus (*P*); **Phosphoreliminierung** *f* ABWASSER phosphorus removal; **Phosphorrücklösung** *f* ABWASSER phosphorus redissolution
phosphorsauer *adj* ANALYTIK phosphorous
***Phosphor*: Phosphorsäure-Brennstoffzelle** *f* LUFT phosphoric acid fuel cell (*PAFC*)
Photo- *in cpds* ALLG photo; **Photoabbau** *m* ABFALL *Kunststoffe* photodegradation
photoabbaubar *adj* ABFALL photodegradable; **~e Kunststoffe** *m pl* ABFALL photodegradable plastics

photoakustisch: **~e Spektroskopie** *f* (*PAS*) ANALYTIK photoacoustic spectroscopy (*PAS*)
photobiologisch *adj* ABFALL photobiological; **~e Reaktion** *f* ABFALL photobiological reaction
photochemisch *adj* ABFALL photodegradable, photochemical, ABWASSER, LUFT photochemical; **~er Abbau** *m* ABFALL *Kunststoffe* photodegradation; **~ abbaubare Kunststoffe** *m pl* ABFALL photodegradable plastics; **~e Behandlung** *f* ABWASSER photochemical treatment; **~es Oxidationsmittel** *nt* LUFT photochemical oxidant; **~e Reaktion** *f* ABFALL, WASSER photochemical reaction; **~er Smog** *m* LUFT photochemical smog
Photo-: **Photoelektronenspektroskopie** *f* ANALYTIK *Thermoplast*, WASSER photoelectron spectroscopy
photographisch: **~er Film** *m* ANALYTIK *Filmdosimeter* photographic film; **~e Infrarotkontrolle** *f* ANALYTIK photographic infrared control
Photo-: **Photoionisation** *f* ANALYTIK photoionization; **Photoionisationsdetektor** *m* (*PID*) ANALYTIK photoionization detector (*PID*); **Photokatalysator** *m* ABFALL photocatalyst
photokatalytisch *adj* ABFALL photocatalytic; **~e Oxidation** *f* ABFALL photocatalytic oxidation
Photolyse *f* ABFALL, BODEN photolysis
photolytisch *adj* ANALYTIK *Konverter*, BODEN photolytic; **~er Abbau** *m* BODEN photolytic degradation; **~ abbaubar** *adj* ABFALL photodegradable; **~ abbaubare Kunstoffe** *m pl* ABFALL photodegradable plastics; **~e Zerstörung** *f* BODEN photolytic destruction
photomechanisch *adj* ABFALL photomechanical; **~es Verfahren** *nt* ABFALL *Aussortieren von Keramik* photomechanical process
Photo-: **Photometer** *nt* ANALYTIK photometer; **Photometrie** *f* ANALYTIK photometry
photometrisch *adj* ANALYTIK photometric; **~e Analyse** *f* ANALYTIK photometric analysis
Photo-: **Photomultiplier** *m* ANALYTIK photomultiplier (*PM*); **Photooxidantien** *nt pl* ABFALL, LUFT photooxidants; **Photooxidation** *f* ABFALL, LUFT photooxidation; **Photophorese** *f* LUFT photophoresis; **Photosensor** *m* ABFALL photosensor; **Photospannung** *f* ERNEU ENER *oszillierende Wassersäule* photoelectric voltage; **Photostrom** *m* ERNEU ENER *Wellenenergie* photoelectric current; **Photosynthese** *f* ABWASSER, ERNEU ENER *Biomasse* photosynthesis
phototroph *adj* ABWASSER *Bakterien* phototrophic
Photo-: **Photovervielfacher** *m* ANALYTIK photomultiplier (*PM*)
Photovoltaik *f* (*PV*) ERNEU ENER photovoltaics (*PV*); **Photovoltaik-Kraftwerk** *nt* ERNEU ENER photovoltaic power plant; **Photovoltaikmodul** *nt* ERNEU ENER photovoltaic module; **Photovoltaikzelle** *f* ERNEU ENER photovoltaic cell
photovoltaisch *adj* ERNEU ENER photovoltaic; **~ betrieben** *adj* ERNEU ENER photovoltaic-powered, photovoltaic-operated, photovoltaic-driven; **~ betriebenes Fahrzeug** *nt* ERNEU ENER photovoltaic-powered vehicle; **~er Effekt** *m* ERNEU ENER photovoltaic effect; **~es Element** *nt* ERNEU ENER photovoltaic element; **~e Energiegewinnung** *f* ERNEU ENER photovoltaic energy generation; **~e Energieumwandlung** *f* ERNEU ENER photovoltaic energy conversion; **~es System** *nt* ERNEU ENER photovoltaic system
pH: **pH-Papierstreifen** *m* ANALYTIK pH-paper strip; **pH-Stabilisierung** *f* ABWASSER pH stabilization; **pH-Wert** *m* ABFALL, ANALYTIK, BODEN, LUFT pH value; **pH-Wert-Erhöhung** *f* ABFALL, ANALYTIK, BODEN, LUFT pH value increase
physikalisch *adj* ALLG physical; **~e Adsorption** *f* LUFT physical adsorption; **~e Belastung** *f* ABFALL *Deponie* physical strain; **~e Enthärtung** *f* ABFALL, WASSER mechanical softening; **~er Parameter** *m* ABWASSER, ANALYTIK, BODEN physical parameter
physikalisch-chemisch *adj* ALLG physical-chemical, physiochemical; **~es Reinigungsverfahren** *nt* ABWASSER physiochemical treatment; **~es Verfahren** *nt* ABFALL, ABWASSER, WASSER physical-chemical process, physiochemical process; **~e Umgebung** *f* ABFALL, LUFT, WASSER physiochemical environment
physikochemisch *adj* ALLG physiochemical; **~e Reaktion** *f* LUFT physiochemical reaction
physiologisch: **~e Einwirkung** *f* SICHERHEIT physiological effect
Physisorption *f* LUFT physisorption
PID *abbr* (*Photoionisationsdetektor*) ANALYTIK PID (*photoionization detector*)
piezoelektrisch *adj* ANALYTIK piezoelectric; **~es Mikrofon** *nt* ANALYTIK piezoelectric microphone
Piezometer *nt* ANALYTIK piezometer
Pigmentschlamm *m* ABWASSER pigment sludge
Pilot- *in cpds* ALLG pilot; **Pilotanlage** *f* ALLG pilot plant; **Pilotanlagemaßstab** *m* ABWASSER, BODEN, LUFT pilot plant scale; **Pilot-Audit** *nt* POLITIK pilot audit; **Pilotprogramm** *nt* ALLG pilot program (*AmE*), pilot programme (*BrE*); **Pilotprojekt** *nt* ALLG pilot project; **Pilotversuch** *m* ANALYTIK pilot test
Pilz *m* ABWASSER fungus
pipettieren *vt* ANALYTIK transfer with a pipette
Pistonfon *nt* ANALYTIK *Kalibrationsgerät für Schallmeßgeräte* pistonphone
Pitot-Rohr *nt* ANALYTIK Pitot tube
PIXE *abbr* (*protoneninduzierte Röntgenemission*) ANALYTIK PIXE (*proton-induced x-ray emission*); **PIXE-Analyse** *f* ANALYTIK PIXE analysis
Planck: **~scher Strahler** *m* ANALYTIK, LUFT blackbody; **~sche Strahlungsformel** *f* LUFT Planck's law; **~sches Wirkungsquantum** *nt* LUFT Planck's constant
Planetengetriebe *nt* ERNEU ENER *Windturbine* planetary gear
Planfeststellungsbeschluß *m* POLITIK planning assessment decision
Planfeststellungsverfahren *nt* POLITIK planning assessment procedure
Planiergerät *nt* MEER grader
Planierraupe *f* ABFALL bulldozer
Plankton *nt* WASSER plankton
Plasmabehandlung *f* ABFALL *von Sonderabfällen* plasma treatment
Plasma-Ultrahochtemperaturverfahren *nt* BODEN high-temperature plasma process
Plastifizieraggregat *nt* ABFALL plasticizing unit
plastifizieren *vt* ABFALL plasticize
Plastifizierung *f* ABFALL plasticizing
Plastikator *m* ABFALL plasticator
Plättchen *nt* ABFALL plate
Platte *f* LÄRM panel, LUFT lamella, MEER panel; **Plattenabscheider** *m* LUFT plate precipitator; **Plattenabsorber** *m* LÄRM *Tiefenabsorber* panel absorber; **Plattenband** *nt* ABFALL apron conveyor; **Plattenelektrofilter** *nt* LUFT plate precipitator, plate-wire ESP; **Plattenförderband** *nt* ABFALL apron conveyor; **Plattenkatalysator** *m* LUFT plate catalyst; **Plattenmodul** *nt* ABWASSER *Ultrafiltration* plate module; **Plattenrasterdecke** *f* LÄRM panel-louvered ceiling (*AmE*), panel-louvred ceiling (*BrE*); **Plattenresonator** *m* LÄRM panel resonator; **Plattenschwinger** *m* LÄRM panel absorber
Plenumkammer *f* LÄRM plenum chamber
Plotter *m* ANALYTIK plotter
Ploughshare-Mischer *m* WASSER ploughshare mixer
Plunger *m* WASSER press plunger; **Plungerpumpe** *f* ABWASSER, WASSER plunger pump
PM *abbr* (*Schwebstoff-Teilchen*) LUFT PM (*particulate matter*)
pneumatisch *adj* ALLG pneumatic; **~er Antrieb** *m* ABFALL,

ABWASSER pneumatic driving gear; **~er Schalldämpfer** *m* LÄRM, MEER pneumatic sound absorber; **~e Sortieranlage** *f* ABFALL pneumatic sorter; **~es Sortieren** *nt* ABFALL pneumatic classification

Polarisation *f* ANALYTIK polarization

Polarität *f* LUFT polarity

Polarograph *m* ANALYTIK polarograph

Polarographie *f* ANALYTIK polarography

polarographisch *adj* ANALYTIK polarographic; **~e Bestimmungsmethoden** *f pl* ANALYTIK polarographic determination methods

Polder *m* ABFALL polder, landfill cell, refuse cell

Polieren *nt* ABFALL polishing

Politikinstrument *nt* POLITIK policy instrument

politisch: **~e Zielsetzungen** *f pl* POLITIK policy objectives

Polizeirecht *nt* POLITIK public order law

Polster *nt* ABFALL cushion

Polyacrylnitril *nt* LUFT *Filter-Fasermaterial* polyacrylnitrile

Polyaddition *f* ABFALL polyaddition

Polyamid *nt* (*PA*) ABFALL, ABWASSER polyamide (*PA*); **Polyamidmembran** *f* ABFALL, ABWASSER polyamide membrane

Polycarbonat *nt* ABFALL, ABWASSER, WASSER polycarbonate (*PC*); **Polycarbonatmembran** *f* ABWASSER polycarbonate membrane

polychloriert: **~es Biphenyl** *nt* (*PCB*) BODEN polychlorinated biphenyl (*PCB*); **~es Dibenzofuran** *nt* (*PCDF*) BODEN polychlorinated dibenzofurane (*PCDF*); **~es Dibenzo-p-Dioxin** *nt* (*PCDD*) BODEN polychlorinated dibenzo-p-dioxin (*PCDD*); **~es Naphthalin** *nt* (*PCN*) BODEN, LUFT polychlorinated naphthalene (*PCN*); **~es Terphenyl** *nt* (*PCT*) BODEN, LUFT polychlorinated terphenyl (*PCT*)

Polyester *m* (*PES*) ABFALL, ABWASSER *Duroplast* polyester (*PES*); **Polyestermembran** *f* ABWASSER polyester membrane; **Polyesterschaumstoff** *m* ABFALL polyester foam

Polyethylen *nt* (*PE*) ABFALL polyethylene (*PE*); **~ mit hoher Dichte** ABFALL high-density polyethylene (*HDPE*); **~ mit niedriger Dichte** ABFALL low-density polyethylene (*LDPE*); **Polyethylenterephthalat** *nt* ABFALL polyethylene terephthalate (*PET*)

Polykondensation *f* ABFALL polycondensation

polykristallin *adj* ERNEU ENER polycrystalline; **~e Siliziumzelle** *f* ERNEU ENER *Deponiegaskomponente* polycrystalline silicon cell

Polymer *nt* ABFALL, ABWASSER, LUFT polymer; **Polymerbeton** *m* ABFALL, WASSER polymer concrete (*PC*); **Polymerflockungsmittel** *nt* ABFALL, ABWASSER, ANALYTIK, LUFT, WASSER flocculant

Polymerisation *f* ABFALL polymerization

Polymeter *nt* ANALYTIK polymeter

polynuklear: **~e Aromaten** *m pl* BODEN polynuclear aromatics (*PNA*)

Polyolefin *nt* ABFALL polyolefin (*PO*)

Polyolefin-Container *m* ABFALL polyolefin container

Polyphosphat *nt* ABWASSER polyphosphate

Polypropylen *nt* (*PP*) ABFALL, LUFT polypropylene (*PP*)

Polysaccharid *nt* ABFALL polysaccharide

polysaprob *adj* ABWASSER polysaprobic

Polystyrol *nt* (*PS*) ABFALL polystyrene (*PS*)

Polytetrafluorethylen *nt* (*PTFE*) ABFALL polytetrafluoroethylene (*PTFE*)

Polyurethan *nt* (*PU*) ABFALL polyurethane (*PU*)

polyvalent *adj* ABFALL multivalent

Polyvinylchlorid *nt* (*PVC*) ABFALL, BODEN polyvinyl chloride (*PVC*)

polyzyklisch: **~er aromatischer Kohlenwasserstoff** *m* (*PAK*) ABFALL, ABWASSER, LUFT, WASSER polycyclic aromatic hydrocarbon (*PAH*)

Pol-Zelle *f* ANALYTIK *Prozeßkonduktometer* electrode cell

Pore *f* ABFALL, ABWASSER, LUFT pore; **Porendurchmesser** *m* LUFT pore diameter; **Porengröße** *f* LUFT pore size; **Porenmembran** *f* ABWASSER pore membrane; **Porenraum** *m* LUFT pore volume, pore space; **Porenvolumen** *nt* ABFALL absorptive capacity, pore volume, LUFT pore volume; **Porenvolumenanteil** *m* LUFT pore volume; **Porenwasser** *nt* ABFALL, WASSER interstitial water

Porometer *nt* ANALYTIK porometer

porös *adj* ABFALL, ABWASSER, BODEN, WASSER porous; **~e Abdeckung** *f* ABFALL porous cover; **~er Absorber** *m* LÄRM treble absorber; **~e Membran** *f* ABFALL, ABWASSER, WASSER porous membrane

Porosität *f* ABFALL, ABWASSER, LUFT porosity

portabel *adj* ANALYTIK portable

positionieren *vt* ANALYTIK position

positiv *adj* ALLG positive; **~e chemische Ionisation** *f* (*PCI*) ANALYTIK positive chemical ionization (*PCI*); **~e Elektrode** *f* LUFT positive electrode; **~es Ion** *nt* ABFALL, ABWASSER positive ion; **~es Kation** *nt* WASSER positive cation

Potential *nt* ANALYTIK, BODEN, ERNEU ENER, SICHERHEIT potential; **Potentialmessung** *f* ANALYTIK potential measurement

potentiell *adj* BODEN, ERNEU ENER, SICHERHEIT potential; **~e Azidität** *f* BODEN potential acidity; **~e Energie** *f* ERNEU ENER potential energy; **~e Gefährdung** *f* SICHERHEIT potential danger

Potentiometer *nt* ANALYTIK potentiometer

Potentiometrie *f* ANALYTIK *Polarographie* potentiometry

potentiometrisch *adj* ANALYTIK potentiometric, electrometric; **~er Sensor** *m* ANALYTIK potentiometric sensor; **~e Titration** *f* ANALYTIK potentiometric titration

Potentiostat *m* ANALYTIK potentiostat

potentiostatisch *adj* ANALYTIK potentiostatic; **~er Elektrodensensor** *m* ANALYTIK potentiostatic electrode sensor

Pottasche *f* ABWASSER potash

POX *abbr* (*ausblasbare organische Halogenverbindungen*) BODEN, WASSER POX (*purgeable organic halogen compounds*)

PP *abbr* (*Polypropylen*) ABFALL, LUFT PP (*polypropylene*)

Prall *m* ABFALL, ABWASSER, LUFT collision, impact; **Pralldüse** *f* LUFT impactor nozzle; **Prallfläche** *f* LUFT baffle; **Prallmühle** *f* ABFALL impact crusher; **Prallplattenwäscher** *m* LUFT impingement plate scrubber; **Prallwand** *f* ABWASSER baffle

Prandtl: **~sches Staurohr** *nt* ANALYTIK pitostatic tube, Prandtl's pitot tube

praxiserprobt *adj* ALLG field-proven

Präzedenzfall *m* POLITIK precedence case

Präzisionsmessung *f* ANALYTIK precision measurement

Präzisionsschallpegelmesser *m* ANALYTIK, LÄRM precision sound level meter

Precoat-Filtration *f* LUFT precoat filtration

Precoat-Schicht *f* ABFALL, LUFT *Schlauchfilter* precoat layer

Preß- *in cpds* ABFALL, ABWASSER, ANALYTIK compressing, press, pressing, pressure

Pressemitteilung *f* POLITIK press communication

pressen *vt* ABFALL press

Pressen *nt* ABFALL compression moulding (*BrE*), compression molding (*AmE*), pressing, ABWASSER, ANALYTIK pressing; **Pressenbeschickungsband** *nt* ABFALL press feeder belt; **Pressenhub** *m* ABFALL press stroke; **Pressenstempel** *m* ABFALL press ram, press plunger; **Pressenstößel** *m* ABFALL press ram, press plunger

Pressesprecher *m* POLITIK spokesperson

Pressestelle *f* POLITIK press office

Preß-: **Preßkolben** *m* ABFALL press ram, press plunger; **Preßkraft** *f* ABFALL pressure force

Preßling *m* ABFALL *Kompostierung* briquette; **Preßlingstechnik** *f* ANALYTIK pellet technique

Preß-: **Preßmüllfahrzeug** *nt* ABFALL compression vehicle, compactor vehicle, packer body; **Preßmüllwagen** *m* ABFALL compression vehicle, compactor vehicle, packer body; **Preßtrocknung** *f* ABWASSER press drying

Pressung *f* ABFALL, ABWASSER, ANALYTIK pressing

Preß-: **Preßvorgang** *m* ABFALL pressing operation

Primär- *in cpds* ALLG primary

primär: **~e Setzung** *f* ABFALL, BODEN, WASSER primary settlement

Primär-: **Primärabfallvermeidung** *f* ABFALL primary waste avoidance; **Primäraluminium** *nt* ABFALL primary aluminium (*BrE*), primary aluminum (*AmE*); **Primärbrennstoff** *m* ABFALL primary fuel; **Primärenergiebedarf** *m* ERNEU ENER primary energy demand; **Primärenergiebilanz** *f* ERNEU ENER primary energy balance; **Primärenergiesubstitution** *f* ERNEU ENER primary energy substitution; **Primärenergieträger** *m* ERNEU ENER primary energy source; **Primärenergieverbrauch** *m* ERNEU ENER primary energy consumption; **Primärfallrohr** *nt* LUFT *Schüttschichtfilter* primary downpipe; **Primärfilter** *nt* ABWASSER primary filter; **Primärluft** *f* LUFT primary air; **Primärrecycling** *nt* ABFALL primary recycling; **Primärschlamm** *m* ABWASSER raw sludge, primary sludge, undigested sludge; **Primärstoff** *m* ABFALL primary material; **Primärverbrennungsluft** *f* ABFALL primary combustion air

Prinzip *nt*: **~ der ergonomischen Gestaltung** SICHERHEIT ergonomic design principle

Priorität *f* BODEN, MANAG, POLITIK priority; **Prioritätensetzung** *f* BODEN, MANAG, POLITIK priority determination; **Prioritätsliste** *f* BODEN, MANAG, POLITIK *von Maßnahmen* priority list

privat: **~es Entsorgungsunternehmen** *nt* ABFALL private disposal company; **~e Sammlung** *f* ABFALL private collection; **~er Trinkwasserverbrauch** *m* ABFALL private drinking water consumption

Probe *f* ANALYTIK sample, test; **eine ~ nehmen** *phr* ANALYTIK sample, take a sample; **Probebelüfter** *m* ABWASSER testing aerator; **Probebeurteilung** *f* ANALYTIK sample assessment; **Probeflasche** *f* ANALYTIK sampling bottle; **Probeglas** *nt* ANALYTIK test tube; **Probehahn** *m* ABWASSER, WASSER gauge cock (*BrE*), gage cock (*AmE*); **Probemenge** *f* ANALYTIK sample amount

Probenahme *f* ANALYTIK sampling, sample collection; **Probenahmegerät** *nt* ANALYTIK sampler, sampling device; **Probenahmehäufigkeit** *f* ANALYTIK sampling frequency; **Probenahmekapazität** *f* ANALYTIK sampling capacity; **Probenahmeprotokoll** *nt* ANALYTIK sample record; **Probenahmepumpe** *f* ANALYTIK sampling pump; **Probenahmepunkt** *m* ANALYTIK, BODEN sample point; **Probenahmeraster** *m* ANALYTIK, BODEN sampling grid; **Probenahmerohr** *nt* ANALYTIK sampling tube; **Probenahmesonde** *f* ANALYTIK sampling probe; **Probenahmestelle** *f* ABFALL sampling bay, ANALYTIK sampling bay, sampling point; **Probenahmesystem** *nt* ANALYTIK sampling system

Probe: **Probenanalyse** *f* ANALYTIK sample analysis; **Probenaufbereitung** *f* ANALYTIK sample treatment, sampling preparation; **Probenaufbewahrung** *f* ANALYTIK sample storage; **Probenbehälter** *m* ANALYTIK snap sample container; **Probendosierung** *f* ANALYTIK sample introduction, sample dosing; **Probendurchsatz** *m* ANALYTIK sample throughput, sample thruput (*AmE*); **Probenehmer** *m* ANALYTIK sampler; **Probenflasche** *f* ANALYTIK sampling bottle; **Probengas** *nt* ANALYTIK sample gas, sampling gas; **Probengasfilter** *nt* ANALYTIK sample gas filter; **Probengaskapillare** *f* ANALYTIK sample gas capillary; **Probengeber** *m* ANALYTIK sample dispenser, sample injector; **Probengefäß** *nt* ANALYTIK snap sample container; **Probengröße** *f* ANALYTIK sample size; **Probenhalter** *m* ANALYTIK sample holder; **Probenheber** *m* ANALYTIK sampling tube; **Probenhöchstmenge** *f* ANALYTIK maximum sample quantity; **Probeninjektor** *m* ANALYTIK sample injector; **Probenkalibrierung** *f* ANALYTIK sample calibration; **Probenkondensation** *f* ANALYTIK sampling condensation; **Probenlösung** *f* ANALYTIK sample solution; **Probenschöpfer** *m* ANALYTIK water sampler

probenspezifisch *adj* ANALYTIK *Daten* sample-specific

Probe: **Probenträger** *m* ANALYTIK sample holder; **Probenumfang** *m* ANALYTIK sample size; **Probenvolumen** *nt* ANALYTIK sample volume; **Probenvorbereitung** *f* ANALYTIK sampling preparation; **Probenweiterleitung** *f* ANALYTIK sampling transmission; **Probenzuführung** *f* ANALYTIK sample introduction; **Probenzusammensetzung** *f* ANALYTIK sample composition; **Probestück** *nt* ANALYTIK specimen; **Probeverarbeitung** *f* ANALYTIK sample processing; **Probezelle** *f* ANALYTIK sampling cell

Probierhahn *m* ABWASSER, WASSER gage cock (*AmE*), gauge cock (*BrE*)

Probiertiegel *m* ANALYTIK test crucible

Produkt *nt* ALLG product; **~ am Ende des Lebenszyklus** ABFALL product at end of life cycle; **Produkthaftung** *f* ABFALL, POLITIK product liability

Produktion *f* ABFALL, ERNEU ENER, WASSER production; **Produktionsabfall** *m* ABFALL process waste; **Produktionspotential** *nt* ERNEU ENER production potential; **Produktionsrückstand** *m* ABFALL production residue

produktionsspezifisch: **~er Abfall** *m* ABFALL process waste, production-specific waste

Produktion: **Produktions-Verbrauchs-Kreislauf** *m* ABFALL production-consumption loop

Produkt: **Produktlebenszyklus** *m* MANAG product life cycle; **Produktlinienanalyse** *f* MANAG product-line-analysis; **Produktnormen** *f pl* POLITIK product standards; **Produktrückgewinnung** *f* LUFT product recovery; **Produktsicherheit** *f* SICHERHEIT product safety; **Produktverantwortung** *f* ABFALL, MANAG, POLITIK product stewardship; **Produktzyklus** *m* MANAG product life cycle

Prognose *f* ABFALL, ANALYTIK, WASSER forecasting; **Prognosemodell** *nt* ANALYTIK forecast model

prognostizieren *vt* ABFALL, ANALYTIK, WASSER forecast

Programm *nt* ANALYTIK, POLITIK program (*AmE*), programme (*BrE*)

programmierbar *adj* ANALYTIK programmable

programmieren *vt* ANALYTIK program

progressiv: **~e Steuer** *f* POLITIK progressive tax

Projektphase *f* ERNEU ENER project phase

Pro-Kopf-Verbrauch *m* WASSER per capita consumption

Promille *nt* ANALYTIK thousandth part

Promotor *m* LUFT *Katalysator* promoter

Propansäure *f* ABWASSER propionic acid

Propeller *m* MEER propeller; **Propellerpumpe** *f* ABWASSER, WASSER axial-flow rotary pump, propeller pump

proportional *adj* ANALYTIK proportional; **~e Probenahme** *f* ANALYTIK proportional sampling

Proportionalprobenehmer *m* ANALYTIK sampler for flow-proportional samples

Proportionalzähler *m* ANALYTIK proportional counter

Protokoll *nt* POLITIK protocol

protokollieren *vt* ANALYTIK record

Protonenabsorptionsfähigkeit *f* LUFT proton absorptive capacity

protoneninduziert: **~e Röntgenemission** *f* (*PIXE*) ANALYTIK proton-induced x-ray emission (*PIXE*)

Protoplasma *nt* ABWASSER protoplasm

Protozoen *nt pl* ABWASSER protozoa

provisorisch: **~e Auswertung** *f* ANALYTIK, POLITIK provisional evaluation

Prozeß *m* ALLG process; **Prozeßabgas** *nt* LUFT process waste gas; **Prozeßdampf** *m* ERNEU ENER process steam; **Prozeßgasstrom** *m* LUFT process gas flow; **Prozeßkonduktometer** *nt* ANALYTIK process conductivity meter

prozeßnachgeschaltet: **~e Maßnahmen** *f pl* ABWASSER, LUFT end-of-pipe technology

Prozeß: **Prozeßstabilität** *f* ABWASSER process stability; **Prozeßwärme** *f* ERNEU ENER process heat; **Prozeßwasser** *nt* ABWASSER, BODEN, LUFT, WASSER process water

Prüf- *in cpds* ALLG test, testing, inspection; **Prüfbedingungen** *f pl* ANALYTIK testing conditions

prüfen *vt* ALLG inspect, test

Prüf-: **Prüfgas** *nt* ANALYTIK test gas; **Prüfgaskalibrator** *m* ANALYTIK test gas calibrator; **Prüfgerät** *nt* ANALYTIK, SICHERHEIT inspection equipment; **Prüfglas** *nt* SICHERHEIT inspection window; **Prüfröhrchen** *nt* ANALYTIK *Gasspürgerät* detector tube; **Prüfsiebung** *f* ANALYTIK test sieving; **Prüfsonde** *f* ANALYTIK test probe

Prüfung *f* ALLG inspection; **~ durch Sachverständige** ANALYTIK, POLITIK, SICHERHEIT inspection by experts

Prüf-: **Prüfwasser** *nt* ANALYTIK test water; **Prüfwert** *m* ANALYTIK test value

PS *abbr* (*Polystyrol*) ABFALL PS (*polystyrene*)

Psychrometer *nt* ANALYTIK psychrometer

psychrophil *adj* ABWASSER psychrophilic; **~er Mikrooganismus** *m* ABWASSER psychrophilic microorganism

PTFE *abbr* (*Polytetrafluorethylen*) ABFALL PTFE (*polytetrafluoroethylene*)

PU *abbr* (*Polyurethan*) ABFALL PU (*polyurethane*)

Puffer *m* ALLG buffer; **Puffereffekt** *m* LUFT buffer effect; **Pufferkapazität** *f* ABWASSER, BODEN buffer capacity; **Pufferlösung** *f* ANALYTIK buffer solution

puffern *vt* ALLG buffer

Puffer: **Pufferreaktion** *f* LUFT buffer reaction; **Puffertank** *m* LUFT buffer tank

Pufferung *f* ALLG buffering

Puffer: **Pufferwirkung** *f* ANALYTIK, LUFT buffering action; **Pufferzone** *f* LÄRM buffer zone

Pulpe *f* ABFALL pulp

Pulper *m* ABFALL pulper

Pulse-Polarographie *f* ANALYTIK pulse-polarography

Puls-Jet-Filter *nt* LUFT reverse-jet filter

Pulveraktivkohle *f* ABWASSER, LUFT powdered activated carbon (*PAC*)

Pump- *in cpds* ALLG pump, pumping; **Pumpbetrieb** *m* **über einen Hochbehälter** ABWASSER, WASSER pump operation via water tower

Pumpe *f* ABWASSER, LUFT, MEER, WASSER pump; **~ mit konstanter Drehzahl** ABWASSER, WASSER constant speed pump

Pump-: **Pumpeinheit** *f* ABFALL, ABWASSER, LUFT, MEER, WASSER pumping unit

pumpen *vt* ABFALL, ABWASSER, LUFT, WASSER lift, pump

Pumpen *nt* ABFALL, ABWASSER, LUFT, WASSER pumping, pumping-out, LUFT lifting; **~ und Behandeln** *nt* WASSER pump and treat; **Pumpenaggregat** *nt* ABWASSER, WASSER pump set; **Pumpenanlage** *f* ABWASSER, WASSER pump station, pumping plant, pumping station; **Pumpenanschluß** *m* ABWASSER, WASSER pump connection; **Pumpenantrieb** *m* ABWASSER, WASSER pump-driving gear; **Pumpenbrunnen** *m* WASSER pump well; **Pumpencharakteristik** *f* ABWASSER, LUFT, WASSER pump characteristics; **Pumpendruckstutzen** *m* ABWASSER, WASSER pump outlet branch; **Pumpeneinrichtung** *f* ABWASSER, WASSER pumping plant; **Pumpenflügelrad** *nt* ABWASSER, WASSER pump impeller; **Pumpengehäuse** *nt* ABWASSER, LUFT, WASSER pump case, pump casing; **Pumpengeschwindigkeit** *f* ABWASSER, LUFT, WASSER pump speed; **Pumpengetriebe** *nt* ABWASSER, WASSER pump gear; **Pumpenhaus** *nt* ABWASSER, WASSER pump house; **Pumpenkavitation** *f* ABWASSER, LUFT, WASSER pump cavitation; **Pumpenkennlinie** *f* ABWASSER, WASSER pump characteristics; **Pumpenkolben** *m* ABWASSER, WASSER pump plunger; **Pumpenkolbenwelle** *f* ABWASSER, WASSER pump plunger shaft; **Pumpenleistung** *f* ABWASSER, LUFT, WASSER pump capacity; **Pumpenraum** *m* ABWASSER, WASSER pump room; **Pumpensatz** *m* ABWASSER pump set; **Pumpensaugstutzen** *m* ABWASSER, WASSER pump inlet branch; **Pumpenschacht** *m* ABWASSER, WASSER pump shaft, pumping shaft; **Pumpenstand** *m* MEER pumping unit; **Pumpenstange** *f* ABWASSER, WASSER pump rod; **Pumpensteuerung** *f* ABWASSER, LUFT, WASSER pump controls; **Pumpensumpf** *m* ABWASSER, LUFT, WASSER pump well; **Pumpentreibmittel** *nt* ABWASSER, WASSER pump fluid; **Pumpenwirkungsgrad** *m* ABWASSER, LUFT, WASSER pump efficiency; **Pumpenzylinder** *m* ABWASSER, LUFT, WASSER pump barrel, pump cylinder

pumpfähig *adj* ABFALL, LUFT, WASSER pumpable; **~e Abfälle** *m pl* ABFALL pumpable wastes; **~e Flüssigkeit** *f* LUFT pumpable liquid; **~e Sonderabfälle** *m pl* ABFALL pumpable hazardous wastes

Pump-: **Pumpmaschine** *f* ABWASSER, WASSER pumping engine; **Pumpstation** *f* ABWASSER, WASSER pump station, pumping station; **Pumpsteuerung** *f* ABWASSER, LUFT, WASSER pump controls; **Pumpversuch** *m* ABWASSER, ANALYTIK, WASSER pump test, pumping test; **Pumpwasser** *nt* WASSER pump water; **Pumpwerk** *nt* ABWASSER, WASSER pump station, pumping station

Punktabsaugsystem *nt* SICHERHEIT point vacuum cleaning system

punktförmig: **~es Fassungselement** *nt* ABFALL single-point collection element

Punktquelle *f* ABWASSER, BODEN, LUFT, WASSER *Verschmutzung* point source

Purge-und-Trap: **~-GC** *f* ANALYTIK *Gaschromatographie* purge-and-trap GC; **~-Sampler** *m* ANALYTIK purge-and-trap sampler; **~-Verfahren** *nt* ANALYTIK purge-and-trap method

Putz *m* LÄRM *Flächenverkleidung* plaster

Puzzolan *nt* BODEN *Bindemittel* puzzolan

PV *abbr* (*Photovoltaik*) ERNEU ENER PV (*photovoltaics*)

PVC *abbr* (*Polyvinylchlorid*) ABFALL, BODEN PVC (*polyvinyl chloride*); **PVC-Einsatz** *m* ABFALL PVC insert fitment; **PVC-Flasche** *f* ABFALL PVC bottle; **PVC-Reststoff** *m* ABFALL PVC residue; **PVC-Rohr** *nt* WASSER PVC pipe

Pyknometer *nt* ANALYTIK pyknometer, density bottle

pyknometrisch *adj* ANALYTIK pyknometric

Pyranometer *nt* ANALYTIK pyranometer

Pyrheliometer *nt* ANALYTIK pyrheliometer

Pyrolyse *f* ABFALL, ABWASSER, ERNEU ENER, LUFT pyrolysis; **~ in Kopplung mit Gaschromatographie** ANALYTIK pyrolysis-gas chromatography; **~ in Kopplung mit Gaschromatographie und Massenspektromet** ANALYTIK pyrolysis-gas chromatography-mass spectrometry; **Pyrolysegas** *nt* ABFALL, ABWASSER pyrolysis gas; **Pyrolyseöl** *nt* ABFALL pyrolysis oil; **Pyrolyseprodukt** *nt* ABFALL pyrolysis product

pyrolysieren *vt* ABFALL, BODEN pyrolize

pyrolytisch *adj* ABFALL, BODEN pyrolytic; **~e Vorbehandlung** *f* ABFALL pyrolytic pretreatment

pyrophor *adj* LUFT pyrophoric; **~er Staub** *m* LUFT pyrophoric dust

Q

QS *abbr* (*Qualitätssicherung*) ALLG QA (*quality assurance*); **QS-System** *nt* ANALYTIK, LUFT, MANAG, SICHERHEIT, WASSER quality assurance system

Quadrupol *m* ANALYTIK quadrupole; **Quadrupol-Massenspektrometer** *nt* ANALYTIK quadrupole mass spectrometer

Qualität *f* ALLG quality

qualitativ *adj* ANALYTIK qualitative; **~e Analyse** *f* ANALYTIK qualitative analysis; **~e Auswertung** *f* ANALYTIK qualitative evaluation; **~e Bestimmung** *f* ANALYTIK qualitative determination; **~e Gasanalyse** *f* ANALYTIK qualitative gas analysis; **~e Geruchsbestimmung** *f* ANALYTIK qualitative odor analysis (*AmE*), qualitative odour analysis (*BrE*); **~er Nachweis** *m* ANALYTIK qualitative detection

Qualität: **Qualitätsanforderungen** *f pl* ALLG quality requirements; **Qualitätsfaktor** *m* ABFALL, LUFT, WASSER quality factor; **Qualitätskontrolle** *f* ALLG quality control; **Qualitätskontrolle** *f* **des Trinkwassers** WASSER drinking water quality control; **Qualitätssicherung** *f* (*QS*) ALLG quality assurance (*QA*); **Qualitätssicherungssystem** *nt* ANALYTIK, LUFT, MANAG, SICHERHEIT, WASSER quality assurance system

quantifizieren *vt* ANALYTIK quantify

Quantifizierung *f* ANALYTIK quantification

quantitativ *adj* ANALYTIK quantitative; **~e Analyse** *f* ANALYTIK quantitative analysis; **~e Auswertung** *f* ANALYTIK quantitative evaluation; **~e Bestimmung** *f* ANALYTIK quantitative determination; **~e Gasanalyse** *f* ANALYTIK quantitative gas analysis; **~e Geruchsbestimmung** *f* ANALYTIK quantitative odor analysis (*AmE*), quantitative odour analysis (*BrE*); **~er Nachweis** *m* ANALYTIK quantitative detection

Quarzrohr *nt* ABWASSER quartz tube

quasisimultan *adj* ANALYTIK quasi-simultaneous

quasitrocken: **~e Abgasbehandlung** *f* LUFT partially dry waste gas treatment; **~er Wäscher** *m* LUFT wet/dry scrubber, semidry scrubber

Quecksilber *nt* (*Hg*) ABFALL, ABWASSER, BODEN mercury (*Hg*); **Quecksilberdampfmischlichtlampe** *f* ABFALL mercury vapor lamp (*AmE*), mercury vapour lamp (*BrE*)

quecksilberfrei *adj* ABFALL, ABWASSER mercury-free; **~e Batterie** *f* ABFALL mercury-free battery

Quecksilber: **Quecksilbergehalt** *m* ABFALL, BODEN mercury content

quecksilberhaltig *adj* ABFALL, ABWASSER mercury-containing; **~es Abwasser** *nt* ABWASSER mercurial wastewater; **~e Lampe** *f* ABFALL mercury-containing lamp; **~es Produkt** *nt* ABFALL mercury-containing product

Quecksilber: **Quecksilberrückgewinnung** *f* ABFALL *von Batterien* mercury recovery

Quell- *in cpds* LUFT, WASSER source; **Quellbereich** *m* LUFT, WASSER source area

Quelle *f* LUFT, WASSER source; **an der ~** *phr* ABWASSER, LUFT at source; **~ von Treibhausgasen** LUFT source of greenhouse gases

quellenbezogen *adj* ALLG source-related

quellenorientiert *adj* ABWASSER, BODEN, WASSER source-oriented; **~e Maßnahmen** *f pl* BODEN source-oriented measures

Quell-: **Quellfassung** *f* WASSER tapping of a spring; **Quellhöhe** *f* LUFT *Schornstein* source height; **Quellwasser** *nt* WASSER spring water

Quenchbehälter *m* LUFT quencher tank, quench tank

Quencher *m* LUFT quencher; **Quenchersuspension** *f* LUFT quencher suspension

Quenchzone *f* LUFT quench zone

querangeströmt *adj* LUFT transverse-entry; **~er Wanderschichtreaktor** *m* LUFT transverse-entry traveling-bed reactor (*AmE*), transverse-entry travelling-bed reactor (*BrE*)

Querdurchstrom *m* ABWASSER *Filtermedium* nonradial flow

Querschnittsprobenahme *f* ANALYTIK cross-section sampling

Querstrom *m* LUFT *Abgasstrom* cross-flow; **Querstromabscheider** *m* LUFT cross-flow separator; **Querstromwäscher** *m* LUFT cross-flow scrubber

R

Rad *abbr* (*radiation absorbed dose*) ANALYTIK rad (*radiation absorbed dose*)

Radarsonde *f* ANALYTIK radar probe

Radial- *in cpds* ABWASSER, LUFT, MEER, WASSER centrifugal, radial; **Radialkreiselpumpe** *f* ABWASSER radial-flow pump; **Radialrechen** *m* ABWASSER horizontal bar screen; **Radialströmung** *f* LUFT *Zyklon* radial flow; **Radialstromwäscher** *m* LUFT radial-flow scrubber; **Radialventilator** *m* LUFT centrifugal fan; **Radialverdichter** *m* LUFT centrifugal compressor

radiation: **~ absorbed dose** *f* (*Rad*) ANALYTIK radiation absorbed dose (*rad*); **~ equivalent man** *f* (*Rem*) ANALYTIK radiation equivalent man (*rem*)

Radikalbildung *f* LUFT radical formation

Radio- *in cpds* ANALYTIK radio

radioaktiv *adj* ALLG radioactive; **~er Abfall** *m* ABFALL, SICHERHEIT nuclear waste, radioactive waste; **~e Abwasser** *nt* ABWASSER radioactive wastewater; **~es Edelgas** *nt* LUFT radioactive rare gas; **~e Emissionen** *f pl* LUFT radioactive emissions; **~es Gas** *nt* LUFT radioactive gas; **~ kontaminiertes Wasser** *nt* ABFALL, ABWASSER, WASSER radioactive contaminated water; **~er Niederschlag** *m* ABWASSER radioactive fallout, LUFT radioactive fallout, rainout; **~er Staub** *m* LUFT radioactive dust; **~er Stoff** *m* ABFALL, ABWASSER, SICHERHEIT radioactive substance; **~e Verschmutzung** *f* ABFALL, ABWASSER, LUFT, MEER, SICHERHEIT radioactive pollution; **~e Verseuchung** *f* SICHERHEIT radioactive pollution; **~er Zerfall** *m* ERNEU ENER *Geothermie* radioactive decay

Radiometer *nt* ANALYTIK radiometer

Radiometrie *f* ANALYTIK radiometry

radiometrisch *adj* ANALYTIK radiometric

Radio-: **Radiophotolumineszenz** *f* ANALYTIK radiophotoluminescence; **Radiopolarographie** *f* ANALYTIK radiopolarography; **Radiowelle** *f* ANALYTIK radiowave

Radlader *m* ABFALL, BODEN wheel loader

Radon *nt* ANALYTIK radon; **Radonkonzentration** *f* ANALYTIK radon concentration

Raffinageprodukt *nt* ABWASSER, LUFT, MEER refined product

Raffination *f* ABFALL, LUFT, MEER refining

Raffineriegas *nt* LUFT refinery gas

Raffinerierückstände *m pl* ABFALL refinery waste

raffinieren *vt* ABFALL, LUFT, MEER refine

raffiniert *adj* ABFALL, LUFT, MEER refined; **~es Produkt** *nt* ABWASSER, LUFT, MEER refined product

Rahmenfilterpresse *f* ABWASSER frame filter press

Rainout *nt* LUFT rainout

Raketensonde *f* ANALYTIK rocket probe

rammen *vt* ABFALL pile drive

Rammkernsondierung *f* ANALYTIK pile core sampling

Rammung *f* ABFALL pile-driving

Rand: **Randbereich** *m* ABFALL, WASSER boundary area; **Randgraben** *m* ABFALL, BODEN peripheral ditch; **Randschrift** *f* POLITIK *des Umweltzeichens* explanatory clause; **Randwall** *m* ABFALL boundary wall

Rankine-Kreislaufprozeß-Dampfmaschine *f* LUFT Rankine cycle steam engine

Rapsöl *nt* ERNEU ENER rape seed oil

Raschig-Ring *m* ABWASSER, LUFT raschig ring

Raster- *in cpds* ANALYTIK scanning; **Rasterdampfsperre** *f* LUFT chevron baffle; **Rasterelektronenmikroskop** *nt* ANALYTIK scanning electron microscope (*SEM*); **Rasterelektronenmikroskopie** *f* ANALYTIK scanning electron microscopy (*SEM*)

rasterförmig *adj* ANALYTIK grid-shaped

Raster-: **Rastermeßnetz** *nt* ANALYTIK scanning network; **Rastertunnelmikroskop** *nt* ANALYTIK scanning tunnel microscope; **Rasterverfahren** *nt* ANALYTIK scanning method

ratifizieren *vt* POLITIK ratify

Rauch *m* LUFT, SICHERHEIT smoke; **Rauchabzugsanlage** *f* LUFT, SICHERHEIT smoke control system; **Rauchabzugskanal** *m* LUFT flue; **Rauchbekämpfung** *f* LUFT smoke abatement; **Rauchdichtemesser** *m* ANALYTIK smoke density meter; **Rauchemissionen** *f pl* LUFT smoke emissions

räuchern *vt* LUFT fumigate

Rauch: **Rauchfahne** *f* LUFT plume, smoke plume, smoke streamer

Rauchgas *nt* LUFT flue gas, fumes; **Rauchgasabkühlung** *f* LUFT flue-gas cooling; **Rauchgasdurchsatz** *m* LUFT flue-gas throughput, flue-gas thruput (*AmE*); **Rauchgaseintritt** *m* LUFT fume inlet; **Rauchgasentschwefelung** *f* LUFT, SICHERHEIT flue-gas desulfurization (*AmE*) (*FGD*), flue-gas desulphurization (*BrE*) (*FGD*); **Rauchgasentschwefelungsanlage** *f* (*REA*) LUFT flue-gas desulfurization plant (*AmE*) (*FGD plant*), flue-gas desulphurization plant (*BrE*) (*FGD plant*); **Rauchgasentstaubung** *f* LUFT dust removal from flue gas, particulate collection; **Rauchgasentstickung** *f* LUFT flue-gas nitrogen removal; **Rauchgasfahne** *f* LUFT flue-gas plume; **Rauchgasfilter** *nt* LUFT smoke filter; **Rauchgasgebläse** *nt* LUFT flue-gas blower; **Rauchgasprüfer** *m* ANALYTIK flue-gas tester; **Rauchgasreinigung** *f* LUFT flue-gas cleaning; **Rauchgasreinigungsanlage** *f* LUFT flue-gas cleaning installation; **Rauchgasreinigungsrückstand** *m* ABFALL, LUFT flue-gas cleaning residue; **Rauchgasrezirkulation** *f* LUFT flue-gas recirculation; **Rauchgasrichtung** *f* LUFT direction of flue gas; **Rauchgasseite** *f* LUFT flue-gas side; **Rauchgasstrom** *m* LUFT flue-gas stream; **Rauchgastemperatur** *f* ABFALL, LUFT waste gas temperature; **Rauchgasverbrennungsofen** *m* LUFT fume incinerator; **Rauchgaswärmetauscher** *m* LUFT flue-gas heat exchanger; **Rauchgaswäsche** *f* LUFT flue-gas scrubbing; **Rauchgaswiederaufheizung** *f* LUFT flue-gas reheating; **Rauchgaswolke** *f* LUFT flue-gas cloud

Rauch: **Rauch- und Hitzeabzugsanlage** *f* SICHERHEIT smoke and heat exhaust installation; **Rauchkonzentration** *f* LUFT smoke concentration; **Rauchschattenverfahren** *nt* ANALYTIK smoke shade method; **Rauchschwaden** *m* LUFT plume; **Rauchwolke** *f* LUFT smoke plume

Raum- *in cpds* ABFALL, ANALYTIK, LÄRM, LUFT space; **Raumakustik** *f* LÄRM room acoustics; **Raumauskleidung** *f* LÄRM room lining

Räumer *m* ABWASSER scraper; **Räumerbrücke** *f* ABWASSER scraper bridge

Raum-: **Raumgeschwindigkeit** *f* LUFT space velocity; **Raumgewicht** *nt* ABFALL *des Mülls* volume weight; **Raumladung** *f* LUFT *Elektrofilter* volume charge; **Raumladungsdichte** *f* LUFT *Elektrofilter* density of volume charge

räumlich: **~e Auflösung** *f* ANALYTIK spatial resolution; **~e Auswertung** *f* ANALYTIK spatial evaluation; **~e Struktur** *f* ANALYTIK spatial pattern; **~e Tendenz** *f* ABFALL, LUFT, WASSER spatial trend; **~e Veränderlichkeit** *f* ANALYTIK,

LUFT spatial variability; **~e Verteilung** *f* ANALYTIK spatial distribution
Raum-: **Raumtemperatur** *f* ANALYTIK room temperature
Räumvorgang *m* ABWASSER scraping operation
Raum-: **Raum-Zeit-Ausbeute** *f* ANALYTIK space-time yield
Raupe *f* BODEN caterpillar
Rauschpegel *m* ANALYTIK, LÄRM background level
Rayleigh: **~sche Streustrahlung** *f* ANALYTIK, LUFT Rayleigh scattering; **Rayleigh-Refraktometer** *nt* ANALYTIK Rayleigh refractometer; **Rayleigh-Scheibe** *f* ANALYTIK Rayleigh disc (*BrE*), Rayleigh disk (*AmE*)
Re *abbr* (*Reynoldsche Zahl*) LUFT, WASSER Re (*Reynold's number*)
REA *abbr* (*Rauchgasentschwefelungsanlage*) LUFT FGD plant (*flue-gas desulfurization plant AmE, flue-gas desulphurization plant BrE*); **REA-Gebläse** *nt* LUFT FGD plant blower
Reagens *nt* ANALYTIK, BODEN, LUFT reagent
Reagenz *nt* ANALYTIK, BODEN, LUFT reagent; **Reagenzglas** *nt* ANALYTIK test tube; **Reagenzlösung** *f* ANALYTIK reagent solution
reagieren *vi* ALLG react
REA: **REA-Gips** *m* LUFT FGD plant gypsum
Reaktand *m* LUFT reactant
Reaktion *f* ALLG reaction; **Reaktionsbecken** *nt* ABWASSER reaction tank
reaktionsfähig *adj* LUFT reactive; **~e Partikel** *f* LUFT reactive particle
Reaktion: **Reaktionsgeschwindigkeit** *f* ABFALL, ABWASSER, ANALYTIK, LUFT reaction rate, reaction velocity; **Reaktionsgleichgewicht** *nt* LUFT reaction equilibrium; **Reaktionsharzbeton** *m* ABFALL, WASSER polymer concrete (*PC*); **Reaktionsindikator** *m* ANALYTIK reaction indicator; **Reaktionskammer** *f* ANALYTIK, LUFT reaction chamber; **Reaktionsmittel** *nt* ABFALL, ANALYTIK solidifying agent; **Reaktionsoberfläche** *f* LUFT reaction surface; **Reaktionspartner** *m* ANALYTIK reacting agent, LUFT reactant; **Reaktionsprodukt** *nt* ANALYTIK, LUFT reaction product; **Reaktionsrohr** *nt* ANALYTIK reaction tube; **Reaktionsverzögerer** *m* ABFALL retarder, retarding agent; **Reaktionsverzögerung** *f* LUFT negative catalysis; **Reaktionswärme** *f* ANALYTIK heat of reaction; **Reaktionszeit** *f* LUFT reaction time; **Reaktionszone** *f* ABWASSER reaction zone
reaktiv *adj* ABFALL, LUFT reactive; **~er Sondermüll** *m* ABFALL reactive hazardous waste
reaktivierbar *adj* ABFALL, BODEN, ERNEU ENER, POLITIK reactivatable
reaktivieren *vt* ABFALL, BODEN, ERNEU ENER, POLITIK reactivate
reaktiviert *adj* ABFALL, BODEN, ERNEU ENER, POLITIK reactivated
Reaktivierung *f* ABFALL, BODEN, ERNEU ENER, POLITIK reactivation
Reaktivität *f* ABFALL, BODEN, ERNEU ENER, POLITIK reactivity
Reaktor *m* ABWASSER, LUFT reactor; **~ mit Katalysator** LUFT reactor with catalyst; **Reaktorboden** *m* ABWASSER reactor bottom; **Reaktordrucksensor** *m* LUFT reactor pressure sensor
Rechen: **Rechenanlage** *f* ABWASSER screening plant; **Rechengefällsverlust** *m* ABWASSER screen loss of head
Rechengut *nt* ABWASSER screenings; **Rechengutklassierer** *m* ABWASSER rakings classifier, screenings classifier; **Rechengutpresse** *f* ABWASSER screenings bale press; **Rechengutzerkleinerer** *m* ABWASSER comminutor
Rechen: **Rechenreiniger** *m* ABWASSER screen cleaner; **Rechenreinigung** *f* ABWASSER screen cleaning; **Rechenreinigungsmaschine** *f* ABWASSER screen cleaner; **Rechenrinne** *f* ABWASSER screen gutter; **Rechenverlust** *m* ABWASSER screen loss; **Rechenverstopfung** *f* ABWASSER screen clogging
rechnergestützt: **~e GC-MS** *f* ANALYTIK gas chromatography-mass spectrometry computer techniques (*GC-MS-COMP techniques*)
rechnerisch: **~ ermitteln** *vt* ANALYTIK determine mathematically
Recht *nt* POLITIK law, legal system; **~ auf Informationen** POLITIK *Umweltinformationen* right-to-know principle
Rechteckbecken *nt* ABWASSER rectangular tank, rectangular clarifier, rectangular basin
rechtlich *adj* POLITIK lawful, legal; **~e Auflagen** *f pl* POLITIK legal requirements
recyceln *vt* ABFALL recycle
recycelt: **~e Flasche** *f* ABFALL recycled bottle; **~es Papier** *nt* ABFALL recycled paper
Recyclat *nt* ABFALL recycled material
Recyclate *nt pl*: **~ aus Mehrschichtfolien** ABFALL multifoil recycled material
Recycling *nt* ABFALL, ABWASSER, MANAG, WASSER recycling; **~ von Multimaterialien** ABFALL multimaterial recycling; **~ von Schlämmen** ABWASSER sludge recycling; **Recyclinganlage** *f* ABFALL recycling plant
recyclingfähig *adj* ABFALL recyclable; **~e Batterie** *f* ABFALL recyclable battery
Recycling: **Recyclingfähigkeit** *f* ABFALL recyclability
recyclingfreundlich *adj* ABFALL recycling-friendly; **~es Produkt** *nt* ABFALL recycling-friendly product
Recycling: **Recyclinggebot** *nt* ABFALL, WASSER obligation to recycle
recyclinggerecht *adj* ABFALL recycling-oriented
Recycling: **Recyclinghindernis** *nt* ABFALL impediment to recycling; **Recyclinghof** *m* ABFALL recycling center (*AmE*), recycling centre (*BrE*); **Recyclingkosten** *pl* MANAG recycling costs; **Recyclingmaterial** *nt* ABFALL recycled material, recycling material; **Recyclingpapier** *nt* ABFALL recycled paper, recycling paper; **Recyclingpotential** *nt* ABFALL recycling potential; **Recyclingquote** *f* ABFALL recycling rate; **Recyclingraupe** *f* ABFALL recycling caterpillar; **Recyclingumfang** *m* ABFALL recycling quantity; **Recyclingverfahren** *nt* ABFALL recycling process; **Recyclingvorgang** *m* ABFALL recycling operation
Redox-Potential *nt* ABWASSER, ANALYTIK, LUFT redox potential, reduction intensity, oxidation reduction potential (*ORP*)
Redox-Prozeß *m* LUFT oxidation reduction process
Reduktion *f* ABFALL, ABWASSER, LUFT reduction; **Reduktionsgas** *nt* LUFT reducing gas; **Reduktionskatalysator** *m* LUFT reduction catalyst; **Reduktionsmittel** *nt* ABWASSER, ANALYTIK, LUFT reducing agent; **Reduktions-Oxidations-Potential** *nt* (*rH-Wert*) ABWASSER, ANALYTIK, LUFT reduction-oxidation potential (*rH-value*); **Reduktionsprodukt** *nt* ABFALL reducing product; **Reduktionsstoff** *m* LUFT reducing substance
reduzieren *vt* ABFALL, ABWASSER, ANALYTIK reduce
reduzierend *adj* ABFALL, LUFT reducing; **~er Katalysator** *m* LUFT reducing catalyst; **~es Katalysatorbett** *nt* LUFT reducing catalyst bed; **~e Substanz** *f* LUFT reducing substance
reduziert *adj* ANALYTIK reduced: **~er Satz** *m* POLITIK reduced rate of charge
Refiltrationsflockung *f* WASSER refiltration flocculation
reflektieren *vt* LÄRM reflect
reflektiert *adj* LÄRM reflected; **~er Schall** *m* LÄRM reflected sound
Reflektometer *nt* ANALYTIK, BODEN reflectometer
Reflex *m* MEER sheen
Reflexion *f* ANALYTIK, LÄRM reflection
reflexionsfrei *adj* LÄRM anechoic
Reflexion: **Reflexionsschall** *m* LÄRM secondary sound; **Reflexionsschalldämpfer** *m* LÄRM resonant-absorption

silencer, baffle silencer; **Reflexionsspektroskopie** *f* ANALYTIK reflection spectroscopy
Reformierungsreaktion *f* LUFT reforming reaction
refraktär: **~er Stoff** *m* ABWASSER refractory substance
REGAVO *abbr* (*regenerativer Gasvorwärmer*) LUFT regenerative gas preheater
Regeleinheit *f* ALLG regulating unit
Regelung *f* ALLG regulation; **Regelungssytem** *nt* ERNEU ENER, LUFT, SICHERHEIT, WASSER control system
Regelventil *nt* WASSER flow-regulating valve
Regenerat *nt* ABFALL reclaimed material, reground material, regrind
Regeneration *f* ALLG regeneration, reactivation; **~ von Lösungsmitteln** BODEN solvent regeneration; **Regenerationsbehälter** *m* ABWASSER regeneration basin; **Regenerationsfähigkeit** *f* ABWASSER regeneration capacity; **Regenerationsmiete** *f* BODEN regeneration pit
regenerativ *adj* ALLG renewable, regenerative; **~e Energie** *f* ERNEU ENER renewable energy; **~e Energiequelle** *f* ERNEU ENER renewable energy source; **~er Energieträger** *m* ERNEU ENER renewable energy source; **~er Gasvorwärmer** *m* (*REGAVO*) LUFT regenerative gas preheater; **~e Heizungstechnik** *f* ERNEU ENER regenerative heating technology; **~er Luftvorwärmer** *m* LUFT regenerative air preheater; **~e Nachverbrennung** *f* (*RNV*) LUFT regenerative incineration; **~es Reagenz** *nt* LUFT regenerative reagent; **~er Wärmeaustauscher** *m* LUFT regenerative heat exchanger
Regenerativ-: **Regenerativgas** *nt* LUFT *Inertgas* regenerative gas; **Regenerativgasgebläse** *nt* LUFT regenerative gas fan; **Regenerativpumpe** *f* ABWASSER regenerative pump
regenerierbar *adj* LUFT regenerative; **~e Rauchgasentschwefelung** *f* LUFT regenerable flue-gas desulfurization (*AmE*), regenerable flue-gas desulphurization (*BrE*); **nicht ~e Rauchgasentschwefelung** *f* LUFT nonregenerable flue-gas desulfurization (*AmE*), nonregenerable flue-gas desulphurization (*BrE*)
regenerieren *vt* ALLG reactivate, regenerate; **mikrobiell ~** *vt* LUFT microbially regenerate
regeneriert *adj* ALLG reactivated, regenerated; **~es Adsorbens** *nt* LUFT regenerated adsorbent; **~e Aktivkohle** *f* LUFT regenerated activated carbon; **~er Aktivkoks** *m* LUFT regenerated activated coke; **~es Lösungsmittel** *nt* LUFT regenerated solvent
Regenerierung *f* ABWASSER, LUFT regeneration, WASSER regeneration; **Regenerierungsmittel** *nt* ABFALL, ABWASSER, BODEN, LUFT, WASSER regenerant
regenfrei: **~e Zeit** *f* LUFT rain-free period
Regen: **Regenklärbecken** *nt* ABWASSER stormwater reservoir, stormwater tank, WASSER stormwater reservoir; **Regenmesser** *m* WASSER rain gage (*AmE*), rain gauge (*BrE*); **Regenrückhaltebecken** *nt* ABFALL, ABWASSER stormwater retention tank; **Regenüberlauf** *m* ABWASSER, WASSER stormwater overflow; **Regenüberlaufbecken** *nt* ABWASSER, WASSER stormwater overflow tank
Regenwasser *nt* ABWASSER, WASSER rainwater, stormwater; **Regenwasserabfluß** *m* ABWASSER, WASSER stormwater flow; **Regenwasserbecken** *nt* ABWASSER, WASSER stormwater tank; **Regenwasserbehandlung** *f* ABWASSER, WASSER stormwater treatment; **Regenwasserkanal** *m* ABWASSER, WASSER stormwater sewer; **Regenwasserkanalisation** *f* ABWASSER, WASSER stormwater main; **Regenwasserleitung** *f* ABWASSER, WASSER stormwater main; **Regenwassernutzung** *f* WASSER rainwater utilization; **Regenwasserpumpe** *f* ABWASSER, WASSER stormwater pump
Regierung *f* POLITIK government; **Regierungsbezirk** *m* POLITIK land district; **Regierungspräsident** *m* POLITIK land district president; **Regierungsprognose** *f* POLITIK governmental projection
Regionalpolitik *f* POLITIK regional policy
registrieren *vt* ANALYTIK, POLITIK record, register
registrierend *adj* ANALYTIK recording; **~es Spektralphotometer** *nt* ANALYTIK recording spectrophotometer
Registriergerät *nt* ANALYTIK recording device
Registrierstreifen *m* ANALYTIK recording strip
Registrierung *f* POLITIK recording, registration
Regranulat *nt* ABFALL regranulate
Regranulation *f* ABFALL regranulation
regranulieren *vt* ABFALL regranulate
Regranulierextruder *m*: **~ für Abfallaufbereitung** ABFALL regranulating extruder
Regressionskurve *f* ANALYTIK regression curve
regressiv *adj* POLITIK regressive; **~e Steuer** *f* POLITIK regressive tax
Regularisation *f* LUFT, WASSER regularization
Regulierapparat *m* ANALYTIK regulator
regulierbar: **~es Grundwehr** *nt* WASSER adjustable submerged weir
regulieren *vt* ANALYTIK, POLITIK regulate
reguliert *adj* ANALYTIK, POLITIK regulated
Reibung *f* ABFALL, ABWASSER, LUFT friction; **Reibungshöhe** *f* ABFALL, ABWASSER friction head; **Reibungskraft** *f* LUFT frictional force; **Reibungsschicht** *f* LUFT friction layer
Reichgasverarbeitung *f* LUFT rich gas processing
Reichweite *f* ANALYTIK range
Reifekompost *m* ABFALL mature compost
Reifung *f* ABWASSER *Schlammkonditionerung* ripening
Rein- *in cpds* LUFT, MEER, WASSER clean, pure
rein *adj* ALLG pure, clean; **~es Wasser** *nt* ABWASSER, WASSER pure water, clean water
Reingas *nt* LUFT clean gas, cleaned gas, scrubbed gas; **Reingasaustritt** *m* LUFT clean gas outlet; **Reingasdeckel** *m* LUFT clean gas lid; **Reingaskanal** *m* LUFT clean gas canal; **Reingaskonzentration** *f* LUFT clean gas concentration; **Reingasraum** *m* LUFT clean gas compartment; **Reingasstaubgehalt** *m* LUFT clean gas dust content; **Reingaswiederaufheizung** *f* LUFT clean gas reheating
Reinhaltung *f* ABFALL, ABWASSER, LUFT, MEER, POLITIK pollution control, SICHERHEIT pollution deposition, WASSER pollution control
Reinheitsgrad *m* WASSER degree of purity
reinigen *vt* ALLG clean, cleanse, purify; **hydraulisch ~** *vt* ABFALL, ABWASSER, WASSER clean hydraulically
Reiniger *m* SICHERHEIT cleaner
Reinigung *f* ALLG cleaning, cleansing, purification; **~ der Wärmetauscher** ABFALL, LUFT cleaning of heat exchanger; **Reinigungsanlage** *f* ABWASSER purification plant; **Reinigungsbehälter** *m* ABWASSER filtration vat; **Reinigungsdüse** *f* LUFT cleaning nozzle; **Reinigungsfällung** *f* LUFT below-cloud scavenging; **Reinigungsgeräte** *nt pl* SICHERHEIT cleaning equipment; **Reinigungsgrad** *m* ABWASSER degree of purification, BODEN degree of decontamination; **Reinigungshahn** *m* WASSER purging cock; **Reinigungsleistung** *f* ABWASSER, LUFT cleaning capacity; **Reinigungsstufe** *f* ABWASSER treatment stage; **Reinigungsverfahren** *nt* MEER clean-up technique; **Reinigungsvermögen** *nt* ABWASSER purification capacity; **Reinigungsvorgang** *m* ABWASSER purification operation, LUFT cleaning operation; **Reinigungswirkungsgrad** *m* BODEN clean-up efficiency
Rein-: **Reinluft** *f* LUFT unpolluted air, clean air; **Reinluftansaugung** *f* LUFT clean air suction; **Reinluftgebiet** *nt* LUFT area of pure air; **Reinraum** *m* ANALYTIK clean room; **Reinregen** *m* LUFT clean rain; **Reinsauerstoff** *m* LUFT, WASSER pure oxygen
Reinstwasser *nt* BODEN ultrapure water
Rein-: **Reinwasser** *nt* ABWASSER, WASSER clear water, pure water; **Reinwasserbehälter** *m* WASSER clean water tank

Reißdehnungsprüfgerät *nt* LUFT *Filter* extension break tester
Reißen *nt* LUFT *Filter* tearing
Reißfestigkeit *f* ABFALL bursting strength, LUFT *Filter* tear strength, MEER resistance to tearing
Reißwiderstand *m* LUFT *Filter* tear strength
Reizgas *nt* LUFT tear gas
Rekalibration *f* ANALYTIK recalibration
rekontaminieren *vt* ABWASSER recontaminate
Rektifikation *f* ABWASSER, LUFT rectification
Rektifizierung *f* ABWASSER, LUFT rectification
rekultivieren *vt* ABFALL, ABWASSER, BODEN reclaim, recultivate, revegetate
rekultiviert *adj* ABFALL, ABWASSER, BODEN, LUFT reclaimed
Rekultivierung *f* ABFALL, BODEN reclamation, revegetation, recultivation; **Rekultivierungsschicht** *f* BODEN recultivation, revegetationrecultivation layer
rekuperativ *adj* LUFT recuperative; **~e Wärmerückgewinnung** *f* LUFT recuperative heat recovery; **~er Wärmetauscher** *m* LUFT recuperative heat exchanger
Rekuperator *m* LUFT heat exchanger, recuperator
relativ: **~e Feuchte** *f* ANALYTIK relative humidity
Relativgeschwindigkeit *f* LUFT *von Partikel zum Tropfen* relative speed
rem *nt* (*radiation equivalent man*) ANALYTIK rem (*radiation equivalent man*)
Remobilisierung *f* WASSER remobilization
Reparierbarkeit *f* ABFALL *der Deponie* reparability
Replastifizierung *f* ABFALL *Recyclingasphalt* replastification
Reproduzierbarkeit *f* ANALYTIK consistency
Reserve *f* MEER stand-by
Reservoir *nt* ABWASSER reservoir, water tank, ERNEU ENER reservoir, WASSER reservoir, water tank; **Reservoirreinigung** *f* ABFALL, WASSER reservoir cleaning
Residuum *nt* MEER residue
resistent *adj* ABFALL, BODEN, WASSER resistant
Resonanz *f* LÄRM resonance; **Resonanzabsorber** *m* LÄRM resonance absorber; **Resonanzabsorption** *f* LÄRM resonance absorption
resonanzaktiv: **~es Ion** *nt* ANALYTIK resonant ion
resonanzfähig *adj* LÄRM able to resonate
Resonanz: **Resonanzschalldämpfer** *m* LÄRM, SICHERHEIT reactive attenuator, reactive resonator, reactive silencer; **Resonanzschallschlucker** *m* LÄRM resonance sound absorber; **Resonanzsieb** *nt* LÄRM resonance screen; **Resonanzwandler** *m* LÄRM resonance transducer
Resonator *m* LÄRM resonator
Ressource *f* ABWASSER, BODEN, MEER, WASSER resource
Ressourcenschonung *f* BODEN, WASSER protection of resources, resource conservation
restabilisiert *adj* ABWASSER restabilized; **~es Teilchen** *nt* ABWASSER restabilized particle
Restabilisierung *f* ABWASSER restabilization
Rest-: **Restaktivität** *f* LUFT *Katalysatort* residual activity; **Restbeladung** *f* LUFT residue load; **Restbelastung** *f* **des Bodens mit Schadstoffen** BODEN residual impact of soils with contaminants; **Restemissionen** *f pl* ABFALL, ABWASSER, LUFT residue emissions; **Restentstaubung** *f* LUFT remaining dust removal; **Restfeuchtigkeit** *f* ABFALL, ANALYTIK residual moisture; **Restgas** *nt* ABFALL, LUFT residual gas, tail gas; **Restkonzentration** *f* ABWASSER, WASSER residue concentration; **Restmengenpumpe** *f* MEER stripping pump; **Restrisiken** *nt pl* MANAG, SICHERHEIT remaining risks; **Restsauerstoff** *m* LUFT residual oxygen; **Restschall** *m* LÄRM residual sound; **Reststaubbeladung** *f* LUFT residue dust loading; **Reststaubgehalt** *m* LUFT residual dust content
Reststoff *m* ABFALL residue; **~e** *m pl* SICHERHEIT residues; **Reststoffanfall** *m* ABFALL residue yield
reststoffarm *adj* ABFALL low-residue; **~es Produktionsverfahren** *nt* ABFALL low-residue production method
Reststoff: **Reststoffdeponie** *f* ABFALL, ABWASSER residue landfill; **Reststoffdeponierung** *f* ABFALL, ABWASSER residue landfilling
reststofffrei *adj* ABFALL residue-free; **~es Produktionsverfahren** *nt* ABFALL residue-free production method
Reststoff: **Reststoffvermeidung** *f* **an der Quelle** ABFALL residue avoidance at source; **Reststoffverwertungsgebot** *nt* POLITIK obligation to reuse residue materials
Rest-: **Restverschmutzung** *f* ABWASSER, BODEN, WASSER residual pollution
Resuspension *f* WASSER resuspension
Retentionsbecken *nt* ABWASSER retention tank
reversibel *adj* ALLG reversible; **reversibles Verfahren** *nt* ABWASSER *Chemisorption*, LUFT *Physisorption* reversible process
Revitalisierung *f* BODEN revitalization
Reynold: **~sche Zahl** *f* (*Re*) LUFT, WASSER Reynold's number (*Re*)
RFA *abbr* (*Röntgenfluoreszenzanalyse*) ANALYTIK XRFA (*x-ray fluorescence analysis*)
Rh *abbr* (*Rhodium*) LUFT Rh (*rhodium*)
Rheometer *nt* ANALYTIK rheometer
Rhodium *nt* (*Rh*) LUFT rhodium (*Rh*)
rH-Wert *m* (*Reduktions-Oxidations-Potential*) ABWASSER, ANALYTIK, LUFT rH-value (*reduction-oxidation potential*)
Richtlinie *f* POLITIK directive, guideline
Richtungsänderung *f* LUFT direction change
Richtungsumkehr *f* LUFT *Zyklon* direction change
Richtwert *m* ANALYTIK approximate reading, approximate value, POLITIK guide value
riechstoffbildend *adj* ABFALL, LUFT, WASSER odorous
Riemenantrieb *m* ABWASSER, WASSER *an Pumpe* belt drive
rieselfähig *adj* ABFALL pourable, ABWASSER irrigable, WASSER pourable
Riesel: **Rieselfeld** *nt* ABWASSER wastewater farm; **Rieselfilm** *m* ABWASSER trickling film, biofilm surface, LUFT *Naßabscheider*, WASSER trickling film; **Rieselfilmoberfläche** *f* ABWASSER, WASSER trickling film surface; **Rieselfilmströmung** *f* ABWASSER, WASSER *Füllkörpersäule* trickling film flow; **Rieselfiltration** *f* ABWASSER dry filtration
rieseln *vi* ABFALL, ABWASSER, LUFT, WASSER trickle
Rigole *f* ABFALL, BODEN trench
Rigolen *nt* ABFALL, BODEN trenching
Ring: **Ringadsorber** *m* LUFT ring adsorber; **Ringdüse** *f* LUFT ring nozzle; **Ringdüsenelement** *nt* LUFT ring jet element
Ringelmann-Skala *f* ANALYTIK, LUFT *Staubmessung* Ringelmann scale
Ring: **Ring-Jet-Aerosol-Abscheider** *m* LUFT ring jet aerosol precipitator; **Ringleitung** *f* WASSER circular main; **Ringraum** *m* LUFT annulus; **Ringschicht** *f* LUFT ringed layer; **Ringspalt** *m* BODEN annular gap, LUFT annular gap, annulus, *Venturidüse* expander; **Ringspaltabscheider** *m* LUFT nozzle separator; **Ringspaltwäscher** *m* LUFT annular-gap scrubber, annular-type scrubber
Rinne *f* ABFALL, ABWASSER flume
rinnen *vi* ABFALL, ABWASSER trickle
Rinnenspülung *f* ABFALL, ABWASSER flume flushing
Rippenrohrluftvorwärmer *m* LUFT finned tube air heater
Rippenrohrluvo *abbr* (*Rippenrohrluftvorwärmer*) LUFT finned tube air heater
Risiko *nt* ALLG hazard, risk; **Risikoabschätzung** *f* ALLG risk assessment, risk estimation; **Risikoanalyse** *f* ALLG risk analysis; **Risikobewertung** *f* ALLG risk assessment, risk estimation; **Risikofaktor** *m* ALLG risk factor; **Risikopotential** *nt* ALLG risk potential
RNV *abbr* (*regenerative Nachverbrennung*) LUFT regenerative incineration

Robustheit *f* ABFALL robustness, ANALYTIK *Analysegeräte* ruggedness

Roh- *in cpds* ALLG raw; **Rohabgas** *nt* LUFT raw waste gas

Rohabwasser *nt* ABWASSER raw wastewater, untreated wastewater; **Rohabwasserqualität** *f* ABWASSER raw wastewater quality; **Rohabwasserversorgung** *f* ABWASSER raw wastewater supply

Roh-: **Rohboden** *m* BODEN raw soil; **Rohbrei** *m* ABFALL crude pulp; **Rohdaten** *pl* ANALYTIK raw data; **Rohdatenspeicherung** *f* ANALYTIK raw data storage

Rohgas *nt* LUFT dirty gas; **Rohgasabkühlung** *f* LUFT dirty gas cooling; **Rohgasbeladung** *f* LUFT dirty gas loading; **Rohgaseintritt** *m* LUFT dirty gas inlet; **Rohgasfracht** *f* LUFT dirty gas load; **Rohgaskanal** *m* LUFT dirty gas canal; **Rohgaskonditionierung** *f* LUFT dirty gas conditioning; **Rohgaskonzentration** *f* LUFT dirty gas concentration; **Rohgasraum** *m* LUFT dirty gas compartment; **Rohgasstaubbeladung** *f* LUFT dirty gas dust loading; **Rohgasstaubgehalt** *m* LUFT dirty gas dust content; **Rohgasstrom** *m* LUFT dirty gas flow; **Rohgasvolumenstrom** *m* LUFT dirty gas volume flow

Roh-: **Rohgips** *m* LUFT raw gypsum; **Rohkompost** *m* ABFALL raw compost; **Rohluftzufuhr** *f* LUFT raw air conveyance; **Rohmeßwert** *m* ANALYTIK raw-measured value; **Rohmüll** *m* ABFALL untreated refuse, crude refuse; **Rohöl** *nt* ABFALL, MEER crude oil; **Rohplanum** *nt* ABFALL *einer Deponie* unfinished support; **Rohprobe** *f* ANALYTIK raw sample

Rohr *nt* ALLG pipe, tube; **Rohranschluß** *m* ABWASSER, WASSER pipe joint; **Rohrbelüfter** *m* WASSER pipe aerator; **Rohrbruch** *m* ABWASSER, WASSER pipe break, pipe breakage; **Rohrbrücke** *f* ABWASSER pipe bridge; **Rohrbündel** *nt* LUFT tube bundle; **Rohrbündelwärmeaustauscher** *m* LUFT shell-and-tube heat exchanger; **Rohrdurchbruch** *m* WASSER pipe bursting

Röhre *f* ANALYTIK tube

Rohr: **Rohrelektrofilter** *nt* LUFT tubular ESP

Röhre: **Röhrendränung** *f* WASSER tubular drainage

Rohr: **Rohrendstrang** *m* WASSER branch end, pipe end

röhrenförmig *adj* LUFT tubular; **~er Staubabscheider** *m* LUFT tubular precipitator

Röhre: **Röhrenkondensator** *m* **mit Mantel** LUFT shell-and-tube condenser; **Röhrensystem** *nt* WASSER piping system

Rohr: **Rohrgraben** *m* ABWASSER sewer trench; **Rohrkäfigrekuperator** *m* LUFT tube-cage recuperator; **Rohrleckage** *f* ABWASSER, WASSER pipe leakage

Rohrleitung *f* ABWASSER, WASSER conduit, pipeline; **Rohrleitungsstandsicherheit** *f* ABFALL pipe stability; **Rohrleitungwiderstand** *m* ABWASSER, WASSER pipeline resistance

Rohr: **Rohrmuffe** *f* ABWASSER, WASSER pipe sleeve, socket

Rohrnetz *nt* WASSER grid of pipes, network of pipes; **Rohrnetzberechnung** *f* WASSER network calculation; **Rohrnetzspülung** *f* ABWASSER, WASSER flushing of pipe network; **Rohrnetzverlust** *m* WASSER water loss within distribution system

Rohr: **Rohrstützblech** *nt* LUFT tube support plate; **Rohrstutzen** *m* LUFT connecting branch; **Rohrsystem** *nt* WASSER piping system; **Rohrwand** *f* WASSER tube wall; **Rohrwärmeaustauscher** *m* **mit Mantel** LUFT shell-and-tube heat exchanger; **Rohrzug** *m* LUFT duct

Roh-: **Rohschlacke** *f* ABFALL untreated slag; **Rohschlamm** *m* ABWASSER raw sludge, primary sludge, untreated sludge, undigested sludge

Rohstoff *m* ABFALL, ERNEU ENER, MANAG raw material; **Rohstoffabbau** *m* ABFALL raw material decomposition; **Rohstoffbedarf** *m* ABFALL demand for raw material; **Rohstoffgewinnung** *f* ERNEU ENER *Biokonversion* raw material generation, MANAG raw material extraction; **Rohstoffindustrie** *f* MANAG primary resources industry, raw material delivering industry

rohstofflich: **~es Recycling** *nt* ABFALL raw material recycling; **~es Recyclingverfahren** *nt* ABFALL raw material recycling process; **~e Verwertung** *f* ABFALL raw material recycling

Rohstoff: **Rohstoffressourcen** *f pl* ABFALL raw material ressources; **Rohstoffrückgewinnung** *f* ABFALL resource recovery; **Rohstoffverbrauch** *m* ABFALL resource consumption

Rohwasser *nt* ABWASSER, WASSER raw water, untreated water; **Rohwasserbecken** *nt* ABWASSER, WASSER untreated water tank; **Rohwasserleitung** *f* ABWASSER, WASSER raw water main; **Rohwasserpumpe** *f* ABWASSER, WASSER raw water pump; **Rohwasserqualität** *f* ABWASSER, WASSER raw water quality, untreated water quality; **Rohwasserversorgung** *f* ABWASSER, WASSER raw water supply

rollend: **~e Fraktion** *f* ABFALL rolling fraction

Rollenförderer *m* ABFALL gravity roller conveyor

Röntgen: **Röntgenbrille** *f* SICHERHEIT x-ray protective glasses; **Röntgenfluoreszenz** *f* ANALYTIK x-ray fluorescence; **Röntgenfluoreszenzanalysator** *m* ANALYTIK x-ray fluorescence analyser (*BrE*), x-ray fluorescence analyzer (*AmE*); **Röntgenfluoreszenzanalyse** *f* (*RFA*) ANALYTIK x-ray fluorescence analysis (*XRFA*)

röntgenographisch *adj* ABFALL radiographic; **~es Verfahren** *nt* ABFALL radiographic process

Röntgen: **Röntgenschutzglas** *nt* SICHERHEIT x-ray protective glass

röntgenstrahlangeregt *adj* ANALYTIK x-ray-induced

Röntgen: **Röntgenstrahlen** *m pl* ANALYTIK x-rays

Rost *m* ABFALL, ABWASSER, WASSER *Eisen* rust, *Feuerrost* grate; **Rostdurchfall** *m* ABFALL, ABWASSER grate screenings; **Rostfeuerung** *f* ABFALL, LUFT grate firing; **Rost-Inhibitor** *m* ABFALL rust preventative; **Rostschutz** *m* ABWASSER, WASSER rust protection; **Rostschutzmittel** *nt* ABFALL rust preventative; **Rostwärmebelastung** *f* ABFALL grate heat exposure

Rotalgen *f pl* WASSER red algae

Rotameter *m* ANALYTIK rotameter

Rotation *f* ABWASSER, ANALYTIK, LUFT rotation; **Rotationsgeschwindigkeit** *f* LUFT rotation velocity; **Rotationsofen** *m* ABWASSER rotary furnace, rotary kiln; **Rotationspumpe** *f* ABWASSER rotary pump; **Rotationsviskosimeter** *nt* ANALYTIK drag-torque viscosimeter; **Rotationszerstäubung** *f* LUFT *Sprühabsorption* rotary atomization

rotierend *adj* ABFALL, LUFT rotating; **~er Adsorber** *m* LUFT rotating adsorber; **mit rotierenden Einbauten** *phr* LUFT *Waschraum* with rotating built-in structure

Rotor *m* ERNEU ENER *Biokonversion* rotor; **Rotorblatt** *nt* ERNEU ENER *Biokonversion* rotor blade; **Rotorbremse** *f* ERNEU ENER *Flüssiggas* rotor brake; **Rotordrehzahl** *f* ERNEU ENER *Windturbine* number of revolutions of rotor; **Rotornabe** *f* ERNEU ENER rotor hub; **Rotornabe** *f* **mit Blattverstellmechanismus** ERNEU ENER *Windturbine* rotor hub with adjustable blade mechanism

Rotschlamm *m* ABWASSER red mud

Rotte *f* ABFALL rot; **Rottebox** *f* ABFALL composting drum, rotting chamber; **Rottecontainer** *m* ABFALL rotting container; **Rottedeponie** *f* ABFALL rot landfill; **Rottegrad** *m* ABFALL degree of rotting process; **Rottehalle** *f* ABFALL rotting hall; **Rotteprozeß** *m* ABFALL rotting process; **Rottestillstand** *m* ABFALL rotting standstill; **Rotteturm** *m* ABFALL rotting tower; **Rottevorgang** *m* ABFALL rotting operation; **Rottezeit** *f* ABFALL rotting time; **Rottezelle** *f* ABFALL composting drum, rotting chamber; **Rottezellenkompostierungsverfahren** *nt* ABFALL rotting chamber composting process

Routine *f* ANALYTIK routine; **Routineanalyse** *f* ANALYTIK routine analysis; **Routinebestimmung** *f* ANALYTIK routine determination; **Routineprüfung** *f* ANALYTIK routine testing; **Routineüberwachung** *f* ANALYTIK routine monitoring

Ru *abbr* (*Ruthenium*) LUFT Ru (*ruthenium*)

Rück: **Rückblasfilter** *nt* LUFT blowring filter; **Rückflußverhinderer** *m* WASSER vacuum breaker; **Rückführung** *f* ABWASSER, WASSER feedback, recirculation, recovery

Rückgabe *f* ALLG return; **Rückgabebehälter** *m* ABFALL returnable container

rückgewinnbar *adj* ABFALL, ABWASSER, LUFT recoverable; **~er Abfall** *m* ABFALL recoverable waste; **~e Materialien** *nt pl* ABFALL recoverable materials

rückgewinnen *vt* ABFALL, ABWASSER reclaim

Rückgewinnung *f* ABFALL, LUFT recovery; **Rückgewinnungsanlage** *f* ABFALL reclamation plant; **Rückgewinnungskessel** *m* ABFALL, LUFT recovery boiler

rückgewonnen *adj* ABFALL, ABWASSER, LUFT, WASSER reclaimed, recovered; **~es Abwasser** *nt* ABWASSER recovered wastewater, reclaimed wastewater; **~es Öl** *nt* ABFALL *Altölregenierung* recovered oil; **~e Pulpe** *f* ABFALL recovered pulp; **~e Wärme** *f* ABFALL, LUFT recovered heat; **~es Wasser** *nt* ABWASSER recovered water, reclaimed water

Rück: **Rückgut** *nt* ABFALL returns; **Rückgut-Granulat** *nt* ABFALL returns granulate; **Rückhaltebecken** *nt* ABWASSER storage basin, WASSER storage basin, storage reservoir; **Rückhaltekäfig** *m* LUFT retention cage; **Rückhalterost** *m* LUFT retaining grate; **Rückhaltevermögen** *nt* LUFT retentivity

Rückholbarkeit *f* ABFALL retrieval capacity

Rücklauf: **Rücklaufautomat** *m* ABFALL reverse vending machine; **Rücklaufquote** *f* ABFALL return rate, *des Rücklaufsystems* recovery rate; **Rücklaufschlamm** *m* ABWASSER return sludge, sludge recycle; **Rücklaufschlammfluß** *m* ABWASSER sludge return flow; **Rücklaufschlammleitung** *f* ABWASSER return sludge pipe; **Rücklaufverhältnis** *nt* ABFALL reflux ratio

Rück: **Rückleimung** *f* ABFALL, ABWASSER reconditioning of paper fiber (*AmE*), reconditioning of paper fibre (*BrE*); **Rücklösung** *f* ABWASSER redissolution

Rücknahme *f* ABFALL take-back; **Rücknahmepflicht** *f* ABFALL take-back obligation; **Rücknahmeverpflichtung** *f* ABFALL take-back obligation

Rück: **Rückoxidation** *f* ABFALL, LUFT reoxidation; **Rückschlammhebewerk** *nt* ABWASSER return sludge elevator

rückspülbar *adj* ABFALL backwashable

Rück: **Rückspülgas** *nt* LUFT backflushing gas; **Rückspülsiebwechsler** *m* ABFALL backflush screen changer; **Rückspülung** *f* ABFALL, ABWASSER, WASSER backwashing, stripping; **Rückspülwasser** *nt* ABFALL, ABWASSER backwash water

Rückstand *m* ALLG residue; **Rückstandanalyse** *f* ANALYTIK analysis of residue

Rückstände *m pl* ABFALL, ABWASSER residues; **~ aus Abwasseranlage** *m pl* ABFALL sewage treatment plant residue; **~ aus der Sickerwasserreinigung** *m pl* ABFALL residues from leachate purification

Rückstand: **Rückstandsbeseitigung** *f* ABWASSER residue disposal; **Rückstandsdeponie** *f* ABFALL residue landfill; **Rückstandsdeponierung** *f* ABFALL, ABWASSER residue landfilling; **Rückstandsentsorgung** *f* ABWASSER residue disposal

rückstandsfrei *adj* ALLG residue-free

Rückstand: **Rückstandsprodukt** *nt* ERNEU ENER *Bioenergieausgangsstoff* residue raw material; **Rückstandsstrom** *m* ABFALL residue stream

Rück: **Rückstau** *m* ABFALL, WASSER backwater; **Rückstauwasser** *nt* ABFALL, WASSER backwater; **Rückstrahlung** *f* LUFT reflection; **Rückstreuung** *f* ANALYTIK back scattering

Rücktitration *f* ANALYTIK retitration, back titration

rückwärts: **~ gekrümmter Schaufel-Verdichter** *m* LUFT backward-curved blade compressor

ruhend: **~e Luftschicht** *f* LUFT static air layer

Rühr: **Rührarm** *m* LUFT agitating arm, WASSER rabble arm; **Rührbehälter** *m* ABWASSER, WASSER impeller-type mixer

Rühren *nt* ALLG mixing

Rühr: **Rührwerk** *nt* BODEN stirrer, agitator, WASSER stirring mechanism; **Rührwerksverdampfer** *m* ABWASSER mixing device evaporator

Rund- *in cpds* ALLG circular, round; **Rundbecken** *nt* ABWASSER circular basin, circular clarifier

runderneuert: **~er Reifen** *m* ABFALL retreaded tire (*AmE*), retreaded tyre (*BrE*)

Rund-: **Rundkolben** *m* ANALYTIK round-bottomed flask; **Rundräumer** *m* ABWASSER circular scraper; **Rundregner** *m* ABFALL, ABWASSER rotating sprayer; **Rundsandfang** *m* ABWASSER circular degritter; **Rundzelle** *f* ABFALL dry battery

Ruß *m* ABFALL, LUFT soot, *schwebstaubhaltig* smoke; **Rußbildung** *f* LUFT smoke buildup; **Rußemissionen** *f pl* LUFT smoke emissions; **Rußfänger** *m* LUFT soot catcher

rußfrei *adj* ABFALL, LUFT soot-free; **~e Verbrennung** *f* ABFALL, LUFT soot-free incineration

Ruß: **Rußkohle** *f* LUFT soot coal

Rüstungsaltlasten *f pl* BODEN military contaminated sites, post-closure military contaminated sites

Ruthenium *nt* (*Ru*) LUFT ruthenium (*Ru*)

Rüttel: **Rüttelfilter** *nt* LUFT shaker baghouse; **Rüttelmechanismus** *m* LUFT shaker mechanism; **Rüttelsieb** *nt* ABWASSER vibarting sieve

rütteln *vt* ABFALL, LUFT shake

Rütteln *nt* ABFALL, LUFT shaking, *Elektrofilterplatte* vibration

Rüttel: **Rüttelplatte** *f* ABFALL tumbling station; **Rüttelsieb** *nt* ABFALL vibrating screen

S

Sabin *nt* ANALYTIK, LÄRM *Einheit der Schallschluckung* sabin
Sachbilanz *f* MANAG life cycle inventory
Sackbildung *f* ABFALL subsidence
saisonal: **~er Wärmespeicher** *m* ERNEU ENER *Solarenergie* seasonal heat storage
salinisch: **~e Quelle** *f* WASSER saline spring
Salpeter: **Salpetersäure** *f* ABFALL, LUFT, WASSER nitric acid; **Salpeter-Schwefelsäure-Verfahren** *nt* LUFT nitric and sulphuric acid process
Salz *nt* ABFALL, LUFT, WASSER salt; **Salzaerosol** *nt* LUFT salt aerosol; **Salzgehalt** *m* ABFALL, WASSER salinity
salzhaltig *adj* ABFALL, WASSER saline; **~es Wasser** *nt* ABFALL, WASSER saline water
salzig *adj* ABFALL, WASSER saline
Salz: **Salzkonzentration** *f* ERNEU ENER saline concentration
Salzmann: **Salzmann-Absorptionslösung** *f* ANALYTIK Salzmann absorption solution; **Salzmann-Methode** *f* ANALYTIK Salzmann method
Salz: **Salzschlacke** *f* ABFALL *aus dem konventionellen Alurecycling* salt slag; **Salzwasser** *nt* WASSER saline water, salt water; **Salzwasserumwandlung** *f* WASSER saline water conversion
Sammel- *in cpds* ABWASSER, ANALYTIK, LUFT collecting, storing; **Sammelbecken** *nt* ABWASSER, WASSER collecting basin, storage reservoir, storage basin; **Sammelbehälter** *m* ABFALL collecting basin, sink, storage tank, ABWASSER receiving tank, LUFT storage tank; **Sammelbehältnis** *nt* ANALYTIK collecting receptacle; **Sammelcontainer** *m* ABFALL collection container; **Sammelgefäß** *nt* ANALYTIK receptacle; **Sammelkanal** *m* WASSER main drain; **Sammelkosten** *pl* ABFALL collection costs
sammeln *vt* ALLG collect
Sammel-: **Sammelschacht** *m* WASSER *Pflanzenklärsystem* collector shaft; **Sammelstelle** *f* ABFALL drop-off center (*AmE*), drop-off centre (*BrE*); **Sammelsystem** *nt* ABFALL receiving system
Sammlung *f* ALLG collection; **~ durch beauftragte Dritte** ABFALL collection by contracted third party; **~ an zentralen Orten** ABFALL collection at central sites; **Sammlungszeitraum** *m* ANALYTIK collection period
Sampler *m* ANALYTIK sampler
Sand *m* WASSER sand; **Sandablagerung** *f* WASSER sand deposition
Sandfang *m* ABWASSER grit chamber, detritor, sand trap, WASSER diffuser; **Sandfanggut** *nt* ABWASSER grit, detritus; **Sandfanggutentfernung** *f* ABWASSER grit removal; **Sandfangräumer** *m* ABWASSER degritter
Sand: **Sandfilter** *nt* ABWASSER *Volumenfilter* sand filter; **Sandfiltergraben** *m* ABWASSER sand filtration ditch; **Sandgrube** *f* ABWASSER grit can; **Sandräumer** *m* ABWASSER grit scraper; **Sandschicht** *f* ABFALL, ABWASSER sand layer; **Sandstrahlen** *nt* MEER sandblasting; **Sandstrahlreinigung** *f* MEER sandblasting; **Sandwaschanlage** *f* ABWASSER grit washer; **Sandwaschung** *f* ABWASSER grit washing
Sandwich: **Sandwich-Konstruktion** *f* LÄRM sandwich construction; **Sandwich-Platte** *f* LÄRM *Schwingungsdämpfung* sandwich panel; **Sandwich-Spritzgießen** *nt* ABFALL sandwich injection molding (*AmE*), sandwich injection moulding (*BrE*); **Sandwich-Struktur** *f* ERNEU ENER sandwich structure
sanft: **~er Tourismus** *m* MANAG soft tourism
sanieren *vt* ABFALL, ABWASSER clean up, redevelop, BODEN decontaminate, remediate, MANAG, WASSER clean up, redevelop
saniert *adj* ABFALL, ABWASSER cleaned up, redeveloped, BODEN decontaminated, remediated, MANAG, WASSER cleaned-up, redeveloped
Sanierung *f* ABFALL, ABWASSER cleaning up, redevelopment, BODEN decontamination, remediation, MANAG, WASSER cleaning up, redevelopment; **Sanierungsbedarf** *m* BODEN remediation need
sanierungsbedürftig *adj* BODEN in need of remediation
Sanierung: **Sanierungsgelände** *nt* BODEN site targeted for decontamination; **Sanierungsgrad** *m* BODEN remediation rate, clean-up rate; **Sanierungskosten** *pl* BODEN, MANAG clean-up costs; **Sanierungsmaßnahme** *f* ABFALL, BODEN measure of redevelopment; **Sanierungstechnik** *f* BODEN decontamination technique; **Sanierungszielwerte** *m pl* BODEN remediation target values
sanitär: **~e Einrichtungen** *f pl* ABFALL, ABWASSER, WASSER sanitation
Sanitärtechnik *f* ABWASSER, WASSER sanitary engineering
Sattelschlepperanhänger *m* ABFALL semitrailer
sättigen *vt* ALLG saturate
Sättigung *f* ALLG saturation; **Sättigungsbeladung** *f* LUFT *Adsorption* saturation load; **Sättigungsdefizit** *nt* LUFT saturation deficit; **Sättigungsdruck** *m* LUFT saturation pressure; **Sättigungsebene** *f* WASSER saturation level; **Sättigungsindex** *m* WASSER saturation index; **Sättigungskonzentration** *f* WASSER saturation concentration; **Sättigungstemperatur** *f* LUFT saturation temperature; **Sättigungszone** *f* ABFALL, BODEN, WASSER saturated zone
sauber *adj* ALLG clean; **~e Energiequelle** *f* ERNEU ENER clean energy source; **~e Technologie** *f* ABFALL, ABWASSER, LUFT, MANAG, WASSER clean technology, non-waste-technology (*NWT*)
sauberhalten *vt* ABFALL, WASSER keep clean
Säuberung *f*: **~ der Küstenlinie** MEER shoreline clean-up
sauer *adj* ALLG acid, acidic; **saurer Abfluß** *m* ABFALL, BODEN acid runoff; **saures Abgas** *nt* ABFALL, LUFT acid waste gas; **saure Ablagerung** *f* LUFT acid deposition, acid deposit; **saures Aerosol** *nt* LUFT acid aerosol; **saurer Anionenaustauscher** *m* ABWASSER acidic anionic exchanger; **saure Erde** *f* BODEN acid earth; **saure Gärung** *f* ABFALL acid fermentation; **saures Gas** *nt* LUFT acid gas, sour gas
Sauergas *nt* LUFT acid gas, sour gas
sauer: **saure Gasschadstoffe** *m pl* LUFT acid gas pollutants; **saurer Kationenaustauscher** *m* ABWASSER acidic cationic exchanger
säuern *vt* ABFALL, ABWASSER, WASSER acidify
sauer: **saurer Niederschlag** *m* BODEN, LUFT acid fallout, acidic rain
sauer-oxidativ *adj* LUFT acid-oxidative
sauer: **saurer Regen** *m* BODEN, LUFT acid fallout, acid rain, acidic rain; **saurer Schnee** *m* BODEN, LUFT acid snow; **saure Sedimentation** *f* LUFT acid deposition; **saurer Smog** *m* LUFT acid smog
Sauerstoff *m* (O_2) ALLG oxygen (O_2); **Sauerstoffangebot** *nt* ABWASSER, LUFT, SICHERHEIT, WASSER oxygen supply; **Sauerstoffanreicherung** *f* WASSER oxygen accumulation
sauerstoffarm *adj* ABFALL, WASSER oxygen-deficient
Sauerstoff: **Sauerstoffatemgerät** *nt* SICHERHEIT oxygen breathing apparatus; **Sauerstoffaufnahme** *f* ABWASSER oxygen uptake; **Sauerstoffausschluß** *m* ABWASSER absence

of oxygen; **Sauerstoffbedarf** *m* ABWASSER, WASSER oxygen demand

sauerstoffbedürftig *adj* ABWASSER, BODEN oxygen-demanding

Sauerstoff: **Sauerstoffdefizit** *nt* ABWASSER, WASSER oxygen deficiency; **Sauerstoffdonator** *m* WASSER oxygen donator; **Sauerstoffeintrag** *m* ABFALL, ABWASSER oxygen charge, oxygen load; **Sauerstoffentzug** *m* ABWASSER deoxygenation; **Sauerstoffflasche** *f* ANALYTIK, LUFT oxygen flask; **Sauerstoffgehalt** *m* ANALYTIK, ERNEU ENER *von Biomasse* oxygen content

sauerstoffgesättigt *adj* ABFALL, ABWASSER, WASSER oxygen-saturated; **~es Wasser** *nt* WASSER oxygen-saturated water

sauerstoffhaltig *adj* ABWASSER oxygenic

Sauerstoff: **Sauerstoffhaushalt** *m* LUFT oxygen budget; **Sauerstoffkonzentration** *f* ABWASSER, ANALYTIK, LUFT, WASSER oxygen concentration; **Sauerstoffkorrosion** *f* LUFT, WASSER oxygen corrosion; **Sauerstoffmangel** *m* ABWASSER, ANALYTIK, WASSER lack of oxygen, oxygen deficiency; **Sauerstoffmeßfühler** *m* ANALYTIK, LUFT, SICHERHEIT oxygen sensor; **Sauerstoffmeßgerät** *nt* ANALYTIK, LUFT, SICHERHEIT oxygen measuring device; **Sauerstoffsättigung** *f* ABWASSER, LUFT oxygen saturation; **Sauerstoffsättigungsfaktor** *m* ABWASSER, LUFT oxygen saturation factor; **Sauerstoffsonde** *f* LUFT lambda probe, *Katalysator* oxygen sensor; **Sauerstofftransport** *m* ABWASSER oxygen transfer; **Sauerstoffübersättigung** *f* ABWASSER oxygen supersaturation; **Sauerstoffüberschuß** *m* LUFT, SICHERHEIT excess oxygen; **Sauerstoffverbrauch** *m* ABWASSER, LUFT oxygen consumption; **Sauerstoffversorgung** *f* ABWASSER, LUFT, SICHERHEIT, WASSER oxygen supply

sauerstoffzehrend *adj* ABFALL, ABWASSER oxygen-depleting, BODEN oxygen-demanding, WASSER oxygen-depleting

Sauerstoff: **Sauerstoffzehrung** *f* ABWASSER, ANALYTIK, LUFT oxygen depletion; **Sauerstoffzufuhr** *f* ABWASSER oxygen transfer, ERNEU ENER *Biomasseumwandlung*, LUFT, WASSER oxygen introduction; **Sauerstoffzufuhrvermögen** *nt* **in Abwasser** ABWASSER oxygen transfer capacity in wastewater; **Sauerstoffzufuhrvermögen** *nt* **in Reinwasser** ABWASSER oxygen transfer capacity in clean water; **Sauerstoffzugabe** *f* ABWASSER oxygen addition

sauer: **saures Teilchen** *nt* LUFT acid particle, acidic particle; **saure Umweltverschmutzung** *f* BODEN, LUFT acid pollution; **saures Wasser** *nt* ABFALL, WASSER acidic water

Saug- *in cpds* ABFALL, ABWASSER, MEER, WASSER suction; **Saugabschöpfgerät** *nt* MEER *für Öl von der Wasseroberfläche* direct suction skimmer; **Saugbagger** *m* WASSER suction dredge; **Saugbrunnen** *m* WASSER suction well; **Saug- und Druckpumpe** *f* ABWASSER, WASSER lift-and-force pump; **Saugdüse** *f* ABWASSER, WASSER suction nozzle

Saugen *nt* ALLG suction

Saug-: **Saugfähigkeit** *f* WASSER absorbency; **Saugfegmaschine** *f* ABFALL vacuum street sweeper; **Saugfilter** *nt* ABWASSER vacuum filter, WASSER suction filter; **Sauggebläse** *nt* LUFT exhaust fan; **Saughilfspumpe** *f* ABWASSER, WASSER priming pump; **Saughöhe** *f* ABWASSER, WASSER suction head, suction lift; **Saugkasten** *m* BODEN suction box; **Saugkessel** *m* WASSER suction tank; **Saugkopf** *m* MEER *eines Abschöpfgerätes* skimming head, WASSER suction strainer; **Saugkorb** *m* MEER strainer, WASSER rose, suction strainer; **Saugkrümmer** *m* LUFT suction box; **Saugleitung** *f* ABWASSER, WASSER suction pipe, suction piping; **Saugmund** *m* ABFALL *am Kehrfahrzeug* suction port; **Saugpumpe** *f* ABWASSER, WASSER lift pump, suction pump; **Saugrohr** *nt* LUFT suction pipe, WASSER suction pipe, tail pipe; **Saugsammler** *m* WASSER suction collector; **Saugschacht** *m* WASSER suction well; **Saugschlauch** *m* WASSER suction hose; **Saugschlauchfilter** *nt* LUFT suction filter; **Saugseite** *f* ABWASSER, WASSER *einer Pumpe* suction side

Saugzug *m* LUFT induced draft (*AmE*), induced draught (*BrE*); **Saugzuggebläse** *nt* LUFT induced-draft fan (*AmE*), induced-draught fan (*BrE*), ID fan; **Saugzugventilator** *m* LUFT induced-draft fan (*AmE*)induced-draught fan (*BrE*), ID fan

Säule *f* ANALYTIK column; **Säulenchromatographie** *f* ANALYTIK column chromatography; **Säuleninjektor** *m* ANALYTIK column injector; **Säulenmaterial** *nt* ABFALL column material; **Säulenmethode** *f* ANALYTIK column method; **Säulenofen** *m* ANALYTIK heated column oven

Säure *f* ABFALL, ABWASSER, BODEN, LUFT acid; **durch ~ belastet** *phr* BODEN acid-stressed; **Säureanteil** *m* LUFT acid content; **Säurebelastung** *f* BODEN acid stress, LUFT acid loading

säurebeständig *adj* BODEN, LUFT, WASSER acid-proof, acid-resistant

Säure: **Säurebeständigkeit** *f* BODEN, LUFT, WASSER acid resistance; **Säurebildner** *m* ABFALL, ABWASSER, BODEN, LUFT acid former; **Säurebildung** *f* ABFALL, ABWASSER, BODEN, LUFT acid formation; **Säurefällung** *f* LUFT acid precipitation

säurefrei *adj* BODEN, LUFT, WASSER acid-free

Säure: **Säuregehalt** *m* BODEN, LUFT, WASSER acidity; **Säuregrad** *m* BODEN, LUFT acidity level; **Säurekapazität** *f* ABWASSER acid capacity; **Säurekonzentration** *f* LUFT acid concentration; **Säurenebel** *m* LUFT acid fog; **Säureneutralisation** *f* LUFT acid neutralizing; **Säureneutralisierungsvermögen** *nt* LUFT acid neutralizing capacity; **Säurenvorstufe** *f* BODEN acidic precursor; **Säurepufferkapazität** *f* ABWASSER acid buffering capacity; **Säureresistenz** *f* LUFT acid resistance; **Säurespaltung** *f* ABWASSER acid cleavage; **Säurespaltung** *f* **mit Flotation** ABWASSER acid cleavage with flotation; **Säureverträglichkeit** *f* BODEN acid tolerance; **Säurewäsche** *f* LUFT acid scrubbing

SAV *abbr* (*Sprühabsorptionsverfahren*) LUFT spray absorption process

Savonius-Rotor *m* ERNEU ENER Savonius rotor

Sb *abbr* (*Antimon*) WASSER Sb (*antimony*)

SBR *abbr* (*Sequential Batch Reactor*) BODEN SBR (*sequential batch reactor*)

Scavenger *m* ABFALL, BODEN, LUFT scavenger

schaben *vt* MEER scrape

Schaber *m* MEER scraper

Schacht *m* ABFALL, LUFT shaft

Schachtel *f* ALLG carton

Schacht: **Schachtfutter** *nt* LUFT stack lining; **Schachtofen** *m* ABFALL, LUFT shaft furnace; **Schachtring** *m* ABFALL ring of tubbings; **Schachtstarrheit** *f* ABFALL shaft rigidity; **Schachtverfüllmaterial** *nt* ABFALL shaft-filling material

Schadensermittlung *f* SICHERHEIT damage assessment

Schadgas *nt* LUFT harmful gas; **Schadgasadsorption** *f* BODEN, LUFT harmful gas adsorption

schädigen *vt* ALLG damage, harm

schädlich *adj* ALLG harmful, injurious; **~e Emissionen** *f pl* LUFT harmful emissions; **~e Wirkung** *f* ABFALL, LUFT, SICHERHEIT, WASSER harmful effect

Schadstoff *m* ALLG contaminant, harmful substance, pollutant, toxic substance; **~ aus Flächenquelle** ABWASSER, BODEN, LUFT area source pollutant; **~ aus Punktquelle** ABWASSER, BODEN, LUFT point source pollutant; **Schadstoffabbau** *m* ABWASSER, LUFT contaminant degradation; **Schadstoffablagerung** *f* ABFALL, BODEN, LUFT, WASSER pollutant deposition; **Schadstoffanalyse** *f* ANALYTIK pollutant analysis; **Schadstoffanalytik** *f* ANALYTIK harmful substances analysis

schadstoffarm *adj* ALLG low in harmful substances, low-pollutant, low-polluting; **~es Fahrzeug** *nt* LUFT low-emission vehicle

Schadstoff: **Schadstoffausbreitung** *f* ABFALL, ABWASSER,

BODEN, LUFT, WASSER dispersion of harmful substances, emission of pollutants, propagation of pollutants; **Schadstoffausstoß** *m* LUFT exhaust emissions; **Schadstoffaustritt** *m* BODEN *ins Erdreich* escape of harmful substances
schadstoffbelastet *adj* ALLG polluted; **~es Erdreich** *nt* ABFALL, ABWASSER, BODEN polluted soil
Schadstoff: **Schadstoffbelastung** *f* ABFALL, ABWASSER, BODEN, LUFT pollution burden
schadstoffbeständig *adj* ABWASSER, ANALYTIK, BODEN, LUFT contaminant-resistant, pollutant-resistant
Schadstoff: **Schadstoffbrüden** *m pl* LUFT contaminant vapors (*AmE*), contaminant vapours (*BrE*)
schadstoffdicht *adj* BODEN *Dichtwände* harmful-substance-sealing
Schadstoff: **Schadstoffdiffusion** *f* BODEN migration of harmful substances; **Schadstoffeigenschaften** *f pl* ABWASSER, LUFT, WASSER contaminant properties; **Schadstoffeintrag** *m* BODEN introduction of harmful substances; **Schadstoffeliminierung** *f* ABWASSER, WASSER contaminant elimination; **Schadstoffemission** *f* ABFALL, ABWASSER, BODEN, LUFT emission of contaminant; **Schadstoffentfernung** *f* LUFT pollutant removal; **Schadstoffentfrachtung** *f* ABFALL pollutant extraction; **Schadstofffracht** *f* ABFALL harmful substance load, LUFT pollutant load
schadstofffrei *adj* ABWASSER, LUFT, WASSER contaminant-free, pollutant-free
Schadstoff: **Schadstoff-Gasstrom** *m* LUFT contaminant gas stream; **Schadstoffgehalt** *m* ABFALL, ABWASSER, BODEN, LUFT, WASSER contaminant content, pollutant content
schadstoffhaltig *adj* ALLG containing contaminants; **~er Abfall** *m* ABFALL contaminated waste, waste containing pollutants, waste contaminated by pollutants; **~e Abgase** *nt pl* LUFT waste gas containing harmful substances; **~er Filterstaub** *m* ABFALL, LUFT filter dust containing pollutants; **~e Produkte** *nt pl* ABFALL products containing pollutants; **~er Reststoff** *m* ABFALL residue containing pollutants
Schadstoff: **Schadstoffinventar** *nt* SICHERHEIT pollutants inventory; **Schadstoffkonzentrat** *nt* BODEN contaminant concentrate; **Schadstoffkonzentration** *f* ABWASSER, LUFT harmful substance concentration; **Schadstoffmenge** *f* ABFALL harmful substance volume, ANALYTIK pollutant amount; **Schadstoffoxidation** *f* LUFT pollutant oxidation; **Schadstoffregister** *nt* POLITIK harmful substance register
schadstoffresistent *adj* ABWASSER, ANALYTIK, BODEN, LUFT contaminant-resistant, pollutant-resistant
Schadstoff: **Schadstoffresistenz** *f* ABWASSER, ANALYTIK, BODEN, LUFT resistance to pollutants; **Schadstoffrestkonzentration** *f* LUFT residue concentration of pollutants
schadstoffspezifisch *adj* ABWASSER, BODEN, LUFT pollutant-specific, contaminant-specific
Schadstoff: **Schadstofftransport** *m* BODEN transport of pollutants, transport of contaminants; **Schadstofftyp** *m* ABWASSER, BODEN, LUFT pollutant type, contaminant type; **Schadstoffumsatz** *m* LUFT pollutant conversion; **Schadstoffumwandlung** *f* ABWASSER, BODEN, LUFT transformation of pollutants; **Schadstoffverfrachtung** *f* BODEN transport of pollutants, transport of contaminants; **Schadstoffverhalten** *nt* ABWASSER, BODEN, LUFT pollutant behavior (*AmE*), pollutant behaviour (*BrE*); **Schadstoffverteilung** *f* LUFT pollutant transmission; **Schadstoffvorstufe** *f* LUFT precursor pollutant; **Schadstoffwandlung** *f* ABFALL, BODEN migration of harmful substances
Schäkel *m* MEER shackle
Schale *f* LÄRM shell; **Schalenkreuzanemometer** *nt* ANALYTIK cup anemometer
Schall *m* LÄRM, MEER, SICHERHEIT sound; **~ in festen Körpern** LÄRM structure-borne noise; **Schallabsorber** *m* LÄRM sound absorber
schallabsorbierend *adj* LÄRM sound-absorbing; **~e Auskleidung** *f* LÄRM sound-absorbing lining; **~es Material** *nt* LÄRM sound-absorbing material; **~e Platte** *f* LÄRM acoustical board; **~er Putz** *m* LÄRM acoustic plaster; **~e Verkleidung** *f* LÄRM acoustic paneling (*AmE*), acoustic panelling (*BrE*), acoustic lining; **~e Wandverkleidung** *f* LÄRM sound-absorbing wall lining
Schall: **Schallabsorptionsmaterial** *nt* LÄRM sound-absorbing material; **Schallabstrahlung** *f* LÄRM sound radiation; **Schallausbreitung** *f* LÄRM noise propagation, spread of noise; **Schallausbreitung** *f* **im Freien** LÄRM outdoor sound propagation; **Schallbelastungsmesser** *m* ANALYTIK, LÄRM sound exposure meter; **Schalldämmaß** *nt* LÄRM sound reduction index (*BrE*) (*SRI*), sound transmission loss (*AmE*); **Schalldämmaterial** *nt* LÄRM sound insulation material
schalldämmend *adj* LÄRM anti-noise, sound-absorbing, sound-dampened, sound-insulating; **~e Haube** *f* LÄRM, MEER anti-noise hood; **~e Kapselung** *f* LÄRM, MEER anti-noise encapsulation; **~e Weichgummibespannung** *f* LÄRM, MEER, SICHERHEIT anti-noise soft rubber lining
Schall: **Schalldämmfolie** *f* LÄRM acoustic sheet; **Schalldämmklasse** *f* LÄRM sound transmission class (*STC*); **Schalldämmnoppenplatte** *f* LÄRM knobbed panel; **Schalldämmplatte** *f* LÄRM, SICHERHEIT acoustic panel, acoustic plate, acoustical slab; **Schalldämmstoff** *m* LÄRM sound insulation material; **Schalldämmung** *f* LÄRM sound isolation, sound insulation, acoustic insulation; **Schalldämmwand** *f* LÄRM sound-insulating wall
schalldämpfend *adj* LÄRM, MEER, SICHERHEIT anti-noise
Schall: **Schalldämpfer** *m* LÄRM muffler (*AmE*), silencer (*BrE*), sound damper, sound deadener; **Schalldämpfung** *f* LÄRM sound damping, sound deadening, sound attenuation
schalldicht *adj* LÄRM sound-insulating, soundproof; **~e Kammer** *f* LÄRM, MEER, SICHERHEIT noise protection booth
Schall: **Schalldose** *f* LÄRM sound box
Schalldruck *m* LÄRM sound pressure; **Schalldruckmessung** *f* ANALYTIK, LÄRM sound pressure measuring; **Schalldruckpegel** *m* LÄRM sound pressure level (*SPL*); **Schalldruck-Schallschnelle-Sonde** *f* ANALYTIK pressure velocity probe; **Schalldruckspektrum** *nt* ANALYTIK, LÄRM sound pressure spectrum
Schall: **Schalldurchgang** *m* LÄRM sound transmission
schalldurchlässig *adj* LÄRM sound-transparent, sound-permeable
Schall: **Schalldurchlässigkeit** *f* LÄRM sound permeability; **Schalleistung** *f* ANALYTIK, LÄRM audibility, acoustic capacity, sound power; **Schalleistungspegel** *m* ANALYTIK audibility level, LÄRM sound power level
schalleitend *adj* LÄRM sound-conducting
Schall: **Schallemission** *f* (*SE*) ANALYTIK, LÄRM acoustic emission (*AE*), sound emission; **Schallemissionsanalyse** *f* (*SEA*) ANALYTIK acoustic emissions analysis (*AE analysis*); **Schallemittent** *m* LÄRM noise emitter, sound emitter; **Schallenergie** *f* LÄRM sound energy; **Schallentstehung** *f* LÄRM sound generation; **Schallfeld** *nt* LÄRM sound field; **Schallfortpflanzung** *f* LÄRM noise propagation
schallgedämmt *adj* LÄRM soundproof
Schall: **Schallgeschwindigkeit** *f* LÄRM speed of sound
schallhart *adj* LÄRM sound-reflecting
Schall: **Schallimmission** *f* LÄRM noise immission, sound immission
Schallintensität *f* LÄRM sound intensity; **Schallintensitätsanalysator** *m* ANALYTIK, LÄRM sound intensity analyser (*BrE*), sound intensity analyzer (*AmE*); **Schallintensitätsmeßgerät** *nt* ANALYTIK, LÄRM sound intensity meter; **Schallintensitätspegel** *m* ANALYTIK, LÄRM sound intensity level

Schall: **Schallisolation** *f* LÄRM sound isolation; **Schallisolationsmaterial** *nt* LÄRM sound insulation material
schallisolieren *vt* LÄRM soundproof
Schall: **Schallisolierung** *f* LÄRM soundproofing; **Schallmauer** *f* LÄRM sound barrier; **Schalloch** *nt* LÄRM acoustic louver (*AmE*), acoustic louvre (*BrE*); **Schallöffnung** *f* LÄRM acoustic louver (*AmE*), acoustic louvre (*BrE*)
Schallpegel *m* LÄRM noise level, sound level; **Schallpegelanalysator** *m* ANALYTIK noise level analyser (*BrE*), noise level analyzer (*AmE*); **Schallpegelkalibrator** *m* ANALYTIK, LÄRM sound level calibrator; **Schallpegelmesser** *m* ANALYTIK, LÄRM volume unit meter (*VU meter*), phonometer, sound level meter (*SLM*); **Schallpegelmeßgerät** *nt* ANALYTIK noise level analyser (*BrE*), noise level analyzer (*AmE*)
Schall: **Schallquelle** *f* LÄRM sound source; **Schallradiometer** *nt* ANALYTIK sound radiometer
schallreflektierend *adj* LÄRM sound-reflecting
Schall: **Schallreflexion** *f* LÄRM sound reflection; **Schallschirm** *m* LÄRM baffle collector, baffle board, acoustic screen; **Schallschluckeigenschaft** *f* LÄRM sound-absorbing property
schallschluckend *adj* LÄRM sound-absorbing; **~e Auskleidung** *f* LÄRM sound-absorbing lining; **~e Decke** *f* LÄRM sound-absorbing ceiling; **~es Glas** *nt* LÄRM, MEER, SICHERHEIT sound-absorbing glass; **~es Material** *nt* LÄRM sound-absorbing material; **~e Platte** *f* LÄRM sound-absorbing panel, acoustical board; **~e Schürze** *f* LÄRM sound-absorbing apron; **~er Stein** *m* LÄRM sound-absorbing brick; **~e Tapete** *f* LÄRM sound-absorbing wall paper; **~e Tür** *f* LÄRM, MEER, SICHERHEIT sound-absorbing door; **~e Verkleidung** *f* LÄRM acoustic paneling (*AmE*), acoustic panelling (*BrE*), acoustic lining; **~e Wand** *f* LÄRM sound-absorbing wall; **~e Wandverkleidung** *f* LÄRM sound-absorbing wall lining; **~e Zone** *f* LÄRM sound absorbent zone
Schall: **Schallschluckgrad** *m* LÄRM sound absorption coefficient
Schallschutz *m* LÄRM, SICHERHEIT noise protection; **Schallschutzfenster** *nt* LÄRM sound-protecting window; **Schallschutzhaube** *f* LÄRM sound-absorbing hood; **Schallschutzkabine** *f* LÄRM acoustic booth; **Schallschutzmaßnahmen** *f pl* LÄRM, SICHERHEIT noise prevention measures; **Schallschutzschirm** *m* LÄRM acoustic shield; **Schallschutztechnik** *f* LÄRM, MEER noise protection technology; **Schallschutzwand** *f* LÄRM acoustic barrier, sound-absorbing barrier
Schall: **Schallschwingung** *f* LÄRM acoustic oscillation, sound vibration, acoustic vibration; **Schallsonde** *f* ANALYTIK, LÄRM sound probe; **Schallspektrum** *nt* LÄRM sound spectrum; **Schallstärke** *f* LÄRM sound intensity; **Schallstrahler** *m* LÄRM sound emitter; **Schallstrahlungsdruck** *m* LÄRM sound radiation pressure
schalltechnisch *adj* LÄRM sonic
schalltot *adj* LÄRM *Raum* anechoic
Schall: **Schall- und Vibrationsisolierung** *f* LÄRM, MEER, SICHERHEIT insulation against sound and vibration; **Schall- und Vibrationsmesser** *m* ANALYTIK, LÄRM, MEER, SICHERHEIT noise and vibration measuring equipment; **Schallvorgang** *m* LÄRM sound occurrence; **Schallwand** *f* LÄRM baffle board, acoustic shield, acoustic screen, acoustic baffle; **Schallwandler** *m* ANALYTIK, LÄRM sound transducer; **Schallwechseldruck** *m* LÄRM excess pressure; **Schallwellen** *f pl* LÄRM sound waves
scharfkantig: **~es Partikel** *nt* LUFT sharp-edged particle
Schattenpreis *m* POLITIK shadow price
Schaufel *f* ABWASSER paddle, LUFT blade, MEER shovel; **Schaufelentnahmegerät** *nt* ABFALL, WASSER scraper extractor; **Schaufellader** *m* MEER front-end loader; **Schaufelrad** *nt* ABWASSER paddle wheel, ERNEU ENER *Wellenenergie, Wasserkraft* blade wheel
Schaum *m* LUFT froth, foam; **Schaumbekämpfung** *f* WASSER foaming abatement
Schaum: **Schaumbildung** *f* LUFT *Sprudelbett-Waschturm* foaming, WASSER foam formation; **Schaumflotation** *f* BODEN foam flotation; **Schaumschalldämmplatte** *f* LÄRM, MEER, SICHERHEIT sound-absorbing foam panel; **Schaumstoffplatte** *f* LÄRM foam panel; **Schaumteppich** *m* WASSER foam film; **Schaumverfahren** *nt* BODEN foam process
Scheiben: **Scheibenabschöpfgerät** *nt* MEER *für Öl von der Wasseroberfläche* disc skimmer (*BrE*), disk skimmer (*AmE*); **Scheibentauchkörper** *m* ABWASSER rotating biological disc contactor (*BrE*) (*RBC*), biodisc (*BrE*), biodisk (*AmE*); **Scheibentauchtropfkörper** *m* ABWASSER disc aerator (*BrE*), disk aerator (*AmE*)
Scheidepresse *f* ABFALL wringer
Scheidetrichter *m* ANALYTIK *Extraktion* separating funnel
Scherbe *f* ABFALL glass fragment
Scherben *f pl* ABFALL cullet, refuse glass; **Scherbenrückführung** *f* ABFALL glass cullet recycling
Scher-: **Scherfestigkeit** *f* ABWASSER *des Klärschlammes* shearing resistance; **Scherkraft** *f* ABWASSER, LUFT shearing force; **Scherwirkung** *f* ABWASSER shearing action; **Scherwirkungsgrad** *m* LUFT shearing efficiency
Schicht *f* ABFALL lamination, ABFALL, BODEN, LUFT layer; **Schichtbildung** *f* LUFT layer formation; **Schichtdicke** *f* ABFALL, ANALYTIK, LÄRM layer thickness
schichten *vt* ABFALL laminate, BODEN, LUFT layer
schichtenförmig *adj* ABFALL, LUFT, WASSER laminar
Schicht: **Schichtenwasser** *nt* ABFALL layer water; **Schichtprobenahme** *f* ANALYTIK layer sampling; **Schichtquelle** *f* WASSER contact spring; **Schichtstärke** *f* ABFALL lamination strength, LUFT layer strength
Schichtungsfestigkeit *f* ABFALL lamination strength
schichtweise *adj* ANALYTIK layered
Schiebeboden *m* ABFALL sliding floor
Schiedsanalyse *f* ANALYTIK arbitrary analysis
Schiedsprobe *f* ANALYTIK arbitrary test
Schiene: **Schienenbeförderung** *f* ABFALL rail transport, railroad transportation (*AmE*), railway transportation (*BrE*); **Schienentransport** *m* ABFALL rail transport, railroad transportation (*AmE*), railway transportation (*BrE*)
Schiff *nt* MEER ship, vessel; **Schiffspropeller** *m* MEER propeller; **Schiffsschraube** *f* MEER propeller
Schillern *nt* MEER iridescence
Schimmer *m* MEER sheen
Schlachthausabfall *m* ABFALL abattoir waste
Schlacke *f* ABFALL clinker, slag; **Schlackenanfall** *m* ABFALL slag generation; **Schlackenaustrag** *m* ABFALL slag discharge
schlackenhaltig *adj* ABFALL slag-containing
Schlacke: **Schlackennachbehandlung** *f* ABFALL final treatment of slag; **Schlackenschmelz** *m* ABFALL molten slag; **Schlackesammelbehälter** *m* ABFALL, BODEN slag tank; **Schlackeverwertung** *f* ABFALL slag recovery
Schlagstock *m* LUFT *Stator-Rotor System* picker
Schlamm *m* ABFALL, ABWASSER, LUFT, WASSER sludge, slurry; **Schlammablagerung** *f* ABFALL, ABWASSER sludge deposit; **Schlammalter** *nt* ABWASSER sludge age; **Schlammanfall** *m* ABFALL, ABWASSER sludge accumulation
Schlämmapparat *m* LUFT elutriator
Schlamm: **Schlammart** *f* ABWASSER sludge type; **Schlammaufbereitung** *f* ABWASSER sludge processing, sludge treatment; **Schlammaustragung** *f* ABFALL, ABWASSER, WASSER sludge extraction, desludging; **Schlammbehandlung** *f* ABWASSER sludge treatment; **Schlammbelebungsverfahren** *nt* ABWASSER activated sludge process; **Schlammbeseitigung** *f* ABWASSER sludge disposal, sludge removal;

Schlammbildung *f* LUFT sludge formation; **Schlammblähung** *f* ABWASSER bulking; **Schlammbüchse** *f* ABWASSER bailer; **Schlammdecke** *f* ABWASSER zoogloeal layer, sludge blanket; **Schlammdeckenfilter** *nt* ABWASSER sludge blanket filter; **Schlammdeponierung** *f* ABWASSER sludge landfilling; **Schlammdichteindex** *m* ABWASSER sludge density index (*SDI*); **Schlammeigenschaften** *f pl* ABWASSER sludge characteristics; **Schlammeindickung** *f* ABWASSER sludge thickening

schlämmen *vt* LUFT elutriate

Schlämmen *nt* LUFT elutriation

Schlamm: **Schlammentsorgung** *f* ABWASSER sludge disposal; **Schlammentstehung** *f* ABWASSER sludge generation; **Schlammentwässerung** *f* ABWASSER sludge dewatering; **Schlammentwässerungsanlage** *f* ABWASSER sludge dewatering equipment; **Schlammfang** *m* ABWASSER sludge sump, silt trap; **Schlammfaulbecken** *nt* ABWASSER stabilization pond; **Schlammfaulung** *f* ABWASSER sludge digestion

schlammförmig *adj* ABFALL, ABWASSER sludge-forming; **~er Rückstand** *m* ABFALL, ABWASSER sludge-forming residue

Schlamm: **Schlammgefrieren** *nt* ABWASSER sludge freezing; **Schlammgehalt** *m* ABWASSER mud content; **Schlammhandhabung** *f* ABWASSER sludge handling; **Schlammheber** *m* ABWASSER sludge siphon; **Schlammindex** *m* ABWASSER sludge index; **Schlammkonditionierung** *f* ABWASSER sludge conditioning; **Schlammkontaktflockungsbecken** *nt* ABWASSER sludge contact clarifier, **Schlammkontaktverfahren** *nt* ABWASSER upflow sludge blanket process; **Schlammkonzentration** *f* ABWASSER sludge concentration; **Schlammkuchen** *m* ABWASSER sludge cake; **Schlammlöffel** *m* ABWASSER bailer; **Schlamm-Öl-Reaktion** *f* ABWASSER sludge-to-oil reaction; **Schlamm-Öl-Reaktor** *m* (*S.-Ö.-Reaktor*) ABWASSER sludge-to-oil reactor (*STOR*); **Schlammpumpwerk** *nt* ABWASSER sludge pump works; **Schlammräumer** *m* ABWASSER scraper; **Schlammröhrenwürmer** *m pl* ABWASSER sludge worms; **Schlammrücklauf** *m* ABWASSER sludge return; **Schlammsammelbehälter** *m* ABWASSER sludge sump; **Schlammsaugwagen** *m* ABWASSER eductor-basin cleaner, gully emptier; **Schlammstabilisierung** *f* ABWASSER sludge stabilization; **Schlammteich** *m* ABWASSER sludge lagoon, sludge pond; **Schlammteichverfahren** *nt* ABWASSER lagooning; **Schlammtrichter** *m* ABWASSER sludge hopper; **Schlammtrockenbett** *nt* ABWASSER sludge drying bed; **Schlammtrocknung** *f* ABWASSER sludge drying

Schlämmung *f* LUFT elutriation

Schlamm: **Schlammverbrennungsanlage** *f* ABWASSER sludge incinerator; **Schlammverklappung** *f* ABWASSER sludge barging; **Schlammverweilzeit** *f* ABWASSER sludge residence time (*SRT*), SRT (*sludge residence time*); **Schlammverwertung** *f* ABWASSER sludge recovery, sludge recycling; **Schlammwasser** *nt* ABWASSER, WASSER sludge liquor

schlank: **~es Ökobilanzieren** *nt* MANAG lean eco-balancing

Schlauch *m* LUFT bag; **Schlauchboden** *m* LUFT bag bottom; **Schlauchfilter** *nt* ABWASSER, BODEN bag filter, LUFT bag filter, baghouse; **Schlauchfilter** *nt* **mit Druckluftabreinigung** LUFT reverse-jet filter

schlauchförmig *adj* LUFT *Filter* bag-shaped

Schlauch: **Schlauchhäuser** *nt pl* LUFT multichamber bag filter; **Schlauchkolonne** *f* LUFT bag column; **Schlauchlänge** *f* LUFT bag length; **Schlauchquetschpumpe** *f* ABWASSER peristaltic pump; **Schlauchreihen** *f pl* LUFT bag rows

Schlaufenreaktor *m* ABWASSER loop-type bubble column

schlecht: **~ abbaubarer Stoff** *m* ABWASSER refractory substance; **~ behandeltes Abwasser** *nt* ABWASSER poorly treated wastewater; **~ benetzbar** *adj* LUFT poorly wettable; **~ benetzbare Stäube** *m pl* LUFT poorly wettable dusts

schleifenförmig: **~e Rauchfahne** *f* LUFT looping smoke plume

Schleifmittel *nt* SICHERHEIT abrasive medium, grinding medium

Schleppboot *nt* MEER tow-boat, tug boat

schleppen *vt* MEER tow

Schlepper *m* MEER tow-boat, tug boat

Schleppgeschirr *nt* MEER towing gear

Schleppleine *nt* MEER towline

Schleppnetz *nt* MEER trawl net

schleudern *vt* LUFT *zur Wandung* fling, *Zyklon* spin out

Schleuse *f* WASSER sluice

Schlick *m* ABWASSER silt; **Schlickablagerung** *f* WASSER silt deposition

schließen *vt* ABFALL *Deponie* close

Schließkraftbereich *m* ABFALL, WASSER *Sandwich-Spritzgießen* clamping force

Schließung *f*: **~ der Deponie** ABFALL landfill closure

Schlinge *f* ABFALL, WASSER sling

Schlitz *m* BODEN cut

schlitzen *vt* ABFALL slit, BODEN cut, WASSER slit

Schlitz: **Schlitzplatte** *f* LÄRM slotted panel; **Schlitzwand** *f* ABFALL, BODEN subterraneous curtain; **Schlitzwandlamelle** *f* ABFALL, BODEN subterraneous curtain plate

schluckbar *adj* ALLG absorbable

Schluckbrunnen *m* ABFALL, WASSER injection well

schlucken *vt* LÄRM absorb

Schmalwand *f* WASSER narrow curtain

schmelzen 1. *vt* ABFALL, ABWASSER, BODEN melt, **2.** *vi* ABFALL, ABWASSER, BODEN melt, SICHERHEIT *Kernkraftwerk* melt down

Schmelzen *nt* ABWASSER melting

Schmelz: **Schmelzenthalpie** *f* LUFT enthalpy of fusion; **Schmelzfilter** *nt* ABFALL *Kunststoffe* melt filter; **Schmelzflußelektrolyse** *f* ABFALL melting flux electrolysis; **Schmelzgranulat** *nt* ABFALL fusion granulate; **Schmelzkammerfeuerung** *f* ABFALL melting chamber combustion firing; **Schmelzmischer** *m* ABFALL melt mixer; **Schmelzpunkt** *m* ABFALL melting point; **Schmelzwanne** *f* ABFALL melting end

Schmiermittel *nt* ABFALL lubricant; **Schmiermittelrückstände** *m pl* ABFALL, ABWASSER waste lubricants

Schmutz *m* ALLG dirt, pollution; **Schmutzbehälter** *m* ABFALL *des Kehrfahrzeugs* dust collector; **Schmutzdecke** *f* ABWASSER mud blanket; **Schmutzfrachtindex** *m* POLITIK pollution index; **Schmutzstoff** *m* ABFALL, ABWASSER, ANALYTIK, BODEN, LUFT, MEER, POLITIK, SICHERHEIT, WASSER pollutant; **Schmutzwasser** *nt* ABWASSER wastewater

Schnecke *f* ABWASSER worm; **Schneckenbohrer** *m* ABFALL gimlet bit; **Schneckendosiereinheit** *f* ABFALL screw feeder; **Schneckendosierer** *m* ABFALL screw feeder; **Schneckenextruder** *m* ABFALL screw extruder; **Schneckenförderer** *m* ABFALL, LUFT screw conveyor; **Schneckenfördergerät** *nt* ABFALL screw conveyor; **Schneckenpumpe** *f* MEER screw pump; **Schneckenverdichter** *m* ABFALL screw compressor

Schneidepresse *f* ABFALL cutting press

Schneidöl *nt* ABFALL, ABWASSER cutting oil

Schneidzerkleinerer *m* ABFALL cutting shredder

Schnell- *in cpds* ABFALL, ABWASSER, ANALYTIK, WASSER quick, rapid

schnell *adj* ALLG accelerated, rapid; **~e Kompostierung** *f* ABFALL accelerated composting; **~ trocknend** *adj* ABFALL quick drying

Schnell-: **Schnellanalyse** *f* ANALYTIK quick analysis; **Schnellbestimmung** *f* ANALYTIK quick determination; **Schnellentkarbonisierung** *f* WASSER rapid decarbonization; **Schnellfilter** *nt* ABWASSER, WASSER rapid sand filter; **Schnellfiltration** *f* ABWASSER rapid filtration; **Schnellflockungsreaktor** *m* ABWASSER sludge contact clarifier

Schnelligkeit *f* ANALYTIK speed

Schnell-: **Schnellkompostierung** *f* ABFALL mechanical composting, rapid fermentation; **Schnellmischbecken** *nt*

ABWASSER rapid-mixing basin; **Schnellmischen** *nt* ABWASSER rapid mixing; **Schnellmischer** *m* ABFALL flash mixer; **Schnellprüfung** *f* ABFALL high-speed inspection; **Schnellrotte** *f* ABWASSER rapid rotting; **Schnellrührbecken** *nt* ABWASSER rapid-mixing basin; **Schnelltest** *m* ANALYTIK rapid test; **Schnelltrockner** *m* LUFT flash dryer
Schnittstelle *f* ANALYTIK *Computer* interface; **Schnittstellenmodul** *nt* ANALYTIK interface module
Schokoladenschaum *m* MEER *Öl-Wasser-Emulsion mit halbfesten Klumpen* chocolate mousse
schonen *vt* BODEN conserve
Schonung *f* BODEN conservation
Schönung *f* ABFALL polishing; **Schönungsteich** *m* ABWASSER tertiary pond
Schöpfgerät *nt* ANALYTIK scoop, scooper
Schornstein *m* ABFALL, ERNEU ENER, LUFT chimney, flue, smoke stack, stack; **Schornsteinauswurf** *m* LUFT chimney discharge, chimney emissions; **Schornsteinbauhöhe** *f* LUFT overall stack height, overall chimney height; **Schornsteindurchmesser** *m* LUFT stack diameter; **Schornsteingas** *nt* ABFALL, LUFT tail gas; **Schornsteinhöhe** *f* LUFT stack height, chimney height; **Schornsteinmündung** *f* LUFT stack orifice; **Schornsteinwand** *f* LUFT stack wall
Schotter *m* ABFALL ballast, LUFT rubble, WASSER ballast; **Schottermaterial** *nt* ABFALL, WASSER ballast material; **Schotterpfahl** *m* ABFALL, WASSER ballasted pile
Schräg- *in cpds* ABFALL inclined; **Schräginfiltration** *f* ABFALL inclined infiltration; **Schrägrohrverdampfer** *m* ABFALL inclined tube evaporator; **Schrägsortiermaschine** *f* ABFALL inclined sorting machine; **Schrägtisch** *m* ABFALL inclined table
Schrapper *m* MEER scraper; **Schrappergefäß** *nt* ABFALL scoop
Schrauben- *in cpds* LUFT helical
schraubenförmig *adj* LUFT helical
Schrauben-: **Schraubenpumpe** *f* MEER screw pump, WASSER screw-type pump; **Schraubenradpumpe** *f* ABWASSER mixed-flow pump; **Schraubenschaufler** *m* ABWASSER, MEER screw pump; **Schraubenwasser** *nt* MEER wake
Schredder *m* ABFALL shredder; **Schredderanlage** *f* ABFALL shredding machine, *für Autos* vehicle shredder; **Schreddermüll** *m* ABFALL shredder fluff, shredder remains, shredder residue, shredder waste
Schreddern *nt* ABFALL shredding
Schrott *m* ABFALL scrap; **Schrottballen** *m* ABFALL scrap bale; **Schrotteinsatz** *m* ABFALL scrap application; **Schrotthändler** *m* ABFALL scrap metal dealer, scrap metal merchant; **Schrottpaketierpresse** *f* ABFALL baling press, scrap baler; **Schrottplatz** *m* ABFALL scrap yard; **Schrottpresse** *f* ABFALL scrap-baling press, junk press, scrap press; **Schrottqualität** *f* ABFALL scrap quality; **Schrottsammlung** *f* ABFALL scrap collection; **Schrottschere** *f* ABFALL scrap shear; **Schrottsortierung** *f* ABFALL scrap sorting; **Schrottverhüttung** *f* ABFALL scrap smelting; **Schrottverwertung** *f* ABFALL scrap processing, scrap reuse
Schrumpfverhalten *nt* ABFALL, BODEN shrinking behavior (*AmE*), shrinking behaviour (*BrE*)
Schub *m* MEER thrust; **Schubkraft** *f* MEER thrust; **Schubspannung** *f* BODEN transverse strain
Schuldenerlaß *m*: ~ **für Naturschutz** POLITIK debt-for-nature swaps
Schürzensperre *f* MEER curtain boom
Schütt: **Schüttbett** *nt* LUFT packed bed; **Schüttdichte** *f* ABFALL apparent density, bulk density, ANALYTIK settled apparent density
Schüttelmethode *f* ANALYTIK shaking technique, vibration technique
schütteln *vt* ANALYTIK shake, vibrate
Schütteln *nt* ANALYTIK shaking, vibration
schüttfähig *adj* ABFALL pourable
Schütt: **Schüttgut** *nt* ABFALL bulk goods, LUFT bulk material, pellet; **Schüttgutkatalysator** *m* LUFT bulk material catalyst, pellet catalyst; **Schüttgutreaktor** *m* ABFALL bulk material reactor; **Schütthöhe** *f* ABFALL dumping height; **Schüttphasen** *f pl* ABFALL dumping phases; **Schüttschicht** *f* LUFT packed bed; **Schüttschichtfilter** *nt* LUFT packed-bed filter, *aus körnigen Schichten* granular-bed filter
Schüttung *f* ABFALL packed bed
Schutz *m* GESCHÄFT conservation, protection; **Schutzanzug** *m* SICHERHEIT protective suit; **Schutzausrüstung** *f* SICHERHEIT protective equipment
schutzbelüftet: **~e Kabine** *f* SICHERHEIT protective-aerated cab
Schutz: **Schutzbelüftung** *f* SICHERHEIT protective aeration
schützen *vt* ALLG protect
schützend: **~es Gehäuse** *nt* LUFT protective housing
Schutz: **Schutz- und Entschädigungsvereinigung** *f* MEER Protection & Indemnity Association; **Schutzfiltration** *f* BODEN protective insulation; **Schutzfunktion** *f* LUFT protective function; **Schutzinfiltrationsbrunnen** *m* BODEN protective insulation well; **Schutzkabine** *f* SICHERHEIT cab enclosure; **Schutzmaßnahmen** *f pl* ABWASSER, LÄRM, LUFT, MEER, SICHERHEIT protective measures; **Schutzmittel** *nt* ABFALL, WASSER protective agent; **Schutzschicht** *f* ABFALL protective layer; **Schutzschild** *m* LUFT protective shield; **Schutzwirkung** *f* LÄRM, MEER protection capacity; **Schutzzone** *f* SICHERHEIT protection zone
schwach: **~e Probe** *f* ANALYTIK weak sample; **~ radioaktiv** *adj* ABFALL low-level radioactive; **~ sauer** *adj* BODEN mildly acidic; **~ verdichtete Deponie** *f* ABFALL lightly compacted landfill
schwachbelastet *adj* ABWASSER low-rate; **~es kommunales Abwasser** *nt* ABWASSER low-contaminated municipal wastewater; **~er Tropfkörper** *m* ABWASSER low-rate trickling filter, standard-rate trickling filter; **~es Verfahren** *nt* ABWASSER low-rate process
schwächen *vt* ANALYTIK, LÄRM, LUFT attenuate, weaken
Schwachstellenanalyse *f* ANALYTIK analysis of weakest points
Schwächung *f* ANALYTIK, LÄRM, LUFT attenuation, weakening
Schwade *f* MEER swath
Schwaden *m* MEER swath; **Schwadenachse** *f* LUFT plume axis; **Schwadenfetzen** *m* LUFT plume shred
schwankend *adj* LUFT variable; **~er Abgasstrom** *m* LUFT variable waste gas flow; **~er Gasdurchsatz** *m* LUFT variable gas throughput; **~e Gasdurchsatzmenge** *f* LUFT variable gas throughput amount, variable gas thruput amount (*AmE*)
schwarz: **~e Liste** *f* ABWASSER *der wassergefährdenden Stoffe*, POLITIK black list
Schweb- *in cpds* ABFALL, ABWASSER, ANALYTIK, LUFT suspended, suspension
Schwebebett *nt* ABWASSER packed suspended bed (*PSB*)
schwebend *adj* ABFALL, ABWASSER suspended, LUFT airborne, WASSER suspended; **~es Grundwasser** *nt* WASSER perched groundwater, vadose water; **~e Tröpfchen** *nt pl* ABFALL, LUFT, WASSER suspended liquid droplets
Schwebetrockner *m* ABWASSER, LUFT suspension dryer
Schwebkörper *m* ANALYTIK suspended solid particle; **Schwebkörperdurchflußmesser** *m* ANALYTIK rotameter; **Schwebkörpermesser** *m* ANALYTIK suspended solid particle meter; **Schwebkörperviskosimeter** *nt* ANALYTIK rotameter viscosimeter
Schweb-: **Schweb- und Sinkstoffe** *m pl* ABWASSER suspended and settable solids; **Schwebstaub** *m* ANALYTIK suspended solid particle, LUFT airborne particulate matter, solid particulates, suspended particulate matter
Schwebstoff *m* ABWASSER suspended matter, LUFT, WASSER suspended particle; **Schwebstoffe** *m pl* ABWASSER, WASSER

suspended solids; **Schwebstoffilter** *nt* LUFT particulate matter filter, *Speicherfilter* submicron particulate filter
schwebstofffrei *adj* ABWASSER free from suspended matter
***Schwebstoff*: Schwebstoffkontakt** *m* ABFALL, ABWASSER solids contact; **Schwebstoffteilchen** *nt* (*PM*) LUFT particulate matter (*PM*)
***Schweb-*: Schwebteilchen** *nt* WASSER suspended particle
Schwefel *m* ABFALL, LUFT, WASSER sulfur (*AmE*), sulphur (*BrE*)
schwefelarm *adj* ABWASSER low-sulfur (*AmE*), low-sulphur (*BrE*)
***Schwefel*: Schwefeldioxid** *nt* (SO_2) LUFT sulfur dioxide (*AmE*) (SO_2), sulphur dioxide (*BrE*) (SO_2); **Schwefeldioxidemissionen** *f pl* LUFT sulfur dioxide emissions (*AmE*), sulphur dioxide emissions (*BrE*); **Schwefeldioxidreduktion** *f* ABFALL, LUFT sulfur dioxide reduction (*AmE*), sulphur dioxide reduction (*BrE*); **Schwefeldioxidvergiftung** *f* LUFT sulphur dioxide poisoning (*BrE*), sulfur dioxide poisoning (*AmE*)
schwefelfrei *adj* LUFT sulfur-free (*AmE*), sulphur-free (*BrE*)
***Schwefel*: Schwefelgehalt** *m* LUFT sulfur content (*AmE*), sulphur content (*BrE*)
schwefelhaltig *adj* ABFALL, LUFT, WASSER sulfurous (*AmE*), sulphurous (*BrE*); **~er Abfall** *m* ABFALL sulfurous waste (*AmE*), sulphurous waste (*BrE*); **~e Brennstoffe** *m pl* ABFALL, LUFT sulfurous combustibles (*AmE*), sulphurous combustibles (*BrE*); **~es Gas** *nt* LUFT sulfurous gas (*AmE*), sulphurous gas (*BrE*)
***Schwefel*: Schwefelhaushalt** *m* LUFT sulfur budget (*AmE*), sulphur budget (*BrE*); **Schwefelkreislauf** *m* LUFT sulfur cycle (*AmE*), sulphur cycle (*BrE*); **Schwefeloxid** *nt* (*SO*) ABFALL, LUFT sulfur oxide (*AmE*) (*SO*), sulphur oxide (*BrE*) (*SO*); **Schwefeloxide** *nt pl* (SO_x) LUFT sulfur oxides (*AmE*) (SO_x), sulphur oxides (*BrE*) (SO_x); **Schwefeloxidwäscher** *m* LUFT sulfur oxides scrubber (*AmE*), sulphur oxides scrubber (*BrE*)
schwefelreich *adj* LUFT sulfur-rich (*AmE*), sulphur-rich (*BrE*)
***Schwefel*: Schwefelrückgewinnungsanlage** *f* ABFALL, LUFT *Claus-Anlage* sulfur recovery plant (*AmE*), sulphur recovery plant (*BrE*); **Schwefelsäure** *f* ABFALL, LUFT sulfuric acid (*AmE*), sulphuric acid (*BrE*); **Schwefelsäureanhydrid** *nt* ABFALL sulfur anhydride (*AmE*), sulphur anhydride (*BrE*); **Schwefelsäurenanhydrid** *nt* LUFT, WASSER sulfuric anhydride (*AmE*), sulphuric anhydride (*BrE*); **Schwefeltetroxid** *nt* ($(SO_4)_x$) LUFT sulfur tetraoxide (*AmE*) ($(SO_4)_x$), sulphur tetraoxide (*BrE*) ($(SO_4)_x$); **Schwefeltrioxid** *nt* (SO_3) ABFALL, LUFT sulfur trioxide (*AmE*) (SO_3), sulphur trioxide (*BrE*) (SO_3); **Schwefelwasserstoff** *m* ABFALL, LUFT hydrogen sulfide (*AmE*), hydrogen sulphide (*BrE*)
schweflig: **~e Säure** *f* ABFALL, LUFT sulfurous acid (*AmE*), sulphurous acid (*BrE*)
Schwelbrennverfahren *nt* ABWASSER *Klärschlammverbrennung* process combining pyrolysis and incineration
Schwellenwert *m* ANALYTIK threshold, threshold value, POLITIK threshold; **Schwellenwertaudiometrie** *f* LÄRM threshold audiometry
Schwelwasser *nt* ABWASSER low-temperature carbonization wastewater
Schwemmkegel *m* WASSER alluvial cone
Schwemmlandebene *f* WASSER alluvial plain
schwer: **~ abbaubar** *adj* ABWASSER nondegradable; **~ abbaubarer Stoff** *m* ABWASSER nondegradable substance; **~ benetzbar** *adj* LUFT damp-resistant; **~ benetzbarer Staub** *m* LUFT damp-resistant dust; **~ entflammbar** *adj* SICHERHEIT flame-resistant; **~es Kohlenwasserstoffgas** *nt* LUFT ethylene; **~es Wasser** *nt* ERNEU ENER heavy water; **~ zersetzbar** *adj* ABFALL, WASSER nondegradable
Schwerkraft *f* ALLG gravitational force, gravity; **Schwerkraftabscheidekammer** *f* LUFT gravity settling chamber; **Schwerkraftabscheider** *m* ABFALL, ABWASSER gravity separator, LUFT gravity settler, gravity separator, WASSER gravity separator; **Schwerkraftabscheidung** *f* ABWASSER *Behandlung von ölhaltigen Abwässern*, BODEN gravity separation, LUFT gravity settling, gravitational settling; **Schwerkraftabsetzung** *f* LUFT gravitational settling; **Schwerkrafteindicker** *m* ABWASSER gravity thickener; **Schwerkrafteindickung** *f* ABWASSER gravity thickening; **Schwerkraftentwässerung** *f* ABWASSER gravity thickening; **Schwerkraftentwässerung** *f* **in Containern** ABWASSER gravity thickening in containers; **Schwerkraftentwässerung** *f* **in Filtersäcken** ABWASSER gravity thickening in filter bags; **Schwerkraftförderer** *m* ABFALL, WASSER gravity roller conveyor; **Schwerkraftsedimentation** *f* ANALYTIK *Meßverfahren für Staube* gravity sedimentation, LUFT gravity settling, gravitational settling; **Schwerkraftsprühturm** *m* LUFT gravity spray tower; **Schwerkrafttrennung** *f* LUFT gravity settling, gravitational settling
schwerlöslich *adj* ABFALL, LUFT, WASSER difficult to dissolve, sparingly soluble; **~e Komponente** *f* ABFALL, LUFT, WASSER difficult-to-dissolve component
Schwermetall *nt* ABFALL, ABWASSER, LUFT, WASSER heavy metal; **Schwermetallablagerung** *f* ABWASSER, BODEN, LUFT heavy metals deposition; **Schwermetallanreicherung** *f* ABWASSER heavy metals accumulation, ANALYTIK heavy metal enrichment, BODEN, LUFT heavy metals accumulation; **Schwermetallbestimmung** *f* ANALYTIK heavy metal determination, heavy metal analysis; **Schwermetallemissionen** *f pl* LUFT heavy metal emissions; **Schwermetallentfernung** *f* ABWASSER, LUFT, WASSER heavy metals removal; **Schwermetallfällung** *f* ABWASSER heavy metals precipitation
schwermetallfrei *adj* ABWASSER, BODEN, LUFT, WASSER free of heavy metal
***Schwermetall*: Schwermetallgehalt** *m* ABFALL, ABWASSER, BODEN, LUFT, WASSER heavy metal content
schwermetallhaltig *adj* ALLG wastewater containing heavy metals
***Schwermetall*: Schwermetallkonzentration** *f* ABFALL, ABWASSER, LUFT, WASSER heavy metal concentration; **Schwermetallkreislauf** *m* ABWASSER, BODEN, LUFT heavy metals circulation; **Schwermetallquelle** *f* ABWASSER, BODEN, LUFT heavy metals source
Schwer: **Schwerölentschwefelung** *f* ABFALL heavy oil desulfurization (*AmE*), heavy oil desulphurization (*BrE*); **Schwerölrückstand** *m* ABFALL heavy oil residue; **Schwerpunktlaboratorium** *nt* ANALYTIK specialist lab (*specialist laboratory*); **Schwerstofffraktion** *f* ABFALL coarse fraction, coarse medium fraction
Schwimm: **Schwimmbagger** *m* MEER dredger; **Schwimmdecke** *f* ABWASSER *in einem Gärungsbottich* supernatant liquid
schwimmend *adj* ALLG floating; **~er Fußboden** *m* LÄRM *federnde Schicht* floating floor; **~er Kunststoffbehälter** *m* MEER dracone; **~es Meereswärmekraftwerk** *nt* ERNEU ENER floating OTEC plant; **~er Rückstand** *m* ABFALL floating refuse
***Schwimm*: Schwimmhaut** *f* MEER webbing; **Schwimmprobenahme** *f* ANALYTIK floating sampling
Schwimmschlamm *m* ABWASSER scum; **Schwimmschlammbildung** *f* ABWASSER scum formation; **Schwimmschlammentfernung** *f* ABWASSER scum removal; **Schwimmschlammräumer** *m* ABWASSER scum skimmer, scum scraper
***Schwimm*: Schwimm-Sink-Flotation** *f* WASSER sink-float flotation; **Schwimmstoffabweiser** *m* ABWASSER scum board; **Schwimmstoffe** *m pl* ABWASSER floating solids
Schwing: **Schwingrinne** *f* WASSER vibrating conveyor; **Schwingsieb** *nt* ABFALL vibrating screen, ABWASSER vibrat-

ing sieve, LUFT *Schüttschichtfilter* shaking screen; **Schwingsviskosimeter** *nt* ANALYTIK vibration viscosimeter
Schwingung *f* ANALYTIK, LÄRM swing, vibration; **Schwingungsaufnehmer** *m* ANALYTIK vibration sensor; **Schwingungsdämpfer** *m* LÄRM vibration dampener; **Schwingungsdämpfung** *f* LÄRM vibration attenuation, vibration damping; **Schwingungsisolation** *f* LÄRM vibration isolation; **Schwingungsisolator** *m* LÄRM isolator, vibration isolator
schwingungsisoliert: **~e Aufstellung** *f* LÄRM antivibration mounting
Schwingung: **Schwingungsschutz** *m* LÄRM vibration isolation; **Schwingungsverhalten** *nt* ERNEU ENER vibration behavior (*AmE*), vibration behaviour (*BrE*); **Schwingungswandler** *m* ANALYTIK vibration transducer
Schwungmasse *f* ERNEU ENER centrifugal mass
Scraper *m* MEER scraper
Scrap-Kaskade *f* WASSER scrap cascade
Screening *nt* ANALYTIK, MEER screening; **Screening-Analyse** *f* ANALYTIK screening analysis; **Screening-Test** *m* ANALYTIK screening test
SdT *abbr* (*Stand der Technik*) ALLG state of the art
SE *abbr* (*Schallemission*) ANALYTIK, LÄRM AE (*acoustic emission*), sound emission
SEA *abbr* (*Schallemissionsanalyse*) ANALYTIK AE analysis (*acoustic emissions analysis*)
SEALOSAFE-Verfahren® *nt* ABFALL SEALOSAFE method® (*BrE*), SEALOSAFE process® (*BrE*)
Sediment *nt* ALLG sediment; **Sedimentablagerung** *f* BODEN *Geologie* deposition, LUFT sedimentation, WASSER sediment discharge
Sedimentation *f* ABWASSER sedimentation, settling out, ANALYTIK, LUFT sedimentation, WASSER sedimentation, settling out; **Sedimentationsbecken** *nt* ABWASSER sedimentation tank
sedimentierbar *adj* ABWASSER, ANALYTIK, LUFT, WASSER sedimentable; **~er Stoff** *m* LUFT sedimentable material
sedimentieren *vt* ABWASSER, LUFT sediment
sedimentierfähig *adj* ABWASSER able to be sedimented
sedimentiert *adj* ABWASSER, ANALYTIK, LUFT, WASSER settled, sedimented; **~er Schlamm** *m* ABWASSER sedimented sludge
Sedimentierzentrifuge *f* ABWASSER sedimentation centrifuge
Sediment: **Sedimentprobenahme** *f* ANALYTIK sediment sampling; **Sedimentzuwachs** *m* WASSER aggradation
See *f* ABWASSER lake, MEER sea, WASSER lake; **Seefassung** *f* WASSER *zur Trinkwassergewinnung* tapping of lake water; **Seegang** *m* ERNEU ENER motion of the sea; **Seeklima** *nt* MEER marine climate; **Seeschiff** *nt* MEER sea-going ship, sea-going vessel; **Seetüchtigkeit** *f* MEER seaworthiness; **Seewasser** *nt* ERNEU ENER, MEER, WASSER seawater; **Seewasser-Rohöl-Emulsion** *f* LUFT seawater-in-crude oil emulsion
Seger-Kegel *m* ANALYTIK Seger cone
Segmentleitblech *nt* LUFT segment baffle
seilgeführt: **~er Bohrlochgreifer** *m* BODEN cable-guided drilling grab
seilverspannt *adj* ERNEU ENER cable-braced; **~e Konstruktion** *f* ERNEU ENER *Solarmodule* cable-braced construction
Seiten: **Seitenbereich** *m* ABFALL, WASSER side area; **Seitenkanal** *m* WASSER lateral canal; **Seitenkanalpumpe** *f* ABWASSER regenerative pump; **Seitenkanalverdichter** *m* BODEN *Bodenluftabsaugung* side-channel compressor; **Seitensichtradar** *nt* MEER sideways-looking airborne radar (*SLAR*)
Sekretariat *nt* POLITIK secretariat
sektoral: **~er Ansatz** *m* POLITIK sectoral approach
Sektorfeldmassenspektrometrie *f* (*Sektorfeld-MS*) ANALYTIK magnetic sector mass spectrometer (*magnetic sector MS*)
Sektorfeld-MS *f* (*Sektorfeldmassenspektrometrie*) ANALYTIK magnetic sector MS (*magnetic sector mass spectrometer*)
Sekundär- *in cpds* ALLG secondary
sekundär *adj* ALLG secondary; **~e Adsorption** *f* ABWASSER secondary adsorption; **~e Setzung** *f* ABFALL, BODEN, WASSER secondary settlement
Sekundär-: **Sekundärabfallvermeidung** *f* ABFALL secondary waste avoidance; **Sekundäradsorber** *m* LUFT secondary adsorber; **Sekundäradsorption** *f* LUFT secondary adsorption; **Sekundäraluminium** *nt* ABFALL secondary aluminium (*BrE*), secondary aluminum (*AmE*); **Sekundärbeladung** *f* LUFT *des Abgases* secondary loading; **Sekundärelektronenvervielfacher** *m* (*SEV*) ANALYTIK photomultiplier tube (*PMT*); **Sekundärenergieträger** *m* ERNEU ENER *Strom* secondary energy source; **Sekundärfallrohr** *nt* LUFT *Schüttschichtfilter* secondary downpipe; **Sekundärionen-Massenspektrometrie** *f* (*SIMS*) ANALYTIK secondary ion mass spectrometry (*SIMS*); **Sekundärluft** *f* ABWASSER, LUFT secondary air; **Sekundärlufteinblasung** *f* LUFT *Katalysator* secondary air injection; **Sekundärrecycling** *nt* ABFALL secondary recycling; **Sekundärrohstoff** *m* ABFALL secondary raw material, waste product capable of being reprocessed; **Sekundärschlamm** *m* ABWASSER secondary sludge; **Sekundärsiebung** *f* ABFALL, ABWASSER secondary sieving; **Sekundärwertstoff** *m* ABFALL recycled material
Selbst- *in cpds* ALLG self
selbstansaugend: **~e Kreiselpumpe** *f* ABFALL, ABWASSER, WASSER self-priming centrifugal pump
selbstdränend *adj* BODEN self-draining
selbstentzündlich *adj* LUFT pyrophoric
selbstladend: **~es Müllfahrzeug** *nt* ABFALL self-loading refuse vehicle
selbstnachstellend: **~es Schwimmwehr** *nt* MEER self-adjusting floating weir
selbstreinigend *adj* ABWASSER self-cleaning; **~es Drehtrommelsieb** *nt* ABWASSER self-cleaning rotary-drum screen; **~er Zentrifugalsortierer** *m* ABWASSER self-cleaning centrifugal screen
Selbst-: **Selbstreinigung** *f* ABWASSER, BODEN, WASSER natural purification, self-purification; **Selbstreinigungskraft** *f* ABWASSER, BODEN, WASSER natural purification force, self-purification force; **Selbstüberwachung** *f* POLITIK, WASSER self-monitoring; **Selbstverdichtung** *f* ABFALL self compaction; **Selbstverpflichtung** *f* POLITIK voluntary commitment
selektiv: **~es Ionenaustauscherharz** *nt* ABFALL, WASSER selective ion exchange resin; **~es nicht-katalytisches Verfahren** *nt* (*SNCR-Verfahren*) WASSER selective non-catalytic reduction (*SNCR*)
selektiv-katalytisch *adj* LUFT selective catalytic; **~e Reduktion** *f* (*SKR*) ABFALL, LUFT selective catalytic reduction (*SCR*); **~er Reduktionsreaktor** *m* (*SKR-Reaktor*) ABFALL, LUFT selective catalytic reduction reactor
semiflüchtig: **~e organische Verbindungen** *f pl* (*SVOC*) BODEN semivolatile organic compounds (*SVOC*)
semipermeabel: **semipermeable Membran** *f* ABFALL, ABWASSER, WASSER *Umkehrosmose* semipermeable membrane
semitragbar *adj* ANALYTIK semiportable
Senkgrube *f* ABWASSER cesspit, cesspool
Senkung *f* ABFALL, WASSER *der Deponiesohle* sinking; **Senkungskegel** *m* ABFALL, WASSER depression cone; **Senkungsquelle** *f* ABFALL, WASSER depression spring
Senkwasser *nt* ABFALL, WASSER percolating water
Sensor *m* ANALYTIK probe, sensor
Sensorik *f* ANALYTIK sensory analysis
Sensor: **Sensormodul** *nt* ANALYTIK sensor module
Separator *m* MEER separator
separierfähig *adj* ABFALL separable
Sequencing-Batch-Reaktor *m* ABWASSER sequencing batch reactor

Sequential: ~ **Batch Reactor** *m* (*SBR*) BODEN sequential batch reactor (*SBR*)
sequentiell *adj* ABWASSER sequential; **~e Dosierung** *f* ABWASSER sequential dosage
servomotorisch *adj* ANALYTIK servomotoric
Setzen *nt* ABFALL, BODEN jigging
Setzung *f* ABFALL, BODEN, WASSER settlement; **Setzungsgeschwindigkeit** *f* ABFALL settlement speed; **Setzungspolster** *nt* ABFALL settlement cushion; **Setzungsstillstand** *m* ABFALL, BODEN, WASSER settlement stoppage
SEV *abbr* (*Sekundärelektronenvervielfacher*) ANALYTIK PMT (*photomultiplier tube*)
sicher: **~er Betrieb** *m* SICHERHEIT safe operation; **~e Entsorgung** *f* ABFALL safe disposal; **~e Lagerung** *f* ABFALL safe keeping
Sicherheit *f* ANALYTIK, SICHERHEIT safety; **Sicherheitsabstand** *m* SICHERHEIT safety distance, safety zone; **Sicherheitsanalyse** *f* SICHERHEIT safety analysis; **Sicherheitsbeschwerden** *f pl* SICHERHEIT safety complaints
sicherheitsbezogen *adj* SICHERHEIT safety-related; **~e Ausrüstung** *f* SICHERHEIT safety-related equipment
Sicherheit: **Sicherheitsfaktor** *m* SICHERHEIT safety factor; **Sicherheitskontrolle** *f* SICHERHEIT safety check; **Sicherheitsmaßnahmen** *f pl* SICHERHEIT safety measures; **Sicherheitsrisiko** *nt* SICHERHEIT safety risk; **Sicherheitsstandard** *m* SICHERHEIT safety standard; **Sicherheitsstrategie** *f* SICHERHEIT safety strategy; **Sicherheitstechnik** *f* SICHERHEIT safety engineering
sicherheitstechnisch *adj* SICHERHEIT safety-related; **~e Anforderung** *f* SICHERHEIT technical safety demand; **~e Maßnahmen** *f pl* SICHERHEIT technical safety measures
Sicherheit: **Sicherheitsverpackung** *f* ABFALL barrier packaging; **Sicherheitsvorschriften** *f pl* SICHERHEIT safety code, safety regulations, safety requirements
Sicherung *f* SICHERHEIT security; **Sicherungsmaßnahmen** *f pl*, SICHERHEIT security measures, safeguards; **Sicherungssystem** *nt* SICHERHEIT security system
Sicht: **Sichtanzeigegerät** *nt* ANALYTIK display unit
sichtbar *adj* ANALYTIK, LUFT visible; **~e Lichtabsorption** *f* ANALYTIK visible optical absorption; **~er Schwaden** *m* LUFT visible plume
Sichtbaren: **im ~ und Ultraviolett** *phr* (*UV/VIS*) ANALYTIK ultraviolet and visible (*UV/VIS*)
Sichter *m* LUFT separator; **Sichteranlage** *f* ABFALL classification station
Sicht: **Sichtgerät** *nt* ANALYTIK display unit; **Sichtglas** *nt* ANALYTIK inspection window; **Sichtscheibe** *f* ANALYTIK inspection glass; **Sichtschutz** *m* ABFALL, WASSER *einer Deponie* screen
Sichtschutz *m* ABWASSER screen
Sicht: **Sichttiefe** *f* ANALYTIK visibility depth
Sichtung *f* ABWASSER classification; **Sichtungsstation** *f* ABFALL classification station; **Sichtungstisch** *m* ABFALL sorting table
Sicker: **Sickerbecken** *nt* ABFALL, ABWASSER, WASSER infiltration basin; **Sickerbrunnen** *m* ABFALL, WASSER percolation well; **Sickergalerie** *f* ABFALL, WASSER infiltration gallery; **Sickergrube** *f* ABWASSER absorbing well
sickern *vt* ABFALL, ABWASSER, WASSER seep
Sicker: **Sickerrohr** *nt* ABFALL drainage pipe; **Sickerschacht** *m* ABWASSER absorbing well; **Sickerschlitz** *m* ABFALL, WASSER infiltration slit; **Sickerstollen** *m* ABFALL, WASSER infiltration gallery
Sickerwasser *nt* ABFALL, WASSER leachate, leak water, leakage water, *tropfend* percolating water, seepage water; **Sickerwasserableitung** *f* ABFALL, WASSER leachate drainage; **Sickerwasseranfall** *m* ABFALL, WASSER leachate yield; **Sickerwasseraufbereitung** *f* ABWASSER percolating water treatment plant; **Sickerwasserbehandlungsanlage** *f* ABFALL *einer Deponie* landfill leachate treatment facility; **Sickerwasserbestimmung** *f* ANALYTIK leachate determination; **Sickerwasserdränage** *f* ABFALL, WASSER leachate drainage; **Sickerwasserdränage** *f* **mit Filterschotter** ABFALL, WASSER leachate drainage with filtering rubble; **Sickerwassererfassung** *f* ABFALL, WASSER leachate collection; **Sickerwasserpumpe** *f* ABFALL leachate pump; **Sickerwasserreinigung** *f* ABFALL leachate purification; **Sickerwassersammlung** *f* ABFALL, WASSER leachate collection; **Sickerwasserverhalten** *nt* ABFALL, WASSER leachate behavior (*AmE*), leachate behaviour (*BrE*); **Sickerwasserverminderung** *f* ABFALL leachate reduction
Sieb *nt* ABFALL screening equipment, sieve, ABWASSER, BODEN, LUFT sieve; **Siebanalyse** *f* ANALYTIK screen analysis, sieve analysis; **Siebanlage** *f* ABFALL sieve installation
siebbar *adj* ABFALL, ABWASSER sievable
Sieb: **Siebblech** *nt* LUFT perforated plate; **Siebboden** *m* LUFT sieve plate; **Siebbodenkolonne** *f* LUFT tray plate column; **Siebdurchgang** *m* ABFALL screen underflow, *Kompostierung* through fraction
sieben *vt* ABFALL, ABWASSER, BODEN, LUFT MEER sieve, screen, sift
Sieben *nt* ALLG screening, sieving, sifting
Sieb: **Siebfilter** *nt* ABFALL strainer; **Siebfraktion** *f* ABFALL sieve fraction; **Siebgewebe** *nt* ABFALL wire gauze; **Siebgröße** *f* ABFALL sieve size; **Siebkette** *f* ANALYTIK bandpass filter; **Siebmaschine** *f* LUFT screening machine; **Siebrest** *m* ABFALL screenings, sieve residue; **Siebrost** *m* LUFT sieve grate; **Siebrückstand** *m* ABFALL screenings; **Siebsatz** *m* ABWASSER sieve set; **Siebüberlauf** *m* ABFALL screen overflow
Siebung *f* ALLG screening, sieving
Sieb: **Siebwechselvorrichtung** *f* ABFALL screen-changing device
Siedeanalyse *f* ANALYTIK analysis boiling
sieden *vt* ANALYTIK boil
siedend *adj* ANALYTIK boiling; **~e Probe** *f* ANALYTIK boiling sample
Siedepunkt *m* ANALYTIK, ERNEU ENER boiling point
Siedlungsabfall *m* ABFALL urban waste, domestic waste, municipal solid waste (*MSW*), human settlement waste; **Siedlungsabfallstrom** *m* ABFALL municipal waste stream (*MWS*)
Signalerfassung *f* ANALYTIK peak detection
Silage *f* ABWASSER silage
silieren *vt* ABFALL convey into silos
Silieren *nt* ABFALL ensilage, silo storage
Silikagel *nt* ABWASSER, LUFT *Adsorbens* silica gel
Silikat *nt* ABFALL silicate
Silo *m* ABFALL, ABWASSER silo; **Silolagerung** *f* ABWASSER silage
SIMS *abbr* (*Sekundärionen-Massenspektrometrie*) ANALYTIK SIMS (*secondary ion mass spectrometry*)
Simulation *f* MANAG simulation; ~ **von Flußsystemen** MANAG simulation of river systems
simultan *adj* ABWASSER, ANALYTIK simultaneous; **~e Anzeige** *f* ANALYTIK simultaneous readout; **~e Denitrifikation** *f* ABWASSER simultaneous denitrification
Simultan: **Simultanbetrieb** *m* ANALYTIK simultaneous operation; **Simultanmessung** *f* ANALYTIK simultaneous measurement
sinkend *adj* ABFALL, WASSER *Grundwasserspiegel* sinking
Sink: **Sinkgeschwindigkeit** *f* LUFT settling speed; **Sinkgut** *nt* MEER sinking agent; **Sinkmaterial** *nt* MEER sinking agent; **Sinkstoffe** *m pl* ABWASSER, LUFT sinking material, sinking matter
Sinterband *nt* BODEN *zum Pelletbrennen* band oven, sinter oven
sintern *vt* ABFALL, BODEN sinter, vitrify
Sintern *nt* ABFALL, BODEN sintering, vitrification
Skala *f* ANALYTIK scale
Skimmer *m* MEER skimmer

Skin-Packung *f* ABFALL blister pack

SKR *abbr* (*selektiv-katalytische Reduktion*) ABFALL, LUFT SCR (*selective catalytic reduction*); **SKR-Reaktor** *m* (*selektiv-katalytischer Reduktionsreaktor*) ABFALL, LUFT SCR reactor

Slug-Bail-Test *m* ANALYTIK slug-bail test

Smog *m* LUFT smog; **Smogalarm** *m* LUFT smog alarm; **Smoggebiet** *nt* LUFT smog area; **Smoggefahr** *f* LUFT smog risk; **Smogperiode** *f* LUFT smog period; **Smogwarnung** *f* LUFT smog warning; **Smogwetterlage** *f* LUFT smog weather conditions

SNCR-Verfahren *nt* (*selektives nicht-katalytisches Verfahren*) WASSER selective non-catalytic reduction (*SNCR*)

SNG *abbr* (*Erdgasaustauschgas*) ERNEU ENER SNG (*synthetic natural gas*)

SO *abbr* (*Schwefeloxid*) ABFALL, LUFT SO (*sulfur oxide AmE, sulphur oxide BrE*)

SO_2 *abbr* (*Schwefeldioxid*) ABFALL, LUFT SO_2 (*sulfur dioxide AmE, sulphur dioxide BrE*); **SO_2/NO_x-Entstickung** *f* (*Desonox*) BODEN, LUFT SO_2/NO_x removal (*desonox*)

SO_3 *abbr* (*Schwefeltrioxid*) ABFALL, LUFT SO_3 (*sulfur trioxide AmE, sulphur trioxide BrE*)

$(SO_4)_x$ *abbr* (*Schwefeltetroxid*) ABFALL, LUFT $(SO_4)_x$ (*sulfur tetraoxide AmE, sulphur tetraoxide BrE*)

SO_x *abbr* (*Schwefeloxide*) ABFALL, LUFT SO_x (*sulfur oxides AmE, sulphur oxides BrE*)

Soda *nt* ABFALL *Glasrecycling*, ABWASSER soda

Sofortmaßnahmen *f pl* POLITIK immediate measures

Software *f* ANALYTIK software

software-gesteuert *adj* ANALYTIK software-controlled

Sog *m* MEER wake

Sohldränage *f* ABFALL, WASSER bottom drainage

Sohleabdichtung *f* BODEN bottom sealing

solar *adj* ERNEU ENER solar; **~e Entsorgung** *f* **chemischer Abfälle** ABFALL solar disposal of chemical wastes; **~e erneuerbare Energiequellen** *f pl* ERNEU ENER solar-based renewable energy sources; **~e Fernwärme** *f* ERNEU ENER solar district heating; **~e Fernwärmelieferung** *f* ERNEU ENER solar district heating supply; **~e Hochtemperaturzerstörung** *f* ABFALL, WASSER *Entgiftung von Abfällen* solar high-temperature destruction; **~es Wärmekraftwerk** *nt* ERNEU ENER solar heat power plant

Solar-: **Solararchitektur** *f* ERNEU ENER solar architecture

solarbeheizt *adj* ERNEU ENER solar-heated; **~er Dampferzeuger** *m* ERNEU ENER solar-heated steam generator

solarbetrieben *adj* ERNEU ENER solar-powered, solar-driven, solar-operated; **~e Brunnenanlage** *f* ERNEU ENER solar-driven well installation

Solar-: **Solardish-Anlage** *f* ERNEU ENER solar dish installation; **Solarenergie** *f* ERNEU ENER solar energy; **Solarenergienutzung** *f* ERNEU ENER solar energy use; **Solarfarm** *f* ERNEU ENER solar farm; **Solarfarmanlage** *f* ERNEU ENER solar farm installation; **Solarfarmleistung** *f* ERNEU ENER solar farm capacity

solargestützt *adj* ERNEU ENER solar-supported; **~es Fernwärmesystem** *nt* ERNEU ENER solar-supported district heating system

Solar-: **Solarheizung** *f* ERNEU ENER solar heating; **Solarheizungsanlage** *f* ERNEU ENER solar heating plant

Solarimeter *nt* ANALYTIK solarimeter

Solar-: **Solarkollektor** *m* ERNEU ENER solar collector; **Solarkraftwerk** *nt* ERNEU ENER solar power plant; **Solarmodul** *nt* ERNEU ENER solar module; **Solarreaktor** *m* ERNEU ENER solar reactor; **Solarstand** *m* ERNEU ENER position of the sun; **Solarstrahlung** *f* ERNEU ENER, LUFT solar radiation; **Solarstrom** *m* ERNEU ENER solar power; **Solartechnik** *f* ERNEU ENER solar technology; **Solarteich** *m* ERNEU ENER solar pond; **Solarteichanlage** *f* ERNEU ENER solar pond installation; **Solarthermie** *f* ERNEU ENER solar-thermal energy

solarthermisch *adj* ERNEU ENER solar-thermal; **~es Kraftwerk** *nt* ERNEU ENER solar power plant

Solar-: **Solarturm** *m* ERNEU ENER solar tower; **Solarturmanlage** *f* ERNEU ENER solar tower installation; **Solarunternehmen** *nt* ERNEU ENER solar energy enterprise; **Solar-Wasserstoff-Technologie** *f* ERNEU ENER solar-hydrogen-technology; **Solarzelle** *f* ERNEU ENER solar cell; **Solarzellenkette** *f* ERNEU ENER array of solar cell submodules; **Solarzellenmodul** *nt* ERNEU ENER solar cell module

Solubilisation *f* ABWASSER solubilization

Solubilisierung *f* ABWASSER solubilization

Solvation *f* ABFALL, ABWASSER, BODEN, WASSER solvation

solvatisieren *vt* ABFALL, ABWASSER, BODEN, WASSER solvate

solvatisiert *adj* ABFALL, ABWASSER, BODEN, WASSER solvated

Solvolyse *f* ABFALL solvolysis

sommerlich: **~er Ozonwert** *m* LUFT summertime ozone value

Sommersmog *m* LUFT photochemical smog

Sondenmikrofon *nt* ANALYTIK probe microphone

Sonderabfall *m* ABFALL, ABWASSER hazardous waste; **Sonderabfalldeponie** *f* ABFALL hazardous waste landfill; **Sonderabfallverbrennung** *f* ABFALL hazardous waste incineration; **Sonderabfallverbrennungsanlage** *f* ABFALL hazardous waste incineration plant; **Sonderabfallzwischenlager** *nt* ABFALL intermediate storage site for hazardous waste

Sondermüll *m* ABFALL, ABWASSER hazardous waste; **Sondermüllanfall** *m* ABFALL generation of hazardous waste; **Sondermüllbehandlung** *f* ABFALL hazardous waste treatment; **Sondermülldeponie** *f* ABFALL hazardous waste landfill; **Sondermülldeponierung** *f* ABFALL hazardous waste landfilling; **Sondermülleinsammlung** *f* ABFALL hazardous waste collection; **Sondermüllkonzentrat** *nt* ABFALL hazardous waste concentrate; **Son-dermüllverbrennung** *f* ABFALL hazardous waste incineration; **Sondermüllverbrennungsanlage** *f* ABFALL hazardous waste incineration plant; **Sondermüllvermeidung** *f* ABFALL hazardous waste avoidance

Sondierbohrung *f* ANALYTIK sounding boring

sondieren *vt* ANALYTIK sound

Sondierung *f* ANALYTIK sounding; **Sondierungsbohrung** *f* ANALYTIK sounding boring

Sone *nt* ANALYTIK *phonometrische Einheit* sone

Sonnen- *in cpds* ERNEU ENER solar; **Sonneneinstrahlung** *f* LUFT global insolation; **Sonnenenenergieressourcen** *f pl* ERNEU ENER solar energy resources; **Sonnenenergie** *f* ERNEU ENER solar energy; **Sonnenenergienutzung** *f* ERNEU ENER solar energy use; **Sonnengürtel** *m* **der Erde** ERNEU ENER earth's sunbelt; **Sonnenhöhewinkel** *m* ERNEU ENER angle of sun's altitude; **Sonnenlicht** *nt* ERNEU ENER sunlight; **Sonnenlichtabstrahlung** *f* LUFT sunlight radiation

sonnenschwach: **~e Zeit** *f* ERNEU ENER weak-sun period

Sonnen-: **Sonnenstand** *m* ERNEU ENER position of the sun

sonnenstark: **~e Zeit** *f* ERNEU ENER strong-sun period

Sonnen-: **Sonnenstrahlung** *f* ERNEU ENER, LUFT solar radiation; **Sonnenstrom** *m* ERNEU ENER solar power; **mit Sonnenstrom geladen** *adj* ERNEU ENER charged with solar power; **mit Sonnenstrom geladene Batterien** *f pl* ERNEU ENER solar-charged batteries

Sorbat *nt* LUFT, MEER sorbate

Sorbens *nt* LUFT sorbent

sorbieren *vt* ALLG sorb

sorbiert: **~es Gas** *nt* LUFT, MEER sorbate

S.-Ö.-Reaktor *m* (*Schlamm-Öl-Reaktor*) ABWASSER STOR (*sludge-to-oil reactor*)

Sorption *f* ABFALL, ABWASSER, LUFT, MEER sorption; **Sorptionsfähigkeit** *f* LUFT sorption capacity; **Sorptionsisotherme** *f* LUFT sorption isotherm; **Sorptionsmittel** *nt* LUFT, MEER sorbent;

Sorptionsuntersuchung *f* ANALYTIK sorption analysis; **Sorptionsvermögen** *nt* ABFALL, WASSER sorption ability
sortenrein *adj* ABFALL type-specific; **~e Erfassung** *f* ABFALL recovery by type; **~e Getrenntsammlung** *f* ABFALL separate collection by type; **~e Kunststoffe** *m pl* ABFALL type-specific plastics
Sortenreinheit *f* ABFALL type specification
Sortier- *in cpds* ABFALL sorting; **Sortieranalyse** *f* ANALYTIK sorting analysis; **Sortieranlage** *f* ABFALL sorting plant
sortieren *vt* ABFALL sort
Sortieren *nt* ABFALL separating, sorting
Sortier-: **Sortierkapazität** *f* ABFALL sorting capacity; **Sortiermaschine** *f* ABFALL sorting machine; **Sortierpersonal** *nt* ABFALL sorting personnel; **Sortierquote** *f* ABFALL sorting rate; **Sortier- und Sammelsystem** *nt* ABFALL collating system; **Sortiersystem** *nt* ABFALL sorting system
sortiert *adj* ABFALL sorted; **~e Fraktion** *f* ABFALL sorted fraction
Sortier-: **Sortiertechnik** *f* ABFALL separation technology; **Sortiertisch** *m* ABFALL sorting table
Sortierung *f* ABFALL sorting; **~ von Kunststoffen nach Farben** ABFALL color sorting of plastics (*AmE*), colour sorting of plastics (*BrE*); **~ mit Röntgenstrahlen** ABFALL sorting via X-rays
Soxhlet-Extraktion *f* ANALYTIK Soxhlet extraction
Soxhlet-Extraktor *m* ANALYTIK Soxhlet extractor
spalten *vt* ABFALL *Rohr* crack
Spalter *m* ABWASSER cleaver
Spaltgas *nt* LUFT reformed gas
Spaltprodukt *nt* ABFALL fission product
Spaltung *f* ABFALL *Kunststoffrecycling* scission, ABWASSER separation; **~ mit organischen Chemikalien** ABWASSER separation with organic chemicals
Spann: **Spannbetonrohr** *nt* ABFALL, ABWASSER, WASSER prestressed concrete pipe, reinforced concrete pipe; **Spannkraft** *f* ABFALL, WASSER clamping force
Spannungskorrosion *f* WASSER stress corrosion; **Spannungskorrosionsriß** *m* WASSER stress corrosion cracking
Spanplatte *f* LÄRM chipboard
Sparauto *nt* LUFT low-energy automobile
sparen *vt* ERNEU ENER, MANAG, POLITIK save, conserve
SPE *abbr* (*Festphasenextraktion*) ANALYTIK SPE (*solid-phase extraction*)
Speicher *m* ABFALL, ABWASSER, LUFT, WASSER reservoir; **~ von Treibhausgasen** LUFT reservoir of greenhouse gases
speicherbar *adj* ERNEU ENER storable
Speicher: **Speicherbecken** *nt* ABWASSER, WASSER storage basin, storage reservoir; **Speicherfilter** *nt* LUFT storage filter
speicherfrei *adj* ERNEU ENER storage-free
Speicher: **Speicherfunktion** *f* ERNEU ENER storage function; **Speicherkapazität** *f* ABFALL, ABWASSER, ERNEU ENER, LUFT, WASSER storage capacity; **Speichermedium** *nt* ERNEU ENER *Wärmepumpe* storage medium
speichern *vt* ALLG store
Speicher: **Speichernutzraum** *m* ABFALL live storage; **Speichertechnologie** *f* ERNEU ENER *Energie* storage technology
Speicherung *f* ALLG storage; **Speicherungskapazität** *f* ABFALL, ABWASSER, ERNEU ENER storage capacity; **Speicherungsvermögen** *nt* ABFALL, ABWASSER, ERNEU ENER, WASSER storage capacity
Speicher: **Speichervolumen** *nt* ERNEU ENER storage volume; **Speicherwasserkraftwerk** *nt* ERNEU ENER storage hydroelectric plant; **Speicherwirkungsgrad** *m* ERNEU ENER storage efficiency
Speisepumpe *f* LUFT feed pump
Speisewasser *nt* WASSER feed water
Spektral- *in cpds* ANALYTIK spectral
spektral *adj* ANALYTIK spectral; **~er Absorptionskoeffizient** *m* ANALYTIK spectral absorption coefficient, LUFT extinction coefficient; **~e Halbwertsbreite** *f* ANALYTIK spectral half-width value
Spektral-: **Spektralanalyse** *f* ANALYTIK spectral analysis; **Spektralbereich** *m* ANALYTIK spectral range; **Spektrallinie** *f* ANALYTIK spectral line; **Spektralphotometer** *nt* ANALYTIK spectrophotometer
Spektrometer *nt* ANALYTIK spectrometer
Spektrometrie *f* ANALYTIK spectrometry
spektrometrisch *adj* ANALYTIK spectrometric
Spektrophotometer *nt* ANALYTIK spectrophotometer
Spektroskopie *f* ANALYTIK spectroscopy
spektroskopisch *adj* ANALYTIK spectroscopic; **~e Methoden** *f pl* ANALYTIK spectroscopic methods
Sperr-: **Sperrbrunnen** *m* BODEN retaining well
Sperre *f* MEER *zum Einschließen von ausgetretenem Öl* boom
sperren *vt* ERNEU ENER dam
Sperre: **Sperrengebinde** *nt* MEER boom pack; **Sperrenrückgewinnung** *f* MEER boom retrieval; **Sperrenschleppe** *f* MEER boom towing
Sperr-: **Sperrflüssigkeit** *f* ANALYTIK sealing liquid
sperrig *adj* ABFALL, ABWASSER bulky; **~e Inhaltsstoffe** *m pl* ABWASSER bulky constituents
Sperrmüll *m* ABFALL bulky waste; **Sperrmüllabfuhr** *f* ABFALL bulky waste collection; **Sperrmüllsammlung** *f* ABFALL bulky waste collection
Sperrschicht *f* ABFALL barrier layer, resistive layer; **Sperrschichtfolie** *f* ABFALL barrier film; **Sperrschichtstoff** *m* ABFALL barrier material; **Sperrschichtverpackung** *f* ABFALL barrier packaging
Sperrsystem *nt* ABFALL barrier system
Sperr-: **Sperrventil** *nt* LUFT shutoff valve
Spezial: **Spezialfahrzeug** *nt* ABFALL dedicated collection vehicle; **Speziallaboratorium** *nt* ANALYTIK specialist lab (*specialist laboratory*)
spezifisch: **~er Brennwert** *m* ABFALL, LUFT upper calorific value (*UCV*); **~e elektrische Leitfähigkeit** *f* ANALYTIK, BODEN, LUFT, WASSER conductivity; **~er Heizwert** *m* ABFALL, LUFT net calorific value (*NCV*), specific conductivity; **~e Koagulation** *f* ABWASSER specific coagulation; **~e Leitfähigkeit** *f* ABFALL, ANALYTIK electrical conductivity; **~e Oberfläche** *f* ABWASSER specific surface area; **~es Sauerstoffzufuhrvermögen** *nt* **in Abwasser** ABWASSER specific oxygen transfer capacity in wastewater; **~es Sauerstoffzufuhrvermögen** *nt* **in Reinwasser** ABWASSER specific oxygen transfer capacity in clean water
Spiegel *m* ERNEU ENER mirror, reflector
Spiral- *in cpds* ABWASSER, BODEN, LUFT spiral
Spirale *f* ABWASSER, BODEN, LUFT spiral
Spiral-: **Spiralklassierer** *m* BODEN spiral classifier; **Spiralmodul** *nt* ABWASSER spiral module; **Spiralplatte** *f* LUFT spiral plate; **Spiralplattenkondensator** *m* LUFT spiral plate condenser; **Spiralplattenwärmetauscher** *m* LUFT spiral plate heat exchanger
Spitze *f* ALLG peak; **Spitzenbedarf** *m* ABFALL, WASSER peak demand; **Spitzendurchfluß** *m* ABWASSER peak flow; **Spitzendurchsatz** *m* ABWASSER peak flow; **Spitzeninfiltration** *f* WASSER peak infiltration; **Spitzenkonzentration** *f* LUFT peak concentration; **Spitzenlast** *f* ERNEU ENER peak load; **Spitzenwasserbedarf** *m* WASSER peak water demand; **Spitzenwasserdurchfluß** *m* WASSER peak water flow
spitzwinklig: **~e Abzweigung** *f* WASSER branching at an acute angle
Splittzuschlagstoff *m* ABFALL grit additive
spontan: **~e Verbrennung** *f* SICHERHEIT spontaneous combustion
Spraydose *f* ABFALL aerosol container
Spreaderkran *m* ABFALL spreader crane
sprengen *vt* ABFALL, LUFT, SICHERHEIT blast, blow up

Sprenger *m* ABWASSER sparger
Sprengstoff *m* ABFALL, LUFT, SICHERHEIT explosive
Springflut *f* ERNEU ENER spring tide
Sprinkleranlage *f* ABWASSER, SICHERHEIT sprinkler
Spritz- *in cpds* ABFALL, ANALYTIK injection moulding (*BrE*), injection molding (*AmE*); **Spritzblasautomat** *m* ABFALL automatic injection blow molder (*AmE*), automatic injection blow moulder (*BrE*); **Spritzdruck** *m* ABFALL injection-molding pressure (*AmE*), injection-moulding pressure (*BrE*); **Spritzdüse** *f* LUFT spray nozzle; **Spritzflasche** *f* ANALYTIK gas-washing bottle; **Spritzgußformung** *f* ABFALL injection molding (*AmE*), injection moulding (*BrE*); **Spritzgußmaschine** *f* ABFALL injection-molding machine (*AmE*), injection-moulding machine (*BrE*); **Spritzkegel** *m* LUFT spraying cone; **Spritzmasse** *f* ABFALL injection-molding compound (*AmE*), injection-moulding compound (*BrE*)
Sprudelbett-Waschturm *m* LUFT foam scrubber
Sprudelschicht *f* LUFT *Sprudelbett-Waschturm* foam bed
Sprüh- *in cpds* ABFALL, ABWASSER, LUFT, MEER spray, spraying; **Sprühabsorber** *m* LUFT spray absorber; **Sprühabsorptionsverfahren** *nt* (*SAV*) LUFT spray absorption process; **Sprühausleger** *m* MEER spray-boom; **Sprühbehälter** *m* ABFALL, LUFT aerosol; **Sprühdraht** *m* LUFT charging electrode wire; **Sprühdüse** *f* LUFT spray nozzle
sprühen *vt* ALLG spray
Sprüh-: **Sprühkanone** *f* MEER spray-gun; **Sprühkolonne** *f* LUFT *Absorptionsapparate* spray column; **Sprühnebel** *m* LUFT spray; **Sprühöffnungen** *f pl* MEER *für Dispergatoren* spray apertures; **Sprühpfad** *m* MEER spray path; **Sprührahmen** *m* LUFT spray frame; **Sprühreaktor** *m* LUFT spray reactor; **Sprühregen** *m* LUFT spray; **Sprühstrom** *m* LUFT *Elektrofilter* ion current; **Sprühtrockner** *m* ABWASSER spray dryer; **Sprühtrocknungswäscher** *m* LUFT spray dry absorber (*SDA*), spray dry scrubber; **Sprühturm** *m* LUFT spray tower; **Sprühwäscher** *m* LUFT spray scrubber, spray tower; **Sprühzone** *f* LUFT spraying zone
Sprungschicht *f* ABWASSER metalimnion, thermocline
spülen *vt* ALLG blow off, scour, rinse, purge, flush
Spül: **Spülgas** *nt* LUFT scavenging gas; **Spülgeschwindigkeit** *f* ABWASSER, WASSER flushing speed, flushing velocity; **Spülhydrant** *m* WASSER flush hydrant; **Spülkraft** *f* ABWASSER, WASSER flushing strength; **Spüllufteinlaß** *m* LUFT purge air inlet; **Spülmittel** *nt* ABFALL, BODEN scavenger; **Spülpumpe** *f* ABWASSER flushing pump; **Spülstange** *f* BODEN kelly
Spülung *f* ALLG flushing, scouring, rinsing, purging
Spül: **Spülwasser** *nt* ABFALL, BODEN rinsing water; **Spülwasseraufbereitung** *f* ABFALL, BODEN rinsing water processing
Spundwand *f* ABFALL sheet piling
Spur *f* ANALYTIK trace; **Spurenanalytik** *f* ANALYTIK trace analysis; **Spurenelement** *nt* ANALYTIK trace element; **Spurenelementbestimmung** *f* ANALYTIK trace element determination; **Spurengas** *nt* LUFT trace gas; **Spurenkonzentration** *f* ANALYTIK trace concentration; **Spurenmeßgerät** *nt* ANALYTIK trace-measuring device; **Spurenmessung** *f* ANALYTIK measurement of trace amounts
Stab *m* ABWASSER *des Rechens* bar
stabil *adj* ABFALL, ANALYTIK, LUFT stabile, stable; **~e Methangärung** *f* ABFALL stable methane fermentation; **~e Temperaturschichtung** *f* LUFT stabile temperature layering
Stabilisator *m* ALLG stabilizer
stabilisieren *vt* ALLG stabilize
stabilisiert *adj* ALLG stabilized; **~er Schlamm** *m* ABWASSER stabilized sludge
Stabilisierung *f* ALLG *Klärschlamm* stabilization; **~ von Treibhausgasen** LUFT stabilization of greenhouse gases
Stabilität *f* ABFALL, ABWASSER, ANALYTIK stability
Stab: **Stabkonzentration** *f* ABWASSER *des Rechens* bar concentration; **Stabrechen** *m* ABWASSER bar screen
Stadtgas *nt* LUFT town gas
städtisch: **~es Einzugsgebiet** *nt* ABFALL, WASSER urban catchment; **~e Kanalisation** *f* ABWASSER urban canalization; **~e Wasserbewirtschaftung** *f* WASSER urban water management
Stadtwasser *nt* WASSER town water
Stadtwerke *nt pl* ABFALL, ABWASSER, WASSER public works
Stagnation *f* ABWASSER, WASSER stagnation
stagnieren *vi* ABWASSER, WASSER stagnate
Stahl *m* ABFALL steel; **Stahlbandumreifung** *f* ABFALL steel band strapping; **Stahlbetonrohr** *nt* ABFALL, ABWASSER, WASSER reinforced concrete pipe; **Stahlfaser** *f* LUFT *Filtermedium* steel fiber (*AmE*), steel fibre (*BrE*); **Stahlkörper** *m* LUFT steel body; **Stahlrohr** *nt* ABFALL, WASSER steel pipe; **Stahlverpackung** *f* ABFALL steel packaging; **Stahlwerkschlacke** *f* ABFALL steelwork slag
Stamm: **Stammabfluß** *m* LUFT stemflow; **Stammdaten** *nt pl* ANALYTIK historical data; **Stammlösung** *f* ANALYTIK parent solution, stock solution
Stampfdichte *f* ANALYTIK tamped density
Stampffußverdichter *m* ABFALL ramming machine
Stampfvolumeter *nt* ANALYTIK tamping volume meter
standardisiert *adj* SICHERHEIT standardized; **~e Ausrüstung** *f* SICHERHEIT standardized equipment
Standard: **Standardlösung** *f* ANALYTIK calibration solution, standard solution; **Standardtrübung** *f* ANALYTIK standard turbidity
Stand-: **Standbehälter** *m* LUFT stationary tank
standfest *adj* BODEN firm; **~er Untergrund** *m* BODEN firm subsurface
ständig: **~ erneuert werden** *vt* ERNEU ENER renew continually
Standort *m* ALLG site; **~ der Deponie** ABFALL landfill site; **Standortbewertung** *f* ANALYTIK *Altlasten* site assessment
standortbezogen *adj* ALLG site-related; **~e Kriterien** *nt pl* ANALYTIK, ERNEU ENER site-related criteria; **~er Lärm** *m* LÄRM site-related noise; **~e Untersuchung** *f* ANALYTIK site-related analysis
Standort: **Standortebene** *f* ABFALL, WASSER site level; **Standortgegebenheiten** *f pl* BODEN site characteristics; **Standortkriterien** *nt pl* BODEN, LUFT, WASSER site criteria; **Standortparameter** *m pl* ANALYTIK site parameters
standortspezifisch *adj* ALLG site-specific; **~e Anlage** *f* ABWASSER, ANALYTIK, BODEN, ERNEU ENER, LÄRM, LUFT site-specific installation; **~e Eigenschaften** *f pl* ERNEU ENER site-specific properties; **~e Kriterien** *nt pl* ERNEU ENER site-specific criteria
Standort: **Standorttyp** *m* BODEN site type
Stand-: **Standzeit** *f* ABWASSER serviceable life, ANALYTIK *einer Probe* exposure time, *der Ausrüstung* serviceable life, BODEN, LÄRM, LUFT, MEER, SICHERHEIT, WASSER serviceable life
Stange *f* ABFALL, WASSER *einer Pumpe* rod
Stapeln *nt*: **~ der Ballen** ABFALL stapling of bales
stark: **~ gefärbtes Abwasser** *nt* ABWASSER heavily colored wastewater (*AmE*), heavily coloured wastewater (*BrE*); **~ getrübt** *adj* ABFALL, ABWASSER, WASSER highly turbid; **~ sauer** *adj* BODEN highly acidic; **~ verdichtete Deponie** *f* ABFALL tightly compacted landfill; **~ veschmutzt** *adj* ABFALL, ABWASSER heavily soiled; **~ wassergefährdend** *adj* ABWASSER strongly harmful to water
Stärke *f* ABFALL *Deponieschichten* thickness
stärkehaltig *adj* ERNEU ENER starchy; **~e Pflanzen** *f pl* ERNEU ENER *Bioenergienausgangsstoff* starchy plants; **~er Rohstoff** *m* ERNEU ENER *Bioenergienausgangsstoff* starchy raw material
Starr-: **Starrflügelflugzeug** *nt* MEER fixed-wing aircraft; **Starrflügelluftfahrzeug** *nt* MEER fixed-wing aircraft; **Starrflügler** *m* MEER fixed-wing aircraft
Start: **Startbrenner** *m* LUFT start burner; **Startkatalysator** *m*

LUFT primary catalyst; **Startpunkt** *m* ANALYTIK starting point

stationär *adj* ALLG stationary; **~er Kompaktor** *m* ABFALL, WASSER stationary compactor; **~es Probenahmegerät** *nt* ANALYTIK stationary sampler; **~e Wirbelschicht** *f* (*SWS*) LUFT stationary fluidized bed; **~er Wirbelschichtofen** *m* BODEN stationary fluidized-bed oven; **~e Zerkleinerungsanlage** *f* ABFALL stationary crushing unit

statisch *adj* ABFALL, ABWASSER, ANALYTIK, WASSER static; **~er Auslaugtest** *m* ANALYTIK static leachability test; **~e Auslenkung** *f* LÄRM static deflection; **~er Druck** *m* LUFT static pressure; **~e Druckhöhe** *f* ABWASSER static head; **~er Eluationstest** *m* ANALYTIK, BODEN static leachability test, static leaching test; **~e Förderhöhe** *f* ABFALL, ABWASSER, WASSER static discharge head; **~e Höhe** *f* ABFALL, WASSER static head; **~er Mischer** *m* ABWASSER static mixer; **~e Saughöhe** *f* ABWASSER, WASSER static suction head, static suction lift

Statistik *f* ANALYTIK statistics

statistisch *adj* ANALYTIK statistical; **~e Auswertung** *f* ANALYTIK statistical evaluation; **~e Fehlermöglichkeit** *f* ANALYTIK statistical possibility of error

Stator-Rotor-System *nt* LUFT stator-rotor system

Stau *m* WASSER backwater, banked-up water level

Staub *m* ABFALL, ANALYTIK, LUFT, SICHERHEIT dust; **~ in der Luft** SICHERHEIT airborne dust; **Staubablagerung** *f* LUFT dust deposit, dust build-up; **Staubabsaugung** *f* LUFT dust extraction; **Staubabscheider** *m* ABFALL, LUFT dust separator; **Staubabscheidung** *f* ABFALL, LUFT particulate collection, dedusting; **Staubabzug** *m* SICHERHEIT dust exhaust fan; **Staubanalyse** *f* ANALYTIK dust analysis; **Staubanteil** *m* LUFT dust content; **Staubart** *f* LUFT dust type; **Staubauffanggerät** *nt* ANALYTIK dust collector; **Staubaustrag** *m* LUFT dust discharge; **Staubaustragsöffnung** *f* LUFT *eines Zyklon* dust discharge opening; **Staubaustritt** *m* LUFT dust outlet

staubbeladen *adj* LUFT, SICHERHEIT dust-laden; **~er Film** *m* LUFT dust-laden film; **~es Gas** *nt* LUFT dust-laden gas; **~e Luft** *f* LUFT dust-laden air

***Staub*: Staubbeladung** *f* LUFT dust loading, dust burden; **Staubbelag** *m* LUFT dust coating, dust deposit, dust build-up; **Staubbelastung** *f* LUFT dust burden; **Staubbindemittel** *nt* LUFT dust-binding agent; **Staubbrand** *m* LUFT dust combustion; **Staubbunker** *m* LUFT dust-collecting hopper, dust receiver

staubdicht *adj* SICHERHEIT dust-proof

***Staub*: Staubdurchbruch** *m* LUFT dust breakthrough

***Stau*: Staubecken** *nt* ABWASSER, WASSER reservoir, storage basin, storage reservoir

***Staub*: Staubeinlagerung** *f* LUFT dust dispersion; **Staubemissionen** *f pl* ERNEU ENER, LUFT dust emissions; **Staubexplosion** *f* LUFT dust explosion; **Staubfänger** *m* LUFT dust catcher, dust trap; **Staubfeinheit** *f* LUFT dust fineness; **Staubfilter** *nt* LUFT dust filter; **Staub-Flüssigkeits-Gemisch** *nt* LUFT dust-liquid mixture; **Staubfraktion** *f* LUFT dust fraction

staubfrei *adj* LUFT dust-free; **~e Abluft** *f* LUFT dust-free waste air; **~es Gas** *nt* LUFT dust-free gas

***Staub*: Staubgas** *nt* LUFT particulate gas; **Staubgehalt** *m* ABFALL, LUFT dust content; **Staubgehalt** *m* **im Rauchgas** LUFT dust content in flue gas; **Staubgehaltsmesser** *m* ANALYTIK konimeter; **Staubgemisch** *nt* LUFT dust mixture

staubhaltig *adj* LUFT, SICHERHEIT dust-laden; **~es Abgas** *nt* LUFT dust-laden flue gas; **~es Gas** *nt* LUFT dust-laden gas; **~e Luft** *f* SICHERHEIT dust-laden air

***Staub*: Staubkatalysator** *m* LUFT dust catalyst; **Staubkomponente** *f* LUFT dust component; **Staubkonzentration** *f* ANALYTIK, LUFT, SICHERHEIT dust concentration; **Staubkonzentration** *f* **in der Luft** SICHERHEIT airborne dust concentration; **Staubkonzentrationsmeßgerät** *nt* LUFT dust concentration measuring device; **Staubkonzentrationsmessung** *f* ANALYTIK measurement of dust concentration; **Staublast** *f* LUFT dust content in flue gas; **Staubmessung** *f* ANALYTIK dust measurement; **Staubniederschlag** *m* BODEN dust deposit; **Staubniederschlagsmessung** *f* ANALYTIK measurement of dust precipitation; **Staubpartikel** *f* ANALYTIK, LUFT dust particle; **Staubrückgewinnung** *f* LUFT dust recovery; **Staubsammelbehälter** *m* LUFT dust receiver; **Staubsammeltrichter** *m* LUFT dust-collecting hopper; **Staubschicht** *f* LUFT *Elektrofilter* dust particulate layer; **Staubschutzhaube** *f* SICHERHEIT dust cover; **Staubschutzmaske** *f* SICHERHEIT dust mask, dust respirator; **Staubsilo** *m* LUFT dust silo; **Staubteilchen** *nt* ANALYTIK, LUFT dust particle; **Staubtransportschnecke** *f* LUFT dust transporting worm; **Staubwaage** *f* ANALYTIK dust scale; **Staubzähler** *m* ANALYTIK dust counter

***Stau*: Staudruck** *m* ANALYTIK dynamic pressure; **Staufläche** *f* WASSER surface area; **Staumauer** *f* ERNEU ENER, WASSER masonry dam; **Stauschwelle** *f* WASSER drowned weir; **Stausee** *m* ABWASSER reservoir, WASSER artificial lake, impounding reservoir, reservoir; **Stauwasser** *nt* WASSER backwater

stehend: **~es Gewässer** *nt* ABWASSER, WASSER stagnant water, still water

steigend: **~e Energiepreise** *m pl* ERNEU ENER rising energy prices

Steiggeschwindigkeit *f* ABWASSER rising rate

Stein- *in cpds* ABFALL, LÄRM, LUFT mineral, rock, stone; **Steinkohle** *f* LUFT hard coal, glance coal; **Steinkohlenasche** *f* ABFALL hard coal ash; **Steinwolle** *f* LÄRM mineral fiber (*AmE*), mineral fibre (*BrE*), rock wool; **Steinzeugrohr** *nt* ABFALL, ABWASSER, WASSER vitrified clay pipe, stoneware pipe

STEL *abbr* (*Kurzzeit-Exposition*) ANALYTIK STEL (*short-term exposure limit*)

stereospezifisch: **~er Katalysator** *m* LUFT stereospecific catalyst

Sternverfahren *nt* POLITIK star-shaped procedure

stetig *adj* ANALYTIK continuous; **~e Messung** *f* ANALYTIK continuous measuring

Steuer *f* POLITIK tax

Steuerbarkeit *f* ALLG controllability

***Steuer*: Steuerbefreiung** *f* POLITIK tax exemption; **Steuerbelastung** *f* POLITIK tax burden; **Steuerbemessungsgrundlage** *f* POLITIK tax base; **Steuereinheit** *f* ABWASSER, ANALYTIK, BODEN, LUFT, WASSER control unit; **Steuerentlastung** *f* POLITIK tax rebate; **Steuergesetze** *nt pl* POLITIK tax laws; **Steuerkürzung** *f* POLITIK tax reduction

steuerlich: **~er Anreiz** *m* POLITIK tax incentive; **~e Regelung** *f* POLITIK tax provision

***Steuer*: Steuerminderung** *f* POLITIK tax relief

steuern *vt* ALLG control

***Steuer*: Steuerquote** *f* POLITIK tax quota; **Steuersatz** *m* POLITIK tax rate; **Steuerstand** *m* ABFALL, LUFT, WASSER control room; **Steuertarif** *m* POLITIK tax tariff; **Steuerüberwälzung** *f* POLITIK tax shift

Steuerung *f* ALLG *Regulierung* control, *Steuervorrichtung* controls

Steuler: **~ Verfahren** *nt* LUFT Steuler process

stichfest *adj* ABFALL compacted, semisolid, ABWASSER watertight; **~er Sonderabfall** *m* ABFALL semisolid hazardous waste

Stichleitung *f* BODEN stub cable

Stichprobe *f* ANALYTIK grab sample, spot sample, random sample, snap sample; **Stichprobenbehälter** *m* ANALYTIK snap sample container; **Stichprobenplan** *m* ANALYTIK sampling plan

Stickgas *nt* ABFALL, LUFT asphyxiating gas

Stickstoff *m* (*N*) ALLG nitrogen (*N*)

stickstoffarm *adj* LUFT low-nitrogen
Stickstoff: **Stickstoffbildung** *f* LUFT nitrogen formation; **Stickstoffdioxid** *nt* (NO_2) ABFALL, LUFT *Schwermetall* nitrogen peroxide (NO_2); **Stickstoffoxide** *nt pl* (NO_x) LUFT nitrogen oxides (NO_x); **Stickstoffpentoxid** *nt* (N_2O_5) LUFT nitrogen pentoxide (N_2O_5); **Stickstoff-Phosphor-Detektor** *m* (*NPD*) ANALYTIK *ultraviolett* nitrogen-phosphorus detector (*NPD*); **Stickstoffwäscher** *m* LUFT nitrogen oxides scrubber
stillegen *vt* ABFALL close, *Deponie* close down
stillgelegt *adj* ABFALL closed, closed down; **~er Deponieabschnitt** *m* ABFALL closed landfill section
Stirlingmotor *m* ERNEU ENER Stirling motor
stochastisch: **~es Rauschen** *nt* LÄRM random noise
Stöchiometrie *f* ABWASSER, ANALYTIK, LUFT stoichiometry
stöchiometrisch *adj* ABWASSER, ANALYTIK, LUFT stoichiometric; **~es Sauerstoffangebot** *nt* ABWASSER, LUFT stoichiometric oxygen supply; **~e Verbrennung** *f* LUFT stoichiometric combustion
Stoff *m* ABWASSER substance, ANALYTIK material, substance, BODEN, LUFT substance; **Stoffaustausch** *m* ABFALL material exchange
stoffbezogen *adj* ABWASSER, LUFT substance-related
Stoff: **Stoffbilanz** *f* MANAG materials balance
stofflich *adj* ABFALL material; **~es Recycling** *nt* ABFALL materials recycling, physical recycling, ABWASSER materials recycling; **~e Verwertung** *f* ABFALL materials recycling, physical recycling, ABWASSER materials recycling
Stoff: **Stoffflußbilanz** *f* POLITIK, WASSER material flow analysis
stoffspezifisch *adj* ABWASSER, ANALYTIK, LUFT substance-specific; **~e Kriterien** *nt pl* ANALYTIK substance-specific criteria; **~e Untersuchung** *f* ANALYTIK substance-specific analysis
Stoff: **Stoffstrom** *m* ABFALL materials stream; **Stofftrennprozeß** *m* ABFALL material separation operation; **Stoffübergang** *m* LUFT mass transfer; **Stoffumsatz** *m* ABFALL, ABWASSER metabolic rate
Stoffwechsel *m* ABFALL, ABWASSER, LUFT metabolism; **Stoffwechselprodukt** *nt* ABFALL metabolic product, ABWASSER metabolite, LUFT metabolic product; **Stoffwechselprozeß** *m* ABFALL metabolic process; **Stoffwechseltätigkeit** *f* ABFALL *Kompostierung*, ABWASSER metabolic activity
Stokes: **~sches Reibungsgesetz** *nt* ANALYTIK, LUFT Stokes's law
stopfen *vt* ABFALL, WASSER *ein Leck* stop
Stopfwerk *nt* WASSER *zur kontinuierlichen Beschickung* densifier
Störfall *m* ANALYTIK accident; **Störfallverordnung** *f* POLITIK directive on major industrial hazards, SICHERHEIT regulation in case of accident; **Störfallvorsorge** *f* SICHERHEIT accident prevention
Stör-: **Störgeräusch** *nt* LÄRM disturbing noise; **Störpegel** *m* ANALYTIK, LÄRM disturbance level
störschallunterdrückend: **~es Mikrophon** *nt* LÄRM anti-noise microphone
Stör-: **Störstoff** *m* ABFALL, WASSER foreign material
Störung *f* SICHERHEIT malfunction; **Störungsüberwachung** *f* SICHERHEIT malfunction detection; **Störungsursache** *f* SICHERHEIT cause of malfunction
Stoß *m* ALLG collision; **Stoßbelastung** *f* ABWASSER shock loading; **Stoßdämpfer** *m* SICHERHEIT resilient isolator; **Stoßheber** *m* ABWASSER hydraulic ram; **Stoßionisation** *f* ANALYTIK, LUFT collision ionization, impact ionization
Strafrecht *nt* POLITIK penal law
Strahl: **Strahlbelüfter** *m* ABWASSER jet aerator, sludge siphon; **Strahlbelüftung** *f* ABWASSER jet aeration; **Strahldüse** *f* ABWASSER, LUFT, WASSER discharge nozzle, jet
strahlen *vi* ALLG radiate
Strahlen *m pl* ALLG radiation; **Strahlendosis** *f* LUFT, SICHERHEIT radiation dose; **Strahlenschutz** *m* LUFT, SICHERHEIT radiation protection; **Strahlenteiler** *m* ANALYTIK beamsplitter; **Strahlentrockner** *m* ABFALL radiation dryer; **Strahlenüberwachung** *f* ANALYTIK radiological survey
Strahler *m* LUFT radiator, WASSER radiation emitter
Strahl: **Strahlpumpe** *f* ABWASSER, WASSER jet pump; **Strahlrohr** *nt* LUFT duct, nozzle pipe
Strahlung *f* ALLG radiation; **Strahlungsabschirmung** *f* SICHERHEIT radiation shield; **Strahlungsabsorption** *f* ANALYTIK radiation absorption, LUFT radiative absorption; **Strahlungsdetektor** *m* ANALYTIK *Strahlungskontrolle* radiation detector; **Strahlungsenergie** *f* ERNEU ENER radiation energy; **Strahlungsgefahr** *f* SICHERHEIT radiation hazard; **Strahlungsgleichgewicht** *nt* LUFT radiation equilibrium; **Strahlungsheizung** *f* BODEN direct heating by radiation; **Strahlungsintensität** *f* ANALYTIK radiation intensity, ERNEU ENER intensity of radiation; **Strahlungskontrolle** *f* ANALYTIK radiation monitoring; **Strahlungsmesser** *m* ANALYTIK actinometer; **Strahlungsmeßgeräte** *nt pl* ANALYTIK radiation-measuring instruments; **Strahlungsnachweisgerät** *nt* ANALYTIK *Strahlungskontrolle* radiation detector; **Strahlungspyrometer** *nt* ANALYTIK radiation pyrometer; **Strahlungsrekuperator** *m* LUFT radiation recuperator; **Strahlungsschutz** *m* SICHERHEIT radiation protection, radiation shielding
Strahl: **Strahlwasserhahn** *m* WASSER jet cock
Strang: **Stranggußanlage** *f* ABFALL continuous casting machine; **Strangpreßwerkzeug** *nt* ABFALL extrusion die
Straße *f* ALLG road; **Straßenaufbruch** *m* ABFALL road construction rubble; **Straßen-Emissionspegel** *m* LÄRM street emissions level; **Straßenkehricht** *m* ABFALL litter, road sweepings, street refuse; **Straßenkehrmaschine** *f* ABFALL mechanical sweeper; **Straßentransport** *m* ABFALL road transportation
Stratosphäre *f* LUFT stratosphere
stratosphärisch *adj* LUFT stratospheric; **~es Ozon** *nt* LUFT stratospheric ozone; **~e Ozonbelastung** *f* LUFT stratospheric ozone pollution; **~e Ozonkonzentration** *f* LUFT stratospheric ozone concentration
Strecke *f*: **~ eines Flußes** ABFALL, WASSER reach of a river
streng: **~e Überwachung** *f* SICHERHEIT close supervision
streuen *vt* ANALYTIK scatter
Streulicht *nt* ANALYTIK light scattering, diffused light; **Streulichtdetektor** *m* ANALYTIK light-scattering detector; **Streulichtfernmessung** *f* ANALYTIK dispersion telemetering; **Streulichtmessung** *f* ANALYTIK diffused light measurement; **Streulichtpartikelzähler** *m* ANALYTIK dispersion particle counter; **Streulichtsensor** *m* ANALYTIK diffused light sensor
Streuung *f* ANALYTIK scattering
Streuwolke *f* ANALYTIK *Streulicht-Fernmessung* scattering cloud
Strichkodierung *f*: **~ zur Identifizierung** MANAG bar coding for identification
Strichplatte *f* ANALYTIK cross-lines
strippen *vt* ABFALL, ABWASSER strip
Strippen *nt* ABFALL, ABWASSER, WASSER stripping
Strippersäule *f* ANALYTIK separation column
Stripping: **Stripping-Purging-Technik** *f* ANALYTIK stripping-purging technique; **Stripping-Test** *m* ANALYTIK stripping test; **Stripping-Turm** *m* BODEN stripping tower
Strippverfahren *nt* ABWASSER, WASSER stripping
Stroboskop *nt* ANALYTIK stroboscope
Stroh: **Strohballen** *m* ERNEU ENER *Bioenergienausgangsstoff* straw bale; **Strohfeuerungsanlage** *f* ERNEU ENER *Biokonversion* straw-burning plant
Strom *m* ABWASSER stream, ERNEU ENER current, power, WASSER stream
stromabwärts *adv* ABFALL, LUFT, WASSER downstream

Strom: **Stromaufnahme** *f* BODEN *Bodenluftabsaugung* current drawing
stromaufwärts *adv* LUFT, WASSER upstream
Strom: **Strombelüftung** *f* ABWASSER stream aeration
strömen *vi* ABWASSER, LUFT, WASSER flow
Strom: **Stromerzeugung** *f* ABFALL power generation, ERNEU ENER electricity generation, power generation
Stromlinie *f* LUFT streamline, *Trägermedium* flow line; **Stromlinienbild** *nt* WASSER flow pattern (*AmE*), flow configuration (*BrE*); **Stromlinien des Gases** *f pl* LUFT gas streamline
Strom: **Strommesser** *m* WASSER current meter; **Stromnetz** *nt* ERNEU ENER power grid, electricity grid; **Stromquelle** *f* ERNEU ENER electricity source, power source; **Stromüberangebot** *nt* ERNEU ENER excess power supply
Strömung *f* ABWASSER, ERNEU ENER, LUFT, WASSER flow; **~ für viskose Flüssigkeiten** ANALYTIK rheometer
strömungsabwärts *adv* LUFT downstream
strömungsaufwärts *adv* LUFT upstream
Strömung: **Strömungsbild** *nt* WASSER flow pattern (*AmE*), flow configuration (*BrE*); **Strömungsfeld** *nt* LUFT *Zyklon* flow field; **Strömungsgeschwindigkeit** *f* ANALYTIK velocity of flow, LUFT, WASSER flow rate; **Strömungskanal** *m* LUFT duct, WASSER *Filter* flow channel; **Strömungskraft** *f* LUFT flow force; **Strömungsmesser** *m* ANALYTIK flowmeter; **Strömungsmeßfühler** *m* LUFT flow sensor; **Strömungsring** *m* WASSER *Pumpe* volute insert; **Strömungsumlenkung** *f* LUFT flow deviation; **Strömungswiderstand** *m* LUFT drag, WASSER flow resistance
Strom: **Stromversorgung** *f* ERNEU ENER power supply; **Stromversorgungsnetz** *nt* ERNEU ENER energy supply grid
Strontiumsulfat *nt* LUFT strontium sulfate (*AmE*), strontium sulphate (*BrE*)
Stropp *m* MEER *Anschlagmittel* sling
Stufen: **Stufenbelüftung** *f* ABWASSER step aeration
stufenlos: **~ einstellbar** *adj* ABFALL continuously adjustable
Stufen: **Stufenverbrennung** *f* LUFT staged incineration; **Stufenvorschubrost** *m* WASSER stepped forward feed grate
stufenweise *adv* ANALYTIK step-by-step
Stunden: **Stundenverbrauch** *m* WASSER hourly consumption; **Stundenwert** *m* ANALYTIK hourly value
stündlich *adj* ANALYTIK hourly; **~er Durchschnittswert** *m* ANALYTIK hourly mean value (*HMV*)
Stütz- *in cpds* ABFALL, ANALYTIK, BODEN, LUFT support, auxiliary; **Stützbalken** *m* LUFT brace; **Stützbrenner** *m* LUFT auxiliary burner; **Stützelement** *nt* LUFT support element; **Stützgewebe** *nt* LUFT support scrim; **Stützkörper** *m* LUFT support body; **Stützmauer** *f* LÄRM retaining wall; **Stützring** *m* LUFT support ring; **Stützsäule** *f* LUFT end support; **Stützsuspension** *f* BODEN supporting suspension
subletal *adj* SICHERHEIT sub-lethal; **~e Dosis** *f* SICHERHEIT sub-lethal dose; **~er Effekt** *m* SICHERHEIT sub-lethal effect
Submikron *nt* LUFT submicron; **Submikron-Schwebstoff-Teilchen** *nt* LUFT submicron particulate matter
Submikropore *f* LUFT submicro-pore
submikroskopisch: **~es Teilchen** *nt* LUFT submicron
Subsidiarität *f* POLITIK subsidiarity; **Subsidiaritätsprinzip** *nt* POLITIK subsidiarity principle
substanzspezifisch *adj* ABWASSER, ANALYTIK, LUFT substance-specific
Substrat *nt* ABFALL, ABWASSER substrate; **Substratverwendung** *f* ABWASSER substrate utilization
Subvention *f* POLITIK subsidy
subventionieren *vt* POLITIK subsidize
subventioniert *adj* POLITIK subsidized
Sulfat *nt* ABFALL, BODEN, LUFT sulfate (*AmE*), sulphate (*BrE*); **Sulfat-Anion** *nt* ABFALL sulfate anion (*AmE*), sulphate anion (*BrE*); **Sulfatatmung** *f* WASSER sulfate respiration (*AmE*), sulphate respiration (*BrE*); **Sulfat-Ion** *nt* ABFALL sulfate ion (*AmE*), sulphate ion (*BrE*); **Sulfatkristallisator** *m* LUFT sulfate crystallizer (*AmE*), sulphate crystallizer (*BrE*); **Sulfatzellulose** *f* ABFALL sulfate cellulose (*AmE*), sulphate cellulose (*BrE*)
Sulfid *nt* LUFT sulfide (*AmE*), sulphide (*BrE*); **Sulfidfällung** *f* ABFALL, ABWASSER, WASSER *Abwasserreinigung bei der MVA* sulfide precipitation (*AmE*), sulphide precipitation (*BrE*)
Sulfit *nt* ABFALL, LUFT sulfite (*AmE*), sulphite (*BrE*); **Sulfitzellstoff** *m* ABFALL, WASSER sulfite cellulose (*AmE*), sulphite cellulose (*BrE*)
Sulzer-Ariqua-Verfahren *nt* LUFT Sulzer-Ariqua process
Summenhäufigkeit *f* ANALYTIK cumulative frequency
Summenparameter *m* ANALYTIK cumulative parameter
Sumpf *m* ABFALL swamp, ABWASSER sump, WASSER swamp; **Sumpfpflanzen** *f pl* WASSER *Pflanzenklärsystem* marsh plants
superstark: **~e faserarmierte Wellfaserplatte** *f* ABFALL heavy-duty corrugated fiberboard (*AmE*), heavy-duty corrugated fibreboard (*BrE*); **~er faserarmierter Wellkarton** *m* ABFALL heavy-duty corrugated fiberboard (*AmE*), heavy-duty corrugated fibreboard (*BrE*)
suspendiert *adj* ABFALL, ABWASSER, WASSER suspended; **~e Bakterien** *pl* ABWASSER suspended bacteria; **~es Mangandioxid** *nt* ABFALL suspended manganese dioxide; **~e organische Stoffe** *m pl* ABWASSER suspended organic solids; **~es Teilchen** *nt* ABFALL, WASSER suspended particle
Suspension *f* ALLG suspension
Süßwasservorrat *m* WASSER freshwater reserve
SVOC *abbr* (*semiflüchtige organische Verbindungen*) BODEN SVOC (*semivolatile organic compounds*)
SWS *abbr* (*stationäre Wirbelschicht*) LUFT stationary fluidized bed
synchron *adj* ANALYTIK, ERNEU ENER synchronous
Synchrongenerator *m* ERNEU ENER *Windturbine* synchronous generator
Synergetik *f* ABFALL, LUFT, WASSER synergetics
Synergismus *m* ABFALL, LUFT, WASSER synergy
Synthesegas *nt* ABFALL synthesis gas; **Synthesegaserzeugung** *f* ABFALL synthesis gas generation; **Synthesegasgewinnung** *f* ABFALL synthesis gas recovery
synthetisch *adj* ALLG synthetic; **~e Faser** *f* LUFT synthetic fiber (*AmE*), synthetic fibre (*BrE*); **~e Materialien** *nt pl* ABFALL synthetic materials; **~es Öl** *nt* ERNEU ENER synthetic oil; **~es Polymer** *nt* ABWASSER synthetic polymer
System *nt* ANALYTIK, MANAG, POLITIK system
systematisch *adj* ANALYTIK, MANAG, POLITIK systematic; **~er Fehler** *m* ANALYTIK systematic error
System: **Systemgrenze** *f* MANAG system boundary; **Systemgröße** *f* ANALYTIK system parameter
Szintillationszähler *m* ANALYTIK scintillation counter

T

TA *abbr* (*technische Anleitung*) POLITIK technical directive; **TA-Abfall** *f* ABFALL, POLITIK German technical directive on waste management
tablettieren *vt* ABFALL pelletize
Tablettiermaschine *f* ABFALL pelletizer
Tablettierung *f* ABFALL pelletizing
Tages- *in cpds* ALLG daily; **Tageslärmdosis** *f* LÄRM daily noise exposure; **Tageslärmexposition** *f* LÄRM daily noise exposure; **Tagesleistung** *f* ABFALL, WASSER daily output; **Tagesmittelwert** *m* ANALYTIK daily average value, daily mean value; **Tagesspitze** *f* ABFALL daily peak, ABWASSER peak of the day, WASSER daily peak; **Tagesverbrauch** *m* WASSER daily consumption; **Tageswasserfluß** *m* WASSER daily water flow
täglich *adj* ALLG daily; **~e Abdeckung** *f* ABFALL, WASSER daily cover; **~e Gesamtbelastung** *f* SICHERHEIT daily total exposure; **~er Trinkwasserverbrauch** *m* WASSER daily drinking water consumption; **~e Verarbeitungskapazität** *f* ABWASSER, BODEN, LUFT, WASSER daily processing capacity; **~er Wasserverbrauch** *m* ABFALL, WASSER daily water consumption
Takt: **Taktmaximalpegel** *m* LÄRM maximum level in consecutive time intervals; **Taktzeit** *f* ANALYTIK time interval
TA: **TA-Lärm** *f* LÄRM, POLITIK German technical directive on noise protection
Taldeponie *f* ABFALL valley landfill
Talsperre *f* ERNEU ENER dam, storage reservoir; **Talsperrenwasser** *nt* WASSER reservoir water
TA: **TA-Luft** *f* POLITIK German technical directive on air quality control
Tandem-Detektor *m* ANALYTIK tandem detector
tangential *adj* LUFT tangential; **~ angeströmter Zyklon** *m* LUFT tangential-entry cyclone
Tank *m* WASSER tank
Tanker *m* MEER tanker
Tank: **Tankfahrzeug** *nt* MEER road tanker; **Tankschiff** *nt* MEER tanker; **Tankwagen** *m* MEER road tanker
Taschenfilter *nt* ABWASSER bag filter, LUFT pocket filter
taschenförmig *adj* LUFT *Filter* pocket-shaped
Taschenmeßgerät *nt* ANALYTIK pocket measuring instrument
Tastversuch *m* ANALYTIK preliminary trial
tatsächlich *adj* LUFT actual; **~er Luftbedarf** *m* LUFT actual air requirement; **~e Luftmenge** *f* LUFT actual air volume
Tauch- *in cpds* WASSER submerged, submersible; **Tauchbrenner** *m* ABWASSER, WASSER submersible burner; **Tauchbrett** *nt* ABWASSER baffle plate; **Tauchkörper** *m* ABFALL submersible plunger pump, WASSER submerged trickling filter, submersible plunger pump; **Tauchpumpe** *f* ANALYTIK submersible pump, WASSER subaqueous pump; **Tauchrohr** *nt* LUFT cylindrical inlet tube, *des Zyklons* vortex finder; **Tauchtropfkörper** *m* WASSER submerged trickling filter; **Tauchwand** *f* ABWASSER baffle plate, scum board
tauen *vt* MEER tow
Taupunkt *m* LUFT dew point; **Taupunktbestimmung** *f* ANALYTIK determination of dew point; **Taupunkthygrometer** *nt* ANALYTIK dew point hygrometer; **Taupunkttemperatur** *f* LUFT dew point temperature; **Taupunktunterschreitung** *f* LUFT dropping below dew point
TC *abbr* ANALYTIK (*gesamter vorhandener Kohlenstoff*) TC (*total carbon*), SICHERHEIT (*toxische Konzentration*) TC (*toxic concentration*)
TD *abbr* (*toxische Dosis*) SICHERHEIT *Flüssigkeitschromatographie* TD (*toxic dose*)
Technik *f* ALLG technology
technisch *adj* ALLG technical; **~e Anleitung** *f* (*TA*) POLITIK technical directive; **~e Anleitung-Abfall** *f* (*TA-Abfall*) ABFALL, POLITIK German technical directive on waste management; **~e Anleitung-Lärm** *f* (*TA-Lärm*) LÄRM, POLITIK German technical directive on noise protection; **~e Anleitung-Luft** *f* (*TA-Luft*) POLITIK German technical directive on air quality control; **~es Gas** *nt* LUFT manufactured gas; **~es Potential** *nt* ERNEU ENER technical potential
Technologie *f* ALLG technology; **Technologiefolgenabschätzung** *f* (*TFA*) POLITIK technology assessment (*TA*)
Teerklumpen *m* MEER tarball
TE/F *abbr* (*Trübungseinheit/Formazin*) ANALYTIK FTU (*formazin turbidity unit*)
Teich *m* WASSER pond
Teil- *in cpds* ALLG partial; **Teil** *m* ANALYTIK part; **Teile pro Billiarde** *phr* ANALYTIK *Flüssigkeitschromatrographie/elektrochemische-Detektion* parts per quadrillion (*ppq*); **Teile pro Billion** *phr* ANALYTIK parts per billion (*ppb*); **Teile pro Million** *phr* ANALYTIK *Atomfluoreszenzspektrometrie* parts per million (*ppm*); **Teile pro Trillion** *phr* ANALYTIK parts per trillion (*ppt*); **Teilbodendekontamination** *f* BODEN partial soil decontamination, partial soil remediation; **Teilbodensanierung** *f* BODEN partial soil remediation
Teilchen *nt* ALLG particle; **Teilchenbahn** *f* LUFT *Abscheidevorgang* particle orbit; **Teilchengrößenanalysator** *m* ANALYTIK particle size analyser (*BrE*), particle size analyzer (*AmE*); **Teilchenstoß** *m* LUFT particle impact; **Teilchenverunreinigung** *f* ABFALL, WASSER particulate pollution; **Teilchenweg** *m* LUFT particle path; **Teilchenzählungsmeßgerät** *nt* ANALYTIK particle-counting device
Teil-: **Teilgasmenge** *f* ANALYTIK partial gas quantity; **Teilgenehmigung** *f* POLITIK partial authorization
teilhalogeniert *adj* LUFT partially halogenated; **~e FCKW** *m pl* LUFT partially halogenated CFCs; **~e Fluorchlorkohlenwasserstoffe** *m pl* (*HFCKW*) LUFT hydrochlorofluorocarbon (*HCFC*)
teilintegriert: **~es System** *nt* ABFALL, WASSER partially integrated system
Teil-: **Teilkapselung** *f* LÄRM partial enclosure
teilstabilisiert *adj* ABWASSER partially stabilized; **~er Schlamm** *m* ABWASSER partially stabilized sludge
Teil-: **Teilstrom** *m* ABWASSER *von Produktionswasser* partial current, LUFT partial stream
teilweise: **~ löschen** *vt* MEER lighten
teilweise: **~ wasserlöslich** *adj* LUFT partially water-soluble
Telemetrie *f* ANALYTIK telemetry
Teleskop *nt* ANALYTIK *LIDAR* telescope
Tellerbürste *f* ABFALL *an der Straßenkehrmaschine* circular broom
Temperatur *f* ALLG temperature; **Temperaturanstieg** *m* LUFT temperature rise; **Temperaturbereich** *m* LUFT temperature range
temperaturbeständig *adj* ABFALL heat-resistant, BODEN, LUFT temperature-resistant
Temperatur: **Temperaturbeständigkeit** *f* ABFALL, BODEN *Filtermaterialien*, LUFT temperature resistance
Temperatureinfluß *m* ABFALL, WASSER influence of tempera-

ture; ~ **auf die Korrosivität** WASSER influence of temperature on corrosivity; ~ **auf die Wassergüte** ABFALL influence of temperature on water quality; ~ **auf die Wasserspeicherung** WASSER influence of temperature on water storage

Temperatur: **Temperaturfühler** *m* ANALYTIK *pH-Meter* temperature probe

temperaturgeregelt *adj* ANALYTIK temperature-controlled

Temperatur: **Temperaturgradient** *m* ABWASSER, LÄRM, LUFT temperature gradient; **Temperaturinversion** *f* LUFT temperature inversion; **Temperaturmeßfühler** *m* ANALYTIK temperature sensor; **Temperaturmessung** *f* ANALYTIK temperature measurement; **Temperaturregelung** *f* LUFT attemperation

temperaturstabil *adj* ANALYTIK temperature-stable

Temperatur: **Temperaturstabilität** *f* ANALYTIK temperature stability

temperieren *vt* ANALYTIK temper

temporär: **~e Härte** *f* ABWASSER temporary hardness

Tender *m* MEER *Marinefahrzeug* tender

Tensid *nt* ABWASSER, MEER surface-active agent, surfactant

Tensiometer *nt* ANALYTIK, BODEN tensiometer

teratogen *adj* SICHERHEIT teratogenic

Teratogen *nt* SICHERHEIT teratogen

tertiär: **~e Abwasserbehandlung** *f* ABWASSER advanced wastewater treatment, tertiary wastewater treatment

Test *m* ANALYTIK test; **Testanlage** *f* ANALYTIK, WASSER test plant

testen *vt* ANALYTIK test

Test: **Testfiltrationsanlage** *f* WASSER test filtration plant; **Testpapier** *nt* ANALYTIK test paper; **Testperson** *f* ANALYTIK test person; **Testphase** *f* ERNEU ENER *von erneuerbaren Energieanlagen* test phase

Tetra: ~ **CDD** *abbr* (*Tetrachlordibenzo-p-dioxin*) ABFALL, BODEN, LUFT Tetra CDD (*tetrachlorinated dibenzo-p-dioxines*)

Tetrachlordibenzo-p-dioxin *nt* (*Tetra CDD*) ABFALL, BODEN, LUFT tetrachlorinated dibenzo-p-dioxines (*Tetra CDD*)

Textilabfall *m* ABFALL textile waste

TFA *abbr* (*Technologiefolgenabschätzung*) POLITIK *Filtermedium* TA (*technology assessment*)

theoretisch: **~es Potential** *nt* ERNEU ENER theoretical potential

Thermalöl *nt* ERNEU ENER thermal oil; **Thermalölerhitzer** *m* LUFT thermal oil heater

Thermalquelle *f* ERNEU ENER thermal spring

thermisch *adj* ANALYTIK, ERNEU ENER, LUFT thermal; **~er Abbau** *m* ABFALL thermal degradation; **~e Alterung** *f* ABFALL thermal ageing; **~e Analyse** *f* ANALYTIK *Biege-, Scher-, Kompressions- und Zugbeanspruchung* thermal analysis; **~e Belastung** *f* ABFALL, LUFT thermal load; ~ **belüftet** *adj* ANALYTIK thermally aerated; **~e Desorption** *f* ANALYTIK thermal desorption; **~e Desorptionskammer** *f* ANALYTIK thermal desorption chamber; **~e Energie** *f* ERNEU ENER thermal energy; **~e Enthärtung** *f* ABFALL, WASSER thermal softening; ~ **isoliert** *adj* ANALYTIK thermally isolated; **~e Leistung** *f* ABFALL, LUFT, WASSER thermal capacity; **~er Nachbrenner** *m* LUFT fume incinerator, thermal incinerator; **~e Nachverbrennung** *f* (*TNV*) LUFT thermal combustion; **~e Nachverbrennungsanlage** *f* LUFT thermal combustion plant; **~e Oxidation** *f* LUFT thermal oxidation; **~es Recycling** *nt* ABFALL *der vierten Stufe* quaternary recycling; **~e Regeneration** *f* ABWASSER thermal regeneration; **~e Resteverwertung** *f* ABFALL thermal residue recovery; **~e Rückgewinnung** *f* ABFALL thermal recovery; **~es Solarkraftwerk** *nt* ERNEU ENER solar heat power plant; ~ **stabilisiert** *adj* ABWASSER thermally stabilized; **~e Stabilität** *f* ABFALL heat stability; **~e Trocknung** *f* ABWASSER thermal drying; **~e Verbrennung** *f* LUFT thermal combustion; **~es Verfahren** *nt* ABWASSER thermal process; **~e Zerlegung** *f* ABFALL thermal decomposition; **~e Zersetzung** *f* ABFALL thermal decomposition; **~e Zerstörung** *f* BODEN thermal destruction

Thermistor *m* ANALYTIK thermistor

Thermo- *in cpds* ANALYTIK, ERNEU ENER, LUFT thermal

thermochemisch: **~e Umwandlung** *f* ERNEU ENER *Biomasse* thermochemical conversion

Thermo-: **Thermoelement** *nt* LUFT *KNV* thermocouple; **Thermogravimetrie** *f* ANALYTIK thermogravimetry; **Thermohygrograph** *m* ANALYTIK thermohygrograph; **Thermoionischer Detektor** *m* ANALYTIK thermionic detector; **Thermolumineszenz** *f* ANALYTIK thermoluminescence; **Thermolumineszenz-Dosimeter** *nt* (*TLD*) ANALYTIK thermoluminescence dosimeter (*TLD*)

thermomagnetisch *adj* ANALYTIK thermomagnetic

Thermometer *nt* ANALYTIK thermometer

thermophil *adj* ABWASSER thermophilic; **~e Bakterien** *pl* ABWASSER thermophilic bacteria; **~e Faulung** *f* ABWASSER *selten angewendet* thermophilic digestion

thermophil-mesophil *adj* ABWASSER thermophilic-mesophilic; **~e Faulung** *f* ABWASSER thermophilic-mesophilic digestion

Thermo-: **Thermophorese** *f* LUFT thermophoresis; **Thermoplast** *m* ABFALL, BODEN thermoplast

thermoplastisch *adj* ABFALL thermoplastic; **~er Stoff** *m* LUFT thermoplastic material

Thermo-: **Thermoreaktor** *m* LUFT thermal reactor; **Thermoschockfestigkeit** *f* LUFT *Katalysator* thermoshock resistance; **Thermosphäre** *f* LUFT thermosphere; **Thermospray-Interface** *nt* ANALYTIK *Zwischenglied für Kopplungen LC-MS* thermospray interface

thermostabil *adj* LUFT thermostabile

Thermostabilität *f* LUFT thermostability

Thermostat *m* ANALYTIK thermostat

thermostatgeregelt: **~es Luftfilter** *nt* LUFT thermal air cleaner

thermostatisch *adj* ANALYTIK thermostatic

Thermostat: **Thermostat-Luftfilter** *nt* LUFT thermal air cleaner

Thioether *m* LUFT organic sulphides

Thiolalkohol *m* ABFALL, LUFT mercaptan, thiol

thixotrop: **~es Verpreßmittel** *nt* BODEN thixotropic grouting material

TIC *abbr* (*gesamter anorganisch gebundener Kohlenstoff*) ANALYTIK TIC (*total inorganic carbon*)

Tidebecken *nt* ERNEU ENER *Gezeitenkraftwerk* tidal basin

Tidengebiet *nt* MEER *Zone zwischen Hoch- und Niedrigwasser* intertidal zone

Tidenhub *m* ERNEU ENER tidal range

Tief- *in cpds* ABWASSER, BODEN deep; **Tiefbagger** *m* MEER backhoe; **Tiefbehälter** *m* WASSER earth reservoir; **Tiefbrunnen** *m* WASSER deep well; **Tiefbrunneneinspritzung** *f* WASSER deep well injection

Tiefenabsorber *m* LÄRM low-frequency absorber

Tiefenfilter *nt* ABWASSER, LUFT *Abscheider* deep-bed filter

Tiefenfiltration *f* ABWASSER, LUFT deep-bed filtration

Tief-: **Tieffiltration** *f* ABWASSER depth filtration; **Tiefkühlzerkleinerung** *f* ABFALL cryogenic process; **Tiefpflügen** *nt* BODEN deep ploughing; **Tiefsandfang** *m* ABWASSER deep sand trap; **Tiefschachtreaktor** *m* ABWASSER deep-shaft reactor; **Tiefseegraben** *m* MEER trench; **Tieftemperaturzerkleinerung** *f* ABFALL cryogenic process, freeze-grinding, *von Festabfällen* cryogrinding, cryogenic crushing

Tierkörperbeseitigungsanlage *f* ABFALL offal destruction plant

Tilling-Verfahren *nt* BODEN tilling process

Tilt-top-Container *m* ABFALL open-top container

Tintenwäsche *f* ABFALL, ABWASSER de-inking

Titer *m* ANALYTIK titer (*AmE*), titre (*BrE*); **Titerlösung** *f* ANALYTIK standard solution

Titrans *nt* ANALYTIK titrant
Titration *f* ANALYTIK titration; **Titrationskalorimeter** *nt* ANALYTIK titration calorimeter; **Titrationsmethode** *f* ANALYTIK titration method
titrieren *vt* ANALYTIK titrate
Titriergefäß *nt* ANALYTIK titration vessel
Titriergerät *nt* ANALYTIK titrometer
titrimetrisch *adj* ANALYTIK titrimetric; **~er Test** *m* ANALYTIK *Wasseranalyse* titrimetric test
TK *abbr* (*Tropfkörper*) ABWASSER, WASSER percolating filter, sprinkling filter, trickling filter
TLD *abbr* (*Thermolumineszenz-Dosimeter*) ANALYTIK TLD (*thermoluminescence dosimeter*)
TNV *abbr* (*thermische Nachverbrennung*) LUFT *Brennstoff aus Müll* thermal combustion
TOC *abbr* (*gesamter organisch gebundener Kohlenstoff*) ABFALL, ANALYTIK *Schwermetall* TOC (*total organic carbon*); **TOC-Analysator** *m* ANALYTIK TOC analyser (*BrE*), TOC analyzer (*AmE*)
tödlich *adj* LUFT lethal; **~e Konzentration** *f* ABFALL, LUFT, WASSER lethal concentration; **~e Wirkung** *f* ANALYTIK lethal effect
Toleranz *f* ABFALL, SICHERHEIT, WASSER tolerance
tolerierbar *adj* ABFALL, SICHERHEIT, WASSER *Restemissionen* tolerable; **~e tägliche Aufnahme** *f* (*ADI-Wert*) SICHERHEIT acceptable daily intake (*ADI*)
Toluol *nt* ABFALL, WASSER toluene
TOMS-Spektrometer *nt* ANALYTIK, LUFT total ozone mapping spectrometer (*TOMS*)
Ton: **Tonfalle** *f* LÄRM sound trap; **Tonlösung** *f* ABFALL, WASSER clay suspension; **Tonmergel** *m* ABFALL, WASSER marly clay; **Tonmineralschicht** *f* BODEN clay mineral layer; **Tonschicht** *f* ABFALL, WASSER clay layer; **Tonsuspension** *f* ABFALL, WASSER clay suspension
tot: **~e Bakterien** *pl* ABWASSER dead bacteria
Totalklärung *f* ABWASSER extended-period aeration
Toxikologie *f* SICHERHEIT toxicology
toxisch *adj* ALLG toxic; **~er Abfall** *m* ABFALL toxic waste; **~es Abfallprodukt** *nt* ABFALL toxic degradation product; **~es Abgas** *nt* LUFT toxic waste gas; **~e Dosis** *f* (*TD*) SICHERHEIT toxic dose (*TD*); **~e Gesamtwirkung** *f* SICHERHEIT cumulative toxic effect; **~e Konzentration** *f* (*TC*) SICHERHEIT toxic concentration (*TC*); **~er Staub** *m* LUFT toxic dust; **~e Substanz** *f* ANALYTIK, BODEN, SICHERHEIT toxic substance; **~e Wirkung** *f* ANALYTIK, SICHERHEIT toxic effect
Toxizität *f* ALLG toxicity; **Toxizitätsäquivalent** *nt* ANALYTIK toxicity equivalent; **Toxizitätspotential** *nt* BODEN toxicity potential; **Toxizitätstest** *m* ANALYTIK toxicity test
TR *abbr* (*Trockenrückstand*) ABFALL, ABWASSER *Abwasserreinigung bei der Müllverbrennungsanlage* dry residue
Tracer *m* ANALYTIK *Bestimmung der Grundwasserfließrichtung*, LUFT tracer; **Tracerversuch** *m* ANALYTIK tracer experiment
traditionell: **~er Energieträger** *m* ERNEU ENER traditional energy source
Trafo *m* (*Transformator*) ABFALL, ERNEU ENER transformer
tragbar *adj* ANALYTIK transportable, *Gasanalysengerät* portable; **~es digitales Kolorimeter** *nt* ANALYTIK portable digital colorimeter; **~er Gaschromatograph** *m* ANALYTIK portable gas chromatograph; **~er GC** *m* ANALYTIK portable GC; **~er Röntgenfluoreszenz-Analysator** *m* ANALYTIK portable x-ray fluorescence analyser (*BrE*), portable x-ray fluorescence analyzer (*AmE*); **~es Spektrophotometer** *nt* ANALYTIK portable spectrophotometer
Trag: **Tragbeutel** *m* ABFALL carrier bag
Träger *m* LUFT carrier; **Trägergas** *nt* ANALYTIK, LUFT carrier gas; **Trägermaterial** *nt* ABFALL, ABWASSER, ANALYTIK, BODEN, LUFT, WASSER carrier material; **Trägermedium** *nt* LUFT carrier medium; **Trägeroberfläche** *f* LUFT *Katalysator* carrier surface; **Trägerschicht** *f* LUFT *Katalysator* carrier layer
Tragetasche *f* ABFALL carrier bag
Trägheit *f* ABFALL, LUFT, WASSER inertia; **Trägheitsabscheider** *m* LUFT inertial separator; **Trägheitskraft** *f* LUFT inertial force
Trag: **Tragschicht** *f* ABFALL bearing layer, *von Recyclingmaterial* carrier layer; **Tragvermögen** *nt* ABFALL carrying capacity
Transaktionskosten *pl* POLITIK transaction costs
Transformator *m* (*Trafo*) ABFALL, ERNEU ENER transformer
Transmission *f* ANALYTIK transmission; **Transmissionsspektrum** *nt* ANALYTIK transmission spectrum
Transport *m* ABFALL, BODEN, MANAG transport
transportabel *adj* ABFALL, ANALYTIK, MANAG transportable
Transport: **Transportband** *nt* ABFALL belt conveyor
transportfähig *adj* ABFALL, BODEN, WASSER transportable
transportierbar *adj* ABFALL, BODEN transportable; **~e Gasflasche** *f* SICHERHEIT transportable gas container
transportieren *vt* ABFALL, BODEN, MANAG transport
Transport: **Transportkapazität** *f* ABFALL, WASSER *des Verteilrohrnetzes* transport capacity; **Transportkiste** *f* ABFALL, MANAG carrier box; **Transportmodell** *nt* MANAG transport model; **Transportquelle** *f* WASSER transportation source; **Transportsicherheit** *f* SICHERHEIT transport safety; **Transportvolumen** *nt* ABFALL transport volume
Transversalwelle *f* LÄRM transverse wave
Trapezmiete *f* ABFALL trapezoid compost heap
treibend: **~es Ölfeld** *nt* MEER oil spill, spill
Treib: **Treibgas** *nt* LUFT propellant; **Treibgut** *nt* ABFALL floating refuse; **Treibhauseffekt** *m* LUFT greenhouse warming, greenhouse effect; **Treibhausgasakkumulierung** *f* LUFT greenhouse gas build-up; **Treibhausgase** *nt pl* LUFT greenhouse gases; **Treibmittel** *nt* ABFALL blowing agent, LUFT propellant
treibmittelhaltig: **~es Thermoplast** *nt* ABFALL expandable thermoplastics
Treib: **Treiböl** *nt* MEER bunkers; **Treibstrahl** *m* LUFT *Ringspaltdüse* propelled jet
Trenn- *in cpds* ALLG separate
trennbar *adj* ABFALL separable
trennen *vt* ALLG separate
Trenn-: **Trennentwässerung** *f* ABWASSER separate dewatering; **Trennfläche** *f* LÄRM *Wände, Decke* partition surface; **Trennflüssigkeit** *f* ANALYTIK separating liquid; **Trenngrad** *m* ABFALL separation rate, LUFT grade efficiency, WASSER separation rate; **Trennkanalisation** *f* ABWASSER separate canalization; **Trennkurve** *f* LUFT grade efficiency curve; **Trennleistung** *f* ANALYTIK separating capacity; **Trennsäule** *f* ANALYTIK separation column; **Trennschärfe** *f* LUFT separation effect; **Trennschicht** *f* ABFALL, WASSER separation layer; **Trennschleuder** *f* ABWASSER centrifuge; **Trennschnitt** *m* BODEN separating cut; **Trennstufen** *f pl* WASSER separation stages
Trennung *f* ABFALL separation, ABWASSER, LUFT *der Schadstoffe* isolation; **~ nach Farben** ABFALL *Altglas* separation by color (*AmE*), separation by colour (*BrE*)
Trenn-: **Trennverfahren** *nt* ABWASSER separate treatment, LUFT separation process; **Trennwand** *f* LÄRM partition
TRI *abbr* (*Trichlorethylen*) LUFT TRI (*trichloroethylene*)
Trichlorethylen *nt* (*TRI*) LUFT trichloroethylene (*TRI*)
Trichlorfluormethan *nt* LUFT *halogenierter Kohlenwasserstoff* trichlorofluoromethane
Trichter *m* ABWASSER hopper; **Trichterbecken** *nt* ABWASSER hopper-bottom tank; **Trichterbildung** *f* LUFT funneling (*AmE*), funnelling (*BrE*)
trichterförmig *adj* ABWASSER funnel-shaped; **~er Absetzraum** *m* ABWASSER funnel-shaped settling chamber; **~er Stoffänger** *m* ABWASSER funnel-shaped settling chamber
Triebkraft *f* LUFT propelling force

Triglycerid *nt* ABWASSER triglyceride

trinkbar *adj* WASSER potable; **~es Wasser** *nt* WASSER drinking water

Trinkwasser *nt* ANALYTIK, WASSER drinking water; **Trinkwasseranalyse** *f* ANALYTIK drinking water analysis; **Trinkwasseraufbereitungsverordnung** *f* POLITIK, WASSER drinking water treatment regulation; **Trinkwasserentnahme** *f* WASSER tapping of drinking water; **Trinkwassergewinnung** *f* WASSER drinking water recovery; **Trinkwassergewinnung** *f* **aus Oberflächengewässern** WASSER drinking water recovery from surface waters; **Trinkwasserqualität** *f* WASSER drinking water quality; **Trinkwasserschwellenwert** *m* WASSER threshold value for drinking water; **Trinkwasserverbrauch** *m* WASSER drinking water consumption; **Trinkwasserversorgung** *f* WASSER assurance of drinking water supply; **Trinkwasserverteilsystem** *nt* WASSER drinking water distribution network; **Trinkwasserverwendung** *f* WASSER drinking water use; **Trinkwasservorrat** *m* WASSER drinking water stock

Trocken- *in cpds* ABFALL, BODEN, LUFT dry

trocken 1. *adj* ABFALL, BODEN, LUFT dry; **2. ~ laufen** *vt* ABFALL, WASSER run dry

trocken: **~e Abgasbehandlung** *f* LUFT dry waste gas treatment; **~e Abreinigung** *f* LUFT *des trockenen Elektrofilters durch Klopfen* dry cleaning process; **~e Abscheidung** *f* LUFT dry gas cleaning; **~er adiabatischer Temperaturgradient** *m* LUFT dry adiabatic lapse rate; **~e Deposition** *f* BODEN, LUFT dry deposition; **~e Destillation** *f* BODEN dry distillation; **~er Dunst** *m* LUFT atmospheric fumes; **~es Elektrofilter** *nt* LUFT dry electrostatic filter; **~e Korrosion** *f* ABFALL, WASSER dry corrosion; **~er saurer Fallout** *m* BODEN, LUFT dry acid deposit; **~er saurer Niederschlag** *m* BODEN, LUFT dry acid fallout, dry acidic fallout

Trocken-: **Trockenabscheider** *m* LUFT dry scrubber; **Trockenabscheidung** *f* LUFT dry scrubbing; **Trockenabsorption** *f* LUFT dry absorption; **Trockenadditivverfahren** *nt* LUFT dry additive process; **Trockenadsorption** *f* LUFT dry adsorption; **Trockenbett** *nt* WASSER drying bed; **Trockendosierer** *m* ABWASSER dry feeder; **Trockeneinrichtung** *f* ANALYTIK dry equipment; **Trockenentschwefelungsprozess** *m* LUFT dry desulfurization process (*AmE*), dry desulphurization process (*BrE*); **Trockenfeuerung** *f* ABFALL dry firing; **Trockenfiltration** *f* ABWASSER dry filtration; **Trockengut** *nt* ABWASSER dry material, dry substance; **Trockengutverladung** *f* ABWASSER loading of dry mass

trockenlegen *vt* ABWASSER, BODEN, WASSER drain

Trocken-: **Trockenlegen** *nt* BODEN, WASSER drainage; **Trockenluftfilter** *nt* LUFT dry air filter; **Trockenmasse** *f* ABFALL, ABWASSER dry mass, dry solid matter; **Trockenmassekonzentration** *f* ABWASSER dry mass concentration; **Trockenmedium** *nt* ANALYTIK drying agent; **Trockenmittel** *nt* ANALYTIK drying agent; **Trockenrost** *m* ABFALL drying grate; **Trockenrückstand** *m* (*TR*) ABFALL, ABWASSER dry residue; **Trockenschlamm** *m* ABWASSER dry sludge; **Trockenschlammdeponie** *f* ABWASSER dry sludge disposal site; **Trockenschrank** *m* ANALYTIK drying closet; **Trockensieb** *nt* ABWASSER dry screen; **Trockensorption** *f* ABFALL dry sorption; **Trockenstaubbeseitigungsanlage** *f* SICHERHEIT dry dust removal installation

Trockensubstanz *f* (*TS*) ABFALL, ABWASSER dry mass, dry solid matter; **Trockensubstanzabscheidung** *f* ABWASSER dry mass separation; **Trockensubstanzgehalt** *m* ABWASSER dry mass content; **Trockensubstanzgehalt** *m* **im Belebungsbecken** (*TSBB*) ABWASSER mixed liquor suspended solids (*MLSS*)

Trocken: **Trockentrennung** *f* ABFALL *von Müll* dry sorting; **Trockentrommel** *f* BODEN rotary dryer; **Trockenturm** *m* LUFT drying tower; **Trockenzerkleinern** *nt* ABFALL dry grinding

trocknen *vt* ABFALL, ABWASSER, BODEN, LUFT dry, desiccate

Trocknung *f* ABFALL, ABWASSER, BODEN, LUFT drying, dessication; **Trocknungsmedium** *nt* LUFT drying medium; **Trocknungszone** *f* LUFT drying zone

Trogkettenförderer *m* ABFALL troughed chain conveyor

Trommel *f* ABWASSER, BODEN, LUFT, MEER drum; **Trommelabschöpfgerät** *nt* MEER drum skimmer; **Trommelfilter** *nt* ABFALL, ABWASSER, WASSER drum filter; **Trommelsieb** *nt* ABFALL trommel screen, drum sieve, ABWASSER drum screen, WASSER drum sieve; **Trommeltrockner** *m* BODEN drum-type dryer

Tröpfchen *nt* LUFT droplet; **Tröpfchendurchmesser** *m* LUFT droplet diameter; **Tröpfchenfänger** *m* LUFT droplet trap; **Tröpfchengröße** *f* LUFT droplet size

Tropfen *m* LUFT drop; **Tropfenabscheider** *m* LUFT demister unit, mist eliminator, moisture eliminator

tropfenbeladen *adj* LUFT drop-laden

Tropfen: **Tropfendispergierung** *f* LUFT droplet dispersion; **Tropfenerzeugung** *f* LUFT droplet generation; **Tropfenschleier** *m* LUFT drop film; **Tropfenverteilung** *f* LUFT drop distribution; **Tropfenzerteilung** *f* LUFT droplet separation

Tropfkörper *m* (*TK*) ABWASSER, WASSER percolating filter, sprinkling filter, trickling filter; **Tropfkörperanlage** *f* ABWASSER bacteria bed; **Tropfkörperboden** *m* ABWASSER trickling filter bottom; **Tropfkörperverfahren** *nt* ABWASSER biofiltration process, trickling filter process, LUFT *Biowäscher* trickling filter process

Troposphäre *f* LUFT troposphere

troposphärisch *adj* LUFT tropospheric; **~es Ozon** *nt* LUFT tropospheric ozone; **~e Ozonbelastung** *f* LUFT tropospheric ozone pollution; **~e Ozonkonzentration** *f* LUFT tropospheric ozone concentration

trübe *adj* ABWASSER, WASSER turbid, cloudy; **~ Abwasserprobe** *f* ANALYTIK turbid wastewater sample; **~s Wasser** *nt* ABFALL, WASSER turbid water

trüben *vt* ABFALL, WASSER cloud

Trübstoff *m* ABWASSER turbid matter

Trübung *f* ABWASSER turbidity, LUFT air turbidity; **Trübungsanalyse** *f* ANALYTIK turbidimetry, *Wasseranalyse* turbidimetric analysis, nephelometric analysis; **Trübungseinheit** *f* ANALYTIK nephelometry turbidity unit (*NTU*); **Trübungseinheit/Formazin** *f* (*TE/F*) ANALYTIK formazin turbidity unit (*FTU*); **Trübungsmesser** *m* ANALYTIK nephelometer, turbidimeter; **Trübungsmessung** *f* ANALYTIK turbidity value; **Trübungswert** *m* ANALYTIK turbidity value

TS *abbr* (*Trockensubstanz*) ABFALL, ABWASSER dry solid matter, *Schwefeloxid* dry mass; **TS-Abscheidung** *f* ABWASSER dry mass separation

TSBB *abbr* (*Trockensubstanzgehalt im Belebungsbecken*) ABWASSER MLSS (*mixed liquor suspended solids*)

TS: **TS-Gehalt** *m* ABWASSER dry mass content; **TS-Konzentration** *f* ABWASSER dry mass concentration

Tubificiden *f pl* ABWASSER sludge worms

Tubularmodul *m* ABWASSER *Ultrafiltration* tubular module

Turbidimeter *nt* ANALYTIK turbidimeter

Turbidimetrie *f* ANALYTIK turbidimetry

turbidimetrisch *adj* ANALYTIK turbidimetric; **~e Titration** *f* ANALYTIK turbidimetric titration

Turbine *f* ERNEU ENER turbine; **Turbinenblatt** *nt* ERNEU ENER turbine blade; **Turbinenpumpe** *f* WASSER turbo-pump

Turbo- *in cpds* ABWASSER, ERNEU ENER, WASSER turbo-; **Turbobelüfter** *m* ABWASSER turbo-aerator; **Turbogenerator** *m* ERNEU ENER turbo-generator; **Turbomaschine** *f* ABFALL turbo-machine, ABWASSER, ERNEU ENER, LUFT, WASSER kinetic-energy pump, turbo-machine; **Turbomischer** *m* ABFALL flash mixer; **Turbopumpe** *f* ABFALL, WASSER turbo-pump; **Turboverdichter** *m* ABWASSER turbo-compressor

turbulent: **~e Strömung** *f* LUFT turbulent flow
Turbulenz *f* ABFALL, ABWASSER turbulence; **Turbulenzkoeffizient** *m* LUFT eddy viscosity coefficient
Turm *m* ERNEU ENER *Windturbine* tower; **Turmbehälter** *m* WASSER elevated water reservoir, water tower; **Turmbehälterhöhe** *f* WASSER water tower height; **Turm-Biofilter** *nt* LUFT *biologisches Sorptionsverfahren* tower biofilter; **Turmhöhe** *f* ERNEU ENER *Windturbine* tower height; **Turmspitze** *f* ERNEU ENER *Biokonversion* solar tower top; **Turmsystem** *nt* ABFALL *Kompostierungsverfahren* tower system; **Turmtropfkörper** *m* ABWASSER tower-type trickling filter, bio-tower; **Turmwäscher** *m* LUFT scrubbing tower, scrubber tower
TWA *abbr* (*zeitlicher Durchschnitt*) ANALYTIK TWA (*time-weighted average*)
Tyndall-Effekt *m* ANALYTIK Tyndall effect
Typenzulassung *f* POLITIK model operation permission

U

übelriechend: **~er Schwaden** *m* LUFT malodorous plume
Überangebot *nt* WASSER excess supply
überbaut: **~e Fläche** *f* BODEN built-up surface
Überbrand *m* BODEN overburning
Überdachung *f* ABFALL *Deponie*, LUFT *des Filters*, WASSER *Deponie* roofing
überdeckt: **~er Abzugsgraben** *m* ABWASSER, WASSER covered drain
Überdestillation *f* ANALYTIK overdistillation
Überdosierung *f* LUFT, SICHERHEIT overdosage, overdosing
Überdosis *f* ANALYTIK, SICHERHEIT overdose (*OD*)
Überdruck *m* ABWASSER excess pressure; **Überdruckbunker** *m* BODEN lock hopper
Übereinkommen *nt* POLITIK *gesetzliches Instrument* convention
Überfall *m* ABFALL, WASSER overfall; **Überfallkante** *f* ABFALL, WASSER crest; **Überfallschwelle** *f* ABFALL, WASSER sill; **Überfallwehr** *f* WASSER overflow weir
Überflurhydrant *m* WASSER post hydrant
Überflurwasserpfosten *m* WASSER post hydrant
überfluten *vt* ABFALL, WASSER submerge
überflutet *adj* ABFALL, WASSER submerged
Überflutung *f* ABWASSER, WASSER *Berieselung von Flächen mit Abwasser* flooding, inundation, overflow
Übergabeort *m* ABFALL transfer site
Übergabestelle *f* ABFALL *Förderband* discharge point
Übergangs: **Übergangsmetall** *nt* ABFALL transition metal; **Übergangsphase** *f* POLITIK transition phase; **Übergangsregelung** *f* POLITIK transition regulation; **Übergangssee** *m* WASSER transition lake
Überhitzer *m* ABWASSER, LUFT superheater
Überhitzung *f* ABWASSER, LUFT overheating
Überhöhung *f*: **~ der Abluftfahne** LUFT plume-rise
Überkornfraktion *f* ABFALL, BODEN oversize fraction
Überkornmaterial *nt* ABFALL, BODEN *bei Sortierung* oversize material
überkritisch *adj* ANALYTIK, BODEN supercritical; **~e Flüssig-Chromatographie** *f* ANALYTIK supercritical-fluid chromatography (*SFC*); **~e Flüssig-Extraktion** *f* ANALYTIK supercritical-fluid extraction (*SFE*); **~es Wasser** *nt* BODEN supercritical water
Überladung *f* LUFT overload
Überlastung *f* BODEN excess load, LUFT overload
Überlauf *m* ABFALL, ABWASSER, WASSER overflow; **Überlaufbauwerk** *nt* ABWASSER, WASSER overflow structure; **Überlaufbehälter** *m* ABWASSER, WASSER overflow tank
überlaufen *vi* ABFALL, ABWASSER, WASSER run over, overflow
Überlauf: **Überlaufmenge** *f* ABFALL, ABWASSER, WASSER overflow rate; **Überlaufquelle** *f* WASSER overflow spring; **Überlaufrohr** *nt* ABFALL, ABWASSER, WASSER overflow pipe; **Überlaufwehr** *nt* ABWASSER effluent weir
übermäßig: **~e Ablagerung** *f* LUFT excessive deposit build-up
Übernutzung *f* BODEN overexploitation
Überproduktion *f* ABFALL, MANAG excess production
überprüfbar *adj* ANALYTIK verifiable
überprüfen *vt* ABFALL, ANALYTIK, SICHERHEIT check, examine, verify
überprüft *adj* ABFALL, ANALYTIK, SICHERHEIT checked, examined, verified
Überprüfung *f* ABFALL, ANALYTIK, SICHERHEIT check, checking, examination, verification
übersättigt *adj* ABFALL, LUFT, WASSER supersaturated; **~e Lösung** *f* ABFALL, WASSER supersaturated solution
Überschallgeschwindigkeit *f* LÄRM supersonic speed
Überschallknall *m* LÄRM supersonic boom
Überschlag *m* LUFT *Elektrofilter* flash-over, spark-over, arc-over; **Überschlagsspannung** *f* LUFT *Elektrofilter* spark-over voltage, arc-over voltage, flash-over voltage
Überschuß *m* ABFALL, ABWASSER surplus; **Überschußbiomasse** *f* ABFALL, ABWASSER excess biomass; **Überschußgas** *nt* ABFALL excess gas
überschüssig: **~e Düngersteuer** *f* POLITIK excess manure tax
Überschuß: **Überschußschlamm** *m* ABWASSER excess sludge; **Überschußschlammvolumen** *nt* ABWASSER excess sludge volume
Überschwemmung *f* ABFALL, ABWASSER, WASSER overflow; **Überschwemmungsgebiet** *nt* WASSER flood plain
Übersprechen *nt* LÄRM crosstalk
Überstaubewässerung *f* WASSER flood irrigation
überstauen *vt* ABFALL, WASSER overdam
Überstauhöhe *f* ABFALL, WASSER submerged bed height
Überstaupflanzenbecken *nt* ABWASSER submerged plant tank
überstehend: **~e Flüssigkeit** *f* ABWASSER supernatant liquid
überströmen *vi* ANALYTIK, WASSER flow over
übertägig *adj* ABFALL, WASSER above-ground; **~e Ablagerung** *f* ABFALL above-ground disposal
übertragen *vt* ABFALL, ABWASSER transfer, ERNEU ENER, LÄRM transmit, WASSER transfer
Übertragungsdämpfung *f* LÄRM transmission loss (*TL*)
Übertragung *f* ERNEU ENER LÄRM transmission
Übertragungsrate *f* ABFALL, ABWASSER, WASSER transfer rate
überwachen *vt* ALLG monitor, supervise
Überwachung *f* ALLG monitoring, supervision, surveillance; **~ von gesundheitsgefährdenden Stoffen** MANAG, SICHERHEIT Control of Substances Hazardous to Health (*BrE*) (*COSHH*); **~ partikulärer Stoffe** ANALYTIK particulates monitoring; **~ der Wasserbeschaffenheit** WASSER water quality monitoring; **~ der Wirksamkeit von Abgasreinigungsanlagen** ANALYTIK monitoring of the effectiveness of waste gas installations
überwachungsbedürftig *adj* ABFALL, SICHERHEIT in need of monitoring
Überwachung: **Überwachungsinstrument** *nt* POLITIK monitoring instrument; **Überwachungsmaßnahme** *f* SICHERHEIT monitoring measure; **Überwachungspflicht** *f* POLITIK, SICHERHEIT monitoring obligation
überwachungspflichtig: **~e Anlage** *f* POLITIK controlled plant
Überwachung: **Überwachungssystem** *nt* ANALYTIK, SICHERHEIT monitoring system
überwälzen *vt* POLITIK *Kosten, Verantwortung* shift
Überwasser *nt* WASSER surplus water; **Überwasserspiegelzone** *f* BODEN zone of aeration
Überweidung *f* BODEN overgrazing
Überzugsdicke *f* ABFALL coating thickness
UCT-Verfahren *nt* ABWASSER *Phosphorelimination* UCT method (*University of Cape Town method*)
UEG *abbr* (*untere Explosionsgrenze*) SICHERHEIT lower explosion limit
UF *abbr* (*Ultrafiltration*) ALLG UF (*ultrafiltration*)
Ufer *nt* ABWASSER, WASSER *eines Flusses* bank
uferanliegend *adj* WASSER riparian, waterfront

Ufer: **Uferfiltrat** *nt* WASSER bank filtrate
uferfiltriert *adj* WASSER bank-filtered; **~es Grundwasser** *nt* WASSER bank-filtered groundwater
Ufer: **Uferlinie** *f* MEER shoreline; **Ufermauer** *f* **eines Flusses** WASSER river wall
UIS *abbr* (*Umweltinformationssystem*) MANAG, POLITIK environmental information system
Ultrafiltration *f* (*UF*) ALLG ultrafiltration (*UF*)
Ultraschall *m* ANALYTIK, LÄRM ultrasound; **Ultraschallextraktion** *f* ANALYTIK ultrasound extraction; **Ultraschallsensor** *m* ANALYTIK ultrasound sensor; **Ultraschallsiebung** *f* ABFALL, ABWASSER, LÄRM sonification; **Ultraschallstrahlung** *f* BODEN ultrasound irradiation; **Ultraschallzerstäuber** *m* ANALYTIK *Analyse geringster Konzentration* ultrasonic nebulizer
Ultraspur *f* ANALYTIK ultratrace; **Ultraspurenanalyse** *f* ANALYTIK ultratrace analysis
ultraviolett *adj* (*UV*) ALLG ultraviolet (*UV*)
Ultraviolett *nt* (*UV*) ALLG ultraviolet (*UV*)
Umfeldanalyse *f* ANALYTIK surrounding field analysis
Umgebung *f* ALLG environment
umgebungsbeeinflußt: **~e Spannungsrißbildung** *f* ABFALL *Kunststoffe* environmental stress cracking
Umgebung: **Umgebungsgeräusch** *nt* LÄRM ambient noise; **Umgebungslärm** *m* LÄRM ambient noise; **Umgebungsluft** *f* ANALYTIK, LUFT ambient air; **Umgebungsluftüberwachung** *f* SICHERHEIT quality monitoring of ambient air
umkehrbar *adj* ALLG reversible; **~es Verfahren** *nt* ABWASSER reversible process
Umkehrosmose *f* (*UO*) ABFALL, ABWASSER, WASSER reverse osmosis (*RO*); **~ mit Konzentrateindampfung** ABFALL, WASSER reverse osmosis with concentrate evaporation; **Umkehrosmoseanlage** *f* ABFALL, WASSER reverse osmosis plant
Umkehrprisma *nt* ANALYTIK *Grubengas-Interferometer* inverting prism
Umladeanlage *f* ABFALL transfer station
Umladehalle *f* ABFALL transfer hall
Umladestation *f* ABFALL transfer station
Umladung *f* ABFALL transfer, LUFT charge reversal
Umlauf *m* ABFALL, ABWASSER, LUFT, WASSER circulation; **Umlaufbecken** *nt* ABWASSER activated sludge tank with circulating flow; **Umlauffilter** *nt* LUFT recirculating filter; **Umlaufpumpe** *f* ABWASSER recirculating pump, circulation pump; **Umlaufverdampfer** *m* ABWASSER, WASSER circulation evaporator, rotary evaporator
umleiten *vt* LUFT bypass
Umlenkabscheider *m* LUFT deflector separator; **Umlenkblech** *nt* LUFT deflector vane
umlenken *vt* LUFT *Gasstrom auf die Oberfläche* deflect
Ummantelung *f* MEER skirt; **~ mit schallschluckenden Stoffen** LÄRM sound-absorbing material casing
Umrechnungsfaktor *m* ANALYTIK conversion factor
umpumpen: **im Kreislauf ~** *vt* ABFALL, ABWASSER, LUFT, WASSER recirculate
Umreifungsgeräte *nt pl* ABFALL strapping equipment
Umrüstanlage *f* ABFALL conversion machinery
Umsatz *m* ALLG conversion; **Umsatzgeschwindigkeit** *f* LUFT conversion force
Umschlagbereich *m* ANALYTIK indicator range
Umschlagpumpe *f* MEER transfer pump
umschmelzen *vt* ABFALL remelt
umsetzbar: **~e Anlage** *f* BODEN relocatable plant, relocatable unit
umsetzen *vt* ABFALL, ABWASSER convert, LUFT *katalytisch* transfer, POLITIK implement, enforce; **chemisch ~** *vt* LUFT transfer chemically
Umsetzer *m* ABFALL, ABWASSER converter
Umsetzung *f* ALLG conversion, ANALYTIK *chemisch* reaction, POLITIK implementation, *des Rechts, der Gesetze* enforcement; **Umsetzungsprodukt** *nt* ABFALL, WASSER solidified waste, CFS-processed waste, CFS-treated waste; **Umsetzungsprozeß** *m* ABFALL, WASSER *in der Deponie* reaction process
Umströmung *f* ABFALL, ABWASSER, LUFT, WASSER circulation
Umverpackung *f* ABFALL secondary packaging, over-packaging, MANAG secondary packaging
umwälzen *vt* ABFALL, ABWASSER, LUFT, WASSER circulate
Umwälzen *nt* ABFALL, ABWASSER, LUFT, WASSER circulation
Umwälzungsbecken *nt* WASSER spiral-flow tank
Umwälzvorrichtung *f* WASSER stirring mechanism
umwandeln *vt* ABFALL *schädliche oder gefährliche Substanzen in Abfällen*, BODEN, ERNEU ENER, LUFT *in elektrische Energie* convert
Umwandler *m* ERNEU ENER transformer
Umwandlung *f* ALLG conversion; **~ in eine nutzbare Energieform** ERNEU ENER conversion into usable energy form; **~ der Wellenenergie** ERNEU ENER wave energy conversion; **Umwandlungsgrad** *m* LUFT conversion rate; **Umwandlungspunkt** *m* ANALYTIK transition point; **Umwandlungsrate** *f* ABFALL, LUFT transformation rate; **Umwandlungstemperatur** *f* ANALYTIK transition temperature; **Umwandlungswirkungsgrad** *m* ERNEU ENER conversion efficiency
Umwelt *f* ALLG environment; **Umweltaktivist** *m* POLITIK environmental activist; **Umweltaktivität** *f* MANAG, POLITIK environmental activity; **Umweltanalytik** *f* ANALYTIK, MANAG environmental analysis; **Umweltatlas** *m* POLITIK environment atlas; **Umwelt-Audit** *nt* MANAG, POLITIK eco-audit; **Umweltausschuß** *m* POLITIK *des Parlaments* environment committee; **Umweltauswirkung** *f* ABWASSER, BODEN, ERNEU ENER, LÄRM, LUFT, MEER environmental effect, environmental impact; **Umweltbeauftragter** *m* POLITIK environmental representative; **Umweltbedingungen** *f pl* POLITIK environmental conditions; **Umweltbelastung** *f* ABFALL, LUFT, POLITIK, WASSER environmental impact, environmental stress; **Umweltbeobachtung** *f* ANALYTIK, MANAG, POLITIK environmental monitoring; **Umweltberater** *m* POLITIK environmental adviser, environmental counsellor (*BrE*), environmental counselor (*AmE*); **Umweltberatung** *f* POLITIK environmental counseling (*AmE*), environmental counselling (*BrE*), environmental advice; **Umweltbericht** *m* POLITIK environmental report; **Umweltberichterstattung** *f* POLITIK environmental reporting
umweltbewußt *adj* MANAG, POLITIK environmentally aware, green; **~es Einkaufen** *nt* MANAG environmentally-aware purchasing; **~es Konsumverhalten** *nt* MANAG green consumerism
Umwelt: **Umweltbewußtsein** *nt* MANAG, POLITIK environmental awareness; **Umweltdelikt** *nt* POLITIK environmental crime; **Umwelteinwirkung** *f* ABWASSER, BODEN, LUFT, MEER environmental impact; **Umweltengagement** *nt* MANAG, POLITIK environmental commitment; **Umwelterziehung** *f* POLITIK environmental education; **Umweltethik** *f* POLITIK environmental ethics; **Umweltfaktor** *m* ABFALL, ABWASSER, BODEN, MEER, POLITIK, WASSER ecological factor; **Umweltfonds** *m* POLITIK environment fund; **Umweltforschung** *f* ANALYTIK, POLITIK environmental research; **Umweltfragen** *f pl* POLITIK environmental issues
umweltfreundlich *adj* ABFALL, MANAG, POLITIK, WASSER environmentally friendly, eco-friendly; **~e Logistik** *f* MANAG environmentally-friendly logistics; **~es Produkt** *nt* MANAG environmentally-friendly product; **~e Technologie** *f* ABFALL, ABWASSER, LUFT, MANAG, WASSER clean technology, non-waste-technology (*NWT*), low-waste technology
Umwelt: **Umweltfreundlichkeit** *f* MANAG, POLITIK, WASSER environmental friendliness
umweltgefährdend *adj* ALLG harmful to the environment
Umwelt: **Umweltgefährdung** *f* ALLG environmental endan-

germent; **Umweltgenehmigung** *f* POLITIK environmental licence (*BrE*), environmental license (*AmE*)

umweltgerecht *adj* MANAG, POLITIK environmentally-oriented; **~es Produkt** *nt* MANAG environmentally-oriented product

Umwelt: **Umweltgerichtshof** *m* POLITIK environmental court; **Umweltgesetz** *nt* POLITIK environmental law; **Umweltgesetzbuch** *nt* POLITIK environment statute book, environmental code; **Umweltgesetzgebung** *f* POLITIK environmental legislation; **Umweltgestaltung** *f* POLITIK environmental planning; **Umweltgesundheit** *f* POLITIK, SICHERHEIT environmental health; **Umweltgipfel** *m* POLITIK environmental summit; **Umweltgutachten** *nt* POLITIK environmental expert opinion; **Umwelthaftung** *f* ABFALL environmental liability, ABWASSER environmental fiability, POLITIK, SICHERHEIT, WASSER environmental liability; **Umweltidee** *f* MANAG environmental idea; **Umweltinformatik** *f* POLITIK environmental communication; **Umweltinformationen** *f pl* MANAG, POLITIK environmental information; **Umweltinformationsbeschaffung** *f* MANAG, POLITIK environmental information collection; **Umweltinformationssystem** *nt* (*UIS*) MANAG, POLITIK environmental information system; **Umweltjury** *f* POLITIK environmental jury; **Umweltkatastrophe** *f* BODEN, LUFT, MANAG, POLITIK environmental catastrophe, environmental disaster; **Umweltkonto** *nt* MANAG ecological account; **Umweltkriminalität** *f* POLITIK environmental criminality; **Umweltlärm** *m* LÄRM, SICHERHEIT environmental noise; **Umweltleitlinie** *f* POLITIK environmental guideline; **Umweltlizenz** *f* POLITIK *environment* emission right; **Umweltlobby** *f* POLITIK environmental lobby; **Umweltmanagement** *nt* MANAG environmental management; **Umweltmedium** *nt* BODEN, LUFT, MANAG, POLITIK, WASSER environmental medium; **Umweltmesse** *f* MANAG environmental trade fair; **Umweltminister** *m* POLITIK environmental minister; **Umweltministerium** *nt* POLITIK ministry of environment; **Umwelt-Monitoring** *nt* ANALYTIK, MANAG, POLITIK environmental monitoring

umweltnachhaltig *adj* POLITIK environmentally sustainable

Umwelt: **Umweltnutzung** *f* MANAG, POLITIK environmental use; **Umweltnutzungsrechte** *nt pl* MANAG, POLITIK environmental use rights; **Umweltökonom** *m* MANAG, POLITIK environmental economist; **Umweltökonomie** *f* MANAG, POLITIK environmental economics

umweltökonomisch *adj* MANAG, POLITIK environmentally-economic

Umwelt: **Umweltorganisation** *f* POLITIK environmental organization

umweltorientiert *adj* MANAG, POLITIK environmentally-oriented; **~es Management** *nt* MANAG environmentally-oriented management; **~es Marketing** *nt* MANAG environmentally-oriented marketing; **~e Wiedergewinnung** *f* ABFALL *Industrie* environmentally-oriented recovery

Umwelt: **Umweltpartner** *m* MANAG, POLITIK environmental partner; **Umweltplanung** *f* MANAG, POLITIK environmental planning; **Umweltpolitik** *f* MANAG, POLITIK environmental policy

umweltpolitisch *adj* POLITIK environmentally political, ecopolitical; **~e Arbeitsgruppe** *f* POLITIK environmentally-political working group; **~e Instrumente** *nt pl* POLITIK instruments of environmental policy; **~e Maßnahmen** *f pl* POLITIK environmental policy measures

Umwelt: **Umweltpreis** *m* POLITIK environmental award, environmental prize; **Umweltprogramm** *nt* POLITIK environmental program (*AmE*), environmental programme (*BrE*); **Umweltprogramm** *nt* **der Vereinten Nationen** POLITIK United Nations Environmental Programme (*UNEP*); **Umweltqualität** *f* POLITIK environmental quality; **Umweltqualitätsnormen** *f pl* POLITIK environmental quality standards; **Umweltrecht** *nt* POLITIK environmental law, environmental protection law

umweltrelevant *adj* BODEN, LÄRM, LUFT environmentally relevant

Umwelt: **Umweltrichtlinie** *f* POLITIK environmental directive, environmental regulation; **Umweltschaden** *m* ALLG environmental damage

umweltschädigend: **~e Subvention** *f* POLITIK environmentally-harmful subsidy

Umwelt: **Umweltschädigung** *f* ALLG environmental damage

umweltschädlich *adj* ALLG ecologically harmful, harmful to the environment

Umwelt: **Umweltschadstoff** *m* ALLG pollutant

umweltschonend *adj* ABFALL eco-friendly, environmentally acceptable, environmentally friendly; **~e Abfallentsorgung** *f* ABFALL environmentally-acceptable waste disposal

Umweltschutz *m* POLITIK environmental protection; **Umweltschutz-Audit** *nt* MANAG, POLITIK eco-audit; **Umweltschutzbehörde** *f* POLITIK environmental protection agency; **Umweltschutzbewegung** *f* POLITIK environmentalism

Umwelt: **Umweltschützer** *m* POLITIK environmentalist

Umweltschutz: **Umweltschutzgesetz** *nt* POLITIK environmental protection law; **Umweltschutzinitiative** *f* POLITIK environmental protection initiative; **Umweltschutzinvestition** *f* POLITIK environmental protection investment; **Umweltschutzmaßnahmen** *f pl* POLITIK environmental protection measures; **Umweltschutznormen** *f pl* POLITIK environmental protection standards

umweltschutzorientiert *adj* POLITIK environmental-protection-oriented

Umweltschutz: **Umweltschutzpolitik** *f* MANAG, POLITIK environmental policy; **Umweltschutzrecht** *nt* POLITIK environmental protection law; **Umweltschutz- und Sicherheitstechnik** *f* MANAG environmental and safety engineering; **Umweltschutzziel** *nt* POLITIK environmental protection aim

Umwelt: **Umweltsektor** *m* POLITIK environmental sector

umweltsicher *adj* ABWASSER, BODEN, LUFT, SICHERHEIT ecologically safe, environmentally safe

Umwelt: **Umweltsicherheit** *f* SICHERHEIT environmental safety; **Umwelt-Sponsoring** *nt* MANAG environmental sponsoring; **Umweltsprecher** *m* POLITIK *einer Parlamentsfraktion* environment spokesperson; **Umweltstatistik** *f* POLITIK environmental statistics; **Umweltsteuer** *f* POLITIK environmental tax; **Umweltstrafrecht** *nt* POLITIK environmental criminal law; **Umweltstreß** *m* ABWASSER, BODEN, LUFT environmental stress; **Umweltsünder** *m* POLITIK environmental polluter; **Umwelttechnik** *f* ABWASSER, ANALYTIK, BODEN, ERNEU ENER, LÄRM, LUFT, MANAG, WASSER environmental technology; **Umwelttechnikmarkt** *m* MANAG environmental technology market; **Umwelttestverfahren** *nt* ANALYTIK, MANAG environmental testing procedure; **Umweltverantwortung** *f* MANAG, POLITIK environmental responsibility; **Umweltverband** *m* POLITIK environmental group; **Umweltverbrecher** *m* POLITIK environmental criminal; **Umweltvereinbarung** *f* POLITIK environmental agreement

Umweltverschmutzung *f* ALLG pollution; **~ durch Wärme** LUFT heat pollution; **Umweltverschmutzungsforschung** *f* ABWASSER, ANALYTIK, LUFT, MANAG, MEER, POLITIK pollution research;

umweltverträglich *adj* ABFALL, MANAG, POLITIK *Produkte, Stoffe* environmentally compatible; **~er Betrieb** *m* MANAG environmentally-compatible operation; **~e Entsorgung** *f* ABFALL environmentally-compatible disposal; **~e Entwicklung** *f* POLITIK environmentally-compatible development; **~es Produkt** *nt* MANAG environmentally-compatible product

Umweltverträglichkeit *f* LUFT, POLITIK, WASSER environmental compatibility; **Umweltverträglichkeitsprüfung** *f* (*UVP*) POLITIK *beste verfügbare Technik* environmental impact assessment (*EIA*); **Umweltverträglichkeitsprüfungsbericht** *m* POLITIK environmental impact assessment report; **Umweltverträglichkeitsprüfungsverfahren** *nt* POLITIK environmental impact assessment procedure

Umwelt: **Umweltverwaltung** *f* MANAG, POLITIK environmental administration; **Umweltvorreiterposition** *f* POLITIK environmental leader position; **Umweltvorsorgeprinzip** *nt* POLITIK environmental prevention principle; **Umweltzeichen** *nt* POLITIK eco-label, environmental label

unaufbereitet *adj* ALLG unprocessed; **~er Abfall** *m pl* ABWASSER unprocessed waste

unbegrenzt: **~e Ablagerung** *f* ABFALL unlimited storage

unbehandelt *adj* ABFALL, ABWASSER, MEER untreated; **~es Abwasser** *nt* ABWASSER, MEER raw wastewater, untreated wastewater

unbeladen *adj* ABFALL, ABWASSER, MEER unladen

unbelüftet *adj* ABWASSER nonaerated; **~er Abwasserteich** *m* ABWASSER nonaerated oxidation pond; **~e Mietenkompostierung** *f* ABFALL unaerated bed composting

unbeschichtet *adj* ABFALL uncoated; **~e Fläche** *f* ABFALL uncoated surface

unbrennbar *adj* SICHERHEIT nonflammable, noncombustible

Unbrennbarkeitstest *m* SICHERHEIT noncombustibility test

undurchlässig *adj* ABFALL, BODEN, SICHERHEIT, WASSER impermeable, impervious; **~e Abdichtungswand** *f* BODEN impermeable barrier; **~e Schicht** *f* WASSER aquiclude, aquitard

undurchsichtig *adj* ABWASSER opaque

unempfindlich *adj* LUFT insensitive

unerschöpfbar: **~er Energiestrom** *m* ERNEU ENER inexhaustible energy stream

unerschöpflich *adj* ERNEU ENER inexhaustible; **~e Energiequelle** *f* ERNEU ENER inexhaustible energy source

Unerschöpflichkeit *f* ERNEU ENER *der Energiequellen* inexhaustibility

Unfall *m* SICHERHEIT accident; **Unfallursache** *f* SICHERHEIT cause of accident

unfiltriert *adj* ANALYTIK unfiltered; **~e Probe** *f* ANALYTIK unfiltered sample

ungedämpft *adj* LÄRM undamped

ungefährlich: **~er nuklearer Niederschlag** *m* BODEN dry deposition

ungeladen *adj* ABWASSER, LUFT uncharged; **~es Teilchen** *nt* LUFT, MEER uncharged particle

ungelöscht *adj* ABWASSER unslaked

ungelöst *adj* ABFALL, ABWASSER undissolved; **~er Stoff** *m* ABFALL, ABWASSER colloidal substance, undissolved substance

ungenießbar *adj* WASSER unpalatable; **~es Wasser** *nt* WASSER unpalatable water

ungeordnet *adj* ABFALL uncontrolled; **~e Ablagerung** *f* ABFALL uncontrolled dumping, uncontrolled disposal, uncontrolled tipping (*BrE*); **~e Deponie** *f* ABFALL fly tipping (*BrE*), open dump, uncontrolled tip (*BrE*)

ungeregelt: **~er Katalysator** *m* LUFT open loop catalytic converter, open loop catalyst

ungereinigt *adj* LUFT uncleaned; **~e Abluft** *f* LUFT uncleaned waste air

ungesättigt *adj* ABFALL, LUFT, WASSER unsaturated; **~e Bodenzone** *f* BODEN unsaturated soil zone; **~e Kohlenwasserstoffe** *m pl* LUFT unsaturated hydrocarbons; **~es Polyesterharz** *nt* ABFALL unsaturated polyester resin; **~e Zone** *f* ABFALL, WASSER unsaturated zone

ungespannt: **~es Grundwasser** *nt* WASSER unconfined groundwater, unconfined water

ungestört *adj* ANALYTIK undisturbed; **~e Probe** *f* ANALYTIK undisturbed sample

Ungeziefer *nt* ABFALL vermin

ungiftig *adj* ALLG nontoxic; **~e Metalle** *nt pl* ABFALL, ABWASSER nontoxic metals

ungleich: **~e Besteuerung** *f* POLITIK differential taxation; **~er Steuersatz** *m* POLITIK differential tax rate

unipolar *adj* ABWASSER homopolar

Universalmeßinstrument *nt* ANALYTIK universal measurement instrument (*UMI*)

Universalzerkleinerer *m* ABFALL universal shredder

unklar *adj* ABWASSER, WASSER cloudy

unkontrolliert *adj* ABFALL, ABWASSER uncontrolled; **~e Denitrifikation** *f* ABWASSER uncontrolled denitrification; **~e Deponie** *m* ABFALL phantom dump

unlöslich *adj* ABFALL, ABWASSER, WASSER insoluble

unrein *adj* ABFALL, ABWASSER, LUFT, WASSER impure; **~e Gewässer** *nt pl* ABFALL, WASSER impure waters

Unreinheit *f* ABFALL, ABWASSER, LUFT, WASSER impurity

unsachgemäß: **~er Ackerbau** *m* BODEN inappropriate cropping method

unschädlich *adj* LUFT innocuous; **~es Kohlendioxid** *nt* LUFT harmless carbon dioxide

unsicher *adj* SICHERHEIT unsafe; **~er Umweltzustand** *m* SICHERHEIT unsafe environmental condition

Unsicherheit *f* SICHERHEIT unsafeness

unsichtbar *adj* LUFT invisible; **~er Schwaden** *m* LUFT invisible plume

unsortiert: **~er Abfall** *m* ABFALL unsorted waste

unspezifisch: **~e Koagulation** *f* ABWASSER unspecific coagulation

Unterabteilung *f* POLITIK *Verwaltungseinheit* subdivision

unterbelüftet *adj* ABWASSER underaerated

Unterboden *m* BODEN B-horizon, C-horizon

Unterdecke *f* LÄRM underfloor

Unterdruck *m* BODEN, LUFT *Bodenluftabsaugung* negative pressure

unterdruckgesteuert *adj* LUFT vacuum-controlled

Unterdruck: **Unterdruckmanometer** *nt* ANALYTIK vacuum gage (*AmE*), vacuum gauge (*BrE*); **Unterdruckmesser** *m* ANALYTIK vacuum gage (*AmE*), vacuum gauge (*BrE*); **Unterdruckverdampferbrunnen** *m* (*UVB*) BODEN underpressure evaporator well

untere: **unteres Bett** *nt* LUFT lower bed; **untere Explosionsgrenze** *f* (*UEG*) SICHERHEIT *Wassergehalt* lower explosion limit; **unterer Heizwert** *m* ABFALL, LUFT net calorific value (*NCV*); **untere Nachweisgrenze** *f* ANALYTIK detection limit, LUFT detection limit, lower limit of detectibility; **untere Zündgrenze** *f* LUFT lower ignition limit

Unterflurhydrant *m* WASSER underground hydrant, sunk hydrant

Unterflurwasserpfosten *m* ABFALL sunk hydrant, underground hydrant, WASSER underground hydrant, sunk hydrant

Untergrund *m* ABFALL, WASSER subsoil, underground, subsurface; **Untergrundabdichtung** *f* ABFALL, WASSER underground sealing, bottom sealing; **Untergrundverformung** *f* ABFALL, WASSER underground deformation

unterirdisch *adj* ABFALL, ABWASSER, BODEN, WASSER subsurface, underground; **~e Belüftung** *f* ABWASSER subsurface diffusion air release; **~e Bewässerung** *f* WASSER subsurface irrigation; **~e Dränage** *f* ABFALL, WASSER underground drainage; **~e Entwässerung** *f* ABWASSER underdrainage; **~e Erosion** *f* ABFALL, WASSER subsurface erosion; **~e Gewässer** *nt pl* WASSER underground water; **~e Infiltrationsanlage** *f* WASSER underground infiltration facility; **~e Speicherung** *f* ABFALL, WASSER underground storage; **~es Wasser** *nt* WASSER subterranean water, underground water; **~er Wasserabzug** *m* ABWASSER underdrain

unterkühlen *vt* LUFT subcool

Unterlagen *f pl* POLITIK documents

Unterlauf *m* WASSER underflow

Unterspülung *f* ABFALL underwashing; **~ der Deponie** ABFALL washout of landfill
untersuchen *vt* ABFALL, ANALYTIK, SICHERHEIT analyse (*BrE*), analyze (*AmE*), test, examine, investigate
untersuchend *adj* ANALYTIK exploratory
untersucht *adj* ABFALL, ANALYTIK, SICHERHEIT analysed (*BrE*), analyzed (*AmE*), examined, investigated, tested; **~e Probe** *f* ANALYTIK analysed sample; **~e Wasserprobe** *f* ANALYTIK analysed water sample
Untersuchung *f* ABFALL, ANALYTIK, SICHERHEIT analysis, examination, investigation, test; **~ von Headspace-Proben** ANALYTIK static headspace method; **Untersuchungsbefund** *m* ANALYTIK analysis result, testing result; **Untersuchungsergebnis** *nt* ANALYTIK analytical finding, analytical result; **Untersuchungshäufigkeit** *f* ANALYTIK analysis frequency; **Untersuchungslaboratorium** *nt* ANALYTIK testing laboratory, testing lab; **Untersuchungsvolumen** *nt* ANALYTIK analysis volume; **Untersuchungszweck** *m* ANALYTIK analysis aim
Untertagedeponie *f* ABFALL underground depot; **~ für Sonderabfall** ABFALL underground landfill for special waste, underground landfill for hazardous waste
untertägig *adj* ABFALL, WASSER underground; **~e Sondermülldeponie** *f* ABFALL underground landfill for special waste
Unterwasser *nt* WASSER tailwater; **Unterwasserpumpe** *f* ANALYTIK submersible pump, WASSER subaqueous pump; **Unterwassersaugleitung** *f* WASSER underwater suction pipe
Unterwind: **Unterwindgebläse** *nt* ABFALL underblast fan; **Unterwindkettenrostfeuerung** *f* ABFALL underblast chain grate stoker; **Unterwindzone** *f* ABFALL, WASSER underblast zone
Unverbranntes *nt* ABFALL, LUFT incombustibles
unverbrennbar *adj* ABFALL, LUFT incombustible; **~er Rückstand** *m* LUFT incombustible residue
Unverbrennbarkeit *f* ABFALL, LUFT incombustibility
Unverbrennliches *nt* ABFALL, LUFT incombustibles
unverdichtet *adj* ABFALL, BODEN uncompacted
unverdünnt *adj* ABFALL, MEER, WASSER neat, undiluted; **~es Sickerwasser** *nt* ABFALL, WASSER undiluted leachate
unvermischbar *adj* ABWASSER immiscible; **~e Flüssigkeiten** *f pl* ABWASSER immiscible liquids
unvermischt *adj* ABFALL undiluted, MEER, WASSER neat, undiluted; **~e Abfallerfassung** *f* ABFALL unmixed waste recovery; **~e Schadstofferfassung** *f* ABFALL unmixed recovery of harmful substances; **~es Sickerwasser** *nt* ABFALL, WASSER undiluted leachate
unverschmutzt: **~e Kunststoffe** *m pl* ABFALL clean plastics; **~e Luft** *f* LUFT unpolluted air
unverwertbar *adj* ABFALL unusable; **~er Reststoff** *m* ABFALL unusable residue
unvollkommen *adj* ABFALL, LUFT incomplete; **~e Verbrennung** *f* ABFALL, LUFT incomplete burning, incomplete combustion
unvollständig *adj* ABFALL, LUFT incomplete; **~e Verbrennung** *f* ABFALL, LUFT incomplete burning, incomplete combustion
unwirtschaftlich *adj* MANAG, POLITIK uneconomical, unprofitable
Unzuverlässigkeit *f* ANALYTIK unreliability
UO *abbr* (*Umkehrosmose*) ABFALL *Verfestigungsverfahren für Sonderabfälle*, ABWASSER, WASSER RO (*reverse osmosis*)
urban: **~es Ökosystem** *nt* ABWASSER, BODEN, LUFT urban ecosystem
UV *abbr* (*ultraviolett, Ultraviolett*) ALLG UV (*ultraviolet*); **UV-Absorption** *f* ABFALL, ANALYTIK UV absorption
UVB *abbr* (*Unterdruckverdampferbrunnen*) BODEN underpressure evaporator well
UV: **UV-Bestrahlung** *f* ABFALL, LUFT UV irradiation; **UV-C-Strahlung** *f* ABFALL, LUFT UV-C radiation; **UV-Desinfektion** *f* ABFALL UV disinfection; **UV-Filter** *nt* LUFT UV filter, SICHERHEIT, UV filter; **UV-Lampe** *f* ANALYTIK UV lamp; **UV-Lichtabsorption** *f* ANALYTIK UV optical absorption; **UV-Oxidation** *f* WASSER UV oxidation
UVP *abbr* (*Umweltverträglichkeitsprüfung*) POLITIK, WASSER EIA (*environmental impact assessment*); **UVP-Bericht** *m* POLITIK EIA report; **UVP-Bewertung** *f* POLITIK EIA evaluation
UV: **UV-Photometrie** *f* ANALYTIK UV photometry
UVP: **UVP-Studie** *f* POLITIK EIA study; **UVP-Untersuchung** *f* POLITIK EIA investigation; **UVP-Verfahren** *nt* POLITIK EIA procedure
UVS: **UVS-Detektor** *m* ANALYTIK UV/VIS detector
UV: **UV-Spektroskopie** *f* ANALYTIK UV spectroscopy
UVS: **UVS-Spektrometer** *nt* ANALYTIK UV/VIS spectrometer; **UVS-Spektrophotometrie** *f* ANALYTIK UV/VIS spectrophotometry; **UVS-Spektroskopie** *f* ANALYTIK UV/VIS spectroscopy
UV: **UV-Stabilität** *f* ABFALL UV resistance; **UV-Strahlen** *m pl* ANALYTIK, SICHERHEIT UV rays; **UV-Strahler** *m* ABWASSER UV radiator
UV/VIS *abbr* (*im Sichtbaren und Ultraviolett*) ANALYTIK UV/VIS (*ultraviolet and visible*); **UV/VIS-Detektor** *m* ANALYTIK UV/VIS detector; **UV/VIS-Spektrometer** *nt* ANALYTIK UV/VIS spectrometer; **UV/VIS-Spektrophotometrie** *f* ANALYTIK UV/VIS spectrophotometry; **UV/VIS-Spektroskopie** *f* ANALYTIK UV/VIS spectroscopy

V

V *abbr* (*Vanadium*) LUFT V (*vanadium*)
vados: **~es Grundwasser** *nt* WASSER perched groundwater, vadose water
Vakuum *nt* ABWASSER, ANALYTIK, BODEN, LUFT vacuum; **Vakuumbandfilter** *nt* LUFT vacuum belt filter; **Vakuumbrunnen** *m* BODEN vacuum well; **Vakuumeindampfung** *f* ABFALL, ABWASSER vacuum concentration; **Vakuumentlüfter** *m* ABWASSER dry feeder; **Vakuumfilter** *nt* ABWASSER, WASSER vacuum filter; **Vakuumfilterentwässerung** *f* ABWASSER vacuum filter dewatering; **Vakuumfilterung** *f* ABWASSER, WASSER vacuum filtration; **Vakuumflotation** *f* ABWASSER vacuum flotation; **Vakuumkollektor** *m* ERNEU ENER mirror evacuated tube; **Vakuumprinzip** *nt* ANALYTIK vacuum principle; **Vakuumpumpe** *f* ANALYTIK, ERNEU ENER, LUFT, MEER vacuum pump; **Vakuumverdampfer** *m* WASSER vacuum evaporator
Valenz *f* ABWASSER, ANALYTIK valence; **Valenzkraft** *f* ABWASSER, LUFT valence force
Vanadium *nt* (*V*) LUFT vanadium (*V*)
Van-der-Waals-Kräfte *f pl* ABWASSER van der Waals forces
variabel *adj* LUFT variable
VC *abbr* (*Vinylchlorid*) LUFT, SICHERHEIT VC (*vinyl chloride*)
Vegetationsschaden *m* ABFALL, BODEN damage to vegetation
Ventil *nt* MEER valve
Ventilator *m* ABWASSER blower, ventilator, LUFT fan, ventilator; **Ventilatorflügel** *m* LÄRMLUFT fan vane; **Ventilatorgehäuse** *nt* LÄRM fan shroud
Venturi *nt* ANALYTIK, LUFT Venturi tube
venturiartig *adj* LUFT Venturi-shaped
Venturi: **Venturi-Brenner** *m* LUFT Venturi burner; **Venturidüse** *f* ANALYTIK, LUFT Venturi nozzle, Venturi tube; **Venturikehle** *f* LUFT Venturi throat; **Venturimesser** *m* LUFT Venturi meter; **Venturiprinzip** *nt* LUFT Venturi principle; **Venturirohr** *nt* ANALYTIK, LUFT Venturi tube; **Venturiwäscher** *m* LUFT Venturi scrubber
verabschieden *vt* POLITIK pass
Verabschiedung *f* POLITIK adoption, *LWB* passing
verändern 1. *vt* ANALYTIK alter, change; **2. sich ~** *v refl* ANALYTIK alter, change
Veränderung *f* ANALYTIK alteration, change
verankern *vt* MEER moor
Verankern *nt* MEER mooring
Verankerungskette *f* MEER mooring chain
verarbeiten *vt* ALLG process, reprocess
Verarbeitung *f* ALLG processing; **Verarbeitungsabfälle** *m pl* ABFALL industrial scrap, process scrap; **Verarbeitungskapazität** *f* ABWASSER, BODEN, LUFT, WASSER processing capacity
Veraschungstest *m* ANALYTIK incineration test
verbinden *vt* ABFALL bond
Verbindungsglied *nt* MEER shackle
Verbindungsleitung *f* BODEN, WASSER connecting pipe
Verblockung *f* ABWASSER interlocking; **Verblockungsgefahr** *f* ABWASSER interlocking risk
Verbot *nt* POLITIK prohibition
verbrannt *adj* ABFALL, ERNEU ENER, LUFT incinerated; **~e Flüssigkeit** *f* ABWASSER incinerated liquid
Verbrauch *m* ABFALL, MANAG, WASSER consumption; **zum ~ bestimmtes Wasser** *nt* WASSER consumption water
verbrauchen *vt* ABFALL, MANAG, WASSER consume
Verbraucher *m* ABFALL, MANAG, WASSER consumer; **Verbraucherabgabe** *f* POLITIK user charge; **Verbraucherverhalten** *nt* MANAG customer behavior (*AmE*), customer behaviour (*BrE*)
Verbrauch: **Verbrauchsspitze** *f* ABFALL, WASSER peak consumption
Verbreitung *f* MEER spreading
Verbrennbarkeit *f* ABFALL, ERNEU ENER, LUFT, SICHERHEIT combustibility
verbrennen *vt* ABFALL, ERNEU ENER, LUFT, SICHERHEIT incinerate
Verbrennung *f* ABFALL, ERNEU ENER, LUFT, SICHERHEIT combustion, incineration; **~ mit Energierückgewinnung** ABFALL incineration with energy recovery; **Verbrennungsanlage** *f* LUFT incineration plant; **Verbrennungsanlage** *f* **ohne Wärmenutzung** ABFALL, ERNEU ENER incineration plant without heat use; **Verbrennungsanlage** *f* **mit Wärmenutzung** ABFALL, ERNEU ENER incineration plant with heat use; **Verbrennungsenthalpie** *f* ABFALL incineration enthalpy, LUFT combustion enthalpy; **Verbrennungsgas** *nt* ABFALL combustion gas, incinerator gas, LUFT combustion gas; **Verbrennungsgase** *nt pl* LUFT flue gas; **Verbrennungsgrad** *m* LUFT degree of incineration; **Verbrennungsgut** *nt* ABFALL incineration stock; **Verbrennungskalorimeter** *nt* ANALYTIK incineration calorimeter; **Verbrennungskraftmaschine** *f* ABFALL, LUFT, WASSER internal-combustion turbine; **Verbrennungsleistung** *f* ABFALL incineration capacity; **Verbrennungsluft** *f* ABFALL, ABWASSER, LUFT combustion air; **Verbrennungsmittel** *nt* MEER *für Öl* burning agent; **Verbrennungsmotor** *m* ABFALL, ABWASSER, LUFT, WASSER internal-combustion engine, internal-combustion turbine; **Verbrennungsprodukt** *nt* ABFALL, LUFT combustion product; **Verbrennungsreaktion** *f* ABFALL, LUFT combustion reaction; **Verbrennungsrohr** *nt* ANALYTIK combustion tube; **Verbrennungsrückstand** *m* ABFALL ash, combustion residue, incineration ash, incineration residue; **Verbrennungstemperatur** *f* ABFALL, LUFT combustion temperature; **Verbrennungsvorgang** *m* ABFALL, LUFT combustion operation, incineration operation; **Verbrennungszone** *f* ABFALL, ABWASSER, LUFT incineration zone
Verbund: **Verbundkarton** *m* ABFALL composite cardboard; **Verbundplatte** *f* LÄRM sandwich panel; **Verbundstoff** *m* ABFALL composite; **Verbundstoffbehälter** *m* ABFALL composite container; **Verbundverpackung** *f* ABFALL composite packaging, laminated packaging; **Verbundwerkstoff** *m* ABFALL composite secondary resource, laminated secondary resource
verdampfen 1. *vt* ABFALL, ABWASSER, ANALYTIK evaporate, BODEN, ERNEU ENER vaporize, WASSER evaporate; **2.** *vi* ABFALL, ABWASSER, ANALYTIK evaporate, BODEN, ERNEU ENER vaporize, WASSER evaporate
Verdampfer *m* ERNEU ENER vaporizer; **Verdampfer-Destillation** *f* BODEN evaporator distillation
Verdampfrückstand *m* ABFALL, ANALYTIK, ERNEU ENER evaporation residue
Verdampfung *f* ABWASSER, ANALYTIK evaporation, ERNEU ENER vaporization, WASSER evaporation; **Verdampfungsemissionen** *f pl* LUFT evaporative emissions; **Verdampfungsenthalpie** *f* LUFT enthalpy of vaporization; **Verdampfungskristallisator** *m* LUFT evaporation crystallizer; **Verdampfungskühlung** *f* LUFT evaporation cooling
Verdeckungsgeräusch *nt* LÄRM masking noise

verdichtbar *adj* ABFALL, ABWASSER, WASSER compressible; **~er Abfall** *m* ABFALL compressible waste
Verdichtbarkeit *f* ABFALL compressibility
verdichten *vt* ABFALL condense, ERNEU ENER compress, LUFT thicken
Verdichten *nt* ABFALL compacting, ERNEU ENER compression, LUFT thickening
Verdichter *m* ABWASSER, ERNEU ENER, LUFT compressor; **Verdichtergehäuse** *nt* LUFT compressor casing
verdichtet *adj* ABFALL compacted, ERNEU ENER compressed, LUFT thickened
Verdichtung *f* ABFALL compacting, compaction, LUFT thickening; **Verdichtungsanlage** *f* ABFALL compacting unit, landfill compactor, packer unit; **Verdichtungsdeponie** *f* ABFALL compaction landfill; **Verdichtungsgerät** *nt* ABFALL compacting device, compaction equipment, BODEN compaction equipment; **Verdichtungsgrad** *m* ABFALL degree of compaction; **Verdichtungszone** *f* ABFALL compaction zone
Verdolung *f* ABFALL dolination
Verdrängerpumpe *f* ABWASSER, WASSER positive displacement pump
Verdrängung *f* WASSER displacement; **Verdrängungskörper** *m* LUFT displacer
verdünnen *vt* ABFALL, ABWASSER, ANALYTIK, LUFT dilute, thin
verdünnt *adj* ABFALL, ABWASSER, ANALYTIK, LUFT thinned, diluted; **~e Lösung** *f* ANALYTIK diluted solution; **~e Probe** *f* ANALYTIK diluted sample
Verdünnung *f* ABFALL, ABWASSER, ANALYTIK, LUFT thinning, dilution; **~ der Abgase** LUFT waste gas dilution; **Verdünnungsanalyse** *f* ANALYTIK dilution analysis; **Verdünnungsprobe** *f* ANALYTIK dilution sample
verdunsten 1. *vt* ABFALL, ABWASSER, WASSER evaporate; **2.** *vi* ABFALL, ABWASSER, WASSER evaporate
Verdunstung *f* WASSER effective evaporation, evaporation; **Verdunstungskessel** *m* ABFALL evaporation pan; **Verdunstungsrate** *f* ABWASSER, WASSER evaporation rate; **Verdunstungsverlust** *m* ABWASSER, WASSER evaporation loss
verdüsen *vt* ABFALL, WASSER atomize
Verdüser *m* ABFALL, WASSER atomizer
vereinheitlichen *vt* ABFALL, WASSER homogenize
Verfahren *nt* ALLG method, process, procedure; **Verfahrenskosten** *pl* MANAG process costs
verfahrensspezifisch *adj* ABFALL, ABWASSER, ERNEU ENER, LUFT process-specific
Verfahren: **Verfahrenstechnik** *f* ABWASSER, ANALYTIK unit operations
verfahrenstechnisch *adj* ABFALL, ABWASSER, ERNEU ENER, LUFT process-related
Verfahren: **Verfahrensweise** *f* ABWASSER, ANALYTIK, LUFT, WASSER procedure
Verfärbung *f* ANALYTIK discoloration (*AmE*), discolouration (*BrE*)
verfaulen *vi* ABFALL putrefy
verfestigen *vt* ABFALL, BODEN solidify
verfestigt *adj* ABFALL, BODEN solidified; **~er Boden** *m* BODEN solidified soil; **~es Endprodukt** *nt* ABFALL solidified end product
Verfestigung *f* ABFALL, BODEN fixation, fixing, solidification; **~ von Abfällen** ABFALL solidification of waste; **~ mit organischen Polymeren** ABFALL solidification with organic polymers; **~ von Sonderabfällen** ABFALL solidification of hazardous waste; **~ mit Zement** ABFALL solidification with cement; **Verfestigungsmittel** *nt* ABFALL, BODEN solidifier, fixative, solidifying agent; **Verfestigungsprodukt** *nt* ABFALL solidified material, solidified product, solidified waste; **Verfestigungsverfahren** *nt* ABFALL fixation technique, solidification process, solidification technique, stabilization process, stabilization technique
verfilzen *vt* LÄRM *Glaswolle* felt
verflüchtigen: sich ~ *v refl* BODEN, LUFT volatilize
Verflüchtigung *f* BODEN, LUFT volatilization; **Verflüchtigungsuntersuchung** *f* ANALYTIK volatility analysis
verflüssigen *vt* ABFALL, ABWASSER, ERNEU ENER liquefy
Verflüssigung *f* ABFALL, ABWASSER, ERNEU ENER *Biokonversion* liquefaction
Verformbarkeit *f* ABFALL ductibility
Verformung *f* ABFALL deformation
verfügbar *adj* ALLG available; **~e Technologie** *f* ANALYTIK, MANAG, SICHERHEIT available technology
Verfüllboden *m* ABFALL, BODEN *Erdbau* backfill
verfüllt: ~e Deponie *f* ABFALL complete fill
Verfüllung *f* ABFALL, BODEN backfilling, filling
Vergabebedingungen *f pl* POLITIK *des Umweltzeichens* award conditions
vergärbar *adj* ABFALL, ABWASSER, ERNEU ENER, WASSER fermentable
Vergärbarkeit *f* ABFALL, ABWASSER, ERNEU ENER, WASSER fermentability
vergären 1. *vt* ABFALL, ABWASSER, ERNEU ENER, WASSER ferment; **2.** *vi* ABFALL, ABWASSER, ERNEU ENER, WASSER ferment
vergärt: ~er Abfall *m* ABFALL fermented waste
Vergärung *f* ABFALL, ABWASSER, ERNEU ENER, WASSER *Biokonversion* fermentation; **~ zu Biogas** ERNEU ENER fermentation to biogas; **Vergärungsgrad** *m* ABFALL degree of fermentation
vergasen *vt* ABFALL, ABWASSER, ERNEU ENER, LUFT gasify
Vergaser *m* ABFALL, ABWASSER, ERNEU ENER gasifier, LUFT carburetor (*AmE*), carburettor (*BrE*), gasifier; **Vergaseremissionen** *f pl* LUFT carburettor emissions (*BrE*); carburetor emissions (*AmE*)
Vergasung *f* ABFALL, ABWASSER, ERNEU ENER, LUFT gasification
Vergiftung *f* ABFALL, LUFT, SICHERHEIT, WASSER poisoning
verglasen *vt* ABFALL, BODEN vitrify
verglast *adj* ABFALL, BODEN vitrified; **~e Schlacke** *f* ABFALL, BODEN vitrified slag
Vergleich *m* ANALYTIK comparison
vergleichen *vt* ANALYTIK compare
Vergleich: **Vergleichsgas** *nt* ANALYTIK reference gas; **Vergleichskammer** *f* ANALYTIK reference chamber; **Vergleichsmessung** *f* ANALYTIK comparative measurement; **Vergleichsprobe** *f* ANALYTIK comparative test; **Vergleichsspektrum** *nt* ANALYTIK reference spectrum; **Vergleichswiderstand** *m* ANALYTIK reference resistor
Vergraben *nt* ABFALL *von Müll* land burial
Verhalten *nt* ALLG *eines Ölteppichs* behavior (*AmE*), behaviour (*BrE*)
verhandelbar *adj* POLITIK tradeable; **~e Emissionsrechte** *nt pl* POLITIK tradeable emission rights; **~e Umweltnutzungsrechte** *nt pl* POLITIK tradeable environmental use rights
Verhandlung *f* POLITIK negotiation
verhindern *vt* ALLG prevent, curb, prohibit, suppress
Verhütung *f* SICHERHEIT prevention; **~ von Luftverschmutzung** LUFT prevention of atmospheric pollution
Verjüngen *nt*: **~ der Proben** ANALYTIK sample diminution
verkalken *vi* WASSER calcify
Verkalkung *f* WASSER calcination
Verkapselung *f* ABFALL encapsulation
Verklammerung *f* ABFALL cramping
verklappen *vt* ABFALL, MEER dump at sea
Verklappung *f* ABFALL, MEER dumping at sea, ocean disposal, ocean dumping
verkleben *vt* ABFALL, LUFT conglutinate
Verkleben *nt* ABFALL, LUFT conglutination
Verklebung *f* ABFALL, LUFT conglutination

verkleiden *vt* LÄRM line
Verkleidung *f* LÄRM cladding
verknüpfen *vt* ABWASSER link
Verkrustung *f* LUFT scaling; **Verkrustungsgefahr** *f* LUFT scaling risk
Verkürzung *f* LUFT reduction; ~ **der Abgasverweildauer** LUFT reduction of waste gas retention time; ~ **der Abgasverweilzeit** LUFT reduction of waste gas retention time
Verlagerung *f* ABFALL, BODEN relocation
Verlärmung *f* LÄRM noise intrusion
verlegen *vt* ABFALL, BODEN, WASSER *Leitungen* lay
Verlegung *f* ABFALL, BODEN, WASSER laying
vermarktbar *adj* MANAG marketable
Vermarktung *f* MANAG marketing; **Vermarktungsfähigkeit** *f* MANAG marketability
vermascht *adj* ABFALL, WASSER meshed; **~es Hauptleitungs- und Versorgungsleitungsnetz** *nt* ABFALL, WASSER meshed mains and supply network
Vermaschung *f* ABFALL, WASSER meshing
vermeiden *vt* ABFALL, LÄRM avoid
Vermeidung *f* ABFALL, ABWASSER, LÄRM prevention, avoidance; ~ **von Lärmbelästigung** LÄRM, SICHERHEIT noise pollution prevention; **Vermeidungsabgabe** *f* POLITIK *Deponie* prevention charge
Vermengung *f* ABFALL commixture
vermessen *vt* MEER survey
vermischt: **~e Kunststoffe** *m pl* ABFALL commingled plastics
vernetzen *vt* ABFALL, LUFT crosslink
vernetzt *adj* ABFALL, LUFT crosslinked
Vernetzung *f* ABFALL, LUFT, WASSER crosslinking; **Vernetzungseffizienz** *f* ABFALL, WASSER crosslinking efficiency; **Vernetzungswirksamkeit** *f* ABFALL, WASSER crosslinking efficiency
Vernichtung *f* ABFALL elimination
Verockerung *f* ABFALL, ABWASSER, BODEN, WASSER *Filter* clogging, plugging, incrustation
Verockerungsgefahr *f* ABFALL, WASSER risk of iron clogging
Veröffentlichung *f* POLITIK publication
verordnen *vt* POLITIK regulate
verordnet *adj* POLITIK regulated; **~e Verbraucherabgabe** *f* POLITIK mandatory user charge
Verordnung *f* POLITIK regulation
Verpackung *f* ABFALL, MANAG packaging; **Verpackungsfraktion** *f* ABFALL packaging fraction; **Verpackungsindustrie** *f* MANAG packaging industry; **Verpackungsmaterial** *nt* ABFALL packaging material; **Verpackungsmüll** *m* ABFALL packaging waste; **Verpackungsoptimierung** *f* ABFALL, MANAG packaging optimization; **Verpackungswertstoff** *m* ABFALL secondary resource for packaging
verpflichten *vt* POLITIK oblige
Verpreßabschnitt *m* BODEN grouting section, injection section
Verpreßmittel *nt* BODEN grouting material
Verrauchung *f* LUFT fumigation
Verregnung *f* ABWASSER sprinkler
verriegeln *vt* ABFALL lock
verriegelt *adj* ABFALL locked
Verriegelungsmechanik *f* ABFALL locking mechanism
Verrieselung *f* ABWASSER wastewater irrigation
verringern *vt* ABFALL, ABWASSER, WASSER reduce
verrottbar: **~er Stoff** *m* ABFALL putrescible matter
verrotten *vi* ABFALL, ABWASSER decay, putrefy
Verrottung *f* ABFALL, ABWASSER decay, putrefaction
versalzen *vt* ABFALL, WASSER salinate
Versalzung *f* ABFALL, WASSER salination
versauern *vi* ALLG acidify
Verschlackung *f* ABFALL, ABWASSER, LUFT slagging; ~ **im Feuerraum** ABFALL slagging in combustion chamber
verschlammen *vi* ABWASSER silt up
Verschlammung *f* ABWASSER silting-up
verschleißbeständig *adj* ALLG wear-resistant
Verschleiß: **Verschleißeigenschaften** *f pl* ABFALL *Kunststoffe* wear characteristics
verschleißfest *adj* ALLG wear-resistant; **~e Auskleidung** *f* LUFT wear-resistant lining
Verschleiß: **Verschleißfestigkeit** *f* ABFALL wear resistance; **Verschleißwiderstand** *m* ABFALL wear resistance
verschließen *vt* ABFALL block, MEER seal
Verschließung *f* ABFALL blockage, MEER sealing; ~ **mit Sickerwasser** ABFALL leachate blockage
verschlossen *adj* MEER sealed
verschmutzen *vt* ALLG contaminate, pollute
verschmutzt *adj* ALLG contaminated, polluted; **~es Abgas** *nt* LUFT polluted process gas; **~er Altkunststoff** *m* ABFALL contaminated plastic; **~e Gewässer** *nt pl* ABWASSER polluted water body; **~e Kunststoffe** *m pl* ABFALL soiled plastics; **~es Probegas** *nt* ANALYTIK contaminated sampling gas; **~es Regenwasser** *nt* ABWASSER polluted rainwater; **~es Wasser** *nt* ABWASSER polluted water
Verschmutzung *f* ALLG contamination, pollution; ~ **des Abwassers abgebaut durch Mikroorganismen** ABWASSER contaminated wastewater purified by microbes; ~ **durch Feststoffe** *f* WASSER material pollution; **Verschmutzungsausmaß** *nt* ABWASSER, BODEN, LUFT, MEER degree of pollution; **Verschmutzungsgrad** *m* ABWASSER, BODEN, LUFT, MEER degree of pollution; **Verschmutzungsgrad** *m* **des Wassers** WASSER pollution index
Verschrottungsgebühr *f* POLITIK scrapping fee
verschütten *vt* ABFALL, WASSER spill
verschüttend *adj* ABFALL, WASSER spilled; **~er Abfall** *m* ABFALL spilled waste
Verschüttung *f* ABFALL, WASSER spillage; ~ **von Sondermüll** WASSER spillage of hazardous waste
Verschwendung *f* MANAG mismanagement
verseuchen *vt* ALLG contaminate
verseucht *adj* ALLG contaminated
Verseuchung *f* ALLG contamination
Versicherungsbecken *nt* WASSER recharge basin
versickern *vt* ABFALL, ABWASSER, MEER, WASSER seep, percolate, infiltrate
Versickern *nt* ABFALL, ABWASSER, MEER, WASSER seepage, percolation, infiltration
Versickerung *f* ABFALL, ABWASSER, MEER, WASSER seepage, percolation, infiltration; **Versickerungsleistung** *f* ABFALL, WASSER percolation capacity
versiegeln *vt* ABFALL, BODEN seal
versiegelt *adj* ABFALL, BODEN sealed
Versiegelung *f* ABFALL, BODEN sealing; **Versiegelungsverschluß** *m* ABFALL tight-fitting cap
Versinterung *f* LUFT fusion
versorgen *vt* WASSER supply
Versorgung *f* WASSER supply; **Versorgungsgebiet** *nt* WASSER supply area; **Versorgungskanal** *m* WASSER delivery race; **Versorgungsleitung** *f* WASSER supply line; **Versorgungsnetz** *nt* ABFALL supply network, WASSER supply system, supply network; **Versorgungspumpe** *f* MEER supply pump; **Versorgungszone** *f* WASSER supply district
verspiegeln *vt* ERNEU ENER reflect
verspiegelt *adj* ERNEU ENER reflected; **~es Paraboloid** *nt* ERNEU ENER reflected paraboloid
verspinnen *vt* ABFALL spin
verspritzen *vt* ABWASSER spray
Versprödung *f* LUFT *Filtermaterial* brittleness
versprühen *vt* BODEN atomize, MEER spray
Versprühen *nt*: ~ **von Dispersionsmitteln** MEER dispersant spraying
Versprüher *m* BODEN atomizer

Versprühung *f* ALLG spraying
verstärken *vt* ABFALL, WASSER reinforce
Verstärker *m* ANALYTIK amplifier
verstärkt: **~e Kunststoffe** *m pl* ABFALL reinforced plastics
Verstärkungsmittel *nt* ABFALL reinforcing material
Versteigerung *f* POLITIK *von Emissionszertifikaten* auctioning
verstellbar: **~es Rotorblatt** *nt* ERNEU ENER *Windturbine* variable-pitch rotor blade; **~er Spülluftkanal** *m* LUFT movable clean air feed
Verstellpropeller *m* MEER variable-pitch propeller
Verstelluftschraube *f* MEER variable-pitch propeller
verstopfen *vt* ABFALL, ABWASSER, BODEN, WASSER clog, *Wasserröhre* foul, block
Verstopfung *f* ABFALL, ABWASSER *Filter, Pumpe*, BODEN, WASSER clogging, blocking, stoppage; **Verstopfungsgefahr** *f* ABFALL, LUFT, WASSER risk of clogging, risk of plugging
verstopfungssicher *adj* ABWASSER nonclogging; **~es Füllmaterial** *nt* ABWASSER nonclogging packing material
Versuch *m* ANALYTIK assay, test; **Versuchsanlage** *f* ABFALL, ABWASSER, BODEN, ERNEU ENER, LUFT, WASSER pilot plant; **Versuchsbecken** *nt* ABFALL, WASSER experimental basin, test basin; **Versuchsbetrieb** *m* MANAG, POLITIK *einer Anlage* trial operation; **Versuchsbrunnen** *m* WASSER experimental well; **Versuchspumpe** *f* WASSER test pump
Versumpfung *f* WASSER marshiness
Vertäuboje *f* MEER mooring buoy
verteilen *vt* ANALYTIK apportion, WASSER distribute
Verteiler *m* ABFALL distributor; **Verteilerboden** *m* LUFT *Naßabscheider* distributing bottom; **Verteilerrohrnetz** *nt* WASSER distribution network
Verteilsystem *nt* WASSER distribution network
Verteilung *f* WASSER distribution; **Verteilungschromatographie** *f* ANALYTIK partition chromatography; **Verteilungskoeffizient** *m* ABFALL, WASSER distribution coefficient; **Verteilungsnetz** *nt* WASSER distribution network; **Verteilungsrinne** *f* ABFALL, WASSER distribution channel; **Verteilungsrohr** *nt* WASSER distributing pipe; **Verteilungssystem** *nt* WASSER distribution network
Vertikal- *in cpds* ERNEU ENER, LUFT, WASSER vertical
vertikal: **~er Brunnen** *m* WASSER vertical well; **~e Entgasung** *f* ABFALL, WASSER vertical degassing; **~er Festbettadsorber** *m* LUFT vertical fixed-bed adsorber; **~es flächiges Fassungselement** *nt* ABFALL vertical collection element; **~er Gasbrunnen** *m* ABFALL vertical gas well; **~e Gaserfassung** *f* ABFALL vertical gas collection; **~es linienförmiges Fassungselement** *nt* ABFALL vertical linear-formed collection element; **~er Sickerbrunnen** *m* WASSER vertical percolation well; **~e Verteilung** *f* BODEN vertical distribution, LUFT vertical dispersion; **~er Wanderbettadsorber** *m* LUFT vertical moving-bed adsorber
Vertikal-: **Vertikalachsenmaschine** *f* ERNEU ENER vertical axis wind turbine (*VAWT*); **Vertikalbrunnen** *m* WASSER vertical well; **Vertikalfilterbrunnen** *m* (*VFB*) WASSER vertical filtering well; **Vertikalverteilung** *f* BODEN, LUFT vertical dispersion, vertical distribution; **Vertikalverteilung** *f* **des Ozons** LUFT vertical distribution of ozone
Verträglichkeitsprüfung *f* MANAG impact study
vertraulich *adj* POLITIK confidential; **~e Informationen** *f pl* POLITIK confidential information; **~e Unterlagen** *f pl* POLITIK confidential documents
Vertriebsagent *m* ABFALL, WASSER distributor
verunreinigen *vt* ALLG contaminate, pollute
verunreinigt *adj* ALLG contaminated, polluted
Verunreinigung *f* ALLG contamination, pollution; **Verunreinigungen** *f pl* ABFALL, ABWASSER, LUFT, WASSER impurities; **Verunreinigungsquelle** *f* ABFALL, ABWASSER, BODEN, LUFT, POLITIK, WASSER pollution source
Verursacher *m* POLITIK tortfeasor, polluter
verursachergerecht *adj* POLITIK tortfeasor-oriented, polluter-oriented
Verursacher: **Verursachergruppe** *f* POLITIK polluter group; **Verursacherprinzip** *nt* POLITIK polluter-pays-principle
verwaltend *adj* MANAG, POLITIK administrative
Verwaltung *f* MANAG, POLITIK administration, management; **Verwaltungseinheit** *f* POLITIK administrative unit; **Verwaltungsgebühr** *f* POLITIK administrative fee; **Verwaltungsverordnung** *f* POLITIK administrative regulation; **Verwaltungsvorschrift** *f* POLITIK administrative regulation
Verwehung *f* ABFALL drift
Verweil: **Verweildauer** *f* ABFALL, ABWASSER, LUFT dwell time, residence time, retention time; **Verweilzeit** *f* ABFALL, ABWASSER, LUFT dwell time, residence time, retention time
verwendbar: **~es Abfallprodukt** *nt* ABFALL usable waste product; **~es Nebenprodukt** *nt* ABFALL, ABWASSER usable by-product
Verwendungsgrad *m* ABFALL, WASSER degree of use
verwertbar *adj* ABFALL, LUFT usable; **~e Fraktion** *f* ABFALL usable fraction; **~er Gips** *m* LUFT reusable gypsum
Verwertbarkeit *f* ABFALL, ABWASSER, LUFT usability
verwerten *vt* ABFALL, ABWASSER, LUFT recycle, use, reprocess
Verwertung *f* ABFALL, ABWASSER, LUFT, WASSER recovery, recycling; **~ der Flugaschen** ABFALL recycling of fly ash; **~ von Klärschlamm als Brennstoff** ABWASSER use of treated sludge as fuel; **~ in der Landwirtschaft** ABWASSER agricultural reuse; **Verwertungsgebot** *nt* ABFALL, WASSER obligation to recycle; **Verwertungskapazität** *f* ABFALL recycling capacity
verwertungsorientiert *adj* ABFALL recycling-oriented
Verwertung: **Verwertungsquote** *f* ABFALL recycling rate; **Verwertungstechnik** *f* ABFALL reprocessing technology; **Verwertungsverfahren** *nt* ABFALL recycling process
Verwirbelung *f* ABFALL *Deponiegas* turbulence
verwittert: **~es Öl** *nt* MEER weathered oil
Verwitterung *f* ABFALL, BODEN, MEER weathering
verzinken *vt* ABFALL galvanize
verzinkt *adj* ABFALL galvanized
verzinnen *vt* ABFALL tin-plate
verzinnt *adj* ABFALL tin-plated
verzögert: **~er Unterdruck** *m* LUFT time-delayed vacuum
Verzögerung *f* LUFT *Filtermedium* deceleration
verzweigt *adj* WASSER branched; **~es Verteilrohrnetz** *nt* WASSER branched distribution network
Verzweigung *f* WASSER branching
VFB *abbr* (*Vertikalfilterbrunnen*) WASSER vertical filtering well
Vibration *f* LÄRM, LUFT vibration
vibrationsdämpfend *adj* LÄRM vibration-damping
Vibration: **Vibrationsschnecke** *f* BODEN vibrating worm; **Vibrationssieb** *nt* BODEN vibrating screen
Vibratorsieb *nt* BODEN vibrating screen
vibrieren *vi* LÄRM, LUFT vibrate
Vibrieren *nt* LÄRM, LUFT vibrating
Vibriersieb *nt* ABFALL vibrator screen
Vibrometer *nt* ANALYTIK vibrometer
Vielflügler *m* ERNEU ENER *Windturbine* multiple-blade machine
vielzellig: **~er Zyklon** *m* LUFT multiunit cyclone
Vier-Liter-Auto *nt* LUFT four-liter automobile (*AmE*), four-litre car (*BrE*)
Vierpol *m* ANALYTIK quadrupole
Viertaktmotor *m* LUFT four-stroke engine (*BrE*), four-cycle engine (*AmE*)
Viertakt-Ottomotor *m* LUFT four-stroke engine (*BrE*), four-cycle engine (*AmE*)
Viertelwelle-Resonator *m* LÄRM quarter wave resonator
Vinylchlorid *nt* (*VC*) LUFT, SICHERHEIT vinyl chloride (*VC*)
viskos *adj* ABFALL, ABWASSER, ANALYTIK, LUFT viscid, viscous; **~e Flüssigkeit** *f* LUFT viscous liquid

Viskosimeter *nt* ANALYTIK viscosity meter
Viskosität *f* ABFALL, ABWASSER, ANALYTIK, LUFT viscosity
visuell: **~e Prüfung** *f* ANALYTIK visual test
Vlies *nt* ABFALL *Abdeckung* fibrous web, LUFT *Filterfaser* shear wool
VOC *abbr* (*flüchtige organische Verbindungen*) LUFT VOC (*volatile organic compounds*)
Völkerrecht *nt* POLITIK international law
voll: **~er Abgabensatz** *m* POLITIK *in bezug auf Emissionen* standard rate of charge
Vollanalyse *f* ANALYTIK full analysis
volldurchgemischt *adj* ABFALL, ABWASSER, WASSER completely mixed; **~es Becken** *nt* ABWASSER completely mixed tank; **~e Belüftung** *f* ABWASSER complete mix aeration
volldurchmischen *vt* ABFALL, WASSER mix completely
vollhalogeniert *adj* LUFT fully halogenated; **~e FCKW** *m pl* LUFT fully halogenated CFCs
vollintegriert: **~es System** *nt* MANAG fully integrated system
Vollkapselung *f* LÄRM complete enclosure
Vollkegeldüse *f* LUFT *Strahlwäscher* two-phase jet
Vollpappe *f* ABFALL solid fiberboard (*AmE*), solid fibreboard (*BrE*)
vollständig: **~es Verfahren** *nt* POLITIK comprehensive procedure
Vollziegel *m* LÄRM solid brick
vollziehen *vt* POLITIK implement
Vollzug *m* POLITIK implementation; **Vollzugsbehörde** *f* POLITIK environmental inspectorate; **Vollzugsdefizit** *nt* POLITIK lack of implementation
Voltameter *nt* ANALYTIK coulometer, voltameter
Voltametrie *f* ANALYTIK coulometry, voltametry
voltametrisch *adj* ANALYTIK *Detektoren* coulometric, voltametric
voltamperemetrisch *adj* ANALYTIK *Detektoren* voltametric
Volumen *nt* ALLG volume; **Volumendurchfluß** *m* ANALYTIK volume flow rate; **Volumenfilter** *nt* ABWASSER volume filter; **Volumenfiltration** *f* ABWASSER volume filtration; **Volumenkonzentration** *f* LUFT volume concentration; **Volumenreduktion** *f* ABFALL, ANALYTIK volume reduction; **Volumenreduzierung** *f* ABFALL, ANALYTIK volume reduction; **Volumenstrom** *m* ABFALL, ABWASSER, LUFT, WASSER volume flow
volumetrisch *adj* ANALYTIK volumetric, titrimetric; **~e Gasanalyse** *f* ANALYTIK volumetric gas analysis; **~e Titriermethode** *f* ANALYTIK volumetric analysis by titration; **~er Verlust** *m* ABFALL, ABWASSER, WASSER volumetric loss
vorabscheiden *vt* LUFT prescrub
Vor- *in cpds* ALLG pre-, preliminary, primary; **Vorabscheider** *m* LUFT prescrubber; **Vorabscheidezyklon** *m* LUFT preseparation cyclone; **Vorabscheidung** *f* ABFALL *von Schadstoffen vor Deponierung* prior separation
vorbehandeln *vt* ALLG pretreat
vorbehandelt *adj* ALLG pretreated
Vor-: **Vorbehandlung** *f* ABFALL, ABWASSER, ANALYTIK, BODEN, LUFT preliminary treatment, pretreatment; **Vorbelüftung** *f* ABWASSER preaeration; **Vorbelüftungsbecken** *nt* ABWASSER preaeration tank; **Vorbescheid** *m* POLITIK preliminary decision; **Vorbrecher** *m* ABFALL primary crusher; **Vorbrenner** *m* BODEN precombustor; **Vorbrennkammer** *f* LUFT precombustion chamber; **Voreindicker** *m* ABWASSER primary thickener; **Voreindickung** *f* ABWASSER prethickening; **Vorentstaubung** *f* LUFT preliminary dust removal; **Vorentwässerung** *f* ABWASSER preliminary dewatering; **Vorerkundung** *f* ANALYTIK preliminary assessment; **Vorfilter** *nt* ABWASSER prefilter, primary filter, ANALYTIK, LUFT prefilter
vorfiltern *vt* ABWASSER, ANALYTIK, LUFT prefilter
Vor-: **Vorfilterung** *f* ABWASSER, ANALYTIK, LUFT prefiltering, prefiltration; **Vorfiltration** *f* ABWASSER preliminary filtration; **Vorfluter** *m* ABWASSER receiving water; **Vorformung** *f* ABFALL premolding (*AmE*), premoulding (*BrE*)
Vorgang *m* ALLG operation
vorgeformt: **~er Sprühwäscher** *m* LUFT preformed spray scrubber
vorgeheizt: **~e Verbrennungsluft** *f* BODEN preheated combustion air
vorgereinigt: **~es Gas** *nt* LUFT precleaned gas
vorgeschaltet *adj* ABWASSER preceding; **mit ~er Denitrifikation** *phr* ABWASSER with preceding denitrification; **~e Denitrifikation** *f* ABWASSER predenitrification
Vorgesetzter *m* MEER supervisor
vorgespannt: **~es Betonrohr** *nt* ABFALL, ABWASSER, WASSER prestressed concrete pipe
vorgewärmt *adj* LUFT heated; **~e Abluft** *f* LUFT heated waste air
Vor-: **Vorhebepumpe** *f* ABFALL, WASSER priming pump
vorheizen *vt* ABFALL, BODEN, LUFT preheat
Vor-: **Vorheizung** *f* ABFALL, BODEN, LUFT preheating
Vorhersagefähigkeit *f* LUFT predictive capability
Vor-: **Vorkammer** *f* ABWASSER receiving chamber
Vorkehrung *f* SICHERHEIT precaution; **Vorkehrungen** *f pl* **gegen Staub** SICHERHEIT precautions against dust; **Vorkehrungen treffen** *vi* SICHERHEIT take precautions
Vor-: **Vorklärbecken** *nt* ABWASSER preliminary settling tank, primary settling tank, primary settling basin, primary clarifier; **Vorklärschlamm** *m* ABWASSER preliminary settled sludge; **Vorkonzentrat** *nt* ANALYTIK preconcentrate; **Vorkonzentrierung** *f* ANALYTIK preconcentration; **Vorkopfschüttung** *f* ABFALL frontally-packed bed; **Vorkühler** *m* LUFT precooler
Vorlage *f* ABWASSER receiver tank, LUFT collecting main
Vorlaufbehälter *m* LUFT feed tank
Vor-: **Vormischer** *m* ABFALL, WASSER premixer
Vor-Ort- *in cpds* ABFALL, ANALYTIK, BODEN, WASSER on-site, on-the-spot; **Vor-Ort-Analyse** *f* ANALYTIK on-site analysis, field screening; **Vor-Ort-Datenspeicherung** *f* ANALYTIK on-site data storage; **Vor-Ort-Einsatz** *m* ANALYTIK on-site assembly; **Vor-Ort-Einsatzleiter** *m* MEER on-scene commander (*OSC*); **Vor-Ort-Testanlage** *f* WASSER on-site test plant
Vor-: **Vorprobe** *f* ANALYTIK preliminary sample, pilot sample
vorprogrammierbar *adj* ANALYTIK *Meßgerät* preprogrammable
vorprogrammieren *vt* ANALYTIK preprogram
vorprogrammiert *adj* ANALYTIK preprogrammed
Vorrat *m* MANAG supply; **Vorratsbehälter** *m* WASSER storage tank; **Vorratsflasche** *f* **für Flüssigkeiten** ANALYTIK winchester; **Vorratstank** *m* WASSER feeding tank
vorreinigen *vt* ABFALL, LUFT, WASSER preclean
Vorreinigung *f* ABFALL preliminary purification, LUFT precleaning, WASSER preliminary purification; **Vorreinigungsbecken** *nt* ABWASSER precleaning basin
Vorrichtung *f* ALLG device, equipment
Vor-: **Vorrotte** *f* ABFALL *Abbau- und Hygienisierungsphase* prerotting treatment
vorsätzlich: **~e Einleitung** *f* ABFALL, ABWASSER *von Abwässern* intentional discharge
Vorsatzschale *f* LÄRM *Flächenverkleidung* attached shell
Vor-: **Vorsäulenrückspülung** *f* ANALYTIK pre-column/backflush operation
Vorschlag *m* POLITIK proposal
Vorschrift *f* POLITIK, SICHERHEIT regulation; **Vorschriften** *f pl* **zur Überwachung von gesundheitsgefährdenden Substanzen** MANAG, SICHERHEIT Control of Substances Hazardous to Health (*BrE*) (*COSHH*)
Vorschub *m* WASSER forward feed; **Vorschubgeschwindigkeit** *f* WASSER feed rate; **Vorschubrost** *m* WASSER forward feed grate
Vor-: **Vorsedimentation** *f* WASSER presedimentation

Vorsicht *f* SICHERHEIT precaution; **Vorsichtsmaßnahmen** *f pl* SICHERHEIT precautionary measures
Vorsitzender *m* POLITIK chairman
Vorsorge *f* ALLG precaution, prevention; **~ treffen** *phr* SICHERHEIT take precautions; **Vorsorgemaßnahmen** *f pl* ABWASSER, LÄRM, LUFT, MANAG, MEER, SICHERHEIT preventative measures
vorsorgend *adj* POLITIK preventative; **~e Umweltpolitik** *f* POLITIK preventative environmental policy; **~er Umweltschutz** *m* MANAG, POLITIK preventative environmental protection
Vorsorge: **Vorsorgepflicht** *f* POLITIK duty of prevention; **Vorsorgeplanung** *f* MEER contingency plan
vorsortieren *vt* ABFALL, BODEN presort
vorsortiert *adj* ABFALL, BODEN presorted; **~er Abfall** *m* ABFALL presorted waste
Vor-: **Vorsortierung** *f* ABFALL, BODEN presorting; **Vorsortierung** *f* **vor Ort** ABFALL on-site presorting
vorübergehend *adj* ABFALL, ABWASSER, BODEN interim; **~e Härte** *f* ABWASSER, WASSER temporary hardness; **~e Sanierung** *f* BODEN temporary remediation
Vor-: **Voruntersuchung** *f* ANALYTIK preliminary analysis; **Vorverrottung** *f* ABFALL *Abbau- und Hygienisierungsphase* prerotting treatment; **Vorverstärker** *m* ANALYTIK preamplifier; **Vorversuch** *m* ANALYTIK pretest; **Vorwärmer** *m* ABFALL, BODEN, LUFT preheater; **Vorwärmerstützfuß** *m* LUFT heater support; **Vorwärmung** *f* ABFALL, BODEN, LUFT preheating
vorwärts: **~ gekrümmter Schaufel-Verdichter** *m* LUFT forward-curved blade compressor
Vorwärtsstreuung *f* ANALYTIK *Schwermetallbestimmung* forward scattering of radiation
Vor-: **Vorwaschbehälter** *m* LUFT *biologische Gaswäsche* prescrubbing tank; **Vorwäsche** *f* ABFALL prewash, LUFT prescrubbing
vorwaschen *vt* ABFALL prewash, LUFT prescrub
Vor-: **Vorwäscher** *m* ABFALL prewasher, LUFT prescrubber
VRS-Verfahren *nt* ABFALL *Verfestigungsverfahren für Sonderabfälle* WPC-VRS process
vulkanisieren *vt* ABFALL vulcanize
vulkanisiert *adj* ABFALL vulcanized

W

Waage *f* ABFALL scales, weighbridge, ANALYTIK scales
wabenförmig: **~er Kanal** *m* LUFT *Katalysator* honeycomb-shaped channel
Wabenkörperelement *nt* LUFT *Katalysator* honeycomb element
Wabenrohrkatalysator *m* LUFT honeycomb catalyst
Wachstum *nt*: **~ des biologischen Rasens** ABWASSER biofilm growth
wägbar *adj* ANALYTIK weighable
wahrnehmbar *adj* ANALYTIK distinguishable
Walkerde *f* LUFT *Adsorbens* Fuller's earth
Walther-Verfahren *nt* LUFT Walther process
Walze *f* ABFALL roller; **Walzenbrecher** *m* WASSER roll crusher; **Walzenextruder** *m* ABFALL roller die extruder; **Walzenkörper** *m* BODEN *Glattwalze* roll body; **Walzenrost** *m* ABFALL rolling grate; **Walzenrostanlage** *f* ABFALL rolling grate installation
Walzwerk *nt* ABFALL rolling mill
Wand *f* ABFALL, WASSER wall
Wander- *in cpds* ABFALL, LUFT travelling (*BrE*), traveling (*AmE*), moving; **Wanderbett** *nt* LUFT moving bed; **Wanderbettadsorber** *m* LUFT moving-bed adsorber; **Wanderbettzone** *f* LUFT moving-bed zone
Wandern *nt* ABFALL, BODEN, LUFT traveling (*AmE*), travelling (*BrE*)
Wander-: **Wanderrost** *m* ABFALL, BODEN traveling grate (*AmE*), travelling grate (*BrE*); **Wanderrostanlage** *f* ABFALL traveling-grate installation (*AmE*), travelling-grate installation (*BrE*); **Wanderschichtadsorber** *m* ABWASSER, LUFT moving-bed adsorber; **Wanderschichtreaktor** *m* LUFT traveling-bed reactor (*AmE*), travelling-bed reactor (*BrE*); **Wandersprenger** *m* WASSER traveling distributor (*AmE*), travelling distributor (*BrE*)
Wanderungsgeschwindigkeit *f* ANALYTIK, LUFT velocity of migration
Wand: **Wandhydrant** *m* ABWASSER, WASSER wall hydrant
Wandtiefe *f* BODEN wall depth, curtain depth, WASSER wall depth
Wanne *f* ABFALL *Glasschmelze* vat
Warentransportsystem *nt* ABFALL conveyor handling system
warm: **~es Meerwasser** *nt* ERNEU ENER warm seawater
Warmbreitband *nt* ABFALL hot-rolled wide strip
Wärme *f* ALLG heat; **Wärmeabgabe** *f* ABFALL heat loss; **Wärmeausdehnung** *f* LUFT *Katalysator* thermal heat expansion; **Wärmeausgleich** *m* LUFT heat balance; **Wärmeaustausch** *m* LUFT heat exchange; **Wärmeaustauscher** *m* ABWASSER heat exchanger, LUFT recuperator, regenerator; **Wärmebelastung** *f* ABFALL, LUFT thermal load; **Wärmebelastungsplan** *m* ABFALL, LUFT, WASSER heat-load plan
wärmebeständig *adj* SICHERHEIT heatproof
Wärme: **Wärmebilanz** *f* ABWASSER thermal balance, LUFT heat balance, thermal balance; **Wärmedämmung** *f* LUFT heat insulation, lagging; **Wärmeemissionen** *f pl* LUFT heat emissions; **Wärmeenergie** *f* LUFT thermal energy; **Wärmeerzeugung** *f* ERNEU ENER heat generation
wärmefest *adj* SICHERHEIT heat-resistant
Wärme: **Wärmeinhalt** *m* ERNEU ENER heat content; **Wärmeisolierung** *f* ABWASSER heat insulation; **Wärmekonvektion** *f* ABWASSER heat transfer via convection, LUFT heat convection, heat transfer via convection; **Wärmekraftmaschine** *f* **mit äußerer Verbrennung** LUFT external-combustion engine; **Wärmeleistung** *f* LUFT thermal output; **Wärmeleitfähigkeit** *f* ANALYTIK heat conductivity, caloric conductibility; **Wärmeleitfähigkeit-Detektor** *m* (*WLD*) ANALYTIK thermal conductivity detector (*TCD*); **Wärmeleitung** *f* LUFT heat conduction, thermal conduction; **Wärmelichtquelle** *f* ANALYTIK incandescent lamp; **Wärmenutzung** *f* ABFALL heat use, ABWASSER *aus stabilisiertem Schlamm, aus der Abluft* thermal use, heat use, LUFT, WASSER heat use; **Wärmepumpe** *f* ERNEU ENER heat pump; **Wärmepumpenanlage** *f* ERNEU ENER heat pump plant; **Wärmequelle** *f* ERNEU ENER heat source; **Wärmerückgewinnung** *f* ABFALL, ABWASSER, LUFT, WASSER heat recovery; **Wärmeschutz** *m* LUFT lagging; **Wärmespeicher** *m* ERNEU ENER heat storage, LUFT heat accumulator; **Wärmestabilität** *f* LUFT heat stability; **Wärmestrahlung** *f* ERNEU ENER heat radiation; **Wärmestrom** *m* LUFT heat flow; **Wärmeströmung** *f* LUFT heat convection; **Wärmetauscher** *m* ABWASSER, ERNEU ENER heat exchanger, LUFT heat exchanger, recuperator, regenerator; **Wärmetauschermaterial** *nt* LUFT heat exchanger material; **Wärmetauscherwirkungsgrad** *m* LUFT heat exchanger efficiency; **Wärmetönung** *f* ANALYTIK *als Meßgröße für Gaskonzentration* heat tone; **Wärmeträger** *m* ABWASSER heat transfer medium; **Wärmeträgermedium** *nt* ERNEU ENER heat carrier medium; **Wärmeübergang** *m* **durch Leitung** LUFT heat transfer via conduction; **Wärmeübertragung** *f* ABWASSER, LUFT, WASSER heat transfer; **Wärmeverlust** *m* ABFALL, LUFT heat loss; **Wärmeverlust** *m* **nach Abstellen des Motors** LUFT hot soak loss
Warmwalzwerk *nt* ABFALL hot rolling work
Warmwasserbereiter *m* ABWASSER, LUFT warm water heater
Warmwassererzeugung *f* ABWASSER, LUFT warm water generation
warnen *vt* SICHERHEIT warn
Warnsignal *nt* SICHERHEIT warning signal
Wartezeit *f* ANALYTIK waiting time
Wartung *f* ALLG maintenance
Wasch- *in cpds* ABFALL, ANALYTIK, LUFT washing
waschaktiv *adj* ABWASSER, BODEN washing-active
Wasch-: **Waschberge** *m pl* WASSER tailings (*BrE*), tails (*AmE*)
Wäsche *f* ABFALL *Brennstoffaufbereitung* scrubbing, BODEN washing, LUFT *Brennstoffaufbereitung* scrubbing
waschen *vt* ABFALL, BODEN, LUFT, WASSER scrub, wash
Wäscher *m* LUFT scrubber; **Wäscherverstopfung** *f* LUFT scrubber plugging, scrubber clogging
Wasch-: **Waschflasche** *f* ANALYTIK gas-washing bottle; **Waschflügel** *m* LUFT scrubbing vane; **Waschflüssigkeitsstrahl** *m* LUFT *Strahlwäscher* scrubbing liquid jet; **Waschflüssigkeitstank** *m* LUFT scrubbing liquid tank; **Waschflüssigkeitstropfen** *m* LUFT scrubbing liquid droplet; **Waschlösung** *f* LUFT washing solution
Waschmittel *nt* LUFT washing agent; **Waschmittelaufbereitung** *f* LUFT washing agent processing; **Waschmittelregenerierung** *f* LUFT washing agent regeneration; **Waschmittelumlauf** *m* LUFT washing agent circulation
Wasch-: **Waschprozeß** *m* LUFT washing process; **Waschraumgestaltung** *f* LUFT scrubbing area design; **Waschstufen** *f pl* LUFT scrubbing stages; **Waschsystem** *nt* LUFT washing system; **Waschvorgang** *m* LUFT washing operation; **Waschwasser** *nt* ABFALL wash water, BODEN washings, LUFT, WASSER wash water; **Waschwasserbehälter** *m* LUFT wash water tank; **Waschwasserkreislauf** *m* LUFT wash water circulation;

Waschzone *f* LUFT scrubbing zone; **Waschzyklus** *m* LUFT washing cycle
Washcoat *nt* LUFT *Katalysator* wash coat
Washout *nt* LUFT washout; **Washout-Rate** *f* LUFT washout rate
Wasser *nt* ALLG water; **auf ~ basierte Beschichtung** *f* ABFALL, WASSER water-based coating; **Wasserabgabe** *f* WASSER water delivery; **Wasserablaßhahn** *m* WASSER priming cock; **Wasserablaufrinne** *f* LUFT water outlet
wasserabstoßend *adj* ABWASSER, LUFT hydrophobic
Wasser: **Wasseranalyse** *f* ANALYTIK water analysis; **Wasseraufbereitung** *f* ABWASSER, WASSER water purification; **Wasseraufbereitungsanlage** *f* ABWASSER, WASSER water treatment plant
wasseraufnehmend *adj* ABWASSER, LUFT hydrophilic
Wasser: **Wasserbecken** *nt* WASSER water tank; **Wasserbedarf** *m* ABFALL, WASSER water demand; **Wasserbehälter** *m* ABWASSER, WASSER reservoir, water tank; **Wasserbehandlungsanlage** *f* WASSER water treatment plant; **Wasserbehörde** *f* POLITIK, WASSER water authority; **Wasserbelieferung** *f* WASSER water delivery; **Wasserbeschaffenheit** *f* ANALYTIK, WASSER water quality
wasserbeständig *adj* WASSER water-resistant
Wasser: **Wasserbewirtschaftung** *f* WASSER water management; **Wasserbilanz** *f* POLITIK, WASSER water balance, hydrological balance; **Wasserbindung** *f* WASSER water retention
wasserbürtig *adj* ABWASSER waterborne; **~er Schadstoff** *m* ABFALL, ABWASSER waterborne contaminant, waterborne pollutant
Wasserdampf *m* ANALYTIK, LUFT water vapor (*AmE*), water vapour (*BrE*); **Wasserdampfdestillation** *f* ANALYTIK steam distillation; **Wasserdampfgehalt** *m* LUFT condensed water content; **Wasserdampfkreislauf** *m* ERNEU ENER steam circulation; **Wasserdampfschwaden** *m* LUFT water vapor plume (*AmE*), water vapour plume (*BrE*)
Wasser: **Wasserdargebot** *nt* WASSER water supply; **Wasserdruck** *m* WASSER water pressure, hydraulic pressure; **Wassereigenschaften** *f pl* WASSER water properties; **Wassereindüsung** *f* LUFT water spraying; **Wassereinlaß** *m* WASSER water inflow; **Wassereinstau** *m* WASSER banking-up of water; **Wassereintritt** *m* WASSER water inlet; **Wasserenthärtung** *f* WASSER water softening; **Wasserfassung** *f* WASSER water catchment
wasserfest *adj* WASSER watertight
Wasser: **Wasserfilm** *m* LUFT water film; **Wasserfilter** *nt* WASSER water filter; **Wasserförderschnecke** *f* MEER screw pump
wasserfrei *adj* WASSER water-free
wasserführend *adj* WASSER aquiferous
wassergefährdend *adj* ABFALL, ABWASSER, WASSER harmful to water, hazardous to water; **~er Abfall** *m* ABFALL, ABWASSER, WASSER waste harmful to water
Wasser: **Wassergehalt** *m* (*WG*) ABWASSER *von Klärschlamm*, ANALYTIK water content
wassergesättigt *adj* WASSER water-saturated
Wasser: **Wassergewinnung** *f* WASSER water recovery; **Wasserglas** *nt* BODEN *Bindemittel* soluble glass; **Wassergüte** *f* ANALYTIK, WASSER water quality; **Wassergütebestimmung** *f* ANALYTIK, WASSER water quality determination; **Wasserhahn** *m* WASSER water tap; **Wasserhärte** *f* WASSER water hardness; **Wasserhaushaltsgesetz** *nt* POLITIK, WASSER water law; **Wasserhochbehälter** *m* WASSER elevated water reservoir, water tower
Wasserkraft *f* ERNEU ENER hydropower, WASSER water power; **Wasserkraftangebot** *nt* ERNEU ENER hydropower supply; **Wasserkraftressourcen** *f pl* ERNEU ENER hydropower resources; **Wasserkrafttalsperre** *f* ERNEU ENER hydroelectric dam; **Wasserkraftwerk** *nt* ERNEU ENER hydroelectric plant, hydropower plant, WASSER hydroelectric plant
Wasser: **Wasserkrankheit** *f* ABWASSER, SICHERHEIT waterborne disease; **Wasserkreislauf** *f* ABFALL hydrological cycle, ABWASSER water circulation, WASSER hydrologic cycle; **Wasserkreislaufführung** *f* ABWASSER water circulation; **Wasserlauf** *m* WASSER watercourse; **Wasserleitungsrohr** *nt* WASSER water pipe; **Wasserlinie** *f* MEER waterline
wasserlöslich *adj* ABFALL, LUFT, WASSER water-soluble; **~e Sonderabfälle** *m pl* ABFALL water-soluble hazardous wastes
Wasser: **Wasserlöslichkeit** *f* ABFALL, LUFT, WASSER water solubility; **Wassermangel** *m* WASSER water deficiency; **Wassermasse** *f* ERNEU ENER body of water, *Gezeitenenergie* water mass; **Wassermenge** *f* WASSER volume of water; **Wassermengendauerlinie** *f* ABWASSER flow duration curve, WASSER final reach; **Wassermesser** *m* WASSER instant flowmeter; **Wasseroberfläche** *f* ERNEU ENER water surface; **Wasserpfennig** *m* POLITIK water penny; **Wasserpflanzenfilter** *nt* ABWASSER aquatic plant filter; **Wasserpreis** *m* WASSER price of water; **Wasserprobe** *f* ANALYTIK water sample; **Wasserpumpe** *f* ERNEU ENER, WASSER water pump; **Wasserqualität** *f* WASSER water quality; **Wasserrad** *nt* ERNEU ENER waterwheel; **Wasserrecht** *nt* POLITIK, WASSER law on water
wasserrechtlich *adj* ABWASSER, POLITIK, WASSER according to the law on water; **~e Anlagengenehmigung** *f* ABWASSER, POLITIK, WASSER plant permit according to the law on water; **~er Widerruf** *m* ABWASSER, POLITIK, WASSER repeal according to law on water
Wasser: **Wasserreinigung** *f* ABWASSER, WASSER water purification; **Wasserreinigungsschlamm** *m* ABFALL water purification sludge; **Wassersäule** *f* ABWASSER water column, water head, ANALYTIK water column, WASSER water column, water head; **Wasserscheide** *f* WASSER watershed; **Wasserschöpfer** *m* ANALYTIK water sampler; **Wasserspeicher** *m* ABWASSER, WASSER reservoir; **Wasserspeicherung** *f* WASSER water storage; **Wasserspiegel** *m* ERNEU ENER, WASSER *Wellenkraftwerk* water level; **Wasserspiegelabsenkung** *f* WASSER lowering of water level; **Wasserspiegelanstieg** *m* WASSER rising of water level; **Wassersprung** *m* WASSER hydraulic jump
Wasserstand *m* ERNEU ENER, WASSER water level; **Wasserstandsanzeiger** *m* WASSER water gage (*AmE*), water gauge (*BrE*), water indicator, water mark; **Wasserstandshahn** *m* ABWASSER, WASSER gage cock (*AmE*), gauge cock (*BrE*); **Wasserstandsmarke** *f* WASSER water mark; **Wasserstandsmesser** *m* WASSER water gage (*AmE*), water gauge (*BrE*), water level indicator; **Wasserstandsregler** *m* ABWASSER, WASSER water level regulator
Wasserstoff *m* ABFALL, LUFT, WASSER hydrogen; **Wasserstoffexponent** *m* (*pH*) ABFALL, ABWASSER, ANALYTIK, BODEN, WASSER potential of hydrogen (*pH*); **Wasserstoffperoxid** *nt* ABWASSER, LUFT hydrogen peroxide
Wasser: **Wasserstrahl** *m* ABWASSER, BODEN, WASSER water jet; **Wasserstrahlpumpe** *f* BODEN water jet pump; **Wassertechnologie** *f* WASSER water technology; **Wassertransport** *m* MEER, WASSER water transport; **Wasserturbine** *f* ERNEU ENER water turbine; **Wasserturm** *m* WASSER elevated water reservoir, water tower
wasserundurchlässig *adj* ABFALL, WASSER tight; **~e Schicht** *f* ABFALL, WASSER tight layer
wasserunlöslich *adj* ABWASSER, LUFT, WASSER insoluble in water, water-insoluble; **~er Sonderabfall** *m* ABFALL, ABWASSER, WASSER water-insoluble hazardous waste
Wasser: **Wasseruntersuchung** *f* ANALYTIK water examination; **Wasserverbrauch** *m* WASSER water consumption; **Wasserverbraucher** *m* WASSER water user, water consumer; **Wasserverdampfung** *f* LUFT water evaporation; **Wasserverlust** *m* ABFALL, ABWASSER, ANALYTIK, MEER, SICHERHEIT, WASSER leakage, water loss; **Wasserverschmutzung** *f* ABFALL, ABWASSER, WASSER water pollution

Wasserversorgung *f* WASSER water supply; **~ im ländlichen Raum** WASSER rural water supply; **Wasserversorgungsleitung** *f* WASSER water supply pipe; **Wasserversorgungsplanung** *f* WASSER water supply planning; **Wasserversorgungsunternehmen** *nt* WASSER water supply company

Wasser: **Wasserverteilung** *f* WASSER water distribution; **Wasserverteilungsnetz** *nt* WASSER water distribution system

wasserverunreinigend: **~er Stoff** *m* ABFALL, ABWASSER, WASSER water pollutant

Wasser: **Wasserverunreinigung** *f* ABFALL, ABWASSER, WASSER water pollution; **Wasserverunreinigungssteuer** *f* POLITIK water pollution tax; **Wasservorkommen** *nt* WASSER water resources; **Wasservorrat** *m* WASSER water reserve; **Wasserwäsche** *f* LUFT water scrubbing; **Wasserwerk** *nt* WASSER waterworks; **Wasserwirtschaft** *f* ANALYTIK water industry; **Wasserzähler** *m* WASSER water meter; **Wasserzapfstelle** *f* WASSER water tap connection; **Wasserzurückhaltung** *f* WASSER water retention

wäßrig *adj* ABFALL, WASSER aqueous

Webbing *nt* MEER webbing

Wechsel- *in cpds* ANALYTIK, ERNEU ENER, LUFT alternating, reciprocal

wechselnd: **~e Belüftung** *f* ABWASSER tapered aeration; **~es Gefälle** *nt* ABFALL changing head, WASSER changing fall of water, changing head

Wechsel-: **Wechselrichter** *m* ERNEU ENER DC/AC converter; **Wechselstrom** *m* ERNEU ENER alternating current; **Wechselstromleistung** *f* ERNEU ENER alternating current capacity; **Wechselwirkung** *f* ANALYTIK, ERNEU ENER reciprocal effect; **Wechselwirkungskräfte** *f pl* LUFT interaction force

Weg *m* LUFT *von Gas* path

wegwaschen *vt* ANALYTIK wash away

Wegwerf- *in cpds* ABFALL one-way, throw-away; **Wegwerferzeugnis** *nt* ABFALL throw-away product; **Wegwerfpackung** *f* ABFALL throw-away pack; **Wegwerfprodukt** *nt* ABFALL throw-away product; **Wegwerfverpackung** *f* ABFALL one-way packaging

Wehr *nt* ERNEU ENER, MEER weir; **Wehrabschöpfgerät** *nt* MEER weir skimmer; **Wehrölsperre** *f* MEER weir boom

weich: **~es Instrument** *nt* POLITIK soft instrument; **~es Wasser** *nt* WASSER soft water

Weichmachen *nt* ABFALL, WASSER plasticizing, softening

Weichmacher *m* ABFALL, WASSER plasticizer, softener

weichmacherhaltig *adj* ABFALL, WASSER plasticized, softened

weiß: **~es Rauschen** *nt* LÄRM white noise

Weißblech *nt* ABFALL tin plate; **Weißblechabfall** *m* ABFALL tin-plate waste; **Weißblechdose** *f* ABFALL tin-plated can; **Weißblechdose** *f* **für Konserven** ABFALL metal can; **Weißblechdosenschrott** *m* ABFALL metal can scrap

Weißglas *nt* ABFALL clear glass

weiterentwickelt: **~e Kohlereinigung** *f* LUFT advanced coal cleaning

Weiterverwertung *f* ABFALL reclamation

Weithalsstandflasche *f* ANALYTIK wide-mouthed laboratory bottle

weitmaschig *adj* ABWASSER wide-meshed; **~er Rechen** *m* ABWASSER wide-meshed screen

Welle *f* ERNEU ENER wave; **Wellenberg** *m* ERNEU ENER wave crest; **Wellenbewegung** *f* ERNEU ENER wave movement; **Wellenenergie** *f* ERNEU ENER wave energy; **Wellenenergienutzung** *f* ERNEU ENER wave energy use; **Wellenenergieressourcen** *f pl* ERNEU ENER wave energy resources; **Wellenfrontbreite** *f* ERNEU ENER wavefront length; **Wellenkraftwerk** *nt* ERNEU ENER wave power plant; **Wellenlänge** *f* ANALYTIK wavelength; **Wellenlängenskala** *f* ANALYTIK wavelength scale; **Wellental** *nt* ERNEU ENER wave hollow

Well: **Wellfaserplatte** *f* ABFALL corrugated fiberboard (*AmE*), corrugated fibreboard (*BrE*)

Wellmann-Lord-Verfahren *nt* LUFT Wellmann-Lord process

Well: **Wellpapier** *nt* ABFALL corrugated paper; **Wellpappe** *f* ABFALL corrugated board; **Wellpappenkarton** *m* ABFALL corrugated board box

Welt- *in cpds* ERNEU ENER, LUFT global; **Weltenergiebedarf** *m* ERNEU ENER global energy demand; **Weltenergieverbrauch** *m* ERNEU ENER global energy consumption; **Weltenergieversorgung** *f* ERNEU ENER global energy supply; **Weltraum** *m* LUFT universal space; **Weltschwefelhaushalt** *m* LUFT global sulfur budget (*AmE*), global sulphur budget (*BrE*)

weltweit: **~e Emissionen** *f pl* LUFT global emissions; **~es Ökosystem** *nt* ABWASSER, BODEN, LUFT, MEER global ecosystem; **~es Potential** *nt* ERNEU ENER worldwide potential; **~e Sonnenstrahlung** *f* ERNEU ENER global solar radiation; **~e Strahlung** *f* LUFT global radiation

Wendelbelüfter *m* ABWASSER spiral aerator

wendelförmig *adj* LUFT spiralling

Werkstoff *m* ABFALL, ABWASSER material

werkstofflich: **~es Recycling** *nt* ABFALL materials recycling

Werkstoff: **Werkstoffnutzungszyklus** *m* ABFALL utilization cycle of materials

Wertigkeit *f* ANALYTIK valence

Wertstoff *m* ABFALL secondary resource; **Wertstoffballen** *m* ABFALL secondary resources bale; **Wertstoffcontainer** *m* ABFALL secondary resource container; **Wertstofffraktion** *f* ABFALL secondary resource fraction; **Wertstoffgehalt** *m* ABFALL secondary resource content; **Wertstoffqualität** *f* ABFALL secondary resource quality; **Wertstoffrückgewinnung** *f* ABFALL resource recovery

Wetterbedingungen *f pl* ERNEU ENER *wichtig für Windkraftanlagen und Solaranlagen* weather conditions

Wetterlage *f* LUFT atmospheric conditions, meteorological conditions

WG *abbr* (*Wassergehalt*) ABWASSER *von Klärschlamm*, ANALYTIK water content

Wheatstone-Brücke *f* ANALYTIK Wheatstone bridge

Widerruf *m* POLITIK repeal

widerrufen *vt* POLITIK repeal

Widerspruch *m* POLITIK appeal; **Widerspruchsbefugnis** *f* POLITIK appeal authorization; **Widerspruchsbescheid** *m* POLITIK appeal communication; **Widerspruchsfrist** *f* POLITIK appeal deadline; **Widerspruchsverfahren** *nt* POLITIK appeal procedure

Widerstand *m* ABFALL, ANALYTIK resistance, LUFT resistance, *Staub* resistivity, WASSER resistance; **Widerstandsbeiwert** *m* LUFT drag coefficient; **Widerstandshöhe** *f* ABWASSER, WASSER friction head; **Widerstandskapazität** *f* ANALYTIK resistance capacity; **Widerstandsofen** *m* ANALYTIK resistance kiln; **Widerstandsthermometer** *nt* ANALYTIK, LUFT resistance thermometer

Wieder- *in cpds* ABFALL, BODEN, LUFT re-; **Wiederanfiltrieren** *nt* LUFT refiltering

wiederaufbereiten *vt* ABFALL regrind, reprocess

wiederaufbereitet *adj* ABFALL reground, reprocessed; **~es Material** *nt* ABFALL reground material, reprocessed material

Wieder-: **Wiederaufbereitung** *f* ABFALL regrinding; **Wiederaufbereitung** *f* **von Lösungsmitteln** BODEN solvent regeneration; **Wiederaufbringung** *f* BODEN *von kulturfähigem unbelastetem Boden* redisposal

wiederauffangen *vt* LUFT recapture

wiederaufheizen *vt* LUFT reheat

Wieder-: **Wiederaufheizung** *f* LUFT reheating

wiederaufladbar *adj* ALLG rechargeable; **~er Akkumulator** *f* ABFALL rechargeable storage battery

wiederauflebend *adj* ABFALL, WASSER resurgent

Wieder-: **Wiederauflösung** *f* ABWASSER redissolution

wiederaufschmelzbar *adj* ABFALL *Thermoplast* remeltable; **nicht ~** *adj* ABFALL nonremeltable
wiederaufwärmen *vt* LUFT reheat
Wieder-: **Wiederaufwirbelung** *f* LUFT rewhirling
wiederausbrechen *vi* SICHERHEIT *Feuer* break out again
wiederbefüllen *vt* ABFALL *Verpackung* regenerate
Wieder-: **Wiederbelüftung** *f* ABWASSER re-aeration; **Wiederbenutzung** *f* LUFT *abgeschiedenen Staubes* reuse; **Wiedereinlagerung** *f* BODEN off-site disposal, *nach Behandlung* redeposition
wiedergewinnbar *adj* ALLG *Rohstoffe* recoverable
wiedergewinnen *vt* ALLG *Rohstoffe* recover
Wieder-: **Wiedergewinnung** *f* ABFALL, ABWASSER recovery, BODEN *von Land* reclamation, LUFT, WASSER *von Rohstoffen* recovery; **Wiedergewinnungsanlage** *f* ABFALL resource recovery plant, reclamation plant
wiedergewonnen *adj* ALLG recovered; **~er Rohstoff** *m* ABFALL secondary raw material
Wieder-: **Wiederholungsmessung** *f* ANALYTIK repeated measurement; **Wiedernutzbarmachung** *f* BODEN recultivation, rehabilitation
wiederurbar: **~ gemachtes Gebiet** *nt* BODEN reclaimed area
Wieder-: **Wiederurbarmachung** *f* BODEN recultivation
wiederverarbeiten *vt* ABFALL reprocess
Wieder-: **Wiederverarbeitung** *f* ABFALL reprocessing
wiederverunreinigen *vt* ABWASSER recontaminate
wiederverwendbar *adj* ABFALL recyclable, reusable; **~er Karton** *m* ABFALL reusable box; **~es Verpackungsmaterial** *nt* ABFALL reusable packaging
Wieder-: **Wiederverwendbarkeit** *f* ABFALL recyclability, reusability
wiederverwenden *vt* ABFALL, LUFT, WASSER reuse
Wieder-: **Wiederverwendung** *f* ABFALL, LUFT, WASSER *abgeschiedenes Staubes* reuse
wiederverwertbar *adj* ABFALL recyclable; **~es Abfallprodukt** *nt* ABFALL usable waste product, LUFT reusable waste product; **~e Verpackung** *f* ABFALL recyclable packaging, reusable packaging
Wieder-: **Wiederverwertbarkeit** *f* ABFALL recyclability
Wiederverwertung *f* ABFALL recycling; **~ von Abfällen** ABFALL recycling of waste
wiegen *vt* ABFALL, ANALYTIK weigh
Wiegung *f* ABFALL, ANALYTIK weighing
wild: **~e Ablagerung** *f* ABFALL uncontrolled dumping, uncontrolled tipping (*BrE*), POLITIK illegal dumping; **~e Deponie** *m* ABFALL phantom dump; **~e Müllablagerung** *f* ABFALL fly tipping (*BrE*), uncontrolled dumping, uncontrolled tip (*BrE*), uncontrolled tipping (*BrE*); **~e Müllkippe** *f* ABFALL illegal dump site
Wind *m* ALLG wind
Wind: **mit dem ~** *adv* LUFT downwind
Wind: **Windbedingungen** *f pl* ERNEU ENER wind conditions; **Winddauer** *f* ERNEU ENER wind duration; **Wind-Diesel-System** *nt* ERNEU ENER wind-diesel system
Windenergie *f* ERNEU ENER wind energy; **Windenergieanlage** *f* ERNEU ENER wind turbine generator (*WTG*); **Windenergiekonverter** *m* ERNEU ENER wind turbine generator (*WTG*); **Windenergieressourcen** *f pl* ERNEU ENER wind power resources; **Windenergieumwandlung** *f* ERNEU ENER wind energy conversion (*WEC*); **Windenergieumwandlungssytem** *nt* ERNEU ENER wind energy conversion system (*WECS*)
Wind: **Windexposition** *f* ABFALL wind exposure; **Windfarm** *f* ERNEU ENER wind farm, wind park; **Windfarmleistung** *f* ERNEU ENER wind farm capacity, wind park capacity; **Windgeschwindigkeit** *f* ERNEU ENER, LUFT wind velocity
windgünstig *adj* ERNEU ENER wind-favorable (*AmE*), wind-favourable (*BrE*); **~es Gebiet** *nt* ERNEU ENER wind-favorable region (*AmE*), wind-favourable region (*BrE*)
Wind: **Windkraft** *f* ERNEU ENER wind energy; **Windkraftanlage** *f* ERNEU ENER wind turbine generator (*WTG*), windmill; **Windkraftanlage** *f* **im Offshore-Bereich** ERNEU ENER offshore wind farm; **Windkraftnutzung** *f* ERNEU ENER wind power use; **Windkraftwerk** *nt* ERNEU ENER wind power plant; **Windmühle** *f* ERNEU ENER windmill; **Windpark** *m* ERNEU ENER wind farm, wind park; **Windparkleistung** *f* ERNEU ENER wind farm capacity, wind park capacity; **Windrad** *nt* ERNEU ENER windmill
windreich *adj* ERNEU ENER windy; **~es Gebiet** *nt* ERNEU ENER windy region
Wind: **Windrichtung** *f* LUFT wind direction; **Windrichtungsnachführung** *f* ERNEU ENER wind direction tracking mechanism; **Windrose** *f* LUFT wind rose; **Windsensor** *m* ANALYTIK wind sensor; **Windsichter** *m* ABFALL, LUFT air classifier, air separator, air classification plant; **Windsichtung** *f* ABFALL, LUFT air classification; **Windstrom** *m* ERNEU ENER wind power; **Windturbine** *f* ERNEU ENER wind turbine; **Windturbinenblatt** *nt* ERNEU ENER wind turbine blade
Windung *f* LUFT winding
Winkler-Methode *f* ANALYTIK Winkler method
Winsch *f* MEER winch
Wintersmog *m* LUFT acid smog
Wirbel *m* ABFALL eddy, LUFT eddy, vortex, WASSER eddy; **Wirbelabschöpfgerät** *nt* MEER vortex skimmer; **Wirbelbettabsorber** *m* LUFT fluidized-bed absorber; **Wirbelbettreaktor** *m* LUFT fluidized-bed reactor
wirbeln *vt* BODEN, LUFT fluidize
Wirbel: **Wirbelrohr** *nt* LUFT cyclone tube
Wirbelschicht *f* ALLG fluidized bed; **Wirbelschichtadsorber** *m* LUFT fluidized-bed adsorber; **Wirbelschichtboden** *m* LUFT fluidized-bed bottom; **Wirbelschichtfeuerung** *f* (*WSF*) LUFT fluidized-bed combustion (*FBC*); **Wirbelschichtofen** *m* ABWASSER fluidized-bed furnace, LUFT fluidized-bed oven; **Wirbelschicht-Pyrolyse** *f* ABFALL fluidized-bed pyrolysis; **Wirbelschichtreaktor** *m* ABFALL, ABWASSER, LUFT fluidized-bed reactor; **Wirbelschichttrockner** *m* LUFT fluidized-bed dryer; **Wirbelschichttrocknung** *f* ABWASSER fluidized-bed drying; **Wirbelschichtverbrennung** *f* ABWASSER, BODEN fluidized-bed incineration; **Wirbelschichtverbrennungsanlage** *f* ABFALL fluidized-bed incinerator, ABWASSER fluidized-bed incineration plant; **Wirbelschichtverfahren** *nt* BODEN *für feinkörnigen kontaminierten Boden* fluidized-bed process
Wirbel: **Wirbelsenke** *f* LUFT eddy sink; **Wirbelstromabscheidung** *f* ABFALL eddy current separation; **Wirbelströmung** *f* LUFT turbulent flow, turbulence, *Zyklon* eddy flow, vortex flow; **Wirbelströmungswäscher** *m* LUFT eddy flow scrubber
Wirbelung *f* ABFALL, ABWASSER turbulence
Wirbel: **Wirbelzähigkeit** *f* LUFT eddy viscosity; **Wirbelzone** *f* LUFT turbulence zone
Wirkpegel *m* ANALYTIK effective level
Wirksamkeit *f* MEER, POLITIK, SICHERHEIT effectiveness
Wirkung *f* ALLG effect; **Wirkungsgrad** *m* ALLG efficiency
wirkungsorientiert *adj* BODEN effect-oriented; **~e Sicherungsmaßnahmen** *f pl* BODEN effect-oriented security measures
wirkungsvoll: **~e Verpackung** *f* ABFALL efficient packaging
Wirkung: **Wirkungsweise** *f* ABWASSER, ANALYTIK, BODEN, LUFT, WASSER action; **Wirkungszone** *f* ABFALL area of influence
wirtschaftlich *adj* MANAG, POLITIK cost-effective, economic, profitable; **~es Potential** *nt* ERNEU ENER *der erneuerbaren Energiequellen* economic potential; **~ vertretbare beste verfügbare Technik** *f* (*BATNEEC*) ABFALL, ABWASSER, BODEN, LUFT, POLITIK, WASSER best available technology not entailing excessive costs (*BATNEEC*)
Wirtschaftlichkeit *f* MANAG *von umweltpolitischen Maßnahmen* economic feasibility, cost-effectiveness, POLITIK,

MANAG cost-effectiveness, profitability, *von umweltpolitischen Maßnahmen* economic feasibility
Wissenschaft *f*: **~ der Gewässerkunde** WASSER hydroscience
Witterung: **Witterungsbeständigkeit** *f* ABFALL *Kunststoffe*, BODEN *Bahnen* weathering resistance; **Witterungseigenschaften** *f pl* ABFALL *Kunststoffe* weathering characteristics; **Witterungsempfindlichkeit** *f* BODEN sensitivity to weathering; **Witterungsstabilität** *f* ABFALL *Kunststoffe* weathering resistance; **Witterungsverhältnisse** *nt pl* LUFT meteorological conditions
WLD *abbr* (*Wärmeleitfähigkeit-Detektor*) ANALYTIK TCD (*thermal conductivity detector*)
Wochenverbrauch *m* WASSER weekly consumption
Wolle *f* LUFT *Filter-Fasermaterial* wool
Woulfe: **~sche Flasche** *f* ANALYTIK Woulfe's bottle
WSF *abbr* (*Wirbelschichtfeuerung*) LUFT *Kunststoffabfall* FBC (*fluidized-bed combustion*)
Würfeldruckfestigkeit *f* BODEN cube compressive strength
Wurzelraumverfahren *nt* ABWASSER root bed method
Wüstengebiet *nt* ERNEU ENER desert region

X Z

Xanthat-Polymer-Komplex *m* ABWASSER xanthat polymer complex
Xe *abbr* (*Xenon*) LUFT Xe (*xenon*)
Xenon *nt* (*Xe*) LUFT xenon (*Xe*)

zähflüssig *adj* ALLG viscous
Zähflüssigkeit *f* ALLG viscosity
zahlenmäßig *adj* ANALYTIK numerical
Zahlenwert *m* ANALYTIK numerical value
Zähler *m* ANALYTIK meter
Zahnradantrieb *m* ABFALL *Pumpe*, ABWASSER, WASSER *Pumpe* gear drive
Zaunölsperre *f* MEER fence boom
Zechenabraum *m* ABFALL coal mining waste, colliery waste
Zehrungsdauer *f* ABWASSER depletion time
Zeigerinstrument *nt* ANALYTIK *Gasmesser* pointer instrument
Zeigerpflanze *f* BODEN bioindicator plant
zeitabhängig *adj* ANALYTIK *Probenahme*, ERNEU ENER time-dependent
zeitintensiv *adj* ALLG time-intensive; **~e Methode** *f* ANALYTIK time-intensive method
Zeitintervall *m* ANALYTIK time interval
zeitlich: **~e Auflösung** *f* ABFALL, LUFT, WASSER temporal resolution; **~es Auflösungsvermögen** *nt* ABFALL, LUFT, WASSER temporal resolution; **~e Auswertung** *f* ANALYTIK temporal evaluation; **~e Dosisverteilung** *f* ABFALL, LUFT, WASSER temporal dose distribution; **~er Durchschnitt** *m* (*TWA*) ANALYTIK time-weighted average (*TWA*); **~ repräsentativ** *adj* ANALYTIK *Probe* temporally representative; **~e Schwankung** *f* ABFALL, LUFT, WASSER temporal fluctuation, temporal variation
Zeittrend *m* ABFALL, WASSER time trend
zeitweilig: **~e Hörschwellenverschiebung** *f* LÄRM temporary threshold shift (*TTS*)
Zellenfilter *nt* LUFT revolving filter
Zellenradschleuse *f* LUFT cellular wheel sluice
Zellenstack *m* LUFT *Brennstoffzelle* cell stack
Zellglasbeschichtung *f* ABFALL film lamination
Zellstoff *m* ABFALL cellulose, paper pulp, pulp; **Zellstoffaser** *f* ABFALL cellulose fiber (*AmE*), cellulose fibre (*BrE*); **Zellstoffbleiche** *f* ABFALL cellulose bleach; **Zellstoff-Preßsystem** *nt* ABFALL pulp molding system (*AmE*), pulp moulding system (*BrE*)
Zelluloid *nt* ABFALL celluloid
Zellulose *f* ABFALL cellulose; **Zelluloseacetat** *nt* (*CA*) ABFALL, ABWASSER cellulose acetate (*CA*); **Zelluloseacetat-Membran** *f* ABFALL, ABWASSER, WASSER cellulose acetate membrane; **Zellulosegehalt** *m* ABFALL cellulose content
zellulosehaltig *adj* ERNEU ENER cellulose-containing; **~er Abfall** *m* ABFALL cellulose waste; **~er Rohstoff** *m* ERNEU ENER *Bioenergienausgangsstoff* cellulose raw material
Zement *m* BODEN cement; **Zement/Tongemisch** *nt* BODEN cement/clay mixture; **Zement/Wassergemisch** *nt* BODEN cement/water mixture
zentral *adj* ALLG central; **~er Brennpunkt** *m* ERNEU ENER central focal point; **~e Leitstelle** *f* ABFALL, LUFT, WASSER central control room; **~er Sammelplatz** *m* ABFALL, WASSER central collection station; **~e Sammelstation** *f* ABFALL, WASSER central collection station; **~e Solarheizungsanlage** *f* ERNEU ENER central solar heating plant; **~er Steuerstand** *m* ABFALL, LUFT, WASSER central control room
Zentraldeponie *f* ABFALL central landfill
zentralisieren *vt* ABFALL, ERNEU ENER, LUFT, WASSER centralize
Zentralisierung *f* ABFALL, ERNEU ENER, LUFT, WASSER centralization
Zentralrohr *nt* ABFALL central pipe, LUFT central well, WASSER central pipe
Zentrifugal- *in cpds* ALLG centrifugal
zentrifugal *adj* ALLG centrifugal
Zentrifugal-: **Zentrifugalabscheider** *m* LUFT centrifugal collector; **Zentrifugalabschöpfgerät** *nt* MEER *für Öl* centrifugal skimmer; **Zentrifugalbeschleunigung** *f* LUFT centrifugal acceleration; **Zentrifugalfeld** *nt* ABFALL *Kunststoffsortierung* centrifugal field; **Zentrifugalkraft** *f* ERNEU ENER *Gezeitenenergie*, LUFT centrifugal force; **Zentrifugalkraftabscheider** *m* LUFT centrifugal force precipitator; **Zentrifugalpumpe** *f* ABWASSER, MEER centrifugal pump; **Zentrifugalsortierer** *m* ABWASSER centrifugal screen
Zentrifugation *f* ALLG centrifugation, centrifuging
Zentrifuge *f* ALLG centrifuge; **Zentrifugenentwässerung** *f* ABWASSER centrifuge dewatering; **Zentrifugenmantel** *m* ABFALL centrifuge casing; **Zentrifugenvorlagebehälter** *m* LUFT centrifuge prestorage tank
zentrifugieren *vt* ALLG centrifuge
Zentrifugieren *nt* ALLG centrifuge
Zentrifugierung *f* ALLG centrifugation
Zeolith *m* ABFALL, LUFT, WASSER zeolite, zeolith
zeolithisch *adj* LUFT zeolitic; **~es Absorbens** *nt* LUFT zeolitic absorbent
Zeolith: **Zeolithkatalysator** *m* LUFT zeolith catalyst
zerbrechlich *adj* ABFALL fragile
Zerfall *m* BODEN disintegration
zerfallen *vt* BODEN disintegrate
zerfasern *vt* ABFALL, ANALYTIK, BODEN beat, *Papierfasern* grind
zerfressen *vt* ABFALL corrode
Zerkleinerer *m* ABFALL crusher, comminutor, grinder, disintegrator, BODEN crusher, grinder
zerkleinern *vt* ABFALL grind, pulverize, shear, crush, ABFALL, ABWASSER comminute, ANALYTIK grind, BODEN grind, crush
Zerkleinern *nt* ABFALL grinding, shearing, ABWASSER, ANALYTIK, BODEN grinding; **~ der Proben** ANALYTIK sample grinding
zerkleinert *adj* ABFALL crushed, ground, shredded, BODEN crushed; **~er Abfall** *m* ABFALL crushed waste; **~er Festabfall** *m* ABFALL shredded solid waste; **~e Grobstoffe** *m pl* ABWASSER ground coarse solids; **~e Scherben** *f pl* ABFALL ground glass cullet
Zerkleinerung *f* ABFALL, ANALYTIK, BODEN *Altpapier* shredding, crushing, pulverization, pulverizing, grinding; **Zerkleinerungsanlage** *f* ABFALL crushing plant, crushing installation, crushing unit; **Zerkleinerungsmaschine** *f* ABFALL *für Altpapier* shredding machine, shredder; **Zerkleinerungsturm** *m* ABFALL shredding tower; **Zerkleinerungsverfahren** *nt* ABFALL crushing system
Zerklüftung *f* BODEN fracturing
Zerlegung *f* ABFALL decomposition
zermahlen *vt* ABFALL, ANALYTIK, BODEN grind, pulverize
Zermahlung *f* ABFALL, ANALYTIK, BODEN pulverization, grinding
zerreißen *vt* LUFT *Flüssigkeitsteilchen in feinste Tröpfchen* tear, shear, break
Zerreißen *nt* LUFT tearing, shearing, breaking

zerschneiden *vt* ABFALL shear
Zerschneiden *nt* ABFALL shearing
zersetzen *vt* ABFALL, ABWASSER, BODEN, WASSER disintegrate, decompose
Zersetzung *f* ABFALL, ABWASSER, BODEN, WASSER disintegration, decomposition
zerspanen *vt* ABFALL chip
Zerspanung *f* ABFALL chipping
zerstäuben *vt* ABFALL nebulize, pulverize, LUFT atomize
Zerstäuber *m* ABFALL nebulizer, pulverizer, LUFT atomizer; **Zerstäuberluftdüse** *f* LUFT atomizer nozzle
Zerstäubung *f* ABFALL, LUFT *Sprühabsorption* nebulization, pulverisation; **Zerstäubungsmittel** *nt* LUFT spray, dispersion medium
zerstören *vt* ALLG destroy
Zerstörung *f* ALLG destruction
Zerstreuen *nt* ANALYTIK scattering
zerteilen *vt* ABFALL, LUFT, WASSER separate
zertrennen *vt* LUFT separate
zertrümmern *vt* ABFALL, LUFT break up
Zeta-Potential *nt* BODEN zeta potential (*ZP*)
Zickzacksichter *m* ABFALL, WASSER zigzag classifier
Zielgruppe *f* POLITIK target group
Zieltröpfchen *nt* LUFT target droplet
Zink *nt* (*Zn*) ALLG zinc (*Zn*)
zinkhaltig *adj* ABFALL, WASSER zinc-containing; **~er Staub** *m* ABFALL dust containing zinc
Zirkulation *f* ABFALL, ABWASSER, LUFT, WASSER circulation
Zirkulationspumpe *f* ABWASSER circulation pump
zirkulieren *vi* ABWASSER circulate
zirkulierend: **~e Wirbelschicht** *f* (*ZWS*) BODEN, LUFT circulating fluidized bed (*CBC*)
Zisterne *f* ABWASSER, WASSER cistern
Zivilrecht *nt* POLITIK civil law
Zn *abbr* (*Zink*) ALLG Zn (*zinc*)
Zubringerbesen *m* ABFALL *am Kehrfahrzeug* feeder broom
Zubringerleitung *f* WASSER delivery main
zuckerhaltig *adj* ERNEU ENER sugar-containing; **~e Pflanzen** *f pl* ERNEU ENER *Bioenergienausgangsstoff* sugar-containing plants; **~er Rohstoff** *m* ERNEU ENER *Bioenergienausgangsstoff* sugar-containing raw material
zudosieren *vt* BODEN *Prozeßwasser* feed in
Zudosierung *f* BODEN dosing
Zufahrtsrampe *f* ABFALL *zu einer Deponie* access ramp
Zufallsrauschen *nt* LÄRM random noise
Zufallsstichprobe *f* ANALYTIK grab sample, spot sample, random sample, snap sample
Zufluß *m* WASSER inflow; **~ von natürlichen Gewässern** WASSER inflow from natural waters; **~ zu Pumpwerken** WASSER inflow into pumping station; **~ zu Speicheranlagen** WASSER inflow into storage facility; **~ zum Verteilrohrnetz** WASSER inflow into distribution system
zuführen *vt* ABFALL feed, *Substanzen* introduce, BODEN *Substanzen* introduce, LUFT *tangential* feed, supply, WASSER feed, *Substanzen* introduce
Zuführungsleitung *f* WASSER supply pipe
Zugang *m* POLITIK access; **~ zu Umweltinformationen** POLITIK access to environmental information
zugbeansprucht: **~es Glied** *nt* MEER tension member
Zugfestigkeit *f* LUFT tensile strength
Zugglied *nt* MEER tension member
zukünftig: **~e Nutzbarkeit** *f* ERNEU ENER future usability
Zulassung *f* POLITIK *einer Anlage* authorization, license (*AmE*), licensing
Zulauf *m* ABWASSER, WASSER influent, inflow; **~ von kommunalen Klärschlammanlagen** (*KKA-Zulauf*) ABWASSER discharge of municipal wastewater sludge plant; **Zulaufkanal** *m* ABWASSER, WASSER influent canal; **Zulaufkonzentration** *f* ABWASSER influent concentration; **Zulaufwehr** *nt* LUFT liquid inlet
Zünd- *in cpds* ABFALL ignition, ANALYTIK ignition, LUFT, WASSER ignition; **Zündempfindlichkeit** *f* ABFALL, LUFT ignitability
zündfähig *adj* ABFALL, LUFT, SICHERHEIT flammable; **~er Staub** *m* LUFT flammable dust
Zünd-: **Zündgas** *nt* LUFT ignition gas; **Zündgeschwindigkeit** *f* LUFT rate of flame propagation; **Zündgewölbe** *nt* LUFT ignition arch; **Zündgrenze** *f* ANALYTIK, LUFT inflammability limit; **Zündpunkt** *m* LUFT ignition point; **Zündtemperatur** *f* LUFT ignition temperature
Zündung *f* ABFALL, ABWASSER, ANALYTIK, LUFT ignition; **Zündungstemperatur** *f* LUFT ignition temperature
zunehmend: **~e Erwärmung** *f* LUFT incremental heating
zurückführen *vt* ABFALL channel back, recirculate, ABWASSER, LUFT, WASSER recirculate
zurückgestrahlt *adj* LUFT reflected
zurückgewinnen *vt* ABFALL *Rohstoffe*, ABWASSER, LUFT recover
zurückprallen *vi* LUFT *Partikeln* rebound
zurückpumpen *vt* ABFALL, ABWASSER, WASSER pump back
zurückstrahlen *vi* LUFT reflect
zurückverwandeln *vt* LUFT reconvert
zusammenballen 1. *vt* LUFT agglomerate, mass together; **2. sich ~** *vr* ABWASSER agglomerate
zusammenfallen *vi* ABFALL *Deponie* subside
Zusammenfallen *nt* ABFALL subsidence
Zusammenfluß *m* WASSER confluence, *eines Kanals* junction
Zusammenprall *m* ABFALL collision, LUFT impact, impaction
zusammenprallen: **1. ~ mit** *vt* LUFT collide with, impact; **2.** *vi* LUFT impinge
zusammenpressen *vt* ABFALL press together
Zusammenstoß *m* SICHERHEIT collision
zusammentreffen *vi* LUFT meet
Zusatz *m* ABFALL, ABWASSER, ERNEU ENER, LUFT addition; **Zusatzbrenner** *m* LUFT auxiliary burner; **Zusatzbrennstoff** *m* LUFT supplementary fuel; **Zusatzfeuerung** *f* ABFALL, ERNEU ENER additional firing; **Zusatzgebläse** *nt* LUFT auxiliary fan
zusätzlich *adj* ALLG additional
Zusatz: **Zusatzluft** *f* LUFT auxiliary air; **Zusatzreinigung** *f* LUFT supplementary purification; **Zusatzstoff** *m* ABFALL, ABWASSER, BODEN, LUFT, WASSER additive; **Zusatzwasser** *nt* LUFT, WASSER make-up water
Zuschlagstoff *m* ABFALL, ABWASSER, BODEN, LUFT additive, aggregate, WASSER additive
zuschneiden *vt* ABFALL, ANALYTIK cut to size
zusetzen *vt* ANALYTIK add
Zusickerung *f* ABFALL, WASSER intrusive filtration
zuständig: **~e Behörde** *f* POLITIK competent authority
Zustandsstörerhaftung *f* POLITIK liability for disturbance of current local conditions
Zuverlässigkeit *f* ANALYTIK, MEER reliability
Zuwachs *m* WASSER accretion
Zwang *m* ABFALL, ABWASSER, BODEN, POLITIK force; **Zwangsbelüftung** *f* ABFALL, ABWASSER, BODEN, WASSER forced aeration; **Zwangsentgasung** *f* ABFALL, WASSER forced degassing; **Zwangskühlung** *f* WASSER forced cooling; **Zwangsumlaufverdampfer** *m* WASSER forced circulation evaporator
Zweikreislaufwäscher *m* LUFT double-cycle scrubber
zweischalig *adj* LÄRM double-leaf, *Kapselung* double-sheet
zweistufig *adj* ABWASSER two-stage; **~e aerob-thermophile Stabilisierung** *f* ABWASSER two-stage aerobic thermophilic stabilization; **~e Belebung** *f* **mit alternierender Nitrifikation und Denitrifikation** ABWASSER two-stage activated sludge process with alternating nitrification and denitrification; **~es Belebungsverfahren** *nt* ABWASSER two-stage activated sludge process; **~e Belebung** *f* **mit vorgeschal-**

teter Denitrifikation ABWASSER two-stage activated sludge process with pre-denitrification; **~er Erhitzer** *m* LUFT two-stage fired heater; **~er Faulraum** *m* ABWASSER two-stage digester; **~er Gaswäscher** *m* LUFT two-stage gas scrubber; **~e Rekarbonisierung** *f* ABWASSER two-stage recarbonation; **~e thermophil-mesophile Faulung** *f* ABWASSER two-stage thermophilic-mesophilic digestion

Zweiweghahn *m* WASSER two-way cock

Zweiwegstück *nt* LUFT breeches pipe

Zweiwellenschneidzerkleinerer *m* ABFALL twin-cutting shredder

Zwillingshahn *m* WASSER twin cock

Zwischen- *in cpds* ABFALL, BODEN temporary, interim; **Zwischenabdeckmaterial** *nt* ABFALL temporary covering material; **Zwischenabdeckung** *f* ABFALL *einer Deponie* temporary cover; **Zwischenbehälter** *m* BODEN interim tank, LUFT temporary tank; **Zwischenentwässerung** *f* ABFALL interim dewatering

zwischengeschaltet: **mit ~er Denitrifikation** *phr* ABWASSER with intermittent denitrification

Zwischen-: **Zwischenklärbecken** *nt* ABWASSER intermediate settling tank; **Zwischenklärung** *f* ABWASSER intermediate settling, intermediate purification; **Zwischenlage** *f* LÄRM intermediate layer; **Zwischenlager** *nt* ABFALL interim store; **Zwischenlagerfläche** *f* ABFALL interim storage area

zwischenlagern *vt* ABFALL store temporarily, BODEN *Aushubmassen* store intermediately

Zwischen-: **Zwischenlagerung** *f* ABFALL, BODEN temporary storage; **Zwischenprodukt** *nt* ABFALL intermediate product; **Zwischenpumpe** *f* ABWASSER, WASSER booster pump; **Zwischenräume** *m pl* LUFT *Gase beim Elektrofilter* flow lanes; **Zwischensilo** *m* LUFT temporary storage silo; **Zwischenspeicher** *m* ANALYTIK trap; **Zwischenzeit** *f* ANALYTIK lag time

ZWS *abbr* (*zirkulierende Wirbelschicht*) LUFT CBC (*circulating fluidized bed*); **ZWS-Reaktor** *m* LUFT circulating fluidized-bed reactor

Zyklon *m* LUFT cyclone; **Zyklonabscheider** *m* LUFT cyclone separator; **Zyklonabscheidung** *f* LUFT cyclone separation; **Zyklonbrennkammer** *f* BODEN cyclone furnace; **Zyklon-innenwand** *f* LUFT inner cyclone wall; **Zyklonmantel** *m* LUFT cyclone housing; **Zyklonradius** *m* LUFT cyclone radius; **Zyklonsinterung** *f* BODEN cyclone vitrification; **Zyklonwand** *f* LUFT cyclone wall

Zylinderentgasungsanlage *f* WASSER barrel-venting system

zylindrisch *adj* LUFT cylindrical; **~er Schneckenextruder** *m* ABFALL, WASSER cylindrical screw extruder; **~er Teil** *m* LUFT *des Zyklons* cylindrical section; **~er Zyklon** *m* LUFT cylindrical cyclone

English–German
Englisch–Deutsch

A

AAS *abbr* ANALYSIS (*atomic absorption spectrometry*) AAS (*Atomabsorptionsspektrometrie*), (*atomic absorption spectrometer*) AAS (*Atomabsorptionsspektrometer*)

abatement *n* AIR Bekämpfung *f*; **~ measures** *n pl* AIR Bekämpfungsmaßnahmen *f pl*

abattoir: **~ waste** *n* WASTE Schlachthausabfall *m*

abiotic *adj* WASTE, WATER POLL, WATER SUPP abiotisch

above: **~-ground** *adj* WASTE, WATER SUPP oberirdisch, übertägig; **~-ground disposal** *n* WASTE übertägige Ablagerung *f*; **~-ground infiltration facility** *n* WATER SUPP oberirdische Infiltrationsanlage *f*

abrasion *n* AIR, WASTE Abrieb *m*; **~ protection** *n* WASTE Abrasionsschutz *m*; **~ resistance** *n* AIR Abriebfestigkeit *f*

abrasive *adj* AIR abschleißend, WASTE abrasiv; **~ dust** *n* AIR abschleißender Staub *m*; **~ medium** *n* SAFETY Schleifmittel *nt*

ABS *abbr* WASTE (*alkylbenzenesulfonate AmE, alkylbenzenesulphonate BrE*) ABS (*Alkylbenzolsulfonat*), (*acrylonitrile butadiene styrene*) ABS (*Acrylnitril-Butadien-Styrol*)

absence: **~ of oxygen** *n* WATER POLL Sauerstoffausschluß *m*

absolute: **~ pressure** *n* WATER POLL, WATER SUPP *pump* Absolutdruck *m*

absorb *vt* GEN absorbieren, schlucken, aufnehmen

absorbable *adj* GEN absorbierbar, schluckbar

absorbate *n* GEN Absorbat *nt*

absorbed *adj* GEN absorbiert, geschluckt, aufgenommen; **~ compound** *n* AIR adsorbierte Komponente *f*; **~ state** *n* AIR absorbierter Zustand *m*

absorbency *n* GEN Absorbierbarkeit *f*, Absorptionsfähigkeit *f*, Aufnahmefähigkeit *f*, Saugfähigkeit *f*

absorbent 1. *adj* GEN absorptionsfähig, saugfähig; **2.** *n* GEN *agent* Absorbens *nt*, Absorptionsmittel *nt*, Absorptiv *nt*, *material* Absorptionsmaterial *nt*

absorbent: **~ cleaning** *n* AIR Absorbensreinigung *f*; **~-lined** *adj* NOISE absorbierend ausgekleidet

absorber *n* GEN Absorber *m*; **~ head** *n* AIR Absorberkopf *m*; **~ motor** *n* AIR Absorbermotor *m*; **~ plate** *n* RENEW ENERG Absorber *m*; **~ sump** *n* AIR Absorbersumpf *m*; **~ suspension** *n* AIR Absorbersuspension *f*; **~ tank** *n* AIR Absorberbehälter *m*; **~ wall** *n* AIR *bubble-cap tray tower* Absorberwand *f*

absorbing *adj* GEN absorbierend; **~ well** *n* WATER POLL Sickergrube *f*, Sickerschacht *m*

absorption *n* GEN Absorption *f*, Aufnahme *f*; **~ apparatus without internal fittings** *n* AIR Absorptionsapparat *m* ohne Einbauten; **~ band** *n* ANALYSIS Absorptionsband *nt*; **~ capacity** *n* GEN Absorptionsfähigkeit *f*, Absorptionsvermögen *nt*, Aufnahmefähigkeit *f*, Aufnahmekapazität *f*; **~ coefficient** *n* GEN Absorptionsfaktor *m*, Absorptionskoeffizient *m*, Absorptionsgrad *m*, Absorptionsfaktor *m*; **~ column** *n* AIR *scrubber column* Absorptionskolonne *f*

absorption/filtration/adsorption: **~ process** *n* (*AFA process*) AIR Absorption/Filtration/Adsorption-Verfahren *nt* (*AFA-Verfahren*)

absorption: **~ isotherm** *n* AIR Absorptionsisotherme *f*; **~ liquid** *n* AIR Absorptionsflüssigkeit *f*; **~ pipette** *n* ANALYSIS Absorptionspipette *f*; **~ plant for the dry separation of noxious gases** *n* SAFETY Absorptionseinrichtung *f* zur Trockenabtrennung giftiger Gase; **~ silencer** *n* NOISE, SAFETY Absorptionsschalldämpfer *m*; **~ solution** *n* ANALYSIS Absorptionslösung *f*; **~ spectrum** *n* ANALYSIS *of IR spectrophotometer* Absorptionsspektrum *nt*; **~ stage** *n* AIR Absorptionsstufe *f*; **~ temperature** *n* AIR Absorptionstemperatur *f*; **~ tower** *n* AIR Absorptionsturm *m*; **~ vessel** *n* ANALYSIS Absorptionsgefäß *nt*

absorptive *adj* GEN absorptionsfähig, saugfähig; **~ attenuator** *n* NOISE, SAFETY Absorptionsschalldämpfer *m*; **~ capacity** *n* WASTE Porenvolumen *nt*; **~ silencer** *n* NOISE, SAFETY Absorptionsschalldämpfer *m*

absorptivity *n* GEN Absorptionsfähigkeit *f*, Aufnahmefähigkeit *f*

abundance: **~ coefficient** *n* ANALYSIS Häufigkeitskoeffizient *m*

accelerate *vt* AIR *particle*, ANALYSIS, WASTE beschleunigen

accelerated *adj* AIR, ANALYSIS, WASTE beschleunigt, schnell; **~ composting** *n* WASTE schnelle Kompostierung *f*; **~ electron** *n* AIR beschleunigtes Elektron *nt*

acceleration *n* AIR, ANALYSIS, WASTE Beschleunigung *f*; **~ due to gravity** *n* (*g*) AIR Erdbeschleunigung *f* (*g*)

accelerometer *n* ANALYSIS Beschleunigungsaufnehmer *m*, Beschleunigungsmesser *m*

acceptable: **~ daily intake** *n* (*ADI*) SAFETY tolerierbare tägliche Aufnahme *f* (*ADI-Wert*)

acceptance: **~ certificate** *n* POLICY Annahmeerklärung *f*

access *n* POLICY Zugang *m*; **~ ramp** *n* WASTE Zufahrtsrampe *f*; **~ to environmental information** *n* POLICY Zugang *m* zu Umweltinformationen

accident *n* SAFETY Störfall *m*, Unfall *m*; **~ prevention** *n* SAFETY Störfallvorsorge *f*

accompanying: **~ document** *n* POLICY Begleitschein *m*; **~ hydraulic measures** *n pl* SOIL *vapour extraction* hydraulische Begleitmaßnahmen *f pl*

ACCR *abbr* (*activated carbon catalytic reduction*) AIR ACCR-Verfahren *nt*

accretion *n* AIR, SOIL, WASTE, WATER POLL Ablagerung *f*, WATER SUPP *in flowing water* Ablagerung *f*, Zuwachs *m*

accumulability *n* AIR *dust*, WASTE, WATER SUPP Akkumulierbarkeit *f*

accumulate 1. *vt* ANALYSIS, MARINE, WASTE, WATER SUPP akkumulieren, GEN anreichern; **2.** *vi* GEN sich anreichern

accumulation *n* GEN Anreicherung *f*, Akkumulation *f*, Ansammlung *f*; **~ of contaminants** *n* AIR, SOIL, WASTE, WATER POLL, WATER SUPP Anreicherung *f* von Schadstoffen; **~ of harmful substances** *n* AIR, SOIL, WASTE, WATER POLL, WATER SUPP Anreicherung *f* von Schadstoffen; **~ of pollutants** *n* AIR, SOIL, WASTE, WATER POLL, WATER SUPP Anreicherung *f* von Schadstoffen

acetate *n* WASTE Acetat *nt*; **~ build-up** *n* WASTE Acetatbildung *f*; **~ formation** *n* WASTE Acetatbildung *f*

acetic: **~ acid** *n* WATER POLL Essigsäure *f*; **~ acid formation** *n* WATER POLL Essigsäurebildung *f*

acetogenesis *n* WATER POLL Acetogenese *f*

acetogenic *adj* WATER POLL acetogen; **~ bacteria** *n pl* WATER POLL acetogene Bakterien *f pl*

acetone *n* AIR, WASTE Aceton *nt*

acid 1. *adj* GEN sauer; **2.** *n* GEN Säure *f*

acid: **~ aerosol** *n* AIR saures Aerosol *nt*; **~ buffering capacity** *n* WATER POLL Säurepufferkapazität *f*; **~ capacity** *n* WATER POLL Säurekapazität *f*; **~ cleavage** *n* WATER POLL Säurespaltung *f*; **~ cleavage with flotation** *n* WATER POLL Säurespaltung *f* mit Flotation; **~ concentration** *n* AIR Säurekonzentration *f*, *activity* Eindampfen *nt* einer Säure; **~ content** *n* AIR Säureanteil *m*; **~ deposit** *n* AIR saure

Ablagerung *f*; ~ **deposition** *n* AIR saure Ablagerung *f*, saure Sedimentation *f*; ~ **earth** *n* SOIL saure Erde *f*; ~ **fallout** *n* AIR, SOIL saurer Niederschlag *m*, saurer Regen *m*; ~ **fermentation** *n* WASTE saure Gärung *f*; ~ **fog** *n* AIR Säurenebel *m*; ~ **formation** *n* AIR, SOIL, WASTE, WATER POLL Absäuerung *f*, Ansäuerung *f*, Säurebildung *f*; ~ **former** *n* AIR, SOIL, WASTE, WATER POLL *stabilization* Säurebildner *m*; ~**-free** *adj* AIR säurefrei; ~ **gas** *n* AIR saueres Gas *nt*, Sauergas *nt*; ~ **gas pollutants** *n pl* AIR saure Gasschadstoffe *m pl*

acidic *adj* GEN sauer; ~ **anionic exchanger** *n* WATER POLL saurer Anionenaustauscher *m*; ~ **cationic exchanger** *n* WATER POLL saurer Kationenaustauscher *m*; ~ **particle** *n* AIR saures Teilchen *nt*; ~ **precursor** *n* SOIL Säurenvorstufe *f*; ~ **rain** *n* AIR, SOIL saurer Niederschlag *m*, saurer Regen *m*; ~ **water** *n* WATER SUPP saures Wasser *nt*

acidification *n* AIR, SOIL, WASTE, WATER POLL Ansäuerung *f*

acidified: ~ **area** *n* SOIL angesäuerte Bodenfläche *f*

acidify 1. *vt* GEN ansäuern; **2.** *vi* GEN säuern, versauern

acidifying *adj* GEN ansäuernd

acidity *n* GEN Säuregehalt *m*; ~ **level** *n* AIR, SOIL Säuregrad *m*

acid: ~ **loading** *n* AIR Säurebelastung *f*; ~ **neutralizing** *n* AIR Säureneutralisation *f*; ~ **neutralizing capacity** *n* AIR Säureneutralisierungsvermögen *nt*

acidogenesis *n* WATER POLL Acidogenese *f*

acid: ~**-oxidative** *adj* AIR sauer-oxidativ; ~ **particle** *n* AIR saures Teilchen *nt*; ~ **pollution** *n* AIR, SOIL saure Umweltverschmutzung *f*; ~ **precipitation** *n* AIR Säurefällung *f*; ~**-proof** *adj* AIR, WATER SUPP säurebeständig; ~ **rain** *n* AIR, SOIL saurer Regen *m*, saurer Niederschlag *m*; ~ **resistance** *n* AIR, SOIL, WATER SUPP Säurebeständigkeit *f*, Säureresistenz *f*; ~**-resistant** *adj* AIR, SOIL, WATER SUPP säurebeständig; ~ **runoff** *n* SOIL, WASTE saurer Abfluß *m*; ~ **scrubbing** *n* AIR Säurewäsche *f*; ~ **smog** *n* AIR Wintersmog *m*, saurer Smog *m*; ~ **snow** *n* AIR, SOIL saurer Schnee *m*; ~ **stress** *n* SOIL Säurebelastung *f*; ~**-stressed** *adj* SOIL durch Säure belastet; ~ **tolerance** *n* SOIL Säureverträglichkeit *f*; ~ **waste gas** *n* AIR, WASTE saures Abgas *nt*

acoustic *adj* ANALYSIS, NOISE, SAFETY akustisch

acoustical *adj* ANALYSIS, NOISE, SAFETY akustisch; ~ **board** *n* NOISE Absorberplatte *f*, schallabsorbierende Platte *f*, schallschluckende Platte *f*; ~ **ceiling** *n* NOISE Akustikdecke *f*; ~ **emissions** *n pl* ANALYSIS, NOISE akustische Emissionen *f pl*; ~ **leak** *n* NOISE akustische Schwachstelle *f*, akustisches Leck *nt*; ~ **slab** *n* NOISE, SAFETY Dämmplatte *f*, Schalldämmplatte *f*; ~ **test** *n* ANALYSIS, NOISE, SAFETY Akustiktest *m*; ~ **tile** *n* NOISE Absorberplatte *f*

acoustic: ~ **baffle** *n* NOISE Schallwand *f*; ~ **barrier** *n* NOISE Schallschutzwand *f*; ~ **booth** *n* NOISE Lärmschutzkabine *f*, Schallschutzkabine *f*; ~ **capacity** *n* ANALYSIS, NOISE Schallleistung *f*; ~ **coupler** *n* ANALYSIS, NOISE Akustikkoppler *m* akustischer Koppler *m*; ~ **emission** *n* (*AE*) ANALYSIS, NOISE Schallemission *f* (*SE*); ~ **emissions analysis** *n* (*AE analysis*) ANALYSIS Schallemissionsanalyse *f* (*SEA*); ~ **insulation** *n* NOISE Schalldämmung *f*; ~ **lining** *n* NOISE schallabsorbierende Verkleidung *f*, schallschluckende Verkleidung *f*; ~ **louver** *AmE*, ~ **louvre** *BrE* *n* NOISE Schalloch *nt*, Schallöffnung *f*; ~ **oscillation** *n* NOISE Schallschwingung *f*; ~ **panel** *n* NOISE, SAFETY Akustikplatte *f*, Schalldämmplatte *f*; ~ **paneling** *AmE*, ~ **panelling** *BrE* *n* NOISE schallabsorbierende Verkleidung *f*, schallschluckende Verkleidung *f*; ~ **plaster** *n* NOISE schallabsorbierender Putz *m*; ~ **plate** *n* NOISE, SAFETY Akustikplatte *f*, Schalldämmplatte *f*

acoustics *n* ANALYSIS, NOISE, SAFETY Akustik *f*

acoustic: ~ **screen** *n* NOISE Schallschirm *m*, Schallwand *f*; ~ **sheet** *n* NOISE Schalldämmfolie *f*; ~ **shield** *n* NOISE Schallschutzschirm *m*, Schallwand *f*; ~ **shielding** *n* NOISE Abschirmung *f* von Schall, akustische Abschirmung *f*; ~ **tile** *n* NOISE Absorberplatte *f*, Dämmplatte *f*; ~ **vibration** *n* NOISE Schallschwingung *f*

acrid: ~ **odor** *AmE*, ~ **odour** *BrE* *n* AIR, WASTE, WATER POLL, WATER SUPP beißender Geruch *m*

acrylic: ~ **plastic** *n* WASTE Acrylglas *nt*

acrylonitrile: ~ **butadiene styrene** *n* (*ABS*) WASTE Acrylnitril-Butadien-Styrol *nt* (*ABS*)

act *n* POLICY Gesetz *nt*

actinometer *n* ANALYSIS Aktinometer *nt*, Strahlungsmesser *m*

action *n* GEN Wirkungsweise *f*

activate *vt* AIR, SOIL, WASTE, WATER POLL aktivieren, WATER POLL *sludge* beleben

activated *adj* AIR, SOIL, WASTE, WATER POLL aktiviert, WATER POLL *sludge* belebt; ~ **alumina** *n* WATER POLL *adsorbent* Aktivtonerde *f*; ~ **aluminium oxide** *BrE*, ~ **aluminum oxide** *AmE* *n* AIR *adsorbent*, WASTE aktiviertes Aluminiumoxid *nt*; ~ **carbon** *n* AIR, SOIL, WASTE, WATER POLL Aktivkohle *f* (*AK*); ~ **carbon adsorber** *n* AIR, WATER POLL Aktivkohleadsorber *m*; ~ **carbon adsorption** *n* AIR, WATER POLL Aktivkohleadsorption *f*; ~ **carbon canister** *n* AIR Aktivkohlefilter *nt*; ~ **carbon catalytic reduction** *n* (*ACCR*) AIR ACCR-Verfahren *nt*; ~ **carbon column** *n* ANALYSIS Aktivkohlesäule *f*; ~ **carbon filter** *n* WASTE, WATER POLL, WATER SUPP *volume filter* AK-Filter *nt*, Aktivkohlefilter *nt*; ~ **carbon filtration** *n* AIR, WASTE, WATER POLL, WATER SUPP Aktivkohlefiltration *f*; ~ **carbon fixed-bed reactor** *n* AIR Aktivkohlefestbettreaktor *m*; ~ **carbon fluidized bed** *n* AIR Aktivkohlewirbelbett *nt*; ~ **carbon moving bed** *n* AIR Aktivkohlewanderbett *nt*; ~ **carbon process** *n* AIR Aktivkohleverfahren *nt*; ~ **carbon suspension** *n* ANALYSIS Aktivkohlesuspension *f*; ~ **carbon treatment** *n* AIR Aktivkohlebehandlung *f*; ~ **coke** *n* AIR, SOIL, WASTE, WATER POLL Aktivkoks *m* (*AK*); ~ **coke adsorber** *n* AIR, WASTE, WATER SUPP Aktivkoksadsorber *m*; ~ **coke process** *n* AIR Aktivkoksverfahren *nt*; ~ **sludge** *n* WATER POLL Belebtschlamm *m*; ~ **sludge content** *n* WATER POLL Belebtschlammgehalt *m*; ~ **sludge floc** *n* WATER POLL Belebtschlammflocke *f*; ~ **sludge plant** *n* WATER POLL Belebtschlammanlage *f*, Belebungsanlage *f*; ~ **sludge process** *n* WATER POLL Schlammbelebungsverfahren *nt*, Belebungsverfahren *nt*, Belebtschlammverfahren *nt*; ~ **sludge return flow** *n* WATER POLL Belebtschlammrücklauf *m*; ~ **sludge tank** *n* WATER POLL Belebtbecken *nt* (*BB*), Belebtschlammbecken *nt*, Belebungsbecken *nt*; ~ **sludge tank with circulating flow** *n* WATER POLL Umlaufbecken *nt*

activation *n* AIR, ANALYSIS, SOIL, WATER POLL Aktivierung *f*, WATER POLL *sludge* Belebung *f*; ~ **analysis** *n* ANALYSIS *heavy metal detection* Aktivierungsanalyse *f*; ~ **energy** *n* AIR Aktivierungsenergie *f*

active *adj* GEN aktiv; ~ **degassing** *n* WASTE *base-surface sealing* Aktiventgasung *f*; ~ **pellistor** *n* ANALYSIS aktiver Pellistor *m*; ~ **solvent** *n* WASTE aktives Lösungsmittel *nt*; ~ **thermal solar energy use** *n* RENEW ENERG aktive thermische Sonnenenergienutzung *f*; ~ **water** *n* WATER SUPP aktives Wasser *nt*

actual: ~ **air requirement** *n* AIR tatsächlicher Luftbedarf *m*; ~ **air volume** *n* AIR tatsächliche Luftmenge *f*

acute: ~ **effect** *n* AIR, WASTE akute Wirkung *f*; ~ **poisoning** *n* SAFETY akute Vergiftung *f*

acyclic *adj* AIR azyklisch, acyclisch

adaptive *adj* WATER POLL adaptiv; ~ **enzyme** *n* WATER POLL adaptives Enzym *nt*

add *vt* GEN zusetzen; ~ **on** *vt* AIR *dust particulate to liquid droplet* anlagern

addition *n* GEN Zusatz *m*

additional *adj* GEN zusätzlich, Neben-; ~ **display** *n* ANALYSIS *measuring instruments* Nebenanzeige *f*; ~ **firing** *n* RENEW ENERG, WASTE Zusatzfeuerung *f*

additive *n* GEN Zusatzstoff *m*, Additiv *nt*, Zuschlagstoff *m*

additives: ~ **silo** *n* AIR Additivsilo *m*

additive: ~ **suspension** *n* AIR Additivsuspension *f*; ~ **system** *n* WASTE, WATER SUPP additives System *nt*
adenosine: ~ **triphosphate** *n* (*ATP*) WATER POLL Adenosintriphosphat *nt* (*ATP*)
adhere *vi* GEN anhaften; ~ **by suction** *vi* AIR sich ansaugen
adherence *n* GEN Anhaftung *f*, Haftfestigkeit *f*, Adhäsion *f*
adherent *adj* GEN anhaftend
adhesion *n* GEN Haftung *f*, Haftungsfestigkeit *f*
adhesive 1. *adj* GEN anhaftend, klebend; **2.** *n* AIR, WASTE Kleber *m*
adhesive: ~ **agent** *n* AIR Haftmittel *nt*; ~ **force** *n* AIR Haftfestigkeit *f*, Haftvermögen *nt*, Haftkraft *f*, ANALYSIS *chromatography* Haftvermögen *nt*, WATER SUPP Adhäsionskraft *f*
adhesiveness *n* GEN Haftfähigkeit *f*
ADI *abbr* (*acceptable daily intake*) NOISE, SAFETY ADI-Wert *m* (*tolerierbare tägliche Aufnahme*)
adiabatic *adj* AIR adiabatisch
adjacent: ~ **bag** *n* AIR Nachbarschlauch *m*
adjust *vt* AIR, ANALYSIS einsetzen, einstellen
adjustable *adj* AIR, ANALYSIS einsetzbar, einstellbar; ~ **submerged weir** *n* WATER SUPP regulierbares Grundwehr *nt*
adjustment *n* SOIL, WASTE, WATER SUPP Ausgleich *m*
administration *n* MANAGE, POLICY Verwaltung *f*
administrative *adj* MANAGE, POLICY verwaltend; ~ **discretion** *n* POLICY Ermessen *nt*; ~ **fee** *n* POLICY Verwaltungsgebühr *f*; ~ **law** *n* POLICY Ordnungsrecht *nt*; ~ **regulation** *n* POLICY Verwaltungsvorschrift *f*, Verwaltungsverordnung *f*; ~ **unit** *n* POLICY Verwaltungseinheit *f*
adoption *n* POLICY *of law* Verabschiedung *f*
adsorb *vt* GEN adsorbieren
adsorbability *n* GEN Adsorbierbarkeit *f*, Adsorptionseffizienz *f*
adsorbable *adj* GEN adsorbierbar; ~ **organic halogen compounds** *n pl* (*AOX*) ANALYSIS, WASTE, WATER POLL, WATER SUPP adsorbierbare organische Halogenverbindungen *f pl* (*AOX*)
adsorbate *n* GEN Adsorbat *nt*
adsorbed *adj* GEN adsorbiert; ~ **component** *n* AIR adsorbierte Komponente *f*; ~ **gas** *n* AIR adsorbiertes Gas *nt*; ~ **substance** *n* AIR adsorbierter Stoff *m*
adsorbent 1. *adj* GEN adsorbierend; **2.** *n* GEN Adsorbens *nt*, Adsorptionsmittel *nt*
adsorbent: ~ **bed** *n* AIR Adsorbensbett *nt*; ~ **density** *n* AIR Adsorbensdichte *f*; ~ **dosage** *n* AIR Adsorbensdosierung *f*; ~ **inlet** *n* AIR Adsorbenseintritt *m*; ~ **outlet** *n* AIR Adsorbensaustritt *m*, ~ **particle** *n* AIR Adsorbenspartikel *f*; ~ **surface** *n* AIR Adsorbensoberfläche *f*
adsorber *n* GEN Adsorber *m*; ~ **filter** *n* AIR Adsorberfilter *nt*; ~ **resin** *n* WATER POLL Adsorberharz *nt*
adsorbing *adj* GEN adsorbierend, adsorptiv; ~ **agent** *n* ANALYSIS Adsorbens *nt*
adsorption *n* GEN Adsorption *f*; ~ **capacity** *n* AIR, WASTE, WATER SUPP Adsorptionsleistung *f*, Adsorptionskapazität *f*, Adsorptionsvermögen *nt*; ~ **chromatography** *n* ANALYSIS Adsorptionschromatographie *f*; ~ **column** *n* ANALYSIS Adsorptionssäule *f*; ~ **duration** *n* AIR Adsorptionsdauer *m*; ~ **efficiency** *n* AIR, SOIL, WASTE, WATER SUPP Adsorptionseffizienz *f*, Adsorptionswirkungsgrad *m*; ~ **filter** *n* WASTE, WATER SUPP Adsorptionsfilter *nt*; ~ **isotherm** *n* AIR, WASTE, WATER POLL, WATER SUPP Adsorptionsisotherme *f*; ~ **potential** *n* AIR Adsorptionspotential *nt*; ~ **power** *n* AIR Adsorptionsfähigkeit *f*, Adsorptionsvermögen *nt*; ~ **stage** *n* AIR Adsorptionsstufe *f*; ~ **temperature** *n* AIR Adsorptionstemperatur *f*; ~ **test** *n* ANALYSIS Adsorptionstest *m*; ~ **tube** *n* ANALYSIS Adsorptionsrohr *nt*; ~ **velocity** *n* AIR Adsorptionsgeschwindigkeit *f*; ~ **zone** *n* AIR Adsorptionszone *f*
adsorptive *adj* GEN adsorptiv, adsorbierend; ~ **agent** *n* AIR, SOIL, WASTE, WATER POLL, WATER SUPP Adsorptiv *nt*
ADSOX: ~ **process** *n* AIR *adsorption* ADSOX-Verfahren *nt*
advanced: ~ **coal cleaning** *n* AIR weiterentwickelte Kohlereinigung *f*; ~ **wastewater treatment** *n* WATER POLL dritte Reinigungsstufe *f*, fortgeschrittene Abwasserbehandlung *f*, tertiäre Abwasserbehandlung *f*
AE *abbr* (*acoustic emission*) ANALYSIS, NOISE SE (*Schallemission*); ~ **analysis** *n* (*acoustic emissions analysis*) ANALYSIS SEA (*Schallemissionsanalyse*)
AED *abbr* (*atomic emission detection*) ANALYSIS AED (*Atomemissionsdetektion*)
aerate *vt* SOIL, WASTE, WATER POLL, WATER SUPP belüften
aerated *adj* SOIL, WASTE, WATER POLL, WATER SUPP belüftet; ~ **bed composting** *n* SOIL, WASTE, WATER POLL belüftete Mietenkompostierung *f*; ~ **compost pile** *n* WATER POLL belüftete Kompostmiete *f*; ~ **grit chamber** *n* WATER POLL belüfteter Sandfang *m*; ~ **lagoon** *n* WATER POLL Oxidationsteich *m*, belüftetes Becken *nt*; ~ **oxidation pond** *n* WATER POLL belüfteter Abwasserteich *m*; ~ **pile composting** *n* (*APC*) SOIL, WASTE, WATER POLL belüftete Mietenkompostierung *f*; ~ **tunneled compost heap** *AmE*, ~ **tunnelled compost heap** *BrE* *n* WASTE belüfteter vertunnelter Komposthaufen *m*
aeration *n* SOIL, WASTE, WATER POLL, WATER SUPP *groundwater treatment* Belüftung *f*; ~ **basin** *n* WATER POLL Belüftungsbecken *nt*, Belüftungsbehälter *m*; ~ **paddle** *n* WATER POLL Belüftungspaddel *nt*, Paddelbelüfter *m*; ~ **system** *n* WATER POLL Belüftungssystem *nt*; ~ **tank** *n* WATER POLL Belüftungsbecken *nt*, Belüftungsbehälter *m*; ~ **test** *n* WATER POLL Belüftungsversuch *m*; ~ **time** *n* WATER POLL Belüftungsdauer *f*, Belüftungszeit *f*
aerator *n* WASTE Belüfter *m*, WATER POLL Belüfter *m*, Belüftungsaggregat *nt*, Belüftungseinrichtung *f*, WATER SUPP Belüfter *m*; ~ **pipe** *n* WATER POLL Belüftungsrohr *nt*
aerial: ~ **conveyor** *n* WASTE, WATER SUPP Luftförderer *m*; ~ **reconnaissance** *n* MARINE Luftaufklärung *f*; ~ **surveillance** *n* MARINE Luftüberwachung *f*; ~ **view evaluation** *n* ANALYSIS Luftbildauswertung *f*
aerobic *adj* RENEW ENERG, WASTE, WATER POLL, WATER SUPP aerob
aerobically: ~ **digested** *adj* WATER POLL aerob stabilisiert; ~**-digested sludge** *n* WATER POLL aerob stabilisierter Schlamm *m*; ~ **stabilized** *adj* WATER POLL aerob stabilisiert; ~**-stabilized sludge** *n* WATER POLL aerob stabilisierter Schlamm *m*
aerobic: ~ **bacteria** *n pl* WASTE, WATER POLL, WATER SUPP aerobe Bakterien *f pl*; ~ **bioreactor** *n* WATER POLL aerobischer Bioreaktor *m*; ~ **conditions** *n pl* WATER POLL aerobe Bedingungen *f pl*; ~ **decomposition** *n* WASTE, WATER POLL, WATER SUPP aerober Abbau *m*; aerobe Zersetzung *f*; ~ **decomposition process** *n* WASTE, WATER POLL, WATER SUPP aerober Abbauprozeß *m*; ~ **degradation** *n* WASTE, WATER SUPP aerober Abbau *m*; ~ **digestion** *n* WATER POLL *biological sludge stabilization* aerobe Ausfaulung *f*, aerobe Faulung *f*, WATER SUPP aerobe Faulung *f*; ~ **fermentation** *n* RENEW ENERG *biomass conversion* aerobe Vergärung *f*, WASTE aerobe Gärung *f*, WATER POLL aerobe Ausfaulung *f*, aerobe Faulung *f*, aerobe Gärung *f*, WATER SUPP aerobe Faulung *f*, aerobe Gärung *f*; ~ **heterotroph** *n* WATER POLL aerobisches Heterotroph *nt*; ~ **lagoon** *n* WATER POLL aerober Teich *m*, aerobes Becken *nt*; ~ **metabolism** *n* WATER POLL aerober Stoffwechsel *m*; ~ **microbe** *n* WATER POLL aerobe Mikrobe *f*; ~ **microorganism** *n* WATER POLL aerobe Mikrobe *f*, aerober Mikroorganismus *m*; ~ **oxidation** *n* WASTE, WATER SUPP aerobe Oxidation *f*; ~ **pond** *n* WATER POLL aerobes Becken *nt*; ~ **stabilization** *n* WATER POLL aerobe Stabilisierung *f*; ~**-thermophilic** *adj* WATER POLL aerob-thermophil; ~**-thermophilic stabilization** *n* WATER POLL aerob-thermophile Stabilisierung *f* (*ATS*); ~**-thermophilic stabilization with digestion** *n* WATER POLL aerob-thermophile

Stabilisierung *f* mit anschließender Faulung; ~ **zone** *n* WATER POLL *of biofilm* aerobe Zone *f*
aerodynamic *adj* AIR, ANALYSIS, RENEW ENERG aerodynamisch; ~ **diameter** *n* AIR, ANALYSIS aerodynamischer Durchmesser *m*; ~ **lift** *n* RENEW ENERG *wind turbine* aerodynamischer Auftrieb *m*
aerometer *n* AIR, ANALYSIS Aerometer *nt*
aerosol *n* AIR, WASTE *substance* Aerosol *nt*, *can* Sprühbehälter *m*, Spraydose *f*, Sprühdose *f*; ~ **container** *n* WASTE Spraydose *f*, Sprühdose *f*, Sprühbehälter *m*; ~ **dispenser** *n* AIR, WASTE Aerosolpackung *f*; ~ **and dust measuring and analysis apparatus** *n* ANALYSIS, SAFETY Meßgerät *nt* zur Analyse von Staub und Aerosolen; ~ **precipitation** *n* AIR Aerosolabscheidung *f*
AES *abbr* (*atomic emission spectrometry*) ANALYSIS AES (*Atomemissionsspektrometrie*)
AFA: ~ **process** *n* (*absorption/filtration/adsorption process*) AIR AFA-Verfahren *nt* (*Absorption/Filtration/Adsorption-Verfahren*)
affect *vt* SOIL beeinträchtigen
affinity *n* WATER SUPP Affinität *f*
AFS *abbr* (*atomic fluorescence spectrometry*) ANALYSIS AFS (*Atomfluoreszenzspektrometrie*)
afterburner *n* AIR *thermal incinerator* Nachbrenner *m*, Nachverbrenner *m*; ~ **chamber** *n* AIR Nachbrennkammer *f*; ~ **housing** *n* AIR Nachverbrennergehäuse *nt*
afterburning *n* AIR *in thermal processes* Nachverbrennung *f*
aftercooler *n* AIR Nachkühler *m*
after: ~-**use** *n* WASTE, WATER SUPP Folgenutzung *f*
aged: ~ **catalyst** *n* AIR gealterter Katalysator *m*
ageing *n* AIR, WASTE, WATER SUPP Alterung *f*; ~ **characteristics** *n pl* AIR, WASTE, WATER SUPP Alterungseigenschaften *f pl*; ~ **resistance** *n* AIR, WASTE, WATER SUPP Alterungsbeständigkeit *f*; ~ **stability** *n* AIR, WASTE, WATER SUPP Alterungsbeständigkeit *f*; ~ **test** *n* AIR, WASTE, WATER SUPP Alterungstest *m*
age: ~-**resistant** *adj* AIR, WASTE, WATER SUPP alterungsbeständig
agglomerate **1.** *n* AIR, ANALYSIS, SOIL, WATER POLL Agglomerat *nt*; **2.** *vt* AIR, ANALYSIS, SOIL, WASTE, WATER POLL, WATER SUPP agglomerieren, zusammenballen; **3.** *vi* AIR, ANALYSIS, SOIL, WATER POLL sich zusammenballen
agglomerating: ~ **dust** *n* AIR *bag filter* agglomerierender Staub *m*
agglomeration *n* AIR *process* Anhäufung *f*, Agglomeration *f*, Agglomerieren *nt*, SOIL *piling up* Anhäufung; Ansammlung *f*, WASTE, WATER POLL, WATER SUPP Agglomeration *f*; ~ **capability** *n* AIR Agglomerationsfähigkeit *f*
aggradation *n* WASTE Aggradation *f*, WATER SUPP Aggradation *f*, Sedimentzuwachs *m*
aggradational: ~ **deposit** *n* WATER SUPP Flußablagerung *f*
aggregate *n* AIR, SOIL, WASTE, WATER POLL Zuschlagstoff *m*
aggregation *n* AIR, SOIL, WASTE, WATER POLL Aggregation *f*
aggressive *adj* AIR, WASTE, WATER SUPP aggressiv; ~ **gas component** *n* AIR aggressive Gaskomponente *f*, aggressiver Gasbestandteil *m*; ~ **groundwater** *n* WATER SUPP aggressives Grundwasser *nt*; ~ **pit water** *n* WATER SUPP aggressives Grubenwasser *nt*; ~ **water** *n* WATER SUPP aggressives Wasser *nt*
aggressivity: ~ **of lime** *n* WATER SUPP Kalkaggressivität *f*
agitating: ~ **arm** *n* AIR Rührarm *m*
agitator *n* SOIL Rührwerk *nt*
AGP *abbr* (*algae growth potential*) WATER POLL Algenwachstumspotential *nt*
agricultural: ~ **reuse** *n* WATER POLL *of wastewater* Verwertung *f* in der Landwirtschaft , landwirtschaftliche Verwertung *f*; ~ **waste** *n* WASTE Abfall *m* aus der Landwirtschaft, landwirtschaftlicher Abfall *m*
A-horizon *n* SOIL A-Horizont *m*, Krume *f*, Oberboden *m*

air *n* GEN Luft *f*; ~ **absorption** *n* NOISE *airborne noise damping* Luftabsorption *f*
airborne *adj* AIR schwebend; ~ **dust** *n* SAFETY Staub *m* in der Luft; ~ **dust concentration** *n* SAFETY Staubkonzentration *f* in der Luft; ~ **noise** *n* NOISE, SAFETY Luftschall *m*; ~ **noise-proofing** *n* NOISE Luftschalldämmung *f*; ~ **noise-reduction** *n* NOISE *air absorption* Luftschalldämpfung *f*; ~ **noise-transmission** *n* NOISE Luftschallübertragung *f*; ~ **particulate matter** *n* AIR Schwebstaub *m*; ~ **pollutant observation model** *n* MANAGE Luftschadstoff-Prognosemodell *nt*; ~ **remote sensing** *n* MARINE Luftbildfernerkundung *f*
air: ~ **bubble boom** *n* MARINE Luftsperre *f*; ~ **charge** *n* AIR Beladeluft *f*; ~ **charge temperature** *n* AIR Beladelufttemperatur *f*; ~ **classification** *n* AIR, WASTE Windsichtung *f*; ~ **classification plant** *n* AIR, WASTE Windsichter *m*; ~ **classifier** *n* AIR, WASTE Windsichter *m*; ~ **compressor** *n* AIR Luftverdichter *m*; ~ **conditioning installation** *n* AIR Klimaanlage *f*; ~ **conduction cylinder** *n* AIR Luftführungszylinder *m*; ~-**cooled** *adj* WASTE, WATER SUPP luftgekühlt
aircraft: ~ **noise tax** *n* NOISE, POLICY Fluglärmsteuer *f*; ~ **waste gas** *n* AIR Flugzeugabgas *nt*
air: ~ **cushion** *n* NOISE Luftpolster *nt*; ~ **dehumidifier** *n* AIR, SAFETY Luftentfeuchter *m*; ~ **diffuser** *n* WATER POLL Luftdiffusor *m*, Luftverteiler *m*; ~ **distribution** *n* AIR *filters* Luftverteilung *f*; ~ **dosage** *n* AIR Luftdosierung *f*; ~ **drying** *n* WATER POLL Lufttrocknung *f*; ~ **emissions** *n pl* AIR Luftemissionen *f pl*; ~ **excess** *n* WASTE, WATER SUPP Luftabzug *m*; ~ **exchange** *n* AIR Luftaustausch *m*; ~ **filter** *n* AIR, SAFETY Luftfilter *nt*; ~ **flotation** *n* AIR Luftblasenflotation *f*; ~ **flow** *n* AIR Luftströmung *f*; ~ **flow sensor** *n* AIR Luftmengenmesser *m*; ~ **friction damping** *n* ANALYSIS *in measuring instruments* Luftdämpfung *f*; ~ **humidifier** *n* AIR, SAFETY Luftbefeuchter *m*; ~ **humidity** *n* AIR Luftfeuchtigkeit *f*; ~ **intake** *n* AIR Lufteinlaß *m*; ~ **introduction** *n* WASTE, WATER SUPP Luftzufuhr *f*; ~ **jet** *n* AIR Luftdüse *f*, *bag filter* Luftstrahl *m*, WATER POLL Luftstrahl *m*; ~-**jet screening method** *n* ANALYSIS *dust analysis* Luftstrahlsieb-Verfahren *nt*; ~ **layer** *n* AIR Luftschicht *f*
airlock *n* SAFETY Luftschleuse *f*
air: ~ **mass** *n* RENEW ENERG *wind power* Luftmasse *f*; ~ **mold process** *AmE*, ~ **mould process** *BrE* *n* WASTE, WATER SUPP Airmould-Verfahren *nt*; ~ **permeability** *n* SOIL Luftdurchlässigkeit *f*; ~ **poisoning** *n* AIR, SAFETY Luftvergiftung *f*, Luftverseuchung *f*; ~ **pollutant** *n* AIR, SAFTEY Luftschadstoff *m*; ~ **pollution** *n* AIR, SAFETY Luftverschmutzung *f*, Luftverunreinigung *f*; ~ **pollution control** *n* AIR, SAFETY Luftverschmutzungskontrolle *f*; ~ **pollution emissions** *n pl* AIR, SAFETY Luftverunreinigungsemissionen *f pl*; ~ **pollution forecast** *n* AIR, SAFETY Luftverschmutzungsvorhersage *f*; ~ **pollution measurement** *n* AIR, ANALYSIS, SAFETY Messung *f* der Luftverschmutzung; ~ **preheater** *n* AIR, WASTE, WATER POLL, WATER SUPP Luftvorwärmer *m* (*LUVO*); ~ **preheating** *n* AIR, WASTE, WATER POLL, WATER SUPP Luftvorwärmung *f*; ~ **purification and deodorization equipment** *n* AIR, SAFETY Luftreinigungs- und Deodorisierungsgerät *nt*; ~ **purity** *n* AIR Luftreinheit *f*; ~ **quality** *n* AIR, ANALYSIS, SAFETY Luftbeschaffenheit *f*, Luftqualität *f*; ~ **quality criteria** *n pl* AIR, ANALYSIS, SAFTEY Luftqualitätskriterien *nt pl*; ~ **quality data** *n pl* AIR, ANALYSIS Luftqualitätsdaten *pl*; ~ **quality measurement** *n* AIR, ANALYSIS Messung *f* der Luftqualität; ~ **quality protection official** *n* AIR, MANAGE, POLICY Betriebsbeauftragter *m* für Immissionsschutz; ~ **quality standards** *n pl* AIR, SAFETY Luftbeschaffenheitsnormen *f pl*; ~ **relief valve** *n* AIR, WASTE Entlüftungsventil *nt*; ~ **sampler** *n* ANALYSIS Luftprobennahmegerät *nt*; ~ **separator** *n* AIR, WASTE Windsichter *m*; ~ **shed** *n* AIR Luftausfällung *f*; ~ **sparging** *n* SOIL Luftverteilung *f*; ~ **speed** *n* AIR Luftgeschwindigkeit

f; ~ **spring** *n* NOISE *vibration isolator* Luftfeder *f*; ~ **stream** *n* AIR, RENEW ENERG, WASTE, WATER SUPP Luftstrom *m*, Luftströmung *f*; ~ **stripping** *n* WASTE, WATER POLL, WATER SUPP Luftstrippen *nt*; ~ **supply** *n* AIR *thermal combustion* Luftzufuhr *f*; ~ **temperature** *n* AIR, WATER POLL Lufttemperatur *f*; ~**-tight** *adj* AIR luftdicht; ~**-to-cloth ratio** *n* AIR *fabric filtration* Filterflächenbelegung *f*; ~ **turbidity** *n* AIR Atmosphärentrübung *f*, Lufttrübung *f*, Trübung *f*; ~ **unfit for respiration** *n* AIR, SAFETY zu Atemzwecken ungeeignete Luft *f*; ~ **velocity** *n* AIR Luftgeschwindigkeit *f*

Al *abbr* (*aluminium BrE*, *aluminum AmE*) SOIL, WASTE, WATER POLL Al (*Aluminium*)

albedo *n* AIR Albedo *f*; ~ **change** *n* RENEW ENERG *effect of renewable energy technology* Albedo-Veränderung *f*

albumen *n* GEN *highly efficient enzyme* Eiweiß *nt*

alcohol *n* AIR, RENEW ENERG *end product of fermentation* Alkohol *m*

alcoholysis *n* WASTE, WATER SUPP Alkoholyse *f*

algae *n pl* WASTE, WATER POLL, WATER SUPP Algen *f pl*; ~ **abatement** *n* WASTE, WATER POLL, WATER SUPP Algenbekämpfung *f*; ~ **growth** *n* WATER POLL Algenwachstum *nt*; ~ **growth potential** *n* (*AGP*) WATER POLL Algenwachstumspotential *nt*; ~ **pond** *n* WASTE, WATER SUPP Algenteich *m*; ~ **test** *n* ANALYSIS Algentest *m*

algal: ~ **biomass** *n* WATER POLL Algenbiomasse *f*; ~ **photosynthesis** *n* WATER POLL Algenphotosynthese *f*

alicyclic *adj* AIR alizyklisch, alicyclisch; ~ **compounds** *n pl* AIR alicyclische Verbindungen *f pl*

aliphatic *adj* AIR, WASTE, WATER POLL aliphatisch; ~ **compounds** *n pl* AIR aliphatische Verbindungen *f pl*; ~ **hydrocarbons** *n pl* AIR, WASTE, WATER POLL aliphatische Kohlenwasserstoffe *m pl*; ~ **polyamid** *n* AIR *filter fibre material* aliphatisches Polyamid *nt*

alkali *n* GEN Alkali *nt*

alkaline *adj* GEN alkalisch; ~ **earths** *n pl* WASTE, WATER SUPP Erdalkalien *f pl*; ~ **fermentation** *n* WATER POLL alkalische Gärung *f*; ~**-oxidative** *adj* AIR alkalisch-oxidativ; ~ **wastewater** *n* WATER POLL alkalisches Abwasser *nt*

alkalinity *n* GEN Alkalinität *f*, Alkalität *f*; ~ **measurement** *n* ANALYSIS Alkalinitätsmessung *f*

alkane *n* AIR Alkan *nt*

alkene *n* AIR Alken *nt*

alkylbenzenesulfonate *AmE see alkylbenzenesulphonate*

alkylbenzenesulphonate *n* (*ABS*) *BrE* WASTE Alkylbenzolsulfonat *nt* (*ABS*)

allochthonous: ~ **matter** *n* SOIL bodenfremde Substanz *f*

alluvial: ~ **bed** *n* WASTE, WATER SUPP Alluvialschicht *f*; ~ **cone** *n* WATER SUPP Schwemmkegel *m*; ~ **deposit** *n* WASTE, WATER SUPP Alluvium *nt*, Anlandung *f*; ~ **plain** *n* WATER SUPP Schwemmlandebene *f*

alpha: ~**-mesosaprobic** *adj* WATER POLL alpha-mesosaprob

alteration *n* ANALYSIS Veränderung *f*

alternate *vi* AIR sich abwechseln

alternating *adj* AIR, ANALYSIS, RENEW ENERG wechselnd; ~ **current** *n* RENEW ENERG Wechselstrom *m*; ~ **current capacity** *n* RENEW ENERG Wechselstromleistung *f*

alum *n* WASTE, WATER SUPP Alaun *m*

aluminate *n* SOIL, WASTE, WATER POLL Aluminat *nt*

aluminium *BrE*, **aluminum** *AmE* *n* (*Al*) SOIL, WASTE, WATER POLL Aluminium *nt* (*Al*); ~ **chloride** *n BrE* WASTE, WATER SUPP Aluminiumchlorid *nt*; ~**-containing** *adj BrE* WASTE aluminiumhaltig; ~**-containing material** *n BrE* WASTE aluminiumhaltiges Material *nt*; ~ **foil** *n BrE* WASTE Alufolie *f*, Aluminiumfolie *f*; ~ **oxide** *n BrE* WASTE, WATER POLL Aluminiumoxid *nt*; ~ **packaging** *n BrE* WASTE Aluminiumverpackung *f*, Aluverpackung *f*; ~ **recycling** *n BrE* WASTE Aluminiumrecycling *nt*; ~ **salts** *n pl BrE* WATER POLL Aluminiumsalze *nt pl*; ~ **scrap** *n BrE* WASTE Aluminiumschrott *m*; ~ **sulphate** *n BrE* WASTE, WATER POLL, WATER SUPP Aluminiumsulfat *nt*

ambient: ~ **air** *n* AIR, ANALYSIS Immission *f*, Umgebungsluft *f*; ~ **air analysis** *n* AIR, ANALYSIS Immissionsanalyse *f*; ~ **air concentration** *n* AIR, ANALYSIS Immissionskonzentration *f*; ~ **air concentration value** *n* AIR, ANALYSIS Immissionsgrenzwert *m*, Luftqualitätswert *m*; ~ **air measurement** *n* AIR, ANALYSIS Immissionsmessung *f*; ~**-air neutral** *adj* AIR, ANALYSIS immissionsneutral; ~ **air quality analyser** *BrE*, ~ **air quality analyzer** *AmE* *n* AIR, ANALYSIS Immissionsmeßgerät *nt*; ~ **air quality standard** *n* AIR, ANALYSIS Immissionsgrenzwert *m* der Luft; ~ **air quality value** *n* AIR, ANALYSIS Immissionswert *m*; ~ **noise** *n* NOISE Umgebungsgeräusch *nt*, Umgebungslärm *m*

amend *vt* POLICY novellieren

amendment *n* POLICY Novelle *f*

ammeter *n* ANALYSIS Amperemeter *nt*

ammonia *n* AIR, SOIL, WASTE, WATER POLL Ammoniak *nt*; ~ **process** *n* AIR *Walther* Ammoniak-Verfahren *nt*; ~ **stripping** *n* WATER POLL Ammoniak-Strippen *nt*; ~ **water** *n* WATER POLL Gaswasser *nt*

ammonium *n* AIR, SOIL, WASTE, WATER POLL Ammonium *nt*: ~**-containing** *adj* WATER POLL ammoniumhaltig; ~**-containing wastewater** *n* WATER POLL ammoniumhaltiges Abwasser *nt*; ~ **load** *n* WATER POLL Ammoniumfracht *f*

amorphous *adj* AIR amorph, gestaltlos, nichtkristallin, RENEW ENERG amorph; ~ **adsorbent** *n* AIR, WATER POLL amorphes Adsorbens *nt*, amorphes Adsorptionsmittel *nt*; ~ **silicon cell** *n* RENEW ENERG *photovoltaic* amorphe Siliziumzelle *f*

amount *n* GEN Menge *f*; ~ **of energy** *n* RENEW ENERG Energiemenge *f*; ~ **of gas** *n* AIR Gasmenge *f*; ~ **of irradiated energy** *n* RENEW ENERG eingestrahlte Energiemenge *f*; ~ **of precipitation** *n* WATER SUPP Niederschlagsmenge *f*

amperometer *n* ANALYSIS Amperemeter *nt*

amperometric *adj* ANALYSIS amperometrisch; ~ **determination methods** *n pl* ANALYSIS amperometrische Bestimmungsmethoden *f pl*; ~ **measuring head** *n* ANALYSIS amperometrische Meßsonde *f*; ~ **titration** *n* ANALYSIS amperometrische Titration *f*

amperometry *n* ANALYSIS Amperometrie *f*

amplifier *n* ANALYSIS Verstärker *m*

anabolism *n* WATER POLL Anabolismus *m*

anaerobic *adj* RENEW ENERG, WASTE, WATER POLL, WATER SUPP anaerob

anaerobically: ~ **stabilized** *adj* WATER POLL anaerob stabilisiert; ~**-stabilized sludge** *n* WATER POLL anaerob stabilisierter Schlamm *m*

anaerobic: ~ **bacteria** *n pl* WATER POLL anaerobe Bakterien *f pl*; ~ **bioreactor** *n* WATER POLL anaerobischer Bioreaktor *m*; ~ **conditions** *n pl* WATER POLL anaerobe Bedingungen *f pl*; ~ **degradation** *n* WASTE, WATER POLL, WATER SUPP anaerober Abbau *m*; ~ **digestion** *n* WASTE, WATER POLL, WATER SUPP anaerobe Ausfaulung *f*, anaerobe Faulung *f*, anaerobe Gärung *f*; ~ **fermentation** *n* RENEW ENERG *biomass conversion* anaerobe Vergärung *f*, WASTE, WATER POLL, WATER SUPP anaerobe Faulung *f*, anaerobe Gärung *f*; ~ **heterotroph** *n* WATER POLL anaerobisches Heterotroph *nt*; ~ **lagoon** *n* WATER POLL anaerober Teich *m*, anaerobes Becken *nt*; ~ **microbe** *n* WATER POLL anaerobe Mikrobe *f*; ~ **microorganism** *n* WATER POLL anaerobe Mikrobe *f*, anaerober Mikroorganismus *m*; ~ **pond** *n* WATER POLL anaerober Teich *m*, anaerobes Becken *nt*; ~ **respiration process** *n* WASTE, WATER SUPP anaerober Atmungsprozeß *m*; ~ **zone** *n* WATER POLL *of biofilm* anaerobe Zone *f*

analog: ~ **channel** *n* ANALYSIS Analogkanal *m*; ~**-controlled** *adj* ANALYSIS analoggesteuert

analysable *adj BrE* ANALYSIS analysierbar

analyse *vt BrE* ANALYSIS analysieren, untersuchen

analysed *adj BrE* ANALYSIS untersucht; ~ **sample** *n* ANALYSIS untersuchte Probe *f*; ~ **water sample** *n* ANALYSIS untersuchte Wasserprobe *f*

analyser *n BrE* ANALYSIS Analysator *f*, Analysengerät *nt*
analysis *n* ANALYSIS, SAFETY Analyse *f*, Bestimmung *f*, Untersuchung *f*; ~ **aim** *n* ANALYSIS Untersuchungszweck *m*; ~ **boiling** *n* ANALYSIS Siedeanalyse *f*; ~ **calibration function** *n* ANALYSIS Analysenkalibrierfunktion *f*; ~ **data** *n pl* ANALYSIS Analysedaten *pl*; ~ **of exhaust gas** *n* ANALYSIS Abgasanalyse *f*; ~ **extent** *n* ANALYSIS Analysenumfang *m*; ~ **frequency** *n* ANALYSIS Untersuchungshäufigkeit *f*; ~ **of iron in wastewater** *n* ANALYSIS Eisenbestimmung *f* von Abwasser; ~ **location** *n* ANALYSIS Analysenort *m*; ~ **measuring techniques** *n pl* ANALYSIS Analysenmeßtechniken *f pl*; ~ **of residue** *n* ANALYSIS Rückstandanalyse *f*; ~ **result** *n* ANALYSIS Analyseergebnis *nt*, Bestimmungsergebnis *nt*, Untersuchungsbefund *m*, Untersuchungsergebnis *nt*; ~ **time** *n* ANALYSIS Analysezeit *f*; ~ **volume** *n* ANALYSIS Untersuchungsvolumen *nt*; ~ **of weakest points** *n* ANALYSIS Schwachstellenanalyse *f*
analyte *n* ANALYSIS Analyt *m*; ~ **separation** *n* ANALYSIS Analyt-Abscheidung *f*
analytical *adj* ANALYSIS analytisch; ~ **error** *n* ANALYSIS Analysenfehler *m*; ~ **finding** *n* ANALYSIS Analysenergebnis *nt*, Bestimmungsergebnis *nt*, Untersuchungsergebnis *nt*
analytically: ~ **pure** *adj* ANALYSIS analysenrein
analytical: ~ **parameter** *n* ANALYSIS Analysenparameter *m*; ~ **procedure** *n* ANALYSIS Analysenvorschrift *f*, Analyseverfahren *nt*; ~ **process** *n* ANALYSIS Analyseverfahren *nt*; ~ **quality control** *n* ANALYSIS analytische Qualitätssicherung *f* (*AQS*); ~ **result** *n* ANALYSIS Analysenergebnis *nt*, Bestimmungsergebnis *nt*, Untersuchungsbefund *m*, Untersuchungsergebnis *nt*; ~ **scale** *n* ANALYSIS Analysenwaage *f*
analyzable *AmE see analysable BrE*
analyze *AmE see analyse BrE*
analyzed *AmE see analysed BrE*
analyzer *AmE see analyser BrE*
anatomical: ~ **waste** *n* WASTE, WATER POLL infektiöser Abfall *m*, pathogener Abfall *m*
anchor: ~ **group** *n* WATER POLL Ankergruppe *f*
Andreasen: ~ **pipette** *n* ANALYSIS *sedimentation of particulates in liquids* Andreasen-Pipette *f*
anechoic *adj* NOISE *room* reflexionsfrei, schalltot
anemometer *n* AIR, ANALYSIS *wind flow gauge* Anemometer *nt*; ~ **height** *n* AIR *stack* Anemometerhöhe *f*
angle: ~ **of sun's altitude** *n* RENEW ENERG Sonnenhöhewinkel *m*
anhydrite *n* AIR *gypsum form* Anhydrit *m*
anion *n* WASTE, WATER POLL, WATER SUPP Anion *nt*; ~ **exchange** *n* SOIL, WASTE, WATER SUPP Anionenaustausch *m*; ~**-exchange chromatography** *n* ANALYSIS Anionenaustausch-Chromatographie *f*; ~ **exchanger** *n* WASTE, WATER POLL, WATER SUPP Anionenaustauscher *m* (*An-IA*)
anionic: ~**-active** *adj* WATER POLL anionenaktiv
anion: ~**-permeable membrane** *n* WATER POLL, WATER SUPP anionendurchlässige Membran *f*; ~**-selective** *adj* WATER POLL, WATER SUPP anionenselektiv; ~**-selective membrane** *n* WATER POLL, WATER SUPP anionenselektive Membran *f*
anisokinetic *adj* ANALYSIS anisokinetisch; ~ **sampling** *n* ANALYSIS anisokinetische Probenahme *f*
annual *adj* GEN jährlich; ~ **drainage** *n* WASTE, WATER SUPP jährliche Entwässerung *f*; ~ **evaporation** *n* WASTE, WATER SUPP jährliche Verdunstung *f*; ~ **flood** *n* WATER SUPP Jahreshochwasser *nt*; ~ **flow** *n* WATER SUPP Jahresdurchfluß *m*; ~ **load** *n* WASTE, WATER SUPP jährliche Last *f*; ~ **mean value** *n* ANALYSIS Jahresmittelwert *m*; ~ **mean water level** *n* WATER SUPP jährlicher Mittelwasserstand *m*; ~ **precipitation** *n* WATER SUPP jährliche Niederschlagshöhe *f*; ~ **refuse volume** *n* WASTE Jahresmüllmenge *f*; ~ **runoff** *n* WATER SUPP Jahresabfluß *m*; ~ **water consumption** *n* WATER SUPP Jahreswasserverbrauch *m*; ~ **wind velocity** *n* RENEW ENERG Jahreswindgeschwindigkeit *f*
annular: ~ **gap** *n* AIR, SOIL Ringspalt *m*; ~**-gap scrubber** *n* AIR Ringspaltwäscher *m*; ~**-type scrubber** *n* AIR Ringspaltwäscher *m*
annulus *n* AIR Ringraum *m*, Ringspalt *m*
anode *n* AIR, SOIL Anode *f*
anorganic: ~ **nitrogen fertilizer** *n* SOIL *conditioner* anorganischer Stickstoffdünger *m*
anoxic *adj* WATER POLL anoxisch; ~ **conditions** *n pl* WATER POLL anoxische Bedingungen *f pl*; ~ **denitrification zone** *n* WATER POLL anoxische Denitrifikationszone *f*
anthropogenic: ~ **acidification** *n* SOIL anthropogen bedingte Übersäuerung *f*, menschlich bedingte Übersäuerung *f*
anthropogenically: ~ **caused** *adj* AIR, SOIL, WATER POLL anthropogen verursacht; ~**-caused soil contamination** *n* SOIL anthropogen verursachte Bodenkontamination *f*
anthropogenic: ~ **climate changes** *n pl* AIR anthropogene Klimaveränderungen *f pl*; ~ **ecosystem** *n* AIR, SOIL, WATER POLL anthropogenes Ökosystem *nt*; ~ **greenhouse effect** *n* AIR anthropogen bedingter Treibhauseffekt *m*; ~ **pollutant** *n* AIR, SOIL, WATER POLL anthropogener Schadstoff *m*; ~ **pollution** *n* AIR, SOIL, WATER POLL anthropogene Belastung *f*; ~ **sources** *n pl* AIR, WATER SUPP anthropogene Quellen *f pl*; ~ **water components** *n pl* WATER SUPP anthropogene Wasserinhaltsstoffe *m pl*; ~ **water pollution** *n* WATER POLL anthropogen bedingte Wasserverschmutzung *f*, anthropogen bedingte Wasserverunreinigung *f*
anthroposphere *n* SOIL Anthroposphäre *f*
anti-fermentative *adj* WATER POLL gärungshemmend
antifouling *adj* WASTE, WATER SUPP bewuchshindernd
antiknock: ~ **additive** *n* AIR, WASTE Antiklopfmittel *nt*, Gegenklopfmittel *nt*; ~ **agent** *n* AIR, WASTE Antiklopfmittel *nt*, Gegenklopfmittel *nt*; ~ **fuel** *n* AIR, WASTE Antiklopfbrennstoff *m*; ~ **value** *n* AIR, WASTE Klopffestigkeitswert *m*, Klopfwert *m*
antimonial *adj* WASTE, WATER SUPP antimonhaltig
antimony *n* (*Sb*) WASTE, WATER SUPP Antimon *nt* (*Sb*)
anti-noise *adj* NOISE, SAFETY geräuschunterdrückend, schalldämmend, schalldämpfend; ~ **encapsulation** *n* NOISE schalldämmende Kapselung *f*; ~ **hood** *n* NOISE schalldämmende Haube *f*; ~ **microphone** *n* NOISE störschallunterdrückendes Mikrophon *nt*; ~ **soft rubber lining** *n* NOISE, SAFETY schalldämmende Weichgummibespannung *f*
antiquated: ~ **plant** *n* AIR Altanlage *f*
antistatic: ~ **protective clothing against chemical aggression** *n* SAFETY antistatische Schutzkleidung *f* gegen aggressive Chemikalien
antivibration: ~ **mounting** *n* NOISE erschütterungsfreie Aufstellung *f*, schwingungsisolierte Aufstellung *f*
AOX *abbr* (*adsorbable organic halogen compounds*) ANALYSIS, WASTE, WATER POLL, WATER SUPP AOX (*adsorbierbare organische Halogenverbindungen*); ~ **analysis** *n* ANALYSIS AOX-Analyse *f*
APC *abbr* (*aerated pile composting*) SOIL, WASTE, WATER POLL belüftete Mietenkompostierung *f*
APCI *abbr* (*atmospheric pressure chemical ionization*) ANALYSIS chemische Atmosphärendruck-Ionisation *f*
apparatus: ~ **calibration** *n* ANALYSIS Geräteeichung *f*
apparent: ~ **density** *n* WASTE Schüttdichte *f*; ~ **water table** *n* WASTE Nebengrundwasserspiegel *m*
appeal *n* POLICY Widerspruch *m*; ~ **authorization** *n* POLICY Widerspruchsbefugnis *f*; ~ **communication** *n* POLICY Widerspruchsbescheid *m*; ~ **deadline** *n* POLICY Widerspruchsfrist *f*; ~ **procedure** *n* POLICY Widerspruchsverfahren *nt*
appearance: ~ **of waste** *n* WASTE Abfallaufkommen *nt*, Müllaufkommen *nt*
applicability *n* GEN *of processes* Anwendbarkeit *f*, Einsetzbarkeit *f*
application *n* GEN *of process, instrument* Anwendung *f*,

Einsatz *m*; ~ **possibilities** *n pl* ANALYSIS, MANAGE, SAFETY *of processes* Anwendungsmöglichkeiten *f pl*, Einsatzmöglichkeiten *f pl*; ~ **rate** *n* WASTE, WATER SUPP Beschickungshöhe *f*; ~**-technological** *adj* ANALYSIS, MANAGE, SAFETY anwendungstechnisch; ~ **technology** *n* AIR, ANALYSIS, MANAGE, SAFETY Anwendungstechnik *f*

applied: ~ **environmental research** *n* ANALYSIS angewandte Umweltforschung *f*; ~ **research** *n* ANALYSIS angewandte Forschung *f*

apportion *vt* ANALYSIS verteilen

approximate: ~ **reading** *n* ANALYSIS Richtwert *m*; ~ **value** *n* ANALYSIS Richtwert *m*

approximation *n* ANALYSIS Annäherung *f*; ~ **test** *n* ANALYSIS Annäherungstest *m*

apron: ~ **conveyor** *n* WASTE Plattenband *nt*, Plattenförderband *nt*

aquaculture *n* WASTE, WATER SUPP Aquakultur *f*

aquatic *adj* MARINE aquatisch; ~ **ecosystem** *n* MARINE aquatisches Ökosystem *nt*; ~ **life** *n* MARINE, WATER SUPP aquatische Lebensform *f*; ~ **plant filter** *n* WATER POLL Wasserpflanzenfilter *nt*

aqueduct *n* WATER SUPP Aquädukt *m*

aqueous *adj* WASTE, WATER SUPP wäßrig

aquiclude *n* WATER SUPP geringpermeabler Grundwasserleiter *m*, undurchlässige Schicht *f*

aquifer *n* WATER POLL, WATER SUPP Aquifer *m*, Grundwasserleiter *m*, Grundwasserstockwerk *nt*; ~ **contamination** *n* WATER POLL, WATER SUPP *from wastewater* Aquiferverunreinigung *f*

aquiferous *adj* WATER SUPP wasserführend

aquifuge *n* WATER SUPP Grundwassernichtleiter *m*, Grundwasserstauer *m*

aquitard *n* WATER SUPP geringpermeabler Grundwasserleiter *m*, undurchlässige Schicht *f*

Ar *abbr* (*argon*) AIR, WASTE Ar (*Argon*)

arable: ~ **land** *n* SOIL Kulturboden *m*

aramide: ~ **fiber reinforced thermoset** *AmE*, ~ **fibre reinforced thermoset** *BrE* *n* WASTE aramidfaserverstärktes Duromer *nt*

arbitrary: ~ **analysis** *n* ANALYSIS Schiedsanalyse *f*; ~ **test** *n* ANALYSIS Schiedsprobe *f*

Archimedean: ~ **screw** *n* WATER POLL, WATER SUPP archimedische Wasserschnecke *f*

Archimedes': ~ **principle** *n* WATER POLL, WATER SUPP archimedisches Prinzip *nt*

arc: ~**-over** *n* AIR *electrostatic precipitator* Überschlag *m*; ~**-over voltage** *n* AIR *electrostatic precipitator* Überschlagsspannung *f*

area: ~ **aeration** *n* WATER POLL Flächenbelüftung *f*; ~ **emission source** *n* AIR Emissionsquelle *f* in einem Gebiet; ~ **of influence** *n* WASTE Wirkungszone *f*, WATER SUPP Einflußgebiet *nt*; ~ **of pure air** *n* AIR Reinluftgebiet *nt*; ~**-related** *adj* NOISE, WATER SUPP flächenbezogen, gebietsbezogen; ~**-related noise** *n* NOISE flächenbezogener Lärm *m*, gebietsbezogener Lärm *m*; ~**-related water demand** *n* WATER SUPP flächenbezogener Wasserbedarf *m*; ~ **source** *n* AIR, SOIL, WATER POLL Flächenquelle *f*; ~ **source pollutant** *n* AIR, SOIL, WATER POLL Schadstoff *m* aus Flächenquelle; ~ **of water filtration** *n* WATER SUPP Infiltrationsfläche *f*, Infiltrationsgebiet *nt*

argon *n* (*Ar*) AIR, WASTE Argon *nt* (*Ar*)

armature *n* WASTE, WATER POLL, WATER SUPP Armatur *f*, Fitting *nt*

aromatic *adj* AIR, WASTE, WATER POLL, WATER SUPP aromatisch; ~ **compounds** *n pl* AIR, WASTE, WATER POLL, WATER SUPP aromatische Verbindungen *f pl*; ~ **hydrocarbons** *n pl* AIR, WASTE, WATER POLL, WATER SUPP aromatische Kohlenwasserstoffe *m pl*; ~ **polyamid** *n* AIR *filter fibre material* aromatisches Polyamid *nt*

array: ~ **of solar cell submodules** *n* RENEW ENERG Solarzellenkette *f*

arsenic *n* (*As*) SOIL, WASTE, WATER POLL Arsen *nt* (*As*)

artesian *adj* WATER SUPP artesisch; ~ **groundwater** *n* WATER SUPP artesisches Grundwasser *nt*; ~ **spring** *n* WATER SUPP artesische Quelle *f*; ~ **well** *n* WATER SUPP artesischer Brunnen *m*

article *n* POLICY *of act* Artikel *m*, Paragraph *m*

artificial *adj* GEN künstlich; ~ **acidification** *n* SOIL künstliche Übersäuerung *f*; ~ **aeration** *n* WATER POLL künstliche Belüftung *f*; ~ **ageing** *n* WASTE, WATER SUPP künstliche Alterung *f*; ~ **groundwater** *n* WATER SUPP künstliches Grundwasser *nt*; ~ **groundwater recharge** *n* WATER SUPP künstliche Grundwasseranreicherung *f*; ~ **lake** *n* WATER SUPP Stausee *m*

artificially: ~ **recharged** *adj* WASTE, WATER SUPP künstlich angereichert; ~ **recharged groundwater** *n* WATER SUPP künstlich angereichertes Grundwasser *nt*; ~ **replenished** *adj* WASTE, WATER SUPP künstlich angereichert; ~ **replenished groundwater** *n* WATER SUPP künstlich angereichertes Grundwasser *nt*

As *abbr* (*arsenic*) SOIL, WASTE, WATER POLL As (*Arsen*)

asbestos *n* AIR, SAFETY, WATER POLL Asbest *nt*; ~ **cement pipe** *n* WASTE, WATER POLL, WATER SUPP Asbestzementrohr *nt*; ~ **exposure** *n* SAFETY, WATER POLL Asbestbelastung *f*; ~**-exposure limit** *n* SAFETY, WATER POLL Asbestbelastungsgrenzwert *m*; ~**-free** *adj* SAFETY asbestfrei; ~**-free insulating plate** *n* SAFETY asbestfreie Isolierplatte *f*

ash *n* AIR, ANALYSIS, WASTE Asche *f*, Verbrennungsrückstand *m*; ~ **and combustion residue** *n* AIR, ANALYSIS, WASTE Asche- und Verbrennungsrückstand *m*; ~ **content** *n* AIR, WASTE Aschegehalt *m*; ~ **discharge** *n* WASTE Aschenaustrag *m*; ~**-free** *adj* AIR, WASTE aschefrei; ~ **removal** *n* WASTE *incineration plant*, WATER POLL Entaschung *f*; ~ **remover** *n* WASTE, WATER POLL Entascher *m*; ~ **vitrification** *n* WASTE Aschesinterung *f*

asphalt *n* SOIL, WASTE Asphalt *m*; ~ **recycling** *n* SOIL, WASTE Asphaltrecycling *nt*; ~ **sealing** *n* SOIL, WASTE Asphaltbetondichtung *f*

asphyxiating: ~ **gas** *n* AIR, WASTE Stickgas *nt*

aspiration: ~ **psychrometer** *n* ANALYSIS Aspirationspsychrometer *nt*

assay *n* ANALYSIS Versuch *m*

assemble *vt* WASTE, WATER POLL, WATER SUPP montieren

assembling: ~ **unit** *n* WASTE Konfektioniereinheit *f*

assembly *n* WASTE, WATER POLL, WATER SUPP Montage *f*

assess *vt* GEN beurteilen, bewerten, abschätzen; ~ **quantitively** *vt* ANALYSIS qualitativ bestimmen

assessed *adj* GEN abgeschätzt, beurteilt; ~ **sample** *n* ANALYSIS beurteilte Probe *f*

assessment *n* GEN Abschätzung *f*, Bewertung *f*; ~ **criteria** *n pl* ANALYSIS Bewertungskriterien *nt pl*; ~ **level** *n* ANALYSIS Beurteilungspegel *m*; ~ **of risks** *n* SAFETY Gefährdungsabschätzung *f*

assimilate *vt* ANALYSIS, WASTE *data, results* assimilieren

assimilation *n* WATER POLL *biological denitrification process* Assimilation *f*

association: ~ **colloid** *n* WASTE, WATER POLL, WATER SUPP Assoziationskolloid *nt*

assurance: ~ **of drinking water supply** *n* WATER SUPP Trinkwasserversorgung *f*

asymptote *n* AIR *stack* Asymptote *f*

atmosphere *n* AIR Atmosphäre *f*; ~ **which affects breathing** *n* AIR, SAFETY Atmung beeinflussende Umgebungsluft *f*

atmospheric *adj* AIR atmosphärisch; ~ **acidity** *n* AIR atmosphärische Säurekapazität *f*; ~ **barrier layer** *n* AIR atmosphärische Sperrschicht *f*; ~ **chemical process** *n* AIR chemischer Prozeß *m* in der Atmosphäre; ~ **chemistry** *n* AIR Luftchemie *f*; ~ **concentration** *n* AIR Konzentration *f* in der Atmosphäre; ~ **conditions** *n pl* AIR Wetterlage *f*;

~ **fallout** *n* AIR atmosphärischer Niederschlag *m*; ~ **fumes** *n pl* AIR trockener Dunst *m*; ~ **gases** *n pl* AIR atmosphärische Gase *nt pl*; ~ **heat trapping** *n* AIR atmosphärische Aufwärmung *f*; ~ **inversion layer** *n* AIR Inversionsschicht *f*; ~ **lifetime** *n* AIR Lebensdauer *f* der Atmosphäre; ~ **loading** *n* AIR atmosphärische Luftbelastung *f*; ~ **mass** *n* AIR Atmosphärengewicht *nt*, Atmosphärenmasse *f*; ~ **obscurity** *n* AIR atmosphärische Verdunklung *f*; ~ **oxygen** *n* AIR atmosphärischer Sauerstoff *m*, Luftsauerstoff *m*; ~ **ozone** *n* AIR atmosphärisches Ozon *nt*; ~ **ozone concentration** *n* AIR atmosphärische Ozonkonzentration *f*; ~ **phenomenon** *n* AIR atmosphärische Erscheinung *f*; ~ **pollutant** *n* AIR atmosphärischer Schadstoff *m*; ~ **pollution** *n* AIR Luftverschmutzung *f*; ~ **precipitation** *n* AIR, WATER SUPP atmosphärischer Niederschlag *m*; ~ **pressure** *n* AIR Atmosphärendruck *m*; ~ **pressure chemical ionization** *n* (*APCI*) ANALYSIS chemische Atmosphärendruck-Ionisation *f* (*APCI*); ~ **reflection** *n* AIR atmosphärische Rückstrahlung *f*; ~ **scrubbing** *n* AIR atmosphärischer Auswaschvorgang *m*; ~ **sulfur** *AmE*, ~ **sulphur** *BrE* *n* AIR atmosphärischer Schwefel *m*; ~ **temperature** *n* AIR Atmosphärentemperatur *f*; ~ **trace gases** *n pl* AIR atmosphärische Spurengase *nt pl*

atom *n* ANALYSIS Atom *nt*

atomic: ~ **absorption spectrometer** *n* (*AAS*) ANALYSIS Atomabsorptionsspektrometer *nt* (*AAS*); ~ **absorption spectrometry** *n* (*AAS*) ANALYSIS Atomabsorptionsspektrometrie *f* (*AAS*); ~ **emission detection** *n* (*AED*) ANALYSIS Atomemissionsdetektion *f* (*AED*); ~ **emission detector** *n* ANALYSIS Atomemissionsdetektor *m*; ~ **emission spectrometry** *n* (*AES*) ANALYSIS Atomemissionsspektrometrie *f* (*AES*); ~ **fluorescence** *n* ANALYSIS Atomfluoreszenz *f*; ~ **fluorescence spectrometry** *n* (*AFS*) ANALYSIS Atomfluoreszenzspektrometrie *f* (*AFS*)

atomize *vt* AIR zerstäuben, SOIL versprühen, WASTE, WATER SUPP verdüsen

atomizer *n* AIR Einspritzdüse *f*, Zerstäuber *m*, SOIL Versprüher *m*, WATER SUPP Verdüser *m*; ~ **nozzle** *n* AIR Zerstäuberluftdüse *f*

atomspectrometric *adj* ANALYSIS atomspektrometrisch

ATP *abbr* (*adenosine triphosphate*) WATER POLL ATP (*Adenosintriphosphat*)

attached: ~ **shell** *n* NOISE *surface lining* Vorsatzschale *f*

attemperation *n* AIR Temperaturregelung *f*

attenuate *vt* AIR, ANALYSIS schwächen, NOISE dämpfen, schwächen

attenuation *n* AIR, ANALYSIS Schwächung *f*, NOISE Dämpfung *f*, Schwächung *f*

attenuator *n* NOISE *silencer* Dämpfungsregler *m*, Lautstärkeregler *m*; ~ **pad** *n* NOISE Dämpfungsglied *nt*

attraction *n* AIR, WATER POLL Anziehung *f*

attrition *n* AIR Abrieb *m*; ~ **resistance** *n* AIR Abriebfestigkeit *f*

auctioning *n* POLICY *of tradeable emissions rights* Versteigerung *f*

audibility *n* NOISE Schalleistung *f*; ~ **level** *n* ANALYSIS Schalleistungspegel *m*; ~ **reference value** *n* ANALYSIS Bezugsschalleistung *f*

audible *adj* NOISE hörbar; ~ **sound** *n* NOISE hörbarer Schall *m*, Hörschall *m*

audiometer *n* ANALYSIS Audiometer *nt*

auger *n* ANALYSIS, SOIL *bioremediation* Bodenbohrer *m*

Auger: ~ **electron spectroscopy** *n* (*AES*) ANALYSIS Auger-Elektronspektroskopie *f* (*AES*)

authority *n* POLICY Behörde *f*; ~ **decision** *n* POLICY Behördenentscheidung *f*

authorization *n* POLICY Erlaubnis *f*, Genehmigung *f*, Zulassung *f*; ~ **application** *n* POLICY Genehmigungsantrag *m*; ~ **impact** *n* POLICY Genehmigungswirkung *f*; ~ **process** *n* POLICY Genehmigungsverfahren *nt*

autochthonous: ~ **matter** *n* SOIL eigenständige Substanz *f*

automate *vt* GEN automatisieren

automated *adj* GEN automatisiert

automatic *adj* GEN automatisch

automatically: ~ **injected** *adj* ANALYSIS automatisch eingespritzt

automatic: ~ **detector tube** *n* ANALYSIS automatisches Prüfröhrchen *nt*; ~ **injection blow molder** *AmE*, ~ **injection blow moulder** *BrE* *n* WASTE Spritzblasautomat *m*; ~ **recognition unit** *n* WASTE automatische Erkennungseinrichtung *f*; ~ **sampling** *n* ANALYSIS automatische Probenahme *f*; ~ **sorting** *n* WASTE automatische Sortierung *f*, automatische Trennung *f*, mechanische Sortierung *f*, mechanische Trennung *f*

automobile: ~ **battery** *n* *AmE* (*cf car battery BrE*) WASTE Autobatterie *f*; ~ **tax** *n* *AmE* (*cf road tax BrE*) POLICY Kfz-Steuer *f*

autosampler *n* ANALYSIS Autosampler *m*, automatischer Probengeber *m*

autosampling *n* ANALYSIS Autosampling *nt*

autothermic *adj* AIR autotherm

autotrophic *adj* WATER POLL *bacteria* autotroph; ~ **bacteria** *n pl* WATER POLL autotrophe Bakterien *f pl*; ~ **denitrification** *n* WATER POLL autotrophe Denitrifikation *f*

auxiliary: ~ **air** *n* AIR Zusatzluft *f*; ~ **burner** *n* AIR Hilfsbrenner *m*, Stützbrenner *m*, Zusatzbrenner *m*; ~ **fan** *n* AIR Zusatzgebläse *nt*; ~ **gas** *n* ANALYSIS Hilfsgas *nt*

available *adj* GEN nutzbar, verfügbar; ~ **capacity** *n* WATER POLL nutzbare Kapazität *f*; ~ **technology** *n* ANALYSIS, MANAGE, SAFETY verfügbare Technologie *f*

average 1. *adj* ANALYSIS durchschnittlich; **2.** *n* ANALYSIS Durchschnitt *m*

average: ~ **monthly dose** *n* ANALYSIS, SAFETY mittlere Monatsdosis *f*; ~ **sample** *n* ANALYSIS Durchschnittsprobe *f*; ~ **sound level** *n* NOISE Mittelungspegel *m*; ~ **value** *n* ANALYSIS Mittelwert *m*

Avogadro: ~**'s constant** *n* AIR Avogadro-Konstante *f*

avoid *vt* NOISE *reflection*, WASTE vermeiden

avoidance *n* NOISE, WASTE Vermeidung *f*

award: ~ **conditions** *n pl* POLICY *of environmental label* Vergabebedingungen *f pl*

A-weighted: ~ **sound level** *n* NOISE A-bewerteter Schallpegel *m*

axial *adj* AIR axial; ~ **deformation** *n* WASTE *landfill* axiale Verformung *f*; ~**-entry cyclone** *n* AIR Axialeinlauf-Zyklon *m*, axial angeströmter Zyklon *m*; ~ **fan** *n* AIR Axialventilator *m*; ~**-flow rotary pump** *n* WATER POLL, WATER SUPP Axialkreiselpumpe *f*, Propellerpumpe *f*

axis *n* ANALYSIS, RENEW ENERG *wind turbine rotor* Achse *f*

B

Babcock: ~ **method** *n* WATER POLL *enzymatic stabilization* Babcock-Verfahren *nt*
Bacharach: ~ **scale** *n* ANALYSIS *dust measurement* Bacharach-Skala *f*
backfill *n* SOIL, WASTE Auffüllung *f*, Verfüllboden *m*
backfilling *n* SOIL, WASTE Verfüllung *f*
backflushing: ~ **gas** *n* AIR Rückspülgas *nt*
backflush: ~ **screen changer** *n* WASTE Rückspülsiebwechsler *m*
background *n* AIR, NOISE Hintergrund *m*; ~ **concentration** *n* AIR Grundkonzentration *f*; ~ **level** *n* AIR natürliche Grundbelastung *f*, ANALYSIS Hintergrundkonzentration *f*, Rauschpegel *m*, NOISE Rauschpegel *m*, SOIL, WATER SUPP Hintergrundkonzentration *f*; ~ **noise** *n* NOISE Grundgeräusch *nt*, Hintergrundgeräusch *nt*; ~ **ozone concentration** *n* AIR Ozonhintergrundkonzentration *f*; ~ **pollution** *n* AIR grundlegende Verschmutzung *f*; ~ **sample** *n* ANALYSIS Hintergrundprobe *f*
backhoe *n* MARINE Löffelbagger *m*, Löffeltiefbagger *m*, Tiefbagger *m*
back: ~ **pressure** *n* AIR Gegendruck *m*; ~ **scattering** *n* ANALYSIS Rückstreuung *f*; ~ **titration** *n* ANALYSIS Rücktitration *f*; ~-**up requirement** *n* POLICY Nachholbedarf *m*
backward: ~-**curved blade compressor** *n* AIR rückwärts gekrümmter Schaufel-Verdichter *m*
backwashable *adj* WASTE rückspülbar
backwashing *n* WASTE, WATER POLL, WATER SUPP Rückspülung *f*
backwash: ~ **water** *n* WASTE, WATER POLL Rückspülwasser *nt*
backwater *n* WASTE Rückstau *m*, WATER SUPP Rückstauwasser *nt*, Stau *m*, Stauwasser *nt*
bacteria *n pl* AIR *biological filter*, SOIL, WASTE, WATER POLL, WATER SUPP Bakterien *f pl*; ~ **activity** *n* AIR *biological filter* Bakterientätigkeit *f*; ~ **bed** *n* WATER POLL Tropfkörperanlage *f*, biologischer Körper *m*; ~-**containing** *adj* WATER POLL bakterienhaltig; ~-**containing sludge** *n* WATER POLL bakterienhaltiger Schlamm *m*; ~ **content** *n* WASTE, WATER SUPP Bakteriengehalt *m*; ~ **count** *n* WASTE, WATER POLL, WATER SUPP Bakterienzahl *f*; ~ **elimination** *n* WASTE, WATER SUPP Bakterieneliminierung *f*; ~ **filter** *n* AIR Bakterienfilter *nt*
bacterial *adj* GEN bakteriell; ~ **activity** *n* AIR, SOIL, WASTE, WATER SUPP *composting* Bakterientätigkeit *f*; ~ **contamination** *n* AIR, SOIL, WASTE, WATER SUPP *composting* Bakterienverseuchung *f*; ~ **count** *n* AIR, SOIL, WASTE, WATER POLL, WATER SUPP Bakterienzahl *f*; ~ **growth** *n* AIR, SOIL, WASTE, WATER POLL, WATER SUPP Bakterienwachstum *nt*; ~ **growth rate** *n* AIR, SOIL, WASTE, WATER POLL, WATER SUPP Bakterienwachstumsrate *f*; ~ **poison** *n* AIR, SOIL, WATER POLL Bakteriengift *nt*
bacteria: ~ **mass** *n* WATER POLL Bakterienmasse *f*; ~ **population** *n* WASTE, WATER POLL, WATER SUPP Bakterienpopulation *f*; ~ **test** *n* ANALYSIS Bakterientest *m*
bacteriological: ~ **purification** *n* WASTE, WATER SUPP bakteriologische Reinigung *f*; ~ **treatment** *n* WASTE, WATER SUPP bakteriologische Reinigung *f*
baffle *n* AIR Prallfläche *f*, WATER POLL Prallwand *f*; ~ **board** *n* NOISE Schallschirm *m*, Schallwand *f*; ~ **chamber** *n* AIR Blendengehäuse *nt*; ~ **collector** *n* NOISE Schallschirm *m*; ~ **plate** *n* WATER POLL Tauchbrett *nt*, Tauchwand *f*; ~ **silencer** *n* NOISE Reflexionsschalldämpfer *m*
bag *n* AIR *of filter* Schlauch *m*, WASTE Beutel *m*; ~ **bottom** *n* AIR *of filter* Schlauchboden *m*; ~ **column** *n* AIR Schlauchkolonne *f*; ~ **filter** *n* AIR, SOIL Schlauchfilter *nt*, WATER POLL Schlauchfilter *nt*, Taschenfilter *nt*
baghouse *n* AIR Schlauchfilter *nt*
bag: ~ **length** *n* AIR Schlauchlänge *f*; ~ **packaging** *n* WASTE Beutelverpackung *f*; ~ **paper** *n* WASTE Beutelpapier *nt*; ~ **rows** *n pl* AIR Schlauchreihen *f pl*; ~-**shaped** *adj* AIR *filter* schlauchförmig
bailer *n* WATER POLL Schlammbüchse *f*, Schlammlöffel *m*
balance 1. *n* AIR, RENEW ENERG, WASTE, WATER SUPP *ecological* Gleichgewicht *nt*; **2.** *vt* AIR, RENEW ENERG, WASTE, WATER SUPP *maintain equilibrium* im Gleichgewicht halten
bale *n* WASTE Ballen *m*; ~ **cording** *n* WASTE Ballenverschnürung *f*; ~ **disintegrator** *n* WASTE Ballenauflöser *m*; ~ **landfill** *n* WASTE Ballendeponie *f*; ~ **storage belt** *n* WASTE Ballenspeicherband *nt*
baling: ~ **press** *n* WASTE Ballenpresse *f*, Schrottpaketierpresse *f*
ballast *n* WASTE, WATER SUPP Ballast *f*, Schotter *m*
ballasted: ~ **pile** *n* WASTE, WATER SUPP Schotterpfahl *m*
ballast: ~ **material** *n* WASTE, WATER SUPP Ballastmaterial *nt*, Schottermaterial *nt*
ballistic *adj* WASTE ballistisch; ~ **separation** *n* WASTE ballistische Sichtung *f*, ballistische Sortierung *f*; ~ **sorter** *n* WASTE ballistische Auslesevorrichtung *f*, ballistischer Sichter *m*; ~ **sorting** *n* WASTE ballistische Sichtung *f*, ballistische Sortierung *f*
balloon: ~ **sonde** *n* ANALYSIS Ballonsonde *f*
band *n* WATER POLL Band *nt*; ~ **filter** *n* WATER POLL Bandfilter *nt*; ~ **filter press** *n* WATER POLL *dewatering* Bandfilterpresse *f*; ~ **oven** *n* SOIL *pellet incineration* Sinterband *nt*
bandpass: ~ **filter** *n* ANALYSIS Bandfilter *nt*, Siebkette *f*
bank *n* WATER POLL, WATER SUPP *of river* Ufer *nt*
banked: ~-**up water level** *n* WATER SUPP Stau *m*
bank: ~-**filtered** *adj* WATER SUPP uferfiltriert; ~-**filtered groundwater** *n* WATER SUPP uferfiltriertes Grundwasser *nt*; ~ **filtrate** *n* WATER SUPP Uferfiltrat *nt*
banking: ~-**up of water** *n* WATER SUPP Wassereinstau *m*
bar *n* ANALYSIS Bar *nt*, WATER POLL *of screen* Stab *m*; ~ **coding for identification** *n* MANAGE Strichkodierung *f* zur Identifizierung; ~ **concentration** *n* WATER POLL *of screen* Stabkonzentration *f*
barometer *n* ANALYSIS Barometer *nt*
barrel: ~-**venting system** *n* WATER SUPP Zylinderentgasungsanlage *f*
barrier *n* SOIL, WASTE Barriere *f*, WATER SUPP Barriere *f*, Grundwasserstauer *m*; ~ **film** *n* WASTE Sperrschichtfolie *f*; ~ **layer** *n* WASTE Sperrschicht *f*; ~ **material** *n* WASTE Sperrschichtstoff *m*; ~ **packaging** *n* WASTE Sicherheitsverpackung *f*, Sperrschichtverpackung *f*; ~ **system** *n* WASTE Sperrsystem *nt*, Barrierensystem *nt*, Dichtungssystem *nt*
bar: ~ **screen** *n* WATER POLL Stabrechen *m*
base: ~ **cation** *n* AIR Basenkation *nt*; ~ **load power station** *n* RENEW ENERG Grundlastkraftwerk *nt*; ~ **material** *n* RENEW ENERG *biomass conversion* Ausgangsstoff *m*, WASTE, WATER POLL, WATER SUPP Ausgangsmaterial *nt*, Ausgangsstoff *m*
bases *n pl* SOIL *additive* Laugen *f pl*
base: ~-**sealed landfill** *n* WASTE basisabgedichtete Deponie *f*; ~ **sealing** *n* SOIL Basisdichtung *f*, WASTE, WATER SUPP Basisabdichtung *f*
basic *adj* GEN basisch; ~ **absorbent** *n* AIR basisches Absorbens *nt*; ~ **anionic exchanger** *n* WATER POLL basischer

Anionenaustauscher *m*; ~ **capacity** *n* WASTE, WATER POLL, WATER SUPP Basizität *f*; ~ **cationic exchanger** *n* WATER POLL basischer Kationenaustauscher *m*

basicity *n* WASTE, WATER POLL, WATER SUPP Basizität *f*

basic: ~ **protection level** *n* POLICY Basis-Schutz-Niveau *nt*

basin *n* ANALYSIS, WASTE, WATER POLL, WATER SUPP Becken *nt*; ~ **bottom** *n* WATER POLL Beckensohle *f*; ~ **floor** *n* WASTE, WATER SUPP Beckensohle *f*

BAT *abbr* (*best available technology*) GEN BAT (*beste verfügbare Technik*)

batch *n* ANALYSIS, SOIL, WATER POLL Charge *f*, Batch *nt*; ~**-bulk** *adj* SOIL abschnittweise; ~**-bulk injection** *n* SOIL abschnittweises Einpressen *nt*; ~ **distillation** *n* SOIL Batchdestillation *f*; ~ **distillation tank** *n* SOIL Batchdestillationstank *m*

batching: ~ **tank** *n* WATER POLL Ansetzbehälter *m*

batch: ~ **neutralization** *n* WATER POLL Chargenneutralisation *f*; ~ **operation** *n* AIR, WATER POLL Chargenbetrieb *m*; ~ **process** *n* AIR Chargenprozeß *m*, Chargenverfahren *nt*, WATER POLL Batchverfahren *nt*, diskontinuierliches Verfahren *nt*; ~**-processed** *adj* AIR, WATER POLL chargenmäßig betrieben; **in ~ quantities** *phr* AIR, SOIL, WASTE, WATER POLL chargenweise; ~ **test** *n* ANALYSIS Batchversuch *m*

BATNEEC *abbr* (*best available technology not entailing excessive costs*) GEN BATNEEC (*wirtschaftlich vertretbare beste verfügbare Technik*)

battery *n* WASTE Batterie *f*; ~**-operated** *adj* ANALYSIS batteriebetrieben; ~**-operated vehicle** *n* AIR batteriebetriebenes Fahrzeug *nt*; ~**-powered** *adj* ANALYSIS batteriebetrieben; ~**-powered gas chromatograph** *n* ANALYSIS batteriebetriebener Gaschromatograph *m*; ~ **recycling** *n* WASTE Batterierecycling *nt*; ~ **residue** *n* WASTE Batteriereststoff *m*

bauxite *n* AIR *adsorbent* Bauxit *m*

bay *n* RENEW ENERG *tidal power plant* Bucht *f*, Meeresbucht *f*

Be *abbr* (*beryllium*) AIR Be (*Beryllium*)

beaker: ~ **glass** *n* ANALYSIS Becherglas *nt*

beamsplitter *n* ANALYSIS Strahlenteiler *m*

bearing: ~ **bridge** *n* AIR *packed-bed filter* Lagerbrücke *f*; ~ **layer** *n* WASTE Tragschicht *f*

beat *vt* WASTE zerfasern

behavior *AmE see behaviour BrE*

behaviour *n BrE* GEN Verhalten *nt*

beheaded: ~ **river** *n* WATER SUPP angezapfter Fluß *m*

below: ~**-cloud scavenging** *n* AIR Reinigungsfällung *f*

belt *n* WATER POLL Band *nt*; ~ **conveyor** *n* WASTE Förderband *nt*, Transportband *nt*; ~ **drive** *n* WATER POLL, WATER SUPP *pump* Riemenantrieb *m*; ~ **dryer** *n* WATER POLL Bandtrockner *m*; ~ **filter** *n* WATER POLL Bandfilter *nt*; ~ **press** *n* WATER POLL *in dewatering* Bandpresse *f*; ~ **skimmer** *n* MARINE Förderband *nt* zur Ölaufnahme

bench: ~**-scale study** *n* ANALYSIS Laboruntersuchung *f*

bend *n* WATER SUPP *in canal* Bogen *m*

bending: ~ **waves** *n pl* NOISE Biegewellen *f pl*

benzene *n* AIR, WASTE *landfill gas component* Benzol *nt*; ~, **toluol, ethyl benzene, xylene** *n* (*BTEX*) WATER POLL *solvent mix* Benzol, Toluol, Ethylbenzol, Xylol *nt* (*BTEX*); ~, **toluol, xylene** *n* (*BTX*) WATER POLL *solvent mix* Benzol, Toluol, Xylol *nt* (*BTX*)

benzofurane *n* SOIL Benzofuran *nt*

berl: ~ **saddles** *n pl* AIR *packing of packed-tower scrubber* Berl-Füllkörper *m pl*

berm: ~ **ditch** *n* WASTE, WATER SUPP Entwässerungsgraben *m* mit Böschung

Bernoulli: ~**'s equation** *n* ANALYSIS Bernoulli-Gleichung *f*, Bernoullische Gleichung *f*

beryllium *n* (*Be*) AIR Beryllium *nt* (*Be*)

best: ~ **available technology** *n* (*BAT*) GEN beste verfügbare Technik *f* (*BAT*); ~ **available technology not entailing excessive costs** *n* (*BATNEEC*) AIR, POLICY, SOIL, WASTE, WATER POLL, WATER SUPP wirtschaftlich vertretbare beste verfügbare Technik *f* (*BATNEEC*); ~ **practical environmental option** *n* (*BPEO*) POLICY bestmögliche Option *f* für die Umwelt (*BPEO*)

beta: ~**-mesosaprobic** *adj* WATER POLL beta-mesosaprob; ~**-ray absorption sampler** *n* ANALYSIS Beta-Absorption-Probenahmemeßgerät *nt*

beverage: ~ **bottle** *n* WASTE Getränkeflasche *f*; ~ **container** *n* WASTE Getränkeverpackung *f*; ~ **packaging** *n* WASTE Getränkeverpackung *f*

B-horizon *n* SOIL B-Horizont *m*, Unterboden *m*

bifurcation: ~ **point** *n* WATER SUPP *of plume stack* Gabelpunkt *m*

bilateral *adj* POLICY bilateral; ~ **environmental agreement** *n* POLICY bilaterale Umweltvereinbarung *f*; ~ **negotiation** *n* POLICY bilaterale Verhandlung *f*

bilge *n* MARINE Bilge *f*; ~ **pump** *n* MARINE Bilgenpumpe *f*; ~ **water** *n* MARINE Bilgenwasser *nt*

binary *adj* ANALYSIS binär

bind *vt* AIR *particles with washing liquid* binden

binder *n* MARINE, SOIL, WASTE, WATER SUPP Binder *m*, Bindemittel *nt*

binding *n* MARINE, SOIL, WASTE, WATER SUPP Einbindung *f*; ~ **agent** *n* MARINE, SOIL, WASTE, WATER SUPP Bindemittel *nt*, Binder *m*; ~ **layer** *n* WASTE Binderschicht *f*; ~ **strength** *n* WATER POLL *ions* Bindefähigkeit *f*

bio- *pref* GEN biologisch, Bio-

bioaccumulate *vi* AIR, SOIL, WATER POLL biologisch akkumulieren

bioaccumulation *n* AIR, SOIL, WATER POLL Bioakkumulation *f*; ~ **capacity** *n* AIR, SOIL, WATER POLL Bioakkumulationsvermögen *nt*; ~ **factor** *n* AIR, SOIL, WATER POLL Bioakkumulationsfaktor *m*; ~ **potential** *n* AIR, SOIL, WATER POLL Bioakkumulationspotential *nt*

bioaccumulative *adj* AIR, SOIL, WATER POLL bioakkumulativ

bioaugmentation *n* SOIL Biorestauration *f*

biocenosis *AmE see biocoenosis BrE*

biocenotic *AmE see biocoenotic BrE*

biochemical *adj* ANALYSIS, WASTE, WATER POLL, WATER SUPP biochemisch; ~ **conversion** *n* RENEW ENERG *biomass* biochemische Umwandlung *f*; ~ **oxygen demand** *n* (*BOD*) ANALYSIS, WASTE, WATER POLL, WATER SUPP biochemischer Sauerstoffbedarf *m* (*BSB*); ~ **oxygen demand in five days** *n* (BOD_5) AIR, ANALYSIS, WASTE, WATER POLL, WATER SUPP biochemischer Sauerstoffbedarf *m* in fünf Tagen (BSB_5); ~ **self-purification** *n* WASTE, WATER POLL, WATER SUPP biochemische Selbstreinigung *f*; ~ **tracer** *n* ANALYSIS, WASTE, WATER SUPP biochemischer Indikator *m*

biocide *n* SOIL, WATER POLL Biozid *nt*

biocoenosis *n BrE* AIR, WATER POLL Biozönose *f*

biocoenotic *adj BrE* AIR, WATER POLL biozönotisch

biodegradability *n* GEN biologische Abbaubarkeit *f*

biodegradable *adj* GEN biologisch abbaubar; ~ **matter** *n* WASTE, WATER POLL, WATER SUPP abbaubare organische Substanz *f* (*AOS*), biologisch abbaubarer Stoff *m*; ~ **organics** *n pl* WATER POLL biologisch abbaubare organische Verbindungen *f pl*; ~ **packaging material** *n* WASTE biologisch abbaubarer Packstoff *m*; ~ **plastics** *n pl* WASTE biologisch abbaubare Kunststoffe *m pl*; ~ **substance** *n* WASTE biologisch abbaubare Substanz *f*; ~ **waste** *n* WASTE biologisch abbaubarer Abfall *m*

biodegradation *n* GEN biologischer Abbau *m*, biologische Zersetzung *f*

Biodenipho: ~ **method** *n* WATER POLL *phosphorous removal* Biodenipho-Verfahren *nt*

biodiesel *n* AIR, RENEW ENERG Biodiesel *m*

biodisc *n BrE* WATER POLL Scheibentauchkörper *m*

biodisk *AmE see biodisc BrE*

bioenergy *n* RENEW ENERG *bioconversion* Bioenergie *f*; ~ **plant** *n* RENEW ENERG Bioenergieanlage *f*, Biokonversionsanlage *f*

biofilm *n* WATER POLL Biofilm *m*, biologischer Rasen *m*;

~ **growth** *n* WATER POLL Biofilmwachstum *nt*, Wachstum *nt* des biologischen Rasens; ~ **surface** *n* WATER POLL Biofilmoberfläche *f*, Rieselfilm *m*

biofilter *n* AIR, WASTE, WATER POLL, WATER SUPP Biofilter *nt*

biofiltration *n* AIR, WASTE, WATER POLL, WATER SUPP Biofiltration *f*; ~ **process** *n* WATER POLL Tropfkörperverfahren *nt*

bioflocculation *n* WATER POLL Bioflockung *f*

biofuel *n* RENEW ENERG Biokraftstoff *m*

biogas *n* RENEW ENERG, WASTE Biogas *nt*, Faulgas *nt*, WATER POLL Faulgas *nt*, Klärgas *nt*; ~ **composition** *n* WASTE Biogaszusammensetzung *f*; ~ **generation** *n* RENEW ENERG Biogasgewinnung *f*; ~ **reactor** *n* WASTE Biogasreaktor *m*; ~ **recovery** *n* RENEW ENERG, WASTE Biogasgewinnung *f*

biogenic: ~ **sediment** *n* WASTE biogenes Sediment *nt*

bioindicator *n* WASTE, WATER SUPP Bioindikator *m*; ~ **plant** *n* SOIL Zeigerpflanze *f*

biological *adj* GEN biologisch; ~ **activated sludge plant** *n* WATER POLL biologische Belebungsanlage *f*; ~ **agent** *n* WASTE, WATER SUPP biologisches Agens *nt*; ~ **bed** *n* SOIL Biobeet *nt*; ~ **catalyst** *n* AIR Biokatalysator *m*; ~ **clarification plant** *n* WATER POLL biologische Kläranlage *f*; ~ **degradation** *n* GEN biologische Zersetzung *f*, biologischer Abbau *m*; ~ **denitrification** *n* WASTE, WATER SUPP biologische Denitrifikation *f*; ~ **elimination** *n* SOIL biologische Eliminierung *f*; ~ **energy conversion** *n* WASTE, WATER SUPP biologische Umwandlung *f*; ~ **filter** *n* AIR, WASTE, WATER SUPP biologisches Filter *nt*, Biofilter *nt*; ~ **filtration** *n* AIR, WASTE, WATER SUPP Biofilterung *f*; ~ **gas scrubbing** *n* AIR *absorption process* Biowäsche *f*, biologische Gaswäsche *f*; ~ **ground filter** *n* AIR, biologisches Erdfilter *nt*; ~ **hazard** *n* SAFETY biologische Gefahr *f*; ~ **indicator** *n* ANALYSIS, WASTE, WATER SUPP Bioindikator *m*

biologically: ~ **catalytic** *adj* AIR biokatalytisch, biologisch katalytisch; ~ **catalytic regeneration** *n* AIR biokatalytische Regeneration *f*, biologisch katalytische Regeneration *f*; ~ **catalytic waste air cleaning** *n* AIR biokatalytische Abluftreinigung *f*, biologisch katalytische Abluftreinigung *f*; ~ **dead** *adj* SOIL biologisch tot; ~ **difficult to decompose** *adj* WATER POLL biologisch schwer abbaubar

biological: ~ **nutrient removal** *n* WATER POLL biologische Nährstoffentfernung *f*, biologischer Nährstoffentzug *m*; ~ **oxygen demand** *n* (*BOD*) ANALYSIS, WASTE, WATER POLL, WATER SUPP biologischer Sauerstoffbedarf *m* (*BSB*); ~ **oxygen demand in five days** *n* (*BOD$_5$*) AIR, ANALYSIS, WASTE, WATER POLL, WATER SUPP biologischer Sauerstoffbedarf *m* in fünf Tagen (*BSB$_5$*); ~ **phosphorus removal** *n* WATER POLL biologische Phosphoreliminierung *f*; ~ **purification** *n* WATER POLL biologische Reinigung *f*; ~ **sorption process** *n* AIR biologisches Sorptionsverfahren *nt*; ~ **stabilization** *n* SOIL biologische Stabilisierung *f*; ~ **strain** *n* WASTE *landfill* biologische Belastung *f*; ~ **test** *n* ANALYSIS Biotest *m*; ~ **treatment** *n* WASTE, WATER SUPP biologische Behandlung *f*; ~ **washing powder** *n* WATER POLL biologisches Waschpulver *nt*; ~ **waste** *n* WASTE Bioabfall *m*; ~ **waste air treatment** *n* AIR biologische Abluftreinigung *f*; ~ **wastewater treatment** *n* WATER POLL biologische Abwasseraufbereitung *f*; ~ **water quality** *n* WATER SUPP biologische Gewässergüte *f*; ~ **water quality determination** *n* ANALYSIS biologische Wassergütebestimmung *f*

biomass *n* GEN Biomasse *f*; ~ **concentration** *n* WATER POLL Biomassekonzentration *f*; ~ **conversion** *n* RENEW ENERG Biokonversion *f*; ~ **curve** *n* WATER POLL Biomassekurve *f*; ~ **potential** *n* RENEW ENERG Biomassepotential *nt*; ~ **production** *n* RENEW ENERG Biomasseproduktion *f*

bio: ~-**oil** *n* RENEW ENERG Bio-Öl *nt*

bioreactor *n* WATER POLL Bioreaktor *m*

bioremediation *n* SOIL *for soil containing organic compounds* biologische Altlastensanierung *f*

biosensor *n* ANALYSIS Biosensor *m*

biostabilizer *n* AIR, WASTE, WATER SUPP Biostabilisator *m*

biotechnological *adj* AIR, SOIL biotechnologisch; ~ **treatment** *n* AIR biotechnologische Behandlung *f*

biotic *adj* WASTE, WATER POLL, WATER SUPP biotisch; ~ **accumulation** *n* WASTE, WATER SUPP biotische Akkumulation *f*

bio: ~-**tower** *n* WATER POLL Turmtropfkörper *m*

biowasher *n* AIR *biological sorption process* Biowäscher *m*

biowaste *n* WASTE Bioabfall *m*, Biomüll *m*, organischer Abfall *m*

blackbody *n* AIR, ANALYSIS Planckscher-Strahler *m*

blacklist *n* POLICY schwarze Liste *f*; ~ **tide** *n* MARINE Ölpest *f*; ~ **water** *n* MARINE öliges Wasser *nt*

blade *n* AIR Schaufel *f*, RENEW ENERG *wind turbine* Blatt *nt*; ~ **angle** *n* RENEW ENERG *wind turbine* Blatteinstellungswinkel *m*; ~ **angle control** *n* RENEW ENERG *wind turbine* Blatteinstellungswinkelregelung *f*; ~ **rotation** *n* RENEW ENERG *wind turbine* Blattumdrehung *f*; ~ **wheel** *n* RENEW ENERG *wave energy, hydropower* Schaufelrad *nt*

blank: ~ **test** *n* ANALYSIS Blindprobe *f*; ~ **value** *n* ANALYSIS Blindwert *m*

blast 1. *vt* AIR, SAFETY, WASTE sprengen; **2.** *vi* AIR, SAFETY, WASTE sprengen

blast: ~ **connection** *n* AIR Düsenstock *m*; ~ **furnace slag** *n* WASTE *industrial by-product* Hochofenschlacke *f*; ~ **furnace sludge** *n* WATER POLL Gichtschlamm *m*

bleaching: ~ **plant wastewater** *n* WATER POLL Bleichereiabwasser *nt*

blend: ~ **into** *vt* WASTE *plastics* beimischen

blender *n* WATER POLL Mischer *m*

blind: ~ **drainage area** *n* WATER SUPP Einzugsgebiet *nt* mit Binnenentwässerung, abflußloses Einzugsgebiet *nt*; ~ **test** *n* ANALYSIS Blindprobe *f*

blister: ~ **pack** *n* WASTE Blisterpackung *f*, Durchdrückverpackung *f*, Skin-Packung *f*

block *vt* WASTE verschließen

blockage *n* AIR *catalyst* Blockierung *f*, WASTE Verschließung *f*

blocking *n* SOIL, WASTE Abriegelung *f*, WATER POLL *filter, pump* Verstopfung *f*, WATER SUPP Abriegelung *f*

block: ~ **polymer** *n* WASTE Blockpolymer *nt*

blow 1. *vt* AIR blasen; ~ **down** WASTE, WATER POLL abschlämmen; ~ **into** AIR, WATER POLL einblasen; ~ **off** WASTE entleeren, spülen, WATER POLL entleeren, WATER SUPP entleeren, spülen; ~ **out** WASTE, WATER SUPP ausblasen; ~ **up** AIR, SAFETY, WASTE sprengen; **2.** *vi* ~ **up** AIR, SAFETY, WASTE sprengen

blowby: ~ **gas** *n* AIR Blowby-Gas *nt*, Kurbelgehäusegas *nt*

blower *n* AIR, WATER POLL Ventilator *m*; ~ **stream velocity** *n* AIR *catalyst* Anströmgeschwindigkeit *f*

blowing: ~ **agent** *n* WASTE Treibmittel *nt*; ~-**off** *n* WASTE, WATER POLL, WATER SUPP Entleeren *nt*, Entleerung *f*; ~-**through** *n* AIR Durchblasen *nt*

blowring: ~ **filter** *n* AIR Rückblasfilter *nt*

blue: ~ **algae** *n pl* WASTE, WATER SUPP Blaualgen *f pl*; ~ **angel** *n* POLICY *eco-label* blauer Engel *m*

BOD *abbr* ANALYSIS, WASTE, WATER POLL, WATER SUPP (*biochemical oxygen demand*), BSB (*biochemischer Sauerstoffbedarf*), (*biological oxygen demand*) BSB (*biologischer Sauerstoffbedarf*)

BOD$_5$ *abbr* (*biological oxygen demand in five days*) ANALYSIS, WASTE, WATER POLL, WATER SUPP BSB$_5$ (*biochemischer Sauerstoffbedarf in fünf Tagen*); ~ **decomposition** *n* WASTE, WATER POLL, WATER SUPP BSB$_5$-Abbau *m*; ~ **degradation** *n* WATER POLL BSB$_5$-Abbau *m*; ~ **determination** *n* ANALYSIS BSB$_5$-Bestimmung *f*

BOD: ~ **determination** *n* ANALYSIS BSB-Bestimmung *f*; ~ **determination in wastewater** *n* ANALYSIS, WATER POLL, WATER SUPP Bestimmung des BSB von Abwasser *f*; ~ **load** *n* WATER POLL BSB-Fracht *f*

body: ~ **of water** *n* MARINE, RENEW ENERG Wassermasse *f*
boil 1. *vt* ANALYSIS, WATER SUPP sieden, kochen; ~ **off** WATER SUPP abkochen; **2.** *vi* ANALYSIS, WATER SUPP sieden
boiler *n* AIR, RENEW ENERG, WASTE, WATER POLL Kessel *m*; ~ **corrosion** *n* WASTE Kesselkorrosion *f*; ~ **feedwater** *n* AIR, WASTE Kesselspeisewasser *nt*; ~ **feedwater pump** *n* AIR, WASTE Kesselspeisewasserpumpe *f*; ~ **makeup water** *n* AIR, WASTE, WATER SUPP Kesselzusatzwasser *nt*; ~ **plant** *n* AIR, RENEW ENERG, WASTE Feuerungsanlage *f*, Kesselanlage *f*; ~ **room ash** *n* WASTE Kesselraumasche *f*; ~ **scale** *n* WASTE Kesselstein *m*
boiling *adj* ANALYSIS, WATER SUPP siedend; ~ **point** *n* ANALYSIS, RENEW ENERG Siedepunkt *m*; ~ **sample** *n* ANALYSIS siedende Probe *f*; ~ **vessel** *n* AIR Aufkocher *m*
bomb: ~ **calorimeter** *n* ANALYSIS Kalorimeterbombe *f*
bond *vt* AIR, WASTE verbinden
bonding *n* AIR *particles on surface*, WASTE Bindung *f*; ~ **agent** *n* AIR Haftmittel *nt*
boom *n* MARINE Sperre *f*; ~ **pack** *n* MARINE Sperrengebinde *nt*, Ölsperrengebinde *nt*; ~ **retrieval** *n* MARINE Sperrenrückgewinnung *f*; ~ **towing** *n* MARINE Sperrenschleppe *f*
booster: ~ **fan** *n* AIR Druckerhöhungsventilator *m*; ~ **pump** *n* WATER POLL, WATER SUPP Boosterpumpe *f*, Druckerhöhungspumpe *f*, Zwischenpumpe *f*; ~ **pump works** *n* WATER SUPP Druckerhöhungspumpwerk *nt*
bore: ~ **through** *vt* WASTE, WATER SUPP durchbohren
bored: ~ **well** *n* WATER SUPP Bohrbrunnen *m*
borehole *n* ANALYSIS *groundwater and soil sampling*, RENEW ENERG *geothermal power plant*, SOIL, WASTE, WATER SUPP Bohrloch *nt*; ~ **pump** *n* WASTE, WATER SUPP Bohrlochpumpe *f*; ~ **wall** *n* SOIL Bohrlochwand *f*
bore: ~ **machine** *n* ANALYSIS, WATER SUPP Bohrgerät *nt*
boring *n* WATER SUPP Bohrung *f*; ~ **pipe** *n* WATER SUPP Bohrrohr *nt*; ~ **tool** *n* WATER SUPP Bohrwerkzeug *nt*
bottle *n* ANALYSIS Flasche *f*; ~ **bank** *n* WASTE Altglasbehälter *m*, Glascontainer *m*; ~ **cap** *n* WASTE Flaschenverschlußkappe *f*; ~ **crate** *n* WASTE Flaschenkasten *m*; ~ **deposit** *n* WASTE *money* Flaschenpfand *nt*; ~ **glass** *n* WASTE Hohlglas *nt*; ~ **jacket** *n* WASTE Flaschenhülle *f*; ~ **leak detector** *n* WASTE Flaschenlecksensor *m*; **~-rinsing machine** *n* WASTE Flaschenwaschmaschine *f*; ~ **sleeve** *n* WASTE Flaschenhülse *f*; ~ **test** *n* ANALYSIS Flaschenversuch *m*
bottom: ~ **culvert** *n* WATER SUPP Bodenabzugskanal *m*; ~ **current** *n* WASTE, WATER SUPP Bodenströmung *f*; ~ **deposit** *n* WATER POLL Bodenschlamm *m*; ~ **drainage** *n* WASTE, WATER SUPP Sohldränage *f*; ~ **layer** *n* WASTE, WATER SUPP Bodenschicht *f*; ~ **outlet** *n* WATER SUPP Grundablaß *m*; ~ **scraper blade** *n* WATER POLL Bodenräumerschild *nt*; ~ **sealing** *n* SOIL Sohleabdichtung *f*, WASTE, WATER SUPP Basisabdichtung *f*, Untergrundabdichtung *f*
boundary *n* AIR Grenze *f*, NOISE Begrenzungsfläche *f*; ~ **area** *n* WASTE, WATER SUPP Randbereich *m*; ~ **layer** *n* AIR Grenzschicht *f*; ~ **layer flow** *n* AIR Grenzschichtströmung *f*; ~ **wall** *n* WASTE Randwall *m*
box *n* WASTE Kasten *m*, Kiste *f*; **~-making machine** *n* WASTE Kartonautomat *m*; ~ **pallet with sidewalls** *n* WASTE Gitterboxpalette *f*
Boyle: **~'s law** *n* ANALYSIS Boyle-Mariottesches Gesetz *nt*
BPEO *abbr* (*best practical environmental option*) POLICY BPEO (*bestmögliche Option für die Umwelt*)
brace *n* AIR *stringer* Stützbalken *m*
brackish *adj* WATER SUPP brackig; ~ **water** *n* WATER SUPP Brackwasser *nt*
branched *adj* WATER SUPP verzweigt; ~ **distribution network** *n* WATER SUPP verzweigtes Verteilrohrnetz *nt*
branch: ~ **end** *n* WATER SUPP Rohrendstrang *m*
branching *n* WATER SUPP Verzweigung *f*, *canal* Abzweigung *f*; ~ **at an acute angle** *n* WATER SUPP spitzwinklige Abzweigung *f*
branch: ~ **pipe** *n* WASTE, WATER SUPP Abzweig *m*, Abzweigrohr *nt*, Nebensammler *m*; ~ **pipe resistance** *n* WATER SUPP Abzweigwiderstand *m*; ~ **sewer** *n* WATER POLL Nebensammler *m*
breach: ~ **of law** *n* POLICY Ordnungswidrigkeit *f*
break 1. *vt* AIR *liquid into finest droplets* zerreißen, SOIL brechen; ~ **down** WASTE herunterwalzen; ~ **open** SAFETY, WASTE, WATER SUPP aufbrechen; ~ **up** AIR, WASTE zertrümmern; **2.** ~ **out** *vi* SAFETY ausbrechen; ~ **out again** SAFETY wiederausbrechen
breaker *n* SOIL Brecher *m*; ~ **board** *n* MARINE Oberflächenbrecher *m*
breaking *n* AIR Zerreißen *nt*; ~ **load** *n* WASTE, WATER SUPP Bruchbelastung *f*, Bruchlast *f*; **~-up** *n* ANALYSIS, WASTE, WATER SUPP Auflockerung *f*
breakpoint: ~ **chlorination** *n* WATER POLL, WATER SUPP Knickpunktchlorung *f*
breath *n* SAFETY Atem *m*
breathable: ~ **air** *n* SAFETY Atemluft *f*
breathing: ~ **apparatus** *n* SAFETY Atemgerät *nt*; ~ **capacity** *n* AIR Atemgrenzwert *m*; ~ **protection** *n* SAFETY Atemschutz *m*; ~ **protection system** *n* SAFETY Atemschutzsystem *nt*
breeches: ~ **pipe** *n* AIR Gabelrohr *nt*, Zweiwegstück *nt*
brewery: ~ **wastewater** *n* WATER POLL Brauereiabwasser *nt*
bridging: ~ **molecule** *n* WATER POLL *flocculation* Brückenmolekül *nt*
Brigg: **~'s logarithm** *n* AIR Briggscher Logarithmus *m*
briquette *n* RENEW ENERG *biomass conversion base material* Brikett *nt*, WASTE Brikett *nt*, Preßling *m*; ~ **press piston** *n* WASTE Brikettpreßstempel *m*
briquetting: ~ **press** *n* WASTE Brikettierpresse *f*
British: ~ **Standards Institute** *n* (*BSI*) ANALYSIS Britisches Institut *nt* für Normung (*BSI*)
brittleness *n* AIR *filter materials* Versprödung *f*
broad: ~ **irrigation with wastewater** *n* WATER POLL Abwasserverrieselung *f*
bromide: **~-containing** *adj* WASTE, WATER SUPP bromidhaltig; **~-containing water** *n* WATER SUPP bromidhaltiges Wasser *nt*
brown: ~ **coal** *n* AIR, SOIL, WASTE Braunkohle *f*; ~ **coal power station** *n* AIR Braunkohlenkraftwerk *nt*; ~ **glass** *n* WASTE Braunglas *nt*
Brownian: ~ **movement** *n* AIR, WATER POLL *stochastic movement of particles* Brownsche Molekularbewegung *f*
brush: ~ **aeration** *n* WATER POLL Bürstenbelüftung *f*
BSI *abbr* (*British Standards Institute*) ANALYSIS BSI (*Britisches Institut für Normung*)
BTEX *abbr* (*benzene, toluol, ethyl benzene, xylene*) WATER POLL *solvent mix* BTEX (*Benzol, Toluol, Ethylbenzol, Xylol*)
BTX *abbr* (*benzene, toluol, xylene*) WATER POLL *solvent mix* BTX (*Benzol, Toluol, Xylol*)
bubble *n* AIR Blase *f*; ~ **cap** *n* AIR *tray tower* Glocke *f*; **~-cap column** *n* AIR Glockenbodenkolonne *f*; **~-cap tray column** *n* AIR Glockenbodenkolonne *f*; **~-cap tray tower** *n* AIR Glockenboden *m*; ~ **column** *n* AIR Blasensäule *f*; ~ **pack** *n* WASTE Blasenverpackung *f*; ~ **washer** *n* AIR Glockenbodenwäscher *m*
bucket *n* MARINE Dispersantbehälter *m*
buffer 1. *n* GEN Puffer *m*; **2.** *vt* GEN puffern
buffer: ~ **capacity** *n* SOIL, WATER POLL Pufferkapazität *f*; ~ **effect** *n* AIR Puffereffekt *m*
buffering *n* GEN Pufferung *f*; ~ **action** *n* AIR, ANALYSIS Pufferwirkung *f*
buffer: ~ **reaction** *n* AIR Pufferreaktion *f*; ~ **solution** *n* ANALYSIS Pufferlösung *f*; ~ **tank** *n* AIR Puffertank *m*; ~ **zone** *n* NOISE Pufferzone *f*
build: ~ **in** *vt* ANALYSIS, SOIL, WASTE, WATER SUPP einbauen
building: ~ **density** *n* MANAGE, POLICY, WATER SUPP Baudichte *f*, Bebauung *f*, Bebauungsdichte *f*; ~ **development plan** *n* POLICY, WATER SUPP Bebauungsplan *m*; ~ **permit** *n* POLICY, WATER SUPP Baugenehmigung *f*; ~ **rubble** *n* WASTE Bau-

schutt *m*; ~ **rubble recycling** *n* WASTE Bauschuttrecycling *nt*; ~ **site** *n* NOISE, SOIL, WASTE Baustelle *f*; ~ **site waste** *n* WASTE Baustellenabfall *m*; ~ **vibration** *n* NOISE Bauwerkschwingung *f*

build: ~**-up** *n* AIR, SOIL, WASTE, WATER POLL, WATER SUPP Ablagerung *f*

built: ~**-in** *adj* ANALYSIS, SOIL, WASTE, WATER SUPP eingebaut; ~**-in measuring equipment** *n* ANALYSIS, SOIL eingebaute Meßeinrichtung *f*; ~**-in monitoring equipment** *n* ANALYSIS, SOIL eingebaute Überwachungseinrichtung *f*; ~**-in obsolescence** *n* WASTE, WATER SUPP eingebaute Alterung *f*; ~**-in regulating equipment** *n* ANALYSIS, SOIL eingebaute Regeleinrichtung *f*; ~**-in roof surface collector** *n* RENEW ENERG dachintegrierter Flachkollektor *m*; ~**-up surface** *n* SOIL überbaute Fläche *f*

bulk: ~ **density** *n* WASTE Schüttdichte *f*; ~ **deposition** *n* WASTE, WATER SUPP lose Ablagerung *f*; ~ **goods** *n pl* WASTE Schüttgut *nt*

bulking *n* WATER POLL *of sludge* Schlammblähung *f*; ~ **agent** *n* SOIL Füllstoff *m*; ~ **sludge** *n* WATER POLL Blähschlamm *m*; ~ **sludge control** *n* WATER POLL Blähschlammbekämpfung *f*; ~ **sludge formation** *n* WATER POLL Blähschlammbildung *f*

bulk: ~ **material** *n* AIR *packed-bed filter* Schüttgut *nt*; ~ **material catalyst** *n* AIR Schüttgutkatalysator *m*; ~ **material reactor** *n* WASTE Schüttgutreaktor *m*; ~ **transport** *n* WASTE, WATER SUPP loser Transport *m*

bulky *adj* WASTE, WATER POLL sperrig; ~ **constituents** *n pl* WATER POLL *in wastewater* sperrige Inhaltsstoffe *m pl*; ~ **waste** *n* WASTE Sperrmüll *m*; ~ **waste collection** *n* WASTE Sperrmüllabfuhr *f*, Sperrmüllsammlung *f*

bulldozer *n* WASTE Planierraupe *f*

bundle *n* WASTE Bündel *nt*; ~**-tying machine** *n* WASTE Bündelmaschine *f*, Paketieranlage *f*, Paketiermaschine *f*

bunkers *n pl* MARINE Treiböl *nt*

bunker: ~ **tank** *n* AIR, WASTE, WATER SUPP Bunkertank *m*

Bunsen: ~ **burner** *n* ANALYSIS Bunsenbrenner *m*; ~ **distribution coefficient** *n* WASTE, WATER SUPP Bunsenscher Verteilungskoeffizient *m*

burden *n* AIR Belastbarkeit *f*

burette *n* ANALYSIS Bürette *f*, Meßgefäß *nt*

burn *vt* AIR, WASTE *solids* brennen; ~ **off** *vt* WASTE, WATER POLL abfackeln; ~ **out** *vt* AIR *flue gases*, SOIL, WASTE ausbrennen

burnable: ~ **gas** *n* AIR Brenngas *nt*

burner *n* AIR *thermal and catalytic combustion*, WATER POLL Brenner *m*

burning: ~ **agent** *n* MARINE Verbrennungsmittel *nt*; ~ **off** *n* WASTE, WATER POLL Abfackeln *nt*; ~**-out** *n* AIR *of pollutants in thermal combustion* Ausbrand *m*

burn: ~**-off rate** *n* AIR Abbrennrate *f*

burnout *n* AIR Ausbrand *m*; ~ **aggregate** *n* WASTE, WATER SUPP Ausbrennaggregat *nt*; ~ **grate** *n* WASTE Ausbrandrost *m*

burnt: ~ **lime** *n* WASTE, WATER SUPP gebrannter Kalk *m*

bursting: ~ **pressure** *n* WASTE, WATER SUPP Berstdruck *m*; ~ **strength** *n* AIR Berstfestigkeit *f*, WASTE Berstfestigkeit *f*, Reißfestigkeit *f*, WATER SUPP Berstfestigkeit *f*; ~ **strength tester** *n* AIR, WASTE, WATER SUPP Berstfestigkeitsprüfgerät *nt*

butadiene *n* AIR Butadien *nt*

butane *n* AIR Butan *nt*

butterfly: ~ **valve** *n* AIR, WATER SUPP Drosselklappe *f*, Absperrklappe *f*

bypass 1. *n* AIR *protection against overheating* Bypass *m*; **2.** *vt* AIR *gas* umleiten

bypass: ~ **chimney** *n* AIR Bypass-Kamin *m*; ~ **duct** *n* AIR Bypass-Leitung *f*; ~ **filter** *n* AIR Bypass-Filter *nt*; ~ **flow** *n* AIR Bypass-Strom *m*; ~ **pressure sensor** *n* AIR Bypass-Drucksensor *m*; ~ **pump** *n* AIR Bypass-Pumpe *f*; ~ **stack** *n* AIR Bypass-Kamin *m*; ~ **stop valve** *n* AIR Bypass-Absperrklappe *f*; ~ **system** *n* AIR, ANALYSIS Bypass-System *nt*

by-product *n* AIR *of combustion* WASTE Nebenprodukt *nt*; ~ **recovery** *n* AIR, RENEW ENERG, WASTE Nebenproduktverwertung *f*

C

CA *abbr* (*cellulose acetate*) WASTE, WATER POLL CA (*Zelluloseacetat*)
cab *n* SAFETY *building machinery* Kabine *f*; **~ enclosure** *n* SAFETY Schutzkabine *f*; **~ filter** *n* SAFETY Kabinenfilter *nt*; **~ with integrated activated carbon filter unit** *n* SAFETY Kabine *f* mit integrierter Aktivkohlefilteranlage
cable: **~-braced** *adj* RENEW ENERG seilverspannt; **~-braced construction** *n* RENEW ENERG *solar modules* seilverspannte Konstruktion *f*; **~ drum** *n* WASTE Kabeltrommel *f*; **~-guided drilling grab** *n* SOIL seilgeführter Bohrlochgreifer *m*
cadmium *n* (*Cd*) SOIL, WASTE, WATER POLL, WATER SUPP Cadmium *nt* (*Cd*)
caisson *n* RENEW ENERG Caisson *m*
cake *vt* AIR, WASTE, WATER POLL, WATER SUPP anbacken
caked: **~ dust** *n* AIR anbackender Staub *m*
caking *n* AIR, WASTE, WATER POLL, WATER SUPP Anbackung *f*
calcareous *adj* WASTE, WATER SUPP kalkhaltig; **~ sinter** *n* WASTE, WATER SUPP Kalksinter *m*
calcify *vi* WATER SUPP verkalken
calcination *n* AIR Kalzinierung *f*, WATER SUPP *in drinking water pipes* Verkalkung *f*
calcine *vt* AIR, WASTE, WATER SUPP kalzinieren
calcium *n* WASTE, WATER POLL, WATER SUPP Calcium *nt*; **~ carbonate** *n* WASTE, WATER SUPP Calciumcarbonat *nt*; **~ hardness** *n* WATER SUPP Kalkhärte *f*; **~ hydroxide** *n* WASTE, WATER POLL, WATER SUPP Calciumhydroxid *nt*, Kalkhydrat *nt*; **~ ion** *n* WASTE, WATER SUPP Calciumion *nt*; **~ oxide** *(CaO) n* AIR Calciumoxid *nt (CaO) n*; **~ phosphate** *n* WATER POLL Calciumphosphat *nt*; **~ sulfate** *AmE*, **~ sulphate** *BrE n* AIR, SOIL, WASTE Gips *m*; **~ sulfate hemihydrated** *AmE*, **~ sulphate hemihydrated** *n BrE* AIR *gypsum form* Calcium-Sulfat-Dihydrat *nt*
calculable *adj* ANALYSIS berechenbar
calculate *vt* ANALYSIS berechnen
calculation *n* ANALYSIS Berechnung *f*
calender *n* AIR *in filter medium* Kalender *m*
calibrate *vt* ANALYSIS eichen, kalibrieren
calibrated *adj* ANALYSIS geeicht, kalibriert; **~ substance** *n* ANALYSIS Eichsubstanz *f*; **~ value** *n* ANALYSIS Eichwert *m*, Kalibrierwert *m*; **~ watershed** *n* ANALYSIS, SOIL *agriculture* geeichtes Wasserrückhaltebecken *nt*
calibrating: **~ gas** *n* AIR Kalibriergas *nt*
calibration *n* ANALYSIS Eichung *f*, Kalibrieren *nt*, Kalibrierung *f*; **~ buffer** *n* ANALYSIS *pH-meter* Kalibrierpuffer *m*; **~ curve** *n* ANALYSIS Eichkurve *f*, Kalibrierkurve *f*; **~ data** *n* ANALYSIS Kalibrierdaten *pl*; **~ device** *n* ANALYSIS Kalibrationsgerät *nt*; **~ gas** *n* ANALYSIS Eichgas *nt*, Kalibriergas *nt*; **~ gas bottle** *n* ANALYSIS Kalibriergasflasche *f*; **~ mark** *n* ANALYSIS Eichmarke *f*; **~ mode** *n* ANALYSIS Kalibriermodus *m*; **~ pulse** *n* ANALYSIS Eichimpuls *m*, Kalibrierimpuls *m*; **~ range** *n* ANALYSIS Kalibrierbereich *m*; **~ record** *n* ANALYSIS Kalibrierprotokoll *nt*; **~ solution** *n* ANALYSIS Standardlösung *f*, Eichlösung *f*; **~ storer** *n* ANALYSIS Kalibrierspeicher *m*; **~ unit** *n* ANALYSIS Kalibriereinheit *f*
calibrator *n* ANALYSIS Kalibrator *m*
caloric: **~ conductibility** *n* ANALYSIS Wärmeleitfähigkeit *f*
calorific: **~ value** *n* GEN Heizwert *m*
calorimeter *n* ANALYSIS Kalorimeter *nt*
calorimetric *adj* ANALYSIS kalorimetrisch; **~ pararosaniline method** *n* ANALYSIS *gas measurement* kalorimetrische Pararosanilinmethode *f*
calorimetry *n* ANALYSIS Kalorimetrie *f*
CA: **~ membrane** *n* WATER SUPP CA-Membran *f*
can *n* WASTE *food container* Büchse *f*, Dose *f*, Konservenbüchse *f*
canal *n* WASTE, WATER POLL, WATER SUPP Kanal *m*; **~ branching** *n* WATER SUPP Kanalabzweigung *f*; **~ constriction** *n* WATER SUPP Kanalverengung *f*; **~ drainage** *n* WATER SUPP Kanalentwässerung *f*
canalization *n* WASTE, WATER POLL, WATER SUPP Kanalisation *f*, Abwasserleitung *f*, *system of pipes* Kanalisationssystem *nt*, Kanalisationsnetz *nt*
canalize *vt* WASTE, WATER POLL, WATER SUPP kanalisieren
canal: **~ junction** *n* WATER SUPP Kanaleinmündung *f*
cancer: **~-causing** *adj* SAFETY krebserzeugend; **~ risk** *n* SAFETY Krebsrisiko *nt*
candle: **~ filter** *n* AIR, WATER SUPP Kerzenfilter *nt*
canister *n* WASTE Kanister *m*
can: **~ scrap** *n* WASTE Dosenschrott *m*
CaO *abbr* (*calcium oxide*) AIR CaO (*Calciumoxid*)
caoutchouc *n* WASTE Kautschuk *m*
cap 1. *n* AIR *bubble-cap tray tower* Glocke *f*; **2.** *vt* SOIL einkapseln
capacity *n* GEN *capability* Fähigkeit *f*, Kapazität *f*, *output* Leistung *f*
capillary 1. *adj* ANALYSIS kapillar; **2.** *n* ANALYSIS Kapillare *f*
capillary: **~ electrophoresis** *n* (*CE*) ANALYSIS Kapillar-Elektrophorese *f*; **~ electrophoresis mass spectrometry** *n* (*CE-MS*) ANALYSIS Kapillar-Elektrophorese *f* in Kopplung mit Massenspektrometrie; **~ flow meter** *n* ANALYSIS Kapillarströmungsmesser *m*; **~ GC** *n* ANALYSIS Kapillar-GC *f*; **~ GC-MS** *n* ANALYSIS Kapillar-GC-MS *nt*; **~ gel electrophoresis** *n* (*CGE*) ANALYSIS Kapillargel-Elektrophorese *f*; **~ isoelectric focusing** *n* (*CIEF*) ANALYSIS kapillarisoelektrische Fokussierung *f* (*CIEF*); **~ rise** *n* SOIL, WASTE, WATER SUPP kapillare Steighöhe *f*; **~ tube** *n* ANALYSIS Kapillarrohr *nt*; **~ viscosimeter** *n* ANALYSIS Kapillarviskosimeter *nt*; **~ water** *n* SOIL, WATER POLL, WATER SUPP Kapillarwasser *nt*; **~ zone electrophoresis** *n* (*CZE*) ANALYSIS Kapillarzone-Elektrophorese *f*
capnograph *n* ANALYSIS *measurement of dust concentration* Kapnograph *m*
capping *n* WASTE Einkapselung *f*
capture 1. *n* GEN Auffangen *nt*, Erfassung *f*; **2.** *vt* GEN *of contaminants* auffangen, erfassen
capture: **~ rate** *n* WASTE, WATER POLL Erfassungsquote *f*
carbon *n* RENEW ENERG Kohlenstoff *m*; **~ absorption** *n* SOIL Kohlenstoff-Absorption *f*
carbonate: **~ balance** *n* WATER SUPP Kalkkohlensäuregleichgewicht *nt*; **~ hardness** *n* (*CH*) WATER POLL, WATER SUPP Carbonathärte *f* (*CH*); **~ hardness determination in wastewater** *n* ANALYSIS Carbonathärtebestimmung *f*
carbon: **~ cycle** *n* RENEW ENERG *biomass* Kohlenstoffkreislauf *m*; **~ dioxide** *n* (CO_2) AIR, WASTE, WATER SUPP Kohlendioxid *nt* (CO_2); **~ dioxide budget** *n* AIR, POLICY Kohlendioxidhaushalt *m*; **~ dioxide greenhouse effect** *n* AIR Kohlendioxidtreibhauseffekt *m*; **~ fiber paper** *AmE*, **~ fibre paper** *BrE n* AIR Kohlefaserpapier *nt*
carbonic *adj* AIR kohlenstoffhaltig, WASTE, WATER SUPP kohlensäurehaltig; **~ acid** *n* WATER SUPP Kohlensäure *f*;

~ **adsorbent** *n* AIR kohlenstoffhaltiges Adsorbens *nt*; ~ **anhydride** *n* WATER SUPP Kohlensäureanhydrid *nt*; ~ **water** *n* WASTE, WATER SUPP kohlensäurehaltiges Wasser *nt*
carboniferous *adj* AIR kohlenstoffhaltig
carbonized: ~ **RDF** *n* WASTE karbonisierter BRAM *m*
carbon: ~ **monoxide** *n* (*CO*) AIR, WASTE Kohlenmonoxid *nt* (*CO*); ~ **monoxide detector** *n* ANALYSIS Kohlenmonoxidanzeiger *m*; ~**-nitrogen-phosphorus ratio** *n* (*C/N/P ratio*) ANALYSIS Kohlenstoff-Stickstoff-Phosphor-Verhältnis *nt* (*C/N/P-Verhältnis*); ~**-nitrogen ratio** *n* (*C/N ratio*) ANALYSIS Kohlenstoff-Stickstoff-Verhältnis *nt* (*C/N-Verhältnis*)
carbonyl: ~ **sulfide** *AmE*, ~ **sulphide** *BrE n* AIR, WASTE, WATER SUPP Carbonylsulfid *nt*
carburetor *AmE see carburettor BrE*
carburettor *n BrE* AIR Vergaser *m*; ~ **emissions** *n pl BrE* AIR *hydrocarbons* Vergaseremissionen *f pl*
carcinogenic *adj* SAFETY karzinogen, krebserzeugend; ~ **substance** *n* SAFETY Karzinogen *nt*, krebserzeugender Stoff *m*
cardboard *n* WASTE Karton *m*, Pappe *f*; ~ **packaging** *n* WASTE Kartonverpackung *f*, Kartonage *f*; ~ **tube** *n* WASTE Kartonröhre *f*; ~ **waste** *n* WASTE Pappeabfall *m*
carded: ~ **packaging** *n* WASTE Kleinware *f* auf Kartonunterlagen, kardierte Produkte *nt pl*
car: ~ **battery** *n BrE* (*cf automobile battery AmE*) WASTE Autobatterie *f*; ~ **fragmentation plant** *n* WASTE Altfahrzeugverschrottungsanlage *f*
cargo: ~ **pump** *n* MARINE Ladepumpe *f*
carousel: ~ **extractor** *n* SOIL Karussellextraktor *m*
carrier *n* AIR Träger *m*; ~ **bag** *n* WASTE Tragbeutel *m*, Tragetasche *f*; ~ **box** *n* MANAGE, WASTE Transportkiste *f*; ~ **gas** *n* AIR, ANALYSIS Trägergas *nt*; ~ **layer** *n* AIR *catalyzer* Trägerschicht *f*, WASTE *of recycled material* Tragschicht *f*; ~ **material** *n* AIR, ANALYSIS, SOIL, WASTE, WATER POLL, WATER SUPP Trägermaterial *nt*; ~ **medium** *n* AIR Trägermedium *nt*; ~ **surface** *n* AIR *catalyst* Trägeroberfläche *f*
carry: ~ **out** *vt* GEN *tests* durchführen
carrying: ~ **capacity** *n* WASTE Kapazitätsbelastung *f*, Tragvermögen *nt*
car: ~ **scrap** *n* WASTE Altfahrzeugschrott *m*, Autoschrott *m*
carton *n* GEN Karton *m*, Kasten *m*, Kiste *f*, Schachtel *f*
cartridge: ~ **filter** *n* AIR Patronenfilter *nt*
cascade *n* ANALYSIS, SOIL, WATER POLL Kaskade *f*; ~ **acidification** *n* SOIL Kaskadensäuerung *f*; ~ **aeration** *n* WATER POLL Kaskadenbelüftung *f*; ~ **denitrification** *n* WATER POLL Kaskadendenitrifikation *f*; ~ **impactor** *n* ANALYSIS Kaskadenimpaktor *m*
cast: ~**-iron pipe** *n* WATER POLL Gußeisenrohr *nt*
catabolism *n* WATER POLL Katabolismus *m*
catalyse *vt BrE* AIR, SOIL, WATER SUPP katalysieren, katalytisch behandeln
catalysis *n* AIR, SOIL, WATER SUPP Katalyse *f*
catalyst *n* AIR, SOIL, WATER SUPP Katalysator *m*; ~ **bed** *n* AIR Katalysatorbett *nt*, Katalysatorschicht *f*; ~ **carrier** *n* AIR, WATER SUPP Katalysatorträger *m*; ~ **charge** *n* AIR Katalysatorfüllung *f*, Katalysatorschüttung *f*; ~ **container** *n* AIR Katalysatorbehälter *m*; ~ **contamination** *n* AIR Katalysatorvergiftung *f*; ~ **deactivation** *n* AIR Katalysatordesaktivierung *f*; ~ **degradation** *n* AIR Katalysatorschädigung *f*; ~ **deterioration** *n* AIR Katalysatorverschlechterung *f*, Katalysatorzerstörung *f*; ~ **dust** *n* AIR Katalysatorstaub *m*; ~ **efficiency** *n* AIR Katalysatorwirksamheit *f*, Katalysatorwirkungsgrad *m*; ~ **element** *n* AIR Katalysatorelement *nt*; ~ **fines** *n pl* AIR Katalysatorstaub *m*; ~ **level** *n* AIR Katalysatorebene *f*; ~ **lifetime** *n* AIR Katalysatorlebensdauer *f*, Katalysatorstandzeit *f*, Katalysatorwirkzeit *f*; ~ **material** *n* AIR Katalysatormaterial *nt*; ~ **module** *n* AIR Katalysatormodul *nt*, Katalysatorpaket *nt*; ~ **plate** *n* AIR Katalysatorplatte *f*; ~ **poison** *n* AIR Katalysatorgift *nt*, Kontaktgift *nt*; ~ **poisoning** *n* AIR Katalysatorvergiftung *f*; ~ **regeneration** *n* AIR Katalysatorregeneration *f*; ~ **step** *n* AIR Katalysatorstufe *f*; ~ **surface** *n* AIR Katalysatoroberfläche *f*; ~ **technology** *n* AIR Katalysatortechnik *f*; ~ **volume** *n* AIR Katalysatorvolumen *nt*
catalytic *adj* AIR, SOIL, WATER SUPP katalytisch; ~ **acceleration** *n* AIR katalytische Beschleunigung *f*; ~ **active layer** *n* AIR katalytisch aktive Schicht *f*; ~ **active metal** *n* AIR katalytisch aktives Metall *nt*; ~ **activity** *n* AIR katalytische Aktivität *f*; ~ **afterburning** *n* WATER POLL katalytische Nachverbrennung *f* (*KNV*); ~ **coating** *n* AIR *of filter material* katalytische Beschichtung *f*; ~ **combustion** *n* AIR katalytische Nachverbrennung *f* (*KNV*); ~ **combustion plant** *n* AIR KNV-Anlage *f*, katalytische Nachverbrennungsanlage *f*; ~ **converter** *n BrE* (*cf catalytic muffler AmE*) AIR Kat *m*, Katalysator *m*; ~ **converter closed loop** *n* AIR Katalysator *f* mit Lambdaregelung; ~ **converter open loop** *n* AIR Katalysator *f* ohne Lambdaregelung; ~ **cracker** *n* AIR katalytisches Krackmittel *nt*; ~ **cracking** *n* AIR katalytisches Kracken *nt*; ~ **cracking unit** *n* AIR Katalysator-Krackanlage *f*; ~ **impregnation** *n* AIR katalytische Imprägnierung *f*; ~ **incineration** *n* AIR katalytische Nachverbrennung *f* (*KNV*); ~ **incineration plant** *n* AIR KNV-Anlage *f*, katalytische Nachverbrennungsanlage *f*; ~ **incinerator** *n* AIR katalytischer Verbrenner *m*; ~ **muffler** *n AmE* (*cf catalytic converter BrE*) AIR Kat *m*, Katalysator *m*; ~ **oxidation** *n* AIR, SOIL, WASTE, WATER SUPP katalytische Oxidation *f*; ~ **reaction** *n* AIR katalytische Reaktion *f*; ~ **reactor** *n* AIR katalytischer Reaktor *m*; ~ **reduction process** *n* AIR katalytisches Reduktionsverfahren *nt*; ~ **tower** *n* AIR Katalysatorturm *m*; ~ **UV oxidation** *n* WASTE, WATER POLL, WATER SUPP katalytische UV-Oxidation *f*; ~ **waste gas cleaner** *n* AIR katalytischer Abgasreiniger *m*
catalyze *AmE see catalyse BrE*
catch: ~ **feeder** *n* WASTE, WATER SUPP Berieselungsgraben *m*, Berieselungskanal *m*, Bewässerungsgraben *m*
catchment: ~ **area** *n* WATER SUPP Einzugsgebiet *nt*, Niederschlagsgebiet *nt*
caterpillar *n* SOIL Raupe *f*
cathode *n* AIR Kathode *f*; ~ **ray tube** *n* SAFETY Kathodenstrahlröhre *f*
cathodic: ~ **protection** *n* RENEW ENERG *tidal power plant* kathodischer Schutz *m*
cation *n* WATER POLL, WATER SUPP Kation *nt*; ~ **exchange** *n* SOIL, WATER POLL, WATER SUPP Kationenaustausch *m*; ~ **exchange capacity** *n* (*CEC*) SOIL Kationenaustauschkapazität *f*; ~ **exchanger** *n* SOIL, WATER SUPP Kationenaustauscher *m*
cationic *adj* WATER POLL, WATER SUPP kationisch; ~**-active** *adj* WATER POLL kationenaktiv; ~ **exchanger** *n* WATER POLL Kationenaustauscher *m*
cation: ~**-permeable** *adj* WATER POLL kationenpermeabel; ~**-permeable membrane** *n* WATER POLL kationenpermeable Membran *f*; ~**-selective** *adj* WATER POLL kationenselektiv; ~**-selective membrane** *n* WATER POLL kationenselektive Membran *f*
cause: ~ **of accident** *n* SAFETY Unfallursache *f*; ~ **of malfunction** *n* SAFETY Störungsursache *f*
caustic: ~ **scrubbing** *n* AIR Laugewäsche *f*; ~ **soda solution** *n* WATER SUPP Natronlauge *f*
cavitation: ~ **determination** *n* WATER POLL, WATER SUPP Kavitationsbestimmung *f*; ~ **susceptibility** *n* WATER POLL, WATER SUPP Kavitationsempfindlichkeit *f*
cavity: ~ **absorber** *n* NOISE Hohlraumabsorber *m*
CBC *abbr* (*circulating fluidized bed*) AIR ZWS (*zirkulierende Wirbelschicht*)
CC *abbr* (*closed cycle*) RENEW ENERG geschlossener Kreislauf *m*

CCGT *abbr* (*combined-cycle gas turbine*) AIR GuD-Anlage *f* (*Gas- und Dampfanlage*)
CCS *abbr* (*controlled combustion system*) AIR CCS-System *nt*
Cd *abbr* (*cadmium*) SOIL, WASTE, WATER POLL, WATER SUPP Cd (*Cadmium*)
CD *abbr* (*chlorine demand*) WASTE, WATER SUPP Chlorbedarf *m*
CE *abbr* (*capillary electrophoresis*) ANALYSIS Kapillar-Elektrophorese *f*
CEC *abbr* (*cation exchange capacity*) SOIL Kationenaustauschkapazität *f*
ceiling: ~ **panel** *n* NOISE Deckenplatte *f*
cell: ~ **method** *n* WASTE, WATER SUPP Fächermethode *f*; ~ **stack** *n* AIR *fuel cell* Zellenstack *m*
cellular: ~ **wheel sluice** *n* AIR *of cyclone* Zellenradschleuse *f*
celluloid *n* WASTE Zelluloid *nt*
cellulose *n* WASTE Zellstoff *m*, Zellulose *f*; ~ **acetate** *n* (*CA*) WASTE, WATER POLL Zelluloseacetat *nt* (*CA*); ~ **acetate membrane** *n* WASTE, WATER POLL, WATER SUPP Zellulose-acetat-Membran *f*; ~ **bleach** *n* WASTE Zellstoffbleiche *f*; **~-containing** *adj* RENEW ENERG zellulosehaltig; ~ **content** *n* WASTE *compost* Zellulosegehalt *m*; ~ **fiber** *AmE*, ~ **fibre** *BrE* *n* WASTE Zellstoffaser *f*; ~ **raw material** *n* RENEW ENERG *biomass conversion base material* zellulosehaltiger Rohstoff *m*; ~ **waste** *n* WASTE zellulosehaltiger Abfall *m*
cement *n* SOIL *binding agent* Zement *m*
cement/clay: ~ **mixture** *n* SOIL *slurry* Zement/Tongemisch *nt*
cement: ~ **with fly ash** *n* WATER SUPP Flugaschezement *m*
cement/water: ~ **mixture** *n* SOIL *slurry* Zement/Wassergemisch *nt*
CE-MS *abbr* (*capillary electrophoresis mass spectrometry*) ANALYSIS Kapillar-Elektrophorese in Kopplung mit Massenspektrometrie *f*
CEN *abbr* (*European Standardization Committee*) ANALYSIS CEN (*Europäisches Normungsinstitut*)
central *adj* GEN zentral; ~ **collection station** *n* WASTE, WATER SUPP zentrale Sammelstation *f*, zentraler Sammelplatz *m*; ~ **control room** *n* AIR, WASTE, WATER SUPP zentrale Leitstelle *f*, zentraler Steuerstand *m*; ~ **focal point** *n* RENEW ENERG zentraler Brennpunkt *m*
centralization *n* AIR, RENEW ENERG, WASTE, WATER SUPP Zentralisierung *f*
centralize *vt* AIR, RENEW ENERG, WASTE, WATER SUPP zentralisieren
central: ~ **landfill** *n* WASTE Zentraldeponie *f*; ~ **pipe** *n* WASTE, WATER SUPP Zentralrohr *nt*; ~ **solar heating plant** *n* RENEW ENERG zentrale Solarheizungsanlage *f*; ~ **well** *n* AIR Zentralrohr *nt*
centrifugal *adj* GEN zentrifugal, Radial-; ~ **acceleration** *n* AIR Zentrifugalbeschleunigung *f*; ~ **collector** *n* AIR *cyclones included* Zentrifugalabscheider *m*; ~ **compressor** *n* AIR Radialverdichter *m*; ~ **fan** *n* AIR Radialventilator *m*; ~ **field** *n* WASTE *plastics sorting* Zentrifugalfeld *nt*; ~ **force** *n* AIR, RENEW ENERG *tidal power* Fliehkraft *f*, Zentrifugalkraft *f*; ~ **force activity** *n* AIR Fliehkraftwirkung *f*; ~ **force precipitator** *n* AIR *cyclone* Zentrifugalkraftabscheider *m*; ~ **force sedimentation** *n* ANALYSIS *measurement for dust* Fliehkraftsedimentation *f*; ~ **mass** *n* RENEW ENERG Schwungmasse *f*; ~ **precipitation** *n* AIR Fliehkraftabscheidung *f*; ~ **pump** *n* MARINE, WATER POLL Kreiselpumpe *f*, Zentrifugalpumpe *f*, WATER SUPP *volute, circular casing and stormwater pumps* Kreiselpumpe *f*; ~ **screen** *n* WATER POLL Zentrifugalsortierer *m*; ~ **skimmer** *n* MARINE Zentrifugalabschöpfgerät *nt*
centrifugation *n* GEN Zentrifugation *f*, Zentrifugierung *f*
centrifuge 1. *n* GEN *machine* Zentrifuge *f*, *process* Zentrifugieren *nt*, WATER POLL *in dewatering* Trennschleuder *f*; **2.** *vt* GEN zentrifugieren
centrifuge: ~ **casing** *n* WASTE Zentrifugenmantel *m*; ~ **dewatering** *n* WATER POLL Zentrifugenentwässerung *f*; ~ **prestorage tank** *n* AIR Zentrifugenvorlagebehälter *m*
centrifuging *n* GEN Zentrifugation *f*
ceramic: ~ **body** *n* AIR *catalyst* Keramikkörper *m*; ~ **carrier** *n* AIR *catalyst* Keramikträger *m*; ~ **catalyst** *n* AIR Keramikkatalysator *m*; ~ **filter** *n* AIR Keramikfilter *nt*; **~-lined combustion chamber** *n* AIR keramisch ausgekleidete Brennkammer *f*; ~ **material** *n* AIR *filter material* keramisches Material *nt*; ~ **monolith** *n* AIR *catalyst* Keramikmonolith *m*; ~ **pellet** *n* SOIL keramisches Pellet *nt*
ceramics *n pl* AIR Keramik *f*
ceramization *n* SOIL *of pellets* Keramisierung *f*
cesspit *n* WATER POLL Absetzgrube *f*, Faulgrube *f*, Klärgrube *f*, Senkgrube *f*
cesspool *n* WATER POLL *sedimentation* Absetzgrube *f*, Faulgrube *f*, *catch pit* Klärgrube *f*, *drainage pit for groundwater* Senkgrube *f*; ~ **overflow** *n* WATER POLL Grubenüberlauf *m*
cetane *n* AIR Cetan *m*; ~ **number** *n* AIR Cetanwert *m*, Cetanzahl *f* (*CaZ*)
CFC *abbr* (*chlorofluorocarbon*) AIR FCKW (*Fluorchlorkohlenwasserstoff*)
CFR *abbr* (*countercurrent regeneration*) WATER POLL Gegenstromregeneration *f*
CFS: **~-processed waste** *n* WASTE, WATER SUPP Umsetzungsprodukt *nt*; **~-treated waste** *n* WASTE, WATER SUPP Umsetzungsprodukt *nt*
CGE *abbr* (*capillary gel electrophoresis*) ANALYSIS Kapillargel-Elektrophorese *f*
CH *abbr* (*carbonate hardness*) WATER POLL, WATER SUPP CH (*Carbonathärte*)
chain *n* WASTE, WATER POLL, WATER SUPP Kette *f*; **~-and-flight collecting mechanism** *n* WATER POLL Kettenkratzer *m*, Kettenräumer *m*; ~ **conveyor** *n* AIR *packed-bed filter* Becherwerk *nt*; ~ **screen** *n* WATER POLL Kettenstabrechen *m*
chairman *n* POLICY Vorsitzender *m*
chamber *n* GEN Kammer *f*; ~ **filter press** *n* SOIL, WASTE, WATER SUPP Kammerfilterpresse *f*; ~ **water level** *n* RENEW ENERG *wave power plant* Kammerwasserspiegel *m*
change 1. *n* ANALYSIS Veränderung *f*; **2.** *vt* ANALYSIS verändern; **3.** *vi* ANALYSIS sich verändern
changing: ~ **fall of water** *n* WATER SUPP wechselndes Gefälle *nt*; ~ **head** *n* WASTE, WATER SUPP wechselndes Gefälle *nt*
channel 1. *n* AIR, WATER SUPP Kanal *m*; **2.** ~ **back** *vt* WASTE zurückführen
channel: ~ **impeller pump** *n* WATER SUPP Kanalradkreiselpumpe *f*
channeling *AmE see channelling BrE*
channelling *n BrE* AIR Bachbildung *f*
channel: ~ **recuperator** *n* AIR Kanalrekuperator *m*
characteristics: ~ **of materials** *n pl* WASTE Materialeigenschaften *f pl*
charge 1. *n* MANAGE *fee* Abgabe *f*, Gebühr *f*, POLICY *tax, rate* Abgabe *f*; **2.** *vt* AIR, WASTE, WATER POLL, WATER SUPP beschicken
charge: ~ **carrier** *n* AIR *electrostatic precipipator* Ladungsträger *m*
charged *adj* AIR, WATER POLL geladen; ~ **gas stream** *n* AIR geladener Gasstrom *m*; ~ **ion** *n* WATER POLL geladenes Ion *nt*; ~ **particle** *n* AIR aufgeladene Partikel *f*, geladenes Teilchen *nt*; ~ **polymer** *n* WATER POLL geladenes Polymer *nt*; ~ **with solar power** *adj* RENEW ENERG mit Sonnenstrom geladen
charge: ~ **exemption** *n* MANAGE, POLICY Abgabenbefreiung *f*; ~ **reduction** *n* MANAGE, POLICY Abgabenkürzung *f*; ~ **reversal** *n* AIR *particulates in ESP* Umladung *f*
charges: ~ **quota** *n* MANAGE, POLICY Abgabenquote *f*; ~ **tariff** *n* MANAGE, POLICY Abgabentarif *m*
charging *n* GEN Beschickung *f*; ~ **chamber** *n* WASTE, WATER SUPP Beschickungskammer *f*; ~ **electrode wire** *n* AIR *in plate-wire ESPs* Sprühdraht *m*; ~ **facility** *n* WASTE, WATER SUPP Beschickungseinrichtung *f*; ~ **material** *n* AIR Aufgabegut *nt*; ~ **zone** *n* AIR aktive Zone *f*

check 1. *n* ANALYSIS, WASTE Kontrolle *f*, Überprüfung *f*; **2.** *vt* ANALYSIS, WASTE kontrollieren, überprüfen
checked *adj* ANALYSIS, WASTE überprüft
checking *n* ANALYSIS, WASTE Kontrolle *f*, Überprüfung *f*
check: **~ valve** *n* WASTE, WATER SUPP Absperrventil *nt*
chelate *n* SOIL *additive*, WATER POLL Chelat *nt*; **~ former** *n* SOIL, WATER POLL Chelatbildner *m*
chelating *adj* WATER POLL chelatbildend; **~ ion exchanger** *n* WATER POLL chelatbildender Ionenaustauscher *m*
CHEMFIX: **~ process**® *n* SOIL, WASTE, WATER SUPP CHEMFIX-Verfahren® *nt*
chemical *adj* GEN chemisch; **~ adsorption** *n* AIR Chemosorption *f*, chemische Adsorption *f*; **~ aggressivity** *n* WASTE, WATER SUPP chemische Aggressivität *f*; **~ conditioning** *n* WASTE, WATER SUPP chemische Konditionierung *f*; **~ disinfection** *n* WASTE, WATER SUPP chemische Desinfektion *f*; **~ dosimeter** *n* ANALYSIS chemisches Dosimeter *nt*; **~ hazard** *n* SAFETY chemische Gefahr *f*; **~ ionization mass spectrometry** *n* (*CI-MS*) ANALYSIS chemische Ionisation-Massenspektrometrie *f* (*CI-MS*)
chemically: **~ aggressive** *adj* AIR chemisch aggressiv; **~ aggressive dust** *n* AIR chemisch aggressiver Staub *m*; **~ qualitative** *adj* ANALYSIS chemisch qualitativ; **~ quantitative** *adj* ANALYSIS chemisch quantitativ; **~ resistant** *adj* AIR, SAFETY, WASTE, WATER POLL chemiebeständig, chemikalienbeständig, chemisch resistent; **~ resistant protective suit** *n* SAFETY chemikalienbeständiger Schutzanzug *m*; **~ stabilized** *adj* WATER POLL chemisch stabilisiert
chemical: **~ oxidation** *n* WASTE, WATER POLL, WATER SUPP chemische Oxidation *f*; **~-oxidative gas scrubbing** *n* AIR chemisch oxidative Gaswäsche *f*; **~ oxygen demand** *n* (*COD*) AIR, WASTE, WATER POLL, WATER SUPP chemischer Sauerstoffbedarf *m* (*CSB*); **~ parameter** *n* SOIL, WATER POLL chemischer Parameter *m*; **~-physical** *adj* AIR, SOIL, WASTE, WATER POLL, WATER SUPP chemisch-physikalisch; **~-physical pretreatment** *n* WASTE, WATER SUPP chemisch-physikalische Vorbehandlung *f*; **~ precipitation** *n* AIR, WASTE, WATER POLL chemische Fällung *f*; **~ reagent** *n* AIR, WATER POLL, WATER SUPP Chemikalie *f*; **~ regeneration** *n* WATER POLL chemische Regeneration *f*; **~ residue** *n* WASTE Chemikalienreste *m pl*; **~ resistance** *n* AIR *filter materials* chemische Beständigkeit *f*
chemicals *n pl* AIR, WATER POLL, WATER SUPP Chemikalien *f pl*
chemical: **~ sediment** *n* WASTE, WATER POLL, WATER SUPP chemisches Sediment *nt*; **~ self-purification** *n* WASTE, WATER POLL, WATER SUPP chemische Selbstreinigung *f*; **~ stabilization** *n* WATER POLL chemische Stabilisierung *f*; **~ strain** *n* WASTE *landfill* chemische Belastung *f*; **~ waste** *n* WASTE chemischer Abfall *m*; **~ wastewater** *n* WATER POLL Chemieabwasser *nt*; **~ wastewater treatment** *n* WATER POLL chemische Abwasseraufbereitung *f*; **~ water quality** *n* WATER SUPP chemische Gewässergüte *f*
chemiluminescence *n* ANALYSIS Chemilumineszenz *f*; **~ detection** *n* (*CLD*) ANALYSIS Chemilumineszenzdetektion *f* (*CLD*); **~ measuring instrument** *n* ANALYSIS *ozone detection* Chemilumineszenzmeßgerät *nt*
chemisorb *vt* AIR chemisorbieren
chemisorption *n* AIR Chemosorption *f*, chemische Adsorption *f*; **~ filter** *n* AIR Chemisorptionsfilter *nt*
chemolysis *n* WATER POLL Chemolyse *f*
chevron: **~ baffle** *n* AIR Chevron-Baffle *nt*, Rasterdampfsperre *f*
chimney *n* AIR, RENEW ENERG, WASTE Kamin *m*, Schornstein *m*; **~ discharge** *n* AIR Kaminauswurf *m*, Schornsteinauswurf *m*; **~ emissions** *n pl* AIR Kaminauswurf *m*, Schornsteinauswurf *m*; **~ height** *n* AIR Kaminhöhe *f*, Schornsteinhöhe *f*
chip *vt* WASTE zerspanen
chipboard *n* NOISE Spanplatte *f*, WASTE Holzspanplatte *f*
chipping *n* WASTE Zerspanung *f*
chlor: **~-alkali electrolysis** *n* WASTE, WATER POLL, WATER SUPP Chlor-Alkali-Elektrolyse *f*
chlorate *n* AIR, SOIL, WASTE, WATER SUPP Chlorat *nt*
chloride *n* AIR, SOIL, WASTE, WATER SUPP Chlorid *nt*
chlorinate *vt* AIR, SOIL, WATER POLL, WATER SUPP chlorieren
chlorinated *adj* AIR, SOIL, WATER POLL, WATER SUPP chloriert; **~ hydrocarbon** *n* AIR Chlorkohlenwasserstoff *m* (*CKW*), chlorierter Kohlenwasserstoff *m*
chlorination *n* AIR, SOIL, WATER POLL, WATER SUPP Chlorung *f*
chlorinator *n* AIR, SOIL, WATER POLL, WATER SUPP Chlorzusatzgerät *nt*
chlorine *n* (*Cl*) AIR, SOIL, WATER POLL, WATER SUPP Chlor *nt* (*Cl*); **~ bleaching** *n* WASTE, WATER SUPP Chlorbleichlauge *f*; **~ demand** *n* (*CD*) WASTE, WATER SUPP Chlorbedarf *m*; **~ feeding** *n* WATER SUPP Chlordosierung *f*; **~-free** *adj* WASTE chlorfrei; **~-free paper** *n* WASTE chlorfreies Papier *nt*; **~ gas** *n* AIR, WASTE Chlorgas *nt*
chlorobenzene *n* AIR Chlorbenzol *nt*
chloroethylene *n* WASTE *vinyl chloride* Chlorethen *nt*
chlorofluorocarbon *n* (*CFC*) AIR Fluorchlorkohlenwasserstoff *m* (*FCKW*)
chlorophenol *n* AIR Chlorphenol *nt*
chocolate: **~ mousse** *n* MARINE Schokoladenschaum *m*, Ölschlamm *m*
chop *vt* WASTE häckseln
chopping *n* WASTE Häckseln *nt*
C-horizon *n* SOIL C-Horizont *m*, Unterboden *m*
CHP *abbr* (*combined heat and power*) AIR, WASTE KWK (*Kraft-Wärme-Kopplung*); **~ station** *n* AIR, RENEW ENERG BHKW (*Blockheizkraftwerk*); **~ station with fuel cell plant** *n* AIR BHKW mit Brennstoffzellen *nt*
chromate: **~-containing** *adj* WATER POLL chromathaltig; **~-containing waste** *n* WASTE chromathaltiger Abfall *m*; **~-containing wastewater** *n* WATER POLL chromathaltiges Abwasser *nt*; **~ reduction** *n* WASTE Chromatreduktion *f*
chromatogram *n* ANALYSIS Chromatogramm *nt*
chromatograph *vt* ANALYSIS chromatographieren
chromatographic *adj* ANALYSIS chromatographisch; **~ column** *n* ANALYSIS chromatographische Säule *f*; **~ data system** *n* ANALYSIS Chromatographiedatensystem *nt*
chromium *n* (*Cr*) AIR, SOIL, WATER POLL Chrom *nt* (*Cr*); **~-containing** *adj* WATER POLL chromhaltig; **~-containing sludge** *n* WATER POLL chromhaltiger Schlamm *m*
chromogen *n* ANALYSIS Chromogen *nt*
chromophoric *adj* WATER POLL farbgebend; **~ water components** *n pl* WATER POLL farbgebende Wasserinhaltsstoffe *m pl*
chronic *adj* SAFETY chronisch; **~ effect** *n* SAFETY, WASTE chronische Wirkung *f*; **~ poisoning** *n* SAFETY chronische Vergiftung *f*
chronopotentiometry *n* ANALYSIS Chronopotentiometrie *f*
chunk: **~-form activated carbon** *n* AIR *adsorbent* Formaktivkohle *f*
CIEF *abbr* (*capillary isoelectric focusing*) ANALYSIS CIEF (*kapillarisoelektrische Fokussierung*)
CI-MS *abbr* (*chemical ionization mass spectrometry*) ANALYSIS CI-MS (*chemische Ionisation-Massenspektrometrie*)
circular: **~ basin** *n* WATER POLL Rundbecken *nt*; **~ broom** *n* WASTE Tellerbürste *f*; **~ clarifier** *n* WATER POLL Rundbecken *nt*; **~ degritter** *n* WATER POLL Rundsandfang *m*; **~ main** *n* WATER SUPP Ringleitung *f*; **~ movement** *n* AIR Kreisbewegung *f*; **~ scraper** *n* WATER POLL Rundräumer *m*
circulate 1. *vt* AIR, WASTE, WATER POLL, WATER SUPP umwälzen; **2.** *vi* WATER POLL zirkulieren
circulating: **~ fluidized bed** *n* (*CBC*) AIR, SOIL zirkulierende Wirbelschicht *f* (*ZWS*); **~ fluidized-bed reactor** *n* AIR ZWS-Reaktor *m*, zirkulierende Wirbelschicht *f*; **~ liquid** *n* AIR Kreislaufflüssigkeit *f*; **~ water** *n* AIR Kreislaufwasser *nt*
circulation *n* AIR, WASTE, WATER POLL, WATER SUPP *circuit*

Kreislauf *m*, *rotation, revolution* Umlauf *m*, Umströmung *f*, Umwälzen *nt*, Zirkulation *f*; ~ **evaporator** *n* WATER POLL, WATER SUPP Umlaufverdampfer *m*; ~ **pump** *n* AIR Kreislaufpumpe *f*, WATER POLL Umlaufpumpe *f*, Zirkulationspumpe *f*
cistern *n* WATER POLL Behälter *m*, Zisterne *f*, WATER SUPP Zisterne *f*
civil: ~ **law** *n* POLICY Zivilrecht *nt*
Cl *abbr* (*chlorine*) AIR, SOIL, WASTE, WATER SUPP Cl (*Chlor*)
cladding *n* NOISE Verkleidung *f*
clamping: ~ **force** *n* WASTE, WATER SUPP *machine* Schließkraftbereich *m*, Spannkraft *f*
clarification: ~ **operation** *n* WATER POLL Klärwirkung *f*; ~ **plant** *n* WATER POLL Abwasserbehandlungsanlage *f*, Kläranlage *f*, Klärwerk *nt*; ~ **tank** *n* WATER POLL Klärbecken *nt*
clarifier *n* WATER POLL *circular unit for wastewater* Klärbecken *nt*
clarify *vt* WASTE abschlämmen, WATER POLL klären, abschlämmen
classical: ~ **recycling sectors** *n pl* MANAGE, WASTE *metal, scrap and paper* klassische Recyclingbranchen *f pl*
classifiable *adj* POLICY, SOIL, WASTE, WATER SUPP klassierfähig
classification *n* POLICY, SOIL, WASTE Klassifizierung *f*, WATER POLL Sichtung *f*, WATER SUPP Klassifizierung *f*; ~ **station** *n* WASTE Sichteranlage *f*, Sichtungsstation *f*
classifier *n* SOIL Klassierer *m*
classify *vt* POLICY, SOIL, WASTE, WATER SUPP absieben, klassieren, klassifizieren
Claus: ~ **plant** *n* AIR, WATER SUPP *sulphur recovery* Claus-Anlage *f*; ~ **process** *n* AIR, WATER SUPP *sulphur recovery* Claus-Verfahren *nt*
clay: ~ **barrier** *n* WASTE, WATER SUPP Lehmwall *m*; ~ **layer** *n* WASTE, WATER SUPP Tonschicht *f*; ~ **mineral layer** *n* SOIL Tonmineralschicht *f*; ~ **suspension** *n* WASTE, WATER SUPP Tonlösung *f*, Tonsuspension *f*
CLD *abbr* (*chemiluminescence detection*) ANALYSIS CLD (*Chemilumineszenzdetektion*)
clean 1. *adj* GEN rein, sauber; **2.** *vt* GEN abreinigen, reinigen; ~ **hydraulically** WASTE, WATER POLL, WATER SUPP hydraulisch reinigen; ~ **out** MANAGE, WASTE, WATER POLL, WATER SUPP ausspülen; ~ **up** MANAGE, WASTE, WATER POLL, WATER SUPP sanieren
clean: ~ **air** *n* AIR Reinluft *f*; ~ **air device** *n* SAFETY Luftreinigung *f*; ~ **air suction** *n* AIR Reinluftansaugung *f*
cleaned *adj* GEN gereinigt; ~ **gas** *n* AIR Reingas *nt*; ~ **glass cullet** *n* WASTE gereinigte Scherben *f pl*; ~ **pulp fiber** *AmE*, ~ **pulp fibre** *BrE n* WASTE gereinigter Faserbrei *m*; ~**-up** *adj* MANAGE saniert; ~ **waste air** *n* AIR gereinigte Abluft *f*
clean: ~ **energy source** *n* RENEW ENERG saubere Energiequelle *f*
cleaner *n* SAFETY Reiniger *m*
clean: ~ **gas** *n* AIR Reingas *nt*, gereinigtes Gas *nt*; ~ **gas canal** *n* AIR Reingaskanal *m*; ~ **gas compartment** *n* AIR Reingasraum *m*; ~ **gas concentration** *n* AIR Reingaskonzentration *f*; ~ **gas dust content** *n* AIR Reingasstaubgehalt *m*; ~ **gas lid** *n* AIR Reingasdeckel *m*; ~ **gas outlet** *n* AIR Reingasaustritt *m*; ~ **gas reheating** *n* AIR Reingaswiederaufheizung *f*
cleaning *n* GEN Reinigung *f*; ~ **capacity** *n* AIR, WATER POLL Reinigungsleistung *f*; ~ **with compressed air** *n* AIR Abreinigung *f* mit Druckluft, Druckluftreinigung *f*; ~ **equipment** *n* SAFETY Reinigungsgeräte *nt pl*; ~ **of heat exchanger** *n* AIR, WASTE Reinigung *f* der Wärmetauscher; ~ **nozzle** *n* AIR Reinigungsdüse *f*; ~ **operation** *n* AIR Reinigungsvorgang *m*, Abreinigungsvorgang *m*; ~ **out** *n* WASTE, WATER SUPP Ausspülen *nt*, Ausspülung *f*; ~ **position** *n* AIR *bag filter* Abreinigungsstellung *f*; ~ **pressure** *n* AIR Abreinigungsdruck *m*; ~ **pulse** *n* AIR *bag filter* Abreinigungsimpuls *m*; ~ **up** *n* MANAGE Sanierung *f*
clean: ~ **plastics** *n pl* WASTE unverschmutzte Kunststoffe *m pl*; ~ **rain** *n* AIR Reinregen *m*; ~ **room** *n* ANALYSIS Reinraum *m*
cleanse *vt* AIR, MARINE, SAFETY, WASTE reinigen, WATER POLL *river* ausschlämmen, reinigen, WATER SUPP reinigen
cleansing *n* GEN Reinigung *f*
clean: ~ **technology** *n* AIR, MANAGE, WASTE, WATER POLL, WATER SUPP saubere Technologie *f*, umweltfreundliche Technologie *f*; ~**-up** *n* MARINE Reinigung *f*; ~**-up of contaminated site** *n* SOIL Altlastaufarbeitung *f*, Altlastensanierung *f*; ~**-up costs** *n pl* MANAGE, SOIL Sanierungskosten *pl*; ~**-up efficiency** *n* SOIL Reinigungswirkungsgrad *m*; ~**-up rate** *n* SOIL Sanierungsgrad *m*; ~**-up technique** *n* MARINE Reinigungsverfahren *nt*; ~ **water** *n* WATER, WATER POLL reines Wasser *nt*
Clean: ~ **Water Act** *n* (*CWA*) MARINE Gesetz *nt* zur Wasserhaushaltung
clean: ~ **water tank** *n* WATER SUPP Reinwasserbehälter *m*
clearance: ~ **slide** *n* WASTE Freiraumschieber *m*, Freiräumer *m*, WATER SUPP Freiraumschieber *m*
clear: ~ **glass** *n* WASTE Weißglas *nt*; ~ **health risk** *n* SAFETY eindeutige Gesundheitsgefahr *f*; ~ **wastewater sample** *n* ANALYSIS klare Abwasserprobe *f*; ~ **water** *n* WATER POLL, WATER SUPP Reinwasser *nt*
cleaver *n* WATER POLL Spalter *m*
climate *n* AIR Klima *nt*; ~ **changes** *n pl* AIR *global warming* Klimaänderungen *f pl*; ~**-damaging** *adj* AIR klimaschädlich; ~**-damaging gases** *n pl* AIR klimaschädliche Gase *nt pl*; ~ **deterioration** *n* AIR Klimaverschlechterung *f*; ~ **factor** *n* AIR Klimafaktor *m*; ~ **fluctuations** *n pl* AIR Klimaschwankungen *f pl*; ~ **protection tax** *n* POLICY Klimaschutzsteuer *f*; ~**-relevant** *adj* AIR klimarelevant; ~**-relevant trace gases** *n pl* AIR klimarelevante Spurengase *nt pl*; ~ **system** *n* AIR Klimasystem *nt*
climatic *adj* AIR klimatisch; ~ **chamber** *n* AIR, WASTE Klimakammer *f*, Klimaraum *m*; ~ **conditions** *n* AIR, WASTE, WATER SUPP klimatische Bedingungen *f pl*; ~ **hazard** *n* AIR Klimagefahr *f*; ~ **protection** *n* AIR, POLICY Klimaschutz *m*; ~ **stress** *n* AIR klimatischer Streß *m*; ~ **test** *n* WASTE Klimaversuch *m*; ~ **variations** *n pl* AIR Klimaschwankungen *f pl*; ~ **zone** *n* AIR Klimazone *f*
climbing: ~ **screen** *n* WATER POLL Kletterrechen *m*
clinker *n* WASTE Schlacke *f*; ~ **pit** *n* WASTE Aschenabzug *m*
clip *vt* WATER SUPP befestigen
clog *vt* SOIL, WASTE, WATER POLL, WATER SUPP verstopfen
clogging *n* SOIL, WASTE, WATER POLL, WATER SUPP Verockerung *f*, Verstopfung *f*; ~ **risk** *n* WASTE, WATER SUPP Verstopfungsgefahr *f*
close *vt* WASTE schließen, stillegen; ~ **down** *vt* WASTE *landfill* stillegen
close: ~**-coupled** *adj* SOIL direktgekoppelt
closed *adj* WASTE stillgelegt; ~ **canal** *n* WASTE, WATER POLL, WATER SUPP geschlossener Kanal *m*; ~ **cycle** *n* (*CC*) RENEW ENERG geschlossener Kreislauf *m*; ~ **down** *adj* WASTE stillgelegt; ~ **landfill section** *n* WASTE stillgelegter Deponieabschnitt *m*; ~ **loop** *n* WATER POLL *direct reuse of wastewater* geschlossener Kreislauf *m*; ~**-loop circulation** *n* AIR Kreislaufführung *f*; ~**-loop system** *n* WASTE, WATER SUPP Kreislaufsystem *nt*; ~**-loop wet scrubbing** *n* AIR *FGD* geschlossener Kreislauf-Naßwäscher *m*; ~ **pit landfill** *n* WASTE geschlossene Grubendeponie *f*; ~ **Rankine cycle** *n* RENEW ENERG *ocean thermal energy* geschlossener Rankine-Kreislauf *m*; ~ **system** *n* WASTE, WATER SUPP geschlossenes System *nt*
close: ~**-meshed** *adj* AIR, WASTE, WATER POLL, WATER SUPP *grate* engmaschig; ~**-meshed sieve plate** *n* AIR engmaschiger Siebboden *m*; ~ **supervision** *n* SAFETY strenge Überwachung *f*
cloud *vt* WATER SUPP trüben
cloud: ~ **amount** *n* AIR *stack* Bedeckungsgrad *m*
cloudy *adj* WATER POLL, WATER SUPP trübe, unklar

clusters *n pl* WATER SUPP Flocken *f pl*
CNG *abbr* (*compressed natural gas*) AIR Druck-Erdgas *nt*
C/N/P: ~ **ratio** *n* (*carbon-nitrogen-phosphorus ratio*) ANALYSIS C/N/P-Verhältnis *nt* (*Kohlenstoff-Stickstoff-Phosphor-Verhältnis*)
C/N: ~ **ratio** *n* (*carbon-nitrogen ratio*) ANALYSIS C/N-Verhältnis *nt* (*Kohlenstoff-Stickstoff-Verhältnis*)
CO *abbr* (*carbon monoxide*) AIR CO (*Kohlenmonoxid*)
Co *abbr* (*cobalt*) SOIL Co (*Kobalt*)
CO_2 *abbr* (*carbon dioxide*) AIR, WASTE, WATER SUPP CO_2 (*Kohlendioxid*); ~ **concentration** *n* AIR CO_2-Konzentration *f*; ~ **emissions reduction** *n* AIR CO_2-Emissionsverminderung *f*
coagulability *n* AIR, ANALYSIS, WATER POLL, WATER SUPP Flockbarkeit *f*
coagulable *adj* AIR, ANALYSIS, WATER POLL, WATER SUPP ausflockbar, flockbar
coagulant *n* GEN Flockungsmittel *nt*, *precipitant* Fällmittel *nt*, Koagulans *nt*, Koagulationsmittel *nt*, anorganisches Flocungsmittel *nt*
coagulate *vi* GEN ausflocken, flocken, koagulieren
coagulation *n* AIR, ANALYSIS, WASTE, WATER SUPP Ausflockung *f*, Flockung *f*, Koagulation *f*; ~ **aid** *n* WATER POLL, WATER SUPP Flockungshilfsmittel *nt*; ~-**flocculation** *n* WATER POLL Koagulationsflockung *f*; ~-**precipitation** *n* WATER POLL Koagulationsabscheidung *f*
coal *n* AIR Kohle *f*; ~ **desulfurization** *AmE*, ~ **desulphurization** *BrE n* AIR Kohle-Entschwefelung *f*
coalesce *vi* AIR, WASTE, WATER POLL, WATER SUPP koaleszieren
coalescence *n* AIR, WASTE, WATER POLL, WATER SUPP Koaleszenz *f*; ~ **method** *n* WATER POLL *physical oil separation* Koaleszenz *f*; ~ **oil separator** *n* WATER POLL Koaleszenzabscheider *m*
coal: ~ **gas** *n* AIR Leuchtgas *nt*; ~ **mining waste** *n* WASTE Zechenabraum *m*; ~ **power station** *n* AIR Kohlenkraftwerk *nt*
coarse *adj* AIR, SOIL, WATER POLL grob; ~ **ash separator** *n* AIR Grobascheabscheider *m*; ~ **filter** *n* AIR *storage filter* Grobfilter *nt*; ~ **flocculent** *adj* WASTE, WATER SUPP grobflockig; ~ **flocs** *n pl* WATER POLL Grobflocken *f pl*; ~ **fraction** *n* WASTE Schwerstofffraktion *f*; ~-**grained** *adj* WASTE grobkörnig; ~ **medium fraction** *n* WASTE Schwerstofffraktion *f*; ~ **particle** *n* AIR grobe Partikel *f*; ~-**particled** *adj* WASTE grobstückig; ~ **refuse** *n* WASTE Grobmüll *m*; ~ **screen** *n* WATER POLL, WATER SUPP Grobrechen *m*; ~ **solids** *n pl* WATER POLL grobe Feststoffe *m pl*
coastal: ~ **deposit** *n* WASTE Küstenschlamm *m*, Küstenschlick *m*; ~ **river** *n* WATER SUPP Küstenfluß *m*
coastline: ~ **form** *n* RENEW ENERG Küstenform *f*
coat *vt* AIR beschichten
coated *adj* AIR, WASTE beschichtet; ~ **ceramic body** *n* AIR beschichteter Keramikkörper *m*; ~ **paper** *n* WASTE gestrichenes Papier *nt*; ~ **surface** *n* AIR, WASTE beschichtete Fläche *f*
coating *n* GEN Beschichtung *f*; ~ **thickness** *n* WASTE Überzugsdicke *f*
cobalt *n* (*Co*) AIR, SOIL Kobalt *nt* (*Co*)
COC *abbr* (*conventional oxidation catalyst*) AIR konventioneller Oxidationskatalysator *m*
co-current: ~ **centrifuge** *n* WATER POLL gleichlaufende Zentrifuge *f*; ~ **flow** *n* AIR *waste gas* Gleichstrom *m*, WATER SUPP Gleichstrom *m*, Gleichstromung *f*; ~ **flow scrubber** *n* AIR Gleichstromwäscher *m*; ~ **scrubber** *n* AIR Gleichstromwäscher *m*; ~ **washing** *n* AIR Gleichstromwäsche *f*
COD *abbr* (*chemical oxygen demand*) AIR, WASTE, WATER POLL, WATER SUPP CSB (*chemischer Sauerstoffbedarf*); ~ **determination** *n* ANALYSIS CSB-Bestimmung *f*
co-deposition *n* WASTE gemischte Ablagerung *f*
coding: ~ **of measured value** *n* ANALYSIS Meßwertcodierung *f*
co-disposal *n* WASTE gemischte Ablagerung *f*; ~ **landfill** *n* WASTE Mischdeponie *f*
COD: ~ **reduction** *n* WATER POLL CSB-Reduktion *f*
coefficient: ~ **of diffusion** *n* WATER SUPP Diffusionsbeiwert *m*, Diffusionskoeffizient *m*; ~ **of haze** *n* AIR Dunstkoeffizient *m*; ~ **of permeability** *n* AIR, SOIL, WASTE, WATER SUPP Durchlässigkeitskoeffizient *m*, Durchlässigkeitswert *m*, Kf-Wert *m*
CO: ~ **emissions** *n pl* AIR CO-Emissionen *f pl*
co-extruded: ~ **film** *n* WASTE gemeinschaftlich stranggepreßte Folie *f*
co-extrusion *n* WASTE *plastics* Co-Extrudierung *f*, Co-Extrusion *f*, Ko-Extrusion *f*
coffer *n* NOISE Kassette *f*
cofferdam *n* RENEW ENERG *tidal power plant* Kofferdamm *m*
cogeneration *n* AIR *waste heat*, WASTE Abwärmeverwertung *f*, Kraft-Wärme-Kopplung *f* (*KWK*)
cohesion *n* AIR Kohäsion *f*
co-injection *n* WASTE *plastics* Co-Injektion *f*
cold: ~ **seawater** *n* RENEW ENERG *OTEC plant* kaltes Meerwasser *nt*; ~-**vapor AAS** *AmE*, ~-**vapour AAS** *BrE n* ANALYSIS Kaltdampf-Atomabsorptionsspektroskopie *f*
coli: ~ **bacillus** *n* WATER SUPP *indicator germ* Escherichia coli *f*
coliform: ~ **bacteria** *n pl* ANALYSIS *indicator* coliforme Bakterien *f pl*
coliforms *n pl* WASTE, WATER SUPP coliforme Keime *m pl*
coli: ~ **test** *n* ANALYSIS Colititer *m*
collapse *n* AIR *of bag in filter* Kollaps *m*
collating: ~ **device** *n* AIR *catalytic incineration* Mischeinrichtung *f*; ~ **system** *n* WASTE Sortier- und Sammelsystem *nt*
collect *vt* GEN auffangen, einsammeln, sammeln, *fetch* abholen
collectable *adj* ANALYSIS erfaßbar
collected *adj* AIR abgeschieden; ~ **dust** *n* AIR abgeschiedener Staub *m*, *wet electrostatic precipitator* niedergeschlagener Staub *m*; ~ **dust particle** *n* AIR abgeschiedene Staubpartikel *f*
collecting: ~ **basin** *n* WASTE Sammelbehälter *m*, WATER POLL, WATER SUPP Sammelbecken *nt*; ~ **electrode** *n* AIR, ANALYSIS Niederschlagselektrode *f*; ~ **main** *n* AIR Vorlage *f*; ~ **receptacle** *n* ANALYSIS Sammelbehältnis *nt*
collection *n* AIR, ANALYSIS Sammlung *f*, MARINE Auffangen *nt*, Sammlung *f*, WASTE *removal* Abfuhr *f*, Sammlung *f*, *picking up* Abholen *nt*, Einsammeln *nt*, WATER POLL Abfangen *nt*, Sammlung *f*, WATER SUPP Sammlung *f*; ~ **at central sites** *n* WASTE Sammlung *f* an zentralen Orten; ~ **behavior** *AmE*, ~ **behaviour** *BrE n* AIR Abscheideverhalten *nt*; ~ **body** *n* WASTE Müllfahrzeug *nt*, Müllwagen *m*; ~ **by contracted third party** *n* WASTE Sammlung *f* durch beauftragte Dritte; ~ **container** *n* WASTE Sammelcontainer *m*; ~ **costs** *n pl* WASTE Sammelkosten *pl*; ~ **device** *n* MARINE, WASTE, WATER SUPP Auffangeinrichtung *f*; ~ **efficiency** *n* AIR Abscheidewirkungsgrad *m*; ~ **element** *n* WASTE *collector* Fassungselement *nt*; ~ **hopper** *n* MARINE, WASTE, WATER SUPP Auffangtrichter *m*; ~ **logistics** *n pl* WASTE Einsammellogistik *f*; ~ **operation** *n* AIR Abscheidevorgang *m*; ~ **period** *n* ANALYSIS Sammlungszeitraum *m*; ~ **point system** *n* WASTE, WATER SUPP Bringsystem *nt*; ~ **rate** *n* WASTE, WATER SUPP Einsammelquote *f*; ~ **route** *n* WASTE Einsammlungsrunde *f*; ~ **sewer** *n* WATER POLL Abfangleitung *f*; ~ **station** *n* WASTE Annahmestelle *f*; ~ **vehicle** *n* WASTE Müllfahrzeug *nt*, Müllsammelfahrzeug *nt*, Müllwagen *m*
collector *n* WASTE Müllabfuhrunternehmen *nt*; ~ **field** *n* RENEW ENERG *solar farm* Kollektorfeld *nt*; ~ **head** *n* WASTE Kollektorkopf *m*; ~ **shaft** *n* WATER SUPP Sammelschacht *m*
collide: **1.** ~ **with** *vt* AIR aufprallen auf, aufschlagen auf, auftreffen auf, zusammenprallen mit; **2.** *vi* AIR *particle* kollidieren

colliery: ~ **waste** *n* WASTE Zechenabraum *m*
collimator *n* ANALYSIS Kollimator *m*
collision *n* AIR Prall *m*, SAFETY Zusammenstoß *m*, WASTE Aufprall *m*, Prall *m*, Stoß *m*, Zusammenprall *m*, WATER POLL Prall *m*; ~ **ionization** *n* AIR, ANALYSIS Stoßionisation *f*
colloid *n* AIR, SOIL, WATER POLL, WATER SUPP Kolloid *nt*
colloidal *adj* AIR, SOIL, WATER POLL, WATER SUPP kolloidal; ~ **concentration** *n* WATER POLL, WATER SUPP Kolloidkonzentration *f*; ~ **particle** *n* AIR, WASTE, WATER POLL, WATER SUPP Kolloidteilchen *nt*, kolloidales Teilchen *nt*; ~ **solution** *n* WATER POLL kolloidale Lösung *f*; ~ **substance** *n* WATER POLL ungelöster Stoff *m*, WATER SUPP *in wastewater* kolloidale Stoffe *m pl*; ~ **suspension** *n* WATER POLL, WATER SUPP Kolloidsuspension *f*
colloid: ~-**disperse** *adj* WATER POLL kolloiddispers; ~-**disperse system** *n* AIR kolloiddisperses System *nt*
color *AmE see colour BrE*
colorfast *AmE see colourfast BrE*
colorfastness *AmE see colourfastness BrE*
colorimeter *n* ANALYSIS Kolorimeter *nt*
colorimetric *adj* ANALYSIS kolorimetrisch; ~ **detection** *n* ANALYSIS kolorimetrischer Nachweis *m*; ~ **determination** *n* ANALYSIS kolorimetrische Bestimmung *f*; ~ **indicator tube** *n* ANALYSIS kolorimetrische Anzeigeröhre *f*; ~ **pH-value determination** *n* ANALYSIS kolorimetrische pH-Wert-Bestimmung *f*
colorimetry *n* ANALYSIS Kolorimetrie *f*
coloring *AmE see colouring BrE*
colorless *AmE see colourless BrE*
colour *n BrE* ANALYSIS, WASTE Farbe *f*; ~ **analysis** *n BrE* ANALYSIS Färbungsbestimmung *f*; ~ **change** *n BrE* ANALYSIS Farbumschlag *m*; ~ **depth** *n BrE* ANALYSIS Farbtiefe *f*; ~ **designation** *n BrE* ANALYSIS Farbkennzeichnung *f*
colourfast *adj BrE* WASTE farbecht
colourfastness *n BrE* WASTE Farbechtheit *f*
colour: ~ **indicator** *n BrE* ANALYSIS Farbindikator *m*
colouring *n BrE* ANALYSIS, WASTE Färbung *f*
colourless *adj BrE* AIR farblos; ~ **gas** *n BrE* AIR farbloses Gas *nt*
colour: ~ **reaction** *n BrE* ANALYSIS *absorption solution* Farbreaktion *f*; ~ **sorting of plastics** *n BrE* WASTE Sortierung von Kunststoffen nach Farben *f*
column *n* ANALYSIS Säule *f*; ~ **chromatography** *n* ANALYSIS Säulenchromatographie *f*; ~ **injector** *n* ANALYSIS *GC* Säuleninjektor *m*; ~ **material** *n* WASTE Säulenmaterial *nt*; ~ **method** *n* ANALYSIS Säulenmethode *f*
combination *n* AIR, WASTE Kombination *f*; ~ **pH-electrode** *n* ANALYSIS *pH-meter* Einstabmeßkette *f*; ~ **scrubber** *n* AIR Mischwäscher *m*; ~ **sealing** *n* WASTE Kombinationsdichtung *f*; ~ **of trickling filter and activated sludge tank** *n* WATER POLL Kombination *f* Belebtbecken-Tropfkörper (*Kombination BB-TK*)
combined *adj* SOIL, WATER POLL kombiniert; ~ **aeration** *n* WATER POLL Klärwerkspumpwerk *nt*, kombinierte Belüftung *f*; ~ **base sealing** *n* SOIL kombinierte Basisdichtung *f*; ~ **base-sealing system** *n* SOIL kombiniertes Basisdichtungssystem *nt*; ~-**cycle gas turbine** *n* (*CCGT*) AIR Gas- und Dampfanlage *f* (*GuD-Anlage*); ~ **heat and power** *n* (*CHP*) AIR, WASTE Kraft-Wärme-Kopplung *f* (*KWK*); ~ **heat and power station** *n* AIR, RENEW ENERG Blockheizkraftwerk *nt* (*BHKW*); ~ **heat and power station with fuel cell plant** *n* AIR Blockheizkraftwerk mit Brennstoffzellen *nt*; ~ **packaging** *n* WASTE kombinierte Verpackung *f*; ~ **sewer** *n* WATER POLL Gemischtwasserleitung *f*; ~ **treatment** *n* WATER POLL *wastewater* Mischverfahren *nt*
combustibility *n* AIR, RENEW ENERG, SAFETY, WASTE, Brennbarkeit *f*, Verbrennbarkeit *f*
combustible *adj* AIR, SAFETY, WASTE, WATER POLL brennbar; ~ **dust** *n* AIR brennbarer Staub *m*; ~ **fossil fuel** *n* AIR fossiler Brennstoff *m*; ~ **gaseous pollutants** *n pl* AIR brennbare gasförmige Schadstoffe *m pl*; ~ **material** *n* AIR, SAFETY, WASTE, WATER POLL brennbares Material *nt*, Brennstoff *m*
combustibles *n pl* AIR, SAFETY, WASTE, WATER POLL Brennstoffe *m pl*
combustible: ~ **solvent** *n* AIR brennbares Lösemittel *nt*; ~ **waste** *n* WASTE brennbarer Abfall *m*
combustion *n* AIR, RENEW ENERG *biomass conversion*, SAFETY, WASTE Feuerung *f*, Verbrennung *f*; ~ **air** *n* AIR, WASTE, WATER POLL Verbrennungsluft *f*; ~ **chamber** *n* AIR, SOIL, WASTE, WATER POLL *incineration* Brennkammer *f*, Feuerraum *m*; ~ **chamber ash silo** *n* AIR Feuerraumaschesilo *m*; ~ **chamber final temperature** *n* AIR Brennkammerendtemperatur *f*; ~ **chamber outlet** *n* AIR Brennkammeraustritt *m*; ~ **chamber slagging** *n* WASTE Feuerraumanbackung *f*; ~ **chamber temperature** *n* WASTE Feuerraumtemperatur *f*; ~ **enthalpy** *n* AIR Verbrennungsenthalpie *f*; ~ **gas** *n* AIR Abgas *nt*, Verbrennungsgas *nt*, WASTE Verbrennungsgas *nt*; ~ **operation** *n* AIR, WASTE Verbrennungsvorgang *m*; ~ **product** *n* AIR, WASTE Verbrennungsprodukt *nt*; ~ **reaction** *n* AIR, WASTE Verbrennungsreaktion *f*; ~ **residue** *n* WASTE Verbrennungsrückstand *m*; ~ **system** *n* AIR Brennersystem *nt*; ~ **temperature** *n* AIR, WASTE Verbrennungstemperatur *f*; ~ **tube** *n* ANALYSIS *nitrogen determination* Verbrennungsrohr *nt*; ~ **waste gas** *n* AIR Feuerungsabgas *nt*
come: ~ **into force** *vi* POLICY *law, agreement* in Kraft treten
commercial *adj* GEN gewerblich, kommerziell; ~ **global energy demand** *n* RENEW ENERG kommerzieller Globalenergiebedarf *m*
commercially: ~ **available** *adj* ANALYSIS, RENEW ENERG handelsüblich; ~ **available installation** *n* RENEW ENERG handelsübliche Anlage *f*; ~ **available measuring device** *n* ANALYSIS handelsübliches Meßgerät *nt*; ~ **available pH-measuring device** *n* ANALYSIS handelsübliches pH-Meßgerät *nt*; ~ **available thermometer** *n* ANALYSIS handelsübliches Thermometer *nt*; ~ **usable** *adj* RENEW ENERG kommerziell nutzbar
commercial: ~ **operation** *n* MANAGE, POLICY, WATER SUPP Dauerbetrieb *m*; ~-**scale** *adj* GEN großtechnisch; ~-**scale application** *n* GEN großtechnische Anwendung *f*, großtechnischer Einsatz *m*; ~-**scale plant** *n* GEN großtechnische Anlage *f*; ~ **source** *n* AIR, WASTE, WATER SUPP gewerbliche Quelle *f*; ~ **waste** *n* WASTE gewerblicher Abfall *m*
commingled: ~ **plastics** *n pl* WASTE *recycling* vermischte Kunststoffe *m pl*
comminute *vt* WASTE, WATER POLL zerkleinern
comminutor *n* WASTE Zerkleinerer *m*, WATER POLL Rechengutzerkleinerer *m*
Commission: ~ **on Sustainable Development** *n* (*CSD*) POLICY CSD-Kommission *f*
committee *n* MANAGE, POLICY Ausschuß *m*
commixture *n* WASTE Vermengung *f*
communal: ~ **waste collection by local authority** *n* WASTE kommunale Müllabfuhr *f*, kommunale Mülleinsammlung *f*
compact *adj* AIR, WATER POLL, WATER SUPP kompakt; ~ **absorber** *n* NOISE Kompaktabsorber *m*
compacted *adj* WASTE stichfest, verdichtet
compact: ~ **filter** *n* AIR *in container* Kompaktfilter *nt*; ~ **flocculation plant** *n* WATER SUPP Kompaktflockungsanlage *f*; ~ **gas scrubber** *n* AIR Kompaktgaswäscher *m*; ~ **heat exchanger** *n* AIR Kompaktwärmeaustauscher *m*
compacting *n* WASTE Kompaktierung *f*, Verdichten *nt*, Verdichtung *f*; ~ **device** *n* WASTE Verdichtungsgerät *nt*; ~ **unit** *n* WASTE Verdichtungsanlage *f*
compaction *n* WASTE Kompaktierung *f*, Verdichtung *f*; ~ **equipment** *n* SOIL *smooth roller*, WASTE Verdichtungsgerät

nt; ~ **landfill** *n* WASTE Verdichtungsdeponie *f*; ~ **zone** *n* WASTE Verdichtungszone *f*
compact: ~ **module** *n* WATER POLL *with membrane* Kompaktmodul *nt*
compactor *n* WASTE *device* Kompaktor *m*; ~ **vehicle** *n* WASTE Preßmüllfahrzeug *nt*, Preßmüllwagen *m*
compact: ~ **reactor** *n* SOIL, WATER POLL Kompaktreaktor *m*
company: ~**-owned treatment plant** *n* WATER POLL industrieeigene Kläranlage *f*; ~**-specific** *adj* POLICY betriebsspezifisch; ~**-specific documents** *n pl* POLICY betriebsspezifische Unterlagen *f pl*; ~**-specific information** *n* POLICY betriebsspezifische Informationen *f pl*
comparative: ~ **measurement** *n* ANALYSIS Vergleichsmessung *f*; ~ **test** *n* ANALYSIS Vergleichsprobe *f*
compare *vt* ANALYSIS *data, results* vergleichen
comparison *n* ANALYSIS Vergleich *m*
compensate *vt* ANALYSIS kompensieren, POLICY, SAFETY *counterbalance* ausgleichen, *indemnify* entschädigen
compensating: ~ **tube** *n* WATER SUPP Kompensationsschlauch *m*
compensation *n* POLICY Ausgleichszahlung *f*, Entschädigung *f*, SAFETY Entschädigung *f*, SOIL, WASTE, WATER SUPP Ausgleich *m*; ~ **fee** *n* POLICY Bußgeld *nt*
competent: ~ **authority** *n* POLICY zuständige Behörde *f*
competitiveness *n* MANAGE, RENEW ENERG Konkurrenzfähigkeit *f*
complete: ~ **enclosure** *n* NOISE Vollkapselung *f*; ~ **fill** *n* WASTE abgeschlossene Deponie *f*, verfüllte Deponie *f*
completely: ~ **mixed** *adj* WASTE, WATER POLL, WATER SUPP volldurchgemischt; ~ **mixed tank** *n* WATER POLL volldurchgemischtes Becken *nt*
complete: ~ **mix aeration** *n* WATER POLL volldurchgemischte Belüftung *f*; ~ **mixing** *n* AIR, WASTE, WATER POLL, WATER SUPP Durchmischung *f*
complexing: ~ **agent** *n* ANALYSIS, WATER POLL Komplexbildner *m*
complexometric *adj* ANALYSIS komplexometrisch; ~ **determination** *n* ANALYSIS komplexometrische Bestimmung *f*; ~ **method** *n* ANALYSIS komplexometrische Methode *f*; ~ **titration** *n* ANALYSIS komplexometrische Titration *f*
component *n* AIR, ANALYSIS, SOIL Bestandteil *m*; ~**-specific** *adj* ANALYSIS, SOIL komponentenspezifisch; ~**-specific gas analyser** *BrE*, ~**-specific gas analyzer** *AmE* *n* ANALYSIS komponentenspezifischer Gasanalysator *m*
composite *n* WASTE Verbundstoff *m*; ~ **cardboard** *n* WASTE Verbundkarton *m*; ~ **container** *n* WASTE Verbundstoffbehälter *m*; ~ **packaging** *n* WASTE Verbundverpackung *f*; ~ **sample** *n* ANALYSIS Mischprobe *f*; ~ **seal** *n* WASTE Kombinationsdichtung *f*; ~ **secondary resource** *n* WASTE Verbundwerkstoff *m*
compost 1. *n* WASTE, WATER POLL Kompost *m*; **2.** *vt* WASTE, WATER POLL kompostieren
compostability *n* WASTE, WATER POLL Kompostierfähigkeit *f*
compostable *adj* WASTE, WATER POLL kompostierbar, kompostierfähig; ~ **fraction** *n* WASTE kompostierbare Fraktion *f*; ~ **sludge** *n* WATER POLL kompostierbarer Schlamm *m*; ~ **waste** *n* WASTE kompostierbarer Abfall *nt*
compost: ~ **aeration** *n* WASTE Kompostbelüftung *f*; ~ **container** *n* WASTE Kompostbehälter *m*, Komposttonne *f*; ~ **filter** *n* AIR *biological filter* Kompostfilter *nt*; ~ **heap** *n* WASTE Komposthalde *f*
composting *n* WASTE, WATER POLL Kompostierung *f*; ~ **drum** *n* WASTE Rottebox *f*, Rottezelle *f*; ~ **material** *n* WASTE Kompostgut *nt*; ~ **operation** *n* WASTE Kompostierungsvorgang *m*; ~ **plant** *n* SOIL, WASTE Kompostwerk *nt*, Kompostierungsanlage *f*; ~ **residue** *n* WASTE Kompostierungsrückstand *m*; ~ **technique** *n* WASTE Kompostierungstechnik *f*; ~ **of waste and sludge** *n* WATER POLL Müll- und Schlammkompostierung *f*; ~ **works** *n pl* SOIL, WASTE Kompostwerk *nt*, Kompostierungsanlage *f*
compost: ~ **material** *n* WASTE Kompostmaterial *nt*; ~ **maturing** *n* WASTE Kompostreifung *f*; ~ **quality** *n* WASTE Kompostqualität *f*; ~ **raw material** *n* WASTE Ausgangsmaterial *nt*, Kompostrohstoff *m*, WATER POLL Ausgangsmaterial *nt*; ~ **reprocessing** *n* WASTE Kompostaufbereitung *f*; ~ **ripening** *n* WASTE Kompostreifung *f*; ~ **sorting plant** *n* WASTE Kompostsortieranlage *f*
compound: ~ **burner** *n* WATER POLL *incineration* Mehrstoffbrenner *m*
compounder *n* WASTE Kompounder *m*
compound: ~ **wall** *n* NOISE mehrschalige Wand *f*
comprehensive: ~ **procedure** *n* POLICY vollständiges Verfahren *nt*; ~ **wastewater treatment plant** *n* WATER POLL mehrstufige Abwasserbehandlungsanlage *f*
compress *vt* RENEW ENERG verdichten
compressed *adj* RENEW ENERG verdichtet; ~ **air** *n* AIR, SAFETY, SOIL, WATER POLL Druckluft *f*; ~**-air blast** *n* AIR *bag filter* Druckluftstoß *m*; ~**-air-cleaned** *adj* AIR druckluftabgereinigt; ~**-air-cleaned jet filter** *n* AIR druckluftgereinigtes Jet-Filter *nt*; ~**-air consumption** *n* AIR Druckluftverbrauch *m*; ~**-air impulse** *n* AIR Druckluftimpuls *m*; ~**-air industrial breathing apparatus** *n* SAFETY Druckluftatemgerät *nt* für industrielle Einsätze; ~**-air lance** *n* SOIL *vapour extraction* Druckluftlanze *f*; ~**-air receiver** *n* AIR Druckluftbehälter *m*; ~**-air storage tank** *n* AIR *pulse jet filter* Druckluftspeicher *m*; ~ **natural gas** *n* (*CNG*) AIR Druck-Erdgas *nt*
compressibility *n* WASTE Verdichtbarkeit *f*, *of refuse* Kompressibilität *f*
compressible *adj* WASTE, WATER POLL, WATER SUPP komprimierbar, verdichtbar; ~ **waste** *n* WASTE komprimierbarer Abfall *m*, verdichtbarer Abfall *m*
compression *n* AIR, ANALYSIS Druck *m*, RENEW ENERG Verdichten *nt*, SOIL, WASTE, WATER SUPP Druck *m*; ~ **damage** *n* WASTE, WATER SUPP Druckverformung *f*; ~ **filtration** *n* ANALYSIS Druckfiltration *f*; ~ **load** *n* WASTE, WATER SUPP Druckbeanspruchung *f*; ~ **molding** *AmE*, ~ **moulding** *BrE* *n* WASTE Formpressen *nt*, Pressen *nt*; ~**-proof** *adj* AIR druckfest; ~ **strength** *n* WASTE, WATER SUPP Druckfestigkeit *f*; ~ **test** *n* WASTE, WATER SUPP Druckversuch *m*; ~ **test machine** *n* ANALYSIS, WATER SUPP Druckprüfmaschine *f*; ~ **vehicle** *n* WASTE Preßmüllfahrzeug *nt*, Preßmüllwagen *m*
compressor *n* AIR, RENEW ENERG, WATER POLL Kompressor *m*, Verdichter *m*; ~ **casing** *n* AIR Verdichtergehäuse *nt*
computable *adj* ANALYSIS berechenbar
computation *n* ANALYSIS Berechnung *f*
compute *vt* ANALYSIS berechnen
computer: ~**-controlled catalytic converter** *n* AIR geregelter Katalysator *m*
computerized *adj* ANALYSIS computergesteuert; ~ **analyser** *BrE*, ~ **analyzer** *AmE* *n* ANALYSIS computergesteuerter Analysator *m*, mikroprozessorgesteuerter Analysator *m*
concentrate 1. *n* AIR, SOIL, WATER POLL, WATER SUPP Konzentrat *nt*; **2.** *vt* AIR entwässern, konzentrieren, RENEW ENERG *rays* bündeln, SOIL, WASTE, WATER POLL konzentrieren, WATER SUPP entwässern
concentrated *adj* AIR, SOIL, WASTE, WATER POLL, WATER SUPP entwässert, konzentriert, RENEW ENERG gebündelt, konzentriert; ~ **amount of sample** *n* ANALYSIS eingedampfte Probenmenge *f*; ~ **sludge** *n* WATER POLL entwässerter Schlamm *m*; ~ **solar energy** *n* RENEW ENERG gebündelte Sonnenenergie *f*
concentrate: ~ **flow** *n* WATER POLL, WATER SUPP Konzentratstrom *m*; ~ **recovery** *n* WATER POLL Konzentratgewinnung *f*
concentrating: ~ **collector** *n* RENEW ENERG *parabolic collectors* konzentrierender Kollektor *m*
concentration *n* GEN Konzentration *f*; ~ **fluctuation** *n* AIR *of pollutants* Konzentrationsschwankung *f*; ~ **of noise to area**

n MARINE, NOISE Konzentrierung *f* von Lärm auf einen Bereich
concrete: **~ pipe** *n* WASTE, WATER SUPP Betonrohr *nt*; **~ ring of tubbings** *n* WASTE, WATER SUPP Betonschachtring *m*
concurrent: **~ centrifuge** *n* WATER POLL gleichlaufende Zentrifuge *f*
condensate *n* AIR, SOIL, WATER POLL Kondensat *nt*; **~ film** *n* AIR Kondensatfilm *m*; **~ flow** *n* AIR Kondensatstrom *m*; **~ formation** *n* AIR Kondensatbildung *f*; **~ pipe** *n* WASTE kondensatführende Leitung *f*; **~ polishing** *n* WATER POLL Kondensatreinigung *f*; **~ separator** *n* WASTE Kondensatabscheider *m*
condensation *n* AIR, SOIL, WATER POLL Kondensation *f*; **~ aerosol** *n* AIR Kondensationsaerosol *nt*; **~-force ESP** *n* AIR Kondensations-E-Filter *nt*; **~ nucleus** *n* AIR Kondensationskern *m*; **~ nucleus counter** *n* AIR Kondensationskernzähler *m*; **~ scrubber** *n* AIR Kondensationswäscher *m*; **~ water** *n* WATER POLL Kondenswasser *nt*; **~ water separator** *n* WATER POLL Kondenswasserabscheider *m*
condense *vt* AIR auskondensieren, kondensieren, WASTE verdichten
condensed: **~ water content** *n* AIR Wasserdampfgehalt *m*
condenser *n* AIR, ANALYSIS, RENEW ENERG, SOIL Kondensator *m*
condition *vt* AIR, WASTE, WATER POLL, WATER SUPP konditionieren
conditioned *adj* AIR, WASTE, WATER POLL, WATER SUPP konditioniert; **~ sludge** *n* WATER POLL konditionierter Schlamm *m*
conditioning *n* AIR, WASTE, WATER POLL, WATER SUPP Konditionierung *f*; **~ process** *n* WASTE, WATER SUPP Aufbereitungsverfahren *nt*; **~ reactor** *n* AIR Konditionierungsreaktor *m*; **~ of waste** *n* WASTE Müllaufbereitung *f*
condition: **~ of matter** *n* AIR Aggregatzustand *m*
conditions *n pl* AIR, ANALYSIS, SOIL, WATER POLL Bedingungen *f pl*; **~ for authorization** *n pl* POLICY Genehmigungsvoraussetzung *f*; **~ for licensing** *n pl* POLICY Genehmigungsvoraussetzung *f*
conduct *vt* AIR durchführen, ANALYSIS *electricity* leiten, *tests* durchführen, MANAGE durchführen, NOISE *structure-borne sound* fortleiten, RENEW ENERG *electricity* leiten, WATER POLL, WATER SUPP *tests* durchführen
conductibility *n* GEN Leitfähigkeit *f*
conductimetry *n* ANALYSIS Konduktometrie *f*, Leitfähigkeitsmessung *f*
conductivity *n* GEN Leitfähigkeit *f*, spezifische elektrische Leitfähigkeit *f*; **~ measuring instrument** *n* ANALYSIS Leitfähigkeitsmeßgerät *nt*; **~ meter** *n* ANALYSIS Leitfähigkeitsmesser *m*, Leitfähigkeitsmeßgerät *nt*
conductometer *n* ANALYSIS Leitungsmesser *m*, Leitwertmesser *m*
conductometric *adj* ANALYSIS konduktometrisch; **~ analysis** *n* ANALYSIS konduktometrische Maßanalyse *f*; **~ method** *n* ANALYSIS konduktometrische Methode *f*, konduktometrisches Verfahren *nt*
conduit *n* AIR Abzug *m*, WATER POLL, WATER SUPP Rohrleitung *f*
cone *n* AIR *of cyclone* Konus *m*; **~ of depression** *n* WASTE, WATER SUPP Entnahmetrichter *m*; **~ drive** *n* WATER POLL, WATER SUPP *pump* Konusgetriebe *nt*
confidential *adj* POLICY vertraulich; **~ documents** *n pl* POLICY vertrauliche Unterlagen *f pl*; **~ information** *n* POLICY vertrauliche Informationen *f pl*
configure *vt* ANALYSIS konfigurieren
confined: **~ aquifer** *n* WATER SUPP gespannter Grundwasserleiter *m*; **~ groundwater** *n* WATER SUPP gespanntes Grundwasser *nt*
confluence *n* WATER SUPP Zusammenfluß *m*
conforming: **~ to the law** *adj* POLICY gesetzeskonform
conglomerate *vt* WATER POLL konglomerieren
conglutinate *vt* AIR, WASTE verkleben
conglutination *n* AIR, WASTE Verkleben *nt*, Verklebung *f*
conical *adj* AIR konisch; **~ base** *n* AIR *of cyclone* konisches Unterteil *nt*; **~ screw extruder** *n* WASTE, WATER SUPP konischer Schneckenextruder *m*
coning: **~ smoke plume** *n* AIR kegelförmige Rauchfahne *f*
conjugate: **~ compound** *n* ANALYSIS konjugierte Verbindung *f*
connect *vt* WASTE ankuppeln, WATER SUPP ankuppeln, *buildings to drinking water supply* anschließen
connecting: **~ branch** *n* AIR Rohrstutzen *m*; **~ pipe** *n* SOIL Verbindungsleitung *f*, WASTE Anschlußleitung *f*, Anschlußrohr *nt*, WATER POLL *between reactors* Verbindungsleitung *f*, WATER SUPP Anschlußleitung *f*, Anschlußrohr *nt*
connection *n* WASTE, WATER POLL, WATER SUPP Anschluß *m*
conservation *n* MANAGE, POLICY, RENEW ENERG Erhaltung *f*, Schutz *m*, SOIL Schonung *f*
conserve *vt* MANAGE, POLICY, RENEW ENERG erhalten, sparen, SOIL schonen
consistence *n* WASTE Konsistenz *f*
consistency *n* ANALYSIS Reproduzierbarkeit *f*, WASTE Konsistenz *f*
constant: **~ speed pump** *n* WATER POLL, WATER SUPP Pumpe *f* mit konstanter Drehzahl; **~ volume sampling** *n* (*CVS*) ANALYSIS CVS-Test
constituent *n* AIR, ANALYSIS, SOIL Bestandteil *m*
constitutive *adj* WATER POLL konstitutiv; **~ enzyme** *n* WATER POLL konstitutives Enzym *nt*
constrained: **~ layer** *n* NOISE *vibration damping* eingezwängter Dämpfungsbelag *m*
construct *vt* WASTE *landfill body* aufbauen
constructed: **~ area** *n* WASTE *of landfill* Einbaufläche *f*; **~ layers** *n pl* WASTE *of landfill* Einbauschichten *f pl*
construction: **~ licence** *BrE*, **~ license** *AmE* *n* MANAGE, POLICY Baugenehmigung *f*; **~ noise** *n* NOISE Lärm *m* im Bauwesen; **~ permission** *n* POLICY Baugenehmigung *f*; **~ site** *n* NOISE, SOIL, WASTE Baustelle *f*; **~ waste** *n* WASTE Bauschutt *m*, Baustellenabfall *m*
consume *vt* MANAGE, WASTE, WATER SUPP konsumieren, verbrauchen
consumer *n* MANAGE, WASTE, WATER SUPP Konsument *m*, Verbraucher *m*; **~ behavior** *AmE*, **~ behaviour** *BrE* *n* AIR, MANAGE, WASTE, WATER SUPP Konsumverhalten *nt*; **~ goods** *n pl* MANAGE, WASTE Konsumgüter *nt pl*; **~ waste** *n* WASTE Hausmüll *m*
consumption *n* MANAGE, WASTE, WATER SUPP Konsum *m*, Verbrauch *m*; **~ water** *n* WATER SUPP zum Verbrauch bestimmtes Wasser *nt*
contact *n* AIR, WASTE, WATER POLL, WATER SUPP Kontakt *m*; **~ bed** *n* WATER POLL Kontaktfilterbett *nt*; **~ catalysis** *n* AIR Kontaktkatalyse *f*; **~-catalytic** *adj* WASTE, WATER SUPP kontaktkatalytisch; **~ condenser** *n* AIR Kontaktkondensator *m*; **~ dryer** *n* WATER POLL Kontakttrockner *m*; **~ filtration** *n* WATER POLL Flockungsfiltration *f*; **~ process** *n* WATER SUPP *recovery of biogas* Kontaktverfahren *nt*; **~ reactor** *n* AIR Kontaktofen *m*; **~ spring** *n* WATER SUPP Schichtquelle *f*; **~ stabilization** *n* WATER POLL Kontaktstabilisierung *f*; **~ surface** *n* AIR Kontaktoberfläche *f*; **~ time** *n* WATER POLL, WATER SUPP Kontaktzeit *f*; **~ tower** *n* AIR Kontaktturm *m*
contained: **~ energy** *n* RENEW ENERG enthaltene Energie *f*
container *n* GEN Behälter *m*, Container *m*; **~ glass** *n* WASTE Behälterglas *nt*
containerizable *adj* SOIL, WASTE containerisierbar
containerization *n* SOIL, WASTE Containerisierung *f*
containerize *vt* SOIL, WASTE containerisieren
containerized *adj* SOIL, WASTE containerisiert
container: **~ press** *n* WASTE Containerpresse *f*; **~ remedia-**

tion installation *n* SOIL Container-Sanierungsanlage *f*, mobile Sanierungsanlage *f*; ~ **slide** *n* WASTE Container-schlitten *m*; ~ **system** *n* WASTE Containersystem *nt*, WATER POLL Behältersystem *nt*; ~ **wastewater treatment plant** *n* WATER POLL Abwasserreinigungsanlage *f* in Containern, Containerkläranlage *f*

contaminant *n* GEN Kontaminant *nt*, Schadstoff *m*; ~ **concentrate** *n* SOIL Schadstoffkonzentrat *nt*; ~ **content** *n* AIR, SOIL, WASTE, WATER POLL, WATER SUPP Schadstoff-gehalt *m*; ~ **degradation** *n* AIR, WATER POLL Schadstoffabbau *m*; ~**-free** *adj* AIR, WATER POLL, WATER SUPP schadstofffrei; ~ **gas stream** *n* AIR Schadstoff-Gas-strom *m*; ~**-laden** *adj* AIR, SOIL, WASTE, WATER POLL, WATER SUPP schadstoffhaltig; ~ **properties** *n pl* AIR, WATER POLL, WATER SUPP Schadstoffeigenschaften *f pl*; ~**-resistant** *adj* AIR, ANALYSIS, SOIL, WATER POLL schadstoffbeständig, schadstoffresistent

contaminants: ~ **bound to the soil** *n pl* SOIL im Boden eingebundene Schadstoffe *m pl*

contaminant: ~**-specific** *adj* AIR, SOIL, WATER POLL schadstoffspezifisch; ~ **type** *n* AIR, SOIL, WATER POLL Schadstofftyp *m*; ~ **vapors** *AmE*, ~ **vapours** *BrE n pl* AIR Schadstoffbrüden *m pl*

contaminate *vt* GEN *burden* belasten, *pollute* verschmutzen, *with radioactivity* verseuchen, verunreinigen, kontaminieren

contaminated *adj* GEN belastet, *polluted* verschmutzt, *radio-actively* verseucht, verunreinigt, kontaminiert; ~ **dust** *n* AIR belasteter Staub *m*; ~ **groundwater** *n* WASTE, WATER POLL, WATER SUPP kontaminiertes Grundwasser *nt*; ~ **plastic** *n* WASTE kontaminierter Kunststoff *m*, verschmutzter Alt-kunststoff *m*; ~ **sampling gas** *n* ANALYSIS verschmutztes Probegas *nt*; ~ **site** *n* SOIL Altlast *f*, WASTE Altlast *f*, kontaminierter Standort *m*; ~ **site generation** *n* SOIL Altlastenentstehung *f*; ~ **site pyrolysis** *n* SOIL Altlast-pyrolyse *f*; ~ **site register** *n* SOIL Altlastenkataster *m*; ~ **sludge** *n* WATER POLL kontaminierter Schlamm *m*; ~ **waste** *n* WASTE schadstoffhaltiger Abfall *m*; ~ **waste-water purified by microbes** *n* WATER POLL Verschmutzung *f* des Abwassers abgebaut durch Mikroorganismen; ~ **water** *n* WATER POLL, WATER SUPP kontaminiertes Wasser *nt*

contamination *n* GEN Belastung *f*, Kontamination *f*, Verschmutzung *f*, Verseuchung *f*, Verunreinigung *f*; ~ **fallout** *n* AIR Fremdstoffniederschlag *m*; ~ **meter** *n* ANA-LYSIS Geiger-Müllerzähler *m* (*GM-Zähler*), Geigerzähler *m*; ~ **pathway** *n* WATER POLL, WATER SUPP Kontaminationspfad *m*; ~ **type** *n* SOIL Kontaminationsart *f*

contingency: ~ **plan** *n* MARINE Vorsorgeplanung *f*

continuous *adj* ANALYSIS kontinuierlich, stetig; ~ **casting machine** *n* WASTE Stranggußanlage *f*; ~ **cleaning** *n* AIR *filter bag*, WATER POLL kontinuierliche Reinigung *f*; ~ **flow-ing film** *n* AIR kontinuierlich strömender Wasserfilm *m*; ~ **hydrolysis** *n* SOIL kontinuierliche Hydrolyse *f*

continuously: ~ **adjustable** *adj* WASTE *machinery* kontinuier-lich regelbar, stufenlos einstellbar

continuous: ~ **measuring** *n* ANALYSIS kontinuierliche Mes-sung *f*, stetige Messung *f*; ~ **neutralization** *n* WATER POLL Durchlaufneutralisation *f*; ~ **press** *n* WASTE kontinuierliche Presse *f*; ~ **press lines** *n pl* WASTE *waste material* kontinuierliche Pressenstraßen *f pl*; ~ **recording** *n* ANALYSIS kontinuierliche Registrierung *f*; ~ **sampling** *n* ANALYSIS kontinuierliche Probenahme *f*; ~ **screen belt press** *n* WA-TER POLL kontinuierliche Siebbandpresse *f*; ~ **ultrafiltration** *n* WATER POLL kontinuierliche UF *f*, kontinuierliche Ultra-filtration *f*

control 1. *n* GEN Kontrolle *f*, Steuerung *f*; **2.** *vt* GEN kontrollieren, steuern

control: ~ **device** *n* WASTE, WATER SUPP Leitapparat *m*

controllability *n* GEN Kontrollierbarkeit *f*, Steuerbarkeit *f*

controlled: ~ **combustion system** *n* (*CCS*) AIR CCS-System *nt*; ~ **dumping** *n* WASTE geordnete Ablagerung *f*, kontrollierte Müllablagerung *f*; ~ **fermentation** *n* WASTE kontrollierte Gärung *f*; ~ **high-level landfill** *n* WASTE geordnete Hochdeponie *f*; ~ **plant** *n* POLICY überwachungspflichtige Anlage *f*; ~ **tipping** *n BrE* WASTE geordnete Ablagerung *f*, kontrollierte Müllablagerung *f*

controlling: ~ **loop** *n* WASTE Controlling-Kreislauf *m*

control: ~ **room** *n* AIR, WASTE, WATER SUPP Leitstelle *f*, Steuerstand *m*

controls *n pl* GEN Steuerung *f*

control: ~ **shaft** *n* WATER SUPP Kontrollschacht *f*

Control: ~ **of Substances Hazardous to Health** *n BrE* (*COSHH*) MANAGE, SAFETY Vorschriften *f pl* zur Über-wachung von gesundheitsgefährdenden Substanzen, Überwachung *f* von gesundheitsgefährdenden Stoffen

control: ~ **system** *n* AIR, RENEW ENERG, SAFETY, WATER SUPP Regelungssytem *nt*; ~ **of toxicity at the workplace** *n* SAFETY Kontrolle *f* der Giftigkeit am Arbeitsplatz; ~ **unit** *n* AIR, ANALYSIS, SOIL, WATER POLL, WATER SUPP Steuereinheit *f*

convection *n* ANALYSIS, RENEW ENERG Konvektion *f*; ~ **dryer** *n* WASTE, WATER POLL Konvektionstrockner *m*; ~ **heat exchanger** *n* AIR Konvektionswärmeaustauscher *m*; ~ **section** *n* AIR Konvektionsteil *nt*

convention *n* POLICY *legal instrument* Abkommen *nt*, Konvention *f*, Übereinkommen *nt*

conventional *adj* GEN herkömmlich, konventionell; ~ **coal cleaning** *n* AIR herkömmliche Kohlereinigung *f*; ~ **drainage system** *n* WASTE, WATER SUPP konventionelles Dränsystem *nt*; ~ **energy supply** *n* RENEW ENERG konventionelle Ener-gieversorgung *f*; ~ **energy supply system** *n* RENEW ENERG konventionelles Energieversorgungssystem *nt*; ~ **energy technology** *n* RENEW ENERG konventionelle Energietechnik *f*, konventionelle Energietechnologie *f*; ~ **filter system** *n* AIR konventionelles Filtersystem *nt*; ~ **groundwater treatment** *n* WASTE, WATER SUPP klassische Grundwasseraufbereitung *f*, konventionelle Grundwasseraufbereitung *f*; ~ **heating technology** *n* RENEW ENERG konventionelle Heizungstech-nik *f*; ~ **oxidation catalyst** *n* (*COC*) AIR konventioneller Oxidationskatalysator *m*; ~ **technology** *n* AIR, ANALYSIS, NOISE, SOIL, WATER POLL konventionelle Technik *f*, konventionelle Technologie *f*; ~ **transportable pallet** *n* WASTE konventionelle Transportpalette *f*; ~ **wastewater clarification** *n* WATER POLL konventionelle Abwasserreini-gung *f*

conversion *n* GEN Konversion *f*, *chemical* Umsatz *m*, Umsetzung *f*, *transformation* Umwandlung *f*; ~ **efficiency** *n* RENEW ENERG Umwandlungswirkungsgrad *m*; ~ **factor** *n* ANALYSIS Umrechnungsfaktor *m*; ~ **force** *n* AIR Umsatz-geschwindigkeit *f*; ~ **into usable energy form** *n* RENEW ENERG Umwandlung *f* in eine nutzbare Energieform; ~ **machinery** *n* WASTE Umrüstanlage *f*; ~ **rate** *n* AIR Umwandlungsgrad *m*; ~ **reaction** *n* AIR *fuel cell* Konvertierungsreaktion *f*

convert *vt* AIR *energy*, RENEW ENERG, SOIL umwandeln, WASTE umsetzen, umwandeln, WATER POLL umsetzen; ~ **to electricity** *vt* RENEW ENERG in elektrische Energie umwan-deln

converter *n* RENEW ENERG Konverter *m*, WASTE Konverter *m*, Umsetzer *m*, WATER POLL Umsetzer *m*

convey: ~ **into silos** *vt* WASTE silieren

conveying: ~ **belt** *n* WASTE Förderband *nt*; ~ **equipment** *n* WASTE, WATER SUPP Förderanlage *f*; ~ **well** *n* SOIL Förder-brunnen *m*

conveyor *n* WASTE, WATER SUPP Förderer *m*; ~**-belt skimmer** *n* WASTE, WATER SUPP Förderbandskimmer *m*; ~ **handling system** *n* WASTE Warentransportsystem *nt*; ~ **pump** *n* SOIL Förderpumpe *f*

cool 1. *vt* GEN abkühlen, kühlen; **2.** *vi* GEN abkühlen; ~ **down** AIR, ANALYSIS, WATER SUPP abkühlen

coolant *n* AIR, WASTE, WATER SUPP Kühlmittel *nt*; ~ **flow rate** *n* AIR Kühlmittelströmungsgeschwindigkeit *f*; ~ **in** *n* AIR

Kühlmitteleintritt *m*; ~ **inlet** *n* AIR Kühlmitteleintritt *m*; ~ **out** *n* AIR Kühlmittelaustritt *m*; ~ **outlet** *n* AIR Kühlmittelaustritt *m*
cooler *n* AIR, ANALYSIS, WASTE, WATER SUPP Kühler *m*
cooling *n* GEN *refrigeration* Abkühlung *f*, Kühlung *f*; ~ **air** *n* AIR Kühlluft *f*; ~ **air inlet** *n* AIR Kühllufteintritt *m*; ~ **air outlet** *n* AIR Kühlluftaustritt *m*; ~ **down** *n* Abkühlung *f*; ~ **drum** *n* SOIL Kühltrommel *f*; ~ **gas fan** *n* AIR Kühlgasgebläse *nt*; ~ **liquid** *n* AIR Kühlflüssigkeit *f*; ~ **medium** *n* AIR, WASTE, WATER SUPP Kühlmittel *nt*; ~ **oil** *n* WASTE, WATER SUPP Kühlöl *nt*; ~ **pond** *n* WATER POLL Kühlbecken *nt*, Kühlteich *m*; ~ **system** *n* AIR, RENEW ENERG, WASTE Kühlsystem *nt*; ~ **tank** *n* AIR Kühlbehälter *m*; ~ **tower** *n* RENEW ENERG Kühlturm *m*; ~ **tower exhaust air** *n* AIR Kühlturmabluft *f*; ~ **water** *n* AIR, WATER POLL Kühlwasser *nt*; ~ **water loop** *n* RENEW ENERG Kühlkreislauf *m*; ~ **zone** *n* AIR Abkühlzone *f*
cooperation: ~ **among administrations** *n* POLICY Amtshilfe *f*
copolymer *n* WASTE Copolymer *nt*
copolymerization *n* WASTE Copolymerisation *f*
copper *n* (*Cu*) GEN Kupfer *nt* (*Cu*); ~ **carbonate** *n* WATER SUPP Kupfer(II)-Carbonat *nt*; **~-indium diselenide cell** *n* RENEW ENERG *photovoltaic* Kupfer-Indium-Diselenid-Zelle *f*; ~ **pipe** *n* WATER SUPP Kupferrohr *nt*
cordierite *n* AIR *ceramic carrier* Cordierit *nt*
core *n* ANALYSIS Kern *m*; ~ **barrel** *n* ANALYSIS *soil* Kernrohr *nt*; ~ **material** *n* ANALYSIS Kernmaterial *nt*; ~ **sample** *n* ANALYSIS Kernprobe *f*
Coriolis: ~ **force** *n* AIR Coriolis-Kraft *f*
corona *n* AIR Korona *f*; ~ **discharge** *n* AIR Koronaentladung *f*; ~ **effect** *n* AIR Koronaeffekt *m*; ~ **resistance** *n* AIR Koronabeständigkeit *f*
corral *vt* MARINE, WATER POLL *oil pollution* eingrenzen
corresponding: ~ **sampling** *n* ANALYSIS korrespondierende Probenahme *f*
corrode 1. *vt* WASTE *metal* zerfressen; **2.** *vi* AIR, WASTE, WATER POLL, WATER SUPP korrodieren
corrosion *n* AIR, WASTE, WATER POLL, WATER SUPP Korrosion *f*; ~ **damage** *n* WATER SUPP Korrosionsschaden *m*; ~ **in fittings** *n* WATER SUPP Korrosion *f* an Armaturen; **~-inhibiting** *adj* WASTE, WATER SUPP korrosionshemmend; ~ **inhibitor** *n* WATER SUPP Korrosionsinhibitor *m*, Korrosionshemmstoff *m*; ~ **in pumping stations** *n* WATER SUPP Korrosion *f* in Pumpwerken; ~ **in pumps** *n* WATER SUPP Korrosion *f* an Pumpen; ~ **in water installations** *n* WATER SUPP Korrosion *f* in Wasserinstallation; ~ **protection** *n* WATER POLL, WATER SUPP Korrosionsschutz *m*; ~ **protection lining** *n* WATER SUPP Korrosionsschutzauskleidung *f*; ~ **resistance** *n* AIR, WATER SUPP *supply network* Korrosionsbeständigkeit *f*; **~-resistant** *adj* AIR, WASTE, WATER SUPP korrosionsbeständig; **~-resistant leachate pump** *n* WASTE korrosionsbeständige Sickerwasserpumpe *f*; ~ **risk** *n* AIR Korrosionsgefahr *f*; ~ **test** *n* ANALYSIS Korrosionstest *m*; ~ **of water pipelines** *n* WATER POLL Korrosion *f* von Wasserleitungen
corrosive *adj* AIR, SAFETY, WASTE, WATER SUPP korrosiv; ~ **gas** *n* AIR korrosives Gas *nt*; ~ **hazardous waste** *n* WASTE korrosiver Sondermüll *m*
corrosiveness *n* AIR, SAFETY, WASTE, WATER SUPP Korrosivität *f*
corrosive: ~ **properties** *n pl* WASTE, WATER SUPP korrosive Eigenschaften *f pl*; ~ **substance** *n* SAFETY korrosiver Stoff *m*
corrosivity *n* AIR, SAFETY, WASTE, WATER SUPP Korrosivität *f*
corrugated: ~ **board** *n* WASTE Wellpappe *f*; ~ **board box** *n* WASTE Wellpappenkarton *m*; ~ **fiberboard** *AmE*, ~ **fibreboard** *BrE* *n* WASTE Wellfaserplatte *f*; ~ **paper** *n* WASTE Wellpapier *nt*
COSHH *abbr BrE* (*Control of Substances Hazardous to Health*) MANAGE, SAFETY Vorschriften *f pl* zur Überwachung von gesundheitsgefährdenden Substanzen, Überwachung *f* von gesundheitsgefährdenden Stoffen
cosmic *adj* AIR kosmisch; ~ **radiation** *n* AIR kosmische Strahlung *f*; ~ **solar radiation** *n* AIR kosmische Solarstrahlung *f*
cost *n* MANAGE Kosten *pl*, Kostenaufwand *m*; ~ **advantage** *n* MANAGE Kostenvorteil *m*; ~ **analysis** *n* MANAGE Kostenanalyse *f*; ~ **appraisal** *n* MANAGE Kostenbewertung *f*; **~-benefit-analysis** *n* MANAGE Kosten-Nutzen-Rechnung *f*; **~-benefit ratio** *n* MANAGE Aufwand-Nutzen-Relation *f*, Kosten-Nutzen-Verhältnis *nt*; **~-benefit study** *n* MANAGE Kosten-Nutzen-Analyse *f*; ~ **calculation** *n* MANAGE Kostenrechnung *f*; **~-effective** *adj* MANAGE kosteneffektiv, wirtschaftlich, POLICY wirtschaftlich; **~-effectiveness** *n* MANAGE Kosteneffizienz *f*, Wirtschaftlichkeit *f*, POLICY Wirtschaftlichkeit *f*; ~ **factor** *n* MANAGE Kostenfaktor *m*; **~-intensive** *adj* AIR, MANAGE, WATER POLL, WATER SUPP kostenintensiv; **~-intensive methods of treatment** *n pl* AIR, MANAGE, WATER POLL, WATER SUPP kostenintensive Behandlungsverfahren *nt pl*; ~ **planning** *n* MANAGE Kostenplanung *f*; **~-service life ratio** *n* MANAGE Kosten-Standzeit-Relation *f*; ~ **structure** *n* MANAGE Kostenstruktur *f*
cotton *n* AIR *filter material* Baumwolle *f*
Cottrel: ~ **process** *n* AIR *electrostatic gas cleaning process* Cottrell-Möller-Verfahren *nt*
coulometer *n* ANALYSIS Coulometer *nt*, Voltameter *nt*
coulometric *adj* ANALYSIS coulometrisch, voltametrisch; ~ **analyser** *BrE*, ~ **analyzer** *AmE* *n* ANALYSIS coulometrischer Analysator *m*; ~ **titration** *n* ANALYSIS coulometrische Titration *f*
coulometry *n* ANALYSIS Coulometrie *f*, Voltametrie *f*
countercurrent: ~ **aerator** *n* WATER POLL Gegenstrombelüfter *m*; ~ **centrifuge** *n* WATER POLL Gegenstromzentrifuge *f*; ~ **extraction** *n* SOIL, WATER POLL Gegenstromextraktion *f*; ~ **extractor** *n* SOIL, WATER POLL Gegenstromextraktor *m*; ~ **regeneration** *n* (*CFR*) WATER POLL Gegenstromregeneration *f*; ~ **screen** *n* WATER POLL Gegenstromrechen *m*; ~ **stripping** *n* SOIL Gegenstromstripping *nt*; ~ **washing** *n* WATER POLL *with filters* Gegenstromspülung *f*, Gegenstromwäsche *f*
counter: **~-ion** *n* WATER POLL, WATER SUPP Gegenion *nt*
coupled: ~ **with** *phr* ANALYSIS *processes* in Kopplung mit; ~ **biological scrubber** *n* AIR, WATER POLL gekoppelter Biowäscher *m*
coupling *n* AIR Kupplung *f*, ANALYSIS Kopplung *f*
CO: **~-value** *n* AIR CO-Wert *m*
cover 1. *vt* GEN abdecken; **2.** ~ **energy demand** *vi* RENEW ENERG Energiebedarf decken
coverage: ~ **rate** *n* AIR Bedeckungsgrad *m*
covered *adj* WASTE *landfill* abgedeckt; ~ **digester** *n* WATER POLL geschlossener Faulbehälter *m*; ~ **drain** *n* WATER POLL, WATER SUPP überdeckter Abzugsgraben *m*; ~ **trickling filter** *n* WATER POLL geschlossener Tropfkörper *m*
covering *n* GEN Abdecken *nt*, Abdeckung *f*; ~ **layer** *n* SOIL Abdeckungsschicht *f*, WASTE Abdeckschicht *f*, Deckschicht *f*, WATER SUPP Abdeckschicht *f*; ~ **material** *n* WASTE, WATER SUPP Abdeckmaterial *nt*; ~ **sheet** *n* NOISE Abdeckfolie *f*; ~ **soil** *n* SOIL, WASTE Abdeckboden *m*
cover: ~ **plate** *n* AIR *bag filter* Deckplatte *f*, Filterdeckel *m*; ~ **tube** *n* SOIL *ground filter* Abdeckrohr *nt*
Cr *abbr* (*chromium*) AIR, SOIL, WATER POLL Cr (*Chrom*)
crack *vt* WASTE *pyrolysis* cracken, spalten
cradle: **~-to-grave analysis** *n* MANAGE, POLICY *life cycle assessment*, WATER SUPP Cradle-to-Grave-Analyse *f*
cramping *n* WASTE Verklammerung *f*
crankcase *n* AIR Kurbelgehäuse *nt*; ~ **emissions** *n pl* AIR *hydrocarbon* Kurbelgehäuseemissionen *f pl*; ~ **ventilation** *n* AIR Kurbelgehäuseentlüftung *f*
crest *n* WASTE, WATER SUPP Überfallkante *f*
criteria *n pl* ANALYSIS Kriterien *nt pl*

critical: ~ **load** *n* SOIL kritische Belastung *f*; ~ **water level** *n* WASTE, WATER SUPP kritische Wasserlinie *f*, kritischer Wasserstand *m*

crop: ~ **waste** *n* WASTE Ernteabfall *m*

cross: ~**-flow** *n* AIR *waste gas* Kreuzstrom *m*, Querstrom *m*; ~**-flow filtration** *n* WATER POLL Crossflow-Filtration *f*; ~**-flow filtration with membranes** *n* WATER POLL Crossflow-Filtration *f* mit Membranen; ~**-flow microfiltration with underlying electrical field** *n* WATER POLL Crossflow-Mikrofiltration *f* mit überlagertem elektrischem Feld; ~**-flow scrubber** *n* AIR Kreuzstromwäscher *m*, Querstromwäscher *m*; ~**-flow separator** *n* AIR Querstromabscheider *m*; ~**-lines** *n pl* ANALYSIS *interferential pit gas tester* Strichplatte *f*

crosslink *vt* AIR *particle*, WASTE *plastics recycling* vernetzen

crosslinked *adj* AIR *particle*, WASTE *plastics recycling* vernetzt

crosslinking *n* AIR *particle*, WASTE *plastics recycling* Vernetzung *f*; ~ **efficiency** *n* WASTE, WATER SUPP Vernetzungseffizienz *f*, Vernetzungswirksamkeit *f*

cross: ~**-sectional area of flow** *n* WATER POLL, WATER SUPP Durchflußquerschnitt *m*; ~**-section sampling** *n* ANALYSIS Querschnittsprobenahme *f*

crosstalk *n* NOISE *transfer of airborne noise* Nebensprechen *nt*, Übersprechen *nt*

crude: ~ **oil** *n* AIR *auxiliary fuel* Heizöl *nt*, MARINE, WASTE Rohöl *nt*; ~ **pulp** *n* WASTE Rohbrei *m*; ~ **refuse** *n* WASTE Rohmüll *m*

crush *vt* SOIL, WASTE zerkleinern

crushed *adj* SOIL, WASTE zerkleinert; ~ **waste** *n* WASTE zerkleinerter Abfall *m*

crusher *n* SOIL, WASTE Zerkleinerer *m*

crushing *n* SOIL, WASTE Zerkleinerung *f*; ~ **installation** *n* WASTE Zerkleinerungsanlage *f*; ~ **plant** *n* WASTE Zerkleinerungsanlage *f*; ~ **system** *n* WASTE Zerkleinerungsverfahren *nt*; ~ **unit** *n* WASTE Zerkleinerungsanlage *f*

cryogenic: ~ **crushing** *n* WASTE Tieftemperaturzerkleinerung *f*; ~ **process** *n* WASTE Tiefkühlzerkleinerung *f*, Tieftemperaturzerkleinerung *f*

cryogrinding *n* WASTE Tieftemperaturzerkleinerung *f*

crystalline *adj* AIR kristallin, kristallinisch, RENEW ENERG kristallin; ~ **silicon cell** *n* RENEW ENERG *photovoltaic* kristalline Siliziumzelle *f*

crystallization *n* AIR Kristallisation *f*; ~ **germ** *n* AIR Kristallisationskeim *m*; ~ **zone** *n* AIR Kristallisationszone *f*

crystallize *vi* AIR auskristallisieren

crystallography *n* ANALYSIS Kristallographie *f*

CSD *abbr* (*Commission on Sustainable Development*) POLICY CSD-Kommission *f*

Cu *abbr* (*copper*) GEN Cu (*Kupfer*)

cube: ~ **compressive strength** *n* SOIL *grouting* Würfeldruckfestigkeit *f*

cullet *n* WASTE Glasbruch *m*, Scherben *f pl*

cumulative: ~ **frequency** *n* ANALYSIS Summenhäufigkeit *f*; ~ **parameter** *n* ANALYSIS Summenparameter *m*; ~ **toxic effect** *n* SAFETY toxische Gesamtwirkung *f*

cup *n* WASTE, WATER SUPP Becher *m*; ~ **anemometer** *n* ANALYSIS Schalenkreuzanemometer *nt*

cupola: ~**-blast heater** *n* AIR Kupolofen-Winderhitzer *m*

curb *vt* MARINE, NOISE verhindern

curb: ~**-side collection system** *n AmE* (*cf kerb-side collection system BrE*) WASTE Holsystem *nt*

current *n* RENEW ENERG Strom *m*; ~ **drawing** *n* SOIL *suction* Stromaufnahme *f*; ~ **meter** *n* WATER SUPP Strommesser *m*

curtain: ~ **boom** *n* MARINE Schürzensperre *f*; ~ **depth** *n* SOIL Wandtiefe *f*

curved: ~ **screen** *n* WATER POLL Bogenrechen *m*; ~ **sewer** *n* WASTE, WATER POLL, WATER SUPP gebogener Abwasserkanal *m*; ~ **sieve** *n* WATER POLL Bogensieb *nt*

cushion 1. *n* WASTE Polster *nt*; **2.** *vt* NOISE dämpfen

customer: ~ **behavior** *AmE*, ~ **behaviour** *BrE* *n* MANAGE Verbraucherverhalten *nt*

cut 1. *n* SOIL Schlitz *m*; **2.** *vt* SOIL schlitzen; ~ **to size** ANALYSIS, WASTE zuschneiden

cutter *n* SOIL Meißel *m*

cutting: ~ **off** *n* SOIL *using walls* Abschottung *f*; ~ **oil** *n* WASTE, WATER POLL Schneidöl *nt*; ~ **press** *n* WASTE Schneidepresse *f*; ~ **shredder** *n* WASTE Schneidzerkleinerer *m*

CVS *abbr* (*constant volume sampling*) AIR *motor vehicles* CVS-Test

CWA *abbr* (*Clean Water Act*) MARINE Gesetz zur Wasserhaushaltung *nt*

cyanide *n* SOIL, WATER POLL Cyanid *nt*; ~**-containing** *adj* WATER POLL cyanidhaltig; ~**-containing sludge** *n* WASTE, WATER POLL, WATER SUPP cyanidhaltiger Schlamm *m*; ~**-containing waste** *n* WASTE, WATER POLL cyanidhaltiger Abfall *m*; ~ **removal** *n* WATER POLL Cyanideliminierung *f*

cyclone *n* AIR Fliehkraftabscheider *m*, Zyklon *m*; ~ **furnace** *n* SOIL Zyklonbrennkammer *f*; ~ **housing** *n* AIR Zyklonmantel *m*; ~ **radius** *n* AIR Zyklonradius *m*; ~ **separation** *n* AIR Zyklonabscheidung *f*; ~ **separator** *n* AIR Zyklonabscheider *m*; ~ **tube** *n* AIR Wirbelrohr *nt*; ~ **vitrification** *n* SOIL Zyklonsinterung *f*; ~ **wall** *n* AIR Zyklonwand *f*

cylinder: ~ **broom** *n* WASTE Kehrwalze *f*

cylindrical *adj* AIR zylindrisch; ~ **cyclone** *n* AIR zylindrischer Zyklon *m*; ~ **graduated measuring glass** *n* ANALYSIS Meßzylinder *m*; ~ **inlet tube** *n* AIR *of cyclone* Tauchrohr *nt*; ~ **screw extruder** *n* WASTE, WATER SUPP zylindrischer Schneckenextruder *m*; ~ **section** *n* AIR *of cyclone* zylindrischer Teil *m*

CZE *abbr* (*capillary zone electrophoresis*) ANALYSIS Kapillarzone-Elektrophorese *f*

D

dacron *n* AIR *filter material* Dacron *nt*
daily *adj* GEN täglich, Tages-; **~ average value** *n* ANALYSIS Tagesmittelwert *m*; **~ consumption** *n* WATER SUPP Tagesverbrauch *m*; **~ cover** *n* WASTE, WATER SUPP tägliche Abdeckung *f*; **~ drinking water consumption** *n* WATER SUPP täglicher Trinkwasserverbrauch *m*; **~ mean value** *n* ANALYSIS Tagesmittelwert *m*; **~ noise exposure** *n* NOISE Tageslärmdosis *f*, Tageslärmexposition *f*; **~ output** *n* WASTE, WATER SUPP Tagesleistung *f*; **~ peak** *n* WASTE, WATER SUPP Tagesspitze *f*; **~ processing capacity** *n* AIR, SOIL, WATER POLL, WATER SUPP tägliche Verarbeitungskapazität *f*; **~ total exposure** *n* SAFETY tägliche Gesamtbelastung *f*; **~ water consumption** *n* WASTE, WATER SUPP täglicher Wasserverbrauch *m*; **~ water flow** *n* WATER SUPP Tageswasserfluß *m*
Dalton: **~'s law of partial pressure** *n* AIR Daltonsches Gesetz *nt*, Partialdruckgesetz *nt* von Dalton
dam 1. *n* RENEW ENERG Talsperre *f*; **2.** *vt* RENEW ENERG sperren
damage *vt* GEN schädigen
damage: **~ assessment** *n* SAFETY Schadensermittlung *f*; **~ to health** *n* SAFETY Gesundheitsschaden *m*; **~ to vegetation** *n* SOIL, WASTE Vegetationsschaden *m*
damp *vt* NOISE dämpfen
damped: **~ layer** *n* NOISE Dämpfungsbelag *m*
damp-resistant *adj* AIR schwer benetzbar; **~ dust** *n* AIR schwer benetzbarer Staub *m*
danger *n* SAFETY *hazard* Gefahr *f*, *threat* Gefährdung *f*; **~ assessment** *n* SAFETY Gefährlichkeitsabschätzung *f*; **~ estimate** *n* ANALYSIS, MANAGE, POLICY, SOIL Gefährdungsabschätzung *f*
dangerous *adj* GEN gefährlich; **~ anthropogenic interference with the climate system** *n* AIR gefährliche anthropogene Störung *f* des Klimasystems; **~ goods** *n pl* WASTE, WATER SUPP Gefahrgüter *nt pl*, gefährliche Güter *nt pl*; **~ hazardous waste** *n* WASTE, WATER SUPP gefährlicher Sonderabfall *m*; **~ loads** *n pl* WASTE, WATER SUPP gefährliche Ladungen *f pl*; **~ material** *n* SAFETY Gefahrstoff *m*; **~ situation** *n* WASTE, WATER SUPP gefährliche Situation *f pl*; **~ substance** *n* SAFETY Gefahrstoff *m*
danger: **~ potential** *n* SAFETY Gefährdungspotential *nt*; **~ prevention** *n* SAFETY Gefahrenvorsorge *f*; **~ symbol** *n* SAFETY Gefahrsymbol *nt*; **~ zone** *n* SAFETY Gefahrenzone *f*
daphnia: **~ test** *n* ANALYSIS Daphnientest *m*
dark: **~ heat-absorbing material** *n* RENEW ENERG *solar pond* dunkle Absorberfolie *f*
Darrieus: **~ rotor** *n* RENEW ENERG *vertical axle* Darrieus-Rotor *m*
data *n pl* ANALYSIS Daten *pl*; **~ acquisition** *n* ANALYSIS Datenerfassung *f*; **~ analysis** *n* ANALYSIS Datenanalyse *f*; **~ base** *n* ANALYSIS Datenbank *f*; **~ interpretation** *n* ANALYSIS Datenauswertung *f*; **~ logger** *n* ANALYSIS Datenlogger *m*; **~ logging** *n* ANALYSIS Datenerfassung *f*, Meßdatenerfassung *f*; **~ logging programme** *n* ANALYSIS *computing* Meßdatenerfassungsprogramm *nt*; **~ network** *n* ANALYSIS Datennetz *nt*; **~ processing** *n* ANALYSIS Datenverarbeitung *f*; **~ security** *n* ANALYSIS Datensicherheit *f*; **~ storage** *n* ANALYSIS Datenspeicherung *f*, Meßwertspeicherung *f*; **~ system** *n* ANALYSIS Datensystem *nt*; **~ transfer** *n* ANALYSIS Datenübertragung *f*
dB *abbr* (*decibel*) ANALYSIS dB (*Dezibel*)
DC/AC: **~ converter** *n* RENEW ENERG Wechselrichter *m*
DCR *abbr* (*dispersion by chemical reaction*) WASTE, WATER SUPP DCR (*Dispersion durch chemische Reaktion*); **~ process** *n* WASTE *hazardous waste treatment* DCR-Verfahren *nt*
DC: **~ voltage field** *n* AIR Gleichspannungsfeld *nt*
DDT *abbr* (*dichlorodiphenyltrichloroethane*) SOIL DDT (*Dichlordiphenyltrichlorethan*)
deactivate *vt* AIR, WASTE desaktivieren
deactivated *adj* AIR, WASTE desaktiviert
deactivation *n* AIR, WASTE Desaktivierung *f*
dead: **~ bacteria** *n pl* WATER POLL tote Bakterien *pl*
deaden *vt* NOISE *intrusive sound* dämpfen
deaerate *vt* WATER POLL entlüften
debt: **~-for-nature swaps** *n pl* POLICY Schuldenerlaß *m* für Naturschutz
decant *vt* GEN abgießen, absetzen, dekantieren
decantation *n* ANALYSIS, SOIL Dekantieren *nt*, WASTE Absetzklärung *f*, Absetzen *nt*, Dekantieren *nt*, WATER POLL *sedimentation* Absetzen *nt*, Dekantieren *nt*, WATER SUPP Absetzklärung *f*, Absetzen *nt*, Dekantieren *nt*
decanter *n* WATER POLL Dekanter *m*
decanting: **~ centrifuge** *n* WASTE, WATER SUPP Dekantierzentrifuge *f*
decarbonization *n* WATER SUPP Entcarbonisieren *nt*, Entcarbonisierung *f*; **~ with lime water** *n* WASTE, WATER SUPP Entcarbonisierung *f* mit Kalkwasser; **~ plant** *n* WATER SUPP Entcarbonisierungsanlage *f*; **~ with slaked lime** *n* WATER SUPP Entcarbonisierung *f* mit Kalkhydrat
decarbonize *vt* WASTE, WATER SUPP entcarbonisieren
decarbonized *adj* WASTE, WATER SUPP entcarbonisiert; **~ groundwater** *n* WASTE, WATER SUPP entcarbonisiertes Grundwasser *nt*
decay 1. *n* WASTE, WATER POLL Verrottung *f*; **2.** *vi* WASTE, WATER POLL verrotten
decay: **~ tank** *n* WATER POLL Absetzbehälter *m*
deceleration *n* AIR *filter medium* Verzögerung *f*
decentral: **~ photoelectric use** *n* RENEW ENERG dezentrale photoelektrische Nutzung *f*
dechlorinate *vt* SOIL, WASTE, WATER SUPP entchloren
dechlorination *n* SOIL, WASTE, WATER SUPP Entchlorung *f*
decibel *n* (*dB*) ANALYSIS Dezibel *nt* (*dB*)
decoding: **~ of measured value** *n* ANALYSIS Meßwertdecodierung *f*
decoloration *AmE see decolouration BrE*
decolouration *n BrE* ANALYSIS Entfärbung *f*
decompact *vt* WASTE, WATER SUPP dekompaktieren
decompose *vti* GEN abbauen
decomposition *n* GEN *biological* Abbau *m*, *disintegration* Zersetzung *f*, *chemical* Zerlegung *f*; **~ capacity** *n* WATER POLL Abbauleistung *f*; **~ of harmful substances** *n* WASTE, WATER POLL, WATER SUPP Abbau *m* von Schadstoffen; **~ potential** *n* SOIL Abbaupotential *nt*
decontaminability *n* GEN Dekontaminierbarkeit *f*
decontaminate *vt* AIR, SAFETY dekontaminieren, entseuchen, SOIL entseuchen, sanieren, WASTE dekontaminieren, WATER POLL, WATER SUPP dekontaminieren, entseuchen
decontaminated *adj* AIR, SAFETY, SOIL, WATER POLL, WATER SUPP entseucht, GEN dekontaminiert, gereinigt, saniert; **~ soil** *n* SOIL dekontaminierter Boden *m*, gereinigter Boden *m*
decontamination *n* GEN Dekontamination *f*, Entseuchung *f*, Sanierung *f*; **~ reactor** *n* SOIL Dekontaminationsreaktor *m*; **~ technique** *n* SOIL Sanierungstechnik *f*

decoupling: ~ **biological scrubber** *n* AIR, WATER POLL entkoppelter Biowäscher *m*
dedicated: ~ **collection vehicle** *n* WASTE Spezialfahrzeug *nt*
deduster *n* AIR Entstauber *m*, Entstaubungsanlage *f*
dedusting *n* AIR Entstaubung *f*, Staubabscheidung *f*
deep: ~**-bed filter** *n* AIR *separator*, WATER POLL Tiefenfilter *nt*; ~**-bed filtration** *n* AIR, WATER POLL Tiefenfiltration *f*; ~ **ploughing** *n* SOIL Tiefpflügen *nt*; ~ **sand trap** *n* WATER POLL Tiefsandfang *m*; ~**-shaft reactor** *n* WATER POLL Tiefschachtreaktor *m*; ~ **well** *n* WATER SUPP Tiefbrunnen *m*; ~ **well injection** *n* WATER SUPP Tiefbrunneneinspritzung *f*
defective: ~ **canal systems** *n pl* SOIL *degradation* defekte Kanalsysteme *nt pl*
deferrization *n* WATER SUPP *groundwater treatment* Enteisenung *f*
deficiency: ~ **of air** *n* AIR, WASTE, WATER SUPP Luftmangel *m*
definable *adj* ANALYSIS definierbar
deflect *vt* AIR *gas flow on surface* umlenken
deflector *n* AIR Ablenker *m*; ~ **separator** *n* AIR Umlenkabscheider *m*; ~ **vane** *n* AIR Umlenkblech *nt*
deforestation *n* AIR, SOIL Abholzung *f*, Entwaldung *f*
deformation *n* WASTE *surface of landfill* Verformung *f*
degas *vt* GEN entgasen
degasification *n* GEN Entgasung *f*; ~ **filter tube** *n* SOIL Entgasungsfilterrohr *nt*
degasify *vt* AIR, SOIL, WASTE, WATER POLL entgasen
degasser *n* AIR, SOIL, WASTE, WATER POLL Entgaser *m*
degassing *n* AIR, SOIL, WASTE, WATER POLL Entgasung *f*; ~ **with ballasted column** *n* WASTE Entgasung *f* über Schottersäule; ~ **chamber** *n* WASTE Entgasungskammer *f*; ~ **facility** *n* WASTE Entgasungseinrichtung *f*; ~ **with gabion** *n* WASTE Entgasung *f* über Gabionen; ~ **with gas wells** *n* WASTE Entgasung *f* über Gasbrunnen; ~ **layer** *n* WASTE Entgasungsschicht *f*; ~ **pipe** *n* WASTE Entgasungsrohr *nt*; ~ **shaft** *n* WASTE Entgasungsschacht *m*; ~ **with sonde** *n* WASTE Entgasung *f* über Sonden
degradability *n* WASTE, WATER POLL, WATER SUPP Abbaubarkeit *f*; ~ **test** *n* ANALYSIS Abbaubarkeitstest *m*
degradable *adj* WASTE, WATER POLL, WATER SUPP abbaubar; ~ **organic matter** *n* WASTE, WATER POLL, WATER SUPP abbaubare organische Substanz *f* (*AOS*)
degradation *n* GEN Abbau *m*; ~ **in lower atmosphere** *n* AIR Abbau *m* in der unteren Atmosphäre; ~ **performance** *n* WATER POLL Abbauleistung *f*
degrade *vt* GEN abbauen, degradieren
degraded *adj* GEN abgebaut, degradiert
degreaser *n* WATER POLL Ölabscheider *m*
degree: ~ **of absorption** *n* AIR Absorptionsgrad *m*; ~ **of collection** *n* WASTE Erfassungsgrad *m*; ~ **of compaction** *n* WASTE Verdichtungsgrad *m*; ~ **of contamination** *n* SOIL Kontaminationsgrad *m*; ~ **of decontamination** *n* SOIL Reinigungsgrad *m*; ~ **of degradability** *n* WASTE, WATER POLL, WATER SUPP Abbaugrad *m*; ~ **of degradation** *n* WASTE, WATER POLL, WATER SUPP Abbaugrad *m*; ~ **of desulfurization** *AmE*, ~ **of desulphurization** *BrE* *n* AIR Entschwefelungsgrad *m*; ~ **of dissociation** *n* ANALYSIS Dissoziationsgrad *m*; ~ **of drainage** *n* WATER POLL Entwässerungsgrad *m*; ~ **of dust removal** *n* AIR Entstaubungsgrad *m*; ~ **of exchange** *n* WASTE, WATER SUPP Austauschgrad *m*; ~ **of fermentation** *n* WASTE Vergärungsgrad *m*; ~ **of fraction collection** *n* AIR Fraktionsabscheidegrad *m*; ~ **of fraction removal** *n* AIR Fraktionsabscheidegrad *m*; ~ **of hardness** *n* ANALYSIS, WATER POLL, WATER SUPP Härtegrad *m*; ~ **of humification** *n* SOIL, WASTE, WATER POLL Humifizierungsgrad *m*; ~ **of incineration** *n* AIR Verbrennungsgrad *m*; ~ **of nitrogen oxides removal** *n* AIR Entstickungsgrad *m*; ~ **of ozone depletion** *n* AIR Ausmaß *nt* des Ozonabbaus; ~ **of pollution** *n* AIR, MARINE, SOIL, WATER POLL Verschmutzungsausmaß *nt*, Verschmutzungsgrad *m*; ~ **of purification** *n* WATER POLL Reinigungsgrad *m*; ~ **of purity** *n* WATER SUPP Reinheitsgrad *m*; ~ **of rotting process** *n* WASTE Rottegrad *m*; ~ **of separation** *n* AIR, WASTE, WATER POLL Abscheidegrad *m*; ~ **of stabilization** *n* WATER POLL Ausfaulgrad *m*; ~ **of thickening** *n* WATER POLL *dewatering* Eindickgrad *m*; ~ **of total removal** *n* AIR Gesamtabscheidegrad *m*; ~ **of use** *n* WASTE, WATER SUPP Verwendungsgrad *m*
degrit *vt* WATER POLL entsanden
degritter *n* WATER POLL Sandfangräumer *m*
Degussa-Plinke: ~ **process** *n* AIR Degussa-Plinke-Verfahren *nt*
dehalogenate *vt* SOIL dehalogenieren
dehumidification *n* ANALYSIS Entfeuchtung *f*; ~ **effect** *n* ANALYSIS Entfeuchtungswirkung *f*
dehumidify *vt* ANALYSIS entfeuchten
dehydrogenase: ~ **test** *n* ANALYSIS Dehydrogenasetest *m*
de-inking *n* WASTE, WATER POLL De-Inking *nt*, Tintenwäsche *f*; ~ **by flotation** *n* WATER POLL De-Inking *nt* durch Flotation; ~ **plant** *n* WASTE, WATER POLL De-Inking-Anlage *f*; ~ **sludge** *n* WATER POLL De-Inking-Schlamm *m*; ~ **unit** *n* WASTE, WATER POLL De-Inking-Anlage *f*
deionization *n* WASTE, WATER POLL, WATER SUPP Deionisation *f*, Deionisierung *f*, Demineralisierung *f*
deionize *vt* WASTE, WATER POLL, WATER SUPP deionisieren, demineralisieren
deionized *adj* WASTE, WATER POLL, WATER SUPP deionisiert, demineralisiert; ~ **water** *n* WATER POLL, WATER SUPP Deionat *nt*
deliver *vt* WASTE, WATER POLL, WATER SUPP *through pump* fördern
delivery *n* WASTE, WATER POLL, WATER SUPP *through pump* Förderleistung *f*, Anlieferung *f*; ~ **channel** *n* WASTE, WATER SUPP Abzugskanal *m*; ~ **head** *n* WATER POLL, WATER SUPP Förderhöhe *f*; ~ **main** *n* WASTE, WATER SUPP Druckleitung *f*, Druckrohrleitung *f*, Förderleitung *f*, Zubringerleitung *f*; ~ **pipe** *n* WASTE, WATER POLL, WATER SUPP Ablaufrohr *nt*, Ausflußrohr *nt*, Ausgußrohr *nt*, Druckleitung *f*; ~ **pressure** *n* WATER SUPP Förderdruck *m*; ~ **program** *AmE*, ~ **programme** *BrE* *n* ANALYSIS Lieferprogramm *nt*; ~ **race** *n* WATER SUPP Versorgungskanal *m*; ~ **time** *n* WASTE Anlieferungszeit *f*; ~ **vehicle** *n* WASTE Anlieferungsfahrzeug *nt*
demand: ~ **for raw material** *n* WASTE Rohstoffbedarf *m*
demands *n pl* MANAGE Anforderungen *f pl*
demanganization *n* WATER SUPP *groundwater treatment* Entmanganung *f*; ~ **filter** *n* WATER SUPP Entmanganungsfilter *nt*
demineralization *n* WASTE, WATER POLL, WATER SUPP Deionisation *f*, Demineralisierung *f*, Deionisierung *f*
demineralize *vt* WASTE, WATER POLL, WATER SUPP deionisieren, demineralisieren
demineralized *adj* WASTE, WATER POLL, WATER SUPP deionisiert, demineralisiert; ~ **water** *n* WATER POLL, WATER SUPP Deionat *nt*
demister *n* AIR Demister *m*; ~ **unit** *n* AIR Tropfenabscheider *m*
demixing *n* WATER POLL Entmischung *f*
demolish *vt* WASTE abbrechen
demolition *n* WASTE Abbruch *m*; ~ **waste** *n* WASTE Abbruchabfall *m*, Bauschutt *m*
demonstration: ~ **power plant** *n* RENEW ENERG Demonstrationskraftwerk *nt*
demulsification *n* WASTE, WATER POLL, WATER SUPP Emulsionsspaltung *f*
demulsifier *n* WASTE, WATER POLL, WATER SUPP Demulgator *m*, Emulsionsspalter *m*
demulsifying: ~ **agent** *n* WATER POLL, WATER SUPP Demulgator *m*; ~ **product** *n* WATER POLL, WATER SUPP Demulgierprodukt *nt*
denitrification *n* GEN Denitrifizierung *f*, Denitrifikation *f*;

~ **zone** *n* AIR, WASTE, WATER POLL, WATER SUPP Denitrifikationszone *f*
denitrify *vt* GEN denitrifizieren
denitrifying: ~ **bacteria** *n pl* AIR, WASTE, WATER POLL, WATER SUPP Denitrifikant *m*
denox *n* (*NO_x removal*) AIR, SOIL Denox *nt* (*NO_x-Entstickung*); ~**-catalyst** *n* AIR Denox-Katalysator *m*
denoxer *n* AIR *selective catalytic reduction* Denoxanlage *f*
densely: ~ **populated area** *n* MANAGE, POLICY, WASTE, WATER SUPP *demand* Ballungsgebiet *nt*
densifier *n* WATER SUPP *feeding* Stopfwerk *nt*
densify *vt* SOIL *into pellets* dichtmachen
densimeter *n* ANALYSIS *density determination of liquids* Aräometer *nt*
density *n* ANALYSIS, WATER SUPP *of water* Dichte *f*; ~ **bottle** *n* ANALYSIS Pyknometer *nt*; ~ **current** *n* WATER SUPP Dichteströmung *f*; ~ **determination** *n* ANALYSIS Dichtebestimmung *f*; ~ **of measurement points** *n pl* ANALYSIS Meßstellendichte *f*; ~ **of volume charge** *n* AIR Raumladungsdichte *f*
deodorization *n* AIR, WASTE, WATER POLL, WATER SUPP Geruchsbekämpfung *f*, Geruchsbeseitigung *f*
deodorize *vt* AIR, WASTE, WATER POLL, WATER SUPP den Geruch beseitigen von, desodorieren
deodorized *adj* AIR, WASTE, WATER POLL, WATER SUPP desodoriert
de-oiling *n* WATER POLL Ölabscheidung *f*
deoxygenation *n* WATER POLL Sauerstoffentzug *m*
depalletization *n* WASTE Palettenentnahme *f*
depalletizer *n* WASTE Entpalettisiermaschine *f*, Palettenentnehmer *m*
department *n* POLICY *administrative unit* Abteilung *f*
dependent: ~ **ecosystem** *n* AIR, SOIL, WATER POLL, WATER SUPP abhängiges Ökosystem *nt*
dephenolating *n* WATER POLL Entphenolung *f*
dephenolizing *n* WATER POLL Entphenolung *f*
deplate *vt* GEN abbauen, WASTE entzinken
depletion *n* GEN Abbau *m*; ~ **of stratospheric ozone layer** *n* AIR Abbau *m* der Ozonschicht in der Stratosphäre; ~ **time** *n* WATER POLL Zehrungsdauer *f*
depolarization *n* ANALYSIS Depolarisation *f*
depolarizer *n* ANALYSIS Depolarisator *m*
depolymerization *n* WASTE Depolymerisation *f*
deposit 1. *n* AIR Ablagerung *f*, SOIL Ablagerung *f*, Bodensatz *m*, WASTE Ablagerung *f*, *money on bottle* Flaschenpfand *nt*, Pfand *nt*, WATER POLL, WATER SUPP Ablagerung *f*, Aufbringung *f*; **2.** *vt* AIR, WATER POLL ablagern
deposit: ~ **bottle** *n* WASTE Mehrwegflasche *f*, Pfandflasche *f*
deposited *adj* WATER POLL, WATER SUPP abgelagert
deposit: ~ **fee system** *n* POLICY Pfandregelung *f*
deposition *n* AIR Ablagerung *f*, SOIL Ablagerung *f*, Sedimentablagerung *f*, WASTE, WATER POLL, WATER SUPP Ablagerung *f*; ~ **rate** *n* SOIL Absetzleistung *f*; ~ **value** *n* SOIL, WASTE Ablagerungswert *m*; ~ **velocity** *n* AIR, WASTE, WATER SUPP Ablagerungsgeschwindigkeit *f*; ~ **of wastewater sludge on grassland** *n* WATER POLL Aufbringung *f* von Klärschlamm auf Grünland
depot: ~ **container** *n* WASTE Depotcontainer *m*
depression: ~ **cone** *n* WASTE, WATER SUPP Senkungskegel *m*; ~ **spring** *n* WASTE, WATER SUPP Senkungsquelle *f*
depth: ~ **filtration** *n* WATER POLL Tieffiltration *f*
derrick *n* RENEW ENERG *tidal power plant* Derrick *m*
desalinate *vt* WASTE, WATER SUPP entsalzen
desalination *n* WASTE, WATER SUPP Entsalzung *f*; ~ **plant** *n* WATER SUPP Entsalzungsanlage *f*; ~ **technique** *n* WASTE, WATER SUPP Entsalzungsverfahren *nt*
desert: ~ **region** *n* RENEW ENERG *typical area of solar farm installation* Wüstengebiet *nt*
desiccate *vt* AIR, ANALYSIS, SOIL, WASTE trocknen
desiccator *n* AIR, ANALYSIS, SOIL, WASTE Exsikkator *m*
design *vt* ANALYSIS, WASTE, WATER SUPP konzipieren
designation *n* ANALYSIS Benennung *f*
design: ~ **flow** *n* WASTE, WATER SUPP Bemessungswassermenge *f*
desludge *vt* WASTE, WATER POLL, WATER SUPP entschlammen
desludger *n* WASTE, WATER POLL, WATER SUPP Entschlämmer *m*
desludging *n* WASTE, WATER POLL, WATER SUPP Entschlammung *f*, Schlammaustragung *f*
desolventizing *n* WASTE, WATER SUPP Abtrennung *f* des Extraktionsmittels
desonox *n* (*SO_2/NO_x removal*) AIR, SOIL Desonox *nt* (*SO_2/NO_x-Entstickung*); ~ **in circulating fluidized bed** *n* AIR Desonox-Verfahren *nt* in der zirkulierenden Wirbelschicht; ~ **process** *n* AIR Desonox-Verfahren *nt*
desorb *vt* AIR, ANALYSIS, SOIL *in thermal processes* desorbieren
desorbed *adj* AIR, ANALYSIS, SOIL *thermal processes* desorbiert; ~ **absorbent** *n* AIR desorbiertes Absorbens *nt*; ~ **activated coke** *n* AIR desorbierter Aktivkoks *m*; ~ **rich gas** *n* AIR desorbiertes Reichgas *nt*
desorbent *n* AIR, ANALYSIS, SOIL Desorptionsmedium *nt*, Desorptionsmittel *nt*
desorber: ~ **feed** *n* AIR Desorberzulauf *m*
desorption *n* AIR *recovery of gas components from washing solution*, ANALYSIS, WATER POLL Desorption *f*; ~ **air** *n* AIR Desorptionsluft *f*; ~ **air heat exchanger** *n* AIR Desorptionsluftwärmetauscher *m*; ~ **column** *n* AIR Desorptionskolonne *f*; ~ **time** *n* AIR Desorptionszeit *f*; ~ **zone** *n* AIR Desorptionszone *f*
dessication *n* AIR, SOIL, WASTE, WATER POLL Trocknung *f*
destabilization *n* WASTE, WATER POLL *of colloids* Destabilisierung *f*, Entstabilisierung *f*, WATER SUPP Destabilisierung *f*
destabilize *vt* WASTE, WATER POLL destabilisieren, entstabilisieren, WATER SUPP destabilisieren
destabilized *adj* WASTE, WATER POLL destabilisiert, entstabilisiert, WATER SUPP destabilisiert; ~ **particle** *n* WATER POLL *flocculation* destabilisiertes Teilchen *nt*
destroy *vt* GEN zerstören
destroyed: ~ **in minerals** *adj* SOIL mineralogisch zerstört
destruction *n* GEN Zerstörung *f*; ~ **process** *n* WASTE, WATER SUPP Abbauprozeß *m*
desulfurization *AmE see desulphurization BrE*
desulfurize *AmE see desulphurize BrE*
desulphurization *n BrE* AIR, WASTE, WATER SUPP Entschwefelung *f*; ~ **with magnesium oxide** *n BrE* AIR Entschwefelung *f* mit Magnesiumoxid; ~ **of process gases** *n BrE* AIR Entschwefelung *f* von Prozeßgasen; ~ **sorbent** *n BrE* AIR Entschwefelungssorbens *nt*
desulphurize *vt BrE* AIR, WASTE, WATER SUPP entschwefeln
detach *vt* WASTE, WATER SUPP ablösen
detect *vt* ANALYSIS ermitteln, nachweisen
detectability *n* ANALYSIS Nachweisbarkeit *f*
detectable *adj* ANALYSIS nachweisbar
detection *n* ANALYSIS Detektion *f*, Ermittlung *f*, Nachweis *m*; ~ **frequency** *n* ANALYSIS Nachweishäufigkeit *f*; ~ **limit** *n* AIR untere Nachweisgrenze *f*, ANALYSIS Nachweisgrenze *f*, untere Nachweisgrenze *f*; ~ **method** *n* ANALYSIS Nachweismethode *f*; ~ **principle** *n* ANALYSIS Detektionsprinzip *nt*; ~ **reaction** *n* ANALYSIS Nachweisreaktion *f*; ~ **sensitivity** *n* ANALYSIS Nachweisempfindlichkeit *f*
detector *n* ANALYSIS Detektor *m*, Meßfühler *m*; ~ **lamp** *n* ANALYSIS Detektorlampe *f*; ~ **tube** *n* ANALYSIS *gas detector* Prüfröhrchen *nt*
detention: ~ **time** *n* WASTE, WATER SUPP Aufenthaltszeit *f*
detergent *n* MARINE, WATER POLL, WATER SUPP Detergens *nt*
determination *n* ANALYSIS *testing* Bestimmung *f*, Ermittlung *f*, SAFETY Bestimmung *f*; ~ **of acid capacity in wastewater** *n* ANALYSIS Bestimmung *f* der Säurekapazität von Abwasser; ~ **of ammonium content in wastewater** *n* ANALYSIS, WATER

POLL, WATER SUPP Bestimmung *f* des Ammoniumgehaltes in Abwasser; **~ of chloride content in wastewater** *n* ANALYSIS, WATER POLL, WATER SUPP Bestimmung *f* des Chloridgehaltes in Abwasser; **~ of chromite content in wastewater** *n* ANALYSIS, WATER POLL, WATER SUPP Bestimmung *f* des Chromitgehaltes in Abwasser; **~ component** *n* ANALYSIS Bestimmungskomponente *f*; **~ of concentration of nitrogen dioxide** *n* SAFETY Bestimmung *f* der Stickstoffdioxidkonzentration; **~ of conductivity of wastewater** *n* ANALYSIS, WATER POLL, WATER SUPP Bestimmung *f* der elektrischen Leitfähigkeit von Abwasser; **~ of dew point** *n* ANALYSIS Taupunktbestimmung *f*; **~ following sample distillation** *n* ANALYSIS Bestimmung *f* nach Probendestillation; **~ of gaseous acid air pollution** *n* SAFETY Bestimmung *f* der Luftverschmutzung durch saure Gase; **~ limit** *n* ANALYSIS Bestimmungsgrenze *f*; **~ of magnesium content in wastewater** *n* ANALYSIS, WATER POLL, WATER SUPP Bestimmung *f* des Magnesiumgehaltes in Abwasser; **~ of mass concentration of sulfur dioxide in ambient air** *AmE*, **~ of mass concentration of sulphur dioxide in ambient air** *BrE* *n* SAFETY Massenbestimmung *f* von Schwefeldioxid in der Umgebungsluft; **~ of nitrate content in wastewater** *n* ANALYSIS, WATER POLL, WATER SUPP Bestimmung *f* des Nitratgehaltes in Abwasser; **~ of nitrite content in wastewater** *n* ANALYSIS, WATER POLL, WATER SUPP Bestimmung *f* des Nitritgehaltes in Abwasser; **~ of particulate lead and lead compounds** *n* SAFETY, WATER POLL, WATER SUPP Bestimmung *f* des Gehaltes an Blei und Bleiverbindungen; **~ of potassium content in wastewater** *n* ANALYSIS, WATER POLL, WATER SUPP Bestimmung *f* des Kaliumgehaltes in Abwasser; **~ of settleable and undissolved matter in wastewater** *n* ANALYSIS, WATER POLL, WATER SUPP Bestimmung *f* der absetzbaren und ungelösten Stoffe von Abwasser; **~ of sodium content in wastewater** *n* ANALYSIS, WATER POLL, WATER SUPP Bestimmung *f* des Natriumgehaltes in Abwasser; **~ of sulfate content in wastewater** *AmE*, **~ of sulphate content in wastewater** *BrE* *n* ANALYSIS, WATER POLL, WATER SUPP Bestimmung *f* des Sulfatgehaltes in Abwasser; **~ of total, evaporation and ignition residue in wastewater** *n* ANALYSIS, WATER POLL Bestimmung *f* des Gesamt-, Abdampf- und Glührückstandes von Abwasser; **~ of total hardness in wastewater** *n* ANALYSIS, WATER POLL, WATER SUPP Gesamthärtebestimmung *f* von Abwasser; **~ of wastewater density** *n* ANALYSIS, WATER POLL, WATER SUPP Bestimmung *f* der Abwasserdichte; **~ of wastewater pH-value** *n* ANALYSIS, WATER POLL, WATER SUPP Bestimmung *f* des pH-Wertes von Abwasser; **~ of wastewater temperature** *n* ANALYSIS, WATER POLL, WATER SUPP Bestimmung *f* der Abwassertemperatur; **~ of wastewater turbidity** *n* ANALYSIS, WATER POLL, WATER SUPP Bestimmung *f* der Trübung von Abwasser; **~ of water demand** *n* WATER SUPP Ermittlung *f* des Wasserbedarfs

determine *vt* ANALYSIS bestimmen, ermitteln; **~ mathematically** *vt* ANALYSIS rechnerisch ermitteln

detin *vt* WASTE, WATER POLL entzinnen

detinning *n* WASTE, WATER POLL Entzinnung *f*

detoxification *n* AIR, SOIL, WASTE, WATER SUPP Entgiftung *f*

detoxify *vt* AIR, SOIL, WASTE, WATER SUPP entgiften

detritor *n* WATER POLL Sandfang *m*

detritus *n* WATER POLL Detritus *m*, Sandfanggut *nt*

develop *vt* GEN *processes* entwickeln, *a pump* entsanden

development *n* GEN *of pump* Entsandung *f*

device *n* GEN Einrichtung *f*, Gerät *nt*, Vorrichtung *f*; **~ for measured value transmission** *n* ANALYSIS Meßwertübermittlungsgerät *nt*

Dewar: **~ vessel** *n* ANALYSIS Dewar-Gefäß *nt*

dew: **~ point** *n* AIR Taupunkt *m*; **~ point hygrometer** *n* ANALYSIS Taupunkthygrometer *nt*; **~ point temperature** *n* AIR Taupunkttemperatur *f*

DHS *abbr* (*dynamic headspace technique*) ANALYSIS DHS (*dynamische Headspace-Technik*)

diagnosis *n* ANALYSIS Diagnose *f*

dial: **~ gage** *AmE*, **~ gauge** *BrE* *n* ANALYSIS Meßuhr *f*

dialog: **~-controlled** *AmE see dialogue-controlled BrE*; **~ display** *AmE see dialogue display BrE*

dialogue: **~-controlled** *adj BrE* ANALYSIS *programming* dialoggesteuert; **~ display** *n BrE* ANALYSIS *on measuring instruments* Dialoganzeige *f*

dialysis *n* ANALYSIS, WATER POLL, WATER SUPP Dialyse *f*

diamagnetic *adj* ANALYSIS diamagnetisch

diameter *n* WATER POLL Durchmesser *m*

diaphragm: **~ pump** *n* ANALYSIS, MARINE, WATER POLL, WATER SUPP Membranpumpe *f*; **~ valve** *n* AIR, WATER POLL, WATER SUPP Membranventil *nt*

diatomaceous: **~ earth** *n* WASTE, WATER SUPP Diatomeenerde *f*; **~ earth filtration** *n* WATER POLL Diatomeenerdefiltration *f*, Kieselgurfiltration *f*

dichlorodifluoromethane *n* (*Freon 12*) AIR *propellant* Dichlordifluormethan *nt* (*Freon 12*)

dichlorodiphenyltrichloroethane *n* (*DDT*) SOIL Dichlordiphenyltrichlorethan *nt* (*DDT*)

dichloromethane *n* AIR *halogenated hydrocarbon* Dichlormethan *nt*

diesel *n* AIR, RENEW ENERG Diesel *m*; **~ engine** *n* AIR Diesel *m*, Dieselmotor *m*, RENEW ENERG Diesel *m*; **~ exhaust gas** *n* AIR Dieselabgas *nt*; **~ fuel** *n* AIR, RENEW ENERG Diesel *m*; **~ generator** *n* RENEW ENERG Dieselgenerator *m*

differential *n* ANALYSIS Differential *nt*; **~ calorimetry** *n* ANALYSIS Differenzkalorimetrie *f*; **~ gas chromatogram** *n* ANALYSIS Differential-Gaschromatogramm *nt*; **~ manometer** *n* AIR Differentialmanometer *nt*; **~ pressure** *n* AIR *in filter layer* Druckdifferenz *f*; **~ pressure gage** *AmE*, **~ pressure gauge** *BrE* *n* AIR Druckdifferenzanzeige *f*; **~ taxation** *n* POLICY ungleiche Besteuerung *f*; **~ tax rate** *n* POLICY *on leaded and unleaded gasoline* ungleicher Steuersatz *m*; **~ thermal analysis** *n* (*DTA*) ANALYSIS Differenzthermoanalyse *f* (*DTA*)

diffraction *n* ANALYSIS Beugung *f*, Diffraktion *f*; **~ of light** *n* ANALYSIS Lichtzerlegung *f*

diffuse 1. *vt* AIR, WASTE, WATER SUPP diffundieren; **2.** *vi* AIR, MARINE, WASTE sich ausbreiten

diffused: **~ air** *n* AIR, SAFETY, SOIL, WATER POLL Druckluft *f*; **~-air aeration** *n* WATER POLL Druckluftbelüftung *f*; **~ air tank** *n* WATER POLL Druckluftbecken *nt*; **~ light** *n* ANALYSIS Lichtstreuung *f*, Streulicht *nt*; **~ light measurement** *n* ANALYSIS *turbidity evaluation* Streulichtmessung *f*; **~ light sensor** *n* ANALYSIS Streulichtsensor *m*

diffuse: **~ field** *n* NOISE diffuses Schallfeld *nt*

diffuser *n* AIR, ANALYSIS, NOISE, WASTE Diffusor *m*, WATER SUPP Diffusor *m*, Sandfang *m*

diffuse: **~ radiation** *n* RENEW ENERG diffuse Einstrahlung *f*

diffusion *n* AIR, WASTE, WATER SUPP Ausbreitung *f*, Diffusion *f*; **~ analysis** *n* ANALYSIS Diffusionsuntersuchung *f*; **~ charging** *n* AIR Diffusionsaufladung *f*; **~ test** *n* ANALYSIS Diffusionsversuch *m*

dig *vt* WASTE, WATER SUPP graben; **~ out** *vt* SOIL ausgraben; **~ up** *vt* SOIL *contaminated soil* ausgraben

digest *vt* SOIL, WATER POLL aufschließen

digested: **~ sludge** *n* WATER POLL Faulschlamm *m*

digester *n* WATER POLL Faulbehälter *m*, Faulraum *m*, Faulturm *m*; **~ disruption** *n* WATER POLL Faulraumstörung *f*; **~ gas** *n* WASTE, WATER POLL, WATER SUPP Biogas *nt*, Faulgas *nt*, Klärgas *nt*; **~ gas quantity** *n* WATER POLL Faulgasausbeute *f*, Faulgasmenge *f*; **~ gas utilization** *n* WATER POLL Faulgasverwertung *f*; **~ gas yield** *n* WATER POLL Faulgasertrag *m*

digestion *n* WASTE Faulung *f*, WATER POLL, WATER SUPP Faulung *f*, Gärung *f*, Zersetzung *f*; **~ with added composting** *n* WATER POLL Faulung *f* mit anschließender

Kompostierung; ~ **chamber heating** *n* WATER POLL Faulraumheizung *f*; ~ **period** *n* WATER POLL Faulzeit *f*; ~ **tank** *n* WATER POLL *septic tank, sludge digestor* Faulbecken *nt*, Faulbehälter *m*, Faulraum *m*; ~ **tank heating** *n* WATER POLL Faulraumbeheizung *f*; ~ **time** *n* WATER POLL Faulzeit *f*
digital: ~ **channel** *n* ANALYSIS Digitalkanal *m*; ~ **readout** *n* ANALYSIS Digitalanzeige *f*
dike *n* RENEW ENERG *tidal power plant* Damm *m*
dilute *vt* AIR, ANALYSIS, WASTE, WATER POLL verdünnen
diluted *adj* AIR, ANALYSIS, WASTE, WATER POLL verdünnt; ~ **sample** *n* ANALYSIS verdünnte Probe *f*; ~ **solution** *n* ANALYSIS verdünnte Lösung *f*
dilution *n* AIR, ANALYSIS, WASTE, WATER POLL Dilution *f*, Verdünnung *f*; ~ **analysis** *n* ANALYSIS Verdünnungsanalyse *f*; ~ **capacity** *n* WATER POLL *of waters* Dilutionskapazität *f*, Dilutionsvermögen *nt*; ~ **sample** *n* ANALYSIS Verdünnungsprobe *f*
DIN *abbr* (*German Institute for Standards*) ANALYSIS DIN (*Deutsches Institut für Normung*)
diode *n* ANALYSIS Diode *f*; ~ **array detector** *n* ANALYSIS Dioden-Array-Detektor *m*; ~ **laser** *n* ANALYSIS *measurement of gaseous pollutants* Diodenlaser *m*
dioxins *n pl* SOIL Dioxine *nt pl*
direct *adj* GEN direkt; ~ **air sampler** *n* ANALYSIS direkter Luftprobenehmer *m*; ~ **contact condenser** *n* AIR *contact condenser* Kontaktkondensator *m*; ~ **desulfurization process** *AmE*, ~ **desulphurization process** *BrE* *n* AIR Direktentschwefelungsprozeß *m* (*DEP*); ~ **discharger** *n* WATER POLL Direkteinleiter *m*; ~ **drying** *n* WATER POLL direkte Trocknung *f*
directed: ~ **polypropylene** *n* (*DPP*) WASTE orientiertes Polypropylen *nt* (*DPP*)
direct: ~ **extrusion** *n* WASTE Direktextrusion *f*; ~ **groundwater recharge** *n* WASTE, WATER SUPP direkte Grundwasseranreicherung *f*; ~**-heated rotary kiln** *n* SOIL direkt beheizter Drehrohrofen *m*; ~ **heating** *n* SOIL direkte Beheizung *f*; ~ **heating by radiation** *n* SOIL *thermal destruction* Strahlungsheizung *f*; ~ **industrial wastewater discharge** *n* WATER POLL direkte Industrieabwassereinleitung *f*; ~ **injection** *n* ANALYSIS Direktinjektion *f*, SOIL Direkteinspritzung *f*; ~ **interception** *n* AIR direktes Auffangen *nt*
direction: ~ **change** *n* AIR *cyclone* Richtungsumkehr *f*, Richtungsänderung *f*; ~ **of flue gas** *n* AIR Rauchgasrichtung *f*; ~ **of gas stream** *n* AIR Gasströmungsrichtung *f*
directive *n* POLICY *legal* Richtlinie *f*; ~ **on major industrial hazards** *n* POLICY Störfallverordnung *f*
direct: ~ **liquid introduction** *n* ANALYSIS *chromatography* direkter Flüssigkeitseinlaß *m*; ~ **measurement method** *n* ANALYSIS direkte Meßmethode *f*
directorate *n* POLICY *administrative unit* Direktorat *nt*
direct: ~ **poisoning** *n* AIR direkte Verseuchung *f*; ~ **precipitation** *n* WASTE, WATER SUPP abgesetzter Niederschlag *m*; ~ **pulse filter** *n* AIR Direktimpulsfilter *nt*; ~ **radiation** *n* RENEW ENERG direkte Einstrahlung *f*; ~**-reading** *adj* ANALYSIS mit direkter Ablesung; ~ **reading** *n* ANALYSIS Direktablesung *f*; ~ **sound** *n* NOISE Direktschall *m*; ~ **sound level** *n* NOISE direkter Schallpegel *m*; ~ **suction skimmer** *n* MARINE Saugabschöpfgerät *nt*; ~ **voltage** *n* ANALYSIS Gleichspannung *f*; ~ **voltage recorder** *n* ANALYSIS Gleichspannungsschreiber *m*
dirt *n* GEN Schmutz *m*
dirty: ~ **gas** *n* AIR Rohgas *nt*; ~ **gas canal** *n* AIR Rohgaskanal *m*; ~ **gas compartment** *n* AIR Rohgasraum *m*; ~ **gas concentration** *n* AIR Rohgaskonzentration *f*; ~ **gas conditioning** *n* AIR Rohgaskonditionierung *f*; ~ **gas cooling** *n* AIR Rohgasabkühlung *f*; ~ **gas dust content** *n* AIR Rohgasstaubgehalt *m*; ~ **gas dust loading** *n* AIR Rohgasstaubbeladung *f*; ~ **gas flow** *n* AIR Rohgasstrom *m*; ~ **gas inlet** *n* AIR Rohgaseintritt *m*; ~ **gas load** *n* AIR Rohgasfracht *f*; ~ **gas loading** *n* AIR Rohgasbeladung *f*; ~ **gas volume flow** *n* AIR Rohgasvolumenstrom *m*
disc: ~ **aerator** *n* *BrE* WATER POLL Scheibentauchtropfkörper *m*
discard *vt* WASTE abwerfen
discharge 1. *n* AIR *of waste gases from chimneystack* Ableitung *f*, MARINE, WASTE, WATER POLL, WATER SUPP *of river* Ausfließen *nt*, *outflow* Ausfluß *m*, *into sea* Einleiten *nt*, Einleitung *f*; **2.** *vt* AIR ableiten, *emissions* ausstoßen, MARINE einleiten, *pump* liefern, fördern, WATER POLL, WATER SUPP ableiten, einleiten, *pump* fördern, liefern
dischargeable: ~ **battery** *n* WASTE entladbare Batterie *f*
discharge: ~ **canal** *n* WATER POLL Ableitungskanal *m*; ~ **conditions** *n pl* AIR *stack* Austrittsbedingungen *f pl*; ~ **flume** *n* WATER POLL Entlastungsgerinne *nt*; ~ **head** *n* WATER POLL Druckhöhe *f*, Förderhöhe *f*, WATER SUPP *pumping* Förderhöhe *f*; ~ **hopper** *n* WASTE Ablauftrichter *m*, WATER POLL Ablauftrichter *m*, Kratzerarm *m*, WATER SUPP Ablauftrichter *m*; ~ **nozzle** *n* WATER POLL, WATER SUPP Strahldüse *f*; ~ **pipe** *n* WASTE, WATER POLL, WATER SUPP Abflußleitung *f*, Ablaufleitung *f*; ~ **piping** *n* WASTE Druckleitung *f*, WATER SUPP Druckleitung *f*, Förderleitung *f*; ~ **point** *n* WASTE *conveyor band* Übergabestelle *f*
discharger *n* MARINE, WASTE, WATER POLL, WATER SUPP Austragseinheit *f*, Einleiter *m*
discharge: ~ **sluice** *n* WATER POLL Entlastungsschleuse *f*; ~ **system** *n* MARINE, WATER POLL *for wastewater* Einleitungsanlage *f*; ~ **worm** *n* AIR Austragschnecke *f*
discoloration *AmE see discolouration BrE*
discolouration *n* *BrE* ANALYSIS Verfärbung *f*
discontinuous *adj* GEN *processes* diskontinuierlich; ~ **fixed-bed adsorber** *n* AIR diskontinuierlicher Festbettadsorber *m*; ~ **hydrolysis** *n* SOIL diskontinuierliche Hydrolyse *f*; ~ **measuring** *n* ANALYSIS diskontinuierliche Messung *f*, diskrete Messung *f*; ~ **ultrafiltration** *n* WATER POLL diskontinuierliche UF *f*, diskontinuierliche Ultrafiltration *f*
discretionary: ~ **decision** *n* POLICY Ermessensentscheidung *f*; ~ **error** *n* POLICY Ermessensfehler *m*
disc: ~ **skimmer** *n* *BrE* MARINE Scheibenabschöpfgerät *nt*
disinfect *vt* SAFETY desinfizieren, AIR, SAFETY, WASTE, WATER POLL, WATER SUPP desinfizieren, entkeimen, entseuchen
disinfectant *n* SAFETY, WASTE, WATER POLL, WATER SUPP Desinfektionsmittel *nt*
disinfected *adj* SAFETY, WASTE, WATER POLL, WATER SUPP desinfiziert, entkeimt
disinfection *n* SAFETY, WASTE, WATER POLL, WATER SUPP *groundwater treatment* Desinfektion *f*, Entkeimung *f*, Entseuchung *f*; ~ **effect** *n* WATER POLL, WATER SUPP Desinfektionswirkung *f*; ~ **efficiency** *n* WATER POLL, WATER SUPP Desinfektionswirkung *f*; ~ **protection** *n* WATER SUPP Desinfektionsschutz *m*
disintegrate *vt* SOIL zerfallen, zersetzen, WASTE, WATER POL, WATER SUPP zersetzen
disintegration *n* SOIL Zerfall *m*, Zersetzung *f*, WASTE, WATER POL, WATER SUPP Zersetzung *f*
disintegrator *n* AIR Desintegrator *m*, WASTE Zerkleinerer *m*
disinterment: ~ **of waste** *n* WASTE Abfallausgrabung *f*, Abfallrückholung *f*, WATER SUPP Abfallrückholung *f*
disk: ~ **aerator** *AmE see disc aerator BrE*; ~ **skimmer** *AmE see disc skimmer BrE*
dispersant *n* GEN *dispersing additive* Dispersant *nt*, *oil* Dispergator *m*, Dispersionsmittel *nt*; ~ **spraying** *n* MARINE Versprühen *nt* von Dispersionsmitteln
disperse *vt* AIR, SOIL, MARINE, WATER POLL dispergieren; AIR, MARINE, WASTE sich ausbreiten
dispersed *adj* AIR, SOIL dispergiert, WATER POLL feinstverteilt, dispergiert
disperse: ~ **phase** *n* AIR, ANALYSIS *in liquids* disperse Phase *f*
dispersing: ~ **agent** *n* AIR, MARINE, SOIL, WATER POLL

Dispergator *m*, Dispergiermittel *nt*; **~ equipment** *n* MARINE Dispersionseinrichtung *f*
dispersion *n* AIR, MARINE, SOIL Dispergierung *f*, Dispersion *f*, WASTE *of drops* Feinstzerteilung *f*, Ausbreitung *f*, Dispersion *f*, WATER POLL, WATER SUPP Dispergierung *f*, Dispersion *f*, Feinstzerteilung *f*, Ausbreitung *f*; **~ aerosol** *n* AIR Dispersionsaerosol *nt*; **~ behavior** *AmE*, **~ behaviour** *BrE* *n* WASTE, WATER SUPP Ausbreitungsverhalten *nt*; **~ by chemical reaction** *n* (*DCR*) WASTE Dispersion *f* durch chemische Reaktion (*DCR*); **~ colloid** *n* WASTE, WATER POLL, WATER SUPP Dispersionskolloid *nt*; **~ force** *n* AIR *van der Waals forces* Dispersionskraft *f*; **~ of harmful substances** *n* AIR, SOIL, WASTE, WATER POLL, WATER SUPP Schadstoffausbreitung *f*; **~ medium** *n* AIR Zerstäubungsmittel *nt*; **~ particle counter** *n* ANALYSIS Streulichtpartikelzähler *m*; **~ telemetering** *n* ANALYSIS Streulichtfernmessung *f*
displacement *n* WATER SUPP Verdrängung *f*
displacer *n* AIR Verdrängungskörper *m*
display: **~ unit** *n* ANALYSIS Sichtanzeigegerät *nt*, Sichtgerät *nt*
disposable: **~ paper bag** *n* WASTE Einwegpapiertüte *f*; **~ plastic bag** *n* WASTE Einwegkunststofftüte *f*; **~ product** *n* WASTE Einwegprodukt *nt*
disposal *n* MARINE, SOIL, WATER POLL Beseitigung *f*, Entsorgung *f*; **~ channel** *n* WASTE Entsorgungsweg *m*; **~ costs** *n pl* MANAGE Entsorgungskosten *pl*; **~ fee** *n* WASTE Entsorgungsgebühr *f*; **~-oriented** *adj* WASTE ablagerungsorientiert; **~-oriented treatment of waste** *n* WASTE ablagerungsorientierte Abfallbehandlung *f*; **~ of plastics** *n* WASTE, WATER POLL Kunststoffentsorgung *f*; **~ proof** *n* POLICY Entsorgungsnachweis *m*; **~-related** *adj* WATER POLL abwassertechnisch; **~ route** *n* WASTE Entsorgungsweg *m*; **~ site** *n* WASTE Deponie *f*; **~ zone** *n* WASTE Endlager *nt*, Endlagerungsstätte *nt*
dispose: **~ of** *vt* GEN beseitigen, entsorgen
disposed: **~ of** *adj* GEN beseitigt, entsorgt; **~-of residues** *n pl* WASTE, WATER POLL entsorgte Rückstände *m pl*; **~-of wastes** *n pl* WASTE entsorgte Abfälle *m pl*
dissimilation *n* WATER POLL *biological denitrification process* Dissimilation *f*
dissociable *adj* ANALYSIS, WATER POLL dissoziierbar; **~ substance** *n* ANALYSIS, WATER POLL dissoziierbarer Stoff *m*
dissociate *vt* ANALYSIS, WATER POLL dissoziieren
dissociation *n* ANALYSIS, WATER POLL Dissoziation *f*
dissolution *n* WASTE, WATER POLL, WATER SUPP Auflösung *f*
dissolve *vt* AIR *gas into liquid* auflösen, ANALYSIS lösen, auflösen, WASTE, WATER POLL auflösen; **~ away** *vt* SOIL herauslösen
dissolved *adj* GEN aufgelöst, gelöst; **~ air flotation** *n* WATER POLL Druckflotation *f*, Entspannungsflotation *f*, Mikroluftblasenflotation *f*; **~ inorganic carbon** *n* AIR, ANALYSIS, WASTE, WATER SUPP gelöster anorganischer Kohlenstoff *m*; **~ organic carbon** *n* (*DOC*) AIR, ANALYSIS, WASTE, WATER SUPP gelöster organischer Kohlenstoff *m* (*DOC*); **~ organic matter** *n* AIR, ANALYSIS, WASTE, WATER SUPP gelöste organische Substanz *f*; **~ oxygen** *n* (*DO*) AIR, WASTE, WATER SUPP gelöster Sauerstoff *m*; **~ solids** *n pl* WASTE, WATER POLL, WATER SUPP gelöste Stoffe *m pl*; **~ substances** *n pl* WATER POLL gelöste Stoffe *m pl*
dissolving: **~ away** *n* SOIL *of contaminants* Herauslösung *f*
distant: **~ source** *n* AIR, WASTE, WATER SUPP entfernte Quelle *f*; **~ water supply** *n* WATER SUPP Fernwasserversorgung *f*
distil *vt BrE* AIR, ANALYSIS, WASTE destillieren
distill *AmE see distil BrE*
distillate *n* ANALYSIS Destillat *nt*; **~ flask** *n* ANALYSIS Destillatkolben *m*
distillation *n* ANALYSIS, SOIL, WASTE, WATER SUPP Destillation *f*; **~ column** *n* WATER SUPP Destillationskolonne *f*; **~ residue** *n* WASTE Destillationsrückstand *m*; **~ water** *n* WASTE Destillationswasser *nt*
distilled: **~ water** *n* ANALYSIS *sampling* destilliertes Wasser *nt*
distillery: **~ wastewater** *n* WATER POLL Brennereiabwasser *nt*
distilling: **~ column** *n* AIR Destillationskolonne *f*
distinguishable *adj* ANALYSIS wahrnehmbar
distribute *vt* WATER SUPP verteilen
distributing: **~ bottom** *n* AIR *wet precipitator* Verteilerboden *m*; **~ pipe** *n* WATER SUPP Verteilungsrohr *nt*
distribution *n* WATER SUPP Verteilung *f*; **~ channel** *n* WASTE, WATER SUPP Verteilungsrinne *f*; **~ coefficient** *n* WASTE, WATER SUPP Verteilungskoeffizient *m*; **~ network** *n* WATER SUPP *of pipes* Verteilerrohrnetz *nt*, Verteilsystem *nt*, Verteilungsnetz *nt*, Verteilungssystem *nt*; **~ reservoir** *n* WASTE, WATER SUPP Ausgleichsbehälter *m*
distributor *n* WASTE Vertriebsagent *m*, *for commercial goods* Verteiler *m*, WATER SUPP Vertriebsagent *m*
disturbance: **~ level** *n* ANALYSIS, NOISE Störpegel *m*
disturbed: **~ sample** *n* ANALYSIS gestörte Probe *f*
disturbing: **~ noise** *n* NOISE Störgeräusch *nt*
ditch *n* WASTE, WATER SUPP Graben *m*
ditcher *n* SOIL *for installing retention walls* Löffelbagger *m*
ditch: **~ irrigation** *n* WASTE, WATER SUPP Grabenbewässerung *f*
divert *vt* WATER SUPP ableiten
division *n* POLICY *administrative unit* Abteilung *f*
DO *abbr* (*dissolved oxygen*) AIR, WASTE, WATER SUPP gelöster Sauerstoff *m*
DOC *abbr* (*dissolved organic carbon*) ANALYSIS, WATER SUPP DOC (*gelöster organischer Kohlenstoff*)
documentation *n* WASTE, WATER SUPP Dokumentation *f*
documents *n pl* POLICY *permit application* Unterlagen *f pl*
dolination *n* WASTE Verdolung *f*
dolomite: **~ meal** *n* AIR *additive* Dolomitmehl *nt*
domestic: **~ emissions** *n pl* AIR Binnenemissionen *f pl*, Inlandsemissionen *f pl*; **~ waste** *n* WASTE Haushaltsabfall *m*, Hausmüll *m*, Siedlungsabfall *m*; **~ waste analysis** *n* ANALYSIS Hausmüllanalyse *f*; **~ waste collection** *n* WASTE Hausmüllabfuhr *f*, Hausmülleinsammlung *f*; **~ waste composition** *n* WASTE Hausmüllzusammensetzung *f*; **~ waste incineration** *n* WASTE Hausmüllverbrennung *f*; **~ waste incineration plant** *n* WASTE Hausmüllverbrennungsanlage *f*; **~ waste pyrolysis** *n* WASTE Hausmüllpyrolyse *f*; **~ waste slag** *n* WASTE Hausmüllschlacke *f*; **~ wastewater** *n* WATER POLL Abwasser *nt* aus Haushaltungen, Haushaltsabwasser *nt*, häusliches Abwasser *nt*; **~ water supply** *n* WATER SUPP Haushaltswasserversorgung *f*
dope *vt* AIR, RENEW ENERG *semiconductor material* dotieren
Dorr: **~ grit chamber** *n* WATER POLL Dorr-Sandfang *m*
Dortmund: **~ tank** *n* WATER POLL Dortmundbecken *nt*, Dortmundbrunnen *m*
dosage *n* GEN Beschickung *f*, Dosierung *f*; **~ device** *n* ANALYSIS Dosiergerät *nt*
dose 1. *n* ANALYSIS, SAFETY, WASTE, WATER POLL Dosis *f*; **2.** *vt* ANALYSIS, SAFETY, WASTE, WATER POLL, WATER SUPP beschicken, dosieren
dose: **~ response** *n* ANALYSIS, WASTE, WATER SUPP Dosiswirkung *f*; **~-response relationship** *n* ANALYSIS, WASTE, WATER SUPP Dosis-Wirkungs-Beziehung *f*
dosimeter *n* ANALYSIS Dosisleistungsmesser *m*, Dosismeßgerät *nt*
dosing *n* ANALYSIS Dosierung *f*, SOIL Dosierung *f*, Zudosierung *f*, WASTE, WATER POLL, WATER SUPP Beschickung *f*, Dosierung *f*; **~ apparatus** *n* ANALYSIS, WASTE, WATER SUPP Dosiervorrichtung *f*; **~ feeder** *n* WASTE, WATER SUPP Dosierer *m*; **~ machine** *n* WASTE, WATER SUPP Abfüllmaschine *f*, Dosiermaschine *f*; **~ pump** *n* WASTE Dosierpumpe *f*, WATER POLL Beschickungspumpe *f*, Dosierpumpe *f*, WATER SUPP Dosierpumpe *f*; **~ scales** *n pl* AIR Dosierwaage *f*
double *adj* GEN Doppel-; **~-beam photometer** *n* ANALYSIS Doppelstrahlphotometer *nt*; **~-belt press** *n* WASTE Doppel-

bandpresse *f*; **~-bucket collector** *n* WASTE, WATER SUPP Doppelschaufelabscheider *m*; **~-cycle scrubber** *n* AIR Zweikreislaufwäscher *m*; **~ detector GC** *n* ANALYSIS Doppeldetektor-GC *m*; **~ eco-balance system** *n* MANAGE Ökobilanz *f* als Doppelsystem; **~-leaf** *adj* NOISE zweischalig; **~ pipe** *n* AIR Doppelrohr *nt*; **~-pipe condenser** *n* AIR Doppelrohrkondensator *m*; **~-pipe heat exchanger** *n* AIR Doppelrohrwärmetauscher *m*; **~-sheet** *adj* NOISE *encapsulation* zweischalig; **~ wall** *n* NOISE *for level reduction* Doppelwand *f*

downpipe *n* WATER POLL Fallrohr *nt*

downspout *n* AIR *tray tower* Fallschacht *m*

downstream *adv* AIR stromabwärts, strömungsabwärts, WASTE, WATER SUPP stromabwärts

downward: **~ surface installation** *n* WASTE, WATER SUPP Abwärtsflächeneinbau *m*

downwind *adv* AIR mit dem Wind

downwind: **~ power plant** *n* RENEW ENERG Fallwindkraftwerk *nt*

DPP *abbr* (*directed polypropylene*) WASTE DPP (*orientiertes Polypropylen*)

dracone *n* MARINE Kunststoffbehälter *m*, schwimmender Kunststoffbehälter *m*

draft: **~ a convention** *vi* POLICY ein Abkommen entwerfen

drag *n* AIR Strömungswiderstand *m*, RENEW ENERG *wind turbine* Luftwiderstand *m*; **~ chain** *n* WASTE Kratzerkette *f*; **~ coefficient** *n* AIR Widerstandsbeiwert *m*; **~ conveyor ash remover** *n* WASTE Kratzerkettenentschlacker *m*; **~-torque viscosimeter** *n* ANALYSIS Rotationsviskosimeter *nt*

drain 1. *n* SOIL, WASTE, WATER SUPP Drän *m*; **2.** *vt* SOIL trockenlegen, WASTE *pond pump* entleeren, leermachen, WATER POLL, WATER SUPP *pond* trockenlegen, *pump* entleeren, leermachen; **~ off** WATER SUPP *fluid* entleeren, leermachen; **3.** *vi* WATER POLL *fluid* ablaufen; **~ off** WASTE auslaufen, leerlaufen

drainability *n* AIR, WATER POLL Entwässerbarkeit *f*

drainable *adj* AIR, WATER POLL entwässerbar; **~ crystal** *n* AIR entwässerbarer Kristall *m*

drainage *n* AIR Entwässerung *f*, SOIL, WASTE *of container area* Dränage *f*, Entleerung *f*, Entwässerung *f*, WATER POLL, WATER SUPP *of container area* Trockenlegen *nt*, Dränage *f*, *of fluid* Abfluß *m*, Abfließen *nt*; **~ area** *n* WATER SUPP Einzugsgebiet *nt*, Niederschlagsgebiet *nt*; **~ basin** *n* WATER SUPP Einzugsgebiet *nt*, Niederschlagsgebiet *nt*; **~ by freezing** *n* WATER POLL Entwässerung *f* durch Gefrieren; **~ canal** *n* WASTE, WATER SUPP Entwässerungsgraben *m*, Entwässerungskanal *m*; **~ channel** *n* WASTE, WATER SUPP Entwässerungskanal *m*; **~ ditch** *n* WASTE, WATER SUPP Abzugsgraben *m*, Entwässerungsgraben *m*; **~ layer** *n* WASTE, WATER SUPP Dränschicht *f*, Entwässerungsschicht *f*; **~ line** *n* WASTE Dränleitung *f*; **~ material** *n* WASTE Entwässerungsmaterial *nt*, Dränmaterial *nt*; **~ pipe** *n* WASTE Sickerrohr *nt*, Dränleitung *f*, Dränrohr *nt*; **~ possibility** *n* WASTE Entwässerungsmöglichkeit *f*; **~ pump** *n* WASTE, WATER SUPP Entwässerungspumpe *f*; **~ shaft** *n* WASTE *of landfill site* Entwässerungsschacht *m*; **~ system** *n* WASTE Kanalisation *f*, Dränagesystem *nt*, WATER SUPP Kanalnetz *nt*, Dränagesystem *nt*; **~ wall** *n* WASTE Dränagewand *f*

drained: **~ sludge** *n* WATER POLL entwässerter Schlamm *m*

draining *n* WASTE, WATER SUPP Leerlauf *m*; **~ engine** *n* WASTE Entwässerungsmaschine *f*; **~ pump** *n* WASTE, WATER SUPP Entwässerungspumpe *f*

drain: **~ pipe spout** *n* WATER SUPP Mündung *f* der Ablaufleitung; **~ pump** *n* WATER SUPP Faßpumpe *f*

draw: **~ off** *vt* SOIL abziehen; **~ water from** *vt* WASTE, WATER SUPP anzapfen

drawdown *n* WASTE, WATER SUPP Absinken *nt*

Drechsel: **~ gas-washing bottle** *n* ANALYSIS Drechsel-Gaswaschflasche *f*, Gaswaschflasche *f* nach Drechsel

dredge *vt* MARINE naßbaggern

dredger *n* MARINE Naßbagger *m*, Schwimmbagger *m*

dried *adj* GEN getrocknet; **~ sludge** *n* WATER POLL getrockneter Schlamm *m*

drift *n* MARINE Abdrift *f*, WASTE Verwehung *f*; **~ velocity** *n* AIR Driftgeschwindigkeit *f*

drill 1. *n* SOIL Bohrer *m*; **2.** *vt* ANALYSIS, RENEW ENERG, SOIL bohren

drilling *n* ANALYSIS, SOIL *suction of soil atmosphere* Bohrung *f*; **~ core analysis** *n* ANALYSIS Bohrkernuntersuchung *f*; **~ grab** *n* SOIL Bohrlochgreifer *m*; **~ rod assembly** *n* SOIL Bohrgestänge *nt*; **~ shaft** *n* SOIL Bohrwelle *f*

drinking: **~ water** *n* ANALYSIS, WATER SUPP Trinkwasser *nt*, trinkbares Wasser *nt*; **~ water analysis** *n* ANALYSIS Trinkwasseranalyse *f*; **~ water consumption** *n* WATER SUPP Trinkwasserverbrauch *m*; **~ water disinfection** *n* WATER SUPP Desinfektion *f* von Trinkwasser; **~ water distribution network** *n* WATER SUPP Trinkwasserverteilsystem *nt*; **~ water quality** *n* WATER SUPP Trinkwasserqualität *f*; **~ water quality control** *n* WATER SUPP Qualitätskontrolle *f* des Trinkwassers; **~ water recovery** *n* WATER SUPP Trinkwassergewinnung *f*; **~ water recovery from surface waters** *n* WATER SUPP Trinkwassergewinnung *f* aus Oberflächengewässern; **~ water softening** *n* WATER SUPP Enthärtung *f* des Trinkwassers; **~ water standards** *n pl* WATER SUPP Gütenormen *f pl* für Trinkwasser; **~ water stock** *n* WATER SUPP Trinkwasservorrat *m*; **~ water treatment regulation** *n* POLICY, WATER SUPP Trinkwasseraufbereitungsverordnung *f*; **~ water use** *n* WATER SUPP Trinkwasserverwendung *f*

drinks: **~ packaging** *n* WASTE Getränkeverpackung *f*

drive: **~ train** *n* RENEW ENERG Getriebe *nt*; **~ units** *n pl* WASTE, WATER SUPP Antriebselemente *nt pl*

drop 1. *n* AIR Tropfen *m*; **2.** *vi* WATER SUPP absinken

drop: **~ distribution** *n* AIR Tropfenverteilung *f*; **~ film** *n* AIR Tropfenschleier *m*; **~-laden** *adj* AIR tropfenbeladen

droplet *n* AIR Flüssigkeitströpfchen *nt*, Tröpfchen *nt*; **~ diameter** *n* AIR Tröpfchendurchmesser *m*; **~ dispersion** *n* AIR Tropfendispergierung *f*; **~ generation** *n* AIR Tropfenerzeugung *f*; **~ separation** *n* AIR Tropfenzerteilung *f*; **~ size** *n* AIR Tröpfchengröße *f*; **~ trap** *n* AIR Tröpfchenfänger *m*

drop: **~-off center** *AmE*, **~-off centre** *BrE n* WASTE Sammelstelle *f*

dropping: **~ below dew point** *n* AIR Taupunktunterschreitung *f*

drowned: **~ weir** *n* WATER SUPP Grundwehr *nt*, Stauschwelle *f*

drum *n* AIR *waste heat boiler*, MARINE, SOIL, WATER POLL Trommel *f*; **~ filter** *n* WASTE, WATER POLL, WATER SUPP Trommelfilter *nt*; **~ screen** *n* WATER POLL Drehsieb *nt*, Trommelsieb *nt*; **~ sieve** *n* WASTE, WATER SUPP Trommelsieb *nt*; **~ skimmer** *n* MARINE Trommelabschöpfgerät *nt*; **~-type dryer** *n* SOIL Trommeltrockner *m*

dry 1. *adj* AIR, SOIL, WASTE trocken, Trocken-; **2.** *vt* AIR, SOIL, WASTE trocknen

dry: **~ absorption** *n* AIR Trockenabsorption *f*; **~ acid deposit** *n* AIR, SOIL trockener saurer Fallout *m*; **~ acid fallout** *n* AIR, SOIL trockener saurer Niederschlag *m*; **~ acidic fallout** *n* AIR, SOIL trockener saurer Niederschlag *m*; **~ additive process** *n* AIR Trockenadditivverfahren *nt*; **~ adiabatic lapse rate** *n* AIR trockener adiabatischer Temperaturgradient *m*; **~ adsorption** *n* AIR Trockenadsorption *f*; **~ air filter** *n* AIR Trockenluftfilter *nt*; **~ battery** *n* WASTE Rundzelle *f*; **~ cleaning process** *n* AIR trockene Abreinigung *f*; **~ corrosion** *n* WASTE, WATER SUPP trockene Korrosion *f*; **~ deposition** *n* AIR nuklear ungefährlicher Niederschlag *m*, trockene Deposition *f*, SOIL trockene Deposition *f*, ungefährlicher nuklearer Niederschlag *m*; **~ desulfurization process** *AmE*, **~ desulphurization process** *BrE n* AIR Trockenentschwefelungsprozess *m*; **~ distillation** *n* SOIL trockene Destillation *f*; **~ dust removal installation** *n* SAFETY Trockenstaubbeseitigungsanlage *f*; **~ electrostatic filter** *n* AIR trockenes Elektrofilter *nt*; **~ equipment** *n* ANA-

LYSIS Trockeneinrichtung *f*; ~ **feeder** *n* WATER POLL Trockendosierer *m*, Vakuumentlüfter *m*; ~ **filtration** *n* WATER POLL Rieselfiltration *f*, Trockenfiltration *f*; ~ **firing** *n* WASTE Trockenfeuerung *f*; ~ **gas cleaning** *n* AIR trockene Abscheidung *f*; ~ **grinding** *n* WASTE Trockenzerkleinern *nt*

drying *n* AIR, SOIL, WASTE Trocknung *f*; ~ **agent** *n* ANALYSIS Trockenmedium *nt*, Trockenmittel *nt*; ~ **bed** *n* WATER SUPP Trockenbett *nt*; ~ **closet** *n* ANALYSIS Trockenschrank *m*; ~ **grate** *n* WASTE Trockenrost *m*; ~ **medium** *n* AIR Trocknungsmedium *nt*; ~ **tower** *n* AIR Trockenturm *m*; ~ **up** *n* WASTE, WATER SUPP *of landfill body* Austrocknung *f*; ~ **zone** *n* AIR Trocknungszone *f*

dry: ~ **mass** *n* WASTE Trockenmasse *f*, Trockensubstanz *f* (*TS*), WATER POLL Feststoffgehalt *m*, Trockenmasse *f*, Trokkensubstanz *f* (*TS*); ~ **mass concentration** *n* WATER POLL TS-Konzentration *f*, Trockenmassekonzentration *f*; ~ **mass content** *n* WATER POLL TS-Gehalt *m*, Trockensubstanzgehalt *m*; ~ **mass separation** *n* WATER POLL TS-Abscheidung *f*, Trockensubstanzabscheidung *f*; ~ **material** *n* WATER POLL Trockengut *nt*; ~ **residue** *n* WASTE, WATER POLL Trockenrückstand *m* (*TR*); ~ **screen** *n* WATER POLL Fangsieb *nt*, Trockensieb *nt*; ~ **scrubber** *n* AIR Trockenabscheider *m*; ~ **scrubbing** *n* AIR Trockenabscheidung *f*; ~ **sludge** *n* WATER POLL Trockenschlamm *m*; ~ **sludge disposal site** *n* WATER POLL Trockenschlammdeponie *f*; ~ **solid matter** *n* WASTE, WATER POLL Trockenmasse *f*, Trockensubstanz *f* (*TS*); ~ **sorption** *n* WASTE Trockensorption *f*; ~ **sorting** *n* WASTE Trockentrennung *f*; ~ **substance** *n* WATER POLL Trockengut *nt*; ~ **waste gas treatment** *n* AIR trockene Abgasbehandlung *f*

DTA *abbr* (*differential thermal analysis*) ANALYSIS DTA (*Differenzthermoanalyse*)

dual *adj* GEN Doppel-; ~ **alkaline scrubbing** *n* AIR Doppelalkaliwäsche *f*; ~-**bed bulk material catalyst** *n* AIR Doppelbettschüttgutkatalysator *m*; ~-**bed catalyst** *n* AIR Doppelbettkatalysator *m*; ~-**bed monolithic catalyst** *n* AIR Doppelbettmonolithkatalysator *m*; ~ **system** *n* MANAGE, WASTE *waste management* Dualsystem *nt*

duct *n* AIR Leitung *f*, Rohrzug *m*, Strahlrohr *nt*, Strömungskanal *m*, NOISE Kanal *m*

ductibility *n* WASTE Verformbarkeit *f*

duct: ~ **lining** *n* NOISE Kanalauskleidung *f*

dump *vt* AIR ablagern, MARINE *at sea* verklappen, WASTE, WATER POLL, WATER SUPP abkippen

dumpable *adj* WASTE ablagerungsfähig

dump: ~ **ground** *n* WASTE Müllabladeplatz *m*

dumping: ~ **area** *n* WASTE Ablagerungsbereich *m*; ~ **at sea** *n* MARINE, WASTE Verklappung *f*; ~ **height** *n* WASTE Schütthöhe *f*; ~ **location** *n* WASTE, WATER POLL, WATER SUPP Abkippstelle *f*; ~ **method** *n* WASTE Ablagerungsverfahren *nt*; ~ **phases** *n pl* WASTE Schüttphasen *f pl*; ~ **plan** *n* WASTE Ablagerungsplan *m*

durable *adj* ANALYSIS *sample* haltbar

duration: ~ **of cleaning** *n* AIR *filter bag* Abreinigungsdauer *m*

dust 1. *n* AIR, ANALYSIS, SAFETY, WASTE Staub *m*; **2.** *vt* AIR entstauben

dust: ~ **analysis** *n* ANALYSIS Staubanalyse *f*

dustbin *n BrE* (*cf garbage can AmE*) WASTE Müllbehälter *m*, Mülleimer *m*, Mülltonne *f*

dust: ~-**binding agent** *n* AIR Staubbindemittel *nt*

dustbin: ~ **liner** *n BrE* (*cf garbage can bag AmE*) WASTE Beutel *m*, Müllbeutel *m*, Müllsack *m*; ~ **lorry** *n BrE* (*cf garbage truck AmE*) WASTE Müllfahrzeug *nt*, Müllwagen *m*

dust: ~ **breakthrough** *n* AIR Staubdurchbruch *m*; ~ **build-up** *n* AIR Staubablagerung *f*, Staubbelag *m*; ~ **burden** *n* AIR Staubbeladung *f*, Staubbelastung *f*; ~ **catalyst** *n* AIR Staubkatalysator *m*; ~ **catcher** *n* AIR Staubfänger *m*; ~ **coating** *n* AIR Staubbelag *m*; ~-**collecting hopper** *n* AIR Staubbunker *m*, Staubsammeltrichter *m*; ~ **collector** *n* ANALYSIS Staubauffanggerät *nt*, WASTE Schmutzbehälter *m*; ~ **combustion** *n* AIR Staubbrand *m*; ~ **component** *n* AIR Staubkomponente *f*; ~ **concentration** *n* AIR, ANALYSIS, SAFETY Staubkonzentration *f*; ~ **concentration measuring device** *n* AIR Staubkonzentrationsmeßgerät *nt*; ~ **containing zinc** *n* WASTE zinkhaltiger Staub *m*; ~ **content** *n* AIR Staubgehalt *m*, Staubanteil *m*, WASTE Staubgehalt *m*; ~ **content in flue gas** *n* AIR Staubgehalt *m* im Rauchgas, Staublast *f*; ~ **counter** *n* ANALYSIS Staubzähler *m*; ~ **cover** *n* SAFETY Staubschutzhaube *f*; ~ **deposit** *n* AIR Staubablagerung *f*, Staubbelag *m*, SOIL Staubniederschlag *m*; ~ **discharge** *n* AIR Staubaustrag *m*; ~ **discharge opening** *n* AIR *of cyclone* Staubaustragsöffnung *f*; ~ **dispersion** *n* AIR Staubeinlagerung *f*; ~ **emissions** *n pl* AIR, RENEW ENERG Staubemissionen *f pl*; ~ **exhaust fan** *n* SAFETY Staubabzug *m*; ~ **explosion** *n* AIR Staubexplosion *f*; ~ **extraction** *n* AIR Entstaubung *f*, Staubabsaugung *f*; ~ **extraction plant** *n* AIR Entstaubungsanlage *f*; ~ **filter** *n* AIR Staubfilter *nt*; ~ **fineness** *n* AIR Staubfeinheit *f*; ~ **fraction** *n* AIR Staubfraktion *f*; ~-**free** *adj* AIR staubfrei; ~-**free gas** *n* AIR staubfreies Gas *nt*; ~-**free waste air** *n* AIR staubfreie Abluft *f*

dusting *n* AIR Bestaubung *f*

dust: ~-**laden** *adj* AIR, SAFETY staubbeladen, staubhaltig; ~-**laden air** *n* AIR staubbeladene Luft *f*, SAFETY staubhaltige Luft *f*; ~-**laden film** *n* AIR staubbeladener Film *m*; ~-**laden flue gas** *n* AIR staubhaltiges Abgas *nt*; ~-**laden gas** *n* AIR staubbeladenes Gas *nt*, staubhaltiges Gas *nt*; ~-**liquid mixture** *n* AIR Staub-Flüssigkeits-Gemisch *nt*; ~ **loading** *n* AIR Staubbeladung *f*; ~ **mask** *n* SAFETY Staubschutzmaske *f*; ~ **measurement** *n* ANALYSIS Staubmessung *f*; ~ **mixture** *n* AIR Staubgemisch *nt*; ~ **outlet** *n* AIR Staubaustritt *m*; ~ **particle** *n* AIR, ANALYSIS Staubpartikel *f*, Staubteilchen *nt*; ~ **particle size distribution** *n* ANALYSIS Korngrößenverteilung *f* von Staub; ~ **particulate layer** *n* AIR Staubschicht *f*; ~-**proof** *adj* SAFETY staubdicht; ~ **receiver** *n* AIR Staubbunker *m*, Staubsammelbehälter *m*; ~ **recovery** *n* AIR Staubrückgewinnung *f*; ~ **removal** *n* AIR Entstaubung *f*; ~ **removal efficiency** *n* AIR Entstaubungswirkungsgrad *m*; ~ **removal from flue gas** *n* AIR Rauchgasentstaubung *f*; ~ **removal installation** *n* SAFETY Entstaubungsanlage *f*; ~ **removal plant** *n* AIR Entstaubungsanlage *f*; ~ **removal technology** *n* AIR Entstaubungstechnik *f*; ~-**removed** *adj* AIR entstaubt; ~-**removed dirty gas** *n* AIR entstaubtes Rohgas *nt*; ~ **remover** *n* AIR *cyclone, filter* Entstauber *m*; ~ **respirator** *n* SAFETY Staubschutzmaske *f*; ~ **scale** *n* ANALYSIS Staubwaage *f*; ~ **separator** *n* AIR Staubabscheider *m*, SAFETY *cleaning equipment* Entstauber *m*, WASTE Entstauber *m*, Staubabscheider *m*; ~ **silo** *n* AIR Staubsilo *m*; ~ **transporting worm** *n* AIR Staubtransportschnecke *f*; ~ **trap** *n* AIR Staubfänger *m*; ~ **type** *n* AIR Staubart *f*

Dutch: ~ **list** *n* SOIL *of harmful substance concentrations* Niederländische Liste *f*, Niederländischer Leitfaden *m*

duties: ~ **of operator** *n pl* MANAGE, POLICY, WATER SUPP Betreiberpflichten *f pl*

duty: ~ **of authorities** *n* POLICY Behördenpflicht *f*; ~ **of prevention** *n* POLICY Vorsorgepflicht *f*

dwell: ~ **time** *n* AIR, WASTE, WATER POLL Verweildauer *f*, Verweilzeit *f*

dynamic *adj* ANALYSIS, MARINE, WATER SUPP dynamisch; ~ **headspace** *n* ANALYSIS dynamisches Headspace *nt*; ~ **headspace technique** *n* (*DHS*) ANALYSIS dynamische Headspace-Technik *f* (*DHS*); ~ **leachability test** *n* ANALYSIS dynamischer Auslaugtest *m*, dynamischer Eluationstest *m*; ~ **pressure** *n* ANALYSIS Staudruck *m*, dynamischer Druck *m*; ~ **skimmer** *n* MARINE dynamisches Abschöpfgerät *nt*; ~ **viscosity** *n* WASTE, WATER SUPP dynamische Viskosität *f*, dynamische Zähigkeit *f*

dynamometer *n* ANALYSIS Dynamometer *nt*

E

EA *abbr* (*elementary analysis, elemental analysis*) ANALYSIS EA (*Elementaranalyse*)
ear: ~ **muffs** *n pl* NOISE Gehörschutzkappe *f*
earplugs *n pl* NOISE Gehörschutzstöpsel *m pl*, Ohrstöpsel *m pl*
ear: ~ **protector** *n* NOISE Gehörschützer *m*
earth: ~ **bank** *n* WASTE, WATER SUPP *waste dump* Aufschüttung *f*
earthed: ~ **wall** *n BrE* (*cf grounded wall AmE*) AIR *collection anode* geerdete Wandung *f*
earth: ~-**moving machinery** *n* SOIL Baumaschinen *f pl*; ~ **radiation** *n* AIR Erdstrahlung *f*; ~ **reservoir** *n* WASTE Erdbehälter *m*, WATER SUPP Tiefbehälter *m*; ~**'s atmosphere** *n* AIR Erdatmosphäre *f*; ~**'s crust** *n* RENEW ENERG *geothermal energy* Erdkruste *f*; ~**'s interior** *n* RENEW ENERG *geothermal energy* Erdinnere *nt*; ~**'s rotation** *n* RENEW ENERG *source of tidal power* Erdumdrehung *f*; ~**'s sunbelt** *n* RENEW ENERG Sonnengürtel *m* der Erde; ~**'s surface** *n* AIR, RENEW ENERG, WASTE Erdoberfläche *f*
ebb: ~ **and flow** *n* RENEW ENERG *tidal power* Ebbe *f* und Flut *f*; ~-**tide** *n* RENEW ENERG *tidal power* Ebbe *f*
EBT *abbr* (*eriochrome black*) ANALYSIS *indicator* Eriochromschwarz *nt*
ECD *abbr* (*electron capture detector*) ANALYSIS ECD (*Elektroneneinfangdetektor*)
eco- *pref* GEN Öko-; ~-**audit** *n* MANAGE, POLICY *assessment of environmental impacts in industry* Umwelt-Audit *nt*, Umweltschutz-Audit *nt*, Öko-Audit *nt*; ~-**automobile** *n* AIR Öko-Auto *nt*; ~-**balance** *n* MANAGE, POLICY Ökobilanz *f*; ~-**balance design** *n* MANAGE Ökobilanzsystematik *f*; ~-**car** *n* AIR Öko-Auto *nt*; ~-**controlling** *n* MANAGE Öko-Controlling *nt*; ~-**friendly** *adj* WASTE umweltfreundlich, umweltschonend; ~-**label** *n* POLICY Umweltzeichen *nt*; ~-**labeling** *AmE*, ~-**labelling** *BrE n* POLICY, WATER SUPP Öko-Labeling *nt*
ecological *adj* GEN ökologisch; ~ **account** *n* MANAGE Umweltkonto *nt*; ~ **balance** *n* POLICY, WASTE, WATER SUPP ökologisches Gleichgewicht *nt*; ~ **buying guideline** *n* MANAGE Einkaufsrichtlinie *f*, Einkaufsvorgabe *f*; ~ **collapse** *n* POLICY ökologischer Zusammenbruch *m*; ~ **conditions** *n pl* POLICY ökologische Randbedingungen *f pl*; ~ **damage** *n* AIR, SOIL, WATER POLL ökologischer Schaden *m*; ~ **database** *n* MANAGE Ökodatenbank *f*; ~ **factor** *n* GEN Umweltfaktor *m*; ~ **filter** *n* WASTE Ökofilter *nt*
ecologically: ~ **harmful** *adj* GEN umweltschädlich; ~ **relevant** *adj* POLICY *tax base* ökologierelevant; ~ **safe** *adj* AIR, SOIL, WATER POLL umweltsicher
ecological: ~ **pyramid** *n* POLICY, WASTE, WATER SUPP ökologische Pyramide *f*; ~ **restrictions** *n pl* POLICY ökologische Restriktionen *f pl*; ~ **structural change** *n* POLICY ökologischer Strukturwandel *m*; ~ **tax reform** *n* POLICY ökologische Steuerreform *f*; ~ **tax system** POLICY ökologisches Steuersysystem *nt*
ecology *n* GEN Ökologie *f*; ~-**oriented** *adj* MANAGE ökologieorientiert; ~-**oriented management** *n* MANAGE ökologieorientiertes Management *nt*; ~-**oriented marketing** *n* MANAGE ökologieorientiertes Marketing *nt*
economic *adj* MANAGE, POLICY wirtschaftlich, ökonomisch; ~ **efficiency** *n* MANAGE, POLICY ökonomische Effizienz *f*; ~ **feasibility** *n* MANAGE, POLICY *policy measures* Wirtschaftlichkeit *f*; ~ **potential** *n* RENEW ENERG wirtschaftliches Potential *nt*
eco-: ~-**political** *adj* POLICY umweltpolitisch; ~-**quality criteria** *n* MANAGE Ökoqualitätskriterien *nt pl*; ~-**rating** *n* MANAGE Öko-Rating *nt*; ~-**retailer** *n* MANAGE Ökohändler *m*
ecosphere *n* SOIL Ökosphäre *f*
ecosystem *n* GEN Ökosystem *nt*; ~ **stress** *n* AIR, MARINE, SOIL, WATER POLL Ökosystemstreß *m*
eco-: ~-**tax** *n* MANAGE, POLICY Ökosteuer *f*; ~-**tax package** *n* MANAGE Ökosteuerpaket *nt*, POLICY ökologisches Steuerpaket *nt*, Ökosteuerpaket *nt*; ~-**tax reform** *n* POLICY ökologische Steuerreform *f*, Ökosteuerreform *f*; ~-**tax revenue** *n* POLICY Ökosteueraufkommen *nt*; ~-**tax system** *n* POLICY ökologisches Steuersystem *nt*; ~-**tourism** *n* MANAGE Ökotourismus *m*
ecotoxicity *n* WATER POLL Ökotoxizität *f*
ecotoxicology *n* POLICY, SAFETY, WATER POLL, WATER SUPP Ökotoxikologie *f*
eco-: ~-**window** *n* MANAGE Ökomaske *f*
ED *abbr* (*electrodialysis*) ANALYSIS, WATER POLL, WATER SUPP ED (*Elektrodialyse*)
eddy *n* AIR, WASTE, WATER SUPP Wirbel *m*; ~ **current separation** *n* WASTE Wirbelstromabscheidung *f*; ~ **flow** *n* AIR *cyclone* Wirbelströmung *f*; ~ **flow scrubber** *n* AIR Wirbelströmungswäscher *m*; ~ **sink** *n* AIR Wirbelsenke *f*; ~ **viscosity** *n* AIR Wirbelzähigkeit *f*; ~ **viscosity coefficient** *n* AIR Turbulenzkoeffizient *m*
edge: ~ **pressure** *n* WASTE Kantenpressung *f*; ~-**sealed landfill** *n* WASTE flankengedichtete Deponie *f*
EDTA *abbr* (*ethylene diamine-tetracetic acid*) ANALYSIS Ethylendiamintetraessigsäure *f*
eductor: ~-**basin cleaner** *n* WATER POLL Schlammsaugwagen *m*
effect *n* GEN Wirkung *f*
effective: ~ **energy use** *n* POLICY effektive Energienutzung *f*; ~ **evaporation** *n* WASTE effektive Evaporation *f*, WATER SUPP Verdunstung *f*, effektive Evaporation *f*; ~ **level** *n* ANALYSIS Wirkpegel *m*
effectiveness *n* MARINE, POLICY, SAFETY Wirksamkeit *f*
effective: ~ **sound pressure** *n* NOISE effektiver Schalldruck *m*; ~ **stack height** *n* AIR effektive Kaminhöhe *f*, effektive Schornsteinhöhe *f*
effect: ~-**oriented** *adj* SOIL wirkungsorientiert; ~-**oriented security measures** *n pl* SOIL wirkungsorientierte Sicherungsmaßnahmen *f pl*
efficiency *n* GEN Wirkungsgrad *m*, *supply network* Leistungsfähigkeit *f*, Leistungsvermögen *nt*; ~ **analysis** *n* ANALYSIS Nutzwertanalyse *f*
efficient: ~ **packaging** *n* WASTE wirkungsvolle Verpackung *f*
effluent *n* POLICY, WASTE, WATER POLL, WATER SUPP Abfluß *m*, Ablauf *m*, Abwasser *nt*; ~ **channel** *n* WASTE Ablaufkanal *m*, WATER POLL Abflußkanal *m*, Ablaufkanal *m*, WATER SUPP Ablaufkanal *m*; ~ **charge** *n* POLICY Abwasserabgabe *f*; ~ **concentration** *n* WASTE, WATER POLL, WATER SUPP Ablaufkonzentration *f*; ~ **disposal** *n* WATER POLL Abwasserbeseitigung *f*; ~-**related** *adj* WATER POLL abwassertechnisch; ~ **sewer** *n* WASTE, WATER POLL, WATER SUPP Abflußleitung *f*, Ablaufkanal *m*, Ablaufleitung *f*; ~ **side wall** *n* WASTE, WATER POLL, WATER SUPP Ablaufrinne *f*; ~ **sludge** *n* WATER POLL Abwasserschlamm *m*; ~ **trough** *n* WASTE, WATER POLL, WATER SUPP Abflußrinne *f*; ~ **weir** *n* WATER POLL Überlaufwehr *nt*
EGR *abbr* (*exhaust gas recirculation*) AIR AGR (*Abgasrückführung*)
EIA *abbr* (*environmental impact assessment*) POLICY, WATER

SUPP UVP (*Umweltverträglichkeitsprüfung*); ~ **evaluation** *n* POLICY UVP-Bewertung *f*; ~ **investigation** *n* POLICY UVP-Untersuchung *f*; ~ **procedure** *n* POLICY UVP-Verfahren *nt*; ~ **report** *n* POLICY UVP-Bericht *m*; ~ **study** *n* POLICY UVP-Studie *f*

EI-MS *abbr* ANALYSIS (*electron impact mass spectrometer*) Elektronenstoß-Massenspektrometer *nt*, (*electron impact mass spectrometry*) Elektronenstoß-Massenspektrometrie *f*

EINCS *abbr* (*European inventory of new chemical substances*) POLICY, WATER SUPP EINCS (*Europäisches Neustoffverzeichnis*)

EINECS *abbr* (*European inventory of existing chemical substances*) POLICY, WATER SUPP EINECS (*Europäisches Altstoffverzeichnis*)

EKC *abbr* (*electrokinetic capillary chromatography*) ANALYSIS EKC (*elektrokinetische Kapillarchromatrographie*)

elastic: ~ **temporary layer** *n* AIR elastische Zwischenlage *f*

elastomer *n* NOISE *resilient material*, WASTE Elastomer *nt*

elbow *n* AIR *separator, scrubber* Krümmer *m*

electret: ~ **microphone** *n* ANALYSIS *noise measurement* Elektretmikrofon *nt*

electric *adj* GEN elektrisch

electrical *adj* GEN elektrisch; ~ **breakdown** *n* AIR Durchschlagen *nt*; ~ **conductivity** *n* ANALYSIS, WASTE spezifische Leitfähigkeit *f*; ~ **dust removal installation** *n* AIR, SAFETY elektrische Entstaubungsanlage *f*; ~ **energy** *n* RENEW ENERG elektrische Energie *f*; ~ **field** *n* AIR *ESP*, RENEW ENERG elektrisches Feld *nt*; ~ **force** *n* AIR elektrische Kraft *f*; ~ **hazard** *n* SAFETY elektrische Gefahr *f*

electrically: ~ **driven** *adj* AIR, RENEW ENERG, SAFETY elektrisch betrieben; ~ **driven heat pump** *n* RENEW ENERG elektrisch betriebene Wärmepumpe *f*

electrical: ~ **moisture measurement** *n* ANALYSIS elektrische Feuchtemessung *f*; ~ **motor** *n* WATER POLL Elektromotor *m*; ~ **power** *n* RENEW ENERG elektrischer Strom *m*

electric: ~ **car** *n* AIR Elektroauto *nt*; ~ **drive** *n* AIR Elektroantrieb *m*; ~ **energy** *n* RENEW ENERG elektrische Energie *f*; ~ **field** *n* AIR elektrisches Feld *nt*

electricity: ~ **generation** *n* RENEW ENERG Stromerzeugung *f*; ~ **grid** *n* RENEW ENERG Stromnetz *nt*; ~ **source** *n* RENEW ENERG Stromquelle *f*

electric: ~ **measuring apparatus** *n* SAFETY elektrisches Meßgerät *nt*; ~ **resistance** *n* ANALYSIS elektrischer Widerstand *m*

electroacoustic *adj* SOIL elektroakustisch

electroanalytic *adj* ANALYSIS elektroanalytisch

electrochemical *adj* ANALYSIS elektrochemisch; ~ **detector** *n* ANALYSIS *gas measuring device* elektrochemische Meßzellen *f pl*; ~ **methods** *n pl* ANALYSIS elektrochemische Methoden *f pl*; ~ **ozone measurement** *n* ANALYSIS elektrochemische Ozonmessung *f*; ~ **ozone sonde** *n* ANALYSIS elektrochemische Ozonsonde *f*; ~ **stripping analysis** *n* (*ESA*) ANALYSIS Auflösungsanalyse *f*

electrode *n* ANALYSIS Elektrode *f*, Meßkette *f*, WASTE Elektrode *f*; ~ **cell** *n* ANALYSIS *process conductivity meter* Pol-Zelle *f*; ~ **impedance** *n* ANALYSIS Meßkettenimpedanz *f*; ~ **potential** *n* ANALYSIS *pH-meter span* Meßkettenspannung *f*; ~ **sensor** *n* ANALYSIS Elektrodensensor *m*; ~ **statistics** *n pl* ANALYSIS *pH-measurement* Meßkettenstatistik *f*

electrodialysis *n* (*ED*) ANALYSIS, WASTE, WATER POLL Elektrodialyse *f* (*ED*)

electroflotation *n* WATER POLL Elektroflotation *f*

electrokinetic *adj* SOIL elektrokinetisch; ~ **capillary chromatography** *n* (*EKC*) ANALYSIS elektrokinetische Kapillarchromatrographie *f* (*EKC*); ~ **remediation** *n* SOIL *in-situ* elektrokinetische Sanierung *f*

electrolysis *n* SOIL, WASTE, WATER POLL, WATER SUPP Elektrolyse *f*; ~ **cell** *n* ANALYSIS Elektrolysezelle *f*; ~ **installation** *n* RENEW ENERG Elektrolyseanlage *f*; ~ **residue** *n* WASTE Elektrolyserückstand *m*

electrolyte *n* ANALYSIS, WASTE, WATER SUPP Elektrolyt *m*; ~ **matrix** *n* AIR *fuel cell* Elektrolytmatrix *f*

electrolytic: ~ **cell** *n* WASTE elektrolytische Zelle *f*; ~ **concentration** *n* WATER POLL Elektrolytkonzentration *f*

electromagnet *n* AIR, WASTE, WATER POLL Elektromagnet *m*

electromagnetic *adj* AIR, WASTE, WATER POLL, WATER SUPP elektromagnetisch; ~ **pump** *n* WASTE, WATER POLL, WATER SUPP elektromagnetische Pumpe *f*; ~ **radiation** *n* AIR elektromagnetische Abstrahlung *f*

electromechanical: ~ **separation** *n* WASTE elektromechanische Trennung *f*

electrometric *adj* ANALYSIS elektrometrisch, potentiometrisch; ~ **pH-value determination** *n* ANALYSIS elektrometrische pH-Wert-Bestimmung *f*

electron *n* AIR, ANALYSIS, WATER SUPP Elektron *nt*; ~ **accelerator** *n* AIR Elektronenbeschleuniger *m*; ~ **acceptor** *n* WATER SUPP Elektronenakzeptor *m*; ~ **amelioration** *n* SOIL Elektromelioration *f*; ~ **beam** *n* AIR Elektronenstrahl *m*; ~ **beam process** *n* AIR *flue-gas cleaning* Elektronenstrahlverfahren *nt* (*ESV*); ~ **capture** *n* ANALYSIS Elektroneneinfang *m*; ~ **capture detector** *n* (*ECD*) ANALYSIS Elektroneneinfangdetektor *m* (*ECD*); ~ **diffraction** *n* ANALYSIS Elektronbeugung *f*; ~ **diffraction pattern** *n* ANALYSIS Elektronbeugungsbild *nt*; ~ **donator** *n* WATER SUPP Elektronendonator *m*

electronic *adj* ANALYSIS *measuring devices* elektronisch; ~ **sorting** *n* WASTE elektronische Sortierung *f*

electron: ~ **impact** *n* ANALYSIS Elektronenstoß *m*; ~ **impact ionization** *n* ANALYSIS *ionisation mode* Elektronenstoß-Ionisation *f*; ~ **impact mass spectrometer** *n* (*EI-MS*) ANALYSIS Elektronenstoß-Massenspektrometer *nt*; ~ **impact mass spectrometry** *n* (*EI-MS*) ANALYSIS Elektronenstoß-Massenspektrometrie *f*; ~ **impact method** *n* ANALYSIS Elektronenstoßmethode *f*; ~ **irradiation** *n* WATER SUPP Elektronenbestrahlung *f*; ~ **lens** *n pl* ANALYSIS *quadrupole ion trap* Elektronenlinse *f*; ~ **microprobe** *n* (*EM*) ANALYSIS Elektronenstrahlmikrosonde *f*; ~ **microscope** *n* ANALYSIS *analysis of dust* Elektronenmikroskop *nt*; ~ **microscopy** *n* ANALYSIS Elektronenmikroskopie *f*; ~ **spin resonance** *n* (*ESR*) ANALYSIS Elektronenspinresonanz *f* (*ESR*)

electronvolt *n* (*eV*) ANALYSIS Elektronenvolt *nt* (*eV*)

electroosmosis *n* SOIL Elektroosmose *f*

electroosmotic *adj* SOIL elektroosmotisch

electrophoresis *n* ANALYSIS, SOIL Elektrophorese *f*

electrophoretic *adj* ANALYSIS, SOIL elektrophoretisch

electrostatic *adj* AIR, WASTE, WATER POLL, WATER SUPP elektrostatisch; ~ **attraction** *n* AIR elektrostatische Anziehung *f*; ~ **precipitator** *n* (*ESP*) AIR, WASTE, WATER SUPP Elektrofilter *nt* (*E-Filter*), elektrostatischer Abscheider *m*, elektrostatischer Staubabscheider *m*; ~ **repulsion** *n* WATER POLL elektrostatische Abstoßung *f*; ~ **repulsive force** *n* WATER POLL *colloidal particle*, WATER SUPP Abstoßungskraft *f*, Abstoßungsvermögen *nt*

electrostatics *n pl* AIR Elektrostatik *f*

electrostatic: ~ **separation** *n* WASTE, WATER SUPP elektrostatische Abscheidung *f*

element *n* ANALYSIS Element *nt*

elemental: ~ **analysis** *n* (*EA*) ANALYSIS Elementaranalyse *f* (*EA*); ~ **composition** *n* ANALYSIS Elementarzusammensetzung *f*

elementary: ~ **analysis** *n* (*EA*) ANALYSIS Elementaranalyse *f* (*EA*)

element: ~**-specific** *adj* ANALYSIS elementbezogen, elementspezifisch; ~**-specific analysis** *n* ANALYSIS elementbezogene Analytik *f*, elementspezifische Analytik *f*

elevated: ~ **water reservoir** *n* WATER SUPP Turmbehälter *m*, Wasserhochbehälter *m*, Wasserturm *m*

elimination: **~ of residues** *n* WASTE Vernichtung *f*; **~ of contaminants** *n* WATER POLL, WATER SUPP Schadstoffeliminierung *f*; **~ of halogenated hydrocarbons** *n* WATER SUPP *drinking water treatment* HKW-Eliminierung *f*
eluant *n* ANALYSIS Extraktionsmittel *nt*
eluate *n* ANALYSIS, WATER POLL Eluat *nt*; **~ control** *n* ANALYSIS Eluatkontrolle *f*; **~ preparation** *n* ANALYSIS Eluatherstellung *f*
elute *vt* ANALYSIS eluieren, herauslösen, WASTE, WATER POLL eluieren
elution *n* ANALYSIS Eluierung *f*, Herauslösung *f*, WASTE, WATER POLL, WATER SUPP Eluierung *f*; **~ chromatography** *n* ANALYSIS Elutionschromatographie *f*; **~ test** *n* ANALYSIS Elutionstest *m*
elutriate *vt* AIR schlämmen, ANALYSIS dekantieren, SOIL auswaschen, dekantieren, WASTE, WATER POLL, WATER SUPP dekantieren, auslaugen
elutriation *n* AIR Schlämmen *nt*, Schlämmung *f*, ANALYSIS *in chromatography* Dekantieren *nt*, Elutriation *f*, SOIL Auswaschen *nt*, Dekantieren *nt*, WASTE, WATER POLL, WATER SUPP Dekantieren *nt*, Auslaugung *f*
elutriator *n* AIR Schlämmapparat *m*
eluvial: **~ horizon** *n* SOIL E-Horizont *m*
EM *abbr* (*electron microprobe*) ANALYSIS Elektronenstrahlmikrosonde *f*
embankment *n* WASTE Grubenwand *f*
embed *vt* WASTE einbetten
emergency *n* GEN Notfall *m*; **~ by-pass** *n* WATER POLL Notumlauf *m*; **~ gate** *n* WATER POLL Notverschluß *m*; **~ measures** *n pl* AIR, SAFETY, SOIL, WATER POLL Notfallmaßnahmen *f pl*; **~ overflow** *n* WATER POLL Notüberlauf *m*; **~ signal** *n* ANALYSIS Alarmsignal *nt*; **~ spillway** *n* WATER SUPP Nothochwasserentlastungsanlage *f*; **~ treatment** *n* AIR, SAFETY, SOIL, WATER POLL Notbehandlung *f*
emerging: **~ technology** *n* GEN neuere Technologie *f*
emission *n* AIR, SOIL, WASTE, WATER POLL Emission *f*; **~ of contaminants** *n* AIR, SOIL, WASTE, WATER POLL Schadstoffemission *f*; **~ of pollutants** *n* AIR, SOIL, WASTE, WATER POLL Schadstoffausbreitung *f*; **~ right** *n* POLICY Umweltlizenz *f*
emissions: **~ analysis** *n* ANALYSIS Emissionsanalyse *f*; **~ avoidance** *n* AIR, SOIL Emissionsvermeidung *f*; **~ banking** *n* POLICY Emission-Banking *nt*; **~ charge** *n* POLICY Emissionsabgabe *f*; **~ control** *n* AIR, SOIL Emissionskontrolle *f*; **~ control system** *n* AIR Emissionskontrollsystem *nt*; **~ credit** *n* POLICY Emissionsguthaben *nt*; **~ data** *n pl* AIR, ANALYSIS Emissionsdaten *pl*; **~ during maintenance** *n* AIR Emissionen *f pl* während der Wartung; **~ inventory** *n* ANALYSIS Emissionskataster *m*, *air quality measurement* Emissionsverzeichnis *nt*; **~ level** *n* NOISE *of sound* Emissionspegel *m*; **~ limit** *n* AIR, POLICY Emissionsbegrenzung *f*; **~ limitation** *n* AIR, POLICY Emissionsbegrenzung *f*; **~ limit for cars** *n* AIR, POLICY Emissionsbegrenzung *f* für Kraftfahrzeuge; **~ limit value** *n* AIR, POLICY Emissionsgrenzwert *m*; **~ measurement** *n* ANALYSIS Emissionsmessung *f*; **~ minimalization** *n* AIR, POLICY, WATER POLL, WATER SUPP Emissionsminimierung *f*; **~ monitoring** *n* ANALYSIS Emissionsüberwachung *f*
emission: **~ source** *n* AIR, POLICY, SOIL, WASTE, WATER SUPP Emissionsquelle *f*, Emittent *m*
emissions: **~ pathway** *n* AIR, WATER POLL Emissionenpfad *m*; **~ point** *n* AIR Emissionsort *m*; **~ potential** *n* WASTE Emissionspotential *nt*; **~ rate** *n* AIR Emissionsrate *f*; **~ reduction** *n* AIR, SOIL, WATER POLL Emissionsverminderung *f*; **~ register** *n* POLICY, WATER SUPP Emissionskataster *m*; **~-related** *adj* AIR emissionsbezogen; **~-related maintenance** *n* AIR emissionsbezogene Wartung *f*; **~ requirements** *n pl* POLICY Emissionsauflagen *f pl*; **~ spectrometer** *n* ANALYSIS Emissionsspektrometer *nt*; **~ spectrum** *n* AIR, ANALYSIS Emissionsspektrum *nt*; **~ standard** *n* AIR, ANALYSIS, POLICY Emissionsgrenzwert *m*; **~ statement** *n* POLICY, WATER SUPP Emissionserklärung *f*; **~-targeted** *adj* POLICY emissionsorientiert; **~-targeted automobile tax** *AmE*, **~-targeted road tax** *BrE* *n* POLICY emissionsorientierte Kfz-Steuer *f*; **~ tax** *n* POLICY Emissionssteuer *f*; **~ value** *n* AIR Emissionswert *m*
emissivity *n* AIR Emissionsfähigkeit *f*, Emissionsvermögen *nt*
emit *vt* AIR, NOISE, WATER SUPP ausstoßen, emittieren
EMIT *abbr* (*enzyme-mediated immunoassay test*) ANALYSIS enzymatischer Immunoassaytest *m*
emitted *adj* AIR, NOISE, WATER SUPP emittiert; **~ particle** *n* AIR emittierte Partikel *f*; **~ substance** *n* AIR emittierter Stoff *m*
emitter *n* AIR, POLICY, WATER SUPP Emittent *m*
emplaced: **~ waste** *n* WASTE eingelagerter Abfall *m*
empties *n pl* WASTE Leergut *nt*
empty 1. *adj* WASTE, WATER POLL, WATER SUPP leer; **2.** *vt* WASTE, WATER POLL, WATER SUPP entleeren
emptying *n* WASTE, WATER POLL, WATER SUPP Entleeren *nt*, Entleerung *f*
empty: **~ trip** *n* WASTE Leerfahrt *f*
Emscher: **~ tank** *n* WATER POLL Emscherbrunnen *m*; **~-type settling tank** *n* WATER POLL Emscherbecken *nt*
emulsified *adj* WASTE, WATER POLL, WATER SUPP emulgiert; **~ oil** *n* WATER POLL emulgiertes Öl *nt*
emulsify *vt* WASTE, WATER POLL, WATER SUPP emulgieren
emulsifying: **~ agent** *n* SAFETY, SOIL, WATER POLL, WATER SUPP Emulgator *m*, Emulgierungsmittel *nt*; **~ liquid** *n* SAFETY, SOIL, WATER POLL, WATER SUPP Emulgatorflüssigkeit *f*
emulsion *n* MARINE, WASTE, WATER POLL, WATER SUPP Emulsion *f*; **~ breaker** *n* MARINE, WASTE, WATER SUPP Demulgator *m*, Emulsionsspalter *m*; **~-coated** *adj* WASTE, WATER SUPP emulsionbeschichtet; **~ composition** *n* WATER POLL Emulsionszusammensetzung *f*; **~ separation** *n* WASTE, WATER POLL, WATER SUPP Emulsionsspaltung *f*, Emulsionstrennung *f*
enameling *AmE see enamelling BrE*
enamelling *n BrE* WASTE, WATER SUPP Beschichtung *f*
encapsulate *vt* NOISE, SOIL einkapseln
encapsulated *adj* NOISE, SOIL eingekapselt
encapsulation *n* NOISE, SOIL Einkapselung *f*, WASTE Einkapselung *f*, Verkapselung *f*; **~ measures** *n pl* SOIL Einkapselungsmaßnahmen *f pl*; **~ using inert materials** *n* WASTE Einkapselung *f* durch Inertmaterial
enclosure *n* NOISE *to prevent sound propagation* Kapseln *nt*, Kapselung *f*
endanger *vt* SAFETY gefährden
endangered *adj* SAFETY gefährdet
end: **~ gas purity** *n* AIR Endgasreinheit *f*; **~-of-life vehicle** *n* WASTE, WATER SUPP Altfahrzeug *nt*; **~-of-line packaging** *n* WASTE Endserienverpackung *f*; **~-of-pipe technology** *n* AIR, POLICY, WASTE, WATER POLL prozeßnachgeschaltete Maßnahmen *f pl*, End-of-pipe-Technologie *f*; **~-of-pipe measures** *n pl* MANAGE nachgeschaltete Maßnahmen *f pl*
endogenous: **~ catabolism** *n* WATER POLL endogener Katabolismus *m*
endothermic *adj* AIR, WASTE, WATER SUPP endothermisch; **~ reaction** *n* AIR, WASTE, WATER SUPP endothermische Reaktion *f*
end: **~ product** *n* AIR *gypsum* Endprodukt *nt*; **~ product generation** *n* AIR Endprodukterzeugung *f*; **~ support** *n* AIR Stützsäule *f*; **~ user** *n* WASTE Endverbraucher *m*
energy *n* GEN Energie *f*; **~ absorption** *n* RENEW ENERG Energieabsorption *f*; **~ application possibilities** *n pl* RENEW ENERG *of processes* Energieanwendungsmöglichkeiten *f pl*, Energieeinsatzmöglichkeiten *f pl*; **~ balance** *n* RENEW ENERG, WASTE, WATER POLL, WATER SUPP Energiebilanz *f*; **~ barrier** *n* WASTE Energieschwelle *f*, WATER POLL Energiebarriere *f*; **~ budget** *n* AIR Energiehaushalt *m*; **~ carrier** *n* WASTE, WATER SUPP Energieträger *m*; **~ consumption** *n* AIR, RENEW ENERG, WATER POLL, WATER SUPP *of processes* Energieverbrauch *m*; **~-consumptive** *adj* WATER POLL

aerobic digestion unter Energieverbrauch; ~ **content** *n* RENEW ENERG Energiegehalt *m*, Energieinhalt *m*; ~ **conversion** *n* AIR Energieumwandlung *f*; ~ **costs** *n pl* RENEW ENERG Energiekosten *pl*; ~ **demand** *n* RENEW ENERG Energiebedarf *m*; ~ **density** *n* RENEW ENERG, WASTE Energiedichte *f*; ~ **efficiency** *n* AIR, RENEW ENERG Energieeffizienz *f*, SOIL energetischer Wirkungsgrad *m*; ~ **expenditure** *n* AIR, RENEW ENERG, WATER POLL, WATER SUPP *of processes* Energieaufwand *m*; ~ **extraction** *n* RENEW ENERG Energiegewinnung *f*; ~ **farming** *n* RENEW ENERG Energie-Farming *nt*; ~ **from biomass** *n* RENEW ENERG Energie *f* aus Biomasse; ~ **gain** *n* RENEW ENERG Energiegewinn *m*; ~ **generation** *n* RENEW ENERG Energiegewinnung *f*; ~ **import** *n* RENEW ENERG Energieimport *m*; ~ **industry** *n* RENEW ENERG Energiewirtschaft *f*; ~ **intensity** *n* RENEW ENERG Energieintensität *f*; ~**-intensive** *adj* AIR, WATER POLL, WATER SUPP energieintensiv; ~**-intensive method** *n* AIR, RENEW ENERG, SOIL, WATER POLL, WATER SUPP energieintensives Verfahren *nt*; ~ **loss** *n* RENEW ENERG Energieverlust *m*; ~ **management** *n* RENEW ENERG Energiewirtschaft *f*; ~ **material stream** *n* RENEW ENERG energetischer Stoffstrom *m*; ~ **output** *n* RENEW ENERG Energieleistung *f*; ~ **pay-back time** *n* RENEW ENERG Energieamortisationszeit *f*; ~ **potential** *n* RENEW ENERG Energiepotential *nt*; ~ **productivity** *n* RENEW ENERG Energieproduktivität *f*; ~ **purpose** *n* RENEW ENERG Energiezweck *m*; ~ **raw materials** *n pl* RENEW ENERG Energierohstoffe *m pl*; ~ **recovery** *n* AIR, WASTE Energierückgewinnung *f*, WATER POLL *sludge incineration* Energiegewinnung *f*; ~**-related** *adj* AIR, RENEW ENERG, SOIL, WATER POLL, WATER SUPP energetisch; ~ **reserves** *n pl* RENEW ENERG Energiereserven *f pl*; ~ **resources** *n* AIR, WASTE, WATER SUPP Energieressourcen *f pl*; ~**-saving** *adj* AIR energiesparend; ~**-saving method** *n* AIR energiesparendes Verfahren *nt*; ~**-saving technology** *n* AIR, WASTE, WATER SUPP energiesparende Technologie *f*; ~ **sensitivity** *n* ANALYSIS Energieempfindlichkeit *f*; ~ **source** *n* RENEW ENERG Energieträger *m*, Energiequelle *f*, Energiespender *m*; ~ **storage** *n* RENEW ENERG Energiespeicherung *f*; ~ **supply** *n* AIR Energiezufuhr *m*, RENEW ENERG Energieversorgung *f*; ~ **supply grid** *n* RENEW ENERG Energieversorgungsnetz *nt*, Stromversorgungsnetz *nt*; ~ **tax** *n* POLICY Energiesteuer *f*; ~ **tax revenue** *n* POLICY Energiesteueraufkommen *nt*; ~ **technology** *n* AIR, RENEW ENERG Energietechnik *f*, Energietechnologie *f*; ~ **transport** *n* RENEW ENERG Energietransport *m*; ~ **use** *n* AIR, RENEW ENERG, WASTE, WATER SUPP Energienutzung *f*; ~ **yield** *n* RENEW ENERG Energieausbeute *nt*

enforce *vt* POLICY umsetzen

enforcement *n* POLICY *of legislation* Umsetzung *f*

enrich *vt* ANALYSIS anreichern

enrichment *n* ANALYSIS Anreicherung *f*

ensilage *n* WASTE *plastics* Silieren *nt*

enthalpy *n* AIR, ANALYSIS Enthalpie *f*; ~ **of fusion** *n* AIR Schmelzenthalpie *f*; ~ **of vaporization** *n* AIR Verdampfungsenthalpie *f*

entrain *vt* AIR, SOIL, WATER POLL mitreißen, aufströmen, mitnehmen

entrained *adj* AIR, SOIL, WATER POLL mitgerissen; ~ **liquid** *n* AIR mitgerissene Flüssigkeit *f*; ~ **liquid droplet** *n* AIR mitgerissener Flüssigkeitstropfen *m*; ~ **soil particle** *n* SOIL mitgerissenes Bodenteilchen *nt*; ~ **solid particle** *n* AIR mitgerissenes Feststoffteilchen *nt*

entrainment *n* AIR, SOIL, WATER POLL Mitnehmen *nt*, Mitreißen *nt*; ~ **factor** *n* AIR Mitnahmefaktor *m*; ~ **point** *n* AIR, SOIL, WATER POLL Mitreißgrenze *f*

entrance: ~ **area** *n* WASTE *delivery point* Eingangsbereich *m*; ~ **shaft** *n* WASTE, WATER SUPP Einstiegschacht *m*

entrap *vt* MARINE einschließen

entry: ~ **inlet** *n* AIR *of cyclone* Einlaufschacht *m*

environment *n* GEN Umgebung *f*, Umwelt *f*

environmental: ~ **activist** *n* POLICY Umweltaktivist *m*; ~ **activity** *n* MANAGE, POLICY Umweltaktivität *f*; ~ **administration** *n* MANAGE, POLICY Umweltverwaltung *f*; ~ **advice** *n* POLICY Umweltberatung *f*; ~ **adviser** *n* POLICY Umweltberater *m*; ~ **agreement** *n* POLICY Umweltvereinbarung *f*; ~ **analysis** *n* ANALYSIS, MANAGE Umweltanalytik *f*; ~ **award** *n* POLICY Umweltpreis *m*; ~ **awareness** *n* MANAGE, POLICY Umweltbewußtsein *nt*; ~ **catastrophe** *n* AIR, MANAGE, POLICY, SOIL Umweltkatastrophe *f*; ~ **code** *n* POLICY Umweltgesetzbuch *nt*; ~ **collapse** *n* AIR, SOIL ökologischer Zusammenbruch *m*; ~ **commitment** *n* MANAGE, POLICY Umweltengagement *nt*; ~ **communication** *n* POLICY Umweltinformatik *f*; ~ **compatibility** *n* AIR, POLICY, WATER SUPP Umweltverträglichkeit *f*; ~ **conditions** *n pl* POLICY Umweltbedingungen *f pl*; ~ **counseling** *AmE*, ~ **counselling** *BrE* *n* POLICY Umweltberatung *f*; ~ **counsellor** *BrE*, ~ **counselor** *AmE* *n* POLICY Umweltberater *m*; ~ **court** *n* POLICY Umweltgerichtshof *m*; ~ **crime** *n* POLICY Umweltdelikt *nt*; ~ **criminal** *n* POLICY Umweltverbrecher *m*; ~ **criminality** *n* POLICY Umweltkriminalität *f*; ~ **criminal law** *n* POLICY Umweltstrafrecht *nt*; ~ **damage** *n* GEN Umweltschädigung *f*, Umweltschaden *m*; ~ **directive** *n* POLICY Umweltrichtlinie *f*; ~ **disaster** *n* AIR, MANAGE, POLICY, SOIL Umweltkatastrophe *f*; ~ **economics** *n pl* MANAGE, POLICY Umweltökonomie *f*; ~ **economist** *n* MANAGE, POLICY Umweltökonom *m*; ~ **education** *n* POLICY Umwelterziehung *f*; ~ **effect** *n* GEN Umweltauswirkung *f*; ~ **endangerment** *n* AIR, MARINE, POLICY, SOIL, WASTE, WATER POLL Umweltgefährdung *f*; ~ **ethics** *n pl* POLICY Umweltethik *f*; ~ **expert opinion** *n* POLICY Umweltgutachten *nt*; ~ **fiability** *n* WATER POLL Umwelthaftung *f*; ~ **friendliness** *n* MANAGE, POLICY, WATER SUPP Umweltfreundlichkeit *f*; ~ **group** *n* POLICY Umweltverband *m*; ~ **guideline** *n* POLICY Umweltleitlinie *f*; ~ **health** *n* POLICY, SAFETY Umweltgesundheit *f*; ~ **idea** *n* MANAGE Umweltidee *f*; ~ **impact** *n* GEN Umweltauswirkung *f*, Umwelteinwirkung *f*, *harmful* Umweltbelastung *f*; ~ **impact assessment** *n* (*EIA*) POLICY Umweltverträglichkeitsprüfung *f* (*UVP*); ~ **impact assessment procedure** *n* POLICY Umweltverträglichkeitsprüfungsverfahren *nt*; ~ **impact assessment report** *n* POLICY Umweltverträglichkeitsprüfungsbericht *m*; ~ **information** *n* MANAGE, POLICY Umweltinformationen *f pl*; ~ **information collection** *n* MANAGE, POLICY Umweltinformationsbeschaffung *f*; ~ **information system** *n* MANAGE, POLICY Umweltinformationssystem *nt* (*UIS*); ~ **inspectorate** *n* POLICY Vollzugsbehörde *f*

environmentalism *n* POLICY Umweltschutzbewegung *f*

environmental: ~ **issues** *n pl* POLICY Umweltfragen *f pl*

environmentalist *n* POLICY Umweltschützer *m*

environmental: ~ **jury** *n* POLICY Umweltjury *f*; ~ **label** *n* POLICY Umweltzeichen *nt*; ~ **law** *n* POLICY *body of regulations* Umweltgesetz *nt*, Umweltrecht *nt*; ~ **leader position** *n* POLICY Umweltvorreiterposition *f*; ~ **legislation** *n* POLICY Umweltgesetzgebung *f*; ~ **liability** *n* POLICY, SAFETY, WASTE, WATER SUPP Umwelthaftung *f*; ~ **licence** *BrE*, ~ **license** *AmE* *n* POLICY Umweltgenehmigung *f*; ~ **lobby** *n* POLICY Umweltlobby *f*

environmentally: ~ **acceptable** *adj* WASTE umweltschonend; ~**-acceptable waste disposal** *n* WASTE umweltschonende Abfallentsorgung *f*; ~ **aware** *adj* MANAGE, POLICY umweltbewußt; ~**-aware purchasing** *n* MANAGE umweltbewußtes Einkaufen *nt*; ~ **compatible** *adj* MANAGE, POLICY, WASTE umweltverträglich; ~**-compatible development** *n* POLICY umweltverträgliche Entwicklung *f*; ~**-compatible disposal** *n* WASTE umweltverträgliche Entsorgung *f*; ~**-compatible operation** *n* MANAGE umweltverträglicher Betrieb *m*; ~**-compatible product** *n*

MANAGE umweltverträgliches Produkt *nt*; ~ **friendly** *adj* GEN umweltfreundlich, umweltschonend; ~-**friendly logistics** *n* MANAGE umweltfreundliche Logistik *f*; ~-**friendly product** *n* MANAGE umweltfreundliches Produkt *nt*; ~-**harmful subsidy** *n* POLICY umweltschädigende Subvention *f*; ~-**oriented** *adj* MANAGE, POLICY umweltgerecht, umweltorientiert; ~-**oriented management** *n* MANAGE umweltorientiertes Management *nt*; ~-**oriented marketing** *n* MANAGE umweltorientiertes Marketing *nt*; ~-**oriented product** *n* MANAGE umweltgerechtes Produkt *nt*; ~-**oriented recovery** *n* WASTE *by industry* umweltorientierte Wiedergewinnung *f*; ~ **political** *adj* POLICY umweltpolitisch; ~-**political working group** *n* POLICY umweltpolitische Arbeitsgruppe *f*; ~ **relevant** *adj* AIR, NOISE, SOIL umweltrelevant; ~ **safe** *adj* AIR, SOIL, WATER POLL umweltsicher; ~ **sustainable** *adj* POLICY umweltnachhaltig

environmental: ~ **management** *n* MANAGE Umweltmanagement *nt*; ~ **medium** *n* GEN Umweltmedium *nt*; ~ **medium compartment** *n* SOIL *subsoil, groundwater* Emissionsraum *m*; ~ **minister** *n* POLICY Umweltminister *m*; ~ **monitoring** *n* ANALYSIS, MANAGE, POLICY Umweltbeobachtung *f*, Umwelt-Monitoring *nt*; ~ **noise** *n* NOISE, SAFETY Umweltlärm *m*; ~ **noise measurement** *n* NOISE, SAFETY Messung *f* von Umweltlärm; ~ **organization** *n* POLICY Umweltorganisation *f*; ~ **partner** *n* MANAGE, POLICY Umweltpartner *m*; ~ **planning** *n* MANAGE Umweltplanung *f*, POLICY Umweltplanung *f*, Umweltgestaltung *f*; ~ **policy** *n* MANAGE, POLICY Umweltpolitik *f*, Umweltschutzpolitik *f*; ~ **policy measures** *n pl* POLICY umweltpolitische Maßnahmen *f pl*; ~ **polluter** *n* POLICY Umweltsünder *m*; ~ **prevention principle** *n* POLICY Umweltvorsorgeprinzip *nt*; ~ **prize** *n* POLICY Umweltpreis *m*; ~ **program** *AmE*, ~ **programme** *BrE* *n* POLICY Umweltprogramm *nt*; ~ **protection** *n* POLICY Umweltschutz *m*; ~ **protection agency** *n* POLICY Umweltschutzbehörde *f*; ~ **protection aim** *n* POLICY Umweltschutzziel *nt*; ~ **protection initiative** *n* POLICY Umweltschutzinitiative *f*; ~ **protection investment** *n* POLICY Umweltschutzinvestition *f*; ~ **protection law** *n* POLICY *body of regulations* Umweltrecht *nt*, Umweltschutzrecht *nt*, *ruling* Umweltschutzgesetz *nt*; ~ **protection measures** *n pl* POLICY Umweltschutzmaßnahmen *f pl*; ~-**protection-oriented** *adj* POLICY umweltschutzorientiert; ~ **protection standards** *n pl* POLICY Umweltschutznormen *f pl*; ~ **quality** *n* POLICY Umweltqualität *f*; ~ **quality standards** *n pl* POLICY Umweltqualitätsnormen *f pl*; ~ **regulation** *n* POLICY Umweltrichtlinie *f*; ~ **report** *n* POLICY Umweltbericht *m*; ~ **reporting** *n* POLICY Umweltberichterstattung *f*; ~ **representative** *n* POLICY Umweltbeauftragter *m*; ~ **research** *n* ANALYSIS, POLICY Umweltforschung *f*; ~ **responsibility** *n* MANAGE, POLICY Umweltverantwortung *f*; ~ **safety** *n* SAFETY Umweltsicherheit *f*; ~ **and safety engineering** *n* MANAGE Umweltschutz- und Sicherheitstechnik *f*; ~ **sector** *n* POLICY Umweltsektor *m*; ~ **sponsoring** *n* MANAGE Umwelt-Sponsoring *nt*; ~ **statistics** *n pl* POLICY Umweltstatistik *f*; ~ **stress** *n* GEN Umweltbelastung *f*, Umweltstreß *m*; ~ **stress cracking** *n* WASTE *plastics* umgebungsbeeinflußte Spannungsrißbildung *f*; ~ **summit** *n* POLICY Umweltgipfel *m*; ~ **tax** *n* POLICY Umweltsteuer *f*; ~ **technology** *n* GEN Umwelttechnik *f*; ~ **technology market** *n* MANAGE Umwelttechnikmarkt *m*; ~ **testing procedure** *n* ANALYSIS, MANAGE Umwelttestverfahren *nt*; ~ **trade fair** *n* MANAGE Umweltmesse *f*; ~ **use** *n* MANAGE, POLICY Umweltnutzung *f*; ~ **use rights** *n pl* MANAGE, POLICY Umweltnutzungsrechte *nt pl*

environment: ~ **atlas** *n* POLICY Umweltatlas *m*; ~ **committee** *n* POLICY *of parliament* Umweltausschuß *m*; ~-**economic** *adj* MANAGE, POLICY umweltökonomisch; ~ **fund** *n* POLICY Umweltfonds *m*; ~ **spokesperson** *n* POLICY *of parliamentary party* Umweltsprecher *m*; ~ **statute book** *n* POLICY Umweltgesetzbuch *nt*

enzymatic *adj* WATER POLL enzymatisch; ~ **stabilization** *n* WATER POLL enzymatische Stabilisierung *f*

enzyme *n* AIR *biological scrubber*, WATER POLL Enzym *nt*; ~ **activity** *n* WATER POLL Enzymtätigkeit *f*; ~-**mediated immunoassay test** *n* (*EMIT*) ANALYSIS enzymatischer Immunoassaytest *m*

EOCL *abbr* (*extractable organic chlorides*) AIR, ANALYSIS, SOIL, WATER POLL EOCL (*extrahierbare organische Chloride*)

EOX *abbr* (*extractable organic halogen compounds*) ANALYSIS, WASTE, WATER SUPP EOX (*extrahierbare organische Halogenverbindungen*)

epilimnion *n* WATER POLL Epilimnion *nt*

EPM *abbr* (*equivalent per million*) AIR EPM (*Äquivalent je Million*)

epoxy: ~ **resin** *n* WASTE Epoxidharz *nt*

equalization *n* WATER POLL Ausgleichung *f*

equalizing: ~ **layer** *n* WASTE Ausgleichsschicht *f*; ~ **tank** *n* WASTE, WATER SUPP Ausgleichbecken *nt*

equilibrium: ~ **concentration** *n* WASTE Gleichgewichtskonzentration *f*; ~ **constant** *n* AIR Gleichgewichtskonstante *f*; ~ **load** *n* AIR *adsorption* Gleichgewichtsbeladung *f*

equip *vt* GEN ausrüsten

equipment *n* GEN Einrichtung *f*, Gerät *nt*, Vorrichtung *f*

equipotential: ~ **surface** *n* WASTE, WATER SUPP Äquipotentialfläche *f*

equivalent: ~ **diameter** *n* AIR Äquivalentdurchmesser *m*; ~ **per million** *phr* (*EPM*) AIR Äquivalent je Million (*EPM*); ~ **standard smoke** *n* AIR Äquivalenznormalruß *m*

erect *vt* GEN errichten, installieren

erection *n* GEN Errichtung *f*, Installation *f*

ergonomic: ~ **design principle** *n* SAFETY Prinzip *nt* der ergonomischen Gestaltung

eriochrome: ~ **black** *n* (*EBT*) ANALYSIS *indicator* Eriochromschwarz *nt*

Erlenmeyer: ~ **flask** *n* ANALYSIS Erlenmeyerkolben *m*

erosion *n* SOIL, WATER SUPP Erosion *f*; ~ **protection** *n* SOIL Erosionsschutz *m*

erosive *adj* AIR *dust property*, SOIL, WATER SUPP erosiv

ESA *abbr* (*electrochemical stripping analysis*) ANALYSIS Auflösungsanalyse *f*

escape: ~ **of harmful substances** *n* SOIL Schadstoffaustritt *m*

ESP *abbr* (*electrostatic precipitator*) AIR, WASTE, WATER SUPP elektrostatischer Abscheider *m*, elektrostatischer Staubabscheider *m*

ESR *abbr* (*electron spin resonance*) ANALYSIS ESR (*Elektronenspinresonanz*)

established: ~ **technologies** *n pl* AIR, RENEW ENERG, WATER POLL, WATER SUPP anerkannte Technik *f*

estimable *adj* ANALYSIS bestimmbar

estimate: ~ **of danger** *n* ANALYSIS, MANAGE, SOIL Gefährdungsabschätzung *f*; ~ **of risks** *n* POLICY Gefährdungsabschätzung *f*

estimation *n* ANALYSIS, MANAGE, POLICY, WATER SUPP Abschätzung *f*

ethanol: ~ **generation** *n* RENEW ENERG *biomass conversion* Ethanolerzeugung *f*

ether: ~ **evaporation** *n* ANALYSIS *dew point hygrometer* Etherverdunstung *f*

ethylene *n* AIR Ethylen *nt*, schweres Kohlenwasserstoffgas *nt*; ~ **diamine-tetracetic acid** *n* (*EDTA*) ANALYSIS Ethylendiamintetraessigsäure *f*

ethylmercaptane *n* WASTE, WATER SUPP Ethylmercaptan *nt*

European: ~ **inventory of existing chemical substances** *n* (*EINECS*) POLICY Europäisches Altstoffverzeichnis *nt* (*EINECS*); ~ **inventory of new chemical substances** *n* (*EINCS*) POLICY Europäisches Neustoffverzeichnis *nt* (*EINCS*); ~ **Standardization Committee** *n* (*CEN*) ANALYSIS Europäisches Normungsinstitut *nt* (*CEN*)

eutrophic *adj* WASTE, WATER POLL, WATER SUPP eutroph

eutrophication *n* WASTE, WATER POLL, WATER SUPP Eutrophierung *f*
eV *abbr* (*electronvolt*) ANALYSIS eV (*Elektronenvolt*)
evacuate *vt* WASTE, WATER SUPP evakuieren
evaluate *vt* GEN auswerten, bewerten
evaluation *n* GEN Auswertung *f*, Bewertung *f*; **~ criteria** *n* POLICY *environmental policy* Bewertungskriterien *nt pl*, Bewertungsmaßstäbe *m pl*, Kriterien *nt pl* zur Bewertung; **~ of effects on human beings** *n* SAFETY Auswertung *f* der Effekte auf Menschen; **~ software** *n* ANALYSIS Auswertesoftware *f*; **~ using standard samples** *n* ANALYSIS leitprobengebundene Auswertung *f*; **~ without standard samples** *n* ANALYSIS leitprobenfreie Auswertung *f*
evaporate 1. *vt* GEN verdampfen, verdunsten; **2.** *vi* GEN verdampfen, verdunsten
evaporation *n* GEN Verdampfung *f*, Verdunstung *f*; **~ cooling** *n* AIR Verdampfungskühlung *f*; **~ crystallizer** *n* AIR Verdampfungskristallisator *m*; **~ loss** *n* WATER POLL, WATER SUPP Verdunstungsverlust *m*; **~ pan** *n* WASTE Verdunstungskessel *m*; **~ rate** *n* WATER POLL, WATER SUPP Verdunstungsrate *f*; **~ residue** *n* ANALYSIS, RENEW ENERG, WASTE Abdampfrückstand *m*, Verdampfrückstand *m*; **~ of waters** *n pl* WATER SUPP Gewässerverdunstung *f*
evaporative: **~ emissions** *n pl* AIR Verdampfungsemissionen *f pl*
evaporator: **~ distillation** *n* SOIL Verdampfer-Destillation *f*
evapotranspiration *n* WATER SUPP Evapotranspiration *f*
examination *n* ANALYSIS, SAFETY, WASTE Untersuchung *f*, Überprüfung *f*
examine *vt* ANALYSIS, SAFETY, WASTE untersuchen, überprüfen
examined *adj* ANALYSIS, SAFETY, WASTE untersucht, überprüft
excavate *vt* SOIL ausheben, auskoffern, WASTE ausheben
excavated *adj* SOIL ausgehoben, ausgekoffert, WASTE ausgehoben; **~ mass** *n* SOIL Aushubmasse *f*; **~ material** *n* SOIL ausgekoffertes Material *nt*, WASTE *inert material for monolandfill* Abraum *m*; **~ soil** *n* SOIL Bodenaushub *m*, ausgekofferter Boden *m*, WASTE *waste type* Bodenaushub *m*, Erdaushub *m*
excavation *n* SOIL, WASTE Aushub *m*; **~ system** *n* WASTE Aushubstation *f*; **~ well** *n* SOIL Entnahmebrunnen *m*
excavator *n* SOIL Bagger *m*
excess: **~ air** *n* AIR, WASTE, WATER SUPP Luftüberschuß *m*; **~ biomass** *n* WASTE, WATER POLL Überschußbiomasse *f*; **~ chlorine** *n* WATER SUPP Chlorüberschuß *m*; **~ chlorine gas** *n* AIR, WASTE Chlorgasüberschuß *m*; **~ fuel** *n* AIR Brennstoffüberschuß *m*; **~ gas** *n* WASTE Überschußgas *nt*
excessive: **~ deposit build-up** *n* AIR übermäßige Ablagerung *f*
excess: **~ load** *n* SOIL Überlastung *f*; **~ manure tax** *n* POLICY überschüssige Düngersteuer *f*; **~ oxygen** *n* AIR, SAFETY Sauerstoffüberschuß *m*; **~ power supply** *n* RENEW ENERG Stromüberangebot *nt*; **~ pressure** *n* NOISE Schallwechseldruck *m*, WATER POLL Überdruck *m*; **~ production** *n* MANAGE, WASTE Überproduktion *f*; **~ sludge** *n* WATER POLL Überschußschlamm *m*; **~ sludge volume** *n* WATER POLL Überschußschlammvolumen *nt*; **~ supply** *n* WATER SUPP Überangebot *nt*
exchange 1. *n* AIR, WASTE, WATER POLL, WATER SUPP Austausch *m*; **2.** *vt* AIR, WASTE, WATER POLL, WATER SUPP austauschen
exchangeable *adj* WATER POLL austauschbar; **~ cation** *n* AIR auswechselbares Kation *nt*
exchange: **~ capacity** *n* WASTE, WATER SUPP Austauschfähigkeit *f*; **~ operation** *n* WASTE, WATER SUPP Austauschvorgang *m*
exchanger: **~ capacity** *n* WATER POLL, WATER SUPP Austauscherkapazität *f*; **~ column** *n* WATER POLL Austauschersäule *f*
exchange: **~ resin** *n* WATER POLL Austauscherharz *nt*
exchanger: **~ material** *n* WASTE, WATER POLL, WATER SUPP Austauschermaterial *nt*
exclusion *n* POLICY *of public participation* Ausschluß *m*; **~ of air** *n* WASTE Luftabschluß *m*
exhaust 1. *n* AIR, NOISE Auspuff *m*; **2.** *vt* RENEW ENERG, WATER SUPP *filter absorption capacity* erschöpfen
exhaust: **~ casing** *n* AIR Abgasgehäuse *nt*; **~ cleaning installation** *n* AIR Abluftreiniger *m*; **~ emissions** *n pl* AIR Abgas *nt*, Auspuffgas *nt*, Schadstoffausstoß *m*; **~ emissions cleaning system** *n* AIR Abgasüberwachungsanlage *f*, Abgasreinigungsanlage *f*; **~ emissions control** *n* AIR Abgasüberwachung *f*
exhauster: **~ filter** *n* AIR Absaugfilter *nt*
exhaust: **~ fan** *n* AIR Sauggebläse *nt*; **~ gas** *n* AIR, WASTE, WATER SUPP Abgas *nt*; **~ gas cleaning following waste incineration** *n* AIR, WASTE Abgasreinigung *f* nach Müllverbrennung; **~ gas flow** *n* AIR, ANALYSIS Abgasstrom *m*; **~ gas limit value** *n* AIR, POLICY, WATER SUPP Abgasgrenzwert *m*; **~ gas recirculation** *n* (*EGR*) AIR Abgasrückführung *f* (*AGR*); **~ heat** *n* AIR Abgaswärme *f*; **~ housing** *n* AIR Abgasgehäuse *nt*
exhaustibility *n* RENEW ENERG, WATER SUPP Erschöpflichkeit *f*
exhaustible *adj* RENEW ENERG, WATER SUPP erschöpflich
exhaust: **~ manifold** *n* AIR Auspuffkrümmer *m*; **~ manifold catalyst** *n* AIR Auspuffkrümmerkatalysator *m*, motornaher Katalysator *m*; **~ manifold reactor** *n* AIR Auspuffkrümmerreaktor *m*; **~ muffler** *n* NOISE Auspuffschalldämpfer *m*; **~ noise** *n* NOISE Auspuffgeräusch *nt*; **~ silencer** *n* NOISE Auspuffschalldämpfer *m*; **~ silencing** *n* NOISE Auspuffschalldämpfung *f*; **~ smoke** *n* AIR Abgasrauch *m*; **~ valve** *n* AIR Auslaßventil *nt*, Auspuffventil *nt*; **~ vapors** *AmE*, **~ vapours** *BrE n pl* WATER POLL Brüden *m pl*; **~ vent installation** *n* AIR, SAFETY Abluftsystem *nt*
existing: **~ landfill site** *n* WASTE Altdeponie *f*; **~ plant** *n* POLICY, WASTE Altanlage *f*
exit: **~ percolation** *n* WASTE, WATER SUPP Aussickerung *f*
exoelectron: **~ dosimeter** *n* ANALYSIS Exoelektronendosimeter *nt*
exoenzyme *n* AIR Exo-Enzym *nt*
exothermic *adj* AIR, WASTE, WATER POLL, WATER SUPP exotherm, exothermisch; **~ process** *n* WATER POLL *wastewater incineration* exothermer Prozeß *m*; **~ reaction** *n* AIR, WASTE, WATER SUPP exothermische Reaktion *f*
expandable: **~ thermoplastics** *n pl* WASTE aufgeschäumtes Thermoplast *nt*, treibmittelhaltiges Thermoplast *nt*
expanded: **~ polystyrene** *n* WASTE geschäumtes Polysterol *nt*; **~ polythene packaging** *n* WASTE Einweg-PT-Verpackung *f*
expander *n* AIR *Venturi tube* Ringspalt *m*
expansion: **~ chamber silencer** *n* NOISE Druckausgleichskammer-Schalldämpfer *m*; **~ thermometer** *n* ANALYSIS Ausdehnungsthermometer *nt*
expendable: **~ pallet** *n* WASTE Einwegpalette *f*
experimental: **~ basin** *n* WATER SUPP Versuchsbecken *nt*; **~ catchment basin** *n* WATER SUPP Kleineinzugsgebiet *nt*; **~ well** *n* WATER SUPP Versuchsbrunnen *m*
explanatory: **~ clause** *n* POLICY *of environmental label* Randschrift *f*
exploratory *adj* ANALYSIS untersuchend
explosibility *n* AIR Explosibilität *f*; **~ limit** *n* AIR, ANALYSIS Explosionsgrenze *f*
explosion *n* AIR, ANALYSIS, SAFETY, SOIL Explosion *f*; **~-proof** *adj* AIR, ANALYSIS, SAFETY, WASTE explosionsgeschützt, explosionssicher; **~ protection** *n* AIR, ANALYSIS Explosionsschutz *m*; **~ risk** *n* ANALYSIS, SAFETY, WASTE Explosionsgefahr *f*; **~ safety** *n* ANALYSIS, SAFETY Explosionssicherheit *f*; **~ suppression** *n* AIR *of filter apparatus* Explosionsunterdrückung *f*; **~ warning device** *n* ANALYSIS, SAFETY Explosionswarngerät *nt*
explosive 1. *adj* AIR, SAFETY, WASTE, WATER SUPP explosiv, explosionsfähig; **2.** *n* AIR, SAFETY, WASTE Sprengstoff *m*
explosive: **~ atmosphere** *n* SAFETY explosive Luft *f*; **~ gas** *n*

AIR explosionsfähiges Gas *nt*; ~ **mixture** *n* AIR *gases* explosionsfähiges Gemisch *nt*; ~ **particles** *n pl* AIR explosible Stäube *m pl*; ~ **waste** *n* SAFETY, WASTE explosiver Abfall *m*

exposure *n* SAFETY Exposition *f*, *to hazardous substance* Belastung *f*, *to radiation* Bestrahlung *f*; ~ **limit to hazardous substances** *n* POLICY, SAFETY Grenzwert *m* für Belastung, Grenzwert *m* für Gefahrstoffe; ~ **path** *n* SAFETY Belastungspfad *m*; ~ **time** *n* ANALYSIS Standzeit *f*; ~ **to fumes** *n* SAFETY Belastung *f* durch Dämpfe

extended: ~ **deposition** *n* WATER POLL Langzeitlagerung *f*; ~**-period aeration** *n* WATER POLL Langzeitbelebung *f*, Langzeitbelüftung *f*, Totalklärung *f*

extension: ~ **break tester** *n* AIR Reißdehnungsprüfgerät *nt*; ~ **of tax base** *n* POLICY Erweiterung *f* einer Steuerbemessungsgrundlage

extent *n* SOIL *of pollution* Ausmaß *nt*

external *adj* AIR, WASTE, WATER POLL, WATER SUPP äußerlich; ~**-combustion engine** *n* AIR Wärmekraftmaschine *f* mit äußerer Verbrennung; ~ **costs** *n pl* POLICY externe Kosten *pl*; ~ **effect** *n* POLICY externer Effekt *m*; ~ **input** *n* AIR, WASTE, WATER SUPP Fremdzufuhr *f*

externality *n* POLICY externer Effekt *m*

external: ~ **validation** *n* MANAGE externe Begutachtung *f*, externe Validierung *f*; ~ **water** *n* WATER POLL Fremdwasser *nt*

extinction *n* ANALYSIS Extinktion *f*; ~ **coefficient** *n* AIR Extinktionskoeffizient *m*, spektraler Absorptionskoeffizient *m*, ANALYSIS Extinktionskoeffizient *m*

extract 1. *n* ANALYSIS, WASTE, WATER POLL, WATER SUPP Extrakt *m*; **2.** *vt* AIR, ANALYSIS, SOIL, WASTE, WATER POLL extrahieren

extractable: ~ **organic chlorides** *n pl* (*EOCL*) AIR, ANALYSIS, SOIL, WATER POLL extrahierbare organische Chloride *nt pl* (*EOCL*); ~ **organic halogen compounds** *n pl* (*EOX*) ANALYSIS, WASTE, WATER SUPP extrahierbare organische Halogenverbindungen *f pl* (*EOX*)

extractant *n* ANALYSIS, SOIL, WATER POLL Extraktionsmittel *nt*; ~ **processing** *n* SOIL Extraktionsmittelaufbereitung *f*

extracted *adj* GEN extrahiert, extrahierend; ~ **component** *n* WATER POLL extrahierende Komponente *f*; ~ **soil** *n* SOIL Bodenaushub *m*; ~ **vapor** *AmE*, ~ **vapour** *BrE n* SOIL extrahierter Dampf *m*

extracting: ~ **agent** *n* ANALYSIS, SOIL, WATER POLL Extraktionsmittel *nt*; ~ **agent processing** *n* SOIL Extraktionsmittelaufbereitung *f*

extraction *n* GEN Extraktion *f*; ~ **agent** *n* ANALYSIS, SOIL, WATER POLL Extraktionsmittel *nt*; ~ **agent processing** *n* SOIL Extraktionsmittelaufbereitung *f*; ~ **centrifuge** *n* WATER POLL Extraktionszentrifuge *f*; ~ **gas** *n* ANALYSIS Extraktionsgas *nt*; ~ **operation** *n* WATER POLL Extraktionsvorgang *m*; ~ **solvent** *n* ANALYSIS, SOIL Extraktionslösungsmittel *nt*; ~ **tank** *n* SOIL Extraktionstank *m*; ~ **test** *n* ANALYSIS Absaugversuch *m*; ~ **vessel** *n* WATER POLL Extraktionsturm *m*

extractive: ~ **agent** *n* ANALYSIS, SOIL, WATER POLL Extraktionsmittel *nt*

extractor *n* SOIL, WASTE Entnahmegerät *nt*, WATER POLL, WATER SUPP Extraktor *m*, Entnahmegerät *nt*

extract: ~ **residue** *n* WASTE, WATER SUPP Extraktionsmittelrest *m*

extraneous: ~ **noise** *n* NOISE Fremdgeräusch *nt*

extraterrestrially: ~**-operating satellite** *n* ANALYSIS extraterrestrisch operierender Satellit *m*

extruding *n* WASTE Extrudieren *nt*

extrusion *n* WASTE Extrusion *f*; ~ **blow molding** *AmE*, ~ **blow moulding** *BrE n* WASTE Extrusionsblasen *nt*, Extrusionsblasverfahren *nt*; ~ **die** *n* WASTE Strangpreßwerkzeug *nt*; ~ **head** *n* WASTE Extrusionsspritzkopf *m*; ~ **line** *n* WASTE *processing* Extrusionsanlage *f*; ~ **plant** *n* WASTE *processing* Extrusionsanlage *f*; ~ **with regranulation** *n* WASTE Extrusion *f* mit Regranulierung; ~ **stretch-flow molding machine** *AmE*, ~ **stretch-flow moulding machine** *BrE n* WASTE Extrusionsstreckblasformmaschine *f*

F

F *abbr* (*fluorine*) SOIL F (*Fluor*)
fabric: ~ **filter** *n* AIR *baghouse* Faserfilter *nt*, Gewebefilter *nt*, Gewebeschlauchfilter *nt*
face: ~ **velocity** *n* AIR, WATER POLL Anströmgeschwindigkeit *f*
facility *n* GEN Anlage *f*
facultative *adj* WATER POLL fakultativ; ~-**aerobic** *adj* WATER POLL fakultativ aerob; ~ **anaerobic bacteria** *n pl* WATER POLL fakultativ anaerobe Bakterien *f pl*; ~ **heterotroph** *n* WATER POLL fakultatives Heterotroph *nt*; ~ **lagoon** *n* WATER POLL fakultativer Teich *m*; ~ **microorganisms** *n pl* WATER POLL fakultative Mikroben *f pl*; ~ **zone** *n* WATER POLL fakultative Zone *f*
faecal *adj BrE* WATER POLL Fäkal-; ~ **matter** *n BrE* WATER POLL Fäkalschlamm *m*; ~ **sludge** *n BrE* WATER POLL Fäkalschlamm *m*; ~ **solids sludge composition** *n BrE* WATER POLL Fäkalschlammzusammensetzung *f*; ~ **solids sludge from domestic clarification plants** *n BrE* WATER POLL Fäkalschlamm *m* aus Hauskläranlagen; ~ **solids sludge treatment** *n BrE* WATER POLL Fäkalschlammbehandlung *f*; ~ **wastewater** *n BrE* WATER POLL Fäkalabwasser *nt*
fall *n* WATER SUPP *between two water levels* Gefälle *nt* des Wasserspiegels
falling: ~ **film absorber** *n* AIR Fallfilmabsorber *m*; ~ **sphere viscosimeter** *n* ANALYSIS Fallkörperviskosimeter *nt*
fallout *n* AIR, SOIL Fallout *m*, Niederschlag *m*
fall: ~ **of water** *n* WATER SUPP Gefälle *nt*
fan *n* AIR, NOISE Gebläse *nt*, Lüfter *m*, Ventilator *m*
fanning: ~ **smoke plume** *n* AIR fächerförmige Rauchfahne *f*
fan: ~ **noise** *n* NOISE Lüfterlärm *m*; ~ **shroud** *n* NOISE *casing* Lüftergehäuse *nt*, Ventilatorgehäuse *nt*; ~ **silencing** *n* NOISE Lüfterdämpfung *f*; ~ **vane** *n* AIR Ventilatorflügel *m*
Faraday: ~**'s law of electromagnetic induction** *n* ANALYSIS Faraday-Gesetz *nt* der elektromagnetischen Induktion
farm: ~ **waste** *n* WASTE landwirtschaftlicher Abfall *m*
fat *n* AIR Fett *nt*; ~-**free** *adj* AIR fettfrei; ~-**free waste air** *n* AIR fettfreie Abluft *f*
fatigue 1. *n* GEN *of materials* Ermüdung *f*; **2.** *vi* GEN ermüden
fatty: ~ **acid** *n* WATER POLL Fettsäure *f*; ~ **alcohol** *n* WATER POLL Fettalkohol *m*
faulty: ~ **measurement** *n* ANALYSIS Fehlmessung *f*
FBC *abbr* (*fluidized-bed combustion*) AIR WSF (*Wirbelschichtfeuerung*)
FD *abbr* (*field desorption*) ANALYSIS FD (*Felddesorption*)
feasibility *n* GEN Durchführbarkeit *f*, Machbarkeit *f*; ~ **assessment** *n* MANAGE Machbarkeitsbewertung *f*; ~ **study** *n* MANAGE Machbarkeitsstudie *f*
fecal *AmE see faecal BrE*
feed *vt* AIR beschicken, einbringen, zuführen, ANALYSIS dosieren, SAFETY dosieren, WASTE, WATER POLL, WATER SUPP beschicken, dosiren, einbringen, zuführen; ~ **in** *vt* SOIL *process water* zudosieren
feedback *n* WATER POLL, WATER SUPP Rückführung *f*
feed: ~ **drum** *n* WASTE, WATER SUPP Aufgabetrommel *f*
feeder *n* WASTE, WATER SUPP Aufgabevorrichtung *f*, Beschickungsvorrichtung *f*; ~ **broom** *n* WASTE Zubringerbesen *m*; ~ **worm** *n* SOIL Eintragschnecke *f*
feeding *n* GEN Beschicken *nt*, Beschickung *f*, Dosierung *f*; ~ **hopper** *n* SOIL Beschickungstrichter *m*; ~-**in of liquid** *n* AIR *absorber* Flüssigkeitsaufgabe *f*; ~ **tank** *n* WATER SUPP Vorratstank *m*
feed: ~ **pump** *n* AIR Speisepumpe *f*; ~ **rate** *n* WATER SUPP Vorschubgeschwindigkeit *f*; ~ **shaft** *n* WASTE, WATER SUPP Aufgabeschacht *m*
feedstock: ~ **recycling** *n* WASTE Feedstock-Recycling *nt*, chemische Verwertung *f*, chemisches Recycling *nt*
feed: ~ **tank** *n* AIR Filterzelle *f*, Vorlaufbehälter *m*, WATER POLL Dosierbehälter *m*; ~ **water** *n* WATER SUPP Speisewasser *nt*
feel *vt* ANALYSIS fühlen
FEL *abbr* (*Frank effect level*) SAFETY FEL (*Frank effect level*)
felt 1. *n* NOISE Filz *m*; **2.** *vt* NOISE *fibreglass* verfilzen
fence: ~ **boom** *n* MARINE Zaunölsperre *f*
fender *n* MARINE Fender *m*
ferment 1. *vt* RENEW ENERG fermentieren, gären, *biomass conversion* vergären, WASTE, WATER POLL, WATER SUPP gären, fermentieren, vergären; **2.** *vi* RENEW ENERG, WASTE, WATER POLL, WATER SUPP gären, vergären
fermentability *n* RENEW ENERG, WASTE, WATER POLL, WATER SUPP Vergärbarkeit *f*
fermentable *adj* RENEW ENERG, WASTE, WATER POLL, WATER SUPP gärfähig, vergärbar
fermentation *n* RENEW ENERG *biomass conversion*, WASTE, WATER POLL, WATER SUPP Vergärung *f*; ~ **gas** *n* WASTE, WATER POLL, WATER SUPP Biogas *nt*, Faulgas *nt*, Gärungsgas *nt*; ~ **of refuse** *n* WASTE Abfallgärung *f*; ~ **test** *n* ANALYSIS Gärprobe *f*; ~ **to biogas** *n* RENEW ENERG Vergärung *f* zu Biogas; ~ **tower** *n* WATER POLL Gärturm *m*
fermentative *adj* RENEW ENERG, WASTE, WATER POLL, WATER SUPP gärungserregend
fermented: ~ **waste** *n* WASTE vergärter Abfall *m*
ferric: ~ **chloride** *n* WASTE, WATER SUPP Eisenchlorid *nt*; ~ **salt** *n* WATER POLL Eisen(III)-Salz *nt*, WATER SUPP *coagulant* Ferrisalz *nt*; ~ **sulfate** *AmE*, ~ **sulphate** *BrE n* WASTE, WATER SUPP Eisen(III)-Sulfat *nt*
ferromagnetic *adj* ANALYSIS ferromagnetisch
ferrous: ~ **metals** *n pl* WASTE Eisenmetalle *nt pl*; ~ **scrap** *n* WASTE Eisenschrott *m*
fertilization *n* SOIL Düngung *f*
fertilizer *n* SOIL, WATER POLL Dünger *m*
fertilizing: ~ **value** *n* SOIL, WATER POLL Düngewert *m*
FGD *abbr* (*flue-gas desulfurization AmE*, *flue-gas desulphurization BrE*) AIR, SAFETY Rauchgasentschwefelung *f*; ~ **plant** *n* AIR Rauchgasentschwefelungsanlage *f* (*REA*); ~ **plant blower** *n* AIR REA-Gebläse *nt*; ~ **plant gypsum** *n* AIR REA-Gips *m*
FI *abbr* (*field ionization*) ANALYSIS FI (*Feldionisation*)
FIA *abbr* (*flow injection analysis*) ANALYSIS FIA (*Fließinjektionsanalyse*)
FIAS *n* (*flow injection analysis system*) ANALYSIS FIAS (*Fließinjektionsanalysensystem*)
fiber *AmE see fibre BrE*
fiberglass *AmE see fibreglass BrE*
fibre *n BrE* AIR, WASTE Faser *f*, Faserstoff *m*; ~ **composition** *n BrE* AIR, WASTE Faserstoffzusammensetzung *f*, Faserzusammensetzung *f*; ~ **diameter** *n BrE* AIR Faserdurchmesser *m*; ~ **drum** *n BrE* WASTE Leichtfaß *nt*, Pappfaß *nt*
fibreglass *n BrE* AIR, NOISE, WASTE Glasfaser *f*; ~ **fleece** *n BrE* NOISE *absorption material* Glasfaservlies *nt*; ~-**reinforced** *adj BrE* NOISE, WASTE glasfaserverstärkt; ~-**reinforced plastic** *n BrE* (*FRP*) WASTE glasfaserverstärkter Kunststoff *m* (*GFK*)
fibre: ~ **layer** *n BrE* AIR *filter medium* Faserschicht *f*; ~ **layer**

filter *n BrE* AIR Faserschichtfilter *nt*; **~ material** *n BrE* AIR *for filter* Fasermaterial *nt*; **~ mixture** *n BrE* AIR Fasermischung *f*; **~ surface** *n BrE* AIR *filtering separators* Faseroberfläche *f*; **~ suspension** *n BrE* WATER SUPP *deinking process* Fasersuspension *f*

fibrous *adj* AIR *particle* faserartig, faserförmig, WASTE faserig; **~ constituents** *n pl* WASTE faserige Inhaltsstoffe *m pl*; **~ mat** *n* AIR Fasermatte *f*; **~ mat filter** *n* AIR Fasermattenfilter *nt*; **~ web** *n* WASTE *shear wool sealing* Vlies *nt*

Fick: **~'s law** *n* AIR *of diffusion* Ficksches Gesetz *nt*

FID *abbr* (*flame ionization detector*) ANALYSIS FID (*Flammenionisationsdetektor*)

field: **~ analysis** *n* ANALYSIS Feldanalyse *f*; **~ capacity of landfill** *n* WASTE Feldkapazität *f* der Deponie; **~ charging** *n* AIR Feldaufladung *f*; **~ data** *n pl* ANALYSIS Felddaten *pl*; **~ desorption** *n* (*FD*) ANALYSIS Felddesorption *f* (*FD*); **~ ionization** *n* (*FI*) ANALYSIS Feldionisation *f* (*FI*); **~ of measurement** *n* ANALYSIS Meßnetz *nt*; **~ performance** *n* ANALYSIS Feldwirksamkeit *f*; **~-proven** *adj* AIR, ANALYSIS, RENEW ENERG, SOIL, WATER POLL, WATER SUPP *processes* praxiserprobt; **~ screening** *n* ANALYSIS On-site-Analyse *f*, Vor-Ort-Analyse *f*; **~ strength** *n* AIR, ANALYSIS Feldstärke *f*; **~ test** *n* ANALYSIS Feldversuch *m*; **~ trial** *n* ANALYSIS Feldversuch *m*

fill *vt* WASTE, WATER SUPP füllen; **~ up** *vt* ANALYSIS auffüllen, SOIL aufschütten

filler *n* SOIL Füllstoff *m*

filling *n* ANALYSIS *of tubes in sampler* Abfüllung *f*, SOIL, WASTE Verfüllung *f*; **~ operation** *n* WASTE Füllvorgang *m*; **~ plant** *n* WASTE, WATER SUPP Abfüllbetrieb *m*

fill: **~ mass** *n* SOIL Lagergut *nt*, WASTE Deponiegut *nt*, Lagergut *nt*; **~ material** *n* SOIL *from decontaminated soils* Füllstoff *m*

film *n* WATER SUPP Film *m*; **~ badge** *n* ANALYSIS *radiation dosimeter* Filmplakette *f*; **~ coating** *n* WATER SUPP Film-Coating *nt*; **~ extrusion equipment** *n* WASTE Folienextrudieranlage *f*; **~ lamination** *n* WASTE Zellglasbeschichtung *f*

filter 1. *n* AIR Filter *nt*, Gewebe *nt*, ANALYSIS, WASTE, WATER POLL, WATER SUPP Filter *nt*; **2.** *vt* AIR, ANALYSIS, WASTE, WATER POLL, WATER SUPP filtern, filtrieren; **~ out** AIR, ANALYSIS, WASTE, WATER POLL, WATER SUPP ausfiltern, abfiltern

filterable *adj* GEN filterfähig, filtrierbar; **~ flocs** *n pl* WATER POLL filtrierbare Flocken *f pl*; **~ solids** *n pl* WASTE, WATER POLL, WATER SUPP filtrierbare Feststoffe *m pl*

filter: **~ aid** *n* AIR, WASTE, WATER POLL, WATER SUPP Filterhilfsmittel *nt*, Filterhilfsstoff *m*; **~ aid efficiency** *n* AIR, WASTE, WATER POLL, WATER SUPP Filterhilfsstoffwirkungsgrad *m*; **~ area** *n* WATER POLL Filterfläche *f*; **~ ash** *n* WASTE Filterasche *f*; **~ bag** *n* AIR Filterschlauch *m*, Filtertasche *f*; **~ bed** *n* AIR, WASTE, WATER POLL, WATER SUPP Filterbett *nt*, Filterschicht *f*; **~ bed depth** *n* AIR, WATER POLL Filterschichthöhe *f*; **~ bed height** *n* AIR, WATER POLL Filterbetthöhe *f*; **~ bed porosity** *n* AIR Filterbettporosität *f*; **~ belt** *n* ANALYSIS Filterband *nt*; **~ cake** *n* AIR, WATER POLL Filterkuchen *m*; **~ cloth** *n* AIR, WASTE, WATER POLL, WATER SUPP Filterstoff *m*, Filtertuch *nt*; **~ column** *n* WASTE, WATER SUPP Filtersäule *f*; **~ compartment** *n* AIR, WATER POLL Filterraum *m*; **~ cooler** *n* AIR Filterkühler *m*; **~ dimensioning** *n* AIR Filterdimensionierung *f*; **~ dust** *n* AIR, WASTE Filterstaub *m*; **~ dust containing pollutants** *n* AIR, WASTE schadstoffhaltiger Filterstaub *m*; **~ dust from incineration plant** *n* WASTE Müllverbrennungsfilterstaub *m*

filtered *adj* GEN filtriert, gefiltert; **~ matter** *n* WASTE, WATER POLL, WATER SUPP Filtratgut *nt*; **~ sample** *n* ANALYSIS filtrierte Probe *f*

filter: **~ efficacy** *n* AIR Filterwirksamkeit *f*; **~ efficiency** *n* AIR, WATER POLL Filterwirkungsgrad *m*; **~ element** *n* AIR, SAFETY, WASTE, WATER POLL, WATER SUPP Filterelement *nt*; **~ fabric** *n* AIR Filtergewebe *nt*; **~ felt** *n* AIR Filterfilz *m*; **~ film** *n* AIR *biowasher*, WATER POLL biologischer Rasen *m*; **~ gallery** *n* WATER POLL, WATER SUPP Filterkammer *f*; **~ gravel** *n* WATER POLL Filterkies *m*; **~ housing** *n* AIR, WATER SUPP Filtergehäuse *nt*

filtering *n* GEN Filtern *nt*, Filtration *f*, Filtrierung *f*; **~ agent** *n* AIR Filtermittel *nt*; **~ operation** *n* AIR, WATER POLL Filtrationsvorgang *m*, Filtriervorgang *m*; **~ plant** *n* WATER SUPP Filteranlage *f*; **~ position** *n* AIR *bag filter* Filtrierstellung *f*; **~ precipitator** *n* AIR filternder Abscheider *m*; **~ rate** *n* AIR, ANALYSIS, WATER POLL, WATER SUPP Filtrationsgeschwindigkeit *f*; **~ tank** *n* WATER POLL Klärbehälter *m*, WATER SUPP Filterbecken *nt*; **~ time** *n* AIR, WATER POLL, WATER SUPP Filtrierdauer *f*, Filtrierzeit *f*; **~ unit** *n* AIR, WATER SUPP Filteranlage *f*; **~ well** *n* WATER POLL, WATER SUPP Filterbrunnen *m*

filter: **~ layer** *n* AIR Filterschicht *f*; **~ layout** *n* AIR Filterauslegung *f*; **~ material** *n* AIR, WASTE, WATER POLL, WATER SUPP Filtermaterial *nt*, Filtermedium *nt*; **~ material circulation** *n* AIR Filtermaterialkreislauf *m*; **~ medium** *n* AIR, WASTE, WATER POLL, WATER SUPP Filtermaterial *nt*, Filtermedium *nt*; **~ medium surface** *n* AIR Filtermediumoberfläche *f*; **~ nozzle** *n* WATER POLL, WATER SUPP Filterdüse *f*; **~ operation** *n* WASTE, WATER SUPP Filterbetrieb *m*; **~ paper** *n* ANALYSIS Filterpapier *nt*; **~ plant** *n* SAFETY Filteranlage *f*; **~ plant for dust and fibers** *AmE*, **~ plant for dust and fibres** *BrE n* SAFETY *textile industry* Filteranlage *f* gegen Staub und Fasern; **~ pore** *n* WATER POLL Filterpore *f*; **~ press** *n* WATER POLL, WATER SUPP *surface filter* Filterpresse *f*; **~ regeneration** *n* AIR, WATER POLL Filterregenierung *f*; **~ replacement** *n* AIR, WATER POLL Filteraustausch *m*; **~ residues** *n pl* AIR, WATER POLL Filterrückstände *m pl*; **~ resistance** *n* AIR, WATER POLL, WATER SUPP Filterwiderstand *m*; **~ stage** *n* AIR Filterstufe *f*; **~ surface** *n* AIR, WATER POLL, WATER SUPP Filteroberfläche *f*; **~ test** *n* ANALYSIS Filtertest *m*; **~ type** *n* AIR, WATER POLL Filterart *f*, Filtertyp *m*; **~ unit** *n* AIR *bag filters*, WASTE, WATER POLL, WATER SUPP Filtereinheit *f*; **~ velocity** *n* WATER SUPP Filtergeschwindigkeit *f*; **~ wall** *n* AIR Filterwandung *f*; **~ washing** *n* AIR, WATER POLL, WATER SUPP Filterspülung *f*; **~ worm** *n* AIR Filterschnecke *f*

filtrate 1. *n* GEN Filtrat *nt*; **2.** *vt* AIR, ANALYSIS, WASTE, WATER POLL, WATER SUPP filtern

filtrate: **~ contamination** *n* WATER POLL, WATER SUPP Filtratkontamination *f*, Filtratverschmutzung *f*

filtrated *adj* GEN gefiltert

filtrate: **~ flow** *n* WATER POLL, WATER SUPP Filtratstrom *m*; **~ post-chlorination** *n* WATER SUPP Filtratnachchlorung *f*; **~ quality** *n* WATER POLL, WATER SUPP Filtratqualität *f*; **~ tank** *n* WASTE, WATER SUPP Filtratbehälter *m*

filtration *n* GEN Filtern *nt*, Filterung *f*, Filtration *f*, Filtrierung *f*; **~ capacity** *n* AIR, WATER POLL, WATER SUPP Filtrationsleistung *f*; **~ curve** *n* AIR, WATER POLL, WATER SUPP Filtrationskurve *f*; **~ cycle** *n* AIR Filtrationszyklus *m*; **~ direction** *n* AIR Filtrationsrichtung *f*; **~ effect** *n* AIR, WATER POLL, WATER SUPP Filtrationseffekt *m*, Filtriereffekt *m*; **~ plant** *n* AIR, WATER SUPP Filteranlage *f*, Filtrationsanlage *f*; **~ pressure** *n* AIR Filtrationsdruck *m*; **~ putrefaction process** *n* WATER POLL Gärfaulverfahren *nt*; **~ rate** *n* AIR, ANALYSIS, WATER POLL, WATER SUPP Filtrationsgeschwindigkeit *f*; **~ residue** *n* ANALYSIS Filtrationsrückstand *m*; **~ and residue water return** *n* WATER POLL Filtrat- und Restwasserrücklauf *m*; **~ sludge** *n* WATER POLL gefilterter Schlamm *m*; **~ vat** *n* WATER POLL Kläranlagebehälter *m*, Reinigungsbehälter *m*; **~ velocity** *n* AIR Filtrationsgeschwindigkeit *f*

final: **~ clarification** *n* WATER POLL Nachklärung *f*; **~ cover** *n* WASTE *of a landfill* Endabdeckung *f*, Oberflächenversiegelung *f*; **~ covering** *n* WASTE Endabdeckung *f*; **~ energy** *n* RENEW ENERG Endenergie *f*; **~ energy**

consumption *n* RENEW ENERG Endenergieverbrauch *m*; **~ energy substitution** *n* RENEW ENERG Endenergiesubstitution *f*; **~ filter** *n* AIR, WATER POLL, WATER SUPP Nachfilter *nt*; **~ filtering** *n* AIR, WATER POLL, WATER SUPP Nachfiltern *nt*, Nachfilterung *f*; **~ filtration** *n* AIR, WATER POLL, WATER SUPP Nachfiltern *nt*, Nachfilterung *f*

finally: **~-sealed landfill** *n* WASTE endabgedichtete Deponie *f*; **~-sealed landfill section** *n* WASTE endabgedichteter Deponieabschnitt *m*

final: **~ reach** *n* WASTE Endhaltung *f*, WATER SUPP Wassermengendauerlinie *f*; **~ remediation** *n* SOIL endgültige Sanierung *f*; **~ settling tank** *n* WATER POLL Nachklärbecken *nt*; **~ storage** *n* WASTE Endlagerung *f*; **~ thickener** *n* WATER POLL Nacheindicker *m*; **~ treatment** *n* WASTE Nachbehandlung *f*; **~ treatment of slag** *n* WASTE Schlackennachbehandlung *f*

finance *vt* MANAGE, POLICY finanzieren

fine 1. *adj* AIR fein; **2.** *n* POLICY *penalty* Bußgeld *nt*; **3.** *vt* POLICY Bußgeld verhängen gegen

fine: **~-bubble aeration** *n* WATER POLL feinblasige Belüftung *f*; **~ dust** *n* AIR *smaller than 10*m* Feinstaub *m*; **~ dust precipitation** *n* AIR Feinstaubabscheidung *f*; **~-fibered needled felt** *AmE*, **~-fibred needled felt** *BrE* *n* AIR Feinfaser-Nadelfilz *m*; **~ flocs** *n pl* WATER POLL Feinflocken *f pl*; **~ fraction** *n* WASTE Leichtfraktion *f*, Leichtstofffraktion *f*; **~-grained** *adj* WASTE feinkörnig; **~-grained additive** *n* WATER SUPP Feinzuschlagstoff *m*

fineness: **~ of particle size** *n* WASTE Kornfeinheit *f*

fine: **~ particles** *n pl* AIR feine Partikeln *f pl*; **~ refuse** *n* WASTE Feinmüll *m*; **~ screen** *n* AIR, WASTE, WATER POLL Feinrechen *m*

finest: **~ drop precipitator** *n* AIR Feinsttropfenabscheider *m*; **~ dust** *n* AIR Feinststaub *m*

finished: **~ product** *n* WASTE *reprocessing* Endprodukt *nt*, Fertigprodukt *nt*

finishing: **~ device** *n* WASTE, WATER SUPP Nachbearbeitungsvorrichtung *f*

finned: **~ tube air heater** *n* AIR Rippenrohrluftvorwärmer *m* (*Rippenrohrluvo*)

fire *n* GEN Brand *m*, Feuer *nt*

fired: **~ heater** *n* AIR, WATER SUPP Erhitzer *m*

fire: **~ hazard** *n* SAFETY Brandgefahr *f*, Entzündungsgefahr *f*; **~ monitor** *n* MARINE Feuerlöschmonitor *m*

fireproof 1. *adj* SAFETY feuerfest; **2.** *vt* SAFETY feuerfest machen

fireproofing *n* SAFETY Brandschutz *m*

fire: **~-resistant** *adj* SAFETY feuerbeständig; **~-retardant** *adj* SAFETY feuerhemmend; **~ risk** *n* SAFETY, WASTE Brandgefahr *f*, Entzündungsgefahr *f*; **~ safety** *n* SAFETY Brandsicherheit *f*

firewood *n* RENEW ENERG *biomass conversion base material* Brennholz *nt*

firing: **~ drum** *n* SOIL Brenntrommel *f*

firm *adj* SOIL standfest; **~-specific** *adj* POLICY betriebsspezifisch; **~-specific documents** *n pl* POLICY betriebsspezifische Unterlagen *f pl*; **~-specific information** *n* POLICY betriebsspezifische Informationen *f pl*; **~ subsurface** *n* SOIL standfester Untergrund *m*

first: **~ assessment** *n* SOIL *of risks* Erstbewertung *f*

fish: **~ test** *n* ANALYSIS Fischtest *m*; **~ toxicity** *n* WATER POLL Fischgiftigkeit *f*, Fischtoxizität *f*

fission: **~ product** *n* WASTE Spaltprodukt *nt*

fitting *n* ANALYSIS, WASTE, WATER POLL, WATER SUPP Armatur *f*, Fitting *nt*

fix *vt* ANALYSIS befestigen

fixation *n* SOIL, WASTE Verfestigung *f*, Fixierung *f*; **~ technique** *n* WASTE Verfestigungsverfahren *nt*

fixative *n* SOIL, WASTE Verfestigungsmittel *nt*

fixed: **~ bed** *n* AIR Festbett *nt*; **~-bed adsorber** *n* AIR, WATER POLL Festbettadsorber *m*; **~-bed cascade reactor** *n* WATER POLL Festbettkaskadenreaktor *m* (*FBK*); **~-bed catalyst** *n* AIR Festbettkatalysator *m*; **~-bed electrolysis** *n* WATER POLL Festbettelektrolyse *f*; **~-bed furnace** *n* WATER POLL *incineration of wastewater sludge* Festbettofen *m*; **~-bed ion exchanger** *n* WATER POLL Festbettionenaustauscher *m*; **~-bed reactor** *n* SOIL, WASTE, WATER SUPP Festbettreaktor *m*; **~ carbon** *n* RENEW ENERG *biomass* fester Kohlenstoff *m*; **~ catalyst** *n* AIR Feststoffkatalysator *m*; **~ film** *n* WATER POLL *attached growth* biologischer Rasen *m*; **~ group** *n* WATER POLL Ankergruppe *f*; **~-pitch rotor blade** *n* RENEW ENERG *wind turbine* nichtverstellbares Rotorblatt *nt*; **~ rotor blade** *n* RENEW ENERG *wind turbine* festes Rotorblatt *nt*; **~-wing aircraft** *n* MARINE Starrflügelflugzeug *nt*, Starrflügelluftfahrzeug *nt*, Starrflügler *m*

fixing *n* SOIL Fixierung *f*, Verfestigung *f*

flaking *n* WATER POLL *slow process filter* Abschälen *nt*

flaky *adj* AIR *particle* geschichtet

flame *n* AIR, ANALYSIS Flamme *f*; **~-AAS** *n* ANALYSIS Flammen-AAS *f*; **~ coloration** *AmE*, **~ colouration** *BrE* *n* ANALYSIS Flammenfärbung *f*; **~ cooling** *n* AIR Flammenabkühlung *f*; **~ gas** *n* AIR Flammgas *nt*; **~ ionization** *n* ANALYSIS Flammenionisation *f*; **~ ionization detector** *n* (*FID*) ANALYSIS Flammenionisationsdetektor *m* (*FID*); **~ photometer** *n* ANALYSIS Flammenphotometer *nt*; **~-photometric** *adj* ANALYSIS flammenphotometrisch; **~-photometric detector** *n* (*FPD*) ANALYSIS flammenphotometrischer Detektor *m* (*FPD*); **~-photometric determination** *n* ANALYSIS flammenphotometrische Bestimmung *f*; **~ photometry** *n* ANALYSIS *heavy metal detection* Flammenphotometrie *f*; **~ point** *n* AIR Flammpunkt *m*

flameproof *adj* AIR, ANALYSIS, SAFETY, WASTE flammenfest, flammensicher, nicht entflammbar, nichtentflammbar

flame: **~-resistant** *adj* SAFETY flammbeständig, schwer entflammbar; **~-spectrometric** *adj* ANALYSIS flammenspektrometrisch; **~-spectrometric determination** *n* ANALYSIS flammenspektrometrische Bestimmung *f*; **~ spectrometry** *n* ANALYSIS Flammenspektrometer *nt*, Flammenspektrometrie *f*

flammability *n* AIR, SAFETY, WASTE Entflammbarkeit *f*

flammable *adj* AIR, SAFETY, WASTE entflammbar, zündfähig; **~ dust** *n* AIR entzündlicher Staub *m*, zündfähiger Staub *m*; **~ liquid** *n* SAFETY entflammbare Flüssigkeit *f*; **~ material** *n* SAFETY entflammbarer Werkstoff *m*; **~ vapor** *AmE*, **~ vapour** *BrE* *n* SAFETY entflammbarer Dampf *m*; **~ waste** *n* WASTE entflammbarer Abfall *m*

flange *n* AIR, MARINE Flansch *m*

flanged *adj* AIR, MARINE geflanscht; **~ heat exchanger** *n* AIR geflanschter Wärmeaustauscher *m*

flanking: **~ path** *n* NOISE *around sound-insulating wall* Flankenweg *m*; **~ sealing** *n* SOIL Flankenabdichtung *f*; **~ transmission** *n* NOISE Flankenwegübertragung *f*, Flankenübertragung *f*, Nebenwegübertragung *f*

flare 1. *vt* AIR, WATER SUPP abfackeln; **2.** *vi* AIR, WASTE, WATER POLL, WATER SUPP aufflammen

flaring *n* WATER POLL, WATER SUPP Abfackeln *nt*, Abfackelung *f*; **~ installation** *n* WASTE Abfackelungsanlage *f*

flash: **~ distillation** *n* WASTE Entspannungsverdampfung *f*; **~ dryer** *n* AIR Schnelltrockner *m*; **~ evaporator** *n* WATER POLL Entspannungsverdampfer *m*; **~ mixer** *n* WASTE Schnellmischer *m*, Turbomischer *m*; **~-over** *n* AIR Überschlag *m*; **~-over voltage** *n* AIR Überschlagsspannung *f*; **~ point determination** *n* ANALYSIS Flammpunktbestimmung *f*; **~ point tester** *n* ANALYSIS Flammpunktprüfer *m*

flask *n* ANALYSIS Glaskolben *m*

flat *adj* SOIL, WASTE, WATER SUPP flach; **~-bed biofilter** *n* AIR *biological sorption process* Flachbettbiofilter *nt*; **~ bunker** *n* WASTE Flachbunker *m*; **~ fraction** *n* WASTE flache Fraktion *f*; **~-plate collector** *n* RENEW ENERG Flachkollektor *m*;

~ **sieve** *n* WASTE, WATER SUPP Flachsieb *nt*; ~ **subsurface** *n* SOIL flacher Untergrund *m*

flavor *AmE see flavour BrE*

flavour: ~ **substance** *n BrE* WATER SUPP Geschmacksstoff *m*

flexible *adj* NOISE biegeweich; ~ **hose** *n* MARINE flexibler Schlauch *m*; ~ **plate** *n* NOISE biegeweiche Platte *f*; ~ **plate absorber** *n* NOISE biegeweicher Plattenabsorber *m*; ~ **shell** *n* NOISE biegeweiche Schale *f*

flight: ~ **scraper** *n* WATER POLL Bandräumer *m*

fling *vt* AIR *to the wall* schleudern

floating 1. *adj* GEN schwimmend; **2.** *n* GEN Aufschwimmen *nt*

floating: ~ **flexible tank** *n* MARINE flexibler Schwimmtank *m*; ~ **floor** *n* NOISE *resilient layer* schwimmender Fußboden *m*; ~ **OTEC plant** *n* RENEW ENERG schwimmendes Meereswärmekraftwerk *nt*; ~ **refuse** *n* WASTE Treibgut *nt*, schwimmender Rückstand *m*; ~ **sampling** *n* ANALYSIS Schwimmprobenahme *f*; ~ **solids** *n pl* WATER POLL Schwimmstoffe *m pl*, aufschwimmende Stoffe *m pl*

floc *n* GEN Flocke *f*

flocculant *n* GEN Flokkulant *m*, Flockungsmittel *nt*, Polymerflockungsmittel *nt*

flocculate *vi* GEN ausflocken

flocculation *n* GEN Ausflockung *f*, Flockung *f*; ~ **rate** *n* WATER POLL Flockungsgeschwindigkeit *f*; ~ **tank** *n* WATER POLL Flockungsbecken *nt*; ~ **test** *n* ANALYSIS Flockungstest *m*; ~ **zone** *n* WATER POLL Flockungszone *f*

flocculator *n* GEN *agent* Flockungsmittel *nt*, *device* Flokkungsbecken *nt*, Flockungsreaktor *m*; ~ **basin** *n* WATER POLL Flockungsbecken *nt*; ~ **with cylindrical mixer** *n* WATER POLL Flockungsreaktor *m* mit Zylinderrührer; ~ **with mechanically-operated paddles** *n* WATER POLL mechanisch gerührter Flockungsreaktor *m*

flocculent *adj* GEN flockig

floc: ~ **filtration** *n* WATER SUPP Flockungsfiltration *f*; ~ **formation** *n* WATER POLL Flockenbildung *f*; ~ **fragments** *n pl* WATER POLL Flockenfragmente *nt pl*

flood: ~ **control** *n* WATER SUPP Hochwasserschutz *m*, Hochwasserüberwachung *f*

flooding *n* WATER POLL, WATER SUPP Flutung *f*, Überflutung *f*

flood: ~ **irrigation** *n* WATER SUPP Überstaubewässerung *f*; ~ **plain** *n* SOIL, WATER SUPP Hochwasserüberschwemmungsgebiet *nt*, Überschwemmungsgebiet *nt*; ~ **prevention** *n* WATER SUPP Hochwasserschutz *m*; ~ **spillway** *n* WATER SUPP Hochwasserentlastungsanlage *f*, Oberflächenentlastungsanlage *f*; ~ **tide** *n* RENEW ENERG *tidal power plant* Flut *f*; ~ **water** *n* MARINE, WATER SUPP Hochwasser *nt*

flotation *n* ANALYSIS, MARINE, WATER POLL, WATER SUPP *hydrocyclone process for plastics* Flotation *f*; ~ **agent** *n* SOIL, WATER POLL Flotationsmittel *nt*; ~ **cell** *n* SOIL Flotationszelle *f*; ~ **chamber** *n* MARINE Flotationskammer *f*; ~ **with dispersed gas** *n* WATER POLL Flotation *f* mit dispergiertem Gas; ~ **with electrolytically generated gas** *n* WATER POLL Flotation *f* mit elektrolytisch erzeugtem Gas; ~ **with integrated filtration** *n* WATER POLL Flotation *f* mit integrierter Filtration; ~ **plant** *n* WATER POLL Flotationsanlage *f*; ~ **with submerged aerator** *n* WATER POLL Flotation *m* mit Tauchbelüfter; ~ **tank** *n* WATER POLL Flotationsbecken *nt*; ~ **thickening** *n* WATER POLL Flotationseindickung *f*

flow 1. *n* AIR, RENEW ENERG, WATER POLL, WATER SUPP Strömung *f*; **2.** *vi* AIR *gas* strömen, RENEW ENERG *electricity*, WATER POLL, WATER SUPP *liquid* fließen, strömen; ~ **over** ANALYSIS, WATER SUPP überströmen; ~ **through** AIR, WASTE, WATER POLL, WATER SUPP durchströmen

flow: ~ **channel** *n* WATER SUPP *filter* Strömungskanal *m*; ~ **configuration** *n BrE* (*cf flow pattern AmE*) WATER SUPP Stromlinienbild *nt*, Strömungsbild *nt*; ~ **deviation** *n* AIR Strömungsumlenkung *f*; ~ **duration curve** *n* ANALYSIS Abflußmengendauerlinie *f*, WATER POLL, WATER SUPP Abflußmengendauerlinie *f*, Wassermengendauerlinie *f*; ~ **field** *n* AIR *cyclone* Strömungsfeld *nt*; ~ **force** *n* AIR Strömungskraft *f*

flowing: ~-**out viscosimeter** *n* ANALYSIS Ausflußviskosimeter *nt*; ~ **water** *n* WATER SUPP Fließwasser *nt*, fließendes Wasser *nt*

flow: ~ **injection analysis** *n* (*FIA*) ANALYSIS Fließinjektionsanalyse *f* (*FIA*); ~ **injection analysis spectrometry** *n* ANALYSIS Fließinjektionsanalyse *f* für Atomspektrometrie; ~ **injection analysis system** *n* (*FIAS*) ANALYSIS Fließinjektionsanalysensystem *nt* (*FIAS*); ~ **lanes** *n pl* AIR *of gas in electrostatic precipitator* Zwischenräume *m pl*; ~ **line** *n* AIR *carrier medium* Stromlinie *f*; ~ **measurement** *n* ANALYSIS, WATER SUPP Abflußmessung *f*; ~-**measuring flume** *n* WATER POLL, WATER SUPP Meßgerinne *nt*, Meßkanal *m*

flowmeter *n* ANALYSIS Durchflußmesser *m*, Strömungsmesser *m*, Durchflußzähler *m*, Flowmeter *nt*

flow: ~ **parameter** *n* ANALYSIS Flußgröße *f*; ~ **pattern** *n AmE* (*cf flow configuration BrE*) WATER SUPP Stromlinienbild *nt*, Strömungsbild *nt*; ~-**proportional** *adj* ANALYSIS mengenproportional; ~-**proportional sampler** *n* ANALYSIS mengenproportionaler Probenehmer *m*; ~ **rate** *n* AIR Strömungsgeschwindigkeit *f*, ANALYSIS *in measuring devices* Durchflußmenge *f*, Durchflußrate *f*, WATER POLL Durchflußmenge *f*, Durchflußrate *f*, WATER SUPP Durchflußmenge *f*, Durchflußrate *f*, Strömungsgeschwindigkeit *f*; ~ **rate measuring device** *n* ANALYSIS Durchflußmengenmeßgerät *nt*; ~-**regulating valve** *n* WATER SUPP Regelventil *nt*; ~ **regulation** *n* WATER SUPP Abflußregelung *f*; ~ **resistance** *n* WATER SUPP Fließwiderstand *m*, Strömungswiderstand *m*; ~ **sensor** *n* AIR Strömungsmeßfühler *m*; ~ **velocity** *n* WATER POLL, WATER SUPP Fließgeschwindigkeit *f*

flue *n* AIR Abzug *m*, Kamin *m*, Rauchabzugskanal *m*, Schornstein *m*, RENEW ENERG Kamin *m*, WASTE Kamin *m*, Schornstein *m*; ~ **gas** *n* AIR Rauchgas *nt*, Verbrennungsgase *nt pl*; ~-**gas blower** *n* AIR Rauchgasgebläse *nt*; ~-**gas cleaning** *n* AIR Rauchgasreinigung *f*; ~-**gas cleaning installation** *n* AIR Rauchgasreinigungsanlage *f*; ~-**gas cleaning residue** *n* AIR, WASTE Rauchgasreinigungsrückstand *m*; ~-**gas cloud** *n* AIR Rauchgaswolke *f*; ~-**gas cooling** *n* AIR Rauchgasabkühlung *f*; ~-**gas desulfurization** *AmE see flue-gas desulphurization BrE*; ~-**gas desulfurization plant** *AmE see flue-gas desulphurization plant BrE*; ~-**gas desulphurization** *n BrE* (*FGD*) AIR, SAFETY Rauchgasentschwefelung *f*; ~-**gas desulphurization plant** *n BrE* (*FGD plant*) AIR Rauchgasentschwefelungsanlage *f* (*REA*); ~-**gas heat exchanger** *n* AIR Rauchgaswärmetauscher *m*; ~-**gas nitrogen removal** *n* AIR Rauchgasentstickung *f*; ~-**gas plume** *n* AIR Rauchgasfahne *f*; ~-**gas recirculation** *n* AIR Rauchgasrezirkulation *f*; ~-**gas reheating** *n* AIR Rauchgaswiederaufheizung *f*; ~-**gas scrubbing** *n* AIR *incineration plant* Rauchgaswäsche *f*; ~-**gas side** *n* AIR Rauchgasseite *f*; ~-**gas stream** *n* AIR Rauchgasstrom *m*; ~-**gas tester** *n* ANALYSIS Rauchgasprüfer *m*; ~-**gas throughput** *n* AIR Rauchgasdurchsatz *m*; ~-**gas thruput** *AmE see flue-gas throughput*

fluffy *adj* AIR *dust property* flaumig, flockig

fluid: ~ **catalyst** *n* AIR Fließbettkatalysator *m*

fluidize *vt* AIR, SOIL wirbeln

fluidized: ~ **bed** *n* AIR *adsorption process*, SOIL, WASTE, WATER POLL, WATER SUPP Wirbelschicht *f*; ~-**bed absorber** *n* AIR Wirbelbettabsorber *m*; ~-**bed adsorber** *n* AIR Wirbelschichtadsorber *m*; ~-**bed bottom** *n* AIR Wirbelschichtboden *m*; ~-**bed combustion** *n* (*FBC*) AIR Wirbelschichtfeuerung *f* (*WSF*); ~-**bed dryer** *n* AIR, WATER POLL Wirbelschichttrockner *m*; ~-**bed drying** *n* WATER POLL Wirbelschichttrocknung *f*; ~-**bed furnace** *n* WATER POLL

Wirbelschichtofen *m*; **~-bed incineration** *n* SOIL, WASTE, WATER POLL Wirbelschichtverbrennung *f*; **~-bed incineration plant** *n* WASTE, WATER POLL Wirbelschichtverbrennungsanlage *f*; **~-bed incinerator** *n* WASTE, WATER POLL Wirbelschichtverbrennungsanlage *m*; **~-bed oven** *n* AIR Wirbelschichtofen *m*; **~-bed process** *n* SOIL *for fine-grained contaminated soil* Wirbelschichtverfahren *nt*; **~-bed pyrolysis** *n* WASTE Wirbelschicht-Pyrolyse *f*; **~-bed reactor** *n* AIR, WASTE, WATER POLL Wirbelschichtreaktor *m*, Wirbelbettreaktor *m*; **~ sand** *n* WATER SUPP Fließsand *m*

flume *n* WASTE Rinne *f*, WATER SUPP Gerinne *nt*; **~ flushing** *n* WASTE, WATER POLL Rinnenspülung *f*

fluorescence *n* ANALYSIS Fluoreszenz *f*; **~ analysis** *n* ANALYSIS Fluoreszenzanalyse *f*; **~ detection** *n* ANALYSIS Fluoreszenzdetektion *f*; **~ polarization** *n* ANALYSIS Fluoreszenzpolarisation *f*; **~ spectrometry** *n* ANALYSIS Fluoreszenzspektrometrie *f*; **~ spectrum** *n* ANALYSIS *IR spectrometer* Fluoreszenzspektrum *nt*

fluorescent: **~ compound** *n* AIR fluoreszierender Stoff *m*; **~ indicator adsorption** *n* ANALYSIS Fluoreszenzindikatoradsorption *f*; **~ spectroscopy** *n* ANALYSIS Fluoreszenzspektroskopie *f*; **~ tube** *n* WASTE Leuchtstofflampe *f*

fluoride *n* WATER SUPP Fluorid *nt*; **~ analysis in wastewater** *n* ANALYSIS Fluoridbestimmung *f* in Abwässern

fluorimetric *adj* ANALYSIS fluorimetrisch

fluorimetry *n* ANALYSIS *heavy metal detection* Fluorimetrie *f*

fluorine *n* (*F*) SOIL Fluor *nt* (*F*)

fluorohydrogen *n* AIR Fluorwasserstoff *m*

flush *vt* GEN spülen, WASTE, WATER SUPP ausspülen, durchspülen

flush: **~ hydrant** *n* WATER SUPP Spülhydrant *m*

flushing *n* WATER SUPP Ausspülen *nt*, MARINE, WATER POLL, WATER SUPP Spülung *f*; **~ of pipe network** *n* WATER POLL, WATER SUPP Rohrnetzspülung *f*; **~ pump** *n* WATER POLL Spülpumpe *f*; **~ speed** *n* WATER POLL, WATER SUPP Spülgeschwindigkeit *f*; **~ strength** *n* WATER POLL, WATER SUPP Spülkraft *f*; **~ velocity** *n* WATER POLL, WATER SUPP Spülgeschwindigkeit *f*

fluvial: **~ alluvium** *n* SOIL Auenböden *m pl*

fluxing: **~ agent** *n* WATER SUPP Flußmittel *nt*

fly: **~ ash** *n* AIR *containing solid particulates* Flugasche *f*, Flugstaub *m*; **~ ash conditioning** *n* AIR Flugstaubkonditionierung *f*; **~ ash silo** *n* AIR Flugaschesilo *m*; **~ ash yield** *n* WATER SUPP Flugascheanfall *m*

flying: **~ sparks** *n pl* AIR Funkenflug *m*

fly: **~ stream method** *n* SOIL Flugstromverfahren *nt*; **~ tipping** *n* *BrE* WASTE ungeordnete Deponie *f*, wilde Müllablagerung *f*

foam *n* AIR Schaum *m*; **~ bed** *n* AIR *scrubber* Sprudelschicht *f*; **~ film** *n* WATER SUPP *de-inking process* Schaumteppich *m*; **~ flotation** *n* SOIL Schaumflotation *f*; **~ formation** *n* WATER SUPP Schaumbildung *f*

foaming *n* AIR *scrubber* Schaumbildung *f*; **~ abatement** *n* WATER SUPP Schaumbekämpfung *f*; **~ tendency** *n* AIR Neigung *f* zur Schaumbildung

foam: **~ panel** *n* NOISE Schaumstoffplatte *f*; **~ process** *n* SOIL Schaumverfahren *nt*; **~ scrubber** *n* AIR Sprudelbett-Waschturm *m*

focal: **~ line** *n* RENEW ENERG *on solar energy plant reflector* Brennlinie *f*

focusing *n* RENEW ENERG *of sunlight* Fokussierung *f*

fog *n* AIR Nebel *m*

foil: **~ particle** *n* WASTE, WATER SUPP Folienpartikel *f*; **~ polarometer** *n* ANALYSIS *for measuring oxygen in water* Membranpolarometer *nt*

folded: **~ filter** *n* WATER POLL Faltenfilter *nt*

foliar: **~ surface** *n* AIR Blattoberfläche *f*

food: **~ packaging** *n* WASTE Lebensmittelverpackung *f*

force *n* POLICY, SOIL, WASTE, WATER POLL Zwang *m*; **~ of attraction** *n* RENEW ENERG, WATER POLL Anziehungskraft *f*, Anziehungsvermögen *nt*

forced: **~ aeration** *n* SOIL, WASTE, WATER POLL, WATER SUPP *composting* Zwangsbelüftung *f*; **~ circulation evaporator** *n* WATER SUPP Zwangsumlaufverdampfer *m*; **~ cooling** *n* WATER SUPP Zwangskühlung *f*; **~ degassing** *n* WASTE, WATER SUPP Zwangsentgasung *f*; **~ draft fan** *AmE*, **~ draught fan** *BrE* *n* AIR Druckgebläse *nt*

forecast *vt* ANALYSIS prognostizieren

forecasting *n* ANALYSIS Prognose *f*

forecast: **~ model** *n* ANALYSIS Prognosemodell *nt*

foreign: **~ emissions** *n pl* AIR Fremdemissionen *f pl*; **~ material** *n* WASTE *in recycling* Störstoff *m*, Fremdteil *m*, Fremdstoff *m*; **~ source** *n* AIR Fremdquelle *f*; **~ substance** *n* WASTE *in recycling* Fremdstoff *m*, Störstoff *m*, Fremdteil *m*

fork: **~ lift truck** *n* WASTE Gabelstapler *m*

form *n* GEN Form *f*

formaldehyde *n* AIR Formaldehyd *m*, Methanal *nt*

formazin: **~ turbidity unit** *n* (*FTU*) ANALYSIS Trübungseinheit/Formazin *f* (*TE/F*)

forming: **~ temperature** *n* WASTE Formungstemperatur *f*

form: **~ stability** *n* AIR *of filter* Formstabilität *f*

forward: **~-curved blade compressor** *n* AIR vorwärts gekrümmter Schaufel-Verdichter *m*; **~ feed** *n* WATER SUPP Vorschub *m*; **~ feed grate** *n* WATER SUPP Vorschubrost *m*; **~ scattering of radiation** *n* ANALYSIS *heavy metal detection* Vorwärtsstreuung *f*

fossil: **~ energy source** *n* RENEW ENERG fossiler Energieträger *m*; **~ fuel** *n* RENEW ENERG fossiler Brennstoff *m*; **~ fuel combustion** *n* AIR fossile Verbrennung *f*; **~-fuel-driven** *adj* RENEW ENERG fossil betrieben; **~-fuel-driven heat pump** *n* RENEW ENERG fossil betriebene Wärmepumpe *f*; **~-fuel-fired** *adj* RENEW ENERG fossil befeuert; **~-fuel-fired steam generator** *n* RENEW ENERG fossil befeuerter Dampferzeuger *m*

foul *vt* WASTE, WATER SUPP *pipes* verstopfen

fouling *n* MARINE, WASTE, WATER SUPP Bewuchs *m*

four: **~-cycle engine** *n* *AmE* (*cf four-stroke engine BrE*) AIR Viertakt-Ottomotor *m*, Viertaktmotor *m*

Fourier: **~-transform** *n* ANALYSIS Fourier-Transformation *f*; **~-transform infrared spectrometry** *n* (*FTIR*) ANALYSIS Infrarotspektrometrie *f* mit Fourier-Transformation (*FTIR*)

four: **~-liter automobile** *n* *AmE* (*cf four-liter car BrE*) AIR Vier-Liter-Auto *nt*; **~-litre car** *BrE* *n* (*cf four-liter automobile AmE*) AIR Vier-Liter-Auto *nt*; **~-stroke engine** *n* *BrE* (*cf four-cycle engine AmE*) AIR Viertakt-Ottomotor *m*, Viertaktmotor *m*

FPD *abbr* (*flame-photometric detector*) ANALYSIS FPD (*flammenphotometrischer Detektor*)

fraction *n* AIR, ANALYSIS, SOIL Fraktion *f*

fractional: **~ distillation** *n* ANALYSIS fraktionierende Destillation *f*

fractionate *vt* AIR, ANALYSIS, SOIL fraktionieren

fractionation *n* AIR, ANALYSIS, SOIL Fraktionierung *f*, WATER POLL Rektifikation *f*

fraction: **~ separation** *n* WASTE Fraktionsabscheidung *f*, Fraktionstrennung *f*; **~ separator** *n* WASTE Fraktionsabscheider *m*

fracture: **~ zone of continental plates** *n* RENEW ENERG *geothermal energy potential* Bruchzone *f* der Kontinentalplatten

fracturing *n* SOIL *pneumatic* Zerklüftung *f*

fragile *adj* WASTE zerbrechlich

frame: **~ filter press** *n* WATER POLL Rahmenfilterpresse *f*

Frank: **~ effect level** *n* (*FEL*) SAFETY Frank effect level *nt* (*FEL*)

Fraunhofer: **~ lines** *n pl* ANALYSIS *diffraction of laserlight* Fraunhofersche Linien *f pl*

free: ~ **cyanide** *n* WATER POLL freies Cyanid *nt*; ~ **of discretionary error** *adj* ANALYSIS ermessensfehlerfrei; ~ **electron** *n* AIR freies Elektron *nt*; ~ **field** *n* NOISE *without obstruction* Freifeld *nt*; ~ **from suspended matter** *adj* WATER POLL schwebstofffrei; ~ **of heavy metal** *adj* AIR, SOIL, WATER POLL, WATER SUPP schwermetallfrei; ~ **inflow** *n* WATER SUPP *directly into distribution system* freier Zufluß *m*; ~ **inflow via an elevated reservoir** *n* WATER SUPP freier Zufluß *m* über einen Hochbehälter; ~ **inflow via a water tower** *n* WATER SUPP freier Zufluß *m* über einen Wasserturm; ~ **radical** *n* AIR *electron beam process* freies Radikal *nt*; ~**-standing tower** *n* RENEW ENERG *wind turbine* freistehender Turm *m*
freeze: ~**-grinding** *n* WASTE Tieftemperaturzerkleinerung *f*
freezer *n* WASTE Gefriergerät *nt*
freight: ~ **rate** *n* WASTE Frachtkosten *pl*, Frachttarif *m*
Freon: ~ **12** *n* (*dichlorodifluoromethane*) AIR *propellant* Freon 12 *nt* (*Dichlordifluormethan*)
frequency *n* ANALYSIS, NOISE *number of cycles* Frequenz *f*, *occurrence* Häufigkeit *f*; ~ **analyser** *BrE*, ~ **analyzer** *AmE n* ANALYSIS Frequenzanalysator *m*; ~ **distribution** *n* ANALYSIS Häufigkeitsverteilung *f*; ~ **of rain** *n* WATER SUPP Niederschlagshäufigkeit *f*; ~ **range** *n* NOISE Frequenzbereich *m*; ~ **spectrum** *n* NOISE Frequenzspektrum *nt*
fresh: ~ **coke** *n* AIR Frischkoks *m*; ~ **compost** *n* WASTE Frischkompost *m*; ~ **water** *n* WATER SUPP Frischwasser *nt*
freshwater: ~ **reserve** *n* WATER SUPP Süßwasservorrat *m*; ~ **supply** *n* WATER SUPP Frischwasserzufuhr *m*
Fresnel: ~ **lens** *n* RENEW ENERG *solar collector* Fresnellinse *f*; Fresnelsche Linse *f*
friction *n* AIR, WASTE, WATER POLL Reibung *f*
frictional: ~ **force** *n* AIR Reibungskraft *f*
friction: ~ **head** *n* WASTE, WATER POLL, WATER SUPP Reibungshöhe *f*, Widerstandshöhe *f*; ~ **layer** *n* AIR Reibungsschicht *f*, dichte Atmosphärenschicht *f*
Friedrichs: ~**'s gas-washing bottle** *n* ANALYSIS Friedrichs-Gaswaschflasche *f*, Gaswaschflasche *f* nach Friedrichs
Friends: ~ **Of The Earth** *n* POLICY Freunde der Erde *m pl*
frontally: ~**-packed bed** *n* WASTE Vorkopfschüttung *f*
front: ~**-end loader** *n* MARINE Frontschaufellader *m*, Schaufellader *m*
froth *n* AIR Schaum *m*
frothing: ~ **tendency** *n* AIR Neigung *f* zur Schaumbildung
FRP *abbr* (*fiberglass-reinforced plastics AmE*, *fibreglass-reinforced plastics BrE*) WASTE, WATER SUPP GFK (*glasfaserverstärkter Kunststoff*)
FTIR *abbr* (*Fourier-transform infrared spectrometry*) ANALYSIS FTIR (*Infrarotspektrometrie mit Fourier-Transformation*); ~ **spectrometer** *n* ANALYSIS FTIR-Spektrometer *nt*; ~ **spectrometry** *n* ANALYSIS FTIR-Spektrometrie *f*
FTU *abbr* (*formazin turbidity unit*) ANALYSIS TE/F (*Trübungseinheit/Formazin*)
fuel *n* AIR, RENEW ENERG, WASTE Brennstoff *m*, Kraftstoff *m*; ~**-air ratio** *n* AIR Luftzahl *f*; ~**-air ratio control unit** *n* AIR Gemischregler *m*; ~ **cell** *n* AIR, RENEW ENERG Brennstoffzelle *f*; ~ **combustion** *n* AIR, WASTE Kraftstoffverbrennung *f*; ~ **composition** *n* WASTE Brennstoffzusammensetzung *f*; ~ **consumption** *n* AIR, RENEW ENERG Kraftstoffverbrauch *m*; ~ **gas** *n* AIR Abgas *nt* als Brennstoff; ~ **layer** *n* WASTE Brennstoffschicht *f*; ~ **oil** *n* AIR Heizöl *nt*; ~ **replacement** *n* AIR Kraftstoffersatz *m*; ~ **from biomass** *n* RENEW ENERG Brennstoff *m* aus Biomasse; ~ **tank** *n* AIR Kraftstofftank *m*; ~ **tank emissions** *n pl* AIR *hydrocarbon* Kraftstofftankemissionen *f pl*
fugitive: ~ **emissions** *n pl* AIR flüchtige Emissionen *f pl*
full: ~ **analysis** *n* ANALYSIS Vollanalyse *f*
Fuller: ~**'s earth** *n* AIR *adsorbent* Fullererde *f*, Walkerde *f*
fully: ~ **halogenated** *adj* AIR vollhalogeniert; ~ **halogenated CFCs** *n pl* AIR vollhalogenierte FCKW *m pl*; ~ **integrated system** *n* MANAGE vollintegriertes System *nt*
fumarole *n* RENEW ENERG *geothermal energy* Fumarole *f*
fumehood *n* ANALYSIS Abzug *m*
fume: ~ **incinerator** *n* AIR *afterburner* Rauchgasverbrennungsofen *m*, thermischer Nachbrenner *m*; ~ **inlet** *n* AIR Rauchgaseintritt *m*
fumes *n pl* AIR *containing solid particulates* Abgas *nt*, Rauchgas *nt*, SAFETY *from paints, lacquers, adhesives* Ausdünstungen *f pl*
fumigate *vt* AIR räuchern
fumigation *n* AIR Verrauchung *f*
function *n* SAFETY Funktion *f*; ~ **monitoring** *n* SAFETY Funktionsüberwachung *f*; ~ **reliability** *n* WASTE *of landfill sealing systems* Funktionstüchtigkeit *f*; ~ **safety** *n* SAFETY Funktionssicherheit *f*
fungus *n* WATER POLL Pilz *m*
funneling *AmE see funnelling BrE*
funnelling *n BrE* AIR, WATER POLL Trichterbildung *f*
funnel: ~**-shaped** *adj* WATER POLL trichterförmig; ~**-shaped settling chamber** *n* WATER POLL trichterförmiger Absetzraum *m*, trichterförmiger Stoffänger *m*
furring *n* WASTE Kalkablagerung *f*
fuse *vt* AIR durchschlagen
fusion *n* AIR Versinterung *f*; ~ **granulate** *n* WASTE Schmelzgranulat *nt*; ~ **process** *n* RENEW ENERG *solar energy* Fusionsprozeß *m*
future: ~ **usability** *n* RENEW ENERG zukünftige Nutzbarkeit *f*

G

g *abbr* (*acceleration due to gravity*) AIR g (*Erdbeschleunigung*)

GAC *abbr* (*granulated activated carbon*) AIR, WATER POLL, WATER SUPP GAK (*granulierte Aktivkohle*)

gage *AmE see gauge BrE*

galvanic: **~ sludge** *n* WASTE, WATER POLL, WATER SUPP Galvanikschlamm *m*; **~ wastes** *n pl* WASTE *hazardous* Galvanikabfälle *m pl*

galvanize *vt* WASTE verzinken

galvanized *adj* WASTE verzinkt

galvanometer *n* ANALYSIS Galvanometer *nt*

gamma: **~-ray irradiation** *n* WASTE Gammastrahlenbestrahlung *f*, WATER POLL Gammabestrahlung *f*

garbage *n AmE* (*cf rubbish BrE, refuse*) WASTE Abfall *m*, Müll *m*; **~ bag** *n AmE* WASTE Müllbeutel *m*, Müllsack *m*; **~ bag collection** *n AmE* WASTE Einsammeln *nt* von Müllsäcken; **~ can** *n AmE* WASTE Müllbehälter *m*, Mülleimer *m*, Mülltonne *f*; **~ can bag** *n AmE* WASTE Beutel *m*, Müllbeutel *m*, Müllsack *m*; **~ chute** *n AmE* WASTE Müllabwurfschacht *m*, Müllschlucker *m*; **~ collection** *n AmE* WASTE Müllabfuhr *f*, Mülleinsammlung *f*, Müllsammlung *f*; **~ collection truck** *n AmE* WASTE Müllfahrzeug *nt*, Müllwagen *m*; **~ composition** *n AmE* WASTE Müllzusammensetzung *f*; **~ compost** *n AmE* WASTE Müllkompost *m*; **~ composting** *n AmE* WASTE Müllkompostierung *f*; **~ disposal** *n AmE* POLICY Abfallentsorgung, WASTE Abfallbeseitigung *f*, Hausmülleinsammlung *f*, Müllabfuhr *f*; **~ disposal site** *n AmE* WASTE Deponie *f*, Müllkippe *f*; **~ grinder** *n AmE* WASTE Müllzerkleinerer *m*; **~ incineration** *n AmE* WASTE Müllverbrennung *f*; **~ incineration plant** *n AmE* WASTE Müllverbrennungsanlage *f* (*MVA*); **~ incinerator** *n AmE* AIR, WASTE Abfallverbrennungsanlage *f*; **~ mass** *n AmE* WASTE Müllmenge *f*; **~ recovery** *n AmE* WASTE Müllerfassung *f*; **~ separation plant** *n AmE* WASTE Abfallsortieranlage *f*, Müllsortieranlage *f*; **~ slide** *n AmE* WASTE Müllabwurfschacht *m*, Müllschlucker *m*; **~ truck** *n AmE* WASTE Müllfahrzeug *nt*, Müllwagen *m*; **~ volume** *n AmE* WASTE Abfallmenge *f*, Müllvolumen *nt*

gas *n* GEN Gas *nt*; **~ absorbent** *n* AIR Gasabsorptionsmittel *nt*; **~ absorption** *n* AIR Gasabsorption *f*; **~ acceleration** *n* AIR *Venturi nozzle* Gasbeschleunigung *f*; **~ analyser** *n BrE* ANALYSIS Gasanalysator *m*, Gasanalysengerät *nt*; **~ analysis** *n* ANALYSIS Gasanalyse *f*; **~ analyzer** *AmE see gas analyser BrE*; **~ bubble** *n* WATER POLL Gasblase *f*; **~ build-up** *n* WASTE *of landfill* Gasbildung *f*; **~ build-up potential** *n* WASTE Gasbildungspotential *nt*; **~ build-up rate** *n* WASTE, WATER SUPP Gasbildungsrate *f*; **~ burette** *n* ANALYSIS Gasbürette *f*; **~ capture** *n* AIR, NOISE Gaserfassung *f*; **~ chromatograph** *n* (*GC*) ANALYSIS Gaschromatograph *m* (*GC*); **~-chromatographic** *adj* ANALYSIS gaschromatographisch; **~-chromatographic separation** *n* ANALYSIS gaschromatographische Trennung *f*; **~ chromatography** *n* (*GC*) ANALYSIS Gaschromatographie *f* (*GC*); **~ chromatography coupled with electron capture detection** *n* (*GC-ECD*) ANALYSIS Gaschromatographie *f* in Kopplung mit Elektroneneinfangsdetektoren (*GC-ECD*); **~ chromatography coupled with flame ionization detectors** *n* ANALYSIS Gaschromatographie *f* in Kopplung mit Flammenionisationsdetektoren; **~ chromatography coupled with mass spectrometers** *n* ANALYSIS Gaschromatographie *f* in Kopplung mit Massenspektrometern; **~ chromatography detector** *n* (*GCD*) ANALYSIS Gaschromatographie-Detektor *m* (*GCD*); **~ chromatography with flame ionization detection** *n* (*GC-FID*) ANALYSIS Gaschromatographie *f* mit Flammenionisationsdetektion (*GC-FID*); **~ chromatography with flame photometric detection** *n* (*GC-FPD*) ANALYSIS Gaschromatographie *f* mit flammenphotometrischer Detektion (*GC-FPD*); **~ chromatography with mass spectrometer** *n* (*GC-MS*) ANALYSIS Gaschromatographie *f* mit Massenspektrometer (*GC-MS*); **~ chromatography with mass spectrometry** *n* (*GC-MS*) ANALYSIS Gaschromatographie *f* mit Massenspektrometrie (*GC-MS*); **~ chromatography-mass spectrometry computer techniques** *n pl* (*GC-MS-COMP techniques*) ANALYSIS rechnergestützte GC-MS *f*; **~ chromatography nitrogen/phosphorus detection** *n* (*GC-NPD*) ANALYSIS Gaschromatographie-Stickstoff/Phosphor-Detektion *f* (*GC-NPD*); **~ cloud** *n* AIR Gaswolke *f*; **~ collection outside of landfill** *n* WASTE Gaserfassung *f* außerhalb der Deponie; **~ collector** *n* WASTE Gaskollektor *m*; **~ compartment** *n* AIR Gasraum *m*; **~ component** *n* AIR, WASTE, WATER SUPP Gaskomponente *f*; **~ composition** *n* AIR Gaszusammensetzung *f*; **~ concentration** *n* WASTE Gaskonzentration *f*; **~ container** *n* WATER POLL Gasbehälter *m*; **~ cooler** *n* AIR Gaskühler *m*; **~ cooling** *n* AIR Gaskühlung *f*, Gasabkühlung *f*; **~ desulfurization** *AmE*, **~ desulphurization** *BrE n* AIR, WASTE Gasentschwefelung *f*; **~ detector** *n* AIR Gasprüfer *m*, ANALYSIS Gasprüfer *m*, Gasspürgerät *nt*, Gaswarngerät *nt*; **~ discharge** *n* WASTE *from landfill surface* Gasaustritt *m*; **~ discharge lamp** *n* WASTE Entladungslampe *f*; **~ dispersion** *n* AIR Gasdispergierung *f*; **~ distribution** *n* AIR Gasverteilung *f*; **~ drainage** *n* WASTE Entgasung *f*; **~ drainage layer** *n* WASTE Gasdränschicht *f*; **~ emissions** *n pl* AIR, RENEW ENERG, WASTE *from landfill* Gasemissionen *f pl*; **~ entrainment** *n* AIR Gaseinschlüsse *m pl*; **~ entrance possibility** *n* WASTE, WATER SUPP Gaszutrittsmöglichkeit *f*

gaseous *adj* AIR, WASTE, WATER SUPP gasförmig; **~ air pollution** *n* AIR gasförmige Luftverunreinigung *f*; **~ combustion product** *n* AIR, WASTE gasförmiges Verbrennungsprodukt *nt*; **~ contaminants** *n pl* AIR gasförmige Schadstoffe *m pl*; **~ emissions** *n pl* AIR, WASTE gasförmige Emissionen *f pl*; **~ fuel** *n* AIR gasförmiger Brennstoff *m*; **~ medium** *n* AIR, WASTE, WATER SUPP gasförmiges Medium *nt*; **~ pollutants** *n pl* AIR gasförmige Schadstoffe *m pl*; **~ state of aggregation** *n* AIR gasförmiger Aggregatzustand *m*; **~ working medium** *n* RENEW ENERG *heat pump* gasförmiges Arbeitsmittel *nt*

gases: **~ and fumes** *n pl* SAFETY *safety hazards* Gase *nt pl* und Dämpfe *m pl*

gas: **~ evolution** *n* WASTE, WATER SUPP Ausgasen *nt*, *of waste dump* Ausgasung *f*; **~-filled** *adj* ANALYSIS gasgefüllt; **~-filled detector** *n* ANALYSIS gasgefüllter Detektor *m*; **~ flow** *n* AIR Gasdurchfluß *m*, Gasstrom *m*, ANALYSIS Gasstrom *m*; **~ flow counter** *n* ANALYSIS Gasdurchflußzähler *m*; **~ flow meter** *n* ANALYSIS Gasflußmesser *m*; **~ flow velocity** *n* AIR Gasdurchflußgeschwindigkeit *f*; **~ formation** *n* WASTE Gasaufkommen *nt*; **~ formation potential** *n* ANALYSIS *landfill gas* Gasbildungspotential *nt*; **~ heat exchanger** *n* AIR Gaswärmeaustauscher *m*; **~ hood** *n* WATER POLL Gashaube *f*

gasification *n* AIR, RENEW ENERG, WASTE, WATER POLL Vergasung *f*

gasifier *n* AIR, RENEW ENERG *biomass conversion*, WASTE, WATER POLL Vergaser *m*

gasify *vt* AIR, RENEW ENERG, WASTE, WATER POLL vergasen

gas: ~ **indicator** *n* ANALYSIS Gaswarngerät *nt*; ~ **inlet** *n* AIR Gaseintritt *m*; ~ **interferometer** *n* ANALYSIS Gasinterferometer *nt*; ~ **mask** *n* SAFETY Gasmaske *f*; ~ **measurement** *n* ANALYSIS Gasmessung *f*; ~ **measurement device** *n* ANALYSIS Gasmeßgerät *nt*; ~ **measuring** *n* ANALYSIS Gasmessung *f*; ~ **measuring device** *n* ANALYSIS Gasmeßgerät *nt*; ~ **melt** *n* WASTE Gasschmelze *f*; ~ **meter** *n* ANALYSIS Gasmesser *m*, Gaszähler *m*; ~ **migration** *n* SOIL, WASTE *underground* Gasmigration *f*; ~ **mixer** *n* ANALYSIS Gasmischer *m*; ~-**mixing chamber** *n* ANALYSIS Gasmischkammer *f*; ~ **mixture** *n* ANALYSIS Gasgemisch *nt*; ~ **molecule** *n* AIR Gasmolekül *nt*; ~ **non-specific** *adj* ANALYSIS *measurement principle* gasunspezifisch

gasoline: ~ **engine vehicle** *n AmE* (*cf petrol engine vehicle BrE*) WASTE Fahrzeug *nt* mit Vergasermotor; ~ **vapor recovery plant** *n AmE* (*cf petrol vapour recovery plant BrE*) AIR, WASTE Benzindampfrückgewinnungsanlage *f*

gasometer *n* ANALYSIS Gasbehälter *m*, Gasometer *nt*, WATER POLL Gasometer *nt*

gasometric *adj* ANALYSIS gasvolumetrisch; ~ **method** *n* ANALYSIS gasvolumetrische Methode *f*

gasometry *n* ANALYSIS Gasvermessung *f*

gas: ~ **outlet** *n* AIR Gasaustritt *m*; ~ **partition chromatography** *n* ANALYSIS Gasverteilungs-chromatographie *f*; ~ **permeability** *n* WASTE *in landfill* Gasdurchtritt *m*; ~-**permeable** *adj* WASTE, WATER SUPP gasdurchlässig; ~ **permeation** *n* AIR Gaspermeation *f*; ~ **phase** *n* AIR Gasphase *f*; ~-**phase oxidation** *n* WATER POLL *wastewater incineration* Gasphasenoxidation *f*; ~-**phase titration** *n* (*GPT*) ANALYSIS Gasphasentitration *f* (*GPT*); ~-**phase titrator** *n* ANALYSIS Gasphasentitrator *m*; ~ **preheater** *n* AIR Gasvorwärmer *m* (*GAVO*); ~ **purifier** *n* AIR Gasreiniger *m*; ~ **quantity** *n* ANALYSIS Gasmenge *f*; ~ **recovery** *n* AIR, WASTE Gasrückgewinnung *f*; ~ **release** *n* ANALYSIS *from sample* Ausgasung *f*; ~ **reuse plant** *n* WASTE Gasverwertungsanlage *f*; ~ **sample** *n* ANALYSIS Gasprobe *f*; ~ **sampling** *n* ANALYSIS Gasprobe *f*, Gasprobenehmen *nt*; ~ **sampling bag** *n* ANALYSIS Gassammelbeutel *m*; ~ **sampling pipette** *n* ANALYSIS Gaspipette *f*, Gassammelröhre *f*; ~ **sampling tube** *n* ANALYSIS Gasprobenahmerohr *nt*; ~ **sampling valve** *n* ANALYSIS Gasprobenahmeventil *nt*; ~ **saturation** *n* ANALYSIS Gassättigung *f*; ~ **saturation method** *n* ANALYSIS *vapour pressure determination* Gassättigungsmethode *f*; ~ **scrubbing** *n* AIR Gasabscheidung *f*; ~-**sensitive** *adj* ANALYSIS gassensitiv; ~-**sensitive electrode** *n* ANALYSIS gassensitive Elektrode *f*, *in measuring cell* gasempfindliche Elektrode *f*; ~-**sensitive electrode with gas permeable membrane** *n* ANALYSIS gassensitive Elektrode *f* mit gaspermeabler Membrane; ~ **shaft** *n* WASTE Gasschacht *m*; ~ **soil pipe** *n* SOIL Gasleitung *f*; ~-**solid chromatography** *n* (*GSC*) ANALYSIS Gas-Feststoff-Chromatographie *f* (*GSC*); ~ **solubility** *n* AIR Gaslöslichkeit *f*; ~ **storage tank** *n* WATER POLL Gasspeicher *m*; ~ **stream** *n* AIR, ANALYSIS Gasstrom *m*; ~ **streamline** *n* AIR Stromlinien des Gases *f pl*; ~ **suction** *n* WASTE Gasabsaugung *f*; ~ **temperature** *n* AIR Gastemperatur *f*; ~ **throughput** *n BrE* AIR Gasdurchsatz *m*; ~ **thruput** *AmE see gas throughput BrE*; ~ **trace analyser** *BrE*, ~ **trace analyzer** *AmE n* ANALYSIS Gasspurenanalysator *m*; ~ **transfer** *n* WASTE, WATER POLL, WATER SUPP Gasaustausch *m*; ~ **turbine** *n* WATER POLL Gasturbine *f*; ~ **use plant** *n* WASTE Gasnutzungsanlage *f*; ~ **velocity** *n* AIR Gasgeschwindigkeit *f*; ~ **velocity of flow** *n* AIR Gasströmungsgeschwindigkeit *f*; ~-**wash bottle** *n* ANALYSIS *gas* Gaswaschflasche *f*; ~-**washing bottle** *n* ANALYSIS Gaswaschflasche *f*, Spritzflasche *f*, Waschflasche *f*; ~-**washing bottle with sintered head** *n* ANALYSIS Gaswaschflasche *f* mit Glasfritte; ~ **water** *n* WATER POLL Gaswasser *nt*; ~ **well** *n* WASTE Gasbrunnen *m*; ~ **well chain** *n* WASTE Gasbrunnenkette *f*

gate: ~ **electrode** *n* ANALYSIS *quadrupole ion trap* Gate-Elektrode *f*

gauge *vt BrE* ANALYSIS abmessen, eichen, kalibrieren, messen, SOIL, WASTE, WATER POLL, WATER SUPP *flow of pump* abmessen

gauge: ~ **cock** *n BrE* WATER POLL, WATER SUPP Probehahn *m*, Probierhahn *m*, Wasserstandshahn *m*; ~ **wire** *n BrE* ANALYSIS Meßdraht *m*

Gaussian: ~ **filter** *n* ANALYSIS Gauß-Filter *nt*; ~ **model** *n* ANALYSIS Gauß-Modell *nt*; ~ **plume** *n* AIR Kühlturmschwaden *m*

GC *abbr* ANALYSIS (*gas chromatograph*) GC (*Gaschromatograph*), (*gas chromatography*) GC (*Gaschromatographie*)

GCD *abbr* (*gas chromatography detector*) ANALYSIS GCD (*Gaschromatographie-Detektor*)

GC-ECD *abbr* (*gas chromatography coupled with electron capture detection*) ANALYSIS GC-ECD (*Gaschromatographie in Kopplung mit Elektroneneinfangsdetektoren*)

GC-FID *abbr* (*gas chromatography with flame ionization detection*) ANALYSIS GC-FID (*Gaschromatographie mit Flammenionisations-Detektion*)

GC-FPD *abbr* (*gas chromatography with flame photometric detection*) ANALYSIS GC-FPD (*Gaschromatographie mit flammenphotometrischer Detektion*)

GC: ~-**injector** *n* ANALYSIS GC-Injektor *m*

GC-MS *abbr* ANALYSIS (*gas chromatography with mass spectrometer*) GC-MS (*Gaschromatographie mit Massenspektrometer*), (*gas chromatography with mass spectrometry*) GC-MS (*Gaschromatographie mit Massenspektrometrie*)

GC-MS-COMP: ~ **techniques** *n pl* (*gas chromatography-mass spectrometry computer techniques*) ANALYSIS rechnergestützte GC-MS *f*

GC-NPD *abbr* (*gas chromatography nitrogen/phosphorus detection*) ANALYSIS GC-NPD (*Gaschromatographie-Stickstoff/Phosphor-Detektion*)

GC: ~-**thermodesorption** *n* ANALYSIS GC-Thermodesorption *f*

gear: ~ **drive** *n* WASTE, WATER POLL, WATER SUPP *pump* Zahnradantrieb *m*

GEF *abbr* (*Global Environment Facility*) POLICY globaler Umweltfonds *m*

Geiger: ~ **counter** *n* ANALYSIS Geiger-Müllerzähler *m* (*GM-Zähler*), Geigerzähler *m*

Geiger-Müller: ~ **counter** *n* (*GM counter*) ANALYSIS Geiger-Müllerzähler *m* (*GM-Zähler*), Geigerzähler *m*

gel: ~ **chromatography** *n* ANALYSIS Gelchromatographie *f*; ~ **electrophoresis** *n* ANALYSIS Gelelektrophorese *f*; ~ **electrophoretic** *adj* ANALYSIS gelelektrophoretisch

gelling: ~ **agent** *n* MARINE Geliermittel *nt*

gel: ~ **permeation chromatography** *n* (*GPC*) ANALYSIS Gelpermeations-Chromatographie *f* (*GPC*)

generally: ~ **accepted technical procedures** *n pl* POLICY, WATER SUPP allgemein anerkannte Regeln der Technik *f pl* (*aaRdT*)

generate *vt* NOISE erzeugen, RENEW ENERG *energy* entstehen, gewinnen, *electricity, power, steam* erzeugen, WASTE, WATER SUPP entstehen, erzeugen

generated *adj* NOISE, RENEW ENERG, WASTE, WATER SUPP erzeugt; ~ **electricity** *n* RENEW ENERG erzeugter Strom *m*; ~ **power** *n* RENEW ENERG erzeugter Strom *m*

generation *n* RENEW ENERG *of electrical energy*, WASTE, WATER SUPP Gewinnung *f*; ~ **of hazardous waste** *n* WASTE Sondermüllanfall *m*

generator *n* ANALYSIS, RENEW ENERG Generator *m*; ~ **house** *n* NOISE Generatorgehäuse *nt*

genetic *adj* ANALYTIC, SAFETY genetisch

genetically: ~ **significant dose** *n* AIR, SAFETY genetisch wichtige Dosis *f*

geological: ~ **barrier** *n* WASTE geologische Barriere *f*;

~ **conditions** *n pl* RENEW ENERG geologische Bedingungen *f pl*
geophysical: ~ **site analysis** *n* ANALYSIS geophysikalische Standortuntersuchung *f*
geothermal *adj* RENEW ENERG geothermisch; ~ **deposit** *n* RENEW ENERG geothermische Lagerstätte *f*; ~ **electrical power generation** *n* RENEW ENERG geothermische Stromerzeugung *f*; ~ **energy** *n* AIR geothermale Energie *f*, RENEW ENERG geothermale Energie *f*, Erdwärme *f*, Geothermie *f*, geothermische Energie *f*, WASTE geothermale Energie *f*; ~ **energy exploitation** *n* AIR Ausbeutung *f* der geothermalen Energie; ~ **energy potential** *n* RENEW ENERG geothermisches Energiepotential *nt*; ~ **energy resources** *n pl* RENEW ENERG geothermische Energieressourcen *f pl*; ~ **gradient** *n* RENEW ENERG geothermische Tiefenstufe *f*; ~ **heating plant** *n* RENEW ENERG geothermisches Heizwerk *nt*; ~ **power plant** *n* RENEW ENERG geothermisches Kraftwerk *nt*; ~ **reservoir** *n* RENEW ENERG geothermisches Reservoir *nt*; ~ **resources** *n pl* RENEW ENERG geothermische Ressourcen *f pl*
geothermic *adj* RENEW ENERG geothermisch
geothermics *n pl* RENEW ENERG Geothermik *f*
germ *n* SOIL, WATER POLL, WATER SUPP Keim *m*
German: ~ **Institute for Standards** *n* (*DIN*) ANALYSIS Deutsches Institut *nt* für Normung (*DIN*); ~ **technical directive on air quality control** *n* POLICY technische Anleitung-Luft *f* (*TA-Luft*); ~ **technical directive on noise protection** *n* NOISE, POLICY technische Anleitung-Lärm *f* (*TA-Lärm*); ~ **technical directive on waste management** *n* POLICY, WASTE technische Anleitung-Abfall *f* (*TA-Abfall*); ~ **unit processes for water, wastewater and sludge analysis** *n pl* ANALYSIS Deutsche Einheitsverfahren *nt pl* zur Wasser-, Abwasser- und Schlammanalyse
germ: ~**-free** *adj* ANALYSIS, WASTE keimfrei; ~**-free wastewater sample** *n* ANALYSIS, WASTE keimfreie Abwasserprobe *f*; ~ **reduction** *n* WATER SUPP Keimverminderung *f*
geyser *n* RENEW ENERG *geothermal energy* Geysir *m*
GF *abbr* (*graphite furnace*) ANALYSIS GF (*Graphitrohrofen*)
gimlet: ~ **bit** *n* WASTE Schneckenbohrer *m*
glance: ~ **coal** *n* AIR Steinkohle *f*
glass: ~ **bottle** *n* ANALYSIS Glasflasche *f*; ~ **consumption** *n* WASTE Glasverbrauch *m*; ~ **container** *n* WASTE Glas *nt*, Glasbehälter *m*; ~ **crusher** *n* WASTE Glaszerkleinerer *m*; ~ **cullet recycling** *n* WASTE Scherbenrückführung *f*; ~ **fabric** *n* WASTE Glasgewebe *nt*; ~ **fiber** *AmE*, ~ **fibre** *BrE n* AIR *filter medium*, NOISE *absorbing material*, WASTE Glasfaser *f*; ~ **fibre absorber** *n BrE* NOISE Glasfaserabsorber *m*; ~ **fibre filter** *n BrE* ANALYSIS Glasfaserfilter *nt*; ~ **fibre laminate** *n BrE* WASTE, WATER SUPP GFK-Laminat *nt*, Glasfaserlaminat *nt*, Glasfaserschichtkunststoff *m*; ~ **fibre-mat-reinforced thermoplast** *n BrE* WASTE glasfasermattenverstärktes Thermoplast *nt*; ~**-fibre-reinforced** *adj BrE* NOISE, WASTE, WATER SUPP glasfaserverstärkt; ~ **fragment** *n* WASTE Scherbe *f*; ~ **jar** *n* WASTE Glas *nt*, Glasgefäß *nt*; ~**-mat-reinforced thermoplast** *n* WASTE glasmattenverstärktes Thermoplast *nt*; ~ **melting process** *n* WASTE *treatment of waste incineration slag* Glasschmelzverfahren *nt*; ~ **pane** *n* RENEW ENERG *solar collector* Glasscheibe *f*; ~ **pulverizer** *n* WASTE Glaszerkleinerer *m*; ~ **recycling** *n* WASTE Altglasverwertung *f*, Glasrecycling *nt*, WATER SUPP Altglasverwertung *f*; ~ **roof** *n* RENEW ENERG *solar power plant* Glasdach *nt*; ~ **tube** *n* ANALYSIS Glasrohr *nt*; ~ **wool** *n* WASTE Glaswolle *f*
glazing *n* RENEW ENERG *solar collector* Glasscheibe *f*
GLC *abbr* (*ground-level concentration*) SOIL Konzentration *f* auf Bodenhöhe
global *adj* AIR, RENEW ENERG Welt-, global; ~ **climate modeling** *AmE*, ~ **climate modelling** *BrE n* AIR globales Klimamodell *nt*, globales Klimamodellieren *nt*; ~ **ecosystem** *n* AIR, MARINE, SOIL, WATER POLL weltweites Ökosystem *nt*; ~ **emissions** *n pl* AIR weltweite Emissionen *f pl*; ~ **energy consumption** *n* RENEW ENERG Weltenergieverbrauch *m*; ~ **energy demand** *n* RENEW ENERG Weltenergiebedarf *m*; ~ **energy supply** *n* RENEW ENERG Weltenergieversorgung *f*
Global: ~ **Environment Facility** *n* (*GEF*) POLICY globaler Umweltfonds *m*; ~ **Environment Monitoring System** *n* ANALYSIS, POLICY globales Umweltüberwachungssystem *nt*
global: ~ **insolation** *n* AIR Sonneneinstrahlung *f*; ~ **irradiance** *n* AIR Globalstrahlung *f*; ~ **radiation** *n* AIR weltweite Strahlung *f*; ~ **solar radiation** *n* RENEW ENERG weltweite Sonnenstrahlung *f*; ~ **sulfur budget** *AmE*, ~ **sulphur budget** *BrE n* AIR Weltschwefelhaushalt *m*; ~ **warming** *n* AIR Erwärmung *f* der Erde
GLP *abbr* (*good laboratory practice*) ANALYSIS GLP (*gute Laborpraxis*)
glutinous *adj* AIR *dust property* klebrig, leimig
GM: ~ **counter** *n* (*Geiger-Müller counter*) ANALYSIS GM-Zähler *m* (*Geiger-Müllerzähler*), Geigerzähler *m*
gondola *n* RENEW ENERG Gondel *f*
good: ~ **laboratory practice** *n* (*GLP*) ANALYSIS gute Laborpraxis *f* (*GLP*)
government *n* POLICY Regierung *f*
governmental: ~ **projection** *n* POLICY Regierungsprognose *f*
GPC *abbr* (*gel permeation chromatography*) ANALYSIS GPC (*Gelpermeations-Chromatographie*)
GPT *abbr* (*gas-phase titration*) ANALYSIS GPT (*Gasphasentitration*)
grab *n* WASTE Greifbagger *m*, Greifer *m*; ~ **boring** *n* SOIL Greiferbohrung *f*; ~ **sample** *n* ANALYSIS Stichprobe *f*, Zufallsstichprobe *f*; ~ **system** *n* WASTE Greifersystem *nt*
grade *vt* POLICY, SOIL, WASTE, WATER SUPP klassifizieren
grade: ~ **efficiency** *n* AIR Trenngrad *m*; ~ **efficiency curve** *n* AIR Trennkurve *f*
grader *n* MARINE Planiergerät *nt*
grading *n* POLICY, SOIL, WASTE, WATER SUPP Klassifizierung *f*
graduate *vt* ANALYSIS graduieren, kalibrieren
grain *n* ANALYSIS, SOIL Korn *nt*; ~ **encapsulation** *n* WASTE *of hazardous waste* Einbindung *f*, Einschluß *m*, Mikroeinkapselung *f*; ~ **size** *n* SOIL Korngröße *f*; ~ **size analysis** *n* ANALYSIS Korngrößenanalyse *f*; ~ **size analysis of particulates** *n* ANALYSIS Korngrößenanalyse *f* von Schwebestoffen; ~ **size distribution** *n* ANALYSIS, SOIL Korngrößenverteilung *f*
grandfathering: ~ **principle** *n* POLICY Grandfathering-Prinzip *nt*
granting: ~ **of permit** *n* POLICY Genehmigungserteilung *f*
granular: ~**-bed filter** *n* AIR *of granulate layers* Schüttschichtfilter *nt*; ~ **catalyst** *n* AIR Katalysatorgranulat *nt*
granulate 1. *n* AIR *granular bed filter* Granulat *nt*, Granulatmaterial *nt*; **2.** *vt* WASTE granulieren
granulate: ~ **body** *n* AIR *granular bed filter* Granulatkörper *m*
granulated: ~ **activated carbon** *n* (*GAC*) AIR, WATER POLL, WATER SUPP granulierte Aktivkohle *f* (*GAK*); ~ **layer** *n* AIR *filter medium* körnige Schicht *f*
granulate: ~ **layer** *n* AIR *granular-bed filter* Granulatschicht *f*; ~ **tubular filter** *n* AIR Granulatröhrenfilter *nt*
granulating: ~ **head** *n* WASTE Granulierkopf *m*; ~ **machine** *n* WASTE Granulator *m*
granulator *n* WASTE Granulator *m*
graphite *n* WASTE Graphit *m*; ~ **furnace** *n* (*GF*) ANALYSIS Graphitrohrofen *m* (*GF*); ~ **furnace atomic absorption spectrometry** *n* ANALYSIS Graphitrohrofen-atomabsorptionspektrometrie *f* (*GF-AAS*)
grappler *n* WASTE Greifarm *m*
grate *n* WASTE, WATER POLL, WATER SUPP Rost *m*; ~ **firing** *n* AIR, WASTE Rostfeuerung *f*; ~ **heat exposure** *n* WASTE Rostwärmebelastung *f*; ~ **screenings** *n pl* WASTE, WATER POLL Rostdurchfall *m*
gravel *n* WASTE, WATER POLL, WATER SUPP Kies *m*; ~**-bed filter** *n* AIR Kiesbettfilter *nt*; ~ **column** *n* WASTE Kiessäule *f*;

~ **deposition** *n* WASTE, WATER SUPP Kiesablagerung *f*; ~ **filter** *n* WATER SUPP Kiesfilter *nt*; ~ **filter layer** *n* WATER SUPP Kieselfilterschicht *f*; ~ **layer** *n* WATER SUPP *on floor of filtering tanks* Kiesschicht *f*; ~**-sheathed drainage line** *n* WASTE kiesummantelte Dränleitung *f*
gravimetric *adj* ANALYSIS gewichtsanalytisch, gravimetrisch; ~ **analysis** *n* ANALYSIS Gewichtsanalyse *f*, Gravimetrie *f*, gravimetrische Analyse *f*; ~ **determination** *n* ANALYSIS gravimetrische Bestimmung *f*; ~ **gas analysis** *n* ANALYSIS Gasgravimetrie *f*; ~ **method** *n* ANALYSIS gravimetrisches Verfahren *nt*
gravitation *n* AIR, WATER POLL Gravitation *f*
gravitational: ~ **force** *n* AIR, ANALYSIS, RENEW ENERG, WATER POLL *dewatering*, WATER SUPP Schwerkraft *f*, Gravitationskräfte *f pl*; ~ **settling** *n* AIR Schwerkraftabscheidung *f*, Schwerkraftsedimentation *f*, Schwerkrafttrennung *f*, Schwerkraftabsetzung *f*
gravity *n* GEN Schwerkraft *f*, Gravitation *f*; ~ **roller conveyor** *n* WASTE Rollenförderer *m*, Schwerkraftförderer *m*, WATER SUPP Schwerkraftförderer *m*; ~ **sedimentation** *n* ANALYSIS *measurement for dust* Schwerkraftsedimentation *f*; ~ **separation** *n* SOIL, WATER POLL *treatment of oily wastewaters* Schwerkraftabscheidung *f*; ~ **separator** *n* AIR, WASTE, WATER POLL, WATER SUPP Schwerkraftabscheider *m*; ~ **settler** *n* AIR Schwerkraftabscheider *m*; ~ **settling** *n* AIR Schwerkraftabscheidung *f*, Schwerkraftsedimentation *f*, Schwerkrafttrennung *f*; ~ **settling chamber** *n* AIR Schwerkraftabscheidekammer *f*; ~ **sewer** *n* WATER POLL Freispiegelleitung *f*, Gefälleleitung *f*; ~ **spray tower** *n* AIR Schwerkraftsprühturm *m*; ~ **thickener** *n* WATER POLL Schwerkrafteindicker *m*; ~ **thickening** *n* WATER POLL Schwerkrafteindickung *f*, Schwerkraftentwässerung *f*; ~ **thickening in containers** *n* WATER POLL Schwerkraftentwässerung *f* in Containern; ~ **thickening in filter bags** *n* WATER POLL Schwerkraftentwässerung *f* in Filtersäcken
gray *n* (*Gy*) ANALYSIS Gray *nt* (*Gy*); ~ **container** *AmE see grey container BrE*; ~ **filter** *AmE see grey filter BrE*
grease *n* WASTE, WATER POLL, WATER SUPP Fett *nt*; ~**-free** *adj* WASTE, WATER SUPP fettfrei; ~ **separator** *n* WATER SUPP Fettabscheider *m*; ~ **trap** *n* WATER POLL Fettabscheider *m*, Fettfang *m*, Ölfang *m*, WATER SUPP Fettabscheider *m*
green *adj* MANAGE, POLICY umweltbewußt; ~ **consumerism** *n* MANAGE umweltbewußtes Konsumverhalten *nt*
Green: ~ **Container** *n* WASTE Biotonne *f*
green: ~ **glass** *n* WASTE Grünglas *nt*
greenhouse: ~ **effect** *n* AIR Treibhauseffekt *m*; ~ **gas build-up** *n* AIR Treibhausgasakkumulierung *f*; ~ **gases** *n pl* AIR Treibhausgase *nt pl*; ~ **warming** *n* AIR Treibhauseffekt *m*
green: ~ **product** *n* MANAGE Ökoprodukt *nt*; ~ **revolution** *n* AIR *damaging to the atmosphere* grüne Revolution *f*
Greens: **the** ~ *n pl* POLICY *political party* die Grünen *pl*
greensand *n* WASTE, WATER SUPP *zeolite* Grünsand *m*
green: ~ **taxes** *n pl* MANAGE, POLICY Ökosteuern *f pl*; ~ **tax reform** *n* POLICY ökologische Steuerreform *f*, Ökosteuerreform *f*; ~ **tax system** *n* POLICY ökologisches Steuersysystem *nt*; ~ **tourism** *n* MANAGE Ökotourismus *m*; ~ **waste** *n* WASTE Grünabfall *m*
grey: ~ **container** *n BrE* WASTE *German collection container for residual waste* graue Tonne *f*; ~ **filter** *n BrE* ANALYSIS Neutralglasfilter *nt*
grid *n* ANALYSIS, WASTE, WATER SUPP Gitter *nt*; ~ **connection** *n* RENEW ENERG Netzanschluß *m*; ~**-coupled** *adj* RENEW ENERG netzgekoppelt; ~**-coupled photovoltaic modules** *n pl* RENEW ENERG netzgekoppelte Photovoltaikmodule *m pl*; ~ **crusher** *n* WASTE, WATER SUPP Gitterzerkleinerer *m*; ~ **of pipes** *n* WATER SUPP Rohrnetz *nt*; ~ **point** *n* ANALYSIS Gitterpunkt *m*; ~**-shaped** *adj* ANALYSIS rasterförmig
grind *vt* ANALYSIS, SOIL, WASTE zerfasern, zerkleinern, zermahlen
grinder *n* ANALYSIS, SOIL, WASTE Zerkleinerer *m*
grinding *n* ANALYSIS, SOIL, WASTE, WATER POLL Zerkleinern *nt*, Zerkleinerung *f*; ~ **medium** *n* SAFETY Schleifmittel *nt*
gripper: ~ **robot** *n* WASTE Greifroboter *m*
grit *n* AIR *greater than 10*m* Grobstaub *m*, WATER POLL Grieß *m*, Sandfanggut *nt*; ~ **additive** *n* WASTE Splittzuschlagstoff *m*; ~ **can** *n* WATER POLL Sandgrube *f*; ~ **chamber** *n* WATER POLL Sandfang *m*; ~ **precipitation** *n* AIR Grobstaubabscheidung *f*; ~ **removal** *n* WATER POLL Sandfanggutentfernung *f*; ~ **scraper** *n* WATER POLL Sandräumer *m*; ~ **washer** *n* WATER POLL Sandwaschanlage *f*; ~ **washing** *n* WATER POLL Sandwaschung *f*
gross: ~ **calorific value** *n* WASTE oberer Heizwert *m*
ground 1. *adj* WASTE zerkleinert; **2.** *n* WASTE Erdboden *m*, Erdreich *nt*
ground: ~ **coarse solids** *n pl* WATER POLL zerkleinerte Grobstoffe *m pl*
grounded: ~ **wall** *n AmE* (*cf earthed wall BrE*) AIR geerdete Wandung *f*
ground: ~ **glass cullet** *n* WASTE zerkleinerte Scherben *f pl*; ~ **glass stopper** *n* ANALYSIS *bottle* Glasschliffstopfen *m*
grounding *n* MARINE Aufgrundlaufen *nt*
ground: ~**-level concentration** *n* (*GLC*) SOIL Konzentration *f* auf Bodenhöhe; ~**-level contaminants** *n pl* AIR bodennahe Schadstoffe *m pl*; ~ **stock** *n* WASTE Mahlgut *nt*
groundwater *n* SOIL, WATER SUPP Grundwasser *nt*; ~ **basin** *n* WATER SUPP Grundwasserbecken *nt*; ~ **bottom** *n* WATER SUPP Grundwassersohle *f*; ~ **capture** *n* WATER SUPP Grundwassererfassung *f*, Grundwassererschließung *f*; ~ **catchment** *n* WATER SUPP Grundwassereinzugsgebiet *nt*; ~ **contamination** *n* SOIL Grundwasserverschmutzung *f*, Grundwasserverunreinigung *f*, WATER SUPP Grundwasserkontamination *f*; ~ **depth** *n* WATER SUPP Grundwassertiefe *f*; ~ **extraction tax** *n* POLICY Grundwasserentnahmesteuer *f*; ~ **formation** *n* WATER SUPP Grundwasserbildung *f*; ~ **level** *n* WATER SUPP Grundwasserstand *m*; ~ **management** *n* WATER SUPP Grundwasserbewirtschaftung *f*; ~ **measuring point** *n* ANALYSIS Grundwassermeßstelle *f*; ~ **method** *n* SOIL Grundwasserverfahren *nt* (*GW-Verfahren*); ~ **model** *n* ANALYSIS, MANAGE Grundwassermodell *nt*; ~ **monitoring well** *n* ANALYSIS Grundwasserbeobachtungsbrunnen *m*; ~ **probe** *n* ANALYSIS Grundwassersonde *f*; ~ **problem** *n* WATER SUPP Grundwasserproblem *nt*; ~ **properties** *n pl* WATER SUPP Grundwassereigenschaften *f pl*; ~ **prospecting** *n* WATER SUPP Grundwasserschürfung *f*; ~ **protection** *n* WATER SUPP Grundwasserschutz *m*; ~ **protection area** *n* WATER SUPP Grundwasserschutzgebiet *nt*, Grundwasserschutzzone *f*; ~ **pumping station** *n* WATER SUPP Grundwasserpumpwerk *nt*; ~ **recharge** *n* WATER SUPP Grundwasseranreicherung *f*; ~ **reserve** *n* WATER SUPP Grundwasserbestand *m*, Grundwasservorrat *m*; ~ **resources** *n pl* WATER SUPP Grundwasservorkommen *nt*; ~ **risk** *n* WATER SUPP Grundwassergefährdung *f*; ~ **sampling** *n* ANALYSIS Grundwasserprobeentnahme *f*; ~ **source** *n* WATER SUPP Grundwasserquelle *f*; ~ **supply** *n* WATER SUPP Grundwasserdargebot *nt*; ~ **table** *n* WATER SUPP Grundwasserspiegel *m*; ~ **tapping** *n* WATER SUPP Grundwassererfassung *f*, Grundwassererschließung *f*; ~ **treatment** *n* WATER SUPP Grundwasseraufbereitung *f*; ~ **treatment plant** *n* WATER SUPP Grundwasseraufbereitungsanlage *f*; ~ **use** *n* WATER SUPP Grundwassernutzung *f*; ~ **well** *n* WATER SUPP Grundwasserbrunnen *m*; ~ **withdrawal charge** *n* POLICY Grundwasserentnahmeabgabe *f*; ~ **withdrawal tax** *n* POLICY Grundwasserentnahmesteuer *f*
group: ~ **clarification plant** *n* WATER POLL Gruppenkläranlage *f*; ~ **collective dose** *n* AIR, ANALYSIS gruppenkollektive Dosis *f*; ~ **parameter** *n* ANALYSIS Gruppenparameter *m*
grouting *n* SOIL Einpreßverfahren *nt*; ~ **fluid** *n* SOIL Injektionsflüssigkeit *f*; ~ **material** *n* SOIL Verpreßmittel *nt*;

~ pressure *n* SOIL Einpreßdruck *m*; **~ section** *n* SOIL Einpreßabschnitt *m*, Verpreßabschnitt *m*

GSC *abbr* (*gas-solid chromatography*) ANALYSIS GSC (*Gas-Feststoff-Chromatographie*)

guaranteed: **~ disposal** *n* WASTE Entsorgungssicherheit *f*

guideline *n* POLICY Richtlinie *f*

guide: **~ value** *n* POLICY Richtwert *m*; **~ vane** *n* AIR Leitschaufel *f*, RENEW ENERG Leitflügel *m*, WATER SUPP *pump* Leitschaufel *f*; **~ wall** *n* WATER POLL Leitwand *f*

guillotine *n* WASTE Papierschneidemaschine *f*

gully: **~ emptier** *n* WATER POLL Schlammsaugwagen *m*; **~ sucker** *n* MARINE Bilgenpumpe *f*

Gy *abbr* (*gray*) ANALYSIS Gy (*Gray*)

gypsum *n* AIR, SOIL, WASTE *binding agent* Gips *m*; **~ crystal** *n* AIR Gipskristall *m*; **~ dewatering** *n* AIR Gipsentwässerung *f*; **~ drying** *n* AIR Gipstrocknung *f*; **~ formation** *n* AIR Gipsbildung *f*; **~-generating** *adj* AIR gipserzeugend; **~ particle** *n* AIR Gipspartikel *f*; **~ plaster board** *n* NOISE *wall* Gipskartonplatte *f*; **~ processing** *n* AIR Gipsaufbereitung *f*; **~ separation** *n* AIR Gipsabtrennung *f*; **~ silo** *n* AIR Gipssilo *m*; **~ suspension** *n* AIR Gipssuspension *f*; **~ treatment** *n* AIR Gipsbehandlung *f*; **~ use** *n* AIR Gipsverwertung *f*; **~ yield** *n* AIR Gipsanfall *m*

H

hack: ~ **up** *vt* WASTE häckseln
hair: ~ **hygrometer** *n* ANALYSIS Haarhygrometer *nt*, Haarfeuchtigkeitsmesser *m*
half: ~**-hourly average value** *n* ANALYSIS Halbstundenmittelwert *m*; ~**-hourly mean value** *n* ANALYSIS Halbstundenmittelwert *m*; ~**-life** *n* AIR Halbwertzeit *f*; ~**-width value** *n* ANALYSIS Halbwertsbreite *f*
halide *n* AIR Halid *nt*, Halogenid *nt*
haloform: ~ **compounds** *n pl* WASTE, WATER POLL, WATER SUPP Haloformverbindungen *f pl*
halogenated *adj* WASTE, WATER SUPP halogenhaltig; ~ **gas** *n* AIR halogenhaltiges Gas *nt*; ~ **hydrocarbon** *n* AIR, WASTE Halogenkohlenwasserstoff *m*; ~ **hydrocarbons** *n pl* WATER SUPP halogenierte Kohlenwasserstoffe *f pl* (*HKW*); ~ **organic solvents** *n pl* WASTE, WATER SUPP *hazardous waste* halogenhaltige organische Lösungsmittel *nt pl*
halogen: ~**-free organic solvents** *n pl* WASTE, WATER SUPP *in hazardous waste* halogenfreie organische Lösungsmittel *nt pl*
halon *n* AIR Halon *nt*
hammermill *n* WASTE Hammerbrecher *m*, Hammermühle *f*, Hammerzerkleinerer *m*; ~ **crusher** *n* WASTE Hammerbrecher *m*, Hammermühle *f*, Hammerzerkleinerer *m*
hand: ~ **augering** *n* ANALYSIS, SOIL Handbohrung *f*
handbook *n* ANALYSIS Fibel *f*
hand: ~**-driven** *adj* ANALYSIS handbetrieben; ~**-driven gas detector** *n* ANALYSIS handbetriebenes Gasspürgerät *nt*; ~**-held** *adj* ANALYSIS handgehalten; ~**-held centrifuge** *n* ANALYSIS Handzentrifuge *f*; ~**-held spectroscope** *n* ANALYSIS Handspektroskop *nt*; ~**-loaded compactor** *n* WASTE, WATER SUPP handbeladener Kompaktor *m*; ~**-operated** *adj* ANALYSIS handbedient; ~ **scanner** *n* WASTE, WATER SUPP *waste separation* Handscanner *m*; ~**-sorted fraction** *n* WASTE handsortierte Fraktion *f*; ~ **sorting** *n* WASTE Handauslese *f*, Handsortierung *f*; ~ **sorting line** *n* WASTE Handsortierstrecke *f*
hard: ~ **coal** *n* AIR Steinkohle *f*; ~ **coal ash** *n* WASTE Steinkohlenasche *f*
harden *vt* SOIL erhärten
hardening *n* SOIL Erhärtung *f*
hardness *n* WATER SUPP Härte *f*; ~ **detection** *n* ANALYSIS Härtenachweis *m*; ~ **determination** *n* ANALYSIS Härtebestimmung *f*; ~ **former** *n* WATER SUPP Härtebildner *m*
hard: ~ **water** *n* WASTE, WATER SUPP hartes Wasser *nt*, kalkhaltiges Wasser *nt*
harm *vt* GEN schädigen
harmful *adj* GEN schädlich; ~ **effect** *n* AIR, SAFETY, WASTE, WATER SUPP schädliche Wirkung *f*; ~ **emissions** *n pl* AIR schädliche Emissionen *f pl*; ~ **gas** *n* AIR Schadgas *nt*; ~ **gas adsorption** *n* AIR, SOIL Schadgasadsorption *f*; ~ **substance** *n* GEN Schadstoff *m*; ~ **substance concentration** *n* AIR, WATER POLL Schadstoffkonzentration *f*; ~ **substance load** *n* WASTE Schadstofffracht *f*; ~ **substance register** *n* POLICY Schadstoffregister *nt*; ~ **substances analysis** *n* ANALYSIS Schadstoffanalytik *f*; ~**-substance-sealing** *adj* SOIL *retention wall* schadstoffdicht; ~ **substance volume** *n* WASTE Schadstoffmenge *f*; ~ **to air** *adj* AIR luftgefährdend; ~ **to the environment** *adj* AIR, MANAGE, MARINE, POLICY, SOIL, WASTE, WATER POLL umweltgefährdend, umweltschädlich; ~ **to health** *adj* SAFETY gesundheitsschädlich; ~ **to water** *adj* WASTE, WATER POLL, WATER SUPP wassergefährdend; **not** ~ **to water** *adj* WASTE, WATER POLL, WATER SUPP nichtwassergefährdend
harmless: ~ **carbon dioxide** *n* AIR unschädliches Kohlendioxid *nt*
harmonization *n* POLICY Harmonisierung *f*; ~ **of environmental protection** *n* POLICY Harmonisierung *f* des Umweltschutzes
harmonize *vt* POLICY harmonisieren
harmonized *adj* POLICY harmonisiert; ~ **environmental protection** *n* POLICY harmonisierter Umweltschutz *m*
harness *vt* RENEW ENERG nutzen
harvest 1. *n* RENEW ENERG Ernte *f*; **2.** *vt* RENEW ENERG ernten
hauled: ~ **container system** *n AmE* (*HCS*) WASTE, WATER SUPP Abholcontainersystem *nt*
hauling: ~ **capacity** *n* WATER POLL, WATER SUPP *of transport vehicle* Förderleistung *f*, Fördermenge *f*
HAWT *abbr* (*horizontal axis wind turbine*) RENEW ENERG Horizontalachsenturbine *f*
hazard *n* GEN Risiko *nt*
hazardous: ~ **liquid waste** *n* WASTE flüssiger Sonderabfall *m*; ~ **substance** *n* SAFETY Gefahrstoff *m*, WATER POLL gefährlicher Stoff *m*; ~ **to air** *adj* AIR luftgefährdend; ~ **to water** *adj* WASTE, WATER POLL, WATER SUPP wassergefährdend; **not** ~ **to water** *adj* WASTE, WATER POLL, WATER SUPP nichtwassergefährdend; ~ **waste** *n* WASTE, WATER POLL Sonderabfall *m*, Sondermüll *m*, gefährlicher Abfall *m*, WATER SUPP gefährlicher Abfall *m*; ~ **waste avoidance** *n* WASTE Sondermüllvermeidung *f*; ~ **waste collection** *n* WASTE Sondermülleinsammlung *f*; ~ **waste concentrate** *n* WASTE Sondermüllkonzentrat *nt*; ~ **waste incineration** *n* WASTE Sondermüllverbrennung *f*, Sonderabfallverbrennung *f*; ~ **waste incineration plant** *n* WASTE Sondermüllverbrennungsanlage *f*, Sonderabfallverbrennungsanlage *f*; ~ **waste landfill** *n* WASTE Sondermülldeponie *f*, Sonderabfalldeponie *f*; ~ **waste landfilling** *n* WASTE Sondermülldeponierung *f*; ~ **waste surface landfill** *n* WASTE oberirdische Sonderabfalldeponie *f*; ~ **waste treatment** *n* WASTE Sondermüllbehandlung *f*
hazard: ~ **prevention** *n* MANAGE, SAFETY *in buildings* Gefahrenverhütung *f*
haze *n* AIR Dunst *m*, leichter Nebel *m*
HC *abbr* (*hydrocarbon*) AIR, MARINE, WATER POLL, WATER SUPP KW-Stoff *m* (*Kohlenwasserstoff*)
HCB *abbr* (*hexachlorobenzene*) SOIL HCB (*Hexachlorbenzol*)
HC: ~**-emissions** *n pl* AIR *unburned hydrocarbon* HC-Emissionen *f pl*
HCFC *abbr* (*hydrochlorofluorocarbon*) AIR HFCKW (*teilhalogenierte Fluorchlorkohlenwasserstoffe*)
HCH *abbr* (*hexachlorocyclohexane isomer*) SOIL HCH (*Hexachlorcyclohexan-Isomer*)
HCS *abbr AmE* (*hauled container system*) WASTE, WATER SUPP Abholcontainersystem *nt*
HC: ~ **value** *n* AIR HC-Wert *m* (*Kohlenwasserstoffwert*)
HDPE *abbr* (*high-density polyethylene*) WASTE Hochdruck-Polyethylen *nt*, Polyethylen mit hoher Dichte *nt*
HDR *abbr* (*hot dry rock*) RENEW ENERG Hot-dry-rock *nt*
He *abbr* (*helium*) AIR He (*Helium*)
head *n* AIR Kopf *m*, *of reactor* Kopfseite *f*, NOISE Kopf *m*, WASTE, WATER SUPP Gefälle *nt*; ~ **loss** *n* AIR *filter bag*, WATER POLL, WATER SUPP Druckverlust *m*
headphones *n pl* NOISE Kopfhörer *m*

head: ~ **pipe** *n* WATER SUPP *of pump* Hauptleitung *f*; ~ **plate** *n* AIR Kopfplatte *f*

headspace: ~ **analysis** *n* ANALYSIS *efficient for extraction of VOC from water samples* Dampfraumanalyse *f*; ~ **gas chromatography** *n* (*HSGC*) ANALYSIS Headspace-Gaschromatographie *f* (*HSGC*); ~ **sampler** *n* ANALYSIS Headspace-Sampler *m*; ~ **technique** *n* ANALYSIS Headspace-Technik *f*

health *n* SAFETY Gesundheit *f*; ~ **care waste** *n* WASTE Krankenhausabfall *m*, krankenhausspezifischer Abfall *m*; ~ **hazard** *n* SAFETY Gesundheitsgefahr *f*, Gesundheitsrisiko *nt*; ~**-related** *adj* SAFETY gesundheitsbezogen; ~**-related sampling** *n* SAFETY gesundheitsbezogene Probenentnahme *f*; ~ **risk** *n* SAFETY Gesundheitsgefahr *f*, Gesundheitsrisiko *nt*; ~ **and safety requirements** *n pl* SAFETY Gesundheits- und Sicherheitsbedingungen *f pl*; ~**-threatening** *adj* SAFETY gesundheitsgefährdend

heap *n* WASTE Miete *f*

hearing *n* NOISE Gehör *nt*; ~ **damage** *n* NOISE Gehörschädlichkeit *f*; ~ **protection** *n* NOISE Gehörschutz *m*; ~ **protection helmet** *n* NOISE Gehörschutzhelm *m*; ~ **protection padding** *n* NOISE Gehörschutzwatte *f*; ~ **protection plugs** *n pl* NOISE Gehörschutzstöpsel *m pl*

heat 1. *n* GEN Wärme *f*; **2.** *vt* GEN erhitzen, erwärmen; ~ **up** AIR, ANALYSIS aufheizen

heat: ~ **accumulator** *n* AIR Wärmespeicher *m*; ~ **balance** *n* AIR Wärmeausgleich *m*, Wärmebilanz *f*; ~ **carrier medium** *n* RENEW ENERG Wärmeträgermedium *nt*; ~ **conduction** *n* AIR Wärmeleitung *f*; ~ **conductivity** *n* ANALYSIS *of gases* Wärmeleitfähigkeit *f*; ~ **content** *n* RENEW ENERG Wärmeinhalt *m*; ~ **convection** *n* AIR Wärmekonvektion *f*, Wärmeströmung *f*; ~ **drying** *n* WATER POLL *sludge disposal* Heißtrocknung *f*

heated *adj* AIR vorgewärmt, ANALYSIS, WATER POLL beheizt; ~ **column oven** *n* ANALYSIS Säulenofen *m*; ~ **digestion chamber** *n* WATER POLL beheizter Faulraum *m*; ~ **digestor** *n* WATER POLL beheizter Faulraum *m*; ~ **injector** *n* ANALYSIS *liquid or gas sampling* beheizter Injektor *m*; ~ **kiln** *n* ANALYSIS *Gaschromatographie* beheizter Ofen *m*; ~ **waste air** *n* AIR vorgewärmte Abluft *f*

heat: ~ **emissions** *n pl* AIR Wärmeemissionen *f pl*

heater: ~ **support** *n* AIR Vorwärmerstützfuß *m*

heat: ~ **exchange** *n* AIR Wärmeaustausch *m*; ~ **exchanger** *n* AIR Katalysator *m*, Rekuperator *m*, Wärmetauscher *m*, RENEW ENERG Wärmetauscher *m*, WATER POLL Wärmeaustauscher *m*, Wärmetauscher *m*; ~ **exchanger efficiency** *n* AIR Wärmetauscherwirkungsgrad *m*; ~ **exchanger material** *n* AIR Wärmetauschermaterial *nt*; ~ **flow** *n* AIR Wärmestrom *m*; ~ **generation** *n* RENEW ENERG Wärmeerzeugung *f*

heating *n* GEN Erwärmung *f*; ~ **cylinder** *n* WATER POLL Heizzylinder *m*; ~ **filament** *n* AIR Heizer *m*; ~ **furnace** *n* WASTE, WATER SUPP Aufwärmofen *m*; ~ **installation** *n* RENEW ENERG Heizanlage *f*; ~ **purpose** *n* RENEW ENERG Heizzweck *m*; ~ **system** *n* RENEW ENERG Heizsystem *nt*; ~ **up** *n* AIR, ANALYSIS Aufheizung *f*; ~ **zone** *n* AIR Heizzone *f*

heat: ~ **insulation** *n* AIR Wärmedämmung *f*, WATER POLL *of reactor* Wärmeisolierung *f*; ~**-load plan** *n* AIR, WASTE, WATER SUPP Wärmebelastungsplan *m*; ~ **loss** *n* AIR, WASTE *due to incomplete combustion* Wärmeabgabe *f*, Wärmeverlust *m*; ~ **pollution** *n* AIR Umweltverschmutzung *f* durch Wärme

heatproof *adj* SAFETY hitzeabweisend, wärmebeständig

heat: ~ **pump** *n* RENEW ENERG Wärmepumpe *f*; ~ **pump plant** *n* RENEW ENERG Wärmepumpenanlage *f*; ~ **radiation** *n* RENEW ENERG Wärmestrahlung *f*; ~ **radiation losses** *n pl* WASTE Abstrahlungsverluste *m pl*; ~ **of reaction** *n* ANALYSIS Reaktionswärme *f*; ~ **recovery** *n* AIR, WASTE, WATER POLL, WATER SUPP Wärmerückgewinnung *f*; ~**-resistant** *adj* SAFETY hitzebeständig, wärmefest, hitzeabweisend, WASTE hitzebeständig, temperaturbeständig; ~ **source** *n* RENEW ENERG Wärmequelle *f*; ~ **stability** *n* AIR Wärmestabilität *f*, thermische Stabilität *f*; ~ **storage** *n* RENEW ENERG Wärmespeicher *m*; ~ **tone** *n* ANALYSIS *as measurable variable for gas concentration* Wärmetönung *f*; ~ **transfer** *n* AIR, WATER POLL, WATER SUPP Wärmeübertragung *f*; ~ **transfer medium** *n* WATER POLL Wärmeträger *m*; ~ **transfer medium in** *n* RENEW ENERG *solar collector* Eintritt *m* des Wärmeübertragermediums; ~ **transfer via conduction** *n* AIR *indirect* Wärmeübergang *m* durch Leitung; ~ **transfer via convection** *n* AIR *direct heat transfer*, WATER POLL *direct* Wärmekonvektion *f*; ~ **transition** *n* AIR Durchgang *m*; ~ **use** *n* AIR, WASTE, WATER POLL, WATER SUPP Wärmenutzung *f*

heavily: ~ **colored wastewater** *AmE*, ~ **coloured wastewater** *BrE* *n* WATER POLL stark gefärbtes Abwasser *nt*; ~ **soiled** *adj* WASTE, WATER POLL stark veschmutzt

heavy: ~**-duty corrugated fiberboard** *AmE*, ~**-duty corrugated fibreboard** *BrE* *n* WASTE superstarke faserarmierte Wellfaserplatte *f*, superstarker faserarmierter Wellkarton *m*; ~ **fuel** *n* AIR Heizöl *nt* S; ~ **goods vehicle** *n* (*HGV*) WASTE Lastkraftwagen *m* (*LKW*); ~ **metal** *n* AIR, WASTE, WATER POLL, WATER SUPP Schwermetall *nt*; ~ **metal analysis** *n* ANALYSIS Schwermetallbestimmung *f*; ~ **metal concentration** *n* AIR, WASTE, WATER POLL, WATER SUPP Schwermetallkonzentration *f*; ~ **metal content** *n* AIR, SOIL, WASTE, WATER POLL, WATER SUPP Schwermetallgehalt *m*; ~ **metal determination** *n* ANALYSIS Schwermetallbestimmung *f*; ~ **metal emissions** *n pl* AIR Schwermetallemissionen *f pl*; ~ **metal enrichment** *n* ANALYSIS Schwermetallanreicherung *f*; ~ **metals accumulation** *n* AIR, SOIL, WATER POLL Schwermetallanreicherung *f*; ~ **metals circulation** *n* AIR, SOIL, WATER POLL Schwermetallkreislauf *m*; ~ **metals deposition** *n* AIR, SOIL, WATER POLL Schwermetallablagerung *f*; ~ **metals precipitation** *n* WATER POLL Schwermetallfällung *f*; ~ **metals removal** *n* AIR, WATER POLL, WATER SUPP Schwermetallentfernung *f*; ~ **metals source** *n* AIR, SOIL, WATER POLL Schwermetallquelle *f*; ~ **oil desulfurization** *AmE*, ~ **oil desulphurization** *BrE* *n* WASTE Schwerölentschwefelung *f*; ~ **oil residue** *n* WASTE Schwerölrückstand *m*; ~ **water** *n* RENEW ENERG *deuterium oxide* schweres Wasser *nt*

helical *adj* AIR schraubenförmig

heliostat *n* RENEW ENERG Heliostat *m*

helium *n* (*He*) AIR Helium *nt* (*He*); ~**-neon laser** *n* ANALYSIS *FTIR* Helium-Neon-Laser *m*

Henry-Dalton: ~ **law** *n* WASTE Henry-Daltonsches Gesetz *nt*

Henry: ~**'s law** *n* AIR *of absorption*, SOIL, WATER POLL Henrysches Gesetz *nt*

heptane *n* WASTE Heptan *nt*

heteroatom *n* ANALYSIS *nitrogen, halogene, sulphur* Heteroatom *nt*

heterogeneous: ~ **catalyst** *n* AIR heterogener Katalysator *m*; ~ **wastes** *n pl* WASTE heterogene Abfälle *m pl*

heterotroph *n* WATER POLL Heterotroph *nt*

heterotrophic *adj* WATER POLL *bacteria* heterotroph; ~ **denitrification** *n* WATER POLL heterotrophe Denitrifikation *f*

hexachloro- *pref* SOIL Hexachlor-

hexachlorobenzene *n* (*HCB*) SOIL Hexachlorbenzol *nt* (*HCB*)

hexachlorocyclohexane: ~ **isomer** *n* (*HCH*) SOIL Hexachlorcyclohexanisomer *nt* (*HCH*)

hexane *n* AIR Hexan *nt*

Hg *abbr* (*mercury*) SOIL, WASTE, WATER POLL, WATER SUPP Hg (*Quecksilber*)

HGV *abbr* (*heavy goods vehicle*) WASTE Lastkraftwagen *m* (*LKW*)

high: ~ **concentration** *n* AIR, ANALYSIS, WASTE, WATER SUPP hohe Konzentration *f*; ~**-density polyethylene** *n* (*HDPE*) WASTE Hochdruck-Polyethylen *nt*, Polyethylen *nt* mit hoher Dichte; ~ **dirty gas load** *n* WASTE hohe Rohgasbelastung *f*;

~-**efficiency cyclone** *n* AIR Hochleistungszyklon *m*; ~-**frequency heating** *n* SOIL Hochfrequenzerwärmung *f*; ~-**gloss coating** *n* WASTE *recycling hindrance* Hochglanzanstrich *m*; ~-**grade** *adj* WASTE, WATER SUPP hochwertig; ~-**grade recycling product** *n* WASTE hochwertiges Recyclingprodukt *nt*; ~-**grade scrap** *n* WASTE hochwertiger Schrott *m*; ~-**intensity magnet separator** *n* SOIL, WASTE Magnetabscheider *m* mit hoher Intensität; ~-**level radioactive** *adj* WASTE hoch radioaktiv

highly: ~ **absorbable** *adj* AIR gut absorbierbar; ~ **absorbent** *adj* NOISE hochabsorbierend; ~**absorbent noise protection system** *n* MARINE, NOISE hochabsorbierendes Lärmschutzsystem *nt*; ~ **acidic** *adj* SOIL *soil* stark sauer; ~ **alkaline** *adj* WASTE, WATER SUPP hochalkalisch; ~ **contaminated** *adj* SOIL hochkontaminiert; ~ **contaminated municipal wastewater** *n* WATER POLL hochbelastetes kommunales Abwasser *nt*; ~ **inflammable** *adj* SAFETY *as danger symbol* leichtentzündlich; ~ **polluted** *adj* WASTE, WATER POLL, WATER SUPP hochbelastet, hochverschmutzt; ~ **polluted faecal solids sludge** *n BrE* WATER POLL hochbelasteter Fäkalschlamm *m*; ~ **polluted fecal solids sludge** *AmE see highly polluted faecal solids sludge BrE*; ~ **polluted wastewater** *n* WATER POLL hochbelastetes Abwasser *nt*; ~ **polluted waters** *n pl* WATER POLL hochbelastete Gewässer *nt pl*, hochverschmutzte Gewässer *nt pl*; ~ **sensitive** *adj* ANALYSIS hochempfindlich; ~ **sensitive detector** *n* ANALYSIS hochempfindlicher Detektor *m*; ~ **sensitive measuring device** *n* ANALYSIS hochempfindliches Meßgerät *nt*; ~ **sensitive scale** *n* ANALYSIS *gas weighing* hochempfindliche Waage *f*; ~ **sensitive test** *n* ANALYSIS *water analysis* hochempfindlicher Test *m*; ~ **toxic** *adj* SOIL, WASTE, WATER SUPP hochtoxisch; ~ **toxic hazardous waste** *n* WASTE hochtoxischer Sonderabfall *m*; ~ **turbid** *adj* WASTE, WATER SUPP stark getrübt; ~ **turbulent** *adj* WATER POLL hochturbulent

high: ~ **molecular weight polymer** *n* WASTE hochmolekulares Polymer *nt*; ~-**performance aeration** *n* WASTE, WATER SUPP Hochleistungsbelüftung *f*; ~-**performance cyclone** *n* AIR Hochleistungszyklon *m*; ~-**performance dust removal** *n* AIR Hochleistungsentstaubung *nt*; ~-**performance filter** *n* AIR Hochleistungsfilter *nt*; ~-**performance filtration** *n* AIR, WATER POLL, WATER SUPP Hochleistungsfilterung *f*, Hochleistungsfiltration *f*; ~-**performance fixed-bed reactor** *n* WATER SUPP Hochleistungsfestbettreaktor *m*; ~-**performance liquid chromatography** *n* (*HPLC*) ANALYSIS Hochleistungs-Flüssigkeitschromatographie *f* (*HPLC*); ~-**performance liquid chromatography mass spectrometry** *n* (*HPLC-MS*) ANALYSIS Hochleistungs-Flüssigkeitschromatographie *f* in Kopplung mit Massenspektrometrie; ~-**performance precipitator** *n* AIR Hochleistungsabscheider *m*; ~-**performance scrubber** *n* AIR Hochleistungswäscher *m*; ~-**performance thin-layer chromatography** *n* ANALYSIS Hochleistungs-Dünnschichtchromatographie *f* (*HPDC*); ~-**performance well** *n* WATER SUPP Hochleistungsbrunnen *m*; ~ **position of the sun** *n* RENEW ENERG *summer* hoher Solarstand *m*, hoher Sonnenstand *m*; ~-**pressure cleaning** *n* AIR Hochdruckspülung *f*; ~-**pressure cog** *n* RENEW ENERG *wave energy* Hochdruckkamm *m*; ~-**pressure filter** *n* WATER POLL Hochdruckfilter *nt*; ~-**pressure flushing** *n* MARINE Hochdruckspülen *nt*; ~-**pressure hydrogenation** *n* WASTE Hochdruckhydrierung *f*; ~-**pressure press** *n* WASTE, WATER SUPP Hochleistungspresse *f*; ~-**pressure pump** *n* WASTE, WATER POLL, WATER SUPP Hochdruckpumpe *f*; ~-**pressure radial flow compressor** *n* SOIL *soil vapour extraction* Hochdruckradialverdichter *m*; ~-**pressure reactor** *n* RENEW ENERG *biomass conversion* Hochdruckreaktor *m*; ~-**pressure side** *n* AIR Hochdruckseite *f*; ~-**pressure spray nozzle** *n* AIR Hochdruck-Spritzdüse *f*, Hochdruck-Sprühdüse *f*; ~-**pressure Venturi scrubber** *n* AIR Hochdruck-Venturiwäscher *m*; ~-**pressure water blasting** *n* MARINE Hochdruckreinigung *f*; ~-**pressure water jet** *n* SOIL *soil washing* Hochdruckwasserstrahl *m* (*HDW-Strahl*); ~-**pressure water jet method** *n* SOIL *soil washing* Hochdruckwasserstrahlverfahren *nt*; ~-**pressure water jet pipe** *n* SOIL Hochdruckwasserstrahlrohr *nt*; ~-**pressure water jet torch** *n* WATER SUPP *cleaning of heat exchanger* Hochdruckwasserstrahlgerät *nt*; ~-**pressure wet oxidation** *n* WATER POLL Hochdruck-Naßoxidation *f*; ~-**rate** *adj* WASTE, WATER POLL, WATER SUPP hochbelastet; ~-**rate activated sludge tank** *n* WATER POLL hochbelastetes Belebungsbecken *nt*; ~-**rate coalescence oil separator** *n* WATER POLL Hochleistungskoaleszer *m*; ~-**rate digester** *n* WATER POLL Hochleistungsfaulraum *m*, hochbelasteter Faulraum *m*; ~-**rate filtration** *n* AIR, WATER POLL, WATER SUPP Hochleistungsfilterung *f*, Hochleistungsfiltration *f*; ~-**rate process** *n* WATER POLL *activated sludge* hochbelastetes Verfahren *nt*; ~-**rate trickling filter** *n* WATER POLL hochbelasteter Tropfkörper *m*; ~ **security** *n* SAFETY, SOIL *of site* Hochsicherheit *f*; ~-**security landfill** *n* WASTE Hochsicherheitsdeponie *f*; ~ **soil permeability** *n* SOIL hohe Bodendurchlässigkeit *f*; ~-**speed inspection** *n* WASTE Schnellprüfung *f*; ~-**speed mixer** *n* WASTE, WATER SUPP High-Speed-Mischer *m*; ~-**tech scrap** *n* WASTE High-Tech-Schrott *m*; ~-**temperature corrosion** *n* WASTE Hochtemperaturkorrosion *f*; ~-**temperature gasification** *n* AIR, WASTE Hochtemperaturvergasung *f* (*HTV*); ~-**temperature heat** *n* RENEW ENERG *from solar energy* Hochtemperaturwärme *f*; ~-**temperature incineration** *n* WASTE, WATER POLL Hochtemperaturverbrennung *f*; ~-**temperature plasma process** *n* SOIL *originally for destruction of chemical weapons* Plasma-Ultrahochtemperaturverfahren *nt*; ~-**temperature-resistant oil** *n* RENEW ENERG hochtemperaturbeständiges Öl *nt*; ~-**velocity gas stream** *n* AIR Hochgeschwindigkeitsgasstrom *m*; ~-**voltage aggregate** *n* AIR Hochspannungsaggregat *nt*; ~-**voltage discharge electrode** *n* AIR Hochspannungs-Ausströmelektrode *f*, Hochspannungs-Entladungselektrode *f*; ~-**voltage ESP** *n* AIR Hochspannungselektrofilter *nt*; ~-**voltage generator** *n* AIR Hochspannungserzeuger *m*; ~-**voltage line** *n* RENEW ENERG Hochspannungsleitung *f*; ~-**volume sampler** *n* ANALYSIS großvolumiges Luftprobenahmegerät *nt*; ~ **water** *n* MARINE, WATER SUPP Hochwasser *nt*; ~ **water consumption** *n* WATER SUPP hoher Wasserverbrauch *m*; ~-**water overflow** *n* WATER SUPP Hochwasserüberlauf *m*

hindered: ~ **settling** *n* WATER POLL gehemmtes Absetzen *nt*

hindrance *n* SAFETY Behinderung *f*

historical: ~ **data** *n pl* ANALYSIS Stammdaten *nt pl*

HMV *abbr* (*hourly mean value*) ANALYSIS stündlicher Durchschnittswert *m*

hoist *n* SAFETY Flaschenzug *m*, Hebezug *m*

holding: ~ **capacity** *n* WASTE, WATER SUPP Aufnahmefähigkeit *f*

hollow: ~ **cylinder** *n* AIR *rotating adsorber* Hohlzylinder *m*

homogeneous: ~ **catalyst** *n* AIR homogener Katalysator *m*; ~ **wastes** *n pl* WASTE homogener Abfall *m*

homogenization *n* SOIL, WASTE, WATER SUPP Homogenisierung *f*

homogenize *vt* SOIL homogenisieren, WASTE, WATER SUPP homogenisieren, vereinheitlichen

homopolar *adj* WATER POLL unipolar

homopolymer *n* WASTE Homopolymer *nt*, Homopolymerisat *nt*

honeycomb: ~ **catalyst** *n* AIR Wabenrohrkatalysator *m*; ~ **element** *n* AIR *catalyst* Wabenkörperelement *nt*; ~-**shaped channel** *n* AIR *catalyst* wabenförmiger Kanal *m*

hood *n* AIR Abzug *m*, Haube *f*

hopper *n* WATER POLL Trichter *m*; ~-**bottom tank** *n* WATER

POLL Trichterbecken *nt*; **~-furnace feed chute** *n* WASTE Müllaufgabetrichter *m*
horizontal *adj* GEN horizontal; **~ axis wind turbine** *n* (*HAWT*) RENEW ENERG Horizontalachsenturbine *f*; **~ bar screen** *n* WATER POLL Radialrechen *m*; **~ blanket-type collection element** *n* WASTE horizontalflächiges Fassungselement *nt*; **~ degassing** *n* WASTE, WATER SUPP horizontale Entgasung *f*; **~ feeder ram** *n* WASTE, WATER SUPP horizontaler Dosierstößel *m*; **~ filtering well** *n* WATER SUPP Horizontalfilterbrunnen *m* (*HFB*); **~ filter well** *n* WATER SUPP Horizontalfilterbrunnen *m* (*HFB*); **~ fixed-bed adsorber** *n* AIR horizontaler Festbettadsorber *m*; **~ gas drainage** *n* SOIL, WASTE, WATER SUPP horizontale Gasdränage *f*; **~ gas migration** *n* SOIL, WASTE, WATER SUPP *through soil* horizontale Gaswanderung *f*; **~ linear-shaped collection element** *n* WASTE horizontales linienförmiges Fassungselement *nt*; **~ moving-bed adsorber** *n* AIR horizontaler Wanderbettadsorber *m*; **~ percolation well** *n* WASTE, WATER SUPP horizontaler Sickerbrunnen *m*; **~ surface installation** *n* WASTE, WATER SUPP horizontaler Flächeneinbau *m*; **~ well** *n* WATER SUPP horizontaler Brunnen *m*
hospital: **~ waste** *n* WASTE Krankenhausabfall *m*, krankenhausspezifischer Abfall *m*
hot: **~-air radiation heating system** *n* SAFETY Heißluftheizung *f*; **~-dipped** *adj* WASTE, WATER SUPP feuerverzinkt; **~ dry rock** *n* (*HDR*) RENEW ENERG Hot-dry-rock *nt*; **~-galvanized** *adj* WASTE, WATER SUPP feuerverzinkt; **~-galvanized steel** *n* WASTE feuerverzinkter Stahl *m*; **~-galvanized steel pipe** *n* WASTE, WATER SUPP feuerverzinktes Stahlrohr *nt*; **~ gas** *n* AIR, SOIL Heißgas *nt*; **~ gas compressor** *n* AIR Heißgasverdichter *m*; **~ gas dust removal** *n* AIR *incinerator gas* Heißgasentstaubung *f*; **~ gas filter** *n* AIR Heißgasfilter *nt*; **~ gas injection** *n* SOIL Heißgasinjektion *f*; **~-rolled wide strip** *n* WASTE Warmbreitband *nt*; **~ rolling work** *n* WASTE Warmwalzwerk *nt*; **~ sampling gas** *n* ANALYSIS heißes Probegas *nt*; **~ soak loss** *n* AIR Wärmeverlust *m* nach Abstellen des Motors; **~-tubing** *n* (*ht*) AIR Hot-tubing *nt*; **~ waste gas** *n* AIR heißes Abgas *nt*; **~ water heater** *n* AIR *thermal combustion* Heißwassererhitzer *m*; **~ water source** *n* RENEW ENERG Heißwasserquelle *f*; **~ water washing** *n* MARINE Heißwasserwaschen *nt*; **~-wire anemometer** *n* ANALYSIS Hitzdrahtanemometer *nt*
hourly *adj* ANALYSIS stündlich; **~ consumption** *n* WATER SUPP Stundenverbrauch *m*; **~ mean value** *n* (*HMV*) ANALYSIS stündlicher Durchschnittswert *m*; **~ value** *n* ANALYSIS Stundenwert *m*
house: **~ collection** *n* WASTE Haussammlung *f*
household: **~ wastewater** *n* WATER POLL Abwasser *nt* aus Haushaltungen
housing *n* RENEW ENERG *solar collector* Gehäuse *nt*
HPLC *abbr* (*high-performance liquid chromatography*) ANALYSIS HPLC (*Hochleistungs-Flüssigkeitschromatographie*)
HPLC-MS *abbr* (*high-performance liquid chromatography mass spectrometry*) ANALYSIS HPLC-MS (*Hochleistungs-Flüssigkeitschromatographie in Kopplung mit Massenspektrometrie*)
HSGC *abbr* (*headspace gas chromatography*) ANALYSIS HSGC (*Headspace-Gaschromatographie*)
ht *abbr* (*hot-tubing*) AIR Hot-tubing *nt*
human: **~-induced** *adj* SOIL menscheninduziert; **~-induced soil degradation** *n* SOIL menscheninduzierte Bodendegradation *f*; **~ settlement waste** *n* WASTE Siedlungsabfall *m*
humic: **~ acid** *n* SOIL, WASTE, WATER POLL Huminsäure *f*; **~ mass** *n* SOIL, WASTE, WATER POLL Huminstoff *m*; **~ mass formation** *n* SOIL, WASTE, WATER POLL Huminstoffbildung *f*
humidity *n* AIR, ANALYSIS Feuchtigkeit *f*, Feuchte *f*; **~ content** *n* ANALYSIS *of gases* Feuchtgehalt *m*; **~ content determination** *n* ANALYSIS Feuchtgehaltsbestimmung *f*; **~ measurement** *n* ANALYSIS Feuchtemessung *f*
humification *n* SOIL, WASTE, WATER POLL Humifizierung *f*
humus *n* WASTE, WATER POLL Humus *m*; **~-containing soils** *n pl* SOIL humushaltiger Boden *m*; **~ formation** *n* WATER POLL Humusbildung *f*; **~ tank** *n* WATER POLL Humusbehälter *m*
hundred: **~ points calibration** *n* ANALYSIS 100-Punkt-Kalibrierung *f*
Huygen-Fresnel: **~ principle** *n* ANALYSIS Huygens-Fresnelsches Prinzip *nt*
hybrid: **~ scrubber** *n* AIR Kombinationswäscher *m*, Kombiwäscher *m*
hydrated: **~ lime** *n* WATER POLL Kalkhydrat *nt*
hydrate: **~ process** *n* WATER POLL *demineralization* Hydratverfahren *nt*
hydraulic *adj* GEN hydraulisch
hydraulically: **~ controlled** *adj* AIR hydraulisch gesteuert; **~ controlled flap** *n* AIR hydraulisch gesteuerte Klappe *f*; **~ mixed flocculator** *n* WATER POLL hydraulisch durchmischter Flockungsreaktor *m*; **~ operated** *adj* WASTE, WATER SUPP hydraulisch betrieben; **~ operated feeder ram** *n* WASTE, WATER SUPP hydraulisch betriebener Dosierstößel *m*; **~ operated tipping ramp** *n* WASTE hydraulisch betriebene Kippbühne *f*
hydraulic: **~ classifier** *n* SOIL hydraulischer Klassierer *m*; **~ diameter** *n* WASTE, WATER SUPP hydraulischer Durchmesser *m*; **~ fluid** *n* RENEW ENERG Hydraulikflüssigkeit *f*; **~ jump** *n* WATER SUPP Wassersprung *m*; **~ loading** *n* WATER POLL hydraulische Belastung *f*; **~ loss** *n* WASTE, WATER POLL, WATER SUPP hydraulischer Verlust *m*; **~ press** *n* WASTE hydraulische Presse *f*; **~ pressure** *n* WATER SUPP Wasserdruck *m*; **~ radius** *n* WASTE, WATER SUPP hydraulischer Radius *m*; **~ ram** *n* WATER POLL Stoßheber *m*, hydraulischer Widder *m*; **~ transmission** *n* RENEW ENERG hydraulischer Antrieb *m*
hydrocarbon *n* (*HC*) AIR, MARINE, WATER POLL, WATER SUPP Kohlenwasserstoff *m* (*KW-Stoff*)
hydrocarbonate *n* WATER SUPP Hydrogencarbonat *nt*
hydrocarbon: **~ catalyst** *n* AIR Kohlenwasserstoff-Katalysator *m*; **~ slick** *n* WATER SUPP Kohlenwasserstoffteppich *m*
hydrochlorofluorocarbon *n* (*HCFC*) AIR teilhalogenierte Fluorchlorkohlenwasserstoffe *m pl* (*HFCKW*)
hydrocyclone *n* AIR, SOIL, WATER POLL, WATER SUPP Hydrozyklon *m*; **~ battery** *n* AIR Hydrozyklonbatterie *f*
hydrodynamic *adj* SOIL hydrodynamisch; **~ behavior in subsoil** *AmE*, **~ behaviour in subsoil** *BrE* *n* SOIL hydrodynamische Verhältnisse *nt pl* im Untergrund
hydrodynamics *n pl* WATER SUPP Hydrodynamik *f*
hydroelectric: **~ dam** *n* RENEW ENERG Wasserkrafttalsperre *f*; **~ plant** *n* RENEW ENERG, WATER SUPP Wasserkraftwerk *nt*
hydrogen *n* AIR, WASTE, WATER SUPP Wasserstoff *m*
hydrogenate *vt* AIR, RENEW ENERG, WASTE hydrieren
hydrogenation *n* AIR, RENEW ENERG, WASTE *plastics, solid waste* Hydrierung *f*
hydrogen: **~ fluoride** *n* AIR Fluorwasserstoff *m*; **~ peroxide** *n* AIR *liquid oxidant*, WATER POLL *wastewater disinfection* Wasserstoffperoxid *nt*; **~ sulfide** *AmE*, **~ sulphide** *BrE* *n* AIR, WASTE Schwefelwasserstoff *m*
hydrogeological: **~ analysis** *n* SOIL, WATER SUPP hydrogeologische Untersuchung *f*
hydrogeology *n* SOIL, WATER SUPP Hydrogeologie *f*
hydrokinetic *adj* WATER POLL hydrokinetisch
hydrolift: **~ reactor** *n* AIR Hydroliftreaktor *m*
hydrological *adj* WASTE, WATER SUPP hydrologisch; **~ balance** *n* POLICY, WASTE, WATER SUPP hydrologische Bilanz *f*, Wasserbilanz *f*; **~ cycle** *n* WASTE, WATER SUPP Wasserkreislauf *m*, hydrologischer Zyklus *m*
hydrology *n* WATER SUPP Gewässerkunde *f*, Hydrologie *f*
hydrolysable *adj* *BrE* RENEW ENERG, WASTE, WATER POLL, WATER SUPP hydrolisierbar

hydrolyse *vt BrE* RENEW ENERG, WASTE, WATER POLL, WATER SUPP hydrolysieren

hydrolysed *adj BrE* RENEW ENERG, WASTE, WATER POLL, WATER SUPP hydrolysiert; ~ **raw sludge** *n BrE* WATER POLL hydrolysierter Rohschlamm *m*

hydrolysis *n* RENEW ENERG *biomass conversion*, WASTE, WATER POLL, WATER SUPP Hydrolyse *f*

hydrolytic *adj* RENEW ENERG, WASTE, WATER POLL, WATER SUPP hydrolytisch

hydrolyzable *AmE see hydrolysable BrE*

hydrolyze *AmE see hydrolyse BrE*

hydrolyzed *AmE see hydrolysed BrE*

hydrometry *n* WATER SUPP Hydrometrie *f*

hydrophilic *adj* AIR, WATER POLL hydrophil, wasseraufnehmend

hydrophobic *adj* AIR, WATER POLL hydrophob, wasserabstoßend

hydropower *n* RENEW ENERG Wasserkraft *f*; ~ **plant** *n* RENEW ENERG Wasserkraftwerk *nt*; ~ **resources** *n pl* RENEW ENERG Wasserkraftressourcen *f pl*; ~ **supply** *n* RENEW ENERG Wasserkraftangebot *nt*

hydroscience *n* WATER SUPP Gewässerkunde *f*, Wissenschaft *f* der Gewässerkunde

hydrosphere *n* SOIL, WATER SUPP Hydrosphäre *f*

hydrostatic *adj* WATER POLL hydrostatisch; ~ **barrier** *n* WATER POLL *of wastewater from saltwater intrusion* hydrostatische Barriere *f*; ~ **pressure** *n* WASTE, WATER POLL, WATER SUPP hydrostatischer Druck *m*; ~ **pressure difference** *n* WATER POLL hydrostatische Druckdifferenz *f*

hydrostatics *n pl* WATER SUPP Hydrostatik *f*

hydrothermal *adj* RENEW ENERG *geothermal energy resource* hydrothermal

hydroxide: ~ **sludge** *n* WATER POLL, WATER SUPP Hydroxidschlamm *m*

hygrometer *n* AIR, ANALYSIS, WASTE, WATER SUPP Hygrometer *nt*

hygrometry *n* AIR, ANALYSIS, WASTE, WATER SUPP Hygrometrie *f*

hygroscope *n* AIR, ANALYSIS, WASTE, WATER SUPP Hygroskop *nt*

hygroscopic *adj* AIR, ANALYSIS, WASTE, WATER SUPP hygroskopisch

hygroscopicity *n* AIR, ANALYSIS, WASTE, WATER SUPP Hygroskopizität *f*

hygroscopic: ~ **water** *n* WASTE, WATER SUPP hygroskopisches Wasser *nt*

hypersonic: ~ **range** *n* NOISE Hyperschallbereich *m*

hypochlorination *n* WATER POLL Hypochlorung *f*

hypolimnion *n* WATER POLL Hypolimnion *nt*

I

IC *abbr* (*ion chromatography*) ANALYSIS IC (*Ionenchromatographie*)
ICP-MS *abbr* (*inductively coupled plasma mass spectrometry*) ANALYSIS ICP-MS (*induktiv gekoppeltes Plasma in Kopplung mit Massenspektrometrie*)
ICP-OES *abbr* (*inductively coupled plasma with optical emission spectrometry*) ANALYSIS ICP-OES (*Emissionsspektrometrie mit induktiv gekoppeltem Plasma*)
ID (*induced draft AmE, induced draught BrE*) *abbr* AIR Saugzug *m*
ideal: ~ **gas** *n* AIR ideales Gas *nt*, perfektes Gas *nt*
identifiability *n* ANALYSIS Nachweisbarkeit
identifiable *adj* ANBALYSIS nachweisbar
identification *n* ANALYSIS Nachweis *m*, *of sample* Identifizierung *f*, SAFETY *labelling* Kennzeichnung *f*; ~ **of contents** *n* SAFETY *industrial gas containers* Inhaltskennzeichnung *f*; ~ **frequency** *n* ANALYSIS Nachweishäufigkeit *f*; ~ **limit** *n* ANALYSIS Nachweisgrenze *f*; ~ **method** *n* ANALYSIS Nachweismethode *f*; ~ **reaction** *n* ANALYSIS Nachweisreaktion *f*
identify *vt* ANALYSIS identifizieren, nachweisen, SAFETY kennzeichnen
ID: ~ **fan** *n* AIR Saugzuggebläse *nt*, Saugzugventilator *m*
IE *abbr* (*ion exchange*) ANALYSIS, WATER POLL, WATER SUPP IA (*Ionenaustausch*)
ignitability *n* AIR, WASTE Zündempfindlichkeit *f*
ignite *vt* AIR, ANALYSIS, WASTE, WATER POLL entzünden, glühen
ignition *n* AIR, ANALYSIS WASTE, WATER POLL Glühzündung *f*, Zündung *f*; ~ **arch** *n* AIR Zündgewölbe *nt*; ~ **gas** *n* AIR *thermal combustion* Zündgas *nt*; ~ **loss** *n* AIR, WASTE Glühverlust *m*; ~ **point** *n* AIR Zündpunkt *m*; ~ **residue** *n* ANALYSIS, WATER POLL Glührückstand *m*; ~ **temperature** *n* AIR *thermal waste gas treatment* Zündtemperatur *f*, Zündungstemperatur *f*
illegal *adj* POLICY, WASTE, WATER POLL illegal, ordnungswidrig; ~ **barging of spent acid** *n* WASTE, WATER SUPP illegale Dünnsäureverklappung *f*; ~ **dumping** *n* POLICY wilde Ablagerung *f*; ~ **dumping of spent acid** *n* MARINE illegale Dünnsäureverklappung *f*; ~ **dump site** *n* WASTE wilde Müllkippe *f*
image: ~ **analysis** *n* ANALYSIS Bildanalyse *f*
Imhoff: ~ **sedimentation cone** *n* WATER POLL Imhofftrichter *m*; ~ **tank** *n* WATER POLL *Emscher tank* Imhoffbrunnen *m*
immediate: ~ **measures** *n pl* POLICY Sofortmaßnahmen *f pl*
immersion: ~ **stirrer** *n* ANALYSIS *pH-meter* Eintauchrührer *m*
immiscible *adj* WATER POLL unvermischbar; ~ **liquids** *n pl* WATER POLL unvermischbare Flüssigkeiten *f pl*
immobilization *n* ANALYSIS, SOIL, WASTE, WATER POLL Immobilisierung *f*
immobilize *vt* ANALYSIS, SOIL, WASTE, WATER POLL immobilisieren
immobilized *adj* ANALYSIS, SOIL, WASTE, WATER POLL immobilisiert; ~ **microorganism** *n* WATER POLL immobilisierter Mikroorganismus *m*
immunoassay *n* ANALYSIS Immunversuch *m*; ~ **test** *n* ANALYSIS Immunoanalyse *f*
impact 1. *n* AIR Aufprall *m*, Prall *m*, *collision* Zusammenprall *m*, MANAGE Auswirkung *f*, WASTE Prall *m*, *harmful* Belastung *f*, WATER POLL, WATER SUPP *adverse* Belastung *f*, Prall *m*; **2.** *vt* AIR aufprallen auf, aufschlagen auf, auftreffen auf, zusammenprallen mit
impact: ~ **analysis** *n* MANAGE Auswirkungsanalyse *f*, Bewertung *f* von Auswirkungen; ~ **crusher** *n* WASTE Prallmühle *f*
impaction *n* AIR Aufprall *m*, Zusammenprall *m*, ANALYSIS *measurement for dust* Impaktion *f*
impact: ~ **ionization** *n* AIR, ANALYSIS Stoßionisation *f*; ~ **on groundwater** *n* WATER SUPP Grundwasserbelastung *f*
impactor *n* AIR, ANALYSIS Impaktor *m*; ~ **nozzle** *n* AIR Pralldüse *f*
impact: ~ **study** *n* MANAGE Verträglichkeitsprüfung *f*
impair *vt* GEN beeinträchtigen
impairment *n* GEN Beeinträchtigung *f*
impediment: ~ **to recycling** *n* WASTE Recyclinghindernis *nt*
impeller *n* AIR Laufrad *nt*; ~ **anemometer** *n* ANALYSIS Flügelradanemometer *nt*; ~**-type mixer** *n* WATER POLL, WATER SUPP Rührbehälter *m*
impermeable *adj* SAFETY, SOIL, WASTE, WATER SUPP undurchlässig; ~ **barrier** *n* SOIL undurchlässige Abdichtungswand *f*
impervious *adj* SAFETY, SOIL, WASTE, WATER SUPP undurchlässig
impinge *vi* AIR aufprallen, aufschlagen, auftreffen, zusammenprallen
impingement *n* AIR Aufprall *m*; ~ **plate scrubber** *n* AIR Prallplattenwäscher *m*; ~ **separator** *n* AIR Aufschlag-Abscheider *m*
impinge: ~ **method** *n* ANALYSIS *measurement of air quality* Aufprallmethode *f*
impinger *n* ANALYSIS Impaktor *m*
implement *vt* POLICY umsetzen, vollziehen
implementation *n* POLICY Umsetzung *f*, Vollzug *m*; ~ **procedure** *n* POLICY *in court* Normenkontrollverfahren *nt*
import *vt* RENEW ENERG importieren
impounding: ~ **reservoir** *n* WATER SUPP Stausee *m*
impregnate *vt* AIR, ANALYSIS imprägnieren
impregnated *adj* AIR, ANALYSIS imprägniert; ~ **filter paper** *n* ANALYSIS imprägniertes Filterpapier *nt*
impregnation *n* AIR, ANALYSIS Imprägnierung *f*
impulse: ~ **noise** *n* NOISE Impulslärm *m*; ~ **sound level meter** *n* ANALYSIS Impulsschallpegelmesser *m*
impure *adj* AIR, WASTE, WATER POLL, WATER SUPP unrein; ~ **waters** *n pl* WASTE, WATER SUPP unreine Gewässer *nt pl*
impurities *n pl* AIR, WASTE, WATER POLL, WATER SUPP Verunreinigungen *f pl*
impurity *n* AIR, WASTE, WATER POLL, WATER SUPP Unreinheit *f*
inactivate *vt* WASTE, WATER SUPP inaktivieren
inactivation *n* WASTE, WATER SUPP Inaktivierung *f*
inappropriate: ~ **cropping method** *n* SOIL unsachgemäßer Ackerbau *m*
incandescent: ~ **lamp** *n* ANALYSIS Glühfadenlampe *f*, Wärmelichtquelle *f*
incentive *n* MANAGE, POLICY Anreiz *m*
incident: ~**-dependent** *adj* ANALYSIS ereignisabhängig
incinerate *vt* GEN verbrennen
incinerated *adj* AIR, RENEW ENERG, WASTE verbrannt; ~ **liquid** *n* WATER POLL verbrannte Flüssigkeit *f*
incineration *n* AIR, RENEW ENERG, SAFETY, WASTE Verbrennung *f*; ~ **ash** *n* WASTE Verbrennungsrückstand *m*; ~ **calorimeter** *n* ANALYSIS Verbrennungskalorimeter *nt*; ~ **capacity** *n* WASTE Verbrennungsleistung *f*; ~ **with energy recovery** *n* WASTE Verbrennung *f* mit Energierückgewinnung; ~ **enthalpy** *n* WASTE Verbrennungsenthalpie *f*; ~ **operation** *n* AIR, WASTE Verbrennungsvorgang *m*; ~ **plant** *n* AIR

Verbrennungsanlage *f*; ~ **plant with heat use** *n* RENEW ENERG, WASTE Verbrennungsanlage *f* mit Wärmenutzung; ~ **plant without heat use** *n* RENEW ENERG, WASTE Verbrennungsanlage *f* ohne Wärmenutzung; ~ **residue** *n* WASTE Verbrennungsrückstand *m*; ~ **stock** *n* WASTE Verbrennungsgut *nt*; ~ **test** *n* ANALYSIS Veraschungstest *m*; ~ **of wastewater sludge in cement furnaces** *n* WATER POLL Klärschlammverbrennung *f* in Zementöfen; ~ **of wastewater sludge in coal furnaces** *n* WATER POLL Klärschlammverbrennung *f* in Steinkohlefeuerungen; ~ **of wastewater sludge in lignite furnaces** *n* WATER POLL Klärschlammverbrennung *f* in Braunkohlefeuerungen; ~ **zone** *n* AIR, WASTE, WATER POLL Verbrennungszone *f*

incinerator *n* AIR, RENEW ENERG, WASTE Abfallverbrennungsofen *m*; ~ **charge** *n* WASTE Brenngut *nt*; ~ **gas** *n* WASTE Verbrennungsgas *nt*

inclination *n* WASTE, WATER SUPP Neigung *f*

inclined: ~ **infiltration** *n* WASTE Schräginfiltration *f*; ~ **sorting machine** *n* WASTE Schrägsortiermaschine *f*; ~ **spring** *n* WATER SUPP Hangquelle *f*; ~ **table** *n* WASTE Schrägtisch *m*; ~ **tube evaporator** *n* WASTE Schrägrohrverdampfer *m*

incombustibility *n* AIR, WASTE Unverbrennbarkeit *f*

incombustible *adj* AIR, WASTE unverbrennbar; ~ **residue** *n* AIR unverbrennbarer Rückstand *m*, WASTE inertisierte Rückstände *m pl*

incombustibles *n pl* AIR, WASTE *after incineration* Unverbranntes *nt*, Unverbrennliches *nt*

income: ~**-neutral** *adj* POLICY einkommensneutral; ~**-neutral eco-tax reform** *n* POLICY einkommensneutrale ökologische Steuerreform *f*

incomplete *adj* AIR, WASTE unvollkommen, unvollständig; ~ **burning** *n* AIR, WASTE unvollkommene Verbrennung *f*, unvollständige Verbrennung *f*; ~ **combustion** *n* AIR, WASTE unvollkommene Verbrennung *f*, unvollständige Verbrennung *f*

incremental: ~ **heating** *n* AIR zunehmende Erwärmung *f*

incrustation *n* SOIL Inkrustation *f*, WASTE, WATER SUPP Inkrustation *f*, Verockerung *f*

incubate *vt* ANALYSIS inkubieren

incubated *adj* ANALYSIS inkubiert

incubation *n* ANALYSIS Inkubation *f*; ~ **period** *n* ANALYSIS Inkubationszeit *f*

indicating: ~ **electrode** *n* ANALYSIS Indikatorelektrode *f*; ~ **instrument** *n* ANALYSIS Anzeigeinstrument *nt*, Anzeigevorrichtung *f*

indication: ~ **error** *n* ANALYSIS Fehlanzeige *f*

indicator *n* ANALYSIS, WATER POLL, WATER SUPP Indikator *m*; ~ **organism** *n* WATER POLL Indikatororganismus *nt*; ~ **paper** *n* ANALYSIS Indikatorpapier *nt*; ~ **range** *n* ANALYSIS Umschlagbereich *m*; ~ **solution** *n* ANALYSIS Indikatorlösung *f*; ~ **species** *n* ANALYSIS, WASTE, WATER SUPP Kontrollart *f*

indirect *adj* AIR, ANALYSIS, WATER POLL indirekt; ~ **condenser** *n* AIR *surface condenser* Oberflächenkondensator *m*; ~ **discharge** *n* WATER POLL Indirekteinleitung *f*; ~ **discharger** *n* WATER POLL Indirekteinleiter *m*; ~ **drying** *n* WATER POLL indirekte Trocknung *f*; ~ **emission sources** *n pl* AIR Indirekteinleiterquellen *f pl*; ~ **groundwater recharge** *n* WASTE, WATER SUPP indirekte Grundwasseranreicherung *f*; ~**-heated rotary kiln** *n* SOIL indirekt beheizter Drehrohrofen *m*; ~ **heating** *n* SOIL indirekte Beheizung *f*; ~ **industrial wastewater discharge** *n* WATER POLL indirekte Industrieabwassereinleitung *f*; ~ **measurement method** *n* ANALYSIS indirekte Meßmethode *f*

individual: ~ **droplet** *n* AIR Einzeltröpfchen *nt*; ~ **dust removal apparatus** *n* SAFETY Einzelstaubabscheider *m*; ~ **sample** *n* ANALYSIS Einzelprobe *f*; ~ **substance analysis** *n* ANALYSIS Einzelstoffanalyse *f*

indoor: ~ **air pollution** *n* AIR Innenraumluftverunreinigung *f*; ~ **shielding** *n* NOISE Abschirmung *f* in Räumen

induced: ~ **draft** *AmE*, ~ **draught** *BrE* *n* AIR Saugzug *m*; ~**-draft fan** *AmE*, ~**-draught fan** *BrE* *n* AIR Saugzuggebläse *nt*, Saugzugventilator *m*

induction *n* AIR, WATER POLL Induktion *f*; ~ **generator** *n* RENEW ENERG *wind turbine* Asynchrongenerator *m*; ~ **heating** *n* AIR Induktionserwärmung *f*, Induktionsheizung *f*

inductively: ~ **coupled plasma mass spectrometry** *n* (*ICP-MS*) ANALYSIS induktiv gekoppeltes Plasma *nt* in Kopplung mit Massenspektrometrie (*ICP-MS*); ~ **coupled plasma with optical emission spectrometry** *n* (*ICP-OES*) ANALYSIS Emissionsspektrometrie *f* mit induktiv gekoppeltem Plasma (*ICP-OES*)

industrial *adj* GEN gewerblich, industriell; ~ **accident** *n* SAFETY Industrieunfall *m*; ~ **bulk container system** *n* WASTE industrielles Schüttgutcontainersystem *nt*; ~ **collection** *n* WASTE gewerbliche Sammlung *f*; ~ **effluent** *n* WATER POLL *discharge*, WATER SUPPgewerbliches Abwasser *nt*, Industrieabwasser *nt*; ~ **emissions** *n pl* AIR industrielle Emissionen *f pl*; ~ **insurance** *n* POLICY, SAFETY Industrieversicherung *f*; ~ **landfill** *n* WASTE Industriemülldeponie *f*; ~ **packing** *n* WASTE Industrieverpackung *f*; ~ **process** *n* WASTE, WATER SUPP Betriebsablauf *m*; ~ **recycling** *n* WASTE industrielles Recycling *nt*; ~ **scrap** *n* WASTE Ausschuß *m*, Verarbeitungsabfälle *m pl*; ~ **site** *n* SOIL Industriestandort *m*; ~ **source** *n* AIR, WASTE, WATER SUPP gewerbliche Quelle *f*; ~ **standard** *n* GEN Industrienorm *f*; ~ **waste** *n* WASTE Gewerbeabfall *m*, Gewerbemüll *m*, Industrieabfall *m*, gewerblicher Abfall *m*; ~ **wastewater** *n* WATER POLL Industrieabwasser *nt*, gewerbliches Abwasser *nt*, WATER SUPP Industrieabwasser *nt*; ~ **wastewater purification** *n* WATER POLL, WATER SUPP Industrieabwasserreinigung *f*; ~ **wastewater sludge** *n* WATER POLL gewerblicher Klärschlamm *m*, industrieller Klärschlamm *m*; ~ **water** *n* WASTE Brauchwasser *nt*, WATER POLL Betriebswasser *nt*, Brauchwasser *nt*, WATER SUPP Brauchwasser *nt*; ~ **water consumption** *n* WATER SUPP gewerblicher Wasserverbrauch *m*; ~ **water supply** *n* WATER SUPP Industriewasserversorgung *f*

industry *n* GEN Industrie *f*; ~**-specific** *adj* GEN industriespezifisch; ~**-specific purification** *n* GEN industriespezifische Reinigung *f*

inert: ~ **closure** *n* ANALYSIS *bottle* inerter Verschluß *m*; ~ **fraction** *n* WASTE Inertfraktion *f*; ~ **gas** *n* AIR Inertgas *nt*; ~ **gas combustion chamber** *n* AIR Inertgasbrennkammer *f*; ~ **gas flow** *n* AIR Inertgasstrom *m*

inertia *n* AIR, WASTE, WATER SUPP Trägheit *f*

inertial: ~ **force** *n* AIR Trägheitskraft *f*; ~ **separator** *n* AIR Inertialabscheider *m*, Trägheitsabscheider *m*

inerting *n* AIR, WASTE Inertisierung *f*

inert: ~ **landfill** *n* WASTE Inertdeponie *f*; ~ **material** *n* AIR Inertmaterial *nt*, WASTE Ballastmaterial *nt*, Inertmaterial *nt*, WATER SUPP Ballastmaterial *nt*; ~ **waste** *n* WASTE, WATER POLL inerter Abfall *m*

inexhaustibility *n* RENEW ENERG *of energy sources* Unerschöpflichkeit *f*

inexhaustible *adj* RENEW ENERG unerschöpflich; ~ **energy source** *n* RENEW ENERG unerschöpfliche Energiequelle *f*; ~ **energy stream** *n* RENEW ENERG unerschöpfbarer Energiestrom *m*

infectious: ~ **hospital waste** *n* WASTE infektiöser Krankenhausabfall *m*; ~ **waste** *n* MARINE, WASTE, WATER POLL infektiöser Abfall *m*, pathogener Abfall *m*

infiltrate *vt* WASTE, WATER SUPP eindringen, einsickern

infiltrated: ~ **raw water** *n* WASTE, WATER SUPP infiltriertes Rohwasser *nt*

infiltration *n* WASTE, WATER POLL, WATER SUPP Infiltration *f*, Versickerung *f*, Versickern *nt*; ~ **basin** *n* WASTE Sickerbecken *nt*, WATER POLL Infiltrationsbecken *nt*, Sickerbecken *nt*, WATER SUPP Sickerbecken *nt*; ~ **gallery** *n* WASTE, WATER

SUPP Sickergalerie *f*, Sickerstollen *m*; ~ **ground** *n* WATER SUPP *water works* Infiltrationsgelände *nt*; ~ **rate** *n* WATER SUPP Infiltrationsrate *f*; ~ **slit** *n* WASTE, WATER SUPP Sickerschlitz *m*; ~ **tank** *n* WATER POLL Infiltrationsbecken *nt*; ~ **waters** *n pl* WATER SUPP Infiltrationsgewässer *nt pl*; ~ **well** *n* SOIL, WATER SUPP Infiltrationsbrunnen *m*
inflammability: ~ **limits** *n pl* ANALYSIS Zündgrenze *f*
inflammable *adj* AIR, SAFETY, WASTE entflammbar
inflatable *adj* GEN aufblasbar
inflate *vt* AIR, MARINE aufblähen, aufblasen, aufpumpen
inflation *n* AIR, MARINE Aufblasen *nt*, Aufpumpen *nt*; ~ **cuff** *n* MARINE Aufblasanschluß *m*
inflow *n* WATER POLL Zulauf *m*, WATER SUPP Zufluß *m*; ~ **from natural waters** *n* WATER SUPP Zufluß *m* von natürlichen Gewässern; ~ **into distribution system** *n* WATER SUPP Zufluß *m* zum Verteilrohrnetz; ~ **into pumping station** *n* WATER SUPP Zufluß *m* zu Pumpwerken; ~ **into storage facility** *n* WATER SUPP Zufluß *m* zu Speicheranlagen
influence: ~ **of temperature** *n* WASTE, WATER SUPP Temperatureinfluß *m*; ~ **of temperature on corrosivity** *n* WATER SUPP Temperatureinfluß *m* auf die Korrosivität; ~ **of temperature on water quality** *n* WASTE Temperatureinfluß *m* auf die Wassergüte; ~ **of temperature on water storage** *n* WATER SUPP Temperatureinfluß *m* auf die Wasserspeicherung; ~ **zone of noise** *n* NOISE Lärmzone *f*
influent *n* WATER POLL, WATER SUPP Einlauf *m*, Zulauf *m*; ~ **canal** *n* WATER POLL, WATER SUPP Zulaufkanal *m*; ~ **concentration** *n* WATER POLL Zulaufkonzentration *f*; ~ **pump works** *n* WATER POLL Einlaufhebewerk *nt*
information: ~ **instruments** *n pl* POLICY Informationsinstrumente *nt pl*
infrared *n* (*IR*) AIR, ANALYSIS Infrarot *nt* (*IR*); ~ **continuous furnace** *n* WASTE Infrarot-Durchlaufofen *m*; ~ **kiln** *n* ANALYSIS Infrarotofen *m*; ~ **radiation** *n* AIR IR-Strahlung *f*, Infrarotstrahlung *f*, WASTE Infrarotstrahlung *f*; ~ **region** *n* ANALYSIS Infrarotbereich *m*; ~ **spectrometry** *n* ANALYSIS Infrarot-Spektrometrie *f*; ~ **spectrum** *n* ANALYSIS Infrarotspektrum *nt*; ~ **thermal destruction** *n* SOIL infrarotthermische Zerstörung *f*; ~ **through-type furnace** *n* WASTE Infrarot-Durchlaufofen *m*
ingest *vt* WASTE, WATER SUPP aufnehmen
ingested *adj* WASTE, WATER SUPP aufgenommen
ingestion *n* WASTE, WATER SUPP Aufnahme *f*
inhabitant: ~ **equivalent** *n* WASTE, WATER POLL Einwohnergleichwert *m* (*EGW*)
inhibitor *n* AIR Antikatalysator *m*, Inhibitor *m*, negativer Inhibitor *m*, WATER SUPP Inhibitor *m*
in-house: ~ **recycling** *n* WASTE betriebsinternes Recycling *nt*, innerbetriebliche Verwertung *f*; ~ **water recirculation system** *n* WATER POLL innerbetrieblicher Wasserkreislauf *m*
initial: ~ **adsorption at optimum polymer dosage** *n* WATER POLL *flocculation* Anfangsadsorption *f* bei optimalem Flokkulantenzusatz
initiation: ~ **time** *n* WATER POLL Anfahrzeit *f*
inject *vt* AIR einblasen, AIR, SOIL, WASTE, WATER SUPP eindüsen, ANALYSIS, SOIL injizieren, *microorganisms and nutrients in soil* eindüsen, einspritzen, *inoculate* impfen
injection *n* AIR Einblasen *nt*, Eindüsung *f*, ANALYSIS, SOIL Eindüsung *f*, *inoculation* Einimpfung *f*, Einspritzen *nt*, Injektion *f*; ~ **drilling** *n* SOIL Injektionsbohrung *f*; ~ **fluid** *n* SOIL Injektionsflüssigkeit *f*; ~ **lance** *n* SOIL Injektionslanze *f*; ~ **molding** *AmE see injection moulding BrE*; ~ **moulding** *n BrE* WASTE Spritzgußformung *f*; ~-**moulding compound** *n BrE* WASTE Spritzmasse *f*; ~-**moulding machine** *n BrE* WASTE Spritzgußmaschine *f*; ~-**moulding pressure** *n BrE* WASTE Spritzdruck *m*; ~ **nozzle** *n* AIR Einspritzdüse *f*; ~ **plant** *n* SOIL *grouting* Injektionsanlage *f*; ~ **pressure** *n* SOIL Einpreßdruck *m*; ~ **section** *n* SOIL Einpreßschnitt *m*, Verpreßabschnitt *m*; ~ **spraying** *n* ANALYSIS *dosing of gas, solutions* Injektionsspritzen *nt*; ~ **system** *n* ANALYSIS Injektionssystem *nt*; ~ **well** *n* SOIL Injektionsbrunnen *m*, WASTE, WATER SUPP Schluckbrunnen *m*
injector *n* ANALYSIS Injektor *m*; ~ **action** *n* AIR *bag filter* Injektorwirkung *f*
injurious *adj* GEN schädlich
inland: ~ **waters** *n pl* WATER SUPP Binnengewässer *nt pl*
inlet *n* AIR Einlaß *m*, Eintritt *m*, WATER POLL, WATER SUPP Einlauf *m*, Einlaß *m*; ~ **dust burden** *n* AIR Eintrittstaubbelastung *f*; ~ **emissions** *n pl* AIR Einlaßemissionen *f pl*; ~ **structure** *n* WASTE Entnahmebauwerk *nt*, WATER POLL Einlaufbauwerk *nt*, Entnahmebauwerk *nt*, WATER SUPP Einlaufbauwerk *nt*; ~ **velocity** *n* AIR *cyclone* Eintrittsgeschwindigkeit *f*
inner: ~ **cyclone wall** *n* AIR Zykloninnenwand *f*; ~ **vortex** *n* AIR Innentrombe *f*
innocuous *adj* AIR unschädlich
innovative: ~ **technologies** *n pl* AIR, RENEW ENERG, WATER POLL, WATER SUPP innovative Technik *f*
inorganic *adj* AIR, WASTE, WATER POLL, WATER SUPP anorganisch; ~ **acid gas** *n* AIR anorganisches Säuregas *nt*; ~ **base material** *n* AIR *biofilter* anorganisches Trägermaterial *nt*; ~ **chemical** *n* WATER POLL, WATER SUPP anorganische Chemikalie *f*; ~ **contaminants** *n pl* AIR, WATER POLL anorganische Kontaminanten *nt pl*; ~ **gas** *n* AIR anorganisches Gas *nt*; ~ **pollutants** *n pl* AIR, WATER POLL anorganische Schadstoffe *m pl*; ~ **roughage** *n* WASTE, WATER SUPP *glass, ceramics, slag* anorganischer Ballaststoff *m*; ~ **salts** *n pl* WASTE, WATER SUPP anorganische Salze *nt pl*; ~ **solids** *n pl* WASTE, WATER SUPP anorganische Feststoffe *m pl*
input *n* ANALYSIS *reading into an instrument* Eingang *m*; ~ **of rotor** *n* RENEW ENERG *wind turbine* Leistungsaufnahme *f* des Rotors
inputs *n pl* MANAGE Eingangsgrößen *f pl*
insensitive *adj* AIR unempfindlich
insert *n* WASTE, WATER SUPP Einsatz *m*
insertion: ~ **loss** *n* NOISE *noise level reduction* Einfügungsdämpfung *f*
in-situ *adj* GEN in-situ; ~ **bioremediation** *n* SOIL *for hydrocarbons, chlorinated solvents, metals* In-situ-biologische Altlastensanierung *f*; ~ **denitrification** *n* WATER SUPP In-situ-Denitrifikation *f*; ~ **extraction** *n* SOIL In-situ-Extraktion *f*; ~ **method** *n* SOIL In-situ-Verfahren *nt*; ~ **solidification** *n* SOIL In-situ-Verfestigung *f*; ~ **thermal desorption** *n* SOIL In-situ thermische Desorption *f*; ~ **vitrification** *n* (*ISV*) SOIL In-situ-Verglasung *f*
insoluble *adj* WASTE, WATER POLL, WATER SUPP insolubel, unlöslich; ~ **in water** *adj* AIR, WATER POLL, WATER SUPP wasserunlöslich
inspect *vt* GEN kontrollieren, prüfen
inspection *n* GEN Inspektion *f*, Prüfung *f*; ~ **by experts** *n* ANALYSIS, POLICY, SAFETY Prüfung *f* durch Sachverständige; ~ **equipment** *n* ANALYSIS, SAFETY Kontrollvorrichtung *f*, Prüfgerät *nt*; ~ **glass** *n* ANALYSIS Sichtscheibe *f*; ~ **window** *n* ANALYSIS Prüfglas *nt*, Sichtglas *nt*
instability *n* WASTE Instabilität *f*
install *vt* AIR einbauen, installieren ANALYSIS *measuring station* einrichten, MARINE, NOISE, RENEW ENERG, SAFETY, SOIL, WATER POLL, WATER SUPP einbauen, installieren
installation *n* GEN Anlage *f*, *plant* Einbau *m*, Installation *f*; ~ **area** *n* RENEW ENERG *of solar farm* Einsatzgebiet *nt*; ~-**related** *adj* AIR anlagebezogen, ANALYSIS anlagebezogen, anlagentechnisch, GEN anlagebezogen; ~-**specific** *adj* GEN anlagenspezifisch
instantaneous: ~ **concentration** *n* WATER SUPP Grundkonzentration *f*
instant: ~ **flowmeter** *n* WATER SUPP Durchflußmesser *m*, Wassermesser *m*
institution *n* POLICY Institution *f*

instrumental: ~ **analysis** *n* ANALYSIS instrumentelle Analytik *f*
instrumentation *n* POLICY Anwendung *f*
instrument: ~ **self test** *n* ANALYSIS Geräteselbsttest *m*
instruments: ~ **of environmental policy** *n pl* POLICY umweltpolitische Instrumente *nt pl*
insulate *vt* GEN isolieren
insulated *adj* GEN isoliert
insulating: ~ **board** *n* NOISE Dämmplatte *f*; ~ **brick** *n* NOISE Dämmstein *m*; ~ **compound** *n* NOISE Isoliermasse *f*; ~ **glass** *n* NOISE Isolierglas *nt*; ~ **layer** *n* NOISE Dämmschicht *f*; ~ **material** *n* NOISE Isolierstoff *m*; ~ **properties** *n pl* NOISE Isoliereigenschaften *f pl*
insulation *n* GEN Isolation *f*, Isolierung *f*; ~ **against sound and vibration** *n* MARINE, NOISE, SAFETY Schall- und Vibrationsisolierung *f*
insulator *n* AIR Isolator *m*, ANALYSIS *semiconductor detector* Isolator *m*, Nichtleiter *m*
intake *n* AIR, WATER POLL, WATER SUPP Einlaß *m*, Einlauf *m*, *process* Entnahme *f*; ~ **air** *n* AIR Ansaugluft *f*; ~ **main** *n* WASTE, WATER SUPP Entnahmeleitung *f*; ~ **manifold** *n* AIR Einlaßkrümmer *m*; ~ **pressure** *n* WATER POLL Ansaugdruck *m*; ~ **sill** *n* WATER SUPP Einlaufschwelle *f*; ~ **structure** *n* WASTE, WATER POLL, WATER SUPP Einlaufbauwerk *nt*, Entnahmebauwerk *nt*; ~ **tunnel** *n* WASTE, WATER SUPP Entnahmestollen *m*; ~ **volume** *n* AIR, WATER POLL Ansaugvolumen *nt*; ~ **well** *n* WASTE, WATER SUPP Entnahmebrunnen *m*
intalox: ~ **saddle** *n* AIR *packing of packed-tower scrubber* Novalox-Sattelkörper *m*
integral *adj* ANALYSIS integral
integrated *adj* AIR, MANAGE, POLICY integriert; ~ **approach** *n* POLICY integrierter Ansatz *m*; ~ **environmental protection** *n* MANAGE, POLICY integrierter Umweltschutz *m*; ~ **heat exchanger system** *n* AIR integriertes Wärmeaustauschersystem *nt*; ~ **limit values** *n pl* POLICY integrierte Grenzwerte *m pl*; ~ **pollution control** *n* (*IPC*) MANAGE, POLICY integrierte Vermeidung *f* der Umweltverschmutzung (*IVU*); ~ **pollution prevention** *n* MANAGE integrierte Vorsorge *f* der Umweltverschmutzung; ~ **prevention and pollution control** *n* (*IPPC*) MANAGE, POLICY integrierte Vorsorge *f* und Vermeidung *f* der Umweltverschmutzung; ~ **system** *n* WASTE integriertes System *nt*; ~ **waste management** *n* (*IWM*) MANAGE integrierte Abfallbewirtschaftung *f*
integrating: ~ **motor** *n* ANALYSIS Meßmotor *m*
integrator *n* ANALYSIS *for quick determination* Integrator *m*
intensity *n* ANALYSIS Intensität *f*; ~ **of radiation** *n* RENEW ENERG Strahlungsintensität *f*; ~ **of UV radiation** *n* AIR Intensität *f* der UV-Strahlung
intensive: ~ **farming** *n* SOIL intensiver Ackerbau *m*; ~ **rotting** *n* WASTE Intensivrotte *f*
intentional: ~ **discharge** *n* WASTE, WATER POLL *of sewage* vorsätzliche Einleitung *f*
interaction: ~ **force** *n* AIR Wechselwirkungskräfte *f pl*
intercept *vt* AIR *particle* auffangen, WATER POLL abfangen
intercepting: ~ **sewer** *n* WATER POLL *interceptor* Abfangleitung *f*
interception *n* AIR *particle* Auffangen *nt*, WATER POLL Abfangen *nt*
interchangeable *adj* ANALYSIS austauschbar; ~ **detector** *n* ANALYSIS *gas chromatography* austauschbarer Detektor *m*; ~ **lamp** *n* ANALYSIS austauschbare Lampe *f*
interface *n* AIR Grenzfläche *f*, ANALYSIS Grenzfläche *f*, *computer* Schnittstelle *f*, MARINE Grenzfläche *f*; ~ **module** *n* ANALYSIS Schnittstellenmodul *nt*
interfacial: ~ **renewal** *n* AIR Grenzflächenerneuerung *f*; ~ **tension** *n* ANALYSIS Grenzflächenspannung *f*
interference *n* ANALYSIS Interferenz *f*; ~ **filter** *n* ANALYSIS Interferenzfilter *nt*, Interferenzlichtfilter *nt*; ~ **refractometer** *n* ANALYSIS Interferenzrefraktometer *nt*
interferential: ~ **pit gas tester** *n* ANALYSIS Grubengasinterferometer *nt*
interferogram *n* ANALYSIS Interferogramm *nt*
interferometer *n* ANALYSIS Interferometer *nt*
interferometry *n* ANALYSIS Interferenzmeßverfahren *nt*
interim *adj* SOIL, WASTE, WATER POLL vorübergehend; ~ **dewatering** *n* WASTE Zwischenentwässerung *f*; ~ **storage area** *n* WASTE Zwischenlagerfläche *f*; ~ **store** *n* WASTE Zwischenlager *nt*; ~ **tank** *n* SOIL Zwischenbehälter *m*
interior: ~ **electrode** *n* ANALYSIS Innenelektrode *f*; ~**-polished cyclone** *n* AIR innen polierter Zyklon *m*
interlocking *n* WATER POLL *of membrane* Verblockung *f*; ~ **risk** *n* WATER POLL *of membrane* Verblockungsgefahr *f*
intermediate: ~ **fraction** *n* WASTE Mittelfraktion *f*; ~ **layer** *n* NOISE *in encapsulation* Zwischenlage *f*; ~ **product** *n* WASTE Zwischenprodukt *nt*; ~ **purification** *n* WATER POLL Zwischenklärung *f*; ~ **settling** *n* WATER POLL Zwischenklärung *f*; ~ **settling tank** *n* WATER POLL Zwischenklärbecken *nt*; ~ **storage site for hazardous waste** *n* WASTE Sonderabfallzwischenlager *nt*; ~ **trickling filter** *n* WATER POLL mittelbelasteter Tropfkörper *m*
intermittent *adj* GEN diskontinuierlich, intermittierend; ~ **cleaning** *n* AIR *filter bag* diskontinuierliche Reinigung *f*, intermittierende Reinigung *f*; **with ~ denitrification** *phr* WATER POLL mit zwischengeschalteter Denitrifikation; ~ **spring** *n* WATER SUPP Karstquelle *f*
internal: ~**-combustion engine** *n* AIR, WASTE, WATER POLL *in large pumping stations* Verbrennungsmotor *m*; ~**-combustion turbine** *n* AIR, WASTE, WATER POLL Verbrennungskraftmaschine *f*, Verbrennungsmotor *m*; ~ **eddy** *n* WASTE, WATER SUPP *centrifugal field* Innenwirbel *m*
internalization *n* POLICY Internalisierung *f*; ~ **of external costs** *n* POLICY Internalisierung *f* externer Kosten; ~ **of external effects** *n* POLICY Internalisierung *f* externer Effekten; ~ **of externality** *n* POLICY Internalisierung *f* externer Effekten
internal: ~ **pressure** *n* AIR *bag filter* Innendruck *m*; ~ **recirculation** *n* WATER POLL interne Rezirkulation *f*; ~ **source** *n* WATER SUPP Binnenquelle *f*, Inlandsquelle *f*
international: ~ **law** *n* POLICY Völkerrecht *nt*
internationally: ~ **tradeable emissions rights** *n pl* POLICY international verhandelbare Emissionsrechte *nt pl*
international: ~ **negotiation** *n* POLICY internationale Verhandlung *f*
International: ~ **Referral System** *n* (*IRS*) POLICY Internationales Referenzsystem *nt* (*IRS*); ~ **Register of Potentially Toxic Chemicals** *n* (*IRPTC*) POLICY Internationales Verzeichnis *nt* für potentiell toxische Chemikalien (*IRPTC*); ~ **Standards Organization** *n* (*ISO*) ANALYSIS, POLICY Internationale Normungsorganisation *f* (*ISO*)
interphase *n* AIR Phasengrenzfläche *f*
interstitial: ~ **water** *n* WASTE, WATER SUPP Porenwasser *nt*
intertidal: ~ **zone** *n* MARINE Tidengebiet *nt*
introduce *vt* POLICY *eco-tax* einführen, SOIL, WASTE, WATER SUPP eintragen, *substances* zuführen
introduction *n* POLICY *of eco-tax* Einführung *f*; ~ **of harmful substances** *n* SOIL Schadstoffeintrag *m*
intrude *vi* SAFETY eindringen in
intrusion *n* SAFETY *of polluted air* Eindringen *nt*, WASTE, WATER SUPP *specially designed for mixed plastics* Intrusion *f*
intrusive: ~ **filtration** *n* WASTE, WATER SUPP Zusickerung *f*
inundation *n* WASTE, WATER POLL, WATER SUPP Überflutung *f*
inventory: ~ **of hazardous substances** *n* GEN Gefahrstoffkataster *nt*
inverse: ~ **voltammetry** *n* ANALYSIS Inversvoltammetrie *f*
inversion: ~ **atmospheric conditions** *n pl* AIR Inversionswetterlage *f*; ~ **layer** *n* AIR Inversionsschicht *f*

inverting: ~ **prism** *n* ANALYSIS *interferential pit gas tester* Umkehrprisma *nt*
investigate *vt* ANALYSIS, SAFETY erkunden, untersuchen
investigated *adj* ANALYSIS, SAFETY erkündet, untersucht
investigation *n* ANALYSIS Erkundung *f*, SAFETY Untersuchung *f*
invisible *adj* AIR unsichtbar; ~ **plume** *n* AIR unsichtbarer Schwaden *m*
involutional *adj* AIR involutorisch
iodometric *adj* ANALYSIS iodometrisch; ~ **determination** *n* ANALYSIS iodometrische Bestimmung *f*; ~ **method** *n* ANALYSIS iodometrisches Verfahren *nt*
ion *n* ANALYSIS, WATER POLL, WATER SUPP Ion *nt*; ~ **activity** *n* ANALYSIS Ionenaktivität *f*; ~ **budget** *n* AIR, SOIL, WATER SUPP Ionenhaushalt *m*; ~ **chromatography** *n* (*IC*) ANALYSIS Ionenchromatographie *f* (*IC*); ~ **current** *n* AIR *ESP* Ionenfluß *m*, Ionenstrom *m*, Sprühstrom *m*; ~ **exchange** *n* (*IE*) ANALYSIS, WATER POLL, WATER SUPP Ionenaustausch *m* (*IA*); ~ **exchanger** *n* WATER POLL, WATER SUPP Ionenaustauscher *m*; ~ **exchange resin** *n* SOIL, WATER SUPP Ionenaustauschharz *nt*; ~ **exchanger membrane** *n* WATER POLL Ionaustauschermembran *f*; ~ **exchange softener** *n* ANALYSIS, WATER SUPP Ionenaustauschenthärter *m*; ~ **exchange softening** *n* WATER POLL, WATER SUPP Ionenaustauschenthärtung *f*
ionic *adj* GEN ionisch; ~ **concentration** *n* WATER POLL, WATER SUPP Ionenkonzentration *f*; ~ **form** *n* ANALYSIS, WATER POLL, WATER SUPP Ionenform *f*; ~ **migration** *n* SOIL Ionenwanderung *f*; ~ **mobility** *n* ANALYSIS Ionenbeweglichkeit *f*; ~ **valence** *n* WATER POLL Ionenwertigkeit *f*
ionization *n* AIR, ANALYSIS, SOIL, WATER SUPP Ionisation *f*, Ionisierung *f*; ~ **chamber** *n* ANALYSIS Ionisationskammer *f*; ~ **potential** *n* ANALYSIS Ionisierungsenergie *f*; ~ **spectrometer** *n* ANALYSIS Bragg-Spektrometer *nt*, Braggsches Spektrometer *nt*
ionize *vt* GEN ionisieren
ionized *adj* GEN ionisiert; ~ **gas molecule** *n* AIR ionisiertes Gasmolekül *nt*
ionizing *adj* GEN ionisierend; ~ **radiation** *n* AIR, SOIL, WASTE, WATER SUPP ionisierende Strahlung *f*; ~ **wet scrubber** *n* (*IWS*) AIR ionisierender Naßwäscher *m*; ~ **zone** *n* AIR Ionisationszone *f*
ionogenic *adj* ANALYSIS, WATER POLL ionogen
ion: ~ **pair** *n* (*IP*) ANALYSIS Ionenpaar *nt*; ~**-selective** *adj* ANALYSIS, WASTE, WATER POLL, WATER SUPP ionenselektiv; ~**-selective electrode** *n* (*ISE*) ANALYSIS ionenselektive Elektrode *f* (*ISE*); ~**-selective electrode with glass membrane** *n* ANALYSIS ionenselektive Elektrode *f* mit Glasmembran; ~**-selective membrane** *n* WASTE, WATER POLL, WATER SUPP ionenselektive Membran *f*; ~**-sensitive** *adj* ANALYSIS ionensensitiv; ~**-sensitive electrode** *n* ANALYSIS ionensensitive Elektrode *f*; ~ **source** *n* ANALYSIS *quadrupole mass analyser* Ionenquelle *f*; ~ **trap** *n* ANALYSIS *with MSD* Ionenfalle *f*; ~ **trap detector** *n* ANALYSIS Ionenfalle-Detektor *m*
IP *abbr* (*ion pair*) ANALYSIS Ionenpaar *nt*
IPC *abbr* (*integrated pollution control*) MANAGE, POLICY IVU (*integrierte Vermeidung der Umweltverschmutzung*)
IP: ~ **chromatography** *n* ANALYSIS Ionenpaarchromatographie *f*
IPPC *abbr* MANAGE, POLICY (*integrated prevention and pollution control*) integrierte Vorsorge *f* und Vermeidung *f* der Umweltverschmutzung (*IVU*)
IR *abbr* (*infrared*) AIR, ANALYSIS, WATER SUPP IR (*Infrarot*); ~ **drying** *n* ANALYSIS *moisture content determination* IR-Trocknung *f*
iridescence *n* MARINE Irisieren *nt*, Schillern *nt*
IR: ~ **laser back-scattering** *n* ANALYSIS IR-Laser-Rückstreuung *f*
iron *n* WASTE, WATER SUPP Eisen *nt*; ~**-containing** *adj* WASTE, WATER SUPP eisenhaltig; ~**-containing material** *n* WASTE eisenhaltiges Material *nt*; ~ **scrap** *n* WASTE Eisenschrott *m*; ~ **sulfate** *AmE*, ~ **sulphate** *BrE* *n* WASTE, WATER POLL, WATER SUPP Eisensulfat *nt*
IR: ~ **optical absorption** *n* ANALYSIS IR-Lichtabsorption *f*
IRPTC *abbr* (*International Register of Potentially Toxic Chemicals*) POLICY IRPTC (*Internationales Verzeichnis für potentiell toxische Chemikalien*)
irradiated: ~ **energy** *n* RENEW ENERG eingestrahlte Energie *f*
irradiation: ~ **dosage** *n* WASTE, WATER SUPP Bestrahlungsdosis *f*; ~ **intensity** *n* WASTE, WATER SUPP *of microorganisms* Bestrahlungsstärke *f*; ~ **time** *n* WASTE, WATER SUPP Bestrahlungsdauer *f*
irreversible *adj* AIR, WATER POLL irreversibel; ~ **process** *n* AIR, WATER POLL irreversibles Verfahren *nt*
irrigable *adj* WATER POLL rieselfähig
irrigation *n* WASTE, WATER POLL, WATER SUPP Berieselung *f*, Bewässerung *f*; ~ **ditch** *n* WASTE, WATER SUPP Berieselungsgraben *m*, Berieselungskanal *m*, Bewässerungsgraben *m*; ~ **well** *n* WASTE, WATER SUPP Berieselungsbrunnen *m*, Bewässerungsbrunnen *m*
irritant *adj* SAFETY *as danger symbol* ätzend
IRS *abbr* (*International Referral System*) POLICY IRS (*Internationales Referenzsystem*)
IR: ~ **spectrometry** *n* ANALYSIS *infrared spectrometry* IR-Spektrometrie *f*; ~ **spectrophotometer** *n* ANALYSIS IR-Spektralphotometer *nt*, IR-Spektrophotometer *nt*; ~ **thermometer** *n* ANALYSIS IR-Thermometer *nt*
IR-VOC: ~ **analyser** *n* *BrE* ANALYSIS *infrared volatile organic compounds analyser* IR-VOC Analysator *m*; ~ **analyzer** *AmE see IR-VOC analyser BrE*
ISE *abbr* (*ion-selective electrode*) ANALYSIS ISE (*ionenselektive Elektrode*)
ISO *abbr* (*International Standards Organization*) ANALYSIS, POLICY ISO (*Internationale Normungsorganisation*)
isokinetic *adj* ANALYSIS isokinetisch; ~ **sampling** *n* ANALYSIS isokinetische Probenahme *f*, isokinetisches Probenehmen *nt*; ~ **suction** *n* ANALYSIS isokinetisches Ansaugen *nt*
isolated: ~ **thermal use** *n* RENEW ENERG dezentrale thermische Nutzung *f*
isolation *n* AIR *of pollutants* Trennung *f*, ANALYSIS *of sample* Isolierung *f*, WATER POLL *of pollutants* Trennung *f*; ~ **pad** *n* NOISE Dämpfungsglied *nt*
isolator *n* NOISE Schwingungsisolator *m*
isotherm *n* AIR, ANALYSIS Isotherme *f*
isothermal *adj* AIR, ANALYSIS isotherm, isothermisch; ~ **gas chromatograph** *n* ANALYSIS isothermer Gaschromatograph *m*; ~ **gas chromatography** *n* ANALYSIS isotherme Gaschromatographie *f*
ISV *abbr* (*in-situ vitrification*) SOIL In-Situ-Verglasung *f*
iteration *n* AIR Iteration *f*
IWM *abbr* (*integrated waste management*) MANAGE integrierte Abfallbewirtschaftung *f*
IWS *abbr* (*ionizing wet scrubber*) AIR ionisierender Naßwäscher *m*

J

Jamin: ~ **interferometer** *n* ANALYSIS Jamin-Interferometer *nt*
jar: ~ **test** *n* ANALYSIS, WATER SUPP Jar-Test *m*
jaw: ~ **crusher** *n* WASTE Backenbrecher *m*
jet *n* AIR WATER POLL, WATER SUPP *nozzle* Düse *f*, Strahldüse *f*, *stream* Düsenstrahl *m*; ~ **aeration** *n* WATER POLL Ejektorbelüftung *f*, Strahlbelüftung *f*; ~ **aerator** *n* WATER POLL Ejektorbelüfter *m*, Strahlbelüfter *m*; ~ **cock** *n* WATER SUPP Strahlwasserhahn *m*; ~ **filter** *n* AIR Druckstoßfilter *nt*; ~ **injection** *n* SOIL Düsenstrahlinjektion *f*; ~ **pump** *n* WATER POLL, WATER SUPP Strahlpumpe *f*
jigger *n* SOIL, WASTE, WATER SUPP Jigger *m*
jigging *n* SOIL, WASTE Setzen *nt*
joint: ~ **and several liability** *n* POLICY, WASTE, WATER SUPP Haftung *f* zur ungeteilten Hand, gesamtschuldnerische Haftung *f*
joule: ~ **heating** *n* SOIL *thermal desorption* Joule-Heizung *f*
junction *n* WATER SUPP *of canal* Einmündung *f*, Zusammenfluß *m*
junk *n* WASTE Abfall *m*; ~ **iron** *n* WASTE Eisenschrott *m*; ~ **press** *n* WASTE Eisenschrottpresse *f*, Schrottpresse *f*
juridical *adj* POLICY juristisch
juvenile: ~ **groundwater** *n* WASTE, WATER SUPP juveniles Grundwasser *nt*; ~ **water** *n* WASTE, WATER SUPP juveniles Wasser *nt*

K

Kaplan: ~ **turbine** *n* RENEW ENERG *tidal power plant* Kaplanturbine *f*
karst *n* WATER SUPP Karst *m*; ~ **groundwater** *n* WATER SUPP Karstgrundwasser *nt*; ~ **hydrology** *n* WATER SUPP Karsthydrologie *f*
karstic: ~ **spring** *n* WATER SUPP Karstquelle *f*
karst: ~ **water resources** *n pl* WATER SUPP Karstwasservorkommen *nt*
keep: ~ **clean** *vt* WASTE, WATER SUPP sauberhalten; ~ **cool** *vt* ANALYSIS *sample* kühl halten; ~ **in dark** *vt* ANALYSIS *sample* dunkel halten; ~ **sealed** *vt* ANALYSIS *sample* dichthalten
kelly *n* SOIL *for soil injection* Spülstange *f*
kerb: ~-**side collection system** *n BrE* (*cf curb-side collection system AmE*) WASTE Holsystem *nt*
kilojoule: ~ **per kilogram**, ~ **per kilogramme** *n* (*kJ/kg*) WASTE Kilojoule pro Kilogramm (*kJ/kg*)
kilonewton: ~ **per square meter** *AmE*, ~ **per square metre** *BrE phr* WASTE Kilonewton pro Quadratmeter (*kN/m²*)
kilowatt *n* (*kW*) RENEW ENERG Kilowatt *nt* (*kW*); ~ **hour** *n* RENEW ENERG *heat pump* Kilowattstunde *f*; ~-**hour heat** *n* RENEW ENERG *heat pump* Kilowattstundenwärme *f*; ~ **plant** *n* RENEW ENERG Kilowattanlage *f*, kW-Anlage *f*
kinematic: ~ **viscosity** *n* WASTE, WATER SUPP kinematische Viskosität *f*, kinematische Zähigkeit *f*
kinetically: ~ **induced buoyancy** *n* WASTE, WATER SUPP kinetisch erzeugter Auftrieb *m*
kinetic: ~ **energy** *n* AIR kinetische Energie *f*, RENEW ENERG *wind power* Bewegungsenergie *f*, kinetische Energie *f*; ~-**energy pump** *n* AIR, RENEW ENERG, WATER POLL, WATER SUPP Turbomaschine *f*
kJ/kg *abbr* (*kilojoule per kilogram, kilojoule per kilogramme*) WASTE kJ/kg (*Kilojoule pro Kilogramm*)
Kn *abbr* (*Knudsen number*) AIR Kn (*Knudsen-Zahl*)
kneader *n* WASTE Kneter *m*
kN/m² *abbr* (*kilonewton per square meter AmE, kilonewton per square metre BrE*) WASTE kN/m² (*Kilonewton pro Quadratmeter*)
knobbed: ~ **panel** *n* NOISE Schalldämpf-Noppenplatte *f*
knock *vt* AIR klopfen
knock: ~ **filter** *n* AIR Klopffilter *nt*
knocking *n* AIR Klopfung *f*, Klopfen *nt*
knock: ~ **limit** *n* AIR Klopfgrenze *f*; ~ **resistance** *n* AIR Klopffestigkeit *f*
Knudsen: ~ **number** *n* (*Kn*) AIR Knudsen-Zahl *f* (*Kn*)
konimeter *n* ANALYSIS Konimeter *nt*, Staubgehaltsmesser *m*
Kr *abbr* (*krypton*) AIR Kr (*Krypton*)
krypton *n* (*Kr*) AIR Krypton *nt* (*Kr*)
kW *abbr* (*kilowatt*) RENEW ENERG kW (*Kilowatt*)

L

lab *n* (*laboratory*) ANALYSIS Labor *nt*, Laboratorium *nt*
label *vt* WASTE kennzeichnen
labeling *AmE see labelling BrE*
labelling *n BrE* WASTE Kennzeichnung *f*; **~ obligation** *n BrE* WASTE *of plastics* Kennzeichnungspflicht *f*; **~ system for recyclable plastic containers** *n BrE* WASTE *numbering system from one to seven* Kennzeichnungssystem *nt* für wiederverwertbare Plastikcontainer
labile *adj* AIR *waste gas plume propagation class* instabil, labil; **~ temperature layering** *n* AIR labile Temperaturschichtung *f*
lab: **~ method** *n* ANALYSIS Labormethode *f*
laboratory *n* (*lab*) ANALYSIS Labor *nt*, Laboratorium *nt*; **~ experiment** *n* ANALYSIS Laborexperiment *nt*; **~ extruder** *n* ANALYSIS Laborextruder *m*; **~ findings** *n pl* ANALYSIS Laborbefunde *m pl*; **~ mesocosms** *n* ANALYSIS *simulation* Labormesokosmen *nt pl*; **~ method** *n* ANALYSIS Labormethode *f*; **~ practice** *n* ANALYSIS Laborpraxis *f*; **~ results** *n pl* ANALYSIS Laborbefunde *m pl*; **~ sample** *n* ANALYSIS Laborprobe *f*; **~ scale** *n* ANALYSIS Labormaßstab *m*; **~ technician** *n* ANALYSIS Labortechniker *m*, Labortechnikerin *f*; **~ test** *n* ANALYSIS Labortest *m*, Laborversuch *m*; **~-tested** *adj* ANALYSIS laborgeprüft
labor *AmE see labour BrE*
labour: **~-intensive** *adj BrE* GEN arbeitsintensiv; **~-intensive method** *n BrE* AIR, ANALYSIS, MARINE, NOISE, SOIL, WATER POLL, WATER SUPP arbeitsintensive Methode *f*
lab: **~ practice** *n* ANALYSIS Laborpraxis *f*; **~ results** *n pl* ANALYSIS Laborbefunde *m pl*; **~ sample** *n* ANALYSIS Laborprobe *f*; **~ technician** *n* ANALYSIS Labortechniker *m*, Labortechnikerin *f*; **~ test** *n* ANALYSIS Labortest *m*, Laborversuch *m*; **~-tested** *adj* ANALYSIS laborgeprüft
lack: **~ of fuel** *n* AIR Brennstoffmangel *m*; **~ of implementation** *n* POLICY *of competent authority* Vollzugsdefizit *nt*; **~ of oxygen** *n* ANALYSIS, WATER POLL, WATER SUPP Sauerstoffmangel *m*
laden *adj* AIR, WASTE, WATER SUPP beladen
lagging *n* AIR Wärmedämmung *f*, Wärmeschutz *m*
lagoon *n* RENEW ENERG Meeresbucht *f*
lagooning *n* WATER POLL Schlammteichverfahren *nt*
lag: **~ time** *n* ANALYSIS *between sampling* Zwischenzeit *f*
lake *n* WATER POLL, WATER SUPP See *m*
lambda *n* AIR Lambda *nt*; **~ probe** *n* AIR *catalyst* Lambdasonde *f*, Sauerstoffsonde *f*, ANALYSIS Lambdasonde *f*; **~ window** *n* AIR Lambdafenster *nt*
Lambert-Beer: **~ law** *n* ANALYSIS Lambert-Beersches Gesetz *nt*
lamella *n* AIR Lamelle *f*, Platte *f*, WATER POLL *plates within a settling tank* Blättchen *nt*, Lamelle *f*; **~ bundle** *n* AIR Lamellenbündel *nt*
lamellar *adj* AIR lamellenförmig; **~ heat exchanger** *n* AIR Lamellenwärmeaustauscher *m*; **~ precipitator** *n* AIR Lamellenabscheider *m*
laminar *adj* AIR lamellenförmig, laminar, schichtenförmig, SOIL laminar, WASTE, WATER SUPP lamellenförmig, laminar, schichtenförmig; **~ boundary layer** *n* AIR *cyclone* laminare Grenzschicht *f*; **~ boundary layer flow** *n* AIR laminare Grenzschichtströmung *f*; **~ fixed bed** *n* SOIL laminares Festbett *nt*; **~ flow** *n* AIR Laminarströmung *f*, laminare Strömung *f*, WASTE laminare Strömung *f*, WATER SUPP Laminarströmung *f*, laminare Strömung *f*; **~ flow velocity** *n* AIR Anströmungsgeschwindigkeit *f*
laminate *vt* WASTE laminieren, schichten
laminated: **~ packaging** *n* WASTE Verbundverpackung *f*; **~ secondary resource** *n* WASTE Verbundwerkstoff *m*
lamination *n* WASTE Schicht *f*; **~ strength** *n* WASTE Schichtstärke *f*, Schichtungsfestigkeit *f*; **~ of surface** *n* WASTE Kaschierung *f* der Oberflächen
LAMS *abbr* (*laser mass spectrometry*) ANALYSIS LAMS (*Laser-Massenspektrometrie*)
lance *n* AIR Lanze *f*
land: **~ application** *n* WATER POLL Berieselung *f* von Flächen mit Abwasser, landwirtschaftliche Nutzung *f*; **~-based** *adj* RENEW ENERG von Land aus; **~-based OTEC plant** *n* RENEW ENERG Meereswärmekraftwerk *nt* am Land; **~ burial** *n* WASTE Vergraben *nt*; **~ degradation** *n* WASTE Devastierung *f*; **~ district** *n* POLICY *German Länder* Regierungsbezirk *m*; **~ district president** *n* POLICY *German Länder* Regierungspräsident *m*; **~ disturbance** *n* WASTE Devastierung *f*
landfarming: **~ method** *n* SOIL Landfarming-Methode *f*
landfill *n* GEN Mülldeponie *f*; **~ age** *n* WASTE Deponiealter *nt*; **~ area** *n* WASTE Deponiefläche *f*, Deponieraum *m*; **~ area conservation** *n* WASTE Deponieraumschonung *f*; **~ base sealing** *n* WASTE Deponiebasisabdichtung *f*; **~ behavior** *AmE*, **~ behaviour** *BrE n* WASTE Deponieverhalten *nt*; **~ binder** *n* WASTE Deponiebinder *m*; **~ body** *n* SOIL Deponiekörper *m*; **~ bottom** *n* WASTE Deponiesohle *f*; **~ building** *n* WASTE Deponiebauwerk *nt*; **~ capacity** *n* WASTE Deponiekapazität *f*; **~ cell** *n* WASTE Kassette *f*, Polder *m*; **~ charge** *n* POLICY Deponieabgabe *f*; **~ closure** *n* WASTE Schließung *f* der Deponie; **~ compactor** *n* WASTE Kompaktor *m*, Müllverdichter *m*, Verdichtungsanlage *f*; **~ design** *n* WASTE Deponieart *f*; **~ disposal** *n* WASTE Deponieentsorgung *f*
landfilled *adj* WASTE deponiert
landfill: **~ fee** *n* WASTE Deponiegebühr *f*; **~ foundation** *n* WASTE Deponiebasis *f*; **~ gas** *n* AIR *ozone damage*, WASTE Deponiegas *nt*; **~ gas analysis** *n* ANALYSIS Deponiegasuntersuchung *f*; **~ gas collection** *n* AIR, WASTE Deponiegaserfassung *f*; **~ gas collector** *n* AIR, WASTE Deponiegasfassung *f*; **~ gas composition** *n* WASTE Deponiegaszusammensetzung *f*; **~ gas development** *n* WASTE Deponiegasentwicklung *f*; **~ gas drainage** *n* AIR, WASTE Deponiegasableitung *f*; **~ gas extraction** *n* WASTE Gasgewinnung *f* aus Mülldeponien; **~ gas limit** *n* WASTE Deponiegasgrenzwert *m*; **~ gas measurement** *n* ANALYSIS Deponiegasmessung *f*; **~ gas migration** *n* AIR, WASTE Deponiegasmigration *f*; **~ gas problem** *n* WASTE Deponiegasproblem *nt*; **~ gas production** *n* WASTE Deponiegasproduktion *f*; **~ gas quantity** *n* WASTE Deponiegasmenge *f*; **~ gas treatment** *n* WASTE Deponiegasbehandlung *f*; **~ gas use** *n* WASTE Deponiegasnutzung *f*; **~ gas yield** *n* AIR, WASTE Deponiegasanfall *m*
landfilling *n* WASTE Deponierung *f*, geordnetes Ablagern *nt*; **~ method** *n* WASTE Deponierungsverfahren *nt*; **~ practice** *n* WASTE Deponierungspraxis *nt*; **~ suitability** *n* WASTE Deponierbarkeit *f*, Deponierfähigkeit *f*, WATER POLL *wastewater sludge ash* Deponierbarkeit *f*
landfill: **~ installation** *n* WASTE Deponieeinbau *m*; **~ leachate** *n* WASTE, WATER POLL Deponiesickerwasser *nt*; **~ leachate purification** *n* WASTE, WATER POLL, WATER SUPP Deponiesickerwasserreinigung *f*; **~ leachate tank** *n* WASTE, WATER POLL Deponiesickerwasserbecken *nt*; **~ leachate treatment** *n* WASTE, WATER POLL Deponiesickerwasser-

behandlung *f*; **~ leachate treatment facility** *n* WASTE *of landfill* Sickerwasserbehandlungsanlage *f*; **~ location** *n* WASTE Deponiestandort *m*; **~ location security** *n* SAFETY, WASTE Deponiesicherheit *f*; **~ material** *n* WASTE Deponiematerial *nt*; **~ operation** *n* WASTE Deponiebetrieb *m*; **~ operator** *n* WASTE Deponiebetreiber *m*; **~ perimeter** *n* WASTE Deponiegrenze *f*; **~ phase** *n* WASTE Deponiephase *f*; **~ register** *n* POLICY Deponiekataster *m*; **~ requirement** *n* WASTE Deponieerfordernis *nt*; **~ roads** *n pl* WASTE Deponiestraßen *f pl*; **~ sealing** *n* WASTE Deponieabdichtung *f*; **~ section** *n* WASTE Deponieabschnitt *m*; **~ site** *n* MARINE Deponiestandort *m*, WASTE Standort *m* der Deponie, Deponiegelände *nt*; **~ site technique** *n* WASTE Deponiestandortstechnik *f*; **~ slope** *n* WASTE Deponiehang *m*; **~ structure** *n* WASTE Deponieaufbau *m*; **~-suitable** *adj* WASTE deponiefähig, deponiegeeignet, deponierbar; **~-suitable vehicle** *n* WASTE deponiefähiges Fahrzeug *nt*; **~ support** *n* WASTE Deponieplanum *nt*; **~ technique** *n* WASTE Deponietechnik *f*; **~ thickness** *n* WASTE Deponiedichte *f*; **~ type** *n* WASTE Deponieart *f*; **~ underground** *n* WASTE Deponieuntergrund *m*; **~ volume** *n* WASTE Deponievolumen *nt*

land: **~ pollutants** *n pl* SOIL, WASTE bodenverschmutzende Stoffe *m pl*, bodenverunreinigende Stoffe *m pl*; **~ use** *n* SOIL Bodennutzung *f*

large: **~-area** *adj* RENEW ENERG großflächig; **~-area use of solar energy** *n* RENEW ENERG großflächige Nutzung *f* der Sonnenenergie; **~-bubble aeration** *n* WATER POLL grobblasige Belüftung *f*; **~ collection station** *n* WASTE Großsammelstelle *f*; **~ container** *n* WASTE Großbehälter *m*; **~ refuse container** *n* WASTE Müllgroßbehälter *m* (*MGB*); **~-scale clarification plant** *n* WATER POLL Großkläranlage *f*; **~-scale installation** *n* RENEW ENERG *wind, solar* Großanlage *f*

laser *n* ANALYSIS, SAFETY Laser *m*; **~ beam** *n* ANALYSIS *in light detection and ranging*, WASTE Laserstrahl *m*; **~-enhanced ionization** *n* (*LEI*) ANALYSIS Ionisation *f* durch Laser (*LEI*); **~-excited** *adj* ANALYSIS laserangeregt; **~-excited atomic fluorescence spectrometry** *n* (*LEAFS*) ANALYSIS laserangeregte Atomfluoreszenzspektrometrie *f* (*LEAFS*); **~ mass spectrometry** *n* (*LAMS*) ANALYSIS Laser-Massenspektrometrie *f* (*LAMS*); **~ microprobe mass analysis** *n* ANALYSIS Laser-Mikrospektralanalyse *f*; **~ radar** *n* ANALYSIS *measurement of pollutant concentration* Laser-Radar *nt*; **~ radiation** *n* SAFETY Laserstrahlung *f*; **~ radiation hazard** *n* SAFETY Gefahr *f* durch Laserstrahlung; **~ Raman spectroscopy** *n* ANALYSIS Laser-Raman-Spektroskopie *f*; **~ sampler** *n* ANALYSIS *solids sampling analysis* Laser-Sampler *m*; **~ spectroscopy** *n* ANALYSIS Laserspektroskopie *f*

lateral: **~ canal** *n* WATER SUPP Kanalabzweig *m*, Seitenkanal *m*; **~ sewer** *n* WASTE, WATER POLL *wastewater* Nebensammler *m*

law *n* POLICY *body of law* Gesetz *nt*, *ordinance* Recht *nt*

lawful *adj* POLICY rechtlich

law: **~ on ambient air** *n* AIR, POLICY Immissionsrecht *nt*; **~ on waste** *n* POLICY, WASTE Abfallrecht *nt*; **~ on water** *n* POLICY, WATER SUPP Wasserrecht *nt*

lay *vt* SOIL *liners*, WASTE, WATER SUPP *mains* verlegen

layer 1. *n* AIR, SOIL, WASTE Lage *f*, Schicht *f*; **2.** *vt* AIR schichten

layered *adj* ANALYSIS schichtweise

layer: **~ formation** *n* AIR Schichtbildung *f*; **~ sampling** *n* ANALYSIS Schichtprobenahme *f*; **~ strength** *n* AIR Schichtstärke *f*; **~ thickness** *n* ANALYSIS, NOISE, WASTE Schichtdicke *f*; **~ water** *n* WASTE Schichtenwasser *nt*

laying *n* SOIL *of liners*, WASTE, WATER SUPP *of mains* Verlegung *f*

LBG *abbr* (*low thermal value biogas*) RENEW ENERG LBG (*niedrigkaloriges Biogas*)

LC_0 *abbr* (*lethal concentration zero*) SAFETY LC_0 (*letale Konzentration Null*)

LC_{50} *abbr* (*lethal concentration fifty*) SAFETY LC_{50} (*letale Konzentration, bei der 50% der Probanden überleben*)

LCA *abbr* MANAGE, WASTE (*life cycle analysis*) LWB (*Lebenszyklusanalyse*), (*life cycle assessment*) LWB (*Lebenswegbilanz*)

LCD *abbr* (*liquid crystal display*) ANALYSIS LCD (*Flüssigkristallsichtanzeige*); **~ monitor** *n* ANALYSIS LCD-Monitor *m*; **~ screen** *n* ANALYSIS LCD-Bildschirm *m*

LC-UV: **~ detection** *n* ANALYSIS LC-UV-Detektion *f*

LC_x *abbr* (*lethal concentration*) SAFETY LC_x (*letale Konzentration*)

LD *abbr* (*lethal dose*) ANALYSIS LD (*letale Dosis*)

LD_{50} *abbr* (*lethal dose fifty*) SAFETY LD_{50} (*letale Dosis, bei der 50% der Probanden überleben*)

LD_{Lo} *abbr* (*lethal dose low*) SAFETY LD_{Lo} (*niedrigste letale Dosis*)

LDPE *abbr* (*low-density polyethylene*) WASTE Niederdruck-Polyethylen *nt*, Polyethylen mit niedriger Dichte *nt*

LD_x *abbr* (*lethal dose*) SAFETY LD_x (*letale Dose*)

leach *vt* MARINE, SOIL, WASTE, WATER SUPP auslaugen

leachability *n* SOIL Auslaugbarkeit *f*

leachate *n* WASTE, WATER SUPP Sickerwasser *nt*; **~ behavior** *AmE*, **~ behaviour** *BrE* *n* WASTE, WATER SUPP Sickerwasserverhalten *nt*; **~ blockage** *n* WASTE Verschließung *f* mit Sickerwasser; **~ collection** *n* WASTE, WATER SUPP Sickerwassererfassung *f*, Sickerwassersammlung *f*; **~ determination** *n* ANALYSIS Sickerwasserbestimmung *f*; **~ drainage** *n* WASTE, WATER SUPP Sickerwasserdränage *f*, Sickerwasserableitung *f*; **~ drainage with filtering rubble** *n* WASTE, WATER SUPP Sickerwasserdränage *f* mit Filterschotter; **~ pump** *n* WASTE Sickerwasserpumpe *f*; **~ purification** *n* WASTE Sickerwasserreinigung *f*; **~ reduction** *n* WASTE Sickerwasserverminderung *f*; **~ yield** *n* WASTE, WATER SUPP Sickerwasseranfall *m*

leaching *n* MARINE Auslaugen *nt*, SOIL Auslaugen *nt*, Eluation *f*, WASTE Auslaugung *f*, WATER POLL Auslaugen *nt*, Auslaugung *f*, WATER SUPP Auslaugung *f*; **~ behavior** *AmE*, **~ behaviour** *BrE* *n* SOIL Auslaugverhalten *nt*; **~ control** *n* SOIL Kontrolle *f* der Bodenauslaugung; **~ process** *n* WASTE, WATER SUPP Auslaugverfahren *nt*; **~ property** *n* SOIL Auslaugverhalten *nt*; **~ test** *n* SOIL Auslaugtest *m*, Eluationsversuch *m*; **~ trench** *n* WASTE, WATER SUPP Auslaugungsgraben *m*

lead *n* (*Pb*) GEN Blei *nt* (*Pb*); **~ accumulator** *n* WASTE Bleiakkumulator *m*, Pb-Akku *m*; **~ additives** *n pl* AIR, WASTE Bleizusatzstoffe *m pl*; **~-containing** *adj* AIR, WASTE bleihaltig; **~-containing dust** *n* AIR, WASTE bleihaltiger Staub *m*; **~-free gasoline** *n AmE* (*cf lead-free petrol BrE*) AIR, WASTE bleifreies Benzin *nt*; **~-free petrol** *n BrE* (*cf lead-free gasoline AmE*) AIR, WASTE bleifreies Benzin *nt*; **~ poisoning** *n* SAFETY *occupational disease* Bleivergiftung *f*

leaf: **~-area index** *n* AIR, WASTE Blattindex *m*

LEAFS *abbr* (*laser-excited atomic fluorescence spectrometry*) ANALYSIS LEAFS (*laserangeregte Atomfluoreszenzspektrometrie*)

leak 1. *n* GEN *of groundwater* Leck *nt*, Leckage *f*; **2.** *vi* MARINE auslaufen, SAFETY lecken

leakage *n* GEN Auslaufen, *nt* Leck *nt*, Leckage *f*, Wasserverlust *m*; **~ indicator system** *n* SAFETY Leckwarnsystem *nt*; **~ warning** *n* SAFETY Leckalarm *m*; **~ water** *n* WASTE, WATER SUPP Sickerwasser *nt*

leak: **~ detection** *n* ANALYSIS, SOIL Leckdetektion *f*, WASTE, WATER SUPP Leckbestimmung *f*; **~ detector** *n* ANALYSIS Leckanzeiger *m*

leaking *adj* SOIL *of chemical into soil* leckgeworden

leak: **~ water** *n* WASTE, WATER SUPP Sickerwasser *nt*

lean: **~ eco-balancing** *n* MANAGE schlankes Ökobilanzieren *nt*; **~-mix engine** *n* AIR Magermotor *m*; **~-mix engine with oxidation catalyst** *n* AIR Magermotor *m* mit Oxidationskatalysator

leasing *n* POLICY Leasing *nt*; ~ **of certificates** *n* POLICY Leasing *nt* von Zertifikaten
LED *abbr* (*light-emitting diode*) ANALYSIS LED (*lichtemittierende Diode*); ~ **display** *n* ANALYSIS LED-Anzeige *f*
legal *adj* POLICY gesetzlich, rechtlich; ~ **instrument** *n* POLICY gesetzliches Instrument *nt*; ~ **requirements** *n pl* POLICY gesetzliche Auflagen *f pl*, rechtliche Auflagen *f pl*; ~ **standards** *n pl* POLICY nachbarschützende Rechtsnormen *f pl*; ~ **system** *n* POLICY Recht *nt*
legislator *n* POLICY Gesetzgeber *m*
LEI *abbr* (*laser-enhanced ionization*) ANALYSIS LEI (*Ionisation durch Laser*)
lens *n* RENEW ENERG *solar energy plant* Linse *f*
lethal *adj* AIR tödlich, ANALYSIS giftig, SAFETY letal; ~ **concentration** *n* (LC_x) AIR tödliche Konzentration *f*, ANALYSIS Letalkonzentration *f*, SAFETY letale Konzentration *f*, (LC_x), WASTE, WATER SUPP tödliche Konzentration *f*; ~ **concentration fifty** *n* (LC_{50}) SAFETY letale Konzentration *f* bei der 50% der Probanden überleben (LC_{50}); ~ **concentration zero** *n* (LC_0) SAFETY letale Konzentration *f* Null (LC_0); ~ **dose** *n* AIR, ANALYSIS (*LD*) Letaldosis *f*, SAFETY, WATER SUPP Letaldosis *f*; ~ **dose fifty** *n* (LD_{50}) SAFETY letale Dosis *f*, bei der 50% der Probanden überleben (LD_{50}); ~ **dose low** *n* (LD_{Lo}) SAFETY niedrigste letale Dosis *f* (LD_{Lo}); ~ **effect** *n* ANALYSIS tödliche Wirkung *f*
level: ~ **indicator** *n* ANALYSIS, NOISE, WATER POLL Pegel *m*; ~ **recorder** *n* ANALYSIS, NOISE, WATER POLL Pegelschreiber *m*; ~ **reduction** *n* MARINE, NOISE Pegelminderung *f*
levy *n* MANAGE, POLICY Abgabe *f*
liability *n* POLICY Haftung *f*; ~ **for disturbance of current local conditions** *n* POLICY Zustandsstörerhaftung *f*
licence *BrE*, **license** *AmE n* AIR, POLICY, WASTE, WATER SUPP Genehmigung *f*, Erlaubnis *f*
licensed *adj* POLICY, WASTE, WATER SUPP lizenziert
licensing *n* POLICY Erlaubnis *f*, Zulassung *f*; ~ **application** *n* POLICY, WATER SUPP Genehmigungsantrag *m*; ~ **authority** *n* POLICY Genehmigungsbehörde *f*; ~ **conference** *n* POLICY Genehmigungskonferenz *f*; ~ **impact** *n* POLICY Genehmigungswirkung *f*; ~ **law** *n* POLICY Genehmigungsrecht *nt*; ~ **mark** *n* MANAGE Lizenzzeichen *nt*; ~ **process** *n* POLICY Genehmigungsverfahren *nt*
lid *n* AIR *meteorology* Inversionskappe *f*
LIDAR *abbr* (*light detection and ranging*) ANALYSIS LIDAR (*lichtmaserbetriebenes Radar*)
life: ~ **cycle analysis** *n* (*LCA*) MANAGE Lebenswegbilanz *f* (*LWB*), Lebenszyklusanalyse *f* (*LWB*), WASTE Lebenswegbilanz *f* (*LWB*); ~ **cycle assessment** *n* (*LCA*) MANAGE, WASTE Lebenswegbilanz *f* (*LWB*), Lebenszyklusanalyse *f* (*LWB*); ~ **cycle inventory** *n* MANAGE Sachbilanz *f*
lifespan *n* SOIL *of remedial measure*, WASTE Lebensdauer *f*
lift 1. *n* RENEW ENERG *wind turbine* Auftrieb *m*; **2.** *vt* AIR pumpen, WASTE, WATER POLL, WATER SUPP heben, pumpen
lift: ~**-and-force pump** *n* WATER POLL, WATER SUPP Saug- und Druckpumpe *f*
lifting *n* AIR Pumpen *nt*, WASTE, WATER POLL, WATER SUPP Hub *m*
lift: ~ **pump** *n* WATER POLL, WATER SUPP Saugpumpe *f*
light *n* ANALYSIS, WASTE Licht *nt*; ~ **beam** *n* WASTE Lichtstrahl *m*; ~**-density fraction** *n* WASTE Leichtfraktion *f*, Leichtstofffraktion *f*; ~ **detection and ranging** *n* (*LIDAR*) ANALYSIS lichtmaserbetriebenes Radar *nt* (*LIDAR*); ~**-emitting** *adj* ANALYSIS lichtemittierend; ~**-emitting diode** *n* (*LED*) ANALYSIS lichtemittierende Diode *f* (*LED*)
lighten *vt* MARINE ableichtern, teilweise löschen
lightening *n* MARINE Leichtern *nt*; ~ **vessel** *n* MARINE Leichter *m*
lighter *n* MARINE Leichter *m*
light: ~**-induced pulse** *n* ANALYSIS *LIDAR* Lichtimpuls *m*
lighting: ~ **installation** *n* RENEW ENERG *with solar energy* Beleuchtungsanlage *f*
lightly: ~ **compacted landfill** *n* WASTE schwach verdichtete Deponie *f*
light: ~ **measuring** *n* ANALYSIS Lichtmessung *f*; ~**-off temperature** *n* AIR *catalyst* Anspringtemperatur *f*; ~**-optical microscope** *n* ANALYSIS Lichtmikroskop *nt*; ~ **scattering** *n* ANALYSIS *dispersion* Lichtstreuung *f*, Streulicht *nt*; ~**-scattering detector** *n* ANALYSIS Streulichtdetektor *m*
lightweight: ~ **panel** *n* NOISE Leichtbauplatte *f*
lignite *n* AIR, SOIL, WASTE Braunkohle *f*; ~ **ash** *n* WASTE Braunkohlenflugasche *f*; ~ **power station** *n* AIR Braunkohlenkraftwerk *nt*
lime *n* SOIL *soil conditioner, binding agent*, WASTE *in glass recycling*, WATER SUPP Kalk *m*; ~**-aggressive** *adj* WASTE, WATER POLL, WATER SUPP kalkangreifend; ~ **concentration** *n* WATER SUPP Kalkkonzentration *f*; ~**-slaked** *adj* WASTE kalklöschend; ~ **slaking** *n* WASTE Kalklöschen *nt*; ~ **sludge** *n* WATER SUPP Kalkschlamm *m*; ~ **slurry** *n* WASTE Kalkschlamm *m*
limestone: ~ **ballast** *n* WASTE, WATER SUPP Kalksteinschotter *m*; ~ **meal** *n* WASTE, WATER SUPP *additive* Kalksteinmehl *nt*; ~ **suspension** *n* WASTE, WATER SUPP Kalksteinsuspension *f*
lime: ~ **water** *n* WATER SUPP Kalkwasser *nt*; ~ **water treatment** *n* WATER SUPP Kalkwasserbereitung *f*
liming *n* WATER SUPP Kalkung *f*
limit *vt* AIR *dispersion of hazardous substances*, SOIL, WASTE begrenzen
limit: ~ **concentration** *n* SAFETY Grenzkonzentration *f*; ~ **of inflammability** *n* AIR Zündgrenze *f*
limiting: ~ **concentration** *n* AIR Grenzkonzentration *f*
limit: ~ **value** *n* POLICY, SAFETY, Höchstwert *m*, WATER SUPP Grenzwert *m*; ~ **value control** *n* ANALYSIS Grenzwertkontrolle *f*, Grenzwert-Monitoring *nt*
limnology *n* ANALYSIS Limnologie *f*
line *vt* NOISE *with sound-absorbing material* abdichten, auskleiden, verkleiden, SOIL *bioremediation pit*, WASTE, WATER POLL, WATER SUPP abdichten, auskleiden
linear: ~ **absorption coefficient** *n* ANALYSIS Extinktionskoeffizient *m*
liner: ~ **sheet** *n* WASTE Dichtungsbahn *f*
line: ~ **source** *n* AIR Linienquelle *f*
lining *n* AIR Ausmauerung *f*, NOISE, WASTE, WATER POLL, WATER SUPP *of basin, tank* Auskleidung *f*
link *vt* WATER POLL verknüpfen
lipophilic *adj* WATER POLL lipophil
lipophobic *adj* WATER POLL lipophob
liposolubility *n* WATER POLL Fettlöslichkeit *f*
liquefaction *n* RENEW ENERG, WASTE, WATER POLL *biomass conversion* Verflüssigung *f*
liquefied: ~ **natural gas** *n* (*LNG*) AIR Flüssigerdgas *nt*
liquefy *vt* RENEW ENERG, WASTE, WATER POLL verflüssigen
liquid 1. *adj* GEN flüssig; **2.** *n* GEN Flüssigkeit *f*
liquid: ~ **chromatography** *n* ANALYSIS Flüssigkeitschromatographie *f* (*LC*); ~ **chromatography detector** *n* ANALYSIS Detektor *m* für Flüssigkeitschromatographie; ~ **chromatography-electrochemical detection** *n* ANALYSIS Flüssigkeitschromatographie *f* mit elektrochemischer Detektion; ~ **chromatography-fluorescence detection** *n* ANALYSIS Flüssigkeitschromatographie *f* mit Fluoreszenzdetektion; ~ **chromatography-liquid chromatography** *n* ANALYSIS Flüssigkeit-Flüssigkeitschromatographie *f*; ~ **chromatography-ultraviolet detection** *n* ANALYSIS Flüssigkeitschromatographie *f* mit Ultraviolettdetektion; ~ **column** *n* ANALYSIS Flüssigkeitssäule *f*; ~ **composting** *n* WATER POLL Flüssigkompostierung *f*; ~ **crystal display** *n* (*LCD*) ANALYSIS Flüssigkristallsichtanzeige *f* (*LCD*); ~ **discharge** *n* AIR *absorber* Flüssigkeitsabzug *m*; ~ **distributor** *n* AIR Flüssigkeitsverteiler *m*; ~ **film** *n* AIR Flüssigkeitsfilm *m*; ~ **glass thermometer** *n* ANALYSIS Flüssigkeitsglasthermometer *nt*; ~ **inlet** *n* AIR *bubble-cap*

tray tower Zulaufwehr *nt*; ~ **layer** *n* AIR Flüssigkeitsschicht *f*; ~**-liquid extraction** *n* ANALYSIS, WATER POLL Flüssig-Flüssig-Extraktion *f*; ~ **manure** *n* WATER POLL, WATER SUPP Gülle *f*, Jauche *f*; ~ **petroleum gas** *n* (*LPG*) AIR, RENEW ENERG Flüssiggas *nt* (*LPG*); ~ **phase** *n* AIR Flüssigphase *f*; ~**-reacting agent** *n* ANALYSIS flüssiger Reaktionspartner *m*; ~ **residue** *n* AIR flüssiger Rückstand *m*; ~ **sampling** *n* ANALYSIS Flüssigkeitsprobenahme *f*; ~ **sludge** *n* WATER POLL Flüssigschlamm *m*, Naßschlamm *m*; ~ **sludge conditioning** *n* WATER POLL Flüssigschlammkonditionierung *f*; ~**-solid extraction** *n* ANALYSIS Flüssig-Fest-Extraktion *f*; ~ **spring thermometer** *n* ANALYSIS Flüssigkeitsfederthermometer *nt*; ~ **state of aggregation** *n* AIR flüssiger Aggregatzustand *m*; ~ **surface** *n* AIR Flüssigkeitsoberfläche *f*; ~ **waste** *n* WASTE flüssiger Abfall *m*; ~ **waste substance** *n* AIR, WATER POLL flüssiger Abfallstoff *m*; ~ **working medium** *n* RENEW ENERG *heat pump* flüssiges Arbeitsmittel *nt*

litter *n* WASTE Straßenkehricht *m*

live: ~ **bacteria** *n pl* WATER POLL lebende Bakterien *f pl*; ~ **storage** *n* WASTE Speichernutzraum *m*, nutzbarer Speicherraum *m*

lixiviation *n* WATER POLL Auslaugen *nt*, Auslaugung *f*

LNB *abbr* (*low-*NO_x *burner*) AIR NOx-armer Brenner *m*

LNG *abbr* (*liquefied natural gas*) AIR Flüssigerdgas *nt*

load 1. *n* ANALYSIS Belastung *f*, WATER POLL Fracht *f*; **2.** *vt* AIR beladen, beschicken, einbringen, WASTE, WATER POLL beladen

loadability *n* AIR Belastbarkeit *f*

load: ~ **capacity test** *n* ANALYSIS Belastungsprobe *f*, Belastungsversuch *m*

loaded *adj* AIR, WASTE, WATER SUPP beladen; ~ **activated coke** *n* AIR beladener Aktivkoks *m*; ~ **adsorbent** *n* AIR beladenes Adsorbens *nt*; ~ **filter** *n* AIR, WASTE, WATER SUPP beladenes Filter *nt*; ~ **solvent** *n* AIR beladenes Lösungsmittel *nt*; ~ **suspension** *n* AIR beladene Suspension *f*; ~ **washing agent** *n* AIR beladenes Waschmittel *nt*

load: ~**-induced damage** *n* WASTE auflastbedingte Beschädigung *f*

loading *n* AIR Beladung *f*, Ladung *f*, ANALYSIS Belastung *f*, SOIL, WASTE Beladung *f*, WATER POLL *ion exchanger* Ladung *f*, WATER SUPP Beladung *f*; ~ **capacity** *n* AIR Belastbarkeit *f*, Beladefähigkeit *f*, Ladefähigkeit *f*; ~ **distribution** *n* AIR Ladungsverteilung *f*; ~ **of dry mass** *n* WATER POLL Trockengutverladung *f*; ~ **mechanism** *n* WASTE, WATER SUPP Beschickungseinrichtung *f*; ~ **tractor** *n* SOIL Laderaupe *f*

load: ~ **of landfill** *n* WASTE Auflast *f* der Deponie

LOAEL *abbr* (*lowest observed adverse effect level*) SAFETY LOAEL (*geringste Dosis, die bereits zu nachteiligen Wirkungen führt*)

loam *n* WASTE Lehm *m*

lobbying *n* POLICY *of environmental groups* Lobby-Arbeit *f*

local: ~ **emission source** *n* AIR, RENEW ENERG, WATER POLL örtliche Emissionsquelle *f*

localization *n* SOIL *of contaminated site* Lokalisierung *f*

locally: ~ **representative** *adj* ANALYSIS *sample* örtlich repräsentativ

lock *vt* WASTE verriegeln

locked *adj* WASTE verriegelt

lock: ~ **hopper** *n* SOIL Überdruckbunker *m*

locking: ~ **mechanism** *n* WASTE Verriegelungsmechanik *f*

LOEC *abbr* (*lowest observed effect concentration*) SAFETY LOEC (*niedrigste Konzentration eines Schadstoffes bei der Wirkung*)

logarithmic *adj* ANALYSIS logarithmisch; ~ **calibration** *n* ANALYSIS logarithmische Eichung *f*

logistics: ~ **of disposal** *n pl* WASTE Entsorgungslogistik *f*

long: ~**-distance water main** *n* WATER SUPP Fernwasserleitung *f*; ~**-distance water supply** *n* WATER SUPP Fernwasserversorgung *f*

longevity *n* AIR *of substances like CFCs*, SOIL *of contaminants* Langlebigkeit *f*

longitudinal: ~ **scraper** *n* WATER POLL Längsräumer *m*; ~ **waves** *n pl* NOISE Longitudinalwellen *f pl*

long: ~**-life radioactive waste** *n* WASTE Atommüll *m* mit langer Halbwertzeit; ~**-path monitor** *n* ANALYSIS Langwegmonitor *m*; ~**-running** *adj* AIR, ANALYSIS, MARINE, NOISE, POLICY, RENEW ENERG, SOIL, WASTE, WATER POLL, WATER SUPP langzeitig; ~**-term** *adj* WATER SUPP langzeitig; ~**-term behavior** *AmE*, ~**-term behaviour** *BrE n* WASTE *of landfill* Langzeitverhalten *nt*; ~**-term control** *n* AIR, NOISE, WASTE, WATER POLL Langzeitkontrolle *f*; ~**-term detector tube** *n* ANALYSIS Langzeitprüfröhrchen *nt*; ~**-term dosimeter** *n* ANALYSIS Langzeitmeter *nt*; ~**-term durability** *n* AIR, MARINE, NOISE, WATER POLL, WATER SUPP Langzeitbeständigkeit *f*; ~**-term effect** *n* AIR, MARINE, NOISE, POLICY, WATER POLL, WATER SUPP Langzeiteffekt *m*, Langzeitwirkung *f*; ~**-term gas analysis** *n* ANALYSIS Langzeitgasanalyse *f*; ~**-term measures** *n pl* AIR, MARINE, NOISE, SOIL, WASTE, WATER POLL, WATER SUPP Langzeitmaßnahmen *f pl*; ~**-term monitoring** *n* AIR, ANALYSIS, MARINE, NOISE, WASTE, WATER POLL, WATER SUPP Langzeitüberwachung *f*; ~**-term safety** *n* WASTE *of landfill site* Langzeitabsicherung *f*; ~**-term sampling** *n* ANALYSIS Langzeitprobenahme *nt*; ~**-term security** *n* WASTE Langzeitsicherheit *f*; ~**-term storage** *n* RENEW ENERG Langzeitspeicher *m*, WASTE Langzeitlager *nt*; ~**-throw sprinkler** *n* WATER POLL Großfeldberegner *m*; ~**-wave** *adj* AIR langwellig; ~**-wave heat radiation** *n* AIR langwellige Wärmestrahlung *f*; ~**-wave radiation from the earth** *n* AIR langwellige terrestrische Ausstrahlung *f*

looping: ~ **smoke plume** *n* AIR schleifenförmige Rauchfahne *f*

loop: ~**-type bubble column** *n* WATER POLL Schlaufenreaktor *m*

loosen *vt* SOIL, WASTE *part* ablösen, *surface* auflockern; ~ **up** *vt* SOIL auflockern

loosening: ~ **up** *n* ANALYSIS, WASTE Auflockerung *f*

lorry *n BrE* (*cf truck AmE*) WASTE Lastkraftwagen *m* (*LKW*); ~ **trailer** *n BrE* WASTE LKW-Anhänger *m*, Lastwagenanhänger *m*; ~ **with trailer** *n BrE* (*cf tractor trailer AmE*) WASTE Lastzug *m*

loss: ~ **of head** *n* AIR, WATER POLL, WATER SUPP Druckverlust *m*, Gefällsverlust *m*

loudness *n* NOISE Lautheit *f*; ~ **level** *n* NOISE Lautstärkepegel *m*; ~ **measurement** *n* ANALYSIS, NOISE Lautstärkemessung *f*; ~ **perception** *n* NOISE Lautstärkeempfindung *f*

low *adj* GEN Niedrig-; ~ **alkalinity** *n* WATER SUPP Niedrigalkalität *f*; ~**-chlorine bleached paper** *n* WASTE chlorarm gebleichtes Papier *nt*; ~ **concentration** *n* ANALYSIS niedrige Konzentration *f*; ~**-contaminated municipal wastewater** *n* WATER POLL schwachbelastetes kommunales Abwasser *nt*; ~**-density polyethylene** *n* (*LDPE*) WASTE Niederdruck-Polyethylen *nt*, Polyethylen *nt* mit niedriger Dichte; ~ **dirty gas load** *n* WASTE niedrige Rohgasbelastung *f*; ~**-emission vehicle** *n* AIR schadstoffarmes Fahrzeug *nt*; ~**-energy** *adj* AIR energiearm; ~**-energy automobile** *n* AIR Sparauto *nt*

lower 1. *vt* WATER SUPP absenken; **2.** *vi* AIR *temperature* absinken

lower: ~ **bed** *n* AIR *catalyst* unteres Bett *nt*; ~ **explosion limit** *n* SAFETY untere Explosionsgrenze *f* (*UEG*); ~ **ignition limit** *n* AIR untere Zündgrenze *f*

lowering: ~ **of groundwater** *n* WATER SUPP Grundwasserabsenkung *f*; ~ **of water level** *n* WATER SUPP Wasserspiegelabsenkung *f*

lower: ~**-level administration** *n* POLICY nachgeordnete Behörde *f*; ~ **limit of detectibility** *n* AIR untere Nachweisgrenze *f*; ~**-quality recycled paper** *n* WASTE minderwertiges Altpapier *nt*; ~**-temperature heat distribution** *n* RENEW ENERG Niedertemperaturwärmeverteilung *f*

lowest: ~ **achievable emission rate** *n* AIR niedrigste erreichbare Emissionsrate *f*; ~ **observed effect concentration** *n* (*LOEC*) SAFETY niedrigste Konzentration *f* eines Schadstoffes bei der Wirkung (*LOEC*)

low: ~-**frequency absorber** *n* NOISE Niederfrequenzabsorber *m*, Tiefenabsorber *m*; ~-**grade** *adj* WASTE niederwertig; ~-**grade paper** *n* WASTE niederwertiges Papier *nt*; ~ **in harmful substances** *adj* AIR, SOIL, WASTE, WATER POLL, WATER SUPP schadstoffarm; ~-**intensity magnet separator** *n* SOIL, WASTE Magnetabscheider *m* mit niedriger Intensität; ~-**level radioactive** *adj* WASTE schwach radioaktiv; ~-**lime-content rubble** *n* WASTE kalkarmer Schotter *m*; ~-**nitrogen** *adj* AIR stickstoffarm; ~-**noise** *adj* NOISE lärmarm; ~-**noise building machinery** *n pl* NOISE lärmarme Baumaschinen *f pl*; ~-**noise construction** *n* NOISE lärmarme Konstruktion *f*; ~-**noise equipment** *n* MARINE, NOISE lärmarme Geräte *nt pl*; ~-**noise machine** *n* NOISE lärmarme Maschine *f*; ~-**noise working methods** *n pl* NOISE lärmarme Arbeitsverfahren *nt pl*

low-NO_x: ~ **burner** *n* (*LNB*) AIR NO_x-armer Brenner *m*; ~ **combustor** *n* AIR NO_x-armer Brenner *m*

low: ~-**pollutant** *adj* AIR, SOIL, WASTE, WATER POLL, WATER SUPP schadstoffarm; ~-**polluting** *adj* AIR, SOIL, WASTE, WATER POLL, WATER SUPP schadstoffarm; ~ **position of the sun** *n* RENEW ENERG *in winter* niedriger Solarstand *m*, niedriger Sonnenstand *m*; ~-**pressure** *adj* AIR, MARINE, SOIL, WASTE, WATER POLL, WATER SUPP Niederdruck-, druckschwach; ~-**pressure backwashing** *n* AIR Niederdruckrückspülung *f*; ~-**pressure cleaning** *n* AIR Niederdruckspülung *f*; ~-**pressure flushing** *n* MARINE Niederdruckspülen *nt*; ~-**pressure pump** *n* WATER POLL, WATER SUPP Niederdruckpumpe *f*; ~-**pressure side** *n* AIR Niederdruckseite *f*; ~-**pressure steam** *n* SOIL Niederdruckdampf *m*; ~-**pressure supply** *n* WATER SUPP druckschwache Versorgung *f*; ~-**pressure Venturi scrubber** *n* AIR Niederdruck-Venturiwäscher *m*; ~-**pressure wet oxidation** *n* WATER POLL Niederdruck-Naßoxidation *f*; ~-**rate** *adj* WATER POLL schwachbelastet; ~-**rate process** *n* WATER POLL *activated sludge* schwachbelastetes Verfahren *nt*; ~-**rate trickling filter** *n* WATER POLL schwachbelasteter Tropfkörper *m*; ~-**residue** *adj* WASTE reststoffarm; ~-**residue production method** *n* WASTE reststoffarmes Produktionsverfahren *nt*; ~-**settled apparent density** *n* ANALYSIS *non-metallic wastes* niedrige Schüttdichte *f*; ~-**soil permeability** *n* SOIL geringe Bodendurchlässigkeit *f*; ~-**sulfur** *AmE*, ~-**sulphur** *BrE adj* WATER POLL schwefelarm; ~-**temperature** *adj* RENEW ENERG Niedertemperatur-; ~-**temperature carbonization wastewater** *n* WATER POLL Schwelwasser *nt*; ~-**temperature conversion** *n* WATER POLL *wastewater sludge incineration* Niedertemperaturkonvertierung *f*; ~-**temperature heat** *n* RENEW ENERG *from solar energy* Niedertemperaturwärme *f*; ~-**temperature solar collector plant** *n* RENEW ENERG Niedertemperaturkollektoranlage *f*; ~ **thermal value biogas** *n* (*LBG*) RENEW ENERG niedrigkaloriges Biogas *nt* (*LBG*); ~ **turbidity** *n* WATER SUPP Niedrigtrübung *f*; ~ **voltage** *n* AIR, ANALYSIS Niederspannung *f*; ~-**voltage ESP** *n* AIR Niederspannungs-elektrofilter *nt*; ~-**voltage lamp** *n* ANALYSIS Niederspannungslampe *f*; ~-**waste production method** *n* AIR, WASTE abfallarmes Produktionsverfahren *nt*; ~-**waste technology** *n* AIR abfallarme Technologie *f*, WASTE abfallarme Technologie *f*, umweltfreundliche Technologie *f*, WATER SUPP abfallarme Technologie *f*; ~ **water** *n* MARINE, WATER SUPP Niedrigwasser *nt*; ~ **water consumption** *n* WATER SUPP Niedrigwasserverbrauch *m*; ~-**water discharge** *n* WASTE Niedrigwasserabfluß *m*; ~-**water level** *n* WATER SUPP Niedrigwasserstand *m*

LPG *abbr* (*liquid petroleum gas*) AIR, RENEW ENERG LPG (*Flüssiggas*)

lubricant *n* WASTE Schmiermittel *nt*

luminescent: ~ **bacteria test** *n* ANALYSIS Leuchtbakterientest *m*

luminophore *n* WASTE Leuchtstoff *m*

luminous: ~ **intensity** *n* ANALYSIS Lichtintensität *f*; ~ **substance** *n* WASTE Leuchtstoff *m*

lung: ~-**irritant** *adj* SAFETY lungenreizend

lysimeter *n* ANALYSIS *leachate determination* Lysimeter *nt*

M

MAC *abbr* (*maximum allowable concentration*) AIR, ANALYSIS, SAFETY, WATER SUPP HZK (*höchstzulässige Konzentration*); **~ at the workplace** *n* SAFETY HZK am Arbeitsplatz *f*
macerate *vt* WASTE, WATER SUPP aufweichen
maceration *n* WASTE, WATER SUPP Aufweichen *nt*
machine: **~ enclosure** *n* NOISE Maschinenkapsel *f*; **~ failure** *n* NOISE, SOIL, WATER POLL Maschinenausfall *m*
Mach-Zehnder: **~ interferometer** *n* ANALYSIS Mach-Zehnder-Interferometer *nt*
macro: **~-climate** *n* AIR Makroklima *nt*
macroencapsulation *n* SOIL Makroeinkapselung *f*
macromolecular *adj* WASTE makromolekular
macropore *n* AIR *biological scrubber*, WATER POLL *activated carbon* Makropore *f*
macroscopic: **~ biological water quality determination** *n* ANALYSIS makroskopisch-biologische Wassergütebestimmung *f*
macrowaste *n* WASTE Makroabfall *m*
magma *n* RENEW ENERG *geothermal energy resource* Magma *nt*
magnesium *n* WASTE, WATER SUPP Magnesium *nt*; **~ hardness** *n* WATER SUPP Magnesiumhärte *f*; **~ hydroxide** *n* WATER SUPP Magnesiumhydroxid *nt*; **~ ion** *n* WATER SUPP Magnesiumion *nt*; **~ noncarbonate hardness** *n* WATER SUPP Magnesium-Nichtcarbonathärte *f*; **~ oxide scrubbing** *n* AIR *wet flue-gas desulphurization* Magnesiumoxid-Wäsche *f*; **~ sulfate** *AmE*, **~ sulphate** *BrE n* WASTE, WATER SUPP Magnesiumsulfat *nt*
magnet *n* ANALYSIS, SOIL, WASTE Magnet *m*
magnetic *adj* ANALYSIS, SOIL, WASTE magnetisch; **~ coupling** *n* WATER POLL Magnetkupplung *f*; **~ field** *n* ANALYSIS Magnetfeld *nt*; **~ oxygen analyser** *n BrE* ANALYSIS magnetischer Sauerstoffanalysator *m*; **~ oxygen analyzer** *AmE see magnetic oxygen analyser BrE*; **~ sector mass spectrometer** *n* (*magnetic sector MS*) ANALYSIS Sektorfeldmassenspektrometrie *f* (*Sektorfeld-MS*); **~ sector MS** *n* (*magnetic sector mass spectrometer*) ANALYSIS Sektorfeld-MS *f* (*Sektorfeldmassenspektrometrie*); **~ separation** *n* SOIL, WASTE Magnetabscheidung *f*, magnetische Abscheidung *f*, magnetische Sortierung *f*; **~ separator** *n* SOIL, WASTE, WATER SUPP Magnetabscheider *m*; **~ stirrer** *n* WATER POLL Magnetrührstab *m*; **~ susceptibility** *n* ANALYSIS magnetische Suszeptibilität *f*
main: **~ analysis** *n* ANALYSIS Hauptuntersuchung *f*; **~ catalyst** *n* AIR Hauptkatalysator *m*; **~ collector** *n* WASTE, WATER SUPP Hauptsammler *m*; **~ combustion chamber** *n* AIR, WASTE Hauptbrennkammer *f*; **~ display** *n* ANALYSIS *on measuring instruments* Hauptanzeige *f*; **~ drain** *n* WATER SUPP Sammelkanal *m*; **~ flow** *n* AIR Hauptströmung *f*; **~ gas stream** *n* AIR Hauptgasstrom *m*; **~ pipe** *n* WASTE, WATER SUPP Hauptsammler *m*; **~ pumping station** *n* WATER SUPP Hauptpumpwerk *nt*
mains *n* WATER SUPP Hauptwasserleitung *f*; **~ connection** *n* RENEW ENERG Netzanschluß *m*
main: **~ sealing layer** *n* WATER SUPP Hauptdichtung *f*; **~ stream** *n* AIR Hauptstrom *m*; **~ supply area** *n* WATER SUPP Hauptversorgungsgebiet *nt*
maintain *vt* GEN *ecological balance* im Gleichgewicht halten
maintenance *n* GEN Instandhaltung *f*, Wartung *f*
main: **~ wash circulation** *n* AIR *main stream oxidation* Hauptwaschkreislauf *m*; **~ wash circulation loop** *n* AIR *main stream oxidation* Hauptwaschkreislauf *m*; **~ waste air blower** *n* AIR Hauptabluftgebläse *nt*
major: **~ source** *n* AIR, WATER SUPP Hauptquelle *f*
make: **~-up water** *n* AIR, WATER SUPP Zusatzwasser *nt*
malfunction *n* SAFETY Störung *f*; **~ detection** *n* SAFETY Störungsüberwachung *f*
malodorous: **~ plume** *n* AIR übelriechender Schwaden *m*
mammoth: **~ pump** *n* WATER POLL Mammutpumpe *f*; **~ rotor** *n* WATER POLL Mammutrotor *m*
management *n* MANAGE, POLICY Verwaltung *f*
mandatory: **~ user charge** *n* POLICY verordnete Verbraucherabgabe *f*
manganese *n* AIR Mangan *nt*; **~ bacteria** *n pl* WATER SUPP Manganbakterien *f pl*; **~ hardness** *n* WATER SUPP Manganhärte *f*; **~ hydroxide** *n* WASTE, WATER SUPP Manganhydroxid *nt*
manhole *n* WATER POLL Mannloch *nt*
manmade *adj* GEN *caused by humans* anthropogen verursacht, *synthetic* künstlich; **~ dust** *n* AIR künstlicher Staub *m*; **~ earthquake** *n* SOIL anthropogen verursachtes Erdbeben *nt*; **~ fiber** *AmE*, **~ fibre** *BrE n* WASTE Kunstfaser *f*
manometer *n* ANALYSIS Manometer *nt*; **~ liquid** *n* ANALYSIS Manometerflüssigkeit *f*; **~ value** *n* ANALYSIS Manometerwert *m*
manometric *adj* ANALYSIS, WASTE, WATER POLL, WATER SUPP manometrisch; **~ discharge head** *n* WASTE, WATER POLL, WATER SUPP manometrische Förderhöhe *f*; **~ measurement** *n* ANALYSIS manometrische Messung *f*; **~ suction head** *n* WASTE, WATER POLL, WATER SUPP manometrische Saughöhe *f*; **~ switch** *n* AIR Druckschalter *m*
manual *adj* ANALYSIS, WASTE manuell, von Hand, WATER SUPP von Hand
manually: **~ loaded compactor** *n* WASTE manuell beladender Kompaktor *m*; **~-operated screen** *n* WATER POLL mechanischer Rechen *m*
manual: **~ processing** *n* WASTE manuelle Aufbereitung *f*; **~ separation** *n* WASTE Handsortierung *f*; **~ sorting** *n* WASTE Aussortierung *f* von Hand, Handsortierung *f*, manuelle Aussortierung *f*
manufactured: **~ gas** *n* AIR Industriegas *nt*, technisches Gas *nt*
marine *adj* MARINE marin, Meeres-; **~ climate** *n* MARINE Seeklima *nt*; **~ dredge** *n* MARINE Hochseebagger *m*; **~ ecology** *n* MARINE Meeresökologie *f*; **~ ecosystem** *n* MARINE marines Ökosystem *nt*, Ökosystem *nt* des Meeres; **~ pollution** *n* MARINE Meeresverschmutzung *f*
market *n* MANAGE Markt *m*
marketability *n* MANAGE *of recycled products* Vermarktungsfähigkeit *f*
marketable *adj* MANAGE absatzfähig, vermarktbar, *recycled products* marktgängig
market: **~-dependent** *adj* MANAGE marktabhängig; **~-economy** *adj* MANAGE, POLICY marktwirtschaftlich; **~-economy incentive** *n* MANAGE marktwirtschaftlicher Anreiz *m*; **~-economy instrument** *n* POLICY marktwirtschaftliches Instrument *nt*; **~ forces** *n pl* MANAGE, POLICY Marktkräfte *f pl*
marketing *n* MANAGE *of recycled products* Vermarktung *f*
market: **~ price** *n* POLICY Marktpreis *m*; **~ waste** *n* WASTE *waste type* Marktabfall *m*
marl *n* WASTE Mergel *m*
marly: **~ clay** *n* WASTE Mergelton *m*, Tonmergel *m*, WATER SUPP Tonmergel *m*
marsh *n* WATER SUPP Marsch *f*

marshiness *n* WATER SUPP Versumpfung *f*
marshland *n* WATER SUPP Marsch *f*
marsh: ~ **plants** *n pl* WATER SUPP Sumpfpflanzen *f pl*
masking: ~ **noise** *n* NOISE Verdeckungsgeräusch *nt*
masonry: ~ **dam** *n* RENEW ENERG, WATER SUPP Staumauer *f*
mass: ~ **together** *vt* AIR zusammenballen
mass: ~ **concentration** *n* AIR *gas, dust* Massenkonzentration *f*; ~ **flow** *n* ANALYSIS Mengenfluß *m*; ~ **flow controller** *n* ANALYSIS Mengenflußregelgerät *nt*; ~ **flow rate meter** *n* ANALYSIS Massendurchflußmesser *m*; ~ **flow regulator** *n* ANALYSIS Massenflußregler *m*; ~ **fragmentography** *n* ANALYSIS Massenfragmentographie *f*; ~**-selective** *adj* ANALYSIS massenselektiv; ~**-selective detector** *n* ANALYSIS massenselektiver Detektor *m*; ~ **selective detector** *n* ANALYSIS manuselektiver Detektor *m*; ~ **spectrometer** *n* ANALYSIS Massenspektrometer *nt*; ~ **spectrometry** *n* (*MS*) ANALYSIS Massenspektrometrie *f* (*MS*); ~ **spectroscope** *n* ANALYSIS *mass spectrometer or spectrograph* Massenspektroskop *nt*; ~ **spectroscopy** *n* ANALYSIS *process type within mass spectrometry* Massenspektroskopie *f*; ~ **spectrum** *n* ANALYSIS Massenspektrum *nt*; ~ **transfer** *n* AIR *biological gas scrubbing* Stoffübergang *m*; ~ **waste** *n* WASTE Massenabfall *m*
material 1. *adj* WASTE stofflich; **2.** *n* ANALYSIS Stoff *m*, POLICY Material *nt*, WASTE Werkstoff *m*, Material *nt*, WATER POLL Werkstoff *m*
material: ~ **exchange** *n* WASTE Stoffaustausch *m*; ~ **flow analysis** *n* POLICY, WATER SUPP Stoffflußbilanz *f*; ~ **landfill** *n* WATER SUPP Inertstoffdeponie *f*; ~ **pollution** *n* WATER SUPP Verschmutzung *f* durch Feststoffe
materials: ~ **balance** *n* MANAGE Stoffbilanz *f*, POLICY Materialbilanz *f*
material: ~ **separation operation** *n* WASTE Stofftrennprozeß *m*
materials: ~ **recycling** *n* WASTE stoffliche Verwertung *f*, stoffliches Recycling *nt*, werkstoffliches Recycling *nt*, WATER POLL stoffliche Verwertung *f*, stoffliches Recycling *nt*; ~ **saving** *n* WASTE Materialeinsparung *f*; ~ **stream** *n* WASTE Stoffstrom *m*; ~ **stream distributor** *n* WASTE Materialstromverteiler *m*
mathematical: ~ **climate model** *n* AIR mathematisches Klimamodell *nt*
matrix *n* WASTE Matrize *f*
matting *n* NOISE *surface lining* Matte *f*
mature: ~ **compost** *n* WASTE Reifekompost *m*
maximum: ~ **acceptable toxicant concentration** *n* (*MATC*) SAFETY MATC; ~ **allowable concentration** *n* (*MAC*) AIR, ANALYSIS, SAFETY, WATER SUPP höchstzulässige Konzentration *f* (*HZK*); ~ **allowable concentration at the workplace** *n* SAFETY höchstzulässige Konzentration *f* am Arbeitsplatz (*HAK*); ~ **allowable concentration in the free environment** *n* AIR, ANALYSIS maximale zulässige Immissionskonzentration *f* (*MIK-Wert*); ~ **allowable emissions concentration** *n* ANALYSIS MEK-Wert *m*, maximal zulässige Emissionskonzentration *f*; ~ **allowable organ concentration** *n* ANALYSIS MOK-Wert *m*, maximal zulässige Organkonzentration *f*; ~ **credible accident** *n* (*MCA*) SAFETY, WASTE größter anzunehmender Unfall *m* (*GAU*); ~ **daily demand** *n* WATER SUPP maximaler Tagesbedarf *m*; ~ **emissions concentration** *n* AIR, WASTE maximale Emissionskonzentration *f* (*MEK*); ~ **exposure** *n* SAFETY Höchstbelastung *f*; ~ **exposure limit** *n* SAFETY Höchstbelastungsgrenze *f*; ~ **exposure limits to hazardous substances** *n pl* SAFETY Höchstbelastungsgrenzen *f pl* für Gefahrstoffe; ~ **flow velocity** *n* WASTE, WATER SUPP maximale Fließgeschwindigkeit *f*; ~ **immission concentration** *n* AIR, WASTE maximale Immissionskonzentration *f*; ~ **level in consecutive time intervals** *n* NOISE Taktmaximalpegel *m*; ~ **sample quantity** *n* ANALYSIS Probenhöchstmenge *f*; ~ **tolerated dose** *n* (*MTD*) SAFETY maximal verträgliche Dosis *f* (*MTD*); ~ **water demand** *n* WATER SUPP maximaler Wasserbedarf *m*
MBG *abbr* (*middle thermal value biogas*) RENEW ENERG MBG (*mittelkaloriges Biogas*)
MCA *abbr* (*maximum credible accident*) SAFETY, WASTE GAU (*größter anzunehmender Unfall*)
MCFA *abbr* (*molten carbonate fuel cell*) AIR Karbonatschmelze-Brennstoffzelle *f*
mean: ~ **seasonal runoff** *n* WASTE, WATER SUPP mittlerer jahreszeitlicher Abfluß *m*; ~ **value** *n* ANALYSIS Mittelwert *m*
measurability *n* ANALYSIS Meßbarkeit *f*
measurable *adj* ANALYSIS meßbar
measure *vt* ANALYSIS messen, abmessen
measured: ~ **value** *n* ANALYSIS Meßwert *m*; ~ **value distortion** *n* ANALYSIS Meßwertverfälschung *f*; ~ **value processing** *n* ANALYSIS Meßwertverarbeitung *f*; ~ **value recording device** *n* ANALYSIS Meßwertregistriergerät *nt*; ~ **variable** *n* ANALYSIS *infrared absorption or caloric conductibility* Meßgröße *f*
measurement *n* AIR Messung *f*, ANALYSIS Abmessung *f*, Messung *f*, SAFETY Messung *f*; ~ **component** *n* ANALYSIS Meßkomponente *f*; ~ **conditions** *n pl* ANALYSIS Meßbedingungen *f pl*; ~ **of dust concentration** *n* ANALYSIS Staubkonzentrationsmessung *f*; ~ **of dust precipitation** *n* ANALYSIS Staubniederschlagsmessung *f*; ~ **error** *n* ANALYSIS Meßfehler *m*; ~ **of flow rate** *n* ANALYSIS Messung *f* der Strömungsgeschwindigkeit; ~ **of gaseous pollutants** *n* ANALYSIS Messung *f* von gasförmigen Schadstoffen; ~ **of gas flow velocity** *n* ANALYSIS Messung *f* der Gasstromgeschwindigkeit; ~ **of gas quantities** *n pl* ANALYSIS Messung *f* von Gasmengen; ~ **of intensity of scattered light** *n* ANALYSIS Messung *f* der Streulichtintensität; ~ **of oxygen concentration** *n* ANALYSIS Messung *f* der Sauerstoffkonzentration; ~ **plan** *n* ANALYSIS Meßplan *m*; ~ **prerequisites** *n pl* ANALYSIS Meßvoraussetzungen *f pl*; ~ **program** *AmE*, ~ **programme** *BrE n* ANALYSIS Meßprogramm *nt*; ~ **result** *n* ANALYSIS Meßergebnis *nt*; ~ **sensitivity control** *n* ANALYSIS Meßempfindlichkeitskontrolle *f*; ~ **task** *n* ANALYSIS Meßaufgabe *f*; ~ **technique** *n* ANALYSIS Meßtechnik *f*; ~ **of trace amounts** *n* ANALYSIS Spurenmessung *f*; ~ **vehicle** *n* ANALYSIS Meßfahrzeug *nt*, Meßwagen *m*
measure: ~ **of redevelopment** *n* SOIL, WASTE Sanierungsmaßnahme *f*
measuring *n* GEN Messung *f*; ~ **accuracy** *n* ANALYSIS Meßgenauigkeit *f*; ~ **amplifier** *n* ANALYSIS Meßverstärker *m*; ~ **bridge** *n* ANALYSIS Meßbrücke *f*; ~ **cell** *n* ANALYSIS *gauge head* Meßzelle *f*; ~ **chamber** *n* ANALYSIS *sensor chamber* Meßkammer *f*; ~ **channel** *n* ANALYSIS Meßkanal *m*; ~ **container** *n* ANALYSIS Meßbehälter *m*, Meßgefäß *nt*; ~ **cup** *n* ANALYSIS Meßbecher *m*, Meßkelch *m*; ~ **cuvette** *n* ANALYSIS Meßküvette *f*; ~ **data filtering** *n* ANALYSIS Meßdatenfilterung *f*; ~ **data scanning** *n* ANALYSIS Meßdatenabtastung *f*; ~ **device** *n* ANALYSIS Meßapparat *m*, Meßinstrument *nt*, Meßgerät *nt*; ~ **electrode** *n* ANALYSIS Meßelektrode *f*; ~ **element** *n* ANALYSIS Meßelement *nt*; ~ **equipment** *n* ANALYSIS Meßausstattung *f*, Meßeinrichtung *f*; ~ **error** *n* ANALYSIS Meßfehler *m*; ~ **fluid** *n* ANALYSIS Meßflüssigkeit *f*; ~ **flume** *n* WATER SUPP Meßgerinne *nt*, Meßkanal *m*; ~ **gage** *AmE see measuring gauge BrE*; ~ **gas** *n* ANALYSIS Meßgas *nt*; ~ **gas cooler** *n* ANALYSIS Meßgaskühler *m*; ~ **gauge** *n BrE* ANALYSIS Meßlehre *f*; ~ **head** *n* ANALYSIS Meßkopf *m*, Meßsonde *f*; ~ **head cable** *n* ANALYSIS Meßsondekabel *nt*; ~ **instrument** *n* ANALYSIS Meßapparat *m*, Meßinstrument *nt*, Meßgerät *nt*; ~ **instrument based on paramagnetism** *n* ANALYSIS *measurement of oxygen concentration* Meßgerät *nt* auf der Grundlage des Paramagnetismus; ~ **jar** *n* ANALYSIS Mensur *f*, Meßzylinder *m*; ~ **location** *n* ANALYSIS Meßort *m*, Meßstelle *f*; ~ **means** *n pl* ANALYSIS Meßmittel *nt*; ~ **method** *n* ANALYSIS Meßmethode *f*; ~ **nozzle** *n* ANALYSIS Meßdüse *f*; ~ **orifice** *n*

ANALYSIS Meßblende *f*; ~ **panel** *n* ANALYSIS Meßplatte *f*; ~ **period** *n* ANALYSIS Meßperiode *f*; ~ **pipette** *n* ANALYSIS Meßpipette *f*; ~ **point** *n* ANALYSIS Meßort *m*, Meßpunkt *m*, Meßstelle *f*; ~ **point of groundwater** *n* ANALYSIS Grundwassermeßstelle *f*; ~ **position** *n* ANALYSIS Meßplatz *m*; ~ **precision** *n* ANALYSIS Meßgenauigkeit *f*; ~ **principle** *n* ANALYSIS Meßprinzip *nt*; ~ **probe** *n* ANALYSIS Meßsonde *f*; ~ **procedure** *n* ANALYSIS Meßvorgang *m*; ~ **purpose** *n* ANALYSIS Meßzweck *m*; ~ **quantity** *n* ANALYSIS Meßgröße *f*; ~ **range** *n* ANALYSIS Meßbereich *m*; ~ **range deviation** *n* ANALYSIS Meßbereichsüberschreitung *f*; ~ **report** *n* ANALYSIS Meßprotokoll *nt*; ~ **resistance** *n* ANALYSIS Meßwiderstand *m*; ~ **resistor** *n* ANALYSIS Meßwiderstand *m*; ~ **result** *n* ANALYSIS Meßergebnis *nt*; ~ **rod** *n* ANALYSIS Maßstab *m*; ~ **scoop** *n* ANALYSIS Meßlöffel *m*; ~ **section** *n* ANALYSIS Meßstrecke *f*; ~ **sensitivity** *n* ANALYSIS Meßempfindlichkeit *f*; ~ **set-up** *n* ANALYSIS Meßanordnung *f*; ~ **shaft** *n* ANALYSIS Meßschacht *m*; ~ **signal** *n* ANALYSIS Meßsignal *nt*; ~ **sonde** *n* ANALYSIS Meßkopf *m*, Meßsonde *f*; ~ **station** *n* ANALYSIS Meßstation *f*; ~ **surface** *n* ANALYSIS Meßfläche *f*; ~ **switch** *n* ANALYSIS Meßumschalter *m*; ~ **time** *n* ANALYSIS Meßdauer *f*, Meßzeit *f*; ~ **transducer** *n* ANALYSIS *pH-measurement* Meßumformer *m*, Meßwertgeber *m*; ~ **tube** *n* ANALYSIS Meßrohr *nt*; ~ **with ultrasound** *n* ANALYSIS Messung *f* mit Ultraschall; ~ **unit** *n* ANALYSIS Maßeinheit *f*; ~ **vessel** *n* ANALYSIS Meßbehälter *m*, Meßgefäß *nt*; ~ **weir** *n* WATER SUPP Meßwehr *nt*

mechanical *adj* GEN maschinell, mechanisch; ~ **cleaning** *n* WATER POLL *of screen* mechanische Reinigung *f*; ~ **collector** *n* AIR mechanischer Abscheider *m*; ~ **composting** *n* WASTE Schnellkompostierung *f*, beschleunigte Kompostierung *f*, geschlossene Kompostierung *f*; ~ **energy** *n* RENEW ENERG mechanische Energie *f*; ~ **filtration** *n* WASTE, WATER SUPP mechanische Filtration *f*; ~ **hazard** *n* SAFETY mechanische Gefahr *f*; ~ **loss** *n* WASTE, WATER POLL, WATER SUPP mechanischer Verlust *m*

mechanically: ~ **assisted** *adj* GEN maschinell; ~ **assisted hand separation** *n* WASTE maschinell gestützte Handsortierung *f*; ~ **cleaned** *adj* WATER POLL mechanisch gereinigt; ~ **cleaned bar screen** *n* WATER POLL mechanisch gereinigter Rechen *m*; ~ **dewatered** *adj* WATER POLL maschinell entwässert, mechanisch entwässert; ~-**dewatered sludge** *n* WATER POLL maschinell entwässerter Schlamm *m*, mechanisch entwässerter Schlamm *m*; ~ **expand** *vt* SOIL mechanisch ausdehnen; ~ **thickened** *adj* WATER POLL maschinell entwässert, mechanisch entwässert; ~-**thickened sludge** *n* WATER POLL maschinell entwässerter Schlamm *m*, mechanisch entwässerter Schlamm *m*

mechanical: ~ **processing** *n* WASTE, WATER SUPP maschinelle Aufbereitung *f*, mechanische Aufbereitung *f*; ~ **purification** *n* WATER POLL mechanische Reinigung *f*; ~ **raking** *n* WATER POLL mechanischer Rechen *m*; ~ **recycling** *n* WASTE, WATER POLL mechanische Verwertung *f*, mechanisches Recycling *nt*; ~ **separation** *n* WASTE mechanische Trennung *f*; ~ **sludge-thickening method** *n* WATER POLL *vacuum filtration, centrifugation* mechanisches Entwässerungsverfahren *nt*; ~ **softening** *n* WASTE, WATER SUPP physikalische Enthärtung *f*; ~ **sweeper** *n* WASTE Straßenkehrmaschine *f*; ~ **waste separation** *n* WASTE automatische Müllsortierung *f*, mechanische Müllsortierung *f*, mechanische Mülltrennung *f*; ~ **wastewater treatment** *n* WATER POLL mechanische Abwasseraufbereitung *f*

median: ~ **lethal time** *n* AIR, ANALYSIS mittlere Letalzeit *f*

medium *n* MARINE, WASTE, WATER POLL Mittel *nt*; ~-**grained** *adj* WASTE *compost* mittelkörnig; ~-**pressure cleaning** *n* AIR Mitteldruckspülung *f*; ~-**scale** *adj* RENEW ENERG Mittel-; ~-**scale installation** *n* RENEW ENERG *wind, solar* mittlere Anlage *f*; ~-**size bubble aeration** *n* WATER POLL mittelblasige Belüftung *f*; ~-**sized refuse** *n* WASTE *grain size 8 to 40 mm* mittelgroßer Müll *m*; ~-**specific** *adj* ANALYSIS medienspezifisch ~-**specific component** *n* ANALYSIS medienspezifischer Bestandteil *m*

meet *vi* AIR *drops and particles* zusammentreffen

megawatt: ~ **plant** *n* RENEW ENERG MW-Anlage *f*, Megawatt-Anlage *f*

melt 1. *vt* SOIL, WASTE, WATER POLL aufschmelzen, schmelzen; ~ **down** ANALYSIS, WASTE einschmelzen; **2.** *vi* SOIL, WASTE, WATER POLL aufschmelzen, schmelzen; ~ **down** SAFETY *nuclear reactor core* schmelzen

melt: ~ **filter** *n* WASTE *plastics* Schmelzfilter *nt*

melting *n* WATER POLL Ausschmelzen *nt*, Schmelzen *nt*; ~ **chamber combustion firing** *n* WASTE Schmelzkammerfeuerung *f*; ~ **end** *n* WASTE Schmelzwanne *f*; ~ **flux electrolysis** *n* WASTE Schmelzflußelektrolyse *f*; ~ **point** *n* WASTE Schmelzpunkt *m*

melt: ~ **mixer** *n* WASTE Schmelzemischer *m*

membrane *n* GEN Membran *f*; ~ **biological reactor** *n* AIR Membranbioreaktor *m*; ~ **distillation** *n* WATER POLL Membrandestillation *f*; ~ **filter** *n* ANALYSIS, WASTE, WATER POLL, WATER SUPP Membranfilter *nt*; ~ **filter press** *n* SAFETY, WASTE, WATER POLL, WATER SUPP Membranfilterpresse *f*; ~ **filtration** *n* ANALYSIS, WATER POLL, WATER SUPP Membranfiltration *f*; ~ **layer** *n* WATER POLL, WATER SUPP Membranschicht *f*; ~ **permeability** *n* WATER POLL, WATER SUPP Membrandurchlässigkeit *f*; ~ **process** *n* SAFETY, WATER POLL *electrodialysis, ultrafiltration, reverse osmosis* Membranverfahren *nt*; ~ **reactor** *n* AIR Membranreaktor *m*; ~ **resistance** *n* WATER POLL, WATER SUPP Membranwiderstand *m*; ~ **separation process** *n* WATER POLL, WATER SUPP Membrantrennverfahren *nt*; ~ **surface** *n* ANALYSIS, WATER POLL, WATER SUPP Membranoberfläche *f*

mercaptan *n* AIR, WASTE Thiolalkohol *m*

mercurial: ~ **wastewater** *n* WATER POLL quecksilberhaltiges Abwasser *nt*

mercury *n* (*Hg*) SOIL, WASTE, WATER POLL Quecksilber *nt* (*Hg*); ~ **battery** *n* WASTE Knopfzelle *f*; ~-**containing** *adj* WASTE, WATER POLL Hg-haltig, quecksilberhaltig; ~-**containing lamp** *n* WASTE Hg-haltige Lampe *f*, quecksilberhaltige Lampe *f*; ~-**containing product** *n* WASTE quecksilberhaltiges Produkt *nt*; ~ **content** *n* SOIL, WASTE Quecksilbergehalt *m*; **mercury-free** *adj* WASTE, WATER POLL quecksilberfrei; ~-**free battery** *n* WASTE quecksilberfreie Batterie *f*; ~ **recovery** *n* WASTE *from batteries* Quecksilberrückgewinnung *f*; ~ **vapor lamp** *AmE*, ~ **vapour lamp** *BrE* *n* WASTE Quecksilberdampfmischlichtlampe *f*

mesh *n* WATER POLL Masche *f*

meshed *adj* WASTE, WATER SUPP vermascht; ~ **mains and supply network** *n* WASTE, WATER SUPP vermaschtes Hauptleitungs- und Versorgungsleitungsnetz *nt*

meshing *n* WASTE, WATER SUPP Vermaschung *f*

mesh: ~ **size** *n* WATER POLL *of screen* Maschengröße *f*

mesophilic *adj* WATER POLL mesophil; ~ **bacteria** *n pl* WATER POLL mesophile Bakterien *f pl*; ~ **digestion** *n* WATER POLL *anaerobic digestion* mesophile Faulung *f*

mesopore *n* AIR *biological scrubber* Mesopore *f*

mesoscale: ~ **circulation** *n* AIR Meso-Scale-Zirkulation *f*

mesosphere *n* AIR Mesosphäre *f*

metabolic: ~ **activity** *n* WASTE, WATER POLL *composting* Stoffwechseltätigkeit *f*; ~ **process** *n* WASTE Stoffwechselprozeß *m*; ~ **product** *n* AIR, WASTE Stoffwechselprodukt *nt*; ~ **rate** *n* WASTE *composting*, WATER POLL Stoffumsatz *m*

metabolism *n* AIR, WASTE, WATER POLL Stoffwechsel *m*

metabolite *n* WATER POLL Stoffwechselprodukt *nt*

metabolites: ~ **test** *n* ANALYSIS Metabolitentest *m*

metal *n* AIR, SOIL, WASTE, WATER POLL Metall *nt*; ~ **can** *n* WASTE Weißblechdose *f* für Konserven; ~ **can scrap** *n* WASTE Weißblechdosenschrott *m*; ~ **carrier** *n* AIR *catalyst* Metallträger *m*; ~ **coating** *n* WASTE Metallisierung *f*; ~ **complexes** *n pl* SOIL Metallverbindungen *f pl*; ~ **drum** *n*

WASTE Blechtrommel *f*; ~ **fungicides** *n pl* SOIL metallhaltiges Pflanzenschutzmittel *nt*; ~ **housing** *n* AIR Metallgehäuse *nt*
metalimnion *n* WATER POLL Metalimnion *nt*, Sprungschicht *f*
metallic: ~ **dust** *n* AIR metallischer Staub *m*; ~ **salts** *n pl* WATER POLL Metallsalze *nt pl*; ~ **sludge** *n* WASTE, WATER POLL Metallschlamm *m*
metalliferous *adj* WATER POLL metallhaltig; ~ **wastewater** *n* WATER POLL metallhaltiges Abwasser *nt*
metallurgical: ~ **slag** *n* WASTE metallurgische Schlacke *f*
metal: ~ **pipe** *n* AIR Metallrohr *nt*; ~ **plate** *n* AIR Metallplatte *f*; ~ **scrap** *n* WASTE Metallschrott *m*; ~ **separator** *n* WASTE Metallsortieranlage *f*; ~ **waste** *n* WASTE *production-specific waste* Metallabfall *m*, Metallschrott *m*
meteorological: ~ **conditions** *n pl* AIR Wetterlage *f*, Witterungsverhältnisse *nt pl*
meter 1. *n AmE see metre BrE*; **2.** ANALYSIS *measuring device* Meter *m*, Zähler *m*, Messer *m*, Meßgerät *nt*; **3.** *vt* ANALYSIS messen
meter: ~ **housing** *n* ANALYSIS Meßgerätegehäuse *nt*
metering: ~ **equipment** *n* ANALYSIS, WASTE, WATER SUPP Dosiervorrichtung *f*; ~ **machine** *n* WASTE, WATER SUPP Dosiermaschine *f*; ~ **pump** *n* WATER POLL Beschickungspumpe *f*, Dosierpumpe *f*; ~ **unit** *n* ANALYSIS, WASTE, WATER SUPP Dosiereinheit *f*, Dosiervorrichtung *f*; ~ **worm** *n* WATER POLL Dosierschnecke *f*
meter: ~ **terminal** *n* ANALYSIS Meßgeräteklemme *f*
methane *n* AIR *trace gas*, WASTE *landfill gas component*, WATER POLL Methan *nt*; ~ **build-up** *n* WASTE, WATER POLL Methanbildung *f*; ~ **fermentation** *n* WASTE Methangärung *f*; ~ **former** *n* WASTE, WATER POLL *stabilization* Methanbildner *m*; ~ **reactor** *n* WASTE Methanreaktor *m*; ~ **recovery** *n* WASTE Methangewinnung *f*
methanogenesis *n* WASTE Methanbildung *f*, Methanerzeugung *f*, WATER POLL Methanbildung *f*, Methanogenese *f*
methanogenic *adj* WATER POLL methanogen; ~ **bacteria** *n pl* WATER POLL methanogene Bakterien *f pl*
methanolysis *n* WASTE Methanolyse *f*
method *n* GEN Methode *f*, Verfahren *nt*; ~ **of application** *n* AIR, ANALYSIS, SAFETY, SOIL, WATER POLL, WATER SUPP Anwendungsverfahren *nt*; ~ **of determination** *n* ANALYSIS Bestimmungsmethode *f*
methylene: ~ **blue test** *n* ANALYSIS Methylenblauprobe *f*, Methylenblautest *m*; ~ **chloride** *n* AIR Methylenchlorid *nt*
metre *n BrE* ANALYSIS *unit of length* Meter *m*
metrology: ~**-related** *adj* ANALYSIS meßtechnisch
mg/kg *abbr* (*milligram per kilogram, milligramme per kilogramme*) ANALYSIS mg/kg (*Milligramm pro Kilogramm*)
Michelson: ~ **interferometer** *n* ANALYSIS *FTIR* Michelson-Interferometer *nt*
micro *n* ANALYSIS Mikrometer *nt* (*μm*)
microbe *n* WATER POLL Mikroorganismus *m*, Mikrobe *f*
microbial *adj* SOIL, WATER POLL mikrobiell; ~ **activity** *n* SOIL, WATER POLL Mikrobenaktivität *f*; ~ **community** *n* SOIL, WATER POLL Mikrobenpopulation *f*; ~ **decomposition** *n* SOIL mikrobieller Abbau *m*; ~ **degradation** *n* SOIL mikrobieller Abbau *m*; ~ **leaching** *n* SOIL mikrobielle Auslaugung *f*; ~ **population** *n* SOIL, WATER POLL Mikrobenpopulation *f*; ~ **respiration** *n* WATER POLL mikrobielle Atmung *f*
microbially: ~ **regenerate** *vt* AIR regenerieren mikrobiell
microbiological *adj* GEN mikrobiologisch; ~ **decomposition** *n* WATER POLL mikrobiologischer Abbau *m*; ~ **metabolic process** *n* AIR mikrobiologischer Stoffwechselprozeß *m*; ~ **processes** *n pl* SOIL *soil treatment* mikrobiologische Verfahren *nt pl*; ~ **testing** *n* ANALYSIS mikrobiologische Untersuchung *f*
microclimate *n* AIR Mikroklima *nt*
microclimatic: ~ **changes** *n pl* RENEW ENERG Mikroklimaveränderungen *f pl*
microcoagulation *n* WATER POLL Mikroflockung *f*
microcoulometric *adj* ANALYSIS mikrocoulometrisch; ~ **titration** *n* ANALYSIS mikrocoulometrische Titration *f*
microcoulometry *n* ANALYSIS Mikrocoulometrie *f*
microdipper *n* ANALYSIS *dosing of liquids* Mikrodipper *m*
microencapsulation *n* SOIL Mikroeinkapselung *f*
microfilter *n* WATER POLL Feinfilter *nt*, Mikrosiebfilter *nt*
microfiltration *n* WATER POLL Mikrofiltration *f*, Feinfiltration *f*
microfloc *n* WATER SUPP Mikroflocke *f*
micromachined *adj* ANALYSIS mikromaterialbearbeitet
μm *abbr* (*micrometer AmE, micrometre BrE*) ANALYSIS μm (*Mikrometer*)
micrometer 1. *n AmE unit of length see micrometre BrE*; **2.** *measuring device* Meßschraube *f*, Mikrometerschraube *f*
micrometre *n* (*μm*) *BrE* ANALYSIS *unit of length* Mikrometer *nt* (*μm*)
micronutrient *n* WATER POLL Mikronährstoff *m*
microorganism *n* AIR *biological sorption process*, SOIL, WASTE, WATER POLL, WATER SUPP Mikroorganismus *m*, Mikrobe *f*
microphone *n* ANALYSIS Mikrofon *nt*
micropore *n* AIR *biological scrubber*, WATER POLL *activated carbon* Mikropore *f*
microporous *adj* AIR, WASTE, WATER POLL, WATER SUPP feinporig, mikroporös; ~ **membrane** *n* WASTE, WATER POLL, WATER SUPP mikroporöse Membran *f*
microprocessor *n* ANALYSIS *built-in in analyser* Mikroprozessor *m*; ~**-controlled PID** *n* ANALYSIS mikroprozessorgesteuerter PID *m*
microscope *n* ANALYSIS Mikroskop *nt*
microscopic *adj* ANALYSIS mikroskopisch; ~ **dust** *n* SAFETY mikroskopischer Staub *m*; ~ **qualitative** *adj* ANALYSIS mikroskopisch qualitativ; ~ **qualitative detection method** *n* ANALYSIS mikroskopisch qualitative Nachweismethode *f*
microscreening *n* WATER POLL Mikrosiebung *f*
microwave: ~ **process** *n* WASTE *plastics sorting* Mikrowellenverfahren *nt*
middle: ~ **thermal value biogas** *n* (*MBG*) RENEW ENERG mittelkaloriges Biogas *nt* (*MBG*)
mid: ~**-temperature heat** *n* RENEW ENERG *from solar energy* Mitteltemperaturwärme *f*; ~**-temperature incineration** *n* AIR Mitteltemperaturverbrennung *f*
Mie: ~ **scattering** *n* AIR, ANALYSIS Mie-Streuung *f*
migrate *vi* SOIL, WASTE *of landfill gas* migrieren
migrating: ~ **gas** *n* SOIL, WASTE migrierendes Gas *nt*
migration: ~ **of harmful substances** *n* SOIL Migration *f* der Schadstoffe, Schadstoffdiffusion *f*, Schadstoffwandlung *f*, WASTE Migration *f* der Schadstoffe, Schadstoffwandlung *f*
mildly: ~ **acidic** *adj* SOIL schwach sauer
military: ~ **contaminated sites** *n pl* SOIL Rüstungsaltlasten *f pl*
milligramme *see milligram*
milligram: ~ **per kilogram** *phr* ANALYSIS Milligramm pro Kilogramm (*mg/kg*)
mined: ~ **space** *n* WASTE bergmännisch hergestellter Hohlraum *m*
mineral: ~ **aggregate landfill** *n* WASTE Mineralstoffdeponie *f*; ~ **dust** *n* AIR mineralischer Staub *m*; ~ **fiber** *AmE see mineral fibre BrE*; ~ **fibre** *BrE n* AIR Mineralfaser *f*, NOISE *material for airborne noise damping* Mineralfaser *f*, Mineralwolle *f*, Steinwolle *f*; ~ **fibre absorber** *n BrE* NOISE Mineralfaserabsorber *m*; ~ **fibre panel** *n BrE* MARINE, NOISE Mineralwolleplatte *f*
mineralization *n* AIR Mineralisierung *f*, SOIL *biological treatment* Inertisierung *f*, Mineralisierung *f*, WASTE, WATER POLL Mineralisierung *f*, Mineralisation *f*; ~ **level** *n* WASTE Mineralisierungsgrad *m*
mineralize *vt* GEN mineralisieren
mineral: ~**-matter-free** *adj* WASTE, WATER SUPP mineralstoffrei; ~ **oil** *n* WASTE Mineralöl *nt*; ~ **oil product**

n WASTE Mineralölprodukt *nt*; **~ oil tax** *n* POLICY Mineralölsteuer *f*; **~ oil waste** *n* WASTE *production-specific waste* Mineralölabfall *m*; **~ sealing** *n* WASTE *base-sealing* mineralische Dichtung *f*, *out of mineral mixed clay and coarse clay* mineralische Abdichtung *f*; **~ sealing layer** *n* WASTE mineralische Dichtungsschicht *f*; **~ spring** *n* WATER SUPP Heilbrunnen *m*, Mineralquelle *f*; **~ wool** *n* NOISE Mineralwolle *f*

minimal: **~ daily demand** *n* WATER SUPP minimaler Tagesbedarf *m*; **~ water demand** *n* WATER SUPP minimaler Wasserbedarf *m*

minimization *n* AIR, WASTE Minimierung *f*

minimize *vt* AIR *emissions*, MANAGE *costs*, WASTE minimieren

minimum: **~ flow** *n* WATER SUPP niedrigster Niedrigwasserabfluß *m*; **~ flow velocity** *n* WASTE, WATER SUPP minimale Fließgeschwindigkeit *f*; **~ low water level** *n* WATER SUPP niedrigster Niedrigwasserstand *m*; **~ sample quantity** *n* ANALYSIS Mindestprobemenge *f*

ministry: **~ of environment** *n* POLICY Umweltministerium *nt*

mirror *n* RENEW ENERG *solar tower* Spiegel *m*; **~ evacuated tube** *n* RENEW ENERG *non-concentrating collector* Vakuumkollektor *m*

miscibility *n* AIR, WATER POLL Mischbarkeit *f*

miscible: **~ substance** *n* AIR, WASTE, WATER POLL, WATER SUPP mischbare Substanz *f*

mismanagement *n* MANAGE, POLICY Verschwendung *f*

missing: **~ data** *n* MANAGE *in eco-balance* fehlende Angaben *f pl*

mist *n* AIR *containing liquid particulates* leichter Nebel *m*; **~ drops** *n pl* AIR *from stack* Nebeltropfen *m*; **~ eliminator** *n* AIR Tropfenabscheider *m*

mitigation *n* POLICY Linderung *f*, Minderung *f*

mix: **~ completely** *vt* WASTE, WATER SUPP volldurchmischen; **~ thoroughly** *vt* ANALYSIS durchmischen

mixed *adj* GEN gemischt; **~-bed ion exchanger** *n* WATER POLL Mischbettionenaustauscher *m*; **~ canalization** *n* WATER POLL Mischkanalisation *f*; **~-color** *AmE see mixed-colour BrE*; **~-color glass cullet** *AmE see mixed-colour glass cullet BrE*; **~-colour** *BrE adj* WASTE farblich gemischt; **~-colour glass cullet** *n BrE* WASTE farblich gemischtes Bruchglas *nt*; **~ culture** *n* WATER POLL Mischkultur *f*; **~-flow pump** *n* WATER POLL Schraubenradpumpe *f*; **~ gas** *n* AIR Mischgas *nt*; **~ glass** *n* WASTE Mischglas *nt*; **~ glass cullet** *n* WASTE Mischglasscherben *f pl*; **~ liquor suspended solids** *n pl* (*MLSS*) WATER POLL Trockensubstanzgehalt *m* im Belebungsbecken (*TSBB*); **~ plastics** *n pl* WASTE gemischte Kunststoffe *m pl*; **~ primary sludge** *n* WATER POLL gemischter Primärschlamm *m*; **~ salt** *n* WASTE Mischsalz *nt*; **~ sludge** *n* WATER POLL Mischschlamm *m*; **~ water** *n* WATER SUPP Mischwasser *nt*

mixer: **~-settler extractor** *n* WATER POLL Mixer-Settler-Extraktor *m*

mixing *n* AIR Rühren *nt*, SOIL Mischen *nt*, Rühren *nt*, WASTE, WATER POLL, WATER SUPP Mischung *f*, Rühren *nt*; **~ with aggregates** *n* SOIL Mischen *nt* mit Zuschlagsstoffen; **~ device** *n* WATER POLL Mischer *m*; **~ device evaporator** *n* WATER POLL Rührwerksverdampfer *m*; **~ tank** *n* WATER POLL, WATER SUPP Mischbecken *nt*; **~ worm** *n* WATER POLL Mehrkammerabsetzbecken *nt*, Mischschnecke *f*

MLSS *abbr* (*mixed liquor suspended solids*) WATER POLL TSBB (*Trockensubstanzgehalt im Belebungsbecken*)

mobile *adj* GEN mobil, beweglich; **~ component** *n* AIR, WASTE bewegliche Komponente *f*; **~ crushing unit** *n* WASTE mobile Zerkleinerungsanlage *f*; **~ emission sources** *n pl* AIR bewegliche Emissionsquellen *f pl*; **~ field-measuring technique** *n* ANALYSIS mobile Feldmeßtechnik *f*; **~ lab** *n* ANALYSIS mobiles Labor *nt*; **~ laboratory** *n* ANALYSIS Mobillabor *nt*, mobiles Labor *nt*; **~ laboratory-measuring technique** *n* ANALYSIS mobile Labormeßtechnik *f*; **~ measuring unit** *n* ANALYSIS mobile Meßstation *f*; **~ pumping unit** *n* WATER POLL bewegliche Pumpanlage *f*; **~ sampler** *n* ANALYSIS mobiles Probenahmegerät *nt*; **~ unit** *n* SOIL *for soil treatment* mobile Anlage *f*

mobilization *n* SOIL *of heavy metals*, WASTE *of pollutants*, WATER SUPP *of harmful substances* Mobilisierung *f*

mobilize *vt* SOIL, WASTE, WATER SUPP mobilisieren

mobilized *adj* SOIL, WASTE, WATER SUPP mobilisiert

model: **~ calculation** *n* AIR Modellberechnung *f*; **~ calibration** *n* ANALYSIS Modelleichung *f*

modeling: **~ of water quality** *AmE see modelling of water quality BrE*

modelling: **~ of water quality** *n BrE* WATER SUPP Modellierung *f* der Gewässerqualität

model: **~ operation permission** *n* POLICY Typenzulassung *f*

moderately: **~ acidic** *adj* SOIL *soil* mäßig sauer

modernization *n* POLICY *of tax system* Modernisierung *f*; **~ of existing plants** *n* WATER POLL Altanlagenmodernisierung *f*

modernize *vt* POLICY *tax system*, WASTE, WATER POLL, WATER SUPP modernisieren

modified: **~ soil materials** *n pl* SOIL, WATER SUPP *landfill sealing* modifizierte Bodenmaterialien *nt pl*

modifier *n* GEN Mittel *nt*

modular *adj* ANALYSIS *analysis system* modular

module: **~ sound level meter** *n* ANALYSIS Modulschallpegelmesser *m*

Mohr: **~ balance** *n* ANALYSIS Mohrsche Waage *f*

moisture *n* ANALYSIS Feuchte *f*, WASTE Feuchtigkeit *f*; **~ content** *n* AIR *dust*, RENEW ENERG *biomass*, WASTE, WATER POLL Feuchtgehalt *m*, Feuchtigkeitsgehalt *m*; **~ content determination** *n* ANALYSIS Feuchtgehaltsbestimmung *f*; **~ determination** *n* WASTE Feuchtigkeitsbestimmung *f*; **~ eliminator** *n* AIR Tropfenabscheider *m*; **~ of installation** *n* WASTE Einbaufeuchte *f*; **~-proof** *adj* WASTE, WATER SUPP feuchtigkeitsbeständig; **~-resistant** *adj* WASTE, WATER SUPP feuchtigkeitsfest; **~ sensor** *n* ANALYSIS Feuchtesensor *m*; **~ test** *n* WASTE Feuchtigkeitsprobe *nt*

molar: **~ volume** *n* AIR Molvolumen *nt*

mold *AmE see mould BrE*

molded *AmE see moulded BrE*

mole *n* (*gram-molecule*) AIR Mol *nt*

molecular *adj* GEN Molekular-; **~ colloid** *n* WATER POLL, WATER SUPP Molekülkolloid *nt*; **~ force** *n* AIR *adsorption* Molekularkraft *f*; **~ sieve** *n* AIR, ANALYSIS, SOIL, WATER POLL, WATER SUPP Molekularsieb *nt*; **~ sieving** *n* AIR, ANALYSIS, SOIL, WATER POLL, WATER SUPP Molekularsiebung *f*; **~ spectrometry** *n* ANALYSIS Molekülspektrometrie *f*; **~ weight** *n* WASTE, WATER POLL, WATER SUPP Molekulargewicht *nt*

molecule *n* GEN Molekül *nt*

mol: **~ ratio** *n* AIR Molverhältnis *nt*

molten: **~ carbonate fuel cell** *n* (*MCFA*) AIR Karbonatschmelze-Brennstoffzelle *f*; **~ metal** *n* SAFETY geschmolzenes Metall *nt*; **~ slag** *n* WASTE Schlackenschmelz *m*

monitor **1.** *n* ANALYSIS Monitor *m*; **2.** *vt* AIR, ANALYSIS, MARINE, SAFETY, SOIL, WASTE, WATER SUPP beobachten, überwachen, kontrollieren

monitoring *n* GEN Beobachtung *f*, *after site closure* Nachsorge *f*, Monitoring *nt*, Überwachung *f*; **~ of the effectiveness of waste gas installations** *n* ANALYSIS Überwachung *f* der Wirksamkeit von Abgasreinigungsanlagen; **~ instrument** *n* POLICY Überwachungsinstrument *nt*; **~ measure** *n* SAFETY Überwachungsmaßnahme *f*; **~ network** *n* ANALYSIS Meßstellennetz *nt*; **~ obligation** *n* POLICY, SAFETY Überwachungspflicht *f*; **~ system** *n* ANALYSIS, SAFETY Überwachungssystem *nt*; **~ time** *n* ANALYSIS Beobachtungszeit *f*; **~ of toxicity at the workplace** *n* SAFETY Kontrolle *f* der Giftigkeit am Arbeitsplatz; **~ well** *n* ANALYSIS, WASTE, WATER SUPP Beobachtungsbrunnen *m*, Kontrollbrunnen *m*

monitor: **~ tube** *n* ANALYSIS *mass spectroscopy* Monitorröhre *f*

monochlorobenzene *n* AIR Monochlorbenzol *nt*
mono: **~-dump** *n* WASTE, WATER POLL Monodeponie *f*
monolith *n* AIR *catalyst* Monolith *m*
monolithic *adj* AIR monolithisch
monomer *n* WASTE Monomer *nt*
monthly: **~ mean value** *n* ANALYSIS Monatsmittelwert *m*
moor 1. *n* WATER SUPP Moor *nt*; **2.** *vt* MARINE befestigen, verankern
mooring *n* MARINE Mooring *nt*, Verankern *nt*; **~ bracket** *n* MARINE Befestigungsgeschirr *nt*; **~ buoy** *n* MARINE Ankerboje *f*, Vertäuboje *f*; **~ chain** *n* MARINE Verankerungskette *f*
moorland *n* WATER SUPP Moor *nt*
morainic: **~ filter layer** *n* WASTE Moränenfilterschicht *f*
motion: **~ of the sea** *n* RENEW ENERG Seegang *m*
motor *n* AIR, MARINE, WASTE, WATER SUPP Motor *m*; **~ pump** *n* MARINE, WATER SUPP Motorpumpe *f*; **~ vehicle** *n* AIR Kraftfahrzeug *nt* (*Kfz*); **~ vehicle exhaust gas** *n* AIR Kfz-Abgas *nt*, Kraftfahrzeugabgas *nt*
mould 1. *n BrE* WASTE Form *f*; **2.** *vt BrE* WASTE *reprocessed plastics* formen
moulded: **~ parts using recycled plastics for core material** *n pl BrE* WASTE Formteile *nt pl* unter Verwendung von Recyclingrohstoffen
mould: **~-filling technique** *n BrE* WASTE Formfülltechnik *f*
movable: **~ blade** *n* AIR drehbarer Flügel *m*; **~ bulkhead** *n* WASTE bewegliche Trennwand *f*; **~ clean air feed** *n* AIR verstellbarer Spülluftkanal *m*; **~ fittings** *n pl* AIR bewegliche Einbauten *m pl*
movement: **~ of groundwater** *n* WATER SUPP Bewegung *f* des Grundwassers
moving: **~ bed** *n* AIR, WATER SUPP Fließbett *nt*, Wanderbett *nt*; **~-bed adsorber** *n* AIR Fließbettadsorber *m*, Wanderbettadsorber *m*, Wanderschichtadsorber *m*, WATER POLL Wanderschichtadsorber *m*; **~-bed dryer** *n* WATER POLL Fließbetttrockner *m*; **~-bed reactor** *n* SOIL, WASTE Fließbettreaktor *m*; **~-bed sand filter** *n* WATER POLL, WATER SUPP Fließbettsandfilter *nt*; **~-bed zone** *n* AIR Wanderbettzone *f*; **~ text** *n* WATER POLL Fließbett *nt*
MS *abbr* (*mass spectrometry*) ANALYSIS MS (*Massenspektrometrie*)
MSW *abbr* (*municipal solid waste*) WASTE Hausmüll *m*, Siedlungsabfall *m*
MTD *abbr* (*maximum tolerated dose*) SAFETY MTD (*maximal verträgliche Dosis*)
mud: **~ blanket** *n* WATER POLL Schmutzdecke *f*; **~ content** *n* WATER POLL Schlammgehalt *m*
muffle *vt* NOISE dämpfen
muffler *n AmE* (*cf silencer BrE*) NOISE Dämpfer *m*, Schalldämpfer *m*
mullite *n* AIR *ceramic carrier* Mullit *m*
multibarrier: **~ principle** *n* WASTE Mehrschichtprinzip *nt*
multichamber *adj* AIR, WASTE Mehrkammer-; **~ bag filter** *n* AIR Mehrkammerschlauchfilter *nt*, Schlauchhäuser *nt pl*
multicompartment *adj* AIR, WASTE Mehrkammer-; **~ container** *n* WASTE Mehrkammer-Container *m*
multifoil: **~ recycled material** *n* WASTE Recyclate *nt pl* aus Mehrschichtfolien
multilateral *adj* POLICY mehrseitig; **~ negotiation** *n* POLICY mehrseitige Verhandlung *f*
multilayer *adj* WASTE mehrschichtig
multilayered: **~ construction** *n* WASTE mehrschichtiger Aufbau *m*; **~ container** *n* WASTE mehrschichtiger Behälter *m*; **~ filtration** *n* WASTE, WATER POLL, WATER SUPP mehrschichtige Filtration *f*
multilayer: **~ filter** *n* WATER POLL Mehrschichtfilter *nt*
multimaterial: **~ recycling** *n* WASTE Recycling *nt* von Multimaterialien; **~ system** *n* WASTE Mehrstoffsystem *nt*
multiple: **~-blade machine** *n* RENEW ENERG *wind turbine* Vielflügler *m*; **~ material container** *n* WASTE Combicontainer *m*; **~ measuring stations** *n pl* ANALYSIS Mehrfachmeßstellen *f pl*
multirange *adj* ANALYSIS Mehrbereich-; **~ instrument** *n* ANALYSIS Mehrbereichsinstrument *nt*
multisheet *adj* NOISE *noise prevention walls* mehrschalig
multistage *adj* GEN mehrstufig; **~ activated sludge process** *n* WATER POLL mehrstufiges Belebungsverfahren; **~ cyclone** *n* ANALYSIS mehrstufiger Zyklon *m*; **~ evaporator** *n* WATER POLL Mehrstufenverdampfer *m*; **~ fluidized bed** *n* AIR mehrstufige Wirbelschicht *f*; **~ fluidized-bed adsorber** *n* AIR mehrstufiger Wirbelschichtadsorber *m*; **~ peripheral pump** *n* WASTE, WATER POLL, WATER SUPP mehrstufige Peripheralpumpe *f*
multitube: **~ cyclone** *n* AIR Mehrröhrenzyklon *m*
multiunit: **~ container** *n* WASTE Mehreinheit-Container *m*; **~ cyclone** *n* AIR Multizyklon *m*, vielzelliger Zyklon *m*
multivalent *adj* WASTE mehrwertig, polyvalent; **~ cation** *n* WASTE, WATER SUPP mehrwertiges Kation *nt*
municipal *adj* POLICY, WATER POLL, WATER SUPP kommunal; **~ environmental protection** *n* POLICY kommunaler Umweltschutz *m*; **~ solid waste** *n* (*MSW*) WASTE Haushaltsabfall *m*, Hausmüll *m*, Siedlungsabfall *m*; **~ waste collection** *n* WASTE Hausmüllabfuhr *f*; **~ waste collection load** *n* WASTE Müllabfuhrlast *f*; **~ waste composting** *n* WASTE Hausmüllkompostierung *f*; **~ waste landfill** *n* WASTE Hausmülldeponie *f*; **~ waste stream** *n* (*MWS*) WASTE Siedlungsabfallstrom *m*; **~ wastewater** *n* WATER POLL kommunales Abwasser *nt*; **~ wastewater discharge** *n* WATER POLL kommunale Abwassereinleitung *f*; **~ wastewater plant** *n* WATER POLL kommunale Kläranlage *f* (*KKA*); **~ wastewater sludge** *n* (*MWS*) WASTE, WATER POLL kommunaler Abwasserschlamm *m*, kommunaler Klärschlamm *m*
mutagen *n* SAFETY mutagene Substanz *f*
mutagenic *adj* SAFETY mutagen
MWS *abbr* WASTE (*municipal wastewater sludge*) kommunaler Klärschlamm *m*, (*municipal waste stream*) Abwasserschlamm *m*, Siedlungsabfallstrom *m*, WATER POLL (*municipal wastewater sludge*) kommunaler Abwasserschlamm *m*, kommunaler Klärschlamm *m*

N

N *abbr* AIR (*nitrogen*) N (*Stickstoff*), ANALYSIS (*newton*) N (*Newton*), WASTE, WATER POLL (*nitrogen*) N (*Stickstoff*)
Na *abbr* (*sodium*) AIR, WATER SUPP Na (*Natrium*)
nanogram *n* ANALYSIS Nanogramm *nt* (*ng*)
nanogramme *see nanogram*
nanometer *n* ANALYSIS Nanometer *nt* (*nm*)
narrow: **~ curtain** *n* WATER SUPP Schmalwand *f*
national: **~ environmental protection** *n* POLICY nationaler Umweltschutz *m*
natural *adj* GEN natürlich; **~ acidification** *n* AIR, SOIL, WASTE, WATER SUPP natürliche Säuerung *f*; **~ convection** *n* WATER POLL natürliche Konvektion *f*; **~ dewatering system** *n* WATER POLL *sludge lagoon, sand drying bed etc* natürliches Entwässerungssystem *nt*; **~ drainage** *n* WASTE, WATER SUPP natürliche Entwässerung *f*; **~ ecosystem** *n* AIR, SOIL, WATER SUPP natürliches Ökosystem *nt*; **~ energy balance** *n* RENEW ENERG natürliche Energiebilanz *f*; **~ energy source** *n* RENEW ENERG natürliche Energiequelle *f*; **~ environment** *n* AIR, POLICY, WATER SUPP natürliche Umwelt *f*; **~ greenhouse effect** *n* AIR natürlicher Treibhauseffekt *m*; **~ groundwater recharge** *n* WATER SUPP natürliche Grundwasseranreicherung *f*, natürliche Grundwasserneubildung *f*; **~ gypsum** *n* AIR Naturgips *m*; **~ hydrological cycle** *n* WASTE, WATER SUPP natürlicher Wasserkreislauf *m*; **~ infiltration** *n* WASTE, WATER SUPP natürliche Infiltration *f*; **~ lining** *n* WASTE natürliche Abdichtung *f*
naturally: **~ acid lake** *n* WATER SUPP natursaurer See *m*
natural: **~ ozone content** *n* AIR natürlicher Ozongehalt *m*; **~ percolation** *n* WASTE, WATER SUPP natürliche Versickerung *f*; **~ pollutant** *n* AIR, SOIL, WATER POLL natürlicher Schadstoff *m*; **~ purification** *n* SOIL, WATER POLL, WATER SUPP Selbstreinigung *f*; **~ purification force** *n* SOIL, WATER POLL, WATER SUPP Selbstreinigungskraft *f*; **~ sedimentation** *n* WASTE, WATER POLL, WATER SUPP natürliche Sedimentation *f*; **~ sedimentation process** *n* WASTE, WATER POLL, WATER SUPP natürlicher Sedimentationsvorgang *m*; **~ self-purification** *n* WASTE, WATER POLL, WATER SUPP natürliche Selbstreinigung *f*; **~ soil materials** *n pl* WASTE natürliche Bodenmaterialien *nt pl*; **~ sources** *n pl* AIR natürliche Quellen *f pl*
NC: **~-curve** *n* (*noise criterion curve*) ANALYSIS Lärmgrenzwertkurve *f*, Lärmkriteriumkurve *f*
NCH *abbr* (*noncarbonate hardness*) WASTE bleibende Härte *f*, WATER POLL, WATER SUPP NCH (*Nichtkarbonathärte*), bleibende Härte *f*, permanente Härte *f*
NCI *abbr* (*negative chemical ionization*) ANALYSIS NCI (*negative chemische Ionisation*)
NCV *abbr* (*net calorific value*) AIR, WASTE spezifischer Heizwert *m*, unterer Heizwert *m*
NDIR *abbr* (*nondispersive infrared*) ANALYSIS NDIR (*nichtstreuendes Infrarot*)
NDIRA *abbr* (*nondispersive infrared absorption*) ANALYSIS NDIRA (*nichtstreuende Infrarot-Absorption*)
NDIR: **~ spectrometry** *n* ANALYSIS NDIR Spektrometrie *f*
Ne *abbr* (*neon*) AIR Ne (*Neon*)
neap: **~ tides** *n pl* RENEW ENERG *tidal energy* Nipptiden *f pl*
near: **~ infrared spectroscopy** *n* ANALYSIS Nah-Infrarot-Spektroskopie *f*
neat *adj* MARINE, WASTE, WATER SUPP unverdünnt, unvermischt
nebulization *n* AIR, WASTE Zerstäubung *f*
nebulizer *n* AIR, WASTE Zerstäuber *m*
needled: **~ felt** *n* AIR *common early filter medium* Nadelfilz *m*
NEF *abbr* (*noise exposure forecast*) NOISE Lärmimmissionsprognosewert *m*
negative *adj* AIR, WASTE negativ; **~ catalysis** *n* AIR Antikatalyse *f*, Reaktionsverzögerung *f*; **~ chemical ionization** *n* (*NCI*) ANALYSIS negative chemische Ionisation *f* (*NCI*); **~ electrode** *n* AIR Kathode *f*, negative Elektrode *f*; **~ pressure** *n* AIR, SOIL *suction of soil atmosphere* Unterdruck *m*; **~ skin friction** *n* WASTE negative Mantelreibung *f*
negotiation *n* POLICY Abstimmung *f*, Verhandlung *f*
neon *n* (*Ne*) AIR Neon *nt* (*Ne*)
nephelometer *n* ANALYSIS Trübungsmesser *m*
nephelometric *adj* ANALYSIS nephelometrisch; **~ analysis** *n* ANALYSIS *of water* Trübungsanalyse *f*
nephelometry *n* ANALYSIS Nephelometrie *f*; **~ turbidity unit** *n* (*NTU*) ANALYSIS Trübungseinheit *f*
Nessler: **~'s solution** *n* ANALYSIS Neßlers-Reagens *nt*
net: **~ calorific value** *n* (*NCV*) AIR, WASTE spezifischer Heizwert *m*, unterer Heizwert *m*; **~ donator** *n* WASTE, WATER SUPP Nettodonator *m*; **~ positive suction head** *n* (*NPSH*) WASTE, WATER POLL NPSH-Wert *m*
network: **~ calculation** *n* WATER SUPP Rohrnetzberechnung *f*; **~ of pipes** *n* WATER SUPP Rohrnetz *nt*
neutral *adj* GEN neutral; **~ gas molecule** *n* AIR neutrales Gasmolekül *nt*
neutralization *n* GEN Entsäuerung *f*, Neutralisation *f*, Neutralisierung *f*; **~ of acid wastewaters** *n* WATER POLL Neutralisation *f* saurer Abwässer; **~ of alkaline wastewater** *n* WATER POLL Neutralisation *f* alkalischer Abwässer; **~ with carbon dioxide** *n* WATER POLL Neutralisation *f* mit Kohlendioxid; **~ with lime** *n* WATER POLL Neutralisation *f* mit Kalk; **~ with mineral acids** *n* WATER POLL Neutralisation *f* mit Mineralsäuren; **~ point** *n* ANALYSIS Neutralisationspunkt *m*
neutralize *vt* GEN entsäuern, neutralisieren
neutralized *adj* GEN neutralisiert, entsäuert
neutralizer *n* GEN Neutralisationsmittel *nt*
neutralizing *n* GEN Neutralisierung *f*; **~ agent** *n* AIR Neutralisationsmittel *nt*, SAFETY Neutralisierungsmittel *nt*, WATER POLL Neutralisationsmittel *nt*
neutral: **~ tax** *n* POLICY neutrale Steuer *f*; **~ temperature layering** *n* AIR neutrale Temperaturschichtung *f*; **~ wastewater sample** *n* ANALYSIS neutrale Abwasserprobe *f*
new: **~ plant** *n* AIR, POLICY, WASTE, WATER SUPP neue Betriebsanlage *f*, neue Anlage *f*
newton *n* (*N*) ANALYSIS Newton *nt* (*N*)
ng *abbr* (*nanogram, nanogramme*) ANALYSIS ng (*Nanogramm*)
Ni *abbr* (*nickel*) AIR, SOIL, WATER POLL, WATER SUPP Ni (*Nickel*)
nickel *n* (*Ni*) AIR, SOIL, WATER POLL, WATER SUPP Nickel *nt* (*Ni*)
NIPTS *abbr* (*noise-induced permanent threshold shift*) NOISE lärmbedingte dauerhafte Hörschwellenverschiebung *f*
nitrate *n* AIR, SOIL, WATER POLL, WATER SUPP Nitrat *nt*; **~ concentration** *n* WATER POLL Nitratkonzentration *f*; **~-containing** *adj* WATER POLL nitrathaltig; **~-containing effluent** *n* WATER POLL nitrathaltiges Abwasser *nt*; **~-containing wastewater stream** *n* WATER POLL nitrathaltiger Abwasserstrom *m*; **~ load** *n* WATER POLL Nitratfracht *f*; **~ removal** *n* WATER POLL Nitrateliminierung *f*
nitric: **~ acid** *n* AIR, WASTE, WATER SUPP Salpetersäure *f*; **~ bacteria** *n pl* WATER POLL Nitrifikanten *pl*; **~ and sul-**

phuric acid process *n* AIR Salpeter-Schwefelsäure-Verfahren *nt*
nitrification *n* WASTE, WATER POLL, WATER SUPP Nitrifikation *f*; **~-denitrification** *n* WATER POLL Nitrifikation-Denitrifikation *f*; **~ zone** *n* WATER POLL Nitrifikationszone *f*
nitrify *vt* WASTE, WATER POLL, WATER SUPP nitrifizieren
nitrifying *adj* WASTE, WATER POLL, WATER SUPP nitrifizierend; **~ bacteria** *n pl* WASTE, WATER POLL, WATER SUPP nitrifizierende Bakterien *f pl*
nitrite *n* WASTE, WATER POLL, WATER SUPP Nitrit *nt*; **~ concentration** *n* WATER POLL Nitritkonzentration *f*; **~ rich sludge** *n* WASTE nitritreicher Schlamm *m*
nitrogen *n* (*N*) AIR, ANALYSIS, WASTE, WATER POLL Stickstoff *m* (*N*); **~ formation** *n* AIR Stickstoffbildung *f*; **~ oxides** *n pl* (NO_x) AIR Stickstoffoxide *nt pl* (NO_x); **~ oxides scrubber** *n* AIR Stickstoffwäscher *m*; **~ pentoxide** *n* (N_2O_5) AIR Stickstoffpentoxid *nt* (N_2O_5); **~ peroxide** *n* (NO_2) AIR, WASTE Stickstoffdioxid *nt* (NO_2); **~-phosphorus detector** *n* (*NPD*) ANALYSIS Stickstoff-Phosphor-Detektor *m* (*NPD*)
nitrometer *n* ANALYSIS Azotometer *nt*
nitrosomas: **~ bacteria** *n pl* WATER POLL *oxidizing ammonia to nitrate* Nitrosomas-Bakterien *f pl*
nitrous: **~ oxide** *n* (N_2O) AIR Distickstoffoxid *nt* (N_2O), Lachgas *nt*, WASTE, WATER SUPP Distickstoffoxid *nt* (N_2O)
nm *abbr* (*nanometer AmE, nanometre BrE*) ANALYSIS nm (*Nanometer*)
NO_2 *abbr* (*nitrogen peroxide*) AIR NO_2 (*Stickstoffdioxid*)
NO_x *abbr* (*nitrogen oxides*) AIR NO_x (*Stickstoffoxide*)
N_2O *abbr* (*nitrous oxide*) AIR, WASTE, WATER SUPP N_2O (*Distickstoffoxid*)
N_2O_5 *abbr* (*nitrogen pentoxide*) AIR N_2O_5 (*Stickstoffpentoxid*)
NOAEL *abbr* (*no observed adverse effect level*) SAFETY NOAEL (*höchste Dosis, die noch nicht zu nachteiligen Wirkungen führt*)
noble: **~ metal** *n* AIR, WASTE, WATER SUPP Edelmetall *nt*; **~ metal catalyst** *n* AIR Edelmetallkatalysator *m*
nodulize *vt* AIR agglomerieren
nodulizing *n* AIR Agglomerieren *nt*
NOEC *abbr* (*no observed effect concentration*) SAFETY NOEC (*höchste Konzentration ohne eine beobachtete nachteilige Wirkung*)
no-effect: **~ level** *n* SAFETY No-Effect-Level *m*
NO_x: **~-emissions** *n pl* AIR NO_x-Emissionen *f pl*
noise *n* NOISE, SAFETY Geräusch *m*, Lärm *m*; **~ abatement** *n* NOISE Geräuschbekämpfung *f*, Lärmbekämpfung *f*; **~ abatement zone** *n* NOISE Lärmschutzbereich *m*, Lärmschutzzone *f*; **~-absorbing** *adj* NOISE lärmschluckend; **~-absorbing structural element** *n* NOISE lärmschluckendes Bauelement *nt*; **~ ambient impairment** *n* NOISE Lärmimmissionen *f pl*; **~ analysis** *n* NOISE Geräuschanalyse *f*; **~ area** *n* NOISE Lärmzone *f*; **~ area map** *n* NOISE Lärmzonenkarte *f*; **~ associated with construction** *n* NOISE Baulärm *m*; **~ at the workplace** *n* NOISE Lärm *m* am Arbeitsplatz, Lärmimmissionen *m* am Arbeitsplatz; **~ ban** *n* NOISE Lärmverbot *nt*; **~ carpet** *n* NOISE Lärmteppich *m*; **~ check** *n* NOISE Lärmüberwachung *f*; **~-contaminated** *adj* NOISE lärmverseucht; **~ contamination** *n* NOISE Lärmverseuchung *f*; **~ control** *n* NOISE, SAFETY Lärmbegrenzung *f*, Lärmkontrolle *f*; **~ criterion curve** *n* (*NC-curve*) ANALYSIS Lärmgrenzwertkurve *f*, Lärmkriteriumkurve *f*; **~ deafness** *n* NOISE Lärmschwerhörigkeit *f*; **~ dosimeter** *n* ANALYSIS Lärmdosimeter *nt*; **~ emissions** *n pl* NOISE Geräuschemissionen *f pl*, Lärmemissionen *f pl*, RENEW ENERG *from wind turbines* Lärmemissionen *f pl*; **~ emitter** *n* NOISE Schallemittent *m*, Geräuschemittent *m*; **~-excluding** *adj* NOISE lärmdämmend; **~ exclusion helmet** *n* NOISE Gehörschutzhelm *m*; **~ exposure** *n* NOISE Lärmexposition *f*, Lärmeinwirkung *f*; **~ exposure forecast** *n* (*NEF*) NOISE Lärmimmissionsprognosewert *m*; **~ exposure meter** *n* ANALYSIS Lärmexposimeter *nt*; **~ generation** *n* NOISE Lärmentstehung *f*, Lärmerzeugung *f*; **~ guards** *n pl* NOISE Lärmschutzmittel *nt*; **~ guiding value** *n* NOISE Lärmrichtwert *m*; **~ immission** *n* NOISE Schallimmission *f*; **~-induced** *adj* MARINE durch Lärm verursacht, NOISE durch Lärm verursacht, lärmbedingt, SAFETY durch Lärm verursacht; **~-induced deafness** *n* NOISE Lärmtaubheit *f*; **~-induced hearing damage** *n* NOISE, SAFETY Lärmschaden *m*; **~-induced hearing impairment** *n* MARINE, NOISE, SAFETY durch Lärm verursachter Hörschaden *m*; **~-induced permanent threshold shift** *n* (*NIPTS*) NOISE lärmbedingte dauerhafte Hörschwellenverschiebung *f*; **~-insulating barrier** *n* NOISE *on streets* Lärmschutzwand *f*; **~-insulating wall** *n* NOISE *earthen* Lärmschutzwand *f*; **~ insulation wall** *n* NOISE *earthen* Lärmschutzwall *m*; **~-intensive** *adj* NOISE geräuschintensiv, lärmintensiv; **~ intrusion** *n* NOISE Lärmstörung *f*, Verlärmung *f*; **~ labeling** *AmE*, **~ labelling** *BrE* *n* NOISE Lärmemissionskennzeichnung *f*
noiseless *adj* GEN geräuschlos; **~ motor** *n* NOISE, SAFETY geräuschloser Motor *m*
noise: **~ level** *n* NOISE Geräuschpegel *m*, Lärmpegel *m*, Schallpegel *m*; **~ level analyser** *n BrE* ANALYSIS Schallpegelanalysator *m*, Schallpegelmeßgerät *nt*; **~ level analyzer** *AmE see noise level analyser BrE*; **~ level indicator** *n* NOISE Geräuschpegelanzeiger *m*; **~ limit** *n* ANALYSIS, NOISE Lärmgrenzwert *m*; **~ limit indicator** *n* ANALYSIS, NOISE Lärmgrenzwertmelder *m*; **~ limiting value** *n* ANALYSIS, NOISE Lärmgrenzwert *m*; **~ limit value** *n* ANALYSIS, NOISE Lärmgrenzwert *m*; **~ map** *n* NOISE Lärmkarte *f*; **~ measurement** *n* ANALYSIS Lärmmessung *f*; **~ meter** *n* ANALYSIS Lärmmeßgerät *nt*; **~ monitor** *n* ANALYSIS, NOISE Lärmüberwachungsgerät *nt*; **~ monitoring** *n* ANALYSIS, NOISE Lärmüberwachung *f*; **~ nuisance** *n* NOISE Lärmbelastung *f*; **~ peak** *n* NOISE Lärmspitze *f*; **~ pollution** *n* NOISE, SAFETY Lärmbelästigung *f*; **~ pollution prevention** *n* NOISE, SAFETY Vermeidung *f* von Lärmbelästigung; **~ prevention** *n* NOISE, SAFETY Lärmverhinderung *f*, Lärmvorsorge *f*; **~ prevention measures** *n pl* NOISE, SAFETY Schallschutzmaßnahmen *f pl*; **~ propagation** *n* NOISE Fortpflanzung *f* von Schall, Schallausbreitung *f*, Schallfortpflanzung *f*; **~ protection** *n* NOISE, SAFETY Lärmschutz *m*, Schallschutz *m*; **~ protection area** *n* NOISE Lärmschutzbereich *m*, Lärmschutzgebiet *nt*, SAFETY Lärmschutzbereich *m*; **~ protection booth** *n* MARINE, NOISE, SAFETY schalldichte Kammer *f*; **~ protection for compressors** *n* SAFETY Kompressorenschalldämpfer *m*; **~ protection law** *n* NOISE, SAFETY Lärmschutzgesetz *nt*, Lärmschutzrecht *nt*; **~ protection measures** *n pl* NOISE Lärmschutzmaßnahmen *f pl*; **~ protection system** *n* NOISE Lärmschutzsystem *nt*; **~ protection technology** *n* MARINE, NOISE Schallschutztechnik *f*; **~ protection tunnel** *n* NOISE Lärmschutztunnel *m*; **~ protection wall** *n* NOISE Lärmschutzwand *f*; **~ protection wall made from plastic waste** *n* NOISE Lärmschutzwand *f* aus Kunststoffmüll; **~ protection wall made from waste** *n* NOISE Lärmschutzwand *f* aus Abfall; **~ protection zone** *n* NOISE Lärmschutzbereich *m*, Lärmschutzzone *f*; **~ radiation** *n* NOISE Lärmabstrahlung *f*; **~ rating** *n* (*NR*) ANALYSIS, NOISE Lärmbewertung *f*, Lärmbeurteilung *f*; **~ rating curve** *n* (*NR curve*) ANALYSIS, NOISE Lärmbewertungskurve *f*; **~ rating index** *n* (*NR index*) ANALYSIS, NOISE Lärmbewertungsindex *m*, Lärmbewertungszahl *f*; **~ rating level** *n* NOISE Beurteilungspegel *m*; **~-reducing** *adj* NOISE lärmgedämpft, lärmmindernd; **~ reduction** *n* NOISE Geräuschminderung *f*, Lärmminderung *f*; **~ register** *n* NOISE Lärmkataster *m*; **~ screen** *n* NOISE Lärmschirm *m*; **~ shield** *n* NOISE Lärmschutzschirm *m*, Lärmschirm *m*; **~ source** *n* NOISE Lärmentstehung *f*, Lärmquelle *f*; **~ standard value** *n* NOISE Lärmrichtwert *m*; **~ suppression** *n* NOISE Lärmabwehr *f*;

~ **type** *n* NOISE Geräuschart *f*; ~ **and vibration measuring equipment** *n* ANALYSIS, MARINE, NOISE, SAFETY Schall- und Vibrationsmesser *m*; ~ **weighting** *n* NOISE Lärmbeurteilung *f*
nonaerated *adj* WATER POLL unbelüftet; ~ **oxidation pond** *n* WATER POLL unbelüfteter Abwasserteich *m*
nonbiodegradable *adj* GEN biologisch nicht abbaubar; ~ **matter** *n* WASTE, WATER POLL, WATER SUPP biologisch nicht abbaubarer Stoff *m*; ~ **organics** *n pl* AIR, SOIL, WATER POLL biologisch nicht abbaubare organische Verbindungen *f pl*; ~ **packaging material** *n* WASTE biologisch nicht abbaubarer Packstoff *m*; ~ **substance** *n* WASTE, WATER SUPP biologisch nicht abbaubare Substanz *f*; ~ **waste** *n* WASTE, WATER SUPP biologisch nicht abbaubarer Abfall *m*
noncarbonate: ~ **hardness** *n* (*NCH*) WASTE, WATER POLL, WATER SUPP Nichtkarbonathärte *f* (*NCH*), bleibende Härte *f*, permanente Härte *f*
noncatalytic *adj* AIR nichtkatalytisch
nonclogging *adj* WATER POLL verstopfungssicher; ~ **packing material** *n* WATER POLL verstopfungssicheres Füllmaterial *nt*; ~ **pump** *n* WATER POLL, WATER SUPP Kanalradpumpe *f*
nonclog: ~ **pump** *n* WATER POLL Kanalradpumpe *f*
noncombustibility: ~ **test** *n* SAFETY Unbrennbarkeitstest *m*
noncombustible *adj* SAFETY unbrennbar
noncompostable *adj* WASTE nichtkompostierbar; ~ **waste** *n* WASTE nichtkompostierbarer Abfall *m*
nonconcentrating: ~ **collector** *n* RENEW ENERG *flat-plate collector* nichtkonzentrierender Kollektor *m*
nonconing: ~ **smoke plume** *n* AIR nichtkegelförmige Rauchfahne *f*
noncorrosive *adj* AIR, WASTE, WATER SUPP korrosionsbeständig
noncrystalline *adj* AIR amorph, gestaltlos, nichtkristallin
nondegradable *adj* WATER POLL schwer abbaubar, schwer zersetzbar; ~ **substance** *n* WATER POLL schwer abbaubarer Stoff *m*
nondispersive *adj* ANALYSIS nichtdispersiv, nichtstreuend; ~ **infrared** *n* (*NDIR*) ANALYSIS nichtstreuendes Infrarot *nt* (*NDIR*); ~ **infrared absorber** *n* ANALYSIS nichtdispersives Absorptionsgerät *nt*; ~ **infrared absorption** *n* (*NDIRA*) ANALYSIS nichtstreuende Infrarotabsorption *f* (*NDIRA*); ~ **infrared analyser** *n BrE* ANALYSIS nichtdispersiver Infrarotanalysator *m*, nichtstreuender Infrarotanalysator *m*; ~ **infrared analyzor** *AmE see nondispersive infrared analyser BrE*; ~ **infrared spectrometry** *n* ANALYSIS nichtstreuende Infrarotspektrometrie *f*
nonferrous: ~ **metals** *n pl* WASTE NE-Metalle *nt pl*, Nichteisenmetalle *nt pl*; ~ **scrap** *n* WASTE Nichteisenschrott *m*
nonfilterable *adj* WATER POLL nichtfiltrierbar; ~ **residues** *n pl* WATER POLL nichtfiltrierbare Rückstände *m pl*
nonflammable *adj* AIR, SAFETY, WASTE nicht entflammbar, unbrennbar
noninflammable *adj* AIR, ANALYSIS, SAFETY, WASTE nicht entflammbar
nonionic *adj* WATER POLL nichtionisch
nonionogenic *adj* WATER POLL nichtionogen; ~ **polymer** *n* WATER POLL nichtionogenes Polymer *nt*
nonleachable *adj* SOIL nichtauslaugbar
nonpaper: ~ **component** *n* WASTE papierfremder Bestandteil *m*
nonporous *adj* WASTE, WATER POLL, WATER SUPP nichtporös
nonputrefactive *adj* WASTE, WATER POLL fäulnisunfähig
nonputrescible *adj* WASTE, WATER POLL fäulnisunfähig, haltbar
nonradial: ~ **flow** *n* WATER POLL *of settling tank* Querdurchstrom *m*
nonrecoverable *adj* WASTE nichtrückgewinnbar; ~ **waste** *n* WASTE nichtrückgewinnbare Abfälle *m pl*
nonregenerable: ~ **flue-gas desulfurization** *AmE*, ~ **flue-gas desulphurization** *BrE n* AIR nicht regenerierbare Rauchgasentschwefelung *f*
nonremeltable *adj* WASTE nicht wiederaufschmelzbar
nonresonant: ~ **ion** *n* ANALYSIS *quadrupole mass analyser* nichtresonanzaktives Ion *nt*
nonreturnable: ~ **bottle** *n* WASTE Einwegflasche *f*, Flasche *f* ohne Pfand; ~ **container** *n* WASTE Einwegverpackung *f*; ~ **packaging** *n* WASTE Einwegverpackung *f*; ~ **pallet** *n* WASTE Einwegpalette *f*
non-solar-based: ~ **renewable energy sources** *n pl* RENEW ENERG *tidal and geothermal energy* nichtsolare erneuerbare Energiequellen *f pl*
nontoxic *adj* GEN nichttoxisch, ungiftig; ~ **metals** *n pl* WASTE, WATER POLL ungiftige Metalle *nt pl*
nonvolatile *adj* AIR *absorbent* nichtflüchtig
nonvulcanized *adj* WASTE nichtvulkanisiert
non-waste-technology *n* (*NWT*) GEN abfallose Technologie *f*, saubere Technologie *f*, umweltfreundliche Technologie *f*
nonweighable *adj* ANALYSIS nichtwägbar; ~ **lysimeter** *n* ANALYSIS nichtwägbares Lysimeter *nt*
nonwetting *adj* AIR nichtnetzend
nonwoven *adj* AIR nichtgewebt; ~ **fabric** *n* AIR *filter fibre* Faservlies *nt*
normal: ~ **operation** *n* MANAGE, POLICY *of a plant* Normalbetrieb *m*; ~ **pressure** *n* ANALYSIS Normdruck *m*
north: ~**-south relationship** *n* POLICY Nord-Süd-Beziehung *f*
note *n* POLICY Umweltgesetzbuch *nt*
notification *n* MARINE Benachrichtigung *f*, Meldung *f*
nozzle *n* AIR Düse *f*; ~ **atomization** *n* AIR *spray absorption* Düsenzerstäubung *f*; ~ **atomizer** *n* AIR *spray absorber* Düsenzerstäuber *m*; ~ **blade** *n* AIR *cartridge filter* Düsenflügel *m*; ~ **pipe** *n* AIR Strahlrohr *nt*; ~ **separator** *n* AIR Ringspaltabscheider *m*; ~ **tube** *n* AIR Düsenrohr *nt*
NPD *abbr* (*nitrogen-phosphorus detector*) ANALYSIS NPD (*Stickstoff-Phosphor-Detektor*)
NPSH *abbr* (*net positive suction head*) WASTE, WATER POLL NPSH-Wert *m*
NR *abbr* (*noise rating*) ANALYSIS, NOISE Lärmbewertung *f*; ~ **curve** *n* ANALYSIS, NOISE Lärmbewertungskurve *f*; ~ **index** *n* ANALYSIS, NOISE Lärmbewertungsindex *m*, Lärmbewertungszahl *f*
NTU *abbr* (*nephelometry turbidity unit*) ANALYSIS Trübungseinheit *f*
nuclear *adj* RENEW ENERG, SAFETY, WASTE nuklear; ~ **energy source** *n* RENEW ENERG nuklearer Energieträger *m*; ~ **safety** *n* SAFETY Nuklearsicherheit *f*; ~ **waste** *n* SAFETY, WASTE radioaktiver Abfall *m*
null: ~ **balance** *n* ANALYSIS *nullification* Nullabgleich *m*
number: ~ **of revolutions** *n* RENEW ENERG *wind turbine* Drehzahl *f*; ~ **of revolutions of rotor** *n* RENEW ENERG *wind turbine* Rotordrehzahl *f*
numerical *adj* ANALYSIS numerisch, zahlenmäßig; ~ **grid model** *n* ANALYSIS numerisches Gittermodell *nt*; ~ **value** *n* ANALYSIS Zahlenwert *m*
nutrient *n* MARINE, SOIL, WATER POLL, WATER SUPP Nährstoff *m*; ~**-poor** *adj* WASTE, WATER SUPP nährstoffarm, oligotroph; ~**-poor water** *n* WATER SUPP nährstoffarmes Wasser *nt*; ~ **removal** *n* WATER POLL Nährstoffentfernung *f*, Nährstoffentzug *m*; ~**-rich** *adj* WASTE, WATER POLL, WATER SUPP eutroph, nährstoffreich; ~**-rich water** *n* WATER POLL, WATER SUPP nährstoffreiches Wasser *nt*; ~ **solution tank** *n* AIR *biological gas scrubbing* Nährlösungsbehälter *m*; ~ **supply** *n* AIR, SOIL, WASTE, WATER POLL, WATER SUPP Nährstoffversorgung *f*
NWT *abbr* (*non-waste-technology*) GEN abfallose Technologie *f*, saubere Technologie *f*, umweltfreundliche Technologie *f*
nylon: ~**-reinforced** *adj* WASTE, WATER SUPP nylonverstärkt

O

O_2 *abbr* (*oxygen*) GEN O_2 (*Sauerstoff*)
object: **~-related** *adj* MARINE, NOISE objektbezogen; **~-related noise** *n* MARINE, NOISE objektbezogener Lärm *m*
obligation *n* POLICY Gebot *nt*; **~ to recycle** *n* WASTE, WATER SUPP Recyclinggebot *nt*, Verwertungsgebot *nt*; **~ to reuse residue materials** *n* POLICY Reststoffverwertungsgebot *nt*
oblige *vt* POLICY verpflichten
observable *adj* ANALYSIS beobachtbar
observation *n* ANALYSIS, WASTE, WATER SUPP Beobachtung *f*; **~ time** *n* ANALYSIS Beobachtungszeit *f*; **~ well** *n* ANALYSIS, WASTE, WATER SUPP Beobachtungsbrunnen *m*, Kontrollbrunnen *m*
observed: **no ~ adverse effect level** *n* (*NOAEL*) SAFETY höchste Dosis *f*, die noch nicht zu nachteiligen Wirkungen führt (*NOAEL*); **no ~ effect concentration** *n* (*NOEC*) SAFETY höchste Konzentration *f* ohne eine beobachtete nachteilige Wirkung (*NOEC*)
OC *abbr* (*open cycle*) RENEW ENERG offener Kreislauf *m*
occupational *adj* SAFETY beruflich; **~ disease** *n* SAFETY Berufskrankheit *f*; **~ dose limit** *n* SAFETY Dosisgrenzwert *m* für berufliche Bestrahlung; **~ exposure limits** *n pl* (*OEL*) SAFETY berufsbedingte Belastungsgrenzen *f pl* (*OEL*); **~ MAC** *n* (*occupational maximum allowable concentration*) WASTE maximal zulässige Arbeitsplatzkonzentration *f* (*MAK-Wert*); **~ maximum allowable concentration** *n* (*occupational MAC*) WASTE maximal zulässige Arbeitsplatzkonzentration *f* (*MAK-Wert*); **~ threshold limit value** *n* (*occupational TLV*) WASTE maximal zulässige Arbeitsplatzkonzentration *f* (*MAK-Wert*); **~ TLV** *n* (*occupational threshold limit value*) WASTE MAK-Wert *m* (*maximal zulässige Arbeitsplatzkonzentration*)
ocean: **~ basin** *n* MARINE Ozeanbecken *nt*; **~ current** *n* RENEW ENERG Meeresströmung *f*; **~ current energy** *n* RENEW ENERG Meeresströmungsenergie *f*; **~ current power** *n* RENEW ENERG Energie *f* der Meeresströmung; **~ current power plant** *n* RENEW ENERG Meeresströmungskraftwerk *nt*; **~ current resources** *n pl* RENEW ENERG Meeresströmungsressourcen *f pl*; **~ disposal** *n* MARINE, WASTE Verklappung *f*, WATER POLL *of wastewater* Meereseinleitung *f*; **~ dumping** *n* MARINE, WASTE Verklappung *f*; **~ energy** *n* RENEW ENERG *wave, ocean, thermal* Meeresenergie *f*
oceanic *adj* MARINE ozeanisch
ocean: **~ thermal energy** *n* RENEW ENERG Meereswärmeenergie *f*; **~ thermal energy conversion** *n* (*OTEC*) RENEW ENERG Meereswärmeumwandlung *f*; **~ thermal energy conversion plant** *n* RENEW ENERG (*OTEC plant*) Meereswärmekraftwerk *nt*; **~ water** *n* MARINE, RENEW ENERG, WATER SUPP Meerwasser *nt*
octane: **~ number** *n* AIR Klopfwert *m*, Octanzahl *f*
ocular *n* ANALYSIS *interferential pit gas tester* Okular *nt*
OD *abbr* (*overdose*) SAFETY Überdosis *f*
odor *AmE see odour BrE*
odoriferous: **~ substance** *n* AIR Geruchsstoff *m*; **~ substance emissions** *n pl* AIR Geruchsstoffemissionen *f pl*
odorimetry *n* ANALYSIS Geruchsmessung *f*
odoriphore *n* AIR Geruchsträger *m*
odorless *AmE see odourless BrE*
odorous *adj* AIR, WASTE, WATER SUPP riechstoffbildend; **~ substance** *n* WASTE, WATER SUPP Geruchsstoff *m*
odour *n BrE* GEN Geruch *m*; **~ analysis** *n BrE* ANALYSIS Geruchsbestimmung *f*; **~ analysis in wastewater** *n BrE* ANALYSIS Geruchsbestimmung *f* bei Abwässern; **~ annoyance** *n BrE* AIR, WASTE, WATER POLL, WATER SUPP Geruchsbelästigung *f*; **~-annoying substance** *n BrE* AIR geruchsbelästigender Stoff *m*; **~-annoying wastewater** *n BrE* AIR, WATER POLL geruchsbelästigendes Abwasser *nt*; **~ control** *n BrE* AIR, WASTE, WATER POLL, WATER SUPP Geruchsbekämpfung *f*; **~ determination** *BrE n* ANALYSIS Geruchsbestimmung *f*; **~ emission** *n BrE* AIR, WASTE, WATER POLL, WATER SUPP Geruchsemission *f*; **~ formation** *n BrE* WASTE, WATER POLL, WATER SUPP Geruchsbildung *f*; **~-forming** *adj BrE* WASTE, WATER POLL, WATER SUPP geruchsbildend; **~-forming substance** *n BrE* WATER POLL geruchsbildende Substanz *f*; **~-free** *adj BrE* WASTE, WATER SUPP geruchsfrei; **~ impairment** *n BrE* AIR, WASTE, WATER POLL, WATER SUPP Geruchsbelästigung *f*; **~-intensive** *adj BrE* AIR, SOIL, WASTE, WATER POLL, WATER SUPP geruchsintensiv; **~-intensive solvent** *n BrE* AIR geruchsintensives Lösungsmittel *nt*; **~-intensive substance** *n BrE* AIR geruchsintensiver Stoff *m*; **~-laden** *adj BrE* AIR geruchsbeladen; **~-laden dirty gas** *n BrE* AIR geruchsbeladenes Rohgas *nt*; **~-laden waste air** *n BrE* AIR geruchsbeladene Abluft *f*
odourless *adj BrE* GEN geruchlos; **~ gas** *n BrE* AIR geruchloses Gas *nt*
odour: **~ nuisance** *n BrE* AIR, WASTE, WATER POLL, WATER SUPP Geruchsbelästigung *f*; **~ pollution** *n BrE* AIR, WASTE, WATER POLL, WATER SUPP Geruchsbelästigung *f*; **~ problem** *n BrE* WATER POLL Geruchsproblem *nt*; **~ quotient** *n BrE* ANALYSIS Geruchszahl *f*; **~ removal** *n BrE* AIR Geruchsentfernung *f*, WATER POLL Geruchsbeseitigung *f*; **~ source** *n BrE* AIR, WATER POLL Geruchsentstehung *f*; **~-substance-laden air** *n BrE* AIR geruchsstoffbeladene Luft *f*; **~ test** *n BrE* ANALYSIS Geruchsprüfung *f*
ODP *abbr* (*ozone depletion potential*) AIR Ozonabbaupotential *nt*, Ozonschutz-Zerstörungspotential *nt*
OEL *abbr* (*occupational exposure limits*) SAFETY OEL (*berufsbedingte Belastungsgrenzen*)
OES *abbr* (*optical emission spectrometry*) ANALYSIS optische Emissionsspektrometrie *f*
offal: **~ destruction plant** *n* WASTE Tierkörperbeseitigungsanlage *f*
offer: **~ incentives** *vi* MANAGE *to recycle* Anreiz setzen
off-gas *n* AIR, SOIL Abgas *nt*
office: **~ waste** *n* WASTE Büroabfall *m*
official: **~ journal** *n* POLICY *of EU* Amtsblatt *nt*; **~ language** *n* POLICY Amtssprache *f*; **~ monitoring** *n* POLICY behördliche Überwachung *f*; **~ supervision** *n* POLICY *of plants* behördliche Überwachung *f*
off-line *adj* ANALYSIS off-line; **~ operation** *n* ANALYSIS Off-line-Betrieb *m*
offshore: **~ potential** *n* RENEW ENERG Off-shore-Potential *nt*; **~ potential of wind energy** *n* RENEW ENERG Off-shore-Windenergiepotential *nt*; **~ wind farm** *n* RENEW ENERG Off-shore-Windpark *m*, Windkraftanlage *f* im Offshore-Bereich
off-site: **~ analysis** *n* ANALYSIS Off-site-Analyse *f*, Off-site-Untersuchung *f*; **~ disposal** *n* SOIL Wiedereinlagerung *f*; **~ incineration** *n* SOIL Off-site-Verbrennung *f*
offtake *n* WASTE, WATER POLL, WATER SUPP Ableitung *f*, Förderung *f*
OFS *abbr AmE* (*oil from sludge*) WATER POLL Öl aus Schlamm *nt*
oil *n* MARINE, RENEW ENERG, WATER POLL Öl *nt*; **~ aerosol**

separator *n* WASTE Ölnebelabscheider *m*; ~ **boom** *n* MARINE Ölsperre *f*; ~**-concentrating agent** *n* WASTE Ölverdichtungsmittel *nt*; ~ **concentration** *n* WATER POLL Ölkonzentration *f*; ~**-containing** *adj* RENEW ENERG, WASTE ölhaltig; ~**-containing plants** *n pl* RENEW ENERG *biomass conversion* ölhaltige Pflanzen *f pl*; ~ **contamination** *n* MARINE Ölverschmutzung *f*; ~ **droplet** *n* WATER POLL Öltröpfchen *nt*; ~**-filled receiving tube** *n* RENEW ENERG *in solar power plant* mit Öl gefüllte Empfängerröhre *f*; ~ **film** *n* MARINE Ölfilm *m*, Ölschicht *f*, Ölteppich *m*; ~ **from sludge** *n AmE* (*OFS*) WATER POLL Öl *nt* aus Schlamm; ~ **leakage** *n* MARINE Ölaustritt *m*, Ölleck *nt*; ~ **mop** *n* MARINE adhäsiv wirkendes Ölaufnahmeband *nt*; ~ **patch** *n* MARINE Ölfleck *m*; ~**-polluted wastewater** *n* WATER POLL ölverschmutztes Abwasser *nt*; ~ **pollution** *n* MARINE Ölverschmutzung *f*, WASTE, WATER POLL Ölpest *f*; ~**-pollution combatting ship** *n* MARINE Ölbekämpfungsschiff *nt*; ~ **pollution emergency** *n* MARINE, WASTE Ölverschmutzungsnotfall *m*; ~**-recovery barge** *n* WASTE *ship* Ölrückgewinnungsschute *f*; ~**-recovery skimmer** *n* WASTE Ölrückgewinnungsskimmer *m*; ~**-recovery vessel** *n* WASTE Ölrückgewinnungsschiff *nt*; ~ **regeneration plant** *n* WASTE Altölaufbereitungsbetrieb *m*, Ölaufbereitungsanlage *f*; ~ **removal** *n* WATER POLL Entölen *nt*, Ölabscheidung *f*; ~ **separation** *n* WATER POLL Entölen *nt*, Ölabscheidung *f*; ~ **separator** *n* SOIL, WASTE, WATER POLL Ölabscheider *m*; ~ **slick** *n* MARINE Ölfilm *m*, Ölschicht *f*, Ölteppich *m*, Ölfleck *m*; ~ **spill** *n* MARINE *leakage* Ölverschüttung *f*, ausgeflossenes Öl *nt*, *slick* Öllache *f*, treibendes Ölfeld *nt*; ~ **spillage** *n* MARINE Ölverschüttung *f*; ~ **spill combatting equipment** *n* MARINE Ölteppichbekämpfungsausrüstung *f*; ~ **spill disaster** *n* MARINE, WASTE, WATER POLL Ölkatastrophe *f*; ~ **spill identification** *n* MARINE Ölteppichidentifikation *f*; ~ **spill identification system** *n* MARINE Ölspill-Identifikationssystem *nt*; ~**-spill response** *n* MARINE Ölbekämpfung *f*; ~ **trap** *n* WASTE Ölfang *m*, WATER POLL Fettabscheider *m*, Fettfang *m*, Ölfang *m*; ~ **waste** *n* WASTE Ölabfall *m*, Ölrückstände *m pl*

oil/water: ~ **interface** *n* MARINE, SOIL Öl/Wasser-Berührungsfläche *f*; ~ **separation** *n* SOIL Öl/Wasser-Abscheidung *f*, WATER POLL Öl/Wasser-Abscheidung *f*, Öl/Wasser-Trennung *f*; ~ **separator** *n* SOIL, WATER POLL Öl/Wasser-Abscheider *m*

oily: ~ **wastewater** *n* WATER POLL ölhaltiges Abwasser *nt*, fetthaltiges Abwasser *nt*

old: ~ **oil product** *n* SOIL Altölprodukt *nt*

oleaginous *adj* RENEW ENERG, WASTE, WATER POLL ölhaltig; ~ **wastes** *n pl* WASTE ölhaltige Abfälle *m pl*; ~ **wastewater** *n* WASTE ölhaltiges Abwasser *nt*

oleophilic *adj* MARINE, WATER POLL oleophil, ölanziehend, ölaufnehmend; ~ **belt** *n* MARINE ölaufnehmendes Band *nt*; ~ **belt skimmer** *n* MARINE ölaufnehmendes Förderband-Abschöpfgerät *nt*

olfactometry *n* ANALYSIS Geruchsmessung *f*, Olfaktometrie *f*

oligodynamic *adj* WASTE oligodynamisch

oligosaprobic *adj* WASTE, WATER POLL, WATER SUPP oligosaprob

oligotrophic *adj* WASTE, WATER POLL, WATER SUPP oligotroph

on-board: ~ **computer** *n* ANALYSIS on-board Mikroprozessor *m*

one: ~**-way bottle** *n* WASTE Einwegflasche *f*; ~**-way container** *n* WASTE Einwegbehälter *m*; ~**-way pack** *n* WASTE Einwegpackung *f*; ~**-way packaging** *n* WASTE Einwegverpackung *f*, Wegwerfverpackung *f*; ~**-way pallet** *n* WASTE Einwegpalette *f*

on-line *adj* ANALYSIS On-line; ~ **analyser** *BrE*, ~ **analyzer** *AmE* *n* ANALYSIS On-line-Analysengerät *nt*; ~ **measurement** *n* ANALYSIS On-line-Messung *f*; ~ **operation** *n* ANALYSIS On-line-Betrieb *m*; ~ **storage tank** *n* WATER SUPP Gegenbehälter *m*

on-scene: ~ **commander** *n* (*OSC*) MARINE Vor-Ort-Einsatzleiter *m*

on-site: ~ **analysis** *n* ANALYSIS On-site-Analyse *f*, Vor-Ort-Analyse *f*; ~ **application** *n* SOIL On-site-Anwendung *f*, On-site-Einsatz *m*; ~ **assembly** *n* ANALYSIS Vor-Ort-Einsatz *m*; ~ **data storage** *n* ANALYSIS Vor-Ort-Datenspeicherung *f*; ~ **excavation** *n* SOIL On-site-Aushub *m*; ~ **extraction** *n* SOIL On-site-Aushub *m*; ~ **incineration** *n* SOIL In-Situ-Verbrennung *f*, On-site-Verbrennung *f*; ~ **presorting** *n* WASTE On-site-Vorsortierung *f*, Vorsortierung *f* vor Ort; ~ **storage** *n* WASTE On-site-Lagerung *f*; ~ **test plant** *n* WATER SUPP Vor-Ort-Testanlage *f*; ~ **treatment** *n* WASTE On-site-Behandlung *f*

opaque *adj* WATER POLL *degree of turbidity* undurchsichtig

open: ~ **air measurement** *n* ANALYSIS Freilandmessung *f*; ~ **cycle** *n* (*OC*) RENEW ENERG offener Kreislauf *m*; ~**-cycle OTEC plant** *n* RENEW ENERG Offener-Kreislauf-Meereswärmekraftwerk *nt*; ~ **drain** *n* WASTE, WATER POLL offener Abzugsgraben *m*; ~ **dump** *n* WASTE ungeordnete Deponie *f*; ~ **landfill** *n* WASTE offene Deponie *f*; ~ **landfill section** *n* WASTE offener Deponieabschnitt *m*; ~ **loop catalyst** *n* AIR ungeregelter Katalysator *m*; ~ **loop catalytic converter** *n* AIR ungeregelter Katalysator *m*; ~**-ocean ecosystem** *n* MARINE Ökosystem *nt* des offenen Ozeans; ~ **pit landfill** *n* WASTE offene Deponie *f*; ~ **Rankine cycle** *n* RENEW ENERG *ocean thermal energy* offener Rankine-Kreislauf *m*; ~ **sewer** *n* WATER POLL offener Abzugsgraben *m*; ~ **system** *n* WASTE offenes System *nt*; ~**-top container** *n* WASTE Open-Top-Container *m*, Tilt-top-Container *m*; ~ **trickling filter** *n* WATER POLL offener Tropfkörper *m*

operate *vt* MANAGE, POLICY, RENEW ENERG, WATER SUPP betreiben

operating: ~ **authorization** *n* POLICY Betriebsgenehmigung *f*; ~ **error** *n* ANALYSIS Fehlbedienung *f*; ~ **face** *n* WASTE Deponieoberfläche *f*; ~ **licence** *BrE*, ~ **license** *AmE* *n* POLICY Betriebsgenehmigung *f*; ~ **material** *n* AIR Betriebsmittel *nt*; ~ **permission** *n* POLICY Betriebsgenehmigung *f*; ~ **speed** *n* RENEW ENERG *wind turbine* Betriebsdrehzahl *f*; ~ **standards** *n pl* POLICY Betriebsnormen *f pl*; ~ **voltage** *n* AIR Betriebsspannung *f*

operation *n* GEN *process* Vorgang *m*, *business* Betrieb *m*

operational: ~ **error** *n* MANAGE, WASTE, WATER SUPP Betriebsfehler *m*, betriebsbedingter Fehler *m*; ~ **landfill section** *n* WASTE betriebener Deponieabschnitt *m*

operations: ~ **safety** *n* SAFETY Betriebssicherheit *f*

operation: ~ **time** *n* RENEW ENERG Betriebsdauer *f*

operative: ~ **range** *n* SOIL Einsatzbereich *m*

operator *n* MANAGE, POLICY, WATER SUPP Betreiber *m*

optical *adj* ANALYSIS *analyzers* optisch; ~ **absorption** *n* ANALYSIS Lichtabsorption *f*; ~ **absorption in gases** *n* ANALYSIS Lichtabsorption *f* in Gasen; ~ **emission spectrometry** *n* (*OES*) ANALYSIS optische Emissionsspektrometrie *f*; ~ **sorter** *n* WASTE optischer Abscheider *m*; ~ **sorting** *n* WASTE optische Sortierung *f*

optimum: ~ **polymer dosage** *n* WATER POLL optimaler Flokkulantenzusatz *m*

optoelectric *adj* WASTE elektronenoptisch; ~ **process** *n* WASTE *form of binding* elektronenoptisches Verfahren *nt*

orbit *n* AIR *raw gas in cyclone* Kreisbahn *f*

ore *n* WASTE Erz *nt*

organic *adj* AIR, WASTE, WATER POLL, WATER SUPP organisch; ~ **acid gas** *n* AIR organisches Sauergas *nt*; ~ **base material** *n* AIR *biofilter* organisches Trägermaterial *nt*; ~ **biomass** *n* RENEW ENERG *biomass conversion* organische Biomasse *f*; ~ **chemical** *n* WATER POLL, WATER SUPP organische Chemikalie *f*; ~ **chlorine compounds** *n pl* WASTE, WATER SUPP organische Chlorverbindungen *f pl*; ~ **compounds** *n pl* WATER POLL organische Verbindungen *f pl*; ~ **contaminant** *n* AIR, WATER POLL organischer Kontaminant *m*, organischer Schadstoff *m*; ~ **dry mass** *n* WASTE, WATER POLL, WATER SUPP organische Trockensubstanz *f*; ~ **fraction** *n* WASTE, WATER POLL organische Fraktion *f*; ~ **gas** *n* AIR organisches Gas *nt*; ~ **harmful substance** *n* AIR, WATER POLL organischer Schadstoff *m*; ~ **loading** *n* WATER POLL organische Belastung *f*; ~ **matter** *n* AIR, WASTE, WATER SUPP organischer Stoff *m*; ~ **pollutant** *n* AIR, WATER POLL organischer Kontaminant *m*, organischer Schadstoff *m*; ~ **roughage** *n* WASTE organischer Ballaststoff *m*; ~ **solids** *n pl* WASTE, WATER POLL organische Feststoffe *m pl*; ~ **solvent** *n* AIR organisches Lösungsmittel *nt*; ~ **sulphides** *n pl* AIR Thioether *m*; ~ **waste** *n* WASTE, WATER SUPP Bioabfall *m*, Biomüll *m*, organischer Abfall *m*; ~ **waste fraction** *n* WASTE, WATER SUPP Bioabfallfraktion *f*, Biomüllfraktion *f*, organische Müllfraktion *f*; ~ **waste recovery** *n* WASTE Biomüllerfassung *f*; ~ **waste use** *n* WASTE Biomüllverwertung *f*

oriented *adj* MANAGE, POLICY orientiert, WASTE *plastics* gereckt; ~ **plastics** *n pl* WASTE gereckte Kunststoffe *m pl*; ~ **polypropylene film** *n* WASTE DPP-Film *m*, gereckte Polypropylenfolie *f*

original: ~ **sample** *n* ANALYSIS Originalprobe *f*; ~ **soil** *n* SOIL Originalboden *m*

orlon *n* AIR *filter material* Orlon *nt*

ORP *abbr* (*oxidation reduction potential*) ANALYSIS Redox-Potential *nt*

Orsat: ~ **apparatus** *n* ANALYSIS Gasanalyseapparat *m* nach Orsat, Orsatapparat *m*; ~ **burette** *n* ANALYSIS Orsatbürette *f*; ~ **gas analyser** *n BrE* ANALYSIS Gasanalyseapparat *m* nach Orsat, Orsatapparat *m*; ~ **gas analyzer** *AmE see Orsat gas analyser BrE*

orthokinetic *adj* WASTE orthokinetisch; ~ **flocculation** *n* WASTE, WATER SUPP orthokinetische Flockung *f*

orthophosphate *n* WATER POLL Orthophosphat *nt*, o-Phosphat *nt*

OSC *abbr* (*on-scene commander*) MARINE Vor-Ort-Einsatzleiter *m*

oscillate 1. *vt* RENEW ENERG oszillieren; **2.** *vi* NOISE oszillieren

oscillating: ~ **water column** *n* (*OWC*) RENEW ENERG oszillierende Wassersäule *f* (*OWC*)

oscillator *n* NOISE Oszillator *m*

oscilloscope *n* ANALYSIS *mass spectroscopy* Oszilloskop *nt*

osmophore *n* AIR Geruchsträger *m*

osmosis *n* WATER POLL, WATER SUPP Osmose *f*

osmotic *adj* WASTE, WATER POLL, WATER SUPP osmotisch; ~ **pressure** *n* WASTE, WATER POLL, WATER SUPP osmotischer Druck *m*

OTEC *abbr* (*ocean thermal energy conversion*) RENEW ENERG Meereswärmeumwandlung *f*; ~ **plant** *n* RENEW ENERG Meereswärmekraftwerk *nt*

outdoor: ~ **noise** *n* NOISE Außenlärm *m*; ~ **noise-free field** *n* NOISE lärmfreies Feld *nt* im Freien; ~ **shielding** *n* NOISE Abschirmung *f* im Freien; ~ **sound propagation** *n* NOISE Schallausbreitung *f* im Freien

outer: ~ **air** *n* AIR Außenluft *f*; ~ **eddy** *n* WASTE, WATER SUPP *centrifugal field* Außenwirbel *m*; ~ **electrode** *n* ANALYSIS Außenelektrode *f*; ~ **filter** *n* ANALYSIS Außenfilter *nt*; ~ **shell** *n* NOISE *in encapsulation* Außenhaut *m*; ~ **vortex** *n* AIR Außentrombe *f*

outflow *n* WATER POLL Abfließen *nt*, Ablauf *m*, Ausfluß *m*, WATER SUPP Abfließen *nt*

outlet *n* AIR Austritt *m*, WATER POLL Ablaß *m*, Auslaß *m*; ~ **channel** *n* WATER POLL Ablaßgraben *m*; ~ **flow control** *n* WATER POLL Auslaßdurchflußregelung *f*; ~ **pipe** *n* WATER POLL Ablaßrohr *nt*

output *n* ANALYSIS *instrument reading* Ausgang *m*, MANAGE Ausgangsgrößen *f pl*, Förderung *f*

overall: ~ **chimney height** *n* AIR Kaminbauhöhe *f*, Schornsteinbauhöhe *f*; ~ **stack height** *n* AIR Kaminbauhöhe *f*, Schornsteinbauhöhe *f*; ~ **water quality** *n* WATER SUPP Gesamtgewässergüte *f*

overburning *n* SOIL Überbrand *m*

overcrowded: ~ **area** *n* MANAGE, POLICY, WASTE, WATER SUPP Ballungsgebiet *nt*

overdam *vt* WASTE, WATER SUPP überstauen

overdistillation *n* ANALYSIS Überdestillation *f*

overdosage *n* AIR, SAFETY Überdosierung *f*

overdose *n* (*OD*) ANALYSIS, SAFETY Überdosis *f*

overdosing *n* AIR, SAFETY Überdosierung *f*

overexploitation *n* SOIL *of soil* Übernutzung *f*

overfall *n* WASTE, WATER SUPP Überfall *m*

overflow *n* WASTE, WATER POLL, WATER SUPP Überflutung *f*, Überlauf *m*, Überschwemmung *f*; ~ **pipe** *n* WASTE, WATER POLL, WATER SUPP Überlaufrohr *nt*; ~ **rate** *n* WASTE, WATER POLL, WATER SUPP Überlaufmenge *f*; ~ **spring** *n* WATER SUPP Überlaufquelle *f*; ~ **structure** *n* WATER POLL, WATER SUPP Überlaufbauwerk *nt*; ~ **structure for settled combined water** *n* WATER POLL Klärüberlauf *m*; ~ **tank** *n* WATER POLL, WATER SUPP Überlaufbehälter *m*; ~ **weir** *n* AIR *bubble-cap tray tower*, WASTE, WATER POLL Ablaufwehr *nt*, WATER SUPP Ablaufwehr *nt*, Überfallwehr *f*

overgrazing *n* SOIL Überweidung *f*

overhead: ~ **irrigation** *n* SOIL Beregnung *f*; ~ **sound absorber** *n* NOISE abhängbares Schallschluckelement *nt*

overheating *n* AIR, WATER POLL Überhitzung *f*

overload *n* AIR Überladung *f*, Überlastung *f*

over-packaging *n* WASTE Umverpackung *f*

oversampling *n* ANALYSIS Oversampling *nt*

oversize: ~ **fraction** *n* SOIL, WASTE Überkornfraktion *f*; ~ **material** *n* SOIL, WASTE *in sorting* Überkornmaterial *nt*

OWC *abbr* (*oscillating water column*) RENEW ENERG OWC (*oszillierende Wassersäule*); ~ **operation** *n* RENEW ENERG *wave energy* OWC-Betrieb *m*

oxidability *n* ANALYSIS Oxidierbarkeit *f*

oxidant *n* AIR, WATER POLL, WATER SUPP Oxidationsmittel *nt*

oxidation *n* GEN Oxidation *f*; ~ **air** *n* AIR Oxidationsluft *f*; ~ **ditch** *n* WASTE, WATER POLL, WATER SUPP Oxidationsgraben *m*; ~ **pond** *n* WATER POLL Abwasserteich *m*, *maturation pond* Oxidationsteich *m*; ~ **reduction** *n* WATER POLL Oxidationsreduktion *f*; ~ **reduction potential** *n* (*ORP*) ANALYSIS Redox-Potential *nt*; ~ **reduction process** *n* AIR *for waste air cleaning* Redox-Prozeß *m*; ~ **tank** *n* AIR Oxidationsbehälter *m*; ~ **zone** *n* AIR Oxidationszone *f*

oxidative *adj* WATER POLL oxidativ; ~ **decomposition** *n* WATER POLL oxidativer Abbau *m*

oxidator *n* AIR Oxidator *m*

oxidimetric *adj* ANALYSIS oxidimetrisch; ~ **determination** *n* ANALYSIS oxidimetrische Bestimmung *f*; ~ **titration** *n* ANALYSIS oxidimetrische Titration *f*

oxidizable *adj* GEN oxidierbar; ~ **substance** *n* AIR oxidierbarer Stoff *m*

oxidize 1. *vt* SOIL *contaminants in extract* oxidieren; **2.** *vi* AIR, ANALYSIS, WASTE, WATER POLL, WATER SUPP oxidieren

oxidizing *adj* GEN oxidierend; ~ **agent** *n* AIR, WATER POLL, WATER SUPP Oxidationsmittel *nt*; ~ **catalyst** *n* AIR oxidierender Katalysator *m*; ~ **catalyst bed** *n* AIR oxidierendes Katalysatorbett *nt*; ~ **substance** *n* AIR, SAFETY oxidierender Stoff *m*

oxygen *n* (O_2) GEN Sauerstoff *m* (O_2); ~ **accumulation** *n* WATER SUPP Sauerstoffanreicherung *f*; ~ **addition** *n* WATER POLL Sauerstoffzugabe *f*

oxygenate *vt* WATER POLL oxygenieren

oxygen: ~ **breathing apparatus** *n* SAFETY Sauerstoffatemgerät*nt*; ~ **budget** *n* AIR Sauerstoffhaushalt *m*; ~ **charge** *n* WASTE, WATER POLL Sauerstoffeintrag *m*; ~ **concentration** *n* AIR, ANALYSIS, WATER POLL, WATER SUPP Sauerstoffkonzentration *f*; ~ **consumption** *n* AIR, WATER POLL Sauerstoffverbrauch *m*; ~ **content** *n* ANALYSIS, RENEW ENERG *of biomass* Sauerstoffgehalt *m*; ~ **corrosion** *n* AIR, WATER SUPP Sauerstoffkorrosion *f*; ~ **deficiency** *n* ANALYSIS Sauerstoffmangel *m*, WATER POLL, WATER SUPP Sauerstoffmangel *m*, Sauerstoffdefizit *nt*; ~**-deficient** *adj* WASTE, WATER

SUPP sauerstoffarm; ~ **demand** *n* WATER POLL, WATER SUPP Sauerstoffbedarf *m*; ~**-demanding** *adj* SOIL sauerstoffbedürftig, sauerstoffzehrend, WATER POLL sauerstoffbedürftig; ~**-depleting** *adj* WASTE, WATER POLL, WATER SUPP sauerstoffzehrend; ~ **depletion** *n* AIR, ANALYSIS, WATER POLL Sauerstoffzehrung *f*; ~ **donator** *n* WATER SUPP Sauerstoffdonator *m*; ~ **flask** *n* AIR, ANALYSIS Sauerstoffflasche *f*

oxygenic *adj* WATER POLL sauerstoffhaltig

oxygen: ~ **introduction** *n* AIR, RENEW ENERG, WATER SUPP Sauerstoffzufuhr *f*; ~ **load** *n* WASTE, WATER POLL Sauerstoffeintrag *m*; ~ **measuring device** *n* AIR, ANALYSIS, SAFETY Sauerstoffmeßgerät *nt*; ~**-saturated** *adj* WASTE, WATER POLL, WATER SUPP sauerstoffgesättigt; ~**-saturated water** *n* WATER SUPP sauerstoffgesättigtes Wasser *nt*; ~ **saturation** *n* AIR, WATER POLL Sauerstoffsättigung *f*; ~ **saturation factor** *n* AIR, WATER POLL Sauerstoffsättigungsfaktor *m*; ~ **sensor** *n* AIR Sauerstoffmeßfühler *m*, *catalyst* Lambdasonde *f*, Sauerstoffsonde *f*, ANALYSIS, SAFETY Sauerstoffmeßfühler *m*; ~ **supersaturation** *n* WATER POLL Sauerstoffübersättigung *f*; ~ **supply** *n* AIR, SAFETY, WATER POLL, WATER SUPP Sauerstoffangebot *nt*, Sauerstoffversorgung *f*; ~ **transfer** *n* WATER POLL Sauerstofftransport *m*, Sauerstoffzufuhr *f*; ~ **transfer capacity in clean water** *n* WATER POLL Sauerstoffzufuhrvermögen *nt* in Reinwasser; ~ **transfer capacity in wastewater** *n* WATER POLL Sauerstoffzufuhrvermögen *nt* in Abwasser; ~ **uptake** *n* WATER POLL Sauerstoffaufnahme *f*

ozone *n* AIR, ANALYSIS, SAFETY, WASTE Ozon *nt*; ~ **analyser** *BrE*, ~ **analyzer** *AmE* *n* ANALYSIS Ozonanalysator *m*; ~ **budget** *n* AIR Ozonhaushalt *m*; ~ **capillary** *n* AIR Ozonkapillare *f*; ~ **concentration** *n* AIR, WASTE Ozonkonzentration *f*; ~ **content** *n* AIR Ozongehalt *m*; ~ **damage** *n* AIR Ozonschaden *m*; ~**-damaging** *adj* AIR ozonschädlich; ~ **deficit** *n* AIR Ozondefizit *nt*; ~**-depleting** *adj* AIR ozonabbauend; ~ **depletion** *n* AIR Ozonabbau *m*; ~ **depletion potential** *n* (*ODP*) AIR Ozonabbaupotential *nt*, Ozonschutzzerstörungspotential *nt*; ~ **destroyer** *n* AIR Ozonzerstörer *m*; ~ **destruction** *n* AIR Ozonzerstörung *f*; ~ **detection** *n* ANALYSIS Ozonbestimmung *f*; ~ **formation** *n* AIR Ozonbildung *f*; ~**-friendly** *adj* AIR ozonfreundlich; ~**-generating** *adj* AIR ozonerzeugend; ~ **generator** *n* AIR, ANALYSIS Ozongenerator *m*; ~ **hole** *n* AIR Ozonloch *nt*; ~ **immissions** *n pl* AIR Ozonimmissionen *f pl*; ~ **impact** *n* AIR Ozonbelastung *f*; ~ **layer** *n* AIR Ozonschicht *f*; ~ **layer destruction** *n* AIR Ozonschichtzerstörung *f*; ~ **level** *n* AIR Ozonpegel *m*; ~ **loss** *n* AIR Ozonverlust *m*; ~ **measuring** *n* ANALYSIS Ozonmessung *f*; ~ **plant** *n* WASTE Ozonanlage *f*; ~**-resistant** *adj* AIR ozonresistent; ~ **shield** *n* AIR Ozonschicht *f*; ~ **shrinkage** *n* AIR Ozonschwund *m*; ~ **smog** *n* AIR Ozonsmog *m*; ~**-tolerant** *adj* AIR ozontolerant; ~ **value** *n* AIR Ozonwert *m*; ~ **warning** *n* AIR, SAFETY Ozonwarnung *f*

ozonization *n* SOIL, WASTE, WATER POLL, WATER SUPP Ozonierung *f*, Ozonisierung *f*

P

P *abbr* (*phosphorus*) AIR, WATER POLL P (*Phosphor*)
Pa *abbr* (*pascal*) ANALYSIS *SI unit*, SAFETY Pa (*Pascal*)
PA *abbr* (*polyamide*) WASTE, WATER POLL PA (*Polyamid*)
packaging *n* MANAGE, WASTE Verpackung *f*; **~ fraction** *n* WASTE Verpackungsfraktion *f*; **~ industry** *n* MANAGE Verpackungsindustrie *f*; **~ material** *n* WASTE Packstoff *m*, Verpackungsmaterial *nt*; **~ material producing industry** *n* MANAGE Packstoffindustrie *f*; **~ optimization** *n* MANAGE, WASTE Verpackungsoptimierung *f*; **~ waste** *n* WASTE Verpackungsmüll *m*
packed: **~ bed** *n* AIR Schüttbett *nt*, Schüttschicht *f*, WASTE Schüttung *f*; **~-bed column** *n* AIR Füllkörperkolonne *f*; **~-bed filter** *n* AIR Schüttschichtfilter *nt*; **~-bed height** *n* AIR Füllkörperhöhe *f*; **~-bed scrubber** *n* AIR Füllkörperwäscher *m*; **~ column** *n* ANALYSIS gepackte Säule *f*, WATER SUPP Füllkörperkolonne *f*; **~-column scrubber** *n* AIR Füllkörperkolonne-Abscheider *m*, Füllkörpersäulenabscheider *m*; **~ suspended bed** *n* (*PSB*) WATER POLL Schwebebett *nt*; **~-tower scrubber** *n* AIR Füllkörperkolonnenabscheider *m*, Füllkörpersäulenabscheider *m*
packer *n* MANAGE, WASTE Packer *m*; **~ body** *n* WASTE Preßmüllfahrzeug *nt*, Preßmüllwagen *m*; **~ unit** *n* WASTE Kompaktor *m*, Verdichtungsanlage *f*
packing *n* AIR Füllkörper *m*, Füllkörperschüttung *f*, WASTE Füllkörper *m*; **~ press** *n* WASTE Emballagenpresse *f*
paddle *n* WATER POLL Paddel *nt*, Schaufel *f*; **~ mixer** *n* WATER SUPP Paddelmischer *m*; **~ wheel** *n* WATER POLL Schaufelrad *nt*; **~-wheel aeration** *n* WATER POLL Paddelradbelüftung *f*
PAFC *abbr* (*phosphoric acid fuel cell*) AIR Phosphorsäure-Brennstoffzelle *f*
PAH *abbr* (*polycyclic aromatic hydrocarbon*) AIR, WASTE, WATER POLL, WATER SUPP PAK (*polyzyklischer aromatischer Kohlenwasserstoff*)
paint *n* WASTE Farbe *f*; **~ sludge** *n* WATER POLL Lackschlamm *m*
palladium *n* (*Pd*) AIR Palladium *nt* (*Pd*)
pallet *n* WASTE Palette *f*
pall: **~ ring** *n* AIR *packed-tower scrubbers* Pallring *m*
PAN *abbr* (*peroxoacetyl nitrate*) AIR, WATER SUPP PAN (*Peroxoacetylnitrat*)
panel *n* MARINE, NOISE Platte *f*; **~ absorber** *n* NOISE *bass absorber* Plattenabsorber *m*, Plattenschwinger *m*; **~-louvered ceiling** *AmE*, **~-louvred ceiling** *BrE* *n* NOISE Plattenrasterdecke *f*; **~ resonator** *n* NOISE Plattenresonator *m*
paper *n* ANALYSIS, WASTE Papier *nt*; **~ bag** *n* WASTE Papierbeutel *m*, Papiertasche *f*, Papiertüte *f*; **~ board waste** *n* WASTE Pappeabfall *m*; **~ chromatography** *n* (*pc*) ANALYSIS Papierchromatographie *f* (*PC*); **~ collection** *n* WASTE Altpapiersammlung *f*; **~ consumption** *n* WASTE Papierverbrauch *m*; **~ fiber** *AmE*, **~ fibre** *BrE* *n* WASTE Papierfaser *f*, Papierfaserstoff *m*; **~ filter** *n* ANALYSIS Papierfilter *nt*; **~ pulp** *n* WASTE Zellstoff *m*; **~ quality** *n* WASTE Papierqualität *f*; **~ raw material** *n* WASTE Papierrohstoff *m*; **~ sack** *n* WASTE Papiersack *m*; **~ wrapping** *n* WASTE Papierverpackung *f*
parabolic *adj* RENEW ENERG Parabol-, parabolisch
parabolically: **~ formed gutter** *n* RENEW ENERG *collector* parabolisch geformte Rinne *f*
parabolic: **~ dish** *n* RENEW ENERG Parabolschüssel *f*; **~ dish installation** *n* RENEW ENERG Parabolschüsselanlage *f*; **~ reflector** *n* RENEW ENERG *solar power plant* Parabolspiegel *m*
parallel *adj* ANALYSIS *of measuring* parallel; **~ pump** *n* WATER POLL, WATER SUPP Parallelpumpe *f*
paramagnetic *adj* ANALYSIS paramagnetisch; **~ fluid analytical method** *n* ANALYSIS paramagnetische Flüssigkeitsanalyse *f*
paramagnetism *n* ANALYSIS Paramagnetismus *m*
parameter *n* GEN Kennwert *m*, Parameter *m*; **~ determination** *n* ANALYSIS Parameterbestimmung *f*
pararosaniline *n* ANALYSIS Pararosanilin *nt*; **~ method** *n* ANALYSIS *West-Gaeke method* Pararosanilinmethode *f*
parent: **~ solution** *n* ANALYSIS Stammlösung *f*
part *n* ANALYSIS *component* Teil *nt*
partial: **~ authorization** *n* POLICY Teilgenehmigung *f*; **~ current** *n* WATER POLL *of production water, etc* Teilstrom *m*; **~ differential pressure** *n* AIR Partialdruckdifferenz *f*; **~ enclosure** *n* NOISE Teilkapselung *f*; **~ gas quantity** *n* ANALYSIS Teilgasmenge *f*
partially: **~ automated** *adj* WASTE, WATER SUPP halbautomatisch; **~ continuous** *adj* ANALYSIS, WATER POLL halbkontinuierlich; **~ dry waste gas treatment** *n* AIR quasitrockene Abgasbehandlung *f*; **~ halogenated** *adj* AIR teilhalogeniert; **~ halogenated CFCs** *n pl* AIR teilhalogenierte FCKW *m pl*; **~ integrated system** *n* WASTE, WATER SUPP teilintegriertes System *nt*; **~ stabilized** *adj* WATER POLL teilstabilisiert; **~ stabilized sludge** *n* WATER POLL teilstabilisierter Schlamm *m*; **~ water-soluble** *adj* AIR teilweise wasserlöslich
partial: **~ pressure** *n* AIR Partialdruck *m*; **~ pressure difference** *n* WATER POLL Partialdruckdifferenz *f*; **~ soil decontamination** *n* SOIL Teilbodendekontamination *f*; **~ soil remediation** *n* SOIL Teilbodendekontamination *f*, Teilbodensanierung *f*; **~ stream** *n* AIR Teilstrom *m*
particle *n* GEN Korn *nt*, Partikel *f*, Teilchen *nt*; **~ beam liquid chromatography-mass spectrometry** *n* ANALYSIS Partikelstrahl-Flüssigchromatographie-Massenspektrometrie *f*; **~ charging** *n* AIR Partikelaufladung *f*; **~ counter** *n* ANALYSIS Partikelzähler *m*; **~-counting device** *n* ANALYSIS Teilchenzählungsmeßgerät *nt*; **~ diameter** *n* WATER POLL Partikeldurchmesser *m*; **~-drop agglomerate** *n* AIR Partikel-Tropfen-Agglomerat *nt*; **~ emissions** *n pl* AIR Partikelemissionen *f pl*; **~ filtration** *n* WATER POLL Partikelfiltration *f*; **~ friction** *n* AIR Partikelreibung *f*; **~ impact** *n* AIR Teilchenstoß *m*; **~ orbit** *n* AIR *collection operation* Teilchenbahn *f*; **~ path** *n* AIR Teilchenweg *m*; **~ separation** *n* AIR Partikelabscheidung *f*; **~ size** *n* AIR, WASTE Korngröße *f*, Partikelgröße *f*; **~ size analyser** *BrE*, **~ size analyzer** *AmE* *n* ANALYSIS Teilchengrößenanalysator *m*; **~ size class** *n* WASTE Korngrößenklasse *f*; **~ size distribution** *n* AIR, WATER POLL Korngrößenverteilung *f*; **~ size separation** *n* WASTE Korngrößentrennung *f*; **~ surface** *n* AIR *wet precipitator* Oberfläche *f* der Partikeln, Oberfläche *f* der Teilchen; **~ thickness** *n* AIR Partikeldichte *f*; **~ wetting** *n* AIR Partikelbenetzung *f*
particulate *adj* AIR partikelförmig; **~ air pollution** *n* AIR partikelförmige Luftverunreinigung *f*; **~ collection** *n* AIR Rauchgasentstaubung *f*, Staubabscheidung *f*, WASTE Staubabscheidung *f*; **~ emissions** *n pl* AIR Feststoffausstoß *m*; **~ gas** *n* AIR Staubgas *nt*; **~-loaded** *adj* AIR partikelbeladen; **~-loaded gas stream** *n* AIR partikelbeladener Gasstrom *m*; **~-loaded waste gas** *n* AIR partikelbeladenes Abgas *nt*;

~ **matter** *n* (*PM*) AIR Schwebstoffteilchen *nt* (*PM*); ~ **matter filter** *n* AIR Schwebstofffilter *nt*; ~ **pollutants** *n* AIR *in dust, fumes, mist, smoke, spray* Partikelschadstoffe *m pl*, partikelförmige Schadstoffe *m pl*; ~ **pollution** *n* WASTE, WATER SUPP Teilchenverunreinigung *f*; ~ **size distribution** *n* AIR Partikelgrößenverteilung *f*; ~ **smoke emissions** *n pl* AIR partikelförmige Rußemissionen *f pl*
particulates: ~ **monitoring** *n* ANALYSIS Überwachung *f* partikulärer Stoffe
particulate: ~ **solid** *n* AIR partikelförmiger Feststoff *m*
partition *n* NOISE Trennwand *f*; ~ **chromatography** *n* ANALYSIS Verteilungschromatographie *f*; ~ **surface** *n* NOISE *walls, ceilings* Trennfläche *f*
partometer *n* ANALYSIS Partometer *nt*
parts: ~ **per billion** *phr* (*ppb*) ANALYSIS Teile pro Billion; ~ **per million** *phr* (*ppm*) ANALYSIS Teile pro Million; ~ **per quadrillion** *phr* (*ppq*) ANALYSIS Teile pro Billiarde; ~ **per trillion** *phr* (*ppt*) ANALYSIS Teile pro Trillion
PAS *abbr* (*photoacoustic spectroscopy*) ANALYSIS PAS (*photoakustische Spektroskopie*)
pascal *n* (*Pa*) ANALYSIS *SI unit*, SAFETY Pascal *nt* (*Pa*)
pass *vt* POLICY *law* verabschieden; ~ **through** *vt* AIR durchlaufen
passing *n* POLICY *of law* Verabschiedung *f*
passive *adj* GEN passiv; ~ **degassing** *n* WASTE *natural process* Passiventgasung *f*; ~ **pellistor** *n* ANALYSIS passiver Pellistor *m*; ~ **sound protection** *n* NOISE *earphones, headphones* passiver Schallschutz *m*; ~ **thermal solar energy use** *n* RENEW ENERG passive thermische Sonnenenergienutzung *f*
pasteurization *n* WATER POLL Pasteurisierung *f*
pasty *adj* WASTE pastös; ~ **hazardous waste** *n* WASTE pastöser Sonderabfall *m*; ~ **waste** *n* WASTE pastöser Abfall *m*
paternoster: ~ **pump** *n* WATER POLL, WATER SUPP Kettenpumpe *f*
path *n* AIR *of gas* Weg *m*
pathogen *n* WASTE Pathogen *nt*, WATER POLL, WATER SUPP Krankheitserreger *m*, Pathogen *nt*; **~-free** *adj* WASTE pathogen-frei
pathogenic *adj* WASTE, WATER POLL, WATER SUPP krankheitserregend, pathogen; ~ **bacteria** *n pl* WASTE pathogene Bakterien *pl*; ~ **germs** *n pl* WASTE, WATER SUPP pathogene Keime *pl*; ~ **hazard** *n* SAFETY pathogene Gefahr *f*; ~ **microorganism** *n* WASTE, WATER POLL pathogener Mikroorganismus *m*
pathological: ~ **waste** *n* MARINE, WASTE, WATER POLL infektiöser Abfall *m*, pathogener Abfall *m*
path: **~-specific** *adj* SOIL pfadspezifisch
patrol: ~ **boat** *n* MARINE Patrouillenboot *nt*
pay: **~-back time** *n* RENEW ENERG Amortisationszeit *f*
payload *n* MARINE Nutzlast *f*, Nutzmasse *f*
Pb *abbr* (*lead*) GEN Pb (*Blei*)
pc *abbr* (*paper chromatography*) ANALYSIS PC (*Papierchromatographie*)
PC *abbr* WASTE, WATER POLL, WATER SUPP (*polycarbonate*) Polycarbonat *nt*, (*polymer concrete*) Polymerbeton *m*, Reaktionsharzbeton *m*
PCB *abbr* (*polychlorinated biphenyl*) SOIL PCB (*polychloriertes Biphenyl*); **~-capacitor** *n* AIR, WASTE PCB-Kondensator *m*
PC: ~ **connection** *n* (*personal computer connection*) ANALYSIS *to measuring devices* PC-Anschluß *m* (*Personalcomputer-Anschluß*)
PCDD *abbr* (*polychlorinated dibenzo-p-dioxin*) SOIL PCDD (*polychloriertes Dibenzo-p-Dioxin*)
PCDF *abbr* (*polychlorinated dibenzofurane*) SOIL PCDF (*polychloriertes Dibenzofuran*)
PCI *abbr* (*positive chemical ionization*) ANALYSIS PCI (*positive chemische Ionisation*)
PCN *abbr* (*polychlorinated naphthalene*) AIR, SOIL PCN (*polychloriertes Naphthalin*)
PCP *abbr* (*pentachlorophenol*) AIR, SOIL PCP (*Pentachlorphenol*)
PCT *abbr* (*polychlorinated terphenyl*) AIR, SOIL PCT (*polychloriertes Terphenyl*)
PCV *abbr* (*positive crankcase ventilation*) AIR Kurbelgehäuseentlüftung *f*
Pd *abbr* (*palladium*) AIR Pd (*Palladium*); ~ **II/Cu II catalyst** *n* AIR PdII/Cu II-Katalysator *m*
PE *abbr* (*polyethylene*) WASTE PE (*Polyethylen*)
peak *n* GEN Spitze *f*, Peak *nt*; ~ **concentration** *n* AIR Spitzenkonzentration *f*; ~ **consumption** *n* WASTE, WATER SUPP Verbrauchsspitze *f*; ~ **of the day** *n* WATER POLL Tagesspitze *f*; ~ **demand** *n* WASTE, WATER SUPP Spitzenbedarf *m*; ~ **detection** *n* ANALYSIS Signalerfassung *f*; ~ **flow** *n* WATER POLL Spitzendurchfluß *m*, Spitzendurchsatz *m*; ~ **infiltration** *n* WATER SUPP Spitzeninfiltration *f*; ~ **load** *n* RENEW ENERG Spitzenlast *f*; ~ **ozone damage** *n* AIR Ozonspitzenbelastung *f*; ~ **velocity** *n* ANALYSIS, SAFETY, WASTE, WATER SUPP Höchstgeschwindigkeit *f*; ~ **water demand** *n* WATER SUPP Spitzenwasserbedarf *m*; ~ **water flow** *n* WATER SUPP Spitzenwasserdurchfluß *m*, Anfallspitze *f*
PE: ~ **bottle** *n* WASTE PE-Flasche *f*; ~ **container** *n* WASTE PE-Container *m*; ~ **film** *n* WASTE PE-Film *m*
pellet *n* AIR Schüttgut *nt*, SOIL *vitrification*, WATER POLL *sludge cake* Pellet *nt*; ~ **catalyst** *n* AIR Schüttgutkatalysator *m*; ~ **incineration** *n* AIR, SOIL Pelletbrennen *nt*
pelletization *n* SOIL, WASTE, WATER POLL Pelletisierung *f*
pelletize *vt* SOIL pelletisieren, WASTE granulieren, pelletisieren, tablettieren, WATER POLL pelletisieren
pelletizer *n* WASTE *machine* Granuliervorrichtung *f*, Tablettiermaschine *f*
pelletizing *n* WASTE Granulierung *f*, Tablettierung *f*
pellet: ~ **layer** *n* SOIL Pelletschicht *f*; ~ **mill** *n* WASTE Pelletmühle *f*, Pelletpresse *f*; ~ **technique** *n* ANALYSIS *in IR spectroscopy* Preßlingstechnik *f*
pellistor *n* ANALYSIS Pellistor *m*
Peltier: ~ **element** *n* ANALYSIS *for cooling* Peltier-Element *nt*
penal: ~ **law** *n* POLICY Strafrecht *nt*
penetrate *vt* AIR eindringen in; ~ **by radiation** *vt* ANALYSIS durchstrahlen
penetration *n* AIR *of dust particle* Eindringen *nt*
penetrometer *n* ANALYSIS Penetrometer *nt*
penstock *n* WASTE Druckleitung *f*, Druckrohrleitung *f*, WATER SUPP Druckleitung *f*, Druckrohrleitung *f*, Förderleitung *f*
pentaCDF *abbr* (*pentachlorodibenzofurane*) SOIL PentaCDF (*Pentachlordibenzofuran*)
pentachlorodibenzofurane *n* (*pentaCDF*) SOIL Pentachlordibenzofuran *nt* (*PentaCDF*)
pentachlorophenol *n* (*PCP*) AIR, SOIL Pentachlorphenol *nt* (*PCP*)
PER *abbr* (*perchloroethylene*) AIR PER (*Perchlorethylen*)
per: ~ **capita consumption** *n* WATER SUPP Pro-Kopf-Verbrauch *m*; ~ **capita energy consumption** *n* RENEW ENERG Energie-Pro-Kopf-Verbrauch *m*
perceived: ~ **noise level** *n* (*PNdB*) NOISE Lästigkeitspegel *m*
percentile: ~ **value** *n* ANALYSIS Perzentilwert *m*
perched: ~ **groundwater** *n* WASTE vadoses Grundwasser *nt*, WATER SUPP schwebendes Grundwasser *nt*, vadoses Grundwasser *nt*
perchloroethylene *n* (*PER*) AIR Perchlorethylen *nt* (*PER*)
percolate 1. *vt* ANALYSIS perkolieren, WASTE, WATER POLL, WATER SUPP *strain* durchfiltern, perkolieren, versickern; ~ **through** WASTE, WATER SUPP eindringen in, einsickern in; **2.** *vi* MARINE, WATER POLL durchsickern
percolating: ~ **filter** *n* WATER POLL, WATER SUPP Tropfkörper *m* (*TK*); ~ **water** *n* WASTE, WATER SUPP Senkwasser *nt*, Sickerwasser *nt*; ~ **water treatment plant** *n* WATER POLL Sickerwasseraufbereitung *f*
percolation *n* GEN *straining* Durchsickern *nt*, Perkolation *f*, Versickerung *f*, Versickern *nt*; ~ **capacity** *n* WASTE, WATER

SUPP Versickerungsleistung *f*; ~ **test** *n* ANALYSIS Perkolationsversuch *m*; ~ **well** *n* WASTE, WATER SUPP Sickerbrunnen *m*

percolative: ~ **infiltration** *n* WASTE, WATER SUPP perkolative Infiltration *f*

perfect: ~ **gas** *n* AIR ideales Gas *nt*, perfektes Gas *nt*

perforate *vt* GEN *pipe* lochen

perforated *adj* GEN *pipe* gelocht, perforiert; ~ **panel** *n* NOISE Lochblech *nt*, Lochplatte *f*; ~ **panel absorber** *n* NOISE Lochblechabsorber *m*, Lochplattenabsorber *m*; ~ **plate** *n* AIR Lochblech *nt*, Lochplatte *f*, Siebblech *nt*, WATER POLL *of screen* Lochblech *nt*, WATER SUPP Lochplatte *f*

perikinetic *adj* WASTE, WATER SUPP perikinetisch; ~ **flocculation** *n* WASTE, WATER SUPP perikinetische Flockung *f*

periodic *adj* AIR, ANALYSIS periodisch; ~ **cleaning** *n* AIR *filter bag* periodische Reinigung *f*; ~ **flushing** *n* WASTE, WATER SUPP periodische Spülung *f*

peripheral: ~ **ditch** *n* SOIL, WASTE Randgraben *m*; ~ **pump** *n* WATER POLL, WATER SUPP Peripheralpumpe *f*

periphery *n* NOISE *areas to be lined* Begrenzungsfläche *f*; ~ **particle orbit** *n* AIR *collection operation* Grenzteilchenbahn *f*

peristaltic: ~ **pump** *n* WASTE peristaltische Pumpe *f*, WATER POLL Schlauchquetschpumpe *f*, peristaltische Pumpe *f*

permanent: ~ **audibility threshold shift** *n* (*PTS*) NOISE dauerhafte Hörschwellenverschiebung *f*; ~ **hardness** *n* WASTE, WATER POLL, WATER SUPP Nichtkarbonathärte *f* (*NCH*), bleibende Härte *f*, permanente Härte *f*; ~ **infiltration** *n* WASTE, WATER SUPP permanente Infiltration *f*; ~ **magnet** *n* ANALYSIS, WASTE Permanentmagnet *m*; ~ **magnetic** *adj* WASTE permanentmagnetisch; ~ **storage** *n* WASTE Endlagerung *f*

permeability *n* ANALYSIS Permeabilität *f*, SOIL Durchlässigkeit *f*, WATER POLL Durchlässigkeit *f*, Permeabilität *f*; ~ **measuring instrument** *n* ANALYSIS Permeabilitätsmeßgerät *nt*, Permeameter *nt*

permeable *adj* GEN durchlässig, permeabel

permeameter *n* ANALYSIS *permeability measuring instrument* Permeabilitätsmeßgerät *nt*, Permeameter *nt*

permeate 1. *n* AIR, WATER POLL, WATER SUPP Permeat *nt*; **2.** *vt* AIR, SOIL, WASTE, WATER POLL, WATER SUPP durchdringen

permeate: ~ **recovery** *n* WATER POLL Permeatgewinnung *f*

permeation: ~ **membrane** *n* AIR *membrane reactor* Permeationsmembran *f*; ~ **velocity** *n* AIR Permeationsgeschwindigkeit *f*

permission *n* AIR, POLICY, WASTE, WATER SUPP Erlaubnis *f*, *licence, authorization* Genehmigung *f*

permit *n* AIR Genehmigung *f*, POLICY Genehmigung *f*, Erlaubnis *f*, WASTE, WATER SUPP Genehmigung *f*; ~ **application** *n* POLICY, WATER SUPP Genehmigungsantrag *m*; ~ **impact** *n* POLICY Genehmigungswirkung *f*; ~ **process** *n* POLICY Genehmigungsverfahren *nt*

peroxoacetyl: ~ **nitrate** *n* (*PAN*) AIR, WATER SUPP Peroxoacetylnitrat *nt* (*PAN*)

persistence *n* AIR, SOIL Persistenz *f*; ~ **potential** *n* SOIL Persistenzpotential *nt*

persistent: ~ **oil** *n* SOIL, WASTE persistentes Öl *nt*

personal: ~ **computer connection** *n* (*PC connection*) ANALYSIS *to measuring devices* Personalcomputer-Anschluß *m* (*PC-Anschluß*)

pervious *adj* SOIL durchlässig; ~ **area** *n* WATER POLL durchlässige Fläche *f*

PES *abbr* (*polyester*) WASTE PES (*Polyester*)

pesticide *n* WATER POLL Pestizid *nt*

PET *abbr* (*polyethylene terephthalate*) WASTE Polyethylenterephthalat *nt*

PETRIFIX: ~ **process** *n* WASTE PETRIFIX-Verfahren *nt*

petrol: ~ **engine vehicle** *n* *BrE* (*cf gasoline engine vehicle AmE*) WASTE Fahrzeug mit Vergasermotor *nt*

petroleum *n* WASTE Erdöl *nt*; ~ **product** *n* WASTE Erdölerzeugnis *nt*

petrol: ~ **vapour recovery plant** *n* *BrE* (*cf gasoline vapor recovery plant AmE*) AIR, WASTE Benzindampfrückgewinnungsanlage *f*

pH *abbr* (*potential of hydrogen*) GEN pH (*Wasserstoffexponent*)

phantom: ~ **dump** *n* WASTE unkontrollierte Deponie *m*, wilde Deponie *m*

pharmaceutical: ~ **industry waste** *n* WASTE Abfall *m* der Pharmaindustrie

phase *n* AIR, ANALYSIS Phase *f*; ~ **analysis** *n* ANALYSIS *of sample* Phasenanalyse *f*; ~ **break vessel** *n* AIR Phasentrenngefäß *nt*; ~ **separation** *n* AIR, ANALYSIS, WASTE Phasentrennung *f*; ~**-specific** *adj* WASTE phasenspezifisch; ~**-specific sequential extraction** *n* WASTE phasenspezifische sequentielle Extraktion *f*

pH: ~ **depression** *n* AIR, ANALYSIS, WATER POLL Herabsetzung des pH-Wertes *f*; ~ **drop** *n* AIR, ANALYSIS pH-Abnahme *f*, WATER POLL Absinken des pH-Wertes *nt*; ~ **electrode** *n* ANALYSIS pH-Elektrode *f*

phenol *n* WATER POLL Phenol *nt*; ~**-formaldehyde-resin** *n* WASTE Phenol-Formaldehyd-Harz *nt* (*PF*)

phenolic: ~ **resin** *n* WASTE, WATER SUPP Phenolharz *nt*

pheromone *n* SOIL Pheromon *nt*

pH: ~ **measurement** *n* ANALYSIS pH-Messung *f*; ~ **meter** *n* ANALYSIS pH-Meßgerät *nt*

phon *n* ANALYSIS *unit of noise level* Phon *nt*

phonometer *n* ANALYSIS, NOISE Lautstärkemesser *m*, Lautstärkepegelmesser *m*, Phonmeter *nt*, Schallpegelmesser *m*

Phoredox: ~ **method** *n* WATER POLL *phosphorus removal* Phoredox-Verfahren *nt*

phosphate *n* WATER POLL Phosphat *nt*; ~ **of calcium** *n* WATER POLL Calciumphosphat *nt*; ~ **concentration** *n* WATER POLL Phosphatkonzentration *f*; ~**-containing** *adj* WASTE, WATER POLL, WATER SUPP phosphathaltig; ~ **elimination** *n* WATER POLL Phosphateliminierung *f*; ~**-free** *adj* WATER POLL phosphatfrei; ~**-free detergent** *n* WATER POLL phosphatfreies Waschmittel *nt*; ~**-free washing powder** *n* WATER POLL phosphatfreies Waschmittel *nt*; ~ **precipitate** *n* WATER POLL Phosphatfällungsmittel *nt*; ~ **precipitation** *n* WATER POLL Phosphatfällung *f*; ~ **removal** *n* WATER POLL Phosphateliminierung *f*; ~ **sludge** *n* WATER POLL Phosphatschlamm *m*; ~ **stripping** *n* WATER POLL Phosphatstrippen *nt*

phosphoric: ~ **acid fuel cell** *n* (*PAFC*) AIR Phosphorsäure-Brennstoffzelle *f*

phosphorous *adj* ANALYSIS phosphorsauer

phosphorus *n* (*P*) AIR, WATER POLL Phosphor *f* (*P*); ~ **redissolution** *n* WATER POLL Phosphorrücklösung *f*; ~ **removal** *n* WATER POLL Phosphoreliminierung *f*

photoacoustic: ~ **spectroscopy** *n* (*PAS*) ANALYSIS photoakustische Spektroskopie *f* (*PAS*)

photobiological *adj* WASTE photobiologisch; ~ **reaction** *n* WASTE photobiologische Reaktion *f*

photocatalyst *n* WASTE Photokatalysator *m*

photocatalytic *adj* WASTE photokatalytisch; ~ **oxidation** *n* WASTE photokatalytische Oxidation *f*

photochemical *adj* AIR, WASTE, WATER POLL, WATER SUPP photochemisch; ~ **oxidant** *n* AIR photochemisches Oxidationsmittel *nt*; ~ **reaction** *n* WASTE, WATER SUPP photochemische Reaktion *f*; ~ **smog** *n* AIR Sommersmog *m*, photochemischer Smog *m*; ~ **treatment** *n* WATER POLL photochemische Behandlung *f*

photodegradable *adj* WASTE photoabbaubar, photochemisch, photolytisch abbaubar; ~ **plastics** *n pl* WASTE photoabbaubare Kunststoffe *m pl*, photochemisch abbaubare Kunststoffe *m pl*, photolytisch abbaubare Kunstoffe *m pl*

photodegradation *n* WASTE Photoabbau *m*, photochemischer Abbau *m*

photoelectric: ~ **current** *n* RENEW ENERG *solar cell* Photostrom *m*; ~ **voltage** *n* RENEW ENERG *solar cell* Photospannung *f*
photoelectron: ~ **spectroscopy** *n* ANALYSIS, WATER SUPP Photoelektronenspektroskopie *f*
photographic: ~ **film** *n* ANALYSIS photographischer Film *m*; ~ **infrared control** *n* ANALYSIS photographische Infrarotkontrolle *f*
photoionization *n* ANALYSIS Photoionisation *f*; ~ **detector** *n* (*PID*) ANALYSIS Photoionisationsdetektor *m* (*PID*); ~ **gas chromatography** *n* ANALYSIS Gaschromatographie mit Photoionisationsdetektion *f* (*GC-PID*)
photolysis *n* SOIL, WASTE Photolyse *f*
photolytic *adj* ANALYSIS *converter*, SOIL photolytisch; ~ **degradation** *n* SOIL photolytischer Abbau *m*; ~ **destruction** *n* SOIL photolytische Zerstörung *f*
photomechanical *adj* WASTE photomechanisch; ~ **process** *n* WASTE photomechanisches Verfahren *nt*
photometer *n* ANALYSIS Lichtmesser *m*, Photometer *nt*
photometric *adj* ANALYSIS photometrisch; ~ **analysis** *n* ANALYSIS photometrische Analyse *f*
photometry *n* ANALYSIS Photometrie *f*
photomultiplier *n* (*PM*) ANALYSIS Photovervielfacher *m*, Photomultiplier *m*; ~ **tube** *n* (*PMT*) ANALYSIS Photovervielfacher *m*, Sekundärelektronenvervielfacher *m* (*SEV*)
photooxidants *n pl* AIR, WASTE Photooxidantien *nt pl*
photooxidation *n* AIR, WASTE Photooxidation *f*
photophoresis *n* AIR Photophorese *f*
photosensor *n* WASTE Photosensor *m*
photosynthesis *n* RENEW ENERG *biomass*, WATER POLL Photosynthese *f*
phototrophic *adj* WATER POLL *bacteria* phototroph
photovoltaic *adj* RENEW ENERG photovoltaisch; ~ **cell** *n* RENEW ENERG Photovoltaikzelle *f*; ~**-driven** *adj* RENEW ENERG photovoltaisch betrieben; ~ **effect** *n* RENEW ENERG photovoltaischer Effekt *m*; ~ **element** *n* RENEW ENERG photovoltaisches Element *nt*; ~ **energy conversion** *n* RENEW ENERG photovoltaische Energieumwandlung *f*; ~ **energy generation** *n* RENEW ENERG photovoltaische Energiegewinnung *f*; ~ **module** *n* RENEW ENERG Photovoltaikmodul *nt*; ~**-operated** *adj* RENEW ENERG photovoltaisch betrieben; ~**-powered** *adj* RENEW ENERG photovoltaisch betrieben; ~**-powered vehicle** *n* RENEW ENERG photovoltaisch betriebenes Fahrzeug *nt*; ~ **power plant** *n* RENEW ENERG Photovoltaik-Kraftwerk *nt*
photovoltaics *n pl* (*PV*) RENEW ENERG Photovoltaik *f* (*PV*)
photovoltaic: ~ **system** *n* RENEW ENERG photovoltaisches System *nt*
pH: ~**-paper strip** *n* ANALYSIS Lackmuspapier *nt*, pH-Papierstreifen *m*; ~ **range** *n* WATER POLL pH-Bereich *m*; ~**-sensitive** *adj* WASTE, WATER POLL, WATER SUPP pH-empfindlich; ~ **stabilization** *n* WATER POLL pH-Stabilisierung *f*; ~ **value** *n* AIR, ANALYSIS, SOIL, WASTE pH-Wert *m*; ~ **value increase** *n* AIR, ANALYSIS, SOIL, WASTE pH-Wert-Erhöhung *f*
physical *adj* GEN physikalisch; ~ **adsorption** *n* AIR physikalische Adsorption *f*; ~ **agent** *n* WATER SUPP Naturkraft *f*; ~**-chemical** *adj* AIR, SOIL, WASTE, WATER POLL, WATER SUPP physikalisch-chemisch; ~**-chemical process** *n* WASTE, WATER POLL, WATER SUPP physikalisch-chemisches Verfahren *nt*; ~ **condition** *n* AIR Aggregatzustand *m*; ~ **parameter** *n* ANALYSIS, SOIL, WATER POLL physikalischer Parameter *m*; ~ **recycling** *n* WASTE stoffliche Verwertung *f*, stoffliches Recycling *nt*; ~ **stabilization** *n* WASTE mechanischer Einschluß *m*; ~ **strain** *n* WASTE *landfill* physikalische Belastung *f*; ~ **water treatment** *n* WASTE, WATER SUPP mechanische Wasseraufbereitung *f*
physiochemical *adj* GEN physikalisch-chemisch, physikochemisch; ~ **environment** *n* AIR, WASTE, WATER SUPP physikalisch-chemische Umgebung *f*; ~ **phosphorus removal** *n* WATER POLL chemisch-physikalishe Phosphorelimінierung *f*; ~ **process** *n* WASTE, WATER POLL, WATER SUPP physikalisch-chemisches Verfahren *nt*; ~ **reaction** *n* AIR physikochemische Reaktion *f*; ~ **treatment** *n* WATER POLL physikalisch-chemisches Reinigungsverfahren *nt*
physiological: ~ **effect** *n* SAFETY physiologische Einwirkung *f*
physisorption *n* AIR Physisorption *f*
pick: ~ **up** *vt* WASTE abholen
picker *n* AIR *stator-rotor system* Schlagstock *m*
pickup: ~ **mechanism** *n* WASTE Greifer *m*
PID *abbr* (*photoionization detector*) ANALYSIS PID (*Photoionisationsdetektor*)
piezoelectric *adj* ANALYSIS piezoelektrisch; ~ **microphone** *n* ANALYSIS piezoelektrisches Mikrofon *nt*
piezometer *n* ANALYSIS Piezometer *nt*
pigment: ~ **sludge** *n* WATER POLL Pigmentschlamm *m*
pile: ~ **drive** *vt* WASTE einrammen, rammen
pile: ~ **core sampling** *n* ANALYSIS Rammkernsondierung *f*; ~**-driving** *n* WASTE Rammung *f*
pilot: ~ **audit** *n* POLICY Pilot-Audit *nt*; ~ **plant** *n* AIR, RENEW ENERG, SOIL, WASTE, WATER POLL Pilotanlage *f*, Versuchsanlage *f*, WATER SUPP Versuchsanlage *f*; ~ **plant scale** *n* AIR, SOIL, WATER POLL Pilotanlagemaßstab *m*; ~ **program** *AmE*, ~ **programme** *BrE* *n* AIR, RENEW ENERG, SOIL, WASTE, WATER POLL, WATER SUPP Pilotprogramm *nt*; ~ **project** *n* AIR, RENEW ENERG, WASTE, WATER POLL, WATER SUPP Pilotprojekt *nt*; ~ **sample** *n* ANALYSIS Vorprobe *f*; ~ **test** *n* ANALYSIS Pilotversuch *m*
pipe *n* GEN Rohr *nt*; ~ **aerator** *n* WATER SUPP Rohrbelüfter *m*; ~ **break** *n* WATER POLL, WATER SUPP Rohrbruch *m*; ~ **breakage** *n* WATER POLL, WATER SUPP Rohrbruch *m*; ~ **bridge** *n* WATER POLL Rohrbrücke *f*; ~ **bursting** *n* WATER SUPP Rohrdurchbruch *m*; ~ **end** *n* WATER SUPP Rohrendstrang *m*; ~ **fitting** *n* WASTE, WATER SUPP Formstück *nt*; ~ **joint** *n* WATER POLL, WATER SUPP Rohranschluß *m*; ~ **leakage** *n* WATER POLL, WATER SUPP Rohrleckage *f*
pipeline *n* WATER POLL, WATER SUPP Rohrleitung *f*; ~ **resistance** *n* WATER POLL, WATER SUPP Rohrleitungwiderstand *m*
pipe: ~ **sleeve** *n* WATER POLL, WATER SUPP Muffe *f*, Rohrmuffe *f*; ~ **stability** *n* WASTE Rohrleitungsstandsicherheit *f*; ~**-to-pipe recycling** *n* WATER POLL *direct reuse of wastewater* geschlossener Kreislauf *m*
piping: ~ **system** *n* WATER SUPP Röhrensystem *nt*, Rohrsystem *nt*
piston *n* AIR, WATER POLL, WATER SUPP Kolben *m*; ~ **capacity** *n* WASTE, WATER SUPP *of plunger* Hubraum *m*; ~ **compressor** *n* WASTE Hubkolbenverdichter *m*, WATER POLL Hubkolbenverdichter *m*, Kolbenkompressor *m*, WATER SUPP Kolbenkompressor *m*
pistonphone *n* ANALYSIS *calibration device* Pistonfon *nt*
piston: ~ **pump** *n* WATER POLL, WATER SUPP Kolbenpumpe *f*
pit *n* SOIL Miete *f*, WASTE Grube *f*; ~ **landfill** *n* WASTE Grubendeponie *f*
pitostatic: ~ **tube** *n* ANALYSIS Prandtlsches Staurohr *nt*
Pitot: ~ **tube** *n* ANALYSIS Pitot-Rohr *nt*
pitting *n* WATER SUPP Lochfraß *m*
PIXE *abbr* (*proton-induced x-ray emission*) ANALYSIS PIXE (*protoneninduzierte Röntgenemission*); ~ **analysis** *n* ANALYSIS PIXE-Analyse *f*
placement: ~ **of wastes** *n* WASTE *in landfill* Einbau *m* der Abfälle
Planck: ~**'s constant** *n* AIR Plancksches Wirkungsquantum *nt*; ~**'s law** *n* AIR Plancksche Strahlungsformel *f*
plane *n* WASTE hobeln
planetary: ~ **gear** *n* RENEW ENERG *wind turbine* Planetengetriebe *nt*
planking *n* NOISE *of sound absorbing panels* Beplankung *f*
plankton *n* WATER SUPP Plankton *nt*
planning: ~ **assessment decision** *n* POLICY

Planfeststellungsbeschluß *m*; ~ **assessment procedure** *n* POLICY Planfeststellungsverfahren *nt*

plant *n* GEN *factory* Anlage *f*; ~ **authorization** *n* POLICY Anlagengenehmigung *f*; ~ **dust** *n* AIR pflanzlicher Staub *m*; ~ **licence** *BrE n*, ~ **license** *AmE* POLICY Anlagengenehmigung *f*; ~ **oils** *n pl* RENEW ENERG *end product of extraction, biomass conversion* Pflanzenöle *nt pl*; ~ **permit** *n* POLICY Anlagengenehmigung *f*

Plant: ~ **Protection Act** *n* POLICY Pflanzenschutzgesetz *nt*

plant: ~-**related** *adj* GEN anlagebezogen, anlagentechnisch; ~ **representative** *n* MANAGE, POLICY, WATER SUPP Betriebsbeauftragte *f*, Betriebsbeauftragter *m*; ~ **requiring licence** *BrE*, ~ **requiring license** *AmE n* POLICY genehmigungsbedürftige Anlage *f*; ~ **requiring permit** *n* POLICY genehmigungsbedürftige Anlage *f*; ~ **safety** *n* SAFETY Anlagensicherheit *f*; ~-**specific** *adj* GEN anlagenspezifisch, anlagentechnisch

plasma: ~ **emission detection** *n* ANALYSIS Emissionsdetektion *f* mit Plasma

plaster *n* NOISE *surface lining* Putz *m*

plastic *n* WASTE Kunststoff *m*

plasticator *n* WASTE Plastikator *m*

plastic: ~ **bottle** *n* ANALYSIS *in sampling*, WASTE Kunststofflasche *f*; ~ **container** *n* WASTE Kunststoffgebinde *nt*, Kunststoffverpackung *f*

plasticize *vt* WASTE plastifizieren

plasticized *adj* WASTE, WATER SUPP weichmacherhaltig

plasticizer *n* WASTE, WATER SUPP Weichmacher *m*

plasticizing *n* WASTE *plastics* Plastifizierung *f*, Weichmachen *nt*, WATER SUPP Weichmachen *nt*; ~ **unit** *n* WASTE Plastifizieraggregat *nt*

plastic: ~ **liner** *n* SOIL Kunststoffbahn *f*; ~ **matrix** *n* WASTE Kunststoffmatrize *f*; ~ **module** *n* WATER POLL *trickling filter packing material* Kunststoffeinbauten *m pl*; ~ **packaging** *n* WASTE Kunststoffverpackung *f*; ~ **packaging material** *n* WASTE Kunststoffverpackung *f*; ~ **pipe** *n* RENEW ENERG *heat pump* Kunststoffrohr *nt*

plastics: ~ **application** *n* WASTE Kunststoffeinsatz *m*

plastic: ~ **sealing** *n* WASTE Kunststoffabdichtung *f*; ~ **separation** *n* WASTE Kunststoffabtrennung *f*

plastics: ~ **extruder** *n* WASTE Kunststoffverarbeitungsextruder *m*

plastic: ~ **sheet** *n* SOIL Kunststoffbahn *f*, Kunststoffdichtungsbahn *f*, WASTE Kunststoffolie *f*

plastics: ~ **molding** *AmE*, ~ **moulding** *BrE n* WASTE Kunststofformen *nt*; ~-**processing** *adj* WASTE kunststoffverarbeitend; ~-**processing industry** *n* WASTE kunststoffverarbeitende Industrie *f*; ~ **recycling** *n* WASTE Kunststoffrecycling *nt*, Kunststoffverwertung *f*; ~ **recycling fluidized-bed plant** *n* WASTE Kunststoffrecycling-Wirbelschichtanlage *f*; ~ **recycling plant** *n* WASTE Kunststoffrecyclinganlage *f*, Kunststoffverwertungsanlage *f*; ~ **sheeting** *n* WASTE Kunststoffilm *m*, Kunststoffolie *f*; ~ **sorting plant** *n* WASTE Kunststoffsortieranlage *f*; ~ **waste** *n* WASTE Kunststoffabfall *m*

plate *n* AIR *bubble-cap tray tower* Bodenblech *nt*, WASTE Plättchen *nt*; ~ **catalyst** *n* AIR Plattenkatalysator *m*; ~ **column** *n* AIR Bodenkolonne *f*; ~-**column scrubber** *n* AIR *wet scrubber* Bodensäulenabscheider *m*; ~ **module** *n* WATER POLL *ultrafiltration* Plattenmodul *nt*; ~ **precipitator** *n* AIR Plattenabscheider *m*, Plattenelektrofilter *nt*; ~ **tower** *n* AIR *wet scrubber* Bodensäulenabscheider *m*; ~-**wire ESP** *n* AIR *dry electrofilter* Plattenelektrofilter *nt*

platinum: ~ **metal** *n* AIR *catalyst* elementares Platin *nt*

plenum: ~ **chamber** *n* NOISE *type of silencer* Plenumkammer *f*

plot *vt* ANALYSIS *chromatogram* auftragen

plotter *n* ANALYSIS Plotter *m*

ploughshare: ~ **mixer** *n* WATER SUPP Ploughshare-Mischer *m*

plugging *n* WASTE, WATER SUPP Verockerung *f*

plumbiferous *adj* AIR, WASTE bleihaltig

plume *n* AIR Kaminaustritt *m*, Rauchfahne *f*, Rauchschwaden *m*; ~ **axis** *n* AIR Schwadenachse *f*; ~-**rise** *n* AIR Überhöhung *f* der Abluftfahne; ~ **shred** *n* AIR Schwadenfetzen *m*

plunger: ~ **pump** *n* WATER POLL, WATER SUPP Plungerpumpe *f*

PM *abbr* AIR (*particulate matter*) PM (*Schwebstoff-Teilchen*), ANALYSIS (*photomultiplier*) Photomultiplier *m*, Photovervielfacher *m*

PMT *abbr* (*photomultiplier tube*) ANALYSIS Photovervielfacher *m*, SEV (*Sekundärelektronenvervielfacher*)

PNA *abbr* (*polynuclear aromatics*) SOIL polynukleare Aromaten *m pl*

PNdB *abbr* (*perceived noise level*) NOISE Lästigkeitspegel *m*

pneumatic *adj* ANALYSIS *sample*, SOIL, WASTE, WATER POLL, WATER SUPP pneumatisch; ~ **classification** *n* WASTE pneumatisches Sortieren *nt*; ~ **driving gear** *n* WASTE, WATER POLL pneumatischer Antrieb *m*; ~ **flotation** *n* WATER SUPP Flotation *f* durch Lufteinblasen; ~ **injector** *n* WATER POLL Druckluftinjektor *m*; ~ **sorter** *n* WASTE pneumatische Sortieranlage *f*; ~ **sound absorber** *n* MARINE, NOISE pneumatischer Schalldämpfer *m*

PO *abbr* (*polyolefin*) WASTE Polyolefin *nt*

pocket: ~ **filter** *n* AIR Taschenfilter *nt*; ~ **measuring instrument** *n* ANALYSIS Taschenmeßgerät *nt*; ~-**shaped** *adj* AIR *filter* taschenförmig

pointer: ~ **instrument** *n* ANALYSIS *gas meter* Zeigerinstrument *nt*

point: ~ **source** *n* AIR, SOIL, WATER POLL, WATER SUPP *pollution* Punktquelle *f*; ~ **source pollutant** *n* AIR, SOIL, WATER POLL Schadstoff *m* aus Punktquelle; ~ **vacuum cleaning system** *n* SAFETY Punktabsaugsystem *nt*

poison *n* AIR, SAFETY, WASTE, WATER SUPP Gift *nt*

poisoning *n* AIR, SAFETY, WASTE, WATER SUPP Vergiftung *f*

poisonous *adj* SAFETY *as danger symbol* giftig; ~ **waste** *n* WASTE Giftmüll *m*

polarity *n* AIR *positive or negative* Polarität *f*

polarization *n* ANALYSIS Polarisation *f*

polarograph *n* ANALYSIS Polarograph *m*

polarographic *adj* ANALYSIS *detectors* polarographisch; ~ **determination methods** *n pl* ANALYSIS polarographische Bestimmungsmethoden *f pl*

polarography *n* ANALYSIS *potentiometry* Polarographie *f*

polder *n* WASTE Polder *m*

policy: ~ **instrument** *n* POLICY Politikinstrument *nt*; ~ **objectives** *n pl* POLICY politische Zielsetzungen *f pl*

polishing *n* WASTE Schönung *f*, *plastics recycling* Polieren *nt*

pollutant *n* GEN Kontaminant *nt*, Schadstoff *m*, Schmutzstoff *m*, Umweltschadstoff *m*; ~ **amount** *n* ANALYSIS Schadstoffmenge *f*; ~ **analysis** *n* ANALYSIS Schadstoffanalyse *f*; ~ **behavior** *AmE*, ~ **behaviour** *BrE n* AIR, SOIL, WATER POLL Schadstoffverhalten *nt*; ~ **content** *n* AIR, SOIL, WASTE, WATER POLL, WATER SUPP Schadstoffgehalt *m*; ~ **conversion** *n* AIR Schadstoffumsatz *m*; ~ **deposition** *n* AIR, SOIL, WASTE, WATER SUPP Schadstoffablagerung *f*; ~ **extraction** *n* WASTE Schadstoffentfrachtung *f*; ~-**free** *adj* AIR, WATER POLL, WATER SUPP schadstofffrei; ~ **load** *n* AIR Schadstofffracht *f*; ~ **oxidation** *n* AIR Schadstoffoxidation *f*; ~-**resistant** *adj* AIR, ANALYSIS, SOIL, WATER POLL schadstoffbeständig, schadstoffresistent

pollutants: ~ **inventory** *n* SAFETY Schadstoffinventar *nt*

pollutant: ~-**specific** *adj* AIR, SOIL, WATER POLL schadstoffspezifisch; ~ **transmission** *n* AIR Schadstoffverteilung *f*; ~ **type** *n* AIR, SOIL, WATER POLL Schadstofftyp *m*

pollute *vt* GEN belasten, verschmutzen, verunreinigen

polluted *adj* GEN belastet, schadstoffbelastet, verschmutzt, verunreinigt; ~ **dust** *n* AIR belasteter Staub *m*; ~ **process gas** *n* AIR verschmutztes Abgas *nt*; ~ **rainwater** *n* WATER POLL belastetes Regenwasser *nt*, verschmutztes Regenwasser *nt*; ~ **soil** *n* SOIL, WASTE, WATER POLL schadstoffbelastetes

Erdreich *nt*; ~ **water** *n* WATER POLL verschmutztes Wasser *nt*; ~ **water body** *n* WATER POLL belastete Gewässer *nt pl*, verschmutzte Gewässer *nt pl*

polluter *n* POLICY Verursacher *m*; ~ **group** *n* POLICY Verursachergruppe *f*; **~-oriented** *adj* POLICY verursachergerecht; **~-pays-principle** *n* POLICY Verursacherprinzip *nt*

polluting: ~ **agent** *n* GEN belastender Stoff *m*

pollution *n* GEN Belastung *f*, Schmutz *m*, Umweltverschmutzung *f*, Verschmutzung *f*, Verunreinigung *f*

pollutional: ~ **index** *n* WATER SUPP Verschmutzungsgrad *m* des Wassers

pollution: ~ **burden** *n* AIR, SOIL, WASTE, WATER POLL Schadstoffbelastung *f*; ~ **control** *n* GEN Reinhaltung *f*; ~ **deposition** *n* SAFETY Reinhaltung *f*; ~ **emitter** *n* AIR, POLICY, SOIL, WASTE, WATER SUPP Emissionsquelle *f*; ~ **index** *n* POLICY Schmutzfrachtindex *m*; ~ **research** *n* AIR, ANALYSIS, MANAGE, MARINE, POLICY, WATER POLL Umweltverschmutzungsforschung *f*; ~ **source** *n* AIR, POLICY, SOIL, WASTE, WATER POLL, WATER SUPP Verunreinigungsquelle *f*

polyacrylnitrile *n* AIR *filter fibre material* Polyacrylnitril *nt*

polyaddition *n* WASTE Polyaddition *f*

polyamide *n* (*PA*) WASTE, WATER POLL Polyamid *nt* (*PA*); ~ **membrane** *n* WASTE Polyamidmembran *f*, WATER POLL PA-Membran *f*, Polyamidmembran *f*

polycarbonate *n* (*PC*) WASTE, WATER POLL Polycarbonat *nt*; ~ **membrane** *n* WATER POLL Polycarbonatmembran *f*

polychlorinated: ~ **biphenyl** *n* (*PCB*) SOIL polychloriertes Biphenyl *nt* (*PCB*); ~ **dibenzofurane** *n* (*PCDF*) SOIL polychloriertes Dibenzofuran *nt* (*PCDF*); ~ **dibenzo-p-dioxin** *n* (*PCDD*) SOIL polychloriertes Dibenzo-p-Dioxin *nt* (*PCDD*); ~ **naphthalene** *n* (*PCN*) AIR, SOIL polychloriertes Naphthalin *nt* (*PCN*); ~ **terphenyl** *n* (*PCT*) AIR, SOIL polychloriertes Terphenyl *nt* (*PCT*)

polycondensation *n* WASTE Polykondensation *f*

polycrystalline *adj* RENEW ENERG polykristallin; ~ **silicon cell** *n* RENEW ENERG *photovoltaic* polykristalline Siliziumzelle *f*

polycyclic: ~ **aromatic hydrocarbon** *n* (*PAH*) AIR, WASTE, WATER POLL, WATER SUPP polyzyklischer aromatischer Kohlenwasserstoff *m* (*PAK*)

polyester *n* (*PES*) WASTE, WATER POLL Polyester *m* (*PES*); ~ **foam** *n* WASTE Polyesterschaumstoff *m*; ~ **membrane** *n* WATER POLL PES-Membran *f*, Polyestermembran *f*

polyethylene *n* (*PE*) WASTE Polyethylen *nt* (*PE*); ~ **container** *n* WASTE PE-Container *m*; ~ **terephthalate** *n* (*PET*) WASTE Polyethylenterephthalat *nt*

polymer *n* AIR *adsorbent*, WASTE, WATER POLL Polymer *nt*; ~ **concrete** *n* (*PC*) WASTE, WATER SUPP Polymerbeton *m*, Reaktionsharzbeton *m*; **~-concrete composite material** *n* WASTE Kunstharzbeton *m*

polymerization *n* WASTE Polymerisation *f*

polymeter *n* ANALYSIS Polymeter *nt*

polynuclear: ~ **aromatics** *n pl* (*PNA*) SOIL polynukleare Aromaten *m pl*

polyolefin *n* (*PO*) WASTE Polyolefin *nt*; ~ **container** *n* WASTE Polyolefin-Container *m*

polyphosphate *n* WATER POLL Polyphosphat *nt*

polypropylene *n* (*PP*) AIR, WASTE *filter fibre material* Polypropylen *nt* (*PP*)

polysaccharide *n* WASTE Polysaccharid *nt*

polysaprobic *adj* WATER POLL polysaprob

polystyrene *n* (*PS*) WASTE Polystyrol *nt* (*PS*)

polytetrafluoroethylene *n* (*PTFE*) WASTE *filter fibre material* Polytetrafluorethylen *nt* (*PTFE*)

polyurethane *n* (*PU*) WASTE Polyurethan *nt* (*PU*)

polyvinyl: ~ **chloride** *n* (*PVC*) SOIL, WASTE Polyvinylchlorid *nt* (*PVC*)

pond *n* WATER SUPP Teich *m*

population *n* AIR, ANALYSIS, SAFETY, WATER SUPP Bevölkerung *f*; ~ **dose** *n* AIR *nuclear*, SAFETY Bevölkerungsgesamtdosis *f*; ~ **growth** *n* SOIL *massive pressure on soil function* Bevölkerungswachstum *nt*; **~-related** *adj* WATER SUPP einwohnerbezogen; **~-related water consumption** *n* WATER SUPP einwohnerbezogener Wasserverbrauch *m*; **~-specific** *adj* WATER SUPP einwohnerbezogen, einwohnerspezifisch; **~-specific water demand** *n* WATER SUPP einwohnerbezogener Wasserverbrauch *m*

pore *n* AIR, WASTE, WATER POLL Pore *f*; ~ **diameter** *n* AIR Porendurchmesser *m*; ~ **membrane** *n* WATER POLL Porenmembran *f*; ~ **size** *n* AIR Porengröße *f*; ~ **space** *n* AIR Porenraum *m*; ~ **volume** *n* AIR Porenvolumenanteil *m*, Porenraum *m*, Porenvolumen *nt*, WASTE Porenvolumen *nt*

porometer *n* ANALYSIS Porometer *nt*

porosity *n* AIR, WASTE, WATER POLL, WATER SUPP Durchlässigkeit *f*, Porosität *f*

porous *adj* AIR engporig, SOIL, WASTE, WATER POLL, WATER SUPP porös, durchlässig; ~ **activated carbon** *n* AIR engporige Aktivkohle *f*; ~ **adsorbent** *n* AIR engporiges Absorbens *nt*; ~ **cover** *n* WASTE *of landfill* poröse Abdeckung *f*; ~ **membrane** *n* WASTE, WATER POLL, WATER SUPP poröse Membran *f*

portable *adj* ANALYSIS *gas analyser* beweglich, portabel, tragbar; ~ **gas chromatograph** *n* ANALYSIS tragbarer Gaschromatograph *m*; ~ **GC** *n* ANALYSIS tragbarer GC *m*; ~ **spectrophotometer** *n* ANALYSIS tragbares Spektrophotometer *nt*; ~ **x-ray fluorescence analyser** *BrE*, ~ **x-ray fluorescence analyzer** *AmE n* ANALYSIS tragbarer Röntgenfluoreszenz-Analysator *m*

pose: ~ **a chronic health risk** *vi* SAFETY etwas chronisch belasten

position *vt* ANALYSIS positionieren

position: ~ **of the sun** *n* RENEW ENERG Solarstand *m*, Sonnenstand *m*

positive *adj* GEN positiv; ~ **cation** *n* WATER SUPP Kation *nt*, positives Kation *nt*; ~ **chemical ionization** *n* (*PCI*) ANALYSIS positive chemische Ionisation *f* (*PCI*); ~ **crankcase ventilation** *n* (*PCV*) AIR Kurbelgehäuseentlüftung *f*; ~ **displacement pump** *n* WATER POLL, WATER SUPP Verdrängerpumpe *f*; ~ **electrode** *n* AIR Anode *f*, positive Elektrode *f*; ~ **ion** *n* WASTE positives Ion *nt*, WATER POLL Kation *nt*, positives Ion *nt*

possibility: ~ **of error** *n* ANALYSIS Fehlermöglichkeit *f*; ~ **of wastewater deposition** *n* WATER POLL Aufbringfähigkeit *f*

post: **~-closure care** *n* WASTE, WATER SUPP Nachsorge *m*; **~-closure expenditure** *n* WASTE Nachsorgeaufwand *f*; **~-closure measures** *n pl* WASTE Nachsorgemaßnahmen *f pl*; **~-closure military contaminated sites** *n pl* SOIL Rüstungsaltlasten *f pl*; **~-closure monitoring** *n* SOIL, WASTE Nachsorgeüberwachung *f*; **~-coagulation** *n* WASTE nachträgliche Flockung *f*; **~-consumer plastics waste** *n* WASTE Kunststoffabfall *m* nach Gebrauch; **~-denitrification** *n* WATER POLL nachgeschaltete Denitrifikation *f*; **~-filtration** *n* WATER POLL Nachfiltration *f*; ~ **hydrant** *n* WATER SUPP Überflurhydrant *m*, Überflurwasserpfosten *m*; **~-precipitation** *n* WATER POLL Nachfällung *f*; **~-reaction** *n* AIR Nachreaktion *f*; **~-reaction time** *n* AIR Nachreaktionszeit *f*; **~-rotting treatment** *n* WASTE, WATER POLL Nachrotte *f*, Nachverrottung *f*; **~-treatment** *n* SOIL Nachbehandlung *f*

potable *adj* WATER SUPP trinkbar

potash *n* WATER POLL Pottasche *f*

potential 1. *adj* RENEW ENERG, SAFETY, SOIL potentiell; **2.** *n* ANALYSIS, RENEW ENERG, SAFETY, SOIL Potential *nt*

potential: ~ **acidity** *n* SOIL potentielle Azidität *f*; ~ **danger** *n* SAFETY potentielle Gefährdung *f*; ~ **energy** *n* RENEW ENERG potentielle Energie *f*; ~ **of hydrogen** *n* (*pH*) ANALYSIS, SOIL, WASTE, WATER POLL, WATER SUPP Wasserstoffexponent *m* (*pH*); ~ **measurement** *n* ANALYSIS Potentialmessung *f*; ~ **safety risk** *n* SAFETY mögliches Sicherheitsrisiko *nt*

potentiometer *n* ANALYSIS Potentiometer *nt*
potentiometric *adj* ANALYSIS potentiometrisch, elektrometrisch; **~ sensor** *n* ANALYSIS potentiometrischer Sensor *m*; **~ titration** *n* ANALYSIS elektrometrische Titration *f*, potentiometrische Titration *f*
potentiometry *n* ANALYSIS *polarography* Potentiometrie *f*
potentiostat *n* ANALYSIS Potentiostat *m*
potentiostatic *adj* ANALYSIS potentiostatisch; **~ electrode sensor** *n* ANALYSIS potentiostatischer Elektrodensensor *m*
pourable *adj* WASTE rieselfähig, schüttfähig, WATER SUPP rieselfähig
powdered: **~ activated carbon** *n* (*PAC*) AIR, WATER POLL Pulveraktivkohle *f*
power *n* RENEW ENERG *electricity* Strom *m*, *strength* Kraft *f*; **~ capacity** *n* AIR, RENEW ENERG, WATER POLL, WATER SUPP Nennleistung *f*; **~ frequency** *n* RENEW ENERG Netzfrequenz *f*; **~ generation** *n* RENEW ENERG Stromerzeugung *f*, *of renewable energy* Energieerzeugung *f*, WASTE *via biogas* Stromerzeugung *f*; **~ grid** *n* RENEW ENERG Stromnetz *nt*; **~ house** *n* RENEW ENERG Kraftwerk *nt*; **~ output** *n* RENEW ENERG Leistungsabgabe *f*; **~ plant** *n* AIR Kraftwerk *nt*; **~ source** *n* RENEW ENERG Stromquelle *f*; **~ station** *n* RENEW ENERG Kraftwerk *nt*; **~ supply** *n* RENEW ENERG Stromversorgung *f*
POX *abbr* (*purgeable organic halogen compounds*) SOIL, WATER SUPP POX (*ausblasbare organische Halogenverbindungen*)
PP *abbr* (*polypropylene*) AIR, WASTE PP (*Polypropylen*)
ppq *abbr* (*parts per quadrillion*) ANALYSIS Teile pro Billiarde
Prandtl: **~'s pitot tube** *n* ANALYSIS Prandtlsches Staurohr *nt*
preacidification: **~ alkalinity** *n* AIR Alkalität *f* vor der Ansäuerung
preaeration *n* WATER POLL Vorbelüftung *f*; **~ tank** *n* WATER POLL Vorbelüftungsbecken *nt*
preamplifier *n* ANALYSIS Vorverstärker *m*
precaution *n* AIR, MANAGE, MARINE, NOISE Vorsorge *f*, SAFETY Vorsicht *f*, Vorsorge *f*, *measure* Vorkehrung *f*
precautionary: **~ measures** *n pl* SAFETY Vorsichtsmaßnahmen *f pl*
precautions: **~ against dust** *n pl* SAFETY Vorkehrungen *f pl* gegen Staub
precedence: **~ case** *n* POLICY Präzedenzfall *m*
preceding *adj* WATER POLL vorgeschaltet; **with ~ denitrification** *phr* WATER POLL mit vorgeschalteter Denitrifikation
precipitable: **~ organics** *n pl* WATER POLL fällbare organische Verbindungen *f pl*
precipitant *n* WATER POLL Fällmittel *nt*, Fällungsmittel *nt*
precipitate 1. *n* AIR Kondensat *nt*, Niederschlag *m*, WATER POLL Fallmittel *nt*, Fallungsmittel *nt*; **2.** *vt* AIR *from fluid* abscheiden, ausfällen, entstauben, niederschlagen, ANALYSIS ausfällen, SOIL, WASTE, WATER POLL fällen, ausfällen, WATER SUPP ausfällen, fällen; **3.** *vi* AIR *condense* kondensieren
precipitate: **~ feeding** *n* WATER POLL Fällmitteldosierung *f*
precipitation *n* AIR *process* Ausfällung *f*, *rain, snow, sleet* Niederschlag *m*, *atmospheric* atmosphärischer Niederschlag *m*, ANALYSIS Ausfällung *f*, SOIL Niederschlag *m*, WATER POLL, WATER SUPP *chemical* Fällung *f*, Ausfällung *f*; **~ with aluminium salts** *n BrE* WATER POLL Fällung *f* mit Aluminiumsalzen; **~ with aluminum salts** *AmE see precipitation with aluminium salts BrE*; **~ area** *n* WATER SUPP Niederschlagsgebiet *nt*; **~ collector** *n* AIR Niederschlagssammler *m*; **~ event** *n* AIR Niederschlagsvorfall *m*; **~ with ferric salt** *n* WATER POLL Fällung *f* mit Eisen(III)-Salzen; **~ gage** *AmE*, **~ gauge** *BrE n* WATER SUPP Niederschlagsmesser *m*; **~ with lime** *n* WATER POLL Kalkfällung *f*; **~ operation** *n* WATER POLL Fällungsvorgang *m*; **~ sludge** *n* WATER POLL Fällungsschlamm *m*; **~ unit** *n* AIR Abscheideeinheit *f*
precision: **~ measurement** *n* ANALYSIS Präzisionsmessung *f*; **~ sound level meter** *n* ANALYSIS, NOISE Präzisionsschallpegelmesser *m*
preclean *vt* AIR, WASTE, WATER SUPP vorreinigen
precleaned: **~ gas** *n* AIR vorgereinigtes Gas *nt*
precleaning *n* AIR Vorreinigung *f*; **~ basin** *n* WATER POLL Vorreinigungsbecken *nt*
precoat: **~ filtration** *n* AIR Precoat-Filtration *f*; **~ layer** *n* AIR *bag filter*, WASTE Precoat-Schicht *f*
pre-column/back-flush: **~ operation** *n* ANALYSIS Vorsäulenrückspülung *f*
precombustion: **~ chamber** *n* AIR Vorbrennkammer *f*
precombustor *n* SOIL Vorbrenner *m*
preconcentrate *n* ANALYSIS Vorkonzentrat *nt*
preconcentration *n* ANALYSIS Vorkonzentrierung *f*
precooler *n* AIR Vorkühler *m*
precursor: **~ pollutant** *n* AIR Schadstoffvorstufe *f*
predenitrification *n* WATER POLL vorgeschaltete Denitrifikation *f*
predictive: **~ capability** *n* AIR Vorhersagefähigkeit *f*
predrying: **~ of wastewater sludge** *n* WATER POLL *before incineration* KS-Vortrocknung *f*, Klärschlammvortrocknung *f*
prefilter 1. *n* AIR, ANALYSIS, WATER POLL Vorfilter *nt*; **2.** *vt* AIR, ANALYSIS, WATER POLL vorfiltern
prefiltering *n* AIR, ANALYSIS, WATER POLL Vorfilterung *f*
prefiltration *n* AIR, ANALYSIS, WATER POLL Vorfilterung *f*
preformed: **~ spray scrubber** *n* AIR vorgeformter Sprühwäscher *m*
preheat *vt* AIR, SOIL, WASTE vorheizen
preheated: **~ combustion air** *n* SOIL vorgeheizte Verbrennungsluft *f*
preheater *n* AIR, SOIL, WASTE Vorwärmer *m*
preheating *n* AIR, SOIL, WASTE Vorheizung *f*, Vorwärmung *f*
preliminary: **~ analysis** *n* ANALYSIS Voruntersuchung *f*; **~ assessment** *n* ANALYSIS Vorerkundung *f*; **~ decision** *n* POLICY Vorbescheid *m*; **~ dewatering** *n* WATER POLL Vorentwässerung *f*; **~ dust removal** *n* AIR Vorentstaubung *f*; **~ evaluation** *n* ANALYSIS *for soil remediation* Erstbewertung *f*; **~ filtration** *n* WATER POLL Vorfiltration *f*; **~ purification** *n* WASTE, WATER SUPP Vorreinigung *f*; **~ sample** *n* ANALYSIS Vorprobe *f*; **~ settled sludge** *n* WATER POLL Vorklärschlamm *m*; **~ settlement** *n* WASTE, WATER SUPP Anfangssetzung *f*; **~ settling tank** *n* WATER POLL Vorklärbecken *nt*; **~ treatment** *n* AIR, ANALYSIS, SOIL, WASTE, WATER POLL Vorbehandlung *f*; **~ trial** *n* ANALYSIS Tastversuch *m*
premixer *n* WASTE, WATER SUPP Vormischer *m*
premolding *AmE see premoulding BrE*
premoulding *n BrE* WASTE Vorformung *f*
preparation: **~ plant** *n* WASTE Aufbereitungsanlage *f*; **~ process** *n* WASTE Aufbereitungsverfahren *nt*
preprogram *vt* ANALYSIS vorprogrammieren
preprogrammable *adj* ANALYSIS *measuring instrument* vorprogrammierbar
preprogrammed *adj* ANALYSIS vorprogrammiert
prerotting: **~ treatment** *n* WASTE *degradation and sanitation phase* Vorrotte *f*, Vorverrottung *f*
prescrub *vt* AIR vorwaschen, vorabscheiden
prescrubber *n* AIR Vorabscheider *m*, Vorwäscher *m*
prescrubbing *n* AIR Vorwäsche *f*; **~ tank** *n* AIR *biological gas scrubbing* Vorwaschbehälter *m*
presedimentation *n* WATER SUPP Vorsedimentation *f*
presence: **~ of groundwater** *n* WATER SUPP Grundwasservorkommen *nt*
presentation: **~ of documents for public inspection** *n* POLICY Auslage *f* von Unterlagen, Auslegen *nt* von Unterlagen
preseparation: **~ cyclone** *n* AIR Vorabscheidezyklon *m*
presort *vt* SOIL, WASTE vorsortieren
presorted *adj* SOIL, WASTE vorsortiert; **~ waste** *n* WASTE vorsortierter Abfall *m*
presorting *n* SOIL, WASTE Vorsortierung *f*
press *vt* WASTE pressen; **~ down** *vt* WASTE herunterdrücken;

~ **in** *vt* WASTE eindrücken, einpressen; ~ **together** *vt* WASTE zusammenpressen

press: ~ **communication** *n* POLICY Pressemitteilung *f*; ~ **drying** *n* WATER POLL *dewatering by pressure* Preßtrocknung *f*; ~ **drying of waste with wastewater sludge** *n* WATER POLL Müll-Klärschlamm-Preßtrocknung *f*; ~ **feeder belt** *n* WASTE Pressenbeschickungsband *nt*

pressing *n* ANALYSIS, WASTE, WATER POLL Pressen *nt*, Pressung *f*; ~ **operation** *n* WASTE Preßvorgang *m*

press: ~ **office** *n* POLICY Pressestelle *f*; ~ **plunger** *n* WASTE Pressenstempel *m*, Pressenstößel *m*, Preßkolben *m*, WATER SUPP Plunger *m*; ~ **ram** *n* WASTE Pressenstempel *m*, Pressenstößel *m*, Preßkolben *m*; ~ **stroke** *n* WASTE Pressenhub *m*

pressure *n* GEN Druck *m*; ~ **chamber** *n* AIR *ground filter* Druckkammer *f*; ~ **compensation** *n* WASTE, WATER SUPP Druckausgleich *m*; ~ **difference** *n* ANALYSIS, WASTE, WATER SUPP Druckdifferenz *f*; ~ **drainage** *n* WATER POLL Druckentwässerung *f*; ~ **drop** *n* AIR *filter bag* Druckabfall *m*; ~ **energy** *n* AIR Druckenergie *f*; ~ **filter** *n* AIR, WASTE, WATER POLL, WATER SUPP Druckfilter *nt*; ~ **flotation** *n* WATER POLL *dissolved air flotation* Entspannungsflotation *f*; ~ **force** *n* WATER SUPP Preßkraft *f*; ~ **gage** *AmE*, ~ **gauge** *BrE n* AIR Druckmesser *m*; ~ **governor** *n BrE* (*cf pressure regulator AmE*) AIR, ANALYSIS *trace analyser* Druckregler *m*; ~ **gradient** *n* SOIL Druckgradient *m*; ~ **head** *n* WATER SUPP Druckhöhe *f*; ~ **increase** *n* AIR Druckerhöhung *f*; ~ **loss** *n* AIR *filter bag*, WATER POLL, WATER SUPP Druckverlust *m*; ~ **loss in distribution network** *n* WATER SUPP Druckverlust *m* in Verteilerrohrnetzen; ~ **loss of pipeline** *n* WATER SUPP Druckverlust *m* von Rohrleitungen; ~ **oscillation** *n* WATER SUPP Druckschwingung *f*; ~ **pipe** *n* WATER POLL, WATER SUPP Druckrohr *nt*; ~ **pump** *n* WATER SUPP Druckpumpe *f*; ~ **regulator** *n AmE* (*cf pressure governor BrE*) AIR, ANALYSIS *trace analyser* Druckregler *m*; ~ **release** *n* AIR *of filtering apparatus* Druckentlastung *f*; ~ **sensor** *n* AIR Drucksensor *m*; ~ **sounding** *n* ANALYSIS Drucksondierung *f*; ~ **velocity probe** *n* ANALYSIS *p-u probe* Schalldruck-Schallschnelle-Sonde *f*; ~ **well** *n* WATER SUPP Druckschacht *m*, Fallschacht *m*; ~ **zone** *n* WATER SUPP Druckzone *f*

pressurized: ~ **liquid** *n* AIR Druckflüssigkeit *f*; ~ **liquid stream** *n* AIR Druckflüssigkeitsstrom *m*

prestressed: ~ **concrete pipe** *n* WASTE, WATER POLL, WATER SUPP Spannbetonrohr *nt*, vorgespanntes Betonrohr *nt*

pretest *n* ANALYSIS Vorversuch *m*

prethickening *n* WATER POLL Voreindickung *f*

pretreat *vt* GEN vorbehandeln

pretreated *adj* GEN vorbehandelt

pretreatment *n* GEN Vorbehandlung *f*

prevent *vt* GEN verhindern

preventative *adj* POLICY vorsorgend; ~ **environmental policy** *n* POLICY vorsorgende Umweltpolitik *f*; ~ **environmental protection** *n* MANAGE, POLICY vorsorgender Umweltschutz *m*; ~ **measures** *n pl* AIR, MANAGE, MARINE, NOISE, SAFETY, WATER POLL Vorsorgemaßnahmen *f pl*

prevention *n* NOISE Vermeidung *f*, POLICY Vorsorge *f*, SAFETY Verhütung *f*, WASTE, WATER POLL Vermeidung *f*; ~ **of atmospheric pollution** *n* AIR Verhütung *f* von Luftverschmutzung; ~ **charge** *n* POLICY *waste* Vermeidungsabgabe *f*

prewash 1. *n* WASTE Vorwäsche *f*; **2.** *vt* WASTE vorwaschen

prewasher *n* WASTE Vorwäscher *m*

price: ~ **of water** *n* WATER SUPP Wasserpreis *m*

primary: ~ **air** *n* AIR Primärluft *f*; ~ **aluminium** *n BrE* WASTE Primäraluminium *nt*; ~ **aluminum** *AmE see primary aluminium BrE*; ~ **catalyst** *n* AIR Startkatalysator *m*; ~ **clarifier** *n* WATER POLL Vorklärbecken *nt*; ~ **combustion air** *n* WASTE Primärverbrennungsluft *f*; ~ **cosmic radiation** *n* AIR kosmische Primärstrahlung *f*; ~ **crusher** *n* WASTE Vorbrecher *m*; ~ **downpipe** *n* AIR *packed-bed filter* Primärfallrohr *nt*; ~ **energy balance** *n* RENEW ENERG Primärenergiebilanz *f*; ~ **energy consumption** *n* RENEW ENERG Primärenergieverbrauch *m*; ~ **energy demand** *n* RENEW ENERG Primärenergiebedarf *m*; ~ **energy source** *n* RENEW ENERG Primärenergieträger *m*; ~ **energy substitution** *n* RENEW ENERG Primärenergiesubstitution *f*; ~ **filter** *n* WATER POLL Primärfilter *nt*, Vorfilter *nt*; ~ **fuel** *n* WASTE Primärbrennstoff *m*; ~ **material** *n* WASTE Primärstoff *m*; ~ **plastics recycling** *n* WASTE Kunststoffverwertung *f* der ersten Stufe; ~ **production of paper** *n* WASTE Neuproduktion *f* von Papier; ~ **recycling** *n* WASTE Primärrecycling *nt*; ~ **resources industry** *n* MANAGE Rohstoffindustrie *f*; ~ **settlement** *n* SOIL, WASTE, WATER SUPP primäre Setzung *f*; ~ **settling basin** *n* WATER POLL Vorklärbecken *nt*; ~ **settling tank** *n* WATER POLL Vorklärbecken *nt*; ~ **sludge** *n* WATER POLL Primärschlamm *m*, Rohschlamm *m*; ~ **sound** *n* NOISE Direktschall *m*; ~ **thickener** *n* WATER POLL Voreindicker *m*; ~ **waste avoidance** *n* WASTE Primärabfallvermeidung *f*; ~ **wastewater treatment** *n* WATER POLL *physical process* Abwasseraufbereitung *f* der ersten Stufe, mechanische Abwasserreinigung *f*

prime *vt* WASTE, WATER POLL, WATER SUPP *pump* zum Ansaugen bringen

priming *n* WATER POLL, WATER SUPP Ansaugenlassen *nt*; ~ **cock** *n* WATER SUPP Wasserablaßhahn *m*; ~ **pipe** *n* WASTE, WATER POLL, WATER SUPP *of a pump* Ansaugrohr *nt*; ~ **pump** *n* WASTE Vorhebepumpe *f*, WATER POLL, WATER SUPP Saughilfspumpe *f*, Vorhebepumpe *f*

principal: ~ **rotting treatment** *n* WASTE Hauptrotte *f*

printer: ~ **connection** *n* ANALYSIS *to measuring devices* Druckeranschluß *m*

priority *n* POLICY Priorität *f*; ~ **determination** *n* POLICY Prioritätensetzung *f*; ~ **list** *n* MANAGE, POLICY, SOIL *of measures, etc* Prioritätsliste *f*

prior: ~ **separation** *n* WASTE *of harmful substances* Vorabscheidung *f*

private: ~ **collection** *n* WASTE private Sammlung *f*; ~ **disposal company** *n* WASTE privates Entsorgungsunternehmen *nt*; ~ **drinking water consumption** *n* WASTE privater Trinkwasserverbrauch *m*

probe *n* ANALYSIS Meßfühler *m*, Sensor *m*; ~ **microphone** *n* ANALYSIS Sondenmikrofon *nt*

problem: ~ **definition** *n* ANALYSIS Aufgabenstellung *f*; ~ **site** *n* SOIL kontaminierter Standort *m*

procedure *n* GEN Arbeitsweise *f*, Verfahrensweise *f*

process 1. *n* GEN Prozeß *m*, Verfahren *nt*; **2.** *vt* AIR, ANALYSIS, RENEW ENERG, SOIL *contaminated*, WASTE, WATER POLL aufbereiten, verarbeiten; **3.** ~ **raw material** *vi* MANAGE Grundstoffen verarbeiten

process: ~ **combining pyrolysis and incineration** *n* WATER POLL *wastewater sludge incineration* Schwelbrennverfahren *nt*; ~ **conductivity meter** *n* ANALYSIS Prozeßkonduktometer *nt*; ~ **costs** *n pl* MANAGE Verfahrenskosten *pl*

processed: ~ **slag** *n* WASTE aufbereitete Schlacke *f*; ~ **wastes** *n pl* WASTE aufbereitete Abfälle *m pl*

process: ~ **effectiveness** *n* AIR, ANALYSIS, RENEW ENERG, SOIL, WATER POLL, WATER SUPP Effektivität *f* der Verfahren; ~ **gas flow** *n* AIR Prozeßgasstrom *m*; ~ **heat** *n* RENEW ENERG Prozeßwärme *f*

processing *n* GEN Aufbereitung *f*, Verarbeitung *f*; ~ **capacity** *n* AIR, SOIL, WATER POLL, WATER SUPP Verarbeitungskapazität *f*; ~ **center for recyclable solid waste materials** *AmE*, ~ **centre for recyclable solid waste materials** *BrE n* WASTE Aufarbeitungszentrum *nt* für wiederverwertbare feste Altstoffe; ~ **plant** *n* WASTE Aufbereitungsanlage *f*; ~ **unit** *n* WASTE Aufbereitungsaggregat *nt*, Aufbereitungseinheit *f*

process: ~ **materials** *n pl* AIR, WASTE, WATER SUPP Hilfsstoffe *m pl*; ~ **materials stream** *n* AIR Hilfsstoffstrom *m*; ~**-related** *adj* AIR, RENEW ENERG, WASTE, WATER POLL verfahrenstechnisch; ~ **scrap** *n* WASTE Verarbeitungsabfälle *m pl*; ~**-specific** *adj* AIR, RENEW ENERG, WASTE, WATER POLL verfahrensspezifisch; ~ **stability** *n* WATER POLL Prozeßstabilität *f*; ~ **steam** *n* RENEW ENERG Prozeßdampf *m*; ~ **waste** *n* WASTE Gewerbeabfall *m*, Produktionsabfall *m*, gewerblicher Abfall *m*, produktionsspezifischer Abfall *m*; ~ **waste gas** *n* AIR Prozeßabgas *nt*; ~ **water** *n* AIR, SOIL, WASTE, WATER POLL, WATER SUPP Betriebswasser *nt*, Brauchwasser *nt*, Prozeßwasser *nt*

produce *vt* MANAGE, WASTE herstellen

producer *n* MANAGE, WASTE Hersteller *m*

product *n* GEN Produkt *nt*; ~ **at end of life cycle** *n* WASTE Produkt *nt* am Ende des Lebenszyklus

production *n* RENEW ENERG, WASTE, WATER SUPP Produktion *f*; ~**-consumption loop** *n* WASTE Produktions-Verbrauchs-Kreislauf *m*; ~ **line** *n* WASTE Fertigungsstraße *f*; ~ **potential** *n* RENEW ENERG *of renewable energy sources* Produktionspotential *nt*; ~ **of raw material** *n* MANAGE Herstellung *f* des Werkstoffes; ~ **residue** *n* WASTE Produktionsrückstand *m*; ~**-specific waste** *n* WASTE produktionsspezifischer Abfall *m*; ~ **water** *n* WASTE, WATER POLL, WATER SUPP Betriebswasser *nt*, Brauchwasser *nt*

product: ~ **liability** *n* POLICY, WASTE Produkthaftung *f*; ~ **life cycle** *n* MANAGE Lebensweg *m* eines Produktes, Produktlebenszyklus *m*, Produktzyklus *m*; ~**-line-analysis** *n* MANAGE Produktlinienanalyse *f*; ~ **recovery** *n* AIR Produktrückgewinnung *f*; ~ **safety** *n* SAFETY Produktsicherheit *f*

products: ~ **containing pollutants** *n pl* WASTE schadstoffhaltige Produkte *nt pl*

product: ~ **standards** *n pl* POLICY Produktnormen *f pl*; ~ **stewardship** *n* MANAGE, POLICY, WASTE Produktverantwortung *f*

profitability *n* MANAGE, POLICY Wirtschaftlichkeit *f*

profitable *adj* MANAGE, POLICY wirtschaftlich

program 1. *AmE see programme BrE*; **2.** *vt* ANALYSIS *computer* programmieren

programmable *adj* ANALYSIS *computer* programmierbar

programme *n BrE* ANALYSIS, POLICY *schedule, procedure* Programm *nt*

progressive: ~ **tax** *n* POLICY progressive Steuer *f*

prohibit *vt* POLICY verhindern

prohibition *n* POLICY Verbot *nt*

project: ~ **phase** *n* RENEW ENERG *of renewable energy installations* Projektphase *f*

promoter *n* AIR *catalyst* Promotor *m*

propagate *vt* AIR, MARINE, WASTE sich ausbreiten

propagation *n* AIR Ausbreitung *f*; ~ **class** *n* AIR *waste gas plume of stack* Ausbreitungsklasse *f*; ~ **deadening** *n* NOISE Ausbreitungsdämpfung *f*; ~ **path** *n* SOIL *of harmful substances in soil* Ausbreitungsweg *m*; ~ **of pollutants** *n* AIR, SOIL, WASTE, WATER POLL, WATER SUPP Ausbreitung *f* von Schadstoffen, Schadstoffausbreitung *f*

propellant *n* AIR Treibgas *nt*, Treibmittel *nt*

propelled: ~ **jet** *n* AIR *annular gap nozzle* Treibstrahl *m*

propeller *n* MARINE Propeller *m*, Schiffspropeller *m*, Schiffsschraube *f*; ~ **pump** *n* WATER POLL, WATER SUPP Propellerpumpe *f*

propelling: ~ **force** *n* AIR Triebkraft *f*

proper: ~ **disposal** *n* WASTE geordnete Ablagerung *f*

propionic: ~ **acid** *n* WATER POLL Propansäure *f*

proportional *adj* ANALYSIS proportional; ~ **counter** *n* ANALYSIS *gas-filled detectors that measure ionization in gases* Proportionalzähler *m*; ~ **sampling** *n* ANALYSIS proportionale Probenahme *f*

proposal *n* POLICY Vorschlag *m*

protect *vt* GEN schützen

protection *n* GEN Schutz *m*; ~ **against fire** *n* SAFETY Brandschutz *m*

Protection: ~ **& Indemnity Association** *n* MARINE Schutz- und Entschädigungsvereinigung *f*

protection: ~ **capacity** *n* MARINE, NOISE *from noise* Schutzwirkung *f*; ~ **of neighbors** *AmE*, ~ **of neighbours** *BrE n* POLICY Nachbarschaftsschutz *m*; ~ **of resources** *n* SOIL, WATER SUPP Ressourcenschonung *f*; ~ **zone** *n* SAFETY Schutzzone *f*

protective: ~**-aerated cab** *n* SAFETY *of building machinery* schutzbelüftete Kabine *f*; ~ **aeration** *n* SAFETY Schutzbelüftung *f*; ~ **agent** *n* WASTE, WATER SUPP Schutzmittel *nt*; ~ **equipment** *n* SAFETY Schutzausrüstung *f*; ~ **function** *n* AIR Schutzfunktion *f*; ~ **housing** *n* AIR *filter material* schützendes Gehäuse *nt*; ~ **insulation** *n* SOIL Schutzfiltration *f*; ~ **insulation well** *n* SOIL Schutzinfiltrationsbrunnen *m*; ~ **layer** *n* WASTE Schutzschicht *f*; ~ **measures** *n pl* AIR, MARINE, NOISE, SAFETY, WATER POLL Schutzmaßnahmen *f pl*; ~ **shield** *n* AIR *against UV radiation* Schutzschild *m*; ~ **suit** *n* SAFETY Schutzanzug *m*

protocol *n* POLICY *legal instrument* Protokoll *nt*

proton: ~ **absorptive capacity** *n* AIR Protonenabsorptionsfähigkeit *f*; ~**-induced x-ray emission** *n* (*PIXE*) ANALYSIS protoneninduzierte Röntgenemission *f* (*PIXE*)

protoplasm *n* WATER POLL Protoplasma *nt*

protozoa *n pl* WATER POLL Protozoen *nt pl*

provisional: ~ **evaluation** *n* ANALYSIS, POLICY provisorische Auswertung *f*

proximate: ~ **analysis** *n* ANALYSIS *in which only main constituents are determined* Kurzanalyse *f*

proximity: ~ **pick-up** *n* ANALYSIS Annäherungssensor *m*; ~ **sensor** *n* ANALYSIS Annäherungssensor *m*

PS *abbr* (*polystyrene*) WASTE PS (*Polystyrol*)

PSB *abbr* (*packed suspended bed*) WATER POLL Schwebebett *nt*

psychrometer *n* ANALYSIS Psychrometer *nt*

psychrophilic *adj* WATER POLL psychrophil; ~ **microorganism** *n* WATER POLL psychrophiler Mikrooganismus *m*

PTFE *abbr* (*polytetrafluoroethylene*) WASTE *filter fibre material* PTFE (*Polytetrafluorethylen*)

PTS *abbr* (*permanent audibility threshold shift*) NOISE dauerhafte Hörschwellenverschiebung *f*

PU *abbr* (*polyurethane*) WASTE PU (*Polyurethan*)

publication *n* POLICY Veröffentlichung *f*

public: ~ **canalization** *n* WATER POLL öffentliche Kanalisation *f*; ~ **collection station** *n* WASTE öffentliche Sammelstation *f*, öffentlicher Sammelplatz *m*; ~ **drinking water consumption** *n* WATER SUPP öffentlicher Trinkwasserverbrauch *m*; ~ **drinking water supply** *n* WATER SUPP öffentliche Trinkwasserversorgung *f*; ~ **health** *n* POLICY, WATER SUPP öffentliche Gesundheit *f*; ~ **hearing** *n* POLICY Erörterung *f*; ~ **hearing date** *n* POLICY Erörterungstermin *m*; ~ **order law** *n* POLICY Polizeirecht *nt*; ~ **participation** *n* POLICY Öffentlichkeitsbeteiligung *f*; ~ **utilities** *n pl* WATER SUPP öffentliche Versorgungsunternehmen *nt pl*; ~ **water supply** *n* WATER POLL, WATER SUPP öffentliche Wasserversorgung *f*; ~ **works** *n pl* WASTE, WATER POLL, WATER SUPP Stadtwerke *nt pl*

pug: ~ **mill** *n* SOIL Mörtelmischer *m*

pulp *n* WASTE Brei *m*, *from waste paper* Pulpe *f*, Zellstoff *m*

pulper *n* WASTE Pulper *m*

pulp: ~ **molding system** *AmE*, ~ **moulding system** *BrE n* WASTE Zellstoff-Preßsystem *nt*; ~ **wood** *n* WASTE Faserholz *nt*, Papierholz *nt*

pulse: ~**-jet-cleaned filter** *n* AIR druckstoßgereinigtes Filter *nt*; ~**-jet cleaning** *n* AIR Druckstoßreinigung *f*; ~**-polarography** *n* ANALYSIS Pulse-Polarographie *f*; ~ **stroke** *n* AIR Druckluftstoß *m*

pulverization *n* WASTE Zerstäubung *f*, Zerkleinerung *f*, Zermahlung *f*

pulverize *vt* WASTE zerstäuben, mahlen, zerkleinern, zermahlen

pulverizer *n* WASTE Zerstäuber *m*

pulverizing *n* WASTE Zerkleinerung *f*

pump 1. *n* AIR Pumpe *f*, *hydraulic* Hydraulikpumpe *f*, MARINE

Hydropumpe *f*, Pumpe *f*, WASTE, WATER POLL WATER SUPP Pumpe *f*, *hydraulic* Hydraulikpumpe *f*; **2.** *vt* AIR, WASTE, WATER POLL, WATER SUPP pumpen; ~ **back** WASTE, WATER POLL, WATER SUPP zurückpumpen; ~ **out** SOIL, WASTE, WATER POLL, WATER SUPP abpumpen, auspumpen, leerpumpen

pumpable *adj* AIR, WASTE, WATER SUPP pumpfähig; ~ **hazardous wastes** *n pl* WASTE pumpfähige Sonderabfälle *m pl*; ~ **liquid** *n* AIR pumpfähige Flüssigkeit *f*; ~ **wastes** *n pl* WASTE pumpfähige Abfälle *m pl*

pump: ~ **barrel** *n* AIR, WATER POLL, WATER SUPP Pumpenzylinder *m*; ~ **capacity** *n* AIR Pumpenleistung *f*, WASTE Fördermenge *f*, WATER POLL Fördermenge *f*, Pumpenleistung *f*, WATER SUPP Pumpenleistung *f*, Fördermenge *f*; ~ **case** *n* AIR, WATER POLL, WATER SUPP Pumpengehäuse *nt*; ~ **casing** *n* AIR, WATER POLL, WATER SUPP Pumpengehäuse *nt*; ~ **cavitation** *n* AIR, WATER POLL, WATER SUPP Pumpenkavitation *f*; ~ **characteristics** *n pl* AIR Pumpencharakteristik *f*, WATER POLL, WATER SUPP Pumpencharakteristik *f*, Pumpenkennlinie *f*; ~ **connection** *n* WATER POLL, WATER SUPP Pumpenanschluß *m*; ~ **controls** *n pl* AIR, WATER POLL, WATER SUPP Pumpsteuerung *f*, Pumpensteuerung *f*; ~ **cylinder** *n* AIR, WATER POLL, WATER SUPP Pumpenzylinder *m*; ~**-driving gear** *n* WATER POLL, WATER SUPP Pumpenantrieb *m*

pumped *adj* WASTE, WATER SUPP gepumpt; ~ **water flow** *n* WASTE, WATER SUPP gepumpter Wasserstrom *m*

pump: ~ **efficiency** *n* AIR, WATER POLL, WATER SUPP Pumpenwirkungsgrad *m*; ~ **fluid** *n* WATER POLL, WATER SUPP Pumpentreibmittel *nt*; ~ **gear** *n* WATER POLL, WATER SUPP Pumpengetriebe *nt*; ~ **house** *n* WATER POLL, WATER SUPP Pumpenhaus *nt*; ~ **impeller** *n* WATER POLL, WATER SUPP Pumpenflügelrad *nt*

pumping *n* AIR, WASTE, WATER POLL, WATER SUPP Auspumpen *nt*, Pumpen *nt*; ~ **engine** *n* WATER POLL, WATER SUPP Pumpmaschine *f*; ~**-out** *n* AIR, SOIL, WASTE, WATER POLL, WATER SUPP Abpumpen *nt*, Auspumpen *nt*, Pumpen *nt*; ~ **plant** *n* WATER POLL, WATER SUPP Pumpenanlage *f*, Pumpeneinrichtung *f*; ~ **shaft** *n* WATER POLL, WATER SUPP Pumpenschacht *m*; ~ **station** *n* WATER POLL, WATER SUPP Pumpstation *f*, Pumpwerk *nt*, Pumpenanlage *f*; ~ **test** *n* WASTE Entnahmeversuch *m*, WATER POLL, WATER SUPP Entnahmeversuch *m*, Pumpversuch *m*; ~ **unit** *n* AIR Pumpeinheit *f*, MARINE Pumpeinheit *f*, Pumpenstand *m*, WASTE, WATER POLL, WATER SUPP Pumpeinheit *f*

pump: ~ **inlet** *n* AIR, WASTE, WATER POLL, WATER SUPP Ansaugöffnung *f*; ~ **inlet branch** *n* WATER POLL, WATER SUPP Pumpensaugstutzen *m*; ~ **operation via water tower** *n* WATER POLL, WATER SUPP Pumpbetrieb *m* über einen Hochbehälter; ~ **outlet branch** *n* WATER POLL, WATER SUPP Pumpendruckstutzen *m*; ~ **plunger** *n* WATER POLL, WATER SUPP Pumpenkolben *m*; ~ **plunger shaft** *n* WATER POLL, WATER SUPP Pumpenkolbenwelle *f*; ~ **rod** *n* WATER POLL, WATER SUPP Kolbenstange *f* der Pumpe, Pumpenstange *f*; ~ **room** *n* WATER POLL, WATER SUPP Pumpenraum *m*; ~ **set** *n* WATER POLL, WATER SUPP Pumpenaggregat *nt*, Pumpensatz *m*; ~ **shaft** *n* WATER POLL, WATER SUPP Pumpenschacht *m*; ~ **speed** *n* AIR, WATER POLL, WATER SUPP Pumpengeschwindigkeit *f*; ~ **station** *n* WATER POLL, WATER SUPP Pumpstation *f*, Pumpwerk *nt*, Pumpenanlage *f*; ~ **test** *n* ANALYSIS Pumpversuch *m*; ~ **and treat** *n* WATER SUPP Pumpen *nt* und Behandeln *nt*; ~ **water** *n* WATER SUPP Pumpwasser *nt*; ~ **well** *n* AIR, WATER POLL, WATER SUPP Pumpenbrunnen *m*, Pumpensumpf *m*

puncturability *n* AIR, SAFETY, WASTE Durchstoßfestigkeit *f*

puncture: ~ **resistance** *n* AIR, SAFETY, WASTE Durchstoßfestigkeit *f*

punitive: ~ **fee** *n* POLICY Bußgeld *nt*

pure *adj* GEN rein; ~ **oxygen** *n* AIR, WATER SUPP Reinsauerstoff *m*; ~ **water** *n* WATER POLL, WATER SUPP Reinwasser *nt*, reines Wasser *nt*

purge *vt* GEN spülen

purgeable: ~ **organic halogen compounds** *n pl* (*POX*) SOIL, WATER SUPP ausblasbare organische Halogenverbindungen *f pl* (*POX*)

purge: ~ **air inlet** *n* AIR *activated carbon filter* Spüllufteinlaß *m*; ~**-and-trap GC** *n* ANALYSIS *gas chromatography* Purge-und-Trap-GC *f*; ~**-and-trap method** *n* ANALYSIS Purge-und-Trap-Verfahren *nt*; ~**-and-trap sampler** *n* ANALYSIS Purge-und-Trap - Sampler *m*

purging *n* GEN Spülung *f*; ~ **cock** *n* WATER SUPP Reinigungshahn *m*

purification *n* GEN Reinigung *f*; ~ **capacity** *n* WATER POLL *self-purification* Reinigungsvermögen *nt*; ~ **operation** *n* WATER POLL Reinigungsvorgang *m*; ~ **plant** *n* WATER POLL *wastewater* Kläranlage *f*, Reinigungsanlage *f*; ~ **system using plants** *n* WASTE, WATER POLL Pflanzenklärsystem *nt*

purified: ~ **wastewater** *n* WATER POLL geklärtes Abwasser *nt*

purify *vt* AIR, MARINE, SAFETY, WASTE reinigen, WATER POLL klären, *wastewater, water* reinigen, WATER SUPP reinigen

putrefaction *n* WASTE, WATER POLL Faulen *nt*, Fäulnis *f*

putrefactive *adj* WASTE, WATER POLL fäulniserregend, fäulnisfähig

putrefy *vi* WASTE *compost* verfaulen, verrotten, WATER POLL verrotten

putrescibility *n* WASTE Faulfähigkeit *f*

putrescible *adj* WASTE, WATER POLL fäulniserregend, fäulnisfähig; ~ **matter** *n* WASTE fäulnisfähiger Stoff *m*, verrottbarer Stoff *m*; ~ **sludge** *n* WATER POLL fäulnisfähiger Schlamm *m*; ~ **waste** *n* WASTE, WATER SUPP Bioabfall *m*, Biomüll *m*, organischer Abfall *m*

puzzolan *n* SOIL *binding agent* Puzzolan *nt*

PV *abbr* (*photovoltaics*) RENEW ENERG PV (*Photovoltaik*)

PVC *abbr* (*polyvinyl chloride*) SOIL, WASTE PVC (*Polyvinylchlorid*); ~ **bottle** *n* WASTE PVC-Flasche *f*; ~ **insert fitment** *n* WASTE PVC-Einsatz *m*; ~ **pipe** *n* WATER SUPP PVC-Rohr *nt*; ~ **residue** *n* WASTE PVC-Reststoff *m*

pyknometer *n* ANALYSIS Pyknometer *nt*

pyknometric *adj* ANALYSIS pyknometrisch

pyranometer *n* ANALYSIS *global radiation* Pyranometer *nt*

pyrheliometer *n* ANALYSIS *measures direct solar radiation* Pyrheliometer *nt*

pyrolize *vt* GEN *in thermal processes* pyrolysieren

pyrolysis *n* GEN *sludge disposal* Pyrolyse *f*; ~ **gas** *n* WASTE, WATER POLL Pyrolysegas *nt*; ~**-gas chromatography** *n* ANALYSIS Pyrolyse *f* in Kopplung mit Gaschromatographie; ~**-gas chromatography-mass spectrometry** *n* ANALYSIS Pyrolyse *f* in Kopplung mit Gaschromatographie und Massenspektromet; ~ **oil** *n* WASTE Pyrolyseöl *nt*; ~ **product** *n* WASTE Pyrolyseprodukt *nt*

pyrolytic *adj* GEN pyrolytisch; ~ **pretreatment** *n* WASTE pyrolytische Vorbehandlung *f*

pyrophoric *adj* AIR *dust property* luftentzündlich, pyrophor, selbstentzündlich; ~ **dust** *n* AIR pyrophorer Staub *m*

Q

QA *abbr* (*quality assurance*) GEN QS (*Qualitätssicherung*)

quadrupole *n* ANALYSIS Quadrupol *m*, Vierpol *m*; **~ mass spectrometer** *n* ANALYSIS Quadrupol-Massenspektrometer *nt*

qualitative *adj* ANALYSIS qualitativ; **~ analysis** *n* ANALYSIS qualitative Analyse *f*; **~ detection** *n* ANALYSIS qualitativer Nachweis *m*; **~ determination** *n* ANALYSIS qualitative Bestimmung *f*; **~ evaluation** *n* ANALYSIS qualitative Auswertung *f*; **~ gas analysis** *n* ANALYSIS *by hand-driven gas detectors* qualitative Gasanalyse *f*; **~ odor analysis** *AmE*, **~ odour analysis** *BrE n* ANALYSIS qualitative Geruchsbestimmung *f*

quality *n* GEN Qualität *f*; **~ assurance** *n* (*QA*) GEN Qualitätssicherung *f* (*QS*); **~ assurance system** *n* AIR, ANALYSIS, MANAGE, SAFETY, WATER SUPP QS-System *nt*, Qualitätssicherungssystem *nt*; **~ control** *n* AIR, WASTE, WATER SUPP Qualitätskontrolle *f*; **~ factor** *n* AIR, WASTE, WATER SUPP Qualitätsfaktor *m*; **~ monitoring of ambient air** *n* SAFETY Umgebungsluftüberwachung *f*; **~ requirements** *n pl* AIR, ANALYSIS, MANAGE, NOISE, SAFETY, SOIL, WATER POLL, WATER SUPP Qualitätsanforderungen *f pl*

quantification *n* ANALYSIS Quantifizierung *f*

quantify *vt* ANALYSIS quantifizieren, qualitativ bestimmen

quantitative *adj* ANALYSIS quantitativ; **~ analysis** *n* ANALYSIS quantitative Analyse *f*; **~ detection** *n* ANALYSIS quantitativer Nachweis *m*; **~ determination** *n* ANALYSIS quantitative Bestimmung *f*; **~ evaluation** *n* ANALYSIS quantitative Auswertung *f*; **~ gas analysis** *n* ANALYSIS *infrared absorption spectrometry* quantitative Gasanalyse *f*; **~ odor analysis** *AmE*, **~ odour analysis** *BrE n* ANALYSIS quantitative Geruchsbestimmung *f*

quantity: **~-dependent** *adj* ANALYSIS *sample* mengenabhängig; **~ of precipitate** *n* WATER POLL Fällmittelaufwand *m*; **~ to be measured** *n* ANALYSIS Meßgröße *f*

quarter: **~ wave resonator** *n* NOISE Viertelwelle-Resonator *m*

quartz: **~ tube** *n* WATER POLL *UV reactor* Quarzrohr *nt*

quasi: **~-simultaneous** *adj* ANALYSIS quasisimultan

quaternary: **~ recycling** *n* WASTE thermisches Recycling *nt*

quench *vt* AIR löschen, SOIL abkühlen, auslöschen, WASTE, WATER SUPP löschen

quencher *n* AIR Quencher *m*, SOIL Löscher *m*; **~ suspension** *n* AIR Quenchersuspension *f*; **~ tank** *n* AIR Löschbehälter *m*, Quenchbehälter *m*

quenching: **~ tower** *n* AIR Löschturm *m*

quench: **~ tank** *n* AIR Quenchbehälter *m*; **~ zone** *n* AIR Quenchzone *f*

quick: **~ analysis** *n* ANALYSIS Schnellanalyse *f*; **~ determination** *n* ANALYSIS Schnellbestimmung *f*; **~ drying** *adj* WASTE schnell trocknend

quicklime *n* MARINE Branntkalk *m*, Ätzkalk *m*, SOIL Ätzkalk *m*, WATER POLL Branntkalk *m*

R

rabble: ~ **arm** *n* WASTE Krählarm *m*, WATER SUPP Rührarm *m*; ~ **rake** *n* WATER POLL Krählwerk *nt*
rad *abbr* (*radiation absorbed dose*) ANALYSIS Rad (*radiation absorbed dose*)
radar: ~ **probe** *n* ANALYSIS Radarsonde *f*
radial *adj* AIR, MARINE, WATER POLL, WATER SUPP Radial-; ~ **flow** *n* AIR *cyclone* Radialströmung *f*; ~**-flow pump** *n* WATER POLL Radialkreiselpumpe *f*; ~**-flow scrubber** *n* AIR Radialstromwäscher *m*
radiate 1. *vt* AIR, NOISE abstrahlen; **2.** *vi* AIR, NOISE strahlen
radiation *n* GEN Bestrahlung *f*, Strahlen *m pl*, Strahlung *f*; ~ **absorbed dose** *n* (*rad*) ANALYSIS radiation absorbed dose *f* (*Rad*); ~ **absorption** *n* ANALYSIS *dust concentration* Strahlungsabsorption *f*; ~ **detector** *n* ANALYSIS *radiation monitoring* Strahlungsdetektor *m*, Strahlungsnachweisgerät *nt*; ~ **dose** *n* AIR, SAFETY Strahlendosis *f*; ~ **dryer** *n* WASTE Strahlentrockner *m*; ~ **emitter** *n* WATER SUPP *UV disinfection* Strahler *m*; ~ **energy** *n* RENEW ENERG *solar energy plant* Strahlungsenergie *f*; ~ **equilibrium** *n* AIR Strahlungsgleichgewicht *nt*; ~ **equivalent man** *n* (*rem*) ANALYSIS radiation equivalent man *f* (*Rem*); ~ **hazard** *n* SAFETY Strahlungsgefahr *f*; ~ **intensity** *n* ANALYSIS Strahlungsintensität *f*; ~**-measuring instruments** *n pl* ANALYSIS Strahlungsmeßgeräte *nt pl*; ~ **monitoring** *n* ANALYSIS Strahlungskontrolle *f*; ~ **protection** *n* AIR Strahlenschutz *m*, SAFETY Strahlungsschutz *m*, Strahlenschutz *m*; ~ **pyrometer** *n* ANALYSIS Strahlungspyrometer *nt*; ~ **recuperator** *n* AIR Strahlungsrekuperator *m*; ~ **shield** *n* SAFETY Strahlungsabschirmung *f*; ~ **shielding** *n* SAFETY Strahlungsschutz *m*
radiative: ~ **absorption** *n* AIR Strahlungsabsorption *f*
radiator *n* AIR *electron beam process* Strahler *m*
radical: ~ **formation** *n* AIR *electron beam process* Radikalbildung *f*
radio *adj* ANALYSIS Radio-
radioactive *adj* GEN radioaktiv; ~ **contaminated water** *n* WASTE, WATER POLL, WATER SUPP radioaktiv kontaminiertes Wasser *nt*; ~ **decay** *n* RENEW ENERG *geothermal energy* radioaktiver Zerfall *m*; ~ **dust** *n* AIR radioaktiver Staub *m*; ~ **emissions** *n pl* AIR radioaktive Emissionen *f pl*; ~ **fallout** *n* AIR, WATER POLL radioaktiver Niederschlag *m*; ~ **gas** *n* AIR radioaktives Gas *nt*; ~ **pollution** *n* AIR, MARINE, SAFETY radioaktive Verseuchung *f*, radioaktive Verschmutzung *f* WASTE, WATER POLL radioaktive Verschmutzung *f*; ~ **rare gas** *n* AIR radioaktives Edelgas *nt*; ~ **substance** *n* SAFETY, WASTE, WATER POLL radioaktiver Stoff *m*; ~ **waste** *n* SAFETY, WASTE Atommüll *m*, radioaktiver Abfall *m*; ~ **wastewater** *n* WATER POLL radioaktive Abwasser *nt*
radio: ~ **frequency heating** *n* SOIL *thermal desorption* Hochfrequenzerwärmung *f*
radiograph *vt* ANALYSIS durchstrahlen
radiographic *adj* WASTE röntgenographisch; ~ **process** *n* WASTE röntgenographisches Verfahren *nt*
radiological: ~ **survey** *n* ANALYSIS Strahlenüberwachung *f*
radiometer *n* ANALYSIS Radiometer *nt*
radiometric *adj* ANALYSIS radiometrisch
radiometry *n* ANALYSIS Radiometrie *f*
radiophotoluminescence *n* ANALYSIS Radiophotolumineszenz *f*
radiopolarography *n* ANALYSIS Radiopolarographie *f*
radiowave *n* ANALYSIS Radiowelle *f*; ~ **hazard** *n* SAFETY Gefahr *f* durch Hochfrequenzwellen
radon *n* ANALYSIS Radon *nt*; ~ **concentration** *n* ANALYSIS Radonkonzentration *f*
railroad: ~ **transportation** *n AmE* (*cf railway transportation BrE*) WASTE Beförderung *f* über die Schiene, Schienenbeförderung *f*, Schienentransport *m*
rail: ~ **transport** *n* WASTE Beförderung *f* über die Schiene, Schienenbeförderung *f*, Schienentransport *m*
railway: ~ **transportation** *n BrE* (*cf railroad transportation AmE*) WASTE Beförderung *f* über die Schiene, Schienenbeförderung *f*, Schienentransport *m*
rainfall: ~ **area** *n* WATER SUPP Niederschlagsgebiet *nt*
rain: ~**-free period** *n* AIR regenfreie Zeit *f*; ~ **gage** *AmE*, ~ **gauge** *BrE n* WATER SUPP Niederschlagsmesser *m*, Regenmesser *m*
rainout *n* AIR Rainout *nt*, radioaktiver Niederschlag *m*
rainwater *n* WATER POLL, WATER SUPP Niederschlagswasser *nt*, Regenwasser *nt*; ~ **tank** *n* WATER SUPP Durchlaufbecken *nt*; ~ **utilization** *n* WATER SUPP Regenwassernutzung *f*
rakings: ~ **classifier** *n* WATER POLL Rechengutklassierer *m*
ramming: ~ **machine** *n* WASTE Stampffußverdichter *m*
ramp: ~ **landfill** *n* WASTE, WATER SUPP Böschung *f*
random: ~ **noise** *n* NOISE Zufallsrauschen *nt*, stochastisches Rauschen *nt*; ~ **sample** *n* ANALYSIS Stichprobe *f*, Zufallsstichprobe *f*
range *n* ANALYSIS Reichweite *f*; ~ **of audibility** *n* NOISE Hörbereich *m*; ~ **of capacity** *n* RENEW ENERG Leistungsbereich *m*
Rankine: ~ **cycle steam engine** *n* AIR Rankine-Kreislaufprozeß-Dampfmaschine *f*
rap *vt* AIR klopfen
rape: ~ **seed oil** *n* RENEW ENERG *fuel* Rapsöl *nt*
rapid *adj* ANALYSIS, WATER POLL, WATER SUPP schnell; ~ **decarbonization** *n* WATER SUPP Schnellentkarbonisierung *f*; ~ **fermentation** *n* WASTE Schnellkompostierung *f*, beschleunigte Kompostierung *f*, geschlossene Kompostierung *f*; ~ **filtration** *n* WATER POLL Schnellfiltration *f*; ~ **mixing** *n* WATER POLL Schnellmischen *nt*; ~**-mixing basin** *n* WATER POLL Schnellmischbecken *nt*, Schnellrührbecken *nt*; ~ **rotting** *n* WATER POLL Intensivrotte *f*, Schnellrotte *f*; ~ **sand filter** *n* WATER POLL, WATER SUPP Schnellfilter *nt*; ~ **test** *n* ANALYSIS Schnelltest *m*
rapper *n* AIR Klopfvorrichtung *f*
rapping *n* AIR Klopfen *nt*, Klopfung *f*; ~ **of electrodes** *n* AIR *collected dust* Klopfung *f* des Staubes zur Lösung von der Niederschlagselektrode; ~ **interval** *n* AIR *filter* Klopfintervall *nt*
raschig: ~ **ring** *n* AIR *packing of wet scrubber*, WATER POLL Raschig-Ring *m*
rate: ~ **of absorption** *n* WASTE, WATER SUPP Absorptionsrate *f*; ~ **of ascent** *n* AIR *of gases* Aufstiegsgeschwindigkeit *f*; ~ **of charge** *n* MANAGE, POLICY Abgabensatz *m*
rated: ~ **capacity** *n* AIR, RENEW ENERG, WATER POLL, WATER SUPP Nennlast *f*, Nennleistung *f*
rate: ~ **of decomposition** *n* ANALYSIS, WASTE, WATER POLL, WATER SUPP Abbaugeschwindigkeit *f*
rated: ~ **load** *n* WATER SUPP Nennlast *f*
rate: ~ **of flame propagation** *n* AIR Zündgeschwindigkeit *f*; ~ **of flow** *n* ANALYSIS, WATER POLL, WATER SUPP Abflußmenge *f*, Durchflußmenge *f*, Durchflußrate *f*; ~ **of utilization** *n* WASTE, WATER POLL, WATER SUPP Auslastungsgrad *m*
ratify *vt* POLICY ratifizieren
raw: ~ **air conveyance** *n* AIR Rohluftzufuhr *f*; ~ **compost** *n*

WASTE Rohkompost *m*; ~ **data** *n pl* ANALYSIS Rohdaten *pl*; ~ **data storage** *n* ANALYSIS Rohdatenspeicherung *f*; ~ **gypsum** *n* AIR Rohgips *m*; ~ **material** *n* MANAGE, RENEW ENERG, WASTE Rohstoff *m*; ~ **materialdecomposition** *n* WASTE Rohstoffabbau *m*; ~ **material delivering industry** *n* MANAGE Rohstoffindustrie *f*; ~ **material extraction** *n* MANAGE Rohstoffgewinnung *f*; ~ **material generation** *n* RENEW ENERG *biomass conversion* Rohstoffgewinnung *f*; ~ **material recycling** *n* WASTE rohstoffliche Verwertung *f*, rohstoffliches Recycling *nt*; ~ **material recycling process** *n* WASTE rohstoffliches Recyclingverfahren *nt*; ~ **material resources** *n pl* WASTE Rohstoffressourcen *f pl*; ~**-measured value** *n* ANALYSIS Rohmeßwert *m*; ~ **sample** *n* ANALYSIS Rohprobe *f*; ~ **sludge** *n* WATER POLL Primärschlamm *m*, Rohschlamm *m*; ~ **soil** *n* SOIL Rohboden *m*; ~ **waste gas** *n* AIR Rohabgas *nt*; ~ **wastewater** *n* MARINE unbehandeltes Abwasser *nt*, WATER POLL Rohabwasser *nt*, unbehandeltes Abwasser *nt*; ~ **wastewater quality** *n* WATER POLL *untreated* Rohabwasserqualität *f*; ~ **wastewater supply** *n* WATER POLL *untreated* Rohabwasserversorgung *f*; ~ **water** *n* WATER POLL, WATER SUPP Rohwasser *nt*; ~ **water main** *n* WATER SUPP Rohwasserleitung *f*; ~ **water pump** *n* WATER SUPP Rohwasserpumpe *f*; ~ **water quality** *n* WATER POLL, WATER SUPP Rohwasserqualität *f*; ~ **water supply** *n* WATER POLL Rohwasserversorgung *f*

Rayleigh: ~ **disc** *BrE*, ~ **disk** *AmE n* ANALYSIS *measurement of sound intensity* Rayleigh-Scheibe *f*; ~ **refractometer** *n* ANALYSIS Rayleigh-Refraktometer *nt*; ~ **scattering** *n* AIR, ANALYSIS Rayleighsche Streustrahlung *f*

RBC *abbr* (*rotating biological disc contactor*) WATER POLL Scheibentauchkörper *m*

RDF *abbr* (*refuse-derived fuel*) WASTE BRAM (*Brennstoff aus Müll*); ~ **from organic mass** *n* WASTE BRAM aus organischer Masse *m* (*BROM*); ~ **from paper** *n* WASTE BRAM aus Papier *m* (*BRAP*); ~ **incineration** *n* WASTE BRAM-Verbrennung *f*

re- *pref* AIR, SOIL, WASTE Wieder-

Re *abbr* (*Reynold's number*) AIR, WATER SUPP Re (*Reynoldsche Zahl*)

re-treat *vt* AIR, WASTE nachbehandeln

reach: ~ **of a river** *n* WASTE, WATER SUPP Strecke *f* eines Flus ses

react *vi* GEN reagieren; ~ **exothermally** *vi* WASTE exotherm reagieren

reactant *n* AIR Reaktand *m*, Reaktionspartner *m*

reacting: ~ **agent** *n* ANALYSIS Reaktionspartner *m*

reaction *n* AIR Reaktion *f*, ANALYSIS *chemical* Reaktion *f*, Umsetzung *f*, SOIL, WASTE, WATER POLL Reaktion *f*; ~ **chamber** *n* AIR, ANALYSIS Reaktionskammer *f*; ~ **equilibrium** *n* AIR Reaktionsgleichgewicht *nt*; ~ **indicator** *n* ANALYSIS Reaktionsindikator *m*; ~ **process** *n* WASTE, WATER SUPP *in landfill* Umsetzungsprozeß *m*; ~ **product** *n* AIR, ANALYSIS Reaktionsprodukt *nt*; ~ **rate** *n* AIR, ANALYSIS, WASTE, WATER POLL Reaktionsgeschwindigkeit *f*; ~ **surface** *n* AIR Reaktionsoberfläche *f*; ~ **tank** *n* WATER POLL Reaktionsbecken *nt*; ~ **time** *n* AIR Reaktionszeit *f*; ~ **tube** *n* ANALYSIS Reaktionsrohr *nt*; ~**-type axial fan** *n* AIR Axialüberdruckventilator *m*; ~ **velocity** *n* AIR, ANALYSIS, WASTE, WATER POLL Reaktionsgeschwindigkeit *f*; ~ **zone** *n* WATER POLL Reaktionszone *f*

reactivatable *adj* GEN reaktivierbar

reactivate *vt* GEN reaktivieren, regenerieren

reactivated *adj* GEN reaktiviert, regeneriert

reactivation *n* GEN Reaktivierung *f*, Regeneration *f*

reactive *adj* AIR, WASTE reaktionsfähig, reaktiv; ~ **attenuator** *n* NOISE, SAFETY Resonanzschalldämpfer *m*; ~ **hazardous waste** *n* WASTE reaktiver Sondermüll *m*; ~ **particle** *n* AIR reaktionsfähige Partikel *f*; ~ **resonator** *n* NOISE, SAFETY Resonanzschalldämpfer *m*; ~ **silencer** *n* NOISE, SAFETY Resonanzschalldämpfer *m*

reactivity *n* POLICY, RENEW ENERG, WASTE Reaktivität *f*

reactor *n* AIR, WATER POLL Reaktor *m*; ~ **bottom** *n* WATER POLL Reaktorboden *m*; ~ **with catalyst** *n* AIR Reaktor *m* mit Katalysator; ~ **pressure sensor** *n* AIR Reaktordrucksensor *m*

read *vt* ANALYSIS *figure, value* ablesen; ~ **out digitally** *vt* ANALYSIS digital anzeigen

readable *adj* ANALYSIS ablesbar, lesbar

readied: ~ **compost** *n* WASTE Fertigkompost *m*

reading *n* ANALYSIS *taking measurement* Ablesung *f*, *value* Meßwert *m*; ~ **accuracy** *n* ANALYSIS Ablesegenauigkeit *f*; ~ **constancy** *n* ANALYSIS Meßwertkonstanz *f*; ~ **error** *n* ANALYSIS Ablesefehler *m*; ~ **precision** *n* ANALYSIS Ablesegenauigkeit *f*

read: ~**-out display** *n* ANALYSIS Meßwertanzeige *f*, Meßwertausgabe *f*; ~**-out of measured values** *n* ANALYSIS ausgegebene Meßwerte *m pl*

re-aeration *n* WATER POLL Wiederbelüftung *f*

reagent *n* AIR, ANALYSIS, SOIL Reagens *nt*, Reagenz *nt*; ~ **solution** *n* ANALYSIS Reagenzlösung *f*

real: ~**-time analyser** *BrE*, ~**-time analyzer** *AmE n* ANALYSIS Echtzeit-Analysator *m*; ~**-time chromatogram** *n* ANALYSIS Echtzeit-Chromatogramm *nt*; ~**-time operation** *n* ANALYSIS Echtzeitbetrieb *m*; ~**-time sampling** *n* ANALYSIS Echtzeit-Stichprobenahme *f*; ~**-time signal processing** *n* ANALYSIS Echtzeit-Signalverarbeitung *f*

rebound *vi* AIR *particle* zurückprallen, *particles on fibre* abprallen

recalibration *n* ANALYSIS Nachkalibrierung *f*, Rekalibration *f*

recapture *vt* AIR wiederauffangen

receiver *n* RENEW ENERG *solar power plant* Empfängerröhre *f*; ~ **tank** *n* WATER POLL Vorlage *f*

receiving: ~ **bunker** *n* WASTE Aufnahmebunker *m*, Lagerbunker *m*, Müllbunker *m*; ~ **chamber** *n* WATER POLL Vorkammer *f*; ~ **system** *n* WASTE Sammelsystem *nt*; ~ **tank** *n* WATER POLL Sammelbecken *nt*, Sammelbehälter *m*; ~ **telescope** *n* ANALYSIS Empfangsteleskop *nt*; ~ **tower** *n* RENEW ENERG *solar power plant* Empfängerturm *m*; ~ **water** *n* WATER POLL Vorfluter *m*

receptacle *n* ANALYSIS Auffanggefäß *nt*, Sammelgefäß *nt*

receptor *n* AIR Empfänger *m*; ~ **region** *n* AIR Empfängerbereich *m*

recharge *vt* GEN anreichern

rechargeable *adj* GEN wiederaufladbar; ~ **storage battery** *n* WASTE wiederaufladbarer Akkumulator *f*

recharge: ~ **basin** *n* WASTE Anreicherungsbecken *nt*, WATER SUPP Anreicherungsbecken *nt*, Versicherungsbecken *nt*

reciprocal: ~ **effect** *n* ANALYSIS, RENEW ENERG Wechselwirkung *f*

reciprocating: ~ **piston engine** *n* AIR Kolbenkraftmaschine *f*; ~ **pump** *n* WATER POLL, WATER SUPP Kolbenpumpe *f*

recirculate *vt* AIR, WASTE, WATER, WATER POLL im Kreislauf umpumpen, zurückführen

recirculated: ~ **heating** *n* WATER POLL Kreislaufheizung *f*

recirculating: ~ **filter** *n* AIR Umlauffilter *nt*; ~ **pump** *n* WATER POLL Umlaufpumpe *f*; ~ **washing agent** *n* AIR Kreislaufwaschmittel *nt*; ~ **water** *n* WATER POLL Kreislaufwasser *nt*; ~ **water management** *n* WATER SUPP Kreislaufwasserführung *f*

recirculation *n* WATER POLL, WATER SUPP Rückführung *f*; ~ **system** *n* WATER POLL Kreislaufsystem *nt*

reclaim *vt* SOIL gewinnen, rekultivieren, WASTE, WATER POLL rückgewinnen, rekultivieren

reclaimed *adj* AIR, SOIL, WASTE, WATER POLL rekultiviert, rückgewonnen; ~ **area** *n* SOIL wiederurbar gemachtes Gebiet *nt*; ~ **material** *n* WASTE Regenerat *nt*; ~ **wastewater** *n* WATER POLL rückgewonnenes Abwasser *nt*; ~ **water** *n* WATER POLL rückgewonnenes Wasser *nt*

reclamation *n* SOIL *of land* Gewinnung *f*, Rekultivierung *f*, Wiedergewinnung *f*, WASTE Weiterverwertung *f*,

Rekultivierung *f*; ~ **plant** *n* WASTE Wiedergewinnungsanlage *f*, Rückgewinnungsanlage *f*
recommendation *n* POLICY Empfehlung *f*
reconditioning: ~ **of paper fiber** *AmE*, ~ **of paper fibre** *BrE* *n* WASTE, WATER POLL Rückleimung *f*
recontaminate *vt* WATER POLL *treated wastewater* rekontaminieren, wiederverunreinigen
reconvert *vt* AIR zurückverwandeln
record *vt* ANALYSIS protokollieren, registrieren, POLICY registrieren
recording 1. *adj* ANALYSIS registrierend; **2.** *n* POLICY Registrierung *f*; ~ **device** *n* ANALYSIS Registriergerät *nt*; ~ **spectrophotometer** *n* ANALYSIS registrierendes Spektralphotometer *nt*; ~ **strip** *n* ANALYSIS Registrierstreifen *m*
recover *vt* AIR wiedergewinnen, zurückgewinnen, WASTE *capture* erfassen, wiedergewinnen, zurückgewinnen, *metal from scrap* gewinnen, WATER POLL, WATER SUPP erfassen, gewinnen, wiedergewinnen, zurückgewinnen; ~ **in liquid form** *vt* WASTE in flüssiger Form zurückgewinnen
recoverable *adj* AIR rückgewinnbar, wiedergewinnbar, WASTE erfaßbar, gewinnbar, rückgewinnbar, wiedergewinnbar, WATER POLL, WATER SUPP erfaßbar, gewinnbar, wiedergewinnbar; ~ **materials** *n pl* WASTE rückgewinnbare Materialien *nt pl*; ~ **volume of water** *n* WASTE, WATER POLL, WATER SUPP gewinnbare Wassermenge *f*; ~ **waste** *n* WASTE rückgewinnbarer Abfall *m*
recovered *adj* AIR, WASTE, WATER POLL, WATER SUPP rückgewonnen, wiedergewonnen; ~ **heat** *n* AIR, WASTE rückgewonnene Wärme *f*; ~ **oil** *n* WASTE rückgewonnenes Öl *nt*, Ölregenerat *nt*; ~ **pulp** *n* WASTE rückgewonnene Pulpe *f*; ~ **wastewater** *n* WATER POLL rückgewonnenes Abwasser *nt*; ~ **water** *n* WATER POLL rückgewonnenes Wasser *nt*
recovery *n* AIR Wiedergewinnung *f*, *reconditioning* Aufarbeitung *f*, Rückgewinnung *f*, *reuse* Verwertung *f*, WASTE Wiedergewinnung *f*, *of metal from scrap* Gewinnung *f*, *reconditioning* Aufarbeitung *f*, *reuse* Verwertung *f*, Rückgewinnung *f*, *capture* Erfassung *f*, WATER POLL, WATER SUPP *salvage* Rückführung *f*, Wiedergewinnung *f*, *of raw materials* Gewinnung *f*, *reuse* Verwertung *f*; ~ **area** *n* WATER SUPP Gewinnungsgebiet *nt*; ~ **boiler** *n* AIR, WASTE Rückgewinnungskessel *m*; ~ **by type** *n* WASTE sortenreine Erfassung *f*; ~ **device** *n* AIR Aufarbeitungsvorrichtung *f*, MARINE Ölabschöpfgerät *nt*, WASTE, WATER SUPP Aufarbeitungsvorrichtung *f*; ~ **of landfill space** *n* WASTE Deponierückbau *m*; ~ **rate** *n* WASTE Erfassungsquote *f*, *of take-back system* Rücklaufquote *f*, WATER POLL Erfassungsquote *f*; ~ **system** *n* MARINE Ölabschöpfsystem *nt*
rectangular: ~ **basin** *n* WATER POLL Längsbecken *nt*, Rechteckbecken *nt*; ~ **clarifier** *n* WATER POLL Längsbecken *nt*, Rechteckbecken *nt*; ~ **tank** *n* WATER POLL Längsbecken *nt*, Rechteckbecken *nt*
rectification *n* AIR Rektifikation *f*, Rektifizierung *f*, WATER POLL Rektifikation *f*
rectifier *n* AIR Gleichrichter *m* (*GR*)
recultivate *vt* SOIL, WASTE rekultivieren
recultivation *n* SOIL Rekultivierung *f*, Wiedernutzbarmachung *f*, Wiederurbarmachung *f*; ~ **layer** *n* SOIL *tillable soil* Rekultivierungsschicht *f*
recuperative *adj* AIR rekuperativ; ~ **heat exchanger** *n* AIR rekuperativer Wärmetauscher *m*; ~ **heat recovery** *n* AIR rekuperative Wärmerückgewinnung *f*
recuperator *n* AIR Rekuperator *m*, Wärmeaustauscher *m*, Wärmetauscher *m*
recyclability *n* WASTE Recyclingfähigkeit *f*, Wiederverwendbarkeit *f*, Wiederverwertbarkeit *f*
recyclable *adj* WASTE recyclingfähig, wiederverwendbar, wiederverwertbar; ~ **battery** *n* WASTE recyclingfähige Batterie *f*; ~ **packaging** *n* WASTE Mehrwegverpackung *f*, wiederverwertbare Verpackung *f*
recycle *vt* AIR verwerten, WASTE, WATER POLL recyceln, verwerten, wiederverwerten
recycled: ~ **bottle** *n* WASTE Mehrwegflasche *f*, recycelte Flasche *f*; ~ **clothing** *n* WASTE Altkleider *nt pl*; ~ **glass** *n* WASTE Altglas *nt*; ~ **glass application** *n* WASTE Altglaseinsatz *m*; ~ **glass cullet** *n* WASTE Altglasscherben *f pl*; ~ **material** *n* WASTE Recyclat *nt*, Recyclingmaterial *nt*, Sekundärwertstoff *m*; ~ **paper** *n* WASTE Recyclingpapier *nt*, recyceltes Papier *nt*, Altpapier *nt*; ~ **paper application** *n* WASTE Altpapiereinsatz *m*; ~ **paper application rate** *n* WASTE Altpapiereinsatzquote *f*; ~ **paper use** *n* WASTE Altpapierverwertung *f*; ~ **plastic** *n* WASTE, WATER SUPP Altkunststoff *m*
recycling *n* MANAGE Recycling *nt*, WASTE *use of packaging waste for other productive uses* Recycling *nt*, Verwertung *f*, Wiederverwertung *f*, WATER POLL, WATER SUPP Recycling *nt*; ~ **capacity** *n* WASTE Verwertungskapazität *f*; ~ **caterpillar** *n* WASTE Recyclingraupe *f*; ~ **center** *AmE*, ~ **centre** *BrE* *n* WASTE Recyclinghof *m*; ~ **costs** *n pl* MANAGE Recyclingkosten *pl*; ~ **economy** *n* POLICY, WATER SUPP Kreislaufwirtschaft *f*; ~ **economy law** *n* POLICY *Germany*, WATER SUPP Kreislaufwirtschaftsgesetz *nt*; ~ **of fly ash** *n* WASTE Verwertung *f* der Flugaschen; ~**-friendly** *adj* WASTE recyclingfreundlich; ~**-friendly product** *n* WASTE recyclingfreundliches Produkt *nt*; ~ **gas** *n* AIR Kreislaufgas *nt*; ~ **of inoculated compost** *n* WASTE Impfkompostrückführung *f*; ~ **material** *n* WASTE Recyclingmaterial *nt*; ~ **operation** *n* WASTE Recyclingvorgang *m*; ~**-oriented** *adj* WASTE recyclinggerecht, verwertungsorientiert, WATER SUPP recyclinggerecht; ~ **paper** *n* WASTE Recyclingpapier *nt*; ~ **plant** *n* WASTE Recyclinganlage *f*; ~ **potential** *n* WASTE Recyclingpotential *nt*; ~ **process** *n* WASTE Recyclingverfahren *nt*, Verwertungsverfahren *nt*; ~ **quantity** *n* WASTE Recyclingumfang *m*; ~ **rate** *n* WASTE Recyclingquote *f*, Verwertungsquote *f*; ~ **of waste** *n* WASTE Wiederverwertung *f* von Abfällen
red: ~ **algae** *n* WATER SUPP Rotalgen *f pl*
redeposit *vt* SOIL *after treatment* wieder einlagern
redeposition *n* SOIL *after treatment* Wiedereinlagerung *f*
redevelop *vt* WASTE, WATER SUPP sanieren
redeveloped *adj* WASTE, WATER SUPP saniert
redevelopment *n* WASTE, WATER SUPP Sanierung *f*
redisposal *n* SOIL *of unpolluted arable soil* Wiederaufbringung *f*
redissolution *n* WATER POLL Rücklösung *f*, Wiederauflösung *f*
red: ~ **mud** *n* WATER POLL Rotschlamm *m*
redox: ~ **potential** *n* AIR, ANALYSIS, WATER POLL Redox-Potential *nt*
reduce *vt* ANALYSIS reduzieren, NOISE *sound level* mindern, WASTE, WATER POLL, WATER SUPP entspannen, reduzieren, verringern
reduced *adj* ANALYSIS reduziert, NOISE gemindert, WASTE, WATER POLL, WATER SUPP entspannt; ~ **rate of charge** *n* POLICY gekürzter Satz *m*, reduzierter Satz *m*
reducing *adj* AIR, WASTE reduzierend; ~ **agent** *n* AIR, ANALYSIS, WATER POLL Reduktionsmittel *nt*; ~ **catalyst** *n* AIR reduzierender Katalysator *m*; ~ **catalyst bed** *n* AIR reduzierendes Katalysatorbett *nt*; ~ **gas** *n* AIR Reduktionsgas *nt*; ~ **product** *n* WASTE Reduktionsprodukt *nt*; ~ **substance** *n* AIR Reduktionsstoff *m*, reduzierende Substanz *f*
reduction *n* AIR, ANALYSIS Reduktion *f*, *in time* Verkürzung *f*, WASTE, WATER POLL *in pressure* Entspannung *f*, Reduktion *f*; ~ **catalyst** *n* AIR Reduktionskatalysator *m*; ~ **intensity** *n* AIR, ANALYSIS, WATER POLL Redox-Potential *nt*; ~**-oxidation potential** *n* (*rH-value*) AIR, ANALYSIS, WATER POLL Reduktions-Oxidations-Potential *nt* (*rH-Wert*); ~ **of waste gas retention time** *n* AIR Verkürzung *f* der Abgasverweildauer, Verkürzung *f* der Abgasverweilzeit

reel: ~ **in** *vt* MARINE eintrommeln; ~ **out** *vt* MARINE austrommeln
reference: ~ **chamber** *n* ANALYSIS Vergleichskammer *f*; ~ **electrode** *n* ANALYSIS *pH-measurement* Bezugselektrode *f*; ~ **gas** *n* ANALYSIS Vergleichsgas *nt*; ~ **resistor** *n* ANALYSIS Vergleichswiderstand *m*; ~ **sound pressure** *n* NOISE Bezugsschalldruck *m*; ~ **spectrum** *n* ANALYSIS Vergleichsspektrum *nt*; ~ **value** *n* ANALYSIS Bezugsgröße *f*
refiltering *n* AIR Wiederanfiltrieren *nt*
refiltration: ~ **flocculation** *n* WATER SUPP Refiltrationsflockung *f*
refine *vt* AIR, MARINE, WASTE raffinieren
refined *adj* AIR, MARINE, WASTE raffiniert; ~ **product** *n* AIR, MARINE, WATER POLL Raffinageprodukt *nt*, raffiniertes Produkt *nt*
refinery: ~ **gas** *n* AIR Raffineriegas *nt*; ~ **waste** *n* WASTE Raffinerierückstände *m pl*
refining *n* AIR, WASTE Raffination *f*
reflect *vt* AIR zurückstrahlen, NOISE reflektieren, RENEW ENERG verspiegeln
reflected *adj* AIR zurückgestrahlt, NOISE reflektiert, RENEW ENERG verspiegelt; ~ **paraboloid** *n* RENEW ENERG verspiegeltes Paraboloid *nt*; ~ **sound** *n* NOISE reflektierter Schall *m*
reflection *n* AIR Rückstrahlung *f*, ANALYSIS, NOISE Reflexion *f*; ~ **spectroscopy** *n* ANALYSIS Reflexionsspektroskopie *f*
reflectometer *n* ANALYSIS, SOIL Reflektometer *nt*
reflector *n* RENEW ENERG *solar tower* Spiegel *m*
reflux: ~ **ratio** *n* WASTE Rücklaufverhältnis *nt*
reformed: ~ **gas** *n* AIR Spaltgas *nt*
reforming: ~ **reaction** *n* AIR Reformierungsreaktion *f*
refractive: ~ **index** *n* ANALYSIS *for detector* Brechungsindex *m*
refractory: ~ **substance** *n* WATER POLL refraktärer Stoff *m*, schlecht abbaubarer Stoff *m*
refrigerant *n* AIR *containing CFCs*, WASTE Kältemittel *nt*
refueling *AmE see refuelling BrE*
refuelling *n BrE* MARINE Auftanken *nt*, Kraftstoffaufnahme *f*
refuse *n* (*cf garbage AmE, rubbish BrE*) WASTE Abfall *m*, Müll *m*; ~ **body** *n* WASTE Müllfahrzeug *nt*, Müllwagen *m*; ~ **bunker** *n* WASTE Aufnahmebunker *m*, Lagerbunker *m*; ~ **can** *n* WASTE Mülleimer *m*, Müllbehälter *m*, Mülltonne *f*; ~ **cell** *n* WASTE Kassette *f*, Polder *m*; ~ **chute** *n* WASTE Müllabwurfschacht *m*, Müllschlucker *m*; ~ **collection** *n* WASTE Müllabfuhr *f*, Mülleinsammlung *f*, Müllsammlung *f*; ~ **collection service** *n* WASTE Müllabfuhr *f*; ~ **collection vehicle** *n* WASTE Müllfahrzeug *nt*, Müllwagen *m*; ~ **composition** *n* WASTE Müllzusammensetzung *f*; ~ **compost** *n* WASTE Müllkompost *m*; ~ **composting** *n* WASTE Müllkompostierung *f*; ~**-derived fuel** *n* (*RDF*) WASTE Brennstoff *m* aus Müll (*BRAM*); ~ **disposal** *n* POLICY Abfallentsorgung *f*, WASTE Abfallentsorgung *f*, Hausmülleinsammlung *f*, Müllabfuhr *f*; ~ **disposal site** *n* WASTE Deponie *f*, Müllkippe *f*; ~ **extractor** *n* WASTE Austragregler *m*; ~ **glass** *n* WASTE Scherben *f pl*; ~ **grinder** *n* WASTE Müllzerkleinerer *m*; ~ **incineration** *n* WASTE Müllverbrennung *f*; ~ **incineration plant** *n* WASTE Müllverbrennungsanlage *f* (*MVA*); ~ **incinerator** *n* AIR, WASTE Abfallverbrennungsanlage *f*; ~ **mass** *n* WASTE Müllmenge *f*; ~ **recovery** *n* WASTE Müllerfassung *f*; ~ **sack** *n* WASTE Müllbeutel *m*, Müllsack *m*; ~ **sack collection** *n* WASTE Einsammeln *nt* von Müllsäcken; ~ **separation plant** *n* WASTE Abfallsortieranlage *f*, Müllsortieranlage *f*; ~ **volume** *n* WASTE Abfallmenge *f*, Müllvolumen *nt*; ~ **worm** *n* WASTE Austragschnecke *f*
regenerable: ~ **flue-gas desulfurization** *AmE*, ~ **flue-gas desulphurization** *BrE n* AIR regenerierbare Rauchgasentschwefelung *f*
regenerant *n* GEN Regenerierungsmittel *nt*
regenerate *vt* AIR, SOIL regenerieren, WASTE regenerieren, *packaging* wiederbefüllen, WATER POLL, WATER SUPP regenerieren
regenerated *adj* GEN regeneriert; ~ **activated carbon** *n* AIR regenerierte Aktivkohle *f*; ~ **activated coke** *n* AIR regenerierter Aktivkoks *m*; ~ **adsorbent** *n* AIR regeneriertes Adsorbens *nt*; ~ **solvent** *n* AIR regeneriertes Lösungsmittel *nt*
regeneration *n* AIR, SOIL, WASTE WATER POLL, WATER SUPP Regeneration *f*, Regenerierung *f*; ~ **basin** *n* WATER POLL Regenerationsbehälter *m*; ~ **capacity** *n* WATER POLL Regenerationsfähigkeit *f*; ~ **pit** *n* SOIL Regenerationsmiete *f*
regenerative *adj* AIR regenerativ, regenerierbar, SOIL, RENEW ENERG, WASTE, AIR regenerativ; ~ **air preheater** *n* AIR regenerativer Luftvorwärmer *m*; ~ **gas** *n* AIR *inert gas* Regenerativgas *nt*; ~ **gas fan** *n* AIR Regenerativgasgebläse *nt*; ~ **gas preheater** *n* AIR regenerativer Gasvorwärmer *m* (*REGAVO*); ~ **heat exchanger** *n* AIR regenerativer Wärmeaustauscher *m*; ~ **heating technology** *n* RENEW ENERG regenerative Heizungstechnik *f*; ~ **incineration** *n* AIR *exhaust gas* regenerative Nachverbrennung *f* (*RNV*); ~ **pump** *n* WATER POLL Regenerativpumpe *f*, Seitenkanalpumpe *f*; ~ **reagent** *n* AIR regeneratives Reagenz *nt*
regenerator *n* AIR Wärmeaustauscher *m*, Wärmetauscher *m*
regional: ~ **policy** *n* POLICY Regionalpolitik *f*
register *vt* ANALYSIS, POLICY registrieren
register: ~ **of hazardous substances** *n* AIR, POLICY, SOIL, WASTE, WATER SUPP Gefahrstoffkataster *nt*
registration *n* POLICY Registrierung *f*
regranulate 1. *n* WASTE Regranulat *nt*; **2.** *vt* WASTE regranulieren
regranulating: ~ **extruder** *n* WASTE Regranulierextruder *m* für Abfallaufbereitung
regranulation *n* WASTE Regranulation *f*
regression: ~ **curve** *n* ANALYSIS Regressionskurve *f*
regressive *adj* POLICY regressiv; ~ **tax** *n* POLICY regressive Steuer *f*
regrind 1. *n* WASTE Regenerat *nt*; **2.** *vt* WASTE wiederaufbereiten
regrinding *n* WASTE Wiederaufbereitung *f*
reground: ~ **material** *n* WASTE Regenerat *nt*, wiederaufbereitetes Material *nt*
regularization *n* AIR, WATER SUPP Regularisation *f*
regulate *vt* ANALYSIS regulieren, POLICY regulieren, verordnen
regulated *adj* ANALYSIS reguliert, POLICY reguliert, verordnet; ~ **deposition** *n* WASTE geregelte Ablagerung *f*
regulating: ~ **unit** *n* GEN Regeleinheit *f*
regulation *n* ANALYSIS Regelung *f*, POLICY *legal* Verordnung *f*, Vorschrift *f*, SAFETY Vorschrift *f*; ~ **in case of accident** *n* SAFETY Störfallverordnung *f*
regulator *n* ANALYSIS Regulierapparat *m*
regulatory: ~ **tax** *n* POLICY *eco-tax* Lenkungssteuer *f*
rehabilitation *n* SOIL Wiedernutzbarmachung *f*
reheat *vt* AIR wiederaufwärmen, wiederaufheizen
reheating *n* AIR Wiederaufheizung *f*
reinforce *vt* WASTE, WATER SUPP armieren, verstärken
reinforced: ~ **concrete pipe** *n* WASTE, WATER POLL, WATER SUPP Spannbetonrohr *nt*, Stahlbetonrohr *nt*; ~ **plastics** *n pl* WASTE verstärkte Kunststoffe *m pl*
reinforcing: ~ **material** *n* WASTE *plastics reprocessing* Verstärkungsmittel *nt*
relative: ~ **humidity** *n* ANALYSIS relative Feuchte *f*; ~ **speed** *n* AIR *of particle to drop* Relativgeschwindigkeit *f*
relax *vt* WASTE, WATER POLL, WATER SUPP entspannen
relaxation *n* WASTE, WATER POLL, WATER SUPP Entspannung *f*
relaxed *adj* WASTE, WATER POLL, WATER SUPP entspannt
release 1. *n* GEN *of harmful substances* Freisetzung *f*; **2.** *vt* GEN *harmful substances* freisetzen
reliability *n* ANALYSIS *of sample* Zuverlässigkeit *f*, MARINE Betriebszuverlässigkeit *f*, Zuverlässigkeit *f*
relief: ~ **well** *n* WASTE, WATER SUPP Entlastungsbrunnen *m*

relocatable: ~ **plant** *n* SOIL umsetzbare Anlage *f*; ~ **unit** *n* SOIL umsetzbare Anlage *f*
relocation *n* SOIL, WASTE Verlagerung *f*
rem *abbr* (*radiation equivalent man*) ANALYSIS rem *nt* (*radiation equivalent man*)
remaining: ~ **dust removal** *n* AIR Restentstaubung *f*; ~ **risks** *n pl* MANAGE, SAFETY Restrisiken *nt pl*
remediate *vt* SOIL dekontaminieren, sanieren
remediated *adj* SOIL saniert, dekontaminiert
remediation: ~ **need** *n* SOIL Sanierungsbedarf *m*; ~ **rate** *n* SOIL Sanierungsgrad *m*; ~ **target values** *n pl* SOIL Sanierungszielwerte *m pl*
remelt *vt* WASTE umschmelzen
remeltable *adj* WASTE *thermoplasts* wiederaufschmelzbar
remobilization *n* WATER SUPP Remobilisierung *f*
remote: ~ **control system** *n* WATER POLL, WATER SUPP Fernsteueranlage *f*; ~ **control unit** *n* WASTE, WATER POLL Fernsteuereinheit *f*; ~ **laser measuring** *n* ANALYSIS Laserfernmessung *f*; ~ **sensing** *n* ANALYSIS Fernerkundung *f*, MARINE Fernerkundung *f*, Fernmessung *f*
removal *n* GEN *of harmful substances* Entfernung *f*, Entfernen *nt*, Beseitigen *nt*; ~ **behavior** *AmE*, ~ **behaviour** *BrE n* AIR Abscheideverhalten *nt*; ~ **by suction** *n* WATER POLL Absaugung *f*; ~ **of contaminated sites** *n* SOIL Beseitigung *f* der Altlasten; ~ **of contaminated soils** *n* SOIL Beseitigung *f* von kontaminierten Böden; ~ **of dissolved organic substances** *n* WATER SUPP Entfernung *f* gelöster organischer Stoffe; ~ **efficiency** *n* AIR Abscheidewirkungsgrad *m*; ~ **of oil by separators** *n* WASTE Ölbeseitigung *f* durch Trennmittel; ~ **operation** *n* AIR Abscheidevorgang *m*; ~ **of organic matter** *n* AIR, WASTE, WATER SUPP Beseitigung *f* von organischen Bestandteilen; ~ **of pollutants** *n* AIR Schadstoffentfernung *f*; ~ **quality** *n* AIR Abscheidegüte *f*; ~ **of suspended solids by sedimentation** *n* AIR Beseitigung *f* von Schwebstoffen durch Sedimentablagerung, WASTE, WATER SUPP Beseitigung *f* von Schwebstoffen durch Sedimentation
remove *vt* AIR *particles from gas* abscheiden, beseitigen, entfernen, SOIL *contaminants*, WASTE, WATER POLL *solid material*, WATER SUPP beseitigen; ~ **by suction** *vt* SOIL, WASTE, WATER POLL, WATER SUPP absaugen; ~ **color from** *AmE*, ~ **colour from** *BrE vt* WASTE, WATER POLL, WATER SUPP entfärben; ~ **nitrogen from** *vt* AIR entsticken
render: ~ **harmless** *vt* AIR, WATER POLL *contaminants* neutralisieren, unschädlich machen; ~ **inert** *vt* WATER POLL *dewatered sludge* inertisieren
rendered: ~ **harmless** *phr* AIR, WATER POLL neutralisiert; ~ **inert** *phr* WATER POLL *dewatered sludge* inertisiert
renew *vt* RENEW ENERG erneuern; ~ **continually** *vt* RENEW ENERG ständig erneuert werden
renewable *adj* AIR erneuerbar, RENEW ENERG erneuerbar, regenerativ, SOIL, WASTE, WATER SUPP regenerativ; ~ **energy** *n* RENEW ENERG erneuerbare Energie *f*, regenerative Energie *f*; ~ **energy form** *n* RENEW ENERG erneuerbare Energieform *f*; ~ **energy source** *n* AIR erneuerbare Energiequelle *f*, RENEW ENERG erneuerbare Energiequelle *f*, erneuerbarer Energieträger *m*, regenerative Energiequelle *f*, regenerativer Energieträger *m*; ~ **energy supply** *n* RENEW ENERG erneuerbare Energieversorgung *f*; ~ **energy supply system** *n* RENEW ENERG erneuerbares Energieversorgungssystem *nt*; ~ **energy technology** *n* RENEW ENERG erneuerbare Energietechnik *f*, erneuerbare Energietechnologie *f*; ~ **raw materials** *n pl* RENEW ENERG nachwachsende Rohstoffe *m pl*; ~ **resources** *n pl* RENEW ENERG erneuerbare Ressourcen *f pl*; ~ **technology** *n* RENEW ENERG erneuerbare Technologie *f*
reoxidation *n* AIR, WASTE Rückoxidation *f*
reparability *n* WASTE *of landfill* Reparierbarkeit *f*
repeal 1. *n* POLICY Widerruf *m*; **2.** *vt* POLICY widerrufen
repeal: ~ **of authorization** *n* POLICY Genehmigungswiderruf *m*; ~ **of licence** *BrE*, ~ **of license** *AmE n* POLICY Genehmigungswiderruf *m*; ~ **of permit** *n* POLICY Genehmigungswiderruf *m*
repeated: ~ **measurement** *n* ANALYSIS Wiederholungsmessung *f*
repellency: ~ **effect** *n* AIR Abperleffekt *m*
replastification *n* WASTE *recycled asphalt* Replastifizierung *f*
replenish *vt* GEN anreichern
reporting *n* MANAGE Berichterstattung *f*
repository *n* SOIL Lagerstätte *f*, WASTE Deponie *f*, Lagerstätte *f*; ~ **for radioactive waste** *n* WASTE Endlagerstätte *f* für radioaktiven Abfall
reprocess *vt* ANALYSIS verarbeiten, WASTE wiederverarbeiten, aufbereiten, verwerten, *nuclear waste* wiederaufbereiten
reprocessed *adj* WASTE *nuclear waste* wiederaufbereitet; ~ **material** *n* WASTE wiederaufbereitetes Material *nt*
reprocessing *n* WASTE Wiederverarbeitung *f*; ~ **costs** *n pl* WASTE, WATER SUPP Aufbereitungskosten *pl*; ~ **technology** *n* WASTE Verwertungstechnik *f*
repulpable: ~ **adhesive** *n* WASTE einstampfbarer Klebstoff *m*
repulping: ~ **equipment** *n* WASTE Einstampfgerät *nt*
repulsion *n* WASTE, WATER SUPP Abstoßung *f*
repulsive: ~ **power** *n* WATER POLL, WATER SUPP *colloidal particle* Abstoßungskraft *f*, Abstoßungsvermögen *nt*
rescrubber *n* AIR Nachwäscher *m*
reservation: ~ **of permission** *n* POLICY Erlaubnisvorbehalt *m*, Genehmigungsvorbehalt *m*
reservoir *n* AIR Speicher *m*, RENEW ENERG Reservoir *nt*, WASTE *storage basin* Speicher *m*, WATER POLL, WATER SUPP *artificial lake* Staubecken *nt*, Reservoir *nt*, Speicher *m*, Wasserbehälter *m*, Wasserspeicher *m*, Stausee *m*; ~ **cleaning** *n* WASTE, WATER SUPP Behälterreinigung *f*, Reservoirreinigung *f*; ~ **of greenhouse gases** *n* AIR Speicher *m* von Treibhausgasen; ~ **lining** *n* WASTE, WATER POLL, WATER SUPP Beckenauskleidung *f*; ~ **water** *n* WATER SUPP Talsperrenwasser *nt*
residence: ~ **time** *n* AIR, WASTE, WATER POLL Verweildauer *f*, Verweilzeit *f*; ~ **time of solids** *n* WATER SUPP Feststoffverweildauer *m*, Feststoffverweilzeit *f*; ~ **time of waste** *n* WASTE Müllverweildauer *m*, Müllverweilzeit *f*
residual: ~ **activity** *n* AIR *catalyst* Restaktivität *f*; ~ **dust content** *n* AIR Reststaubgehalt *m*; ~ **gas** *n* AIR, WASTE Restgas *nt*, Gasrückstand *m*; ~ **impact of soils with contaminants** *n* SOIL Restbelastung *f* des Bodens mit Schadstoffen; ~ **moisture** *n* ANALYSIS, WASTE Restfeuchtigkeit *f*; ~ **oil** *n* WASTE Altöl *nt*, Ölabfall *m*, Ölrückstände *m pl*; ~ **oxygen** *n* AIR Restsauerstoff *m*; ~ **pollution** *n* SOIL, WATER POLL, WATER SUPP Restverschmutzung *f*; ~ **solvent removal** *n* SOIL Beseitigung *f* von Lösemittelresten; ~ **sound** *n* NOISE Restschall *m*
residue *n* AIR Rückstand *m*, MARINE Residuum *nt*, Rückstand *m*, WASTE Reststoff *m*, Rückstand *m*, WATER POLL Rückstand *m*; ~ **avoidance at source** *n* WASTE Reststoffvermeidung *f* an der Quelle; ~ **concentration** *n* WATER POLL, WATER SUPP Restkonzentration *f*; ~ **concentration of pollutants** *n* AIR Schadstoffrestkonzentration *f*; ~ **containing pollutants** *n* WASTE schadstoffhaltiger Reststoff *m*; ~ **disposal** *n* WATER POLL Rückstandsbeseitigung *f*, Rückstandsentsorgung *f*; ~ **dust loading** *n* AIR Reststaubbeladung *f*; ~ **emissions** *n pl* AIR, WASTE, WATER POLL Restemissionen *f pl*; **~-free** *adj* AIR *spray absorption process*, WASTE reststofffrei, rückstandsfrei, WATER SUPP rückstandsfrei; **~-free production method** *n* WASTE reststoffreies Produktionsverfahren *nt*; ~ **landfill** *n* WASTE Reststoffdeponie *f*, Rückstandsdeponie *f*, WATER POLL Reststoffdeponie *f*; ~ **landfilling** *n* WASTE, WATER POLL Reststoffdeponierung *f*, Rückstandsdeponierung *f*; ~ **load** *n* AIR Restbeladung *f*; ~ **raw material** *n* RENEW ENERG *biomass conversion base material* Rückstandsprodukt *nt*
residues *n pl* SAFETY Reststoffe *m pl*, WASTE, WATER POLL

Rückstände *m pl*; ~ **from leachate purification** *n* WASTE Rückstände aus der Sickerwasserreinigung *m pl*; ~ **to be disposed of** *n* WASTE zu entsorgende Rückstände *m pl*
residue: ~ **stream** *n* WASTE Rückstandsstrom *m*; ~ **yield** *n* WASTE Reststoffanfall *m*
resilient *adj* NOISE federnd; ~ **bearing** *n* NOISE federndes Lager *nt*; ~ **isolator** *n* SAFETY Stoßdämpfer *m*; ~ **layer** *n* NOISE federnde Schicht *f*; ~ **mounting** *n* NOISE *of machines for the limiting of noise propagation* federnde Aufstellung *f*
resin *n* WASTE Harz *nt*, Kunstharz *nt*
resistance *n* AIR, ANALYSIS, WASTE, WATER SUPP Widerstand *m*; ~ **capacity** *n* ANALYSIS Widerstandskapazität *f*; ~ **kiln** *n* ANALYSIS *determination* Widerstandsofen *m*; ~ **thermometer** *n* AIR, ANALYSIS Widerstandsthermometer *nt*; ~ **to oxidation** *n* WASTE, WATER SUPP *recycled materials* Oxidationsbeständigkeit *f*; ~ **to pollutants** *n* AIR, ANALYSIS, SOIL, WATER POLL Schadstoffresistenz *f*; ~ **to tearing** *n* MARINE Reißfestigkeit *f*
resistant *adj* ANALYSIS beständig, SOIL *retention wall* resistent, WASTE, WATER SUPP resistent, beständig
resistive: ~ **layer** *n* WASTE Sperrschicht *f*
resistivity *n* AIR *dust* Widerstand *m*
resonance *n* NOISE Resonanz *f*; ~ **absorber** *n* NOISE Resonanzabsorber *m*; ~ **absorption** *n* NOISE Resonanzabsorption *f*; ~ **screen** *n* NOISE Resonanzsieb *nt*; ~ **sound absorber** *n* NOISE Resonanzschallschlucker *m*; ~ **transducer** *n* NOISE Resonanzwandler *m*
resonant: ~**-absorption silencer** *n* NOISE Reflexionsschalldämpfer *m*; ~ **ion** *n* ANALYSIS *quadrupole mass analyser* resonanzaktives Ion *nt*
resonator *n* NOISE Resonator *m*
resource *n* AIR Betriebsmittel *nt*, MARINE, SOIL, WATER POLL, WATER SUPP Ressource *f*; ~ **conservation** *n* SOIL, WATER SUPP Ressourcenschonung *f*; ~ **consumption** *n* WASTE Rohstoffverbrauch *m*; ~ **recovery** *n* WASTE Rohstoffrückgewinnung *f*, Wertstoffrückgewinnung *f*; ~ **recovery plant** *n* WASTE Wiedergewinnungsanlage *f*
respirable *adj* SAFETY lungengängig; ~ **residues** *n pl* SAFETY lungengängige Reststoffe *m pl*
respiratory: ~ **filter** *n* SAFETY Atemschutzfilter *nt*; ~ **mask** *n* SAFETY Atemschutzmaske *f*; ~ **protection apparatus** *n* SAFETY Atemschutzgerät *nt*; ~ **protection system** *n* SAFETY Atemschutzsystem *nt*
responsible: ~ **authority** *n* POLICY federführende Behörde *f*
restabilization *n* WATER POLL Restabilisierung *f*
restabilized *adj* WATER POLL restabilisiert; ~ **particle** *n* WATER POLL restabilisiertes Teilchen *nt*
restriction: ~ **of toxicity at the workplace** *n* SAFETY Beschränkung *f* der Giftigkeit am Arbeitsplatz
resurgent *adj* WASTE, WATER SUPP wiederauflebend
resuspension *n* WATER SUPP Resuspension *f*
retailer *n* MANAGE Händler *m*
retaining: ~ **grate** *n* AIR Rückhalterost *m*; ~ **wall** *n* NOISE *noise-insulating barriers* Stützmauer *f*; ~ **well** *n* SOIL Sperrbrunnen *m*
retarder *n* WASTE Reaktionsverzögerer *m*
retarding: ~ **agent** *n* WASTE Reaktionsverzögerer *m*
retention: ~ **cage** *n* AIR *fluidized-bed absorber* Rückhaltekäfig *m*; ~ **tank** *n* WATER POLL Retentionsbecken *nt*; ~ **time** *n* AIR, WASTE Aufenthaltszeit *f*, Verweildauer *f*, Verweilzeit *f*, WATER POLL Verweildauer *f*, Verweilzeit *f*, Aufenthaltszeit *f*; ~ **wall** *n* SOIL Abdichtungswand *f*, Dichtungswand *f*, Dichtwand *f*
retentivity *n* AIR Rückhaltevermögen *nt*
retitration *n* ANALYSIS Rücktitration *f*
retreaded: ~ **tire** *AmE*, ~ **tyre** *BrE n* WASTE runderneuerter Reifen *m*
retrieval: ~ **capacity** *n* WASTE Rückholbarkeit *f*
retrofit *vt* GEN nachrüsten
retrofitting *n* GEN Nachrüstung *f*
return *n* GEN Rückgabe *f*
returnable *adj* WASTE Mehrweg-; ~ **bottle** *n* WASTE Mehrwegflasche *f*, Pfandflasche *f*; ~ **container** *n* WASTE Rückgabebehälter *m*, Mehrweggebinde *nt*
return: ~ **rate** *n* WASTE Mehrwegquote *f*, Rücklaufquote *f*
returns *n pl* WASTE Rückgut *nt*; ~ **granulate** *n* WASTE Rückgut-Granulat *nt*
return: ~ **sludge** *n* WATER POLL Rücklaufschlamm *m*; ~ **sludge elevator** *n* WATER POLL Rückschlammhebewerk *nt*; ~ **sludge pipe** *n* WATER POLL Rücklaufschlammleitung *f*
reusability *n* WASTE Wiederverwendbarkeit *f*
reusable *adj* WASTE wiederverwendbar; ~ **box** *n* WASTE Mehrwegkarton *m*, Pfandkasten *m*, wiederverwendbarer Karton *m*; ~ **dust** *n* AIR Nutzstaub *m*; ~ **gypsum** *n* AIR *product of flue-gas cleaning* verwertbarer Gips *m*; ~ **packaging** *n* WASTE Mehrwegverpackung *f*, wiederverwertbare Verpackung *f*, wiederverwendbares Verpackungsmaterial *nt*; ~ **waste product** *n* AIR wiederverwertbares Abfallprodukt *nt*
reuse 1. *n* AIR *of collected dust*, WASTE, WATER SUPP Wiederbenutzung *f*, Wiederverwendung *f*; **2.** *vt* AIR, WASTE, WATER SUPP wiederbenutzen, wiederverwenden
revegetate *vt* SOIL, WASTE rekultivieren
revenue *n* POLICY *from taxes* Aufkommen *nt*, Einnahme *f*; ~ **from eco-taxes** *n* POLICY Einnahme *f* aus Ökosteuern; ~**-neutral** *adj* POLICY aufkommensneutral; ~ **neutrality** *n* POLICY Aufkommensneutralität *f*
reverberant *adj* NOISE nachhallend
reverberation *n* NOISE Nachhall *m*; ~ **chamber** *n* NOISE Hallraum *m*; ~ **time** *n* NOISE *to reduce in noise control* Nachhallzeit *f*
reverse: ~**-air-flow filter** *n* AIR Gegenstromfilter *nt*; ~ **filtration** *n* WASTE, WATER SUPP Aufwärtsfiltration *f*; ~ **flow** *n* AIR *gas flow* Gegenstrom *m*; ~**-flow cleaning** *n* AIR Gegenstromspülung *f*; ~**-flow duct** *n* AIR *bag filter* Gegenstromröhrenleitung *f*; ~**-flow filter** *n* WATER POLL Gegenstromfilter *nt*; ~**-flow packed-bed filter** *n* AIR Gegenstromschüttschichtfilter *nt*; ~**-flow precipitator** *n* AIR Gegenstromabscheider *m*; ~**-flow scrubber** *n* AIR Gegenstromwäscher *m*; ~**-jet filter** *n* AIR Druckstoßfilter *nt*, Puls-Jet-Filter *nt*, Schlauchfilter *nt* mit Druckluftabreinigung; ~ **osmosis** *n* (*RO*) WASTE, WATER POLL, WATER SUPP Umkehrosmose *f* (*UO*); ~ **osmosis with concentrate evaporation** *n* WASTE, WATER SUPP Umkehrosmose *f* mit Konzentrateindampfung; ~ **osmosis plant** *n* WASTE, WATER SUPP Umkehrosmoseanlage *f*; ~ **vending machine** *n* WASTE Rücklaufautomat *m*
reversible *adj* GEN reversibel, umkehrbar; ~ **process** *n* AIR *physical adsorption* reversibles Verfahren *nt*, WATER POLL *chemisorption* reversibles Verfahren *nt*, umkehrbares Verfahren *nt*
revitalization *n* SOIL *of treated soils* Revitalisierung *f*
revolving: ~ **drum** *n* SOIL *thermal soil treatment*, WASTE Drehtrommel *f*; ~ **filter** *n* AIR *nuclear technology* Zellenfilter *nt*
rewhirling *n* AIR Wiederaufwirbelung *f*
Reynold: ~**'s number** *n* (*Re*) AIR, WATER SUPP Reynoldsche Zahl *f* (*Re*)
Rh *abbr* (*rhodium*) AIR Rh (*Rhodium*)
rheometer *n* ANALYSIS Rheometer *nt*, Strömung *f* für viskose Flüssigkeiten
rhodium *n* (*Rh*) AIR Rhodium *nt* (*Rh*)
rH: ~**-value** *n* (*reduction-oxidation potential*) AIR, ANALYSIS, WATER POLL rH-Wert *m* (*Reduktions-Oxidations-Potential*)
rich: ~ **gas processing** *n* AIR Reichgasverarbeitung *f*
ridge: ~**-and-furrow technique** *n* WATER POLL *land application of wastewater* Furchentechnik *f*
right: ~**-to-know principle** *n* POLICY *environmental information* Recht *nt* auf Informationen

rigid: ~ **plate absorber** *n* NOISE biegesteifer Plattenabsorber *m*; ~ **PVC** *n* WASTE Hart-PVC *nt*
ring: ~ **adsorber** *n* AIR *fixed bed adsorber* Ringadsorber *m*
ringed: ~ **layer** *n* AIR Ringschicht *f*
Ringelmann: ~ **scale** *n* AIR, ANALYSIS *dust measurement* Ringelmann-Skala *f*
ring: ~ **jet aerosol precipitator** *n* AIR Ring-Jet-Aerosol-Abscheider *m*; ~ **jet element** *n* AIR Ringdüsenelement *nt*; ~ **nozzle** *n* AIR Ringdüse *f*; ~ **of tubbings** *n* WASTE Schachtring *m*
rinse *vt* GEN spülen
rinsing *n* GEN Spülung *f*; ~ **water** *n* SOIL, WASTE Spülwasser *nt*; ~ **water processing** *n* SOIL, WASTE Spülwasseraufbereitung *f*
riparian *adj* WATER SUPP uferanliegend
ripening *n* WATER POLL *sludge conditioning* Reifung *f*
rise: ~ **of pH-value** *n* WATER POLL Erhöhung *f* des pH-Wertes
rising: ~ **energy prices** *n pl* RENEW ENERG steigende Energiepreise *m pl*; ~ **rate** *n* WATER POLL Steiggeschwindigkeit *f*; ~ **of water level** *n* WATER SUPP Wasserspiegelanstieg *m*
risk *n* GEN Risiko *nt*; ~ **analysis** *n* AIR, MANAGE, POLICY, SAFETY, SOIL, WASTE, WATER POLL Risikoanalyse *f*; ~ **assessment** *n* AIR, ANALYSIS, MANAGE, MARINE, POLICY, SAFETY, SOIL *of contaminated sites*, WASTE, WATER POLL Gefährdungsabschätzung *f*, Risikoabschätzung *f*, Risikobewertung *f*; ~ **of clogging** *n* AIR Verstopfungsgefahr *f*; ~ **estimate** *n* ANALYSIS, MANAGE, POLICY, SOIL Gefährdungsabschätzung *f*; ~ **estimation** *n* MANAGE *eco-audit* Risikoabschätzung *f*, Risikobewertung *f*; ~ **factor** *n* AIR, MANAGE, POLICY, SAFETY, SOIL, WASTE, WATER POLL Risikofaktor *m*; ~ **of iron clogging** *n* WASTE, WATER SUPP Verockerungsgefahr *f*; ~ **limit** *n* SAFETY Grenzrisiko *nt*; ~ **of plugging** *n* AIR Verstopfungsgefahr *f*; ~ **potential** *n* AIR Risikopotential *nt*, MANAGE Gefährdungspotential *nt*, Risikopotential *nt*, POLICY, SAFETY, SOIL, WASTE, WATER POLL Risikopotential *nt*
river *n* POLICY, WATER POLL, WATER SUPP Fluß *m*; ~ **authority** *n* MANAGE, WASTE, WATER SUPP Aufsichtsamt *nt*, Flußaufsichtsbehörde *f*; ~ **basin** *n* WASTE Flußbecken *f*, WATER SUPP Flußbecken *nt*, Flußeinzugsgebiet *nt*, Flußgebiet *nt*; ~ **basin authority** *n* POLICY Flußeinzugsgebietsbehörde *f*; ~ **basin management plan** *n* POLICY Flußverwaltungsplan *m*; ~ **bed** *n* WATER SUPP Flußbett *nt*; ~ **floor** *n* WATER SUPP Flußsohle *f*; ~ **tapping** *n* WATER SUPP Flußfassung *f*; ~ **training** *n* WATER SUPP Flußbau *m*, Flußkanalisierung *f*, Flußregulierung *f*; ~ **wall** *n* WATER SUPP Flußufermauer *f*, Ufermauer *f* eines Flusses; ~ **water** *n* WATER SUPP Flußwasser *nt*; ~ **water purification** *n* WATER SUPP Flußwasserreinigung *f*; ~ **water quality** *n* WATER SUPP Flußwasserqualität *f*; ~ **water tapping** *n* WATER SUPP Flußwasserentnahme *f*, Flußwasserfassung *f*; ~ **works** *n* WATER SUPP Flußbau *m*
RO *abbr* (*reverse osmosis*) WASTE, WATER POLL, WATER SUPP UO (*Umkehrosmose*)
road *n* NOISE, WASTE Straße *f*; ~ **construction rubble** *n* WASTE Straßenaufbruch *m*; ~**-sweeping lorry** *n BrE* (*cf road-sweeping truck AmE*) WASTE Kehrfahrzeug *nt*; ~ **sweepings** *n pl* WASTE Straßenkehricht *m*; ~**-sweeping truck** *n AmE* (*cf road-sweeping lorry BrE*) WASTE Kehrfahrzeug *nt*; ~ **tanker** *n* MARINE Tankfahrzeug *nt*, Tankwagen *m*; ~ **tax** *n BrE* (*cf automobile tax AmE*) POLICY Kfz-Steuer *f*; ~ **transportation** *n* WASTE Beförderung *f* über die Straße, Straßentransport *m*
robustness *n* WASTE Robustheit *f*
rock: ~ **bed** *n* SOIL Gesteinsschicht *f*
rocket: ~ **probe** *n* ANALYSIS Raketensonde *f*
rocking: ~ **grate** *n* WASTE Kippstufenrost *m*
rock: ~ **wool** *n* NOISE Steinwolle *f*
rod *n* WASTE, WATER SUPP *of pump* Stange *f*
roll: ~ **body** *n* SOIL *smooth roller* Walzenkörper *m*; ~ **crusher** *n* WATER SUPP Walzenbrecher *m*
roller *n* WASTE Walze *f*; ~ **die extruder** *n* WASTE Walzenextruder *m*
rolling: ~ **fraction** *n* WASTE rollende Fraktion *f*; ~ **grate** *n* WASTE Walzenrost *m*; ~ **grate installation** *n* WASTE Walzenrostanlage *f*; ~ **mill** *n* WASTE Walzwerk *nt*
roll: ~**-out waste container** *n* WASTE Müllcontainer *m*
roofing *n* AIR *of filter*, WASTE, WATER SUPP *of landfill* Überdachung *f*
room: ~ **acoustics** *n pl* NOISE Raumakustik *f*; ~ **lining** *n* NOISE *insulation* Raumauskleidung *f*; ~ **temperature** *n* ANALYSIS Raumtemperatur *f*
root: ~ **bed method** *n* WATER POLL Wurzelraumverfahren *nt*
rope: ~ **skimmer** *n* MARINE Bandabschöpfgerät *nt*
rose *n* WATER SUPP Saugkorb *m*
rot *n* WASTE Faulfähigkeit *f*, Faulen *nt*, Rotte *f*, WATER POLL Faulen *nt*
rotameter *n* ANALYSIS Durchflußschwimmermesser *m*, Rotameter *m*, Schwebkörperdurchflußmesser *m*; ~ **viscosimeter** *n* ANALYSIS Schwebkörperviskosimeter *nt*
rotary: ~ **air heater** *n* AIR Drehluftvorwärmer *m*, Drehluvo *m*; ~ **atomization** *n* AIR *spray absorption* Rotationszerstäubung *f*; ~ **current frequency** *n* RENEW ENERG *wind turbine* Drehstromfrequenz *f*; ~ **distributor** *n* WATER POLL *trickling filtration* Drehsprenger *m*; ~ **drum furnace** *n* WATER POLL *incineration* Drehtrommelofen *m*; ~ **dryer** *n* SOIL Trockentrommel *f*; ~ **evaporator** *n* WATER POLL, WATER SUPP Umlaufverdampfer *m*; ~ **filter** *n* WASTE, WATER SUPP Drehfilter *nt*; ~ **furnace** *n* AIR, SOIL, WASTE Drehofen *m*, Drehrohrofen *m*, WATER POLL Drehrohrofen *m*, Rotationsofen *m*; ~ **kiln** *n* AIR, SOIL, WASTE Drehofen *m*, Drehrohrofen *m*, WATER POLL *rotary furnace* Drehrohrofen *m*, Rotationsofen *m*; ~ **kiln incinerator** *n* AIR Drehofenverbrenner *m*; ~ **piston compressor** *n* AIR Drehkolbengebläse *nt*, Drehkolbenverdichter *m*; ~ **pump** *n* WASTE Drehkolbenpumpe *f*, WATER POLL Drehkolbenpumpe *f*, Rotationspumpe *f*; ~ **screen** *n* WATER POLL Drehsieb *nt*; ~ **vacuum filter** *n* WATER POLL Drehvakuumfilter *nt*
rotatable: ~ **arm** *n* MARINE Dreharm *m*
rotating *adj* AIR, WASTE rotierend, WATER SUPP Dreh-; ~ **adsorber** *n* AIR rotierender Adsorber *m*; ~ **biological disc contactor** *n BrE* WATER POLL Scheibentauchkörper *m*; **with** ~ **built-in structure** *phr* AIR *scrubbing area* mit rotierenden Einbauten; ~ **gas preheater** *n* AIR Dreh-Gavo *m*, Drehgasvorwärmer *m*; ~ **sprayer** *n* WASTE, WATER POLL *wastewater* Rundregner *m*
rotation *n* AIR, ANALYSIS, WATER POLL Rotation *f*; ~ **movement** *n* AIR *cyclone* Drehbewegung *f*; ~ **velocity** *n* AIR Rotationsgeschwindigkeit *f*
rot: ~ **landfill** *n* WASTE Rottedeponie *f*
rotor *n* RENEW ENERG *wind turbine* Rotor *m*; ~ **blade** *n* RENEW ENERG *wind turbine* Rotorblatt *nt*; ~ **brake** *n* RENEW ENERG *wind turbine* Rotorbremse *f*; ~ **hub** *n* RENEW ENERG *wind turbine* Rotornabe *f*; ~ **hub with adjustable blade mechanism** *n* RENEW ENERG *wind turbine* Rotornabe *f* mit Blattverstellmechanismus
rotors: ~ **with fixed number of revolutions** *n pl* RENEW ENERG *wind turbine* drehzahlfeste Rotoren *m pl*; ~ **with variable number of revolutions** *n pl* RENEW ENERG *wind turbine* drehzahlvariable Rotoren *m pl*
rotting: ~ **body** *n* WASTE Kompostkörper *m*; ~ **chamber** *n* WASTE Rottebox *f*, Rottezelle *f*; ~ **chamber composting process** *n* WASTE Boxenkompostierung *f*, Rottezellenkompostierungsverfahren *nt*; ~ **container** *n* WASTE Rottecontainer *m*; ~ **hall** *n* WASTE Rottehalle *f*; ~ **operation** *n* WASTE Rottevorgang *m*; ~ **process** *n* WASTE Rotteprozeß *m*; ~ **standstill** *n* WASTE Rottestillstand *m*; ~ **time** *n* WASTE Rottezeit *f*; ~ **tower** *n* WASTE Rotteturm *m*

roughage *n* WASTE, WATER SUPP Ballaststoff *m*; **~-free** *adj* WASTE, WATER SUPP ballaststofffrei
roughing: **~ tank** *n* WATER POLL Grobklärbecken *nt*, Grobstoffänger *m*
round *adj* ANALYSIS, WASTE, WATER POLL, WATER SUPP Rund-; **~-bottomed flask** *n* ANALYSIS Rundkolben *m*
routine *n* ANALYSIS Routine *f*; **~ analysis** *n* ANALYSIS Routine-Analyse *f*; **~ determination** *n* ANALYSIS Routinebestimmung *f*; **~ monitoring** *n* ANALYSIS Routineüberwachung *f*; **~ testing** *n* ANALYSIS Routineprüfung *f*
Ru *abbr* (*ruthenium*) AIR Ru (*Ruthenium*)
rubbish *n BrE* (*cf garbage AmE, refuse*) WASTE Abfall *m*, Müll *m*; **~ bag** *n BrE* WASTE Müllbeutel *m*, Müllsack *m*; **~ bin** *n BrE* WASTE Müllbehälter *m*, Mülleimer *m*, Mülltonne *f*; **~ chute** *BrE* WASTE Müllschlucker; **~ collection** *n BrE* Müllabfuhr *f*, Mülleinsammlung *f*, Müllsammlung *f*; **~ tip** *n BrE* Müllkippe *f*, Deponie *f*
rubble *n* AIR *ground filter* Schotter *m*
ruggedness *n* ANALYSIS *of analysis devices* Robustheit *f*
run 1. *n* ANALYSIS *measurements* Meßreihe *f*; **2. ~ through** *vt* ANALYSIS durchströmen; **3. ~ dry** *vi* WASTE, WATER SUPP trocken laufen; **~ off** WASTE, WATER SUPP ablaufen; **~ off the surface** WASTE *precipitation* oberflächig abfließen; **~ out** WASTE auslaufen, WATER SUPP ausfließen; **~ over** WASTE, WATER POLL, WATER SUPP überlaufen
running: **~ water** *n* WASTE fließendes Wasser *nt*
runoff *n* MARINE Abfluß *m*, Ablauf *m*, Ausfluß *m*, WASTE Ablauf *m*, WATER POLL, WATER SUPP Abfließen *nt*, Abfluß *m*, Ablauf *m*; **~ capacity** *n* WASTE Abflußvermögen *nt*, WATER POLL Abflußleistung *f*, WATER SUPP Abflußleistung *f*, Abflußvermögen *nt*; **~ coefficient** *n* WATER POLL, WATER SUPP Abflußkoeffizient *m*, Abflußverhältnis *nt*; **~ cycle** *n* WATER SUPP Abflußkreislauf *m*; **~ duration** *n* WATER POLL Ablaufdauer *f*
rupture *n* WASTE, WATER SUPP *of filter bag* Durchbruch *m*; **~ of floc** *n* WATER POLL Aufbrechen *nt* der Flocke
rural: **~ water supply** *n* WATER SUPP Wasserversorgung *f* im ländlichen Raum
rust *n* WASTE, WATER POLL, WATER SUPP Rost *m*; **~ preventative** *n* WASTE Rostschutzmittel *nt*, Rost-Inhibitor *m*; **~ protection** *n* WATER POLL, WATER SUPP Rostschutz *m*
ruthenium *n* (*Ru*) AIR Ruthenium *nt* (*Ru*)

S

sabin *n* ANALYSIS *unit of sound absorption*, NOISE Sabin *nt*
safe: ~ **against leaching** *adj* SOIL, WASTE, WATER SUPP auslaugsicher; ~ **disposal** *n* WASTE geordnete Beseitigung *f*, sichere Entsorgung *f*
safeguards *n pl* SAFETY Sicherungsmaßnahmen *f pl*
safe: ~ **keeping** *n* WASTE sichere Lagerung *f*; ~ **operation** *n* SAFETY sicherer Betrieb *m*
safety *n* ANALYSIS, SAFETY Sicherheit *f*; ~ **analysis** *n* SAFETY Sicherheitsanalyse *f*; ~ **check** *n* SAFETY Sicherheitskontrolle *f*; ~ **code** *n* SAFETY Sicherheitsvorschriften *f pl*; ~ **complaints** *n pl* SAFETY Sicherheitsbeschwerden *f pl*; ~ **distance** *n* SAFETY Sicherheitsabstand *m*; ~ **engineering** *n* SAFETY Sicherheitstechnik *f*; ~ **factor** *n* SAFETY Sicherheitsfaktor *m*; ~ **measures** *n pl* SAFETY Sicherheitsmaßnahmen *f pl*; ~ **regulations** *n pl* SAFETY Sicherheitsvorschriften *f pl*; ~**-related** *adj* SAFETY sicherheitsbezogen, sicherheitstechnisch; ~**-related equipment** *n* SAFETY sicherheitsbezogene Ausrüstung *f*; ~ **requirements** *n pl* SAFETY Sicherheitsvorschriften *f pl*; ~ **risk** *n* SAFETY Sicherheitsrisiko *nt*; ~ **standard** *n* SAFETY Sicherheitsstandard *m*; ~ **strategy** *n* SAFETY Sicherheitsstrategie *f*; ~ **zone** *n* SAFETY Sicherheitsabstand *m*
salinate *vt* WASTE, WATER SUPP versalzen
salination *n* WASTE, WATER SUPP Aufsalzung *f*, Versalzung *f*
saline *adj* WASTE, WATER SUPP salzhaltig, salzig; ~ **concentration** *n* RENEW ENERG *solar pond* Salzkonzentration *f*; ~ **spring** *n* WATER SUPP salinische Quelle *f*; ~ **water** *n* WASTE, WATER SUPP Salzwasser *nt*, salzhaltiges Wasser *nt*; ~ **water conversion** *n* WATER SUPP Salzwasserumwandlung *f*
salinity *n* WASTE, WATER SUPP Salzgehalt *m*
salt *n* GEN Salz *nt*; ~ **aerosol** *n* AIR Salzaerosol *nt*; ~ **slag** *n* WASTE Salzschlacke *f*; ~ **water** *n* WATER SUPP Salzwasser *nt*; ~ **water plant** *n* WASTE, WATER SUPP Entsalzungsanlage *f*
Salzmann: ~ **absorption solution** *n* ANALYSIS Salzmann-Absorptionslösung *f*; ~ **method** *n* ANALYSIS Salzmann-Methode *f*
sample 1. *n* ANALYSIS Probe *f*; **2.** *vt* ANALYSIS eine Probe nehmen
sample: ~ **amount** *n* ANALYSIS Probemenge *f*; ~ **analysis** *n* ANALYSIS Probenanalyse *f*; ~ **assessment** *n* ANALYSIS Probebeurteilung *f*; ~ **calibration** *n* ANALYSIS Probenkalibrierung *f*; ~ **collection** *n* ANALYSIS Probenahme *f*; ~ **components** *n pl* ANALYSIS Inhaltsstoffe der Proben *m pl*; ~ **composition** *n* ANALYSIS Probenzusammensetzung *f*; ~ **diminution** *n* ANALYSIS Verjüngen *nt* der Proben; ~ **dispenser** *n* ANALYSIS Probengeber *m*; ~ **dosing** *n* ANALYSIS Probendosierung *f*; ~ **gas** *n* ANALYSIS Probengas *nt*; ~ **gas capillary** *n* ANALYSIS Probengaskapillare *f*; ~ **gas filter** *n* ANALYSIS Probengasfilter *nt*; ~ **grinding** *n* ANALYSIS Zerkleinern *nt* der Proben; ~ **holder** *n* ANALYSIS Probenhalter *m*, Probenträger *m*; ~ **injector** *n* ANALYSIS Probengeber *m*, Probeninjektor *m*; ~ **introduction** *n* ANALYSIS Probendosierung *f*, Probenzuführung *f*; ~ **point** *n* ANALYSIS, SOIL Probenahmepunkt *m*; ~ **processing** *n* ANALYSIS Probeverarbeitung *f*; ~ **pump** *n* ANALYSIS Meßgaspumpe *f*
sampler *n* ANALYSIS Probenahmegerät *nt*, Sampler *m*, *person* Probenehmer *m*
sample: ~ **record** *n* ANALYSIS Probenahmeprotokoll *nt*
sampler: ~ **for flow-proportional samples** *n* ANALYSIS Proportionalprobenehmer *m*, mengenproportionaler Probenehmer *m*
sample: ~ **size** *n* ANALYSIS *statistically* Probengröße *f*, Probenumfang *m*; ~ **solution** *n* ANALYSIS Probenlösung *f*; ~**-specific** *adj* ANALYSIS *data* probenspezifisch; ~ **storage** *n* ANALYSIS Probenaufbewahrung *f*; ~ **throughput**, ~ **thruput** *AmE n* ANALYSIS Probendurchsatz *m*; ~ **treatment** *n* ANALYSIS Probenaufbereitung *f*; ~ **volume** *n* ANALYSIS Probenvolumen *nt*
sampling *n* ANALYSIS Beprobung *f*, Probenahme *f*; ~ **aid** *n* ANALYSIS Entnahmehilfsmittel *nt*; ~ **bay** *n* ANALYSIS, WASTE Probenahmestelle *f*; ~ **bottle** *n* ANALYSIS Probenflasche *f*, Probeflasche *f*; ~ **capacity** *n* ANALYSIS Beprobungskapazität *f*, Probenahmekapazität *f*; ~ **cell** *n* ANALYSIS Probezelle *f*; ~ **condensation** *n* ANALYSIS Probenkondensation *f*; ~ **device** *n* ANALYSIS Probenahmegerät *nt*; ~ **frequency** *n* ANALYSIS Probenahmehäufigkeit *f*; ~ **gas** *n* ANALYSIS Probengas *nt*; ~ **grid** *n* ANALYSIS, SOIL Probenahmeraster *m*; ~ **instrument** *n* ANALYSIS Entnahmegerät *nt*, Entnahmevorrichtung *f*; ~ **method** *n* ANALYSIS Entnahmemethode *f*; ~ **plan** *n* ANALYSIS *inspection* Stichprobenplan *m*; ~ **point** *n* ANALYSIS, SOIL, WATER POLL Probenahmestelle *f*, Entnahmestelle *f* von Proben; ~ **preparation** *n* ANALYSIS Probenaufbereitung *f*, Probenvorbereitung *f*; ~ **probe** *n* ANALYSIS Probenahmesonde *f*; ~ **pump** *n* ANALYSIS Probenahmepumpe *f*; ~ **station** *n* ANALYSIS *for water control* Meßstation *f*; ~ **system** *n* ANALYSIS Probenahmesystem *nt*; ~ **transmission** *n* ANALYSIS Probenweiterleitung *f*; ~ **tube** *n* ANALYSIS Probenahmerohr *nt*, Probenheber *m*; ~ **vessel** *n* ANALYSIS Entnahmegefäß *nt*
sand *n* WATER SUPP *glass recycling* Sand *m*
sandblasting *n* MARINE Sandstrahlen *nt*, Sandstrahlreinigung *f*
sand: ~ **deposition** *n* WATER SUPP Sandablagerung *f*; ~ **filter** *n* WATER POLL *volume filter* Sandfilter *nt*; ~ **filtration ditch** *n* WATER POLL Sandfiltergraben *m*; ~ **layer** *n* WASTE, WATER POLL *filtering tanks* Sandschicht *f*; ~ **trap** *n* WATER POLL Sandfang *m*
sandwich: ~ **construction** *n* NOISE Mehrschichtenkonstruktion *f*, Sandwichkonstruktion *f*; ~ **injection molding** *AmE*, ~ **injection moulding** *BrE n* WASTE Sandwich-Spritzgießen *nt*; ~ **panel** *n* NOISE *vibration damping* Sandwichplatte *f*, Verbundplatte *f*, mehrschichtige Platte *f*; ~ **structure** *n* RENEW ENERG Sandwich-Struktur *f*
sanitary: ~ **engineering** *n* WATER POLL, WATER SUPP Sanitärtechnik *f*; ~ **landfill** *n* WASTE Mülldeponie *f*, geordnete Deponie *f*, kontrollierte Müllablagerung *f*; ~ **landfilling** *n* WASTE geordnete Deponierung *f*; ~ **wastewater** *n* WATER POLL Abwasser *nt* aus Sanitäranlagen
sanitation *n* WASTE sanitäre Einrichtungen *f pl*, *compost* Hygienisierung *f*, WATER POLL, WATER SUPP sanitäre Einrichtungen *f pl*
saturate *vt* GEN beladen, sättigen
saturated *adj* GEN beladen, gesättigt; ~ **activated carbon** *n* WATER POLL beladene Aktivkohle *f*; ~ **adsorbent** *n* AIR beladenes Adsorbens *nt*; ~ **gas** *n* ANALYSIS gesättigtes Gas *nt*; ~ **resin** *n* WATER POLL *ion exchanger* beladenes Harz *nt*; ~ **soil zone** *n* SOIL gesättigte Bodenzone *f*; ~ **zone** *n* WASTE, WATER SUPP Sättigungszone *f*, *underground* gesättigte Zone *f*
saturation *n* GEN Sättigung *f*; ~ **concentration** *n* WATER SUPP Sättigungskonzentration *f*; ~ **deficit** *n* AIR Sättigungsdefizit

nt; ~ **index** *n* WATER SUPP Sättigungsindex *m*; ~ **level** *n* WATER SUPP Sättigungsebene *f*; ~ **load** *n* AIR *adsorption* Sättigungsbeladung *f*; ~ **pressure** *n* AIR Sättigungsdruck *m*; ~ **temperature** *n* AIR Sättigungstemperatur *f*

save *vt* RENEW ENERG *energy* sparen

Savonius: ~ **rotor** *n* RENEW ENERG *vertical axle* Savonius-Rotor *m*

Sb *abbr* (*antimony*) WATER SUPP Sb (*Antimon*)

SBR *abbr* (*sequential batch reactor*) SOIL SBR (*Sequential Batch Reactor*)

SC *abbr* (*standard conditions*) ANALYSIS Normzustand *m*

scale *n* ANALYSIS Skala *f*, WASTE Kesselstein *m*; ~ **formation** *n* WASTE Kesselsteinablagerung *f*, Kesselsteinbildung *f*

scales *n pl* ANALYSIS, WASTE Waage *f*

scaling *n* AIR Verkrustung *f*; ~ **risk** *n* AIR Verkrustungsgefahr *f*

scanner *n* ANALYSIS Meßstellenabtaster *m*, Meßstellenumschalter *m*

scanning: ~ **electron microscope** *n* (*SEM*) ANALYSIS Rasterelektronenmikroskop *nt*; ~ **electron microscopy** *n* (*SEM*) ANALYSIS Rasterelektronenmikroskopie *f*; ~ **method** *n* ANALYSIS Rasterverfahren *nt*; ~ **network** *n* ANALYSIS Rastermeßnetz *nt*; ~ **tunnel microscope** *n* ANALYSIS Rastertunnelmikroskop *nt*

scarce *adj* GEN knapp; ~ **environmental resources** *n pl* AIR, POLICY, RENEW ENERG, SOIL, WATER POLL knappe Umweltressourcen *f pl*

scatter *vt* ANALYSIS streuen

scattering *n* ANALYSIS Streuung *f*, Zerstreuen *nt*; ~ **cloud** *n* ANALYSIS *dispersion telemetering* Streuwolke *f*

scavenge *vt* AIR *gases* ausspülen

scavenger *n* AIR, SOIL, WASTE Scavenger *m*, Spülmittel *nt*

scavenging *n* AIR Ausspülen *nt*, Ausspülung *f*, Spülung *f*; ~ **gas** *n* AIR Spülgas *nt*

SCC *abbr* (*secondary combustion chamber*) WASTE Nachverbrennungskammer *f*

scintillation: ~ **counter** *n* ANALYSIS Szintillationszähler *m*

scission *n* WASTE *plastics recycling* Spaltung *f*

scoop *n* ANALYSIS Schöpfgerat *nt*, MARINE Behälter *m*, Container *m*, Frachtbehälter *m*, WASTE Schrappergefäß *nt*

scooper *n* ANALYSIS Schöpfgerat *nt*

scour *vt* AIR, WASTE, WATER SUPP abschwemmen, spülen, ausspülen

scouring *n* AIR, WASTE, WATER SUPP Abschwemmung *f*, Spülung *f*, Ausspülen *nt*, Ausspülung *f*; ~ **outlet with grate** *n* WASTE, WATER SUPP Ausmündungsstück *nt*

SCR *abbr* (*selective catalytic reduction*) AIR, WASTE SKR (*selektiv-katalytische Reduktion*)

scrap *n* WASTE Schrott *m*; ~ **application** *n* WASTE Schrotteinsatz *m*; ~ **bale** *n* WASTE Schrottballen *m*; ~ **baler** *n* WASTE Schrottpaketierpresse *f*; ~-**baling press** *n* WASTE Eisenschrottpresse *f*, Schrottpresse *f*; ~ **cascade** *n* WATER SUPP Scrap-Kaskade *f*; ~ **collection** *n* WASTE Schrottsammlung *f*

scrape *vt* MARINE schaben

scraper *n* MARINE Schaber *m*, Schrapper *m*, Scraper *m*, WATER POLL *sludge rake* Kratzer *m*, Räumer *m*, Schlammräumer *m*; ~ **bridge** *n* WATER POLL Räumerbrücke *f*; ~ **extractor** *n* WASTE, WATER SUPP Schaufelentnahmegerät *nt*

scraping: ~ **operation** *n* WATER POLL Räumvorgang *m*

scrap: ~ **iron** *n* WASTE Eisenschrott *m*; ~ **lead** *n* WASTE Altblei *nt*; ~ **metal** *n* WASTE Altmetall *nt*; ~ **metal dealer** *n* WASTE Altmetallhändler *m*, Schrotthändler *m*; ~ **metal merchant** *n* WASTE Altmetallhändler *m*, Schrotthändler *m*; ~ **metal separation** *n* WASTE Entschrottung *f*; ~ **motorcar** *n* WASTE Autowrack *nt*

scrapped: ~ **plant** *n* WASTE abgehende Anlage *f*

scrapping: ~ **fee** *n* POLICY *automotive* Verschrottungsgebühr *f*

scrap: ~ **press** *n* WASTE Schrottpresse *f*; ~ **processing** *n* WASTE Schrottverwertung *f*; ~ **quality** *n* WASTE Schrottqualität *f*; ~ **reuse** *n* WASTE Schrottverwertung *f*; ~ **shear** *n* WASTE Schrottschere *f*; ~ **smelting** *n* WASTE Schrottverhüttung *f*; ~ **sorting** *n* WASTE Schrottsortierung *f*; ~ **yard** *n* WASTE Altmetallverwertungsanlage *f*, Schrottplatz *m*

screen 1. *n* NOISE Abschirmung *f*, WASTE, WATER POLL, WATER SUPP Sichtschutz *m*; **2.** *vt* GEN absieben, *size* klassieren, *separate out* aussondern, durchsieben, *for impurities* aussieben

screen: ~ **analysis** *n* ANALYSIS Siebanalyse *f*; ~-**changing device** *n* WASTE Siebwechselvorrichtung *f*; ~ **cleaner** *n* WATER POLL Rechenreiniger *m*, Rechenreinigungsmaschine *f*; ~ **cleaning** *n* WATER POLL Rechenreinigung *f*; ~ **clogging** *n* WATER POLL Rechenverstopfung *f*; ~ **for cups** *n* WASTE, WATER SUPP *in sorting plant* Bechersieb *nt*; ~ **gutter** *n* WATER POLL Rechenrinne *f*

screening *n* GEN Sieben *nt*, Siebung *f*, *pretreatment* Durchsieben *nt*, Screening *nt*; ~ **analysis** *n* ANALYSIS Screening-Analyse *f*; ~ **equipment** *n* WASTE Sieb *nt*; ~ **machine** *n* AIR Siebmaschine *f*; ~ **magnet** *n* WASTE, WATER SUPP Aussonderungsmagnet *m*; ~ **plant** *n* WATER POLL Rechenanlage *f*

screenings *n pl* WASTE Siebrest *m*, Siebrückstand *m*, WATER POLL Rechengut *nt*; ~ **bale press** *n* WATER POLL Rechengutpresse *f*; ~ **classifier** *n* WATER POLL Rechengutklassierer *m*

screening: ~ **test** *n* ANALYSIS Screening-Test *m*; ~ **wall** *n* NOISE Abschirmwand *f*

screen: ~ **loss** *n* WATER POLL Rechenverlust *m*; ~ **loss of head** *n* WATER POLL Rechengefällsverlust *m*; ~ **overflow** *n* WASTE Siebüberlauf *m*; ~ **underflow** *n* WASTE Siebdurchgang *m*

screw: ~ **compressor** *n* WASTE Schneckenverdichter *m*; ~ **conveyor** *n* AIR, SOIL, WASTE Förderschnecke *f*, Schneckenförderer *m*; ~ **extruder** *n* WASTE Schneckenextruder *m*; ~ **feeder** *n* WASTE Schneckendosiereinheit *f*, Schneckendosierer *m*; ~ **pump** *n* MARINE Schneckenpumpe *f*, Schraubenpumpe *f*, Schraubenschaufler *m*, Wasserförderschnecke *f*, WATER POLL Schraubenschaufler *m*; ~-**type pump** *n* WATER SUPP Schraubenpumpe *f*

SCR: ~ **reactor** *n* AIR, WASTE SKR-Reaktor *m* (*selektiv-katalytischer Reduktionsreaktor*)

scrub *vt* AIR berieseln, waschen, SOIL, WASTE, WATER SUPP waschen

scrubbed: ~ **gas** *n* AIR Reingas *nt*

scrubber *n* AIR Wäscher *m*; ~ **clogging** *n* AIR Wäscherverstopfung *f*; ~ **plugging** *n* AIR Wäscherverstopfung *f*; ~ **tower** *n* AIR Gaswaschturm *m*, Turmwäscher *m*

scrubbing *n* AIR, WASTE *nuclear fuel reprocessing* Wäsche *f*; ~ **area design** *n* AIR *without fittings, with random movable fitted structure* Waschraumgestaltung *f*; ~ **liquid droplet** *n* AIR Waschflüssigkeitstropfen *m*; ~ **liquid jet** *n* AIR Waschflüssigkeitsstrahl *m*; ~ **liquid tank** *n* AIR Waschflüssigkeitstank *m*; ~ **stages** *n pl* AIR Waschstufen *f pl*; ~ **tower** *n* AIR Gaswaschturm *m*, Turmwäscher *m*; ~ **vane** *n* AIR Waschflügel *m*; ~ **zone** *n* AIR Waschzone *f*

SCS *abbr AmE* (*stationary container system*) WASTE, WATER SUPP Bringbehältersystem *nt*

scum *n* WATER POLL Schwimmschlamm *m*; ~ **board** *n* WATER POLL Schwimmstoffabweiser *m*, Tauchwand *f*; ~ **formation** *n* WATER POLL Schwimmschlammbildung *f*; ~ **removal** *n* WATER POLL Schwimmschlammentfernung *f*; ~ **scraper** *n* WATER POLL Schwimmschlammräumer *m*; ~ **skimmer** *n* WATER POLL Schwimmschlammräumer *m*

SDA *abbr* (*spray dry absorber*) AIR Sprühtrocknungswäscher *m*

SDI *abbr* (*sludge density index*) WATER POLL Schlammdichteindex *m*

sea *n* MARINE See *f*; ~-**based** *adj* MARINE meeresbürtig; ~-**based pollution** *n* MARINE meeresbürtige Verschmutzung *f*; ~-**going ship** *n* MARINE Hochseeschiff *nt*, Seeschiff *nt*; ~-**going vessel** *n* MARINE Hochseeschiff *nt*, Seeschiff *nt*

seal 1. *n* SOIL, WASTE Dichtung *f*; **2.** *vt* MARINE, NOISE, SOIL, WASTE, WATER POLL abdecken, abdichten, versiegeln

seal: ~-**bearing surface** *n* WASTE Dichtungsauflager *nt*

sealed *adj* MARINE, SOIL, WASTE *stopped-up* abgedichtet, *locked* abgeschlossen, versiegelt, *barred, bolted* verschlossen; ~ **ceiling** *n* NOISE geschlossene Decke *f*; ~ **wall** *n* NOISE geschlossene Wand *f*

sea: ~ **level** *n* MARINE Meeresspiegel *m*

sealing *n* GEN Abdeckung *f*, Abdichtung *f*, Versiegelung *f*, *of hazardous waste* Einkapselung *f*; ~ **of the bottom** *n* SOIL Sohleabdichtung *f*; ~ **layer** *n* WASTE, WATER SUPP Abdeckschicht *f*, Abdeckungsschicht *f*; ~ **liquid** *n* ANALYSIS Sperrflüssigkeit *f*; ~ **measures** *n pl* WASTE, WATER SUPP Abdichtungsmaßnahmen *f pl*; ~ **properties** *n pl* WASTE Dichtwirkung *f*; ~ **sheeting** *n* SOIL Dichtungsfolie *f*

SEALOSAFE: ~ **method**® *n BrE* WASTE SEALOSAFE-Verfahren® *nt*; ~ **process**® *n BrE* WASTE SEALOSAFE-Verfahren® *nt*

seasonal: ~ **heat storage** *n* RENEW ENERG *solar energy* saisonaler Wärmespeicher *m*

seawater *n* MARINE, RENEW ENERG, WATER SUPP Meerwasser *nt*, Seewasser *nt*; ~**-in-crude oil emulsion** *n* AIR Seewasser-Rohöl-Emulsion *f*; ~ **intrusion** *n* MARINE Meerwassereinbruch *m*; ~ **recovery** *n* WATER SUPP Meereswassergewinnung *f*

seaweed *n* MARINE Algen *f pl*

seaworthiness *n* MARINE Seetüchtigkeit *f*

secondary *adj* GEN nachgeschaltet, sekundär; ~ **activated sludge tank** *n* WATER POLL nachgeschaltetes Belebtbecken *nt*; ~ **adsorber** *n* AIR Sekundäradsorber *m*; ~ **adsorption** *n* AIR Sekundäradsorption *f*, WATER POLL *of polymer in flocculation* sekundäre Adsorption *f*; ~ **air** *n* AIR, WATER POLL Sekundärluft *f*; ~ **air injection** *n* AIR *catalyst* Sekundärlufteinblasung *f*; ~ **aluminium** *BrE*, ~ **aluminum** *AmE n* WASTE Sekundäraluminium *nt*; ~ **catalyst** *n* AIR nachgeschalteter Katalysator *m*; ~ **combustion chamber** *n* (*SCC*) WASTE Nachverbrennungskammer *f*; ~ **cosmic radiation** *n* AIR kosmische Sekundärstrahlung *f*; ~ **digestion tank** *n* WATER POLL Nachfaulbecken *nt*, Nachfaulraum *m*; ~ **downpipe** *n* AIR *packed-bed filter* Sekundärfallrohr *nt*; ~ **energy source** *n* RENEW ENERG *electricity* Sekundärenergieträger *m*; ~ **ion mass spectrometry** *n* (*SIMS*) ANALYSIS Sekundärionen-Massenspektrometrie *f* (*SIMS*); ~ **loading** *n* AIR *of waste gas* Sekundärbeladung *f*; ~ **packaging** *n* MANAGE, WASTE Umverpackung *f*; ~ **particle separator** *n* AIR nachgeschalteter Partikelabsorber *m*, nachgeschalteter Staubabsorber *m*; ~ **plastics recycling** *n* WASTE Kunststoffverwertung *f* der zweiten Stufe; ~ **polishing pond** *n* WATER POLL nachgeschalteter Schönungsteich *m*; ~ **precipitator** *n* AIR Nachabscheider *m*; ~ **raw material** *n* WASTE Sekundärrohstoff *m*, wiedergewonnener Rohstoff *m*; ~ **reaction time** *n* WATER POLL Nachreaktionszeit *f*; ~ **recycling** *n* WASTE Sekundärrecycling *nt*; ~ **resource** *n* WASTE Wertstoff *m*; ~ **resource container** *n* WASTE Wertstoffcontainer *m*; ~ **resource content** *n* WASTE Wertstoffgehalt *m*; ~ **resource for packaging** *n* WASTE Verpackungswertstoff *m*; ~ **resource fraction** *n* WASTE Wertstofffraktion *f*; ~ **resource quality** *n* WASTE Wertstoffqualität *f*; ~ **resources bale** *n* WASTE Wertstoffballen *m*; ~ **settlement** *n* SOIL, WASTE, WATER SUPP sekundäre Setzung *f*; ~ **settling tank** *n* WATER POLL Nachklärbecken *nt*; ~ **sieving** *n* WASTE, WATER POLL Sekundärsiebung *f*; ~ **sludge** *n* WATER POLL Sekundärschlamm *m*; ~ **sound** *n* NOISE Reflexionsschall *m*; ~ **tertiary pond** *n* WATER POLL nachgeschalteter Schönungsteich *m*; ~ **treatment** *n* AIR Nachbehandlung *f*; ~ **trickling filter** *n* WATER POLL nachgeschalteter Tropfkörper *m*; ~ **waste avoidance** *n* WASTE Sekundärabfallvermeidung *f*; ~ **waste gas treatment** *n* AIR nachgeschaltete Abgasreinigung *f*; ~ **wastewater treatment** *n* WATER POLL *biological* Abwasseraufbereitung *f* der zweiten Stufe, biologische Nachreinigung *f*

secretariat *n* POLICY Sekretariat *nt*

sector *n* POLICY Fach *nt*

sectoral: ~ **approach** *n* POLICY sektoraler Ansatz *m*; ~ **authority** *n* POLICY Fachbehörde *f*; ~ **division** *n* POLICY Fachabteilung *f*; ~ **unit** *n* POLICY Fachbehörde *f*

sector: ~ **sampler** *n* ANALYSIS Sektor-Sampler *m*

security *n* SAFETY Sicherung *f*; ~ **measures** *n pl* SAFETY Sicherungsmaßnahmen *f pl*; ~ **system** *n* SAFETY Sicherungssystem *nt*

sediment 1. *n* GEN Ablagerung *f*, Sediment *nt*; **2.** *vt* AIR, WATER POLL sedimentieren

sedimentable *adj* AIR, ANALYSIS, WATER POLL, WATER SUPP sedimentierbar; ~ **material** *n* AIR sedimentierbarer Stoff *m*

sedimentation *n* AIR, ANALYSIS, WATER POLL, WATER SUPP Sedimentation *f*, Sedimentablagerung *f*; ~ **centrifuge** *n* WATER POLL Sedimentierzentrifuge *f*; ~ **curve** *n* WATER POLL Absetzkurve *f*; ~ **glass** *n* ANALYSIS Absetzglas *nt*; ~ **pond** *n* WASTE, WATER SUPP Absetzteich *m*; ~ **tank** *n* WASTE Absetzbecken *nt*, WATER POLL *wastewater clarification* Absetzbecken *nt*, Sedimentationsbecken *nt*, WATER SUPP Absetzbecken *nt*

sediment: ~ **discharge** *n* WATER SUPP Sedimentablagerung *f*

sedimented *adj* AIR, ANALYSIS, WATER POLL, WATER SUPP sedimentiert; ~ **sludge** *n* WATER POLL sedimentierter Schlamm *m*

sediment: ~ **sampling** *n* ANALYSIS Sedimentprobenahme *f*

seeding: ~ **sludge** *n* WASTE, WATER SUPP Impfschlamm *m*

seep *vt* MARINE, WASTE, WATER POLL, WATER SUPP aussickern, durchsickern, sickern, versickern

seepage *n* MARINE, WASTE, WATER POLL, WATER SUPP Aussickern *nt*, Durchsickern *nt*, Versickerung *f*, Versickern *nt*; ~ **warning device** *n* WATER POLL Leckwarngerät *nt*; ~ **warning installation** *n* WATER POLL Leckwarnanlage *f*; ~ **water** *n* WASTE, WATER SUPP Sickerwasser *nt*

Seger: ~ **cone** *n* ANALYSIS Seger-Kegel *m*

segment: ~ **baffle** *n* AIR Segmentleitblech *nt*

segregate *vt* AIR, WASTE, WATER POLL, WATER SUPP abscheiden

segregation *n* AIR, WASTE, WATER POLL, WATER SUPP Abscheidung *f*; ~ **berm** *n* WASTE, WATER SUPP Berme *f*

selective: ~ **catalytic** *adj* AIR selektiv-katalytisch; ~ **catalytic reduction** *n* (*SCR*) AIR, WASTE selektiv-katalytische Reduktion *f* (*SKR*); ~ **catalytic reduction reactor** *n* AIR, WASTE selektiv-katalytischer Reduktionsreaktor *m* (*SKR-Reaktor*); ~ **collection** *n* WASTE getrennte Müllabfuhr *f*, getrennte Müllsammlung *f*; ~ **ion exchange resin** *n* WASTE, WATER SUPP selektives Ionenaustauscherharz *nt*; ~ **non-catalytic reduction** *n* (*SNCR*) WATER SUPP SNCR-Verfahren *nt* (*selektives nicht-katalytisches Verfahren*); ~ **refuse collection** *n* WASTE Getrenntsammlung *f*, getrennte Müllabfuhr *f*, getrennte Müllsammlung *f*, WATER SUPP getrennte Müllsammlung *f*

self: ~**-adjusting floating weir** *n* MARINE selbstnachstellendes Schwimmwehr *nt*; ~**-cleaning** *adj* WATER POLL selbstreinigend; ~**-cleaning centrifugal screen** *n* WATER POLL selbstreinigender Zentrifugalsortierer *m*; ~**-cleaning rotary-drum screen** *n* WATER POLL selbstreinigendes Drehtrommelsieb *nt*; ~ **compaction** *n* WASTE Selbstverdichtung *f*; ~**-draining** *adj* SOIL selbstdränend; ~**-loading refuse vehicle** *n* WASTE automatisch ladendes Müllfahrzeug *nt*, selbstladendes Müllfahrzeug *nt*; ~**-loading vehicle with internal compactor** *n* WASTE automatisch ladendes Müllfahrzeug *nt* mit internem Kompaktor; ~**-monitoring** *n* POLICY, WATER SUPP Selbstüberwachung *f*; ~**-priming centrifugal pump** *n* WASTE, WATER POLL, WATER SUPP selbstansaugende Kreiselpumpe *f*; ~**-propelled skimmer** *n* MARINE motorbetriebenes Abschöpfgerät *nt*; ~**-purification** *n* SOIL, WATER POLL, WATER SUPP Selbstreinigung *f*; ~**-purification force** *n* SOIL, WATER POLL, WATER SUPP *of geological underground* Selbstreinigungskraft *f*

SEM *abbr* ANALYSIS (*scanning electron microscopy*) Raster-

elektronenmikroskopie *f*, (*scanning electron microscope*) Rasterelektronenmikroskop *nt*
semiconductor *n* ANALYSIS, RENEW ENERG *solar cell* Halbleiter *m*; **~ component** *n* RENEW ENERG *solar cell* Halbleiterbauelement *nt*; **~ detector** *n* ANALYSIS *measuring solid state ionization* Halbleiterdetektor *m*
semidry: **~ scrubber** *n* AIR quasitrockener Wäscher *m*
semi-impermeable: **~ layer** *n* WASTE, WATER SUPP halbdurchlässige Schicht *f*
semipermeable *adj* WASTE, WATER POLL, WATER SUPP halbdurchlässig; **~ membrane** *n* WASTE, WATER POLL, WATER SUPP halbdurchlässige Membran *f*, *reverse osmosis* semipermeable Membran *f*
semiportable *adj* ANALYSIS semitragbar
semirotary: **~ pump** *n* WATER SUPP Flügelpumpe *f*
semisolid *adj* WASTE stichfest; **~ combustible waste** *n* WASTE halbfester brennbarer Abfall *m*; **~ hazardous waste** *n* WASTE stichfester Sonderabfall *m*
semitrailer *n* WASTE Sattelschlepperanhänger *m*
semivolatile: **~ chlorinated hydrocarbons** *n pl* SOIL leichtflüchtige chlorierte Kohlenwasserstoffe *m pl*; **~ organic compounds** *n pl* (*SVOC*) SOIL semiflüchtige organische Verbindungen *f pl* (*SVOC*)
semi-wet: **~ sorting** *n* WASTE Halbfeuchttrennung *f*
sensing: **~ device** *n* ANALYSIS Meßfühler *m*, Meßwertgeber *m*, MARINE Meßgerät *nt*
sensitive *adj* AIR, ANALYSIS *methods* empfindlich; **~ detection** *n* ANALYSIS empfindlicher Nachweis *m*
sensitivity *n* AIR, ANALYSIS *of detectors* Empfindlichkeit *f*; **~ to noise** *n* NOISE Lärmempfindlichkeit *f*; **~ to weathering** *n* SOIL *liners* Witterungsempfindlichkeit *f*
sensor *n* ANALYSIS Meßfühler *m*, Sensor *m*; **~ module** *n* ANALYSIS Sensormodul *nt*
sensory: **~ analysis** *n* ANALYSIS Sensorik *f*
separability: **~ test** *n* ANALYSIS, WATER POLL Abscheidbarkeitstest *m*
separable *adj* AIR abscheidbar, SOIL, WASTE abtrennbar, auslesbar, separierfähig, trennbar, WATER POLL abscheidbar, abtrennbar WATER SUPP abtrennbar; **~ roughage** *n* WASTE auslesbarer Ballaststoff *m*
separate 1. *adj* GEN getrennt; **2.** *vt* AIR *disengage* abscheiden, *divide* zerteilen, zertrennen, SOIL abtrennen, WASTE *harmful substances* abtrennen, *detach* ablösen, trennen, *divide* zerteilen, WATER POLL abscheiden, abtrennen, trennen, WATER SUPP *divide* zerteilen, *detach* ablösen; **~ electrolytically** WASTE *dissolved metal in battery* elektrolytisch abscheiden
separate: **~ canalization** *n* WATER POLL Trennkanalisation *f*; **~ collection** *n* WASTE getrennte Müllabfuhr *f*, getrennte Müllsammlung *f*; **~ collection by type** *n* WASTE sortenreine Getrenntsammlung *f*; **~ composting of dewatered sludges** *n* WATER POLL Alleinkompostierung *f* entwässerter Klärschlämme; **~ dewatering** *n* WATER POLL Trennentwässerung *f*; **~ disposal** *n* WASTE getrennte Entsorgung *f*; **~ refuse collection** *n* WASTE Getrenntsammlung *f*, getrennte Müllabfuhr *f*, getrennte Müllsammlung *f*; **~ sewerage system** *n* WATER POLL Einzelkläranlage *f*, getrennte Klärung *f* von Abwässern; **~ solid matter from gas** *phr* AIR Feststoffe und Gase voneinander trennen; **~ storage** *n* WASTE getrennte Lagerung *f*; **~ treatment** *n* WATER POLL Trennverfahren *nt*
separating *n* WASTE Sortieren *nt*; **~ capacity** *n* ANALYSIS Trennleistung *f*; **~ cut** *n* SOIL Trennschnitt *m*; **~ efficiency** *n* AIR Abscheidewirkungsgrad *m*; **~ funnel** *n* ANALYSIS *extraction* Scheidetrichter *m*; **~ liquid** *n* ANALYSIS Trennflüssigkeit *f*
separation *n* AIR, WASTE, WATER POLL, WATER SUPP Abscheidung *f*, Abtrennung *f*, Trennung *f*; **~ by color** *AmE*, **~ by colour** *BrE* *n* WASTE *recycled glass* Trennung *f* nach Farben, farbliche Sortierung *f*, farbliche Trennung *f*; **~ capacity** *n* AIR Abscheideleistung *f*; **~ column** *n* ANALYSIS Strippersäule *f*, Trennsäule *f*; **~ effect** *n* AIR Trennschärfe *f*; **~ layer** *n* WASTE, WATER SUPP Trennschicht *f*; **~ with organic chemicals** *n* WATER POLL Spaltung *f* mit organischen Chemikalien; **~ of phases** *n* AIR, ANALYSIS, WASTE Phasentrennung *f*; **~ process** *n* AIR Trennverfahren *nt*; **~ properties** *n pl* AIR Abscheideeigenschaften *f pl*; **~ rate** *n* WASTE, WATER SUPP Trenngrad *m*; **~ stages** *n pl* WATER SUPP Trennstufen *f pl*; **~ technology** *n* WASTE Sortiertechnik *f*
separator *n* AIR Sichter *m*, MARINE, WASTE, WATER POLL, WATER SUPP Abscheider *m*, Separator *m*
sequencing: **~ batch reactor** *n* WATER POLL Sequencing-Batch-Reaktor *m*
sequential *adj* WATER POLL sequentiell; **~ batch reactor** *n* (*SBR*) SOIL Sequential Batch Reactor *m* (*SBR*); **~ dosage** *n* WATER POLL sequentielle Dosierung *f*
serviceable: **~ life** *n* GEN Standzeit *f*, *of plant* Nutzungsdauer *f*
service: **~ load** *n* WASTE, WATER POLL Nutzlast *f*; **~ mains** *n pl* WASTE, WATER SUPP Anschlußleitung *f*, Hausanschlußleitung *f*; **~ water** *n* RENEW ENERG *heat pump* Brauchwasser *nt*
servomotoric *adj* ANALYSIS servomotorisch
set 1. *vt* ANALYSIS einstellen, SOIL erhärten; **2.** *vi* WASTE, WATER SUPP erstarren
setting *n* GEN Einstellung *f*, SOIL *separating wall* Erhärtung *f*, WASTE Erstarren *nt*; **~ time** *n* WASTE Abbindezeit *f*
settle *vt* WASTE, WATER POLL, WATER SUPP absetzen, klären; **~ out** *vt* WATER POLL, WATER SUPP absetzen
settleability *n* WASTE, WATER POLL, WATER SUPP Absetzbarkeit *f*
settleable *adj* WASTE, WATER POLL, WATER SUPP absetzbar, absetzfähig; **~ macroflocs** *n pl* WATER POLL absetzfähige Makroflocken *f pl*; **~ sludge** *n* WATER POLL absetzbarer Schlamm *m*; **~ solids** *n pl* WATER POLL absetzbare Feststoffe *m pl*; **~ suspended solids** *n pl* WATER POLL absetzbare Schwebstoffe *m pl*
settled *adj* WASTE, WATER POLL, WATER SUPP abgesetzt, sedimentiert; **~ apparent density** *n* ANALYSIS Schüttdichte *f*; **~ sludge** *n* WATER POLL abgesetzter Schlamm *m*; **~ wastewater** *n* WATER POLL abgesetztes Abwasser *nt*
settlement *n* SOIL, WASTE, WATER SUPP Setzung *f*; **~ cushion** *n* WASTE Setzungspolster *nt*; **~ speed** *n* WASTE Setzungsgeschwindigkeit *f*; **~ stoppage** *n* SOIL, WASTE, WATER SUPP Setzungsstillstand *m*
settling *n* AIR, WASTE, WATER POLL, WATER SUPP Absetzen *nt*; **~ basin** *n* WASTE, WATER POLL, WATER SUPP Absetzbecken *nt*; **~ chamber** *n* WASTE, WATER SUPP Absetzkammer *f*; **~ characteristics** *n pl* WASTE, WATER POLL, WATER SUPP Absetzeigenschaften *f pl*; **~ conditions** *n pl* WATER POLL Absetzbedingungen *f pl*; **~ lagoon** *n* WATER POLL Absetzteich *m*; **~ out** *n* AIR, WASTE, WATER POLL, WATER SUPP Absetzen *nt*, Sedimentation *f*; **~ properties** *n pl* WASTE, WATER POLL, WATER SUPP Absetzeigenschaften *f pl*; **~ speed** *n* AIR Sinkgeschwindigkeit *f*; **~ tank** *n* AIR *biological gas scrubbing* Absetzbehälter *m*, WASTE, WATER POLL, WATER SUPP Absetzbecken *nt*; **~ tank with central inlet** *n* WATER POLL Absetzbecken *nt* mit zentralem Einlauf; **~ tank with peripheral inlet** *n* WATER POLL Absetzbecken *nt* mit peripherem Einlauf; **~ test** *n* ANALYSIS Absetzprobe *f*; **~ time** *n* WASTE, WATER POLL, WATER SUPP Absetzzeit *f*; **~ velocity** *n* WASTE, WATER POLL Absetzgeschwindigkeit *f*; **~ vessel** *n* WATER POLL Absetzgefäß *nt*; **~ zone** *n* WATER POLL Absetzzone *f*
sewage *n* WATER POLL, WATER SUPP Abwasser *nt*; **~ lagoon** *n* WATER POLL Abwasserteich *m*; **~ sludge** *n* WASTE, WATER SUPP Klärschlamm *m* (*KS*); **~ treatment plant residue** *n* WASTE Rückstände aus Abwasseranlage *m pl*
sewer *n* WATER POLL Abwasserkanal *m*, Entwässerungskanal *m*
sewerage *n* WATER POLL Kanalisation *f*; **~ connection duty** *n*

POLICY Anschlußgebühr *f*; ~ **system** *n* WATER POLL Kanalnetz *nt*, Kanalisation *f*
sewer: ~ **cleaning** *n* WATER POLL Abwasserkanalreinigung *f*; ~ **system** *n* WATER POLL Kanalnetz *nt*, Kanalisation *f*; ~ **trench** *n* WATER POLL Rohrgraben *m*
SFC *abbr* (*supercritical-fluid chromatography*) ANALYSIS überkritische Flüssig-Chromatographie *f*
SFE *abbr* (*supercritical-fluid extraction*) ANALYSIS überkritische Flüssig-Extraktion *f*
shackle *n* MARINE Bügel *m*, Schäkel *m*, Verbindungsglied *nt*
shade: ~ **of color** *AmE*, ~ **of colour** *BrE n* ANALYSIS Farbton *m*
shadow: ~ **price** *n* POLICY Schattenpreis *m*
shaft *n* AIR, WASTE Schacht *m*; ~**-filling material** *n* WASTE Schachtverfüllmaterial *nt*; ~ **furnace** *n* AIR, WASTE Schachtofen *m*; ~ **rigidity** *n* WASTE Schachtstarrheit *f*
shake *vt* AIR rütteln, ANALYSIS schütteln; ~ **in** *vt* SOIL einrütteln
shaker: ~ **baghouse** *n* AIR Rüttelfilter *nt*; ~ **mechanism** *n* AIR Rüttelmechanismus *m*
shaking *n* AIR Rütteln *nt*, ANALYSIS Schütteln *nt*, WASTE Rütteln *nt*; ~ **screen** *n* AIR *packed-bed filter* Schwingsieb *nt*; ~ **technique** *n* ANALYSIS Schüttelmethode *f*
shallow: ~ **well** *n* WATER SUPP Flachbrunnen *m*
shape *n* WASTE, WATER SUPP Form *f*; ~ **recognition methods** *n pl* WASTE *sorting of plastics* Formerkennungsverfahren *nt*
share: ~ **of different colored glass** *AmE*, ~ **of different coloured glass** *BrE n* WASTE Fremdfarbanteil *m*; ~ **of surface waters** *n* WATER SUPP Oberflächengewässeranteil *m*
sharp: ~**-edged particle** *n* AIR scharfkantiges Partikel *nt*
shear *vt* AIR *liquid into finest droplets* zerreißen, WASTE zerkleinern, zerschneiden; ~ **off** *vt* WASTE abscheren
shearing *n* WASTE Zerkleinern *nt*, Zerschneiden *nt*, Zerreißen *nt*; ~ **action** *n* WATER POLL Scherwirkung *f*; ~ **efficiency** *n* AIR Scherwirkungsgrad *m*; ~ **force** *n* AIR, WATER POLL Scherkraft *f*; ~ **resistance** *n* WATER POLL *of wastewater sludge* Scherfestigkeit *f*
shear: ~ **wool** *n* AIR *filter fibre* Vlies *nt*
sheen *n* MARINE Reflex *m*, Schimmer *m*
sheet: ~ **absorber** *n* NOISE Folienabsorber *m*; ~ **glass** *n* WASTE Flachglas *nt*
sheeting *n* SOIL Folie *f*
sheet: ~ **metal** *n* NOISE Blech *nt*; ~ **piling** *n* WASTE Spundwand *f*
shell *n* AIR Mantel *m*, NOISE Schale *f*; ~**-and-tube condenser** *n* AIR Röhrenkondensator *m* mit Mantel; ~**-and tube heat exchanger** *n* AIR Rohrbündelwärmeaustauscher *m*, Rohrwärmeaustauscher *m* mit Mantel
shielding: ~ **wall** *n* NOISE Abschirmwand *f*
shift *vt* POLICY *tax* überwälzen
ship *n* MARINE Schiff *nt*
shock: ~ **loading** *n* WATER POLL Stoßbelastung *f*
shoreline *n* MARINE Küstenlinie *f*, Uferlinie *f*; ~ **clean-up** *n* MARINE Säuberung *f* der Küstenlinie
shore: ~ **protection** *n* WATER SUPP Küstenschutz *m*; ~ **reception facility** *n* MARINE Entsorgungsanlage *f* an Land
short: ~**-glass-fiber reinforced thermoplast** *AmE*, ~**-glass-fibre reinforced thermoplast** *BrE n* WASTE, WATER SUPP kurzglasfaserverstärktes Thermoplast *nt*; ~**-term** *adj* WATER SUPP kurzzeitig; ~**-term exposure limit** *n* (*STEL*) ANALYSIS Kurzzeit-Exposition *f* (*STEL*); ~**-term measures** *n pl* AIR, MARINE, NOISE, SOIL, WATER POLL, WATER SUPP Kurzzeitmaßnahmen *f pl*; ~**-term monitoring** *n* ANALYSIS Kurzzeitüberwachung *f*; ~**-term sampling** *n* ANALYSIS Kurzzeitprobenahme *f*; ~**-term storage** *n* RENEW ENERG Kurzzeitspeicher *m*; ~ **and thick cyclone** *n* AIR kurzgebauter dicker Zyklon *m*; ~**-wave** *adj* AIR kurzwellig; ~**-wave incoming radiation** *n* AIR kurzwellige Sonnenstrahlung *f*; ~**-wave listener** *n* (*SWL*) NOISE Kurzwellenhörer *m*; ~**-wave sunlight** *n* AIR kurzwelliges Sonnenlicht *nt*
shot: ~ **cleaning system** *n* AIR Kugelregensystem *nt*
shovel *n* MARINE Löffel *m*, Schaufel *f*
shredded *adj* WASTE geschreddert, zerkleinert; ~ **plastics** *n pl* WASTE Kunststoffschnitzel *nt*, geschredderte Kunststoffe *m pl*; ~ **solid waste** *n* WASTE zerkleinerter Festabfall *m*
shredder *n* WASTE Schredder *m*, Zerkleinerungsmaschine *f*; ~ **fluff** *n* WASTE Schreddermüll *m*; ~ **remains** *n* WASTE Schreddermüll *m*; ~ **residue** *n* WASTE Schreddermüll *m*; ~ **waste** *n* WASTE Schreddermüll *m*
shredding *n* WASTE Schreddern *nt*, Zerkleinerung *f*; ~ **machine** *n* WASTE Schredderanlage *f*, Zerkleinerungsmaschine *f*; ~ **tower** *n* WASTE Zerkleinerungsturm *m*
shrinking: ~ **behavior** *AmE*, ~ **behaviour** *BrE n* SOIL, WASTE Schrumpfverhalten *nt*
shroud *n* SOIL Deckband *nt*
shutoff: ~ **device** *n* WASTE, WATER SUPP Absperrorgan *nt*; ~ **valve** *n* AIR Sperrventil *nt*
side: ~ **area** *n* WASTE, WATER SUPP Seitenbereich *m*; ~**-channel compressor** *n* SOIL *soil vapour extraction* Seitenkanalverdichter *m*; ~ **reaction** *n* AIR Nebenreaktion *f*
sideways: ~**-looking airborne radar** *n* (*SLAR*) MARINE Seitensichtradar *nt*
sievable *adj* WASTE, WATER POLL siebbar
sieve 1. *n* GEN Sieb *nt*; **2.** *vt* GEN sieben
sieve: ~ **analysis** *n* ANALYSIS Siebanalyse *f*; ~ **fraction** *n* WASTE Siebfraktion *f*, Korngrößenbereich *m*; ~ **grate** *n* AIR Siebrost *m*; ~ **installation** *n* WASTE Siebanlage *f*; ~ **plate** *n* AIR Siebboden *m*; ~ **residue** *n* WASTE Siebrest *m*; ~ **set** *n* WATER POLL Siebsatz *m*; ~ **size** *n* WASTE Siebgröße *f*
sieving *n* GEN Sieben *nt*, Siebung *f*
sift *vt* MARINE, WASTE, WATER POLL, WATER SUPP absieben, durchsieben, sieben
sifting *n* GEN Sieben *nt*, MARINE, WATER POLL Durchsieben *nt*; ~ **belt** *n* MARINE Lochband *nt*
silage *n* WATER POLL Silage *f*, Silolagerung *f*
silencer *n BrE* (*cf muffler AmE*) NOISE Dämpfer *m*, Schalldämpfer *m*
silica: ~ **dust** *n* SAFETY Kieselerdestaub *m*; ~ **gel** *n* AIR, WATER POLL *adsorbent* Kieselgel *nt*, Silikagel *nt*
silicate *n* WASTE Silikat *nt*
sill *n* WASTE, WATER SUPP Grundschwelle *f*, Überfallschwelle *f*
silo *n* WASTE, WATER POLL Silo *m*; ~ **storage** *n* WASTE *plastics* Silieren *nt*
silt 1. *n* WATER POLL Schlick *m*; **2.** ~ **up** *vi* WATER POLL verschlammen
silt: ~ **deposition** *n* WATER SUPP Schlickablagerung *f*
silting: ~**-up** *n* WATER POLL Verschlammung *f*
silt: ~ **trap** *n* WATER POLL Schlammfang *m*
simplified: ~ **procedure** *n* POLICY einfaches Verfahren *nt*
SIMS *abbr* (*secondary ion mass spectrometry*) ANALYSIS SIMS (*Sekundärionen-Massenspektrometrie*)
simulation *n* MANAGE Simulation *f*; ~ **of river systems** *n* MANAGE Simulation *f* von Flußsystemen
simultaneous *adj* ANALYSIS, RENEW ENERG, WATER POLL gleichzeitig, simultan; ~ **denitrification** *n* WATER POLL simultane Denitrifikation *f*; ~ **measurement** *n* ANALYSIS Simultanmessung *f*; ~ **operation** *n* ANALYSIS Simultanbetrieb *m*; ~ **readout** *n* ANALYSIS simultane Anzeige *f*
single: ~**-blade machine** *n* RENEW ENERG *wind turbine* Einflügler *m*; ~**-crystalline** *adj* RENEW ENERG monokristallin; ~**-crystalline silicon cell** *n* RENEW ENERG *photovoltaic* monokristalline Siliziumzelle *f*; ~**-cycle scrubber** *n* AIR Einkreislaufwäscher *m*; ~**-event noise level** *n* NOISE Ereignispegel *m*; ~**-leaf** *adj* NOISE einschalig; ~**-material system** *n* WASTE Einstoffsystem *nt*; ~ **measurement** *n* ANALYSIS Einzelmessung *f*; ~**-phase jet** *n* AIR *jet scrubber* Einstoffdüse *f*; ~**-point collection element** *n* WASTE punktförmiges Fassungselement *nt*; ~**-screw extruder** *n* WASTE Einschneckenextruder *m*; ~**-sheet** *adj* NOISE *encapsulation* einschalig; ~**-stage** *adj*

AIR *axial fans*, WASTE, WATER POLL, WATER SUPP einstufig; **~-stage activated sludge process** *n* WATER POLL einstufiges Belebungsverfahren *nt*; **~-stage digestor** *n* WATER POLL einstufiger Faulraum *m*; **~-stage fluidized-bed adsorber** *n* AIR einstufiger Wirbelschichtadsorber *m*; **~-stage peripheral pump** *n* WASTE, WATER POLL, WATER SUPP einstufige Peripheralpumpe *f*; **~-unit cyclone** *n* AIR Einzelzyklon *m*, einzelliger Zyklon *m*; **~-wall corrugated fiberboard** *AmE*, **~-wall corrugated fibreboard** *BrE n* WASTE Einzelschichtwellfaserplatte *f*

sink 1. *n* WASTE *sump* Sammelbehälter *m*, *drainhole* Einlaufschacht *m*, Ablauf *m*, *in kitchen* Ausguß *m*, WATER POLL Ablauf *m*, WATER SUPP *drainhole* Einlaufschacht *m*, Ausguß *m*, Gully *m*, Ablauf *m*; **2.** *vt* WASTE *landfill gas well* abteufen

sink: **~-float flotation** *n* WATER SUPP Schwimm-Sink-Flotation *f*

sinking 1. *adj* WASTE, WATER SUPP *groundwater level* sinkend; **2.** *n* WASTE, WATER SUPP *of landfill base* Senkung *f*

sinking: **~ agent** *n* MARINE Sinkgut *nt*, Sinkmaterial *nt*; **~ material** *n* AIR, WATER POLL *water-borne sediments* Sinkstoffe *m pl*; **~ matter** *n* AIR Sinkstoffe *m pl*, WATER POLL Absinkstoff *m*, Sinkstoffe *m pl*

sinter *vt* SOIL, WASTE sintern

sintering *n* SOIL, WASTE Sintern *nt*

sinter: **~ oven** *n* SOIL *pellet incineration* Sinterband *nt*

siphon: **~ off** *vt* MARINE, WASTE, WATER POLL, WATER SUPP abschöpfen

siphon: **~ head** *n* WASTE, WATER SUPP Drückerkopf *m*

siphoning: **~ off** *n* MARINE, WASTE, WATER POLL Abschöpfen *nt*

site *n* GEN Standort *m*; **~ assessment** *n* ANALYSIS *contaminated site* Standortbewertung *f*; **~ characteristics** *n pl* SOIL Standortgegebenheiten *f pl*; **~ characterization** *n* RENEW ENERG, WASTE, WATER POLL Ortskennzeichnung *f*; **~ closure** *n* WASTE Deponieverschluß *m*; **~ criteria** *n* AIR, SOIL, WATER SUPP Standortkriterien *nt pl*; **~-dependent** *adj* RENEW ENERG *renewable energy technologies* ortsabhängig; **~-dependent emission source** *n* AIR ortsfeste Emissionsquelle *f*; **~ level** *n* WASTE, WATER SUPP Standortebene *f*; **~ operation** *n* WASTE Deponiebetrieb *m*; **~ parameters** *n pl* ANALYSIS Standortparameter *m pl*; **~-related** *adj* GEN standortbezogen; **~-related analysis** *n* ANALYSIS standortbezogene Untersuchung *f*; **~-related criteria** *n pl* ANALYSIS, RENEW ENERG standortbezogene Kriterien *nt pl*; **~-related noise** *n* NOISE standortbezogener Lärm *m*; **~-specific** *adj* GEN standortspezifisch; **~-specific criteria** *n pl* RENEW ENERG standortspezifische Kriterien *nt pl*; **~-specific installation** *n* AIR, ANALYSIS, NOISE, RENEW ENERG, SOIL, WATER POLL standortspezifische Anlage *f*; **~-specific properties** *n pl* RENEW ENERG standortspezifische Eigenschaften *f pl*; **~ targeted for decontamination** *n pl* SOIL Sanierungsgelände *nt*; **~ type** *n* SOIL Standorttyp *m*

skim: **~ off** *vt* WASTE, WATER POLL, WATER SUPP abschöpfen

skimmer *n* MARINE Abschöpfeinrichtung *f*, Skimmer *m*, WATER POLL Abschöpfer *m*

skimming: **~ barge** *n* MARINE Abschöpfbarke *f*; **~ barrier** *n* MARINE Abschöpfölsperre *f*; **~ head** *n* MARINE Abschöpfkopf *m*, Saugkopf *m*; **~ off** *n* MARINE, WATER POLL Abschöpfen *nt*; **~ tank** *n* MARINE Fettabscheider *m*, Fettfänger *m*, WATER POLL Ölabscheider *m*

skip *n* MARINE Kippkübel *m*, WASTE Kipper *m*, Abfuhrwagen *m*; **~ lorry** *n BrE* (*cf skip truck AmE*) MARINE Muldenkippwagen *m*; **~ truck** *n AmE* (*cf skip lorry BrE*) MARINE Muldenkippwagen *m*

skirt *n* MARINE Einfassung *f*, Ummantelung *f*

slag *n* WASTE Schlacke *f*; **~-containing** *adj* WASTE schlackenhaltig; **~ discharge** *n* WASTE Schlackenaustrag *m*; **~ from waste incineration plant** *n* WASTE MVA-Schlacke *f*, Müllverbrennungsschlacke *f*; **~ generation** *n* WASTE Schlackenanfall *m*

slagging *n* AIR, WASTE, WATER POLL Verschlackung *f*; **~ in combustion chamber** *n* WASTE Verschlackung *f* im Feuerraum

slag: **~ recovery** *n* WASTE Schlackeverwertung *f*; **~ removal** *n* AIR, WASTE Entschlackung *f*; **~ removal equipment** *n* WASTE Entschlackungseinrichtung *f*; **~-removing plant** *n* WATER POLL Entschlackungseinrichtung *f*; **~ tank** *n* SOIL, WASTE Schlackesammelbehälter *m*

slaked: **~ lime** *n* AIR, SOIL, WASTE Gips *m*, WATER POLL Kalkhydrat *nt*; **~ lime mixing tank** *n* AIR Kalkmilch-Anrührbehälter *m*; **~ lime suspension** *n* AIR Kalkmilchsuspension *f*

SLAR *abbr* (*sideways-looking airborne radar*) MARINE Seitensichtradar *nt*

slash: **~-and-burn agriculture** *n* AIR *global carbon dioxide budget* Brandrodungslandwirtschaft *f*

sliding: **~ floor** *n* WASTE Schiebeboden *m*

sling *n* MARINE Stropp *m*, WASTE, WATER SUPP Schlinge *f*

slit *vt* WASTE, WATER SUPP schlitzen

SLM *abbr* (*sound level meter*) ANALYSIS, NOISE Schallpegelmesser *m*

slope *n* WASTE, WATER SUPP Böschung *f*, Gefälle *nt*, Hang *m*; **~ landfill** *n* WASTE, WATER SUPP Böschung *f*; **~ leachate** *n* WASTE Hangsickerwasser *nt*; **~ leachate emission** *n* WASTE, WATER SUPP Hangsickerwasseraustritt *m*; **~ recess** *n* WASTE Hangeinschnitt *m*; **~ sealing** *n* SOIL Böschungsdichtung *f*; **~ water** *n* WASTE, WATER SUPP Hangwasser *nt*

slop: **~ tank** *n* WASTE Altöltank *m*

slotted: **~ panel** *n* NOISE Schlitzplatte *f*

sloughed: **~ biomass** *n* WATER POLL Ausschwemmung von Biomassen *f*

sloughing *n* WATER POLL Ausschwemmung *nt*; **~ of biomass** *n* WATER POLL Ausschwemmung *f* von Biomassen

slow: **~ composting** *n* WASTE langsame Kompostierung *f*; **~ fermentation** *n* WASTE, WATER SUPP langsame Gärung *f*; **~-process filter** *n* WATER SUPP Langsamfilter *nt*; **~-process filtrate** *n* WATER SUPP Langsamfiltrat *nt*; **~-process sand filter** *n* WATER SUPP Langsamsandfilter *nt*; **~-process sand filter bed** *n* WATER SUPP Langsamsandfilterschicht *f*; **~-sand filter** *n* WATER POLL Langsamsandfilter *nt*; **~-sand filtration** *n* WATER POLL Langsamsandfiltration *f*

sludge *n* AIR, WASTE, WATER POLL, WATER SUPP Schlamm *m*; **~ accumulation** *n* WASTE, WATER POLL Schlammanfall *m*; **~ age** *n* WATER POLL Schlammalter *nt*; **~ barging** *n* WATER POLL Schlammverklappung *f*; **~ blanket** *n* WATER POLL Schlammdecke *f*; **~ blanket filter** *n* WATER POLL Schlammdeckenfilter *nt*; **~ cake** *n* WATER POLL Schlammkuchen *m*; **~ characteristics** *n pl* WATER POLL Schlammeigenschaften *f pl*, Klärschlammeigenschaften *f pl*; **~ concentration** *n* WATER POLL Schlammkonzentration *f*; **~ conditioning** *n* WATER POLL Schlammkonditionierung *f*; **~ contact clarifier** *n* WATER POLL Schlammkontaktflockungsbecken *nt*, Schnellflockungsreaktor *m*; **~ containing ash** *n* WATER POLL Aschenschlamm *m*; **~ density index** *n* (*SDI*) WATER POLL Schlammdichteindex *m*; **~ deposit** *n* WASTE, WATER POLL Schlammablagerung *f*; **~ dewatering** *n* WATER POLL Schlammentwässerung *f*; **~ dewatering equipment** *n* WATER POLL Schlammentwässerungsanlage *f*; **~ digester** *n* WATER POLL Faulbecken *nt*, Faulbehälter *m*, Faulraum *m*; **~ digestion** *n* WATER POLL Schlammfaulung *f*; **~ disposal** *n* WATER POLL Schlammbeseitigung *f*, Schlammentsorgung *f*; **~ drying** *n* WATER POLL Schlammtrocknung *f*; **~ drying bed** *n* WATER POLL Schlammtrockenbett *nt*; **~ extraction** *n* WATER POLL Schlammaustragung *f*; **~ formation** *n* AIR Schlammbildung *f*; **~-forming** *adj* WASTE, WATER POLL schlammförmig; **~-forming residue** *n* WASTE, WATER POLL schlammförmiger Rückstand *m*; **~ freezing** *n* WATER POLL Schlammgefrieren *nt*; **~ gas** *n* WASTE, WATER POLL, WATER SUPP Faulgas *nt*; **~ generation** *n* WATER POLL Schlammentstehung *f*; **~ handling** *n* WATER POLL

Schlammhandhabung f; ~ **hopper** n WATER POLL Schlammtrichter m; ~ **incinerator** n WATER POLL Schlammverbrennungsanlage f; ~ **index** n WATER POLL Schlammindex m; ~ **lagoon** n WATER POLL Schlammteich m; ~ **landfilling** n WATER POLL Schlammdeponierung f; ~ **liquor** n WATER POLL Faulschlammwasser nt, Schlammwasser nt, WATER SUPP Schlammwasser nt; ~ **pond** n WATER POLL Schlammteich m; ~ **processing** n WATER POLL Schlammaufbereitung f; ~ **pump works** n WATER POLL Schlammpumpwerk nt; ~ **recovery** n WATER POLL Schlammverwertung f; ~ **recycle** n WATER POLL Rücklaufschlamm m; ~ **recycling** n WATER POLL Recycling nt von Schlämmen, Schlammverwertung f; ~ **removal** n WATER POLL Schlammbeseitigung f; ~ **residence time** n (*SRT*) WATER POLL Schlammverweilzeit f; ~ **return** n WATER POLL Schlammrücklauf m; ~ **return flow** n WATER POLL Rücklaufschlammfluß m; ~ **siphon** n WATER POLL Ejektorbelüfter m, Schlammheber m, Strahlbelüfter m; ~ **stabilization** n WATER POLL Schlammstabilisierung f; ~ **sump** n WATER POLL Schlammfang m, Schlammsammelbehälter m; ~ **thickening** n WATER POLL Schlammeindickung f; **~-to-oil reaction** n WATER POLL Schlamm-Öl-Reaktion f; **~-to-oil reactor** n (*STOR*) WATER POLL Schlamm-Öl-Reaktor m (*S.-Ö.-Reaktor*); ~ **treatment** n WATER POLL Schlammaufbereitung f, Schlammbehandlung f; ~ **type** n WATER POLL Schlammart f; ~ **worms** n pl WATER POLL Schlammröhrenwürmer m pl, Tubificiden f pl

slug: **~-bail test** n ANALYSIS Slug-Bail-Test m

sluice n WATER SUPP Schleuse f

sluiceway n RENEW ENERG *tidal power plant* künstlicher Schleusenkanal m

slurry n WASTE Gülle f, Jauche f, WATER POLL Flüssigschlamm m, Jauche f, Naßschlamm m, Schlamm m, WATER SUPP Gülle f, Jauche f

small: ~ **landfill** n WASTE Kleindeponie f; **~-scale installation** n RENEW ENERG *wind, solar* Kleinanlage f

smog n AIR Smog m; ~ **alarm** n AIR Smogalarm m; ~ **area** n AIR Smoggebiet nt; ~ **period** n AIR Smogperiode f; ~ **risk** n AIR Smoggefahr f; ~ **warning** n AIR Smogwarnung f; ~ **weather conditions** n pl AIR Smogwetterlage f

smoke n AIR Rauch m, *containing solid particulates* Ruß m, SAFETY Rauch m; ~ **abatement** n AIR Rauchbekämpfung f; ~ **buildup** n AIR Rußbildung f; ~ **concentration** n AIR Rauchkonzentration f; ~ **control system** n AIR, SAFETY Rauchabzugsanlage f; ~ **density meter** n ANALYSIS Rauchdichtemesser m; ~ **emissions** n pl AIR Rußemissionen f pl, Rauchemissionen f pl; ~ **filter** n AIR Rauchgasfilter nt; ~ **and heat exhaust installation** n SAFETY Rauch- und Hitzeabzugsanlage f; ~ **plume** n AIR Rauchfahne f, Rauchwolke f; ~ **shade method** n ANALYSIS *particulate concentration* Rauchschattenverfahren nt; ~ **stack** n AIR Kamin m, Schornstein m, RENEW ENERG Kamin m, WASTE Kamin m, Schornstein m; ~ **streamer** n AIR Rauchfahne f

snap: ~ **sample** n ANALYSIS Stichprobe f, Zufallsstichprobe f; ~ **sample container** n ANALYSIS Probenbehälter m, Probengefäß nt, Stichprobenbehälter m

SNCR abbr (*selective non-catalytic reduction*) WATER SUPP SNCR-Verfahren nt (*selektives nicht-katalytisches Verfahren*)

SNG abbr (*synthetic natural gas*) RENEW ENERG SNG (*Erdgasaustauschgas*)

SO abbr (*sulfur oxide AmE, sulphur oxide BrE*) AIR, WASTE SO (*Schwefeloxid*)

SO_2 abbr (*sulfur dioxide AmE, sulphur dioxide BrE*) AIR, WASTE SO_2 (*Schwefeldioxid*)

SO_3 abbr (*sulfur trioxide AmE, sulphur trioxide BrE*) AIR, WASTE SO_3 (*Schwefeltrioxid*)

$(SO_4)_x$ abbr (*sulfur tetraoxide AmE, sulphur tetraoxide BrE*) AIR, WASTE $(SO_4)_x$ (*Schwefeltetroxid*)

SO_x abbr (*sulfur oxides AmE, sulphur oxides BrE*) AIR, WASTE SO_x (*Schwefeloxide*)

socket n WATER POLL, WATER SUPP *pipe end extension* Muffe f, Rohrmuffe f

soda n WASTE *in glass recycling*, WATER POLL Soda nt; ~ **ash** n WATER SUPP Natriumcarbonatanhydrid nt

sodium n (*Na*) AIR, WATER SUPP Natrium nt (*Na*); ~ **chloride** n MARINE, WATER POLL Natriumchlorid nt; ~ **ion** n WATER SUPP Natriumion nt; ~ **trioxocarbonate** n MARINE, WATER POLL Natriumchlorid nt, WATER SUPP Natriumcarbonatanhydrid nt

SOFC abbr (*solid oxide fuel cell*) AIR Oxidkeramik-Brennstoffzelle f

soften vt WASTE, WATER SUPP enthärten

softened adj WASTE, WATER SUPP enthärtet, weichmacherhaltig; ~ **water** n WATER SUPP enthärtetes Wasser nt

softener n WASTE Weichmacher m, WATER SUPP Enthärter m, Enthärtungsmittel nt, Weichmacher m

softening n WASTE Weichmachen nt, WASTE, WATER SUPP Enthärtung f, Weichmachen nt; ~ **plant** n WATER SUPP Enthärtungsanlage f

soft: ~ **instrument** n POLICY *information, labelling* weiches Instrument nt; ~ **tourism** n MANAGE sanfter Tourismus m

software n ANALYSIS Software f; **~-controlled** adj ANALYSIS softwaregesteuert

soft: ~ **water** n WATER SUPP enthärtetes Wasser nt, weiches Wasser nt

soil n SOIL, WASTE, WATER SUPP Boden m, Erdboden m, Erdreich nt; ~ **acidity** n SOIL Bodensäure f; ~ **aeration** n SOIL Bodendurchlüftung f; ~ **ameliorant** n SOIL *soil conditioner* Bodenverbesserer m, Bodenverbesserungsmittel nt; ~ **analysis** n ANALYSIS, SOIL Bodenanalyse f, Bodenuntersuchung f; ~ **area** n SOIL Bodenbereich m; ~ **break-up** n SOIL Bodenumbruch m; ~ **characteristics** n pl SOIL, WASTE, WATER SUPP Bodeneigenschaften f pl; ~ **clean-up** n SOIL Bodensanierung f; ~ **column** n SOIL Bodensäule f; ~ **component** n SOIL *soil constituent* Bodenbestandteil m; ~ **condition** n SOIL Bodenzustand m; ~ **conditioner** n WATER POLL *sludge* Bodenverbesserungsmittel nt; ~ **conditioning** n SOIL Bodenverbesserung f; ~ **constituent** n SOIL Bodenbestandteil m; ~ **contamination** n SOIL Bodenkontamination f; ~ **content** n SOIL Bodengehalt m; ~ **covering** n SOIL Bodenbedeckung f; ~ **decontamination** n SOIL *radioactive* Bodenentseuchung f; ~ **degradation following military action** n SOIL Bodendegradation f infolge militärischer Einwirkungen; ~ **degradation through mechanized farming** n SOIL Bodendegradation f durch industrielle Landwirtschaft; ~ **degradation through mining and prospecting** n SOIL Bodendegradation f durch Bergbau und Prospektierung; ~ **degradation through tourism** n SOIL Bodendegradation f durch Tourismus; ~ **degradation through uncontrolled urbanization** n SOIL Bodendegradation f durch ungeregelte Urbanisierung; ~ **deposit sampler** n ANALYSIS Bodenschlammgreifer m; ~ **dust** n SOIL Bodenstaub m

soiled: ~ **plastics** n pl WASTE verschmutzte Kunststoffe m pl

soil: ~ **environment** n SOIL Bodenmilieu nt; ~ **exchange** n SOIL Bodenaustausch m; ~ **filter** n WASTE, WATER SUPP Bodenfilter nt; ~ **filter layer** n WASTE, WATER SUPP Bodenfilterschicht f; ~ **flushing** n SOIL Bodenspülung f; ~ **freezing** n SOIL Bodenvereisung f; ~ **function** n SOIL Bodenfunktion f; ~ **gas** n SOIL Bodengas nt; ~ **horizon** n SOIL Bodenhorizont m; ~ **humidity measuring** n ANALYSIS Bodenfeuchtemessung f; ~ **layer** n SOIL Bodenschicht f; **~-like material** n SOIL *from decontaminated earth* bodenähnliches Material nt; ~ **matrix** n SOIL, WASTE, WATER SUPP Bodenmatrix f; ~ **matrix sample** n ANALYSIS Bodenmatrixprobe f; **~-mechanical** adj WASTE bodenmechanisch; ~ **mechanics** n pl SOIL Bodenmechanik f; ~ **particle** n SOIL Bodenteilchen nt; ~ **passage** n WASTE, WATER SUPP Bodenpassage *f*; ~ **permeability** *n* SOIL, WASTE, WATER SUPP Bodendurchlässigkeit *f*; **~-physical** *adj* SOIL bodenphysikalisch; **~-physical parameter** *n* SOIL bodenphysikalischer Parameter *m*; ~ **pipe** *n* WATER POLL, WATER SUPP Fallrohr *nt*; ~ **pollutants** *n pl* SOIL bodenverschmutzende Stoffe *m pl*, bodenverunreinigende Stoffe *m pl*; ~ **pollution** *n* SOIL Bodenverunreinigung *f*; ~ **probe** *n* ANALYSIS Bodensonde *f*; ~ **processing** *n* SOIL Bodenaufbereitung *f*; ~ **properties** *n pl* SOIL, WASTE, WATER SUPP Bodeneigenschaften *f pl*; **~-protecting** *adj* SOIL bodenschützend; ~ **sample** *n* ANALYSIS, SOIL Bodenprobe *f*; ~ **sample tube** *n* ANALYSIS, SOIL Erdbohrer *m*; ~ **sampling** *n* ANALYSIS, SOIL Bodenprobenahme *nt*; ~ **saturation** *n* SOIL Bodensättigung *f*; ~ **and sediment washing** *n* SOIL Boden- und Sedimentationswäsche *f*; ~ **slurry** *n* SOIL Bodenschlamm *m*; **~-specific** *adj* SOIL bodenspezifisch; ~ **texture** *n* SOIL Bodentextur *f*; ~ **use** *n* SOIL Bodennutzung *f*; ~ **vapor** *AmE*, ~ **vapour** *BrE n* SOIL Bodenluft *f*; ~ **vapour analysis** *n BrE* ANALYSIS Bodenluftanalyse *f*; ~ **vapour concentration** *n BrE* SOIL Bodenluftkonzentration *f*, Melioration *f*; ~ **vapour extraction** *n BrE* SOIL *pneumatic measure* Absaugen *nt* der Bodenluft; ~ **vapour measurement** *n BrE* ANALYSIS Bodenluftmessung *f*; ~ **vapour probe** *n BrE* ANALYSIS Bodenluftsensor *m*; ~ **vapour sampling** *n BrE* ANALYSIS Bodenluftprobenehmen *nt*; ~ **washer** *n* SOIL Bodenwäscher *m*; ~ **water** *n* WASTE Bodenwasser *nt*; ~ **water balance** *n* WATER SUPP Bodenwasserbilanz *f*; ~ **water sampling** *n* ANALYSIS Bodenwasserprobenahme *f*; ~ **zone** *n* SOIL Biobeet *nt*, Bodenzone *f*

solar *adj* RENEW ENERG solar; ~ **architecture** *n* RENEW ENERG Solararchitektur *f*; **~-based renewable energy sources** *n pl* RENEW ENERG solare erneuerbare Energiequellen *f pl*; ~ **cell** *n* RENEW ENERG Solarzelle *f*; ~ **cell module** *n* RENEW ENERG Solarzellenmodul *nt*; **~-charged batteries** *n pl* RENEW ENERG mit Sonnenstrom geladene Batterien *f pl*; ~ **collector** *n* RENEW ENERG Solarkollektor *m*; ~ **dish installation** *n* RENEW ENERG Solardishanlage *f*; ~ **disposal of chemical wastes** *n* WASTE solare Entsorgung *f* chemischer Abfälle; ~ **district heating** *n* RENEW ENERG solare Fernwärme *f*; ~ **district heating supply** *n* RENEW ENERG solare Fernwärmelieferung *f*; **~-driven** *adj* RENEW ENERG solarbetrieben; **~-driven well installation** *n* RENEW ENERG solarbetriebene Brunnenanlage *f*; ~ **energy** *n* RENEW ENERG Solarenergie *f*, Sonnenenergie *f*; ~ **energy enterprise** *n* RENEW ENERG Solarunternehmen *nt*; ~ **energy resources** *n pl* RENEW ENERG Sonnenenenergieressourcen *f pl*; ~ **energy use** *n* RENEW ENERG Nutzung *f* der Solarenergie, Solarenergienutzung *f*, Sonnenenergienutzung *f*; ~ **farm** *n* RENEW ENERG Solarfarm *f*; ~ **farm capacity** *n* RENEW ENERG Solarfarmleistung *f*; ~ **farm installation** *n* RENEW ENERG Solarfarmanlage *f*; **~-heated** *adj* RENEW ENERG solarbeheizt; **~-heated steam generator** *n* RENEW ENERG solarbeheizter Dampferzeuger *m*; ~ **heating** *n* RENEW ENERG Solarheizung *f*; ~ **heating plant** *n* RENEW ENERG Solarheizungsanlage *f*; ~ **heat power plant** *n* RENEW ENERG solares Wärmekraftwerk *nt*, thermisches Solarkraftwerk *nt*; ~ **high-temperature destruction** *n* WASTE, WATER SUPP *detoxification of wastes* solare Hochtemperaturzerstörung *f*; **~-hydrogen-technology** *n* RENEW ENERG Solar-Wasserstoff-Technologie *f*

solarimeter *n* ANALYSIS Solarimeter *nt*

solar: ~ **module** *n* RENEW ENERG Solarmodul *nt*; **~-operated** *adj* RENEW ENERG solarbetrieben; ~ **pond** *n* RENEW ENERG Solarteich *m*; ~ **pond installation** *n* RENEW ENERG Solarteichanlage *f*; ~ **power** *n* RENEW ENERG Solarstrom *m*, Sonnenstrom *m*; **~-powered** *adj* RENEW ENERG solarbetrieben; ~ **power plant** *n* RENEW ENERG Solarkraftwerk *nt*, solarthermisches Kraftwerk *nt*; ~ **radiation** *n* AIR, RENEW ENERG Solarstrahlung *f*, Sonnenstrahlung *f*; ~ **reactor** *n* RENEW ENERG Solarreaktor *m*; **~-supported** *adj* RENEW ENERG solargestützt; **~-supported district heating system** *n* RENEW ENERG solargestütztes Fernwärmesystem *nt*; ~ **technology** *n* RENEW ENERG Solartechnik *f*; **~-thermal** *adj* RENEW ENERG solarthermisch; **~-thermal energy** *n* RENEW ENERG Solarthermie *f*; ~ **tower** *n* RENEW ENERG Solarturm *m*; ~ **tower installation** *n* RENEW ENERG Solarturmanlage *f*; ~ **tower top** *n* RENEW ENERG Turmspitze *f*

solenoid: ~ **valve** *n* AIR Magnetventil *nt*

solid: **~-borne noise** *n* NOISE Körperschall *m*; ~ **brick** *n* NOISE *shielding walls* Vollziegel *m*; **~-containing waste air** *n* AIR feststoffhaltige Abluft *f*; ~ **fiberboard** *AmE*, ~ **fibreboard** *BrE n* WASTE Vollpappe *f*; ~ **fuel** *n* AIR Festbrennstoff *m*, fester Brennstoff *m*

solidification *n* SOIL Verfestigung *f*, WASTE Erstarren *nt*, Verfestigung *f*; ~ **with cement** *n* WASTE Verfestigung *f* mit Zement; ~ **of hazardous waste** *n* WASTE Verfestigung *f* von Sonderabfällen; ~ **with organic polymers** *n* WASTE Verfestigung *f* mit organischen Polymeren; ~ **process** *n* WASTE Verfestigungsverfahren *nt*; ~ **technique** *n* WASTE Verfestigungsverfahren *nt*; ~ **of waste** *n* WASTE Verfestigung *f* von Abfällen

solidified *adj* SOIL, WASTE verfestigt; ~ **end product** *n* WASTE *ash, slag* verfestigtes Endprodukt *nt*; ~ **material** *n* WASTE Verfestigungsprodukt *nt*; ~ **product** *n* WASTE Umsetzungsprodukt *nt*, Verfestigungsprodukt *nt*; ~ **soil** *n* SOIL verfestigter Boden *m*; ~ **waste** *n* WASTE Umsetzungsprodukt *nt*, Verfestigungsprodukt *nt*

solidifier *n* SOIL, WASTE Verfestigungsmittel *nt*

solidify *vt* SOIL, WASTE verfestigen

solidifying: ~ **agent** *n* ANALYSIS Reaktionsmittel *nt*, SOIL Verfestigungsmittel *nt*, WASTE Verfestigungsmittel *nt*, Reaktionsmittel *nt*

solidity *n* RENEW ENERG *wind turbine* Ausfüllungsgrad *m*

solid: **~-liquid extraction** *n* SOIL Fest-Flüssig-Extraktion *f*; ~ **material** *n* AIR, WASTE, WATER POLL, WATER SUPP Feststoff *m*; ~ **matter** *n* AIR, WASTE, WATER POLL, WATER SUPP Feststoff *m*; **~-matter loading** *n* AIR Feststoffbeladung *f*; **~-matter separation** *n* AIR Feststoffabtrennung *f*; ~ **oxide fuel cell** *n* (*SOFC*) AIR Oxidkeramik-Brennstoffzelle *f*; ~ **particle** *n* WATER POLL feste Partikel *f*; ~ **particles** *n pl* WASTE, WATER SUPP Feststoffteilchen *nt*; ~ **particulates** *n pl* AIR Schwebstaub *m*; **~-phase extraction** *n* (*SPE*) ANALYSIS Festphasenextraktion *f* (*SPE*); **~-reacting agent** *n* ANALYSIS fester Reaktionspartner *m*; ~ **residue** *n* WASTE, WATER SUPP fester Rückstand *m*

solids: ~ **contact** *n* WASTE, WATER POLL Schwebstoffkontakt *m*; ~ **contents** *n pl* WASTE, WATER POLL Feststoffgehalt *m*; ~ **separator** *n* WASTE, WATER POLL Feststoffabscheider *m*

solid: ~ **state of aggregation** *n* AIR fester Aggregatzustand *m*; ~ **waste** *n* WASTE Feststoffabfall *m*, fester Abfall *m*

solubility *n* AIR, WATER POLL, WATER SUPP Löslichkeit *f*

solubilization *n* SOIL Löslichmachung *f*, WATER POLL Solubilisation *f*, Solubilisierung *f*, Löslichmachung *f*

solubilize *vt* AIR, ANALYSIS, WASTE, WATER POLL auflösen

solubilized *adj* SOIL, WATER POLL aufgelöst; ~ **organics** *n pl* WATER POLL aufgelöste organische Verbindungen *f pl*

soluble *adj* GEN lösbar, löslich; ~ **gas** *n* AIR lösliches Gas *nt*; ~ **glass** *n* SOIL *binding agent* Wasserglas *nt*

solute *n* AIR, ANALYSIS, WATER POLL *absorbate* Absorbat *nt*, Absorptiv *nt*

solution *n* ANALYSIS, WATER POLL Lösung *f*; ~ **tank** *n* WATER POLL Lösebehälter *m*, Lösungsbecken *nt*

solvability *n* SOIL, WASTE, WATER POLL, WATER SUPP *of contaminant* Löslichkeit *f*

solvate *vt* SOIL, WASTE, WATER POLL, WATER SUPP solvatisieren

solvated *adj* SOIL, WASTE, WATER POLL, WATER SUPP solvatisiert

solvation *n* SOIL, WASTE, WATER POLL, WATER SUPP Solvation *f*

solvent *n* GEN Lösemittel *nt*, Lösungsmittel *nt*; ~ **accumulation** *n* AIR Lösungsmittelanreicherung *f*;

~-based coating *n* WASTE auf Lösungsmittel basierte Beschichtung *f*; **~ concentration** *n* AIR Lösungsmittelkonzentration *f*; **~-containing waste air** *n* AIR lösemittelhaltige Abluft *f*; **~ emissions** *n pl* AIR Lösemittelemissionen *f pl*, Lösungsmittelemissionen *f pl*; **~-free** *adj* AIR, WASTE, WATER SUPP lösemittelfrei, lösungsmittelfrei; **~-free paint** *n* WASTE lösungsmittelfreie Lacke *f*; **~-free waste air** *n* AIR lösemittelfreie Abluft *f*, lösungsmittelfreie Abluft *f*; **~-laden** *adj* AIR lösungsmittelhaltig; **~-laden air** *n* AIR lösemittelhaltige Luft *f*; **~-laden waste air** *n* AIR lösungsmittelhaltige Abluft *f*; **~ loading** *n* AIR Lösungsmittelbeladung *f*; **~ recovery** *n* AIR, WASTE Lösungsmittelrückgewinnung *f*; **~ recovery plant** *n* AIR, WASTE Lösungsmittelrückgewinnungsanlage *f*; **~ recovery unit** *n* AIR, WASTE Lösungsmittelrückgewinnungsanlage *f*; **~ regeneration** *n* SOIL Regeneration *f* von Lösungsmitteln, Wiederaufbereitung *f* von Lösungsmitteln; **~ residue** *n* SOIL, WASTE, WATER SUPP Lösemittelrest *m*

solvents: **~ reactor** *n* WATER SUPP Lösereaktor *m*

solvent: **~ tank** *n* AIR Lösebehälter *m*; **~ vaporization** *n* AIR Lösungsmittelzerstäubung *f*

solvolysis *n* WASTE *plastics* Solvolyse *f*

sone *n* ANALYSIS *phonometric unit* Sone *nt*

sonic *adj* NOISE schalltechnisch

sonification *n* NOISE, WASTE, WATER POLL Ultraschallsiebung *f*

SO_2/NO_x: **~ removal** *n* (*desonox*) AIR, SOIL SO_2/NO_x-Entstickung *f* (*Desonox*)

soot *n* AIR Kaminruß *m*, Ruß *m*; **~ catcher** *n* AIR Rußfänger *m*; **~ coal** *n* AIR Rußkohle *f*; **~-free** *adj* AIR, WASTE rußfrei; **~-free incineration** *n* AIR, WASTE rußfreie Verbrennung *f*

sorb *vt* AIR, MARINE sorbieren

sorbate *n* AIR, MARINE Sorbat *nt*, sorbiertes Gas *nt*

sorbent *n* AIR, MARINE Sorbens *nt*, Sorptionsmittel *nt*

sorption *n* AIR, MARINE, WASTE, WATER POLL Sorption *f*; **~ ability** *n* WASTE, WATER SUPP Sorptionsvermögen *nt*; **~ analysis** *n* ANALYSIS Sorptionsuntersuchung *f*; **~ capacity** *n* AIR Sorptionsfähigkeit *f*; **~ isotherm** *n* AIR Sorptionsisotherme *f*

sort *vt* WASTE sortieren; **~ according to size** *vt* WASTE nach Größe sortieren

sorted *adj* WASTE sortiert; **~ fraction** *n* WASTE sortierte Fraktion *f*

sorting *n* WASTE Sortieren *nt*, Sortierung *f*; **~ analysis** *n* ANALYSIS Sortieranalyse *f*; **~ based on density** *n* WASTE Dichtesortierung *f*; **~ belt** *n* WASTE Leseband *nt*; **~ capacity** *n* WASTE Sortierkapazität *f*; **~ of garbage** *n AmE* (*cf sorting of rubbish BrE*) WASTE Müllsortierung *f*; **~ machine** *n* WASTE Sortiermaschine *f*; **~ personnel** *n* WASTE Sortierpersonal *nt*; **~ plant** *n* WASTE Sortieranlage *f*; **~ rate** *n* WASTE Sortierquote *f*; **~ of refuse** *n* WASTE Müllsortierung *f*; **~ of rubbish** *n BrE* (*cf sorting of garbage AmE*) WASTE Müllsortierung *f*; **~ system** *n* WASTE Sortiersystem *nt*; **~ table** *n* WASTE Sichtungstisch *m*, Sortiertisch *m*; **~ via X-rays** *n* WASTE Sortierung *f* mit Röntgenstrahlen; **~ of waste** *n* WASTE Müllsortierung *f*

sound 1. *n* MARINE, NOISE, SAFETY Schall *m*; **2.** *vt* ANALYSIS sondieren

sound: **~-absorbent damper** *n* NOISE, SAFETY Absorptionsschalldämpfer *m*; **~-absorbent material filler** *n* NOISE Füllung *f* mit schallschluckenden Stoffen; **~ absorbent zone** *n* NOISE schallschluckende Zone *f*; **~ absorber** *n* NOISE Schallabsorber *m*; **~-absorbing** *adj* NOISE schallabsorbierend, schalldämmend, schallschluckend; **~-absorbing apron** *n* NOISE schallschluckende Schürze *f*; **~-absorbing barrier** *n* NOISE Schallschutzwand *f*; **~-absorbing brick** *n* NOISE schallschluckender Stein *m*; **~-absorbing ceiling** *n* NOISE schallschluckende Decke *f*; **~-absorbing door** *n* MARINE, NOISE, SAFETY schallschluckende Tür *f*; **~-absorbing foam panel** *n* MARINE, NOISE, SAFETY Schaumschalldämmplatte *f*; **~-absorbing glass** *n* MARINE, NOISE, SAFETY schallschluckendes Glas *nt*; **~-absorbing hood** *n* NOISE Schallschutzhaube *f*; **~-absorbing lining** *n* NOISE schallabsorbierende Auskleidung *f*, schallschluckende Auskleidung *f*; **~-absorbing material** *n* NOISE Schallabsorptionsmaterial *nt*, schallabsorbierendes Material *nt*, schallschluckendes Material *nt*; **~-absorbing material casing** *n* NOISE Ummantelung *f* mit schallschluckenden Stoffen; **~-absorbing panel** *n* NOISE schallschluckende Platte *f*, Absorberplatte *f*; **~-absorbing property** *n* NOISE Schallschluckeigenschaft *f*; **~-absorbing wall** *n* NOISE schallschluckende Wand *f*; **~-absorbing wall lining** *n* NOISE schallabsorbierende Wandverkleidung *f*, schallschluckende Wandverkleidung *f*; **~-absorbing wall paper** *n* NOISE schallschluckende Tapete *f*; **~ absorption coefficient** *n* NOISE Dämmzahl *f*, Schallschluckgrad *m*; **~ attenuation** *n* NOISE Schalldämpfung *f*; **~ barrier** *n* NOISE Schallmauer *f*; **~ box** *n* NOISE Schalldose *f*; **~-conducting** *adj* NOISE schalleitend; **~-dampened** *adj* NOISE schalldämmend; **~ damper** *n* NOISE Schalldämpfer *m*; **~ damping** *n* NOISE Schalldämpfung *f*; **~ deadener** *n* NOISE Schalldämpfer *m*; **~ deadener material** *n* NOISE Entdröhnungsmaterial *nt*, Entdröhnungsmittel *nt*; **~ deadening** *n* NOISE Dämpfung *f*, Entdröhnung *f*, Schalldämpfung *f*; **~ disposal** *n* WASTE geordnete Ablagerung *f*; **~ emission** *n* ANALYSIS, NOISE Schallemission *f* (*SE*); **~ emitter** *n* NOISE Schallemittent *m*, Schallstrahler *m*; **~ energy** *n* NOISE Schallenergie *f*; **~ exposure meter** *n* ANALYSIS, NOISE Schallbelastungsmesser *m*; **~ field** *n* NOISE Schallfeld *nt*; **~ generation** *n* NOISE Schallentstehung *f*; **~ immission** *n* NOISE Schallimmission *f*

sounding *n* ANALYSIS Sondierung *f*; **~ boring** *n* ANALYSIS Sondierbohrung *f*, Sondierungsbohrung *f*

sound: **~-insulating** *adj* NOISE schalldicht, schalldämmend; **~-insulating wall** *n* NOISE Schalldämmwand *f*; **~ insulation** *n* NOISE Schalldämmung *f*; **~ insulation material** *n* NOISE Schalldämmaterial *nt*, Schalldämmstoff *m*, Schallisolationsmaterial *nt*; **~ intensity** *n* NOISE Schallintensität *f*, Schallstärke *f*; **~ intensity analyser** *BrE*, **~ intensity analyzer** *AmE* *n* ANALYSIS, NOISE Schallintensitätsanalysator *m*; **~ intensity level** *n* ANALYSIS, NOISE Schallintensitätspegel *m*; **~ intensity meter** *n* ANALYSIS, NOISE Schallintensitätsmeßgerät *nt*; **~ isolation** *n* NOISE Schalldämmung *f*, Schallisolation *f*; **~ level** *n* NOISE Schallpegel *m*; **~ level calibrator** *n* ANALYSIS, NOISE Schallpegelkalibrator *m*; **~ level meter** *n* (*SLM*) ANALYSIS, NOISE Schallpegelmesser *m*; **~ occurrence** *n* NOISE Schallvorgang *m*; **~ permeability** *n* NOISE Schalldurchlässigkeit *f*; **~-permeable** *adj* NOISE schalldurchlässig; **~ pollution** *n* NOISE Lärmbelästigung *f*; **~ power** *n* ANALYSIS, NOISE Schalleistung *f*; **~ power level** *n* NOISE *short-wave listener* Schalleistungspegel *m*; **~ pressure** *n* NOISE Schalldruck *m*; **~ pressure level** *n* (*SPL*) NOISE Schalldruckpegel *m*; **~ pressure measuring** *n* ANALYSIS, NOISE Schalldruckmessung *f*; **~ pressure reference value** *n* ANALYSIS, NOISE Bezugsschalldruck *m*; **~ pressure spectrum** *n* ANALYSIS, NOISE Schalldruckspektrum *nt*; **~ probe** *n* ANALYSIS, NOISE Schallsonde *f*

soundproof *adj* NOISE schalldicht, schallgedämmt;

soundproofing *n* NOISE Schallisolierung *f*

sound: **~-protecting window** *n* NOISE Schallschutzfenster *nt*; **~ radiation** *n* NOISE Schallabstrahlung *f*; **~ radiation pressure** *n* NOISE Schallstrahlungsdruck *m*; **~ radiometer** *n* ANALYSIS Schallradiometer *nt*; **~ reduction index** *n BrE* (*cf sound transmission loss AmE*) NOISE Schalldämmaß *nt*; **~-reflecting** *adj* NOISE schallhart, schallreflektierend; **~ reflection** *n* NOISE Schallreflexion *f*; **~ source** *n* NOISE Schallquelle *f*; **~ spectrum** *n* NOISE Schallspektrum *nt*; **~ transducer** *n* ANALYSIS, NOISE Schallwandler *m*; **~ transmission** *n* NOISE Schalldurchgang *m*; **~ transmis-**

sion class *n* (*STC*) NOISE Schalldämmklasse *f*; **~ transmission loss** *n AmE* (*cf sound reduction index BrE*) NOISE Schalldämmaß *nt*; **~-transparent** *adj* NOISE schalldurchlässig; **~ trap** *n* NOISE Tonfalle *f*; **~ vibration** *n* NOISE Schallschwingung *f*; **~ waves** *n pl* NOISE Schallwellen *f pl*

source *n* AIR, WATER SUPP Quelle *f*; **at ~** *phr* AIR, WATER POLL an der Quelle; **~ area** *n* AIR, WATER SUPP Quellbereich *m*; **~ of greenhouse gases** *n* AIR Quelle *f* von Treibhausgasen; **~ height** *n* AIR *of stack* Quellhöhe *f*; **~-oriented** *adj* SOIL, WATER POLL, WATER SUPP quellenorientiert; **~-oriented measures** *n pl* SOIL quellenorientierte Maßnahmen *f pl*; **~-related** *adj* AIR, NOISE, SOIL, WATER POLL, WATER SUPP quellenbezogen; **~ of revenue** *n* POLICY Einnahmequelle *f*; **~ separation** *n* WASTE Abfallsortierung *f* am Anfallsort; **~ of waste generation** *n* WASTE Anfallstelle *f*

sour: **~ gas** *n* AIR Sauergas *nt*, saures Gas *nt*

Soxhlet: **~ extraction** *n* ANALYSIS Soxhlet-Extraktion *f*; **~ extractor** *n* ANALYSIS Soxhlet-Extraktor *m*

space: **~-time yield** *n* ANALYSIS Raum-Zeit-Ausbeute *f*; **~ velocity** *n* AIR *catalyst* Raumgeschwindigkeit *f*

sparger *n* WATER POLL Sprenger *m*

sparingly: **~ soluble** *adj* AIR, WASTE, WATER SUPP schwerlöslich

spark *n* AIR Funken *m*; **~-ablation atomic absorption spectrometry** *n* ANALYSIS Funken-AAS *f*; **~ arrestor** *n* AIR Funkenableiter *m*, Funkenfänger *m*; **~ ignition engine** *n* AIR Motor *m* mit Fremdzündung, Ottomotor *m*; **~-over** *n* AIR Überschlag *m*; **~-over voltage** *n* AIR Überschlagsspannung *f*

spatial: **~ distribution** *n* ANALYSIS räumliche Verteilung *f*; **~ evaluation** *n* ANALYSIS räumliche Auswertung *f*; **~ pattern** *n* ANALYSIS räumliche Struktur *f*; **~ resolution** *n* ANALYSIS *nuclear technology* räumliche Auflösung *f*; **~ trend** *n* AIR, WASTE, WATER SUPP räumliche Tendenz *f*; **~ variability** *n* AIR, ANALYSIS räumliche Veränderlichkeit *f*

SPE *abbr* (*solid-phase extraction*) ANALYSIS SPE (*Festphasenextraktion*)

specialist: **~ lab** *n* (*specialist laboratory*) ANALYSIS Schwerpunktlaboratorium *nt*, Speziallaboratorium *nt*; **~ laboratory** *n* (*specialist lab*) ANALYSIS Schwerpunktlaboratorium *nt*, Speziallaboratorium *nt*

special: **~ waste** *n* WASTE, WATER POLL Sonderabfall *m*, Sondermüll *m*; **~ waste surface landfill** *n* WASTE oberirdische Sonderabfalldeponie *f*

specific: **~ coagulation** *n* WATER POLL spezifische Koagulation *f*; **~ conductivity** *n* AIR, WASTE spezifischer Heizwert *m*; **~ oxygen transfer capacity in clean water** *n* WATER POLL spezifisches Sauerstoffzufuhrvermögen *nt* in Reinwasser; **~ oxygen transfer capacity in wastewater** *n* WATER POLL spezifisches Sauerstoffzufuhrvermögen *nt* in Abwasser; **~ surface area** *n* WATER POLL spezifische Oberfläche *f*

specimen *n* ANALYSIS Probestück *nt*

spectral *adj* ANALYSIS spektral; **~ absorption coefficient** *n* ANALYSIS spektraler Absorptionskoeffizient *m*; **~ analysis** *n* ANALYSIS Spektralanalyse *f*; **~ half-width value** *n* ANALYSIS spektrale Halbwertsbreite *f*; **~ line** *n* ANALYSIS Spektrallinie *f*; **~ range** *n* ANALYSIS Spektralbereich *m*

spectrometer *n* ANALYSIS Spektrometer *nt*

spectrometric *adj* ANALYSIS spektrometrisch

spectrometry *n* ANALYSIS Spektrometrie *f*

spectrophotometer *n* ANALYSIS Spektralphotometer *nt*, Spektrophotometer *nt*

spectroscopic *adj* ANALYSIS spektroskopisch; **~ methods** *n pl* ANALYSIS spektroskopische Methoden *f pl*

spectroscopy *n* ANALYSIS Spektroskopie *f*

speed *n* ANALYSIS Geschwindigkeit *f*, Schnelligkeit *f*; **~ of degradation** *n* WASTE, WATER POLL, WATER SUPP *composting* Abbaugeschwindigkeit *f*

speedometer *n* ANALYSIS Geschwindigkeitsmesser *m*

speed: **~ of sound** *n* NOISE Schallgeschwindigkeit *f*

spent: **~ acid recovery** *n* WASTE Dünnsäurerückgewinnung *f* (*DSR*); **~ lye** *n* WASTE, WATER POLL, WATER SUPP Ablauge *f*; **~ oil** *n* WASTE, WATER POLL Altöl *nt*

spill 1. *n* MARINE ausgeflossenes Öl *nt*; **2.** *vt* WASTE, WATER SUPP verschütten

spillage *n* WATER SUPP Verschüttung *f*; **~ of hazardous waste** *n* WATER SUPP Verschüttung *f* von Sondermüll; **~ tank** *n* WATER SUPP Leckbehälter *m*

spilled *adj* WASTE, WATER SUPP verschüttend; **~ waste** *n* WASTE verschüttender Abfall *m*

spin *vt* WASTE verspinnen; **~ off** *vt* WASTE aufhaspeln; **~ out** *vt* AIR *cyclone* schleudern

spin: **~ dryer** *n* WATER POLL Dralltrockner *m*

spiral *n* AIR, SOIL, WATER POLL Spirale *f*; **~ aerator** *n* WATER POLL Wendelbelüfter *m*; **~ classifier** *n* SOIL Spiralklassierer *m*; **~-flow tank** *n* WATER SUPP Umwälzungsbecken *nt*

spiralling *adj* AIR wendelförmig

spiral: **~ module** *n* WATER POLL Spiralmodul *nt*; **~ plate** *n* AIR Spiralplatte *f*; **~ plate condenser** *n* AIR Spiralplattenkondensator *m*; **~ plate heat exchanger** *n* AIR Spiralplattenwärmetauscher *m*

SPL *abbr* (*sound pressure level*) NOISE Schalldruckpegel *m*

splitting: **~ off** *n* WASTE Abspaltung *f*

spokesperson *n* POLICY Pressesprecher *m*

spontaneous: **~ combustion** *n* SAFETY spontane Verbrennung *f*

spot: **~ analysis** *n* ANALYSIS Einzeluntersuchung *f*; **~ measurement** *n* ANALYSIS Einzelmessung *f*; **~ sample** *n* ANALYSIS Stichprobe *f*, Zufallsstichprobe *f*

spotter: **~ plane** *n* MARINE Luftaufklärungsflugzeug *nt*

spout *n* WASTE, WATER SUPP *of pump* Ausguß *m*

spray 1. *n* AIR Sprühnebel *m*, Sprühregen *m*, *atomizer* Zerstäubungsmittel *nt*; **2.** *vt* AIR, MARINE, SOIL, WASTE, WATER POLL, WATER SUPP eindüsen, sprühen, *drizzle* besprühen, *atomize* versprühen

spray: **~ absorber** *n* AIR Sprühabsorber *m*; **~ absorption process** *n* AIR Sprühabsorptionsverfahren *nt* (*SAV*); **~ apertures** *n pl* MARINE Sprühöffnungen *f pl*; **~-boom** *n* MARINE Sprühausleger *m*; **~ column** *n* AIR *absorption device* Sprühkolonne *f*; **~ dry absorber** *n* (*SDA*) AIR Sprühtrocknungswäscher *m*; **~ dryer** *n* WATER POLL Sprühtrockner *m*; **~ dry scrubber** *n* AIR Sprühtrocknungswäscher *m*; **~ frame** *n* AIR Sprührahmen *m*; **~-gun** *n* MARINE Sprühkanone *f*

spraying *n* GEN Bedüsung *f*, Versprühung *f*; **~ cone** *n* AIR Spritzkegel *m*; **~ zone** *n* AIR Sprühzone *f*

spray: **~ nozzle** *n* AIR Spritzdüse *f*, Sprühdüse *f*; **~ path** *n* MARINE Sprühpfad *m*; **~ reactor** *n* AIR Sprühreaktor *m*; **~ scrubber** *n* AIR Sprühwäscher *m*; **~ tower** *n* AIR Sprühturm *m*, Sprühwäscher *m*

spread *vi* AIR, MARINE, WASTE sich ausbreiten

spreader: **~ crane** *n* WASTE Spreaderkran *m*

spreading *n* MARINE Ausbreitung *f*, Verbreitung *f*; **~ grounds** *n pl* WATER POLL Einstaufelder *nt pl*, Einstauflächen *f pl*

spread: **~ of noise** *n* NOISE Schallausbreitung *f*

spring: **~ tide** *n* RENEW ENERG Springflut *f*; **~ water** *n* WATER SUPP Quellwasser *nt*

sprinkle *vt* WASTE, WATER SUPP besprengen

sprinkler *n* SAFETY Sprinkleranlage *f*, WATER POLL Beregnungsanlage *f*, Sprinkleranlage *f*, Verregnung *f*

sprinkling *n* WATER POLL *land application of wastewater* Beregnung *f*, Berieselung *f*; **~ filter** *n* WATER POLL, WATER SUPP Tropfkörper *m* (*TK*)

squirrel: **~-cage induction motor** *n* RENEW ENERG *wind turbine* Käfigläufermotor *m*, WATER POLL, WATER SUPP *pump driving gear* Kurzschlußläufermotor *m*

SRI *abbr* (*sound reduction index*) NOISE Schalldämmaß *nt*

SRT *abbr* (*sludge residence time*) WATER POLL Schlammverweilzeit *f*

stabile *adj* AIR, ANALYSIS stabil; **~ temperature layering** *n* AIR stabile Temperaturschichtung *f*
stability *n* AIR, ANALYSIS, WASTE, WATER POLL Stabilität *f*
stabilization *n* GEN Stabilisierung *f*; **~ of greenhouse gases** *n* AIR Stabilisierung *f* von Treibhausgasen; **~ pond** *n* WATER POLL Abwasserbecken *nt*, Schlammfaulbecken *nt*, Abwasserteich *m*; **~ process** *n* WASTE Verfestigungsverfahren *nt*; **~ technique** *n* WASTE Verfestigungsverfahren *nt*
stabilize *vt* GEN stabilisieren
stabilized *adj* GEN stabilisiert; **~ sludge** *n* WATER POLL stabilisierter Schlamm *m*
stabilizer *n* GEN Stabilisator *m*
stable *adj* AIR, ANALYSIS, WASTE, WATER POLL stabil; **~ methane fermentation** *n* WASTE stabile Methangärung *f*
stack *n* AIR, RENEW ENERG, WASTE *chimney* Kamin *m*, Schornstein *m*; **~ diameter** *n* AIR Kamindurchmesser *m*, Schornsteindurchmesser *m*; **~ gas** *n* AIR Essengas *nt*, Gichtgas *nt*; **~ height** *n* AIR Kaminhöhe *f*, Schornsteinhöhe *f*; **~ lining** *n* AIR Schachtfutter *nt*; **~ orifice** *n* AIR *opening* Kaminmündung *f*, Schornsteinmündung *f*; **~ wall** *n* AIR Kaminwand *f*, Schornsteinwand *f*
stage: **~ of decompostion** *n* WASTE Abbaustufe *f*
staged: **~ fuel feeding** *n* AIR gestufte Brennstoffdosierung *f*; **~ incineration** *n* AIR Stufenverbrennung *f*
stagnant: **~ water** *n* WATER POLL, WATER SUPP stehendes Gewässer *nt*
stagnate *vi* WATER POLL, WATER SUPP stagnieren
stagnation *n* WASTE, WATER POLL, WATER SUPP Stagnation *f*
stall *n* RENEW ENERG *wind turbine* aerodynamische Strömungsabriß *f*
standard: **~ conditions** *n pl* (*SC*) ANALYSIS Normzustand *m*; **~ curve-time temperature** *n* ANALYSIS Normbrandkurve - Zeit-Temperatur *f*
standardized *adj* SAFETY standardisiert; **~ equipment** *n* SAFETY standardisierte Ausrüstung *f*
standard: **~ rate of charge** *n* POLICY voller Abgabensatz *m*; **~-rate trickling filter** *n* WATER POLL schwachbelasteter Tropfkörper *m*
standards: **~ implementation monitoring** *n* POLICY Normenkontrolle *f*
standard: **~ solution** *n* ANALYSIS Standardlösung *f*, Titerlösung *f*; **~ temperature and pressure** *n* (*STP*) ANALYSIS Normalbedingungen *f pl*; **~ turbidity** *n* ANALYSIS Standardtrübung *f*
stand: **~-by** *n* MARINE Bereitschaft *f*, Reserve *f*; **~-by duty** *n* MARINE Bereitschaftsdienst *m*
standing: **~ wave tube** *n* NOISE Kundtsches Rohr *nt*
standpipe *n* AIR Fallrohr *nt*
stapling: **~ of bales** *n* WASTE Stapeln *nt* der Ballen
starchy *adj* RENEW ENERG stärkehaltig; **~ plants** *n pl* RENEW ENERG *biomass conversion* stärkehaltige Pflanzen *f pl*; **~ raw material** *n* RENEW ENERG stärkehaltiger Rohstoff *m*
star: **~-shaped procedure** *n* POLICY *administrative units* Sternverfahren *nt*
start: **~ burner** *n* AIR Startbrenner *m*
starting: **~ operation** *n* AIR Anfahrbetrieb *m*; **~ point** *n* ANALYSIS Startpunkt *m*
state: **~ of aggregation** *n* AIR Aggregatzustand *m*; **~-of-the-art** *adj* GEN auf dem neuesten Stand der Technik
static *adj* ANALYSIS, WASTE, WATER POLL, WATER SUPP statisch; **~ air layer** *n* AIR ruhende Luftschicht *f*; **~ deflection** *n* NOISE statische Auslenkung *f*; **~ discharge head** *n* WASTE, WATER POLL, WATER SUPP statische Förderhöhe *f*; **~ head** *n* WASTE statische Höhe *f*, WATER POLL statische Druckhöhe *f*, WATER SUPP statische Höhe *f*; **~ headspace method** *n* ANALYSIS Untersuchung *f* von Headspace-Proben; **~ leachability test** *n* ANALYSIS statischer Auslaugtest *m*, statischer Eluationstest *m*; **~ leaching test** *n* SOIL statischer Eluationstest *m*; **~ mixer** *n* WATER POLL statischer Mischer *m*; **~ pressure** *n* AIR statischer Druck *m*; **~ suction head** *n* WATER POLL, WATER SUPP statische Saughöhe *f*; **~ suction lift** *n* WATER POLL, WATER SUPP statische Saughöhe *f*
stationary *adj* AIR, ANALYSIS stationär, RENEW ENERG ortsabhängig; **~-bed adsorber** *n* AIR *adsorption process* Festbettadsorber *m*; **~ compactor** *n* WASTE, WATER SUPP stationärer Kompaktor *m*; **~ container system** *n* (*SCS*) WASTE, WATER SUPP Bringbehältersystem *nt*; **~ crushing unit** *n* WASTE stationäre Zerkleinerungsanlage *f*; **~ emission source** *n* AIR ortsfeste Emissionsquelle *f*; **~ fluidized bed** *n* AIR stationäre Wirbelschicht *f* (*SWS*); **~ fluidized-bed oven** *n* SOIL stationärer Wirbelschichtofen *m*; **~ sampler** *n* ANALYSIS stationäres Probenahmegerät *nt*; **~ tank** *n* AIR Standbehälter *m*
statistical *adj* ANALYSIS *data* statistisch; **~ evaluation** *n* ANALYSIS statistische Auswertung *f*; **~ possibility of error** *n* ANALYSIS statistische Fehlermöglichkeit *f*
statistics *n pl* ANALYSIS Statistik *f*
stator: **~-rotor system** *n* AIR Stator-Rotor-System *nt*
statute *n* POLICY Gesetz *nt*; **~ book** *n* POLICY Gesetzbuch *nt*
statutory *adj* POLICY gesetzlich; **~ power** *n* POLICY, WATER SUPP Ermächtigungsgrundlage *f*; **~ powers** *n pl* POLICY *of administration* Ermächtigung *f*; **~ regulation** *n* POLICY, SAFETY gesetzliche Vorschrift *f*
STC *abbr* (*sound transmission class*) NOISE Schalldämmklasse *f*
steam *n* GEN Dampf *m*; **~ air preheater** *n* AIR Dampfluftvorwärmer *m*, Dampfluvo *m*; **~ boiler** *n* RENEW ENERG, WASTE, WATER POLL Dumpfkessel *m*; **~ circulation** *n* RENEW ENERG Wasserdampfkreislauf *m*; **~ cleaning** *n* WASTE Dampfstrahlen *nt*; **~ distillation** *n* ANALYSIS Wasserdampfdestillation *f*; **~ generator** *n* WATER POLL Dampferzeuger *m*; **~ jet** *n* WATER SUPP Dampfstrahl *m*; **~ jet pump** *n* WASTE Dampfstrahlpumpe *f*; **~ out** *n* AIR Ausdämpfen *nt*; **~ pressure** *n* AIR, ANALYSIS, SOIL *vapour extraction*, WASTE, WATER POLL, WATER SUPP Dampfdruck *m*; **~ pressure thermometer** *n* ANALYSIS Dampfdruckfederthermometer *nt*; **~ process** *n* RENEW ENERG Dampfprozeß *m*; **~ stripping** *n* SOIL, WASTE Dampfstrippen *nt*, WATER POLL Dampfstripping *f*, WATER SUPP Dampfstrippen *nt*; **~ turbine** *n* RENEW ENERG Dampfturbine *f*; **~-vacuum stripping** *n* SOIL Dampfvakuum Strippen *nt*
steel *n* WASTE Stahl *m*; **~ band strapping** *n* WASTE Stahlbandumreifung *f*; **~ body** *n* AIR Stahlkörper *m*; **~ fiber** *AmE*, **~ fibre** *BrE* *n* AIR *filter medium* Stahlfaser *f*; **~ packaging** *n* WASTE Stahlverpackung *f*; **~ pipe** *n* WASTE, WATER SUPP Stahlrohr *nt*
steelwork: **~ slag** *n* WASTE Stahlwerkschlacke *f*
STEL *abbr* (*short-term exposure limit*) ANALYSIS STEL (*Kurzzeit-Exposition*)
stemflow *n* AIR Stammabfluß *m*
step: **~ aeration** *n* WATER POLL Stufenbelüftung *f*
step: **~-by-step** *adv* ANALYSIS stufenweise
stepped: **~ forward feed grate** *n* WATER SUPP Stufenvorschubrost *m*
stereospecific: **~ catalyst** *n* AIR stereospezifischer Katalysator *m*
sterilization *n* WASTE, WATER POLL, WATER SUPP Entkeimung *f*
sterilize *vt* WASTE, WATER POLL, WATER SUPP entkeimen
sterilized *adj* WASTE, WATER POLL, WATER SUPP entkeimt
Steuler: **~ process** *n* AIR Steuler Verfahren *nt*
stickiness *n* AIR, WASTE, WATER SUPP Klebrigkeit *f*
sticky: **~ deposit** *n* AIR klebrige Ablagerung *f*; **~ dust** *n* AIR klebriger Staub *m*
still: **~ water** *n* WATER POLL, WATER SUPP stehendes Gewässer *nt*
Stirling: **~ motor** *n* RENEW ENERG Stirlingmotor *m*
stirrer *n* SOIL Rührwerk *nt*
stirring: **~ mechanism** *n* WATER SUPP Rührwerk *nt*, Umwälzvorrichtung *f*
stock: **~ solution** *n* ANALYSIS Stammlösung *f*

stoichiometric *adj* AIR, ANALYSIS, WATER POLL stöchiometrisch; ~ **combustion** *n* AIR stöchiometrische Verbrennung *f*; ~ **oxygen supply** *n* AIR, WATER POLL stöchiometrisches Sauerstoffangebot *nt*
stoichiometry *n* AIR, ANALYSIS, WATER POLL Stöchiometrie *f*
stoker *n* AIR Heizer *m*
Stokes: ~**'s law** *n* AIR, ANALYSIS Stokessches Reibungsgesetz *nt*
stoneware: ~ **pipe** *n* WASTE, WATER POLL, WATER SUPP Steinzeugrohr *nt*
stop *vt* WASTE, WATER SUPP *leak* stopfen
stoppage *n* WATER POLL, WATER SUPP *sewer* Verstopfung *f*
STOR *abbr* (*sludge-to-oil reactor*) WATER POLL S.-Ö.-Reaktor *m* (*Schlamm-Öl-Reaktor*)
storable *adj* RENEW ENERG speicherbar, SOIL, WASTE, WATER POLL, WATER SUPP lagerfähig
storage *n* MARINE, RENEW ENERG, SOIL Einlagerung *f*, Lagerung *f*, Speicherung *f*, WASTE Einlagerung *f*, Lagerung *f*, WATER POLL Einlagerung *f*, Lagerung *f*, Speicherung *f*, WATER SUPP Speicherung *f*; ~ **area** *n* SOIL Lagerraum *m*, WASTE Ablagerungsplatz *m*; ~ **basin** *n* WATER POLL, WATER SUPP Rückhaltebecken *nt*, Sammelbecken *nt*, Speicherbecken *nt*, Staubecken *nt*; ~ **battery** *n* WASTE Akkumulator *m* (*Akku*); ~ **behavior** *AmE*, ~ **behaviour** *BrE n* WASTE Einlagerungsverhalten *nt*; ~ **capacity** *n* AIR *of filters* Speicherkapazität *f*, RENEW ENERG *energy* Speicherkapazität *f*, Speicherungskapazität *f*, Speicherungsvermögen *nt*, WASTE, WATER POLL Speicherungskapazität *f*, Speicherungsvermögen *nt*, *of ground* Speicherkapazität *f*, WATER SUPP Speicherungsvermögen *nt*, *of ground* Speicherkapazität *f*; ~ **chamber** *n* WASTE Einlagerungskammer *f*, Lagerkammer *f*; ~ **of dangerous materials** *n* SAFETY Gefahrstofflager *nt*; ~ **durability** *n* WASTE Lagerbeständigkeit *f*; ~ **efficiency** *n* RENEW ENERG Speicherwirkungsgrad *m*; ~ **facility** *n* AIR Lagermöglichkeit *f*, SOIL, WASTE Lagermöglichkeit *f*, Lagerstätte *f*, WATER SUPP Lagermöglichkeit *f*; ~ **filter** *n* AIR Speicherfilter *nt*; ~**-free** *adj* RENEW ENERG speicherfrei; ~ **function** *n* RENEW ENERG *energy* Speicherfunktion *f*; ~ **hydroelectric plant** *n* RENEW ENERG Speicherwasserkraftwerk *nt*; ~ **medium** *n* RENEW ENERG *heat pump* Speichermedium *nt*; ~ **reservoir** *n* RENEW ENERG *hydropower* Talsperre *f*, WATER POLL Speicherbecken *nt*, Staubecken *nt*, WATER SUPP Rückhaltebecken *nt*, Sammelbecken *nt*, Speicherbecken *nt*, Staubecken *nt*; ~ **site** *n* WASTE Deponie *f*; ~ **space** *n* WASTE Lagerraum *m*; ~ **tank** *n* AIR, WASTE Sammelbehälter *m*, WATER SUPP Vorratsbehälter *m*; ~ **technology** *n* RENEW ENERG Speichertechnologie *f*; ~ **volume** *n* RENEW ENERG Speichervolumen *nt*
store *vt* ANALYSIS *sample* aufbewahren, MARINE, RENEW ENERG, SOIL speichern, WASTE, WATER POLL, WATER SUPP einlagern, speichern; ~ **intermediately** *vt* SOIL *excavated mass* zwischenlagern; ~ **temporarily** *vt* WASTE zwischenlagern
stored *adj* RENEW ENERG gespeichert; ~ **geothermal energy** *n* RENEW ENERG gespeicherte Erdwärme-Energie *f*; ~ **heat** *n* RENEW ENERG gespeicherte Wärme *f*; ~ **solar energy** *n* RENEW ENERG gespeicherte Sonnenenergie *f*; ~ **water** *n* RENEW ENERG gespeichertes Wasser *nt*
storm: ~ **wastewater** *n* WATER SUPP Niederschlagsabwasser *nt*
stormwater *n* WATER POLL, WATER SUPP Niederschlagswasser *nt*, Regenwasser *nt*; ~ **flow** *n* WATER SUPP Regenwasserabfluß *m*; ~ **main** *n* WATER POLL, WATER SUPP Regenwasserkanalisation *f*, Regenwasserleitung *f*; ~ **overflow** *n* WATER POLL, WATER SUPP Regenüberlauf *m*; ~ **overflow tank** *n* WATER POLL, WATER SUPP Regenüberlaufbecken *nt*; ~ **pump** *n* WATER POLL, WATER SUPP Regenwasserpumpe *f*; ~ **reservoir** *n* WATER POLL, WATER SUPP Regenklärbecken *nt*; ~ **retention tank** *n* WASTE, WATER POLL Regenrückhaltebecken *nt*; ~ **sewer** *n* WASTE, WATER POLL, WATER SUPP Regenwasserkanal *m*; ~ **tank** *n* WATER POLL Regenklärbecken *nt*, Regenwasserbecken *nt*, WATER SUPP Regenwasserbecken *nt*; ~ **treatment** *n* WATER POLL, WATER SUPP Regenwasserbehandlung *f*
STP *abbr* (*standard temperature and pressure*) ANALYSIS Normalbedingungen *f pl*
strainer *n* MARINE Saugkorb *m*, WASTE Siebfilter *nt*
strapping: ~ **equipment** *n* WASTE Umreifungsgeräte *nt pl*; ~ **machine** *n* WASTE Banderoliermaschine *f*
stratosphere *n* AIR Stratosphäre *f*
stratospheric *adj* AIR stratosphärisch; ~ **ozone** *n* AIR stratosphärisches Ozon *nt*; ~ **ozone concentration** *n* AIR stratosphärische Ozonkonzentration *f*; ~ **ozone pollution** *n* AIR stratosphärische Ozonbelastung *f*
straw: ~ **bale** *n* RENEW ENERG *biomass conversion* Strohballen *m*; ~**-burning plant** *n* RENEW ENERG *biomass conversion* Strohfeuerungsanlage *f*
stream 1. *n* WATER POLL, WATER SUPP Strom *m*; **2.** ~ **in** *vi* AIR einströmen
stream: ~ **aeration** *n* WATER POLL Flußbelüftung *f*, Strombelüftung *f*
streamline *n* AIR Stromlinie *f*; ~ **flow** *n* AIR laminare Strömung *f*
street: ~ **cleaner** *n* WASTE *vehicle* Kehrfahrzeug *nt*; ~**-cleaning lorry** *n BrE* (*cf street-cleaning truck AmE*) WASTE Kehrfahrzeug *nt*; ~**-cleaning truck** *n AmE* (*cf street-cleaning lorry BrE*) WASTE Kehrfahrzeug *nt*; ~ **emissions level** *n* NOISE Straßen-Emissionspegel *m*; ~ **refuse** *n* WASTE Straßenkehricht *m*
strength *n* WASTE Festigkeit *f*
stress *n* WASTE Beanspruchung *f*, WATER SUPP Beanspruchung *f*, mechanische Spannung *f*; ~ **capacity** *n* SOIL Belastbarkeit *f*; ~ **corrosion** *n* WATER SUPP Spannungskorrosion *f*; ~ **corrosion cracking** *n* WATER SUPP Spannungskorrosionsriß *m*
strict: ~ **liability** *n* POLICY Gefährdungshaftung *f*
strip *vt* WASTE, WATER POLL strippen
stripping *n* WASTE, WATER POLL, WATER SUPP Rückspülung *f*, Strippen *nt*, Strippverfahren *nt*; ~ **pump** *n* MARINE Restmengenpumpe *f*; ~**-purging technique** *n* ANALYSIS Stripping-Purging-Technik *f*; ~ **test** *n* ANALYSIS Stripping-Test *m*; ~ **tower** *n* SOIL Stripping-Turm *m*
stroboscope *n* ANALYSIS Stroboskop *nt*
strong: ~**-sun period** *n* RENEW ENERG *solar power* sonnenstarke Zeit *f*
strontium: ~ **sulfate** *AmE*, ~ **sulphate** *BrE n* AIR *adsorbent* Strontiumsulfat *nt*
structural: ~ **engineering** *n* RENEW ENERG, WATER POLL, WATER SUPP Bautechnik *f*; ~ **engineering technology** *n* RENEW ENERG, WATER POLL, WATER SUPP bautechnische Verfahrenstechnik *f*; ~ **insulation board** *n* NOISE Isolierplatte *f*
structurally: ~ **engineered** *adj* RENEW ENERG, WATER POLL, WATER SUPP bautechnisch; ~**-engineered barrier** *n* WASTE, WATER SUPP bautechnische Barriere *f*
structure *n* WASTE, WATER SUPP Aufbau *m*; ~**-borne absorber** *n* NOISE Körperschallabsorber *m*; ~**-borne noise** *n* NOISE Körperschall *m*, Schall *m* in festen Körpern; ~**-borne-noise proofing** *n* NOISE Körperschalldämmung *f*; ~**-borne-noise-reducing** *adj* NOISE körperschalldämpfend; ~**-borne-noise reduction** *n* NOISE Körperschalldämpfung *f*; ~**-borne-noise transmission** *n* NOISE Körperschallübertragung *f*
stub: ~ **cable** *n* SOIL Stichleitung *f*
subaqueous: ~ **pump** *n* WATER SUPP Tauchpumpe *f*, Unterwasserpumpe *f*
subcell *n* WASTE Kassette *f*
subcool *vt* AIR unterkühlen
subdivision *n* POLICY *administrative unit* Unterabteilung *f*
subject: ~ **to calibration** *adj* ANALYSIS eichbedürftig; ~ **to obligatory disposal** *adj* WASTE entsorgungspflichtig
sub-lethal *adj* SAFETY subletal; ~ **dose** *n* SAFETY subletale Dosis *f*; ~ **effect** *n* SAFETY subletaler Effekt *m*

submains *n pl* WATER SUPP Nebenzweigung *f*
submerge *vt* WASTE, WATER SUPP überfluten
submerged *adj* WASTE, WATER SUPP überflutet; ~ **bed height** *n* WASTE, WATER SUPP Überstauhöhe *f*; ~ **plant tank** *n* WATER POLL Überstaupflanzenbecken *nt*; ~ **trickling filter** *n* WATER SUPP Tauchkörper *m*, Tauchtropfkörper *m*
submersible: ~ **burner** *n* WATER POLL, WATER SUPP Tauchbrenner *m*; ~ **plunger pump** *n* WASTE, WATER SUPP Tauchkörper *m*; ~ **pump** *n* ANALYSIS Tauchpumpe *f*, Unterwasserpumpe *f*
submicron *n* AIR *particle* Submikron *nt*, submikroskopisches Teilchen *nt*; ~ **particulate filter** *n* AIR *storage filter* Schwebstofffilter *nt*; ~ **particulate matter** *n* AIR Submikron-Schwebstoff-Teilchen *nt*
submicro: ~**-pore** *n* AIR Submikropore *f*
subpopulation: ~ **collective dose** *n* AIR, SAFETY Bevölkerungsteildosis *f*
subsequent: ~ **dewatering** *n* SOIL nachträgliche Entwässerung *f*; ~ **exhaust gas cleaning** *n* AIR, WATER POLL nachgeschaltete Abgasreinigung *f*
subside *vi* AIR einsinken, zusammenfallen
subsidence *n* AIR Einsinken *nt*, Zusammenfallen *nt*, WASTE Sackbildung *f*, Zusammenfallen *nt*
subsidiarity *n* POLICY Subsidiarität *f*; ~ **principle** *n* POLICY Subsidiaritätsprinzip *nt*
subsidize *vt* POLICY subventionieren
subsidized *adj* POLICY subventioniert
subsidy *n* POLICY Subvention *f*
subsoil *n* WASTE, WATER SUPP Untergrund *m*
substance *n* AIR, ANALYSIS, SOIL, WATER POLL Stoff *m*; ~**-related** *adj* AIR, WATER POLL stoffbezogen; ~**-specific** *adj* AIR, ANALYSIS, WATER POLL stoffspezifisch, substanzspezifisch; ~**-specific analysis** *n* ANALYSIS stoffspezifische Untersuchung *f*; ~**-specific criteria** *n pl* ANALYSIS stoffspezifische Kriterien *nt pl*
substrate *n* WASTE, WATER POLL Substrat *nt*; ~ **utilization** *n* WATER POLL Substratverwendung *f*
subsurface 1. *adj* SOIL, WASTE, WATER POLL, WATER SUPP unterirdisch; **2.** *n* WASTE, WATER SUPP Untergrund *m*
subsurface: ~ **diffusion air release** *n* WATER POLL unterirdische Belüftung *f*; ~ **erosion** *n* WASTE, WATER SUPP unterirdische Erosion *f*; ~ **irrigation** *n* WATER SUPP unterirdische Bewässerung *f*
subterranean: ~ **water** *n* WATER SUPP Grundwasser *nt*, unterirdisches Wasser *nt*
subterraneous: ~ **curtain** *n* SOIL, WASTE Schlitzwand *f*; ~ **curtain plate** *n* SOIL, WASTE Schlitzwandlamelle *f*
suck *vt* GEN absaugen
suction *n* GEN Absaugen *nt*, Absaugung *f*, Saugen *nt*; ~ **box** *n* AIR Saugkrümmer *m*, SOIL Saugkasten *m*; ~ **collector** *n* WATER SUPP Saugsammler *m*; ~ **dredge** *n* WATER SUPP Saugbagger *m*; ~ **extraction line** *n* WASTE Absaugleitung *f*; ~ **filter** *n* AIR Saugschlauchfilter *nt*, WATER SUPP Saugfilter *nt*; ~ **and filter installation for dust and chippings** *n* AIR, SAFETY, SOIL, WASTE Absaug- und Filtervorrichtung *f* für Staub und Späne; ~ **head** *n* WATER POLL, WATER SUPP Saughöhe *f*; ~ **hose** *n* WATER SUPP Saugschlauch *m*; ~ **level** *n* SOIL Absaugpegel *m*; ~ **lift** *n* WATER POLL, WATER SUPP Saughöhe *f*; ~ **nozzle** *n* WATER POLL, WATER SUPP Saugdüse *f*; ~ **pipe** *n* AIR Saugrohr *nt*, WASTE Ansaugrohr *nt*, WATER POLL Saugleitung *f*, WATER SUPP Ansaugrohr *nt*, Saugleitung *f*, Saugrohr *nt*; ~ **piping** *n* WATER POLL, WATER SUPP Saugleitung *f*; ~ **port** *n* WASTE Saugmund *m*; ~ **pump** *n* ANALYSIS *for quick determination* Absaugpumpe *f*, WATER POLL, WATER SUPP Saugpumpe *f*; ~ **side** *n* WATER POLL, WATER SUPP *of pump* Saugseite *f*; ~ **strainer** *n* WATER SUPP *of pump* Saugkopf *m*, Saugkorb *m*; ~ **tank** *n* WATER SUPP Saugkessel *m*; ~ **well** *n* WATER SUPP Saugbrunnen *m*, Saugschacht *m*
sugar: ~**-containing** *adj* RENEW ENERG zuckerhaltig; ~**-containing plants** *n pl* RENEW ENERG *biomass conversion* zuckerhaltige Pflanzen *f pl*; ~**-containing raw material** *n* RENEW ENERG *biomass conversion* zuckerhaltiger Rohstoff *m*
sulfate *AmE see sulphate BrE*
sulfide *AmE see sulphide BrE*
sulfite *AmE see sulphite BrE*
sulfur *AmE see sulphur BrE*
sulfuric: ~ **acid** *AmE see sulphuric acid BrE*; ~ **anhydride** *AmE see sulphuric anhydride BrE*
sulfurous *AmE see sulphurous BrE*
sulphate *n BrE* AIR, SOIL, WASTE Sulfat *nt*; ~ **anion** *n BrE* WASTE Sulfat-Anion *nt*; ~ **cellulose** *n BrE* WASTE Sulfatzellulose *f*; ~ **crystallizer** *n BrE* AIR Sulfatkristallisator *m*; ~ **ion** *n BrE* WASTE Sulfat-Ion *nt*; ~ **respiration** *n BrE* WATER SUPP Sulfatatmung *f*
sulphide *n BrE* AIR Sulfid *nt*; ~ **precipitation** *n BrE* WASTE, WATER POLL, WATER SUPP *wastewater purification* Sulfidfällung *f*
sulphite *n BrE* AIR, WASTE Sulfit *nt*; ~ **cellulose** *n BrE* WASTE, WATER SUPP Sulfitzellstoff *m*
sulphur *n BrE* AIR, WASTE, WATER SUPP Schwefel *m*; ~ **anhydride** *n BrE* WASTE Schwefelsäureanhydrid *nt*; ~ **budget** *n BrE* AIR Schwefelhaushalt *m*; ~ **content** *n BrE* AIR Schwefelgehalt *m*; ~ **cycle** *n BrE* AIR Schwefelkreislauf *m*; ~ **dioxide** *n BrE* (SO_2) AIR Schwefeldioxid *nt* (SO_2); ~ **dioxide emissions** *n pl BrE* AIR Schwefeldioxidemissionen *f pl*; ~ **dioxide poisoning** *n BrE* AIR Schwefeldioxidvergiftung *f*; ~ **dioxide reduction** *n BrE* AIR, WASTE Schwefeldioxidreduktion *f*; ~**-free** *adj BrE* AIR schwefelfrei
sulphuric: ~ **acid** *n BrE* AIR, WASTE Schwefelsäure *f*; ~ **anhydride** *n BrE* AIR, WATER SUPP Schwefelsäurenanhydrid *nt*
sulphurous *adj BrE* AIR, WASTE, WATER SUPP schwefelhaltig; ~ **acid** *n BrE* AIR, WASTE schweflige Säure *f*; ~ **combustibles** *n pl BrE* AIR, WASTE schwefelhaltige Brennstoffe *m pl*; ~ **gas** *n BrE* AIR schwefelhaltiges Gas *nt*; ~ **waste** *n BrE* WASTE schwefelhaltiger Abfall *m*
sulphur: ~ **oxide** *n BrE* (SO) AIR, WASTE Schwefeloxid *nt* (SO); ~ **oxides** *n pl BrE* (SO_x) AIR Schwefeloxide *nt pl* (SO_x); ~ **oxides scrubber** *n BrE* AIR Schwefeloxidwäscher *m*; ~ **recovery plant** *n BrE* AIR, WASTE *Claus plant* Schwefelrückgewinnungsanlage *f*; ~**-rich** *adj BrE* AIR schwefelreich; ~ **tetraoxide** *n BrE* ($(SO_4)_x$) AIR Schwefeltetroxid *nt* ($(SO_4)_x$); ~ **trioxide** *n BrE* (SO_3) AIR, WASTE Schwefeltrioxid *nt* (SO_3)
Sulzer-Ariqua: ~ **process** *n* AIR Sulzer-Ariqua-Verfahren *nt*
summertime: ~ **ozone value** *n* AIR sommerlicher Ozonwert *m*
sump *n* WATER POLL Sumpf *m*
sunk: ~ **hydrant** *n* WASTE Unterflurwasserpfosten *m*, WATER SUPP Unterflurhydrant *m*, Unterflurwasserpfosten *m*
sunlight *n* RENEW ENERG Sonnenlicht *nt*; ~ **radiation** *n* AIR Sonnenlichtabstrahlung *f*
supercritical *adj* ANALYSIS, SOIL überkritisch; ~**-fluid chromatography** *n* (*SFC*) ANALYSIS überkritische Flüssig-Chromatographie *f*; ~**-fluid extraction** *n* (*SFE*) ANALYSIS überkritische Flüssig-Extraktion *f*; ~ **water** *n* SOIL überkritisches Wasser *nt*
superheated: ~ **boiler** *n* WASTE Heißdampfkessel *m*; ~ **steam** *n* AIR, RENEW ENERG Heißdampf *m*
superheater *n* AIR, WATER POLL *incineration* Überhitzer *m*
supernatant: ~ **liquid** *n* WATER POLL überstehende Flüssigkeit *f*, *in fermentation vessel* Schwimmdecke *f*; ~ **liquor** *n* WATER POLL Faulraumwasser *nt*
supersaturated *adj* AIR, WASTE, WATER SUPP übersättigt; ~ **solution** *n* WASTE, WATER SUPP übersättigte Lösung *f*
supersonic: ~ **boom** *n* NOISE Überschallknall *m*; ~ **speed** *n* NOISE Überschallgeschwindigkeit *f*
supervise *vt* GEN überwachen

supervision *n* GEN Aufsicht *f*, Überwachung *f*
supervisor *n* GEN Inspektor *m*, MARINE Vorgesetzter *m*
supplementary: ~ **fuel** *n* AIR *thermal and catalytic combustion* Zusatzbrennstoff *m*; ~ **purification** *n* AIR Zusatzreinigung *f*
supplied: ~**-air breathing apparatus** *n* SAFETY Atemgerät *nt* mit externer Luftversorgung
supply 1. *n* MANAGE Vorrat *m*, WASTE, WATER POLL Anschluß *m*, WATER SUPP Versorgung *f*, Anschluß *m*; **2.** *vt* AIR zuführen, WATER SUPP versorgen
supply: ~ **area** *n* WATER SUPP Versorgungsgebiet *nt*; ~ **district** *n* WATER SUPP Versorgungszone *f*; ~ **line** *n* WATER SUPP Versorgungsleitung *f*; ~ **network** *n* WASTE Versorgungsnetz *nt*, WATER SUPP Leitungsnetz *nt*, Versorgungsnetz *nt*; ~ **pipe** *n* WATER SUPP Zuführungsleitung *f*; ~ **potential** *n* RENEW ENERG Angebotspotential *nt*; ~ **pump** *n* MARINE Versorgungspumpe *f*; ~ **system** *n* WATER SUPP Versorgungsnetz *nt*
support *n* POLICY Förderung *f*; ~ **body** *n* AIR Stützkörper *m*; ~ **element** *n* AIR *filter bag* Stützelement *nt*
supporting: ~ **grate** *n* AIR Auflagerost *m*; ~ **suspension** *n* SOIL Stützsuspension *f*
support: ~ **ring** *n* AIR *bag filter* Stützring *m*; ~ **scrim** *n* AIR *of needled felt* Stützgewebe *nt*; ~ **wire cage** *n* AIR *bag filter* Drahtstützkorb *m*
suppress *vt* MARINE, NOISE verhindern
surface 1. *adj* WASTE, WATER SUPP oberirdisch; **2.** *n* GEN Fläche *f*, Oberfläche *f*
surface: ~**-active** *adj* AIR grenzaktiv, SOIL oberflächenaktiv; ~**-active agent** *n* AIR grenzaktiver Stoff *m*, MARINE, WATER POLL, WATER SUPP Detergens *nt*, Tensid *nt*, oberflächenaktiver Stoff *m*; ~**-active substances** *n pl* SOIL *additive* oberflächenaktive Substanzen *f pl*, WATER POLL oberflächenaktive Stoffe *m pl*; ~ **aeration** *n* WASTE, WATER POLL Oberflächenbelüftung *f*; ~ **aerator** *n* WASTE Oberflächenbelüfter *m*; ~ **application** *n* WATER POLL Auftragung *f* von Klärschlamm auf Flächen; ~ **area** *n* SOIL Oberflächenbereich *m*, WATER SUPP *of reservoir* Staufläche *f*; ~ **catalysis** *n* AIR Kontaktkatalyse *f*; ~ **collector** *n* RENEW ENERG Flachkollektor *m*; ~ **condenser** *n* AIR Oberflächenkondensator *m*; ~ **conductivity** *n* AIR Oberflächenleitfähigkeit *f*; ~ **cover** *n* SOIL Oberflächenabdeckung *f*, Off-site-Verfahren *nt*; ~ **evaporation** *n* WATER SUPP Oberflächenverdunstung *f*; ~ **film** *n* WATER POLL Oberflächenfilm *m*; ~ **filter** *n* AIR Oberflächenfilter *nt*, WATER SUPP Flächenfilter *nt*; ~ **filtration** *n* AIR *deep-bed filtration*, WATER POLL Oberflächenfiltration *f*, WATER SUPP Flächenfiltration *f*, Oberflächenfiltration *f*; ~ **filtration with membranes** *n* WATER POLL Oberflächenfiltration *f* mit Membranen; ~ **installation** *n* WASTE, WATER SUPP Flächeneinbau *m*; ~ **irradiation** *n* SOIL Oberflächenausstrahlung *f*; ~ **irrigation** *n* SOIL Berieselung *f*; ~ **landfill** *n* WASTE oberirdische Deponie *f*; ~ **layer** *n* WASTE Oberflächenschicht *f*; ~ **lining** *n* NOISE *sound-absorbent panelling* Flächenverkleidung *f*; ~ **load** *n* AIR Flächenbelastung *f*; ~ **loading** *n* WASTE Flächenbelastung *f*; ~ **method** *n* WASTE, WATER SUPP Oberflächenmethode *f*; ~ **refurbishing** *n* WASTE, WATER SUPP Oberflächenneubildung *f*; ~ **renewal** *n* WASTE, WATER SUPP Oberflächenneubildung *f*; ~ **requirement** *n* RENEW ENERG Flächenbedarf *m*; ~ **reservoir** *n* WATER POLL *indirect reuse of wastewater* Oberflächensammelbecken *nt*; ~ **roughness** *n* WASTE Oberflächenrauhigkeit *f*; ~ **runoff** *n* WASTE, WATER POLL, WATER SUPP Abfluß *m* im Oberflächenbereich, Oberflächenabfluß *m*, oberirdischer Abfluß *m*; ~ **sealing** *n* WASTE Oberflächenabdichtung *f*; ~ **seawater** *n* RENEW ENERG *ocean thermal energy* Oberflächenmeerwasser *nt*; ~ **sensor** *n* ANALYSIS Oberflächenmeßfühler *m*; ~ **smoothing** *n* SOIL *of surface sealing* Oberflächenglättung *f*; ~ **tension modifier** *n* MARINE Mittel *nt* zur Modifizierung der Oberflächenspannung; ~ **water** *n* WATER SUPP Oberflächenwasser *nt*; ~ **waters** *n pl* WATER SUPP Oberflächengewässer *nt*; ~ **waters management** *n* WATER SUPP Oberflächengewässerbewirtschaftung *f*; ~ **waters treatment** *n* WATER SUPP Oberflächengewässeraufbereitung *f*; ~ **water temperature** *n* RENEW ENERG Oberflächenwassertemperatur *f*
surfactant *n* AIR grenzflächenaktiver Stoff *m*, MARINE Tensid *nt*, WASTE grenzflächenaktiver Stoff *m*, WATER POLL Tensid *nt*, WATER SUPP grenzflächenaktiver Stoff *m*
surge: ~ **hopper** *n* SOIL Ausgleichsbunker *m*
surplus *n* WASTE, WATER POLL Überschuß *m*; ~ **water** *n* WATER SUPP Freiwasser *nt*, Überwasser *nt*
surrounding: ~ **field analysis** *n* ANALYSIS Umfeldanalyse *f*
surveillance *n* GEN Überwachung *f*
survey *vt* MARINE vermessen
suspended *adj* AIR dispergiert, WASTE, WATER POLL, WATER SUPP feinverteilt, schwebend, suspendiert; ~ **bacteria** *n* WATER POLL suspendierte Bakterien *pl*; ~ **ceiling** *n* NOISE *resilient layer* abgehängte Decke *f*; ~ **frame** *n* AIR *bag filter* Hängerahmen *m*; ~ **liquid droplets** *n pl* AIR, WASTE, WATER SUPP schwebende Tröpfchen *nt pl*; ~ **liquid phase** *n* AIR dispergierte Flüssigphase *f*; ~ **manganese dioxide** *n* WASTE suspendiertes Mangandioxid *nt*; ~ **matter** *n* WATER POLL Schwebstoff *m*; ~ **organic solids** *n pl* WATER POLL suspendierte organische Stoffe *m pl*; ~ **particle** *n* AIR Schwebstoff *m*, WASTE suspendiertes Teilchen *nt*, WATER SUPP Schwebteilchen *nt*, suspendiertes Teilchen *nt*, Schwebstoff *m*; ~ **particulate matter** *n* AIR Schwebstaub *m*; ~ **and settable solids** *n pl* WATER POLL Detritus *m*, Schweb- und Sinkstoffe *m pl*; ~ **solid particle** *n* AIR dispergiertes Feststoffteilchen *nt*, ANALYSIS Schwebkörper *m*, Schwebstaub *m*; ~ **solid particle meter** *n* ANALYSIS Schwebkörpermesser *m*; ~ **solids** *n pl* WATER POLL, WATER SUPP Schwebstoffe *m pl*
suspension *n* GEN Suspension *f*; ~ **dryer** *n* AIR, WATER POLL Schwebetrockner *m*
sustain *vt* SAFETY *injuries* erleiden
sustainability *n* MANAGE, POLICY, SOIL Nachhaltigkeit *f*
sustainable *adj* MANAGE, POLICY, SOIL nachhaltig; ~ **business** *n* MANAGE nachhaltiges Wirtschaften *nt*; ~ **development** *n* MANAGE, POLICY nachhaltige Entwicklung *f*; ~ **economy** *n* POLICY nachhaltiges Wirtschaften *nt*; ~ **forestry** *n* POLICY nachhaltige Forstwirtschaft *f*; ~ **recycling economy** *n* POLICY nachhaltige Kreislaufwirtschaft *f*; ~ **soil use** *n* SOIL nachhaltige Bodennutzung *f*
SVOC *abbr* (*semivolatile organic compounds*) SOIL SVOC (*semiflüchtige organische Verbindungen*)
swamp *n* WASTE, WATER SUPP Moor *nt*, Morast *m*, Sumpf *m*
swath *n* MARINE Schwade *f*, Schwaden *m*
sweep: ~ **coagulation** *n* WATER POLL *flocculation* Einschlußflockung *f*
sweeper *n* WASTE *machine* Kehrmaschine *f*
swing *n* ANALYSIS, NOISE Schwingung *f*
swinging: ~ **scraper** *n* WATER POLL Pendelräumer *m*
SWL *abbr* (*short-wave listener*) NOISE Kurzwellenhörer *m*
synchronous *adj* ANALYSIS, RENEW ENERG gleichzeitig, synchron; ~ **generator** *n* RENEW ENERG *wind turbine* Synchrongenerator *m*
synergetics *n* AIR, WASTE, WATER SUPP Synergetik *f*
synergy *n* AIR, WASTE, WATER SUPP Synergismus *m*
synthesis: ~ **gas** *n* WASTE Synthesegas *nt*; ~ **gas generation** *n* WASTE *plastics* Synthesegaserzeugung *f*; ~ **gas recovery** *n* WASTE Synthesegasgewinnung *f*
synthetic *adj* GEN künstlich, synthetisch; ~ **fiber** *AmE*, ~ **fibre** *BrE* *n* AIR *filter medium* synthetische Faser *f*; ~ **lining** *n* WASTE, WATER SUPP künstliche Abdichtung *f*; ~ **materials** *n pl* WASTE *landfill sealing* synthetische Materialien *nt pl*; ~ **natural gas** *n* (*SNG*) RENEW ENERG Erdgasaustauschgas *nt* (*SNG*); ~ **oil** *n* RENEW ENERG synthetisches Öl *nt*; ~ **polymer** *n* WATER POLL synthetisches Polymer *nt*

system *n* ANALYSIS, MANAGE, POLICY System *nt*

systematic *adj* ANALYSIS, MANAGE, POLICY systematisch; ~ **error** *n* ANALYSIS systematischer Fehler *m*

system: ~ **boundary** *n* MANAGE Systemgrenze *f*; ~ **parameter** *n* ANALYSIS Systemgröße *f*

T

TA *abbr* (*technology assessment*) POLICY TFA (*Technologiefolgenabschätzung*)
tablet: ~ **press** *n* ANALYSIS Pastillenpresse *f*
tacky: ~ **particulate components** *n pl* AIR klebende Staubbestandteile *m pl*
tail: ~ **gas** *n* AIR, WASTE Abgas *nt*, Endgas *nt*, Restgas *nt*, Schornsteingas *nt*
tailings *n pl BrE* (*cf tails AmE*) WATER SUPP Waschberge *m pl*; ~ **pond** *n* WATER POLL Becken *nt* für Rückstände
tail: ~ **pipe** *n* AIR Abgasrohr *nt*, Auspuffrohr *nt*, WATER SUPP *of pump* Fußröhre *f*, Saugrohr *nt*; ~ **pipe emissions** *n pl* AIR Abgasrohremissionen *f pl*
tails *n pl AmE* (*cf tailings BrE*) WATER SUPP Waschberge *m pl*
tailwater *n* WATER SUPP Unterwasser *nt*
take: ~ **precautions** *vi* SAFETY Vorkehrungen treffen, Vorsorge treffen; ~ **a sample** *vi* ANALYSIS eine Probe nehmen; ~ **a sample automatically** *vi* ANALYSIS automatisch eine Probe nehmen
take: ~**-back** *n* WASTE Rücknahme *f*; ~**-back obligation** *n* WASTE Rücknahmepflicht *f*, Rücknahmeverpflichtung *f*; ~**-back obligation for plastics** *n* WASTE Kunststoffrücknahmeverpflichtung *f*
tamped: ~ **density** *n* ANALYSIS Stampfdichte *f*
tamping: ~ **volume meter** *n* ANALYSIS Stampfvolumeter *nt*
tandem: ~ **detector** *n* ANALYSIS Tandem-Detektor *m*; ~ **mounting** *n* AIR *of cyclones* Hintereinanderschaltung *f*
tangential *adj* AIR tangential; ~**-entry cyclone** *n* AIR tangential angeströmter Zyklon *m*
tank *n* WATER SUPP Behälter *m*, Tank *m*; ~ **adsorber** *n* AIR Behälteradsorber *m*; ~ **bottom** *n* WATER POLL Beckensohle *f*
tanker *n* MARINE Tankschiff *nt*, Tanker *m*
tank: ~ **infiltration** *n* WASTE, WATER SUPP Beckeninfiltration *f*
tapered: ~ **aeration** *n* WATER POLL abgestufte Lüftung *f*, konische Staffelbelüftung *f*, wechselnde Belüftung *f*
tapping *n* WATER SUPP Entnahme *f*; ~ **of drinking water** *n* WATER SUPP Trinkwasserentnahme *f*; ~ **of lake water** *n* WATER SUPP *for drinking water recovery* Seefassung *f*; ~ **pipe** *n* WATER SUPP Fassungsrohr *nt*; ~ **of a spring** *n* WATER SUPP Quellfassung *f*
tap: ~ **water** *n* WATER POLL, WATER SUPP Leitungswasser *nt*
tarball *n* MARINE Teerklumpen *m*
target: ~ **droplet** *n* AIR Zieltröpfchen *nt*; ~ **group** *n* POLICY Zielgruppe *f*
task *n* ANALYSIS Aufgabe *f*
tasteless *adj* AIR *gas*, WATER SUPP geschmacklos, geschmacksfrei, geschmacksneutral
tax 1. *n* POLICY Steuer *f*; **2.** *vt* POLICY besteuern
tax: ~ **base** *n* POLICY Steuerbemessungsgrundlage *f*; ~ **burden** *n* POLICY Steuerbelastung *f*; ~ **exemption** *n* POLICY Steuerbefreiung *f*; ~ **incentive** *n* POLICY steuerlicher Anreiz *m*; ~ **laws** *n pl* POLICY Steuergesetze *nt pl*; ~ **provision** *n* POLICY steuerliche Regelung *f*; ~ **quota** *n* POLICY Steuerquote *f*; ~ **rate** *n* POLICY Steuersatz *m*; ~ **rebate** *n* POLICY Steuerentlastung *f*; ~ **reduction** *n* POLICY Steuerkürzung *f*; ~ **relief** *n* POLICY Steuerminderung *f*; ~ **shift** *n* POLICY Steuerüberwälzung *f*; ~ **tariff** *n* POLICY Steuertarif *m*
TC *abbr* ANALYSIS (*total carbon*) TC (*gesamter vorhandener Kohlenstoff*), SAFETY (*toxic concentration*) TC (*toxische Konzentration*)
TCD *abbr* (*thermal conductivity detector*) ANALYSIS WLD (*Wärmeleitfähigkeit-Detektor*)
TD *abbr* (*toxic dose*) SAFETY TD (*toxische Dosis*)
TDS *abbr* (*total dissolved solids*) WATER SUPP Gesamtschwebestoffe *m pl*
tear: ~ **gas** *n* AIR Reizgas *nt*
tearing *n* AIR *of filter* Reißen *nt*, Zerreißen *nt*
tear: ~ **strength** *n* AIR *filter* Reißfestigkeit *f*, Reißwiderstand *m*
technical *adj* GEN technisch; ~ **directive** *n* POLICY technische Anleitung *f* (*TA*); ~ **potential** *n* RENEW ENERG *of energy source* technisches Potential *nt*; ~ **safety demand** *n* SAFETY sicherheitstechnische Anforderung *f*; ~ **safety measures** *n pl* SAFETY sicherheitstechnische Maßnahmen *f pl*
technology *n* GEN Technik *f*, Technologie *f*; ~ **application** *n* ANALYSIS, MANAGE, SAFETY Anwendung *f* der Technik; ~ **assessment** *n* (*TA*) POLICY Technologiefolgenabschätzung *f* (*TFA*)
telecontrol: ~ **system** *n* WATER POLL, WATER SUPP Fernsteuersystem *nt*
telemetry *n* ANALYSIS Meßwertfernübertragung *f*, Telemetrie *f*
telescope *n* ANALYSIS Teleskop *nt*
TEM *abbr* (*transmission electron microscopy*) ANALYSIS Durchstrahlungselektronenmikroskopie *f*
temper *vt* ANALYSIS mäßigen, temperieren
temperature *n* GEN Temperatur *f*; ~**-controlled** *adj* ANALYSIS temperaturgeregelt; ~ **gradient** *n* AIR, NOISE, WATER POLL Temperaturgradient *m*; ~ **inversion** *n* AIR Temperaturinversion *f*; ~ **measurement** *n* ANALYSIS Temperaturmessung *f*; ~ **probe** *n* ANALYSIS Temperaturfühler *m*; ~ **range** *n* AIR Temperaturbereich *m*; ~ **resistance** *n* AIR, SOIL, WASTE Temperaturbeständigkeit *f*; ~**-resistant** *adj* AIR *filter materials*, SOIL, WASTE temperaturbeständig; ~ **rise** *n* AIR Temperaturanstieg *m*; ~ **sensor** *n* ANALYSIS Temperaturmeßfühler *m*; ~ **stability** *n* ANALYSIS Temperaturstabilität *f*; ~**-stable** *adj* ANALYSIS temperaturstabil
temporal: ~ **dose distribution** *n* AIR, WASTE, WATER SUPP zeitliche Dosisverteilung *f*; ~ **evaluation** *n* ANALYSIS zeitliche Auswertung *f*; ~ **fluctuation** *n* AIR, WASTE, WATER SUPP zeitliche Schwankung *f*
temporally: ~ **representative** *adj* ANALYSIS *sample* zeitlich repräsentativ
temporal: ~ **resolution** *n* AIR, WASTE, WATER SUPP zeitliche Auflösung *f*, zeitliches Auflösungsvermögen *nt*; ~ **variation** *n* AIR, WASTE, WATER SUPP zeitliche Schwankung *f*
temporary: ~ **cover** *n* WASTE Zwischenabdeckung *f*; ~ **covering material** *n* WASTE Zwischenabdeckmaterial *nt*; ~ **hardness** *n* WATER POLL, WATER SUPP temporäre Härte *f*, vorübergehende Härte *f*; ~ **remediation** *n* SOIL vorübergehende Sanierung *f*; ~ **storage** *n* SOIL, WASTE Zwischenlagerung *f*; ~ **storage silo** *n* AIR Zwischensilo *m*; ~ **tank** *n* AIR Zwischenbehälter *m*; ~ **threshold shift** *n* (*TTS*) NOISE zeitweilige Hörschwellenverschiebung *f*
tender *n* MARINE Tender *m*
tensile: ~ **strength** *n* AIR Zugfestigkeit *f*
tensiometer *n* ANALYSIS, SOIL Tensiometer *nt*
tension: ~ **member** *n* MARINE Zugglied *nt*, zugbeanspruchtes Glied *nt*
teratogen *n* SAFETY Teratogen *nt*
teratogenic *adj* SAFETY teratogen
tertiary: ~ **plastic recycling** *n* WASTE Feedstock-Recycling *nt*, chemisches Recycling *nt*; ~ **pond** *n* WATER POLL Schönungsteich *m*; ~ **wastewater treatment** *n* WATER POLL dritte

Reinigungsstufe *f*, fortgeschrittene Abwasserbehandlung *f*, tertiäre Abwasserbehandlung *f*

test 1. *n* ANALYSIS, SAFETY Test *m*, Probe *f*, Versuch *m*; **2.** *vt* ANALYSIS, SAFETY prüfen, testen, untersuchen

test: ~ **basin** *n* WATER SUPP Versuchsbecken *nt*; ~ **crucible** *n* ANALYSIS Probiertiegel *m*

tested *adj* ANALYSIS, SAFETY untersucht

test: ~ **environment** *n* ANALYSIS Meßumgebung *f*; ~ **filtration plant** *n* WATER SUPP Testfiltrationsanlage *f*; ~ **gas** *n* ANALYSIS Prüfgas *nt*; ~ **gas calibrator** *n* ANALYSIS Prüfgaskalibrator *m*

testing: ~ **aerator** *n* WATER POLL Probebelüfter *m*; ~ **conditions** *n pl* ANALYSIS Prüfbedingungen *f pl*; ~ **lab** *n* ANALYSIS Untersuchungslaboratorium *nt*; ~ **laboratory** *n* ANALYSIS Untersuchungslaboratorium *nt*; ~ **limit** *n* ANALYSIS Nachweisgrenze *f*; ~ **result** *n* ANALYSIS Untersuchungsbefund *m*

test: ~ **paper** *n* ANALYSIS Testpapier *nt*; ~ **phase** *n* RENEW ENERG *of installation* Testphase *f*; ~ **plant** *n* ANALYSIS, WATER SUPP Testanlage *f*; ~ **probe** *n* ANALYSIS Prüfsonde *f*; ~ **pump** *n* WATER SUPP Versuchspumpe *f*; ~ **sieving** *n* ANALYSIS *dust* Prüfsiebung *f*; ~ **solution** *n* ANALYSIS Meßlösung *f*; ~ **tube** *n* ANALYSIS Probeglas *nt*, Reagenzglas *nt*; ~ **value** *n* ANALYSIS Prüfwert *m*; ~ **water** *n* ANALYSIS Prüfwasser *nt*

Tetra: ~ **CDD** *abbr* (*tetrachlorinated dibenzo-p-dioxine*) AIR, SOIL, WASTE Tetra CDD (*Tetrachlordibenzo-p-dioxin*)

tetrachlorinated: ~ **dibenzo-p-dioxines** *n* (*Tetra CDD*) AIR, SOIL, WASTE Tetrachlordibenzo-p-dioxin *nt* (*Tetra CDD*)

textile: ~ **waste** *n* WASTE Textilabfall *m*

theoretical: ~ **potential** *n* RENEW ENERG *of energy source* theoretisches Potential *nt*

thermal *adj* GEN thermisch, Thermo-; ~ **ageing** *n* WASTE thermische Alterung *f*; ~ **air cleaner** *n* AIR Thermostat-Luftfilter *nt*, thermostatgeregeltes Luftfilter *nt*; ~ **analysis** *n* ANALYSIS thermische Analyse *f*; ~ **balance** *n* AIR, WATER POLL Wärmebilanz *f*; ~ **capacity** *n* AIR, WASTE, WATER SUPP Heizleistung *f*, thermische Leistung *f*; ~ **combustion** *n* AIR thermische Verbrennung *f*, thermische Nachverbrennung *f* (*TNV*); ~ **combustion plant** *n* AIR thermische Nachverbrennungsanlage *f*; ~ **conduction** *n* AIR Wärmeleitung *f*; ~ **conductivity detector** *n* (*TCD*) ANALYSIS Wärmeleitfähigkeit-Detektor *m* (*WLD*); ~ **decomposition** *n* WASTE thermische Zerlegung *f*, thermische Zersetzung *f*; ~ **degradation** *n* WASTE thermischer Abbau *m*; ~ **desorption** *n* ANALYSIS *pretreatment for pollutant analysis* thermische Desorption *f*; ~ **desorption chamber** *n* ANALYSIS thermische Desorptionskammer *f*; ~ **destruction** *n* SOIL *direct heating by radiation* thermische Zerstörung *f*; ~ **discharge** *n* AIR, WASTE, WATER POLL, WATER SUPP Abwärme *f*; ~ **drying** *n* WATER POLL thermische Trocknung *f*; ~ **energy** *n* AIR Wärmeenergie *f*, RENEW ENERG thermische Energie *f*; ~ **heat expansion** *n* AIR *catalyst* Wärmeausdehnung *f*; ~ **incinerator** *n* AIR thermischer Nachbrenner *m*; ~ **load** *n* AIR, WASTE thermische Belastung *f*, Wärmebelastung *f*

thermally: ~ **aerated** *adj* ANALYSIS thermisch belüftet; ~ **isolated** *adj* ANALYSIS thermisch isoliert; ~ **stabilized** *adj* WATER POLL thermisch stabilisiert

thermal: ~ **oil** *n* AIR, RENEW ENERG Thermalöl *nt*; ~ **oil heater** *n* AIR, RENEW ENERG Thermalölerhitzer *m*; ~ **output** *n* AIR Wärmeleistung *f*; ~ **oxidation** *n* AIR thermische Oxidation *f*; ~ **process** *n* WATER POLL thermisches Verfahren *nt*; ~ **reactor** *n* AIR Thermoreaktor *m*; ~ **recovery** *n* WASTE thermische Rückgewinnung *f*; ~ **regeneration** *n* WATER POLL thermische Regeneration *f*; ~ **residue recovery** *n* WASTE thermische Resteverwertung *f*; ~ **softening** *n* WASTE, WATER SUPP thermische Enthärtung *f*; ~ **spring** *n* RENEW ENERG Thermalquelle *f*; ~ **use** *n* WATER POLL *from stabilized sludge, exhaust gas* Wärmenutzung *f*

thermionic: ~ **detector** *n* ANALYSIS Thermoionischer Detektor *m*

thermistor *n* ANALYSIS Thermistor *m*

thermochemical: ~ **conversion** *n* RENEW ENERG *biomass* thermochemische Umwandlung *f*

thermocline *n* WATER POLL Metalimnion *nt*, Sprungschicht *f*

thermocouple *n* AIR *catalytic incineration* Thermoelement *nt*

thermogravimetry *n* ANALYSIS Thermogravimetrie *f*

thermohygrograph *n* ANALYSIS Thermohygrograph *m*

thermoluminescence *n* ANALYSIS Thermolumineszenz *f*; ~ **dosimeter** *n* (*TLD*) ANALYSIS Thermolumineszenz-Dosimeter *nt* (*TLD*)

thermomagnetic *adj* ANALYSIS thermomagnetisch

thermometer *n* ANALYSIS Thermometer *nt*

thermophilic *adj* WATER POLL thermophil; ~ **bacteria** *n pl* WATER POLL thermophile Bakterien *pl*; ~ **digestion** *n* WATER POLL thermophile Faulung *f*; ~**-mesophilic** *adj* WATER POLL thermophil-mesophil; ~**-mesophilic digestion** *n* WATER POLL thermophil-mesophile Faulung *f*

thermophoresis *n* AIR Thermophorese *f*

thermoplast *n* SOIL *binding agent*, WASTE Thermoplast *m*

thermoplastic *adj* SOIL, WASTE thermoplastisch; ~ **material** *n* AIR thermoplastischer Stoff *m*

thermoset *n* WASTE Duromer *nt*, Duroplast *m*

thermosetting *adj* WASTE duroplastisch

thermoshock: ~ **resistance** *n* AIR *catalyst* Thermoschockfestigkeit *f*

thermosphere *n* AIR Thermosphäre *f*

thermospray: ~ **interface** *n* ANALYSIS Thermospray-Interface *nt*

thermostabile *adj* AIR thermostabil

thermostability *n* AIR Thermostabilität *f*

thermostat *n* ANALYSIS Thermostat *m*

thermostatic *adj* ANALYSIS thermostatisch

thicken *vt* AIR verdichten, WASTE eindicken, WATER POLL *sludge* eindicken

thickened *adj* AIR, SOIL, WASTE, WATER POLL, WATER SUPP entwässert, AIR verdichtet; ~ **sludge** *n* WATER POLL Dickschlamm *m*

thickener *n* WASTE, WATER POLL Eindicker *m*; ~ **overflow tank** *n* WASTE, WATER POLL Eindicker-Überlaufbehälter *m*; ~ **pre-storage tank** *n* WASTE, WATER POLL Eindicker-Vorlagebehälter *m*

thickening *n* AIR Verdichten *nt*, Verdichtung *f*, WASTE Eindickung *f*, WATER POLL *of sludge* Eindickung *f*, Entwässerung *f*; ~ **by freezing** *n* WATER POLL Entwässerung durch Gefrieren *f*

thick: ~ **matter flow** *n* WATER POLL Dickstoffstrom *m*

thickness *n* WASTE *landfill layers* Dichte *f*, Stärke *f*

thin *vt* AIR, ANALYSIS, WASTE, WATER POLL verdünnen

thin: ~**-film evaporator** *n* WATER POLL, WATER SUPP Dünnschichtverdampfer *m*; ~**-film module** *n* RENEW ENERG *solar cell* Dünnschichtmodul *nt*; ~**-film photovoltaic cell** *n* RENEW ENERG Dünnschichtphotovoltaikzelle *f*; ~**-layer chromatography** *n* (*TLC*) ANALYSIS Dünnschichtchromatographie *f* (*DC*); ~**-layer drying** *n* WATER POLL Dünnschichttrocknung *f*; ~**-layer electrophoresis** *n* ANALYSIS Dünnschichtelektrophorese *f*; ~**-layer installation** *n* WASTE, WATER SUPP *surface installation* Dünnschichteinbau *m*

thinning *n* AIR *of ozone layer* Ausdünnung *f*

thin: ~**-walled** *adj* ANALYSIS *sampling* dünnwandig

thiol *n* AIR, WASTE Thiolalkohol *m*

third: ~**-party recycling** *n* WASTE außerbetriebliche Verwertung *f*

thixotropic: ~ **grouting material** *n* SOIL thixotropes Verpreßmittel *nt*

thorough: ~ **mixture** *n* AIR, WASTE, WATER POLL, WATER SUPP Durchmischung *f*

thousandth: ~ **part** *n* ANALYSIS Promille *nt*

three: **~-liter automobile** *n AmE* (*cf three-litre-car BrE*) AIR Drei-Liter-Auto *nt*; **~-litre car** *BrE n* (*cf three-liter automobile AmE*) AIR Drei-Liter-Auto *nt*; **~-way catalyst** *n* (*TWC*) AIR Drei-Wege-Katalysator *m*

threshold *n* ANALYSIS, POLICY Schwellenwert *m*; **~ of audibility** *n* NOISE Hörbarkeitsschwelle *f*; **~ audiometry** *n* NOISE Schwellenwertaudiometrie *f*; **~ limit value** *n* (*TLV*) NOISE, POLICY, SAFETY höchstzulässiger Konzentrationswert *m* (*HZK*); **~ odor** *AmE*, **~ odour** *BrE n* ANALYSIS, WASTE, WATER POLL, WATER SUPP Geruchsschwelle *f*; **~ odour concentration** *n BrE* ANALYSIS, WASTE, WATER POLL, WATER SUPP Geruchsschwellenkonzentration *f*; **~ odour number** *n BrE* (*TON*) ANALYSIS, WASTE, WATER POLL, WATER SUPP Geruchsschwellenwert *m* (*GSW*); **~ value** *n* ANALYSIS Schwellenwert *m*; **~ value for drinking water** *n* WATER SUPP Trinkwasserschwellenwert *m*; **~ value range** *n* ANALYSIS Intervalwert *m*

throttle *n* WATER SUPP Drossel *f*

throughfall *n* AIR Durchfall *m*

through: **~ fraction** *n* WASTE Siebdurchgang *m*

throughput *n* AIR, SAFETY Durchsatz *m*, SOIL *precipitation reaching the ground directly* Durchfall *m*, Durchsatz *m*, WATER POLL, WATER SUPP Durchsatz *m*; **~ fluctuation** *n* AIR Durchsatzschwankung *f*

throw: **~ away** *vt* WASTE abwerfen

throw: **~-away pack** *n* WASTE Einwegverpackung *f*, Wegwerfpackung *f*; **~-away product** *n* WASTE Einwegprodukt *nt*, Wegwerferzeugnis *nt*, Wegwerfprodukt *nt*

thruput *AmE see throughput*

thrust *n* MARINE Druck *m*, Schub *m*, Schubkraft *f*

TIC *abbr* (*total inorganic carbon*) ANALYSIS, WASTE TIC (*gesamter anorganisch gebundener Kohlenstoff*)

tidal: **~ barrage** *n* RENEW ENERG Gezeitensperrwerk *nt*; **~ basin** *n* RENEW ENERG Tidebecken *nt*; **~ energy** *n* RENEW ENERG Gezeitenenergie *f*; **~ energy resources** *n pl* RENEW ENERG Gezeitenenergieressourcen *f pl*; **~ mill** *n* RENEW ENERG *earlier term* Gezeitenkraftwerk *nt*; **~ power** *n* RENEW ENERG Gezeitenenergie *f*, Gezeitenkraft *f*; **~ power plant** *n* RENEW ENERG Gezeitenkraftwerk *nt*; **~ range** *n* RENEW ENERG Tidenhub *m*

tide *n* RENEW ENERG Gezeiten *f pl*

tiered: **~ biofilter** *n* AIR *biological sorption process* Etagenbiofilter *nt*; **~ dryer** *n* WATER POLL Etagentrockner *m*; **~ furnace** *n* WATER POLL Etagenofen *m*

tight *adj* WASTE, WATER SUPP *water-impermeable* wasserundurchlässig; **~-fitting cap** *n* WASTE *on shaft section* Versiegelungsverschluß *m*; **~ layer** *n* WASTE, WATER SUPP wasserundurchlässige Schicht *f*

tightly: **~ compacted landfill** *n* WASTE stark verdichtete Deponie *f*

tilling: **~ process** *n* SOIL Tilling-Verfahren *nt*

time: **~-delayed vacuum** *n* AIR verzögerter Unterdruck *m*; **~-dependent** *adj* ANALYSIS *sample*, RENEW ENERG zeitabhängig; **~-intensive** *adj* ANALYSIS zeitintensiv; **~-intensive method** *n* ANALYSIS zeitintensive Methode *f*; **~ interval** *n* ANALYSIS Taktzeit *f*, Zeitintervall *m*; **~ trend** *n* WASTE, WATER SUPP Zeittrend *m*; **~-weighted average** *n* (*TWA*) ANALYSIS zeitlicher Durchschnitt *m* (*TWA*)

tin 1. *n* WASTE *container* Konservenbüchse *f*, *metal* Zinn *nt*; **2. ~-plate** *vt* WASTE verzinnen

tin: **~ plate** *n* WASTE Weißblech *nt*; **~-plated** *adj* WASTE verzinnt; **~-plated can** *n* WASTE Weißblechdose *f*; **~-plate waste** *n* WASTE Weißblechabfall *m*

tip 1. *n BrE* WASTE Deponie *f*; **2.** *vt BrE* WASTE ablagern

tip: **~ filling method** *n BrE* WASTE Deponierungsverfahren *nt*, Füllart *f* der Deponie

tipped: **~-edge installation** *n* WASTE Kippkanteneinbau *m*

tipper: **~ truck** *n* WASTE Abfuhrwagen *m*, Kipper *m*

tipping: **~ ramp** *n BrE* WASTE Kippbühne *f*; **~ table** *n* WASTE Ablegetisch *m*; **~ trough** *n* WATER POLL Kipprinne *f*

titer *AmE see titre BrE*

titrant *n* ANALYSIS Titrans *nt*

titrate *vt* ANALYSIS titrieren

titration *n* ANALYSIS Titration *f*; **~ calorimeter** *n* ANALYSIS Titrationskalorimeter *nt*; **~ method** *n* ANALYSIS Titrationsmethode *f*; **~ vessel** *n* ANALYSIS Titriergefäß *nt*

titre *n BrE* ANALYSIS Titer *m*

titrimetric *adj* ANALYSIS maßanalytisch, titrimetrisch, volumetrisch; **~ test** *n* ANALYSIS *water analysis* titrimetrischer Test *m*

titrometer *n* ANALYSIS Titriergerät *nt*

TL *abbr* (*transmission loss*) NOISE Übertragungsdämpfung *f*

TLC *abbr* (*thin-layer chromatography*) ANALYSIS DC (*Dünnschichtchromatographie*)

TLD *abbr* (*thermoluminescence dosimeter*) ANALYSIS TLD (*Thermolumineszenz-Dosimeter*)

TLV *abbr* (*threshold limit value*) AIR, NOISE, POLICY, SAFETY HZK (*höchstzulässiger Konzentrationswert*); **~ at the workplace** *n* SAFETY Arbeitsplatzgrenzschwellenwert *m*; **~ in the free environment** *n* SAFETY HZK in der Umgebung *m*

TOC *abbr* (*total organic carbon*) ANALYSIS, WASTE TOC (*gesamter organisch gebundener Kohlenstoff*); **~ analyser** *BrE*, **~ analyzer** *AmE n* ANALYSIS TOC-Analysator *m*

toilet: **~ paper** *n* WASTE Hygienepapier *nt*; **~ tissue** *n* WASTE Hygienepapier *nt*

tolerable *adj* SAFETY tolerierbar, SOIL *harmful substance concentrations* duldbar, WASTE, WATER SUPP *residue emissions* tolerierbar

tolerance *n* SAFETY Toleranz *f*, SOIL *of harmful substance concentrations* Duldbarkeit *f*, WASTE, WATER SUPP Toleranz *f*

toluene *n* WASTE, WATER SUPP Toluol *nt*

TOMS *abbr* (*total ozone mapping spectrometer*) AIR, ANALYSIS TOMS-Spektrometer *nt*

TON *abbr* (*threshold odor number AmE, threshold odour number BrE*) ANALYSIS, WASTE, WATER POLL, WATER SUPP GSW (*Geruchsschwellenwert*)

top *n* AIR Deckel *f*; **~ gas recycling** *n* AIR Gichtgasrückführung *f*; **~ gas sludge** *n* WASTE *hazardous waste* Gichtgasschlamm *m*; **~ height** *n* AIR Mündungshöhe *f*; **~ layer** *n* WASTE Oberboden *m*; **~ soil** *n* WASTE Mutterboden *m*

torque *n* RENEW ENERG Drehmoment *nt*; **~ adjuster** *n* AIR Drallregler *m*

tortfeasor *n* POLICY Verursacher *m*; **~-oriented** *adj* POLICY verursachergerecht

total *adj* GEN gesamt; **~ air quality** *n* AIR Gesamtluftqualität *f*; **~ amount of groundwater tapping** *n* WATER SUPP Grundwasserentnahmesumme *f*; **~ analysis** *n* ANALYSIS Gesamtanalyse *f*; **~ appearance of waste** *n* WASTE Gesamtabfallaufkommen *nt*; **~ bacterial count** *n* WATER SUPP Gesamtbakterienzahl *f*; **~ capacity** *n* RENEW ENERG Gesamtleistung *f*; **~ carbon** *n* (*TC*) ANALYSIS gesamter vorhandener Kohlenstoff *m* (*TC*); **~ chlorine** *n* WASTE *landfill gas component* Gesamtchlor *nt*; **~ collection** *n* AIR Gesamtabscheidung *f*; **~ contamination** *n* WATER POLL Gesamtkontamination *f*; **~ costs** *n pl* MANAGE Gesamtkosten *pl*; **~ deposition** *n* AIR, SOIL, WASTE, WATER SUPP Gesamtablagerung *f*; **~ dissolved solids** *n pl* (*TDS*) WATER SUPP Gesamtschwebestoffe *m pl*; **~ dust** *n* AIR, ANALYSIS Gesamtstaub *m*; **~ dust content** *n* AIR Gesamtstaubgehalt *m*; **~ efficiency** *n* RENEW ENERG Gesamtwirkungsgrad *m*; **~ flow** *n* ANALYSIS *with each spot measurement* Gesamtdurchfluß *m*; **~ fluorine** *n* WASTE *landfill gas component* Gesamtfluor *nt*; **~ gas quantity** *n* ANALYSIS, WASTE Gesamtgasmenge *f*; **~ germ number** *n* WATER SUPP Gesamtkeimzahl *f*; **~ hardness** *n* WATER POLL, WATER SUPP Gesamthärte *f* (*GH*); **~ head** *n* WATER POLL, WATER SUPP Gesamtfallhöhe *f*; **~ inorganic carbon** *n* (*TIC*) ANALYSIS, WASTE gesamter anorganisch gebundener Kohlenstoff *m* (*TIC*); **~ noise** *n*

NOISE Gesamtgeräusch *nt*; ~ **organic carbon** *n* (*TOC*) ANALYSIS, WASTE gesamter organisch gebundener Kohlenstoff *m* (*TOC*); ~ **ozone mapping spectrometer** *n* (*TOMS*) AIR, ANALYSIS TOMS-Spektrometer *nt*; ~ **paper consumption** *n* WASTE Gesamtpapierverbrauch *m*; ~ **pollutant emissions** *n pl* ANALYSIS Gesamtschadstoffmenge *f*; ~ **refuse** *n* WASTE Gesamtmüll *m*; ~ **removal** *n* AIR Gesamtabscheidung *f*; ~ **residue** *n* ANALYSIS Gesamtrückstand *m*; ~ **risk** *n* SAFETY Gesamtrisiko *nt*; ~ **sludge yield** *n* WATER POLL Gesamtschlammanfall *m*; ~ **sulfur** *AmE*, ~ **sulphur** *BrE n* WASTE *landfill gas component* Gesamtschwefel *m*; ~ **undissolved suspended solids** *n pl* (*TSS*) WATER POLL Gesamtmenge *f* ungelöster Schwebstoffe; ~ **volume** *n* ANALYSIS Gesamtvolumen *nt*; ~ **water requirement** *n* WATER SUPP Gesamtwasserbedarf *m*

tow *vt* MARINE schleppen, tauen

tow: ~**-boat** *n* MARINE Schleppboot *nt*, Schlepper *m*

tower *n* RENEW ENERG *wind turbine* Turm *m*; ~ **biofilter** *n* AIR *biological sorption process* Turm-Biofilter *nt*; ~ **height** *n* RENEW ENERG *wind turbine* Turmhöhe *f*; ~ **system** *n* WASTE *composting system* Turmsystem *nt*; ~**-type trickling filter** *n* WATER POLL Turmtropfkörper *m*

towing: ~ **gear** *n* MARINE Schleppgeschirr *nt*

towline *n* MARINE Schleppleine *nt*

town: ~ **gas** *n* AIR Stadtgas *nt*; ~ **water** *n* WATER SUPP Stadtwasser *nt*, kommunales Wasser *nt*

toxic *adj* GEN giftig, toxisch; ~ **agent** *n* AIR, WASTE Giftstoff *m*, *military* chemischer Kampfstoff *m*

toxicant *n* GEN Giftstoff *m*

toxic: ~ **concentration** *n* (*TC*) SAFETY toxische Konzentration *f* (*TC*); ~ **degradation product** *n* WASTE toxisches Abfallprodukt *nt*; ~ **dose** *n* (*TD*) SAFETY toxische Dosis *f* (*TD*); ~ **dust** *n* AIR toxischer Staub *m*; ~ **effect** *n* ANALYSIS, SAFETY toxische Wirkung *f*

toxicity *n* GEN Giftigkeit *f*, Toxizität *f*; ~ **equivalent** *n* ANALYSIS Toxizitätsäquivalent *nt*; ~ **potential** *n* SOIL Toxizitätspotential *nt*; ~ **test** *n* ANALYSIS Toxizitätstest *m*

toxicology *n* GEN Toxikologie *f*

toxic: ~ **substance** *n* GEN Schadstoff *m*, toxische Substanz *f*; ~ **waste** *n* WASTE Giftmüll *m*, giftiger Abfall *m*, toxischer Abfall *m*; ~ **waste disposal plant** *n* WASTE Giftmüllentsorgungsanlage *f*; ~ **waste gas** *n* AIR toxisches Abgas *nt*

toxin *n* GEN Giftstoff *m*

trace *n* ANALYSIS Spur *f*; ~ **analysis** *n* ANALYSIS Spurenanalytik *f*; ~ **concentration** *n* ANALYSIS Spurenkonzentration *f*; ~ **element** *n* ANALYSIS Spurenelement *nt*; ~ **element determination** *n* ANALYSIS Spurenelementbestimmung *f*; ~ **gas** *n* AIR Spurengas *nt*; ~**-measuring device** *n* ANALYSIS Spurenmeßgerät *nt*

tracer *n* ANALYSIS Indikator *m*, Tracer *m*; ~ **experiment** *n* ANALYSIS Tracerversuch *m*

tracker *n* RENEW ENERG Nachführeinrichtung *f*

tracking: ~ **mechanism** *n* RENEW ENERG *solar collectors* Nachführeinrichtung *f*; ~ **reflector** *n* RENEW ENERG *solar tower* nachgeführter Spiegel *m*

tractor: ~ **trailer** *n AmE* (*cf lorry with trailer BrE*) WASTE Lastzug *m*

tradeable *adj* POLICY verhandelbar; ~ **emission rights** *n pl* POLICY verhandelbare Emissionsrechte *nt pl*; ~ **environmental use rights** *n pl* POLICY verhandelbare Umweltnutzungsrechte *nt pl*

trade: ~ **waste** *n* WASTE gewerblicher Abfall *m*

traditional: ~ **energy source** *n* RENEW ENERG traditioneller Energieträger *m*

transaction: ~ **costs** *n pl* POLICY Transaktionskosten *pl*

transboundary *adj* AIR, POLICY, WATER POLL grenzüberschreitend; ~ **air pollution** *n* AIR grenzüberschreitende Luftverunreinigung *f*; ~ **water pollution** *n* WATER POLL grenzüberschreitende Wasserverschmutzung *f*, grenzüberschreitende Wasserverunreinigung *f*

transfer 1. *n* WASTE Umladung *f*; **2.** *vt* WASTE *hazardous substances*, WATER POLL, WATER SUPP übertragen; ~ **catalytically** AIR katalytisch umsetzen; ~ **chemically** AIR chemisch umsetzen; ~ **with a pipette** ANALYSIS pipettieren

transfer: ~ **hall** *n* WASTE Umladehalle *f*; ~ **pump** *n* MARINE Umschlagpumpe *f*; ~ **rate** *n* WASTE, WATER POLL, WATER SUPP Übertragungsrate *f*; ~ **site** *n* WASTE Übergabeort *m*; ~ **station** *n* WASTE Umladeanlage *f*, Umladestation *f*

transformation: ~ **of pollutants** *n* AIR, SOIL, WATER POLL Schadstoffumwandlung *f*; ~ **rate** *n* AIR, WASTE Umwandlungsrate *f*

transformer *n* RENEW ENERG, WASTE *converter* Umwandler *m*, *electrical* Transformator *m* (*Trafo*)

transition: ~ **lake** *n* WATER SUPP Übergangssee *m*; ~ **metal** *n* WASTE Übergangsmetall *nt*; ~ **phase** *n* POLICY Übergangsphase *f*; ~ **point** *n* ANALYSIS Umwandlungspunkt *m*; ~ **regulation** *n* POLICY Übergangsregelung *f*; ~ **temperature** *n* ANALYSIS Umwandlungstemperatur *f*

transmission *n* ANALYSIS Transmission *f*, NOISE, RENEW ENERG Übertragung *f*; ~ **electron microscopy** *n* (*TEM*) ANALYSIS Durchstrahlungselektronenmikroskopie *f*; ~ **loss** *n* (*TL*) NOISE Übertragungsdämpfung *f*; ~ **spectrum** *n* ANALYSIS Transmissionsspektrum *nt*

transmit *vt* ANALYSIS, NOISE, RENEW ENERG übertragen

transmittance: ~ **meter** *n* AIR, ANALYSIS Durchlässigkeitsmeter *nt* Durchlässigkeitsmesser *m*

transparency *n* WASTE, WATER SUPP Durchsichtigkeit *f*; ~ **measurement** *n* WASTE, WATER SUPP Durchsichtigkeitsmessung *f*

transport 1. *n* GEN Beförderung *f*, Transport *m*; **2.** *vt* GEN transportieren, SOIL *off-site method* abtransportieren

transportable *adj* GEN tragbar, transportabel, transportfähig, transportierbar; ~ **gas container** *n* SAFETY transportierbare Gasflasche *f*

transportation *n* GEN Beförderung *f*; ~ **source** *n* WASTE, WATER SUPP Transportquelle *f*

transport: ~ **capacity** *n* WASTE *of distribution network* Transportkapazität *f*, WATER POLL *for sludge* Förderleistung *f*, Fördermöglichkeit *f*, WATER SUPP *of distribution network* Transportkapazität *f*; ~ **container** *n* WASTE, WATER SUPP Abrollcontainer *m*; ~ **of contaminants** *n* SOIL Schadstofftransport *m*, Schadstoffverfrachtung *f*; ~ **of hazardous substances** *n* WASTE Gefahrguttransport *m*; ~ **model** *n* MANAGE Transportmodell *nt*; ~ **of pollutants** *n* SOIL Schadstofftransport *m*, Schadstoffverfrachtung *f*; ~ **safety** *n* SAFETY Transportsicherheit *f*; ~ **volume** *n* WASTE Transportvolumen *nt*

transshipment: ~ **of hazardous goods** *n* WASTE Gefahrgutumschlag *m*

transverse: ~**-entry** *adj* AIR querangeströmt; ~**-entry traveling-bed reactor** *AmE*, ~**-entry travelling-bed reactor** *BrE n* AIR querangeströmter Wanderschichtreaktor *m*; ~ **strain** *n* SOIL Schubspannung *f*; ~ **wave** *n* NOISE Transversalwelle *f*

trap *n* ANALYSIS Zwischenspeicher *m*

trapezoid: ~ **compost heap** *n* WASTE Trapezmiete *f*

trash: ~ **burning** *n AmE* WASTE Müllverbrennung *f*; ~ **can** *n AmE* (*cf rubbish bin BrE*) WASTE Müllbehälter *m*, Mülleimer *m*, Mülltonne *f*

traveling *AmE see travelling BrE*

travelling *n BrE* AIR, SOIL, WASTE Wandern *nt*; ~**-bed reactor** *n BrE* AIR Wanderschichtreaktor *m*; ~ **distributor** *n BrE* WATER SUPP Fahrsprenger *m*, Wandersprenger *m*; ~ **grate** *n BrE* SOIL *for pellet incineration*, WASTE Wanderrost *m*; ~**-grate installation** *n BrE* WASTE Wanderrostanlage *f*

trawl: ~ **net** *n* MARINE Grundnetz *nt*, Grundschleppnetz *nt*, Schleppnetz *nt*

tray *n* AIR Boden *m*, Bodenblech *nt*; ~ **absorber** *n* AIR Bodenabsorber *m*; ~ **plate column** *n* AIR Sieb-

bodenkolonne *f*; **~-tower** *n* AIR *wet scrubber* Bodensäulenabscheider *m*
treat *vt* GEN aufbereiten, behandeln
treated *adj* GEN behandelt; **~ effluent** *n* WATER POLL behandeltes Abwasser *nt*; **~ sludge** *n* WATER POLL behandelter Schlamm *m*
treatment *n* GEN Aufbereitung *f*, Behandlung *f*; **~ cell** *n* SOIL Behandlungsanlage *f*; **~ duration** *n* SOIL Behandlungsdauer *f*; **~ of hazardous wastes with plasma** *n* WASTE Plasmabehandlung *f*; **~ intensity** *n* SOIL Behandlungsintensität *f*; **~ plant** *n* WASTE *for refuse* Behandlungsanlage *f*; **~ process** *n* WASTE, WATER SUPP Aufbereitungsverfahren *nt*; **~ stage** *n* WATER POLL Reinigungsstufe *f*; **~ time** *n* SOIL Behandlungsdauer *f*; **~ unit** *n* WASTE, WATER SUPP Aufbereitungseinheit *f*; **~ of wastewater sludge** *n* WATER POLL Mitbehandlung *f* von Klärschlamm
treble: **~ absorber** *n* NOISE Höhenabsorber *m*, poröser Absorber *m*
trench *n* MARINE Graben *m*, Tiefseegraben *m*, SOIL, WASTE Graben *m*, *drainage ditch* Rigole *f*
trenching *n* SOIL, WASTE Rigolen *nt*
trench: **~ method** *n* WASTE Grabenmethode *f*
TRI *abbr* (*trichloroethylene*) AIR TRI (*Trichlorethylen*)
trial: **~ operation** *n* MANAGE, POLICY *of a plant* Versuchsbetrieb *m*
triangular: **~ compost heap** *n* WASTE Dreiecksmiete *f*
trichloroethylene *n* (*TRI*) AIR Trichlorethylen *nt* (*TRI*)
trichlorofluoromethane *n* AIR *halogenated hydrocarbon* Trichlorfluormethan *nt*
trickle *vi* WATER POLL, WATER SUPP rieseln, rinnen
trickling: **~ film** *n* AIR *wet precipitator*, WATER POLL, WATER SUPP Rieselfilm *m*; **~ film flow** *n* WATER POLL, WATER SUPP *packed column* Rieselfilmströmung *f*; **~ film surface** *n* WATER POLL, WATER SUPP Rieselfilmoberfläche *f*; **~ filter** *n* WATER POLL *biofilter*, WATER SUPP Tropfkörper *m* (*TK*); **~ filter bottom** *n* WATER POLL Tropfkörperboden *m*; **~ filter process** *n* AIR *biowasher*, WATER POLL *biofiltration* Tropfkörperverfahren *nt*
triglyceride *n* WATER POLL Triglycerid *nt*
trommel: **~ screen** *n* WASTE Trommelsieb *nt*
troposphere *n* AIR Troposphäre *f*
tropospheric *adj* AIR troposphärisch; **~ ozone** *n* AIR troposphärisches Ozon *nt*; **~ ozone concentration** *n* AIR troposphärische Ozonkonzentration *f*; **~ ozone pollution** *n* AIR troposphärische Ozonbelastung *f*
troughed: **~ chain conveyor** *n* WASTE Trogkettenförderer *m*
truck *n AmE* (*cf lorry BrE*) WASTE Lastwagen *m*, Lastkraftwagen *m* (*LKW*); **~ trailer** *n AmE* WASTE LKW-Anhänger *m*, Lastwagenanhänger *m*
trunk: **~ main** *n* WATER SUPP Hauptleitung *f*
try: **~ cock** *n* WATER POLL, WATER SUPP Probierhahn *m*
TSS *abbr* (*total undissolved suspended solids*) WATER POLL Gesamtmenge ungelöster Schwebstoffe *f*
TTS *abbr* (*temporary threshold shift*) NOISE zeitweilige Hörschwellenverschiebung *f*
tube *n* GEN Rohr *nt*, Röhre *f*; **~ bundle** *n* AIR *heat exchanger* Rohrbündel *nt*; **~-cage recuperator** *n* AIR Rohrkäfigrekuperator *m*; **~ support plate** *n* AIR Rohrstützblech *nt*; **~ wall** *n* WATER SUPP Rohrwand *f*
tubular *adj* GEN röhrenförmig; **~ drainage** *n* WATER SUPP Röhrendränung *f*; **~ ESP** *n* AIR *dry electrofilter* Rohrelektrofilter *nt*; **~ filter bag** *n* AIR Filterschlauch *m*; **~ module** *n* WATER POLL *ultrafiltration* Tubularmodul *m*; **~ precipitator** *n* AIR röhrenförmiger Staubabscheider *m*
tug: **~ boat** *n* MARINE Schleppboot *nt*, Schlepper *m*
tumbling: **~ station** *n* WASTE Rüttelplatte *f*
turbid *adj* WATER POLL trübe
turbidimeter *n* ANALYSIS Trübungsmesser *m*, Turbidimeter *nt*
turbidimetric *adj* ANALYSIS turbidimetrisch; **~ analysis** *n* ANALYSIS Trübungsanalyse *f*; **~ titration** *n* ANALYSIS turbidimetrische Titration *f*
turbidimetry *n* ANALYSIS Trübungsanalyse *f*, Turbidimetrie *f*
turbidity *n* WATER POLL Trübung *f*; **~ value** *n* ANALYSIS Trübungsmessung *f*, Trübungswert *m*
turbid: **~ matter** *n* WATER POLL Trübstoff *m*; **~ wastewater** *n* WATER POLL getrübtes Abwasser *nt*; **~ wastewater sample** *n* ANALYSIS trübe Abwasserprobe *f*; **~ water** *n* WATER SUPP trübes Wasser *nt*
turbine *n* RENEW ENERG Turbine *f*; **~ blade** *n* RENEW ENERG Turbinenblatt *nt*
turbo: **~-aerator** *n* WATER POLL Turbobelüfter *m*; **~-compressor** *n* WATER POLL Turboverdichter *m*; **~-generator** *n* RENEW ENERG *tidal power plant* Turbogenerator *m*; **~-machine** *n* WASTE, WATER POLL, WATER SUPP Turbomaschine *f*; **~-pump** *n* WASTE, WATER SUPP Turbinenpumpe *f*, Turbopumpe *f*
turbulence *n* AIR Wirbelströmung *f*, WASTE Turbulenz *f*, Wirbelung *f*, *landfill gas* Verwirbelung *f*, WATER POLL Turbulenz *f*, Wirbelung *f*; **~ zone** *n* AIR Wirbelzone *f*
turbulent: **~ flow** *n* AIR Wirbelströmung *f*, turbulente Strömung *f*
TWA *abbr* (*time-weighted average*) ANALYSIS TWA (*zeitlicher Durchschnitt*)
TWC *abbr* (*three-way catalyst*) AIR Drei-Wege-Katalysator *m*
twin: **~ cock** *n* WATER SUPP Zwillingshahn *m*; **~-cutting shredder** *n* WASTE Zweiwellenschneidzerkleinerer *m*; **~-screw extruder** *n* WASTE Doppelschneckenextruder *m*
two: **~-phase jet** *n* AIR *jet scrubber* Vollkegeldüse *f*; **~-stage** *adj* WATER POLL zweistufig; **~-stage activated sludge process** *n* WATER POLL zweistufiges Belebungsverfahren *nt*; **~-stage activated sludge process with alternating nitrification and denitrification** *n* WATER POLL zweistufige Belebung *f* mit alternierender Nitrifikation und Denitrifikation; **~-stage activated sludge process with predenitrification** *n* WATER POLL zweistufige Belebung *f* mit vorgeschalteter Denitrifikation; **~-stage aerobic thermophilic stabilization** *n* WATER POLL zweistufige aerob-thermophile Stabilisierung *f*; **~-stage digester** *n* WATER POLL zweistufiger Faulraum *m*; **~-stage fired heater** *n* AIR zweistufiger Erhitzer *m*; **~-stage gas scrubber** *n* AIR zweistufiger Gaswäscher *m*; **~-stage recarbonation** *n* WATER POLL zweistufige Rekarbonisierung *f*; **~-stage thermophilic-mesophilic digestion** *n* WATER POLL zweistufige thermophil-mesophile Faulung *f*; **~-way cock** *n* WATER SUPP Zweiweghahn *m*
Tyndall: **~ effect** *n* ANALYSIS Tyndall-Effekt *m*
type: **~-specific** *adj* WASTE sortenrein; **~ specification** *n* WASTE Sortenreinheit *f*; **~-specific plastics** *n pl* WASTE sortenreine Kunststoffe *m pl*

U

UCT: ~ **method** *n* (*University of Cape Town method*) WATER POLL *phosphorous removal* UCT-Verfahren *nt*
UCV *abbr* (*upper calorific value*) AIR, WASTE spezifischer Brennwert *m*
UF *abbr* (*ultrafiltration*) ANALYSIS, SOIL UF (*Ultrafiltration*)
ultrafiltration *n* (*UF*) ANALYSIS, SOIL Ultrafiltration *f* (*UF*)
ultrapure: ~ **water** *n* SOIL *extraction* Reinstwasser *nt*
ultrasonic: ~ **hazard** *n* ANALYSIS, NOISE, SAFETY Gefahr *f* durch Ultraschall; ~ **nebulizer** *n* ANALYSIS Ultraschallzerstäuber *m*
ultrasound *n* ANALYSIS, NOISE Ultraschall *m*; ~ **extraction** *n* ANALYSIS Ultraschallextraktion *f*; ~ **irradiation** *n* SOIL Ultraschallstrahlung *f*; ~ **sensor** *n* ANALYSIS Ultraschallsensor *m*
ultratrace *n* ANALYSIS Ultraspur *f*; ~ **analysis** *n* ANALYSIS Ultraspurenanalyse *f*
ultraviolet 1. *adj* (*UV*) GEN ultraviolett (*UV*); **2.** *n* (*UV*) GEN Ultraviolett *nt* (*UV*)
ultraviolet: ~ **and visible** *adj* (*UV/VIS*) ANALYSIS im Sichtbaren und Ultraviolett (*UV/VIS*)
UMI *abbr* (*universal measurement instrument*) ANALYSIS Universalmeßinstrument *nt*
unaerated: ~ **bed composting** *n* WASTE unbelüftete Mietenkompostierung *f*
uncharged *adj* AIR, WATER POLL ungeladen; ~ **particle** *n* AIR, MARINE ungeladenes Teilchen *nt*
uncleaned *adj* AIR ungereinigt; ~ **waste air** *n* AIR ungereinigte Abluft *f*
uncoated *adj* WASTE unbeschichtet; ~ **surface** *n* WASTE unbeschichtete Fläche *f*
uncompacted *adj* SOIL, WASTE nichtverdichtet, unverdichtet; ~ **dump** *n* WATER SUPP nichtverdichtete Deponie *f*; ~ **landfill** *n* WASTE nichtverdichtete Deponie *f*; ~ **tip** *n BrE* WASTE nichtverdichtete Deponie *f*
unconfined: ~ **groundwater** *n* WATER SUPP freies Grundwasser *nt*, ungespanntes Grundwasser *nt*; ~ **water** *n* WATER SUPP freies Grundwasser *nt*, ungespanntes Grundwasser *nt*
uncontrolled *adj* WASTE, WATER POLL unkontrolliert, ungeordnet; ~ **denitrification** *n* WATER POLL unkontrollierte Denitrifikation *f*; ~ **disposal** *n* WASTE ungeordnete Ablagerung *f*; ~ **dumping** *n* WASTE ungeordnete Ablagerung *f*, wilde Ablagerung *f*, wilde Müllablagerung *f*; ~ **tip** *n BrE* WASTE ungeordnete Deponie *f*, wilde Müllablagerung *f*; ~ **tipping** *n BrE* WASTE ungeordnete Ablagerung *f*, wilde Ablagerung *f*, wilde Müllablagerung *f*
undamped *adj* NOISE nichtgedämpft, ungedämpft
underaerated *adj* WATER POLL unterbelüftet
underblast: ~ **chain grate stoker** *n* WASTE Unterwindkettenrostfeuerung *f*; ~ **fan** *n* WASTE Unterwindgebläse *nt*; ~ **zone** *n* WASTE Unterwindzone *f*
underdrain *n* WATER POLL unterirdischer Wasserabzug *m*
underdrainage *n* WATER POLL unterirdische Entwässerung *f*
underfloor *n* NOISE Unterdecke *f*; ~ **catalyst** *n* AIR motorferner Katalysator *m*
underflow *n* WATER SUPP Unterlauf *m*
underground 1. *adj* SOIL, WASTE, WATER SUPP unterirdisch, untertägig; **2.** *n* SOIL, WASTE, WATER SUPP Untergrund *m*
underground: ~ **deformation** *n* WASTE, WATER SUPP Untergrundverformung *f*; ~ **depot** *n* WASTE Untertagedeponie *f*; ~ **drainage** *n* WASTE, WATER SUPP unterirdische Dränage *f*; ~ **hydrant** *n* WATER SUPP Unterflurhydrant *m*, Unterflurwasserpfosten *m*; ~ **infiltration facility** *n* WATER SUPP unterirdische Infiltrationsanlage *f*; ~ **landfill for hazardous waste** *n* WASTE Untertagedeponie *f* für Sonderabfall; ~ **landfill for special waste** *n* WASTE Untertagedeponie *f* für Sonderabfall, untertägige Sondermülldeponie *f*; ~ **sealing** *n* WASTE, WATER SUPP Untergrundabdichtung *f*; ~ **storage** *n* WASTE, WATER SUPP unterirdische Speicherung *f*; ~ **wastewater disposal** *n* WATER POLL Abwassereinleitung *f* in den Untergrund, Abwasserversenkung *f*; ~ **water** *n* WATER SUPP Grundwasser *nt*, unterirdische Gewässer *nt pl*, unterirdisches Wasser *nt*
underhood: ~ **catalyst** *n* AIR motornaher Katalysator *m*
underpressure: ~ **evaporator well** *n* SOIL Unterdruckverdampferbrunnen *m* (*UVB*)
underwashing *n* WASTE Auskolkung *f*, Unterspülung *f*
underwater: ~ **suction pipe** *n* WATER SUPP Unterwassersaugleitung *f*
undigested: ~ **sludge** *n* WATER POLL Primärschlamm *m*, Rohschlamm *m*
undiluted *adj* MARINE, WASTE, WATER SUPP unverdünnt, unvermischt; ~ **leachate** *n* WASTE, WATER SUPP unverdünntes Sickerwasser *nt*, unvermischtes Sickerwasser *nt*
undissolved *adj* WASTE, WATER POLL ungelöst; ~ **substance** *n* WASTE, WATER POLL ungelöster Stoff *m*
undisturbed *adj* ANALYSIS ungestört; ~ **sample** *n* ANALYSIS ungestörte Probe *f*
uneconomical *adj* MANAGE, POLICY unwirtschaftlich
UNEP *abbr* (*United Nations Environmental Programme*) POLICY Umweltprogramm *nt* der Vereinten Nationen
unfiltered *adj* ANALYSIS unfiltriert; ~ **sample** *n* ANALYSIS unfiltrierte Probe *f*
unfinished: ~ **support** *n* WASTE Rohplanum *nt*
uniaxial: ~ **compressive strength** *n* WASTE *landfill* einaxiale Druckfestigkeit *f*
United: ~ **Nations Environmental Programme** *n* (*UNEP*) POLICY Umweltprogramm *nt* der Vereinten Nationen
unit: ~ **operation** *n* ANALYSIS, WATER POLL Einheitsoperation *f*; ~ **operations** *n pl* ANALYSIS, WATER POLL Verfahrenstechnik *f*; ~ **process** *n* ANALYSIS, WATER POLL *contaminant removal* Einheitsverfahren *nt pl*
unity *n* ANALYSIS Einheit *f*
universal: ~ **measurement instrument** *n* (*UMI*) ANALYSIS Universalmeßinstrument *nt*; ~ **shredder** *n* WASTE Universalzerkleinerer *m*; ~ **space** *n* AIR Weltraum *m*
University: ~ **of Cape Town method** *n* (*UCT method*) WATER POLL *phosphorous removal* UCT-Verfahren *nt*
unladen *adj* MARINE, WASTE, WATER POLL nichtbeladen, unbeladen
unlimited: ~ **storage** *n* WASTE unbegrenzte Ablagerung *f*
unload *vt* MARINE löschen, WASTE, WATER SUPP entladen
unloading *n* MARINE, WASTE, WATER SUPP Entladung *f*; ~ **hall** *n* WASTE Entladehalle *f*; ~ **hopper** *n* WASTE Entladebunker *m*
unmixed: ~ **recovery of harmful substances** *n* WASTE unvermischte Schadstofferfassung *f*; ~ **waste recovery** *n* WASTE unvermischte Abfallerfassung *f*
unpalatable *adj* WATER SUPP ungenießbar; ~ **water** *n* WATER SUPP ungenießbares Wasser *nt*
unpolluted: ~ **air** *n* AIR Reinluft *f*, unverschmutzte Luft *f*
unpressurized *adj* WASTE, WATER SUPP drucklos; ~ **supply** *n* WATER SUPP drucklose Versorgung *f*
unprocessed *adj* MARINE, WASTE, WATER POLL, WATER SUPP nichtaufbereitet, unaufbereitet; ~ **waste** *n* WASTE nichtaufbereiteter Abfall *m*, unaufbereiteter Abfall *m*

unprofitable *adj* MANAGE, POLICY unwirtschaftlich
unreliability *n* ANALYSIS *of sample* Unzuverlässigkeit *f*
unsafe *adj* SAFETY unsicher; ~ **environmental condition** *n* SAFETY unsicherer Umweltzustand *m*
unsafeness *n* SAFETY Unsicherheit *f*
unsaturated *adj* AIR, WASTE, WATER SUPP ungesättigt; ~ **hydrocarbons** *n pl* AIR ungesättigte Kohlenwasserstoffe *m pl*; ~ **polyester resin** *n* WASTE *thermoset* ungesättigtes Polyesterharz *nt*; ~ **soil zone** *n* SOIL ungesättigte Bodenzone *f*; ~ **zone** *n* SOIL, WASTE, WATER SUPP ungesättigte Zone *f*
unslaked *adj* WATER POLL ungelöscht
unsorted: ~ **waste** *n* WASTE unsortierter Abfall *m*
unspecific: ~ **coagulation** *n* WATER POLL unspezifische Koagulation *f*
unstable *adj* WASTE, WATER SUPP instabil; ~ **methane fermentation** *n* WASTE instabile Methangärung *f*
unthickened *adj* WATER POLL nichtentwässert; ~ **sludge** *n* WATER POLL nichtentwässerter Schlamm *m*
untreated *adj* GEN unbehandelt, Roh-; ~ **refuse** *n* WASTE Rohmüll *m*; ~ **slag** *n* WASTE Rohschlacke *f*; ~ **sludge** *n* WATER POLL Frischschlamm *m*, Rohschlamm *m*; ~ **wastewater** *n* MARINE, WATER POLL Rohabwasser *nt*, unbehandeltes Abwasser *nt*; ~ **water** *n* WATER POLL, WATER SUPP Rohwasser *nt*; ~ **water quality** *n* WATER POLL, WATER SUPP Rohwasserqualität *f*; ~ **water tank** *n* WATER POLL, WATER SUPP Rohwasserbecken *nt*
unusable *adj* WASTE nichtverwertbar, unverwertbar; ~ **residue** *n* WASTE *from recycling process* nichtverwertbarer Reststoff *m*, unverwertbarer Reststoff *m*
unvulcanized: ~ **rubber** *n* WASTE Kautschuk *m*
up-current: ~ **classifier** *n* WATER SUPP Aufstromklassierer *m*
upflow: ~ **filter** *n* WATER SUPP Aufstromfilter *nt*; ~ **sludge blanket process** *n* WATER POLL Schlammkontaktverfahren *nt*
upgrading *n* AIR Ertüchtigung *f*
upper: ~ **bed** *n* AIR *catalyst* oberes Bett *nt*; ~ **calorific value** *n* (*UCV*) AIR, WASTE spezifischer Brennwert *m*; ~ **surface** *n* WASTE Oberkante *f*
upstream *adv* AIR, WATER SUPP stromaufwärts, strömungsaufwärts
upward: ~ **surface installation** *n* WASTE, WATER SUPP Aufwärtsflächeneinbau *m*
upwind *adv* AIR, RENEW ENERG gegen den Wind
upwind: ~ **power plant** *n* RENEW ENERG Aufwindkraftwerk *nt*
urban: ~ **canalization** *n* WATER POLL städtische Kanalisation *f*; ~ **catchment** *n* WASTE, WATER SUPP städtisches Einzugsgebiet *nt*; ~ **ecosystem** *n* AIR, SOIL, WATER POLL urbanes Ökosystem *nt*; ~ **waste** *n* WASTE Haushaltsabfall *m*, Hausmüll *m*, Siedlungsabfall *m*; ~ **water management** *n* WATER SUPP städtische Wasserbewirtschaftung *f*
urea *n* WATER POLL Harnstoff *m*
usability *n* GEN Verwertbarkeit *f*, Brauchbarkeit *f*, Nutzbarkeit *f*
usable *adj* GEN verwertbar, brauchbar, nutzbar; ~ **by-product** *n* WASTE, WATER POLL verwendbares Nebenprodukt *nt*; ~ **energy form** *n* RENEW ENERG nutzbare Energieform *f*; ~ **fraction** *n* WASTE verwertbare Fraktion *f*; ~ **portion of hydrosphere** *n* AIR *for human use* nutzbarer Anteil *m* der Hydrosphäre; ~ **potential** *n* RENEW ENERG nutzbares Potential *nt*; ~ **waste product** *n* WASTE wiederverwertbares Abfallprodukt *nt*, verwendbares Abfallprodukt *nt*
use **1.** *n* GEN Nutzung *f*; **2.** *vt* GEN nutzen, *recycle* verwerten
used *adj* GEN gebraucht; ~ **aluminium** *BrE*, ~ **aluminum** *AmE n* WASTE gebrauchtes Aluminium *nt*; ~ **battery** *n* WASTE Altbatterie *f*; ~ **bottle** *n* WASTE, WATER SUPP gebrauchte Flasche *f*; ~ **car dump** *n* WASTE Altfahrzeugdeponie *f*; ~ **commercial packaging** *n* WASTE gebrauchte Verkaufsverpackung *f*; ~ **glass** *n* WASTE Altglas *nt*; ~ **material** *n* WASTE Altstoff *m*; ~ **oil** *n* WASTE Altöl *nt*; ~ **product** *n* WASTE gebrauchtes Produkt *nt*; ~ **storage battery** *n* WASTE Altakkumulator *m*; ~ **tire** *AmE*, ~ **tyre** *BrE n* WASTE Altreifen *m*; ~ **tyre carcass** *n BrE* WASTE Altreifenkarkasse *f*; ~ **tyre disposal** *n BrE* WASTE Altreifenbeseitigung *f*; ~ **tyre pyrolysis** *n BrE* WASTE Altreifenpyrolyse *f*
useful: ~ **area** *n* WASTE Nutzfläche *f*
use: ~ **possibilities** *n pl* RENEW ENERG Nutzungsmöglichkeiten *f pl*; ~ **prohibition** *n* POLICY *sites* Nutzungsverbot *nt*
user: ~ **charge** *n* POLICY Verbraucherabgabe *f*; **~'s contract** *n* POLICY *of environmental label* Nutzungsvertrag *m*
use: ~ **recommendation** *n* POLICY *sites* Nutzungsempfehlung *f*; ~ **reduction** *n* POLICY *sites* Nutzungsbeschränkung *f*, Nutzungseinschränkung *f*; ~ **regulation** *n* POLICY Nutzungsregelung *f*; **~-related** *adj* AIR, RENEW ENERG, WATER POLL, WATER SUPP nutzungsbezogen; ~ **safety** *n* SAFETY Anwendungssicherheit *f*; ~ **of treated sludge as fuel** *n* WATER POLL Verwertung *f* von Klärschlamm als Brennstoff; **~-value analysis** *n* MANAGE Nutzwertanalyse *f*
utility: **~-scale** *adj* RENEW ENERG großtechnisch
utilization *n* GEN Nutzbarmachung *f*; ~ **cycle of materials** *n* WASTE Werkstoffnutzungszyklus *m*
utilize *vt* GEN nutzen
UV *abbr* (*ultraviolet*) GEN UV (*ultraviolett, Ultraviolett*); ~ **absorption** *n* ANALYSIS *SO_2, NO_x measurement*, WASTE UV-Absorption *f*
UV-C: ~ **radiation** *n* AIR, WASTE UV-C-Strahlung *f*
UV: ~ **disinfection** *n* WASTE UV-Desinfektion *f*; ~ **filter** *n* AIR, SAFETY UV-Filter *nt*; ~ **irradiation** *n* AIR, WASTE UV-Bestrahlung *f*; ~ **lamp** *n* ANALYSIS *photoionization detector* UV-Lampe *f*; ~ **optical absorption** *n* ANALYSIS UV-Lichtabsorption *f*; ~ **oxidation** *n* WATER SUPP UV-Oxidation *f*; ~ **photometry** *n* ANALYSIS UV-Photometrie *f*; ~ **radiator** *n* WATER POLL UV-Strahler *m*; ~ **rays** *n pl* ANALYSIS, SAFETY UV-Strahlen *m pl*; ~ **resistance** *n* WASTE UV-Stabilität *f*; ~ **spectroscopy** *n* ANALYSIS UV-Spektroskopie *f*
UV/VIS *abbr* (*ultraviolet and visible*) ANALYSIS UV/VIS (*im Sichtbaren und Ultraviolett*); ~ **detector** *n* ANALYSIS UV/VIS-Detektor *m*, UVS-Detektor *m*; ~ **spectrometer** *n* ANALYSIS UV/VIS-Spektrometer *nt*, UVS-Spektrometer *nt*; ~ **spectrophotometry** *n* ANALYSIS UV/VIS-Spektrophotometrie *f*, UVS-Spektrophotometrie *f*; ~ **spectroscopy** *n* ANALYSIS UV/VIS-Spektroskopie *f*, UVS-Spektroskopie *f*

V

V *abbr* (*vanadium*) AIR V (*Vanadium*)
vacuum *n* AIR, ANALYSIS, SOIL, WATER POLL Vakuum *nt*; ~ **belt filter** *n* AIR Vakuumbandfilter *nt*; ~ **breaker** *n* WATER SUPP Rückflußverhinderer *m*; ~ **concentration** *n* WASTE, WATER POLL Vakuumeindampfung *f*; ~-**controlled** *adj* AIR unterdruckgesteuert; ~ **evaporator** *n* WATER SUPP Vakuumverdampfer *m*; ~ **filter** *n* WATER POLL, WATER SUPP Saugfilter *nt*, Vakuumfilter *nt*; ~ **filter dewatering** *n* WATER POLL Vakuumfilterentwässerung *f*; ~ **filtration** *n* WATER POLL, WATER SUPP Vakuumfilterung *f*; ~ **flotation** *n* WATER POLL Vakuumflotation *f*; ~ **gage** *AmE*, ~ **gauge** *BrE n* ANALYSIS Unterdruckmanometer *nt*, Unterdruckmesser *m*; ~ **principle** *n* ANALYSIS *of sampling* Vakuumprinzip *nt*; ~ **pump** *n* AIR, ANALYSIS, MARINE, RENEW ENERG Vakuumpumpe *f*; ~ **street sweeper** *n* WASTE Saugfegmaschine *f*; ~ **well** *n* SOIL Vakuumbrunnen *m*
vadose: ~ **water** *n* WATER SUPP schwebendes Grundwasser *nt*, vadoses Grundwasser *nt*
valence *n* AIR, ANALYSIS, WATER POLL Valenz *f*, Wertigkeit *f*; ~ **force** *n* AIR, WATER POLL Valenzkraft *f*
valley: ~ **landfill** *n* WASTE Taldeponie *f*
valve *n* MARINE, WATER SUPP Ventil *nt*
vanadium *n* (*V*) AIR Vanadium *nt* (*V*)
van der Waals: ~ **forces** *n pl* WATER POLL Van-der-Waals-Kräfte *f pl*
vane *n* WASTE, WATER POLL Flügel *m*; ~ **pump** *n* WATER POLL Flügelzellenpumpe *f*; ~ **shear strength** *n* SOIL, WASTE *landfill* Flügelscherfestigkeit *f*
vapor *AmE see vapour BrE*
vaporization *n* RENEW ENERG, SOIL Verdampfung *f*
vaporize 1. *vt* RENEW ENERG, SOIL *solvents* eindampfen, verdampfen; **2.** *vi* RENEW ENERG, SOIL verdampfen
vaporizer *n* RENEW ENERG, SOIL Verdampfer *m*
vaporous *adj* AIR dampfförmig; ~ **fuel** *n* AIR dampfförmiger Brennstoff *m*; ~ **pollutants** *n pl* AIR dampfförmige Schadstoffe *m pl*; ~ **solvent** *n* AIR dampfförmiges Lösungsmittel *nt*
vapour *n BrE* GEN Dampf *m*; ~ **compressor** *n BrE* AIR Brüdenkompressor *m*; ~ **emissions** *n pl BrE* AIR Dampfemissionen *f pl*; ~ **extraction** *n BrE* SOIL Dampfextraktion *f*; ~-**laden emissions** *n pl BrE* AIR Dampfemissionen *f pl*; ~ **plume** *n BrE* AIR, WATER SUPP Dampffahne *f*; ~ **pressure** *n BrE* AIR, ANALYSIS, WASTE, WATER POLL, WATER SUPP *pump* Dampfdruck *m*; ~ **pressure analysis** *n BrE* ANALYSIS Dampfdruckbestimmung *f*; ~ **pressure determination** *n BrE* ANALYSIS Dampfdruckbestimmung *f*; ~ **pressure measurement** *n BrE* ANALYSIS Dampfdruckmessung *f*; ~ **pressure scale** *n BrE* ANALYSIS Dampfdruckwaage *f*; ~ **riser** *n BrE* AIR *bubble-cap tray tower* Glockenhals *m*; ~ **treatment** *n BrE* WASTE, WATER SUPP Brüdenbehandlung *f*
variable *adj* GEN schwankend, variabel; ~ **gas throughput** *n* AIR schwankender Gasdurchsatz *m*; ~ **gas throughput amount** *n* AIR schwankende Gasdurchsatzmenge *f*; ~ **gas thruput** *AmE see variable gas throughput*; ~-**pitch propeller** *n* MARINE Verstellpropeller *m*, Verstelluftschraube *f*; ~-**pitch rotor blade** *n* RENEW ENERG *wind turbine* verstellbares Rotorblatt *nt*; ~-**speed motor** *n* WATER POLL Motor *m* mit einstellbarer Drehzahl; ~ **waste gas flow** *n* AIR schwankender Abgasstrom *m*
varnish: ~ **waste sludge** *n* WASTE, WATER SUPP Lackschlamm *m*
vat *n* WASTE Wanne *f*
VAWT *abbr* (*vertical axis wind turbine*) RENEW ENERG Vertikalachsenmaschine *f*
VC *abbr* (*vinyl chloride*) AIR, SAFETY VC (*Vinylchlorid*)
vegetable: ~ **waste** *n* WASTE pflanzlicher Abfall *m*
vehicle: ~ **capacity** *n* WASTE Fahrzeugkapazität *f*; ~ **shredder** *n* WASTE Schredderanlage *f*
velocity *n* AIR, WASTE, WATER POLL, WATER SUPP Geschwindigkeit *f*; ~ **of flow** *n* ANALYSIS Strömungsgeschwindigkeit *f*; ~ **head** *n* WASTE, WATER POLL, WATER SUPP Geschwindigkeitshöhe *f*, dynamische Druckhöhe *f*; ~ **of migration** *n* AIR, ANALYSIS Wanderungsgeschwindigkeit *f*
vent *n* AIR, WASTE, WATER POLL, WATER SUPP Entlüftungsrohr *nt*; ~ **condenser** *n* WATER POLL Brüdenkondensator *m*
ventilate *vt* AIR, WASTE, WATER POLL, WATER SUPP entlüften
ventilated: ~ **nappe** *n* WATER SUPP belüftete Wasserlamelle *f*
ventilation *n* AIR, WASTE, WATER POLL, WATER SUPP Luftwechsel *m*, Belüftung *f*, Entlüftung *f*
ventilator *n* AIR, WASTE, WATER POLL, WATER SUPP Gebläse *nt*, Ventilator *m*
venting *n* SOIL *of soil atmosphere* Abzug *m*
vent: ~ **silencer** *n* NOISE Belüftungslochdämpfer *m*
Venturi: ~ **burner** *n* AIR Venturi-Brenner *m*; ~ **meter** *n* AIR Venturimesser *m*; ~ **nozzle** *n* AIR, ANALYSIS Venturidüse *f*; ~ **principle** *n* AIR Venturiprinzip *nt*; ~ **scrubber** *n* AIR Venturiwäscher *m*; ~-**shaped** *adj* AIR venturiartig; ~ **throat** *n* AIR Venturikehle *f*; ~ **tube** *n* AIR, ANALYSIS Venturi *nt*, Venturidüse *f*, Venturirohr *nt*
verifiable *adj* ANALYSIS überprüfbar, nachprüfbar
verification *n* ANALYSIS, SAFETY, WASTE Überprüfung
verified *adj* ANALYSIS, SAFETY, WASTE überprüft
verify *vt* ANALYSIS überprüfen
vermin *n* WASTE Ungeziefer *nt*
vertical: ~ **axis wind turbine** *n* (*VAWT*) RENEW ENERG Vertikalachsenmaschine *f*; ~ **collection element** *n* WASTE vertikales flächiges Fassungselement *nt*; ~ **degassing** *n* WASTE, WATER SUPP vertikale Entgasung *f*; ~ **dispersion** *n* AIR, SOIL vertikale Verteilung *f*, Vertikalverteilung *f*; ~ **distribution** *n* AIR, SOIL vertikale Verteilung *f*, Vertikalverteilung *f*; ~ **distribution of ozone** *n* AIR Vertikalverteilung *f* des Ozons; ~ **filtering well** *n* WATER SUPP Vertikalfilterbrunnen *m* (*VFB*); ~ **fixed-bed adsorber** *n* AIR vertikaler Festbettadsorber *m*; ~ **gas collection** *n* WASTE vertikale Gaserfassung *f*; ~ **gas well** *n* WASTE vertikaler Gasbrunnen *m*; ~ **linear-formed collection element** *n* WASTE vertikales linienförmiges Fassungselement *nt*; ~ **moving-bed adsorber** *n* AIR vertikaler Wanderbettadsorber *m*; ~ **percolation well** *n* WATER SUPP vertikaler Sickerbrunnen *m*; ~ **well** *n* WATER SUPP Vertikalbrunnen *m*, vertikaler Brunnen *m*
vessel *n* AIR, ANALYSIS Behälter *m*, Gefäß *nt*, MARINE Schiff *nt*; ~ **wall** *n* AIR *of cyclone* Behälterwand *f*
viable: ~ **bacteria** *n pl* WASTE, WATER SUPP lebensfähige Bakterien *f pl*
vibrate 1. *vt* AIR, NOISE, WASTE rütteln; **2.** *vi* AIR, NOISE, WASTE rütteln, vibrieren
vibrating *n* AIR, NOISE, WASTE Vibrieren *nt* Rütteln *nt*; ~ **conveyor** *n* WATER SUPP Schwingrinne *f*; ~ **screen** *n* SOIL Vibrationssieb *nt*, Vibratorsieb *nt*, WASTE Rüttelsieb *nt*, Schwingsieb *nt*; ~ **sieve** *n* WATER POLL Rüttelsieb *nt*, Schwingsieb *nt*; ~ **worm** *n* SOIL Vibrationsschnecke *f*
vibration *n* AIR, NOISE, WASTE Rütteln *nt*, Schwingung *f*, Vibration *f*; ~ **attenuation** *n* NOISE Schwingungsdämpfung

f; ~ **behavior** *AmE*, ~ **behaviour** *BrE* *n* RENEW ENERG Schwingungsverhalten *nt*; ~ **dampener** *n* NOISE Schwingungsdämpfer *m*; ~**-damping** *adj* NOISE vibrationsdämpfend; ~ **damping** *n* NOISE Schwingungsdämpfung *f*; ~ **isolation** *n* NOISE Schwingungsisolation *f*, Schwingungsschutz *m*; ~ **isolator** *n* NOISE Schwingungsisolator *m*; ~ **sensor** *n* ANALYSIS Schwingungsaufnehmer *m*; ~ **technique** *n* ANALYSIS Schüttelmethode *f*; ~ **transducer** *n* ANALYSIS Schwingungswandler *m*; ~ **viscosimeter** *n* ANALYSIS Schwingsviskosimeter *nt*

vibrator: ~ **screen** *n* WASTE Vibriersieb *nt*

vibrometer *n* ANALYSIS Vibrometer *nt*

vinyl: ~ **chloride** *n* (*VC*) AIR, SAFETY Vinylchlorid *nt* (*VC*)

viscid *adj* AIR, ANALYSIS, WASTE, WATER POLL viskos

viscidity *n* AIR, ANALYSIS, WASTE, WATER POLL Klebrigkeit *f*

viscosity *n* GEN Viskosität *f*, Zähflüssigkeit *f*; ~ **meter** *n* ANALYSIS Viskosimeter *nt*

viscous *adj* GEN viskos, zähflüssig; ~ **liquid** *n* AIR, WATER POLL viskose Flüssigkeit *f*

visibility: ~ **depth** *n* ANALYSIS Sichttiefe *f*

visible *adj* AIR, ANALYSIS sichtbar; ~ **optical absorption** *n* ANALYSIS sichtbare Lichtabsorption *f*; ~ **plume** *n* AIR sichtbarer Schwaden *m*

visual: ~ **analysis** *n* ANALYSIS Augenscheinanalyse *f*; ~ **test** *n* ANALYSIS visuelle Prüfung *f*

vitrification *n* SOIL, WASTE Sintern *nt*

vitrified *adj* SOIL, WASTE verglast; ~ **clay pipe** *n* WASTE, WATER POLL, WATER SUPP Steinzeugrohr *nt*; ~ **slag** *n* SOIL, WASTE verglaste Schlacke *f*

vitrify *vt* SOIL, WASTE sintern, verglasen

VOC *abbr* (*volatile organic compounds*) AIR VOC (*flüchtige organische Verbindungen*)

void *n* WASTE, WATER POLL Hohlraum *m*

voids: ~ **fraction** *n* AIR Lückengrad *m*

void: ~ **space** *n* WATER POLL *of trickling filter* Hohlraum *m*; ~ **volume** *n* AIR Hohlraumvolumen *nt*

volatile *adj* AIR, SOIL, WATER POLL flüchtig, leichtflüchtig; ~ **chlorinated hydrocarbon** *n* AIR leichtflüchtiger Chlorkohlenwasserstoff *m*; ~ **emissions** *n pl* AIR flüchtige Emissionen *f pl*; ~ **halogenated hydrocarbon** *n* AIR, WASTE, WATER SUPP leichtflüchtiger Halogenkohlenwasserstoff *m*; ~ **heavy metals** *n pl* AIR flüchtige Schwermetalle *nt pl*; ~ **inorganic compounds** *n pl* AIR, SOIL flüchtige anorganische Verbindungen *f pl*; ~ **organic compounds** *n pl* (*VOC*) AIR, SOIL flüchtige organische Verbindungen *f pl* (*VOC*)

volatility: ~ **analysis** *n* ANALYSIS Verflüchtigungsuntersuchung *f*

volatilization *n* AIR, SOIL Verflüchtigung *f*

volatilize *vi* AIR, SOIL sich verflüchtigen

voltameter *n* ANALYSIS Voltameter *nt*

voltametric *adj* ANALYSIS voltametrisch, voltamperemetrisch

voltametry *n* ANALYSIS Voltametrie *f*

volume *n* AIR, ANALYSIS Volumen *nt*, NOISE Lautstärke *f*, SOIL, WATER POLL, WATER SUPP Volumen *nt*; ~ **charge** *n* AIR Raumladung *f*; ~ **concentration** *n* AIR Volumenkonzentration *f*; ~ **filter** *n* WATER POLL Volumenfilter *nt*; ~ **filtration** *n* WATER POLL Volumenfiltration *f*; ~ **flow** *n* AIR, WASTE, WATER POLL, WATER SUPP Volumenstrom *m*; ~ **flow rate** *n* ANALYSIS Volumendurchfluß *m*; ~ **reduction** *n* ANALYSIS, WASTE Mengenreduktion *f*, Volumenreduktion *f*, Volumenreduzierung *f*

volumetric *adj* ANALYSIS maßanalytisch, volumetrisch; ~ **analysis** *n* ANALYSIS Maßanalyse *f*; ~ **analysis by titration** *n* ANALYSIS volumetrische Titriermethode *f*; ~ **determination** *n* ANALYSIS maßanalytische Bestimmung *f*; ~ **factor** *n* ANALYSIS Faktor *m*, Korrekturfaktor *m*; ~ **flask** *n* ANALYSIS Meßkolben *m*; ~ **gas analysis** *n* ANALYSIS Gasvolumetrie *f*, volumetrische Gasanalyse *f*; ~ **loss** *n* WASTE, WATER POLL, WATER SUPP volumetrischer Verlust *m*

volume: ~ **unit meter** *n* (*VU meter*) ANALYSIS, NOISE Schallpegelmesser *m*; ~ **of waste** *n* WASTE Abfallmenge *f*; ~ **of water** *n* WATER SUPP Wassermenge *f*; ~ **weight** *n* WASTE *of refuse* Raumgewicht *nt*

voluntary: ~ **agreement** *n* POLICY freiwillige Vereinbarung *f*; ~ **commitment** *n* POLICY Selbstverpflichtung *f*, freiwillige Selbstverpflichtung *f*

volute: ~ **chamber** *n* WASTE Ausströmraum *m*, Diffusor *m*, WATER SUPP *centrifugal pump* Ausströmraum *m*; ~ **insert** *n* WATER SUPP *pump* Strömungsring *m*

vortex *n* AIR Wirbel *m*; ~ **finder** *n* AIR *cyclone* Tauchrohr *nt*; ~ **flow** *n* AIR *cyclone* Wirbelströmung *f*; ~ **skimmer** *n* MARINE Wirbelabschöpfgerät *nt*

vulcanize *vt* WASTE vulkanisieren

vulcanized *adj* WASTE vulkanisiert

VU: ~ **meter** *n* (*volume unit meter*) ANALYSIS, NOISE Schallpegelmesser *m*

W

Wagner: **~'s bottle method** *n* ANALYSIS Flaschenmethode *f* nach Wagner
wagon: **~ drill** *n* ANALYSIS Bohrwagen *m*
waiting: **~ time** *n* ANALYSIS Wartezeit *f*
wake *n* MARINE Kielwasser *nt*, Schraubenwasser *nt*, Sog *m*
wall *n* WASTE, WATER SUPP Wand *f*; **~ depth** *n* SOIL, WATER SUPP Wandtiefe *f*; **~ hydrant** *n* WATER POLL, WATER SUPP Wandhydrant *m*
Walther: **~ process** *n* AIR Walther-Verfahren *nt*
WAO *abbr* (*wet air oxidation*) SOIL, WASTE, WATER POLL, WATER SUPP Naßoxidation *f*
warm: **~ seawater** *n* RENEW ENERG warmes Meerwasser *nt*; **~ water generation** *n* AIR, WATER POLL Warmwassererzeugung *f*; **~ water heater** *n* AIR, WATER POLL Warmwasserbereiter *m*
warn *vt* SAFETY warnen
warning: **~ signal** *n* SAFETY Warnsignal *nt*
wash *vt* AIR, SOIL, WASTE, WATER SUPP waschen; **~ away** *vt* AIR *collected dust* abschwemmen, ANALYSIS wegwaschen, WASTE *collected dust* abschwemmen, WATER POLL abspülen, WATER SUPP abschwemmen; **~ out** *vt* ANALYSIS, SOIL, WASTE, WATER SUPP auswaschen, WASTE, WATER SUPP, entkrusten
wash: **~ coat** *n* AIR *catalyst* Washcoat *nt*
washing *n* SOIL Wäsche *f*, WASTE, WATER SUPP Abschwemmung; **~-active** *adj* SOIL, WATER POLL waschaktiv; **~ agent** *n* AIR Waschmittel *nt*; **~ agent circulation** *n* AIR Waschmittelumlauf *m*; **~ agent processing** *n* AIR Waschmittelaufbereitung *f*; **~ agent regeneration** *n* AIR Waschmittelregenerierung *f*; **~ cycle** *n* AIR Waschzyklus *m*; **~ operation** *n* AIR Waschvorgang *m*; **~-out** *n* AIR *filter fibres*, ANALYSIS *water determination in gases*, SOIL Auswaschen *nt*; **~ process** *n* AIR Waschprozeß *m*
washings *n pl* SOIL Waschwasser *nt*
washing: **~ solution** *n* AIR Waschlösung *f*; **~ system** *n* AIR Waschsystem *nt*
washout *n* AIR Auswaschung *f*, *nuclear* Washout *nt*, WASTE, WATER SUPP Auswaschung *f*; **~ of landfill** *n* WASTE Unterspülung *f* der Deponie; **~ rate** *n* AIR Washout-Rate *f*
wash: **~ water** *n* AIR, WASTE, WATER SUPP Waschwasser *nt*; **~ water circulation** *n* AIR Waschwasserkreislauf *m*; **~ water tank** *n* AIR Waschwasserbehälter *m*
waste *n* WASTE Abfall *m*, Müll *m*; **~ acid** *n* WASTE Abfallsäure *f*; **~ air concentration** *n* AIR Abluftkonzentration *f*; **~ air constituent** *n* AIR Abluftkomponente *f*; **~ air fan** *n* AIR Abluftventilator *m*; **~ air flow** *n* AIR Abluftstrom *m*; **~ air temperature** *n* AIR Ablufttemperatur *f*; **~ air throughput** *n* AIR Abluftdurchsatz *m*; **~ air thruput** *AmE see waste air throughput*; **~ avoidance** *n* WASTE Abfallvermeidung *f*
Waste: **~ Avoidance and Management Act** *n* POLICY Abfallgesetz *nt*, Gesetz *nt* über Vermeidung und Entsorgung von Abfällen
waste: **~ bag** *n* WASTE Müllbeutel *m*, Müllsack *m*; **~ bale** *n* WASTE Müllballen *m*; **~ body** *n* WASTE Müllkörper *m*; **~ bunker** *n* WASTE Aufnahmebunker *m*, Lagerbunker *m*, Müllbunker *m*; **~ charge** *n* POLICY Abfallabgabe *f*
waste/coal: **~ mixture** *n* WASTE *refuse-derived fuel* Abfall-Kohle-Gemisch *m*
waste: **~ collection** *n* WASTE Abfallabfuhr *f*, Mülleinsammlung *f*; **~ collector** *n* WASTE *business* Müllabfuhrunternehmen *nt*, *person* Müllfahrer *m*, Müllmann *m*; **~ composition** *n* WASTE Abfallzusammensetzung *f*, Müllzusammensetzung *f*; **~ container** *n* WASTE Abfallbehälter *m*, Müllbehälter *m*, Mülleimer *m*, Mülltonne *f*; **~ containing pollutants** *n* WASTE schadstoffhaltiger Abfall *m*; **~ contaminated by pollutants** *n* WASTE schadstoffhaltiger Abfall *m*; **~ dampness** *n* WASTE Müllfeuchte *f*; **~ denitrification** *n* WASTE Denitrierung *f* des Abfalls; **~ detector** *n* WASTE Mülldetektor *m*; **~ disinfection** *n* WASTE Abfalldesinfektion *f*; **~ disintegrator** *n* WASTE Abfallzerkleinerer *m*; **~ disposal** *n* WASTE Abfallbeseitigung *f*, Abfallentsorgung *f*, Müllbeseitigung *f*; **~ disposal company** *n* WASTE Abfallbeseitigungsunternehmen *nt*; **~ disposal firm** *n* WASTE Abfallbeseitigungsunternehmen *nt*; **~ disposal method** *n* WASTE Deponierungsverfahren *nt*; **~ disposal plan** *n* MANAGE, POLICY, WASTE Abfallentsorgungsplan *m*; **~ disposal regulation** *n* POLICY Abfallentsorgungsverordnung *f*; **~ disposal tax** *n* POLICY Abfallbeseitigungssteuer *f*; **~ dump** *n* WASTE Abfalldeponie *f*, Deponie *f*, Müllkippe *f*; **~-economical** *adj* MANAGE, WASTE abfallwirtschaftlich; **~-economical planning** *n* WASTE abfallwirtschaftliche Planung *f*; **~ of energy** *n* AIR, WASTE Energieverschwendung *f*
Waste: **~ Exchange** *n* WASTE Abfallbörse *f*
waste: **~ extraction system** *n* AIR, WASTE, WATER POLL Abfallextraktionssystem *nt*; **~-fired** *adj* WASTE abfallgefeuert; **~-fired plant** *n* WASTE abfallgefeuerte Anlage *f*; **~-free** *adj* WASTE, WATER SUPP abfallfrei; **~ gas** *n* AIR Abgas *nt*, Abluft *f*; **~ gas back pressure** *n* AIR Abgasgegendruck *m*; **~ gas blower** *n* AIR Abgasgebläse *nt*; **~ gas cleaning** *n* AIR Abgasreinigung *f*, Abluftreinigung *f*; **~ gas cleansing** *n* AIR Abgasreinigung *f*, Abluftreinigung *f*; **~ gas component** *n* AIR Abgasinhaltsstoff *m*, Abgaskomponente *f*; **~ gas composition** *n* AIR Abgaszusammensetzung *f*; **~ gas containing harmful substances** *n* AIR schadstoffhaltige Abgase *nt pl*; **~ gas desulfurization** *AmE*, **~ gas desulphurization** *BrE n* AIR, WASTE Abgasentschwefelung *f*; **~ gas dilution** *n* AIR Verdünnung *f* der Abgase; **~ gas emissions** *n pl* AIR Abgasemissionen *f pl*; **~ gas extraction** *n* AIR Abgasabsaugung *f*; **~ gas flow** *n* AIR, ANALYSIS Abgasstrom *m*; **~ gas incineration** *n* AIR Abgasverbrennung *f*; **~ gas loading** *n* AIR Abgasbeladung *f*; **~ gas plume** *n* AIR Abgasfahne *f*, Abgaswolke *f*; **~ gas preheating** *n* AIR Abgasvorwärmung *f*; **~ gas pressure** *n* AIR *stack* Abgasdruck *m*; **~ gas probe** *n* AIR Abgassonde *f*; **~ gas purifier** *n* AIR Abgasreiniger *m*; **~ gas temperature** *n* AIR Rauchgastemperatur *f*, Abgastemperatur *f*; **~ gas throughput** *n* AIR Abgasdurchsatz *m*; **~ gas thruput** *AmE see waste gas throughput*; **~ gas turbidity** *n* ANALYSIS *dust, smoke density* Abgastrübung *f*; **~ gas velocity** *n* AIR *stack* Abgasgeschwindigkeit *f*; **~ gas volume** *n* AIR Abgasvolumen *nt*; **~-generating** *adj* WASTE abfallerzeugend; **~-generating factory** *n* AIR, WASTE, WATER POLL abfallerzeugender Betrieb *m*; **~ generation** *n* WASTE Abfallerzeugung *f*, Müllerzeugung *f*; **~ generator** *n* WASTE Abfallerzeuger *m*; **~ glass** *n* WASTE Altglas *nt*; **~ glass container** *n* WASTE Altglasbehälter *m*; **~ glass cullet** *n* WASTE Glasbruch *m*, Glasscherben *f pl*; **~ harmful to water** *n* WASTE, WATER POLL, WATER SUPP wassergefährdender Abfall *m*; **~ heat** *n* AIR, WASTE, WATER POLL, WATER SUPP *production* Abhitze *f*, Abwärme *f*; **~ heat boiler** *n* AIR, WASTE, WATER POLL *incineration* Abhitzedampfkessel *m*, Abhitzekessel *m*; **~ heat boiler system** *n* AIR, WASTE, WATER POLL Abhitzekesselsystem *nt*; **~ heat recovery** *n* AIR, WASTE Abwärmeverwertung *f*; **~ heat system** *n* AIR, WASTE

Abhitzesystem *nt*; ~ **heat use** *n* AIR, WASTE, WATER SUPP Abwärmenutzung *f*; ~ **heat utilization** *n* AIR, WASTE Abwärmeverwertung *f*; ~ **incineration** *n* WASTE Müllverbrennung *f*; ~ **incineration ash** *n* WASTE Müllverbrennungsfilterstaub *m*; ~ **incineration plant** *n* WASTE Müllverbrennungsanlage *f* (*MVA*); ~ **incineration slag** *n* WASTE MVA-Schlacke *f*, Müllverbrennungsschlacke *f*; ~ **injection** *n* WASTE Abfallbeschickung *f*; ~ **inventory** *n* POLICY, WASTE Abfallkataster *m*; ~ **landfilling** *n* WASTE Abfalldeponierung *f*; ~ **landfill planning** *n* WASTE Deponieplanung *f*; ~ **layer** *n* WASTE Abfallschicht *f*; ~ **lubricants** *n pl* WASTE, WATER POLL Schmiermittelrückstände *m pl*; ~ **lye** *n* WASTE, WATER POLL, WATER SUPP Ablauge *f*; ~ **management** *n* MANAGE Abfallbewirtschaftung *f*, Abfallwirtschaft *f*, WASTE Abfallbehandlung *f*

Waste: ~ **Management Act** *n* MANAGE, POLICY Abfallgesetz *nt*

waste: ~ **management company** *n* MANAGE, WASTE Abfallbeseitigungsunternehmen *nt*; ~ **management concept** *n* MANAGE Abfallwirtschaftskonzept *nt*, abfallwirtschaftliches Konzept *nt*; ~ **management firm** *n* MANAGE, WASTE Abfallbeseitigungsunternehmen *nt*; ~ **management market** *n* MANAGE Abfallwirtschaftsmarkt *m*; ~ **management official** *n* MANAGE, WASTE Betriebsbeauftragter *m* für Abfall; ~ **management planning** *n* MANAGE Abfallwirtschaftsplanung *f*, abfallwirtschaftliche Planung *f*; ~ **management plant** *n* WASTE Abfallentsorgungsanlage *f*; ~ **mass** *n* WASTE Deponiegut *nt*; ~ **material** *n* WASTE Altstoff *m*; ~ **material return rate** *n* WASTE Altstoffrücklaufquote *f*; ~ **material use rate** *n* WASTE Altstoffeinsatzquote *f*; ~ **mixture** *n* WASTE Abfallgemisch *nt*; ~ **oil** *n* WASTE Altöl *nt*

Waste: ~ **Oil Act** *n* POLICY Altölgesetz *nt*

waste: ~ **oil collection** *n* WASTE Altöleinsammlung *f*, Altölsammlung *f*; ~ **oil recovery** *n* WASTE Altölrückgewinnung *f*, Altölwiederverwertung *f*; ~ **paint** *n* WASTE Farbabfälle *m pl*; ~ **paint sludge** *n* WASTE Farbschlamm *m*; ~ **paper** *n* WASTE Altpapier *nt*; ~ **paper collection** *n* WASTE Altpapiersammlung *f*; ~ **paper preparation** *n* WASTE Altpapieraufbereitung *f*; ~ **paper recycling** *n* WASTE Altpapierrecycling *nt*; ~ **plastics** *n pl* WASTE Kunststoffabfälle *m pl*; ~ **pretreatment** *n* WASTE Abfallvorbehandlung *f*; ~ **processing** *n* WASTE Abfallbehandlung *f*, Abfallverwertung *f*; ~ **processing plant** *n* WASTE Abfallverwertungsanlage *f*; ~ **producer** *n* WASTE Abfallerzeuger *m*, Abfallverursacher *m*; ~ **product** *n* WASTE Abfallstoff *m*, nichtverwertbarer Rückstand *m*; ~ **production** *n* WASTE Abfallanfall *m*, Abfallerzeugung *f*, Müllanfall *m*; ~ **pulp** *n* WASTE Abfallzellstoff *m*; ~ **quality** *n* WASTE Abfallqualität *f*; ~ **raw material** *n* RENEW ENERG *biomass conversion* Abfallrohstoff *m*; ~ **removal** *n* WASTE Abfallbeseitigung *f*; ~ **rubber** *n* WASTE Gummiabfall *m*

waste/sludge: ~ **mixture** *n* WASTE Abfall-Klärschlamm-Gemisch *m*

waste: ~ **sorting plant** *n* WASTE Abfallsortieranlage *f*, Müllsortieranlage *f*; ~ **steam** *n* WASTE Abdampf *m*; ~ **storage** *n* WASTE Abfallagerung *f*, Müllbunkerung *f*; ~ **stream** *n* WASTE Abfallstrom *m*; ~ **substance** *n* WASTE *landfill sealing* Abfallstoff *m*; ~ **tax** *n* POLICY Abfallsteuer *f*; ~ **tip** *n BrE* WASTE Deponie *f*, Müllkippe *f*; ~ **tourism** *n* POLICY, WASTE Abfalltourismus *m*, Mülltourismus *m*; ~ **treatment** *n* WASTE *processing* Abfallbehandlung *f*, *recycling* Abfallverwertung *f*, *removal* Abfallbeseitigung *f*; ~ **treatment plant** *n* WASTE *processing* Abfallbehandlungsanlage *f*, *recycling* Abfallverwertungsanlage *f*, *removal* Abfallbeseitigungsanlage *f*; ~ **utilization** *n* WASTE Abfallverwertung *f*; ~ **utilization plant** *n* WASTE Abfallverwertungsanlage *f*

wastewater *n* WASTE *from landfill*, WATER POLL, WATER SUPP Abwasser *nt*, Schmutzwasser *nt*; ~ **analyser** *BrE*, ~ **analyzer** *AmE* *n* ANALYSIS Abwasseranalysegerät *nt*; ~ **bed** *n* WATER POLL Klärfilter *nt*; ~ **clarification** *n* WATER POLL *physical process* Abwasserbehandlung *f*, Abwasserreinigung *f*; ~ **clarification plant** *n* WATER POLL Abwasserkläranlage *f*; ~ **collection** *n* WATER POLL Abwassererfassung *f*; ~ **collection system** *n* WATER POLL Abwassersammelsystem *nt*; ~ **collection tank** *n* WATER SUPP Abwassersammeltank *m*; ~ **collector** *n* WATER POLL Abwassersammler *m*; ~ **coloration** *n* ANALYSIS Abwasserfärbung *f*; ~ **composition** *n* WATER POLL Abwasserzusammensetzung *f*; ~ **containing heavy metals** *n* WATER POLL schwermetallhaltiges Abwasser *nt*; ~ **containing mineral oil** *n* WATER POLL mineralölhaltiges Abwasser *nt*; ~ **control** *n* WATER POLL Abwasserkontrolle *f*, Abwasserüberwachung *f*; ~ **decontamination** *n* WATER POLL Abwasserdekontaminierung *f*, Abwasserentgiftung *f*; ~ **discharge** *n* WATER POLL Abwasserausfluß *m*; ~ **disposal** *n* WATER POLL Abwasserbeseitigung *f*, Abwasserentsorgung *f*; ~ **farm** *n* WATER POLL Rieselfeld *nt*; ~ **fish pond** *n* WATER POLL Abwasserfischteich *m*; ~ **flow** *n* WATER POLL Abwasseranfall *m*, Abwassermenge *f*; ~**-free** *adj* AIR *spray absorption process* abwasserfrei; ~**-free flue-gas cleaning** *n* AIR abwasserfreie Rauchgasreinigung *f*; ~ **from flue-gas scrubbing** *n* AIR Abwasser aus der Rauchgaswäsche *nt*; ~ **from waste incineration plant** *n* WASTE, WATER SUPP Abwasser *nt* aus der Müllverbrennungsanlage; ~ **fungus** *n* WATER POLL Abwasserpilz *m*; ~ **in ash removal** *n* WASTE, WATER SUPP Abwasser *nt* in der Entaschung; ~ **irrigation** *n* WATER POLL Berieselung *f*, Verrieselung *f*; ~ **load** *n* WATER POLL Abwasserabgabe *f*, Abwasserlast *f*; ~ **monitoring** *n* WATER POLL Abwasserkontrolle *f*; ~ **monitoring station** *n* WATER POLL Abwasserkontrollstation *f*; ~ **organisms** *n pl* WATER POLL Abwasserorganismen *m pl*; ~ **outfall** *n* WATER POLL Abwasserausfluß *m*; ~ **outflow** *n* WATER POLL Abwasserausfluß *m*; ~ **production** *n* WATER POLL Abwasserentstehung *f*; ~ **pump** *n* WATER POLL Fäkalienpumpe *f*; ~ **purification** *n* WATER POLL Abwasserklärung *f*, Abwasserreinigung *f*; ~ **purification plant** *n* WATER POLL Abwasserkläranlage *f*; ~ **purification process** *n* WATER POLL Abwasserreinigungsverfahren *nt*; ~ **quality** *n* WATER POLL Abwasserqualität *f*; ~ **renovation** *n* WATER POLL Abwassersanierung *f*; ~ **resulting from drinking water treatment** *n* WATER POLL, WATER SUPP Abwasser *nt* aus der Trinkwasseraufbereitung; ~ **reuse** *n* WATER POLL Abwasserwiederverwertung *f*; ~ **sample** *n* ANALYSIS Abwasserprobe *f*; ~ **sludge** *n* WATER POLL, WATER SUPP Klärschlamm *m* (*KS*); ~ **sludge amount** *n* WATER POLL Klärschlammenge *f*; ~ **sludge catalysis** *n* WATER POLL Klärschlammkatalyse *f*; ~ **sludge composting** *n* WATER POLL Klärschlammkompostierung *f*; ~ **sludge conditioning plant** *n* WATER POLL Klärschlammkonditionierungsanlage *f*; ~ **sludge delivery** *n* WATER POLL Klärschlammabgabe *f*; ~ **sludge deposition ban** *n* WATER POLL Aufbringungsverbot *nt* für Klärschlamm; ~ **sludge dewatering equipment** *n* WATER POLL Klärschlammentwässerungsanlage *f*; ~ **sludge incineration** *n* WATER POLL Klärschlammverbrennung *f*; ~ **sludge incineration plant** *n* WATER POLL Klärschlammverbrennungsanlage *f*; ~ **sludge treatment** *n* WATER POLL Klärschlammbehandlung *f*; ~ **sprinkling** *n* WATER POLL Abwasserverregnung *f*; ~ **storage** *n* WATER POLL Abwasserlagerung *f*, Abwasserspeicherung *f*; ~ **stream** *n* WATER POLL Abwasserstrom *m*; ~ **technology** *n* WATER POLL Abwassertechnik *f*; ~ **temperature** *n* WATER POLL Abwassertemperatur *f*; ~ **toxicity** *n* WATER POLL Abwassertoxizität *f*; ~ **trail** *n* WATER POLL *plume* Abwasserfahne *f*; ~ **treatment** *n*, WATER POLL *processing* Abwasseraufbereitung *f*, Abwasserbehandlung *f*, Abwasserklärung *f*, Abwasserreinigung *f*; ~ **treatment fee** *n* WATER POLL Abwassergebühr *f*; ~ **treatment plant** *n*, WATER POLL *processing* Abwasseraufbereitungsanlage *f*, Abwasserbehandlungsanlage *f*, *purification* Kläranlage *f*, Klärwerk *nt*; ~ **treatment plant installation** *n* WATER POLL Kläranlagenausrüstung *f*; ~ **treatment stage** *n* WATER POLL Abwasserreinigungsstufe *f*; ~ **utilization** *n* WATER POLL Abwasserverwertung *f*; ~ **volume** *n* WATER POLL Abwasseranfall *m*, Abwassermenge *f*; ~ **yield** *n* WATER POLL Abwasseranfall *m*

water *n* GEN Wasser *nt*; ~ **analysis** *n* ANALYSIS Wasseranalyse *f*; ~ **authority** *n* POLICY, WATER SUPP Wasserbehörde *f*; ~ **balance** *n* POLICY, WATER SUPP Wasserbilanz *f*; **~-based coating** *n* WASTE, WATER SUPP auf Wasser basierte Beschichtung *f*; ~ **biology** *n* WATER SUPP Gewässerbiologie *f*

waterborne *adj* WATER POLL wasserbürtig; ~ **contaminant** *n* WASTE, WATER POLL wasserbürtiger Schadstoff *m*; ~ **disease** *n* SAFETY, WATER POLL Wasserkrankheit *f*; ~ **pollutant** *n* WASTE, WATER POLL wasserbürtiger Schadstoff *m*

water: ~ **catchment** *n* WATER SUPP Fassung *f*, Wasserfassung *f*; ~ **circulation** *n* WATER POLL Wasserkreislauf *f*, Wasserkreislaufführung *f*; ~ **clearness** *n* WATER POLL, WATER SUPP Klarheit *f* des Wassers; ~ **column** *n* ANALYSIS, WATER POLL, WATER SUPP Wassersäule *f*; ~ **consumer** *n* WATER SUPP Wasserverbraucher *m*; ~ **consumption** *n* WATER SUPP Wasserverbrauch *m*; ~ **containing pesticides** *n* WATER POLL pestizidhaltiges Wasser *nt*; ~ **content** *n* ANALYSIS, WATER POLL *of wastewater sludge* Wassergehalt m (*WG*)

watercourse *n* WATER SUPP Wasserlauf *m*

water: ~ **deficiency** *n* WATER SUPP Wassermangel *m*; ~ **delivery** *n* WATER SUPP Wasserabgabe *f*, Wasserbelieferung *f*; ~ **demand** *n* WATER SUPP Wasserbedarf *m*; ~ **distribution** *n* WATER SUPP Wasserverteilung *f*; ~ **distribution system** *n* WATER SUPP Wasserverteilungsnetz *nt*; ~ **evaporation** *n* AIR Wasserverdampfung *f*; ~ **examination** *n* ANALYSIS Wasseruntersuchung *f*; ~ **film** *n* AIR Wasserfilm *m*; ~ **filter** *n* WATER SUPP Wasserfilter *nt*; **~-free** *adj* WATER SUPP wasserfrei

waterfront *adj* WATER SUPP uferanliegend

water: ~ **gage** *AmE*, ~ **gauge** *BrE* *n* WATER SUPP Wasserstandsanzeiger *m*, Wasserstandsmesser *m*; ~ **hardness** *n* WATER SUPP Wasserhärte *f*; ~ **head** *n* WATER POLL, WATER SUPP Wassersäule *f*; ~ **hydrant** *n* WATER SUPP Hydrant *m*; ~ **indicator** *n* WATER SUPP Wasserstandsanzeiger *m*; ~ **industry** *n* MANAGE Wasserwirtschaft *f*; ~ **inflow** *n* WATER SUPP Wassereinlaß *m*

watering *n* AIR *of filter* Berieselung *f*, Bewässerung *f*

water: ~ **inlet** *n* WATER SUPP Wassereintritt *m*; **~-insoluble** *adj* AIR, WATER POLL, WATER SUPP wasserunlöslich; **~-insoluble hazardous waste** *n* WASTE, WATER POLL wasserunlöslicher Sonderabfall *m*; ~ **jet** *n* SOIL, WATER POLL, WATER SUPP Wasserstrahl *m*; ~ **jet pump** *n* SOIL, WATER POLL, WATER SUPP Wasserstrahlpumpe *f*; ~ **law** *n* POLICY, WATER SUPP Wasserhaushaltsgesetz *nt*; ~ **level** *n* RENEW ENERG *tidal, wave power plant* Wasserspiegel *m*, WATER SUPP *instrument* Wasserstand *m*; ~ **level indicator** *n* WATER SUPP Wasserstandsmesser *m*; ~ **level regulator** *n* WATER POLL, WATER SUPP Wasserstandsregler *m*

waterline *n* MARINE Wasserlinie *f*

water: ~ **loss** *n* WATER POLL, WATER SUPP Wasserverlust *m*; ~ **loss within distribution system** *n* WATER SUPP Rohrnetzverlust *m*; ~ **main** *n* WATER SUPP Hauptwasserleitung *f*, Hauptwasserrohr *nt*; ~ **management** *n* WATER SUPP Wasserbewirtschaftung *f*; ~ **mark** *n* WATER SUPP Wasserstandsanzeiger *m*, Wasserstandsmarke *f*; ~ **mass** *n* RENEW ENERG *tidal energy* Wassermasse *f*; ~ **meter** *n* WATER SUPP Wasserzähler *m*; ~ **outlet** *n* WATER SUPP Wasserablaufrinne *f*; ~ **penny** *n* POLICY Wasserpfennig *m*; ~ **pipe** *n* WATER SUPP Wasserleitungsrohr *nt*; ~ **pollutant** *n* WATER POLL wasserverunreinigender Stoff *m*; ~ **pollution** *n* WATER POLL Wasserverschmutzung *f*, Wasserverunreinigung *f*, Gewässerbelastung *f*, Gewässerverschmutzung *f*, Gewässerverunreinigung *f*; ~ **pollution tax** *n* POLICY Wasserverunreinigungssteuer *f*; ~ **power** *n* WATER SUPP Wasserkraft *f*; ~ **pressure** *n* WATER SUPP Wasserdruck *m*; ~ **properties** *n pl* WATER SUPP Wassereigenschaften *f pl*; ~ **pump** *n* RENEW ENERG, WATER SUPP Wasserpumpe *f*; ~ **purification** *n* WATER POLL, WATER SUPP Wasseraufbereitung *f*, Wasserreinigung *f*; ~ **purification sludge** *n* WASTE Wasserreinigungsschlamm *m*; ~ **quality** *n* ANALYSIS, WATER SUPP *of drinking water* Wasserqualität *f*, Wasserbeschaffenheit *f*, Wassergüte *f*, *body of water* Gewässerbeschaffenheit *f*, Gewässergüte *f*; ~ **quality class** *n* ANALYSIS, WATER SUPP Gewässergüteklasse *f*; ~ **quality determination** *n* ANALYSIS, WATER SUPP Wassergütebestimmung *f*; ~ **quality monitoring** *n* WATER SUPP Überwachung *f* der Wasserbeschaffenheit; ~ **recovery** *n* WATER SUPP Wassergewinnung *f*; ~ **reserve** *n* WATER SUPP Wasservorrat *m*; **~-resistant** *adj* WATER SUPP wasserbeständig; ~ **resources** *n pl* WATER SUPP Wasservorkommen *nt*; ~ **retention** *n* WATER SUPP Wasserbindung *f*, Wasserzurückhaltung *f*; ~ **sample** *n* ANALYSIS Wasserprobe *f*; ~ **sampler** *n* ANALYSIS Probenschöpfer *m*, Wasserschöpfer *m*; **~-saturated** *adj* WATER SUPP wassergesättigt; ~ **scrubbing** *n* AIR Wasserwäsche *f*

watershed *n* WATER SUPP Wasserscheide *f*

water: ~ **softening** *n* WATER SUPP Wasserenthärtung *f*; ~ **solubility** *n* AIR, WASTE, WATER SUPP Wasserlöslichkeit *f*; **~-soluble** *adj* AIR, WASTE, WATER SUPP wasserlöslich; **~-soluble hazardous wastes** *n pl* WASTE wasserlösliche Sonderabfälle *m pl*; ~ **spraying** *n* AIR Wassereindüsung *f*; ~ **storage** *n* WATER SUPP Wasserspeicherung *f*; ~ **supply** *n* WATER SUPP Wasserversorgung *f*, Wasserdargebot *nt*; ~ **supply company** *n* WATER SUPP Wasserversorgungsunternehmen *nt*; ~ **supply pipe** *n* WATER SUPP Wasserversorgungsleitung *f*; ~ **supply planning** *n* WATER SUPP Wasserversorgungsplanung *f*; ~ **surface** *n* RENEW ENERG *wave energy* Wasseroberfläche *f*; ~ **table** *n* WATER SUPP Grundwasserspiegel *m*; ~ **tank** *n* WATER POLL, WATER SUPP Reservoir *nt*, Wasserbecken *nt*, Wasserbehälter *m*; ~ **tap** *n* WATER SUPP Wasserhahn *m*; ~ **tap connection** *n* WATER SUPP Wasserzapfstelle *f*; ~ **technology** *n* WATER SUPP Wassertechnologie *f*

watertight *adj* WATER POLL, WATER SUPP wasserfest, stichfest

water: ~ **tower** *n* WATER SUPP Turmbehälter *m*, Wasserhochbehälter *m*, Wasserturm *m*; ~ **tower height** *n* WATER SUPP Turmbehälterhöhe *f*; ~ **tower runoff** *n* WATER SUPP Hochbehälterabfluß *m*; ~ **transport** *n* MARINE, WATER SUPP Wassertransport *m*; ~ **treatment plant** *n* WATER POLL, WATER SUPP Wasseraufbereitungsanlage *f*, Wasserbehandlungsanlage *f*; ~ **turbine** *n* RENEW ENERG Wasserturbine *f*; ~ **user** *n* WATER SUPP Wasserverbraucher *m*; ~ **vapor** *AmE*, ~ **vapour** *BrE* *n* AIR *trace gas*, ANALYSIS Wasserdampf *m*; ~ **vapour plume** *n BrE* AIR Wasserdampfschwaden *m*

waterwheel *n* RENEW ENERG *tidal power* Wasserrad *nt*

waterworks *n pl* WATER SUPP Wasserwerk *nt*

wave *n* RENEW ENERG Welle *f*; ~ **crest** *n* RENEW ENERG Wellenberg *m*; ~ **energy** *n* RENEW ENERG Wellenenergie *f*; ~ **energy conversion** *n* RENEW ENERG Umwandlung *f* der Wellenenergie; ~ **energy resources** *n pl* RENEW ENERG Wellenenergieressourcen *f pl*; ~ **energy use** *n* RENEW ENERG Wellenenergienutzung *f*; ~ **frequency** *n* RENEW ENERG Häufigkeit *f* der Wellen

wavefront: ~ **length** *n* RENEW ENERG *wave energy* Wellenfrontbreite *f*

wave: ~ **hollow** *n* RENEW ENERG *wave energy* Wellental *nt*

wavelength *n* ANALYSIS Wellenlänge *f*; ~ **scale** *n* ANALYSIS Wellenlängenskala *f*

wave: ~ **movement** *n* RENEW ENERG Wellenbewegung *f*; ~ **power plant** *n* RENEW ENERG Wellenkraftwerk *nt*

weaken *vt* AIR, ANALYSIS, NOISE schwächen

weakening *n* AIR, ANALYSIS, NOISE Schwächung *f*

weak: ~ **sample** *n* ANALYSIS schwache Probe *f*; ~**-sun period** *n* RENEW ENERG *solar energy* sonnenschwache Zeit *f*
wear: ~ **characteristics** *n pl* WASTE *of plastics* Verschleißeigenschaften *f pl*; ~ **resistance** *n* WASTE *of plastics* Verschleißfestigkeit *f*, Verschleißwiderstand *m*; ~**-resistant** *adj* GEN verschleißbeständig, verschleißfest; ~**-resistant lining** *n* AIR Innenpanzerung *f*, verschleißfeste Auskleidung *f*
weather: ~ **conditions** *n pl* RENEW ENERG Wetterbedingungen *f pl*
weathered: ~ **oil** *n* MARINE verwittertes Öl *nt*
weathering *n* MARINE, SOIL, WASTE *of plastics* Verwitterung *f*; ~ **characteristics** *n pl* WASTE Witterungseigenschaften *f pl*; ~ **resistance** *n* SOIL *liners*, WASTE Witterungsbeständigkeit *f*, Witterungsstabilität *f*
webbing *n* MARINE Schwimmhaut *f*, Webbing *nt*
WEC *abbr* (*wind energy conversion*) RENEW ENERG Windenergieumwandlung *f*
WECS *abbr* (*wind energy conversion system*) RENEW ENERG Windenergieumwandlungssytem *nt*
weekly: ~ **consumption** *n* WATER SUPP Wochenverbrauch *m*
weigh *vt* ANALYSIS wiegen; ~ **out** *vt* ANALYSIS auswiegen; ~ **out analytically** *vt* ANALYSIS analytisch auswiegen
weighable *adj* ANALYSIS wägbar
weighbridge *n* WASTE Waage *f*
weighing *n* ANALYSIS, WASTE Wiegung *f*
weigh: ~ **office** *n* WASTE Eingangskontrolle *f*
weighted: ~ **sound level** *n* NOISE bewerteter Schallpegel *m*
weir *n* MARINE, RENEW ENERG Wehr *nt*; ~ **boom** *n* MARINE Wehrölsperre *f*; ~ **skimmer** *n* MARINE Wehrabschöpfgerät *nt*
well *n* SOIL, WATER SUPP Brunnen *m*; ~ **catchment** *n* WATER SUPP Brunnenfassung *f*; ~ **construction** *n* WATER SUPP Brunnenausbau *m*; ~ **development** *n* WATER SUPP Brunnenentwicklung *f*; ~ **drilling** *n* ANALYSIS *groundwater sampling* Bohren *nt*; ~ **filter** *n* WATER SUPP Brunnenfilter *nt*
Wellmann-Lord: ~ **process** *n* AIR *wet flue-gas desulphurization* Wellmann-Lord-Verfahren *nt*
well: ~ **operation** *n* WATER SUPP Brunnenbetrieb *m*; ~ **pipe** *n* WATER SUPP Brunnenrohr *nt*; ~ **regeneration** *n* WATER SUPP Brunnenregeneration *f*; ~ **shaft** *n* WATER SUPP Brunnenschacht *m*, Brunnenschaft *m*; ~ **sinking** *n* WATER SUPP Brunnenabsenkung *f*; ~ **water** *n* WATER SUPP Brunnenwasser *nt*; ~ **yield** *n* WATER SUPP Brunnenergiebigkeit *f*
WEP *abbr* (*wet electrostatic precipitator*) AIR elektrostatischer Naßabscheider *m*
wet: ~ **acidic fallout** *n* AIR nasser saurer Niederschlag *m*; ~ **air oxidation** *n* (*WAO*) SOIL, WASTE, WATER POLL, WATER SUPP Naßoxidation *f*; ~ **ash remover** *n* WASTE Naßentascher *m*; ~**-chemical methods** *n pl* ANALYSIS naßchemische Methoden *f pl*; ~ **classifier** *n* WATER POLL Naßsortierer *m*; ~ **cleaning** *n* AIR, WASTE, WATER POLL Naßreinigung *f*; ~ **cleaning process** *n* AIR *with liquid or condensate* nasse Abreinigung *f*; ~ **collector** *n* AIR Naßabsorber *m*; ~ **corrosion** *n* WASTE, WATER SUPP nasse Korrosion *f*; ~ **deposition** *n* AIR, WASTE Naßablagerung *f*; ~ **desulfurization process** *AmE*, ~ **desulphurization process** *BrE* *n* AIR, WASTE Naßentschwefelungsprozeß *m*
wet/dry: ~ **scrubber** *n* AIR quasitrockener Wäscher *m*
wet: ~ **dust removal** *n* AIR nasse Entstaubung *f*; ~ **electrostatic filter** *n* AIR nasses Elektrofilter *nt*; ~ **electrostatic precipitator** *n* (*WEP*) AIR elektrostatischer Naßabscheider *m*; ~ **flue-gas desulfurization plant** *AmE*, ~ **flue-gas desulphurization plant** *BrE* *n* AIR Naßrauchgasentschwefelungsanlage *f* (*Naß-REA*); ~ **grind** *n* WASTE Naßzerkleinerung *f*; ~ **grinding** *n* AIR Naßmahlen *nt*; ~ **milling** *n* AIR Naßmahlen *nt*; ~ **mixer** *n* WASTE Naßmischer *m*; ~ **oxidation** *n* SOIL, WASTE, WATER POLL, WATER SUPP Naßoxidation *f*; ~ **period** *n* AIR Naßperiode *f*; ~ **precipitation** *n* AIR Naßabscheidung *f*; ~ **screening** *n* WATER POLL Naßsiebung *f*; ~**-scrubbed waste top gas** *n* AIR naßreinigendes Gichtgas *nt*; ~ **scrubber** *n* AIR, WASTE Naßwäscher *m*, Naßabscheider *m*; ~ **scrubbing device** *n* AIR, WASTE Naßscheider *m*, Naßwäscher *m*; ~ **separation** *n* AIR Naßabscheidung *f*; ~ **sludge** *n* WATER POLL Naßschlamm *m*; ~ **sludge utilization** *n* WATER POLL Flüssigschlammverwertung *f*, Naßschlammverwertung *f*; ~ **sorption** *n* WASTE Naßsorption *f*; ~ **sorting** *n* WASTE Naßtrennung *f*, Naßtrennverfahren *nt*
wettability *n* AIR, SOIL, WATER POLL Benetzbarkeit *f*
wettable *adj* AIR, SOIL, WATER POLL benetzbar, netzbar; ~ **dust** *n* AIR benetzbarer Staub *m*
wetted *adj* AIR, SOIL, WATER POLL benetzt; ~**-filter-type scrubber** *n* AIR benetztes Filter *nt*
wetting *adj* AIR, SOIL, WATER POLL netzend; ~ **agent** *n* AIR, MARINE Benetzungsmittel *nt*, Netzmittel *nt*
wet: ~ **waste gas treatment** *n* AIR nasse Abgasbehandlung *f*
Wheatstone: ~ **bridge** *n* ANALYSIS Wheatstone-Brücke *f*
wheel: ~ **loader** *n* SOIL, WASTE Radlader *m*
white: ~ **noise** *n* NOISE weißes Rauschen *nt*; ~ **product** *n* MARINE farbloses Produkt *nt*
wide: ~**-meshed** *adj* WATER POLL weitmaschig; ~**-meshed screen** *n* WATER POLL weitmaschiger Rechen *m*; ~**-mouthed laboratory bottle** *n* ANALYSIS Weithalsstandflasche *f*
winch *n* MARINE Winsch *f*
winchester *n* ANALYSIS Vorratsflasche *f* für Flüssigkeiten
wind *n* GEN Wind *m*; ~ **conditions** *n pl* RENEW ENERG Windbedingungen *f pl*; ~**-diesel system** *n* RENEW ENERG Wind-Diesel-System *nt*; ~ **direction** *n* AIR *stack* Windrichtung *f*; ~ **direction tracking mechanism** *n* RENEW ENERG *yaw mechanism* Windrichtungsnachführung *f*; ~ **duration** *n* RENEW ENERG Winddauer *f*; ~ **energy** *n* RENEW ENERG Windenergie *f*, Windkraft *f*; ~ **energy conversion** *n* (*WEC*) RENEW ENERG Windenergieumwandlung *f*; ~ **energy conversion system** *n* (*WECS*) RENEW ENERG Windenergieumwandlungssytem *nt*; ~ **exposure** *n* WASTE Windexposition *f*; ~ **farm** *n* RENEW ENERG Windfarm *f*, Windpark *m*; ~ **farm capacity** *n* RENEW ENERG Windfarmleistung *f*, Windparkleistung *f*; ~**-favorable** *AmE*, ~**-favourable** *BrE* *adj* RENEW ENERG windgünstig; ~**-favourable region** *n* *BrE* RENEW ENERG windgünstiges Gebiet *nt*
winding *n* AIR *of cyclone* Windung *f*
windmill *n* RENEW ENERG Windkraftanlage *f*, Windmühle *f*, Windrad *nt*
wind: ~ **park** *n* RENEW ENERG Windfarm *f*, Windpark *m*; ~ **park capacity** *n* RENEW ENERG Windfarmleistung *f*, Windparkleistung *f*; ~ **power** *n* RENEW ENERG Windstrom *m*; ~ **power plant** *n* RENEW ENERG Windkraftwerk *nt*; ~ **power resources** *n pl* RENEW ENERG Windenergieressourcen *f pl*; ~ **power use** *n* RENEW ENERG Windkraftnutzung *f*; ~ **rose** *n* AIR Windrose *f*
windrow: ~ **method** *n* SOIL dynamisches Vorrotteverfahren *nt*
wind: ~ **sensor** *n* ANALYSIS Windsensor *m*; ~ **turbine** *n* RENEW ENERG Windturbine *f*; ~ **turbine blade** *n* RENEW ENERG Windturbinenblatt *nt*; ~ **turbine generator** *n* (*WTG*) RENEW ENERG Windenergieanlage *f*, Windenergiekonverter *m*, Windkraftanlage *f*; ~ **velocity** *n* AIR, RENEW ENERG Windgeschwindigkeit *f*
windy *adj* RENEW ENERG windreich; ~ **region** *n* RENEW ENERG windreiches Gebiet *nt*
Winkler: ~ **method** *n* ANALYSIS Winkler-Methode *f*
wire: ~ **cage** *n* WASTE, WATER SUPP Drahtkäfig *m*; ~ **gauze** *n* WASTE Siebgewebe *nt*; ~**-heated** *adj* ANALYSIS *rotary oven* drahtbeheizt
withdraw: ~ **by pipette** *vt* ANALYSIS abpipettieren
withdrawal *n* WATER SUPP Entnahme *f*
wood: ~ **chips** *n pl* RENEW ENERG *biomass conversion* Holzschnitzel *m pl*; ~ **firing** *n* RENEW ENERG *biomass conversion* Holzfeuerung *f*; ~ **firing plant** *n* RENEW ENERG *biomass*

conversion Holzfeuerungsanlage *f*; **~ waste** *n* WASTE Holzabfall *m*

wool *n* AIR *filter fibre material* Wolle *f*

working: **~ capacity** *n* WATER POLL *of clarification tank* Nutzraum *m*; **~ with dangerous materials** *n* SAFETY Arbeit *f* mit Gefahrstoffen; **~ face** *n* WASTE Deponieoberfläche *f*; **~ with hazardous substances** *n* SAFETY Arbeit *f* mit Gefahrstoffen; **~ medium** *n* RENEW ENERG *heat pump* Arbeitsmedium *nt*, Arbeitsmittel *nt*; **~ temperature** *n* ANALYSIS *of detector* Betriebstemperatur *f*

workplace *n* SAFETY Arbeitsplatz *m*; **at the ~** *phr* SAFETY am Arbeitsplatz; **~-related** *adj* SAFETY arbeitsplatzbezogen; **~-related water demand** *n* WATER SUPP arbeitsplatzbezogener Wasserbedarf *m*

work: **~ safety** *n* SAFETY Arbeitssicherheit *f*

worldwide: **~ potential** *n* RENEW ENERG weltweites Potential *nt*

worm *n* WATER POLL Schnecke *f*

Woulfe: **~'s bottle** *n* ANALYSIS Woulfesche Flasche *f*

wound: **~-rotor induction motor** *n* WATER POLL *pump driving gear* Induktionsmotor *m* mit gewickeltem Läufer

woven: **~-wire cloth** *n* WATER POLL *screen* Drahtgewebe *nt*

WPC-VRS: **~ process** *n* WASTE VRS-Verfahren *nt*

wring *vt* MARINE ansprengen

wringer *n* WASTE Scheidepresse *f*

WTG *abbr* (*wind turbine generator*) RENEW ENERG Windenergieanlage *f*, Windenergiekonverter *m*, Windkraftanlage *f*

X Y Z

xanthat: ~ **polymer complex** *n* WATER POLL Xanthat-Polymer-Komplex *m*
Xe *abbr* (*xenon*) AIR Xe (*Xenon*)
xenon *n* (*Xe*) AIR Xenon *nt* (*Xe*)
x-ray: ~ **fluorescence** *n* ANALYSIS Röntgenfluoreszenz *f*; ~ **fluorescence analyser** *n BrE* ANALYSIS Röntgenfluoreszenz-Analysator *m*; ~ **fluorescence analysis** *n* (*XRFA*) ANALYSIS Röntgenfluoreszenzanalyse *f* (*RFA*); ~ **fluorescence analyzer** *AmE see x-ray fluorescence analyser BrE*; ~**-induced** *adj* ANALYSIS röntgenstrahlangeregt; ~ **protective glass** *n* SAFETY Röntgenschutzglas *nt*; ~ **protective glasses** *n pl* SAFETY Röntgenbrille *f*
x-rays *n pl* ANALYSIS Röntgenstrahlen *m pl*
XRFA *abbr* (*x-ray fluorescence analysis*) ANALYSIS RFA (*Röntgenfluoreszenzanalyse*)

yarn *n* AIR *fabric filtration* Garn *nt*
yearly: ~ **consumption** *n* WATER SUPP Jahresverbrauch *m*
Yellow: ~ **Container** *n* WASTE gelbe Tonne *f*
yellow: ~ **plastic collection bag** *n* WASTE gelber Sack *m*; ~ **plastic liner bag for DSD wastes** *n* WASTE gelber Wertstoffsack *m* für DSD-Abfälle
yield *n* POLICY Aufkommen *nt*, WATER SUPP Ergiebigkeit *f*

zeolite *n* AIR, WASTE, WATER SUPP Zeolith *m*; ~ **catalyst** *n* AIR Zeolithkatalysator *m*
zeolitic *adj* AIR zeolithisch; ~ **absorbent** *n* AIR zeolithisches Absorbens *nt*
zero: ~ **CO_2 emissions** *n pl* AIR ausgeglichene CO_2-Bilanz *f*; ~ **effect compensation** *n* ANALYSIS Nullpunktkorrektur *f*
zeta: ~ **potential** *n* (*ZP*) SOIL Zeta-Potential *nt*
zigzag: ~ **classifier** *n* WASTE, WATER SUPP Zickzacksichter *m*
zinc *n* (*Zn*) GEN Zink *nt* (*Zn*); ~**-containing** *adj* WASTE, WATER SUPP zinkhaltig
Zn *abbr* (*zinc*) GEN Zn (*Zink*)
zone: ~ **of aeration** *n* SOIL Lufteinwirkungszone *f*, Überwasserspiegelzone *f*; ~ **of saturation** *n* SOIL Sättigungszone *f*
zoning: ~ **plan** *n* MANAGE, POLICY, RENEW ENERG, WATER SUPP Flächennutzungsplan *m*
zoogloeal: ~ **layer** *n* WATER POLL Schlammdecke *f*
ZP *abbr* (*zeta potential*) SOIL Zeta-Potential *nt*